Analytical Chemistry

**Edited by
R. Kellner,
J.-M. Mermet,
M. Otto,
H. M. Widmer**

WILEY-V

A selection
of advanced textbooks
in analytical chemistry

W. Funk, V. Dammann, G. Donnevert
Quality Assurance in Analytical Chemistry
1995. 238 pp. with 87 figures and 17 tables.
Hardcover. ISBN 3-527-28668-3

H. Friebolin
**Basic One- and Two-Dimensional NMR
Spectroscopy**
1993. 368 pp. with 161 figures and 48 tables.
Softcover. ISBN 3-527-29059-1

G. Schomburg
Gas Chromatography
A Practical Course
1990. 320 pp. with 125 figures and 6 tables.
Softcover. ISBN 3-527-27879-6

R. Westermeier
Electrophoresis in Practice
1997. 320 pp. with 160 figures and 40 tables.
Hardcover. ISBN 3-527-30070-8

M. Otto
Chemometrics
In preparation. ISBN 3-527-28895-3

H. Günzler, H. M. Heise
IR Spectroscopy
In preparation. ISBN 3-527-28896-1

Analytical Chemistry

The Approved Text
to the FECS Curriculum
Analytical Chemistry

Edited by
R. Kellner,
J.-M. Mermet,
M. Otto,
H. M. Widmer

 **WILEY-VCH**

Weinheim • New York
Chichester • Brisbane
Singapore • Toronto

Robert A. Kellner†
Institute of Analytical Chemistry
Technical University of Vienna
Getreidemarkt 9/151
A-1060 Vienna
Austria

Matthias Otto
Institute of Analytical Chemistry
Freiberg University of Mining
and Technology
Leipziger Str. 29
D-09599 Freiberg
Germany

Jean-Michel Mermet
Laboratory of Analytical Sciences
Université Claude Bernard Lyon – I
43, Bd. du 11 Novembre 1918
F-69622 Villeurbanne Cedex
France

H. Michael Widmer†
Ciba-Geigy Ltd.
CH-4002 Basel
Switzerland

Library of Congress Card No. applied for

A catalogue record for this book is available from the British Library

Deutsche Bibliothek Cataloguing-in-Publication Data:
Analytical chemistry : the authentic text to the FECS curriculum
analytical chemistry / ed. by R. Kellner ... - Weinheim ; Berlin ; New
York ; Chichester ; Brisbane ; Singapore ; Toronto ; Wiley-VCH,
1998
   ISBN 3-527-28881-3   (brosch.)
   ISBN 3-527-28610-1   (Gb.)

Composition: Asco Trade Typesetting Ltd, Hong Kong. Printing and Bookbinding: Aubin Imprimeur, F-86240 Ligugé.
Printed in France.

Foreword

Within the physical sciences, analytical chemistry has probably undergone the most dramatic expansion and development during the past few decades. Consequently, the number of analytical techniques, their degree of sophistication and areas of application have increased tremendously. This poses a problem for the chemical educator: what to include in an undergraduate course and at which level?

Since its establishment a quarter of a century ago, the Federation of European Chemical Societies (FECS) has taken a keen interest in chemical education and analytical chemistry, which is shown by the fact that these two areas are represented as divisions within the FECS structure. After conducting extensive surveys the Division of Analytical Chemistry introduced the concept of a Eurocurriculum which received a favourable response from Europe as well as from overseas. A logical follow-up and tool to implement this Curriculum Analytical Chemistry worldwide is the present textbook.

The late I.M. Kolthoff, one of the great pioneers of analytical chemistry, used the motto *theory guides, experiment decides* in his books. The present textbook aims at imparting analytical insight by a suitable blend of theoretical background, experimental and instrumental details and finally a glimpse from the real world of applications. The ultimate goal is to convey analytical thinking for problem-solving.

Our rapid technological progress and at the same time enhanced concern for the environment are guarantees that the number of analytical problems will certainly increase and become more complex. Let us hope that this timely textbook will make a contribution to the education and training of future analytical chemists in order to have enough analytical problem-solvers qualified to meet these new challenges.

September 1997

Lauri Niinistö
President FECS

Preface

Analytical chemistry is – depending on the point of view – the oldest as well as the youngest branch of chemistry, the science of the transformation of matter. Claiming to be the oldest branch of natural philosophy (as "chemistry" has been described) goes back to the use of the Aristotelian syllogism for argumentation and proof (as shown in his "Analytiken"). Claiming to be the youngest branch became a possibility as "analytics" emerged as an own scientific discipline based on application of modern knowledge theories and information science in chemistry. The emancipation of analytical chemistry from chemistry began with Robert Boyle, continued with the activities of Lavoisier, Berzelius, Wöhler and Liebig, culminated the first time 100 years ago with Wilhem Ostwald (and his work *Die wissenschaftlichen Grundlagen der Analytischen Chemie*) and led to its present autonomy as a separate, very complex and highly attractive branch of science, described in its fundamentals in this book. The development of analytical chemistry is still continuing at a dramatic rate: consider the dynamic impact of Jan Heyrovsky (Electroanalysis), Richard R. Ernst (NMR), Gerhard Binnig and Heinrich Rohrer (STM/AFM) to name just a few of today's great researchers. As a consequence of the early and continuing interest in this field, a wealth of empirical knowledge, both of fundamental and of practical importance, about the material world in and around us has been compiled.

Chemistry as a whole has evolved worldwide into a supporting pillar of human culture, industry and trade, providing numerous goods of urgent daily need for humankind, such as food, clothing, shelter, pharmaceuticals and materials essential for medical use, transport or communication. Purely empirically at first, and mainly a branch of medicine, today's chemistry is a modern experimental science underpinned by physico-chemical and mathematical laws and has itself diversified into organic chemistry, inorganic chemistry, biochemistry, food chemistry, chemical technologies, physical chemistry and lately analytical chemistry.

It is certainly true that modern analytical chemistry, with its plethora of sensitive and selective techniques, has also strongly contributed to the awareness of environmental problems and issues and to the establishment of quality control systems in industrial production, in the health area and in the environment. The world market for analytical instruments has grown significantly over the years into today's impressive 1000 billion US $ size. Under the "responsible care" program of today's chemical industry, misuse of unoptimized technologies will ultimately be banned. The principle of "sustainable development" has been accepted as a basis of the production philosophy of the major chemical companies in Europe and is under consideration elsewhere. Analytical chemistry has been given a decisive role in controling the success of this process and in preserving the ecological balance of our world today.

The Division of Analytical Chemistry (DAC) of the FECS (Federation of European Chemical Societies) defines: "*Analytical chemistry is a scientific discipline that develops and applies methods, instruments and strategies to obtain information on the composition and nature of matter in space and time*".

This textbook – besides its unique role in university education – can also be considered as an interpretation of this DAC definition of analytical chemistry and as a pillar of support of the fact that analytical chemistry is indeed an own branch of science today – an *information science* – answering the theoretically and practically important question, how the material world is composed.

In order to find answers to this question, analytical chemists use chemical, physical and biological reagents to interact with the samples under investigation. This reagent-oriented view is also the scheme adopted for this textbook for practical reasons. In some cases the methods of analysis are clearly chemical, such as acid-base titrations where chemical reagents are used exclusively or purely physical, such as x-ray fluorescence. On the other hand the reader will soon notice that in some cases there are no clear, sharp differentiations, such as in chromatography, listed here under "chemical analysis" or "scanning tunneling microscopy" which has been linked to "physical analysis" although both areas touch both chemical *and* physical aspects.

Analytical chemistry is an *in-between-science* using and depending on the laws of chemistry, physics, mathematics, information science and biology. Its aim is to decipher the information hidden in the sample under investigation, not to change this intrinsic information, hence to tell the *truth* about the composition of the material world. This sounds trivial to a scientist, but it is not in today's complex and complicated technical and environmental matrices where analytical data frequently have to be made available in *real time* and *in situ* (*in the unchanged matrix*). Pressing needs of modern world trade, industry and commerce have led to the creation of national – and more importantly – international bodies for quality assurance, such as EURACHEM and CITAC. These bodies demand that even experienced laboratories prove their technical competence by passing accreditation procedures comprising both technological standards and personal skills from time to time (see Chapter 3).

The worldwide training of analytical chemists in the analytical skills and knowledge required to meet these challenges demands a high level of education using a harmonized scientific language and worldwide agreement on the basic scientific contents to be taught. While English has today the unchallenged role as *lingua franca* in chemistry and physics, the DAC-Curriculum "Analytical Chemistry" is the first broadly accepted attempt to harmonize the basic curricula in analytical chemistry for the benefit of the chemistry students working under this scheme (see *Anal. Chem.* 1994, *66*, 98A for the basic part and *Fres. J. Anal. Chem.* 1997, *357*, 197 for the advanced part of the DAC-Curriculum).

This textbook *Analytical Chemistry* is the authentic version of the DAC-Curriculum "Analytical Chemistry" and has emerged from the former "WPAC-Eurocurriculum". Its concept is based on the balanced mix of traditional methods of chemical analysis (Part II), modern techniques of biological (also Part II) and physical analysis (Part III) as well as chemometrics (Part IV). The textbook chapters in Parts II to IV are preceded by an introductory Part I featuring general topics such as "Aims of analytical chemistry and its importance for society", "The analytical process" and "Quality assurance and quality control". The book is completed by an industrially relevant Part V "Total analysis systems" dealing with more complex "Hyphenated techniques" and "Process analysis systems" of industrial importance today and, in particular, in the future. It is a multiauthored book, in order to guarantee the highest level of competence also in the undergraduate level of modern university education. The book is devoted to the principle of combining solid foundations of scientific knowledge with flexibility towards novel analytical techniques. Its unique concept allows coverage of classical topics, such as acid-base titrations or compleximetry – which are necessary to understand modern chemical sensors technology – while also including recent, trendsetting developments in physical analysis, chemometrics and process analysis. For example, atomic force microscopy and miniaturized total analysis systems (μ-TAS) are introduced at an elementary level. Possible shortcuts in some chapters have to be seen in the light of the editorial decision to limit the size of the book to under 1000 pages and may be overcome by referring to the wealth of specialized textbooks for advanced studies that are part of the DAC-Curriculum scheme.

Besides providing chemistry students with a sound preparation for the requirements of modern industry, university studies in analytical chemistry must also provide fitness in basic academic research. Truth has an essential place in analytical chemistry. The critical student should therefore carefully study Chapters 2 ("The analytical process") and 3 ("Quality assurance and quality control"). Sir

Karl Popper's credo: "The approximation to the truth is in principle possible" can be accepted as a strong philosophical foundation of analytical chemistry as a basic science.

In the applied field, we are confronted with a worldwide flood of analytical data, resulting from the incredible number of 10 000 million analyses per year! It is the attempt of this elementary textbook and its editorial team to make clear that analytical chemistry has today more than ever a sound responsibility for the future development of our society.

Analytical data and model calculations on NO_x production in the stratosphere by supersonic jetliners, for instance, resulted in knowledge which prevented the development of a significant fleet of ozone-killing supersonic jets. This is just one example of our necessity to produce correct analytical data and to transfer them correctly into knowledge that we can use in problem solving. Correct knowledge is also the basis for wisdom, needed for instance for far reaching political decisions in general.

We are convinced, that when taught worldwide, analytical chemistry – based on an educational scheme devoted to the equilibrium between freedom and responsibility, as provided by this textbook and the DAC-Curriculum – can be a key science to provide for a safer future for mankind!

June 1997

R. Kellner, H. Malissa and E. Pungor
Chairmen of WPAC/FECS: 1993–97, 1975–81 and 1981–87
(since Sept. 1996 the WPAC continues as Division of Analytical Chemistry (DAC) of the FECS)

Acknowledgments

We wish to express our deep appreciation to the team of reviewers, who contributed their expertise to this international enterprise by reviewing chapters and by providing countless valuable suggestions. Sincere thanks to:

Professor F. Adams, UIA, Antwerpen, Belgium
Professor K. Ballschmiter, University of Ulm, Germany
Professor M.F. Camoes, University of Lisboa, Portugal
Professor Ch. Ducauze, National Institute of Agronomy, Paris, France
Professor R.R. Ernst, Swiss Federal Institute of Technology, Zurich, Switzerland
Professor S. Gál, Technical University, Budapest, Hungary
Professor D.J. Harrison, University of Alberta, Edmonton, Canada
Professor J.A. de Haseth, University of Georgia, USA
Dr. M. Hoenig, Chemical Research Institute, Tervueren, Belgium
Professor B.R. Kowalski, University of Washington, USA
Professor J. Mink, University of Veszprem, Hungary
Professor N.M.M. Nibbering, University of Amsterdam, The Netherlands
Professor R. Niessner, Technical University of Munich, Germany
Dr. B. te Nijenhuis, Gist-brocades NV, The Netherlands
Professor A. Sanz-Medel, University of Oviedo, Spain
Professor G.R. Scollary, Charles Stuart University, Australia
Professor G. Tölg, University of Dortmund, Germany
Professor M. Valcárcel, University of Córdoba, Spain
Professor C. Wilkins, University of California, Riverside, USA
Professor O.S. Wolfbeis, University of Regensburg, Germany
Professor Yu. A. Zolotov, Academy of Sciences, Russia

This book would not have been published without the strong, efficient and constant support provided by our publisher, Christina Dyllick. The Editors wish to express their gratitude to Dr. Dyllick for her patience, encouragement and helpful advice that were highly beneficial in the preparation of this book. This collaboration was a real pleasure.

Contributors

Holger Becker
Imperial College of Science,
Technology and Medicine
Zeneca Smith Kline Centre for
Analytical Sciences
Department of Chemistry
South Kensington
London SW7 2AY
United Kingdom
Chapter 15

Jack Beconsall
1 St. Tudwals Estate
Myntho
Pwllheli LL53 7RU
Wales
United Kingdom
Section 9.3

Karl Cammann
Institute for Chemical and Biochemical
Sensor Research
Mendelstr. 7
D-48149 Münster
Germany
Section 1.1

Gary D. Christian
Department of Chemistry
University of Washington
Box 351700
Seattle, WA 98195-1700
USA
Chapter 16

Pierre Van Espen
Department of Chemistry
University of Antwerpen
Universiteitsplein 1
B-2610 Antwerpen
Belgium
Section 8.3

Horst Friebolin
Department of Organic Chemistry
University of Heidelberg
Im Neuenheimer Feld 270
D-69120 Heidelberg
Germany
Section 9.3

Gernot Friedbacher
Institute of Analytical Chemistry
Vienna University of Technology
Getreidemarkt 9
A-1060 Vienna
Austria
Chapter 10

Keiichiro Fuwa
3-1-6-402 Sekimae Musashino
Tokyo 180
Japan
Section 1.2

Jeanette G. Grasselli
Ohio University
Department of Chemistry
150 Greentree Road
Chagrin Falls, OH 44022
USA
Section 1.3

Manfred Grasserbauer
Institute of Analytical Chemistry
Vienna University of Technology
Getreidemarkt 9
A-1060 Vienna
Austria
Chapter 10

D. Bernard Griepink
CEC
Rue Montoyer 75
B-1040 Brussels
Belgium
Chapter 3

Elizabeth A.H. Hall
Institute of Biotechnology
University of Cambridge
Tennis Court Road
Cambridge CB2 1QT
United Kingdom
Sections 7.8 and 7.9

Elo H. Hansen
Department of Chemistry
Technical University of Denmark
Building 207
DK-2800 Lyngby
Denmark
Section 7.4

Robert A. Kellner†
Institute of Analytical Chemistry
Technical University of Vienna
Getreidemarkt 9/151
A-1060 Vienna
Austria
Sections 1.1, 7.1, 7.2, 7.6, 9.1, 9.2,
Appendix 5

Viliam Krivan
Sektion Analytik und Höchstreinigung
University of Ulm
Albert-Einstein-Allee 11
D-89081 Ulm
Germany
Section 8.4

Willem E. van der Linden
Faculty of Chemical Technology
University of Twente
P.O. Box 217
NL-7500 AE Enschede
The Netherlands
Sections 2.1–2.3

Eddie A. Maier
European Commission
Standard Measurements
and Testing Programme (MO75)
3rd Floor, Office 7
Rue de la Loi 200
B-1049 Brussels
Belgium
Chapter 3

Andreas Manz
Imperial College of Science,
Technology and Medicine
Zeneca Smith Kline Centre for
Analytical Chemistry
Department of Chemistry
South Kensington
London SW7 2AY
United Kingdom
Chapter 15

Jean-Michel Mermet
Laboratory of Analytical Sciences
Université Claude Bernard Lyon-I
43, Bd. du 11 Novembre 1918
F-69622 Villeurbanne Cedex
France
Sections 8.1, 8.2, 8.5, Appendix 4

Wilfried M.A. Niessen
Hyphen MassSpec Consultancy
De Wetstraat 8
NL-2332 XT Leiden
The Netherlands
Section 9.4

Lauri Niinistö
Helsinki University of Technology
Laboratory of Inorganic and Analytical
Chemistry
P.O. Box 6100
FIN-02015 Espoo
Finland
Section 7.5

Matthias Otto
Institute of Analytical Chemistry
Freiberg University of Mining
and Technology
Leipziger Str. 29
D-09599 Freiberg
Germany
*Sections 5.1–5.5, 7.7, 12.4, 12.5,
Chapter 13, Appendices 1, 3, 6 and 7*

Dolores Pérez Bendito
Department of Analytical Chemistry
University of Córdoba
E-14004 Córdoba
Spain
Chapter 6

Erwin Rosenberg
Institute of Analytical Chemistry
Vienna University of Technology
Getreidemarkt 9
A-1060 Vienna
Austria
Chapter 14

William S. Sheldrick
Department of Analytical Chemistry
University of Bochum
Universitätsstr. 150
D-44801 Bochum
Germany
Chapter 11

Klára Tóth
Institute of General and Analytical
Chemistry
Technical University of Budapest
Gellért tér 4
H-1111 Budapest XI
Hungary
Sections 4.4 and 7.3

Wolfhard Wegscheider
University of Leoben
Franz-Josef-Str. 18
A-8700 Leoben
Austria
Sections 12.2 and 12.3

H. Michael Widmer†
Ciba-Geigy Ltd.
CH-4002 Basel
Switzerland
Sections 4.1–4.3, 4.5, 5.6, 7.1 and 7.2

E. Deniz Yalvac
The Dow Chemical Company
2301 North Brazosport Blvd.
B-3828 Building
Freeport, TX 77541
USA
Chapter 16

Pier Giorgio Zambonin
Department of Chemistry
Campus University
Via E. Orabona, 4
1-70126 Bari
Italy
Sections 2.4 and 12.1

Contents

Symbols

a	Auger yield
A	absorbance
b_0	intercept, background
b_1	sensitivity
$b_{1/2}$	full width at half maximum
c	concentration
C	covariance
CV	coefficient of variation
d	distance, cell thickness lattice spacing, mean deviation
D	diffusion coefficient
e	electron
E	energy, electrode potential, expectation
$E(X^r)$	rth (non-central) moment of X (or of $F(x)$)
$E\{(X - \mu)^r\}$	rth central moment of X (or of $F(x)$)
E_a	activation energy
E_B	binding energy
E_{kin}	kinetic energy
f	frequency
$f(z)$	standard normal density function
F	Fisher ratio, flow rate
$F(z)$	standard normal distribution function
G	free energy
H	plate height, enthalpy
I	intensity, nuclear angular momentum quantum number (spin), current, Kováts retention index, ionic strength
I_0	incident intensity
j	angular moment quantum number
J	coupling constant
k	rate constant, relative sensitivity factor
k'	capacity factor
K	equilibrium constant
K_M	Michaelis-Menten rate constant
l	orbital quantum number
L	distance between sample and observation screen
m	magnetic quantum number
M	multiplicity
n	order of diffraction, rotation axis, refractive index
$\bar{n}$	rotary inversion axis
N	number of counts, areal density, plate number
NA	numerical aperture
P	angular momentum of nucleus, probability
Q	number of incident projectiles
r	distribution correlation coefficient, radius
R	resolution, range of electrons, radius of diffraction, resistance
R_f	retardation factor
R_S	resolution

s	standard deviation (estimate)
S	entropy, similarity
s^2	variance (estimate)
t	time, film thickness, Student factor
$t_{1/2}$	half-life time
t_M	hold-up (mobile) time
t_R	total retention time
T	true value of the measured quantity, transmittance
T_1	spin-lattice relaxation time
T_2	spin-spin relaxation time
$\bar{u}$	average linear velocity of molecules of the mobile phase
U	DC potential, voltage
v	reaction rate, linear velocity
$\bar{v}$	average linear velocity of analyte
v_0	initial reaction rate
V	variance
V_M	hold-up (mobile) volume
V_R	total retention volume
w	peak width
x	scalar variable
$\mathbf{x}$	vector of x-values
$\mathbf{X}$	matrix of x-values
$\bar{x}$	mean, arithmetic mean or average in a set of n observations
y	variable
Y	sputter yield
z	number of elementary charges
Z	standard normal deviate, atomic number
α	significance level, selectivity factor, degree of dissociation
β	yield of ion detector, phase ratio, cumulative stability constant
χ^2	Chi-squared (distribution)
δ	chemical shift
ε	molar absorptivity
Φ	flux density, work function
λ	wavelength, radioactive decay constant
η	efficiency, abundance
γ	gyromagnetic ratio
Γ	Gamma function
μ	population mean, magnetic moment, ionic strength
ν	degrees of freedom, frequency
$\Delta\Omega$	acceptance angle of detector
θ	scatter angle, pulse angle, diffraction angle
ρ	mass density
σ	shielding constant, population standard deviation
σ^2	population variance
τ_p	pulse duration
ω	fluorescence yield, cyclotron frequency
$\varnothing$	dihedral angle

Abbreviations and Acronyms

2D	two-dimensional
AALA	American Association for Laboratory Accreditation
AAS	atomic absorption spectrometry
AC	alternating current
ADC	analog-to-digital converter
AED	atomic emission detector
AEM	analytical electron microscopy
AES	atomic emission spectrometry, Auger electron spectrometry
AFM	atomic force microscopy
AFNOR	Association Française de Normalisation
AFS	atomic fluorescence spectrometry
AOAC	Association of Official Analytical Chemists
AL	atomic layer
APCI	atmospheric-pressure chemical ionization
API	atmospheric pressure ionization
ARM	atomic resolution microscopy
ARUPS	angle resolved UPS
ATR	attenuated total reflectance
BB	broad-band (decoupling)
BCR	Bureau Communautaire de Référence
BE	reversed geometry: magnetic sector + electrostatic analyzer
BIPM	Bureau International des Poids et Mesures
BSI	British Standards Institute
CAMM	computer-aided molecular modelling
CAR	continuous addition of reagent
CCD	charge-coupled device
CCT	constant current topography
CE	capillary electrophoresis
CEN	Comité Européen de Normalisation
CENELEC	Comité Européen de Normalisation Electrotechnique
CFA	continuous flow analysis
CF-FAB	continuous flow fast-atom bombardment
CGC	capillary gas chromatography
CI	chemical ionization
CID	collision-induced dissociation
CID	charge-injection device
CITAC	Cooperation on International Traceability in Analytical Chemistry
COSY	correlation spectrometry
CPAA	charged particle activation analysis
CRM	certified reference material
CTD	charge-transfer device
CV	coefficient of variation
CW	continuous wave
DAD	diode array detection (detector)
DAS	desamino-sulphadimidine
DBE	double-bond equivalent

DC	direct current
DCI	desorption chemical ionization
DCP	direct current plasma
DCPU	dichlorophenyl urea
DEPT	distorsionless enhancement by polarization transfer
DIN	Deutsches Institut für Normung
DLI	direct liquid introduction
DMA	dynamic mechanical analysis
DSC	differential scanning calorimetry
DTA	differential thermal analysis
DTG	differential thermogravimetry
EAL	European Cooperation for Accreditation of Laboratories
EB	forward geometry: electrostatic analyzer + magnetic sector
EC	European Commission
ECD	electron capture detector
ED-XRF	energy-dispersive X-ray fluorescence (spectrometry)
EDL	electrodeless discharge lamp
EELS	electron energy loss spectrometry
EFTA	European Free Trade Association
EGA	evolved gas analysis
EI	electron impact (ionization)
EL	electro luminescence
ELD	electro luminescence display
EN	European Norm
EPA	Environmental Protection Agency (USA)
EPXMA	electron probe X-ray microanalysis
ERD	elastic recoil detection
ESA	electrostatic analyzer
ESP	electrospray
ETA	electrothermal atomizer, emanation thermal analysis
ETSI	European Telecommunication Standard Institute
EU	European Union
EXAFS	extended X-ray absorption fine structure (spectrometry)
FAB	fast atom bombardment
FD	field desorption
FDA	Food and Drug Administration (USA)
FG	functional group (chromatogram)
FIA	flow injection analysis
FID	free induction decay
FID	flame ionization detector
FIM	field ion microscopy
FIR	far infrared (radiation)
FNAA	fast neutron activation analysis
FT	Fourier transform
FT-ICR	Fourier-transform ion cyclotron resonance (spectrometry)
FTIR	Fourier transform infrared (spectrometry)
FT-MS	Fourier-transform mass spectrometry
FWHM	full width at half maximum
GC	gas chromatography
GDL	glow discharge lamp
GDMS	glow discharge mass spectrometry
GF-AAS	graphite furnace atomic absorption spectrometry
GLP	good laboratory practice
GMP	good manufacturing practice
GNP	gross national product
GO	glucose oxidase
GS	Gram-Schmidt (algorithm)
HCL	hollow cathode lamp
HPDE	high-density polyethylene
HPGe	high purity germanium (detector)
HPLC	high performance liquid chromatography
HPTLC	high performance thin-layer chromatography

ICP	inductively coupled plasma
ICP-MS	inductively coupled plasma mass spectrometry
ICTAC	International Confederation of Thermal Analysis and Calorimetry
IDF	International Dairy Federation
IDMS	isotope dilution mass spectrometry
ILAC	International Laboratory Accreditation Cooperation
INAA	instrumental neutron activation analysis
IQR	interquartile range
IR	infrared (radiation)
IRN	indicator radionuclides
ISO	International Organization for Standardization
ISO/REMCO	ISO Council Committee on Reference Materials
ISP	ionspray
ISS	ion scattering spectrometry
IUPAC	International Union for Pure and Applied Chemistry
JCPDS	Joint Committee for Powder Diffraction Standards
KRS-5	thallium-bromide-iodide (ATR crystal material)
LAMMS	laser micro mass spectrometry
LARIS	laser atomization resonance ionization mass spectrometry
LBB	Lambert-Bouguer Beer's law
LC	liquid chromatography
LEED	low energy electron diffraction
LEEM	low energy electron microscopy
LNRI SNMS	laser non-resonant ionization SNMS
LQR	lower quartile range
LRI SNMS	laser resonance ionization SNMS
LRMA	laser Raman micro analysis
m/z	mass-to-charge ratio
MALDI	matrix-assisted laser desorption/ionization
MCA	multichannel analyzer
MCT	mercury cadmium telluride
MEIS	medium energy ion scattering (spectrometry)
MID	multiple-ion detection
MIP	microwave-induced plasma
MIR	middle infrared (radiation)
MS	mass spectrometry
MS-MS	tandem mass spectrometry
M&T	Measuring and Testing Program of the EC
NAA	neutron activation analysis
NATA	National Association of Testing Authorities (Australia)
NBS	National Bureau of Standards (now NIST)
Nd:YAG	neodymium yttrium aluminium garnet (laser)
NEXAFS	near-edge X-ray absorption fine structure
NICI	negative ion chemical ionization
NIR	near infrared (radiation)
NIST	National Institute of Standards and Technology
NMR	nuclear magnetic resonance (spectrometry)
NPD	nitrogen phosphorus detection (detector)
NRA	nuclear reaction analysis
OECD	Organization for Economic Cooperation and Development
OIML	Organisation Internationale de Métrologie Légale
PA	proton affinity
PAA	photon activation analysis
PC	personal computer
PCB	polychlorobiphenyl
PCDD	polychlorodibenzodioxins
PD	plasma desorption
PDF	powder diffraction file
PE	photo electron
PFIA	process flow injection analysis
PFK	perfluorokerosene

PFTBA	perfluorotributylamine
PICI	positive ion chemical ionization
PMT	photomultiplier tube
ppb	parts per billion
ppm	parts per million
ppt	parts per trillion
PSD	position sensitive detector
PTFE	polytetrafluoroethylene
PVD	physical vapor deposition
Q	quadrupole filter
QA	quality assurance
QC	quality control
RBS	Rutherford backscattering spectrometry
REELS	reflection electron energy loss spectrometry
REM	reflection electron microscopy
RHEED	reflection high energy electron diffraction
RIC	reconstructed ion chromatogram
RIMS	resonance ionization mass spectrometry
RM	reference material
RMD	relative mean deviation
RNAA	radiochemical neutron activation analysis
RS	Raman spectrometry
RSC	Royal Society of Chemistry (UK)
RSD	relative standard deviation
SCA	single channel analyzer
SCE	standard calomel electrode
SDM	sulphadimidine
SDS-PAGE	sodium dodecylsulphonate polyacrylamide gel electrophoresis
SEC	size exclusion chromatography
SEM	scanning electron microscopy
SEM	secondary electron multiplier
SERS	surface enhanced Raman scattering
SEXAFS	surface extended X-ray absorption fine structure (spectrometry)
SFC	supercritical fluid chromatography
SHE	standard hydrogen electrode
SI	Système International (d'Unités)
SIM	selected ion monitoring
SIMS	secondary ion mass spectrometry
SIRIS	sputtered initiated resonance ionization mass spectrometry
SNMS	sputtered neutrals mass spectrometry
SOP	standard operating procedure
SPM	scanning probe microscopy
SRM	standard reference material
SRM	selective reaction monotoring
SSMS	spark source mass spectrometry
STM	scanning tunneling microscopy
STS	scanning tunneling spectrometry
TC	Technical Committee (of CEN or ISO)
TCDD	tetrachlorodibenzodioxins
TD	thermodilatometry
TEELS	transmission electron energy loss spectrometry
TG	thermogravimetry
TGA	thermogravimetric analysis
TGS	triglycine sulfate
THEED	transmission high energy electron diffraction
TIC	total ion chromatogram
TIMS	thermo-ionization mass spectrometry
TLC	thin-layer chromatography
TMA	thermomechanical analysis
TMS	tetramethylsilane
TOF	time-of-flight (mass spectrometer)
TSP	thermospray

UHV	ultrahigh vacuum
UPS	ultraviolet photoelectron spectrometry
UQR	upper quartile range
UV	ultraviolet (radiation)
VIM	vocabulaire international de métrologie
VIS	visible (radiation)
VML	vocabulaire de métrologie légale
VOX	volatile organic halogene
WD-XRF	wavelength-dispersive X-ray fluorescence (spectrometry)
WHO	World Health Organization
XAS	X-ray absorption spectroscopy
XPS	X-ray photoelectron spectrometry
XRD	X-ray diffraction
XRF	x-ray fluorescence (spectrometry)
ZAF	Z (for element number) absorption fluorescence

Part I
General Topics

1 Aims of Analytical Chemistry and Its Importance for Society

1.1 Aims of Analytical Chemistry: Its Basic Importance for Society

1.1.1 Introduction: the origins of universe and chemistry

In order to consider the aims of analytical chemistry and its basic importance for present society, we need to first look back briefly at the natural and social history of science and chemistry.

According to one theory, the universe originated fifteen billion years ago when the "Big Bang" took place, and is still expanding today with the materials in it, in the form of stars, cosmic dusts, and various gases, which are composed of the elements in the periodic table starting from hydrogen. Our Earth is a member of the solar system included in one of the galaxies of the universe. That means in a way that chemistry, the central core of materials science, has existed from the beginning. Human beings appeared on the Earth about four million years ago, and since that time have managed to survive the struggle for existence, and have developed civilization in various areas including the natural sciences. Nowadays there is a modern type of chemistry, one branch of which is analytical chemistry [1.1-1].

In the West, Greek philosophers, Plato (428–347 B.C.) and Aristotle (384–322 B.C.), thought that there were four essential elemental substances of matter, namely, earth, water, air, and fire. Later in the Middle Ages (end of the 5th to the middle of the 14th century), attempts to change the base metals into gold was the people's major concern, because of the high value of gold in society. Alchemy was one approach to chemistry at that time.

On the other hand, in the East, mainly in China, certain philosophers thought of five basic substances similar to the Greek philosophers', namely, earth, water, fire, wood, and metal. Besides making gold by alchemy, however, the major effort of chemistry in China was focused on either finding or synthesizing the "Elixir of Life" or the medicine to conquer aging and death. The reason was that the Chinese Emperors in ancient times, including the First Emperor of Ch'in, Shih Huang-ti (221–206 B.C.), badly wanted longevity. Consequently, the old-time "chemists" in China received emperors' orders to produce the medicine of longevity, and had to make many hard efforts, which were often in vain. Thereafter in China, there was certain progress in the fields of producing paper, black-powders for fireworks, and oriental medicine, but virtually no modern chemistry had developed. In Japan, the situation was similar. Modern science including chemistry developed only after the Meiji Evolution in 1868. Prior to that time, only some limited knowledge of chemistry and western medicine was gained, mainly from Dutch visitors [1.1-2]; [1.1-3].

The credit for the first atomic theory should be given to Democritus (~460–370 B.C.), a Greek philosopher, who considered that the universe consisted of matter made out of nondividable and unchangeable particles, atoms. This idea later influenced great scientists and philosophers such as N. Copernicus (1473–1543),

F. Bacon (1561–1626), R. Boyle (1623–1691) and I. Newton (1642–1727). Thereafter, the modern atomic theory was established by the English chemist J. Dalton (1766–1844). A.L. Lavoisier must be credited for his introduction of quantitative measurements in chemistry, which thus gave rise to analytical chemistry.

Lavoisier performed, e.g., important quantitative measurements on the formation of mercuric oxide from mercury and air, or oxygen. He proved that the amount of mercuric oxide formed was exactly the sum of the two components. Lavoisier died as a victim of the Revolution, but he will be remembered as the person who gave modern chemistry its first landmark [1.1-4]. For more examples on the development of chemistry, the reader is referred to F. Szabadvary and other chapters in this textbook.

1.1.2   Aims of analytical chemistry

In order to investigate problems in the field of both natural and social sciences, an analytical approach is in general first used. That is to say, the problem is first broken down into simpler units, each of which is studied, then, after the individual pieces of information have been combined, the problem as a whole can finally be understood. Analysis of matter is carried out in a similar order. The matter is divided first into components – chemical species such as atoms, ions, or molecules (see Part II). Physical analysis now often allows the investigation of matter in its original state, without preseparation or digestion "in situ" (see Part III). The primary purpose or aims of analytical chemistry involves finding out the amount and types of chemical species present in a system. After knowing the constituents of the matter, one can estimate or evaluate the constitution of the original matter in question, just as in the second and the final processes described above.

In the last half of this century, the technological development of instrumental analysis was so wide and rapid that today the field of analytical chemistry has even expanded toward "computer-based analytical chemistry" (see Part IV). In other words, the various characterizations of matter can be carried out nowadays also in total analysis systems (see Part V), and the present day definition of analytical chemistry is found to be: "A scientific discipline that develops and applies methods, instruments, and strategies to obtain information on the composition and nature of matter in space and time." (Edinburgh-Definition of WPAC 1993, [1.1-5]).

1.1.3   Importance of analytical chemistry for society

Human beings are surrounded by natural and man-made environments. Air, water, soil, rocks, plants, and animals compose the natural environment. The man-made environment, on the other hand, consists of houses, clothes, foods, and all other objects such as tools, goods, books, papers, and so forth, which modern society has created for modern people to conduct their lives. These natural and artificial objects are all made out of matter, which is composed of chemical species. In order to recognize and estimate each material, man needs certain methods. Besides the shape, size, hardness, or color, etc., which are physical properties of the material, analytical chemistry is needed to obtain information on the chemical composition and properties of any of these materials. In other words, analytical chemistry is very close to our modern lives and societies.

In order to better understand the detailed relationship between analytical chemistry and society, we need to consider some of the traditional classifications of analytical chemistry.

(a) The first and most direct classification is with the material itself of which a few examples, corresponding to the material to be analyzed, or the sample material, are *water analysis*, *rock analysis*, *food analysis*, and so forth.

In the industrial area, *steel analysis* or *iron analysis* is fundamentally important in order to ensure a high steel quality, which is the basis of other industries. There have been experts of analysis of steel and iron in big companies around the world from the time when wet chemical analysis was in use.

Clinical analysis is vitally important for health, and a modern hospital must be able to perform reliable *blood* and *urine analyses* in order to make a proper diagnosis of the patient. All automated physical analytical instruments are available. *Pharmacological analysis* and *biological analysis* are related fields. In recent years, *environmental analytical chemistry* has become very popular, and air, water, soil, and biological materials are all included for environmental investigation, which is naturally very important for our society.

(b) The classification of chemical species to be analyzed is another item. Again giving only a few examples, *total analysis* is used to find all species in the sample, so that the sum of the weight of each component equals the original sample weight. The total analysis of a rock is a good example. Recently, moon rock analysis received concern from a great number of people around the world. However, often some particular constituent(s), element(s) or molecule(s), are required to be analyzed. *Elemental analysis* is used today to analyze all elements in both organic and inorganic compounds. Analyses such as for NO_x, SO_x, and O_3 give information on air pollution, and *PCB analysis* and *dioxin analysis* are also needed to improve our environmental safety. These molecules, in particular, consist of numerous numbers of isomers, some of which are known to be more toxic than others. *Radioactive analysis* has an obvious importance for our society, as it includes the analysis of ^{90}Sr, ^{137}Cs, ^{235}U, and ^{249}Pu, which are either nuclear fission products or nuclear power materials.

(c) Today *analytical chemistry* encompasses *qualitative analysis* and *quantitative analysis*. The former indicates whether a particular element or compound is in the sample, whereas the latter gives the amount of the species in the sample [1.1-6]. The relation of this concept with the society can be found in various occasions. A quantitative description is better suited to, for instance, the distance between two points, the size of a house, the amount of food, the ability of a child, the seriousness of a crime, and so forth, and such quantitative measures are often also required in chemistry. When the results cannot be described in accurate figures of mg or ppm, but may just be described by such words as "fairly high" or "very little", we call them semiquantitative results. Such semiquantitative descriptions are frequently found in our daily life.

As mentioned in the previous section, physical methods of analytical chemistry such as *spectrochemical analysis* have been developed so widely that they are applicable to all examples listed in this section. That means that the importance of analytical chemistry for our present day society has expanded greatly, and will continue to grow further in the future.

References

[1.1-1] Miller, S.L., Orgel, L.E., *The Origins of Life on the Earth*. New Jersey: Prentice-Hall, 1974.

[1.1-2] Rodzinski, W., *A History of China*. Oxford: Pergamon Press, 1979.

[1.1-3] Kaku, K., Chemistry and Chemical Industry, *The Chem. Soc. of Japan*, 1992, 45, 388.

[1.1-4] Petrucci, R.H., *General Chemistry*. New York: Macmillan Publishing, 1972.

[1.1-5] Kellner, R., Education of Analytical Chemists in Europe., *Anal. Chem.*, 1994, 66, 98A.

[1.1-6] Laitinen, H.A., *Analytical Chemistry, Encyclopedia of Chemistry*, New York: McGraw Hill, 1992, pp72–74.

1.2 Aims of Analytical Chemistry. The Analytical Chemist as a Problem Solver

Learning objectives

■ To introduce students to the concept of the analytical chemist as a "problem solver" using multitechniques and broad, generalized knowledge

■ To explain the unique and special industrial aspects of the environment for the practice of analytical chemistry

■ To provide examples of analytical problem solving in environmental science

1.2.1 Introduction

There is a widespread belief that the decades surrounding the year 2000 will go down in history as the "information age". Considering the breathtaking advances which have been made in collecting, storing, manipulating, transmitting, and presenting information, this view has merit. Massive, parallel computer and data storage, satellite transmission of data, worldwide TV networks, and even "virtual reality" have been built on these breakthroughs.

The science of analytical chemistry has benefited in a number of ways from this burst of technology. Local area networks connecting analytical instruments, the Fourier transform manipulation of spectral information, and the comparative searching of analytical data to identify an "unknown" are only a few examples of this contribution. However, analytical chemistry should be viewed as one of the strongest pillars of the "information age" for several important reasons. While collecting, storing, manipulating, and transmitting information are vital activities, the generation of new, accurate information, which concentrates on enhancing the understanding of critical problem and opportunity areas facing mankind, is the promise of analytical chemistry.

We live in an "information age" which has brought great changes to society and also to the capabilities of analytical chemistry.

In fact, those tasks constitute only one definition of analytical chemistry. The focused approach of defining a problem, addressing it through multiple techniques or interdisciplinary methods if needed, and assuring that results are accurate and reproducible, constitutes analytical science. Past and future applications of this methodology to biotechnology, the chemical industry, materials science, food science, reaction mechanisms, and environmental issues have the potential to produce multiplicative levels of advancement when combined with the other technologies on which the "information age" is being built.

This chapter will present examples of advances which have resulted from the use of analytical chemistry, and will attempt to show how a scientific group can operate as a team in order to define a problem accurately, and to insure that all approaches are considered in its solution. It will discuss additional factors which are vital to our field, such as communicating complex results in a meaningful way to a management or to the public who may lack high levels of scientific literacy.

1.2.2 The industrial environment

To put the importance of analytical chemistry to society in perspective, we need to consider the astounding products which have been developed in the past 50 years which have so altered our lives. Scientists and technicians have produced new drugs, man-made fibers and plastics, new additives to make paint and rubber tougher, new aerospace metal alloys, and transistors and integrated circuits for electronics. At the same time, we have gained a far keener insight into the basic chemistry of life, the control of disease, and the examination of our environment here and in outer space.

Much of this has evolved from initially being able to identify molecules and determine their structure, and to relate structural features to physical properties, to the development of novel methods for tracing the kinetics of new reaction systems under operating conditions, and to the application of intricate and precise trace analysis techniques to samples of environmental concern.

To obtain this kind of information, we rely heavily on analytical methods and instruments. These are the tools that probe deeply into even the smallest sample to unravel the structure of its molecules, or measure a single component that may be even less than a millionth part of the whole.

There is probably nothing that characterizes scientific development and wide applications of novel technology and concepts better than the astonishing developments in analytical instrumentation over the last decade or two. The biggest single change, as noted earlier, has involved computerization, which has allowed not only faster and better data collection, but also the development of many sophisticated data processing methods, which had not been possible with the older, noncomputerized instruments.

But it is the effective use of all of analytical chemistry, including methods and instrumentation, which has had such a tremendous impact on industrial research and development. Since about 80% of analytical chemistry graduates pursue careers in industry, it is important to understand the role of the analytical chemist in industry. The effective use of analytical chemistry is an important part of the operation of any successful industrial laboratory. The rewards for a competent scientist with analytical skills are commensurate with this degree of challenge.

The education or "skill" requirements for analysts will vary directly with the problem under study. "Problem" is really the key word here. Today's industrial analytical chemist is primarily a "problem solver", with the responsibility for applying his or her knowledge of methods and instrumentation as effectively as possible to solve analytical problems of the corporation or industry.

The process of setting goals for industrial analytical chemists must recognize the diversity and scope of the industrial environment. Therefore, it is instructive to consider typical specific objectives for industrial scientists.

1. Develop experimental techniques and knowledge of the theory required to provide consistently accurate results and meaningful information on materials using a variety of established procedures. Fundamental to success in any sphere of analytical chemistry is a good theoretical knowledge of each method employed. Without this basis, the analyst becomes a mechanic rather than a scientist. Nearly equally important is a thorough understanding of the optimum areas of application as well as the limitations inherent in each method. Experimental technique is vital to the analytical chemist. Experiment always leads theory.

   The student should definitely be able to operate an instrument, and even more importantly, know if it is not operating properly, e.g., is it exhibiting the proper signal to noise ratio (S/N) or resolution for the problem under study? Both theory and laboratory work are important since neither can survive without the other. Laboratory work not only develops the capability of the student to think, but also to act independently. This trait will be important regardless of how rapidly the analytical methods change.

2. Develop specialized analytical training for conducting both routine and unconventional analyses, recognize the interaction of time and accuracy in an analysis, develop methods and procedures for handling both routine and unusual samples, and consult with other scientists in other areas of expertise so that the "best" approach to obtaining the required information (as distinguished from the requested information) is utilized.

   Analytical chemistry is not the application of fixed methods to a series of samples. The analytical chemist must be innovative and entrepreneurial in his/her application of the knowledge base in analytical chemistry.

3. Develop the capability to direct the work of other analytical chemists, act as an active participant in the analytical aspects of research programs in the corporation, and do scientific research in uncharted areas of analytical chemistry.

Industry has used analytical chemistry extensively in the development of new products and processes.

The analytical chemist in industry must be a skilled problem solver.

The analytical chemist must have a good theoretical knowledge of analytical methods and instrumentation.

Innovation and team work are important elements for success in the industrial environment.

The ability of any individual to accomplish these goals will depend on the educational level, on-the-job training, further education, and experience that is acquired. Obviously, only a limited portion of the skills required can be obtained in the university. These general comments apply, of course, not only to prospective industrial analytical chemists, but to all analytical chemists – indeed to all chemists.

Let us now turn our attention to some of the factors of the industrial environment which make the industrial analytical chemists somewhat different.

Industrial analytical chemists must balance the constraints between time, cost, and accuracy, which are key elements in every "real world" problem (Fig. 1.2-1). The problem may come from any part of the corporation and the analyst must be skilled in utilizing resources available to him/her to solve the problem.

There are two important aspects of time to the industrial chemist. The first is in the management of one's own time and that of others reporting to you. Productivity of all resources is important in a competitive environment such as industry faces. Although the academic laboratory is becoming more conscious of costs, time management is still not a major concern. It is extremely important that time management becomes a way of life for an industrial analyst, and exposure to this important concept would be a valuable addition to the curriculum.

The second factor of time is the compromise between accuracy and the need for a prompt answer. If a plant operation or a costly experimental program is dependent on analyses for guidance, a good answer on time is often much more valuable than a perfect answer too late. This should not be construed as implying that industrial analytical chemistry is "quick and dirty". Some of the most accurate analytical work is needed and performed in industrial laboratories, for example in pharmaceutical companies seeking clearance from regulators before products can be marketed. In "*Physics Today*" a letter on how to succeed in industrial research [1.2-1] succinctly phrased a relevant rule on accuracy: "Don't spend time and effort on making any one measurement more accurate than the nature of the experiment justifies. If the error in one reading cannot be reduced below $\pm 10\%$ it makes no sense to measure other parameters to within 0.1%."

Industrial laboratories spend large amounts of money on analysis. Industry is less concerned with the cost of instrumentation if equipment is shown to be cost-effective in solving problems. The need for and value of an answer usually outweighs by a substantial margin the cost of obtaining it. Thus, economics dictate that one must consider not only the cost of obtaining the answer to an analytical problem but also the cost of delay to another operation or to a researcher if the answer is not available on time.

Uncertainty is the "name of the game" in industry. Industrial research laboratories, for example, are often engaged in exploratory research looking for new compositions of matter from reactions of feed mixtures over various types of catalyst or from products of organic synthesis. In the early stages of such work product mixtures are complex. Since the work is usually conducted in an area where novelty is important, there is little literature to suggest which products are present. The analyst is faced with a true "unknown". In other instances, such as in plant operating problems, the "problem" can be equally ill-defined. It is not uncommon for a murky, heterogeneous liquid to appear on the analyst's desk labeled in an obscure way, such as "sample taken from surge tank on Isocracker II" or "contamination from unknown source". Contrast this with most samples taken under controlled conditions in an academic laboratory!

In both the laboratory example or the operations case, the analyst must gather together whatever information he/she can on the chemistry surrounding the problem. The analyst must then develop a complete analytical plan to home in on the answer. As mentioned, many industrial samples are truly "unknowns", and sampling in itself presents new uncertainties. Obtaining a representative sample can be almost as hard as the analysis itself. Sample taking and sample preparation should be a vital part of teaching analytical chemistry. Statistics can be a powerful tool, and every student must at least know the difference between accuracy and precision, and how to distinguish between an analysis and a determination (see Chap. 2).

The problems that industry will face at the turn of the century and the methods available at that time for the solution of those problems will be drastically different from those which we are worrying about today. This is the result of rapid change, not only in the development of new analytical techniques (witness the explosive

Fig. 1.2-1. The industrial environment

Cost–benefit and timeliness of analytical results are key industrial factors.

The analyst may be asked to work on problems with little information about sample source or purity.

Problems may change but the problem-solving method will not.

growth of the microscopies, atomic force, etc., in the last five years), but also in the increasingly complex problems of society which we must address. During the next decade, according to projections made by scientists at the National Institute for Standards and Technology (NIST) in the United States, analytical measurement capabilities will experience technical growth equal to that of the previous three decades combined. We must realize that since there is no way to educate today's students in the specific methods of the year 2000, we must concentrate on those parts of learning that will have continuing value. Although the problems will change, the problem-solving method will not. It will still involve:

– an understanding of the chemistry involved
– a knowledge of sampling and sample processing
– the appropriate uses of separation methods
– an appreciation of the importance of proper calibration and the use of standards
– and the ability to select the best method or methods for the measurement step.

As Philip W. West said in his ACS Fisher Award address in 1974, "Analytical chemistry is not spectrometers, polarographs, electron microprobes, etc., it is experimentation, observation, developing facts, and drawing deductions" [1.2-2]. Particularly in industry where instrumentation is more readily available, we must guard against concentration on the measurement step. The answer, not the method, is the goal. The problem and its solution are paramount – all analytical chemists must be problem solvers.

1.2.3 The analytical approach

The analytical chemist must provide information of a sufficiently valid nature so that meaningful decisions can be made by industry or government.

If the analytical scientist is to be effective in solving the problems of society, there is another dimension to his/her role which must be recognized. The successful analytical chemist must be adept at a number of techniques and must be at the forefront of chemical knowledge. The function of the analytical chemist is to provide information of sufficiently valid nature, that is, of the requisite statistical significance so that meaningful decisions can be made about materials or problems. The emphasis here again is that the analytical chemist is essentially a "problem solver", and to do the job most effectively he/she should use a complete analytical approach for solving the problems. This "analytical approach" may be defined by the following steps (see also Chap. 2):

• defining the problem correctly
• ensuring that the samples available are representative of the problem
• interacting with the client to obtain his/her knowledge of the problem and to define the boundary of timeliness and accuracy required
• developing an analytical plan involving an evaluation of the sequence and best methods to be employed
• completing the work using the highest level of expertise and good chemical knowledge
• communicating answers, not data, including the precision and reliability of all numbers and specifying cautions or constraints on the use of the data

The most important responsibilities of an analytical chemist are to interpret results and communicate answers effectively.

• interpreting the information and results in a clear, consistent, and meaningful report that clearly addresses the problem.

Therefore, with the exception of routine, repetitive analytical testing, it is a mistake to consider "measurement" as the primary role of analytical chemists. They must be "problem solvers". The analytical chemist should be an active participant in problem solving teams from the outset. In most industrial laboratories, the analytical chemist acts in a support capacity to research or operating groups. It is important for the analyst to be an active participant in the projects he/she supports rather than to be a remote source of analytical results.

The analytical chemist can make major contributions to research or operations by interacting with expertise in the planning stage of any program. In this way the analyst gains very useful insight into the totality of the problem; including, for example, knowledge of feedstocks, processing conditions, recovery methods, and potentially related by-products. In turn, with this knowledge, the analyst can suggest sampling methods and the combination of analytical techniques best designed to truly decipher the problem.

It seems illogical that the analyst not be involved in establishing the best means for the collection of meaningful and representative samples, and the appropriate order of analysis. In addition, it is important to have the analyst on a problem-solving team take a broad view of the techniques which might be used. In these days of complicated, hyphenated multitechnique analytical methods, the power of combined techniques should be obvious (see Chap. 14). In this way the maximum amount of information can be obtained.

The most important contributions of analytical chemists are from the elucidation of mechanisms rather than from the simple determination of levels of concentration, e.g., understanding how a catalyst works to provide both selectivity and conversion, or of the changes in structure which may result due to the sequence of addition of molecules in a polymer synthesis. In environmental areas, the potential major contribution of analysts could be in spelling out the toxicological mechanism by which a chemical affects human processes.

In summary (Fig. 1.2-2), the analytical scientist must interact professionally with the client in order to properly define the problem. The analytical approach then requires the analyst to interact with other analytical scientists and employ their expertise to utilize combined techniques which provide information to help solve the problem. Finally, the analytical scientist must be at the forefront of analytical advances in all the areas of methodology and instrumentation described in this text. The analytical chemist will then be a true problem solver.

This description of the analytical problem solver in industry suggests that the chemist's academic training should supply a general, broad education rather than excessive specialization. Once in industry it is easier to specialize than to broaden one's analytical background. The foundation of good chemistry and experimental practice should be laid in the university.

Another description of an analytical chemist was presented by deHaseth [1.2-3]. He points out that an analytical chemist must be both an expert and a generalist. It is also necessary, in the industrial world, that the analytical chemist must be a team player. Almost no industrial project is an individual effort. Professor Gary Hieftje, Indiana University, winner of the ACS Award in Analytical Chemistry and of the Outstanding Teaching Award at Indiana, has recognized both these requirements. He has written:

"One of my concerns about graduate research has long been the conflict that a student often encounters when trying to complete a doctoral project in a reasonable length of time while maintaining sufficient breadth to become a good problem solver. Obviously, if the student is to perform original research, a requirement for a doctorate, he/she must become somewhat focused. Unfortunately, this same focus often makes the student a less able problem solver. To combat this problem in my own research group, I employ several tactics. One is to have a large, fairly diverse research effort. It is inevitable that a student's closest contacts, both professional and social, are within the same research group. If individuals in that group are working in a broad number of areas of science, each student assimilates a bit of that breadth and becomes somewhat expert in fields other than his/her own research.

Yet another tactic I employ is to have students work in research teams. By doing so, each student not only becomes broader scientifically, but also learns to work in a team-oriented environment. Interestingly, any student can become a team leader and can recruit into the team anyone else in the group, including visiting scientists or postdocs."

It is also relevant, in discussing the education of analytical chemists, to consider the importance of retraining or upgrading of skills. Short courses are offered by professional societies, by instrument companies or academic institution worldwide.

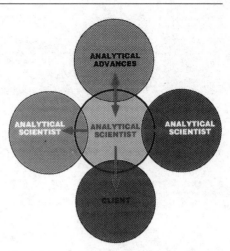

Fig. 1.2-2. Analytical scientist interactions

It is essential for an analytical chemist to keep skills current and upgraded through continued learning and education.

1.2.4 Analytical problem solving in environmental science

As previously stated, there can be no question in anyone's mind that our ability to analyze the structure and composition of materials has had an important impact

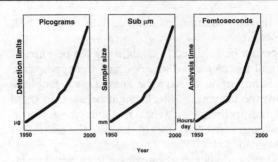

Fig. 1.2-3. Changes in technical progress in analytical science since World War II

in the industrial sector and in society. The analysis of products and the understanding of mechanisms have truly fed the "industrial revolution", and they have been breathtaking. Advances in biotechnology, medicine, electronics, catalysis, chemicals, polymers, and new materials, to name a few areas, would not have been nearly as spectacular without the leverage available from new analytical techniques.

Fig. 1.2-3 illustrates the rapid changes in detection limits, sample size, and analysis time that have occurred in analytical science since World War II [1.2-4]. Detection limits have gone from the microgram to the subpicogram level. A millimeter-sized sample presented a challenge in 1950; today we routinely look at submicrometer units, or even at single atoms which can be resolved in analytical electron microscopes in an almost routine manner. Analysis times have decreased from hours or sometimes even days required for a complete qualitative and quantitative analysis to modern capabilities which include real-time spectroscopy on in situ experiments, giving us data on molecular dynamics in the femtosecond time regime. Analytical science today is indeed the catalyst and the key to our understanding.

Trace analysis, the detection and determination of tiny quantities of elements and molecules in increasingly complex matrices, is readily accomplished today. These are exactly the types of samples usually involved in environmental analysis. Yet, some people tend to view analytical advances as a negative factor in environmental studies because the ability to measure extremely small quantities can focus regulatory restrictions on materials not previously detectable. If so, that is a rare instance where increased knowledge can be considered a setback. This ability to "see" previously invisible components has "good" and "bad" sides. On the "good" side, we can identify some components which are truly deleterious. On the other side, we can initiate "witch-hunts" for no other reason than that a material can be detected or measured. Would it be of more value to industry and to society to direct our efforts in analytical science toward identifying more of the components in environmental samples, thereby providing early warning of impending environmental threats or hazards to human beings and to biota? Analytical science could feed the technological advances which would allow us to mitigate environmental changes or adapt to them.

This brings to a focus the other side of the question; that is, the issues of risk assessment, risk management, and risk communication. These words, along with "detection limits", "selectivity", "accuracy", "analytical variability", "quality assurance", and "total quality management" (see Chaps. 2 and 3) must be defined and made understandable to the public, regulators, and the industrial establishment. We can no longer separate the performance of science from our mandate to communicate the findings and the uncertainties of science to the public.

What role do analytical chemists play in industrial and in environmental science, and how can they be deployed most effectively? How should they influence the quantity and quality of communication between industrial scientists and legislative or public interest groups? It is critical that there be a better understanding of the true risks involved from environmental releases in terms that allow the public to gain a realistic measure of the effects of exposure.

As shown in Fig. 1.2-4 the state-of-the-art in analytical science for the type of trace analysis involved in environmental samples is impressive. But other issues must also be considered. These involve the state-of-the-practice related to the state-of-the-art, the uncertainty in sampling for trace analysis, and the relationship between what can be measured and what is harmful to human health.

Rapid changes in the capabilities of analytical chemists have occurred in the last 50 years which have allowed the lowering of detection limits, working with smaller samples, and dealing with lower concentration levels.

Environmental regulations have followed the ability of analytical chemistry to detect and analyze molecules at lower levels and in more complex matrices.

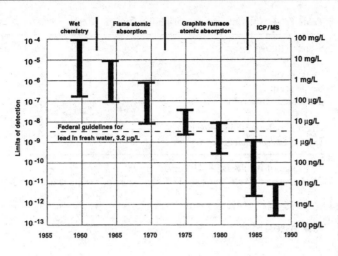

Fig. 1.2-4. Detection limits in routine water quality measurements over time

Chemical toxicity

Most of us are rightly frightened of cancer. We can measure substances, such as those synthetic chemicals determined to be toxic, and regulated to zero level in or on food to extremely low detection limits. It has been pointed out that, during the past two decades, the detection limit has dropped from a concentration near that which produces a measurable dysfunction in a human to a concentration that may be six or more orders of magnitude lower [1.2-5]. In short, there is no scientifically provable way to test the validity of the extrapolation.

New studies [1.2-6] take issue with the animal tests that led to estimates of whether substances cause cancer. They believe that early work was based on inaccurate ideas about how cancer comes about, and, therefore, unwarranted assumptions were made about the degree to which data gained from high doses of chemicals can reveal anything about the effects of low doses. Ames' recent work also shows that there are naturally occurring pesticides present in concentrations a hundred or more times higher in fruits and produce than are any synthetic chemicals.

Because of the difference between animals and humans, there is no way to confidently extrapolate data on toxicity of chemicals between them.

Analytical sources of error

With this perspective, and in the midst of the furor, analytical chemists face their own dilemma with regard to the real-world sources of uncertainty and error in the data they obtain. Although sources of error such as sampling, contamination, and selectivity are recognized, they are extremely difficult to overcome in practice. As a result, at a concentration of 1 ppb, the average uncertainty found for a variety of methods carried out in different labs has been close to ±50%. Furthermore, the uncertainty rapidly grows larger as one goes to lower concentrations (see Chaps. 3 and 12).

The lack of quality control in environmental sample analysis has, for example, been discussed by Hertz [1.2-7], and is illustrated in Table 1.2-1. In this study, conducted by the National Bureau of Standards (now the National Institute of Standards and Technology, or NIST), the results obtained from various laboratories conducting routine tests on round-robin samples fell far short of the accuracy and precision achievable under the best of good laboratory practices by qualified chemists. For example, in measuring polychlorinated biphenyls (PCBs) in oil, one of the six samples sent for the round-robin analyses had no Arochlor (PCBs) added, the NBS analyses confirmed that no PCBs were present at the limit of detertion of the method. Yet, 9 of the 18 laboratories reported the presence of PCBs in all 6 samples! Factors contributing to the variability in results include sample-handling procedures, purity of reagents, and sample contamination from other sources, conditions of facilities and equipment, and the lack of available reference or calibration materials for standardization (see Chap. 3).

The uncertainty in an analytical method grows larger at lower concentrations.

It is difficult to reach agreement between results from different laboratories, even when they are using standard methods.

Table 1.2-1. Interlaboratory variability studies

System studied	No. of samples	No. of labs	Variability (CV)
PCBs in oil	6	18	38–64%
Se in serum		27	(a)
β-Carotene in serum	4	11	100% (per sample)
			35% (grand average)
Organic pollutants in water		5	500%
Leachates from hazardous wastes[b]			–

(a) Range of results is 38–100 µg/L.

(b) For each of 9 elements studied, the range of results was a factor of 10.

Risk assessment

The public is not familiar with risk assessment.

In light of the tremendous current capabilities of analytical science and the inherent difficulties in realizing those capabilities in a reliable manner, how can we reach a balance between our ability to measure and our assessment of the environmental risk? The impact of technological innovation can lead to a greater ability to detect and analyze, or it can bring about wholly new processes or products [1.2-8]. The new analytical understanding helps in the development of processing changes to lower emissions or contaminant; as has been stated: – "technology shapes society and society shapes technology" [1.2-8].

An example is in the history of lead pollution. The presence of lead in air and water was the result of the burning of coal; the metallurgical industry; and, after World War II use of leaded gasoline in automobiles. The toxic effects of lead were well documented long before 1970, when the Clean Water Act in the United States initiated action on determining harmful levels and regulating lead. But links between human activities, environmental lead contamination, and ecological effects had to await more sensitive detection limits.

Fig. 1.2-4 shows the dramatic improvement in our analytical capabilities for lead determination in water. As our ability to detect and measure lead quantitatively improved, society responded with technological innovations to replace lead – for example, unleaded gasoline; substitutes for lead-based paints; replacement material for lead water pipes; and alternatives to lead-containing solder, including seamless aluminum cans. The cycle of awareness, social response, and technological innovation was the direct result of our ability to determine lead in smaller amounts in complex samples.

1.2.5 Standards for analytical laboratory procedures and measurements

The use of traceable standards and approved methods is critical in environmental analysis.

A well-managed and productive analytical laboratory must operate with good and accepted laboratory procedures. There must be quality control for all methods and techniques (see Chap. 3). This includes traceability of all primary and secondary standards, frequent calibration, and transferability of methodology between various labs or parts of the organization (e.g., between research laboratory and plant quality control laboratory). To enhance productivity the use of automated systems [1.2-9] must be employed. National standards laboratories in most countries are resources for accomplishing these goals. Industry demands reliable, state-of-the-art analytical measurement in order to comply with environmental regulations and to improve product and process quality so that competitive advantage may be realized.

An example of an industry/government interaction in this area is the Consortium on Automated Analytical Laboratory Systems (CAALS) established at NIST in the United States. The consortium was formed in 1989 and is establishing guidelines so that industry and NIST can work as partners to develop automated, accredited chemical analysis technologies. Their definition of a chemical analysis system is shown in Table 1.2-2. During the development of an analytical system

Table 1.2-2. Chemical analysis systems

Analyte release	Analyte separation	Analyte detection
Microwave dissolution	Gas chromatography	ICP-OES elemental detection
Fusion	Liquid chromatography	Atomic absorption
Solvation	Matrix modification	ICP/MS elemental detection
Solvent extraction	Liquid–liquid extraction	Flame ionization detection
Supercritical fluid extraction		Spectroscopic detection
Solid-phase extraction		Mass spectrometry

the researchers: (a) build chemical analysis expertise; (b) develop standard reference materials; and (c) build in quality assurance. NIST is also the source of Standard Reference Materials (SRMs) and of all calibration services (ohm, volt, etc.) in the United States.

A very important international organization involved with analytical standards and methods is the International Union of Pure and Applied Chemistry (IUPAC). The Analytical Division is very active but all other Divisions (Physical, Inorganic, Organic, Macromolecular, Clinical, Medicinal, and Applied Chemistry) also have commissions that work on problems of nomenclature and methods in electrochemistry, atomic masses, molecular spectroscopy, food chemistry, polymers, and biotechnology, among others.

1.2.6 Future of analytical science

Many new technologies are on the horizon which will markedly affect the capabilities of the analytical scientist of the future. Fiber optics, lasers, chemometrics, and nanotechnology are just a few of these. The problems of society are more complex and challenging than ever before, especially in areas such as environmental technology, biotechnology, advanced materials, and the entire information age revolution. Some of the future paths of analytical science are shown in Table 1.2-3. To follow those paths the analytical scientist will need to continue innovating in the areas shown in Table 1.2-4. As we continue to employ the analytical approach to new challenges, with new methods and new techniques, the analytical chemist as a problem solver will continue to be a key player in industry, academia and government. Analytical chemistry is a central discipline to all other branches of science.

New technologies continue to improve the capabilities of analytical chemistry.

Table 1.2-3. Future paths of analytical science

Automation and robotics
Networks of instruments
Truly intelligent instruments
More complex methods of data reduction
On-line sensors and miniaturized systems
Advanced remote sensing

Table 1.2-4. Future needs for analytical chemistry

Greater sensitivity / selectivity
More innovative combinations of analytical methods
Advanced three-dimensional micro, nano, and subsurface analysis
More sophisticated understanding of measurement science
Ability to analyze under more rigorous in situ situations
Direct probes of energy localization in molecules, transition states, and reaction dynamics
Interpretation of analytical raw data via expert systems

References

[1.2-1] Sommer, A.H., *Phys. Today*, Sept. 1976, 9.

[1.2-2] West, P.W., *Anal. Chem.* 1974, 46:9, 784A.

[1.2-3] deHaseth, J., *Spectroscopy*, 1990, 5:1, 20.

[1.2-4] Grasselli, J.G., in: *Analytical Applications of Spectroscopy II*: Davies A.M.C., Creaser, C.S. (Eds.). Cambridge, UK: The Royal Society of Chemistry, 1991.

[1.2-5] Harris, W.E., *Anal. Chem.*, 1992, 64, 655A.

[1.2-6] Ames, B.N., Gold, L.S., *Chem. Eng. News*, Jan. 7, 1991, 28.

[1.2-7] Hertz, H.S., *Anal. Chem.*, 1988, 60, 76A.

[1.2-8] White, R.M., Rod, S.R., *Environ. Sci. Technol.*, 1990, 24, 460.

[1.2-9] Salit, M.L., Guenther, F.R., Kramer, G.W., Griesmeyer, J.M., *Anal. Chem.*, 1994, 66, 361A.

1.3.1 Introduction

The purpose of analytical chemistry (see definition, page 4), is often connected with questions to be answered or problems to be solved (see Sec. 1.1, 1.2). Very often important decisions are based directly on the result of chemical measurements. Such experiments sometimes do not necessarily result in quantitative analyses in the traditional way, but cover qualitative aspects as well. There are many more or less standardized tests which are used to determine properties other than concentration (e.g. acid number, chemical oxygen demand, or parameters such as AOX, TOC, leaching rate, biotoxicity tests etc.). In the forensic field the presence or absence of a compound is often essential. In materials science local concentrations (analyte distribution) on surfaces or at locations which show unexpected behavior may be more important than bulk concentrations (see Chap. 10). The quality assurance system applied should guarantee that all chemical measurements are accurate and precise (see Chap. 3). The obtained results must be independent of the operator, place, and time.

In this last respect, reproducibility is more important than the measurement uncertainty obtained by the same person in the same laboratory in a series of measurements (repeatability, see Sec. 2.4). Reasonably good repeatability is essential but not sufficient for total quality control of the work of a chemical laboratory. International harmonization agreements, especially for trade demand bias-free test results which can be checked and agreed upon, independently of time, location and individual. The chemical measurement should result in a true statement, which can be proven (see Chap. 3).

The truth of an analytical result (chemical measurement) of an unknown real-world sample is pragmatically chosen to be the accepted value most experts agree upon. In many cases the expert community defines national or international standards to which the measurements should be related by an unbroken chain (traceability). Measurement methods are often standardized, resulting in operational standards, defined by an exact procedure (see Chap. 3).

In the field of quantitative analysis, primary methods (absolute methods of chemical analysis, needing no calibration) compete with instrumental methods which need calibration with certain standards. In describing the quality elements for the best practice in nonroutine and analytical chemical research and development it is important not to hinder further improvements. Therefore, it is of vital importance that written standard operation procedures (SOPs) are not taken as permanently valid. Even in the case of tests which are defined by common sense operational procedures, analytical expertise must be applied to allow improvement (e.g., in speed, in material and money savings, in waste production, etc.) as long as the results are convincingly demonstrated to be comparable or even better than those obtained with common methods. Accreditation bodies require experts [1.3-1, 1.3-2] to judge the details of a newly developed method.

In order to establish guidelines describing the optimal practice for quality control in nonroutine and research and development-laboratories, it seems appro-

---

*according to CITAC Guide NO. 1 (Quality Assurance in Analytical Chemistry) priate to divide test and measurement procedures into action sequences or "unit operations" which can then be combined as necessary into "unit processes", finally leading to the overall procedure. This splittingup of test procedures into different modules which can be combined, allows one to construct a large variety of testing methods. The idea behind this approach is that there is a high probability for a validable procedure if all unit processes are validated for this kind of analyte measurement, or test goal. In other words: the combination of unit processes, each with its SOP validated for the analyte or test goal in question, must result in a validated overall procedure or method, so that only verification checks for the special combination chosen are needed as a quality control measure.

A guideline on quality assurance in nonroutine and research and development laboratories is being prepared by a combined CITAC/EURACHEM working group as an analytical chemical interpretation of the latest ISO25 guideline to cover all possible unit operations and processes in detail. The trained analyst has to decide what sequence of accurately described actions is considered as a unit operation. This involves knowledge and above all experience. Offered by highly experienced and motivated personnel.

In order to select the optimum combination of unit operations to form a new method, a broad knowledge of the range of available techniques is essential to avoid the trial-and-error approach.

The SOP for each unit process can consist of a decision tree structure. In producing a SOP, it is necessary to bear in mind for what purpose this is done. One is the quality control aspect which demands that laboratory procedures are followed as closely as possible by different persons at different times. Another purpose is a certain traceability so that assessors may evaluate the performance of a certain test unit, or analytical results obtained a long time ago. A further purpose is for the leading analytical chemist (principal investigator) who has to sign the test results so that he/she can follow the history of the sample in an unbroken chain. In this respect the SOPs for the unit operations and processes need not be very lengthy. It is sufficient if a qualified person can understand and follow those descriptions or instructions.

1.3.2   Examples of unit processes in analytical chemistry

The generic approach of defining unit processes demands a splittingup of the general test procedure or method of analysis into more or less closed sequences of actions (unit operations). Natural dividing lines can be points where a test can be interrupted or a sample can be kept or stored without drawbacks before the next step. In qualitative and quantitative analysis it is common practice to divide the general analytical task according to the so-called analytical process (see Chap. 2). The latter can roughly be described by the following main steps:

Analytical task

The exact question to be answered or the exact problem to be solved by the chemical measurement has to be defined in agreement with the user (customer) of the chemical test data. With an experienced analyst who has a broad overview of the analytical techniques, together with their advantages and disadvantages, the exact analytical task can be worked out. This is strongly recommended in order to prevent alteration of the sample by the sampling procedure and storage conditions.

Only the analytical expert understands the general concept of every sound test procedure, since only he/she is aware of the possible uncertainty or reliability of an overall procedure and can take precautions from the very beginning to avoid false planning. Already at this stage a creative thought may result in a preliminary study plan or combination of unit processes describing the conditions for sampling, storage, and sample preparation (see Sec. 2.2). It will become obvious which kind of chemical tests will be needed. It may be an internal or external standardized operationally defined procedure which determines certain properties; it may result in a surface or trace analysis of materials (see Chap. 10) or an environmental, food, or industrial analysis (see Part V).

The final version of the defined analytical task must also include the level of reproducibility, comparability, and truth which is needed for the interpretation of the result.

Specification of the analytical requirements

Reliable and correct analysis (with international comparability) needs a clear and adequate specification of the requirements, especially those concerning the validity of the data interpretation (needed minimum precision, comparability, truth) have to be defined. These quality requirements must be translated into technical requirements.

Given the time and cost constraints, together with the existing measuring and data interpretation uncertainty (reliability) and the possible risk of false results, the analyst has also to formulate the traceability requirements. In the preparation phase of the intended analysis, each chosen unit process has to be judged by an experienced expert to avoid introduction of systematic errors. Counteraction should be considered from the very beginning.

Staff

Chemical analyses on a nonroutine basis, and especially in research and development, must be carried out by, or under the supervision of qualified and experienced analysts, holding a relevant and documented professional qualification, obtained by previous education or by training on the job (see chapt. 3).

Appropriate education for this purpose is a specialization in analytical chemistry at university level. An educational background corresponding to a minimum requirement in this area has been defined by the Working Party on Analytical Chemistry (now Division of Analytical Chemistry) of the Federation of the European Chemical Societies in 1993 under the WPAC EUROCURRICULUM guideline.

This CURRICULUM ANALYTICAL CHEMISTRY is now available worldwide through this textbook for basic education, a series of specialized textbooks and multimedia at graduate level, and by the Eurocourse system and the Internet for the postgraduate education. This, together with a minimum experience of about two years in the field, is definitely required for senior staff positions. The total quality management system must ensure constant retraining and education in a rapidly changing measurement technology.

Laboratory environment

The total quality control management system may require restricted laboratory areas for cleanness, security, and safety. In these cases only certain persons are allowed to work in the restricted areas. The use of such areas has to be properly documented.

Equipment

All equipment used in laboratories should be of a specification sufficient for the intended purpose (validated for the specific test), and kept in a controlled state of maintenance, effectiveness, and calibration consistent with its use.

Equipment normally found in the chemical laboratory can be categorized as

- general service equipment
- volumetric equipment
- chemical and physical measurement standards
- physical measuring instruments (analytical instruments)
- computers and data processors

Reagents

The quality of reagents used in the course of a chemical analysis must be appropriate for the intended purpose. Preferably, only reagents with a guaranteed grade level should be purchased from manufacturers who have a quality assurance system, such as ISO 9000. This becomes even more important as the traceability to the Mol is now demanded in every quantitative analysis[1]. Especially in trace analysis any reagent blank should be determined separately and not automatically compensated for since the latter is uncertain when matrix-effects occur.

Optimal sampling and testing strategy

The sampling and testing strategy guarantee that the final goal (purpose) of the test will be achieved, and meaningful and valid results are produced (see Sec. 2.2). The importance of the sampling stage cannot be overemphasized. If the test portion is not representative of the original material, it will not be possible to relate the analytical result to that in the original material, no matter how good the analytical method is nor how carefully the analysis is performed. Sampling is always an error-generating process, in which the homogeneity of the original sample material plays a major part.

The optimal sampling strategy must be worked out together with the customer or data user. It is strongly connected to the problem to be solved by the chemical test or analysis. At this point creative thinking and/or common sense is very much needed. A possible unit process SOP for a sampling strategy may simply consist of a decision tree in which the required total uncertainty (reliability) is compared with the level of heterogeneity in the number and size of test samples to be drawn.

It is also essential to establish a sampling strategy for the blank values taken in situ (field blanks) to prevent false determination and interpretation of the measurement process. Often this requires comparison samples (e.g., from other areas outside the influence of the test samples, such as pollution-free areas, areas where the material to be checked worked properly, etc.).

A sampling plan may also contain a layout and time plan for accompanying actions: preparation of the right sample containers, including cleaning, preparation of reagents to be added for preventing alterations during storage and transport to the laboratory; and if necessary a sufficient amount of containers for the field, reagent, and container blanks.

Sampling

The actual sampling must be performed exactly as specified in the sampling plan described above. Sampling has to be carried out in accordance with the analytical task and considering the representativeness of the test samples taken to the laboratory. It is important to take into consideration the unavoidable imperfections of sampling in determining the overall uncertainty (reliability) of the chemical test.

Any alteration during the actual sampling process must be documented and justified. The properties of the analyte(s) of interest should be taken into account.

Volatility, sensitivity to light, thermal instability, biodegradability, and chemical reactivity may be important considerations in designing the sampling strategy and the optimal choice of the sampling technique. All equipment and tools used for sampling, subsampling (see below), sample handling, preparation, and extraction should be documented together with the results for the corresponding blanks and control values.

All staff concerned with the administration of the sample handling system should be properly trained and should follow a general unit process procedure describing the minimum quality elements to be fulfilled.

---

[1] P. de Bièvre

Preparation and measures for the preservation of the test samples

Before even thinking of taking subsamples to be analyzed by a certain method, the test samples brought into the laboratory should be viewed as raw samples which may need a further homogenization procedure, since otherwise the representativeness of the subsamples is unknown and no volumetric aliquotization is allowed. In cases of an optically evident heterogeneity of a test sample consisting of a physical mixture of different phases, the quality of the homogenization steps may determine the overall quality. The SOPs for the unit process test-sample preparation should address the variability associated with the test laboratory to guarantee homogenization prior to subsampling. For the total uncertainty balance, the uncertainty (reliability) of the subsampling process is to be determined in a quantitative manner. It determines the number of subsamples to be analyzed in order to achieve the required total uncertainty (reliability).

Subsampling

Taking an exactly known portion of the total test sample brought into the laboratory for a test or analysis is a very important unit operation. Subsamples for a certain chemical analysis and/or operational test are prepared, requiring information on the homogeneity of the sample material.

This portion must resemble the total sample material in every aspect. The uncertainty of subsampling which is determined by the level of homogeneity can be determined by taking more subsamples and determining the uncertainty statistically (see Sec. 2.4).

The measurement of the amount of a testportion is best be determined by a gravimetric procedure (see Sec. 7.2). In the course of the weighing process, the field and test sample container blanks should be treated and prepared in the same manner as the subsample.

All blanks must be processed at the same time as the sample in the same environment and treated exactly like the subsamples. Suspensions such as blood or serum cannot be aliquoted by volumetric means with a high accuracy because of the varying amount of the solid or quasi solid components (e.g., larger protein molecules, liquid layer vessels, etc.).

Subsample treatment

Preparation of the subsample includes all measures needed for an unbiased final measurement, such as solvation, dissolution, digestion, surface cleaning, melting, combustion, etc. Only a few analytical techniques, such as certain spectroscopic methods or NAA allow direct instrumental analysis without any treatment. Any SOP dealing with subsample treatment must contain the variety of treatment methods in the laboratory and deal with the special precautions to be taken for each specific analyte. At this stage in the analytical process, some quality assuring elements are to be introduced in the overall analytical procedure (see Sec. 2.2 and Chap. 3). Besides spiked samples, different blanks (field, container, reagent) and spiked blanks, matrix adopted reference materials (in house) or even certified standard reference material with as nearly as possible the same matrix as the sample can be introduced into the process so that they exactly follow the subsample treatment.

Separation and/or enrichment (see Chap. 5)

The best practice towards quality control in separation and enrichment techniques is difficult to express in detail since it depends upon so many individual and problem-specific factors. Purely formal aspects must be too rigid, cumbersome, or costly, but must deliver a sufficiently low uncertainty (reliability). Depending on the exact analytical task, a high quality standard may require quantitative separation and/or enrichment of the analyte from the sample matrix. It should be noted that the recovery factor may be strongly influenced by the separation or enrichment processes. Every process and matrix has its own recovery function as an indication of analyte gains or losses. The smaller the analyte concentration the higher is the risk that there might be a compound present in the sample matrix or in the reagent blanks which will strongly interfere in the quantification process, directly or indirectly. In any case, the SOP for separation should contain all necessary information on the exact procedure, including the method of determining the recovery factors.

It is important that all blanks, all reference or standard material (spiked samples with analyte or an internal standard behaving like the analyte or question) are also treated by exactly the same separation process within the same time interval (in parallel or shortly afterwards) as the original test samples.

Determination

The presence or absence of a substance (analyte) and/or measuring its amount in the aliquot and/or measuring the outcome of an operational defined procedure is dealt with in the determination (final chemical measuring step). The measurable quantity is defined by the International Vocabulary of Basic and General Terms in Metrology as "an atribute of a ... substance which may be distinguished qualitatively and determined quantitatively". The final measurement process usually results in a signal intensity which does not directly represent the result in its correct dimensions. It is necessary to calibrate the measuring system (except in coulometry, see Sec. 7.3, volumetric titrations or gravimetry, see Sec. 7.1 and 7.2) resulting in a functional relationship between the measured signal intensities and the concentration of the analyte or the amount of substance. This relationship may be quite complicated. A strictly linear relationship is the simplest case, but is not strictly necessary (see Sec. 12.2). In the validation process it has to be shown that the calibration function is also valid during the measuring step of the subsample, with or without its matrix. Additionally in the course of the calibration process it is important to differentiate clearly between background signal, net signal, blank signal, zero signal, etc.

The overall program for the calibration of measuring equipment in the chemical laboratory must be designed to ensure that, where the concept is applicable, all measurements are traceable through certificates held by the laboratory, either to a national or international measurement standard or to a certified reference material.

It is the laboratory's responsibility to use techniques and methods which are appropriate to the required application. The laboratory may use its own judgement, it may select a method in consultation with the customer/data user, or the method may be specified in accepted regulations or by the customer.

Analytical tests may be subdivided into three general classes depending on the type of calibration required:

- metrology in general
- chemical analysis (see Part II)
- physical analysis (see Part III)

Before any more or less matrix-free analytes can be quantified the appropriate instrument has to be checked to function properly. For this "fit for purpose test", the SOP (essentially a simplified instruction manual) should contain a short test to confirm that the instrument meets its specification: mainly selectivity, sensitivity, stability, and freeness from systematic errors. In analytical nonroutine and research and development, awareness of systematic errors is of special importance since less preinformation is available in these fields.

Calculation of the result

Standardization or calibration is defined as determining the response function $S = g(C)$ where S is the measured net signal which is a function g of the given analyte concentration (C). The evaluation function is the inverse of the calibration function. The quality of the function is a direct indicator as to how well a chemical system is understood. Automated instruments must allow a check by the user: which evaluation function, which kind of error correction and which algorithm is applied to yield the final result.

The validity of simple evaluation methods has to be proven. In certain instances of automated total spectra subtraction this quality element is missing nor is there a warning that this makes sense only if the sensitivity (slope of the calibration curve) is exactly the same for different matrices.

Chemometrics should include a performance test of the overall procedure. By this test, such parameters as limit of detection, limit of quantification, dynamic measuring range, precision, accuracy, truth and other terms characterizing an analytical method (see Chap. 4 and 12) should be determined and possibly documented in control charts so that any negative deviation can be noticed and corrective actions can be taken.

The calculation of the result must include the calculation or guess of the overall uncertainty balance of the whole analytical process, including sampling.

Validation

Methods developed in house (research and development) must be adequately validated, documented, and authorized before use. Where they are available, certified reference materials should be used to determine any systematic error (bias), or where this is not possible, results should be compared with those obtained by other techniques, preferably based on different principles. Estimation of uncertainty must be a part of this validation process, and is essential for ongoing quality control. Advice on method validation is given in the corresponding CITAC Guide.

Analytical chemists have a professional duty to ensure that the results they produce are demonstrably fit for the purpose for which they are going to be used. There is a clear responsibility on the part of the test laboratory and its staff to justify the customers 'or data users' trust by providing the right answer to the analytical problem, with results which can be controlled by comparisons made by others.

To ensure that the result of an analysis is fit for its intended purpose it is necessary that the truth of the result is sufficient for any decision based on the result to be taken with confidence. This means that the performance of the method has to be validated and the uncertainty in the results has to be given, so that the user can be assured of the degree of confidence that can be placed on the result. To achieve this, the uncertainty should be evaluated and quoted in a way that is widely accepted, internally consistent, and easy to interpret.

Expressing results

Measurement results should be expressed in such a way that the data user can readily understand them and draw appropriate conclusions. The meaning of the test results should not be distorted by the reporting process. The overall uncertainty (reliability) has to be given (directly or on demand) after its evaluation, or at least estimated according the description above.

Expressed results should include the average measured value and the standard deviation of the statistically random error. In expressing the uncertainty which includes any bias (systematic errors), the number of measurements and the standard deviation should be stated.

Documentation

The reporting process should provide sufficient information to clarify claims made for the results and to allow an overall a posteriori quality assessment by experts, and possibly a repetitive analysis including sampling. Any new methodology must be described in detail, allowing one to follow the procedural practice in a straightforward manner.

1.3.3 Conclusion

This example of the requirements for a typical modern chemical analysis demonstrates to the student how a laboratory can define unit operations leading to unit processes and develop a standard operation procedure (SOP) for this defined sequence of operator actions. It should be mentioned that although each unit operation can be validated for the specific purpose in a separate validation effort, the validation of the final result is sufficient, because the combination of unit operations logically includes the perfection of all single steps involved. However, if the latter is not possible because of a lack of certified or local standard reference material with a similar matrix, and a different combination of unit steps including the final measuring process is not available, the proven validation of each unit process may be the best practice available. Therefore, the unit process approach is a generic device allowing plausible quality judgment and saving a lot of time and effort if they are automatically validated by the overall SOP.

Finally, it is important to underline that this section is not only meant to introduce into the complex structure of modern analytical chemistry schemes in nonroutine (research) laboratories and the absolute necessity to ensure the production of results comparable at an international level. By anticipating issues covered in more detail later in this book, its role is also to stimulate the student's interest in working through the many technical chapters in the light of the great importance of precise, accurate and traceable measurements in analytical chemistry for the benefit of our technology-based society.

References

[1.3-1] Kellner, R. et al., *Fresenius' J. Anal. Chem.* 1997, 357:197.

[1.3-2] Kellner, R. et al., *Anal. Chem.* 1994, 66:98A.

2 The Analytical Process

Learning objectives

■ To discuss the role of analytical chemistry in the solving of problems defined by clients
■ To stress the importance of interaction between client and analyst
■ To discuss the various steps in the total analytical process
■ To specify the basic concepts for characterizing the performance of analytical methods

2.1 Introduction

The objects or test samples to be investigated by methods of analytical chemistry comprise a very broad scope. For instance:

- samples of raw materials and finished products in a (chemical) production process
- soil samples for establishing the need for fertilizers
- freshwater samples to assess the degree of pollution
- tapwater to ascertain its purity
- metal assays for determining the commercial value of ores
- blood samples to determine the cholesterol content
- an ancient painting where the presence of certain pigments may give an indication of its age and origin

These examples serve as an illustration of the very broad scope of analytical chemistry, but also makes clear that we need a common language.

The subject of investigation is called *bulk matter* or *object*. The part that is taken from it for investigation is the *sample*, and the compounds or elements of interest are the *analytes*. They are embedded in the *matrix*, which is not primarily in the interest of investigation but may influence it. This influence is called the *matrix effect*. In order to get reliable results we should also compare the analytical results with that of a *blank sample*.

The analysis of objects or test samples must be performed with a clear objective. The problem is usually formulated by others, in this context further denoted as "clients". In the above examples, the clients are the process engineer, the farmer, a water control authority, a steel-producing industry, a physician, or a museum director, respectively. Very often the general problems are not put in analytical terms such as total composition of a material, content of specific compound in a mixture, or structure of a component. The first task for an analytical chemist is therefore to collect the information necessary to reformulate the problem in analytically manageable statements. It is necessary for a dialog to take place between the client and the analyst in which the objective of the investigation is agreed upon.

Useful analytical results can be obtained only when the general problem that has to be solved is clearly defined.

The tool kit of the analytical chemist is so extensive today that the selection of the appropriate method of analysis is often a difficult task that requires a lot of experience and often a touch of intuition. In general there are no "best" methods for solving analytical problems: each individual problem will ask for a specific approach. However, what can be learned is a feeling for the systematic development of analytical procedures based on a sound knowledge of the most important analytical techniques.

For this purpose, on the one hand, the performance of analytical techniques has to be characterized in an unbiased objective way. The most important parameters for this *performance characteristic* will be discussed in Sec. 2.3.

Performance characteristics should give an unambiguous and unbiased description of analytical techniques.

On the other hand, the problem to be solved should be objectified and described in terms that may have a direct relation to the performance characteristics of the methods (problem description or problem profile).

Table 2.2-1. Scheme of the total analytical process

General statement of problem	Client	Oil spill next to a subsurface soil bank in a garden district
Specific analytical statement of problem and definition of objective	Client ↔ analyst	How far-reaching is the contamination of the soil by mineral oils
Selection of procedure	Analyst	Extraction of the soil, separation, and quantification of the oil
Sampling	Client + analyst	Testing solid samples of representative size (100 g)
Sample preparation	Analyst	Homogenize and take subsamples, and extract with CCl_4 in soxhlet
Measurement	Analyst	Gas chromatography (GC) of an aliquot of the extract
Evaluation of data	Analyst	Identification and quantification of the GC-peaks
Conclusions	Analyst	Are the GC values related to the sample weights below the tolerable limit?
Report	Analyst ↔ client	Recommendations for further action to solve the problem

In practice, other constraints play an important role as well, such as availability of equipment, expertise, and presence of trained personnel. Finding an optimal method of analysis means finding the method or procedure with the best match between performance characteristic and objective problem profile, taking into account the other constraints.

2.2 The Total Analytical Process

The total analytical process can be presented in a flow scheme in which the inter-active role of the client and the analyst is indicated (Table 2.2-1).

All the steps in the scheme affect the final result. Therefore a short discussion of the subsequent steps seems to be in order.

2.2.1 General definition of problem

As indicated above, the general problem will often be related to an issue which is beyond the field of expertise of the analytical chemist (analyst) and often even outside the chemical domain. Moreover the client does not always fully appreciate the opportunities and technological aspects of analytical chemistry. In an intensive discussion between the analyst and the client it should be made clear what is the real basis of the analysis before starting the problem solving procedure.

2.2.2 Specific analytical statement of problem

Client and analyst have to define together the object of study: what *has to be* analyzed and what *can be* analyzed. At this stage, questions have to be answered like:

- what is the nature of the object or sample, and is it the elemental or molecular composition, or is it the presence of functional groups one is interested in?
- Does one need a quantitative analysis or will a qualitative analysis do? What is the accuracy required in the case of a quantitative analysis?
- How much material is available for the analysis, and what is the approximate concentration of the compound of interest (the analyte)?

- What is the gross or "matrix" composition?
- Is a single component analysis sufficient or is it desirable to do a multi-component analysis?
- How much time is available between supplying the sample and the time at which the result has to be available, or in other words: what is the acceptable delay time (waiting time plus analysis time)?
- Will there be a recurrent supply of samples of this kind and, if so, with what frequency; or is it a unique analysis?
- Is it desirable to look for a continuous monitoring system and/or to look for a fully automated system?
- How crucial is the dependability of the analytical system or method?
- Can destruction of the object of study be tolerated, or should the integrity of the sample be preserved?

This survey of questions is by no means exhaustive and only aims at an impression of the kind of dialog needed to get the proper information in order to reach a proper choice of the analytical method.

2.2.3 Selection of procedure

The selection of procedure is of crucial importance. It determines the cost of analysis both in terms of instrumental effort and personnel. But it is also influenced by framework conditions, such as size of sample, time availability, and the information content one gets out of the analytical investigation.

2.2.4 Sampling

Students in classroom experiments are usually faced with homogeneous samples; so they tend to underestimate the importance of the sampling process which is really a crucial step in any analytical procedure. [2.2-1]. In effect, the reliability of an analytical result is often conditioned by the quality of the sample. Sometimes a whole object can be investigated, e.g., an antique ring, by a nondestructive method like X-ray fluorescence, but mostly samples have to be taken, e.g., to assess the iron content of ore transported in a bulk carrier ship. In all cases two operations should be distinguished: (a) the definition of a sampling plan and (b) the taking of the samples as such. An analytical operator should never start an analytical procedure without knowledge of the history of the sample (sampling process, storage of sample, preservation, pretreatment, etc.) and of its representativity of the bulk material to which the analytical results will be related. On the basis of the nature of sampling, the type of analyte, its concentration level, and the chemical composition of the matrix, all the necessary precautions must be taken to avoid any alteration of the composition of the sample.

Solid materials. The samples have to be representative, but for economical reasons should not be larger than strictly necessary. The sample size depends on the precision required, on the heterogeneity of the material, and on the grain size. Sometimes samples of several kilograms or even several hundreds of kilograms have to be taken to meet the demand of representativeness. Such large samples require grinding, sieving, and homogenization before a suitable lab sample is obtained. During grinding, heat is evolved which may cause a loss of volatile components. Exposure of the newly formed grain boundaries to the air may lead to oxidation, thus affecting the valence state of certain elements, e.g., $Fe(II)$ and $Fe(III)$.

Liquids and gases. Liquids and gases are usually sufficiently homogeneous or can be easily homogenized, and small samples will normally do. Because small sample vessels have a large surface-to-volume ratio, losses by adsorption may be relatively large. Thorough rinsing of the vessel with the liquid or gas to be analyzed is therefore a prerequisite for establishing an equilibrium between the sample and the walls of the vessel (this process is called "*equilibration*").

The interaction between test sample and sample container may affect the true composition of the test sample, therefore the walls of the container have to be preconditioned in such a way that no changes in the test sample will occur.

Fig. 2.2-1. Pathways from sample to instrument

1. Direct introduction of sample into the instrument
2. Transfer from gas to liquid: condensation or gas phase extraction
3. Transfer from gas to solid: condensation
4. Transfer from liquid to gas: evaporation
5. Direct introduction into the instrument, solvent–solvent extraction
6. Transfer of liquid into a solid: precipitation, evaporation of solvent, freeze dry.
7. Transfer of a solid into a gas: evaporation
8. Transfer of a solid into a liquid: dissolution, digestion
9. Direct introduction into the instrument

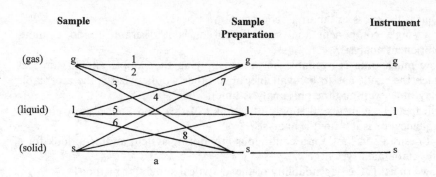

Description of sampling tools is outside the scope of this chapter; specialized publications must be consulted to find the best ways to sample gases, liquids, and solids. Here attention is drawn only to the fact that special care must be taken to ensure that no contamination of the sample occurs from sampling devices and storage containers. Samples should be clearly labeled including information such as source, date and time of sampling, and component to be determined. Specific sampling protocols, if they exist, must be consulted. Finally, it must be emphasized that in some cases sampling can be an extremely hazardous operation and adequate safety procedures should be adopted.

2.2.5 Sample transport and storage

During sample transport the composition of the sample may change due to reactions induced by the vigorous shaking of the sample vessel. The sample may also be brought to the analyzing site by transporting it through tubes. The gas may diffuse through the tubing material thereby depleting or contaminating the sample.

A certain time may elapse between sampling and sample investigation, e.g, ice samples taken at the Antarctic are often analyzed in another part of the world. Again trace compounds may get lost due to absorption on the container walls, or trace compounds from the walls may contaminate the stored sample.

2.2.6 Sample preparation

There are several aspects to sample preparation. One important issue is the adaption of the state of the sample to that accepted by the applied analytical technique (see Fig. 2.2-1). Another point of consideration is to adapt the sample to the optimum concentration range of the chosen technique. In the investigation of liquid samples the most commonly used sample preparation step is the dilution or the enrichment of trace compounds. Other techniques require the careful cleaning of a solid surface.

In the case of solid materials, decomposition or simple dissolution of the sample often has to be performed. In other cases the material must be made electrically conductive. Mixing with graphite or depositing gold by means of sputtering techniques are commonly used techniques.

Separation is often indispensable to eliminate the effect of accompanying components.

The conversion of an accompanying component in another form might help to eliminate the interfering effect of this component.

In trace analysis, concentrations of the analyte are often so low that no method is available for its direct determination [2.2-2]. Methods for enrichment can be applied to increase the fraction of the analyte in the sample. Techniques based on extraction, adsorption, and ion exchange are the most common ones. In general these separation methods can also be used to eliminate compounds that interfere in the determination. Suppression of the contribution of some unwanted components to the measurement result can sometimes be accomplished by the addition of reagents that bind the interferent. This is known as "masking".

2.2.7 Measurement/determination

Most instrumental techniques are typically based on comparison of the signal of a sample of unknown composition with the signal produced by a (series of) standard(s) of exactly known composition: the instrument has to be *calibrated*. In this way the instrument-specific factors can be accounted for. In Sec. 2.4 and in Chap. 12, the statistical aspects of measurement results will be dealt with in more detail. Here it suffices to note that many instruments exhibit unwanted fluctuations in the signal, known as *noise*. Noise is usually described by the standard deviation (see Sec. 2.4) of the signal fluctuation. Optimization of the signal-to-noise ratio is obtained through data acquisition and data processing.

2.2.8 Evaluation of data

Most state-of-the-art instruments are computerized nowadays. This means that the mathematical processing of the data to transfer them in analytically relevant terms, such as concentrations or structural information of the analyte, is an integral part of the analytical system. This automation implies that the analyst doesn't have a direct check on or even access to the raw data. Therefore it is of paramount importance that the software used is properly validated (see Chap. 12).

Computer-based methods are indispensable for the identification of compounds, e.g., on the basis of pattern recognition. Such patterns may be derived from infrared or mass spectra (to be discussed in Secs. 9.2 and 9.3). Procedures based on the comparison of spectra of unknown compounds and spectra of pure compounds stored in the memory of the computer are called "retrieval" or "library search" procedures (see Chap. 13).

Validation of both analytical methods and the software applied is a crucial factor for success in analytical chemistry.

2.2.9 Conclusions and report

The analyst should take full responsibility for the analytical results he/she is going to report. He/she should clearly indicate what reliability can be attributed to the figures and avoid any ambiguity in reporting data. To warrant the quality of the results (see Chap. 3) it is always important to test procedures on certified samples (standards of known composition analyzed by various methods in various laboratories). If such certified samples are not available, results obtained with a proposed procedure can be compared with those obtained with other procedures, preferably accepted reference methods. The analyst should be aware of the fact that his/her results will usually play a role in another context, i.e., the general problem defined by the client.

The conclusions to be drawn on the basis of the analytical results should be the responsibility of the analyst.

Quality of the results must be clearly reported and care should be taken that the client does not jump to unjustified conclusions.

2.3 Performance Characteristics

An analytical method can be characterized by means of criteria based on the "quality" of the analytical result and criteria of more economic origin (not discussed here).

Criteria related to quality of the analytical result	Criteria related to economic aspects
sensitivity	costs of investment
precision	duration of analysis with regard to personnel costs
bias	special safety precautions
accuracy	costs of installation, housing
limit of detection	degree of training of operator because of salary
limit of determination	sample throughput
selectivity	costs of reagents
dynamic range	

Further important characteristics are the kind of required pretreatment of samples; the question whether the method requires destruction of the sample or not; possibility of multicomponent analysis; etc.

Worldwide agreement on the usage of terms forms the basis for proper mutual understanding.

accurate and precise

inaccurate
but precise

accurate
but not precise

inaccurate
and not precise

Fig. 2.3-1. Illustration of the terms accuracy and precision

In the following section the terms mentioned above will be defined. Unfortunately there is not always unanimity in the usage of terms. In this book terms in accordance with recommendations from the International Organization for Standardization (ISO) or the International Union of Pure and Applied Chemistry (IUPAC) are used.

Accuracy
Accuracy is the agreement between the average of the measured values and the accepted reference value (see Sec. 2.4)

Precision
Precision is the scatter of the measured values around the average value (see Sec. 2.4 and also Fig. 2.3-1)

Sensitivity

Sensitivity S is defined as the change in signal Y on change of concentration ($S = \mathrm{d}Y/\mathrm{d}c$) (see Sec. 12.2).

Limit of detection

For an analytical result which is close to the value of the blank, the question arises whether the value corresponds to random values of the blank or to the real presence of analyte. The background signal is that produced by the blank and exhibits noise. The limit of detection corresponds to a signal equal to k times the standard deviation of the background noise (i.e the signal-to-noise ratio is equal to k. Typically k has the value 3; see also Sec. 12.1.) Values above the detection limit can be attributed to the presence of analyte, values below the limit are considered to be indicative for the absence of analyte in detectable amounts.

Limit of determination

For quantitative analysis it must be absolutely clear that only values are used in the further evaluation of data (e.g., averaging of individual results) that can be attributed to the analyte (see also Sec. 12.1). Therefore the condition is more strict; the limit of determination is always higher than the limit of detection.

Selectivity

Selectivity gives an indication of how strongly the result is affected by other components in the sample. In various methods different factors are used to assess this selectivity in a quantitative way. In the chapters concerned this will be indicated (see, e.g., Secs. 5.1 and 7.3). Sometimes the term specificity is used. This suggests that no compound other than the analyte contributes to the result. Hardly any method, except when using so-called hyphenated techniques (coupling of two instruments), is that specific and, in general, the term should be avoided.

2.4 Errors in Analytical Chemistry

Learning objectives

- To provide basic knowledge about: type and sources of errors in chemical analysis, and their propagation
- To stress the significance of bias and accuracy in chemical analysis
- To highlight the meaning of terms such as repeatability, reproducibility, laboratory bias, and method bias
- To provide the basic information necessary for correctly reporting analytical data

2.4.1 Accuracy and precision

Experimentalists are aware that any measurement they make is affected by errors which can seriously affect the experimental result. The actions adopted to reduce these errors as much as possible will depend on the nature of the errors themselves. A fundamental distinction is usually done between three type of errors: *gross, random,* and *systematic*. [2.4-1]

Gross errors do not fit into the usual pattern of errors associated with a particular situation; they should normally be absent but unfortunately they do occasionally occur. Examples are an instrument failure, a macroscopic contamination of a reagent, accidental loss of sample, and so on. When a gross error occurs and it is recognized, there are few alternatives to stop the procedure and start again.

The following discussion will only deal with random and systematic errors; the most relevant features of these two types of errors are summarized and compared in Table 2.4-1.

Accuracy and *precision* are among the most important criteria for defining the quality of an analytical method. Usually precision is evaluated first because systematic errors (affecting accuracy) can be quantified only when the magnitude of random errors is known.

Random errors can be estimated by carrying out replicate measurements; for example, if $X_1, X_2, \ldots X_n$ are the volumes of the standard solution consumed in the case of *n independent* replicate titrations, the given values usually differ from each other because of unavoidable errors occurring in the measuring procedure. The simplest operation one can do is to average the experimental data by calculating, for example, the *(arithmetic) mean*:

Mean, arithmetic mean and *average* are synonyms and may serve as the central value for a set of "*n*" replicate measurements.

$$\bar{X} = \sum_i \frac{X_i}{n} \qquad (2.4\text{-}1)$$

This mean value is considered the best estimate of the true (unknown) value for the measured quantity. The central value in a set of replicate measurements can also be expressed in terms of the *median*. When the replicate data are arranged in order of size the *median* is the data value located halfway between the smallest and the largest value. In the case of an odd number of measurements the *median* is represented by the central value; for instance, the *median* of the following set of 9 results:

In many cases the *median* may be a more realistic measure of the "*central tendency*" than the *mean*.

10.10; 10.20; 10.40; 10.46; 10.50; 10.54; 10.60; 10.80; 10.90

is given by 10.50 – the same as the *mean*. (Note that for a symmetrical distribution – see later – the central location is given by the *median* or the *mean* which, in this particular case only, are identical). For an even number of measurements the *median* is taken as the arithmetic mean of the middle pair; by adding the result 12.80 to the above set

10.10; 10.20; 10.40; 10.46; 10.50; 10.54; 10.60; 10.80; 10.90; **12.80**

Table 2.4-1. Summary of the most salient features of random and systematic errors

Random (or indeterminate) errors
1. Sources include personal, instrumental and methodological uncertainties
2. Not eliminable but reducible by careful working
3. Recognizable through the scatter around the mean
4. Affect precision**
5. Quantifiable through a measure of precision (e.g., the standard deviation)

Systematic (or determinate) errors
1. Sources include personal, instrumental and methodological bias
2. In principle recognizable and reducible (partially or even completely)
3. Recognizable by the lack of agreement between the mean and the true value*
4. Affect accuracy**
5. Quantifiable by the difference between the mean and the true value***

* Valid if the random error is not so large as to obscure any bias.
** Discussed in the next paragraph
*** Valid if a sufficiently high number of measurements are performed to permit a good estimate of the mean.

The *mean* 10.73 is greater than any of the three closely grouped values 10.46, 10.50, and 10.54 and is clearly a less realistic indication of the *central value* than the *median 10.52*. This because in the particular case 12.80 is an "*outlier*".

gives a *median* of $(10.50 + 10.54)/2 = 10.52$ compared to a *mean* of 10.73. The *median* can be used advantageously in a set of data eventually containing an *outlier*, that is to say a result that differs significantly from the rest of the data in the set. As will be shown in more details in Sec. 12.1 (where a more rigorous treatment of outliers is given) an outlying result can affect significantly the *mean* (and the standard deviation) but has no effect on the *median*.

The X_i values in a "*set*" of *size* "*n*", will appear more or less spread around the *mean* value depending on how careful the analytical operator was and how well established was the analytical method adopted.

The closeness of the replicate measurements in a set, i.e., the spreading of the data about the central value (the *mean* $\bar{X}$) is defined as *precision*. The following measures of *precision* within a small set of *n* measurements are widely used:

$$\text{sample standard deviation} \quad s = \left\{ \frac{\sum_i (X_i - \bar{X})^2}{(n-1)} \right\}^{1/2} \tag{2.4-2}$$

$$\text{relative standard deviation} \quad RSD = \frac{s}{\bar{X}} \tag{2.4-3}$$

RSD% is also known as the *Coefficient of Variation* (CV).

$$\text{relative standard deviation \%} \quad RSD\% = \left(\frac{s}{\bar{X}} \right) 100 \tag{2.4-4}$$

$$\text{variance} \quad V = s^2 \tag{2.4-5}$$

In a set of n independent observations there are *n degrees of freedom*. In calculating the mean $\bar{X}$, one *degree of freedom* is used because, with their sign retained, the sum of the individual deviations, or *residuals*, $(X_i - \bar{X})$ must sum to zero. Thus when $(n-1)$ *residuals* have been computed the last one is strictly determined. This means that only $(n-1)$ *residuals* provide an independent measure of the *precision* of the set.

In Eq. 2.4-2 the quantity $(n-1)$ represents the number of degrees of freedom. Other quantities are also defined although they are not to be recommended as measures of precision:

$$\text{mean deviation} \quad d = \frac{\sum_i |X_i - \bar{X}|}{n} \tag{2.4-6}$$

$$\text{relative mean deviation} \quad RMD = \frac{d}{\bar{X}} \tag{2.4-7}$$

$$\text{relative mean deviation \%} \quad RMD\% = \left(\frac{d}{\bar{X}} \right) 100 \tag{2.4-8}$$

The terms "*reproducible*" or "*reproducibility*" are used to indicate the spreading of the results of replicate determinations; so a statement such as "results are highly (poorly) reproducible" means that the scatter of the results around the mean is low (high) so that a low (high) value of *s* or *d* and a high (or low) precision are expected.

Repeatability and reproducibility are different measures of precision

Experience indicates that precision can vary when the replicates are performed in different laboratories or even in the same laboratory in different lapses of time e.g., few days instead of few hours. Then it is necessary to distinguish e.g., between the *within-laboratory precision* (*repeatability*) and the *between-laboratory precision* (*reproducibility*).

According to *ISO* recommendations, *repeatability* indicates the closeness of individual results obtained by the same method on identical test material under the same conditions (operator, apparatus, laboratory) in a short lapse of time. Analogously *reproducibility* indicates the closeness of individual results obtained with the same method on identical test material but under different conditions (e.g., different laboratories, operators, equipments and times). Obviously *repeatability* and *reproducibility* are similar concepts but referred to homogeneous and heterogeneous working conditions, respectively.

Precision can also be seen as the sum of two contributions: random within-laboratories errors and unidentified, normally distributed, systematic errors in individual laboratories i.e., laboratory bias – *vide infra*; according to this view precision has the same meaning of reproducibility as defined above with repeatability as a component. From the above definitions it follows that for a given method, *repeatability* is always better than *reproducibility*.

2.4.2 Bias and accuracy

As seen above, even replicate determinations give different results: X_i. If T is the true value (usually unknown), the difference:

$$E_i = X_i - T \qquad (2.4\text{-}9)$$

is referred to as the *error* of X_i. If a sufficient number of replicates is performed by a given analytical method, a "*stable*" mean, $\overline{X}$, is obtained which is an estimate of the mean μ of an unlimited number of determinations. The difference $\overline{X} - T$ is called bias or *systematic error*. Note that the bias so obtained is only an estimate of the true bias since it is calculated by using $\overline{X}$ which is an estimate itself.

E_i can be rewritten as:

$$E_i = (X_i - \overline{X}) + (\overline{X} - T)$$
$$= E_{\text{random}} + B\text{ias} \qquad (2.4\text{-}10)$$

Bias is a constant or systematic error manifesting as a persistant positive or negative deviation of the mean from the accepted reference (*true*) value.

Systematic errors originate from *laboratory bias* and/or *method bias*. The latter represents the systematic error proper of a particular method and will contribute to the error of any determination performed in all the laboratories using the given method. Then the accuracy in an *inter-laboratory test* is identical to the *method bias* (an *inter-laboratory test* is an exercise in which aliquots of the same sample are analyzed in different, randomly chosen, laboratories by using the same analytical procedure); in Eq. 2.4-10 the mean $\overline{X}$ is the average of the X_i results from the different laboratories (Fig. 2.4-1). *For an individual laboratory the systematic error is the sum of the method bias and laboratory bias.* This quantity is also referred to as *accuracy*. It follows that the meaning of the term accuracy is not univocally stated. Consider also that accuracy is used to indicate both the difference $\overline{X} - T$ (the *absolute accuracy of the mean*) and the difference $X_i - T$ (the *absolute accuracy of an individual measurement*); according to the first definition, *accuracy is a synonym of systematic error*, i.e., method bias plus laboratory bias, while according to the second one, which consider an individual results, indicates a *combination of systematic and random errors*.

From the above it follows that the term *accuracy* can be used only in a colloquial way, a rigorous language requiring the use of unambiguous terms such as *laboratory bias*, *method bias* and *total error* in the case of a combination of method bias and random error.

Error is the sum of the systematic and random parts. If there is only one observation

$$E = X - T.$$

Lack of bias is called *trueness*;
Lack of error is called *accuracy*.

An "*accuracy*" measurement is one that is precise and, at the same time, bias-free.

The term "*accuracy*" should preferably be used not to identify a quantity but to give an indication of the degree to which the requirements of an "*accurate*" measurement can be attained.

2.4.3 Reporting analytical data

As stressed before, an experimental result is useless if it is not accompanied by an estimate of the errors involved in its measurement. An excellent, highly recommendable, approach is to report the *mean* $\overline{X}$ together with the (e.g. 95%) confidence limits $(\overline{X} \pm ts/\sqrt{n})$ as will be described in full details in Sec. 12.1. Note

Non-parametric statistics makes no assumption at all about distribution of data. However it is more concerned with significance testing (see Sec. 12.1) than with estimation.

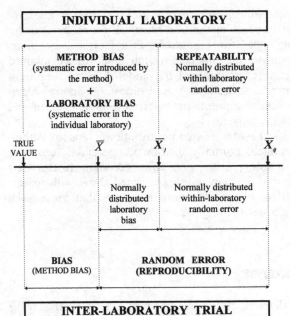

INDIVIDUAL LABORATORY

METHOD BIAS (systematic error introduced by the method)	REPEATABILITY Normally distributed within laboratory random error
+	
LABORATORY BIAS (systematic error in the individual laboratory)	

TRUE VALUE $\overline{X}$ $\overline{X}_j$ $\overline{X}_{ij}$

	Normally distributed laboratory bias	Normally distributed within-laboratory random error

BIAS (METHOD BIAS)	RANDOM ERROR (REPRODUCIBILITY)

INTER-LABORATORY TRIAL

Fig. 2.4-1. Error partitioning scheme: the total error associated with X_{ij}, the ith measurement performed in the jth laboratory is the sum of the *within-laboratory random error*, the *laboratory bias* and the *method bias*. From the point of view of an individual laboratory, accuracy is related to the total bias which is the sum of method bias and laboratory bias. The within-laboratory random error determines the within laboratory precision (*repeatability*). The *laboratory bias*, a component of the systematic error in a within-laboratory exercise, *is assumed normally distributed in an inter-laboratory trial* and then contributes to the inter-laboratory precision (*reproducibility*) which has "*repeatability*" as a component. $\overline{X}_j$ is the mean in the jth laboratory; $\overline{X}$ is the mean in the interlaboratory trial

The median is a robust estimator of the central tendency of a set of analytical data

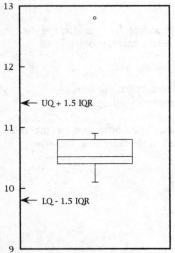

Fig. 2.4-2. Whisker and box plot representation of a set of analytical measurements e.g. 10.10; 10.20; 10.40; 10.46; 10.50; 10.54; 10.60; 10.80; 10.90; 12.80. The *box* encloses 50% of data (i.e., the *interquartile range IQR*) with the *median* displayed as a line. The vertical lines extending outside the box, marks the minimum and maximum values which fall within an acceptable range: e.g., $(UQ + 1.5\ IQR)$ and $(LQ - 1.5\ IQR)$ where UQ and LQ are the upper and lower quartiles. Any values outside this range will be considered an "*outlier*" and is displayed as an individual point

that the above formulation requires a "*normal distribution*" (see Sec. 12.1) of the original data (*classical statistics*); furthermore, an eventual outlier must be recognized by proper tests and removed before extracting the sample mean and the sample standard deviation used in the computation of the confidence interval.

Analytical data often conform only broadly to a normal distribution (e.g., they still possess a unimodal and symmetrical distribution but with severe tailing) and/ or are contaminated by outliers. In such a case (*robust statistics*), the so called "*whisker and box plot*" representation can be used (Fig. 2.4-2). *Median* is used instead of *mean* and, e.g., the *interquartile range* replaces the *sample standard deviation s* as a measure of dispersion. As we have seen, the *median* of measurements set is the data value located halfway between the smallest and the largest values; so the data value located halfway between the *median* and the highest (lowest) value is called *upper* (*lower*) quartile. The *interquartile range* (IQR) is then simply the difference between the upper (UQ) and lower (LQ) quartiles. Another possibility is to report the mean, $\overline{X}$, of n replicates as the estimate of the measured quantity, and the standard deviation, s, as an estimate of the precision; according to this approach the result of the following five replicates, e.g., readings of a 50 mL burette with a 0.1 mL graduated scale:

$$19.14, 19.13, 19.09, 19.21, 19.08$$

should be quoted as $\overline{X} \pm s = 19.11 \pm 0.02$ ($n = 10$). A less satisfactory but often used approach is that known as the "*significant figure convention*". According to this approach the number of significant figures quoted is used, instead of a specific estimate, to indicate the precision of the result; the convention adopted is to quote only the *significant figures*, i.e., all the digits known with certainty plus the first uncertain digit. So each of the burette readings above is correctly reported with four significant figures.

To become more familiar with the meaning of "significant figures" note that the following numbers: 1537, 153.7, 15.37, and 1.537, contain four significant figures. The numbers 1537 and 1.537×10^3 also contain both four significant figures. All the followings numbers 0.001075, 0.01075, 0.1075, 1.075, 10.75, 107.5, 1075, 1.075×10^{-3}, and 10.75×10^{-2}, contain four significant figures. In other words:

- *zeros before the decimal point are not significant*
- *zeros between nonzero digits are significant*

However, *terminal zeros may or may not be significant*. Consider the following record obtained in the measurement of a potential difference: 1750 mV; the situation is now ambiguous because we do not really know if we are dealing with a measuring device that yields 1.75×10^3 mV (three significant figures) or $1.750 \times$

10^3 mV (four significant figures, i.e., a voltmeter with a precision in the order of the millivolt). As a further example consider the different way of quoting the volume of a one litre graduated flask (which typically is given at ±0.4 mL); one could write for example 1.0 L (two significant figures) which, however, means that the volume is known to a few tenths of a litre which is not true. On the other hand if one write 1000 mL the situation is equally confused because it is impossible to decide (without any previous knowledge of the matter) how many zeros are significant; the best way to avoid confusion is to use the scientific notation writing down 1.0×10^3 mL which correctly tell us that the volume is known to a few tenths of a millilitre. The message coming out is twofold:

- *final zeros are not significant unless they follow a decimal point*
- *whenever possible report the result as a number* (containing the right number of significant figures) *expressed as a power of ten*

Care must be exercised in reporting the correct number of significant figures during numerical computation. For example averaging the burette readings given above, one should quote the mean as 19.11 since, as already discussed, all readings are uncertain to the second decimal place (this is also clearly indicated by the magnitude of the standard deviation). However especially when dealing with results which are (or could be) involved in further calculations it is better to avoid an excess of rounding off; so in order to avoid loss of information one can write one digit more than the last significant figure reporting the result as $\bar{X} \pm s = 19.11_2 \pm 0.02_6$ ($n = 10$) and confining the rounding off procedure only at the last stage of calculation.

The above example is also instructive from another point of view. Adopting the "*significant figure convention*" the average of our burette readings is quoted as $\bar{X} = 19.11$; a mistake is often done at the stage of data interpretation since most people assume that the uncertainty in the uncertain figure is ±1. This is of course not true! It must be clear that the "*significant figure convention*" can only indicate the uncertain figure but cannot quantitate the uncertainty; for this purpose we clearly need an estimate of the precision such as the sample standard deviation ($s = 0.02_6$ in the specific case).

Rules of thumb for deciding the number of significant figures in the course of a numerical calculation are summarized in the following.

- *Addition or subtraction*: significant figures after the decimal place not higher than those of that number (being added or subtracted) having the fewest significant figures after the decimal point.

Worked example:

Calculate the molar mass (MW) of HNO_3; atomic masses are: H: 1.00797; N: 14.0067; O: 15.9994.

$$MW = 1.00797 + 14.0067 + 47.9982 = 63.01287 \approx 63.0129 \text{ g mol}^{-1}$$

- *Multiplication or division*: number of significant figures equal to the smallest number carried by any of the value being multiplied and/or divided.

Worked example:

Calculate the molar concentration (M) of a 70% (w/w) HNO_3 solution whose density is 1.413 kg L^{-1}.

$$M = 1.413 \text{ Kg L}^{-1} \; 0.70 \text{ g g}^{-1} \; 1000 \text{ g Kg}^{-1}/63.0129 \text{ g mol}^{-1} = 15.6967 \approx 16 \text{ mol L}^{-1}$$

- *Taking logarithms*: quote the logarithm with the mantissa having as many figures as the significant figures in the original number.

Worked example:

What is the pH of a 1.9×10^{-2} mol L^{-1} HNO_3 solution?

$$pH = -\log[H^+] = -\log 1.9 \times 10^{-2} = 1.7212 \approx 1.72$$

• *Taking anti-logarithms* (*exponentiation*): quote a number of significant figures equal to the number of digits in the mantissa of the logarithm.

Worked example:

What is the $[H^+]$ of a solution whose pH is 4.75?

$$[H^+] = 10^{-4.75} = 1.7782 \times 10^{-5} \approx 1.78 \times 10^{-5}\, mol\, L^{-1}$$

2.4.4   Error propagation

Often the quantity of interest is calculated by combining several *independently* measured parameters each affected by errors which will contribute to the error associated to the final, calculated result. In other words errors are *propagated* in a way which depends on the relationship between the measured and calculated quantities. The procedures used for combining random and systematic errors are different because of the different nature of the errors themselves. Tables 2.4-2 and 2.4-3 summarise the propagation rules for random and systematic errors.

As can be seen from Table 2.4-2 in the case of a linear combination variances are additive; this implies that the largest contributing standard deviation can dominates so that any attempt to reduce σ_y by reducing the other errors can be useless. When calculation involves a difference, variances are still additive so that small relative errors in the measured quantities can be amplified in the difference. Suppose for example that in a three-component X, Z, Y mixture the % content of X and Z has been determined, by two independent methods, to be 80% and 15% respectively with a relative error on both of +2%; it can be easily show that if Y is obtained by difference ($Y = 100 - X - Z$) its relative systematic error is about -38%.

Worked example:

A student is asked to prepare 1 litre of a NaCl (molar mass $58.443\,g\,mol^{-1}$) solution approximately 0.1 molar. He weighs out, by difference, $5.8970\,g$ of the salt; he knows that standard deviations associated to the balance and volumetric flask used are $0.0001\,g$ and $0.4\,mL$ respectively. **What is the relative standard deviation on the molarity of the resulting NaCl solution?**

Molarity (M) can be calculated as:
$M = m/(MW \times V) = 0.10090\,mol\,L^{-1}$ where m is the mass of NaCl, MW is the molar mass (which is assumed to be an exactly known constant) and V is the volume of the solution. According to *case 2* in Table 2.4-2 the relative standard deviation on M is given by:

$$\sigma_M/M = \{(\sigma_m/m)^2 + (\sigma_V/V)^2\}^{1/2}$$

Since the mass is obtained by difference, σ_m is calculated according to *case 1* in Table 2.4-2 i.e.

$$\sigma_m = (\sigma_i^2 + \sigma_f^2)^{1/2} = \{(10^{-4})^2 + (10^{-4})^2\}^{1/2} = 1.4 \times 10^{-4}g$$

then: $\sigma_m/m = 1.4 \times 10^{-4}/5.8970 = 2.3_7 \times 10^{-5}$
The *RSD* on the volume prepared is given by:

$$\sigma_V/V = 0.4/1000.0 = 4 \times 10^{-4} \text{ (or 0.04%)}.$$

then: $\sigma_M/M = \{(2.3_7 \times 10^{-5})^2 + (4 \times 10^{-4})^2\}^{1/2} = 4.01 \times 10^{-4}$
Finally we have:

$$\sigma_M = 4.01 \times 10^{-4} \times 0.10090 = 4.0 \times 10^{-5}$$

i.e., the uncertainty is confined to the 5th decimal place; so the student can write down the molarity as $(1.0090 \pm 0.0004) \times 10^{-1}\,mol\,L^{-1}$. As can be seen the random error associated to the preparation of a standard solution is really very small. Furthermore it can be seen (compare σ_M/M and σ_V/V) that the error originates mainly from the use of graduate flask; the use of a more precise and expensive balance is useless since, in practice, it will not improve the final precision.

Table 2.4-2. Propagation of random errors

Case	Calculated quantity	Random error
1	$y = K + K_A A + K_B B + \ldots$	$\sigma_y^2 = (K_A \sigma_A)^2 + (K_B \sigma_B)^2 + \ldots$
2*	$y = KAB/CD$	$(\sigma_y/y)^2 = (\sigma_A/A)^2 + (\sigma_B/B)^2 + (\sigma_C/C)^2 + (\sigma_D/D)^2$
3**	$y = f(x)$	$\sigma_y = \sigma_x \, dy/dx$

A, B, C, D are individually measured quantities (independent of each other); $K, K_A, K_B \ldots$ are constants whose values are assumed to be known exactly. y can also be calculated from estimates of these quantities; in the above equations $A, B, C, \ldots$ must be replaced by the means $\bar{A}, \bar{B}, \bar{C}, \ldots$ and $\sigma_y, \sigma_A, \sigma_B, \ldots$ by $s_{\bar{y}}, s_{\bar{A}}, s_{\bar{B}}, \ldots$

(*) note that in the case of a quantity raised to a power (e.g. $y = A^2$) the error cannot be calculated as for a multiplicative expression ($y = A \times A$) because the quantities involved are not independent. In the most general case of $y = KA^r$ it can be shown that $\sigma_y/y = r\sigma_A/A$

(**) In the more general case $y = f(x, z, t, \ldots)$ one has the following expression

$$\sigma_y^2 = (\partial f/\partial x)^2 \sigma_x^2 + (\partial f/\partial z)^2 \sigma_z^2 + (\partial f/\partial t)^2 \sigma_t^2$$

which can be used to derive formula for the propagation of random errors such as those of cases 1-3

For example in the case of a logarithmic relationship $y = K \ln x$ (or $y = K \log x$) one has:

$$\sigma_y = K \sigma_x/x \ (\text{or} \ \sigma_y = 0.434 \, K \, \sigma_x/x)$$

Table 2.4-3. Propagation of systematic errors

Case	Calculated quantity	Systematic error
1	$y = K + K_A A + K_B B + \ldots$	$\Delta y = K_A \Delta A + K_B \Delta B + \ldots$
2	$y = KAB/CD$	$\Delta y/y = (\Delta_A/A) + (\Delta_B/B) + (\Delta_C/C) + (\Delta_D/D)$
3*	$y = f(x)$	$\Delta_y = \Delta_x \, dy/dx$

A, B, C, D, are individually measured quantities (independent of each other); $K, K_A, K_B \ldots$ are constants whose values are assumed to be known exactly. $\Delta_A, \Delta_B, \Delta_C, \Delta_D$ are systematic errors in A, B, C, D. Systematic errors are either positive or negative and the sign must be included in the calculation of Δy. The total systematic error can be sometimes zero.

(*) for example in the case of a logarithmic relationship $y = k \log x$ one has:

$$\Delta y = (k/2.303)(\Delta x/x) \ \text{and} \ \Delta y/y = 0.434\Delta x/(x \log x)$$

Worked example:

The H^+ concentration of a solution is quoted $(1.5 \pm 0.1) \ 10^{-3}$ M. What is the standard deviation on the calculated pH?

$$pH = -\log[H^+] = -\log 1.5 \times 10^{-3} = 2.82$$

According to the propagation rules in Table II we have:

$$\sigma_{pH} = 0.434 \, K \, \sigma_{[H^+]}/[H^+] = 0.434 \times (0.1 \times 10^{-3})/1.5 \times 10^{-3} = 0.03$$

Note that in the above example $K = -1$ and that the sign has been dropped since the standard deviation is normally quoted as positive value. Reversing the calculation shown above one can, of course, derive the standard deviation on the $[H^+]$ calculated from a given pH. So if the pH of a solution is 4.70 ± 0.05 one obtain:

$$[H^+] = 10^{-4.70} = 1.7 \times 10^{-5} \ \text{and}$$

$$\sigma_{[H^+]} = 2.303\sigma_{pH}[H^+] = 2.0 \times 10^{-6}; \ \text{then:}$$

$$[H^+] = (1.7 \pm 0.2) \times 10^{-5} M$$

Worked example:

Calculate the error associated to the concentration of an absorbing solution whose transmittance T at a given wavelength is 0.565. The product of the optical pathlength b and the molar absorbivity ε (which are assumed exactly known) is $2.06 \times 10^4 \, dm^3 \, mol^{-1}$. The photometer used has an uncertainty of 0.003 transmittance units.

From the Lambert-Beer law one has

$-\log T = A = \varepsilon b C$ and then

$C = (-\varepsilon b)^{-1} \log T = 1.20 \times 10^{-5} \, mol \, dm^{-3}$

$\sigma_c = 0.434 \, K \, \sigma_T / T = 0.434 \, (-2.06 \times 10^4)^{-1} \, 0.003/0.565$

$\quad = 1.1 \times 10^{-7} \cong 1 \times 10^{-7}$ *(minus sign omitted!)*

then

$C = (1.20 \pm 0.01) \times 10^{-5} \, mol \, dm^{-3}$

The $RSD\%$ is $\sigma_C / C = 0.9\%$

References

[2.2-1] Gy, P.M., Sampling of Heterogeneous and Dynamic Material Systems, *Data Handling in Science and Technology*, Vol. 10, Amsterdam: Elsevier, 1992.

[2.2-2] Dunemann, L., Begerow, J., Bucholski, A., Sample Preparation for Trace Analysis, *Ullmann's Encyclopedia of Industrial Chemistry*; Vol. B5, 65–93. Weinheim: VCH Publishing Group, 1994

[2.4-1] Massart, D.L., Vandeginste, B.G.M., Deming, S.N., Michotte, Y, Kaufmann, L., *Chemometrics: a textbook*. Amsterdam: Elsevier, 1988.

General readings

Anderson R.L., "Practical Statistics for analytical chemistry" Van Nostrand Reinhold, New York, 1987.

Barnett, V., Lewis, T., "Outliers in statistical data" Wiley, New York, 2nd edn., 1984.

Conover, W.J., "Practical non parametric statistics", 2nd Edition, Wiley, New York, 1980.

Green J.R., Margerison D., "Statistical treatment of experimental data" Elsevier Sci. Publishing Co., Amsterdam 1978.

Hawkins, D.M., "Identification of outliers" Chapman & Hall, London, 1980.

Kratochvil, B., Taylor, J.K., "Sampling for chemical Analysis" Anal. Chem., 53 (1981) 924A.

Lindey, D.V., Scott, W.F., "New Cambridge Elementary Statistical Tables" Cambridge University Press, Cambridge, 1984.

Liteanu C., Rica, I., "Statistical Theory and methodology of trace analysis" Ellis Horwood Chichester 1980.

Massart, D.L., Dijkstra, A., Kaufman, L. "Evaluation and optimisation of laboratory methods and analytical procedures" Elsevier Sci. Publishing Co., Amsterdam 1978.

McCormick, D., Roach, A., Chapman, N.B. "Measurement, statistics and computation" John Wiley & Sons Chichester 1987.

Miller, J.C., Miller J.N. "Basic Statistical Methods for Analytical Chemistry. Part I. Statistics of repeated measurements. A review. Analyst 113 (1988) 1351.

Miller J.C., Miller J.N. "Statistics for Analytical Chemistry" Second Edition, Ellis Horwood Chichester 1988.

Miller, J.N., "Outliers in experimental data and their treatment" Analyst 118 (199) 455.

Neave, H.R., "Elementary statistics tables" George Allen & Unwin, London, 1981.

Warnimont, G.T., Spendley, W., "Use of Statistics to develop and evaluate analytical methods" AOAC, Arlington VA, USA, 1985.

Woodget B.W., Cooper D., Chapman, N.B. "Samples and standards" John Wiley & Sons Chichester 1987.

Youden W.J., Steiner E.H., "Statistical manual of the Association of Official Analytical Chemists" AOAC, Arlington VA, USA, 1975.

Questions and problems

1. Explain the difference between random and systematic errors.
2. The determination of nickel in a stainless steel specimen can be carried out gravimetrically using dimethylglyoxime as precipitating agent in a slightly alcaline solution; tartaric acid is used to mask Fe(III) which is the major interferent. The procedure is essentially composed of the following steps:
   - the sample containing Ni is *weighed* and dissolved in a suitable volume of 6 M HCl followed by the addition of a suitable volume of 6 M HNO_3 and boiling;
   - a suitable volume of an aqueous solution of tartaric acid, the masking agent, is then added;
   - a weighed amount of dimethylglioxime, the precipitating agent, is dissolved in ethanol and a suitable volume of the resulting solution is added, followed by addition of 6 M NH_3 until a slight excess is present;
   - the precipitated is digested for a convenient time and the solution allowed to cool;
   - the precipitate so formed is filtered trough a filtering crucible (previously heated at 110 °C to a constant weight) and washed until the washing solution is free from Cl^- ions;
   - the washed precipitated is dried at 110 °C to a constant weight and the necessary calculations, to find out the Ni content in the analyzed sample, are performed.

   Try to identify the category of errors which are likely to be introduced by each of the steps above described. Based on your personal knowledge (theoretical and, hopefully, also practical) of gravimetric methods of analysis try to identify those steps introducing the most significant errors.
3. What is meant by terms "*robust statistics*" and "*nonparametric statistics*"?
4. The following absorbance values were obtained in the course of a spectrophotometric determination of Fe:

   0.390; 0.380; 0.385; 0.381; 0.380; 0.370; 0.375;

   For the above set calculate:
   - median
   - mean
   - standard deviation
   - RSD%
   - the absolute error and the relative error (in parts per thousands) assuming that the true value for the absorbance reading is 0.370.
5. Explain the meaning of terms such as *bias, trueness, accuracy, method bias, laboratory bias*.
6. Explain the meaning of repeatability and reproducibility.
7. What is an *inter-laboratory test*? Why it needs to be performed?
8. For each of the following numerical calculations report the result with the right number of significant figures:
   - calculate the molar concentration of a 37% (w/w) HCl solution (molar mass 36.441 $g\,mol^{-1}$) whose density is 1.201 $kg\,L^{-1}$;
   - calculate the pH of a $2.5 \times 10^{-3}\,mol\,L^{-1}$ HCl solution;
   - calculate the H^+ concentration of a solution having a pH of 2.58
9. Calculate the relative standard deviation on the molarity of a Na_2CO_3 (molar mass 105.99 $g\,mol^{-1}$) solution obtained dissolving 5.3870 g of the salt in a one litre volumetric flask. Assume that weighing has been performed by difference and that the standard deviation associated with the use of the analytical balance and the graduated flask are 0.00012 g and 0.5 mL, respectively.

10. Calculate the RSD% on the concentration of a species giving an absorbance value of 0.248 assuming that the instrument used has an uncertainty of
    a) 0.003 trasmittance units, or
    b) 0.010 trasmittance units
11. Calculate the standard deviation of the pH of a $(1.2 \pm 0.1) \times 10^{-2}\,\mathrm{mol\,L^{-1}}$ HCl solution.
12. Calculate the relative systematic error in the hydrogenionic concentration of a solution whose pH (as determined by a pH-meter having a positive bias of 0.08 pH units) is 3.02.

3 Quality Assurance and Quality Control

Learning objectives

■ To introduce the tools available to analytical chemists for ensuring a quality service
■ To present practical aspects of reliability, precision, trueness, and validation as introduced in Chap. 2
■ To discuss regulatory and legal aspects of quality assurance and quality control
■ To make aware the economical importance of quality systems in analytical work

3.1 Quality and Objectives of Analytical Chemistry

3.1.1 Definitions and general considerations

Analytical chemistry provides much of the basis for:

- assessing the composition and properties, and consequently, the quality of materials and manufactured products
- controlling the manufacturing process
- evaluating the environmental impact of products and process
- guiding the research and development process

Clinical analyses contribute at an ever greater extent in the development of diagnostics. Environmental analyses are an important consideration in making decisions designed to protect the environment.

It is difficult to estimate accurately the real impact of chemical measurements on all fields of economic and social activities. Some authors have made an attempt to give figures based on the importance of measurements in industry or clinical chemistry. G.A. Uriano and C.C. Gravatt of the National Bureau of Standards (NBS) (now National Institute of Standards and Technology – NIST) in the United States stressed that chemical measurements "make a substantial contribution to the gross national product (GNP) of the U.S." [3.1-1].

H.S. Hertz has estimated [3.1-2] that in 1988 250 million chemical analyses were performed every day in the USA. About 10% of these measurements (25 million per day!!) are poor and have to be repeated, which costs an additional $5 billion annually. In some industries, where product performance is tightly linked to chemical composition, up to 30% of the samples must be retested (e.g., electronic components). T.J. Quinn confirmed these figures in a recent evaluation of the BCR-Measurement and Testing Programme of the EC [3.1-3]; at least 5% of the GNP of western-developed countries are devoted to measurements. G. Tölg estimated in 1982 that up to 12 billion DM were spent to repeat unsatisfactory chemical measurements [3.1-4].

All these figures do not include the economical and social side effects of possible wrong measurements. These cannot be evaluated in terms of budget, but may be estimated by taking into account the legal or economic impact of decisions taken on the basis of measured results, for example:

- the closing of factories
- restrictions in the workplace
- waste management
- the discarding of products
- the human consequences following population displacement after industrial accidents

This underlines not only the importance of chemical measurements, but also the need to guarantee their quality to the customers by improving the quality of these

Wrong measurements lead to losses which amount to several billions of US$ every year.

measurements. The word "customer" has to be taken in a very broad sense, since it may mean:

- a client of a testing laboratory
- a colleague in a research laboratory or from the production department of an industry
- a physician requesting biomedical analysis
- the authorities
- the justice system
- the customs

i.e., all those who need answers based on analytical measurements.

This chapter deals with the tools available to analytical chemists for ensuring a quality service to their customers. In other words, the tools which may be applied to avoid certain parts of analytical work are simply a waste of time and money, and drawing wrong conclusions can cause considerable losses of money and adversely affect human well-being.

All the actions undertaken for planning the proper execution of the analytical task represent what is called *quality assurance* (QA). At present the following definitions given by ISO are widely accepted.

Quality Assurance: all those planned and systematic actions necessary to provide adequate confidence that a product, process or service will satisfy given quality requirements [3.1-5].

Quality: The totality of features and characteristics of a product or service that bear on its ability to satisfy stated or implied needs [3.1-5].

Quality is not immediate; it is only achieved if an extensive set of (a priori) measures are taken and complied with. Quality control (QC) involves both the monitoring of the process and the elimination of causes of unsatisfactory performance to achieve quality.

Quality Control: the operational techniques and activities that are used to fulfil requirements for quality [3.1-5].

Quality assurance and quality control are components of the *Quality System* [3.1-6].

Quality System: the organizational structure, responsibilities, procedures, and resources for implementing quality management [3.1-5].

Quality Management: the aspects of the overall management function that determines and implements the quality policy [3.1-5].

Quality Policy: the overall quality intentions and directions of an organization as regards quality as formally expressed by the top management [3.1-5].

3.1.2 Quality of analytical data and adapted information

The analyst has to play a key role in the entire analytical process if the customer wants answers to the problem he is confronted with. By taking customers' requests into account when delivering analytical results, total quality of a service is achieved.

To achieve the required quality the analyst must be involved from the beginning of the process when the needs of the customer are defined, until the end when the answers of the final report are delivered. This means in practice that the analytical chemist has to be consulted on when the selection of the sample is made, and the choice of the parameter(s) to be analyzed, and on the level of accuracy and precision necessary for a proper answer to be given. This will enable the analyst to set up a scientifically and economically adapted and accepted procedure for the intended purpose. Over-sophisticated analytical procedures of involving high costs may be inappropriate for a simple parameter determination as the client may request to reduce the number of determinations, and consequently affect the results of an investigation. On the other hand, less precise or accurate methods do not allow for long-term investigations of slow processes, e.g., natural decontamination processes of polluted areas, contamination of ground water by leaching of waste, etc, or the evaluation of the quality of (high technology) productions. *The answer to the initial question should always be the central aim of the analyst when he/she plans and performs his work.*

Figure 3.1-1 summarizes the major steps which characterize the entire "quality approach". The quality assurance system has to guarantee that all necessary QC

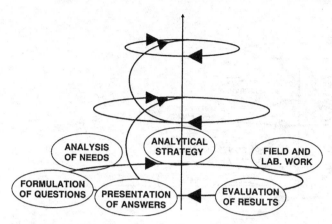

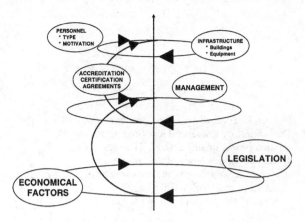

Fig. 3.1-1. Quality assurance – quality control spiral. The various steps for the quality assurance and the quality control of analytical work can be represented on a circle. When all steps lead to adequate results the final evaluation together with an adequate presentation of the answers will give a well-defined answer to the initial question identified by the customer. Alternatively, when the answer is not fully adapted the tandem analyst/consumer may redefine a second approach which leads to a second quality circle. The second circle starts with a greater probability of solving the problem. It is the start of a quality spiral

Fig. 3.1-2. Parameters which influence the quality spiral. Quality assurance and quality control are dependent on various parameters which have to be taken into account when planning the work. These parameters may be internal constraints, e.g., personnel, infrastructure, management, accreditation and certification, or may be dependent on regulatory constraints, e.g., legal requirements or economical aspects

measures are anticipated so that the entire quality circle is under control. If at the end of the chain of actions it is clear that a satisfactory solution cannot be obtained because of some unforeseen circumstances, then the entire circle has to be redone. In view of the information obtained in the first attempt, in most cases it will be possible to come closer to an accurate answer. In Fig. 3.1-1 this is represented as a smaller second circle which starts a "spiral of quality" [3.1-7]. In some very complex situations several quality circles may be necessary to achieve satisfactory answers. For economic reasons the analyst has to keep the number of circles to a minimum. This can only be achieved through close collaboration between the end user or customer and the analyst who desires a proper plan of investigations(s) and the planning of necessary work.

This aspect of global quality assurance, which covers the entire scheme of activities between question and answer, is more and more accepted by analytical chemists [3.1-8] and [3.1-9]. It can be concluded that the task of the analytical chemist is not only limited to the analytical laboratory in the executing of the measurements, but is also vital in the formulation of the questions and the answers.

There is a strong demand on producers from consumers for quality products and services. Parameters like environmentally friendly products and production (free of toxic substances, full recycling possibilities) are developing and require survey. Control of the quality guarantees that they comply with the demands. The analytical chemist is the person who can develop the tools capable of providing quality. In practice, several conditions have to be fulfilled in order to be able to set up a proper quality system. Several persons and factors will intervene and influence the organization and the implementation of the quality system. Figure 3.1-2 schematizes these parameters. Figure 3.1-3 gives the various tools which are at the disposal of the analyst to set up and implement the quality system.

3.2 The Analytical Method

The analytical method is the way in which the analytical chemist obtains the required information. The method has to be fully adapted to the set purpose. Therefore, the analyst has to select a suitable analytical method and should iden-

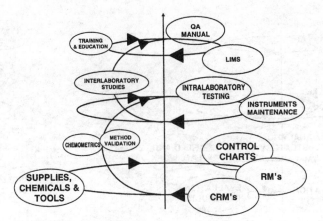

Fig. 3.1-3. Tools to set up the quality spiral. The laboratory has several tools and means to set up a proper quality system. They are applied at various levels of the laboratory structure, personnel or in the quality control of the analytical work itself. Their proper use will lead to the development of the quality objectives

tify all possible sources of errors which might affect its performance. This investigation is called "the validation of the analytical method".

3.2.1 Method selection and definitions

When the analyst has properly defined the problem to be solved he has to select an adapted analytical procedure. When the measurement has never been performed before in the laboratory, the development of the procedure usually starts by investigating the scientific literature. The experience of the analyst and of colleagues on related measurement fields will also influence the choice of the method. This book deals with several of the most important methods applied by analytical chemists. Examples of their application are given and the reader should refer to the relevant chapters.

All analytical methods are composed of a succession of actions (see Chap. 2):

- sampling, sample storage, and preservation of a representative material
- pretreatment of a portion of the sample for the quantitation-calibration
- determination
- calculations and presentation of results

The choice of the measurement method will be governed by the instrumentation available in the laboratory and the experience of the staff. The heart of the procedure consists of the transformation of the sample into the analyte in an accurately determined form compatible with the final detection system available. Selection and development of a method consists of the choice of individual steps to make them compatible, and to develop tools to verify that by the follow-up of the individual steps the entire procedure leads to reliable results. Reliability of analytical data means that they are precise and true. Precision is achieved when random errors are minimized. Trueness is reached when systematic errors are eliminated.

Definitions

Some definitions will help to improve understanding of what is meant by validation of an analytical method which is then treated in Sec. 3.2.2. The basic concepts of these terms and the statistics involved will be introduced in Sec. 12.1. The following definitions are taken from the International Vocabulary of Basic and General Terms in Metrology [3.2-1], ISO 3534 – Statistics – Vocabulary and Symbols [3.2-2], and the Vocabulary of Legal Metrology [3.2-3].

Random error: a component of the error of a measurement which, in the course of a number of measurements of the same measurand, varies in an unpredictable way [3.2-1].

Systematic error: a component of the error of a measurement which, in the course of a number of measurements of the same measurand, remains constant or varies in a predictable way [3.2-1].

Precision: the closeness of agreement between the result obtained by applying the same experimental procedure several times under prescribed conditions [3.2-2].

Trueness: the closeness of agreement between the "true value" and the measured value [3.2-2]. Therefore a method that is both true and precise is described as *accurate*.

True value (of a quantity): a value which characterizes a quantity perfectly defined, in the conditions which exist when that quantity is considered [3.2-1].

The vocabulary of legal metrology gives a definition which makes a direct reference to the measurement:

A value which would be obtained by measurement, if the quantity could be completely defined and if all measurements imperfections could be eliminated [3.2-2].

The selection of an analytical method may also be influenced by external regulatory requirements or by special requests of the customer. The following terms often appear in the analytical jargon, where the classification of methods relies on the purpose of the method or the administrative background [3.2-4].

Official method [3.2-4]: a method required by law or by a regulation issued by an official agency (e.g. EPA, FDA, European Directive, etc).

Reference or standard consensus methods: methods developed by organisations that use interlaboratory studies to validate them (ISO, CEN, DIN, BSI, AFNOR, etc). Their development leads to a known and stated precision or accuracy [3.2-4].

Modified method: reference or standard method which has been modified to simplify or adapt it to the actual state of the art or adapted to other types of samples [3.2-4].

Rapid methods: methods for the rapid determination of large number of samples [3.2-4].

Routine methods: methods used on a routine basis in daily practice. They may be official or standard methods [3.2-4].

Automated methods: methods using automated equipment [3.2-4].

The scientific literature also contains the term definitive method, e.g., in the ISO Guide 35.

Definitive method: method with a high scientific status applied in a laboratory of high proven quality [3.2-5].

In brief, it means that only negligible systematic errors may remain compared to the precision and trueness required for the final result. Isotope dilution mass spectrometry (IDMS) for the determination of traces of multiisotopic elements may be considered as a definitive method when applied in certain laboratories (e.g., laboratories determining the composition of fuels for nuclear fission).

Type of method

The draft ISO Guide 32 on calibration of chemical analysis and the use of certified reference materials [3.2-6], classifies chemical methods into three categories with regards to the calibration procedure.

Calibration: the set of operations which establish, under specified conditions, the relationship between values indicated by a measuring instrument or measuring system, or values represented by a material measure, and the corresponding known values of a measurand [3.2-1].

Calculable methods (absolute methods)

Calculable method: a method that produces the anticipated result by performing a calculation defined on the basis of the laws governing the physical and chemical parameters involved, using measurements taken during the analysis, such as: weight of the test sample, volume of titration reagent, weight of precipitate, volume of titration product generated [3.2-6].

The analyst has to identify every quantity whose measurement is necessary for calculating the end result and to establish the uncertainty of this quantity. Several examples of such types of determinations are known in chemical analysis: e.g., titrimetric, gravimetric, coulometric methods (several examples are given in Secs. 7.1, 7.2, and 7.3).

Relative methods

Relative method: a method which compares the sample to be analyzed with a set of calibration samples of known content, using a detection system for which the response (ideally linear) is recognized in the relevant working area (without necessarily being calculable by theory). The value of the sample is determined by interpolation of the sample signal and with respect to the response curve of the calibration samples [3.2-6].

Differences between sample and calibration sets have no effect or are negligible compared to the uncertainty of the signal. This implies a pretreatment of the sample, matrix matching of the calibration sets, elimination of interferences, etc. Modern spectrometers and chromatographs operate according to this type of method.

Comparative methods

Comparative method: a method where the sample to be analyzed is compared to a set of calibration samples, using a detection system which has to be recognized to be sensitive not only to the content of elements or molecules to be analyzed but also to differences in the matrix [3.2-6].

Ignoring any differences in the matrix will lead to errors. Calibration of such methods requires (Certified) Reference Materials ((C)RM's) with a known matrix composition similar to the matrix of the sample. Such methods are rapid and are often used in the monitoring of manufacturing processes (e.g., WDXRF in the production of metals, alloys and powdered oxides), or for the determination of basic parameters (e.g., viscosity, particle size distribution, etc).

Chemical measurements are usually destructive as they require physical alteration of the sample. These modifications of the sample should not affect the quantitative result of the analysis. Following the entire process step by step is called traceability.

The three categories of methods differ by the way the concentration of the substance is established: in other words, by the way in which the signal of the substance in the sample is linked to the signal of the substance in the calibration material and consequently to its concentration. (Practical examples of method calibration can be found in the various chapters of this book). This link can be made directly to an amount of substance of established purity and stoichiometry in the case of calculable and relative methods (when all steps of the procedure are well-established), or through a (Certified) Reference Material of known matrix composition certified for the concentration of the substance in the case of comparative methods. This link, if established through an unbroken chain of activities to appropriate measurement standards, is called *traceability*.

Traceability

The primary objective of the validation of the analytical method (see Sec. 3.2.2) is to establish the traceability to the recognized reference (pure substance or matrix CRM).

Traceability: the ability to trace the history, application or location of an item or activity, or similar items or activities, by means of recorded identification [3.1-5].

In analytical chemical terms, this could mean that all the steps of the analytical procedure should be performed and recorded in such a way that all essential information is recorded and no wrong information is introduced. In other words the results of the determination and not only the final measurement, should be linked through an unbroken chain of comparisons to appropriate measurement standards, generally international or national standards [3.2-5], e.g., basic SI units, constants, CRM's. This link should be demonstrated.

Conclusions

Depending on the type of analytical method applied and the property of interest to be determined, various steps may be necessary to demonstrate the link between the end signal recorded from the detector and the reference to which it is linked. In principle, the quality of an analytical method can be expressed through two groups of basic and secondary characteristics or figures of merit.

1. Basic characteristics:
   - precision (repeatability and reproducibility)
   - trueness
   - sensitivity
2. Secondary characteristics:
   - specificity
   - range of linear response
   - robustness and ruggedness

Several of these characteristics will be discussed in the section on validation.

The specificity guarantees that the method is really measuring the substance of interest. Interferences of all types affect the specificity of several methods, e.g., chromatographic separations with non-specific detectors such as FID or ECD, etc. The specificity aspects of analytical methods can be appreciated in the various chapters of this book dealing with the description of methods.

Sensitivity is mainly a limiting factor in trace analysis. It is evident that the method applied should be sensitive enough so that the concentration of the substance to be determined is accessible. When the sensitivity of the method of final detection is a limiting factor the analyst may have several possibilities which all will influence the selection, optimization, and validation of the other steps of the procedure.

He may:

- change the method of detection (of course only when available in the laboratory)
- increase the sample intake and adapt the pretreatment steps
- concentrate the determinant at a certain stage of the procedure

The determination of the property of interest in a simple solution will tell the analyst which is his working range in terms of sensitivity of the signal.

3.2.2 Validation of the method

The validation of an analytical method can be defined in the following manner:

validation of an analytical method: the process which allows to demonstrate whether the results produced by this method are reliable, reproducible, and whether the method is suitable for the intended application.

In other words, the method allows the analyst to deliver reliable answers to the initial question. The process to demonstrate that this objective is achieved involves the analyst investigating all possible sources of errors and eliminating them. By doing so, he will show that the method is able to deliver results which are traceable to chosen references. This is only possible within adapted structures, with well-trained and motivated personnel, and with adapted and properly maintained instruments.

Validation of instruments, computer hardware and software

Instrument validation precedes method validation. The same principle applies for computers and the software when automated methods are used. Instrument and computer validation procedures should be provided by the suppliers. All instruments should be validated when delivered by the manufacturers, but on exercising its analytical *responsibility*, an analyst should verify the manufacturer's specifications on performance, sensitivity, stability, ranges of linear response, etc.

A survey done by an international consortium of instrument users – SIREP, WIB and EXERA – tested 126 instruments of various types and arrived at the

A good analyst verifies all the instruments before using them.

Fig. 3.2-1. Compliance of analytical instruments with specifications (EXERA, 1993). (a) The graphical representation indicates the type of instruments tested from 1989 until 1992; (b) The outcome of the tests demonstrate that 75% failed manufacturers' or users' specifications from which 49% failed the manufacturers' own specifications. This is a demonstration for analysts that it is important to validate properly new instruments. (The authors wish to thank EXERA, WIB and SIREP, and in particular Mr. M. Desjardins, for their support and for accepting the publication of the results of their work on instruments' compliance)

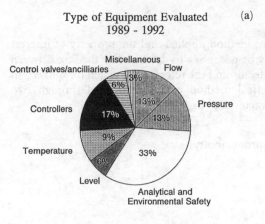

Type of Equipment Evaluated (a)
1989 - 1992

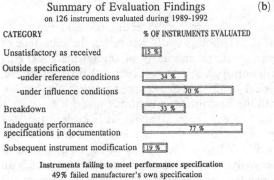

Summary of Evaluation Findings (b)
on 126 instruments evaluated during 1989-1992

CATEGORY	% OF INSTRUMENTS EVALUATED
Unsatisfactory as received	15 %
Outside specification	
-under reference conditions	34 %
-under influence conditions	70 %
Breakdown	33 %
Inadequate performance specifications in documentation	77 %
Subsequent instrument modification	19 %

Instruments failing to meet performance specification
49% failed manufacturer's own specification
75% failed manufacturers' or users' specifications
25% only met all specifications

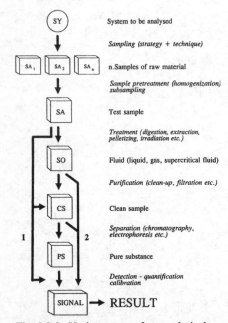

Fig. 3.2-2. Various steps of an analytical procedure. The analytical procedure (field and laboratory work in the QA/QC spiral Fig. 3.1-1) is composed of various steps. Following field sampling, the sample is split into subsamples which may need a physical and/or a chemical treatment before the substance of interest can be measured. Therefore, many methods necessitate a matrix elimination with or without a subsequent purification of the substance(s) to be determined. For some methods one or more steps may be avoided, e.g., sample pretreatment for INAA (route 1), purification for multielemental determinations by ICP-AES (route 2). For organic trace analysis all steps shown are usually necessary

conclusion that 75% of them did not meet the specifications of the manufacturer [3.2-7]. Figure 3.2-1 shows the type of failure and the type of equipment tested by the consortium. The reason for this lack of compliance with specifications is mainly the fact that, especially for the new type of apparatus, the manufacturers and in particular their sales departments exaggerate the potential of their material. Whatever the reasons may be, the analyst should verify specifications of the instrument by not only pure solutions, but also by using his real samples prior to the acceptance of an instrument.

Validation of the analytical method

In chemical analysis the property of interest to be determined, e.g., trace elements or organic substances, is rarely directly measurable, as may be the case in physical analysis (e.g., mass, length, time, etc). To be able to measure the parameter, the chemist has to first select a representative sample and subsample, and then to convert or to separate the analyte in a form that is compatible with the detector. This may imply that he has to change the physical and/or chemical structure of the initial material, but without losing control of this change so that the traceability of the final detection to a predetermined reference (e.g., fundamental units) is not lost. Analytical procedures typically include a pretreatment step, e.g., digestion, extraction, or purification; and/or a separation step; a calibration step in the case of relative and comparative methods; and a final detection step. The classical steps encountered in chemical analysis are shown in Fig. 3.2-2.

In some procedures one or more steps can be avoided, e.g., elimination of the matrix in instrumental neutron activation techniques. The approach consists of stepwise bringing the substance to be determined into a state where it can be detected and quantified accurately. The analyst transforms the initial complex test sample into a simple sample which is compatible with the detection system. Each action undertaken in one of these steps is a potential source of error which adds to the total uncertainty of the determination. The analyst has to identify and possibly

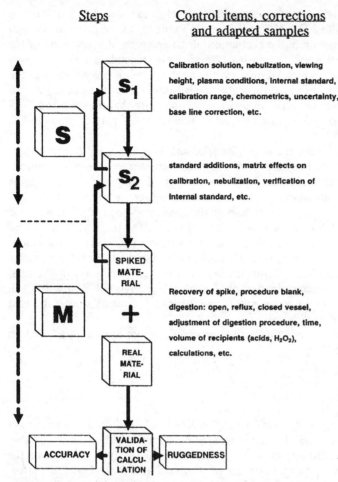

Steps | Control items, corrections and adapted samples

S₁ — Calibration solution, nebulization, viewing height, plasma conditions, internal standard, calibration range, chemometrics, uncertainty, base line correction, etc.

S₂ — standard additions, matrix effects on calibration, nebulization, verification of internal standard, etc.

SPIKED MATE-RIAL — Recovery of spike, procedure blank, digestion: open, reflux, closed vessel, adjustment of digestion procedure, time, volume of recipients (acids, H₂O₂), calculations, etc.

REAL MATE-RIAL

ACCURACY — VALIDA-TION OF CALCU-LATION — RUGGEDNESS

Fig. 3.2-3. Stepwise validation of the determination of Cd and Pb in mussel tissue. The validation of a method for the determination of several trace elements in a matrix requires the investigation of all sources of errors which may exist in each of the steps of the procedure shown in Fig. 3.2-2. The approach consists of starting from the final detection and calibration, and going back stepwise towards the real sample. The various parameters (nonexhaustive list) which can be optimized are listed beside each step

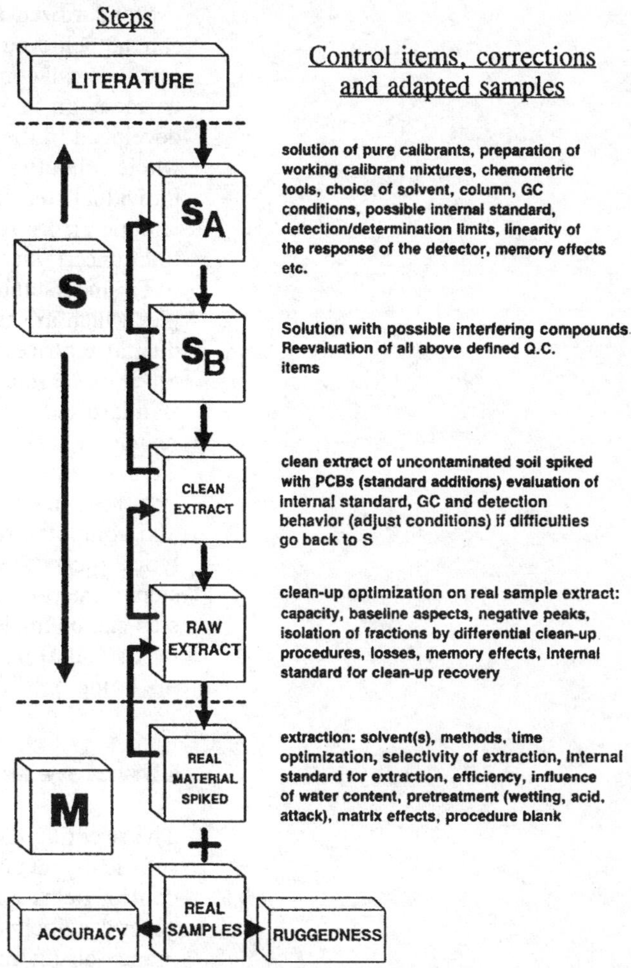

Steps | Control items, corrections and adapted samples

LITERATURE

S_A — solution of pure calibrants, preparation of working calibrant mixtures, chemometric tools, choice of solvent, column, GC conditions, possible internal standard, detection/determination limits, linearity of the response of the detector, memory effects etc.

S_B — Solution with possible interfering compounds. Reevaluation of all above defined Q.C. items

CLEAN EXTRACT — clean extract of uncontaminated soil spiked with PCBs (standard additions) evaluation of internal standard, GC and detection behavior (adjust conditions) if difficulties go back to S

RAW EXTRACT — clean-up optimization on real sample extract: capacity, baseline aspects, negative peaks, isolation of fractions by differential clean-up procedures, losses, memory effects, internal standard for clean-up recovery

REAL MATERIAL SPIKED — extraction: solvent(s), methods, time optimization, selectivity of extraction, internal standard for extraction, efficiency, influence of water content, pretreatment (wetting, acid, attack), matrix effects, procedure blank

ACCURACY — REAL SAMPLES — RUGGEDNESS

Fig. 3.2-4. Stepwise validation of the determination of polychlorobiphenyls in soil. The validation of methods for the determination of traces of organic substances in environmental matrices follows the same principles as those given in Fig. 3.2-3. The analytical methods usually include more steps to be validated than in inorganic trace analysis. The validation of the extraction of the substances from solid matrices is a difficult step in such validations

minimize all these sources of errors. To do so he has to work out a strategy for studying each individual step. One approach consists of taking the analytical procedure described in Fig. 3.2-2 backwards and to study first the most simple system, e.g., the final calibration solutions to be presented to the detector. Having studied the detection and calibration step the analyst may go stepwise towards the real sample. Figures 3.2-3 and 3.2-4 show the main steps to be considered for such an approach for comparative methods, e.g., inorganic and organic trace analysis.

After each new step the analyst evaluates whether the conclusions drawn in the previous step are still valid. If not, he has to come back and reset new conditions. Random errors can be detected and minimized by intralaboratory measures. Systematic errors can be detected and eliminated by internal measures, e.g., use of CRMs, or external means, e.g., interlaboratory studies.

In addition to the reliability of the method, the analyst has to evaluate how it will behave in the real situation of daily "routine measurements" in the hands of the technicians and under economic constraints; he has to estimate the robustness or ruggedness of the method.

Ruggedness, robustness of a method: the ability of a method to be relatively insensitive to minor changes in the procedure, to the quality of reagents, or to the environment [3.2-8].

Before going to the next step of your validation procedure make sure you fully control the one you are working on. This includes prevention of possible problems that may occur in the next step.

Standardized methods are often described in such a manner that a validation can be achieved rapidly. For the individual laboratory, the validation of such methods will consist mainly of the evaluation of the precision, a verification of the trueness, an investigation of the sensitivity and of the ruggedness. For methods developed in the analyst's own laboratory, the analyst himself has to perform the whole validation. In practice this consists of the breakdown of the method into the individual steps mentioned before in order to demonstrate that no sources of systematic errors remain, and to investigate the attribution of various steps to the total uncertainty.

To understand more fully the step-wise approach it is necessary to give examples which are taken from organic and inorganic trace analysis in, e.g., environmental matrices. These analytical fields are typically dealing with very variable types of sample (various matrices), contamination sources and levels. Only few standardized methods exist that can be directly applied without at least extensive revalidation. Figure 3.2-3 shows a validation approach for the determination of Cd and Pb in mussel tissue. Mussels are target organisms which are used to monitor the contamination of the marine environment by various toxic trace elements and compounds. Figure 3.2-4 illustrates the steps for the validation of the analytical procedure for the determination of polychlorobiphenyls (PCB) in soil. Each step of the procedure should provide all the necessary information so that the next step can be made with confidence.

A standardized and validated analytical procedure is also known as a standard operating procedure (SOP).

Literature search

The scientific literature rarely gives descriptions of analytical methods which are sufficiently detailed and validated to be applied directly without risk. Quality control items are usually not presented. Therefore, the analyst has to take the general guidelines given, to build up his own system and to evaluate its performance on his real samples. Standard methods cannot be transferred directly to other types of matrices unless these are fully validated. For example, a standard method for drinking water cannot be transferred to surface or waste water. Similarly a method for a sediment is not suited for a soil or a sludge. Literature search may give indications as to the best possible method, but cannot give more than this. In our examples, one could consider that literature information only defines in broad terms the pretreatment (acid attacks, extraction solvents), the purification, and the final detection systems.

Validation of the final detection (step S in Figs. 3.2-3 and 3.2-4)

The International System of Units (SI) has been set up to avoid confusion in the scientific language. SI should be respected and used fully.

It has to be stressed that errors in the calibration process occur very frequently; this has been concluded by the discussions held in interlaboratory studies conducted over several years within the BCR. The types of mistakes include:

- simple calculation errors (e.g., of concentrations)
- dilution errors
- mistakes during transfer of volatile solvents
- use of impure or nonstoichiometric primary compounds and absence of verification
- contaminations, interferences, unadapted internal standards or unproper introduction of internal standards into sample and calibration solutions
- unsatisfactory background correction and blind trust to integration systems
- abuse and misuse of units (mistakes occur when working on a volume basis for the preparation of calibrant solutions, and when assuming that 1 mL equals 1 g, even for organic solvents or mixtures of acids, mistakes occur in the usage of ppm, ppb, ppt)
- absence of matrix matching of calibration solutions, etc.

A study of the system with a solution of pure compounds is necessary for detecting, evaluating, and solving all problems related to the reliability of the signal produced by the detector: specificity, linearity (several types of detectors present only a limited range where their response is linear, e.g., AAS, ECD), traceability to adequate pure substances, sensitivity, trueness and precision of the calibration. All problems of interference, chromatographic separation, choice of the internal standard for quantification, chemometric tools (see Chap. 12) for a reliable calibration should be investigated, quantified and, if possible, solved at this stage. For this step, the analyst will prepare stock solutions of calibrants with known purity and stoichiometry. Choice, handling, storage, and preparation of such solutions is described in detail by D.E. Wells et al for PCB and other organic compound determinations [3.2-9] and by J.R. Moody et al for inorganic calibration [3.2-10]. Some basic recommendations can be given.

Calibrants and calibration stock solutions should be kept in closed storage facilities (e.g., safe, fridge, deep-freezer), their access should be controlled, and a responsible senior analyst should be designated for their preparation and distribution. For organic compounds which are often dissolved in volatile solvents, sealed vials (e.g., ampoule) should be preferred. Protection from light (e.g., amber glass vials) and increased temperature (refrigerator, deep-freezer) should be guaranteed. Visual inspection of inorganic and organic standard solutions before each use should allow the detection of important precipitates or flocculates. Before and after taking an aliquot of the standard the analyst should weigh the vial so that losses due to evaporation or leakage can be detected. Stock solutions should be replaced regularly, the frequency depending on the stability of the compounds. Errors at the calibration level generally result in differences between laboratories of some orders of magnitude [3.1-7].

Matrix influence

In this step which concerns either an extract or a digest, the matrix is no longer known. All the conclusions obtained in the first step, e.g., optimal settings, have to be verified for the new situation (calibration, linearity, chromatographic conditions and performance, internal standard, etc). With the matrix, new interferences can appear and matrix compounds may influence the system (matrix effect). For the determination of trace organic contaminants, this step is of great importance as it has to ensure that no interfering compounds remain because quantitation is often performed with nonspecific detectors (e.g., ECD, FID, UV). Consequently the reliability of the signal relies on the prior chromatographic separation.

Matrix matching of the calibration solution or standard addition procedure may be necessary. An example of the over-neglected matrix effect is given in Table 3.2-1.

Table 3.2-1. Illustration of the influence of matrix matching for inductively coupled plasma atomic emission as the method of final determination (mass fractions in mg/kg) of trace elements in a light sandy soil – BCR CRM 142R [3.2-11]

Element	Results without matrix matching mean $\pm$ SD	Results with matrix matching mean $\pm$ SD	Certified value mean $\pm$ 95% CI
Co total	6.0 $\pm$ 0.2	7.9 $\pm$ 0.5	5.61 $\pm$ 0.31
Co aqua regia	4.6 $\pm$ 0.5	6.1 $\pm$ 0.5	NC
Cu total	607 $\pm$ 9	667 $\pm$ 18	696 $\pm$ 12
Cu aqua regia	655 $\pm$ 4	745 $\pm$ 12	707 $\pm$ 9
Mn total	139 $\pm$ 2	151 $\pm$ 5	156 $\pm$ 4
Mn aqua regia	122 $\pm$ 2	151 $\pm$ 4	NC
Ni total	216 $\pm$ 5	249 $\pm$ 8	247 $\pm$ 7
Ni aqua regia	207 $\pm$ 3	266 $\pm$ 5	251 $\pm$ 6
Zn total	1826 $\pm$ 34	2072 $\pm$ 47	2122 $\pm$ 23
Zn aqua regia	1856 $\pm$ 15	2238 $\pm$ 26	2137 $\pm$ 50

95% CI: 95% confidence interval
SD: standard deviation of five independent measurements
NC: not certified
For Co and Cu other errors than matrix matching remained unidentified by the laboratory.

These results were obtained in an interlaboratory certification study to demonstrate the importance of matrix matching for the calibration [3.2-11]. Except for Co, for which another source of error remained, the matrix matching of the standard solutions allowed the laboratory to come closer to the certified values and improve the agreement between laboratories.

In organic trace analysis it is usually necessary to clean the extract before the real separation by chromatography can be performed. The clean-up should remove all co-extracted bulk material such as lipids, sulphur, pigments, etc, and other potentially interfering compounds. A proper clean-up will also protect the chromatographic column and the detector from contamination (ionization source of the MS detector, ECD). For organic determinations of traces it is also essential to estimate possible sources of losses due to clean-up. This may be done by spiking a raw extract. Standard additions give an estimate of the capacity of the clean-up and therefore are used in defining the working range in routine. There the analyst usually decides to select an internal standard to follow the clean-up recovery.

Solid material (step M in Figs. 3.2-3 and 3.2-4)

This step optimizes the extraction or digestion procedures and estimates its efficiency. For inorganic analysis, the digestion step can be validated by using a method which can determine the elements directly in the matrix, e.g., WDXRF, INAA. For organic analysis, the validation of the extraction step is difficult and often unsatisfactory. For solids, it is nearly impossible to prove that all the compounds have been extracted. Two approaches are usually applied. The first approach makes use of successive extractions with fresh solvents, and determination of the residual traces of PCBs in the successive extracts. When PCBs can no longer be extracted by any known method, one may assume (but not guarantee) that the extraction is complete. The second approach consists of a standard addition procedure with increasing use of spikes [3.2-12]. After extraction the analyst verifies that he has recovered all added compounds. This method is only valid if the operator allows the spiked substance to be in contact for a sufficient time with the material so that the added and the incurred substance are in the same physicochemical stage in the matrix. By repeating the spiking at each level of enrichment the analysts may estimate the reproducibility of the extraction procedure in addition to its efficiency. When all steps of the method have been developed, optimized, and verified the analyst has to combine the procedure and to work out quality control items for routine use, which may allow him to decide a posteriori that the method is still under control.

Ruggedness/robustness

A method can only be applied in a testing laboratory when it is sufficiently robust. In that case common variations in the method do not affect its reliability; there may be:

- small variations of temperatures during pretreatment, aging of LC or GC columns so decreasing separation power, variations in the water content after drying, etc.
- replacement of parts of the equipment (e.g., Soxhlet, digestion bomb, etc.)
- other technicians
- small variations in the sample matrix, concentration range of the substance (e.g., two soils are never the same), etc.
- variations in the environment, e.g., temperature, humidity, atmospheric pressure.

All parameters fixed during the development and the validation procedure need to be investigated for the effect of variations. The totality of the method ruggedness (including the influence of the operator) needs to be evaluated in conjunction with the total procedure. This is done by modifying some of the parameters which have been identified as critical during the stepwise approach. A method too sensitive towards variations may be inadequate. Ruggedness testing is always part of the development of standardized methods and can be estimated in interlaboratory studies. Chemometric tools [3.2-8] and expert systems have been propsoed [3.2-13] to estimate the ruggedness of a method.

Control points

When the analyst has fully validated all steps of his method he may define a control procedure which will allow him to detect rapidly the origin of errors which can arise when the method is in daily use. The control points are linked to the sensitive items which have been recognized in the validation. Table 3.2-2 lists error sources which may occur in the determination of trace elements in a plant material, and which may be subject to control points. Ideal control points are those which are built in the normal analytical scheme and which can lead to fully automated warning, e.g., solvent purity and/or contamination by blanks. Other control points are required for sensitivity performance checks on the slope of the calibration line, and for investigating the separation power of a chromatographic system for two artificial substances introduced into the samples and calibrants (e.g., internal standards). The result from the critical control points should be included into the standard reporting form delivered to the analysts so that arising problems can be retraced. Each situation and each method needs specially worked-out control points. Some standardized methods or methods recommended by the AOAC [3.2-14] include such control points. More and more standards for difficult analytical tasks contain performance criteria for individual control points.

Control charts

When a fully validated method is available, the analyst can envisage starting a statistical control system including the follow-up of the performance by control charts. The use of these charts helps the analyst to prepare a representative, homogenous, and stable reference material which will be analzed at regular intervals. The quality requirements for such materials are discussed in the paragraph dealing with CRMs (Sec. 3.3.4).

When the laboratory works at a constant level of high quality, only few random errors persist, and fluctuations in the results are small. Only then can the method be considered fully validated, and a statistical control system using control charts can be implemented. This will allow the detection of the introduction of any *new* systematic error, and in some cases, the monitoring of the precision when replicate measurements are performed. At regular intervals, the analyst determines in a reference material the substance to be monitored and reports the result graphically. When starting a chart, the analyst has to determine from several replicate measurements the mean value and the standard deviation which represents the reproducibility of the method. This reproducibility value will allow acceptance limits, e.g., "warning" and "alarm" levels as shown in Figs. 3.2-5 and 3.2-6 to be calculated.

Control charts have been developed for the monitoring of production as a means for statistical process control (SPC). Control charts have also been adapted for analysis as statistical quality control (SQC), where they serve as warning signals for the laboratory. Several situations may be encountered where action on the part of the analyst is necessary. The quality is no longer guaranteed when one measurement exceeds the alarm limit. If the "warning" limit is exceeded once, this should alert the analyst that he/she may have to deal with possible future problems, but immediate action is not necessary. When the "warning" limit is exceeded two consecutive times an investigation into the cause has to be carried out. If results repeatedly, e.g., 10 times, fall on the same side of the mean value the analyst should be alerted to the possible introduction of a systematic error. The action the analyst has to take when the quality runs out-of-control is similar to that taken when there are changes in the analytical procedure between two occasions of measurement of the reference material. In the worst situation he may need to fully revalidate the procedure.

Besides simple Shewart charts, the analyst may also apply X-charts where the

Table 3.2-2. Some possible sources of error and their elimination in trace element analysis of a plant material

Analytical step	Systematic error by	Contribution	Elimination by
A. Methods requiring a preliminary matrix destruction			
Preparation	weighing	+/−	calibrated balance
	volumetric manipulation	+/−	dilution, etc carried out with calibrated glassware, temperature control
Moisture	adsorption/ desorption	+/−	correction to dry mass
Digestion/ oxidation	volatilization	−	for volatile elements (e.g., As, Se ...) treatments carried out in closed systems
	adsorption/ desorption	+/−	acid-washed containers of hard glass. PTFE or HDPE prerinsed surfaces
	incomplete	−	pressurized digestion with oxidizing acids; residue checked to verify the total digestion of the matrix
	reagent contamination	+	reagents of appropriate purity; verification with blank determinations
	contamination by tools/vials	+	acid washing as appropriate: when contents below 1 µg/g to be determined: steaming; verification by blank determinations
	contamination from lab air	+	use of clean benches or clean room; care in performing methods under cover or in closed systems; verification with blank determination
Sample preparation/ clean up/pre- concentration	adsorption/ irreversible precipitation	−	pH-control and/or addition of complexing agents if necessary
	contamination	+	as above for digestion/oxidation
	incomplete conversion	−	excess of reagents; methods verified a priori
Calibration		+/−	reagents of suitable purity and stoichiometry; where necessary verification of stoichiometry and purity of calibrants; different calibration methods when possible: calibration graphs, matrix-matched calibration solutions and standard additions
B. Activation methods			
Counting	peak overlap intrinsic irradiation	+/−	deconvolution; selection of proper decay times; RNAA as an alternative
	high background	+/−	RNAA as an alternative
	geometry	+/−	calibrant and unknowns both in the same form, e.g., solution or powder, measured in identical vials at the same distance of the detector, etc
Irradiation	self-shielding	+/−	verify that for the contents of the elements investigated, shielding does not occur
	changes of flux	+/−	flux monitors added in the irradiation process
Calibration		+/−	same remarks as for destructive methods; additional care to be given to the stability on irradiation of the calibrant

+: overestimation of content
−: underestimation of content

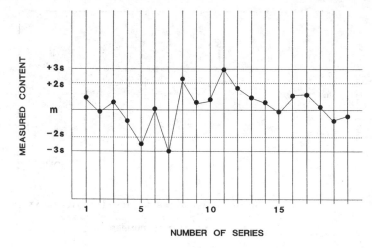

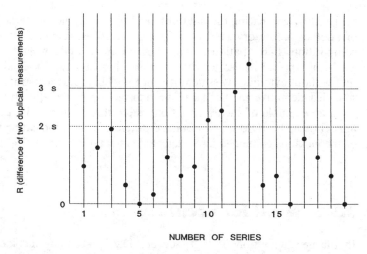

Fig. 3.2-5. Example of an X-chart. X is the value obtained at each occasion of analysis. Warning (2*s*) and Alarm (3*s*) lines correspond respectively to a risk of 5% and 1% that the result does not belong to the population of results. The values for *m* and *s* are determined by several preliminary (at least 10) determinations of the material performed over several days. Rejection criteria are detailed in the text

Fig. 3.2-6. Example of an R-chart. R is the difference of two duplicate determinations. The warning and alarm limits are similar to those of the X-chart and are explained in Fig. 3.2-5 and in the text. R-charts give an indication on the repeatability of methods

mean of several replicate measurements is plotted, or R-charts where the difference of two replicate measurements is plotted. Such X- and R-charts give an indication of the reproducibility of the method. Drifts in analytical procedures, e.g., slow changes in the system caused by the aging of parts of instruments, decalibration in wavelength, aging of calibration stock solutions, etc, may be detected early when applying a Cusum chart (cumulative sum). In Cusum charts, the analyst reports the cumulative sum of the differences of his determination with a reference value. If this value is certified (e.g., by using a CRM) the Cusum chart allows the accuracy of the determination to be monitored. More information on the implementation of control charts (e.g., frequency of measurements, warning and alarm limits, etc.) will be found in Chap. 12.

Chemometric tools

The description of chemometric tools is dealt with in Chap. 12. It should be stated here that these tools can be an important help in the development of a method and in the various steps of validation. They can be applied in calibration, for the optimization of a method, for the evaluation of data, and also for the statistical control of methods over time.

3.3 How to Achieve Accuracy

After having developed the best procedure with a sufficient reproducibility, the accuracy and, in particular, the trueness has to be demonstrated. This can be

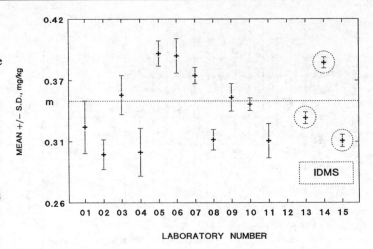

Fig. 3.3-1. Bar graph representation of the results of a BCR interlaboratory study on the determination of Cd in mussel tissue. Three laboratories applied IDMS and found different results. This example demonstrates that a so-called "definitive method" does not guarantee against the presence of systematic errors. The degree of accuracy of results depends, as much as for other techniques, on the quality assurance and quality control system developed within the laboratory. The good precision which can be achieved by this technique allows even the detection of very small systematic errors (between laboratory mean and SD: 0.35 ± 0.04)

achieved by comparing the data obtained on a reference material with the results of another laboratory, or by analyzing a certified reference material of similar composition. Participation in an interlaboratory study where different methods are being applied is also a way to approach accuracy. The analyst has to know that without external assessment it is not possible to demonstrate accuracy. Even, so-called definitive methods can be affected by systematic errors. Figure 3.3-1 gives an example of such a situation where three laboratories determined Cd in mussel tissue by IDMS, and obtained different results, indicating that at least one of the methods contained an undetected systematic error.

3.3.1 Need for accuracy

Accuracy is the basis for comparability of results between laboratories (Sec. 3.3.1)

In the new definition of ISO (see Sec. 3.2.1) accuracy includes precision and trueness. Precision is acquired by working under a total quality management system as discussed above, and by properly validating the analytical method. To demonstrate trueness the analyst is dependent on external help. Trueness is necessary for producing analytical data which are comparable in time and from one laboratory to another. A typical example where trueness/accuracy is necessary is in the monitoring of the state of the quality of the environment. For example, regulations for the reduction of emissions of toxic compounds into the environment have measurable effects on environmental samples, sometimes only after large periods of time. Only very accurate measurements can determine these decreases in the long term. Too many scientists have stated that good reproducibility is sufficient for following trends, and for demonstrating the effects of actions carried out by the authorities for improving the quality of the environment. They have, however, overlooked possible improvements in equipment and methodology.

Reproducibility over time implies minimum changes in the measurement system. In addition the absence of accurate measurements does not allow results issued by other groups in different parts of the world to be compared. Similar problems exist in trade when goods are exported and their quality has to be assessed by measurements. Differences between results of the producer and customer can lead to dramatic commercial conflicts. In clinical chemistry where the measured parameters are used in diagnosis and in the assessment of the patient's status, long-term reproducibility and, in addition, trueness are required, especially, when the patient is changing department or hospital.

Trueness of results can be demonstrated by three different means:

- comparison with a different method
- comparison with other laboratories
- use of certified reference materials

3.3.2 Comparison with a different method

Each method has its own typical sources of errors to which those of the analyst applying it have to be added. For spectrometric methods, e.g., AAS, ICP-AES, and ICP-MS, the error can be the acid-digestion of the matrix, where the material, in which the elements to be determined are trapped, is destroyed by acids (oxidants). The end result is usually a pure acid solution of the elements. This does not play a role in instrumental neutron activation analysis, and therefore, INAA may be used to control results obtained by methods with a pretreatment step, e.g., ICP-AES.

If the results obtained on similar samples by both methods are in good agreement it can be considered that the results from each INAA and ICP-AES are true. This conclusion is most valid when the steps of the procedure of both methods differ widely. If there are similarities in the procedure the comparison of the results may overlook systematic errors due to the common step. A real valid comparison can only be performed when both widely differing methods are under equal control in the laboratory. If the technicians are not sufficiently experienced with the comparison method it may even create confusion or induce additional errors. Therefore, the comparison with another method within the same laboratory is rather difficult and rare when trueness is to be assessed. An intralaboratory comparison may be an efficient tool for controlling the performance of different instruments over time, and may also be helpful for introducing a new method into the laboratory. In the latter case the final accuracy checks will need to pass through one of two means: CRMs or interlaboratory study.

3.3.3 Comparison with other laboratories

An easy way to learn and to improve the quality and the performance of the laboratory is to participate in interlaboratory studies. Interlaboratory studies consist of practical exercises where all participating laboratories have to perform the determination of one or more substances in a sample prepared by the organiser. At the end of the study the results are distributed to the participants. Interlaboratory studies are used for different purposes: to detect the power of the performance of a method in view, for example, of its standardization, to investigate a new or difficult field of chemical analysis, and to certify reference materials. In a report for IUPAC, W. Horwitz [3.3-1] has defined the various types of interlaboratory studies together with the accompanying vocabulary as follows:

You can learn from your mistakes, but you will also learn from other analysts.

Interlaboratory study: a study in which several laboratories analyze one or more identical, homogeneous materials under specified conditions, the results of which are compiled in a single report.

Laboratory performance study: an interlaboratory study that consists of one or more analyzes conducted by a group of laboratories on one or more identical test samples by any method selected or used by the laboratory. The reported results are compared with those from other laboratories or with the known or assigned reference concentration or other characteristic, usually with the objective of evaluating or improving laboratory performance.

Proficiency testing is another term used when the object of the study is appraising laboratory or analyst performance. This term will be encountered in particular in ISO/IEC guide 25 [3.3-2] or in the EN 45001 standard [3.3-3].

Method performance study: an interlaboratory study in which each laboratory uses the same test method on a set of identical test samples to perform analyses by the same protocol. The reported results are used to estimate the performance characteristics of the method. Usually these characteristics are within- and between-laboratory precision. When necessary and possible other pertinent characteristics such as systematic error, internal quality control parameters, sensitivity, and limit of determination, have to be considered as well.

Material certification study: an interlaboratory study that assigns a reference value ("true value") to the analyte concentration (or property) of the test material, usually with a stated uncertainty.

Laboratory performance studies enable participants to evaluate their performance by comparing their own results to those obtained by other, perhaps more experienced, analysts. Often the various participants apply different methods, which can enable them to detect, in their own procedure, unknown sources of systematic errors which affect the trueness of the results. By participating in proper laboratory performance studies it is possible to demonstrate when the laboratory has reached accuracy. This will depend on three major items:

- the motivation and the way the laboratory has prepared itself
- the motivation of the other participants
- the way the study is organized

The latter item is paramount and requires from the organiser a broad knowledge in many fields. The organizer should be able to demonstrate the ability to organize proper exercises so that the participants know that they may obtain the expected answers, i.e., information on their degree of accuracy. Recently, a harmonized approach for the organization of proficiency testing was published by AOAC/ISO-REMCO [3.3-4].

The basic requirements detailed for laboratory performance studies also apply to the other types of interlaboratory studies which are employed in analytical chemistry [3.3-5].

3.3.4   Certified reference materials

Reference material: a material or substance, one or more of whose property values are sufficiently homogeneous and well-established to be used for the calibration of an apparatus, for the assessment of a measurement method, or for assessing values to materials [3.3-6].

Certified reference material: a reference material, accompanied by a certificate, one or more of whose property values are certified by a procedure. This procedure enables the material's traceability to be established in terms of the SI unit (to be understood as: a mole of a substance; the mole is the unit for amount of substance in the SI) in which the property values are expressed. Each certified value is accompanied by an uncertainty at a stated level of confidence [3.3-6].

Types of certified reference materials

The use of certified reference materials is the easiest way to achieve and demonstrate accuracy. Certified reference materials (CRMs) of good suppliers (see Sec. 3.4.2) link the user's results to those of the international analytical chemical community. In addition they enable to verification of the performance of the laboratory at any desired moment. This is a serious advantage over interlaboratory performance studies. CRMs can be of different nature and can fulfil accuracy requirements at several levels of the analytical procedure. CRMs may be:

- pure substances or solutions for calibration and/or identification
- materials of known matrix composition for the calibration of comparative methods (Sec. 3.2.1)
- matrix materials which as far as possible represent the matrix being analyzed by the user and which have a certified content (such materials are used for the verification of a measurement process)
- methodologically defined reference materials for parameters such as: leachable, bio-available, aqua-regia-soluble fractions of elements, chloroform extractable pesticides, etc; the certified value here is defined by the applied method following a strict analytical protocol

Certified reference materials are products whose results are guaranteed by top leading analysts.

The first two categories of CRMs are of particular interest since their chemical composition is artificially enriched and fully known.

Pure substances are usually certified by establishing the maximum amount, in mass fractions, of the impurities which remain in the purified substance. For metals it is possible to estimate the mass fraction of other elements within the limits of sensitivity of the available analytical methods. For inorganic salts or oxides this estimate is already more difficult as the stoichiometry has also to be assessed, e.g., water of crystallization. The determination of the purity and of the stoichiometry of organic substances is far more difficult as it is not always possible to determine the identity of remaining impurities. Mass selective methods have to be applied to determine the apparent mass of the impurities (e.g., GC-FID, GC-MS). Calibrant solutions have to be prepared on a mass basis within specially trained and skilled laboratories. Certification will be based on the metrological weighing procedure, and the purity and stoichiometry assessment.

Materials of known chemical composition (e.g., metals, alloys, powdered oxides and ores, and cements for WDXRF) require also a special knowledge and can be prepared on a mass basis with metals or substances of known purity.

Matrix CRMs are of unknown or only partially known matrix composition in which the amount of a certain number of substances is certified. In such cases, certification can only be based on an analytical approach using complex procedures with limited precision compared to mass determinations. Usually analysts prefer fully natural samples which are as similar as possible to real samples and not artificially enriched samples. In fact, no gravimetric value can be determined in spiked solid samples because, e.g., with organic trace compounds, the inclusion of the spike into the matrix may be only partially achieved as losses due to adsorption on vessel, evaporation, or destruction through homogenization tools, or due to stabilization may occur. The operator, therefore, has to start from a matrix where the substance to be spiked is absent. Alternatively, the operator has to assess whether the possible presence of the substance is negligible compared to the amount added. For solid materials to be analyzed for trace elements or organic compounds, spiked samples behave differently than natural ones. The physical or chemical status of every compound or element is often different, so the pretreatment (extraction or digestion) can be affected. In some circumstances, spiking is the only possibility for carrying out pretreatment as stability or homogeneity of real natural samples cannot be achieved. This may be the case for water samples. The difference in behaviour in the analytical procedure observed for an artificial water sample compared to the procedure for spiked samples may be considered as negligible in terms of verifying the accuracy of the method.

Methodologically defined CRMs are materials which have been certified following a given analytical procedure. This prescribed procedure may represent only a part or the totality of the analytical process. RMs may be certified following a standardized method as the method also defines the parameter. This concerns concepts like the standardizing of compounds such as volatile organic halogens (VOX), etc. Often only a part of the method is prescribed. This involves analytical concepts which try to associate the parameter with a certain property such as mobilizable fractions of elements or compounds in soils or sediments.

Extensive interlaboratory studies have been conducted in the last few years by the BCR on the leaching of trace elements from agricultural soils and sediments. Several QC measures for the good execution of such methods have been worked out through a stepwise approach [3.3-7]. A group of about 20 laboratories has demonstrated that the level of the quality of measurement they have reached authorizes the consideration of the certification of several elements following prescribed leaching procedures, and still allows a free choice of the final determination technique. Such CRMs are now under preparation within the M&T – BCR programme [3.3-8]. Similar concepts are considered for the evaluation of bioavailable toxic compounds, or even nutrients from food or feed the potential leakage of toxic elements from waste and such like.

The certification procedures for matrix CRMs are detailed in the ISO Guide 35 [3.2-5]. Practical examples of certifications and applications of the ISO principles have been published elsewhere [3.3-9] and [3.3-10].

Certification of reference materials implies that the best available analysts and methods under the most strict working conditions have led to the assignment of a value to a substance in a material. Any error in the certification process would mislead the entire analytical community.

Use of CRMs

Calibration, traceability, and accuracy

CRMs are mainly used to demonstrate the accuracy of an analytical method as applied in a laboratory, and consequently to show that the results obtained by this method are linked to the outside world. In other words, CRMs are able to demonstrate the traceability of the results first to the CRM, and consequently to the units in which the certified value is expressed. Some CRMs may also be used to directly calibrate measurements, e.g., certified solutions, alloys, pure metals, etc.

When you use a CRM you will know beforehand the real answer to be found.

Matrix CRMs should be reserved for the verification of the accuracy of methods and to demonstrate their traceability to the fundamental SI units (see Appendix). When validating the accuracy of a relative method, the analyst has to first choose a CRM with a matrix as close as possible to the natural samples he deals with. ISO guide 33 [3.3-11] informs the analyst on the use of matrix CRMs for the validation of the trueness of relative methods. The guide introduces adjustment methods in terms of precision, but also in terms of trueness. CRMs are best suited for this process.

Evaluation of laboratory reference materials

The organizations producing CRMs do not normally produce large quantities of RMs for intralaboratory QC (e.g., for control charts). It is the responsibility of the analysts or other bodies to produce such RMs in or outside his/her own organization. The analyst can compare these RMs with existing similar CRMs and in doing so demonstrate the accuracy of his method and traceability of his results. It must be noticed that this type of traceability of an RM to CRMs is not easy to achieve in one single laboratory, and it is preferable to have the RM analyzed by several laboratories which possibly also apply different methods.

3.4 Regulatory Aspects of QA and QC

Analytical sciences play an increasing role in decision making at official, legal, or private level. Therefore, it has been recognized for a long time by those who need analytical data, and in particular, at political or commercial level that the quality of the delivered data should be guaranteed. Several, sometimes dramatic, accidents in the pharmaceutical area, which were linked to insufficient reliability or traceability of analytical work pushed authorities to take measures to organize evaluation systems.

Pharmaceuticals, the chemical industry producing toxics, and increasingly nowadays, other large fields of chemical industry was the first area to be provided at international level with defined and strict rules, such as "Good Laboratory Practice (GLP)", for the control of the quality of analytical work. For calibration and testing laboratories, the majority of developed countries have set up so-called accreditation systems, some of which have been in force for a very long time (e.g., Australia). There is a tendency nowadays to try to combine EN 45000-ISO/IEC 25 and the GLP requirements into one single and worldwide recognized standard accompanied by guiding principles for all kind of laboratories. This would clarify and simplify the system of laboratory evaluations and sometimes avoid duplication of audits.

3.4.1 Evaluation of laboratories

Good laboratory practice (GLP)

Good laboratory practice is the first attempt to regulate/control industrial practice in the chemical field. GLP mainly verifies that the information on practical activities is traceable.

GLP regulations were first developed in 1978 by the US Food and Drug Administration (FDA). The US Environmental Protection Agency (EPA) issued in 1983 similar regulations for the production of agricultural and industrial toxic chemicals. Following also pioneer work by the WHO, the Organization for Economic Cooperation and Development (OECD) in its C81/30 decision [3.4-1] took over in 1981 the GLP guidelines which then became the rule in all OECD member countries for the quality control of pharmaceutical and chemical toxics (e.g., pesticides) producers. The Council of the European Community adopted several directives for the harmonization (87/18/EEC), the inspection and verification of GLP regulations (88/320/EEC and 90/18/EEC).

In their field of application, GLP are concerned with the organizational process and the conditions under which laboratory studies are planned, performed, monitored, recorded, and reported [3.4-1]. In fact they intend to guarantee that all information and actions undertaken in the course of product development, safety investigations, and toxicity testing are traceable, and at any moment can be found back. W. Merz et al. [3.4-2] resume the objective of GLP as follows: "GLP is therefore only intended to provide a comprehensive insight into the aims, planning, performance, evaluation, and reporting of trials. No more and no less. GLP is not concerned with the requirements or aims of a test or with the interpretation of its results ...".

In other words, following GLP rules does not guarantee the accuracy of test results. Other tools, such as the use of CRMs, and participation into proficiency-testing studies can demonstrate accuracy. GLP as well as accreditation systems only improve the quality of the working environment and the way laboratories function. In doing so, they are able to recommend measures for achieving accuracy.

Accreditation systems

GLP guidelines have been set up for toxicological investigations of newly developed chemicals or pharmaceuticals. Gradually, more and more scientific areas and types of industries have been involved in GLP, especially for safety aspects of new products. General testing laboratories are usually not involved and the GLP rules are mostly not applied to their work. For such laboratories or for other type of industries, the authorities have set up other evaluation systems some of which have been existing for several years. Depending on the field of activity, e.g., food, environment, and biomedical analysis, the control systems took the form of administrative inspections, proficiency testing, or both simultaneously. In several countries the control systems for testing laboratories are based on the ISO/IEC Guide 25 issued in 1978 [3.3-2]. Within the European Union the resolution 90/C/10/01 of the Council of Ministers of 21 December 1989 on "a global approach to conformity assessment" adopted guiding principles mentioned in the European series of norms EN 45000 [3.3-3] for testing laboratories and EN 29000 series (directly derived from ISO/IEC 9000), [3.1-6] for Industrial activities. In fact, the testing laboratory accreditation system involves seven standards:

Accreditation gives a standardized frame for the proper operation of testing laboratories.

- EN 45001: general criteria for the operation of testing laboratories
- EN 45002: general criteria for the assessment of testing laboratories
- EN 45003: general criteria for laboratory accreditation bodies
- EN 45011: general criteria for certification bodies operating product certification
- EN 45012: general criteria for certification bodies operating Quality System certification
- EN 45013: general criteria for certification bodies operating certification of personnel
- EN 45014: general criteria for suppliers' declaration of conformity

The Council of Ministers resolution promotes a mutual recognition principle between the accreditation systems developed in the various member states. The EN 45001 norm lists and recommends a number of quality assurance and quality control items which should be operated by testing laboratories. These include recommendations on the management, infrastructure, and competence of personnel, equipment, and working procedures and, in particular, detailed aspects with regards to calibration and use of reference materials. No statement on a mandatory participation into proficiency testing schemes is given. The organization of proficiency testing is left to the appreciation of the accreditation body. The EN 45002 norm even clearly states that the accreditation of a laboratory cannot be granted or maintained exclusively on the basis of the results obtained by a laboratory in proficiency testing. Also several European countries not belonging to the European Union, e.g., those which ratified the European Free Trade Agreement (EFTA) have adopted the system. All accreditation bodies of Western Europe are cooperating within the European Cooperation for Accreditation of Laboratories (EAL) for a common implementation and development of their systems. In non-European countries similar accreditation systems exist, e.g., NATA in Australia, A2LA in USA; at worldwide level the accreditation bodies are cooperating within the International Laboratory Accreditation Cooperation (ILAC). Several international scientific societies, e.g., IUPAC, AOAC, also associate their efforts to develop common approaches, a common vocabulary, and to set up common guidelines or common means and tools for the improvement of the quality in testing laboratories.

ISO 9000/EN 29000 standards

ISO 9000 is the basic standard for quality systems of industrial activities and services.

GLP requirements aim to ensure public safety for chemical products. The ISO 9000 series of norms [3.1-6] is intended to ensure quality for commercial purposes and for all types of products or services. In 1987, CEN – the European Standards Institution – took over the ISO 9000 norms as the EN 29000 series of norms which were included into the Council resolution 90/C/10/01 of 21 December 1989. The ISO 9000/EN 29000 series of norms include five standards:

- ISO 9000/EN 29000: general guidelines, principles and definitions
- ISO 9001/EN 29001: quality system specifications and requirements for design of production and service
- ISO 9002/EN 29002: quality system specifications and requirements for production and installation
- ISO 9003/EN 29003: quality system specifications and requirements for final inspection and testing
- ISO 9004/EN 29004: guidance on quality management

In ISO 9001, 9002 and 9003, and their EN 29000 counterparts, measurements and quality requirements for measurements or tests are required. Therefore, analytical chemists employed in industries or service laboratories pursuing EN 29000 certification will see their activity audited and controlled.

3.4.2 Accreditation of RM and CRM producers

Reference materials and certified reference materials are essential tools for the assessment of the reliability and comparability of analytical results. Therefore, the quality of the RMs and CRMs, and the scientific and technical ability of the producers should be assessed. The number of producers especially of RMs has increased over the last few years, and quality of the produced materials was not always demonstrated. In order to help producers to perform their task in the most reliable manner and to allow customers to estimate the competence of RM or CRM producers, ISO has started a project on the quality requirements for the production of RMs and CRMs. The work has been conducted under the guidance of the ISO 9000 series of standards and the ISO Guide 25. A draft ISO guide giving the interpretation of ISO 9000 and ISO Guide 25 has been produced. It provides a mechanism for demonstrating that the RM and CRM production is carried out in accordance with the requirements of ISO 9000 and ISO Guide 25. This draft guide covers various requirements and items:

- *Organizational requirements*: management, quality policy and system, staffing and training, contracts and collaboration, storage and long-term monitoring of RM and CRM stability, recording and reporting, after sale service
- *Production control*: planning, preparation of materials, traceability and calibration, measurement equipment and methods, assessment of homogeneity and stability, data treatment and certification

The technical requirements are those set up in the ISO Guides 31 [3.4-3] and 35 [3.2-5]. If successful, such pilot work by ISO could lead to the certification of RM and CRM producers. This would guide customers of RMs and CRMs in their choice of business partner.

3.4.3 Certification of chemists

The first certificate a chemist receives is the scientific degree recognizing successful studies. After this, the analyst has to maintain and improve his knowledge by regularly consulting scientific literature, by participating in scientific events, or by undergoing performance audits within the laboratory. When taking part in inter-laboratory performance studies (see Sec. 3.3.3) the analyst has the opportunity to evaluate his performance and to check the need for training. The chemist should find within his laboratory aids for training, e.g., on new instruments or techniques, on using new analytical parameters, and for investigating new fields of activities. Accreditation of laboratories, in general, requires that the accredited laboratory foresees and implements a programme for training and the continuous education of the staff.

When graduated, the chemist may only get additional recognition from his peers. Very few official postgraduate evaluation systems exist which deliver recognized diplomas. Scientific societies organize postgraduate continuing education programmes in nearly all western countries and deliver certificates of attendance; several universities organize teaching and training courses in specialized fields of analytical chemistry. In the UK, the Royal Society of Chemistry (RSC) has set up since several years ago a Register of Chartered Analytical Chemists; candidature is reserved to members of the RSC and is accepted only under certain conditions, e.g., written support from three sponsors. Similar registers exist in other chemical societies (e.g., clinical chemistry society), and the Federation of European Chemical Societies has a corresponding qualification of European Chemist [3.3-4].

The European Standard EN 45013 [3.3-3] defines criteria for certification bodies operating certification of personnel in the general frame of accreditation of laboratories. This certification may not only be reserved for analytical chemists but also for more managerial or QA officers.

3.4.4 Standardization

Written standards

Written standards (norms) are a way of defining and promoting requirements for product quality and for improving comparability of analytical results. Written standards are also a first step for the introduction of these minimal requirements of quality and comparability into regulatory systems. Regulations at legal or commercial level often refer to standards for the implementation of the regulation. In the field of analytical chemistry, written standards have shown their importance for many applications, such as sampling strategies and techniques, the definition and the measurement of certain global parameters, e.g., total organic carbon, total organic halogens, or fractions of substances such as extractable, leachable, bioavailable fractions of compounds.

In other fields, the analysts often regret the existence of written standards especially when they are bound to regulations. This situation is encountered for standardized analytical methods which are outdated, but still mandatory, as the legislation has not been revised. In situations where the state of the art has improved, the analyst may be obliged to use old-fashioned methods just for legal reasons. Such situations are encountered in the pharmaceutical field, in food control, or in environmental monitoring.

The application of a standardized method is not a guarantee that errors are being avoided by the analyst. In fact, large disagreements have often been noticed in BCR projects between laboratories applying the same standard [3.4-5]. These may be due to systematic errors, improper execution of the method by operators, or due to having ambiguous or insufficient wording of the standard. In fact, standard procedures are often elaborated on by panels of senior analysts who may forget that the procedures will be applied by less qualified personnel in charge of daily measurements. The standard is also often an agreement mainly among senior analysts, so there is no guarantee that the wording is clear to everybody

Written standards or "norms" are the pathway to introducing complex technical items into legislation.

intending to apply the standard. Therefore, ruggedness testing is a must for every standard before adopting it.

Standardization bodies have recognized the difficulty of applying standards and the need to allow progress in analytical sciences to flow into standardized methods. Therefore, in recent written standards more general analytical approaches have been used to give general principles for those methods still requiring performance characteristics (performance standards). Such standards may also recommend to demonstrate that the requested performance is achieved, i.e., upon the analysis of CRMs.

Standardization bodies

Standardization bodies exist at national, regional, and international level. They may depend on public authorities entirely or partly, or they may belong to professional or commercial organizations. For European countries organized in the European Union and the European Free Trade Association, CEN/CENELEC/ETSI are producing European standards which are gradually replacing national standards. At international level more than a hundred countries are collaborating in ISO. ISO and CEN passed an agreement in 1991 in Vienna to avoid an overlapping of tasks. Following this agreement, CEN takes over some written standards developed in ISO when they fulfil the needs of CEN. ISO and CEN cover large fields of activities. Hundreds of technical committees (TCs) exist which support each the work of several working groups. The procedure for the adoption of new standards is very time-consuming, in particular for internationally recognised norms (typically several years). Costs are very high and are mainly supported by industry, public institutions, national standardization bodies, or by the European Commission when the standard is requested to support regulations issued by the European Union.

Professional organizations are also producing written standards which may also be adopted by ISO or CEN afterwards. A typical example is given by the International Dairy Federation (IDF) which produces standards for milk and dairy products and which are adopted by ISO.

3.5 Conclusion

Cheaper is not always better, better is always cheaper (Deming).

Quality assurance and quality control are essential in helping analytical chemists to deliver reliable and meaningful answers to their customers. Analytical results which have been delivered without any quality statements should be considered as counterproductive data: "better to have no result than to have a wrong result". W.E. Deming, a famous American pioneer in the field of quality assurance and quality control in production also stated that: "*cheaper is not always better but better is always cheaper*". As can be deduced from the items which have been briefly addressed in this chapter, to achieve quality the analyst must have a good knowledge of the fundamental sciences, e.g., chemistry, mathematics and statistics, physics, biochemistry, and sometimes biology, of instruments, but also of management, and he must demonstrate economic competencies. What a challenge!

References

[3.1-1] Uriano, G.A., Gravatt, C.C., The role of reference materials and reference methods in chemical analysis. *CRC Critical Review in Analytical Chemistry*, October 1977.

[3.1-2] Hertz, H.S., *Anal. Chem* 1988, 60(2), 75A–80A.

[3.1-3] Quinn, T.J., Bankvall, C., Harrington, M.G., Machado Jorge, H., Repussard, J., Reuter, H.W., *Evaluation of the BCR Programme 1988–1992, Measurement and Testing in Europe*, Report EUR 15041 EN, Commission of the European Communities, Luxembourg, 1992.

[3.1-4] Tölg, G., private communication, 1982.

[3.1-5] ISO, *Quality Vocabulary* (ISO/IEC Standard 8402), International Organization for Standardization, Geneva, CH, 1986.

[3.1-6] ISO, *Quality management and Quality Assurance Standards. Guidelines for Selection and Use* (ISO/IEC Standard 9000), International Organization for Standardization, Geneva, CH, 1987.

[3.1-7] Broderick, B.E., Cofino, W.P., Cornelis, R., Heydorn, K., Horwitz, W., Hunt, D.T.E., Hutton, R.C., Kingston, H.M., Muntau, H., Baudo, R., Rossi, D., van Raaphorst, J.G., Lub, T.T., Schramel, P., Smyth, F.T., Wells, D.E., Kelly, A.G., *Mikrochim. Acta* (Wien), II (1–6), 523–42, 1991.

[3.1-8] Valcarcel, M., Rios, A., *Anal. Chem* 1993, 65(18), 781A–787A.

[3.1-9] Cofino, W.P., in: *Environment Analysis Techniques, Applications and Quality Assurance – Techniques and Instrumentation in Analytical Chemistry*, Barcelo, D. (Ed.). Amsterdam: Elsevier Science Publishers, 1993; Vol. 13, pp. 79–105.

[3.2-1] BIPM/IEC/ISO/OIML, *International Vocabulary of Basic and General Terms in Metrology*, International Organization of Standardization, Geneva, CH, 1984.

[3.2-2] ISO (1977), *Statistics – Vocabulary and Symbols – Part 1: Probability and General Statistical Terms Revision of ISO 3534*: 1977 to be published, International Organization for Standardization, Geneva, CH.

[3.2-3] OIML, *Vocabulary of Legal Metrology, International Organisation for Legal Metrology*, Paris, F., 1978.

[3.2-4] Garfield, F.M., *Quality Assurance Principles for Analytical Laboratories*. 2nd ed. AOAC International Ed., Arlington VA, USA, 1991.

[3.2-5] ISO, *Certification of Reference Materials – General and Statistical Principles*, ISO Guide 35-1985 E, International Organization for Standardization, Geneva, CH, 1985.

[3.2-6] Marschal, A., *Calibration of Chemical Analyses and Use of Certified Reference Materials*, Draft ISO Guide 32, ISO/REMCO N 262 Rev. August 1993, International Organization for Standardization, Geneva, CH, 1993.

[3.2-7] EXERA, SIREP and WIB, *International Instrument Users' Associations, Studies on the compliance of analytical instruments*, EXERA – Parc Technologique ALATA, BP 2, F-60550 Verneuil en Halatte, F, 1993.

[3.2-8] Massart, D.L., Vandeginste, B.G.M., Deming, S.N., Michotte, Y., Kaufmann, L., *Chemometrics: a textbook. Data Handling in science and technology* – Vol. 2, Amsterdam: Elsevier Science Publishers B.V., 1988.

[3.2-9] Wells, D.E., Maier, E.A., Griepink, B, *Intern. J. Environ. Anal. Chem* 1992, 46, 255–64.

[3.2-10] Moody, J.R., Greenberg, R.R., Pratt, K.W., Rains T.C., *Anal. Chem* 1988, 60(21), 1203A–18A.

[3.2-11] Maier, E.A., Griepink B., Muntau, H., Vercoutere, K., *Certification of the total content (mass fractions) of Cd, Co, Cu, Mn, Pb, Ni and Zn and the aqua regia soluble contents (mass fractions) of Cd, Pb, Ni and Zn in a light sandy soil* (CRM 142R). Report EUR 15283 EN, Commission of the European Communities, Luxembourg, 1993.

[3.2-12] Wells, D.E., in: *Environment Analysis Techniques, Applications and Quality Assurance – Techniques and Instrumentation in Analytical Chemistry*, Barcelo, D. (Ed.). Amsterdam: Elsevier Science Publishers B.V., 1993, Vol. 13, pp. 79–105.

[3.2-13] Mullholland, M., Walker, N., van Leuven, J.A., Buydens, L., Maris, F., Hindriks, H., Schoenmakers, P.J., *Mikrochim. Acta* (Wien), 1991, II (1–6), 493–503.

[3.2-14] AOAC (1990), *Official Methods of Analysis*. 15th ed. Association of Official Analytical Chemists, Arlington, VA, USA.

[3.3-1] Horwitz, W., *Nomenclature for interlaboratory studies*, 4th draft, IUPAC, Analytical Chemistry Division, Commission V1, project 27/87.

[3.3-2] ISO, *Guidelines for Assessing the Technical Competence of Testing Laboratories*, ISO/IEC Guide 25-1978 E, International Organization for Standardization, Geneva, CH, 1978.

[3.3-3] CEN, *General Criteria for the Operation of Testing Laboratories (European Standard 45001)*, CEN/CENELEC, Brussels, B, 1989.

[3.3-4] Thompson, M., Wood, R., *The international harmonised protocol for the proficiency testing of (chemical) analytical laboratories*, IUPAC/ISO/AOAC, Pure & Appl. Chem 1993, 65(9), 2123–44.

[3.3-5]   Maier, E.A., Quevauviller, Ph., Griepink, B., Rymen, T., van der Jagt, H., van Rooij, M.A.F.P., *The role of interlaboratory studies in the improvement of the quality of chemical measurements*, first draft, IUPAC, Analytical Chemistry Division, Commission V2, project QA 2/91, 1992.

[3.3-6]   ISO, *Terms and definitions used in connection with reference materials*, ISO Guide 30 revised version 1991, International Organization for Standardization, Geneva, CH, 1991.

[3.3-7]   Griepink, B., *Intern. J. Environ. Anal. Chem* 1993, 51, 123–128.

[3.3-8]   Ure, A., Quevauviller, Ph., Muntau, H., Griepink, B., *Improvement in the determination of extractable contents of trace metals in soil and sediments prior to certification*, Report EUR 14763 EN, Commission of the European Communities, Luxembourg, 1993.

[3.3-9]   Maier, E.A., in: *Environment Analysis Techniques, Applications and Quality Assurance – Techniques and Instrumentation in Analytical Chemistry*, Barcelo D. (Ed.). Amsterdam: Elsevier Science Publishers, 1993, Vol 13, pp 383–401.

[3.3-10] Wise, S.A. (1993), in: *Environment Analysis Techniques, Applications and Quality Assurance – Techniques and Instrumentation in Analytical Chemistry*, Barcelo D. (Ed.). Amsterdam: Elsevier Science Publishers B.V., 1993; Vol. 13, pp. 403–446.

[3.3-11] ISO, *Uses of certified reference materials*, ISO Guide 33, International Organization for Standardization, Geneva, CH, 1989.

[3.4-1]   OECD, *Decision of the Council*: C81/30 (final) Annex 2, *OECD Guidelines for Testing Chemicals, "OECD Principles of Good Laboratory Practice"*, OECD Paris, F, 1981.

[3.4-2]   Merz, W., Weberruss, U., Wittlinger, R., *Fres. J. Anal. Chem* 1992, 342, 779–782.

[3.4-3]   ISO, *Contents of certificates of reference materials*, ISO Guide 31, International Organization for Standardization, Geneva, CH, 1981.

[3.4-4]   Thomas, J.D.R., *Fres. J. Anal. Chem* 1993, 347, 25–28.

[3.4-5]   Marchandise, H., *Fres. Z. Anal. Chem* 1987, 326, 613–16.

Questions and problems

1. What is the objective of the analytical chemist?
2. Who defines the analytical strategy?
3. What are QA and QC?
4. Is it possible/worthwhile/necessary to set up quality control without quality assurance?
5. Who is responsible for the validation of an instrument?
6. Why and when is it necessary to validate an instrument?
7. What are the main quality characteristics (primary and secondary) of an analytical method (give the definitions)?
8. What does the validation of the analytical procedure cover?
9. Why and when is it necessary to validate a method?
10. How is the pretreatment step of a method validated?
11. How is the calibration step validated?
12. What are the aspects/items to be considered in the calibration step?
13. Reproducibility and repeatability: what are the differences?
14. What are the definition and importance of ruggedness and robustness?
15. Is the customer responsible for the results he gets from the analyst?
16. What is an interlaboratory study? What are their different applications?
17. What is a reference material? Give the main applications of RMs?
18. What are the characteristics, properties, and preparation of RMs?
19. How is the homogeneity and the stability of RMs verified?
20. What is the definition and role of certified reference materials?
21. GLP, accreditation and certification of laboratories: what do these terms mean and which activities do they concern?
22. Is accreditation reserved to testing laboratories?
23. Control charts and control points: what are their concepts and places in the validation scheme?
24. How is accuracy achieved?

25. Are precision, trueness, and accuracy necessary for assuring comparability of results between laboratory results?
26. Is quality assurance a luxury or an economical necessity?
27. Can or do written standards (norms) guarantee good results?
28. What should one do when one has delivered poor results in an intercomparison?
29. What are the relations between accreditation and proficiency testing (laboratory performance studies)?
30. Who organizes and is responsible for accreditation?
31. Is trueness of results sufficient to give meaningful answers to customers?
32. Is precision sufficient to draw meaningful conclusions from analytical results?
33. Can you list socioeconomic situations where accuracy is necessary to reach the set objectives?
34. What are the various types of chemical analytical methods?
35. What does traceability of analytical results mean?
36. How is traceability achieved and to what is it traceable?

Part II
Chemical Analysis

In this part of the book, we consider the chemical basis of analytical chemistry, i.e., analytical investigations which depend on chemical reactions. The reason for including chemical reactions is to render compounds accessible to separation techniques (e.g., extraction, chromatography), or to prepare them for qualitative (identification) or quantitative analyses. Many separation procedures are themselves based on chemical reactions, and the appropriate theoretical foundations are a prerequisite for technical developments in the separation sciences.

In order to become an experienced analytical chemist one must have basic knowledge in general, physical, inorganic, and organic chemistry. However, there it is a tendency for modern analytical chemistry to expand into fields that are beyond the boundaries of classical chemistry, and analytical chemistry has therefore become an interdisciplinary science, requiring additional know-how from biology, medicine, materials science, and microtechnology.

Chemical analysis is that part of analytical chemistry where the reagent which reacts with the sample under investigation is a chemical.

4 Fundamentals of Chemical Analysis

The fundamentals of chemical analysis deal with the chemical reactions used in analytical investigations. These may be reactions designed to prepare a sample for a particular analytical step, such as the quantitative analysis of sample components (e.g., precipitation, spectroscopic identification), to exclude matrix effects (e.g., in sensor performance), and to derivatize a certain compound for improved separation and detection (e.g., pre- and post-column derivatization in chromatography, Chap. 5).

The most important principles are equilibria in homogeneous and heterogeneous systems, and steady-state conditions.

4.1 Equilibria in Homogeneous Systems

Learning objectives

■ To give a description of the equilibrium state and the mechanisms leading to it

■ To distinguish between homogeneous and heterogeneous equilibria, and between single-component and multi-component systems

■ To outline the role chemical and physical equilibria play in analytical chemistry. To emphasize the difference between equilibrium and steady-state, and to discuss the relations between the equilibrium constant and thermodynamic properties

4.1.1 Introduction

Most of the analytical procedures used for quantitative determination of chemical species are based on reactions that take place in homogeneous solutions. The resulting chemical equilibrium states are confined to a single phase and are called *homogeneous equilibria*. In contrast, most separation methods make use of reactions in which a second phase is formed (e.g., a precipitate) or where chemical species are distributed between two or more coexisting phases. The resulting equilibrium states are called *heterogeneous equilibria*. Homogeneous and heterogeneous systems may be a single chemical or may consist of several components. Such systems are called *single-* and *multi-component* systems.

In Section 4.1, we discuss homogeneous equilibria, related analytical topics, and examples of homogeneous equilibria important for the analytical chemist, whereas heterogeneous equilibria and related topics are considered in Section 4.5. A homogeneous phase involves a single form of state and is therefore separated from its surroundings by a definite *phase boundary*.

Homogeneous vs. heterogeneous systems.

4.1.2 General considerations of the equilibrium state

In a system in which a multitude of substances interact with each other, certain chemical changes take place until a state is reached that seems to be at rest. This final state is called the *equilibrium state* of the system.

Every chemical equilibrium state can be described by the *equilibrium constant K*. This is a most useful quantity and can be used to describe a vast range of analytically important chemical processes, such as acid–base equilibria, complex formation, and redox reactions. Heterogeneous equilibria play an important role in modern analytical separation techniques, such as extraction and chromatography. The equilibrium constant is the means by which all types of chemical equilibria can be introduced and described explicitly. The equilibrium constant is thus a powerful means of explaining or simulating particular features of chemical processes in systems such as natural and polluted waters, and it is the basis for the introduction and explanation of a variety of analytical methods and separation techniques.

The understanding of a number of thermodynamic concepts is crucial to the description of all kinds of chemical equilibria. In addition, these quantities are extremely useful in equilibrium calculations, showing that physicochemical concepts apply not only to conditions artificially created in the laboratory, but also to natural processes. The equilibrium constant is related to certain thermodynamic quantities such as the *free energy*, *enthalpy* (*heat content*), and *entropy* of the system:

$$\ln K = \frac{\Delta G^\circ}{RT} \tag{4.1-1}$$

and

$$\Delta G° = \Delta H° - T\Delta S° \tag{4.1-2}$$

where K is the equilibrium constant, $\Delta G°$, $\Delta H°$, and $\Delta S°$ are the standard free energy, standard enthalpy, and standard entropy of the reaction, and T is the temperature in Kelvin. R is the *gas constant* (8.314 J mol^{-1} K^{-1}).

4.1.3 Equilibrium systems

If chemicals are mixed together, reactions may occur until an equilibrium state is reached. At this point, no further changes are observed at the macroscopic level. Changes may be of a chemical or physical nature, and phase transitions may be involved. The progress toward equilibrium is best described by an example in which changes of physical parameters occur. Let us consider two equally large compartments, separated by a common wall, each filled with the same amount of a different ideal gas, at the same pressure (e.g., A and B in compartments I and II, respectively). Owing to the thermal motion of the gas molecules, they constantly collide with each other, and with the vessel walls.

If an orifice in the separating wall is opened, the molecules are free to move within the combined volumes (Fig. 4.1-1a–d). Immediately after opening the orifice, there is a net flow of A into compartment II, and of B into compartment I. Flow rates are high immediately after opening, since A molecules that move in the appropriate direction can enter the compartment II (Fig. 4.1-1b). To flow back in the reverse direction they must first collide with other gas molecules or the walls. At a somewhat later time, the left-hand side of the container is still rich in A and the right-hand side still contains mostly B molecules (Fig. 4.1-1c). As time elapses the net flow of A in one direction and B in the other slows down because the molecules move back from where they came, and in the final state (or equilibrium) the two gases are mixed completely and any sample taken from the container contains A and B in a 1:1 ratio (Fig. 4.1-1d).

In the margin: In chemical analysis, "equilibria" describe not a state of reactions at rest but a dynamic state where both reactions (+ and −) proceed at the same velocity.

In the final, equilibrium state, the thermal motion of the molecules continues as before, but the same amount of gas A moves in each direction. These movements are random and cancel each other. Therefore, at the macroscopic level, the system seems to be at rest. Such a system is said to be in *dynamic equilibrium*. In the equilibrium state, there is no net flow of energy or mass within or through the system. All components remain in the system, and they are characterized by a *residence time* τ_R equal to infinity. An equilibrium system is thus a *closed system* in a *time-invariant state*. All chemical and physical equilibria represent dynamic equilibrium states of closed systems. This is demonstrated by a practical example. If carbon dioxide is dissolved in water, carbonic acid is formed, and one can write the following reaction:

$$CO_{2(aq)} + H_2O \rightarrow H_2CO_3 \tag{4.1-3a}$$

The symbol (aq) refers to the fact that dissolved rather than gaseous CO_2 is involved in the reaction. As more CO_2 is dissolved, more H_2CO_3 is formed. The rate of carbonic acid formation is first-order with respect to the concentration of $CO_{2(aq)}$, as expressed by the following equation:

$$\frac{d[H_2CO_3]}{dt_{formation}} = k_1[CO_2]_{aq} \tag{4.1-4}$$

where $k_1 = 0.03$ s^{-1} at 25 °C (forward reaction rate constant). This reaction is accompanied by a structural change, CO_2 is a linear molecule best described by the structure:

$$:O::C::O: \quad \text{or} \quad O=C=O \tag{4.1-5}$$

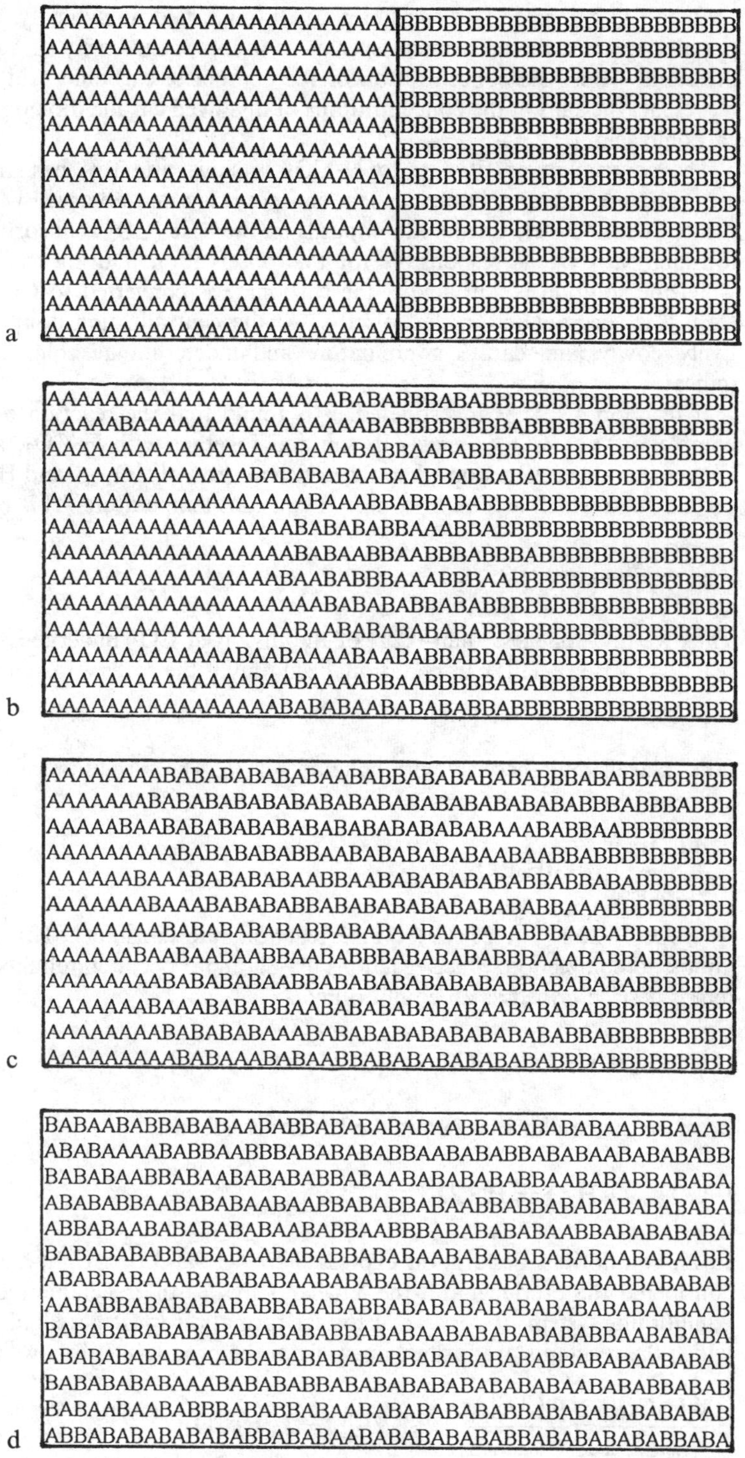

Fig. 4.1-1. Illustration of the equilibrium state a) Two gases A and B, of equal volume and at equal pressure, separated by a common wall; b) The separating wall is removed and the gases spread out into the common volume. The mixing has just started; c) The gases mix through diffusional motion, but areas remote from the former separation wall will be fully mixed later than the center of the container; d) Complete mixing. Diffusion still goes on, but this is not noticed at the macroscopic level

whereas carbonic acid is a planar molecule with new bonds:

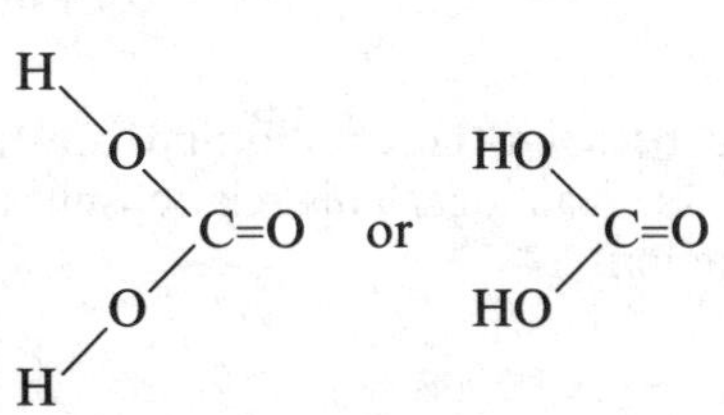

$$
\begin{array}{ccc}
\ce{H-O} & & \ce{HO} \\
\quad\ \ce{C=O} & \text{or} & \quad\ \ce{C=O} \\
\ce{H-O} & & \ce{HO}
\end{array}
\tag{4.1-6}
$$

It is therefore quite clear that $CO_{2(aq)}$ and H_2CO_3 are chemically different species, although it is analytically difficult to distinguish between them. However, the difference can be established from kinetic investigations.

Since $CO_{2(aq)}$ and H_2CO_3 cannot be distinguished analytically, it is common practice to treat these chemical species as a composite entity in equilibrium expressions. Thus, the concentration of the composite carbonic acid, $[H_2CO_3]_{total}$, represents the sum of the concentrations of dissolved carbon dioxide, $[CO_2]_{aq}$, and carbonic acid, $[H_2CO_3]$.

The conversion of $CO_{2(aq)}$ to H_2CO_3 is slow, and this has analytical consequences. A fading phenolphthalein end point is often observed when an acid is titrated with NaOH. Carbonate impurity in the base (CO_2 is absorbed by NaOH, forming Na_2CO_3) is responsible for this observation. During the titration, i.e., when the solution is still acidic, the carbonate is converted to $CO_{2(aq)}$ in which form it is accumulated in the solution. At the neutralization point it reacts relatively slowly, and causes a premature and indeterminate color change of the indicator.

If reaction 4.1.3a is investigated, it is found that the reaction does not go to completion, i.e., $CO_{2(aq)}$ exists in solution together with H_2CO_3, and the interconversion appears to stop when the concentration of $CO_{2(aq)}$ and H_2CO_3 reach a certain ratio which is time-invariant at constant temperature. This ratio is:

$$\frac{[CO_2]_{aq}}{[H_2CO_3]} = 670 \text{ at } 25\,°C \tag{4.1-7}$$

Thus we find that less than 0.2% of the dissolved CO_2 undergoes the hydration reaction (4.1-3a). This can be rationalized only if the reverse reaction:

$$H_2CO_3 \rightarrow CO_{2(aq)} + H_2O \tag{4.1-3b}$$

takes place simultaneously with reaction 4.1-3a. The formation of $CO_{2(aq)}$ can be expressed by:

$$\frac{d[CO_2]_{aq}}{dt_{formation}} = k_2[H_2CO_3] \tag{4.1-8}$$

where $k_2 = 20\,s^{-1}$ at $25\,°C$ (reverse reaction rate constant). This reaction goes almost to completion. Since reactions 4.1-3a and b occur simultaneously, the system is best described as an equilibrium:

$$CO_{2(aq)} + H_2O \underset{k_1}{\overset{k_2}{\rightleftharpoons}} H_2CO_3 \tag{4.1-3c}$$

or, since the equilibrium favors the left-hand side, as:

$$CO_{2(aq)} + H_2O \underset{k_1}{\overset{k_2}{\rightleftharpoons}} H_2CO_3 \tag{4.1-3d}$$

In an equilibrium description expressed by 4.1.3a–d, the right-hand species are considered the products and the left-hand species the reactants (educts). For an equilibrium system, the rate of product formation must be equal to the rate at which the reactants are formed again from the products. It follows that:

$$\frac{d[H_2CO_3]}{dt_{formation}} = \frac{d[CO_2]_{aq}}{dt_{formation}} = k_1[CO_2]_{aq} = k_2[H_2CO_3] \tag{4.1-9}$$

Rearranging, one obtains:

$$\frac{[CO_2]_{aq}}{[H_2CO_3]} = \frac{k_2}{k_1} = \frac{1}{K} = \frac{20\,s^{-1}}{0.03\,s^{-1}} = 670 \text{ at } 25\,°C \tag{4.1-10a}$$

This equation shows the inherent connection between the equilibrium constant and the rate constants of the forward and reverse reactions. This relationship can be represented in general by the reversible reaction:

$$aA + bB + \cdots \rightleftharpoons cC + dD + \cdots \tag{4.1-11}$$

where a, b, c, d represent the stoichiometric coefficients of the reactants A, B, and products C, D. The rate of the forward reaction v_f is given by:

$$v_f = k_1[A]^a[B]^b \cdots \tag{4.1-12}$$

and the rate of the reverse reaction, v_r by:

$$v_r = k_2[C]^c[D]^d \cdots \tag{4.1-13}$$

At equilibrium, v_f and v_r are equal, and therefore:

$$k_1[A]^a[B]^b \cdots = k_2[C]^c[D]^d \cdots \tag{4.1-14}$$

It follows that the equilibrium constant K is:

$$K = \frac{k_1}{k_2} = \frac{[C]^c[D]^d \cdots}{[A]^a[B]^b \cdots} \tag{4.1-15}$$

Equilibrium constant $K = \dfrac{k_+}{k_-}$.

The kinetic approach also demonstrates that the magnitude of the equilibrium constant changes with temperature. The thermal motion of the molecules decreases with decreasing temperature. Accordingly, the reactivity of the reactants and products is decreased and the rate constants k_1 and k_2 assume smaller values. These effects of temperature are not the same for the forward and reverse reaction, but depend on the different activation energies of the two processes. The corresponding rate constants at $0\,°C$ are:

$$k_2 = 2.3\,s^{-1} \tag{4.1-16}$$

and

$$k_1 = 0.0024\,s^{-1} \tag{4.1-17}$$

resulting in an equilibrium ratio:

$$\frac{[CO_2]_{aq}}{[H_2CO_3]} = \frac{1}{K} = 950 \text{ at } 0\,°C \tag{4.1-10b}$$

4.1.4 The steady state

The *steady state* is the *time-invariant state* of an open system in which mass and/or energy flows through the system. To reach a steady state, the input and output of energy and/or mass must be balanced, and remain constant. This condition is satisfied when the flow of energy and/or mass is maintained by large reservoirs connected to the steady-state system. The components of the system flow through it and are characterized by a finite residence time. The resulting time-invariant properties exhibited by the system may be compared with certain equilibrium properties, such as the equilibrium constant, and in many cases the steady state may be approximated by an equilibrium model. Most natural water systems are steady states rather than equilibrium systems. In seawater, components are added to the ocean at the same rate as they are removed, e.g., by precipitation reactions. Owing to the relatively large residence times of seawater components (of the order of 10^3 years), the chemistry of the oceans can be treated by equilibrium models.

Many natural cycles, such as the carbon cycle (aspiration and respiration) and the hydrological cycle (responsible for our freshwater resources) represent steady-state systems which are dependent on the energy input provided by the sun's radiation and the loss of energy from the earth's surface (usually as heat). In this respect the earth is an entropy pump.

Example of a steady state system

The steady state and its quasi-equilibrium properties are well illustrated by a radioactive decay scheme in which parent substance plays the role of the reservoir. There are three natural decay series called the uranium, thorium, and actinium series in which U^{238}, Th^{232}, and U^{235} are the mother substances. They are characterized by extremely large decay constants, comparable to the age of earth.

U^{238}, the mother substance of the uranium series (with mass number $n+2$), is unstable and decays to produce an unstable daughter substance, Th^{234} which

The radioactive decay as an example of steady state.

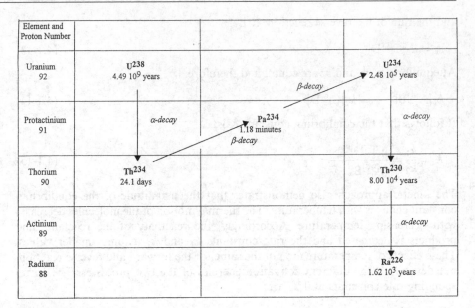

Element and Proton Number			
Uranium 92	U^{238} 4.49 10^9 years		U^{234} 2.48 10^5 years
Protactinium 91	α-decay	Pa^{234} 1.18 minutes β-decay	α-decay
Thorium 90	Th^{234} 24.1 days		Th^{230} 8.00 10^4 years
Actinium 89			α-decay
Radium 88			Ra^{226} 1.62 10^3 years

Fig. 4.1-2. The radioactive decay of U^{238} as an example of a steady state

in turn decays to further daughter substances (Fig. 4.1-2). A decay series may include a large number of different daughter substances, sometimes resulting in different isotopes of the same element (with mass numbers that differ by 4 units), such as Th^{234} and Th^{230}. All elements with atomic number higher than 81 (thallium) possess naturally occurring radioisotopes belonging to the three decay series. Some daughter substances are characterized by very short half-lives (e.g., At^{216} has a half-life, $t_{1/2} = 30\,\mu s$). The half-life and the residence time in such a reaction series are related quantities. The residence or life time τ_R represents the average time an individual chemical or physical unit remains unchanged in a system. The half-life $t_{1/2}$ represents the time during which exactly half of the original material has reacted to form another physical or chemical unit. Since:

$$t_{1/2} = \frac{\ln 2}{k} \quad \text{and} \quad \tau_R = \frac{1}{k} \tag{4.1-18}$$

where k represents the first-order rate constant (in radioactive decay systems it is called the *decay constant* λ), we have:

$$\tau_R = \frac{t_{1/2}}{0.693} = 1.443 t_{1/2} \tag{4.1-19}$$

The first steps of the uranium decay series are shown in Fig 4.1-2. It is observed that there is always the same amount of a particular daughter substance per unit mass of mother substance in the steady state.

4.1.5 Solvent–solute interactions

In a homogeneous equilibrium system, several chemical reactions compete simultaneously with each other and we must learn to classify them. In an aqueous solution the solute is often ionized (at least to some extent) and we refer to it as being an *electrolyte solution*. However, the solute ions are generally not "bare" ions, but exist as the products of specific interactions, which according to their physical nature may be classified as ion–ion, ion–dipole, or covalent interactions (Fig. 4.1-3). Another distinction is based on the chemical nature of the interactions, and we classify them as ion association, hydration (or solvation), and complexation (Fig. 4.1-4).

The two classifications are not synonymous. A complex is generally not the product of a purely covalent interaction. Indeed, in many complexes the ionic contribution may be significant or may even exceed the covalent contribution. Aluminum fluoride AlF_3 is an example of a complex which is held together by predominantly ionic bonds.

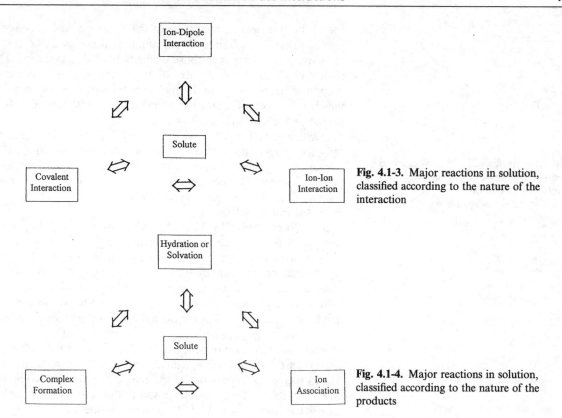

Fig. 4.1-3. Major reactions in solution, classified according to the nature of the interaction

Fig. 4.1-4. Major reactions in solution, classified according to the nature of the products

There is no absolute distinction between complex formation, hydration, solvation, and ion association processes. This can be demonstrated by consideration of the species found in aqueous hydrofluoric acid. The formation of the ion HF_2^- [4.1-1] (which belongs to the same category as $H(H_2PO_4)_2^-$ [4.1-2], $H(IO_3)_2^-$ [4.1-3], and $H(OAc)_2^-$ [4.1-4]) is mainly due to electrostatic interactions, and it can be treated as an ion associate. However, HF_2^- can also be considered as a complex in which the hydrogen ion represents the central atom and the fluoride ions the ligands. Furthermore, HF_2^- can be treated as a solvate. In liquid hydrofluoric acid, characterized by the self-dissociation:

$$HF \rightleftharpoons H^+ + F^- \qquad\qquad (4.1\text{-}20)$$

the formal ions H^+ and F^- do not exist as naked species. They are solvated, forming H_2F^+, HF_2^- or higher solvates. When proton transfer is involved, this reaction is also referred to as a protolysis, and H_2F^+, HF_2^- as protolysis products.

This example demonstrates that there are no firm distinctions possible between a hydrate (or solvate), a complex, and an ion associate, although a formal classification (Figs. 4.1-3 and 4.1-4) is very useful. We first consider the phenomenon of ion hydration.

Ionic hydration

The formation of a hydration shell around an ion can be considered as an ion–dipole interaction, and therefore the arrangements illustrated in Fig. 4.1-5 would be expected for an anion and a cation.

In this arrangement the hydration energy can be calculated on the basis of the Coulomb interaction between the point charge and the dipole (sum of the interactions between the ion and the two point charges of the dipole). The estimated values for many electrolytes are in good agreement with the experimental heats of hydration. This agreement is particularly good for ions possessing the electronic configuration of a noble gas. It is also expected that a small ion will be more strongly hydrated, owing to the greater ion–dipole interaction. Multivalent ions are thus expected to show a stronger hydration than univalent ions.

However, this model does not always give good agreement. For example, it is

Chemical reactions in polar solutions: electrolytes.

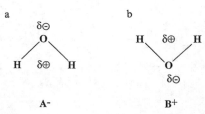

Fig. 4.1-5. Ion–dipole interactions in the hydration of (a) an anion and (b) a cation

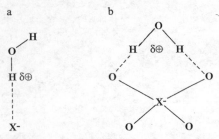

Fig. 4.1-6. Hydration of an anion, based on hydrogen bonds. (a) Small strongly electronegative anion; (b) Large complex anion with electronegative outer groups (oxyanion)

found that the heat of hydration of the silver ion Ag^+ is considerably larger than that of either the sodium or potassium ion, although the crystallographic radius of the silver ion lies between those of the two alkali metal ions. There are also other indications that the hydration of heavy metal ions is caused by complex formation. The color change during the process of dissolution and the slowness with which a number of transition metal compounds (e.g., Cr(III) and Ni(II) compounds) dissolve in aqueous solution indicate complex formation involving covalent interactions. A covalent interaction is possible only with cations, since only the oxygen atom of the water molecule can undergo covalent bonding (the covalently bonded hydrogen atom cannot form a second covalent bond).

The arrangement shown in Fig. 4.1-5a is not generally observed. Small ions, such as F^-, Cl^-, etc., interact with water molecules through hydrogen bonds, as indicated in Fig. 4.1-6a.

It has been shown by Everett and Coulson [4.1-5] (Fig. 4.1-6) that this structural form is characterized by a lower energy than in the dipole–ion interaction (Fig. 4.1-6a). But large complex anions such as the perrhenate ion ReO_4^- are able to stabilize the structure by forming additional hydrogen bonds between the water molecule and the electronegative atom of the complex anion (Fig. 4.1-6b).

We have seen that electrostatic and covalent bonds, in combination with hydrogen bonds, are responsible for the hydration of ions. For the sake of completeness, we must consider yet another kind of hydration, which is not restricted to ionic or dipolar species. Many gases, particularly those with a noble gas electron configuration, are able to form so-called *clathrates*. This is an encapsulation of the dissolved species in the cavities of the bulk water structure. The fact that even noble gases are able to form clathrates shows that the water is not chemically bonded, i.e., the solute is mechanically trapped in the water cavity. Clearly, only substances that possess a diameter smaller than the size of the cavity can be efficiently trapped. Chlorine forms such a clathrate, with the stoichiometric composition $6\,Cl_2 \cdot 46\,H_2O$.

4.1.6 Arrhenius theory of electrolyte dissociation

The Swedish chemist Svante A. Arrhenius (1859–1921) described the dissociation of electrolytes in the following way. An acid, base, or salt dissolved in water has the tendency to dissociate into its ions, so that an equilibrium state is established (the expression shown is valid for all $1:1$ electrolytes):

$$M_mA_a \rightleftharpoons mM^{\nu+} + aA^{\beta-} \tag{4.1-21}$$

The ions $M^{\nu+}$ and $A^{\beta-}$ can move independently in the solution. However, the equilibrium is described by the corresponding dissociation equilibrium constant K_{diss}:

$$K_{diss} = \frac{[M^{\nu+}]^m [A^{\beta-}]^a}{[M_mA_a]} \tag{4.1-22}$$

When an electric field is applied to the solution it exhibits some conductivity, owing to the presence of ions.

If there is complete dissociation, equilibrium 4.1-22 is shifted toward the right-hand side and the predominant species in solution are ions. Each mole of solute then produces a multimolar ion solution. For example, 1 mol {NaCl} produces 1 mol Na^+ and 1 mol Cl^-, whereas 1 mole Na_2SO_4 produces 2 mol Na^+ and 1 mol SO_4^{2-}, and the results are 2 and 3 molar ion solutions, respectively. In this notation and later descriptions { } refers to the solid state, pointing to the fact that it cannot be treated as a monomeric species. {A B} actually means $A_\infty B_\infty$.

The conductance of a completely dissociated electrolyte solution is directly proportional to the concentrations of the ions, which, in turn, are directly proportional to the electrolyte concentration. This is explained by the fact that the ions are the carriers of current through the solution. Therefore, if the concentration is doubled or tripled, the conductance is also doubled or tripled, accordingly.

With incomplete dissociation the predominant solute species is the molecular compound M_mA_a. We can calculate the ion concentration $[M^{\nu+}]$ and $[A^{\alpha-}]$ through the equilibrium constant K_{diss}. Thus:

$$[M^{\nu+}]^m[A^{\beta-}]^a = K_{diss}[M_mA_a] \tag{4.1-23}$$

but since $[M^{\nu+}]/m = [A^{\beta-}]/a$, we can write:

$$\left(\frac{a}{m}\right)^a [M^{\nu+}]^{m+a} = K_{diss}[M_mA_a] \tag{4.1-24}$$

Arrhenius found that, in extremely dilute solutions, all electrolytes are completely dissociated. He also introduced a new concept, called the *degree of dissociation*, α. It is defined as the fraction of solute molecules that are dissociated. Therefore, α assumes values between zero (nonelectrolytes) and 1.0 (strong electrolytes).

In a completely dissociated situation a substance of stoichiometry M_mA_a dissociates into m $M^{\nu+}$ and a $A^{\beta-}$ ions. In this process each solute molecule forms μ ions, where:

$$\mu = m + a \tag{4.1-25}$$

However, if the dissociation is incomplete, then n solute molecules M_mA_a form $n(1 - \alpha)$ undissociated particles and $n\alpha\mu$ ions. The total number of particles is then:

$$n(1 - \alpha + \alpha\mu) \tag{4.1-26}$$

If we define the number of particles formed from one solute molecule as i, where $i = (1 - \alpha + \alpha\mu)$, we obtain the relation:

$$\alpha = \frac{(i-1)}{(\mu-1)} \tag{4.1-27}$$

Since i can be derived experimentally from measurements of osmotic pressure, boiling point elevation, or melting point depression, and since μ can usually be derived from the stoichiometry of the solute, it is possible to determine the value of α experimentally.

On the basis of the α values Arrhenius distinguished between strong and weak electrolytes in the following way:

Strong and weak electrolytes: Arrhenius.

- Weak electrolytes possess α values that change significantly over a wide concentration range. At high electrolyte concentrations they assume values close to zero. Weak electrolytes are only dissociated to a limited degree in dilute solutions. Most organic acids and bases, and many heavy metal compounds belong to this class.
- Strong electrolytes: the α value is not very dependent on the electrolyte concentration, and it is always close to 1.0. Strong electrolytes exhibit high electrolytic conductance. Almost all salts, and all strong mineral acids and bases, belong to this class.

It must be kept in mind that this classification is relative and refers specifically to aqueous systems. Thus, an electrolyte which behaves as a strong electrolyte in an aqueous solution may be a weak electrolyte in an organic solvent of small dipole moment or low dielectric constant.

4.1.7 Ion–ion interactions and ion association

In an electrolyte solution, interactions between ions take place through electrostatic attraction and repulsion (Coulomb forces) and therefore they depend on both ionic charge and the average distance between the individual ions.

These coulomb forces are given by the equation:

$$F = \frac{e_i e_j}{D r^2} \tag{4.1-28}$$

where e_i and e_j denote the ionic charges of the interacting ions i and j, D is the dielectric constant of the medium, and r is the distance between the ions.

The attractive forces are the reason for the formation of ion pairs and higher associates. However, in an aqueous solution, these ion-pairing effects become significant only at extremely high ion concentrations (as the average distance between individual ions decreases). However, in organic solvents, where the dielectric constant is much smaller than in aqueous solutions, the ion-pairing processes become important. Ion-pairing is often observed in solvent–solvent extraction procedures and is used to shift the equilibrium toward the organic phase.

4.1.8 Ion activity

A given solute can exist in more than one form in solution, as illustrated previously. Consider a 0.5 mol/L HCl solution in which a few grams of an iron(III) salt are dissolved. To this solution some reagent X^{3-} is added, which precipitates the iron as FeX. One can now formulate the precipitation process in a number of ways, based on the particular iron species present in the original solution:

$$Fe^{3+} + X^{3-} \rightarrow \{FeX\} \tag{4.1-29a}$$

$$Fe(H_2O)_6{}^{3+} + X^{3-} \rightarrow \{FeX\} + 6\,H_2O \tag{4.1-29b}$$

$$FeCl^{2+} + X^{3-} \rightarrow \{FeX\} + Cl^- \tag{4.1-29c}$$

$$Fe(H_2O)_5\,Cl^{2+} + X^{3-} \rightarrow \{FeX\} + Cl^- + 5H_2O \tag{4.1-29d}$$

These reactions are not the only ones which describe the reaction. In a multitude of possible reactions the question arises as to which of the equations represents the actual events taking place. From a thermodynamic standpoint there is no answer to this question, and only a kinetic study can decide it, by establishing the rate-determining processes. In an equilibrium study, it does not matter which reaction path is chosen, since one ends up with the same equilibrium state. It is therefore unnecessary to write the equations in terms of the predominant species in solution, or in terms of supposed or formal species. We have already seen that bare ions, such as H^+ and Fe^{3+}, are the formal rather than the actual species in solution. However, these questions lead us to be cautious about another thermodynamic concept, namely the widely used parameter of concentration in *mass action law* expressions, for example in process (4.1-29d):

$$K = \frac{[Cl^-][H_2O]^5}{[Fe(H_2O)_5Cl^{2+}][X^{3-}]} \tag{4.1-30}$$

Generally, one does not know the exact concentration of the different species involved in such an expression. Furthermore, the question arises as to whether the chloride ion is hydrated, so that one should really include the concentration of the hydrate. This raises more questions. Does one know the hydration number of the hydrate? How does one compute the concentration of a species such as $Fe(H_2O)_5Cl^{2+}$? What does the water concentration $[H_2O]$ mean, since one knows that water is highly structured, and that only a small fraction of the total water can exist in monomeric form? To answer all these questions and to simplify the computation, one must introduce a concept, called the *activity*, which replaces concentration, but is related to it via the so-called *activity coefficient*.

The concept of activity is a very convenient one, although it may cause some problems for the inexperienced chemist. Owing the fact that there are different concentration scales (e.g., molar concentration c, molal concentration m, and mole fraction n) there are also different scales of activity coefficients. However, in a given system the activity must be independent of the manner in which it was calculated, whether it is derived from the molar, molal, or mole fraction concentration scales. The use of different symbols for the different activity coefficients is therefore necessary. The following symbols are generally used and will be adopted in this book:

- f_i activity coefficient referring to the mole fraction scale
- γ_i activity coefficient referring to the molal concentration scale
- y_i activity coefficient referring to the molar concentration scale

The activity coefficient is a sort of correction term which incorporates a number of chemical and physical effects or computational factors. It must be defined in relation to a standard state, i.e., we must clearly state under what conditions it has the value of unity. A most convenient state is obviously a solution in which there is no hydration or solvation, no complex formation, and no ion association, so that the bare solute exists in solution. Such a standard state, however, does not exist. We cannot avoid interactions between the solvent and the solute, but we can try to avoid complex formation (by keeping the concentration of any potential ligand at zero) and ion association (by keeping ions of opposite charge at infinite distance from each other). This situation exists in an infinitely dilute solution and is approached in extremely dilute solutions. Although the infinitely dilute solution is an ideal case from a physical standpoint, it is nevertheless a most inconvenient one in practice and many reactions cannot be observed in dilute systems, because of experimental constraints.

Lars G. Sillen proposed the adoption of a standard state in a constant ionic medium for very practical reasons. This is a system in which the interaction between the medium and the solute (solvation and complex formation) are not completely absent, but are at least kept constant. Applied to our series of equilibrium equations, this means that the concentration ratios of the naked iron ion (4.1-29a), the hexaquo iron ion (4.1-29b), and the iron chloride complexes (4.1-29c and 4.1-29d) remain constant, so that the equilibrium constants do not change with concentration, and thus comply with the idea of a constant ionic medium. However, the equilibrium constants derived are strictly valid only for this particular medium.

Since the activity relations are much simpler in a constant ionic medium, we treat them before dealing with the Debye–Hückel concept (1923), although they gained full recognition only after the limitations of the Debye–Hückel approach were fully realized.

Activity in a constant ionic medium

It was demonstrated by Lewis and Randall that the value of an equilibrium constant (derived by introducing ion concentrations into the mass action law) does not change as long as the investigations are performed in solutions involving the same ionic medium.

We have already seen that increasing the electrolyte concentration increases the extent of ion association and lowers the "active concentration" of the free ions in solution. This means that more concentrated solutions behave less ideally than dilute ones. However, Lewis pointed out that the same degree of nonideal behavior was observed when the *ionic strength* was kept constant. He defined the ionic strength μ by the equation:

$$\mu = \frac{1}{2} \sum_i \left(c_i z_i^2 \right) \tag{4.1-31}$$

In solution, the ionic interactions based on electrostatic effects are not primarily due to specific properties of the ions involved, but depend on the number of ions present in solution (concentration c_i) and the ionic charges (formal charge on the ion z_i). Lewis observed that, the higher the charge on the ion, the more pronounced the nonideal behavior of the solution became. Thus, if a certain magnitude of a particular "ion-dependent" property is observed for a solution of univalent ions of given concentration, the same effect will be observed in a solution of polyvalent ions at an appreciably smaller concentration. Since the Coulombic forces depend on the product of two ion charges, it is not surprising that the medium effect must also depend on a model in which the ionic charges are raised to the power of two.

In Eq. 4.1-31, c_i denotes the molar concentration of the ion i, and z_i denotes the charge of this ion. Thus, a bivalent ion has a four-fold and a trivalent ion a nine-

Concentration versus activity: analytical dream and measurable reality; an initiation to further studies of Debye-Hückel's theory.

Table 4.1-1. Ionic strength for a number of 1.0 mol/L solutions

Electrolyte	Type	c_i	z_i^2	$\mu = \frac{1}{2}\sum_i(c_i z_i^2)$
NaCl	1:1	$[Na^+] = 1.0$	1.0	1.0
		$[Cl^-] = 1.0$	1.0	
CaCl$_2$	2:1	$[Ca^{2+}] = 1.0$	4.0	3.0
		$[Cl^-] = 2.0$	1.0	
Na$_2$SO$_4$	1:2	$[Na^+] = 2.0$	1.0	3.0
		$[SO_4^{2-}] = 1.0$	4.0	
MgSO$_4$	2:2	$[Mg^{2+}] = 1.0$	4.0	4.0
		$[SO_4^{2-}] = 1.0$	4.0	
LaCl$_3$	3:1	$[La^{3+}] = 1.0$	9.0	6.0
		$[Cl^-] = 3.0$	1.0	
Na$_3$PO$_4$	1:3	$[Na^+] = 3.0$	1.0	6.0
		$[PO_4^{3-}] = 1.0$	9.0	
K$_4$[Fe(CN)$_6$]	1:4	$[K^+] = 4.0$	1.0	10.0
		$[Fe(CN)_6^{4-}] = 1.0$	16.0	
Fe$_3$(PO$_4$)$_2$	2:3	$[Fe^{2+}] = 3.0$	4.0	15.0
		$[PO_4^{3-}] = 2.0$	9.0	
Mg$_2$[Fe(CN)$_6$]	2:4	$[Mg^{2+}] = 2.0$	4.0	12.0
		$[Fe(CN)_6^{4-}] = 1.0$	16.0	
LaPO$_4$	3:3	$[La^{3+}] = 1.0$	9.0	9.0
		$[PO_4^{3-}] = 1.0$	9.0	

fold stronger influence on the ionic strength than a univalent ion. To calculate the ionic strength of a solution, all ions in solution must be considered, and they are regarded as derived from a strong electrolyte. Table 4.1-1 gives a computation of the ionic stength of a number of electrolytes (1 mol/L solutions).

Lewis observed that the activity coefficients of ions remained constant in a solution of constant ionic strength, but of different composition (e.g., when a particular ion, such as Na^+ is replaced by another, such as H^+). This holds only if the medium (inert) electrolyte does not undergo complex reactions with the ions, for which the activity coefficient is supposed to remain constant. It is therefore possible to change individual ion concentrations without altering the activity coefficient of a particular ion.

It is evidently a worthwhile procedure to define an "ionic medium" activity scale in which, by definition, the activity coefficient approaches unity as the solution approaches the pure ionic medium, i.e., when the concentration of all species other than water and the inert electrolyte approach zero. Experience shows that, on this scale, the activity coefficient stays close to unity as long as the concentration of the reacting ion (ligand) is kept low, i.e., does not exceed 10% of the concentration of the inert electrolyte. This approach also enables us to change the pH of the constant ion strength medium without affecting the activity coefficients of the ions under study. Now, the activity of several ions (including the hydrogen ion) may be determined accurately and conveniently by potentiometric methods (e.g., with glass or metal electrodes). The equilibrium constant so derived is thermodynamically as rigidly defined as the constant derived at infinite dilution.

Perchlorates and nitrates are the most commonly used inert electrolytes. They generally have only a small tendency to participate in complex formation. Their alkali salts are very soluble in water (except KClO$_4$) and solutions of ionic strength equal to or greater than 3.0 mol/L are possible with these salts.

Activity at infinite dilution (Debye–Hückel theory)

Peter Debye (1884–1966) and Erich Hückel (1896–1980) calculated the potential energy resulting from the mutual attraction between ions in solution. The calculation was based on a model incorporating certain assumptions, including those of Bjerrum's model of ion association, and some additional restrictions, as follows:

- The interionic forces are exclusively electrostatic; all other interactions are neglected.
- The dielectric constant of the pure solvent is employed in the calculation; all structural changes due to the presence of the dissolved electrolytes are neglected (this is not valid for concentrated solutions, where the dielectric constant of the pure solvent is reduced, owing to dielectric saturation).
- All ions are considered structureless point charges, lacking any polarization (this is not strictly true; polarization *is* observed when the ions approach each other).
- The electric potential resulting from the interionic attraction is small compared with the kinetic energy of thermal motion of the ions (this is true only when the average interionic distance is larger than the critical Bjerrum value *c*).
- Strong electrolytes are considered to be completely dissociated at all concentrations (this is not true in very concentrated solutions).

Since ideal behavior is observed only at infinite dilution, the standard state is the infinitely dilute system, where the activity coefficients are all equal to unity. At any finite concentration, the activity coefficients calculated by Debye–Hückel theory is less than unity.

In their original work, Debye and Hückel proposed the following equation for the calculation of the main activity coefficient:

$$\log y_\pm = -A z_+ z_- \sqrt{\mu} \qquad (4.1\text{-}32)$$

Here, z_+ and z_- denote the charges on the cation and anion, μ is the ionic strength of the solution, and A is a constant, the *Debye–Hückel coefficient*, derived from the Debye–Hückel calculation:

$$A = 1.842 \times 10^6 / (D T^{3/2})$$

where D denotes the dielectric constant of the pure solvent, and T is the temperature in K. Since D is 78.54 for water, the resulting value for A is 0.5091 in pure water at 25 °C.

Equation (4.1-32) is known as the *limiting Debye–Hückel equation*. Its validity is restricted to electrolyte concentrations smaller than 10^{-2}–10^{-3}, depending on the charges involved.

Debye and Hückel later proposed the *extended Debye–Hückel equation*. It extends the validity of Eq. 4.1-33 to 10^{-2}–10^{-1} mol/L solutions:

$$\log y_\pm = \frac{-A z_+ z_- \sqrt{\mu}}{1 + B a \sqrt{\mu}} \qquad (4.1\text{-}33)$$

where a is an ion size parameter (ionic radius), expressed in Å, and B is the *second Debye–Hückel coefficient*, calculated from the relation:

$$B = \frac{50.29}{(DT)^{1/2}} \qquad (4.1\text{-}34)$$

B is 0.32865 in pure water at 25 °C.

It can be seen that the extended Debye–Hückel equation is reduced to the limiting form when $1 \gg B a \sqrt{\mu}$. For most ions a is about 3 Å, so that the product Ba is close to 1.0. The limiting equation therefore replaces the extended form, when $\mu < 0.01$ mol/L.

Many attempts have been made to extend the activity calculation to higher concentrations. However, it is beyond the scope of this book to refer to all the different equations, such as the *Güntelberg equation*.

So far we have treated systems in which the activity coefficients decrease with increasing electrolyte concentration, as found in experimental determinations of activity coefficients performed on dilute solutions, where the nonideal behavior is attributed exclusively to the formation of "ion clouds", i.e., is based on ion–ion interactions. Sometimes there interactions are called *long-range* interactions. There are other interactions (*short-range* effects) in which ions are involved, a very important one being hydration. In dilute systems, the hydration effects are constant since the water activity itself is constant. Therefore, in dilute systems the activity coefficients do not depend on hydration effects, and these are included in the definition of the standard state (although not explicitly). The activity coefficient therefore represents that of the predominant hydrated ion. If the electrolyte concentration is increased, the solvent activity must decrease and thereby must influence the position of the solvation equilibria. Solvation equilibria are shifted toward the lesser solvated or bare ions, and for this reason the activity coefficients of the bare ions start to increase with increasing electrolyte concentration. This is particularly true for small ions such as H^+, Li^+, and Be^{2+}. For example, in mineral acids the activity coefficients may assume values as high as 10^3 or even larger in concentrated solutions. For these systems a general rule can be applied. The activity coefficients deviate from the infinite dilution value of 1.0 at concentrations around 10^{-4}–10^{-3} mol/L. Usually, a minimum is observed for the strongly hydrated ions at a concentration of 0.1–0.5 mol/L. Then the activity coefficients increase again and assume values around 1.0 at concentrations between 1.0 and 2.0 mol/L.

These effects are not accounted for by the Debye–Hückel model. As a matter of fact the effects cannot be calculated from a general model because the hydration of ions is a specific ion–solvent interaction and differs from ion to ion. However, a number of empirical modifications have been suggested with the object of extending the validity of the extended equation to higher concentration ranges. In the general case the following equation is obtained:

$$\log y_\pm = \frac{-Az_+z_-\sqrt{\mu}}{1 + Ba\sqrt{\mu}} + C\mu \tag{4.1-35}$$

In this equation, C is an empirical coefficient, which is determined experimentally.

Robinson and Stokes have proposed yet another equation:

$$\log y_\pm = \frac{-Az_+z_-\sqrt{\mu}}{1 + Ba\sqrt{\mu}} - \frac{H}{2}\log a_w \tag{4.1-36}$$

This expression is the most accurate available, and allows the calculation of activity coefficients in solutions of at least 2–4 mol/L electrolytes. However, the hydration numbers H of ions are needed for this calculation, and these are nearly always unknown.

There is yet another problem connected with the use of activity coefficients. It is not possible to determine the activity coefficient of an individual ion experimentally. This is because each ion always coexists with another, the so-called *counterion*. The individual ion activity coefficient is therefore a hypothetical quantity. Experimentally, the *mean activity coefficient* $y_\pm$ is found, for which the following relationship holds:

$$a_i a_j = c_i c_j y_i y_j = c_i c_j y_\pm{}^2 \tag{4.1-37}$$

where a_i and a_j denote individual ion activities, c_i and c_j the ion concentrations (molar scale), and y_i and y_j the hypothetical activity coefficients of the cation i and anion j; $y_\pm$ is the mean activity coefficient. We can thus conclude that:

$$y_\pm{}^2 = \frac{a_i a_j}{c_i c_j} = (y_i y_j) \tag{4.1-38}$$

Since the electrolyte concentration is calculated from the total amount of solute originally dissolved, and not from the actual ion concentrations, the electrolyte activity must be referred to on the same basis. For example, a 1.0 mol/L HCl solution forms 1 mol/L H^+ and 1 mol/L Cl^-. In this case, the activity coefficient is defined as:

$$y_\pm = \frac{a_{HCl}}{c_{HCl}} = (y_{H^+} y_{Cl^-})^{1/2} \tag{4.1-39}$$

Under these conditions, the activity coefficient is found to be 0.812. Therefore:

$$a_{HCl} = (a_{H^+} a_{Cl^-})^{1/2} = 0.812$$

whereas c_{HCl} is 1.0.

This computation may be illustrated by yet another example. Consider a solution prepared by dissolving 0.2 mol HNO_3 and 0.3 mol $NaCl$ to make 1 L of solution. This solution contains:

$[H^+] = 0.2 \, mol/L$

$[Cl^-] = 0.3 \, mol/L$

We can therefore calculate a formal HCl concentration as:

$$c_{HCl} = (c_{H^+} c_{Cl^-})^{1/2} = (0.2 - 0.3)^{1/2} = \sqrt{0.06} = 0.245 \, mol/L$$

whereas the ionic strength μ is 0.5 mol/L.

References

[4.1-1] Broene, H.H., De Vries, T., J. Amer. Chem. Soc. 69 (1947) 1644.

[4.1-2] Selvaratnam, M., Spiro, M., Trans. Faraday Soc. 61 (1965) 360.

[4.1-3] Pethybridge, A.D., Prue, J.E., Trans. Faraday Soc. 63 (1967) 2019.

[4.1-4] Martin, D.L., Rossotti, F.J.C., Proc. Chem. Soc. 1959, 60.

[4.1-5] Everett, D.H., Coulson, C.A., Trans. Faraday Soc. 36 (1940) 633.

Questions and problems

1. Calculate the ionic strength of a simulated seawater sample containing the following components per liter solution:

   0.478 mol Na^+ 0.550 mol Cl^-

   0.064 mol Mg^{2+} 0.028 mol SO_4^{2-}

   Answer: The ionic strength is:

   $$\mu = \tfrac{1}{2}(0.478 + 4 \times 0.064 + 0.550 + 4 \times 0.028) = 0.698 \, mol/L$$

2. How can you define "Chemical Analysis" as compared to "Physical Analysis"?
3. Which fundamental law describes chemical equilibria?
4. When can we observe a heterogeneous, when a homogeneous equilibrium?
5. Under what circumstances can a weak electrolyte become a strong one?
6. Analytical chemistry is interested in concentration values as result of the analytical measurements. The laws of physical chemistry deliver however activities rather than concentrations. Under what experimental conditions can one bridge that gap and transform activities into concentrations?

4.2 Acid–Base Equilibria

Learning objectives

■ To describe different acid and base concepts, and acidimetric, alkalimetric, proton acceptor, and donor reactions
■ To deal with the pH concept, and the significance of pH indicators and buffers
■ To show the student how to calculate acid and base concentrations, and the pH of solutions and buffers from given concentrations

4.2.1 Introduction

An imaginative idea of acids and bases was expressed by Nicolas Lémery (1645–1715), an apothecary who lived in Paris. In his highly successful book *Cours de Chimie* (1675) he tried to explain the physical and chemical properties of substances in terms of their shape and structure. In this respect his attempts appear extremely modern. Lémary described acids as having sharp spikes on their surfaces, accounting for the prickling sensation they exert on the skin. Bases, called alkalies, were considered to consist of porous bodies into which the spikes of acids penetrate, to be broken or blunted during the production of a neutral salt.

With the discovery and recognition of chemical elements (possible only after the development of atomic and molecular theory) the acid–base concept changed and attempts were made to assign acidic properties to a particular element or molecular group. Antoine L. Lavoisier (1743–1794), defined acids as combinations of radicals and oxygen, and bases as a combination of metals and oxygen. His idea was generally accepted despite the fact that acids such as hydrochloric acid had been known for centuries. It was then believed that chlorine was an oxide rather than an element. In 1809 Joseph L. Gay-Lussac (1778–1850), in collaboration with Louis J. Thénard (1777–1857) found that chlorides, then called muriates, contained no oxygen. However, they were such strong believers in Lavoisier's ideas that they doubted their own results rather than proposing the new concept. It was the English chemist Humphry Davy (1778–1829) who changed acid–base theory by proving that chloride was an element and that hydrochloric acid contained no oxygen. He considered hydrogen as the acid principle (1810) and his theory was accepted shortly afterwards.

A modern approach to the problem was only possible after Arrhenius proposed his theory of electrolyte dissociation, which inluded the concept of strong electrolytes, i.e., salts, acids, and bases. He announced his thoughts to the Swedish Academy of Sciences in 1883 and published his theory in 1887. According to Arrhenius, an acid dissociates into hydrogen ions and some anions, whereas a base dissociates into hydroxide ions and the corresponding cations. The Arrhenius concept also provided an explanation for the different strengths of different acids based on their different degrees of dissociation.

However, further difficulties arose in attempts to explain the acidity and basicity of organic solutions. A solution of ammonia in ether exhibits some basic properties, but it can be shown that the hydroxide ion do not exist in such a solution. Arthur Lapworth (1872–1941), a professor at the University of Manchester, teaching organic as well as physical and inorganic chemistry, described acids as donors of hydrogen ions. T. M. Lowry (1874–1936) was one of his pupils. He and the Danish professor Johann N. Brönsted (1879–1947) independently developed Lapworth's ideas of acid and bases into what is now called the Brönsted–Lowry theory (1923).

4.2.2   The modern concept of acids and bases: Brönsted–Lowry theory

In the Brönsted–Lowry theory, an acid is a proton donor and a base is a proton acceptor. Each acid is related to its conjugate base and vice versa:

$$\text{acid} \rightleftharpoons \text{base} + \text{proton} \tag{4.2-1}$$

A more scientific approach to acid–base theory.

Therefore, the conjugate base of a strong acid must be a weak acid and the conjugate base of a weak acid must be a strong base. Together they form a couple, and an acid without its conjugate base is a meaningless concept. In order to release a proton, the acid must find a base to accept it. In an aqueous solution, the proton, H^+, having an extremely small ionic radius, cannot exist as such. It is hydrated, forming the hydronium ion H_3O^+ and higher hydrates. Thus, an acid–base equilibrium is not a simple dissociation equilibrium, but the result of a proton transfer reaction in which there are at least two reagents and two products. Such a process is also called *protolysis*. The overall reaction is expressed by:

$$HX + H_2O \rightleftharpoons H_3O^+ + X^- \tag{4.2-2}$$

which can be formally split into two single reactions:

$$HX \rightleftharpoons H^+ + X^- \tag{4.2-2a}$$

$$\text{acid 1} \quad \text{proton} \quad \text{base 1}$$

and

$$H^+ + H_2O \rightleftharpoons H_3O^+ \tag{4.2-2b}$$

$$\text{proton} \quad \text{base 2} \quad \text{acid 2}$$

The overall equilibrium constant is:

$$K = \frac{[H_3O^+][X^-]}{[HX][H_2O]} \tag{4.2-3}$$

The concentration of water is unknown, since one should really introduce the concentration of monomeric water. But, whatever this concentration is, it remains constant in dilute solutions. It can therefore be incorporated into the equilibrium constant. This corresponds to an agreed standard state of the system. The resulting equilibrium constant (whether it is calculated in a constant ionic medium or in an infinitely dilute solution) is then:

$$K[H_2O] = K_A = \frac{[H_3O^+][X^-]}{[HX]} \tag{4.2-3a}$$

K_A is called the *acid dissociation constant*. Strictly speaking, K_A reflects not only the *acid strength* of HX, but also the *base strength* of water. This is why different acid dissociation constants are observed for the same acid in different solvents. It must be kept in mind that the actual concentration of the hydronium ion is not known. The symbol $[H_3O^+]$ therefore denotes a practical quantity, namely the concentration representing the sum of all forms of unhydrated and hydrated protons, and the symbol $[H^+]$ could equally well be used for this quantity. For a completely dissociated monobasic acid, i.e., a strong electrolyte, this quantity is equal to the formal acid concentration.

This model, in contrast to the Arrhenius model and earlier definitions of acids and bases, is now applicable to all kinds of acid and base equilibria that are not simply the result of dissociation reactions. Several compounds are known that do not contain hydrogen atoms or hydroxide groups, but form acidic or basic solutions. They do so by reacting with the solvent (water) to form products (hydrates) that subsequently dissociate. Examples of this category are:

$$SO_3 + 2\,H_2O \rightleftharpoons H_3O^+ + HSO_4^- \tag{4.2-4}$$

sulfur trioxide   hydrogen sulfate ion

$$NH_3 + H_2O \rightleftharpoons NH_4^+ + OH^- \tag{4.2-5}$$

ammonia   ammonium ion

$$\{LiH\} + H_2O \rightleftharpoons Li^+ + OH^- + H_2 \qquad (4.2\text{-}6)$$

lithium hydride

$$\{NaNH_2\} + H_2O \rightleftharpoons Na^+ + NH_3 + OH^- \qquad (4.2\text{-}7)$$

sodium amide $\qquad\qquad$ ammonia

We shall use the Brönsted–Lowry concept in our further discussion, although other acid–base theories have been introduced. The Brönsted–Lowry concept is based on a proton transfer reaction in which two pairs of acids and bases are involved. The acid constant K_A is therefore not a straightforward definition of acid strength. Water has a "leveling" property in aqueous solutions, and the relative strengths of very strong acids cannot be determined. So-called *acidity functions* have been defined to overcome such problems. Among other, L.P. Hammett introduced a concept known as the *Hammett acidity function*. However, in discussing only those equilibrium states in which at least one aqueous phase is involved, we restrict ourselves to the Brönsted–Lowry concept.

Another concept, introduced by G.N. Lewis (1875–1946), is mentioned only for the sake of completeness. Lewis introduced the idea in 1916. According to his theory an acid is an electron-pair-deficient species and a base is a compound capable of providing such an electron pair. Consequently, an acid–base reaction is no longer restricted to proton transfer reactions. The theory thus eliminates the special role of the proton, and is replaced by a coordinative complex reaction. BF_3, $AlCl_3$, $FeBr_3$, etc. are *Lewis acids* and NH_3, all amines, and many oxygen compounds, such as ethers, alcohols and ketones, are *Lewis bases*.

Proton acceptor and donor properties in aqueous solution

In aqueous solutions proton donors (acids) produce hydronium ions. In this process water and the conjugate base of the acid compete for the proton according to the equilibrium:

$$HX \;+\; H_2O \rightleftharpoons H_3O^+ + \; X^- \qquad (4.2\text{-}2)$$

acid 1 $\quad$ base 1 $\quad$ acid 2 $\quad$ base 2

Depending on the relative strength of the two acids (or bases) the equilibrium is shifted to one side, or a balanced state is obtained in which all four species are major components. Water is a relatively weak acid, and for strong proton donors such as mineral acids ($HClO_4$, HCl, HBr, etc.), and for a number of complex metal acids such as $HMnO_4$, $HFeCl_4$, etc., the reaction is almost completely shifted to the right-hand side. These strong acids are completely dissociated. Anions such as $FeCl_4^-$ are so weakly basic that they are not considered as bases in the general sense. Furthermore, $FeCl_4^-$ scarcely exists in aqueous solutions, because it requires the presence of extremely high Cl^- concentrations ($HFeCl_4$ exists, however, as a hydrated ion pair in organic solvents, such as methylisobutyl ketone, ether, etc.).

One must keep in mind that an aqueous solution of a molecular gas can produce an acid solution, in no way different from the dissociation of an electrolyte. For example, in the gas phase HCl is a molecular species, but when it is introduced into water hydration takes place producing a considerable amount of heat (comparable with the heat of neutralization produced in the reaction of a strong acid with a moderately strong base, such as NH_3). This underlines the fact that water is a base. A similar reaction takes place when SO_3 is dissolved in water. First the gas is hydrated and then the acid formed dissociates in several steps in the aqueous solution:

$$SO_3 + H_2O \rightleftharpoons H_2SO_4 \qquad\qquad\text{first step: hydration} \qquad (4.2\text{-}8)$$

$$H_2SO_4 + H_2O \rightleftharpoons H_3O^+ + HSO_4^- \quad\text{second step: dissociation} \qquad (4.2\text{-}9)$$

$$HSO_4^- + H_2O \rightleftharpoons H_3O^+ + SO_4^{2-} \quad\text{third step: dissociation} \qquad (4.2\text{-}10)$$

All these equilibria are shifted to the right-hand side, and it is actually more appropriate to write:

$$SO_3 + H_2O \rightarrow H_2SO_4 \qquad\qquad \text{complete hydration} \qquad\qquad (4.2\text{-}8)$$

$$H_2SO_4 + H_2O \rightarrow H_3O^+ + HSO_4^- \quad \text{complete dissociation} \qquad (4.2\text{-}9)$$

$$HSO_4^- + H_2O \rightarrow H_3O^+ + SO_4^{2-} \quad \text{almost complete dissociation} \qquad (4.2\text{-}10)$$

This need not always be the case. A number of oxides are known for which the hydration and dissociation equilibria are only slightly shifted:

$$SO_2 + H_2O \leftarrow H_2SO_3 \qquad\qquad \text{incomplete hydration} \qquad\qquad (4.2\text{-}11)$$

$$H_2SO_3 + H_2O \rightarrow H_3O^+ + HSO_3^- \quad \text{almost complete dissociation} \qquad (4.2\text{-}12)$$

$$HSO_3^- + H_2O \leftarrow H_3O^+ + SO_3^{2-} \quad \text{incomplete dissociation} \qquad (4.2\text{-}13)$$

Carbon dioxide is hydrated in water, but the predominant form of the hydrate is $CO_{2(aq)}$ and not H_2CO_3 (see Sec. 4.1). The molecule H_2CO_3 is a moderately strong acid, but because it is in equilibrium not only with the dissociated form HCO_3^-, but also with the carbon dioxide hydrate CO_2H_2O ($CO_{2(aq)}$), an apparently weak acidity or incomplete dissociation is exhibited.

$$CO_2 + H_2O \rightarrow CO_2 \cdot H_2O \qquad\qquad \text{complete hydration} \qquad\qquad (4.2\text{-}14)$$

The incomplete dissociation of H_2CO_3 results in behavior as a weak acid. The dissociation of a weak electrolyte increases, however, upon dilution (Ostwald's law of dilution).

$$CO_2 \cdot H_2O \leftarrow H_2CO_3 \qquad\qquad \text{molecular rearrangement} \qquad (4.2\text{-}15)$$

$$H_2CO_3 + H_2O \leftarrow H_3O^+ + HCO_3^- \quad \text{dissociation (not complete)} \qquad (4.2\text{-}16)$$

$$HCO_3^- + H_2O \leftarrow H_3O^+ + CO_3^{2-} \quad \text{incomplete dissociation} \qquad (4.2\text{-}17)$$

The equilibrium constant for the second reaction is about 1.5×10^{-3}, resulting in a ratio $[CO_2 \cdot H_2O]/[H_2CO_3] = 670$. The acid dissociation constant is therefore:

$$K_A = \frac{[H_3O^+][HCO_3^-]}{[H_2CO_3]_{total}} = 10^{-6.37} \qquad\qquad (4.2\text{-}18)$$

The symbol $[H_2CO_3]_{total}$ actually stands for $[CO_2 \cdot H_2O] + [H_2CO_3]$. The true acid dissociation constant for the species H_2CO_3 would be about 670 times larger, indicating that it is a rather strong acid, comparable to the second dissociation of HSO_4^-, and is much stronger than acetic acid.

There are some nonmetal oxides which do not dissolve in water and undergo no reaction with it. They are therefore neither acidic nor basic. NO_2 and NO are examples of this class. Owing to their inertness to water they are used in aerosol cans in the food industry.

Several liquid and solid molecular species react with water to form an acidic solution. Pure sulfuric acid exists in molecular form as H_2SO_4. It has low electrical conductance and only becomes a strong electrolyte on addition of water. This reaction produces a considerable amount of heat, indicating a neutralization reaction in which water participates as a base. It is also known that perchloric acid forms a bimolecular species at extremely high acid concentrations. Here, the equilibria involved most probably consist of the following steps:

$$H_3O^+ + ClO_4^- \rightleftharpoons H_3O^+ClO_4^- \qquad\qquad (4.2\text{-}19)$$
$$\text{ion pair}$$

$$H_3O^+ClO_4^- + ClO_4^- \rightleftharpoons H_3O(ClO_4)_2^- \text{ or } H(ClO_4)_2^- \qquad (4.2\text{-}20)$$
$$\text{triple anions}$$

$$H_3O(ClO_4)_2^- + H_3O^+ \rightleftharpoons (H_3O^+ClO_4^-)_2, \text{ or } H_3O^+H(ClO_4)_2^- \qquad (4.2\text{-}21)$$
$$\text{triple anion} \qquad\qquad \text{bimolecular species or quadruple ions}$$

In these equations, $H_3O^+ClO_4^-$ and $H_3O(ClO_4)_2^-$ or $H(ClO_4)_2^-$ stand for the acid ion pair and two forms of anionic triple ions, respectively. In the case of perchloric acid, the ion pair and triple ions lose the hydrating water molecules more readily than the free hydronium ion. Triple ions are commonly observed in media of low dielectric constant, i.e., in solvent extraction systems.

With weak and moderate acids, such as acetic acid, only fractions of a percent exist as dissociated oxonium salt:

$$CH_3COOH + H_2O \leftarrow H_3O^+ + CH_3COO^- \tag{4.2-22}$$

Nonmetal halides exchange the halide with the oxygen of water in aqueous solution and the hydrated cation produced is generally a strong proton donor:

$$PCl_3 + 3\,H_2O \rightleftharpoons P(OH_2)_3{}^{3+} + 3\,Cl^- \tag{4.2-23}$$

$$P(OH_2)_3{}^{3+} + 3\,H_2O \rightleftharpoons 3\,H_3O^+ + H_3PO_3 \tag{4.2-24}$$

$$H_3PO_3 + H_2O \rightleftharpoons H_3O^+ + H_2PO_3{}^- \tag{4.2-25}$$

$$H_2PO_3{}^- + H_2O \rightleftharpoons H_3O^+ + HPO_3{}^{2-} \tag{4.2-26}$$

$$HPO_3{}^{2-} + H_2O \leftarrow H_3O^+ + PO_3{}^{3-} \tag{4.2-27}$$

The first step of this sequence resembles the dissociation process of most heavy metal halides, such as:

$$\{CuCl_2\} + 4\,H_2O \rightleftharpoons Cu(H_2O)_4{}^{2+} + 2\,Cl^- \tag{4.2-28}$$

and

$$SnCl_4 + 6\,H_2O \rightleftharpoons Sn(H_2O)_6{}^{4+} + 4\,Cl^- \tag{4.2-29}$$

However, for the metal halides, the second step is considerably less complete:

$$Cu(H_2O)_4{}^{2+} + H_2O \rightleftharpoons [CuOH(H_2O)_3]^+ + H_3O^+ \tag{4.2-30}$$

$$Sn(H_2O)_6{}^{4+} + 2\,H_2O \rightleftharpoons \{Sn(OH)_4\} + 4\,H_3O^+ \tag{4.2-31}$$

Proton acceptors produce hydroxide ions in aqueous solutions, because water acts as a stronger acid than the protonated base, and the transfer of the proton occurs from the water molecule to the base, to form the corresponding acid. Therefore, water exerts a leveling effect on strong bases. The hydroxide ion concentration of a strong normal base solution is $[OH^-] = 1.0\,mol/L$. Such a solution is obtained from sodium amide $\{NaNH_2\}$ or lithium hydroxide $\{LiOH\}$. The anions of these salts undergo proton transfer reactions with water and produce hydroxide ions.

Weak bases.

That water can act as an acid (proton donor) can be demonstrated by a large number of experiments. The most important is probably the reaction with ammonia:

$$NH_3 + H_2O \leftarrow NH_4{}^+ + OH^- \tag{4.2-32}$$

As indicated, only a small amount of the total ammonia reacts with the water, and more than 99% exists as NH_3. But some $NH_4{}^+$ and OH^- are formed and the aqueous solution assumes a pH value greater than 7. It must be pointed out that the molecular species NH_4OH is not present, but exists rather as the hydrate $NH_3 \cdot H_2O$.

Basic solutions are also formed when metal oxides are dissolved in water. This process is usually accompanied by the production of a considerable amount of heat:

$$\{Na_2O\} + H_2O \rightarrow 2\,Na^+ + 2\,OH^- \tag{4.2-33}$$

In some cases, the metal hydroxide formed is only slightly soluble in the solution, and a precipitate is formed:

$$\{CaO\} + H_2O \rightarrow \{Ca(OH)_2\} \rightleftharpoons Ca^{2+} + 2\,OH^- \tag{4.2-34}$$

Certain anions are strong bases. They combine with a proton from water or another solvent and form a solution of hydroxide ions in an aqueous solution, in combination with the appropriate cation. The above-mentioned class of oxides, and also peroxides, sulfides, phosphates, carbonates, and fluorides belong to this category.

$$F^- + H_2O \rightleftharpoons OH^- + HF \tag{4.2-35}$$

$$O^{2-} + H_2O \rightleftharpoons 2\,OH^- \tag{4.2-36}$$

$$S^{2-} + H_2O \rightleftharpoons OH^- + HS^- \tag{4.2-37}$$

This reaction is followed by a second step in acidic solutions:

$$HS^- + H_3O^+ \rightleftharpoons H_2O + H_2S \tag{4.2-38}$$

$$CO_3^{2-} + H_2O \rightleftharpoons OH^- + HCO_3^- \tag{4.2-39}$$

The last equilibrium is also followed by a second step in acidic solutions:

$$HCO_3^- + H_3O^+ \rightleftharpoons 2\,H_2O + CO_{2(aq)} \tag{4.2-40}$$

These reaction schemes demonstrate that the sulfide ion S^{2-} and the carbonate ion CO_3^{2-} can exist only in basic solutions. It is therefore a physical impossibility to observe NH_4^+ and CO_3^{2-} or S^{2-} as a major species in the same solution.

Of particular interest is the phosphate ion, since it can react with a proton donor in three separate steps:

$$PO_4^{3-} + H_2O \rightleftharpoons HPO_4^{2-} + OH^- \tag{4.2-41}$$

followed by a reaction in neutral solutions:

$$HPO_4^{2-} + H_2O \rightleftharpoons H_2PO_4^- + OH^- \tag{4.2-42}$$

and in acid solutions:

$$H_2PO_4^{2-} + H_3O^+ \rightleftharpoons H_3PO_4 + H_2O \tag{4.2-43}$$

Note that, in all these reactions, the pH of the solution increases. It is obvious that the hydrogen and hydroxide ions cannot exist independently in aqueous solution, i.e., both ions cannot be present at significant concentrations in the same solution. They form an equilibrium with the solvent:

$$OH^- + H^+ \rightleftharpoons H_2O \tag{4.2-44}$$

This can be written as an autoprotolysis reaction:

$$OH^- + H_3O^+ \rightleftharpoons 2\,H_2O \tag{4.2-44}$$

The ion product of water, K_W, and its value at different temperatures.

This equilibrium describes the self-dissociation of water and can be expressed through the so-called *ion product of water*, K_W:

$$K_W = [H^+][OH^-] \quad \text{or} \quad [H_3O^+][OH^-] \tag{4.2-45}$$

The corresponding K_W°, representing the ion product constant derived for an infinitely dilute system (standard state), is:

$$K_W^\circ = a_{H^+}a_{OH^-} = y_\pm^2 [H_3O^+][OH^-] \tag{4.2-46}$$

This quantity is temperature dependent and assumes the values given in Table 4.2-1.

Similar proton transfer reactions exist in all solvents possessing proton donor and acceptor properties. Proton transfer reactions are extremely fast. This makes them very suitable for analytical applications and acid–base reactions have found wide use in volumetric methods and other analytical techniques.

Table 4.2-1. Ion product for water K_W°, as a function of temperature

T (°C)	K_W° (mol²/L²)	T (°C)	K_W° (mol²/L²)
0	$10^{-14.96}$	30	$10^{-13.83}$
10	$10^{-14.53}$	40	$10^{-13.53}$
20	$10^{-14.16}$	50	$10^{-13.26}$
25	$10^{-14.00}$	60	$10^{-13.02}$

4.2.3 The concept of pH

Water has the property of protonating or deprotonating the constituents of an aqueous solution. This tendency is called the *acidity* or *basicity* of the solution. It may be expressed in terms of either a proton pressure or proton vacuum existing in the solution, by the energy required to transfer a proton from the acid to water, or by the energy needed to transfer protons from water to the base. By far the most convenient expression to describe this phenomenon is the so-called *pH value*:

$$pH = -\log[H^+] \tag{4.2-47a}$$

The definition $pH = -\log[H^+]$ was originally given by S.P.L. Sorenson in 1909. In this expression $[H^+]$ denotes the hydrogen ion concentration given by the sum of all hydrated forms of the proton, including H_3O^+, $H_9O_4^+$, etc.

It is a generally accepted convention to use p as symbol for the negative logarithm of a quantity. Therefore pOH and pK_A are given by:

$$pOH = -\log[OH^-] = 14 - pH \tag{4.2-48}$$

$$pK_A = -\log K_A \tag{4.2-49}$$

Thus, for acetic acid at 25 °C:

$$K_A = \frac{[H^+][Ac^-]}{[HOAc]} = 1.745 \times 10^{-5} = 10^{-4.756} \tag{4.2-50}$$

and $pK_A = 4.756$

It is a convention that the proton pressure (excess hydrogen ions) or proton deficiency (hydrogen ion deficiency) in water is expressed in terms of the pH value. Since water as ionic medium exerts a leveling effect, values below $pH = -1$ and above $pH = 15$ cannot be achieved in aqueous solutions. Other solvents have a leveling effect too, but the range is generally different from that of water. This makes acid–base reactions in nonaqueous media an attractive field with special applications. In Fig. 4.2-1 some solvents and their ion products are compared with water.

pH measurements can be made in different ways. They are among the most frequently performed operations in chemistry. Estimates are often made on the basis of color changes of so-called *indicators*, but the most accurate results are obtained from electrochemical methods. Several electrode processes are known to depend accurately on the hydrogen ion activity. The glass electrode is the most popular one for this purpose, but other electrodes, such as the hydrogen electrode (platinum electrode in an H_2 atmosphere), the quinhydrone electrode, etc., are used for special purposes where the glass electrode is not suitable.

At 25 °C, pure water has a pH value of 7 and the hydrogen ion and hydroxide ion concentrations are equal ($[H_3O^+] = [OH^-] = 10^{-7}$ mol/L). At 0 °C, the same water has pH 7.48 and at 60 °C the pH of pure water is 6.51, but again the hydrogen and hydroxide ion concentrations are equal. These solutions are called *neutral*.

An aqueous solution in which the hydronium ion concentration exceeds the hydroxide ion concentration is called *acidic* and its pH at 25 °C is less than 7. An aqueous solution in which the hydroxide ion concentration exceeds the hydrogen ion concentration is called *basic* and its pH is greater than 7 at 25 °C. There is a continuous transition from acidic to neutral to basic solutions.

Owing to the difficulties that arise in more concentrated solutions, where the activity and the concentration of an ion deviate strongly from each other, the pH value must be defined in terms of the ion activity and therefore two different concepts emerge:

- In a constant ionic medium, where the activity coefficients are considered to remain constant at 1.0, the original Sorensen concept still holds. The pH of a sample can easily be related to the measurable pH value of a known standard solution at the same ionic strength. For example, a 0.1 mol/L solution of a strong acid (HCl, $HClO_4$, etc.) at ionic strength of 1.0 mol/L has a pH value of 1.0, and any value measured with a glass electrode or other analytical tool can be related to such a standard value so that the acid concentration of the unknown sample may be derived.

The pH value is a measure for the acidity or basicity of a solution, aqueous or non-aqueous (see Fig. 4.2-1).

Definition of the pH value.

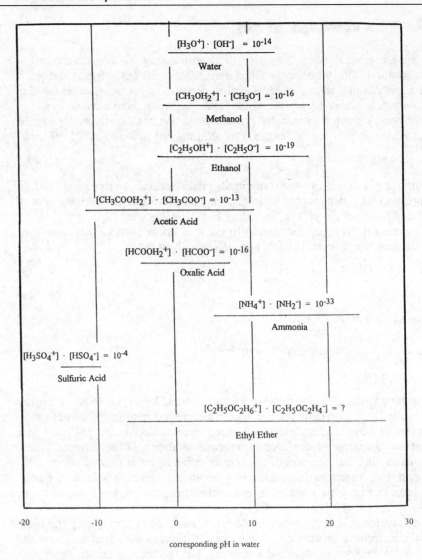

Fig. 4.2-1. The acid–base ion product in different solvents compared with water

- In the infinite dilution region, the pH may be defined in terms of the hydrogen ion activity, according to:

$$p^{a_H} = -\log a_{H^+} = -\log[H^+] - \log y_{H^+} \tag{4.2-47b}$$

However, this is not a practical approach, since we know that the activity coefficient of a particular ion cannot be derived as such, but only in combination with its counterion. This definition is therefore useful only at low concentrations where y_{H^+} does not deviate substantially from unity.

To cope with this difficulty, another more operational definition is endorsed by the International Union of Pure and Applied Chemistry (IUPAC), based on recommendations made by Roger G. Bates. In his definition, the pH is determined relative to that of a standard buffer, for which the pH is calculated on the basis of the infinite dilution system from measurements made on cells with a liquid junction (glass electrode in combination with a calomel electrode). This operational pH is not rigorously identical with p^{a_H}, because the liquid junction potential and the single ion activities cannot be evaluated without nonthermodynamic assumptions.

4.2.4 Acid–base indicators

Acid–base indicators are substances with acid–base properties, having different colors in their protonated and deprotonated forms.

Indicators are chemical substances with acid–base properties, having different colors in their protonated and deprotonated forms. Generally, the acid form is symbolized by I_a and the base form by I_b. The two forms are related to each other in an acid–base equilibrium, since they represent a conjugate pair:

$$K_i = \frac{[H^+][I_b]}{[I_a]} \tag{4.2-51}$$

Note, that the charges have been omitted for simplification. It is possible that either the acid or base form is a positively charged ion (a cation, such as an ammonium ion, R_3NH^+, or $Fe(H_2O)_5OH^{2+}$), a neutral species (e.g., water, which is an acid as well as a base) or a negatively charged species (anions, such as HSO_4^-, which is an acid as well as a base).

The above equation may be rearranged:

$$[H^+] = K_i \frac{[I_a]}{[I_b]} \tag{4.2-51a}$$

so that the pH is derived as:

$$pH = -\log[H^+] = pK_i + \log\frac{[I_b]}{[I_a]} \tag{4.2-52}$$

In an indicator system, the acid form I_a has a different color from the base form I_b. The color change generally takes place over a pH range of about 1–2 pH units, around the pK_i value. This is easily understood by the following calculations:

- When $pH = pK_i$, then $\log[I_b] = \log[I_a]$; both forms are equal in concentration and the color observed is the combination of both colors, i.e., the acid and the base forms.
- When $pH = pK_i - 1$, then $\log[I_b] = \log[I_a] - 1$ and $[I_a]/[I_b] = 10$; this means 90.9% of the indicator is in the acid form. The color of the acid form predominates.
- When $pH = pK_i + 1$, then $\log[I_b] = \log[I_a] + 1$ and $[I_a]/[I_b] = 0.1$; this means 90.9% of the indicator is in the base from and only 9.1% in the acid form. The color of the base form predominates.

Polybasic indicators may exhibit more than one color change (Table 4.2-2).

The pH of a solution can be determined from the concentration ratio of the two indicator forms, when the molar extinction coefficients of the two forms are known. If $pH = pK_i$, then half the indicator exists in the acid and half in the base form. This does not necessarily imply that the color of the solution is a mixture of the two colors. This holds only when both forms have similar molar extinction coefficients.

Table 4.2-2. Properties of the most frequently used indicators

Indicator	Color		pH range of color change	pK_i
	acid I_a	base I_b		
Methyl violet	yellow	blue	0.0–1.6	ca 1.0
Cresol red*	yellow	red	0.4–1.8	ca 1.0
Thymol blue*	red	yellow	1.2–2.8	2.0
Bromophenol blue	colorless	yellow	2.8–4.0	3.3
Congo red	blue	red	3.0–5.0	ca 4.0
Methyl orange	red	yellow	3.1–4.4	3.6
Bromocresol green	yellow	blue	3.8–5.4	4.5
Methyl red	red	yellow	4.4–6.1	5.0
Alizarin*	colorless	yellow	5.5–6.6	6.0
Alizarin**	yellow	red	5.7–7.3	6.5
Bromothymol blue	yellow	blue	6.0–7.6	6.8
Phenol red	yellow	red	6.6–8.0	7.3
Neutral red	red	amber	6.8–8.0	7.4
Cresol red**	red	yellow	7.0–8.8	7.9
Thymol blue**	yellow	blue	8.0–9.6	8.7
Phenolphthalein	colorless	pink	8.2–10.0	9.1
Thymolphthalein	colorless	blue	9.3–10.6	10.0

* Indicator with more than one color change, first acid–base equilibrium
** Indicator with more color changes, second acid–base equilibrium

The use of indicators for accurate determinations is limited by the fact that the indicator itself has acid–base properties and can therefore consume or release protons, acting as a pH buffer. This error may become significant in microtitrations.

4.2.5 Buffers

pH buffers consist of a mixture of a weak acid (base) and its conjugate base (acid).

A most important application of acid–base systems is related to the property of such a system to act as a *buffer*. Many chemical reactions produce either protons (in aqueous solutions hydronium ions) or hydroxide ions. If these products remain in the system, a corresponding pH change is observed. However, if a buffer is present in the solution it reacts with the liberated hydrogen or hydroxide ions so that only a relatively small change of pH occurs. Buffers consist of a mixture of a weak acid and its conjugate base.

Worked example:

The buffer effect is best explained by a practical problem. Phosphoric acid is a polyprotic acid, for which the following dissociation equilibria exist:

$$H_3PO_4 \rightleftharpoons H_2PO_4^- + H^+ \qquad pK_{A1} \simeq 2 \tag{4.2-53}$$

$$H_2PO_4^- \rightleftharpoons HPO_4^{2-} + H^+ \qquad pK_{A2} \simeq 7 \tag{4.2-54}$$

$$HPO_4^{2-} \rightleftharpoons PO_4^{3-} + H^+ \qquad pK_{A3} \simeq 12.5 \tag{4.2-55}$$

Consider the second equilibrium with $pK_{A2} = 7.0$, for which we can write:

$$K_{A2} = \frac{[H^+][HPO_4^{2-}]}{[H_2PO_4^-]} = 1.0 \times 10^{-7} \tag{4.2-56}$$

We prepare the buffer solution by dissolving 10^{-1} mol of {Na_2HPO_4} and 10^{-1} mol of {NaH_2PO_4} in 1 L water. The resulting solution contains 0.1 mol/L $H_2PO_4^-$ and 0.1 mol/L HPO_4^{2-} ions, and the total phosphate concentration is 0.2 mol/L.

To this solution, 10^{-3} mol HCl is added. If this amount of hydrochloric acid were added to 1 L of pure water, the free hydrogen ion concentration would be 10^{-3} mol/L, to give a pH of 3.0. However, in the phosphate buffer solution, the free hydrogen ions react with the monohydrogen phosphate ions, producing dihydrogen phosphate:

$$HPO_4^{2-} + H^+ \rightarrow H_2PO_4^- \tag{4.2-57}$$

This reaction demonstrates *Le Châtelier's principle* applied to the equilibrium. We estimate that 10^{-3} mol H^+ are consumed in this reaction. Therefore, we find that:

$$[HPO_4^{2-}] = 0.100 - 0.001 = 0.099 \, mol/L$$

$$[H_2PO_4^-] = 0.100 + 0.001 = 0.101 \, mol/L$$

and

$$[H^+] = 10^{-7} \frac{0.101}{0.099} = 1.0202 \times 10^{-7} \, mol/L$$

Of the 10^{-3} mol HCl added to the solution only 2.02×10^{-9} mol remain unreacted. The pH of the solution is calculated to be 6.99. The 0.2 mol/L phosphate mixture is an excellent buffer for the quantity of acid added to the solution. A similar calculation, based on the addition of 10^{-3} mol sodium hydroxide {$NaOH$} would yield a solution of pH 7.01.

By analogy with the first phosphate mixture, we now prepare a solution that contains 10^{-2} mol/L HPO_4^{2-} and 10^{-2} mol/L $H_2PO_4^-$, but the same amount of HCl (10^{-3} mol) is added to it. As in the first case, the original solution has a pH of 7. We again estimate that 10^{-3} mol H^+ are consumed in the reaction and:

$$[HPO_4^{2-}] = 0.010 - 0.001 = 0.009 \, mol/L$$

$$[H_2PO_4^-] = 0.010 + 0.001 = 0.011 \, mol/L$$

and

$$[H^+] = 10^{-7} \frac{0.011}{0.009} = 1.222 \times 10^{-7} \, \text{mol/L}$$

Of the 10^{-3} mol HCl only 2.22×10^{-8} mol remain unchanged, i.e., in free ion form. Again our estimate proves to be essentially correct. The pH of the final solution is 6.92.

Again we can see that the 0.02 mol/L phosphate mixture is a good buffer, although a weaker one than the first solution. The *buffer capacity* of the second solution is smaller than that of the first.

By analogy we now prepare a solution that contains only 10^{-3} mol/L HPO_4^{2-} and 10^{-3} mol/L $H_2PO_4^-$, and again add 10^{-3} mol HCl. If we assume that all of the 10^{-3} mol of HCl are used up in the reaction with HPO_4^{2-}, no more HPO_4^{2-} remains. We obtain a system that resembles a solution of 2.0×10^{-3} mol/L NaH_2PO_4 and 10^{-3} mol/L $NaCl$, producing the following concentrations of ions in aqueous solution:

3.0×10^{-3} mol/L Na^+

2.0×10^{-3} mol/L $H_2PO_4^-$

10^{-3} mol/L Cl^-

However, part of the $H_2PO_4^-$ is dissociated into H^+ and HPO_4^{2-}. Obviously, H_3O^+ and HPO_4^{2-} must exist in equal amounts, so we can write:

$$[H_2PO_4^-] = 2.0 \times 10^{-3} \, \text{mol/L}$$

$$[H^+] = [HPO_4^{2-}] < 2.0 \times 10^{-3} \, \text{mol/L}$$

Introducing these quantities into the acid–base equilibrium expression, we obtain:

$$\frac{[H^+][HPO_4^{2-}]}{2.0 \times 10^{-3}} = 10^{-7} \, \text{mol/L}$$

and

$$[H^+][HPO_4^{2-}] = 2.0 \times 10^{-10} \, \text{mol/L}$$

or

$$[H^+] = 1.414 \times 10^{-5} \, \text{mol/L}$$

The pH of the resulting solution is 4.85.

The 0.002 mol/L dihydrogen phosphate mixture is no longer an effective buffer and its capacity is small. However, compared with the pH that results from the addition of 10^{-3} mol/L HCl to pure water, (pH 3) the buffering effect is still appreciable.

Finally, we prepare a mixture of 10^{-4} mol/L HPO_4^{2-} and 10^{-4} mol/L $H_2PO_4^-$ to which 10^{-3} mol/L HCl is added. In the neutralization of the base, HPO_4^{2-}, 10^{-4} mol H^+ are used, so that 9×10^{-4} mol HCl remain unchanged. This cannot be true, because $H_2PO_4^-$ is also a base that can react with hydrogen ions, producing H_3PO_4. However, this is a strong acid, and the relatively small amount of free HCl is insufficient to protonate the $H_2PO_4^-$ fully. We can compare the system to a mixture that contains:

3×10^{-4} mol/L Na^+

2×10^{-4} mol/L $H_2PO_4^-$

9×10^{-4} mol/L H^+

10×10^{-4} mol/L Cl^-

From this mixture x mol H^+ are removed to produce x mol H_3PO_4, to leave in the solution $2 \times 10^{-4} - x$ mol $H_2PO_4^-$. Assuming $pK_{A1} = 2.0$, we can calculate:

$$K_{A1} = 10^{-2} = \frac{[H^+][H_2PO_4^-]}{[H_3PO_4]} = \frac{(9 \times 10^{-4} - x)(2 \times 10^{-4} - x)}{x}$$

or

$$18 \times 10^{-8} - 9 \times 10^{-4}x - 2 \times 10^{-4}x + x^2 - 10^{-2}x = 0$$

rearranging, we obtain:

$$x^2 - 1.11 \times 10^{-2}x + 1.8 \times 10^{-7} = 0$$

this yields:

$$[H_3PO_4] = x = 1.624 \times 10^{-5} \, \text{mol/L}$$

$$[H^+] \qquad\quad = 8.838 \times 10^{-4} \, \text{mol/L}$$

$$[H_2PO_4^-] \quad = 1.838 \times 10^{-4} \, \text{mol/L}$$

and therefore pH 3.05. This solution has essentially no buffer capacity.

As derived earlier, the pH of a buffer solution is given by the relationship:

$$\text{pH} = \text{p}K_A + \log \frac{[B]}{[A]} \tag{4.2-58}$$

where [B] replaces $[I_b]$ and [A] replaces $[I_a]$ of Eq. 4.2-52, denoting the base and acid concentration of the *buffer*, respectively. This equation is sometimes called the *Henderson–Hasselbalch equation*. It can also be expressed as:

$$\text{pOH} = \text{p}K_B + \log \frac{[A]}{[B]} \tag{4.2-59}$$

where K_B denotes the *base constant* $(1/K_A)$ that describes the acid–base equilibrium as follows:

$$K_A = \frac{[H^+][B]}{[HB]} \tag{4.2-60}$$

where HB = A, and:

$$K_B = \frac{[HB]}{[H^+][B]} = \frac{[A]}{[H^+][B]} = \frac{[A][OH^-]}{K_W[B]} \tag{4.2-61}$$

The most efficient buffer for a given pH consists of a 1 : 1 ratio of the protonated and deprotonated forms of a weak acid (with $\text{p}K_A = \text{pH}$). This cannot always be achieved, but if we wish to prepare a solution of a certain pH, we select a weak acid with a $\text{p}K_A$ value close to the desired pH. If the $\text{p}K_A$ and pH values do not match exactly, the [B]/[A] ratio must be adjusted such that $(\text{p}K_A + \log[B] - \log[A])$ is equal to the desired pH value.

A buffer solution resists changes in pH upon addition of strong acids or strong bases. Depending on the relative concentrations of the acid and base forms of the buffer, the system can resist small or large additions of strong acid or base. This *buffer capacity* is defined as the number of moles of strong acid or base required to change the pH of 1 L of buffer solution by one pH unit. It can be seen that the buffer capacity of our first solution is ten times larger than the capacity of the second solution:

$$\text{capacity } 1 = 10^{-3} \, \text{mol/L}/0.0087 = 1.15 \times 10^{-1} \, \text{mol/L}$$

$$\text{capacity } 2 = 10^{-3} \, \text{mol/L}/0.087 = 1.15 \times 10^{-2} \, \text{mol/L}$$

It can also be shown that the largest capacity results from a buffer which has a 1 : 1 ratio of its acid and base forms.

Solutions with high or low pH values, formed as a result of the dissolution of large quantities of a strong base or acid, are characterized by a large buffer capacity, although the electrolyte practically consists of only one of the conjugate forms (e.g., HCl or NaOH solutions).

Table 4.2-3. Standard buffer solutions and their pH values

Solution	pH at*		
	10 °C	25 °C	38 °C
0.05 mol/L K tetroxalate	1.67	1.68	1.69
0.05 mol/L KH_2 citrate	3.820	3.776	3.755
0.05 mol/L KH phthalate	3.998	4.008	4.030
0.025 mol/L $KH_2PO_4^-$, 0.0025 mol/L Na_2HPO_4	6.923	6.865	6.840
0.01 mol/L $Na_2B_4O_7$	9.332	9.180	9.081
0.025 mol/L $NaHCO_3$, 0.0025 mol/L Na_2CO_3	10.179	10.012	9.903
Saturated solution (25 °C) $Ca(OH)_2$	13.00	12.45	12.04

* Values given by R.G. Bates "Determination of pH" (1964)

Standard buffer solutions

The National Bureau of Standards (NBS), now the National Institute of Standards and Technology (NIST), has recommended some standard buffer solutions, for which the pH is accurately known; the most important are summarized in Table 4.2-3.

General reading

Bates, R.G. *Electrometric pH Determinations. Theory and Practice* (1955).

Hammett, L.P. *Physical Organic Chemistry* (1940).

Harned, H.S., Owen, B.B. *Physical Chemistry of Electrolyte Solutions* (1955).

Lowry, T.M. *Historical Introduction to Chemistry* (1936).

Robinson, R.A., Stokes, R.H. *Electrolyte Solutions* (1955).

Stillman, J.M. *The Story of Early Chemistry* (1924).

Questions and problems

1. Given a definition of the pH value!
2. What are acid–base indicators?
3. How can you prepare a pH buffer solution? Give practical examples.
4. Describe the ion product of water and its temperature dependence.

4.3 Complex Formation

Learning objectives

- To review coordination chemistry and its impact on analytical chemistry
- To explain what a coordination compound is, the coordination tendencies of metal ions, the stepwise processes in complex formation, and to characterize chelates and polynuclear complexes
- To provide the basis for an understanding of the role coordination chemistry plays in analytical chemistry, and to introduce compleximetric techniques

4.3.1 Introduction

The theory of *valence states* was developed in the 19th century, largely based on organic chemistry. The product of this theory was *unitary valence*, according to which binary compounds such $NaCl$, $CaCl_2$, $CrCl_3$, and $PtCl_4$ were considered to be in a saturated state. Thus, the stable product of any reaction was considered to be a state which could be described by the stoichiometric ratio, derived from the ratios of the valence states (oxidation numbers) of the different reactants, such as:

$$N_2 + 3H_2 \rightleftharpoons 2NH_3 \tag{4.3-1}$$

However, in the late 19th century, substances were found which could only be classified as compounds of a higher order. They were formed by the combination of ordinary binary compounds, such as:

$$CrCl_3 + 6NH_3 \rightleftharpoons CrCl_3 \cdot 6NH_3 \tag{4.3-2}$$

$$PtCl_4 + 2KCl \rightleftharpoons PtCl_4 \cdot 2KCl \tag{4.3-3}$$

$$PtCl_4 + nNH_3 \rightleftharpoons PtCl_4 \cdot nNH_3 \tag{4.3-4}$$

where $n = 2, 3, 4, 5, 6$.

These compounds were called *complexes*. In the original definition, a complex was considered to be a compound in which the combination of reactants is expressed by a stoichiometry higher than that derived from the formal valence states of the reactants. Of particular interest were transition metal compounds with similar composition, such as:

$$[Co(NH_3)_nCl_{6-n}]Cl_{n-3} \tag{4.3-5}$$

where $n = 3, 4, 5, 6$, and in which the $Co(III)$ ion could be replaced by $Rh(III)$, $Ir(III)$, etc.

Compounds such as these were synthesized by Jorgensen and Werner (1893), who described the compounds in terms of their:

- stoichiometric composition
- color and crystal forms
- behavior in precipitation reactions
- electrolytic behavior

Alfred Werner (1866–1919) was able to show that such compounds had different colors, exhibited differences in their electrolytic conductance, and behaved differently in precipitation reactions.

The yellow compound, later described as $[Co(NH_3)_6]Cl_3$, has the property of a $3:1$ electrolyte (compare with $FeCl_3$), and with silver nitrate, a silver chloride precipitate was obtained corresponding to a $Co:Cl$ ratio of $1:3$.

The purple compound, later described as $[Co(NH_3)_5Cl]Cl_2$, behaved like a $2:1$ electrolyte (compare with $FeCl_2$), and only two of the three chlorides of the compound were precipitated by silver nitrate.

The violet compound, later described as $[Co(NH_3)_3Cl_3]$, behaved as a nonelectrolyte, and gave no precipitation with silver nitrate.

In complexes, atoms of the same species may have different chemical properties such as Cl in $[Co(NH_3)_6]Cl_3$ or in $[Co(NH_3)_5Cl]Cl_2$.

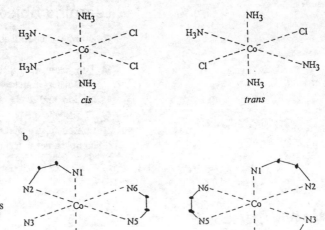

Fig. 4.3-1. Bases of Werner's model of coordination compounds. a) *cis* and *trans* forms of [Co(NH₃)₄Cl₂]⁺; b) Optical isomers of the cobalt tris(ethylenediamine) complex, proving the octahedral structure of the compound

A number of definitions were introduced within the framework of coordination theory:

- *Coordination center* or *central atom*: usually a metal ion
- *Ligands*: groups that surround the central atom; usually anions or neutral basic groups
- *Coordination number*: the number of ligands which are nearest neighbors of the coordination center
- *Structural arrangement* of the ligands around the coordination center, e.g.,
      coordination number 6: octahedral arrangement
      coordination number 4: tetrahedral or square planar arrangement
      coordination number 3: trigonal arrangement
      coordination number 2: linear arrangement
- Square brackets [ ] are used to describe the part of the compound that does not dissociate into smaller parts; the actual complex

Werner first had to prove that a model based on these concepts agreed with the observed properties of these compounds. He was able to show that the anion outside the bracket in the above compounds could easily be exchanged by other anions. The following yellow compound was synthesized with a range of different anions, $[Co(NH_3)_6]X_3$, where $X = F^-$, Cl^-, Br^-, I^-, NO_3^-, ClO_4^-, $\frac{1}{2}C_2O_4^{2-}$, $\frac{1}{2}SO_4^{2-}$, etc.

However, the great success and recognition of Werner's theory was due to findings demonstrating its correctness in a more direct way. Werner, who was awarded the Nobel Prize in 1913, obtained the *cis* and *trans* forms of $[Co(NH_3)_4Cl_2]^+$ (Fig. 4.3-1a) and synthesized the optical isomers of the cobalt *tris*(ethylenediamine) complex, thus proving the octahedral structure of the cobalt complexes (Fig. 4.3-1b).

Werner's concepts are still accepted today. However, in this account we do not distinguish between complexes and binary compounds. There is no need to introduce a separate classification for binary and higher compounds, since the structural aspects of the two are the same in most cases. Simple binary compounds such as $CrCl_3$ are complexes, and such a chemical formula is of value only if the structure is unknown. In the solid phase such compounds as $\{CrCl_3\}$ or $\{FeCl_3\}$ are in fact not characterized by a coordination number of 3 and, in aqueous solution, the molecular species $FeCl_3$ has water molecules in the first coordination sphere. It is therefore logical to describe all members of the stepwise complex formation (e.g., $Fe^{3+} = Fe(H_2O)_6^{3+}$, $FeCl^{2+} = Fe(H_2O)_5Cl^{2+}$, $FeCl_2^+ = Fe(H_2O)_4Cl_2^+$) and the appropriate solid state species as complexes. Any reaction of cations (or Lewis acids) with stable molecular or anionic species containing free electron pairs (Lewis bases) is called complex formation whether or not there is a colvalent or an electrovalent contribution.

4.3.2 Covalent bond formation

Although the covalent bond plays an important role in the formation of complexes, it must be kept in mind that complex formation can also involve electrostatic interactions. We concentrate here on the description and characterization of the covalent bond.

The electronic configuration of the noble gases, i.e., the electron octet, is an energetically favorable situation for most light elements. Not all elements are able to achieve this configuration by direct electron transfer. For many elements the ionization energy is too large for the removal of the excess electrons and the electron affinity is too small to allow addition of the required electrons.

However, an octet arrangement is often achieved when the reacting partners share electron pairs. This type of bonding is very common in organic chemistry, but is neither restricted to carbon chemistry nor to that of nonmetals. The sharing of electrons is characteristic of covalent bond formation; each atom tends toward an electron octet in its outermost electron shell. Hydrogen is an exception, since its outermost shell can incorporate only two electrons. However, the octet rule is not universally observed. For example, in SF_6, the sulfur has twelve electrons in the outermost shell.

According to a modern view, every compound or ion consisting of a central atom and ligands can be considered as a complex (e.g. SO_4^{2-} ... tetraoxo-sulfur VI^+-complex).

A single covalent bond implies the sharing of one electron pair (duplet), a double bond the sharing of two pairs, etc. The formation of bonding duplets may be achieved in different ways. Either each atom contributes a single electron, or the electron pair is supplied by only one of the partners. In the latter case, we have a *Lewis acid–base reaction* (electron-pair donor and acceptor) and consequently the resulting coordinate bond is polar. In other words, an electric dipole is produced and corresponding electrostatic forces are created. The formation of the compound between boron trichloride and phosphorus trichloride represents such a *Lewis acid–base* reaction (neutralization).

Similar polar bonds are formed when each atom contributes an electron to form the bonding duplet, but where, owing to structural arrangements and to the unequal attraction of the electron duplet by the different atomic nuclei, the binding electron pair is shifted from the center of the bond toward one particular nucleus. Such a bond is formed in the water molecule, for which an appreciable ionic interaction is predicted. Polar bonds result when elements with large differences of electronegativity are bonded together. This is expected, particularly for bonds between oxygen or fluorine and certain metal ions (alkali and alkaline earth metals).

Molecular orbitals which describe covalent bonding can be formed by the overlap of the corresponding atomic orbitals. Covalent bonds have characteristic directional properties, which is not the case for ionic bonds.

A number of parameters are characteristic of covalent substances. The enthalpy and entropy values of covalent compounds tend to be negative and the heat content usually has only a slight temperature dependence. These properties, together with other characteristics, are represented in Table 4.3-1, where they are compared with the general properties of ionic compounds.

Table 4.3-1. Properties of covalent and ionic compounds

Property	Ionic compound	Covalent compound
Melting point	high	low
Volatility	low	high
Solubility in water	good	low or insoluble
Solubility in nonpolar solvents	low	good
Electrolyte conductance	conducting solutions	nonconducting
Enthalpy $H°$	positive	negative
Entropy $S°$	positive or negative	negative
$dH°/dT$	positive	close to zero
Bond direction	not directed	directed
Color	composite of single ions	any

4.3.3 Coordination tendencies of metal ions and ligand properties

In coordination chemistry, many attempts have been made to classify the mass of data collected by generations of scientists. The many different aspects of this huge amount of information are often confusing, and it is therefore important to adopt a classification that helps one to extract important general principles.

We shall adopt the classification of Gerold Schwarzenbach (1904–1978), Professor of Inorganic Chemistry at the Swiss Federal Institute of Technology in Zürich. He divided complexing agents into two groups, called *general* and *selective* complexing agents. All complexing agents are classified according to their coordinating ligand atom, and the properties of complicated systems are related to and compared with the properties of the much simpler binary complexes formed by the same metal and the atomic anion or its protonated form as a ligand. Thus hydroxo complexes are representative of all oxygen-containing complexes and sulfide (also called thio) complexes are representative of all sulfur-containing complexes.

The oxygen-containing ligands and the fluoride ion belong to the group of general complexing agents. Almost all metal ions form complexes with these ligands. Owing to the electronegativity of the ligand atoms, one would expect an appreciable ionic contribution to the coordinate bond.

The other halide ions exhibit selective properties, forming weak complexes with some metal ions and extremely strong complexes with others. The selectivity increases with the atomic number of the halides, but is generally smaller than that of special ligands, such as the cyanide ion and phosphorus and sulfur-containing ligands. It should be noted that the electronegativities of these ligand atoms are appreciably lower than those of fluorine and oxygen.

Ralph G. Pearson, Professor of Chemistry at Northwestern University in Evanston, Illinois, has divided the ligand atoms and metal ions into *soft* and *hard* acids and bases, according to their electronegativity and polarizability. It is very attractive to consider complex formation as an extension of the acid–base concept, or alternatively to consider a proton acid (*Brönsted–Lowry acid*) as a hydrogen complex of the corresponding base. The metal ions act as *Lewis acids*, and most ligands are indeed *Lewis bases*.

In Section 4.3.2, we briefly discussed the general stability of the electron octet configuration. There are other electronic configurations which also lead to stability. The electron octet is the configuration of the noble gases. The outermost s and p electron shells are completely filled in this arrangement.

Another stable configuration is found in some metals, namely, $Ni°$, $Pd°$, and $Pt°$. Here, in addition to the filled s and p shells, we find a filled d-electron shell (10 electrons). The inner electron shell is represented by the spherically symmetric noble gas shell (with zero magnetic and electric moments) plus the 10 d-electrons. The individual orbitals of the later are asymmetric and during the filling of the d-shell $(1, 2, \ldots, 9, 10$ electrons$)$, magnetic and electric dipole moments may be superimposed. Although the full d-shell (10 electrons) is again characterized by spherical symmetry and therefore, has zero magnetic moment, the electron shell can easily be polarized by an applied electric or magnetic field.

A similar stable situation is also observed with some of the heavy metals, illustrated by the electronic structure of mercury. Here, in addition to the filled d-shell, the next s-subshell is also filled. This situation is found in the electronic configuration of $[Ni°]ns^2$, $[Pd°]ns^2$, and $[Pt°]ns^2$, the extra stability observed is explained by the so-called inert-pair effect, which is also responsible for the unexpected stability of $Hg°$, Tl^+, Pb^{2+}, and Bi^{3+}. This outer-shell electronic configuration corresponds to that of helium.

Traditionally, the metals are divided into *A- and B-metals*, which are separated by the *transition metals* (*T-metals*) in the periodic table. A-metal ions (or A-cations) have the electronic configuration of a noble gas. The rare earth metal ions M^{3+} are part of this group. A-cations have spherical symmetry and their magnetic and electric dipole moments are zero. They are not easily polarized and Pearson's characterization as hard acids is justified. Obviously, the ionic radius plays an important role in determining the extent of complex formation for these metal ions. Stronger complexes are formed between ions of smaller ionic radius and larger electric charge (Coulomb interactions). For example, the complexes between Be^{2+} and F^- or OH^- are stronger than those between Mg^{2+} and the same anions, and Al^{3+} forms stronger fluoro and hydroxo complexes than Be^{2+} and the alkali metal cations. The complexes formed between hard acids and hard bases have a predominant ionic contribution to the bonding. In aqueous solution, A-cations prefer fluoride ions as ligands and form stronger complexes with H_2O, OH^-, and O^{2-} than with other bases. Indeed, many A-cations form insoluble fluorides and hydroxides (these can be regarded as polynuclear complexes). The basic ligands, such as NH_3, HS^-, CN^-, and even S^{2-} are not able to compete effectively with the more electronegative oxygen and fluoride ligands. Anions which form unsoluble precipitates, such as CO_3^{2-}, $C_2O_4^{2-}$ (oxalate), SO_4^{2-}, PO_4^{3-}, and CrO_4^{2-}, always contain oxygen as a coordinating ligand atom. Organic solvents which are able to solvate these cations also generally contain oxygen (alcohols, ketones, etc.). The following sequence is found for the stability of the halide complexes:

$$F^- \gg Cl^- > Br^- > I^-$$

The stability of complexes generally increases with increasing electronegativity of the ligand atom for a given A-cation.

B-metal ions (B-cations) have the electronic structure of $Ni°$, $Pd°$, and $Pt°$, i.e., they have 10 d-electrons in the outermost electron shell in addition to the noble gas core. We should also include the cations with the inert-pair s-electron configuration ($[Ni°]ns^2$, $[Pd°]ns^2$, and $[Pt°]ns^2$) in this group.

B-cations are also spherically symmetric, but are more readily polarized than A-cations and Pearson's classification follows from this. The ionic radius of the B-cations is not well defined. B-cations form complexes with oxygen ligands (the hydrated ions act as acids and protons and are lost at relatively low pH values, resulting in the corresponding hydroxo complexes). However, ligands such as NH_3, SH^-, S^{2-}, CN^-, and the heavier halide ions are able to compete successfully with oxygen ligand atoms. In particular, fluoride plays only a minor role in complex formation reactions of these ions. In fact, the strength of the complexes increases with *decreasing* electronegativity of the coordinated ligand:

electronegativity: $F > O > N \cong Cl > Br > I \cong C \cong S$ decrease

complex stability: $F < O < N \cong Cl < Br < I < C < S$ increase

A similar reversed trend in the stability of the B-metal complexes (compared with A-metal complexes) is observed in the relation between the stability of the complex and the charge on the ion. This is evident from a comparison of the stabilities of complexes of isoelectronic metal ions. In the series Ag^+, Cd^{2+}, In^{3+}, the complex stability decreases with increasing positive charge of the metal ion. This indicates an increasingly important contribution from covalent bonding, as confirmed by the negative values of enthalpies and entropies of formation. However, it must be recognized that A- or B-character (or, using Pearson's terms, hardness and softness) is not a quantitative parameter, and both properties may overlap. For example, the Pb^{2+} ion exhibits typical hard character toward the electronegative ligands F^- and oxygen, while the color of the iodide precipitate and the insolubility of the sulfide demonstrate that, at least with the heavier halides (Br^- and I^-), the B-character outweighs the A-character, and that covalent bonds are formed in compounds containing the less negative ligands. Generally, one can state that metal ions with an inert-pair electronic configuration possess a hard acid tendency.

Transition metal ions (T-cations) fall between the A- and B-metal classifications. Their properties often suggest a compromise between soft and hard behavior. T-cations have partially filled d-shells ($1, 2, \ldots, 9$ electrons). We may describe them as spherical cores surrounded by a soft and easily deformed outer electronic shell. The electron density is not distributed symmetrically around the core. The magnetic and electric dipole moments may assume different values depending on the number of electrons in the d-shell.

The fluoro complexes are generally weak for this group, and the most stable ones (those with Fe^{3+} and Cr^{3+}) can easily be transformed to polynuclear hydroxo complexes. However, for these two metal ions, NH_3 cannot displace the hydroxide ligands, while Co(II), Ni(II), and Cu(II) form successively stronger complexes with NH_3 (ammines), indicating that the B-character increases as the number of d-electrons increases. Fe^{3+} and Cr^{3+} do not form sulfide complexes in aqueous solution, but Mn^{2+}, Fe^{2+}, Co^{2+}, Ni^{2+}, and Cu^{2+} form insoluble sulfides. Ni^{2+} and Co^{2+} form very strong cyano complexes, $Ni(CN)_4^{2-}$ and $Co(CN)_5^{3-}$, and $Fe(II)$ and $Fe(III)$, as well as $Co(III)$ and $Cr(III)$, form inert hexacyano complexes, $M(CN)_6^{v-6}$. The metal ions far to the right-hand side of the classification (Pd^{2+}, Pt^{2+}, Pt^{4+}, and Au^{3+}) have a strong preference for sulfur and phosphorus ligands over those containing oxygen and fluorine.

From this summary, it can be concluded, all other things being equal, that the ions with the lower charges are those with greatest B-character (softness). On the other hand, a high charge reinforces hard acid behavior. Ions with five or more d-electrons are softer than those with fewer electrons (Ti^{3+}, V^{3+}, and Mn^{3+} form hydrated ions that behave as moderately strong proton acids).

Harry Irving and R.J.P. Williams pointed out that the stability of complexes of the divalent first-row transition metals with a particular ligand follows a sequence (*Irving–Williams order*) [4.3-1]:

$$Mn^{2+} < Fe^{2+} < Co^{2+} < Ni^{2+} < Cu^{2+} < Zn^{2+}$$

Soft and hard central atoms and ligand field theory.

This sequence may be explained in terms of the *crystal field* and *ligand field theories*. The crystal field around the central metal ion is created by the electric field of the ligands arranged around the metal ion. This field affects the energy of the d-orbitals and causes them to split. If the d-electrons can occupy lower energy levels, the complex becomes more stable (ligand field stabilization). Mn^{2+} (5 electrons) and Zn^{2+} (10 electrons) cannot benefit from such stablization, since they are spherically symmetric. Each orbital is singly or doubly occupied in Mn^{2+}_{aq} and Zn^{2+}_{aq}, respectively.

In summary, we note the following:

- Ions of A-metals plus the first few members of the transition metal series (e.g., Cr^{3+}, Mn^{3+}, Fe^{3+}, and Co^{3+}, as well as UO^{2+}, VO^{2+}, and such species as BF_3, SO_3, CO_2, RCO^+, or R_3C^+ (R = alkyl group), are hard acids. They are not very polarizable and they are spherically symmetric. Their affinity for ligands is as follows:

$$N \gg P, O \gg S, F^- \gg Cl^- > Br^- > I^-, PO_4^{3-} \gg SO_4^{2-} \gg ClO_4^-, CO_3^{2-} \gg NO_3^-$$

This indicates that the alkali metal nitrates and perchlorates are the most suitable for use as inert electrolytes in maintaining constant ionic strength.

- Ions of the B-metal group and the latter members of the transition metal series (Pd^{2+}, Pt^{2+}, Pt^{4+}, Au^{3+}, etc.) are soft acids. They are characterized by low electronegativity and high polarizability. The corresponding ligand sequence is observed:

$$P \gg N, S \gg O, S^{2-} > CN^- > I^- > Br^- > Cl^- \cong N < O < F$$

4.3.4 Stepwise complex formation

We shall now discuss complex formation on the basis of what we have established in the previous sections.

Equilibrium in solution-phase complex chemistry involves permanent complex formation and decay:

$$M + L \underset{k_{-1}}{\overset{k_1}{\rightleftharpoons}} ML \tag{4.3-6}$$

If suitable ligands are introduced into a solution of metal ions, we would expect the formation of complexes in which the metal ion plays the role of a coordination center which collects ligand molecules in its coordination sphere. The number and type of ligands coordinated influences the structure of the complex. Sites in the coordination sphere' may be occupied by a single ligand molecule with several coordinated atoms (*multidentate ligands*, forming *chelates*) and the ligand atoms may also be coordinated to more than one metal center (forming *polynuclear complexes*). These aspects will be dealt with in this section.

In complex formation, we must bear in mind that metal ions in solution have an inherent coordination shell. Incoming ligands are not simply incorporated into an empty coordination shell, but rather replace other ligands that already occupy the coordination sites. This replacement or substitution occurs in a stepwise fashion. We know that the free metal ions exist in solution as the hydrated species, in which water plays the role of the ligand (oxygen being the coordinated atom). Therefore, ligand substitution must depend on the concentration of the free ligand species which is to be coordinated. This ligand concentration (ligand pressure) determines how many water molecules may be exchanged for the ligand, and the substitution proceeds in steps, i.e., all the unsubstituted metal ions first exchange one water molecule for the ligand before the next ligand is incorporated into the metal coordination shell, rather than complete substitution of a particular metal ion before the next metal ion is attacked.

It is conventional to refer to the free metal ion when we describe the hydrated metal ion, and therefore stepwise substitution is also expressed in terms of ligand addition. We describe complex formation as follows:

$$M^{v+} + X^{\chi-} \rightleftharpoons MX^{(v-\chi)+}$$

$$MX^{(v-\chi)+} + X^{\chi-} \rightleftharpoons MX_2^{(v-2\chi)+}$$

$$MX_{i-1}^{(v+1-i\chi)+} + X^{\chi-} \rightleftharpoons MX_i^{(v-i\chi)+}$$

but ligand substitution is described as:

$$M(H_2O)_n^{v+} + X^{\chi-} \rightleftharpoons M(H_2O)_{n-j}X^{(v-\chi)+} + jH_2O$$

$$M(H_2O)_{n-j}X^{(v-\chi)+} + X^{\chi-} \rightleftharpoons M(H_2O)_{n-y}X_2^{(v-2\chi)+} + y - jH_2O$$

$$M(H_2O)_{n-g}X_{i-1}^{(v+1-i\chi)+} + X^{\chi-} \rightleftharpoons MX_i^{(v-i\chi)+} + n - gH_2O$$

As mentioned previously, this substitution may involve the replacement of the same number of water molecules or more per incoming ligand.

Negatively charged complexes (anions) generally produce anions with lower coordination numbers (e.g., in $FeCl_4^-$, coordination number four) than in neutral species (e.g., $Fe(H_2O)_2Cl_3$, coordination number five) or in positively charged complexes, such as $Fe(H_2O)_4Cl_2^+$ and $Fe(H_2O)_5Cl^{2+}$ (coordination number six).

This enables us to write the formation of complexes in terms of stepwise ligand addition rather than as ligand substitution, since the liberated water may be excluded from the corresponding mass action law expression. The reason for this is that the solution chemist usually works under conditions of constant ionic strength where the water activity of the solution does not change. The formal complex formation reaction is then expressed in terms of the stepwise complex formation constants K_i or the corresponding cumulative constants β_i or $\beta_{j,i}$. In the use of these constants, the same activity concept must be applied as was discussed in connection with the Sillen approach in constant ionic strength media, the Debye–Hückel theory in infinite dilute systems, and the treatment of acid–base equilibria:

$$K_i = \frac{[MX_i]}{[MX_{i-1}][X]} \quad \beta_i = \frac{[MX_i]}{[M][X]^i} \quad \beta_{j,i} = \frac{[MX_i]}{[MX_j][X]^{i-j}} \tag{4.3-7}$$

The electric charges of the different complexes, ligands, and metal ions have been omitted for simplicity. As a practical example, the complex formation or stability

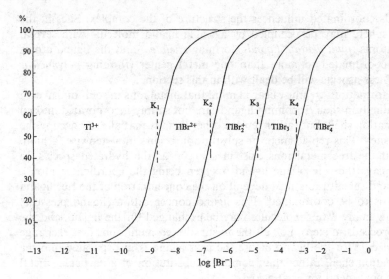

Fig. 4.3-2. Stepwise complex formation of thallium bromide, calculated from the stability constants given in the text

Complex formation is a stepwise process (see Fig. 4.3-2).

constants for bromide complexes of thallium (III) ion are given:

$$K_1 = \frac{[\text{TlBr}^{2+}]}{[\text{Tl}^{3+}][\text{Br}^-]} = 10^{8.15} \qquad (4.3\text{-}8)$$

$$K_2 = \frac{[\text{TlBr}_2^+]}{[\text{TlBr}^{2+}][\text{Br}^-]} = 10^{6.15} \qquad (4.3\text{-}9)$$

$$K_3 = \frac{[\text{TlBr}_3]}{[\text{TlBr}_2^+][\text{Br}^-]} = 10^{4.36} \qquad (4.3\text{-}10)$$

$$K_4 = \frac{[\text{TlBr}_4{}^-]}{[\text{TlBr}_3][\text{Br}^-]} = 10^{2.85} \qquad (4.3\text{-}11)$$

It is common practice to illustrate the stepwise complex formation as a logarithmic plot of the complex species versus free ligand concentration (Fig. 4.3-2).

However, this complex formation can also be expressed in terms of the cumulative constants:

$$\beta_1 = K_1 \qquad (4.3\text{-}8)$$

$$\beta_2 = K_1 K_2 = \frac{[\text{TlBr}_2^+]}{[\text{Tl}^{3+}][\text{Br}^-]^2} = 10^{14.30} \qquad (4.3\text{-}12)$$

$$\beta_3 = K_1 K_2 K_3 = \frac{[\text{TlBr}_3]}{[\text{Tl}^{3+}][\text{Br}^-]^3} = 10^{18.66} \qquad (4.3\text{-}13)$$

$$\beta_4 = K_1 K_2 K_3 K_4 = \frac{[\text{TlBr}_4{}^-]}{[\text{Tl}^{3+}][\text{Br}^-]^4} = 10^{21.51} \qquad (4.3\text{-}14)$$

$$\beta_{1,2} = K_2 = 10^{6.15} \qquad (4.3\text{-}15)$$

$$\beta_{1,3} = K_2 K_3 = \frac{[\text{TlBr}_3]}{[\text{TlBr}^{2+}][\text{Br}^-]^2} = 10^{10.51} \qquad (4.3\text{-}16)$$

$$\beta_{1,4} = K_2 K_3 K_4 = \frac{[\text{TlBr}_4{}^-]}{[\text{TlBr}^{2+}][\text{Br}^-]^3} = 10^{13.36} \qquad (4.3\text{-}17)$$

$$\beta_{2,3} = K_3 = 10^{4.36} \qquad (4.3\text{-}18)$$

$$\beta_{2,4} = K_3 K_4 = \frac{[\text{TlBr}_4{}^-]}{[\text{TlBr}_2^+][\text{Br}^-]^2} = 10^{7.21} \qquad (4.3\text{-}19)$$

On the other hand, if we introduce a complex substance, e.g., a solid-state species, which we denote by a bracket { }, into an aqueous solution in the absence of the free ligand, we must expect the complex to decay. Many forms of complexes are known to exist as solids (e.g., $\{Na_3[TlBr_6]\}$), and these decay in aqueous solution since the complex anion $TlBr_6{}^{3-}$ cannot exist as such in the absence of free Br^-. Ligand substitution now proceeds in the reverse direction, and water molecules replace and liberate the ligand from the coordination shell of the metal ion (similar to an acid dissociation). This process is called *equation* of the complex. The corresponding equilibrium constant is called the *instability constant*. It is simply the reciprocal of the corresponding stability constant.

Complex formation:

$$Tl^{3+} + Br^- \rightarrow TlBr^{2+} \quad K_1 \qquad \text{stability constant} \qquad (4.3\text{-}20)$$

Equation:

$$TlBr^{2+} \rightarrow Tl^{3+} + Br^- \quad K_{i1} \qquad \text{instability constant} \qquad (4.3\text{-}21)$$

The subsequent discussion does not deal specifically with instability constants, but mainly emphasizes the definitions of complex formation.

Complex formation proceeds in distinct steps that generally do not overlap each other. In other words, the different K_i values follow the sequence:

$$K_1 > K_2 > K_3 > K_4 \qquad (4.3\text{-}22)$$

Generally, the ratio K_{i+1}/K_i is smaller than that calculated from statistical considerations. We must keep in mind that, with anions as ligands, the subsequently formed complexes carry a different electric charge, so that each additional ligand is attracted with a smaller Coulomb force, or in the case of a complex anion, must overcome a repulsive Coulomb force. This means that K_1/K_2 is greater than K_2/K_3, etc. Other effects may be involved too, owing to structural changes (change of coordination number) of the complex during stepwise complex formation. This can also occur with covalent bond formation. In order to form such a bond, an orbital of the metal ion must overlap with orbitals of the ligands. This overlap determines the bond angle and structure of the complex. For example, $TlBr_4{}^-$ is tetrahedral. If the higher complex is formed, the coordination number has to increase from four to five or six (as in $TlBr_6{}^{3-}$), accompanied by a corresponding change in the structure. We have already seen that $FeCl_2{}^+$ has a coordination number of six (octahedral) whereas $FeCl_3$ and $FeCl_4{}^-$ have coordination numbers of five and four. The structures most commonly observed are summarized in Table 4.3-2.

Ligand substitution may be either fast or slow. The rate of complex formation is

Table 4.3-2. Most commonly observed structures of complex species

Coordination number	Hybrid orbitals	Structure	Tyical metal ion	Examples
2	sp	linear	H^+	hydrogen bond, H_2F^-
			Ag^+	$\{Ag_2O\}$, $Ag(CN)_2{}^-$, $Ag(SH)_2{}^-$
			Hg^{2+}	$HgCl_2$, $HgBr_2$, HgI_2, $HgS_2{}^{2-}$
			Au^+	$AuCl_2{}^-$
4	dsp^2 or sp^2d	square panar	Pd^{2+}	$Pd(H_2O)_4{}^{2-}$, $Pd(NH_3)_4{}^{2+}$
			Pt^{2+}	$PtCl_4{}^{2-}$
			Ni^{2+}	$Ni(CN)_4{}^{2-}$, Ni-dimethylglyoxime
	sp^3	tetrahedral	Fe^{3+}	$FeCl_4{}^-$, $FeBr_4{}^-$
			Tl^{3+}	$TlCl_4{}^-$, $TlBr_4{}^-$
			Co^{2+}	$CoCl_4{}^{2-}$, $Co(CNS)_4{}^{2-}$
			Li^+	$Li(H_2O)_4{}^+$
			Be^{2+}	$Be(H_2O)_4{}^{2+}$
			$B(III)$	$B(OH)_4{}^-$
			$C(IV)$	organic compounds
6	d^2sp^3 or sp^3d^2	octahedral	most metal ions	$Co(H_2O)_6{}^{2+}$, $Fe(H_2O)_6{}^{2+}$ $Fe(H_2O)_6{}^{3+}$, $Cr(H_2O)_6{}^{3+}$ $Na(H_2O)_6{}^+$ most EDTA complexes

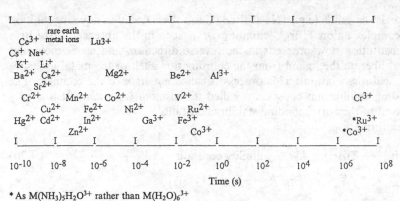

Fig. 4.3-3. Average life times of water molecules in the coordination sphere of different metal ions (aquo complexes)

* As $M(NH_3)_5H_2O^{3+}$ rather than $M(H_2O)_6^{3+}$

best described by the average life time of a particular ligand in the coordination sphere of a metal ion. A good example for evaluating these properties is the rate of water exchange observed with different metal ions. Al^{3+} occupies a position in the middle of the series. Its average half-life for water exchange is about 1 s. Ba^{2+} and Hg^{2+} exchange their coordinated water extremely rapidly (10^{-9} s), while Cr^{3+} and Ru^{3+} exchange their water ligands extremely slowly (exchange half-life $\sim 10^6$ s). Figure 4.3-3 presents a comparison of these times for a number of metal ions.

The half-life of reaction is defined as the time taken for half the species initially present to be transformed to products. In a first-order process the rate law is:

$$\frac{-d[M(H_2O)_5(H_2O)^*]}{dt} = k[M(H_2O)_5(H_2O^*] \tag{4.3-23}$$

where the species $M(H_2O)_5(H_2O)^*$ is exchanging one particular water molecule (marked with an asterisk) and k is the first-order rate constant (s^{-1}). On integration, we obtain:

$$[M(H_2O)_5(H_2O)^*]_t = [M(H_2O)_5(H_2O)^*]_0 e^{-kt} \tag{4.3-24}$$

When:

$$[M(H_2O)_5(H_2O)^*]_t = \tfrac{1}{2}[M(H_2O)_5(H_2O)^*]_0$$

$$t = t_{1/2}$$

where

$$t_{1/2} = \frac{0.693}{k} \tag{4.3-25}$$

The rate constant for ligand exchange is generally influenced by both the nature of the metal ion and that of the ligand. However, there are general properties which can be associated with the nature of particular metal ions. For example, all reactions involving Cr(III), Co(III), Ni(III), and Pt(IV) are relatively slow processes compared with reactions involving other metal ions in the same oxidation state.

4.3.5 Chelates and polynuclear complexes

Several coordinating atoms can be incorporated into single ligand molecules. A ligand of this type can form several bonds to the metal centers. If all the bonds are formed with the same center, then the complex is called a *chelate* and the ligand is said to be *multidentate*. We distinguish uni-, bi-, tri-, tetra-, penta-, and hexadentate ligands.

unidentate ligands: Cl^-, F^-, NH_3, H_2O, etc.
bidentate ligands: oxalate, ethylenediamine ($H_2N–CH_2–CH_2–NH_2$), glycine

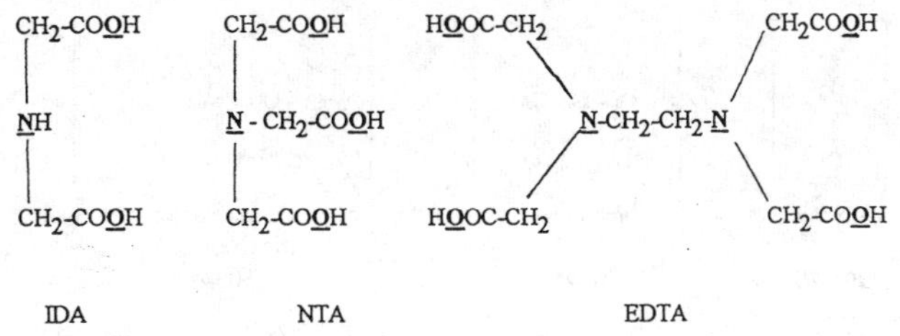

Fig. 4.3-4. Structures of common complexing agents

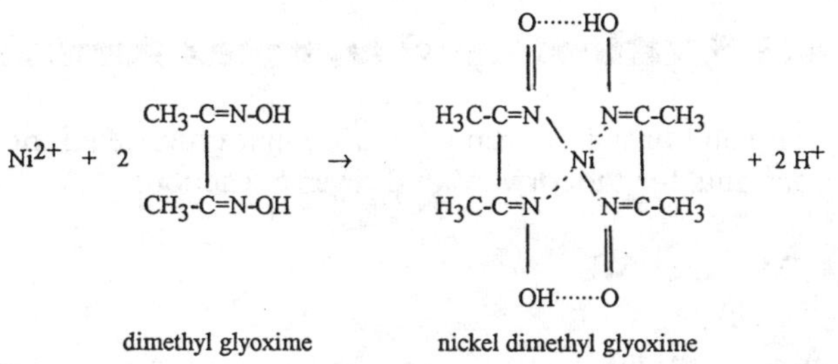

Fig. 4.3-5. Complex formation of Ni^{2+} with dimethylglyoxime to form a neutral five-membered ring complex, which is additionally stabilized by hydrogen bridges

tridentate ligands: iminodiacetic acid (IDA)
tetradentate ligands: nitrilotriacetic acid (NTA)
hexadentate ligands: ethylenediaminetetraacetic acid (EDTA)

Figure 4.3-4 shows the structures of IDA, NTA and EDTA.

Complex formation between a metal ion and a multidentate ligand is called *chelation*. The most obvious feature of the chelated metal center is the presence of a ring system; as for example in complex formation between Ni^{2+} and dimethylglyoxime. Dimethylglyoxime forms a five-membered ring with the nickel ion (Fig. 4.3-5). The corresponding chelate is a neutral species, and it forms an insoluble precipitate.

Chelate complexes are more stable than complexes because the entropy of the system is increased upon chelate formation.

All chelates of importance form either five- or six-membered rings. Smaller or larger rings would create too much strain on the bond angles. Chelates with larger rings can be prepared only with Ag^+, Hg^{2+}, and Au^+, since these metal ions prefer linear coordination ($180°$ bond angle) and are thus stabilized by chelate formation involving rings with six, seven, or eight.

Multidentate ligands forming charged chelates are useful in volumetric methods, since the chelates are water soluble. However, neutral chelates are of ten water insoluble and are therefore widely used in gravimetric procedures and solvent extraction chemistry.

Complexes formed by monodentate ligands are usually less stable than chelates. For example, if a solution of $Cu(NH_3)_4^{2-}$ is diluted with water, aquation is observed. This aquation is much more pronounced for this ammine complex than for the corresponding chelate in which the ammine groups are linked by ethylene groups (e.g., in $H_2N–CH_2–CH_2–NH–CH_2–CH_2–NH_2$ or $H_2N–CH_2–CH_2–NH_2$).

This stabilization of multidentate complexes compared with those with unidentate ligands is called the *chelate effect*. It plays an important role in natural water systems, since the free concentration of unidentate ligands is usually so small that the corresponding complex formation is not observed. However, chelates have such a remarkable stability that they may be found in natural waters.

A ligand atom may be involved in complex formation with more than one coordination center. In this case, *polynuclear complexes* are formed, and the ligand assumes a bridging role. Oxygen is the most important bridging ligand. Examples of such polynuclear complexes are the chromate–bichromate and bichromate–trichromate equilibria (Fig. 4.3-6).

Oxo and hydroxo complexes may be bridged in a similar way, forming polynuclear arrangements of characteristic structure. Lower polymers (oligomers) are usually soluble in aqueous solution, but as the degree of polymerization increases, the species generally become less soluble. Structures shown in Fig. 4.3-7 are quite common.

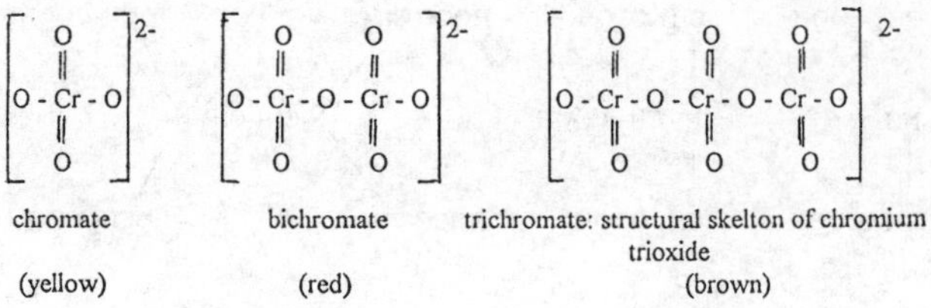

Fig. 4.3-6. Polynuclear complexes of the chromate, bichromate, and trichromate equilibria

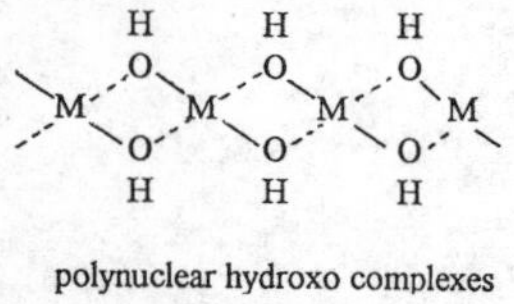

chromate   bichromate   trichromate: structural skelton of chromium
                                                    trioxide
(yellow)   (red)   (brown)

polynuclear hydroxo complexes

Fig. 4.3-7. Polynuclear arrangement with hydroxo groups as bridging ligands

Rapidly formed, water soluble chelates play an important role in modern water analysis (e.g. water hardness determinations).

4.3.6 Kinetics of complex formation

The equilibrium constant for a complexation equilibrium is related to the rate constants for the forward and reverse reactions:

$$M + L \underset{k_{-1}}{\overset{k_1}{\rightleftharpoons}} ML \tag{4.3-26a}$$

At equilibrium:

$$k_1[M][L] = k_{-1}[ML] \tag{4.3-26b}$$

and

$$\frac{[ML]}{[M][L]} = \frac{k_1}{k_{-1}} = K_1 \tag{4.3-26c}$$

Clearly, the magnitude of the equilibrium constant depends on both k_1 and k_{-1}, and the rate constants themselves give much more information about the mechanism of complexation than does K_1, which could, in principle, arise from an infinite combination of different k_1 and k_{-1} values.

As we have seen, the labilities of metal ions (as judged by their relative rates of solvent exchange, Fig. 4.3-3) vary markedly, and from what has been said above, it is obvious that we should not expect thermodynamic stability (represented by K_1) to be a simple function of kinetic lability (represented by the rate constants k_1 and k_{-1}).

Although our present understanding of the mechanisms of inorganic reactions is founded on the analysis of data for complexation of Cr(III), Rh(III) and other kinetically inert metal centers, the metal ions involved in natural water systems are generally much more labile.

It should also be noted that an aquo ion in an aqueuos medium creates three regions in the solution medium: an inner sphere region, where the solvent and other ligand molecules are attached to the metal center in a definite stereochemical arrangement; a second, comparatively disordered region, where the water molecules are sandwiched between the stereochemically rigid sphere; and the differently ordered bulk solvent, which represents the third region.

Although some complexing reactions proceed very slowly, it is important to have fast reactions when complex formation is used in analytical processes, such as compleximetric titrations. The reactions of metal ions with chelating agents such as TNA and EDTA are fast enough for such applications.

Reference

[4.3-1] Irving, H.R., Williams, R.J.P., *Nature* 162, 746 (1948).

4.4 Redox Systems

Learning objectives

- To explain chemical equilibria involving the transfer of electrons from one chemical species to another
- To describe the electrochemical cell
- To explain electrochemical principles of potentiometry, voltammetry, and coulometry.

4.4.1 Fundamentals of electroanalytical methods

Electrochemical methods are important in redox analytical techniques. They are based on electrochemical equilibria and the measurement of electrical quantities, namely the voltage E, current i, charge Q, and resistance R, associated with electrochemical processes in an electrochemical cell. The electrical properties of a solution depend on both the nature of the solution components and their concentration c, and allow qualitative and quantitative methods of analysis to be developed. The electrical quantities can be measured alone, in combination, as a function of time t, or of the reagent addition V_R, hence providing a great variety of electroanalytical techniques. These are surveyed in Table 4.4-1.

Table 4.4-1. Basic electroanalytical methods

Method	Electrical quantity measured	Variable controlled
Direct potentiometry	E	$i = 0$
Potentiometric titration	E vs. V_R	$i = 0$
Voltammetry	i vs. E or i vs. t	E and transport of c
Coulometry	Q	E or i
Direct conductimetry	$1/R$	I_{ac} and transport of c
Conductimetric titration	$1/R$ vs. V_R	I_{ac} and transport of c

4.4.2 Redox reactions

Reduction–oxidation reactions, also called redox reactions, are those in which electrons are exchanged between the reactants. We have learned to describe the proton pressure which is connected with the proton activity of a solution by its pH, even though we know that there are no free protons in a solution. In a similar way we can describe the oxidation–reduction state of a solution, by a hypothetical electron pressure pε, which is connected with the electron activity.

Since there are no free electrons in solution, they must be associated with a carrier, an electron donor. The electron can leave the donor (reductant) only when it is transferred to an electron acceptor (oxidant). Therefore, each oxidation is accompanied by a reduction, and vice versa. Taking the reduction reaction:

Each oxidation in solution is accompanied by a reduction, and vice versa.

$$O_2 + 4H^+ + 4e^- \rightarrow 2H_2O \tag{4.4-1a}$$

which consumes four electrons, we need an oxidation reaction that provides the four electrons, such as:

$$4Fe^{2+} \rightarrow 4Fe^{3+} + 4e^- \tag{4.4-1b}$$

the resulting redox reaction, which is the combination of Eqs. 4.4-1a and 4.4-1b is:

$$O_2 + 4H^+ + 4Fe^{2+} \rightleftharpoons 4Fe^{3+} + 2H_2O \tag{4.4-1c}$$

This represents a chemical equilibrium, with the appropriate equilibrium constant:

$$K_{eq} = \frac{[Fe^{3+}]^4[H_2O]^2}{[Fe^{2+}]^4[H^+]^4[O_2]} \tag{4.4-2}$$

Note that we have chosen both half-reactions such that the electrons are eliminated in the overall reaction. However, we could introduce a hypothetical electron activity a_e, and from that derive the logarithmic function pε, analogous to a_{H^+} and pH.

Equations 4.4-1a and 4.4-1b represent half-reactions, whereas Eq. 4.4-1c is the overall redox reaction. In this system, Fe ions assume different oxidation states, Fe^{3+} being at a higher oxidation state than Fe^{2+}.

The pε-Scale

There are a multitude of such redox reactions, and each provides a particular pε, just as each acid–base equilibrium is connected with a particular pH. Strong acids provide a low pH, and strong bases a high pH. Similarly, strong oxidants provide a low or negative pε, whereas strong reductants are connected with a high pε. We shall show later how pε is related to a convenient, measurable quantity, the redox potential E. A convenient way to describe the oxidation–reduction state of a system is to measure its redox potential in an electrochemical cell.

4.4.3 Electrochemical cells

An electrochemical cell, in general, consists of two electrodes, most commonly electronic conductors, separated by at least one electrolyte, i.e., an ionic conductor (Fig. 4.4-1).

A short-hand notation for an electrochemical cell is given by:

$$Zn \,|\, Zn^{2+}, SO_4^{2-} \,\|\, Cu^{2+}, SO_4^{2-} \,|\, Cu \tag{4.4-3}$$

where a vertical line represents a solid–liquid interface boundary, a double line stands for a solution–solution phase boundary, and a comma separates two components in the same phase.

In general, there is a measurable potential difference between the two electrodes,

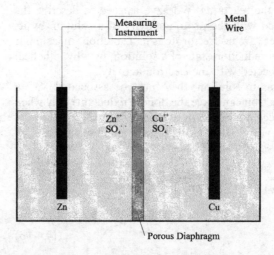

Fig. 4.4-1. Typical electrochemical cell

whether current is passing through the cell or not. The potential profile developed across the whole cell at equilibrium (open circuit, $i = 0$) is shown in Fig. 4.4-2.

In electroanalysis, we are principally concerned with processes at a single interface, the electrode | electrolyte interface. The electrode at which the half-cell reaction of interest takes place is called the indicator- or working electrode, while the other is the reference electrode.

In electroanalytical chemistry, one can measure the potentials at equilibrium (*potentiometry*), control the cell potential (*voltammetry, coulometry*), or measure the solution resistance (*conductimetry*). Depending on the nature of the instrumentation, a redox process in an electrochemical cell can be monitored or controlled (Fig. 4.4-1).

For the reaction:

$$Zn + Cu^{2+} \rightleftharpoons Zn^{2+} + Cu \qquad (4.4-4)$$

if the reactants, and the two half-reactions are physically separated, the electrons have to be transported from the anode to the cathode through the external wire.

If the instrumentation is "passive", it allows or prevents the electron flow until equilibrium is reached, while in many electroanalytical experiments "active" instrumentation is used to control the direction of the electron flow. Thus, by varying the experimental conditions, the electrons can be forced to move in either the spontaneous or the opposite direction between the oxidant and reductant. The unique feature of experimental electrochemistry is that the extent and the direction of the cell reaction can be monitored externally by appropriate choice of measuring instruments.

An electrochemical cell in which the reaction takes place spontaneously is called a *galvanic cell*, while in an *electrolytic cell* the reaction is reversed (Fig. 4.4-3). In other words, electrochemical cells which produce or consume electrical energy are named galvanic and electrolytic cells, respectively.

In a galvanic cell, reactions such as the following occur:

• at the anode (oxidation):

$$Zn \rightleftharpoons Zn^{2+} + 2e^- \qquad (4.4-5)$$

• at the cathode (reduction):

$$Cu^{2+} + 2e^- \rightleftharpoons Cu \qquad (4.4-6)$$

In the electrochemical cell, the potential difference ΔE measured at "open circuit" (without an external controlling instrument) is related to the free energy ΔG of the cell reaction:

$$\Delta E = -\frac{\Delta G}{nF} \qquad (4.4-7)$$

where, the negative sign indicates that the reaction is spontaneous, i.e., the reaction goes to the right, n is the number of electrons exchanged, and F is the Faraday constant (96 486 C/mol, the charge on 1 mol electrons).

In general, if the sign of the cell potential ΔE is negative, the cell reaction goes to the left if the sign is positive, the reaction goes to the right, while at $\Delta E = 0$ the cell reaction is at equilibrium. However, the rate of attaining equilibrium varies from system to system.

The cell potential is equal to the potential of the electrode on the right-hand side

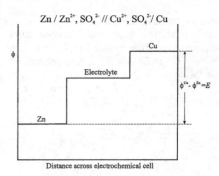

Fig. 4.4-2. Potential profile across the electrochemical cell at "open circuit" condition (where ϕ is the so-called inner potential)

An electrochemical cell (such as in Fig. 4.4-3) can function as a galvanic cell with the chemical reaction following voluntarily towards the equilibrium potential (according to the Nernst equation 4.4-12) or as an electrolytic cell with a potential difference enforced from outside.

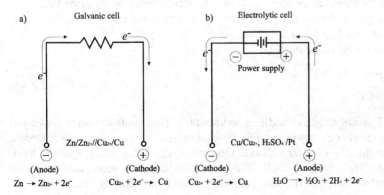

Fig. 4.4-3. (a) Galvanic cell; (b) Electrolytic cell. (From Bard, A.J. and Faulkner, L.R. (1980), *Electrochemical Methods*, New York: Wiley, p. 15, with permission)

Redox reactions can be carried out also outside of electrochemical cells for dissolution purpose or in titrimetry (see Sec. 7.2) e.g. $Ce^{4+} + Fe^{2+} \rightleftharpoons Ce^{3+} + Fe^{3+}$. As a general rule the system with a more positive E° (Ce^{4+}/Ce^{3+} ..... 1.71 V) oxidizes the other system (Fe^{3+}/Fe^{2+} ..... 0.77 V).

with respect to that on the left-hand side in the cell diagram:

$$E_{cell} = E_{right} - E_{left} \qquad (4.4-8)$$

and the sign of the cell potential is in accordance with the IUPAC Stockholm convention, if the left-hand electrode is the anode, and the right-hand electrode is the cathode of the galvanic cell.

Electrode potential

A voltmeter can measure only potential differences; single electrode potentials cannot be measured. Thus, the electrode potential is measured at $i = 0$ in an electrochemical cell against a *standard hydrogen electrode* (SHE) described by:

$$\text{Pt, } H_2(1 \text{ atm}) \,|\, H^+(a = 1) \,\|\, M^{n+}(a = 1) \,|\, M \qquad (4.4-9)$$

$$\text{SHE} \qquad\qquad\qquad\qquad \text{indicator electrode}$$

with a liquid junction potential that arises at the interface of two solutions with constrained mixing.

The standard hydrogen electrode consists of a platinized platinum electrode immersed in an aqueous acidic solution of known constant hydrogen activity ($a_{H^+} = 1$) and saturated with hydrogen by bubbling the gas over the electrode at constant pressure of 1 atm. The platinum does not take part in the electrochemical reaction and serves only as the site where electrons are transferred. The half-reaction responsible for the potential that develops at the electrode is:

$$2\,H^+_{(aq)} + 2e^- \rightleftharpoons H_{2(g)} \qquad (4.4-10)$$

The "electrochemical row" is a collection of half-cell reactions in the order of their E°-values measured in a galvanic cell against the SHE (see Appendix).

The potential of the SHE is defined as zero at all temperatures. It can be used as an anode or a cathode in the electrochemical cell, depending on the nature of the other half-cell to which it is coupled.

The electrode potential is defined as the potential of a cell formed by the electrode of interest acting as cathode and the SHE acting as anode. The corresponding half-reaction at the indicator electrode is:

$$M^{n+} + ne^- \rightleftharpoons M \qquad (4.4-11)$$

The potential E of an electrode for the above oxidation/reduction system is given by the *Nernst equation*:

$$E = E^\circ - \frac{RT}{nF} \ln \frac{1}{a_{M^{n+}}} = E^\circ - \frac{2.3026RT}{nF} \log \frac{1}{a_{M^{n+}}} \qquad (4.4-12)$$

where E° is the standard electrode potential, R is the molar gas constant, T is the absolute temperature, F is the Faraday constant, n is the number of electrons transferred in the electrode reaction, and $a_{M^{n+}}$ is the activity of the metal ion M^{n+}.

If $a_{M^{n+}} = 1$, then $E = E^\circ$, the standard reduction potential of the metal ion. A change of one order of magnitude in the logarithmic term of $a_{M^{n+}}$ changes the value of E by 59.16 mV/n at 25 °C (298 K). It is customary to describe the redox situation in a given chemical equilibrium by its redox potential, and the corresponding values are listed in textbooks.

We can now correlate pϵ with E:

$$p\epsilon = \frac{EF}{2.3026RT} \qquad (4.4-13)$$

Since the SHE is not easy to handle experimentally, potentials are often measured against other reference electrodes. The most commonly used secondary reference electrodes are the silver | silver chloride electrode (Ag | AgCl, KCl), and the calomel electrode, (Hg | Hg$_2$Cl$_2$, KCl). The potentials of these secondary reference electrodes relative to the SHE have been determined very exactly (Eq. 4.4-14).

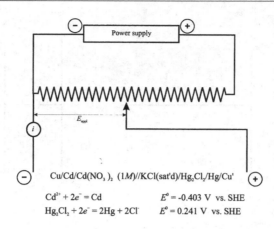

Fig. 4.4-4. Typical electrolytic cell connected to an external power supply. (From Bard, A.J. and Faulkner, L.R. (1980), *Electrochemical Methods*, New York: Wiley, p. 18, with permission)

When, e.g., the saturated calomel electrode (SCE) is coupled to the SHE, the cell potential is equal to the potential of the saturated SCE:

$$Pt, H_2 \mid H^+ (a = 1) \| KCl(sat), Hg_2Cl_2(sat)Hg \qquad (4.4\text{-}14)$$

$E_{SCE} = +241$ mV at 25 °C. The electrode potential of any half-cell is defined as the potential of a cell consisting of the electrode in question and the SHE.

Current–potential relationship

The current observed in an electrolytic cell is carried by electrons in the electrode phase and the external wire, while in the electrolyte solution current is carried by migration of ions in the electrolyte. The current flow requires charge transfer reactions (oxidation–reduction) to occur at the electrode | electrolyte interfaces.

In an electrolytic cell, the energy of the electrons in the electrode phase can be altered by changing its potential by means of external instrumentation (Fig. 4.4-4). If the externally applied potential is more negative than the equilibrium or "open circuit" potential of the electrochemical cell ($E_{appl} < E_{eq}$), redox reactions will occur at the cathode and *reduction current* will flow. On the contrary, when the potential of the electrode is changed in the more positive direction, the energy of the electrons is lowered in the electrode phase and electrons will flow from the electrolyte into the electrode phase, which results in an electrochemical oxidation and in the appearance of the *oxidation current*.

If macroscopic electric current flows across the electrochemical cell, then *faradic processes* must occur at both electrodes. The rate of the electrode reaction v proceeds as predicted by the Faraday law, meaning that nF coulombs of charge require 1 mol of ions to be reduced or oxidized at the relevant electrodes. (One coulomb of electricity is the electric charge passed by a current of one ampere in one second.)

$$v(\text{mol/s}) = \frac{dN}{dt} = \frac{Q}{nF} = \frac{1}{nF} \int_0^t i\, dt \qquad (4.4\text{-}15)$$

where Q is the number of coulombs passed, which is equal to the time integral of the current i; N is the number of moles of substance reacted at the electrode; and n is the number of electrons exchanged.

It follows that the current is a measure of the rate of an electrochemical reaction, but there are exceptions. In electrochemical experiments the electrode | solution interface can exhibit a range of potentials where no charge transfer reactions occur, owing to thermodynamic or kinetic limitations. Under these conditions, adsorption and desorption can occur, and the structure of the electrode solution interface can vary with changes in the potential or the solution composition. These are *nonfaradic processes*, and result in external current flow, called charging or capacitive current. When an electrode reaction takes place, both faradic and nonfaradic processes occur.

An electrode at which there is no charge transfer across the electrode | solution interface, regardless of the external potential applied, is called a polarizable elec-

The electrolytic cell in chemical analysis.

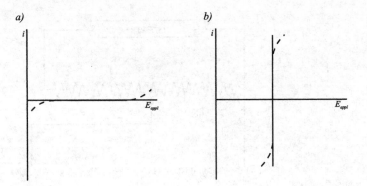

Fig. 4.4-5. Current-potential curves for (a) ideally polarized and (b) non polarized electrodes. Dashed lines show behavior of actual electrodes that approach the ideal behavior over limited current or potential

trode (Fig. 4.4-5a). For example, the mercury electrode in deaerated potassium chloride solution is called an ideally polarized electrode (IPE). The limit of the polarization range is given by the following reaction:

In the anodic region, by the oxidation of Hg:

$$Hg + Cl^- \rightarrow \tfrac{1}{2} Hg_2Cl_2 + e^- \quad (\text{at} \sim +0.25\,\text{V vs NHE}) \tag{4.4-16}$$

in the cathodic region, by the reduction of K^+ at very negative potential:

$$K^+ + Hg + e^- \rightarrow K(Hg) \quad (\text{at} \sim -2.1\,\text{V vs NHE}) \tag{4.4-17}$$

However, in the potential range set by these reactions, no charge transfer reactions occur. The value of the non-faradic current (the charging current) flowing in the cell depends on the value of the polarization potential. This current is required to charge the electrode solution interface, which acts as a capacitor corresponding to a given potential value.

Examples of an ideally nonpolarized electrode are a platinum electrode or a copper electrode immersed in a well-stirred solution of varying concentrations of copper(II) ions. No current flows in the cell as long as the externally applied potential does not exceed the electrode potential at which copper ions are reduced to metallic copper. But when it exceeds that, the following reaction occurs:

$$Cu^{2+} + 2e^- \rightleftharpoons Cu \quad (E^\circ = 0.337\,\text{V}) \tag{4.4-18}$$

The working electrode surface becomes covered with copper, it becomes depolarized, and its potential is given by the following, so-called Nernstian equation at 25 °C:

$$E = E^\circ + \frac{0.05916}{2} \log[Cu^{2+}] \tag{4.4-19}$$

When the electrode is depolarized, the current flows across the cell without altering the electrode potential from its reversible value (Fig. 4.4-5b).

The difference between the applied potential at which the polarizable electrode is depolarized and the equilibrium potential is the *overpotential* η. Both cathodic and anodic processes can exhibit an overpotential which is affected by many factors, such as:

Polarizability: the deviation from the linear relationship of E and i in Ohm's law $E = iR$.

• mass transfer
• electron transfer (charge transfer) at the electrode | solution interface
• chemical reactions preceding or following the electron transfer reaction

The magnitude of the current in an electrochemical cell with a polarized working electrode is often limited by the slowest steps of the electrode reaction. At a given current density, the overpotential term can be affected by different reaction steps; η_{mt} (the mass transfer overpotential or concentration polarization), η_{ct} (the charge transfer overpotential or activation polarization), η_{rxn} (the overpotential associated with a preceding reaction or reaction polarization). The electrode reaction can also be represented by a resistance R composed of a series of resistances representing the various steps of the electrode reaction: R_{mt}, R_{ct}, etc. A fast reaction is represented by a small resistance, while a slow reaction is characterized by a high resistance.

High current density (A/cm^2) favors polarization of an electrode. Small electrodes can be easier polarized than large ones (see polarography).

4.4.4 Charge-transfer controlled reactions

Let us consider a reduction process at which electrons are transferred between oxidant and reductant as:

$$Ox + ne^- \underset{k_r}{\overset{k_f}{\rightleftharpoons}} Red \tag{4.4-20}$$

where k_f is the rate constant of the forward reaction, while k_r is the rate constant of the reversed reaction.

When the reduction is carried out by electrochemical processes, i.e., at an E_{appl} which is more negative than the "open circuit" cell potential, a current will flow across the cell, the rate of which is governed by processes such as the *electron transfer* at the electrode|solution interface or the *mass transport* of the oxidant from the bulk of the solution to the electrode surface, and a new equilibrium will be established:

$$E_{appl} = E^0 - \frac{0.05916}{n} \log \frac{c_{R(x=0)}}{c_{O(x=0)}} \tag{4.4-21}$$

where $c_{O(x=0)}$ and $c_{R(x=0)}$ are the concentration of the oxidant and reductant at the electrode surface ($x = 0$).

The magnitude of the net current flow is:

$$i = nFA(k_f c_O - k_r c_R) \tag{4.4-22}$$

where A is the electrode surface area.

The rate constant of the charge transfer reaction depends on E_{appl}:

$$k_f = k_f^0 \exp\left[-\frac{\alpha n F E_{appl}}{RT}\right] \tag{4.4-23}$$

and

$$k_r = k_r^0 \exp\left[\frac{(1-\alpha)nFE_{appl}}{RT}\right] \tag{4.4-24}$$

where k^0 is the rate constant measured at $E_{appl} = 0$, i.e., the rate constant of the relevant chemical reaction; and α is a factor expressing the fraction of the electrode potential (interfacial potential) contributing to the rate constant of the relevant chemical reaction.

These expressions indicate that the electrode potential affects the rate constants of the relevant chemical reaction via alteration of the activation energies. For example, when the potential of a cathode is made more negative than the equilibrium potential, it facilitates the transfer of the reductant as it raises its potential energy by an amount αnFE, which results in a decrease of the activation energy of the cathode reaction. Conversely, a more negative electrode potential decreases the rate of the anode reaction according to:

$$i = nFA\left\{c_O k_f^0 \exp\left[-\frac{\alpha n F E_{appl}}{RT}\right] - c_R k_r^0 \exp\left[\frac{(1-\alpha)nFE_{appl}}{RT}\right]\right\} \tag{4.4-25}$$

At equilibrium, $E_{appl} = 0$, $i = i_k = i_a = i_0$, $k_f = k_r = k$, and $i_0 = nFkAc$, where i_0 is the exchange current.

Thus, at E_{appl}:

$$i = i_0\left\{\exp\left[-\frac{\alpha n F (E_{appl} - E_{eq})}{RT}\right] - \exp\left[\frac{(1-\alpha)nF(E_{appl} - E_{eq})}{RT}\right]\right\} \tag{4.4-26}$$

where $E_{appl} - E_{eq} = \eta$; and η is the charge transfer overvoltage.

The effect of the current density on the current–potential curve is shown in Fig. 4.4-6.

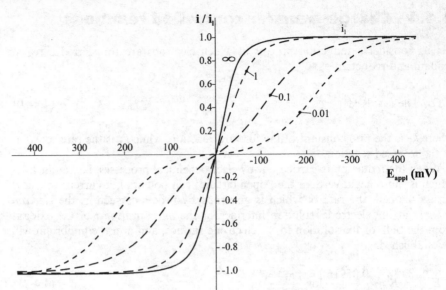

Fig. 4.4-6. Effect of the current density on the shape of the current vs. potential curve. (From Bard, A.J. and Faulkner, L.R. (1980), *Electrochemical Methods*, New York: Wiley, p. 110, with permission) The reaction is $O + ne \leftrightarrow R$ with $\alpha = 0.5$, $n = 1$, $T = 298$ K

4.4.5 Mass-transport-controlled reactions

If the electrode reaction and the associated chemical reaction are fast compared with the mass transfer process, then the rate of the electrode reaction is controlled by the mass transfer rate v_{mt} at which the electroactive species is transported to the surface of the electrode. The mass transfer, i.e., the movement of material from the bulk to the electrode surface, arises by migration of charged species in the electric field, by convection due to the movement of the solution or the electrode, or by diffusion under the influence of a concentration gradient.

The transport of the components to an electrode surface as an effect of migration, convection, and diffusion in the direction of the x-axis can be described by the Nernst–Planck equation:

In quantitative polarography diffusion must be the major mass transport to the electrode and not migration or convection.

$$J_i(x) = -D_i \frac{\delta c_i(x)}{\delta x} - \frac{z_i F}{RT} D_i \cdot c_i \frac{\delta \phi(x)}{\delta x} + c_i v(x) \tag{4.4-27}$$

where $J_i(x)$ is the flux of species i (mol s^{-1} cm^{-2}) at distance x from the electrode surface; D_i is the diffusion coefficient (cm^2/s); $\delta c_i(x)/\delta x$ is the concentration gradient at distance x; $\delta \phi(x)/\delta x$ is the potential gradient; z_i and c_i are the charge and concentration of species i, respectively; and v is the flow rate (mL/s).

In voltammetric experiments, the movement of a species under the influence of the chemical potential gradient, i.e., a concentration gradient, is of primary importance.

Under diffusion mass transport (in the absence of migration, and if the solution velocity is assumed to be zero) the flux and the concentration of the electroactive species can be described by Fick's laws:

a)
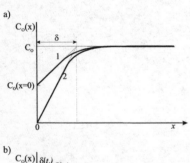

$$v_{mt} = \left(\frac{\delta c_O(x)}{\delta x} \right)_{x=0} \tag{4.4-28}$$

where x is the distance from the electrode surface.

This is approximated by the relation:

b)
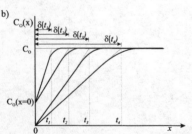

$$v_{mt} = m_O[c_O - c_O(x = 0)] \tag{4.4-29}$$

where c_O is the concentration of species, O in the bulk solution; $c_O(x = 0)$ is the concentration at the electrode surface; m_O is the mass transfer coefficient, (m_O corresponds to D_O/δ_O, where δ_O is the thickness of a hypothetical stagnant, diffusion layer at the electrode surface).

Fig. 4.4-7. The concentration profile at the vicinity of the electrode surface at $E_{appl} > E_{equilibrium}$ (a) at dropping Hg-electrode and (b) at solid electrodes

The variation of the concentration of electroactive species in the vicinity of electrode surfaces of different nature ($x = 0$) at different time intervals after switching on the external voltage source E_{appl} is shown in Fig. 4.4-7.

The largest rate of mass transfer of O occurs when $c_O(x = 0) = 0$. The value of the current under these conditions is called the diffusion controlled limiting current i_d where:

$$i_d = nFAm_Oc_O \qquad (4.4\text{-}30)$$

Questions and problems

1. How can you recognize complexes or complex ions?
2. What are soft or hard central ions?
3. What are *chelates* and why are they more stable than normal complexes?
4. Describe the two major fields of application of chelate complexes in analytical chemistry!
5. Draw the schematics of two principal forms of electrochemical cells (galvanic and electrolytic cell) and describe their application in analytical chemistry for two distinct examples (e.g. for the measurement of pH and for the determination of the oxygen concentration in river water).
6. What is the Standard Hydrogen Electrode (SHE) and what is its role in theory?
7. What is a *polarizable* electrode?
8. Why is it important for the analytical applicability of polarographic methods, that the major mass transport to the electrode takes place by diffusion and not by migration or convection?

Learning objectives

■ To describe equilibrium states involving more than one phase, and the impact such systems have on analytical chemistry

■ To discuss heterogeneous equilibrium concepts, which are the basis of most separation techniques. Knowledge of such systems is fundamental for the understanding of analytical separation processes, although in most cases equilibrium is not established, because the time constants driving the observed changes leading to the equilibrium state are large compared with the time constants of the analytical processes

■ To illustrate the fact that, similar to the situation with homogeneous equilibrium states, the steady state situation in connection with heterogeneous equilibrium states plays an essential role in the practical description of the appropriate analytical processes

4.5.1 Thermodynamic considerations

Introduction

An equilibrium system consisting of more than one phase is called a *heterogeneous equilibrium*. The principles of a multiphase equilibrium have wide applications in separation techniques and processes. Crystallization, extraction, chromatography, and zone refining are only a few examples of such techniques.

Heterogeneous equilibrium states and the kinetic aspects associated with them are of great importance in environmental chemistry, and natural chemical cycles are based on multiphase systems. Thus, the quality of freshwater is determined mainly by the presence of solutes that are introduced from other phases. Examples are dissolved gases from the atmosphere; trace metal compounds from minerals, bedrock material, and industrial processes; and pollutants from solid, liquid, and gaseous wastes. A thorough understanding of heterogeneous equilibrium states is therefore essential for analysis of aquatic systems.

The principles underlying the understanding of heterogeneous equilibrium states were formulated by Josiah Willard Gibbs (1839–1903) in 1876. Before we familiarize ourselves with *Gibbs' phase rule* we must define of the quantities that it introduces. In so doing we expand some earlier ideas developed in connection with the treatment of homogeneous equilibrium states (Sec. 4.1).

A thermodynamic equilibrium exists between different phases when thermal, mechanical, and chemical equilibrium states are established. This statement is summarized in the expression:

$$dG = VdP - SdT + \Sigma\mu_i dn_i = 0 \qquad (4.5\text{-}1)$$

Here, the volume V, entropy S, and mole number n_i of component i are extensive variables, whereas the temperature T, pressure P, and chemical potential μ_i are intensive variables. Thus, V, S, and n_i are additive, and we can derive them for the whole system as sums of the volumes, entropies, and mole numbers of the different phases A, B, and C. For example:

$$V = V^A + V^B + V^C + = \sum_{1}^{N} V \qquad (4.5\text{-}2)$$

where V^A, V^B, and V^C, are the volumes of phases A, B, and C. Similarly we can write:

$$S = \sum_{1}^{N} S \qquad (4.5\text{-}3)$$

and

$$n_i = \sum_1^N n_i \tag{4.5-4}$$

The entire multiphase system is separated from its surroundings by boundaries that do not permit exchange of heat, work, and matter. The system is thermodynamically closed. However, the individual phases within the system are separated by phase boundaries allowing the exchange of heat, work, and matter. This means that if the system is disturbed from its equilibrium it may exchange some or all of these quantities. The question arises whether heat, work, and matter may be exchanged between the phases when equilibrium is already established. In other words, we are investigating the true meaning of thermal, mechanical, and chemical equilibrium states.

We first consider the conditions for a *mechanical equilibrium*. For simplicity we chose a two-phase system with phases A and B, and assume that the pressure in phase A (P^A) is higher than the pressure in phase B (P^B). Therefore, mechanical work δw is done on the system, as one phase expands and the other is compressed. However, this work is done under equilibrium conditions so that the process becomes reversible ($\delta w = \delta w_{rev}$).

We can therefore calculate the volume changes in each phase:

$$\delta V^A = \frac{\delta w}{P^A} \tag{4.5-5}$$

$$\delta V^B = \frac{\delta w}{P^B} \tag{4.5-6}$$

Since the entropy and composition of the entire system do not change, the overall volume change δV of the system must be zero:

$$\delta V = \delta V^A + \delta V^B = 0 \tag{4.5-7}$$

Therefore:

$$\frac{\delta w}{P^A} + \frac{\delta w}{P^B} = 0 \tag{4.5-8}$$

We conclude that for a mechanical equilibrium it is necessary that the pressure of all phases be the same. For our two-phase system we find:

$$P^A = P^B \tag{4.5-9}$$

and therefore, in the equilibrium state the individual phases do not change in volume, and no exchange of work occurs between the phases of the system.

In a similar fashion, we consider a thermal equilibrium. Here, the extensive variable V of the previous discussion is replaced by S and the intensive variable P by T. By a similar procedure, we conclude that for thermal equilibrium it is necessary that the temperature of all phases be the same. For our two-phase system we find:

$$T^A = T^B \tag{4.5-10}$$

In the equilibrium state the individual phases do exchange heat and the entropy stays constant.

Finally, we turn our attention to *chemical equilibrium*, introducing the chemical potential $\Sigma \mu_i$ as intensive variable and the number of moles of component i as the extensive variable n_i. We conclude that for chemical equilibrium no net transport of matter occurs through the phase boundaries, and the chemical potential for any component of the system must be the same in all phases. For our two-phase system we derive:

$$\mu_i^A = \mu_i^B \tag{4.5-11}$$

No exchange of matter occurs.

The Gibbs phase rule

The state of a system consisting of p phases and c components is described when the temperature, pressure, and the composition of each phase are specified. (The symbol p represents the number of phases and not the pressure P.) If the system is in equilibrium, only a limited number of variables is required to describe the system. This number of *independent, intensive variables* is called the *number of degrees of freedom F*. In the following, we derive F for a heterogeneous equilibrium system with c components and p phases. The component i in phase A is characterized by the mole fraction X_i^A, which is obtained according to:

$$X_i^A = \frac{n_i^A}{\sum n_i^A} \tag{4.5-12}$$

where

$$\sum n_i^A = 1 \tag{4.5-13}$$

We need only $c - 1$ mole fractions to describe the composition in phase A, since the last mole fraction can be calculated from Eq. 4.5-13. The phase is therefore fully described by $2 + (c - 1)$ variables. As the temperature and pressure each assume the same value in all phases at equilibrium, the total number of variables in the total systems is $2 + (c - 1)p$. This number still contains a set of dependent variables and therefore, does not represent the number of degrees of freedom.

We have already shown that the chemical potential of each component is the same in every phase. The number of independent variables is therefore reduced by $p - 1$ for each component or by $c(p - 1)$ for the entire system. This leaves the number of independent variables, or degrees of freedom as $2 + (c - 1)p - c(p - 1)$, resulting in:

$$F = c - p + 2 \tag{4.5-14}$$

This equation is called the *Gibbs phase rule* and is generally applicable to equilibrium systems.

Gibbs' phase rule indicates the degree of freedom (number of independent variables) in a heterogeneous system with c components and p phases.

Phase equilibrium states and phase diagrams

The analytical chemist is interested in heterogeneous systems primarily from the viewpoint of developing separation techniques. We approach the thermodynamics of heterogeneous systems from this point of view, and for convenience we distinguish between phase equilibrium and distribution equilibrium.

In *phase equilibrium separation*, the phases are mixtures in which the sample represents a major component. The composition of the phases are therefore conveniently expressed in terms of *mole fractions*.

In *distribution equilibrium*, the phases involve components other than the sample. The sample is not a major component and the phase composition is conveniently expressed in terms of the *sample concentration*, e.g., $mol\ L^{-1}$.

One-component systems

For one-component systems, the phase rule indicates that three different cases are possible. We obtain the relationship:

$$F = 3 - p$$

and distinguish between bivariant one-phase systems, univariant two-phase systems, and invariant three-phase systems.

Since, in a one-phase system, the maximum number of degrees of freedom is two, a two-dimensional phase diagram is the most convenient way to describe the system. Usually T and P are chosen as intensive variables. Figure 4.5-1 shows the phase diagram for water. From this diagram we can conclude the following facts:

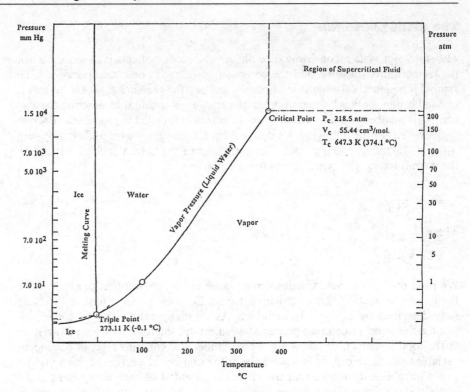

Fig. 4.5-1. Phase diagram for water

- The vapor pressure of ice below 0 °C is smaller than that of liquid water, and therefore ice is the stable form of water in this region. The vapor pressure curve describes the sublimation or its inverse process, condensation. Below a pressure of 4.58 mm Hg (9.87 atm) ice cannot be converted to water, and the only possible route of transformation is sublimation
- The vapor pressure of water is shown between 0 and 218 °C. In an open system exposed to the atmosphere at sea level, we can observe a partial vapor pressure only in the range where liquid water exists, i.e., from 0 to 100 °C. However, in a closed (artificial) system, the two phases remain coexistent until the critical point is reached. This point is characterized by the critical pressure ($P_c = 218.3$ atm) and critical temperature ($T_c = 374.1$ °C).
- Along the vapor pressure curves a two-phase system exists, and the degree of freedom is reduced to $F = 1$, which means that for a given pressure the temperature is fixed and vice versa.
- Beyond P_c and T_c there exists only one phase, the supercritical fluid. Its properties are quite different from liquid water and water vapor. The supercritical region has two degrees of freedom, i.e., any pressure can be applied to the system at a fixed temperature, and at a given pressure there is a free choice of temperature, as long as $T > T_c$ and $P > P_c$.
- At the triple point, all three phases of water coexist and therfore there is no degree of freedom; the temperature (0.01 °C) and pressure (9.87×10^{-4} atm) are fixed.
- The melting point of water is slightly pressure dependent, decreasing with increase of pressure (anomalous behavior of water, compared with other substances).

Supercritical fluids

Supercritical fluid, another analytically interesting state of matter.

At this point we shall briefly describe the properties of supercritical fluids. Table 4.5-1 summarizes the critical pressure, temperature, and density. Such fluids are used in *supercritical fluid extraction* (SFE) and *supercritical fluid chromatography* (SFC).

Figure 4.5-2 shows a phase diagram for CO_2. Since we are interested in the density of the fluid in the supercritical domain, we plot pressure versus density. We recognize that large changes in the density (0.25–0.95 g cm^{-3}) can be achieved in

Table 4.5-1. Properties of fluids at the critical point

Compound	T_c (°C)	P_c (atm)	d_c (g cm^{-3})
Acetonitrile	274.7	47.7	0.24
Ammonia	132.5	112.5	0.24
Argon	−122.3	48.0	
n-Butane	152	37.5	0.23
Carbon dioxide	31.0	72.9	0.47
Carbon disulfide	279	78	
Diethyl ether	192.6	35.6	0.27
Ethane	32.2	48.2	0.20
Helium	−267.9	2.26	
Krypton	−63.8	54.3	
Methane	−82.1	45.8	
Methanol	240	78	0.27
Neon	−228.7	26.9	
Nitrogen dioxide	157.8	100	
Nitrous oxide	36.5	71.7	0.43
1-Pentane	191	39.9	0.24
n-Propane	96	42	0.22
Water	374.1	218.3	0.34

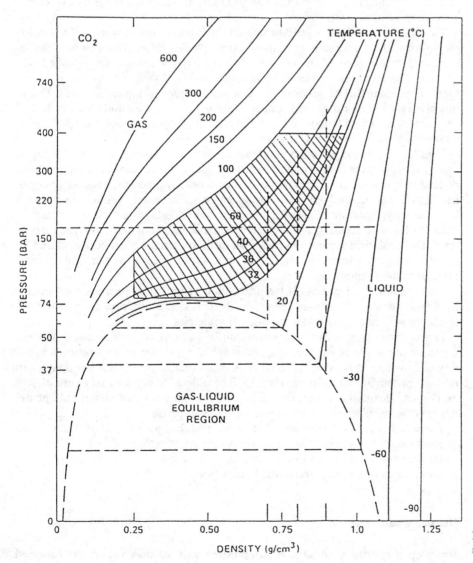

Fig. 4.5-2. Phase Diagram of CO_2 in the supercritical domain

the practical range of the supercritical domain by choosing the appropriate pressure and temperature.

The density of a supercritical fluid assumes values extending over a large range. This is important because the higher the density the better are the solvating properties toward solutes in the supercritical fluid, influencing the solubility of such solutes. The solubility of substances in supercritical fluids often exceeds that in liquids, and supercritical fluids are extraction media with favorable and properties, adjustable to the needs of a particular analytical problem.

In contrast to liquids, supercritical fluids are compressible and therefore the solvation properties and solubility can be manipulated through variations in density (achieved by appropriate variations in pressure and temperature).

4.5.2 Gas–liquid systems

Introduction

Gases are soluble in liquids to a certain degree. Exchange, i.e., the equilibrium process, takes place at the interface between the solvent (e.g., an aqueous solution) and the atmosphere (head space). The solubility of gases in liquids can be calculated from *Henry's Law*.

The transformation of a substance from the liquid to the gaseous state is called *evaporation*. The reverse operation is called *condensation*. The combination of evaporation and subsequent condensation of a chemical species is called *distillation*. The recondensed phase is the *distillate*. The distillation process is particularly important for the separation of larger volumes of liquid mixtures. Often, this process of evaporation and condensation is repeated over and over again, and very sophisticated stills have been designed, which operate over multiple cycles, increasing the efficiency of separation.

Distillation processes remain very popular for the separation of complex liquid mixtures. In process industry, particularly in the petrochemical industry, millions of tons of crude oil are fractionated into commercial products, such as gasoline, heating oil, and natural gas. Distillation processes are used routinely in research laboratories to separate or clean synthesized compounds from their byproducts.

In the last few decades, more efficient separation techniques have been developed, and analytical separations based on distillation have often been replaced by novel techniques. However, distillation still plays a role in preparative applications, including sample clean-up.

Another industrial process is the adsorption of undesired gaseous components by liquids (washing processes). Some of the cleaning procedures for waste gases produced in incinerators make use of such principles.

In gas chromatography, the distribution of gaseous components and a liquid phase, fixed on a solid support (e.g., open tubular wall-coated capillaries in capillary GC, or coated packing material in packed-column gas chromatography), can provide extremely powerful separations. The difference between this kind of separation and distillation is that the packed-column procedure processes bulk phase, whereas the capillary process involves a thin-layer phase.

The words *gas* and *vapor* are often used interchangeably. Vapor is more frequently used for a system, where the compound exists in the liquid or solid form at room temperature, whereas gas is used, when the substance is in the gaseous state at room temperature, i.e., above the boiling point.

Vapor pressure

In an open system, a liquid is in equilibrium with its own vapor; the observed vapor pressure is the result of this equilibrium, depending on the individual substance and the temperature. The vapor pressure reaches atmospheric pressure at the *boiling point*. Figure 4.5-3 shows the vapor pressure of ethanol, water, and hexane, plotted against temperature.

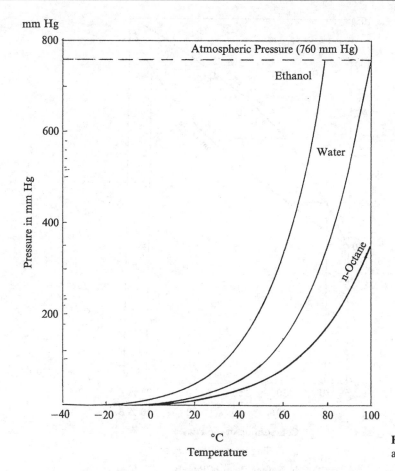

Fig. 4.5-3. Vapor pressure of ethanol, water, and hexane as a function of temperature

Binary mixtures

A binary mixture of two completely miscible liquids exhibits boiling points, that depend on the composition of the liquid phase. The more volatile component has a higher vapor pressure than the less volatile component, and we are able to enrich the more volatile substance in the gas phase relative to that in the liquid phase. If the ratio of the liquid phase concentrations is defined as:

$$\frac{X_A}{X_B} \tag{4.5-15}$$

and the gas phase concentrations as:

$$\frac{Y_A}{Y_B} \tag{4.5-16}$$

where X_A, X_B, Y_A, and Y_B denote the mole fractions of A and B in the liquid and gaseous phase, respectively.

We can now calculate the *relative volatility*, α as follows:

$$\frac{Y_A}{Y_B} = \alpha \frac{X_A}{X_B} \tag{4.5-17}$$

or

$$\alpha = \frac{Y_A X_B}{Y_B X_A} \tag{4.5-17a}$$

α is a measure of the enrichment of substance B in the vapor phase.

Note that A and B are chosen such that $\alpha > 1$, i.e., B is the more volatile component. In practice α- values between 1.0 and 5.0 are commonly observed. α is equal to the ratio of the equilibrium vapor pressure of substance A and B at a given temperature:

$$\alpha = \frac{p^\circ A}{p^\circ B} \tag{4.15-18}$$

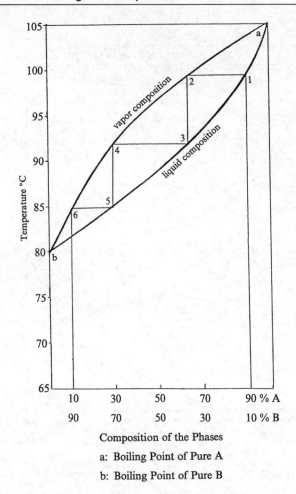

Fig. 4.5-4. Phase diagram with liquid and vapor composition

From Fig. 4.5-3 we can directly determine α- values of the binary mixtures water–ethanol and ethanol–hexane, and obtain, at 70 and 80 °C, the following:

ethanol–water $\alpha = 2.2$ and 2.3, respectively, and

ethanol–hexane $\alpha = 4.4$ and 4.5, respectively

If the vapor phase of a distillation process is condensed, we obtain a liquid phase with a higher concentration of the volatile component compared with the original mixture. This is nicely illustrated in a diagram showing the liquid and gas phase compositions of the mixture (Fig. 4.5-4).

We start by considering a hypothetical binary mixture, consisting of the more volatile substance B and the less volatile component A. In pure form, A and B have boiling points of 105 and 80 °C, respectively. We begin with a mixture of 10% B and 90% A. According to our diagram, this mixture has a boiling point of 99.5 °C (point 1). The appropriate vapor phase has the composition of 36.5% B and 63.5% A (point 2). If this vapor phase is condensed and evaporated in the second cycle, we observe a boiling point of 92 °C (point 3). The composition of this second distillate is 30% A and 70% B (point 4). In the third cycle, we observe a boiling point of 85 °C (point 5) resulting in a vapor composition of 90% A and 10% B (point 6).

The ideas thus far developed are valid only when we consider a large volume of the original mixture, and collect only a small volume of the distillate. Otherwise, the original mixture is changing its composition toward a lower concentration of the more volatile component, and a continuous drift of the boiling point to higher values is observed, indicating that the concentration of the volatile component in the distillate also drifts to smaller values. In our model, where we separated each of the steps, we require three theoretical steps (called *plates*) in order to enrich substance B from 10 to 90%. The system used to explain distillation has been arbitrarily chosen. In general the liquid and vapor composition lines do not show

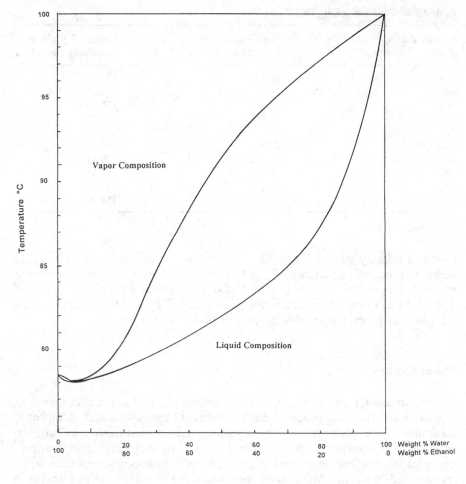

Fig. 4.5-5. Phase diagram of the binary mixture water–ethanol

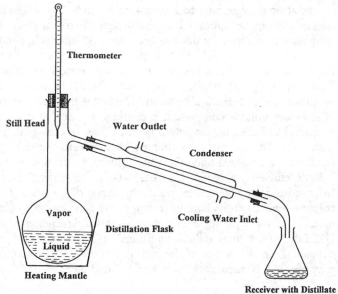

Fig. 4.5-6. Simple laboratory distillation apparatus

the regular curvature depicted in Fig. 4.5-4. In contrast, Fig. 4.5-5 shows the diagram of the binary mixture water–ethanol. The water–ethanol system demonstrates that it is not always possible to obtain the pure liquids. It exhibits a boiling point minimum at 96% ethanol, at this point the vapor and liquid composition are equal and no further separation is achieved. It is said that ethanol and water form a *azeoptropic mixture* at 96% ethanol.

This fact can be used to prepare well-defined solutions. For instance, hydrochloric acid and water form a mixture, called *constant-boiling hydrochloric acid*. It is easily prepared by distillation in simple apparatus (Fig. 4.5-6) and may be

Table 4.5-2. Constant-boiling HCl

Pressure during distillation (mm Hg)	% HCl by weight in distillate	Grams of distillate containing 1 mol HCl
600	20.638	176.55
640	20.507	177.68
680	20.413	178.50
700	20.360	178.96
730	20.293	179.55
740	20.296	179.77
760	20.221	180.19
770	20.197	180.41
780	20.173	180.62
800	20.125	181.05

used as a primary standard for acid–base titrations. It can be stored over a long period of time, with no change of composition. Constant-boiling HCl is prepared by distillation of HCl, specific gravity 1.18 (approximately 38% HCl) at a rate of 3–4 mL/min., discarding the first 75% and the last 5–10% of the distillate. Table 4.5-2 lists the properties of constant-boiling HCl.

Plate theory

In our distillation experiment (Fig. 4.5-4), one step was performed after the other. In sophisticated distillation equipment, the different steps overlap and an improved separation efficient is achieved. The imaginary individual stages are called *theoretical plates*. A theoretical plate can be defined as an imaginary device that produces the same difference in composition as exists at equilibrium between a liquid mixture and its vapor. Any part of the column that produces an enrichment in composition from point 1 to 2, 3 to 4, and 5 to 6 (Fig. 4.5-4) is a theoretical plate. Separation can be improved by the design of special stills. The idea is to build imaginary plates into the device, where the distillates is in equilibrium with its own vapor. In order to enrich the volatile component B from 10 to 90%, we required three steps, and we can imagine plates built into the distillation column, where liquid mixtures with 63.5%, 70.0%, and 90% of the volatile component B are in equilibrium with their liquid phases. The fact is that the variation in concentration of the more volatile component is continuous, ranging over the entire distillation column. Furthermore, the actual height values of each plate are influenced by the reflux rate, i.e., the distillate running back toward the still pot. Theoretical plate numbers are achieved only when the column is operated under *total reflux*.

Plate efficiency is defined as the ratio of theoretical plates to actual plates. In practice plate efficiencies are in the range of 50–75%. The performance of a column depends on its length (or height) and the number of plates it produces. A measure of the separation efficiency is the *height equivalent to a theoretical plate* (HETP). Under total reflux conditions, HETP values of 1–60 cm are achieved, depending on the design of the still. A typical column with overall height 50 cm may produce 10 theoretical plates and would have a HEPT value of 5 cm.

Solubility of gases

Gases can be dissolved in liquids. Most animals living in the oceans or freshwater have a breathing apparatus in which gills can extract the oxygen necessary to support life from the water, in which it is dissolved.

Soda water is drinking water in which CO_2 is dissolved at pressures slightly higher than atmospheric. Exposing sparkling water to the atmosphere produces gas bubbles. From this observation we can derive the most important features of a gas solubility. The solubility of gases in solvents can be calculated from *Henry's Law*.

William Henry (1775–1836), an English physician and chemist, discovered that the amount of gas absorbed by a liquid at a given temperature is proportional to the partial pressure of the gas above the liquid. The concentration of gas [A] is calculated according to:

$$[A] = p_A K_A{}^H \tag{4.5-15}$$

where p_A is partial pressure of the gas and $K_A{}^H$ is a proportionality factor, called *Henry's constant* mol L^{-1} atm^{-1}.

4.5.3 Solid–liquid systems

Precipitation equilibria

The *solubility* of solids in water and other solvents is the visible manifestation of a competition between two different processes. The first, called *crystallization*, is the result of the binding forces in a solid. It causes the growth of a solid crystal exposed to an environment containing its own ions or constituent molecules. The second, called *dissolution*, is the result of the interactions between the solvent molecules and the constituent molecules or ions on the surface of the solid phase. It leads to destruction of the crystal lattice, and produces a more and more concentrated solution until either the whole solid phase is dissolved or an equilibrium is established between the competing processes, leaving a *solid phase* and a *saturated solution*.

A single, "naked" ion or molecule in solution is not in the lowest possible energy state and, therefore tends to undergo aggregation or solvation. Its surface is not saturated and a number of potential bond sites are vacant. Therefore, owing to interionic or intermolecular forces, it undergoes reactions with neighboring groups. The same principle holds for constituent ions and molecules at the surface of the solid phase. Our interest is focused on two processes, associated with crystallization and dissolution, respectively.

- *Formation of bonds between solute species.* This results in the formation of aggregates and, more important, the precipitate. In the process lattice energy is gained. The higher the lattice energy the more solute particles are removed from the solution. It must be kept in mind that the surface of the precipitate remains unsaturated, i.e., the ions or molecules occupying surface positions possess bonding sites that are not used (Fig. 4.5-7). Therefore, particular effects can be observed on the surface (adsorption and other surface phenomena) or more specifically, the solvent molecules interact with solute particles situated on the surface and transfer them from the solid surface to the solution phase.
- *Formation of bonds between the hypothetical "naked" solute particles and water or solvent molecules.* This results in the formation of *hydrates* or *solvates*. Again, energy is released in this process. These interactions may extend over several hydration shells, and small ions in particular can be incorporated into the structure of the solvent.

Fig. 4.5-7. Model demonstrating the lower number of bonds of surface groups compared with groups in the interior of the lattice. The points (·) indicate unsaturated bond sites on the surface of a solid phase

There is competition between crystallization and solvation in solution. In most cases, the hydration or solvation bonds are weaker than the solid–solid interactions, but usually their number outweighs the number of bonds of a molecular or ionic unit in the solid phase. The solution process can therefore be exo- or endothermic. The latter is possible since in the process of dissolution the entropy increases. There is generally an appreciable amount of entropy gained, since the solution represents a more randomly distributed system than the highly ordered solid state. In the process of dissolution, the nature of the solvent plays an important role. Water can exert a large influence on the interionic forces of an ionic solid phase, since in a medium of high dielectric constant the Coulomb forces between the ions are drastically reduced.

The two competing processes establish a thermodynamic equilibrium which is shifted toward the product of lower energy. But a constant exchange of solute particles takes place on the surface of the solid. This can be demonstrated by radiochemical techniques. As an equilibrium state is established, the *mass action law* can be applied and the equilibrium can be expressed by an equilibrium constant. This equilibrium constant must also be related to the solubility of the solid in the liquid phase. It was Walther H. Nernst (1864–1941) who formulated such an equilibrium constant for the first time in 1889. He called it the *solubility product K_{sp}*.

Before we consider this relation in more detail, we have to examine the general effects of ionic and covalent bonds on the solubility. In the ionic crystal lattice, the constituents are ions of opposite charge, held together by electrostatic forces. A pronounced solubility in water, but poor solubility in nonpolar solvents is characteristic of such an ionic crystal. This behavior is called *salt-like* or *saline*. Solids with predominantly ionic interactions are called *salts*. Thus the solubility of many salts is decreased when an organic solvent is added to the aqueous solution. This fact is often used in gravimetric analysis. All truly saline substances are strong electrolytes, and the colors of the solid and its aqueous solutions are the colors of the individual ions.

Solid lattices formed by covalent forces generally show a poor solubility in water. The constituents of the lattice are molecular groups (or complex ions) rather than simple ions. These compounds are often weak electrolytes, and they have a tendency to undergo complex formation reactions in aqueous solution. Often the color of the solid differs from the color of the solution. Most of the heavy metal sulfides ({HgS} red or black; {CdS} yellow, {HgI_2} red) show this behavior.

In the next section the relationship between the solubility and the solubility product will be discussed. However, we shall see that the solubility cannot always be calculated from the solubility product alone. Nernst commented on this fact. He originally formulated the solubility product with some reservations, pointing out that the solubility calculated by means of the solubility product is meaningful only when the solution in equilibrium with the solid phase contains the constituents of the solid phase exclusively in the form of free ions. Svante Arrhenius had already developed his theory of strong and weak electrolytes, and he showed that even weak electrolytes were completely dissociated in dilute solution. Shortly after Nernst published his ideas, Niels Bjerrum introduced the concept of complete dissociation for ionic compounds (salts) in an electrolyte solution and Nernst's reservations were dropped. During the following years, the differentiation between ionic and covalent compounds was not always taken seriously. It was believed that substances with low solubility (covalent substances) were also completely dissociated, since owing to their low solubility only small concentrations were obtained. Therefore, they were also treated as salts. This is not quite correct as we shall see later, and this misconception often causes some confusion, particularly for the inexperienced student.

A knowledge of the characteristics of the solvent and the solute enables the chemist to establish a number of rules that are helpful in the application of precipitation equilibria. Thus it is helpful to know that all nitrates, most of the perchlorates ({$KClO_4$} and {NH_4ClO_4} are exceptions), and almost all alkali metal salts are very soluble in water.

Solubility and solubility product

The *solubility s* of a substance in water is the mass of the solid dissolved in the solution which is in equilibrium with the excess of the solid phase. Such a solution is said to be *saturated*. The solubility is usually temperature dependent. It must be kept in mind that saturated and concentrated solutions are not the same. For example, the poor solubility of silver chloride in water yields a saturated solution of about 10^{-5} mol L^{-1} AgCl (at 25 °C), which is obviously not a concentrated solution.

Solubility may be expressed in different ways. For the analytical chemist, the most convenient one is based on the molar scale, i.e., moles per liter, but other units, such as gram per 100 mL solvent (e.g., the Handbook of Chemistry and Physics uses this scale), are frequently used. If we assume that the solute is completely dissociated (strong electrolytes, salts, etc.) the two-phase system may be described by the following equilibrium, in which sodium chloride is taken as the example of a strong electrolyte:

$$\{NaCl\} \rightleftharpoons Na^+ + Cl^- \qquad (4.5\text{-}19a)$$

The corresponding equilibrium constant is expressed as:

$$K = \frac{[Na^+][Cl^-]}{[\{NaCl\}]} \qquad (4.5\text{-}20)$$

For the solution, the same activity concepts apply as introduced in Section 4.1. The activity of a pure solid phase is always unity, by definition. As a result of this simplification, one obtains the relation known as the *solubility product* K_{sp}:

$$K_{sp} = [Na^+][Cl^-] \qquad (4.5\text{-}21a)$$

If the charges on the ions are not the same, the corresponding equation becomes more complicated. For a salt of general composition M_aB_b, dissociating in an aqueous solution into aM^{v+} and $bB^{\beta-}$ ions, the precipitation equilibrium is expressed by:

$$\{M_aB_b\} \rightleftharpoons aM^{v+} + bB^{\beta-} \qquad (4.5\text{-}19b)$$

and the solubility product by:

$$K_{sp} = [M^{v+}]^a[B^{\beta-}]^b \qquad (4.5\text{-}21b)$$

Under the conditions of complete dissociation, and the requirement that the cation M^{v+} and the anion $B^{\beta-}$ are absent from the original nonequilibrated aqueous phase, each mole of the dissolved salt yields a mol of cations and b mol of anions in the aqueous phase and the solubility and solubility product are simply related to each other. The solubility s expressed in mol of the solid phase per L solution (and not in terms of the individual ion concentrations) is related to the ionic concentration M^{v+} and $B^{\beta-}$ by the following expressions:

$$[M^{v+}] = as \qquad (4.5\text{-}22)$$

and

$$[B^{\beta-}] = bs \qquad (4.5\text{-}23)$$

where s is in (mol/L). The solubility product may be defined as:

$$K_{sp} = [M^{v+}]^a[B^{\beta-}]^b = (as)^a(bs)^b = a^ab^bs^{a+b} \qquad (4.5\text{-}21c)$$

and it yields:

$$s = \frac{(K_{sp})^{1/(a+b)}}{a^ab^b} \qquad (4.5\text{-}24a)$$

Strictly speaking, this equation relates the solubility and the solubility product only under certain conditions (Nernst's reservation) and only if the following considerations are taken in account:

The solubility product is a special case of the mass action law and allows to calculate the solubility of a given compound in water under certain conditions.

- The solubility–solubility product relation holds only if the dissolved compound is completely dissociated. If undissociated solute, or any associated form of it, exists in the solution, or if the cations and anions produced in the dissolution form complexes, then the solubility is larger than the value calculated from the solubility product.
- If the initial aqueous phase contains any of the ions produced by the dissolution of the solid (*common ions*), then the solubility becomes smaller than the value calculated from the solubility product. However, the law of *constant ion product* (the solubility product is a special case of an ion product) is preserved, and thus the final concentration of the common ion becomes larger, and the concentration of its counterion smaller than the calculated values based on the assumption that no common ion effect exists.
- Changing activity coefficients have an effect on the solubility of an electrolyte. Generally, activity coefficients decrease with increase of the total inert electrolyte concentration (ionic strength) and, accordingly, the solubility becomes larger.

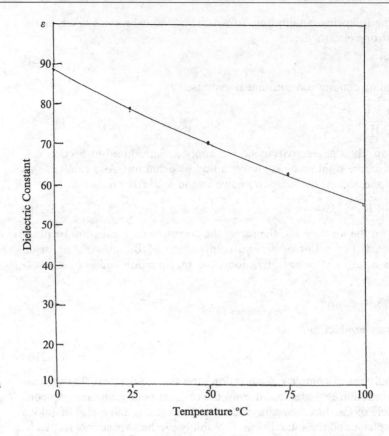

Fig. 4.5-8. Dielectric constant of water as a function of temperature at atmospheric pressure

The solubility is not a thermodynamic quantity, but the solubility product is a thermodynamic expression, provided the proper activity concepts (Debye–Hückel theory, constant ionic strength, etc.) are applied.

It is important to know that the solubility product K_{sp} is not dimensionless. For a 1:1 electrolyte, it has the dimensions $mol^2 L^{-2}$; for a 1:2 electrolyte $mol^3 L^{-3}$, etc. It is therefore meaningless to compare solubility products of electrolytes with different stoichiometry (or different dissociation processes).

The solubility of a substance is influenced by various factors which are discussed in the subsequent sections.

Temperature effects

The solubility and solubility product of a solid are affected by changes of temperature. The bond in the solid lattice, as well as the bonds and structure of the aqueous solution, are weakened by increasing temperature. The dielectric constant of water also decreases with increasing temperature (Fig. 4.5-8). All these phenomena have an effect on the solubility of salts and account for the decreasing solubility of many sulfates (rare earth metals, calcium, and lithium sulfate), acetates (calcium acetate) and carbonates (lithium carbonate). In most systems, however, this effect is outweighed by hydration. The concentration of free "monomeric" water (water activity) increases with the disruption of the solvent structure caused by increasing temperature. It is therefore possible that different hydrates exist in hot and cold solutions. Consequently, the solubility may vary drastically with changes in temperature. Such a case is illustrated in Fig. 4.5-9, represented by $\{Na_2SO_4\}$. Most salt-like electrolytes show a higher solubility in hot than in cold solutions. This is in contrast to the solubility behavior of gases. Some salts exhibit an extreme temperature effect. In some cases, up to 50 times more salt is soluble in hot water (close to the boiling point) than in cold water (close to the freezing point). Borax $\{Na_2B_4O_7 \cdot 10H_2O\}$, an ingredient of many detergents, has such a large temperature effect. However, other substances exhibit a lower solubility in solutions at elevated temperature, behaving like gases. It is interesting that sodium chloride shows about the same solubility over the entire temperature range 0–100 °C. Ad-

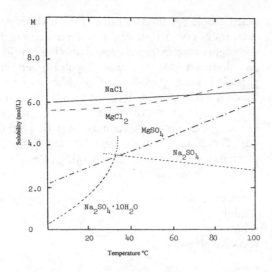

Fig. 4.5-9. Solubility of {Na₂SO₄}, {Na₂SO₄ · 10 H₂O}, {NaCl}, {MgSO₄}, and {MgCl₂} as a function of temperature

vantage may be taken of the different effects of temperature on different electrolytes in a number of analytical and synthetic processes.

Effects of particle size on soulbility

Small solid particles are energetically in a less favorable state than large bodies. This is because their surface is relatively large compared with their volume. It has been shown in Fig. 4.5-7 that the surface of a solid body is in an unsaturated state, and therefore energetically less favorable than the saturated interior of the crystal. In a large body, only an extremely small fraction of the atoms or molecules occupy surface sites. In contrast, extremely small bodies possess a significant fraction of atoms or molecules in surface positions. Therefore, the total energy of a small body is more affected by the surface contribution. Since the energies of small and large bodies are different, they must also exhibit different solubility behavior. One must distinguish between so-called *macro-* and *microsolubility* (Fig. 4.5-10). As a general rule, a constant normal (macro)solubility is observed when the particle size exceeds about 10^{-3} mm in diameter, whereas for smaller crystals the solubility depends on the particle size.

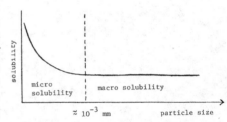

Fig. 4.5-10. Macro- and microsolubility of a crystalline solid

The following example illustrates the problem. Lead chromates {PbCrO₄} is only slightly soluble in water. Its solubility (macrosolubility) is 1.24×10^{-4} mol L^{-1}. This value, is found for a precipitate with an average particle size of 3.0×10^{-2} mm. However, a product with an average particle size of only 9.0×10^{-4} mm is characterized by a solubility value of 2.1×10^{-4} mol L^{-1}. The ratio of micro- and macrosolubility is therefore:

$$\frac{s_{\text{micro}}}{s_{\text{macro}}} = \frac{2.1 \times 10^{-4}}{1.24 \times 10^{-4}} = 1.69$$

A similar calculation can be performed for the solubility product:

$$K_{\text{sp}} = s^2$$

The solubility product of large particles is calculated as:

$$K_{\text{sp}} = (1.24 \times 10^{-4})^2 = 1.54 \times 10^{-8} \, \text{mol}^2 \text{L}^{-2}$$

whereas the solubility product of particles with mm 9×10^{-4} diameter is given by:

$$K_{\text{sp}} = (2.10 \times 10^{-4})^2 = 4.41 \times 10^{-8} \, \text{mol}^2 \, \text{L}^{-2}$$

$$\frac{K_{\text{sp,micro}}}{K_{\text{sp,macro}}} = \frac{4.41 \times 10^{-8}}{1.54 \times 10^{-8}} = 2.86$$

These large differences must be of concern for an analytical chemist who is applying gravimetric methods. It is therefore of important to learn how to produce macroscopic precipitates rather than microcrystalline or colloidal systems. Different sizes of crystals are obtained from saturated and supersaturated solutions.

Systems with extremely low solubility can form large numbers of crystal seeds, and the crystal do not grow beyond a certain size, because each floating solid particle has many close solid neighbors, and only a small solution volume from which it withdraws the ions for its own growth. If the solubility is high, then each of the small number of crystals can grow to an appreciable size. Substances with low solubility products, such as $\{AgCl\}$, $\{BaSO_4\}$, are used in gravimetric methods. Therefore, attention must be paid to the procedures leading to macroscopic precipitates. Generally, the following rules hold. Large crystals are obtained from solutions which are

- not extensively supersaturated (avoid high concentrations)
- dilute
- hot
- cooled slowly

The common ion effect

So far we have treated the solubility of an electrolyte in which the solid phase is equilibrated with pure water. It must be remembered that it does not matter how an equilibrium state is reached. The same result is obtained when a solid body is equilibrated with a certain aqueous solution, or when the solid body is formed in a precipitation reaction by the addition of chemicals. In practice, precipitation is induced by solutions containing a precipitation reagent in excess.

It is clear that the solubility of a solid in an already saturated solution is zero, and that it is small in a solution which already contains the ions of the substance (common ions), but is not yet saturated. Similarly, we would expect a reduced solubility in a system in which only one of the constituent ions is present in the original solution. This result may be derived from an application of *Le Châtelier's principle* to the following equilibrium:

$$\{M_aB_b\} \rightleftharpoons aM^{v+} + bB^{\beta-} \tag{4.5-19b}$$

Le Châtelier's principle states that when M^{v+} or $B^{\beta-}$ exist in excess in the solution, the equilibrium is shifted toward the left-hand side. This results in a smaller solubility, compared with a system in which M^{v+} or $B^{\beta-}$ are not present in the original solution.

From these considerations, we conclude that the common ion effect reduces the solubility. But the solubility product expression is still valid, since it is a thermodynamic constant. After these qualitative conclusions, we should be able to make quantitative statements on the solubility in a system with a common ion. With the exception of systems containing a $1:1$, a $2:2$, or a $3:3$ electrolyte, the mathematical computation is not easy, since it involves an equation of high order. If we denote the concentration of the common ion by c (the concentration of the ion in solution before equilibration), then we are able to calculate the solubility s (identical with the concentration of dissolved solid), with the help of the solubility product expression. We consider a binary electrolyte which contains the cation and anion in a $1:1$ ratio. It does not matter which ion we choose as the common ion, and we assume that the common ion is the anion. The cation concentration in the equilibrium system is therefore equal to s and the anion concentration is equal to $c + s$. Therefore:

$$K_{sp} = [M^{v+}][B^{\beta-}] = s(s+c) = s^2 + sc \tag{4.5-21d}$$

For an $n:n$ electrolyte (where $n = 1, 2, 3$, etc.), $v = \beta$. Rearranging, we obtain:

$$s^2 + sc - K_{sp} = 0$$

or

$$s = \frac{-c \pm \sqrt{c^2 + 4K_{sp}}}{2} \tag{4.5-24b}$$

Solubility calculations

The solubility product of silver chloride {AgCl} is:

$$K_{sp} = 1.0 \times 10^{-10} \, mol^2 \, L^{-2} \tag{4.5-25}$$

In pure water, this gives a solubility of $1.0 \times 10^{-5} \, mol \, L^{-1}$.

Consider the calculation of the silver chloride solubility in solutions which initially contain 10^{-3}, 10^{-5}, and $10^{-7} \, mol \, L^{-1}$ chloride ions, introduced in the form of {NaCl}. It may be assumed that small electrolyte concentrations do not affect the activity coefficients of the silver chloride and that they remain essentially constant.

Case 1

$$[Cl^-] = 10^{-3} \, mol \, L^{-1} = c$$

The calculation yields:

$$K_{sp} = 10^{-10} = s(s + c) = s^2 + sc$$

$$s = \frac{-10^{-3} \pm \sqrt{10^{-6} + 4.0 \times 10^{-10}}}{2} = \frac{(1.00020 - 1.00000) \times 10^{-3}}{2}$$

$$= 1.0 \times 10^{-7} \, mol \, L^{-1}$$

Case 2

$$[Cl^-] = 10^{-5} \, mol \, L^{-1} = c$$

The calculation gives a solubility:

$$s = \frac{10^{-5} \pm \sqrt{10^{-10} - 4.0 \times 10^{-10}}}{2} = \frac{(2.2361 - 1.00000) \times 10^{-5}}{2}$$

$$= 6.1805 \times 10^{-6} \, mol \, L^{-1}$$

Case 3

$$[Cl^-] = 10^{-7} \, mol \, L^{-1} = c$$

The calculation yields a solubility:

$$s = \frac{-10^{-7} \pm \sqrt{10^{-14} + 4.0 \times 10^{-10}}}{2} = \frac{(2.000025 - 0.0100000) \times 10^{-5}}{2}$$

$$= 9.95 \times 10^{-6} \, mol \, L^{-1}$$

Effects of secondary solution equilibria: hydrolysis and complexation

The solubility of a solid is increased, compared with the solubility calculated from the solubility product expression, when one or more of its constituent ions participate in a two-phase distribution, or undergo a second equilibrium process in the aqueous phase. These secondary processes may be either hydrolytic reactions or complex formations.

The problem is best explained by an example. Thallium sulfide {Tl$_2$S} has a solubility product of:

$$K_{sp} = 7 \times 10^{-23} \, mol^3 \, L^{-3} \text{ at } 25\,^\circ C. \tag{4.5-26}$$

From this value we calculate a solubility in pure water of $s = 2.60 \times 10^{-8}$ mol/L. However, a much larger solubility, together with an increase in pH is observed on dissolution. The above calculation is obviously erroneous, because it is based on the assumption that the primary ions, Tl^+ and S^{2-} exist in a neutral aqueous solution. This is not true, as the sulfide ion undergoes a hydrolytic process, according to the following equations:

$$S^{2-} + H_2O \rightleftharpoons HS^- + OH^- \tag{4.5-27}$$

$$HS^- + H_2O \rightleftharpoons H_2S + OH^- \tag{4.5-28}$$

Hydrolysis and complexation may strongly contribute to the actual solubility of a certain substance.

These equilibria are described by the equilibrium constants K_1 and K_2:

$$K_1 = \frac{[HS^-][OH^-]}{[S^{2-}]} \tag{4.5-29}$$

$$K_2 = \frac{[H_2S][OH^-]}{[HS^-]} \tag{4.5-30}$$

Introducing the relation:

$$K_W = [H^+][OH^-] = 10^{-14} \tag{4.2-45}$$

we obtain the familiar acid constants:

$$K_{A,1} = \frac{[H^+][HS^-]}{[H_2S]} = 10^{-7} \tag{4.5-31}$$

$$K_{A,2} = \frac{[H^+][S^{2-}]}{[HS^-]} = 10^{-14} \tag{4.5-32}$$

and we have:

$$K_1 = \frac{K_W}{K_{A,2}} = 1.00 \tag{4.5-29a}$$

and

$$K_2 = \frac{K_W}{K_{A,1}} = 10^{-7} \tag{4.5-30a}$$

K_1 shows that, on formation of hydrogen sulfide ions (HS^-) an equal amount of hydroxide ions (OH^-) is produced, shifting the pH to a higher value. In the pH range 9–12, HS^- is the predominant species and we can neglect the formation of H_2S. We may write:

$$[HS^-] \gg [S^{2-}]$$

and

$$[HS^-] = \tfrac{1}{2}[Tl^+] = [OH^-]$$

Therefore:

$$[H^+] = \frac{2 \times 10^{-14}}{[Tl^+]}$$

Substitution in the expression for $K_{A,2}$ yields:

$$\frac{[HS^-]}{[S^{2-}][H^+]} = 10^{14} = \frac{[Tl^+][Tl^+]}{4 \times 10^{-14}[S^{2-}]} = \frac{[Tl^+]^2}{4 \times 10^{-14}[S^{2-}]}$$

Therefore:

$$[S^{2-}] = \tfrac{1}{4}[Tl^+]^2$$

and

$$[Tl^+][S^{2-}] = K_{sp} = 70 \times 10^{-24}$$

or

$$\tfrac{1}{4}[Tl^+]^4 = 7.0 \times 10^{-23}$$

We calculate:

$$[Tl^+] = 4.09 \times 10^{-6} \, mol \, L^{-1}$$

Around pH $= 9$ the effective solubility of Tl_2S is about 100 times larger than calculated from the solubility product neglecting the hydrolysis equilibrium.

$$[OH^-] = [SH^-] = 2.05 \times 10^{-6} \, mol \, L^{-1}$$

$$[S^{2-}] = 4.18 \times 10^{-12} \, mol \, L^{-1}$$

$$[H^+] = 4.88 \times 10^{-9} \, mol \, L^{-1}$$

The effective solubility ($s = 2.05 \times 10^{-6}\,\mathrm{mol\,L^{-1}}$) is about 100 times larger than that calculated from the solubility product, neglecting the hydrolysis equilibrium. In this calculation we have demonstrated that the free sulfide ion concentration $[S^{2-}]$ is lowered compared with the original value, because this ion is removed from the solution by hydrolytic reactions, and consequently the thallium ion concentration $[Tl^+]$ is greatly increased.

The equilibrium, represented by Eq. 4.5-19b is always shifted toward the right-hand side when the ions in the aqueous solution ($M^{\nu+}$ or $B^{\beta-}$) undergo reactions with the solvent (hydrolysis), the counterion (ion association), or with an additional species in solution (complex formation). If the ions $M^{\nu+}$ or $B^{\beta-}$ are proton acceptors or donors, the solubility becomes pH dependent. It is now understood that the solubility of a salt must increase when its anion has basic properties (proton acceptor), and the solubility increases with decreasing pH. Carbonates, phosphates, borates, oxalates, sulfites, sulfides, fluorides, acetates, and chromates exhibit an increased solubility in solutions of low pH, because the anions CO_3^{2-}, PO_4^{3-}, $B(OH)_4^-$, or BO_2^-, $C_2O_4^{2-}$, SO_3^{2-}, S^{2-}, F^-, OAc^-, and CrO_4^{2-} have basic properties. Chromate, as we know, does not form hydrochromate, but the dichromate ion must be considered an acidic chromate form according to:

$$2\,CrO_4^{2-} + 2\,H^+ \rightleftharpoons 2\,HCrO_4^- \rightleftharpoons Cr_2O_7^{2-} + H_2O \tag{4.5-33}$$

Most of the trivalent cations, such as Fe^{3+} and Al^{3+}, undergo hydrolysis and show proton donor properties. This is understandable if we write the hydrated forms $Fe(H_2O)_6^{3+}$ and $Al(H_2O)_6^{3+}$. These metal ions easily release hydrogen ions and thereby form hydroxo complexes. Consequently, the solubility must increase with increasing pH.

$$M^{3+} + i\,H_2O \rightleftharpoons M(OH)_i^{3-i} + i\,H^+ \tag{4.5-34a}$$

or rearranging:

$$M^{3+} + i\,OH^- \rightleftharpoons M(OH)_i^{3-i} \tag{4.5-34b}$$

However, this solubility behavior is not always observed, since the hydroxo complexes $\{M(OH)_3\}$ often exhibit a lower solubility than the original electrolytes, and the corresponding solid hydroxide or oxides are formed from the original solid. Cr(III), Fe(III), Al(III), Bi(III), and Sb(III) are precipitated as hydroxides at pH values above $5([OH^-] > 10^{-9}\,\mathrm{mol\,L^{-1}})$. But $Cr(OH)_3$, $Al(OH)_3$, $Sb(OH)_3$, and also $Zn(OH)_2$ show an increased solubility at high pH values. This is explained with zinc hydroxide as an example:

$$\underset{\text{precipitate}}{\{Zn(OH)_2\}} + 2\,OH^- \rightleftharpoons \underset{\text{soluble}}{Zn(OH)_4^{2-}} \tag{4.5-35}$$

Some hydroxo complexes also release water, a complication which may confuse the inexperienced student, who does not immediately recognize the pH dependence of these complexes:

$$Cr(OH)_4^- \rightarrow CrO(OH)_2^- + H_2O \tag{4.5-36}$$

$$Al(OH)_4^- \rightarrow AlO_2^- + 2\,H_2O \tag{4.5-37}$$

The solubility of a precipitate is also increased when one of the constituent ions undergoes complex formation with an added reagent. Thus silver chloride is dissolved in a solution containing ammonia or an excess of cyanide or thiosulfate:

$$\{AgCl\} + 2\,NH_3 \rightleftharpoons Ag(NH_3)_2^+ + Cl^- \tag{4.5-38}$$

$$\{AgCl\} + 2\,CN^- \rightleftharpoons Ag(CN)_2^- + Cl^- \tag{4.5-39}$$

$$\{AgCl\} + 2\,S_2O_3^{2-} \rightleftharpoons Ag(S_2O_3)_2^{3-} + Cl^- \tag{4.5-40}$$

In all these systems the free silver ion concentration $[Ag^+]$ is drastically reduced, so that the ion product $[Ag^+][Cl^-]$ becomes smaller than the solubility product K_{sp}. Let us consider one system in more detail.

Worked example:

100 ml 0.5 mol L^{-1} ammonia (NH$_3$) is added to 10^{-4} formula weights of solid *silver chloride* (about 15 mg). The complex Ag(NH$_3$)$_2$$^+$ is thereby formed. The formation constant of this complex is:

$$\beta_2 = \frac{[Ag(NH_3)_2^+]}{[Ag^+][NH_3]^2} = 1.6 \times 10^7 \tag{4.5-41}$$

It is observed that the precipitate dissolves completely and practically all the dissolved silver exists as the complex Ag(NH$_3$)$_2$$^+$(10^{-3} mol L^{-1}). We may calculate the free silver ion concentration:

$$\beta_2 = \frac{10^{-3}}{[Ag^+](0.5)^2} = 1.6 \times 10^7 = 10^{7.20}$$

or

$$[Ag^+] = \frac{10^{-3}}{0.25 \times 1.6 \times 10^7} = 2.5 \times 10^{-10} = 10^{-9.60} \, mol \, L^{-1}$$

The free chloride ion concentration is 10^{-3} mol L^{-1}, and the ion product [Ag$^+$]/[Cl$^-$] assumes a value of 2.5 × 10^{-13} mol^2 L^{-2}, which is far below the solubility product (10^{-10} mol^2 L^{-2}). Therefore, all the silver chloride remains in solution and no solid phase is left.

If the experiment is performed with *silver bromide* the ion product value (10$^{-12.60}$ mol^2 L^{-2}) is of the same order of magnitude as the solubility product:

$$K_{sp} = [Ag^+][Br^-] = 10^{-12.30} \, mol^2 \, L^{-2} \tag{4.5-42}$$

Therefore, the precipitate is barely dissolved. However, it can be seen that the silver bromide is easily dissolved in concentrated ammonia (about 15 mol L^{-1}). For this ammonia concentration, the silver ion concentration is calculated according to:

$$[Ag^+] = \frac{10^{-3}}{(15.0)^2 \times 1.6 \times 10^7} = \frac{10^{-3}}{225 \times 1.6 \times 10^7} = 2.77 \times 10^{-13} = 10^{-12.56} \, mol \, L^{-1}$$

and

$$[Ag^+][Br^-] = 2.77 \times 10^{-16} = 10^{-15.56} \, mol^2 \, L^{-2}$$

This value is well below the solubility product value.

If *silver iodide* is treated with concentrated ammonia, it is found that the precipitate does not dissolve. The silver iodide solubility product is:

$$K_{sp} = [Ag^+][I^-] = 10^{-16.30} \tag{4.5-43}$$

This value is much smaller than the ion product calculated above ([Ag$^+$][I$^-$] = 10$^{-15.56}$ mol^2 L^{-2}).

This kind of complex formation may be used in gravimetric procedures. Consider a mixture of chlorides, bromides, and iodides in solution, accompanied by anions, which need to be separated from each other. The halides are easily separated by precipitation with silver nitrate. By treating the precipitate with dilute ammonia in a second step, only silver chloride is removed, whereas bromide and iodide remain in the solid phase. The resulting dilute ammonia solution may be evaporated, so that the silver chloride is again precipitated (if carefully done by slow evaporation, silver chloride is obtained in a nice crystalline form). In a subsequent treatment of the precipitate with concentrated ammonia, the silver bromide is dissolved and separated from the silver iodide. It may also be obtained in solid form by evaporating the ammonia. Thus it is possible to achieve quantitative separation of the three halides.

Complex formation processes leading to an enhanced solubility are not restricted to the interaction of one constituent of the solid. After all, the same forces which make the precipitates insoluble (strong lattice energy, i.e., strong bonds in the solid, which may be thought of as a polynuclear complex) may also be effective in the solution, forming binary complexes or higher homologs which are soluble (charged complex ions). This is best explained by an example.

Worked example:

In the case of silver chloride, the following complex formation constants are known. They refer to homogeneous equilibrium states (i.e., the species AgCl represents the mononuclear compound and should not be confused with the solid-phase species {AgCl}.

$$Ag^+ + Cl^- \rightleftharpoons AgCl \qquad K_1 = \frac{[AgCl]}{[Ag^+][Cl^-]} = 10^{3.00} \qquad (4.5\text{-}44)$$

$$AgCl + Cl^- \rightleftharpoons AgCl_2^- \qquad K_2 = 10^{2.30} \qquad (4.5\text{-}45)$$

$$AgCl_2^- + Cl^- \rightleftharpoons AgCl_3^{2-} \qquad K_3 = 10^{0.85} \qquad (4.5\text{-}46)$$

$$AgCl_3^{2-} + Cl^- \rightleftharpoons AgCl_4^{3-} \qquad K_4 = 10^{-0.65} \qquad (4.5\text{-}47)$$

These equilibrium constants describe the homogeneous equilibrium, but we are particularly interested in heterogeneous systems in which the solid phase {AgCl} is involved. With the help of K_1 we can calculate the concentration of the aqueous molecular species AgCl, introducing the solubility product of AgCl which is valid when the solid {AgCl} is in equilibrium with the aqueous solution:

$$K_1 = 10^{3.00} = \frac{[AgCl]}{[Ag^+][Cl^-]}$$

where

$$[Ag^+][Cl^-] = 10^{-10.0}$$

Therefore:

$$[AgCl] = 10^{3.00} \times 10^{-10.0} = 10^{-7.0}$$

This is a thermodynamic constant, like K_{sp}, and is independent of the chloride concentration! In other words, if solid chloride is in equilibrium with an aqueous phase, we always have 10^{-7} mol L^{-1} molecular AgCl in the solution. We can now calculate a series of heterogeneous equilibrium constants K_s:

$$\{AgCl\} \rightleftharpoons Ag^+ + Cl^- \qquad (45\text{-}19e)$$

This equilibrium is described by the solubility product:

$$K_{sp} = [Ag^+][Cl^-] = 10^{-10.0} \qquad (4.5\text{-}25)$$

$$\{AgCl\} \rightleftharpoons AgCl$$

This equilibrium is described by the constant K_{s1}:

$$K_{s1} = [AgCl] = 10^{-7.0} \qquad (4.5\text{-}48)$$

$$\{AgCl\} + Cl^- \rightleftharpoons AgCl_2^-$$

This equilibrium is described by K_{s2}:

$$K_{s2} = \frac{[AgCl_2^-]}{[Cl^-]} = K_2[AgCl] = K_2 K_{s1} = 10^{-4.70} \qquad (4.5\text{-}49)$$

$$\{AgCl\} + 2\,Cl^- \rightleftharpoons AgCl_3^{2-}$$

This equilibrium is described by K_{s3}:

$$K_{s3} = \frac{[AgCl_3^{2-}]}{[Cl^-]^2} = K_2 K_3[AgCl] = K_{s2} K_3 = 10^{-3.85} \qquad (4.5\text{-}50)$$

$$\{AgCl\} + 3\,Cl^- \rightleftharpoons AgCl_4^{3-}$$

This equilibrium is described by K_{s4}:

$$K_{s4} = \frac{[AgCl_4^{3-}]}{[Cl^-]^3} = K_2 K_3 K_4[AgCl] = K_{s3} K_4 = 10^{-4.50} \qquad (4.5\text{-}51)$$

We are now able to calculate the concentration of all the different complexes existing in the aqueous solution which is in equilibrium with the solid {AgCl}. Each complex contributes

The solubility curve as a function of pCl for AgCl shows a complex composition (see Fig. 4.5-11).

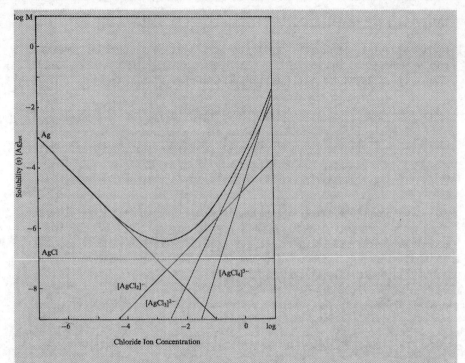

Fig. 4.5-11. Logarithmic plot of the total solubility of silver chloride (bold line) in dependence of the chloride ion concentration in contact with solid {AgCl}

a certain amount to the total solubility. Each of these contributions is represented in Fig. 4.5-11, and is described by a straight line in an appropriate logarithmic plot. We derive:

$$[Ag^+] = \frac{10^{-10.0}}{[Cl^-]} \qquad \frac{d\log[Ag^+]}{d\log[Cl^-]} = -1 \quad (\text{slope} = -1) \tag{4.5-52}$$

$$[AgCl] = 10^{-7.0} \qquad \frac{d\log[AgCl]}{d\log[Cl^-]} = 0 \quad (\text{slope} = 0) \tag{4.5-53}$$

$$[AgCl_2^-] = 10^{-4.70}[Cl^-] \qquad \frac{d\log[AgCl_2^-]}{d\log[Cl^-]} = 1 \quad (\text{slope} = 1) \tag{4.5-54}$$

$$[AgCl_3^{2-}] = 10^{-3.85}[Cl^-]^2 \qquad \frac{d\log[AgCl_3^{2-}]}{d\log[Cl^-]} = 2 \quad (\text{slope} = 2) \tag{4.5-55}$$

$$[AgCl_4^{3-}] = 10^{-4.50}[Cl^-]^3 \qquad \frac{d\log[AgCl_4^{3-}]}{d\log[Cl^-]} = 3 \quad (\text{slope} = 3) \tag{4.5-56}$$

This example is illustrative since it shows that the solubility product does not always lead directly to a correct solubility. The enhanced solubility beyond a certain ligand concentration (in our case, at about $[Cl^-] = 10^{-4}$ mol L^{-1}) is caused by complex formation. This effect is particularly large for substances having covalent bonding, and is therefore observed with substances which are the result of interactions of a soft acid with a soft base. We have already seen that these substances are characterized by a small solubility product. A strong solubility enhancement is expected for the sulfides, selenates, and tellurates of the heavy metal ions. Indeed, the solubility of {HgS} is due to complexes, such as $Hg(SH)_2$, $HgSSH^-$, and HgS_2^{2-}, and over a wide pH region the solubility is more than 10^{20} times larger than that calculated from the solubility product (including the effects of the hydrolysis of S^{2-}). In fact, mercuric sulfide is soluble in macroscopic quantities in alkaline sulfide solutions, in the form of HgS_2^{2-}.

It is also known that silver sulfide forms complexes in hydrogen sulfide and sulfide solutions. Species such as $AgSH$, $Ag(SH)_2^-$, and the binuclear $Ag_2S(SH)_2^{2-}$ are found in these solutions. In Fig. 4.5-12 the silver solubility in a 0.02 mol L^{-1} sulfide solution is given as a function of pH. This is a very complex system since the hydrogen sulfide ion HS$^-$ as well as the sulfide ion S^{2-} are the ligands involved in complex formation. The concentration of these ligands depends on the pH since they represent the acid (HS$^-$) and base form (S^{2-}) of the same acid–base equilibrium. From Fig. 4.5-12 it is seen that at pH = 0, 7, and 13, the total silver solu-

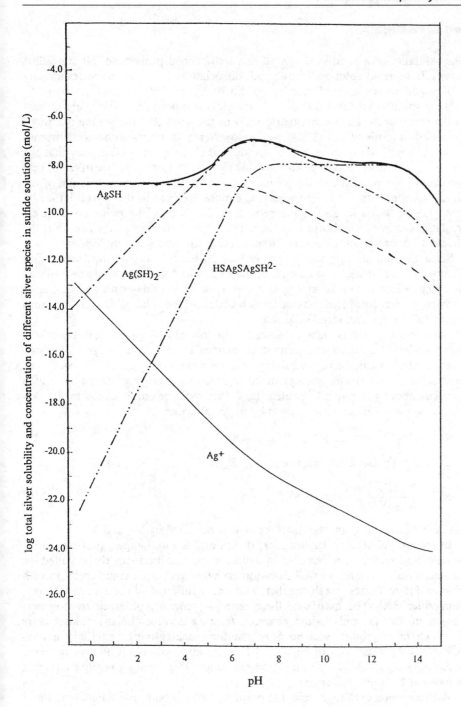

Fig. 4.5-12. Solubility of $\{Ag_2S\}$ in 0.02 mol L^{-1} sulfide solutions as a function of pH. The solubility is not determined by $[Ag^+]$ which predominates only in the unrealistic pH range (pH < -2). However, it is the result of the presence of three silver complexes, $AgSH$, $Ag(SH)^{2-}$, and $HSAgSAgSH^{2-}$

bility is about 10^4, 10^{14}, and 10^{16} times larger than the silver ion concentration. As a matter of fact, the free silver ion never exists as the predominant silver species in the practical pH range. The solid line in Fig. 4.5-12 represents the molar solubility s in terms of the solid phase $\{Ag_2S\}$, and therefore, is the sum of half the values of the concentrations of Ag^+, $AgSH$, and $Ag(SH)_2^-$ and the full concentration of $Ag_2S(SH)_2^{2-}$.

It is known that the sedimentation of heavy metal sulfides from hot, natural sulfide- or hydrogen-sulfide-bearing waters (sulfur springs) does not follow the sequence of the respective solubility products. This is understandable only in terms of the formation of the soluble complexes described above. Such complexes are known for mercury, silver, gold, copper, lead, zinc, cadmium, indium, nickel, iron, arsenic, and antimony sulfides.

Ion exchange

The solubility of a solid in a solvent is a well-defined property of all crystalline bodies. In aqueous solutions, most solids dissociate and exist as the corresponding cations and anions, as explained by Eq. 4.5-20.

If the solution contains no other electrolyte, the anions and cations are present in the solution in the stoichiometric ratio of the solid and the solubility can be expressed in terms of K_{sp} (Eq. 4.5-21). However, this general behavior is not always observed. Many minerals behave quite differently. It is often seen that the anion (e.g., silicates) is part of an insoluble rigid crystal structure (matrix) and that the cations are present only to compensate for the excess negative charge of the rigidly fixed anions. This is so when the cations are held in the crystal lattice by purely electrostatic forces. In the process of dissolution, the polar bonds (electrostatic) can easily be broken by the water dipoles (likes dissolves like), but the covalent bonds are quite resistant to interaction with the water molecules.

Since the cations are held in the crystal lattice by electrostatic forces, they occupy holes, interstices, or cavities of the lattice, and they can easily be replaced by other cations of similar charge and size. However, cations with a large charge and smaller ionic radii are retained much better on the solid surface than cations with small charge and large ionic radii.

Clays belong to this class of minerals. In this case the stoichiometry of the compounds is not fixed and cannot be expressed by simple integral numbers. These materials exhibit low solubility, but they can exchange certain cations in their lattice with cations present in an aqueous solution (e.g., seawater, which contains about 0.7 $mol\,L^{-1}$ electrolytes). This replacement is called an *ion exchange process* and can be expressed by the equilibrium:

$$R^- \cdots X^+_{solid} + Y^+ \rightleftharpoons R^- \cdots Y^+_{solid} + X^+ \tag{4.5-57}$$

and the appropriate equilibrium constant K_{eq}:

$$K_{eq} = \frac{Y^+_R[X^+]}{X^+_R[Y^+]} \tag{4.5-58}$$

where Y^+_R and X^+_R are the fractions of sites occupied by Y^+ and X^+.

In modern laboratory applications, the natural ion exchangers (minerals, such as clays and zeolites) are replaced by synthetic organic products, the so-called *ion exchange resins*. Anion, as well as cation exchange resins are commercially available in different sizes (mesh number) and under different brand names (Dowex, Amberlite, etc.). The matrix of these resins is generally obtained by polymerization of styrene and divinyl benzene, forming a three-dimensional structure to which ionic groups, such as SO_3^- (sulfonic acid group) or tertiary amines ($CH_2-NR_3^+$, where R can be, e.g., CH_3) are attached. If the matrix carries a negative charge, it behaves as a cation exchanger. If it carries a positive charge it behaves as an anion exchanger.

With an anion exchange resin, the positive charge is built into the matrix, and it must be compensated by negatively charged ions that easily exchange with other anions of a solution in contact with the resin. Ion exchange resins can be used in different analytical techniques and for different purposes. Ion exchangers are used in chromatographic methods rather than in batch experiments. Important applications are as suppressors in ion chromatography.

If one uses a mixture of a cation exchange resin in the H^+ form (the negative charge of the matrix is compensated by hydrogen ions) and an anion exchanger in the OH^- form (the positive charge of the matrix is compensated by hydroxide ions), all cations are removed from the solution and replaced by H^+ and all anions are replaced by OH^-. However, the hydrogen ions in solution react with the hydroxide ions according to:

$$H^+ + OH^- \rightleftharpoons H_2O \tag{4.2-44}$$

This equilibrium is described by the ion product for water:

$$K_W = [H^+][OH^-] = 10^{-14} \text{ at } 25°C \tag{4.2-45}$$

Since there must be the same amount of positive and negative charge in the same solution (electroneutrality) we can set:

$$[H^+] = [OH^-] = 10^{-7}\,mol\,L^{-1} \text{ at } 25\,°C$$

Therefore, all electrolytes are removed from the solution and we end up with extremely pure water. This process has practical significance and water prepared this way is called *deionized water*. It is characterized by low electric conductance and is often used in preference over distilled water which always contains small amounts of elctrolytes. However, neutral species, such as sugars, etc., cannot be removed by such an ion exchange process.

4.5.4 Liquid–liquid systems

Introduction

Procedures that involve the application of the phase distribution law to the distribution of a solute between immiscible liquid phases are used extensively in analytical, preparative industrial, and laboratory processes.

In a two-phase system consisting of two immiscible liquid phases, a chemical compound is distributed according to its relative solubility in the individual phases and eventually reaches equilibrium. The distribution of this compound between the two phases is often used to enrich or separate a chemical species or a group of species from each other or from the matrix. This process is called *solvent–solvent extraction*.

The preparative mode of solvent–solvent extractions is still very popular in organic synthesis laboratories. For efficient, industrial applications continuous extraction processes and instrumentation have been developed. The use of dynamic rather than static principles has led to the development of countercurrent extraction systems.

During and after World War II, special solvent extraction procedures were formulated to separate rare earth metal and transuranium compounds. Solvent–solvent extraction was the method of choice before more effective separation techniques, such as chromatography became popular in the 1950s and 1960s.

In the Manhattan Project, a giant scientific research project to produce fissionable material used in the development of the atomic bomb, solvent–solvent extraction played an important role. At that time only one method was available for the production of Pu^{239} in a breeder reactor. Quantity production of this fissionable metal made it necessary to develop chemical extraction procedures which work under extreme conditions.

Today, solvent–solvent extractions play a less important analytical role, because they are time consuming. They have been replaced by more modern and efficient separation techniques (see Sec. 5).

Solvent–solvent extraction is an efficient tool for the enrichment of metal-ion traces from aqueous solutions into organic solvents (e.g. CCl_4) using organic ligands HA forming hydrophobic metal chelates MA_n.

Distribution constant: partition coefficient

The distribution of a solute S, equilibrated between an aqueous phase and an organic solvent may be described by an equilibrium equation:

$$S_{aq} \rightleftharpoons S_{org} \tag{4.5-59a}$$

There is no need to choose an aqueous phase as one of the two phases, and we can describe distributions between any pair of immiscible solvents, liquid 1 and liquid 2, with the appropriate solutes, S_1 and S_2, respectively:

$$S_1 \rightleftharpoons S_2 \tag{4.5-59b}$$

Such systems are described by an equilibrium constant:

$$K_D = \frac{[S]_{org}}{[S]_{aq}} \text{ or } \frac{[S]_2}{[S]_1} \tag{4.5-60}$$

where K_D is called the *partition coefficient*. The determination of the distribution constant K_D' as a thermodynamic quantity requires knowledge of the corresponding activity coefficients. However, these are generally not known, and as a first approximation we use the concentrations. Therefore, the derived quantities depend on the chosen conditions and, strictly speaking, are not constant quantities. The partition is related to the thermodynamic distribution constant according to:

$$K_D' = K_D \frac{y_{org}}{y_{aq}} \tag{4.5-61}$$

where y_{org} and y_{aq} are the activity coefficients in the organic and aqueous phase species, respectively.

Distribution ratio

In the context of practical chemical procedures, we have to go a step further beyond ideal thermodynamic behavior. For example, in the extraction of an acid HA from an aqueous into an organic phase, we are dealing with more than one chemical entity, namely the dissociated and undissociated acid forms. We can define a partition coefficient relating only to the ratio of the undissociated acid forms, according to:

$$K_D = \frac{[HA]_{org}}{[HA]_{aq}} \tag{4.5-62}$$

which does not tell the whole story, because, in the aqueous phase the acid may coexist in the dissociated form, and in the organic phase, higher ion pair products may be formed. In general, total solute concentrations are determined by analytical investigations, and the information gained is used to explain the distribution of the species between the two phases, including speciation. The ratio of the total concentrations of the solute is a practical means of dealing with distribution equilibrium situations, and is called the *distribution ratio D_c*:

$$D_c = \frac{\text{[total concentration of all forms of HA]}_{org}}{\text{[total concentration of all forms of HA]}_{aq}} = \frac{[HA]_{tot,org}}{[HA]_{tot,aq}} \tag{4.5-63}$$

Owing to the differences in speciation in the organic and aqueous phases, D_c is concentration dependent.

Let us pursue the problem of the distribution of an acid between an aqueous and organic phase in more detail, and let us assume that the acid is a moderately strong acid in the aqueous phase and predominantly undissociated in the organic phase. Therefore, we can express our distribution ratio according to Eq. 4.5-63.

In the aqueous phase, we can apply the acid dissociation constant K_A:

$$K_A = \frac{[H^+]_{aq}[A^-]_{aq}}{[HA]_{aq}} \tag{4.2-18}$$

If we assume that the ions cannot exist in the nonpolar organic phase, we postulate the distribution ratio:

$$D_c = \frac{[HA]_{org}}{[HA]_{aq} + [A^-]_{aq}} \tag{4.5-64}$$

Introducing the partition coefficient for the undissociated species:

$$K_D = \frac{[HA]_{org}}{[HA]_{aq}} \tag{4.5-60a}$$

we can combine Eq. 4.5-64, and Eq. 4.5-60a to obtain:

$$D_c = \frac{[HA]_{org}}{\dfrac{[HA]_{org}}{K_D} + \dfrac{K_A}{K_D}\dfrac{[HA]_{org}}{[H^+]_{aq}}} \tag{4.5-65}$$

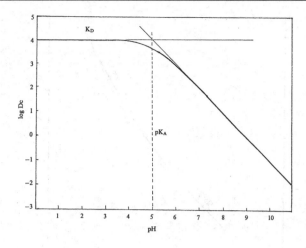

Fig. 4.5-13. Distribution ratio as a function of pH in the partitioning of hypothetical carboxylic acid ($pK_A/5$) between an organic phase and aqueous buffer solution ($K_D = 10^4$)

which can be simplified to yield:

$$D_c = \frac{K_D[H^+]_{aq}}{[H^+]_{aq} + K_A} \tag{4.5-65a}$$

If we plot $\log D_c$ versus pH, we obtain the diagram shown in Fig. 4.5-13, revealing two straight line regions:

When $[H^+]_{aq} > K_A$, then $D_c \approx K_D$, and when $K_A > [H^+]_{aq}$, then:

$$D_c = \frac{K_D[H^+]_{aq}}{K_A} \tag{4.5-65b}$$

In a given solvent–solvent extraction system with a partition coefficient K_D the distribution ratio D_c is strongly dependent on pH of the aqueous solution (for pH > pK_A).

Another interesting case is encountered if the extracted species polymerize in the organic phase. For example, if we extract acetic acid into a nonpolar organic solvent, such as benzene, the organic phase species may be dimerized. We are dealing with two equilibrium states that are linked together. If we keep the pH of the aqueous solution below 2, the predominant aqueous species is the undissociated acetic acid.

In general terms, K_D describes the distribution of the molecular species in the organic and aqueous phases:

$$HAc_{aq} \rightleftharpoons HAc_{org}$$

$$K_D = \frac{[HAc]_{org}}{[HAc]_{aq}} \tag{4.5-60a}$$

and $K_{2,org}$ describes the dimerization in the organic phase:

$$2\,HAc_{org} \rightleftharpoons [HAc]_{2\,org}$$

$$K_{2,org} = \frac{[(HAc)_2]_{org}}{[HAc]_{org}^2} \tag{4.5-66}$$

A plot of $[HAc]_{org}$ versus $[HAc]_{aq}$ is given in Fig. 4.5-14.

We see from these simple examples that variations of D_c cause curvature in the plots of c_{org} (the total concentration of all species in the organic phase) versus c_{aq} (the total concentration of all species in the aqueous phase). Such plots are called *partition isotherms*.

Extraction of molecular species

The distribution of simple molecular species, that do not undergo dissociation or polymerization reactions is straightforward. The corresponding partition isotherm is a straight line with a slope of $+1$, and represents the case where D_c remains constant.

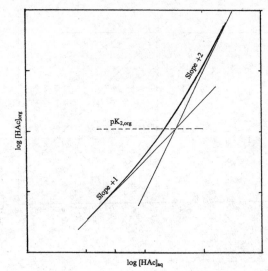

Fig. 4.5-14. Hypothetical extraction of acetic acid from an aqueous phase into an organic solvent in which the acid forms a dimeric species

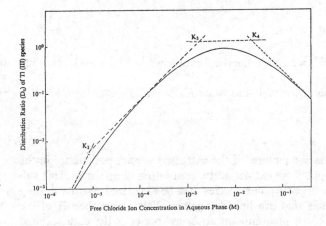

Fig. 4.5-15. Extraction of Tl(III) from aqueous chloride solutions with TBP in hexane at 25°C. From this plot the complex formation constants K_2, K_3, and K_4 of thallium chloride can be determined

Extraction of metal complexes

Metal ions easily form complexes in aqueous solution, when an appropriate ligand is available. Because of their electrical charge, metal ions, even if hydrated, usually do not extract well into an organic phase. The situation may be quite different with neutral complexes, especially if these are covalent in nature. An example is given in Fig. 4.5-15 where the extraction of thallium(III) from aqueous choride solutions with tributyl phosphate (TBP) in hexane is shown. The extracted species is $TlCl_3$. Therefore we can write:

$$K_D = \frac{[TlCl_3]_{org}}{[TlCl_3]_{aq}} = \text{const.} \tag{4.5-60}$$

Consequently, if $TlCl^{2+}$, $TlCl_2^+$, $TlCl_3$ and $TlCl_4^-$ are the predominant species in the aqueous phase, we can simplify to:

$$[TlCl_3]_{org} = \text{const. } [TlCl^{2+}]_{aq}[Cl^-]_{aq}^2 \tag{4.5-67a}$$

where $[TlCl^{2+}]_{aq}$ can be set equal to $[Tl(III)]_{aq,tot}$:

$$[TlCl_3]_{org} = \text{const. } [TlCl_2^+]_{aq}[Cl^-]_{aq} \tag{4.5-67b}$$

where $[TlCl_2^+]_{aq}$ can be set equal to $[Tl(III)]_{aq,tot}$:

$$[TlCl_3]_{org} = \text{const. } [TlCl_3]_{aq} \tag{4.5-67c}$$

where $[TlCl_3]_{aq}$ can be set equal to $[Tl(III)]_{aq,tot}$, and:

$$[TlCl_3]_{org} = \text{const. } [TlCl_4^-]_{aq}[Cl^-]_{aq}^{-1} \tag{4.5-67d}$$

where $[TlCl_4^-]_{aq}$ can be set equal to $[Tl(III)]_{aq,tot}$.

From these relationships, the complex formation constants in aqueous solutions may be derived as:

$$K_2 = \frac{1}{[Cl^-]} \text{ at intercept of } +2 \text{ and slope } +1 \tag{4.5-68a}$$

$$K_3 = \frac{1}{[Cl^-]} \text{ at intercept of } +1 \text{ and slope } 0 \tag{4.5-68b}$$

$$K_4 = \frac{1}{[Cl^-]} \text{ at intercept of } 0 \text{ and slope } -1 \tag{4.5-68c}$$

With certain organic solvents, a number of metal ions are extracted as the anionic complex MX_i^-, where $i = v + 1$ (v being the charge on the free uncomplexed metal ion). This is especially important for tetrahedral complexes, such as $TI(III)X_4^-$, $Fe(III)X_4^-$, ReO_4^-, and ClO_4^- which are well extracted as acids into several organic solvents. Depending on the nature of the organic solvent, the extracted species exist as the free anion over a wide range, showing the behavior of strong acids ($HMeX_4$). An example is given in Fig. 4.5-16. The top diagram (a) shows the experimentally observed distribution of iron(III) between aqueous HCl (2.0 and 1.0 mol L^{-1}) and isobutylmethylketone (hexone, IBMK).

This is a beautiful example to illustrate most of the analytical principles of solution chemistry, involving complex formation, hydration/solvation, and ion association; furthermore, it demonstrates the significance of the common ion effect and proves the interrelationship of one equilibrium with an other. In Fig. 4.5-16a, we observe three straight-line branches in each of the distribution curves. At low concentrations, we have a linear relationship between the aqueous and organic phase iron species (slope $+1$). At concentrations higher than 10^{-5} mol L^{-1} iron(III) species in the organic phase (intercept 1), we observe a slope of $+0.5$ and at concentrations beyond 2.0×10^{-3} mol L^{-1}, the systems again assume a slope of $+1$ (intercept 2). Note that in both systems the intercept 2 is at exactly the same organic phase concentration.

Intercept 1 is caused by the common ion effect. If the extraction experiment is performed with 2.0 mol L^{-1} HCl alone, HCl is extracted into the organic phase to the extent of 8×10^{-4} mol L^{-1} and 2.5×10^{-5} mol L^{-1} as undissociated and dissociated acid, respectively ($pK_{A \, org} = 6.1$). In the iron(III) experiment, the hydrogen ion and chloride ion concentrations cannot be identical as soon as the second anion $FeCl_4^-$ assumes values of the same order of magnitude. The common ion effect then becomes significant, and $[H^+]_{org}$ and $[Cl^-]_{org}$ start to deviate from each other; however, the ion product $[H^+]_{org}[Cl^-]_{org}$ remains constant. Owing to this effect, we also observe a common ion effect for the extracted iron(III) species changing from slope $+1$ to $+0.5$, indicating that $FeCl_4^-{}_{org}$ is the predominant organic phase iron(III) species, coexisting with the same amount of the counterion, $H^+{}_{aq}$. Since a smaller amount of hydrochloric acid is extracted from 1 mol L^{-1} HCl ($[HCl]_{org} = 4 \times 10^{-3}$ and $[H^+]_{org} = [Cl^+]_{org} = 1.8 \times 10^{-5}$ mol L^{-1}), the inflection point must be at a different concentration of the organic phase iron(III) species (1.8×10^{-5} mol L^{-1}), as indicated in the diagram.

The second intercept occurs at the same organic phase iron(III) concentration, indicating that an association reaction is involved. We observe two different additional species. On one hand, the ion-paired or undissociated $HFeCl_{4 \, org}$ becomes a predominant species, overshadowing the dissociated $FeCl_4^-{}_{org}$ form. But in addition, the triple ion $H(FeCl_4)_2^-$ also becomes a significant species in the organic phase. Its slope in this region of Fig. 4.5-16b assumes a value of $+1$, whereas at lower concentration is has slope $+1.5$. This is explained by the fact that the product $[H^+]_{org}[H(FeCl_4)_2^-]_{org}$ also assumes a constant value (ion product relation). Note that $H(FeCl_4)_2^-$ must be considered a dimeric species and one mole corresponds to two moles of iron(III), which must be considered in the definition of total iron(III)$_{org}$ concentration.

We cannot make any statement on the hydration and solvation of the extracted species from the extraction data presented so far. However, in a series of additional experiments it can be demonstrated that the free hydrogen ion in the organic phase is hydrated by four molecules of water ($H_9O_4^+$) and additionally solvated by three IBMK molecules. It is also clear that these solvation numbers

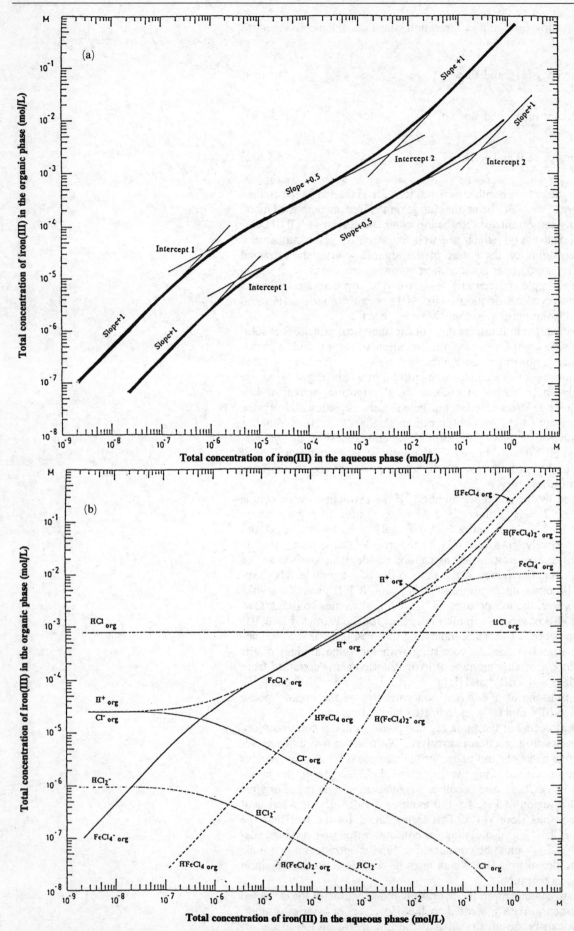

Fig. 4.5-16. Extraction of iron (III) from 1.0 and 2.0 mol L^{-1} aqueous HCl, by IBMK at 20°C. a) Experimental work (distribution data); b) Interpretation: concentration plot of the organic phase species over the observed range for the extraction from 2 mol L^{-1} HCl

are not the same for the hydrogen ion in the triple ion $(H(FeCl_4)_2{}^-)$. This is a nice example, showing that most extraction phenomena can be explained when all the necessary analytical data are available.

Ion-pairing reagents

For organic solvents of low dielectric constant, the phenomenon of ion-pairing is often observed. We have already encountered this in the extraction of iron(III) species from an aqueous hydrochloric acid solution into an IBMK organic phase. Distinct increases in the concentration of organic phase extractants are observed with a number of ion-pairing agents.

As positively charged reagents, the tetraalkyl ammonium compounds (R_4N^+, e.g., tetrabutyl ammonium chloride) are often used. In contrast, the alkylsulfates ($ROSO_3{}^-$) are used as negatively charged ion-pairing agents. The length of the alkyl residue is of importance, and typically shows specific effects for the counterion. The equilibrium states involved in ion-pairing processes are extremely temperature dependent.

4.5.5 Gas–solid systems

Solid surfaces are known to bind molecules of gases with which they are in contact. This process is called *adsorption*; the solid used to adsorb gases is called the *adsorbant*, and the adsorbed species are referred to as *adsorbates*. Adsorption occurs on the external surface of solid and/or its internal surface, i.e. pores, crevices, or capillary channels. The reverse process is called *desorption*.

In the general case, adsorption is based on physical attractive forces, e.g., van der Waals interactions. In contrast, chemical adsorption, also called *chemisorption* occurs, when chemical forces are involved. Chemisorption usually takes place at higher temperature, because it involves an activation energy, and it is generally slower than physical adsorption.

Adsorption isotherms

The amount of adsorbed gas is expressed as the volume of the adsorbed gas at standard temperature and pressure, and the distribution of a chemical species between the solid and gaseous phase is described by *adsorption isotherms* (at constant temperature), in which the volume of the adsorbed gas is plotted against the partial pressure of the gas. Larger volumes are adsorbed at higher vapor pressures, and smaller amounts of gas are adsorbed at higher temperatures. Fig. 4.5-17 shows different types of adsorption isotherms. Fig. 4.5-17a represents the ideal case. Although part of an adsorption isotherm may be linear, it represents a hypothetical case. Fig. 4.5-17b describes saturation. This is the case when only a limited amount of adsorption sites are available. Fig. 4.5-17c illustrates the case where the partial pressure reaches a point where the gas is liquified (reverse process of boiling). Fig. 4.5-17c is a combination of the previously described adsorption isotherms, consisting of a linear range at low partial pressures, saturation at moderate pressures, and an increase due to the effect of condensation.

The adsorption of gases on solid surfaces is of general chemical interest, but there are several applications of analytical significance. The adsorption of gases on activated charcoal is routinely used to clean contaminated air, and plays an important role in the removal of toxic components from the atmosphere (e.g., gas masks) and to clean up gases from incinerators.

Gas adsorption on solid carrier material is used in the analytical determination of pollutants and hazardous material in air (industrial hygiene), and as a sample collector in purge and trap systems. Here, gases are adsorbed on specially prepared material (e.g., silica gel, alumina, porous polymers, polyurethane foam) which are used to enrich gaseous sample components on the adsorbant before they are thermally desorbed (flash evaporation) and introduced into, e.g., a gas chromatograph for further separation and quantitative determination.

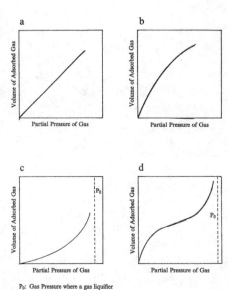

P_0: Gas Pressure where a gas liquifier

Fig. 4.5-17. Adsorption isotherms

Common adsorbant materials

In most analytical applications, the adsorption and desorption processes must be reversible. This is rarely the case when chemisorption is involved, and even with physical adsorption, the processes may not be completely reversible, and may show hysteresis behavior. Therefore, there is a great need for specially prepared adsorbant materials with controlled properties. Enormous efforts have been made to develop procedures that yield large quantities of adsorbant materials with reproducible properties.

The shape of the surface (porous vs. nonporous) and the surface area ($m^2 g^{-1}$ adsorbant) determine the properties of the adsorbant. Useful materials have large surface areas (about $100 m^2 g^{-1}$ and more) with controlled pore sizes. On the other hand, substances with large surface areas are also very effective catalysts.

Activated carbon exists in many forms, differing in surface area, porosity, and surface activity. The surface is generally modified by either physical or chemical means to make it less reactive. Sophisticated procedures exist to modify the properties of *silica gels* and *alumina* to obtain versatile and reproducible adsorbants.

Certain compounds, such as the natural zeolites, have well-defined structures with pores of molecular dimensions, and they are called *molecular sieves*. Gas molecules of dimensions smaller than the cavities can penetrate the interior of the porous structure, and may be efficiently trapped. Larger molecules cannot penetrate, and pass through a column filled with molecular sieve without significant retention. The size of the cavities may be controlled by special preparation procedures.

A number of adsorbant materials are produced synthetically. As an example, porous polymers are obtained from the polymerization of styrene cross-linked with divinylbenzene. The product has an open structure, similar to the molecular sieves. Various types of porous polymers, including Tenax, Porapaks, Chromosorbs, and the XAD materials are commercially available. They are especially useful in purge and trap systems to enrich trace gas constituents from a flow stream. After a certain amount of the constituents are adsorbed they are thermally desorbed and analyzed.

Many sampling procedures depend on such collecting systems, i.e., columns packed with adsorbant materials. They are used either passively, when the adsorption is based on diffusion processes, or actively, with a constant flow of gas.

It is possible to adsorb specific compounds on carrier material impregnated with chemicals (indicators) which change color when a critical amount of the gas or vapor is adsorbed. The German company Dräger has developed a number of devices for the detection of toxic components in air, and they are mainly used in industrial hygiene applications.

Reversible gas-absorbing materials are used in modern trace-gas analysis for enrichment (activated carbon, molecular sieves, porous polymers etc.).

Adsorbing carrier materials impregnated with special chemicals producing color changes as a function of gas concentration are used in rapid screening of toxic gas traces in industry and environment.

Questions and problems

1. Write down and describe *Gibbs' phase rule*!
2. By using the phase diagram of water, explain the status *supercritical fluid*.
3. Under which conditions is it correct to calculate the solubility of a salt in water from the solubility product alone? Compare the salts AgCl and $CaCO_3$.
4. How can gas–solid systems be exploited for rapid analytical measurements?
5. Which side reactions can significantly enhance the solubility of a salt in water:
   (a) in the pH range 1–14?
   (b) in the presence of foreign ions?
   (c) in the presence of a salt forming ion?
6. Write down the simultaneous equilibria which govern the enrichment of metal-ion traces from aqueous solutions into organic solvents using organic chelate forming ligands HA.

5 Chromatography

Learning objectives

■ To understand the basic principles to separate substances by partitioning them between a stationary and mobile phase

■ To learn about the most important chromatographic methods, i.e., gas chromatography (GC), high-performance liquid chromatography (HPLC), thin-layer chromatography (TLC), and ion chromatography (IC)

■ To introduce some of the newer methods, such as supercritical fluid chromatography, capillary electrophoresis, and field flow fractionation

Efficient separation of chemically similar compounds is only possible through repeated use of separation steps. The consecutive use of separation steps – according to the *Craig distribution*, for example – is, however, only suitable in practical terms for preparative purposes.

Efficient multiplicative separations can be undertaken using chromatographic methods. It is estimated today that approximately 60% of all analyses worldwide can be attributed to chromatography.

The Russian botanist, Michail Tswett, introduced column chromatography in 1906. Using a packed column containing finely dispersed calcium carbonate, he was able to separate extracts of leaf pigments, such as chlorophylls and the xanthophyll spirilloxanthin. When doing so, he observed colored zones on the column which prompted him to create the word *chromatography* (Greek *chroma* for "color" and *graphein* for "writing").

5.1.1 Overview

The principle of chromatography is based on the passage of the constituents to be separated between two immiscible phases. For this, the sample is dissolved in the *mobile phase*, which can be a liquid, a gas, or a supercritical phase, and moved across a *stationary phase*, which can either be in a column or on a solid surface. Due to the interactions of the constituents with the stationary phase they separate after sufficient running time.

Depending on the type of development of a chromatogram, a distinction is made between internal and external chromatograms. In the case of an *internal* chromatogram, the various constituents in the sample travel varying distances within the same time. At the end of separation they are still located within the separation bed and are detected there. This type of chromatography is used typically in the *planar* technologies, such as paper and thin-layer chromatography (Sec. 5.3.4). The stationary phase is located on a plate and the mobile phase is moved via capillary forces or by the influence of gravitation across the stationary phase.

External chromatograms are produced in *column chromatography*; that is to say, in gas, high-performance liquid, or in fluid chromatography. Here all the constituents travel the same route through the separation bed and due to the specific interactions with the stationary phase, they appear at the end of the column at different times, where they can be detected externally.

Classification of column chromatographic methods according to the mobile or stationary phase is set out in Table 5.1-1. The two most important separation principles in the passage between the mobile and the stationary phase are *distribution* and *adsorption*. Adsorption chromatography is based on the direct interaction of the analyte with the surface of the stationary backup phase, for instance in GSC or LSC. Distribution chromatography is linked to the presence of an immobilized liquid stationary phase (GLC, LLC).

chlorophyll a

Spirilloxanthin

Table 5.1-1. Classification of column chromatography according to the mobile and stationary phases used

Stationary phase		Mobile phase	

Solid	*gaseous*
	GSC
	(gas-solid
	  chromatography)
		fluid
		SFC
		(supercritical fluid
		  chromatography)
			liquid
			LSC
			(liquid-solid
			  chromatography)

Liquid	GLC
	(gas-liquid
	  chromatography)
			LLC
			(liquid-liquid
			  chromatography)

Liquid chromatography can be undertaken either on a column or on a plane. Thus the physicochemical basic principles are also valid for methods of planar chromatography. Gas chromatographic methods are limited to column chromatography.

The following reflections on the theoretical fundamentals of chromatography concentrate on *column chromatography*.

5.1.2 The development of a chromatogram

Other methods, such as *frontal* and *displacement chromatography* is in addition to elution chromatography. In the frontal technique, the sample is continuously fed in until the break-through of the sample mixture. This can be used for purification processes – to remove hardening constituents from water or to dry solvents, for instance.

The preeminent method in column chromatography is the *elution technique*. In this technique the sample dissolved in the mobile phase is introduced at the head of the column. Then using the mobile phase, elution is undertaken until the substances to be separated can be detected at the end of the column. Figure 5.1-1 clarifies the principle of elution chromatography using the example of separating two substances A and B.

Once the sample has been injected, the constituents distribute themselves between

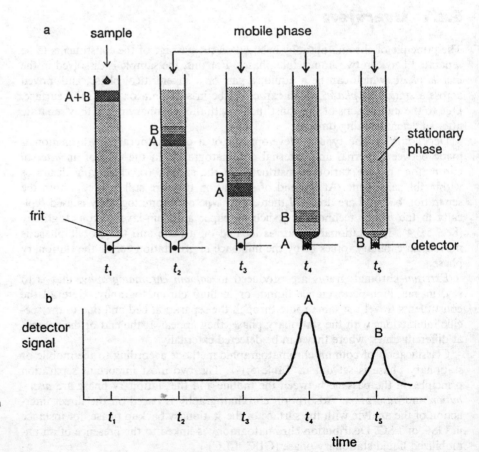

Fig. 5.1-1. Separation of two substances A and B using elution chromatography. (a) Development of the internal chromatogram on the stationary phase. (b) External chromatogram which is monitored by the detector

the mobile and the stationary phase. If the mobile phase is then continuously supplied in the form of the *eluent*, the substances distribute themselves along the column between the new mobile and the stationary phase. Compounds retained more strongly on the stationary phase take longer to be separated than substances which are less strongly interactive. Ideally, the substances are separated after a certain elution time and are individually detected at the end of the column.

In column chromatography, the recording of the detector signal as a function of elution time or elution volume is termed a *chromatogram*. If one follows the zones of substances along the column, then one notices two effects (see Fig. 5.1-1). The distance between the substance peaks increases. However, at the same time, the peaks broaden out, causing some of the separation to be foiled. Separation of the constituents in a chromatogram can accordingly be improved in principle if:
(a) either the *migration rates* of the substances are selectively altered or
(b) the *peak broadening* is kept as low as possible.

The quantities influenced by the migration rate and the peak broadening are dealt with in detail in the sections below.

The term eluent *is synonymous with mobile phase.*

5.1.3 Characteristic values of a chromatogram

Let us start with the *migration rate* with which the particles travel through the column. In the simplest case, the passage of the substances between the mobile (M) and the stationary (s) phase is determined by a partition equilibrium. One can obtain the partition coefficient, K, for this equilibrium using the concentration for the stationary phase, c_s, and for the mobile phase, c_M, corresponding to

$$K = \frac{c_s}{c_M} \tag{5.1-1}$$

The partition coefficient cannot be directly deduced from the chromatogram. The *total retention time*, characterized by the symbol t_R, is immediately readable. We have already used it tacitly in Fig. 5.1-1b. To elucidate the processes a chromatogram is outlined again in Fig. 5.1-2.

The small peak at retention time, t_M, arises from a compound which is not retained at all. Here M stands for mobile; i.e., the retention time, t_M, (*mobile time* or better *hold-up time*) exactly corresponds to the time required for the molecules in the mobile phase to pass through the column. Commonly, it is just described as *dead time*. However, dead time includes the total time from the actual injection point to the actual point of detection.

By using the retention time, the average linear velocity of analyte migration, $\bar{v}$, and u for the molecules of the mobile phase result, corresponding to:

$$\bar{v} = \frac{L}{t_R} \tag{5.1-2}$$

$$u = \frac{L}{t_M} \tag{5.1-3}$$

with L – length of the column packing.

How can the relationship between the retention time and the partition coefficient be established? For this let us look again at the elution of the substances in the column. Since only the mobile phase travels, the retention of a substance

The partition coefficient *describes the equilibrium constant for the distribution of a solute between two immiscible phases. It is also called a* partition ratio.

The difference between total retention and hold-up time is termed adjusted retention time, *i.e.* $t_R' = t_R - t_M$.

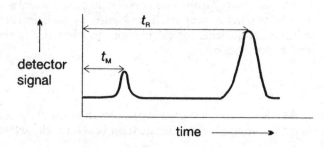

Fig. 5.1-2. Elution chromatogram for a sample consisting of two constituents. The first peak on the left in the chromatogram at retention time, t_M, corresponds to a compound which is not held back at all

corresponds to its *residence time* in the mobile phase. Constituents not retained are located for the whole time in the mobile phase. Substances which interact with the stationary phase remain for only a fraction of the time in the mobile phase compared with the unretained substances. This portion of time can be described using the relationship between the mass of the analyte in the mobile phase and the total mass of the analyte in the column:

$$\bar{v} = u \, \frac{c_M V_M}{c_M V_M + c_s V_s} = u \, \frac{1}{1 + \dfrac{c_s V_s}{c_M V_M}} \tag{5.1-4}$$

In Eq. 5.1-4 the rate of analyte migration is expressed as a *fraction* of the velocity of the mobile phase.

The masses are expressed in Eq. 5.1-4 by the product of the concentration and volume of the mobile or rather the stationary phase.

When the *partition coefficient, K*, is introduced into Eq. 5.1-4 a direct relationship with the migration rate of the substances is obtained:

$$\bar{v} = u \, \frac{1}{1 + K \dfrac{V_s}{V_M}} \tag{5.1-5}$$

The volumes of the stationary and the mobile phase arise from the concrete experimental conditions.

Recently the capacity factor is termed *retention factor* labeled k.

The expression in the denominator in Eq. 5.1-5 is summarized in chromatography as a new quantity, the *capacity factor, k',*:

$$k' = K \, \frac{V_s}{V_M} \quad \text{or} \quad k' = \frac{K}{\beta} \tag{5.1-6}$$

The ratio V_M/V_s is referred to as the *phase ratio β*.

Using the capacity factor k' we can now also establish the connection with the retention time. Insertion of the Eq. 5.1-6 into the equation for migration rate (5.1-5) produces:

$$\bar{v} = u \, \frac{1}{1 + k'} \tag{5.1-7}$$

If the migration rates are replaced by the corresponding equations, (5.1-2) and (5.1-3), then one obtains:

$$\frac{L}{t_R} = \frac{L}{t_M} \, \frac{1}{1 + k'} \tag{5.1-8}$$

After rearrangement the following equation results for the relationship between the capacity factor as a measure for the partition coefficient and the adjusted retention time:

$$k' = \frac{t_R - t_M}{t_M} = \frac{t'_R}{t_M} \tag{5.1-9}$$

The capacity factor can be established directly from the chromatogram based on the total retention times of the constituents of interest and the hold-up time (cf. Fig. 5.1-2). Capacity factors should conveniently be within the range 1 to 5. If the capacity factor is distinctly less than 1, then the compounds are being eluted too quickly, since their retention time hardly differs from the retention time in the mobile phase. If the capacity factor is well over 20, then in practice there are often intolerably long retention times.

The *selectivity factor*, also called *separation factor*, is a measure of the separation of two substances and is indicated by an α in chromatography. Using the partition coefficients for calculating the selectivity factor for two substances, A and B, one obtains:

$$\alpha = \frac{K_B}{K_A} \tag{5.1-10}$$

As we only have the capacity factors from the chromatogram at our disposal, it is better for us to write the equation below for the selectivity factor, by incorporating Eq. 5.1-6:

$$\alpha = \frac{k'_B}{k'_A} \qquad (5.1\text{-}11)$$

Determination of α from an experimental chromatogram is feasible by substitution of the capacity factors in Eq. (5.1-11) by Eq. (5.1-9) revealing the ratio of adjusted retention times:

$$\alpha = \frac{(t'_R)_B}{(t'_R)_A}$$

Quantities derived for determining the resolution of a chromatogram using the capacity and selectivity factors will be dealt with in the sections below.

5.1.4 Chromatographic theory

Let us now look at the second effect – the peak *broadening* along the column. The width of a peak is in direct relation to the separation efficiency or *column efficiency*. This relation results from the classic theory of chromatography.

Column efficiency describes the degree of broadening of a chromatographic peak.

Classic theory

The *classic theory* of chromatography can be understood as the logical transposition of discrete partition steps on to a column. For this purpose, Martin and Synge (Nobel Prize 1952) introduced the *plate height* (i.e., the height equivalent to a theoretical plate) and the *number of theoretical plates*. According to this, on every theoretical plate equilibration for the substance takes place between the stationary and the mobile phase. If the substance moves down the column, this signifies a gradual passage from one separation stage, or rather, equilibrated mobile phase, to the next.

The *number of theoretical plates*, N, results from the height equivalent to the theoretical plate, H, and the length of the column, L, as:

$$N = \frac{L}{H} \qquad (5.1\text{-}12)$$

The relationship between the variance of the peak and the migration distance (column length, L) or retention time, t_R, can be understood as the *height equivalent to the theoretical plate*, H, or shortly *plate height*:

$$H = \frac{\sigma_L^2}{L} \qquad (5.1\text{-}13)$$

$$H = \frac{\sigma_t^2 L}{t_R^2} \qquad (5.1\text{-}14)$$

The variance σ_L^2 relating to the column length is given in cm^2 and the variance σ_t^2 is expressed as the retention time in s^2. Since the length of the column is measured in cm, the dimension of the plate height H is in cm in both equations.

Variance corresponds to a Gaussian distribution as the square of the standard deviation σ.

The smaller the plate height, H, the higher the column efficiency and thus the resolution obtainable.

In practice one uses the chromatogram directly to determine the number of theoretical plates. For this, one approximates the peak width at half-peak height, $b_{1/2}$, using the base-line width, w. The base-line width can be determined from the intersection of the tangents of the inflection points with the base line of the Gaussian curve (Fig. 5.1-3).

We know from the parameters of a Gaussian partition that in a standard deviation of $\pm 2\sigma_t^2$ approximately 96% of the area lies under a Gaussian curve. Measured as retention times, the results for the standard deviation on the time axis corresponds to $\pm 2\sigma_t^2$, or rather just $w = 4\sigma_t^2$ for the basis width. If one inserts the

Chromatographic peaks are also termed chromatographic *bands* or *zones*.

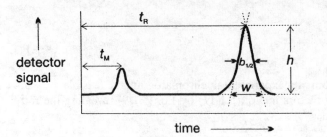

Fig. 5.1-3. Determination of the standard deviation σ as a measure of the peak width from a chromatogram above the base line width $w = 4\,\sigma_t^2$, or the half-band width $b_{1/2}$ (width of peak at half-peak height h)

latter value into the equation for the plate height (5.1-14), then one obtains:

$$H = \frac{w^2}{16} \frac{L}{t_R^2} \tag{5.1-15}$$

Transposition according to the *number of plates* of interest provides:

$$N = 16 \left(\frac{t_R}{w}\right)^2 \tag{5.1-16}$$

Plate height and number of theoretical plates characterize the *efficiency of a column*.

Thus the theoretical number of plates can be determined from a chromatogram by measuring the retention time and the base width of the peak. Better results are often obtained with a modified equation which uses the width at half-maximum, $b_{1/2}$, of the peak (cf. Fig. 5.1-3):

$$N = 5.54 \left(\frac{t_R}{b_{1/2}}\right)^2 \tag{5.1-17}$$

The equation of plate heights and the number of plates are routinely employed in assessing chromatographic separation. However, plate theory merely represents an approximation of the processes actually occurring in the column. Thus the repeated establishment of separate equilibria is often unrealistic, since equilibrium is hardly ever attained due to the movement of the mobile phase. When comparing columns using plate theory and the number of plates, one should always use the *same sample substance*.

Kinetic theory

The peak broadening derives from a *kinetic effect* which is occasioned by the finite rate at which mass-transfer processes occur during migration of the analyte down a column. The extent of this effect depends on the length of possible passages between the mobile and the stationary phase and is thus directly in proportion to the flow rate of the mobile phase.

In order to describe these influences one has to investigate the plate height in dependence on the linear flow rate or simply *linear velocity* $\bar{u}$ in cm s^{-1}. Figure 5.1-4 illustrates this in liquid and gas chromatography. Important independent variables on column efficiency are summarized in Table 5.1-2.

As can be deduced from Fig. 5.1-4, the minimum for the plate height in LC is at

a Liquid chromatograpy   **b Gas chromatography**

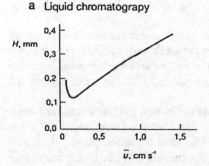
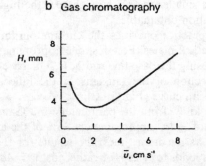

Fig. 5.1-4. Relationship between the number of theoretical plates, H, and the linear velocity, $\bar{u}$, in the mobile phase for liquid chromatography (a) and gas chromatography (b)

Table 5.1-2. Variables that affect column efficiency in chromatography

Variable	Symbol	Dimension
Linear velocity of mobile phase	$\bar{u}$	cm s^{-1}
Diffusion coefficient in the mobile phase	D_M	cm^2 s^{-1}
Diffusion coefficient in the stationary phase	D_s	cm^2 s^{-1}
Diameter of the packing material	d_p	cm
Thickness of the liquid coating on the stationary phase	d_f	cm
Desorption rate of the analyte	t_d	s
Column diameter	d_c	cm

Table 5.1-3. Kinetic influences on peak broadening (for explanation of symbols see Table 5.1-2 and end of table)

Influence	Term in Eq. 5.1-19
Longitudinal diffusion	$\dfrac{B}{\bar{u}} = \dfrac{2k_D D_M}{\bar{u}}$
Mass transfer to and from the liquid stationary phase	$C_s \bar{u} = \dfrac{qk' d_f^2 \bar{u}}{(1+k')^2 D_s}$
Mass transfer to and from the solid stationary phase	$C_s \bar{u} = \dfrac{2t_d k' \bar{u}}{(1+k')^2}$
Mass transfer in the mobile phase	$C_M \bar{u} = \dfrac{f(d_p^2, d_c^2)\bar{u}}{D_M}$

k_D, q – constants; f characterizes a functional dependence.

lower flow rates than in GC. However, this minimum is frequently not observed at all for the customary flow rates in LC. According to the illustration, smaller plate heights can in principle be attained more in LC than in GC. This advantage is however spoiled by the length of the column that can be achieved in practice. While one can only work in LC using columns of between 25 and 50 cm maximum due to the required pressures, columns as long as 100 m can be employed in GC.

Several models have been suggested to mathematically approximate the $H(\bar{u})$-function in Fig. 5.1-4. Originally, an equation developed in the engineering field was transferred to the chromatographic process – the *van Deemter equation*:

$$H = A + \frac{B}{\bar{u}} + C\bar{u} \tag{5.1-18}$$

Here the plate height dependent on the linear flow rate in the mobile phase is described by the constants A as a measurement for the Eddy diffusion, B for the longitudinal diffusion, and C for the mass transfer between mobile and stationary phase. The constant A describes the distance a flowing stream moves before its velocity is seriously changed by the packing. It is also termed the *Eddy diffusion* analogous to the streamlined shape of a turbulent current.

The equation below shows a more detailed model:

$$H = A + \frac{B}{\bar{u}} + C_s\bar{u} + C_M\bar{u} \tag{5.1-19}$$

The constants B, C_M, and C_s correspond to the longitudinal diffusion, to the mass transfer coefficient of and towards mobile phase, as well as to the mass transfer coefficient of and towards the stationary phase, respectively. The contributions of the individual members in Eq. 5.1-19 are explained in more detail in Table 5.1-3 and illustrated in Fig. 5.1-5. In capillary chromatography the A-term is zero since no multitude of flow exists.

Just as in the van Deemter equation (5.1-18), Eq. 5.1-19 provides a minimum for the plate height at an optimum rate. The B term describes the influence of *longitudinal diffusion*. This influence is created by the diffusion of the particles away from the peak centre towards or rather against the direction of flow of the

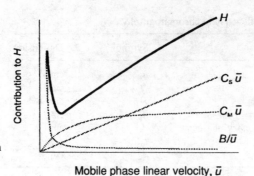

Fig. 5.1-5. Contributions of the diffusion term $B/\bar{u}$ and of the mass transfer terms $C_M\bar{u}$ and $C_s\bar{u}$ to the plate height H

The equilibrium constant for the distribution of a substance between two immiscible solvents is the distribution coefficient.

mobile phase. The B term is the only term which is independent of the particle size of the stationary phase. It is directly proportional to the diffusion coefficient of a substance in the mobile phase and thus increases with decreasing relative molar mass. The B term is hardly of any interest in the rates of flow used in practice in liquid chromatography. It is important, however, in gas chromatography.

The influence of the longitudinal diffusion on the plate height is inversely proportional to the linear velocity; that is to say, the diffusion decreases with increasing flow rate of the mobile phase. By contrast, the plate height increases with increasing linear velocity due to mass transfer effects. This effect arises from the fact that a certain time is required for mass transfer between the phases. Often it is not long enough in the flowing system to reach the state of equilibrium, so that the inhibited mass transfer becomes more evident as the rate of flow increases.

The *mass transfer term in the mobile phase*, $C_M\bar{u}$, represents the convective component of flow dispersion. Under certain limiting conditions it corresponds to the Eddy diffusion in the van Deemter equation. This mass transfer term is inversely proportional to the diffusion coefficient in the mobile phase and is directly dependent on the particle diameter of the packing material, as well as on the column diameter (Table 5.1-3).

For the term, $C_s\bar{u}$, which describes the *mass transfer to and from the stationary phase*, two different cases must be looked at, depending on whether one is dealing with an immobilized liquid or solid stationary phase.

In the case of a liquid phase a distribution equilibrium dominates. The effect increases with the increasing thickness of the liquid film and with decreasing diffusion coefficient in the stationary phase. The thicker the film and the smaller the diffusion coefficient, the slower the substances reach the interface for the mass transfer.

If the stationary phase consists of a solid substance, then the mass transfer coefficient C_s depends on the rate of the adsorption and desorption processes.

In principle, in chromatography, one is striving for *efficient separations in short analysis times*. Following our theoretical reflections, this means that a low plate height (minimum H) should be reached and that the $H(\bar{u})$ function should proceed as evenly as possible, since high separation rates can then be guaranteed without losing efficiency. In principle low H_{min} values and even $H(\bar{u})$ curves can be obtained by:

(a) small particle sizes of solid or low film thicknesses of liquid coating of stationary phases
(b) homogenous packing of the stationary phase using closely distributed packing materials
(c) small column diameter, which over time has led to columns getting increasingly narrower
(d) large diffusion coefficients in the stationary phase and small diffusion coefficients in the mobile phase. In GC diffusion coefficients in the mobile phase can be distinctly reduced by lower temperatures.

Since the diffusion coefficients for various molecular sizes vary, the broadening of a peak also depends on the relative molar mass. Small molar masses favour column efficiency. The molar masses are however determined by the substances to be separated, so that they cannot be freely selected.

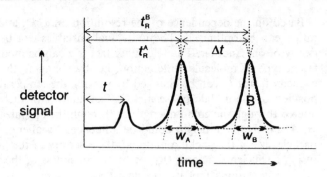

Fig. 5.1-6. Chromatogram for a mixture of the constituents A and B. The chromatographic resolution R_s can be determined according to Eq. 5.1-20 from the data evaluated

5.1.5 The resolution R_s as a measure of peak separation

We are already familiar with the selectivity factor (Eq. 5.1-11) for evaluating the selectivity of two constituents in chromatography. However, the selectivity factor as a ratio of the partition coefficients or of the capacity factors only describes the selectivity of the phase system used. We are aware from chromatographic theory that the efficiency of a column also depends on the number of plates, N, and on the magnitude of the capacity factor.

We therefore use the chromatographic resolution R_s to characterize the selectivity of the whole system. It is calculated for the separation of two Gaussian peaks A and B using their base widths (cf. Fig. 5.1-6):

Resolution characterizes the ability of a chromatographic column to separate two analytes.

$$R_s = \frac{\Delta t}{\dfrac{w_A + w_B}{2}} = \frac{t_R^A - t_R^B}{w} \tag{5.1-20}$$

with $w \cong w_A \cong w_B$

The following extensions of Eq. 5.1-20 are undertaken in order to describe the resolution relationship by including the number of theoretical plates, the capacity factor, and the selectivity factor.

Peaks are frequently asymmetric due to *tailing*

Insertion of Eq. 5.1-16 for the number of theoretical plates gives:

$$R_s = \frac{t_R^A - t_R^B}{t_R^B} \frac{\sqrt{N}}{4} \tag{5.1-21}$$

Use of the capacity factor from Eq. 5.1-9:

or *overloading*

$$R_s = \frac{k_B' - k_A'}{1 + k_B'} \frac{\sqrt{N}}{4} \tag{5.1-22}$$

Introduction of the selectivity factor Eq. 5.1-11 and rearrangement leads to: *Chromatographic resolution*

$$R_s = \frac{\sqrt{N}}{4} \left(\frac{\alpha - 1}{\alpha}\right) \frac{k_B'}{1 + k_B'} \tag{5.1-23}$$

The equation for resolution is often formulated in simplified form in the event of the capacity factors being very similar and thus $k_A' = k_B' = k'$, so to give:

$$R_s = \frac{\sqrt{N}}{4} (\alpha - 1) \frac{k'}{1 + k'} \tag{5.1-24}$$

When the resolution is known, the *number of theoretical plates* can then be calculated from the selectivity factor, and the capacity factor of the constituent, B:

$$N = 16 R_s^2 \left(\frac{\alpha}{\alpha - 1}\right)^2 \left(\frac{1 + k_B'}{k_B}\right)^2 \tag{5.1-25}$$

Based on the dependence of the resolution on α, k', and N (or rather H) separation can be optimized. The individual variables can be varied fairly independently of one another. The selectivity factor α can be modified by choosing a different type of molecular interaction, i.e. by changing the stationary phase. The capacity factor k' can be changed by varying the temperature in GC or the composition of the mobile phase in LC. The number of theoretical plates N can be altered through the column length or through the optimization of the plate height H. Again the plate height could be influenced by alteration of the flow rate, the particle size of the packing material, the viscosity of the phases, and thus the diffusion coefficients D_s and D_M, or by the thickness of the film of an immobilized liquid as a stationary phase (Table 5.1-3).

In the case of equally intensive, symmetric peaks one speaks of a resolution of $R_s = 1$ for base line separation or $4\,\sigma$ – separation. One can easily consider that in very differing intensive peaks or in case of asymmetric peaks higher resolution is necessary for complete separation.

5.1.6 Qualitative information

Chromatographic methods can be used both for qualitative and quantitative analysis and also for preparative purposes. Special features in preparative chromatography are only referred to here in passing.

In an internal chromatogram *qualitative information* is located in the position of a substance on the stationary phase and in external chromatograms in the value for the retention time or the retention volume. Although the reproducibility of retention data is as a rule considerably less than, for instance, the wavelength precision in spectroscopy, through measuring the retention data compared to a standard substance, the presence or absence of a substance can still be relatively well-established. Coupling of chromatographic separation with spectroscopic detectors is undertaken to accurately identify a substance. Thus in GC (gas chromatography) mass spectrometric detectors are used or a UV diode-array detector in HPLC (high-performance liquid chromatography).

In qualitative chromatographic analysis of multicomponent mixtures one must always realize that the peak capacity of a column is limited. *Peak capacity* reflects the number of peaks which can be resolved in a lining-up of peaks on a defined spacing (Fig. 5.1-7). According to Giddings, in elution chromatography the peak capacity, n, can be calculated approximately according to the following equation:

$$n = 1 + \frac{\sqrt{N}}{4} \ln \frac{V_R^n}{V_R^1} \tag{5.1-26}$$

with V_R^1 and V_R^n signifying the retention volumes of the peak eluted first and last, respectively.

If the number of constituents present in a sample exceeds the peak capacity, then overlapping peaks under which two or more constituents elute are unavoidable. Table 5.1-4 shows typical peak capacities for gas chromatography (GC), liquid chromatography (LC), and gel filtration (cf. Fig. 5.1-7).

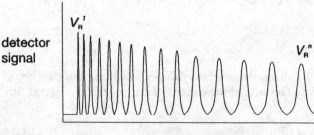

Fig. 5.1-7. Representation of the peak capacity of a column

Table 5.1-4. Typical peak capacities for selected numbers of theoretical plates in GC, LC, and gel chromatography after Giddings

Number of theoretical plates N	Peak capacity n		
	Gel	GC	LC
100	3	11	7
400	5	21	13
1000	7	33	20
2500	11	51	31
10000	21	101	61

5.1.7 Quantitative analysis

The basis of *quantitative analysis* in column chromatography is the evaluation of the height or the area of a peak. In internal chromatograms the complete intensity of a substance spot can be measured, e.g. in thin-layer chromatography (TLC). The chromatographic methods are relative methods; that is to say, a calibration takes place through the analysis of standard substances. It is customary to use both external and internal standards.

When evaluating *peak height* it must be guaranteed that no alterations in peak form arise due to changing chromatographic conditions, otherwise the peak height would then no longer be in direct proportion to the concentration. Variables, such as column temperature, flow rate, and the volume of injection must be precisely controled. In addition, overloading of the column must be avoided.

Peak broadening is of no significance when the *peak area* is being evaluated. Computer aided numerical integration is used for this. Peak areas can be approximated by hand by evaluating the product of the height and the half-band width of the peak (area $= h \, b_{1/2}$, see Fig. 5.1-3). However, problems in area determination can arise if very narrow peaks have to be measured.

Problems with numerical integrators arise from locating the exact positions for the beginning and end of a peak. This is reasoned by the noise of the base-line (cf. Sec. 12.3).

Worked example: Chromatographic parameters

The length of a chromatographic column is to be computed for the retention of two compounds A and B with a selectivity factor of $\alpha = 1.05$, at a resolution of $R_s = 1$, a capacity factor of the second peak of $k'_B = 2.0$ and a plate height of $H = 1$ mm.

From Eq. 5.1-12 we know, how to calculate the length of a column on the basis of the plate height and the number of theoretical plates. Rearranging of this equation leads to:

$$L = N \cdot H$$

In this equation the number of theoretical plates N is not given. However, this can be estimated on the basis of the selectivity factor, the resolution and the peak width as follows. The plate height depends on the average retention time of the two peaks and on their width (Eq. 5.1-16):

$$N = 16 \left(\frac{t_R}{w} \right)^2$$

The average width of the peaks is obtained from Eq. (5.1-20) according to:

$$R_s = \frac{t_R^A - t_R^B}{w}$$

The missing retention time of the first peak A can be derived from the given capacity factor and the selectivity factor α according to Eq. 5.1-11:

$$k'_A = \frac{k'_B}{\alpha}$$

Since no hold-up time is given the ratio of the capacity factors corresponds directly to the ratio of corrected retention times, i.e.

$$\alpha = \frac{k'_B}{k'_A} = \frac{t_R^B - t_M}{t_R^A - t_M}$$

with $t_M = 0$ or at least $t_M \ll t_R^A$. Thus, we can use the capacity factors in the sense of corrected retention times in order to insert those into the above mentioned equations. The detailed calculations reveal:

$$k_A' = \frac{k_B'}{\alpha} = \frac{2}{1.05} = 1.90$$

$$w = \frac{t_R^A - t_R^B}{R_s} = \frac{2.00 - 1.90}{1} = 0.1$$

$$N = 16 \left(\frac{t_R}{w}\right)^2 = 16 \left(\frac{(1.90 + 2.00)/2}{0.1}\right)^2 = 6084$$

$$L = N \cdot H = 6084 \cdot 1\,mm = 6084\,mm \text{ or } 6.084\,m$$

As result one deduces that the length of the column needed should be about 6 m.

Questions and problems

1. Which separation functions are important for chromatographic processes and which fundamental physical and chemical effects are underlying these processes?
2. Explain the effects that cause peak broadening in chromatography.
3. Describe the relationship between the plate height and the velocity of the mobile phase in a chromatographic column.
4. Characterize relationships between the retention time and the actual equilibrium, column length and flow rate.
5. Which quantities can improve the resolution of a chromatographic separation and how can they be determined from a chromatogram?

In gas chromatography the compounds to be analyzed are vaporized and eluted with the aid of a gas as a mobile phase through the column. The mobile phase is used alone as a *carrier gas*, so that interactions of the mobile phase with the analyte are of no significance. A solid substance can serve as a stationary phase on which the constituents to be separated can be adsorbed. In practice, *gas-solid chromatography* (GSC) or adsorption chromatography are especially important for the analysis of air gases and will be explained at the end of the section. The use of a liquid as a stationary phase is preponderant for the analysis of organic compounds. It is termed *gas-liquid chromatography* (GLC) or simply gas chromatography. The predominant separation principle is the partition of substances between the liquid stationary phase and the gaseous mobile phase.

5.2.1 Retention data and partition coefficient

Before we discuss the possibilities of substance separations in the gas phase, we will first look at the factors which determine retention in gas chromatography. In order to be able to take into consideration the influence of the pressure and the temperature, *retention volume* is employed instead of retention time. The total retention volume, V_R and the mobile (hold-up) volume, V_M, is obtained by multiplying the relevant retention times by the carrier gas flow, F, (mL carrier gas per minute), to:

In practice the retention time is predominantly used.

$$V_R = F \cdot t_R \tag{5.2-1}$$

$$V_M = F \cdot t_M \tag{5.2-2}$$

Air or the methane peak can be used to determine the mobile time. In order to be able to correctly describe the retention of a constituent in the stationary phase, the mobile volume must be subtracted from the total retention volume. The corrected quantity is termed *adjusted retention volume*:

$$V_R' = V_R - V_M \tag{5.2-3}$$

The adjusted retention volume is characteristic of a substance and is directly linked via the volume of the stationary phase, V_s, with the partition coefficient, K,:

$$V_R' = KV_s \tag{5.2-4}$$

The resistance of the separating column causes greater gas pressure at the column inlet than at the column outlet. So a pressure difference exists along the column. Due to the compressibility of the carrier gas, the gas flow increases with rising pressure difference. In order to be able to describe the retention volume independently of the fall in pressure, the adjusted retention volume is corrected using the factor j (the Martin factor) and the *net retention volume*, V_N, is obtained:

$$V_N = j \cdot V_R' \tag{5.2-5}$$

with j being $j = 3[(p_i/p_0)^2 - 1]/2[(p_i/p_0)^3 - 1]$ and p_i and p_0 characterizing the inlet and the outlet pressure of the column, respectively.

The *specific retention volume*, V_g, which is only dependent on the net retention volume, V_N, of the mass of the stationary phase (in g), W_s, and on the column temperature T (in K) can be inferred using the net retention volume:

$$V_g = \frac{V_N}{W_s}\frac{273}{T} \tag{5.2-6}$$

The specific retention volume is least influenced by the conditions of analysis. However, it takes a lot of trouble to determine it, so that for comparative examinations the corrected retention volume or relative retention values are in practice considered preferentially.

5.2.2 Separations in the gas phase

The number of theoretical plates N and the plate height H in kinetic theory are used to evaluate the separation efficiency of various columns.

Longitudinal diffusion (the B term in Eq. 5.1-19) was seen to be a peculiarity more pronounced in gas chromatography compared with other chromatographic processes.

For the diffusion coefficients in gases and liquids see Table 5.4-1.

The reasons for this are the high diffusion coefficients in gases which are approximately 10^4 times greater than in liquids. Thus the minimum in the $H(\bar{u})$ function is considerably flatter and is located at higher linear velocities in the mobile phase.

However, the parameters of the plate theory and of the kinetic models in chromatography are not sufficient for deducing the suitability, i.e., the resolving power of a column for a concrete problem of analysis. For this, one must also include as an additional factor the *separation efficiency* of a column. In gas chromatography this depends on the vapour pressure of the compounds and the degree of interaction with the stationary phase.

Separation efficiency can be inferred from the Raoult and Henry laws and is known as the separation formula after Herington:

$$\lg\frac{V_{g_2}}{V_{g_1}} = \lg\frac{p_1^0}{p_2^0} + \lg\frac{\gamma_1^0}{\gamma_2^0} \tag{5.2-7}$$

with V_{g_2} and V_{g_1} signifying the specific retention volume for the constituents 2 and 1, respectively. If one looks at the retention times for two constituents to be separated, then the following applies:

$$\frac{t_{R_2}}{t_{R_1}} \approx \frac{p_1^0\gamma_1^0}{p_2^0\gamma_2^0} \quad \text{or} \quad \lg\frac{t_{R_2}}{t_{R_1}} \approx \lg\frac{p_1^0}{p_2^0} + \lg\frac{\gamma_1^0}{\gamma_2^0} \tag{5.2-8}$$

The relationship between the net retention times of compounds 1 and 2 is proportional to the vapour pressure of the pure constituents, p_1^0 and p_2^0 as well as to the activity coefficients of the constituents, γ_1^0 and γ_2^0, in infinite dilution.

According to this, the separation of two constituents is determined once by their *relative volatility*. Naturally, the vapour pressures are dependent on the temperature. On the other hand, the *activity coefficients* are the expression of the interactions between the constituents of the analyte and the stationary phase. Thus they determine the selective behaviour of the stationary phase for which there are approximately 1000 liquid phases available.

The differences between the vapour pressures form the basis for the separation of chemically related compounds such as, for example, the members of an homologous series. Substances with the same boiling points can be separated based on differing activity coefficients. Unwanted azeotropes from distillatory separations, which we are familiar with, are not of any importance here.

5.2.3 Components of a gas chromatograph

The modules of a gas chrómatograph are illustrated in Fig. 5.2-1. The differences in gas chromatographic systems lie in the type of carrier gas used, in the sample injection system, as well as in the columns and detectors used.

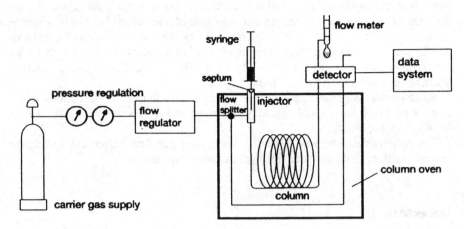

Fig. 5.2-1. Schematic diagram of a gas chromatograph

Carrier gases

Inert gases, such as helium, argon, nitrogen, carbon dioxide, and hydrogen, are used as carrier gases. The choice of carrier gas is in part determined by the detector. The gas can be fed through a molecular sieve to remove traces of water. The flow of gas is ensured by the excess pressure of the gas cylinder, so that one can manage without a pump. The carrier flow must be kept constant for reproducible measurements.

As long as one is working with a column *isotherm*, it is sufficient to set the supply pressure of the column using a two-stage reduction valve. In the case of a method *programmed by temperature*, or in column switching, a flow regulator must be used in addition on account of the changing liquid resistance. A rotometer can be used at the column inlet, or a soap-bubble flow meter at the column outlet to measure the rate of flow. The carrier gas varies for packed columns in the range between 25 and 150 mL min^{-1}, as well as for capillaries between 1 and 25 mL min^{-1}.

Sample injection system

The sample can be injected straight into the carrier gas flow (up to 20 µL sample volume) when analyzing gaseous samples. Liquid and solid samples must first be vaporized in a flash vaporizer port. The injection system can be heated and connects the flow of the carrier gas with the column (cf. Fig. 5.2-1). It is sealed on the outside with a silicone rubber diaphragm called the *septum*.

The sample is injected into the system be puncturing the septum. The injection must be carried out in such a way that a plug of vapour is created. A slow injection causes broad peaks and chromatograms difficult to evaluate. Using packed columns one works with injection volumes of between 0.5 and 20 µL. In capillary GC, volumes of 0.001 µL are directed on to the column via a division of the gas flow in a split injection system. *Sample injection systems* which work automatically guarantee a reproducibility of up to a relative error of 0.5% during sample injection.

The evaporator temperature is typically approximately 50 °C above the boiling point of the least volatile constituent of the sample mixture.

Column packings and column ovens

Separation columns may be manufactured from stainless steel, glass, or fused silica tubes. Fused silica is gaining increasing acceptance as a column material.

The strength of columns is ensured by casing in polyimide. The columns are stored in an oven for preliminary heating. A distinction is drawn between packed columns and open tubular, or capillary columns. In *packed columns* the stationary phase is set up on a granular support material. This is fed by a funnel into the column which has an internal diameter of between 3 and 8 mm and is 1 to 3 m long. *Capillary columns* contain no carrier material. Here the liquid stationary phase is discharged on the wall of the capillaries. For example, the liquid reaches the column via a plug of concentrated solution of the stationary phase which is pressed through the column at an even speed. Another way consists of evaporating the solvent using a vacuum pump from a column completely filled with a very well diluted solution. Capillaries can be up to 100 m long. They have an internal diameter of between 0.15 and 1 mm.

Short packed columns can be stored straight or u-shaped in the column oven. Longer packed columns and capillaries are arranged as helices with a diameter of between 10 and 30 cm.

The column must be well-heated in the carrier gas flow before use to remove solvent residues or to activate silica gel or molecular sieves.

Detectors

The thermal conductivity detector (TCD) and the flame-ionization detector (FID) have gained acceptance as the universal detectors in gas chromatography. The mass spectrometric detector (GC-MS) is increasingly employed for compound specific detection. In addition, other detection principles of interest are those which permit selective detection or show particularly good detection sensitivity. The most important detection principles will be described in more detail in the following text.

Thermal conductivity detector (TCD)

In this detector the thermal conductivity of the carrier gases helium and hydrogen is reduced in the presence of the analyte. The thermal conductivity of helium and hydrogen is approximately 6 to 10 times higher than for organic compounds. Other carrier gases cannot be used in this detection principle, since the difference in the thermal conductivity towards the substances to be detected is too slight. Thermal conductivity is established by measuring the resistance on a heating filament (Fig. 5.2-2). The measurement and the comparative current are compared with one another in a bridge circuit. The TCD works *proportionally to concentration*. Since TCD reacts nonspecifically, it can be used universally for the detection of both organic and inorganic substances.

As can be deduced from Table 5.2-1, the detection capacity and the linear (dynamic) range for TCD are worse compared to other detectors. It is not suitable for capillary GC on account of its low sensitivity. An advantage of TCD arises from its nondestructive character.

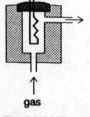

gas

Fig. 5.2-2. Thermal conductivity detector

Table 5.2-1. Detectors in GC

Detector	Species detected	Detection limit	Linear range
TCD	nonspecific	$10^{-8}\,\mathrm{g\,mL^{-1}}$	10^4
FID	compounds containing CH	$10^{-13}\,\mathrm{gs^{-1}}$	10^7
ECD	electronegative groups	$5 \times 10^{-14}\,\mathrm{gs^{-1}}$	$5 \cdot 10^4$
TID	P	$10^{-15}\,\mathrm{gs^{-1}}$	10^5
	N	$10^{-14}\,\mathrm{gs^{-1}}$	
FPD	P	$3 \times 10^{-13}\,\mathrm{gs^{-1}}$	10^5
	S	$2 \times 10^{-11}\,\mathrm{gs^{-1}}$	

TCD – thermal conductivity detector, FID – flame-ionization detector, ECD – electron capture detector, TID – thermionic detector, FPD – flame photometric detector

Flame-ionization detector (FID)

The FID is currently most frequently used. The detection principle is based on the change in the electric conductivity of a hydrogen flame in an electric field when feeding organic compounds. The organic compounds escaping from the separation column are pyrolyzed; that is to say, fragmented. During subsequent oxidation by oxygen which is fed into the flame from outside, ions are formed by the following reaction:

$$CH\cdot + O \rightarrow CHO^+ + e \qquad (5.2\text{-}9)$$

The flow of ions is recorded as a voltage drop across a collecter electrode (Fig. 5.2-3).

The FID is sensitive to all compounds which contain C–C or C–H bondings. Thus it can be generally employed. Considerably less sensitivity up to insensitiveness can be observed for functional groups such as the carbonyl, alcohol, halogen or amino groups. It is also insensitive to the nonflammable gases: H_2O, CO_2, SO_2, or NO_x.

Since the detector reacts to the number of carbon atoms per time unit, it is proportional to the *mass* of the substance detected. Thus changes in the flow rate of the mobile phase have only a slight influence on the detector signal. The FID is distinguished by a very low detection limit and a large linear range (Table 5.2-1). However sample constituents are destroyed.

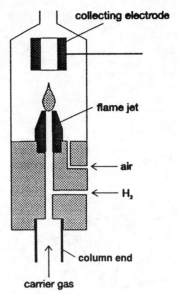

Fig. 5.2-3. Principle structure of a flame-ionization detector

Electron capture detector (ECD)

The electron capture detector (ECD) operates similarly to a proportional counter for measurement of X-ray radiation. Using β emitters, such as ^{63}Ni or tritium, ions and a burst of electrons are created in the carrier gas. In the absence of an analyte a constant standing current flows. This background current decreases in the presence of organic compounds and especially when these are compounds which can capture electrons. Amongst these compounds are those with electrophilic (electronegative) groupings, such as halogens, peroxides, quinones, phthalates, or nitro groups. Thus the ECD is used as a highly sensitive special detector for chlorinated insecticides, for example.

ECD is insensitive to amines, alcohols or hydrocarbons. Other performance characteristic values are listed in Table 5.2-1.

Special detectors

Thermionic detector (TID). The TID is used as a highly specific detector for compounds containing nitrogen and phosphorus (Table 5.2-1). Its sensitivity to both these elements is higher by about a factor of 10 000 compared to carbon. The TID is a flame detector with a very anhydrogenous, no longer flammable, gas mixture. A glass bead containing rubidium is hung on an incandescent platinum wire between the flame jet and the collecting electrode. A plasma is formed around the glass bead in which compounds containing N or P produce radicals, for example:

$$\stackrel{>}{_{>}}C\text{--}\overline{N}\stackrel{\frown}{\rightarrow} \cdot C\equiv N| \qquad (5.2\text{-}10)$$

The radicals react with the rubidium vapour around the glass bead to give:

$$\cdot C\equiv N| + Rb\cdot \rightarrow |C\equiv N|^- + Rb^+ \qquad (5.2\text{-}11)$$

While the alkali ion created is captured again by the negatively charged glass bead, the cyanide ion travels either directly to the collecting electrode or, when released, is burned by electrons.

Figure 5.2-4 shows radicals which play a role in the detection of phosphorus.

$$\cdot \overline{P}=\overline{\underline{O}} \qquad\qquad \overline{\underline{O}}=\dot{P}=\overline{\underline{O}}$$

Fig. 5.2-4. Phosphorus radicals in thermionic detection

Spectroscopic detectors

Flame photometric detector (FPD). The simplest type of spectroscopic detector for the selective indication of phosphorus and sulphur is the FPD. The organic compounds are partially combusted in a hydrogen/air flame, and the emission light thus produced is measured with a photomultiplier at 526 nm for P and at 394 nm for S.

Atomic emission detector (AED). The AED also operates by exploiting emission phenomena. It is an element specific detector and is based on the atomic emission of elements such as N, P, S, C, Si, Hg, Br, Cl, H, D, F or O. Atomization and excitation take place in a helium microwave plasma (cf. Sec. 8.1). Detection of the emission light is undertaken using a diode-array photometer in the wavelength range between 170 and 780 nm.

The precision of elemental analyzers has not yet been attained in AED, since relative errors are between 2% and 20%. Despite this, AED can be used to draw inferences regarding the relation of elements. The sensitivity of the detector is especially high for the elements C, P, and S. The dynamic range is considerably smaller than with the FID, for example.

Element specific detection is especially advantageous for the analysis of complex mixtures. Thus, for instance, alcohols can be selectively detected in a gasoline mixture by measuring on the atomic emission line of oxygen.

Mass spectrometric detector (MSD). Direct coupling of a mass spectrometer with a column is possible in capillary GC due to the slight flow rate of the carrier gas (1 to 25 mL min^{-1}). In the case of packed columns the carrier gas flow must be split across a jet *separator.*

The special features of coupling GC with MS as well as with IR spectroscopy will be discussed in Chap. 14, respectively.

5.2.4 Stationary phases for gas-liquid chromatography

High numbers of theoretical plates, good selectivity in the stationary phase, and a high degree of load-bearing capacity which is obtained by a high coating level with immobilized liquid are all prerequisites for efficient columns.

We have already drawn a distinction between packed and capillary columns in the structure of a gas chromatograph. When GC was introduced *packed columns* were exclusively used. For this the stationary phase is set up on a support. As a rule, packed columns are less than 5 m long. The absolute upper limit is 20 m, which occurs when the columns become cumbersome and the pressure drop, which is contingent on the tight packing, becomes too great. Thus the number of theoretical plates in packed columns is usually less than 10 000.

With the introduction of *capillaries* by Golay (1958) a start was made to overcome these limits. Since the stationary liquid phase is applied to the internal wall in capillaries, they have a free gas passage. So column lengths of up to 100 m and thus numbers of theoretical plates up to 100 000 and beyond can be realized. In addition, the volume of the mobile phase is very large in comparison with that of the stationary phase.

Support materials for packed columns

The ideal support material for the stationary phase consists of small, homogeneous, spherical particles which are chemically inert, thermically and mechanically stable, and have a specific surface area in the range between 0.5 and 4 m^2 g^{-1}. The carrier materials used are particles in size of about 149 to 250 μm (this corresponds to the American particle size measurement of 100 to 60 mesh).

Diatomite is the most frequently used carrier material. It consists of up to 90% amorphous silicic acid. Here one is dealing with specially processed skeleton sediments of diatoms of fossil origin. These algae transport their nutrients through

Diatomaceous earth represents the siliceous skeletons of micellular algae.

their pores by molecular diffusion. Thus the residues are excellently suited for GC, since the same kind of molecular diffusion is required in the carrier material. Depending on the specific surface area, the optimum load-bearing capacity with the liquid phase is between 5% to 30%.

There are other carrier materials on the market based on silica gel (cf. Sec. 5.3.1), porous glass beads, or on polymers (cf. Table 5.2-4).

packed column

thin-film capillary

thin-layer capillary

Capillary columns

The immobilization of the liquid stationary phase in a capillary can occur in two ways. It is either applied to the internal wall of the capillary as a thin film (WCOT – wall coated open tubular) or the liquid is located on a porous support approximately 30 μm in thickness (SCOT – support coated open tubular) (Fig. 5.2-5). Diatomite can again be used as the support material. The thin-layer capillary differs from the thin-film capillary due to a *higher sample capacity*, as considerably more stationary phase is available. However, the *efficiency* of the thin-layer capillary is less and is in-between that of packed and thin-film columns.

Fused-silica is the preponderant column material, which is produced from highly purified quartz with a very low metal oxide content.

A comparison of characteristics of packed columns and capillary columns is listed in Table 5.2-2.

Fig. 5.2-5. Columns in gas liquid chromatography

Table 5.2-2. Typical characteristics of packed and capillary columns in gas chromatography

Parameters	Packed	WCOT	SCOT
Length, m	1–5	10–100	10–100
Inside diameter, mm	2–4	0.1–0.75	0.5
Efficiency, N/m	500–1000	1000–4000	600–1200
Sample size, ng	$10–10^6$	10–1000	10–1000
Relative pressure	high	low	low

Deactivation of surfaces

The untreated carrier materials or capillary columns are a problem for the user, since residual silanol groups are still present on the surface of the silica gel (Fig. 5.2-6) which lead to the physical adsorption of polar compounds, such as alcohols or aromatic hydrocarbons. This makes itself apparent in the chromatogram by peak broadening or on account of delayed desorption by *peak tailing*.

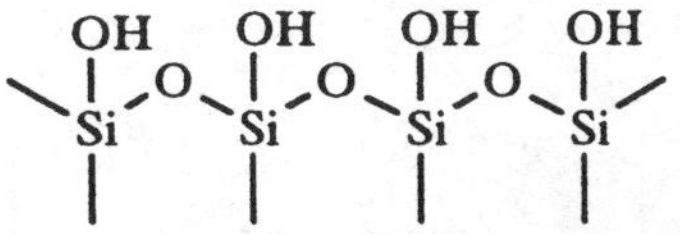

In order to avoid these unwanted interactions the residual silanol groups are deactivated using dimethylsilyl or trimethylsilyl groups. A hydrophobic surface is formed. For this let us observe the reaction of dichlorosilane to silica gel and a subsequent washing stage using methanol:

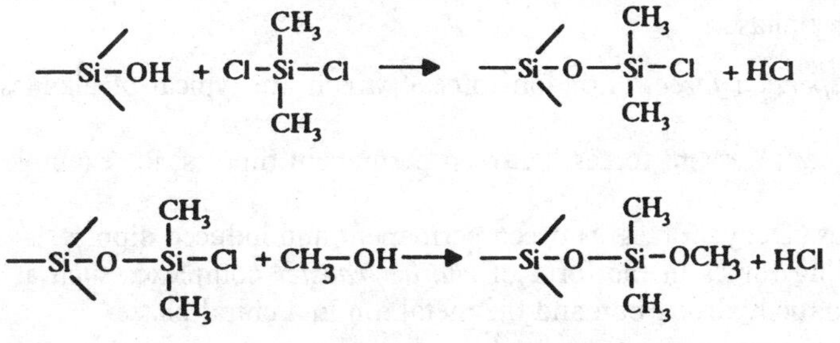

$$—Si—OH + Cl—\underset{\underset{CH_3}{|}}{\overset{\overset{CH_3}{|}}{Si}}—Cl \longrightarrow —Si—O—\underset{\underset{CH_3}{|}}{\overset{\overset{CH_3}{|}}{Si}}—Cl + HCl$$

$$—Si—O—\underset{\underset{CH_3}{|}}{\overset{\overset{CH_3}{|}}{Si}}—Cl + CH_3—OH \longrightarrow —Si—O—\underset{\underset{CH_3}{|}}{\overset{\overset{CH_3}{|}}{Si}}—OCH_3 + HCl$$

Residual active centres derive from contamination with metal oxides. The latter problem does not exist in fused silica capillaries, as they come on the market as highly purified materials.

Fig. 5.2-6. Fully hydrolyzed silica gel surface

Stationary phases

Cross-linking at a stationary phase is performed in-situ after the column is coated with a polymer. The columns become thermally more stable.

Liquids which are to be used as stationary phases in GC must be *thermically and chemically stable*. They must exhibit *low volatility* to prevent bleeding of the column. The boiling point of the separating liquid should be about 100 °C above the required column temperature. The separating liquid must show a certain *selectivity*, i.e., it must generate differing partition coefficients for various analytes. However, the partition coefficients must not be either too great nor too small, otherwise the retention of the compounds will correspondingly take too long or is so short that no separation can occur.

The chemical rule of thumb applies in the choice of separating liquid. Similia similibus solventur (similar material is dissolved by similar material).

Typical *polar phases* contain the functional groupings –CN, –C=O, –OH or polyester. They possess a marked selectivity for alcohols, organic acids, or amines. *Unpolar phases* are hydrocarbons or siloxanes, which are suitable for separating saturated or halogenated hydrocarbons. Analytes of medium polarity, such as ethers, ketones, or aldehydes, must be separated on correspondingly modified phases. Polar phases contain cyano-, trifluoro- or nitrile groups. Table 5.2-3 shows general families for separating liquids with their polarity and the appropriate temperature ranges. The formulas for selected representatives are listed in Fig. 5.2-7.

Table 5.2-3. Liquid phases as stationary phases in GC (selected formulas in Fig. 5.2-7)

Analytes	Phase	Temperature, °C	Polarity
Hydrocarbons	Squalan	20 … 150	unpolar
	Apolan-87	50 … 300	unpolar
Polyglycols	Polyethylene glycol (CARBOWAX)	50 … 225	polar
Ester	Ethylene glycol succinate	100 … 200	high polarity
	Diisodecyl adipate	20 … 125	medium polarity
Compounds containing N	1,2,3-Tris-(2-cyanoethoxy)-propane (CYANO B)	110 … 200	polar
Silicones	Methyl siloxanes (OV-1, SE-30)	20 … 300	unpolar
	Phenyl siloxanes (OV-22)		medium polarity
	Nitrile siloxanes (OE-4178)		high polarity

Chemically bonded phases represent a peculiarity. Here the separating liquid is covalently bonded by a chemical reaction to the surface of the carrier material (column filling or capillary inner wall). They exhibit higher temperature stability and there are no problems with bleeding of the column.

In principle, separating substances can already be produced using a suitable variation of substituents by means of which most separation problems can be solved (cf. Table 5.2-3).

The variety of interactions of analytes with the stationary phase make the choice of a suitable phase for a concrete problem of analysis more difficult. One must take into account the following as typical *interactions* between the analyte and the stationary phase:

- nonspecific *dispersion forces* (London forces), which are typical of alkanes or benzene
- *orientation forces* (Kessom forces) between permanent dipoles, for example, in hydrogen bonds
- *induction forces* (Debye forces) between permanent and induced dipoles
- chemical binding forces in the form of *charge transfer* complexes such as between an aromatic hydrocarbon and the metal ion in a chiral phase

Chiral phases are used to separate optical isomers (enantiomers). They can be derived, e.g., from optically active amino acids.

Rohrschneider, therefore, and later McReynolds, developed characteristic values which permit the typical interactions to be characterized using standard substances (benzene, ethanol, butanone, nitromethane, pyridine) and a phase to be selectively chosen based on this data. The basis for this is the additive assemblage of retention indices from increments for functional groups and for the type of bond.

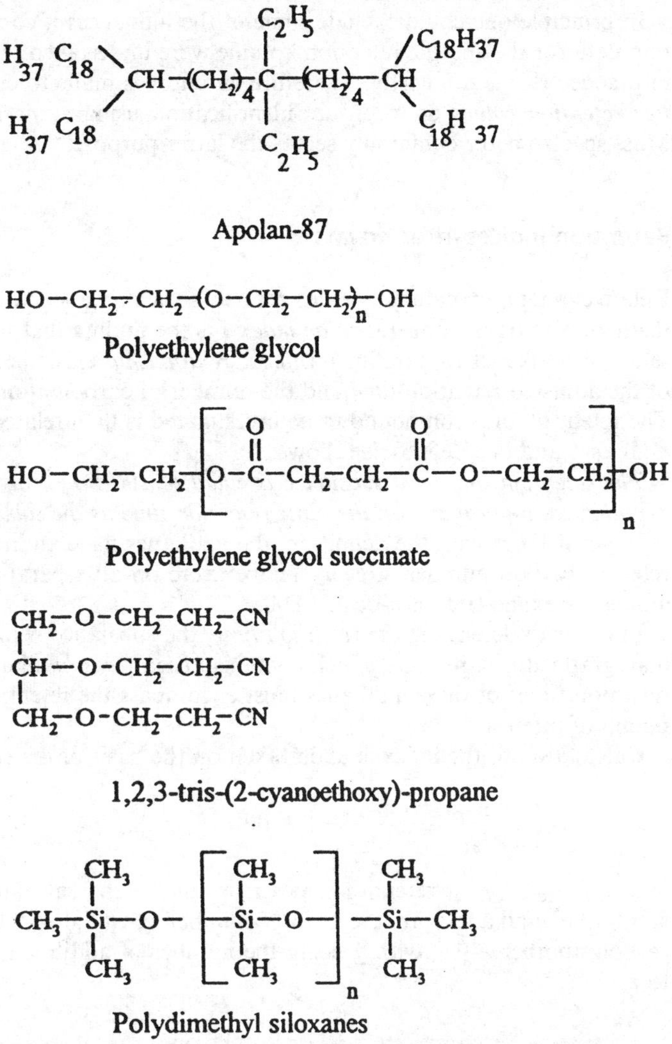

$$H_{37}C_{18}\diagdown CH-(CH_2)_4-\underset{\underset{C_2H_5}{|}}{\overset{\overset{C_2H_5}{|}}{C}}-(CH_2)_4-CH\diagup^{C_{18}H_{37}}_{C_{18}H_{37}}$$

Apolan-87

$$HO-CH_2-CH_2-(O-CH_2-CH_2)_n-OH$$

Polyethylene glycol

$$HO-CH_2-CH_2-\left[O-\overset{\overset{O}{\|}}{C}-CH_2-CH_2-\overset{\overset{O}{\|}}{C}-O-CH_2-CH_2\right]_n-OH$$

Polyethylene glycol succinate

$$\begin{array}{l} CH_2-O-CH_2-CH_2-CN \\ | \\ CH-O-CH_2-CH_2-CN \\ | \\ CH_2-O-CH_2-CH_2-CN \end{array}$$

1,2,3-tris-(2-cyanoethoxy)-propane

$$CH_3-\underset{\underset{CH_3}{|}}{\overset{\overset{CH_3}{|}}{Si}}-O-\left[\underset{\underset{CH_3}{|}}{\overset{\overset{CH_3}{|}}{Si}}-O\right]_n-\underset{\underset{CH_3}{|}}{\overset{\overset{CH_3}{|}}{Si}}-CH_3$$

Polydimethyl siloxanes

Fig. 5.2-7. Formulas for selected liquid phases

5.2.5 Applications of gas-liquid chromatography

The exploitation of GC can, on the one hand, be based on a general separation of compounds. These methods are used to examine the *purity* of a substance or to isolate substances from a mixture *preparatively*. The analytical chemist is primarily interested in the application of GC for *qualitative and quantitative analysis*, which will be discussed in this sequence.

Qualitative analysis

In a chromatogram, the qualitative information is found in the retention data; that is to say, in the corrected retention volume or in the net retention time (cf. Sec. 5.2.1). If an authentic standard substance is available for comparison with the suspected substance, then an unknown substance can be identified due to its appearance in the chromatogram.

However, a simple transfer of absolute retention data to other chromatographic conditions is not permissible. As we already know, the retention data are too dependent on the experimental conditions such as:

- column temperature
- carrier gas speed
- type of carrier gas
- pressure drop in the column
- type and quantity of stationary phase
- column measurement (length and diameter of column)

In principle one could exclude some of the influences if correction of the retention data for the specific retention volume were undertaken (Eq. 5.2-6). However, in practice this is too costly. Therefore an effort is made to either determine *relative retention values* or to ensure identification using a *spectroscopic dimension*. Mass spectrometry commonly serves the latter purpose.

Retention indices after Kovats

Relative retention values are determined in accordance with a Kovats (1958) method. The basis of his *retention index I* is the finding that within a homologous series of *n*-alkanes (*n*-paraffins) a *linear relationship* exists between the logarithm of the adjusted retention time and the number of carbon atoms in the compound. The retention of a compound to be investigated is then related to the retention of *n*-alkanes and one defines as follows:

The retention time of a substance is equal to 100 times the carbon number of a hypothetical n-paraffin with the same retention time as the substance of interest.

In accordance with the definition, the *n*-alkanes have an index of 100 times the relevant carbon number at every temperature on all separation columns, for example, *n*-hexane 600, or *n*-octane 800.

In order to determine the *retention time*, the substance being examined is chromatographed in a mixture which contains at least two *n*-alkanes. In so doing the retention times of these *n*-alkanes must encompass the retention time of the compound of interest.

Calculation of the index is undertaken on the basis of the equation:

$$I = 100y\left(\frac{\log t_{Rx} - \log t_{Rz}}{\log t_{R(z+y)} - \log t_{Rz}}\right) + 100z \qquad (5.2\text{-}12)$$

with t_{Rx}, t_{Rz}, $t_{R(z+y)}$ retention times relevant for the substance being examined, x, for the *n*-alkane with the carbon number z, and for the *n*-alkane with the carbon number z + y, with y being the number of additional C-atoms compared to z.

Worked example:

Figure 5.2-8 shows the gas chromatogram for determining the Kovats index of benzene. In this example the benzene to be tested was chromatographed together with *n*-pentane and *n*-heptane. Using the retention times of *n*-pentane (z = 5), $t_{R5} = 2.0$, of *n*-heptane (z + y = 5 + 2), $t_{R(5+2)} = 2.8$ min, and $t_{Rx} = 2.56$ min for benzene obtained from this, the following retention index is produced:

$$I = 100 \cdot 2\left(\frac{\log 2.56 - \log 2.0}{\log 2.8 - \log 2.0}\right) + 100 \cdot 5 = 640$$

Figure 5.2-9 illustrates the procedure when determining the Kovats index of benzene on a squalan column at 60 °C.

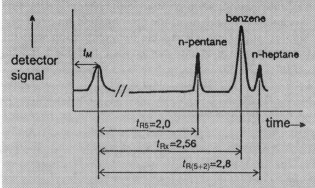

Fig. 5.2-8. Gas chromatogram to determine the Kovats retention index

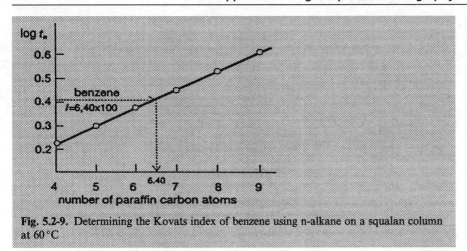

Fig. 5.2-9. Determining the Kovats index of benzene using n-alkane on a squalan column at 60 °C

Retention indices are available for many substances as *data collections* for the most varied stationary phases. They are relatively independent of temperature.

Quantitative analysis

We introduced the height or the area of a peak as a measurement for the concentration of compounds in a chromatogram (Sec. 5.1). Prerequisites for correct analyses are:

- Complete and reproducible vaporization of the sample
- The substances to be quantified are sufficiently separated and correctly identified

Moreover, if the detector reacts in a linear fashion, then both the height and the area of a peak are in a linear relationship with the mass of the analyte (or in case of the TCD with the concentration of it).

Isothermal and temperature programmed GC

The partition coefficient expressed in gas-liquid chromatography by the volumes (Eq. 5.2-4) is temperature dependent – like every equilibrium constant. The retention volumes too depend on the vapour pressure of the compounds (Eq. 5.2-8). An increase in temperature causes greater vapour pressure and thus leads to a higher elution speed. The correlation is given by the *Clausius Clapeyron* relationship. For the integral form the following applies:

$$\log p = \frac{\Delta_v H}{2,303 RT} + \text{const} \tag{5.2-13}$$

with $\Delta_v H$ differential molar enthalpy of vaporization of the pure substance.

According to this, the vapor pressure rises and, as can be shown, the retention time also rises logarithmically with the decrease in the temperature.

In *isothermal* GC the temperature is kept constant in the column. This method suffices if compounds are to be separated in a restricted boiling range. However problems arise for mixtures with a large range of boiling point (>100 °C). If the temperature chosen is too high, then the peaks appear too quickly in the chromatogram and are not fully separated. At too low a temperature, the analysis time is extended and compounds with a high boiling point appear as flat peaks, which are difficult to evaluate, at the end of the chromatogram.

These shortcomings can be avoided using *temperature programmed* GC. For this the temperature during the analysis is, as a rule, increased in stages uniformly or continuously. The starting temperature is so selected in order that the highly volatile constituents can be optimally separated. Compounds with higher boiling

Temperature programming is analogous to gradient elution in LC and to pressure programming in SFC.

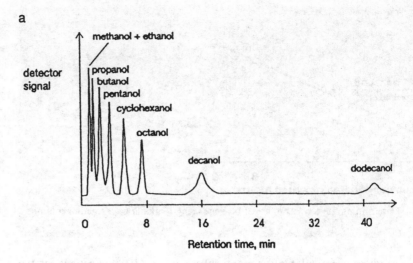

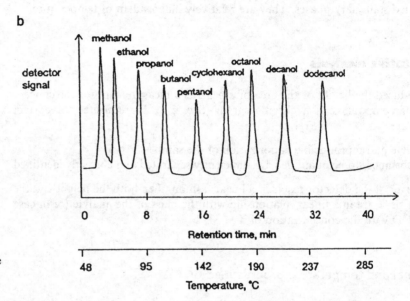

Fig. 5.2-10. Separation of an alcohol mixture using (a) isothermal GC at 175 °C and (b) temperature-programmed GC

packed
column

empty
capillary

thin-layer
capillary

Fig. 5.2-11. Columns for adsorption chromatography

points are first retained in the column head. They only begin to move at a higher temperature.

A comparison between isothermal and temperature programmed GC for the separation of an alcohol mixture is illustrated in Fig. 5.2-10.

Thus by using the temperature-programmed method mixtures with a wide boiling point range can be separated in the course of an analysis. There is the additional advantage in quantification that the detection limits and the precision of the peak evaluation can largely be kept constant throughout the whole chromatogram.

5.2.6 Adsorption chromatography

From a historical perspective adsorption chromatography was the first to be used. The basis of gas-solid chromatography is an adsorption medium as a stationary phase. Separation occurs by adsorption/desorption processes. Separation can take place both in a packed column and in a capillary one. In principle, an empty tube comes into consideration as a capillary, the inner wall of which is activated. The immobilization of the adsorbant on the inner wall of the capillary is more typical. Like SCOT columns, the columns are termed thin-layer capillaries or PLOT columns (PLOT – porous layer open tubular) (Fig. 5.2-11).

Table 5.2-4. Solid column fillings

Family	Trade name	Maximum working temperature, °C	Specific surface area, $m^2\,g^{-1}$	Applications
Diatomite	CHROMOSORB A GASCHROM	400	0.5 to 4	carrier for gas-liquid chromatography
Silica gel	PORASIL	400	1.5 to 500	all separation problems
Activated carbon		400	1300	inorganic gases
Polystyrene copolymers	CHROMOSORB B PORAPAK P,Q,T	275 250	50 to 800 100 to 600	low molecular, polar substances
Teflon	CHROMOSORB T	250	7 to 8	extreme polarity substances

Advantages of adsorption chromatography compared to partition chromatography may be its:

- wide temperature range
- good base line stability which is important for temperature programmed GC or for coupling with MS
- rapid equilibration (express analyses)

Unfortunately there are also substantial disadvantages:

- asymmetric peaks due to the small linear range of the adsorption isotherm
- the large adsorption enthalpies signify long retention times
- heterogeneous surfaces and catalytic activities of many adsorbents
- a smaller number of adsorption media which are more difficult to standardize

Inorganic adsorbents such as molecular sieves (aluminum silicate) or graphitized black, as well as porous polymers such as styrene-divinyl benzene copolymerisate are used as *stationary phases*. The specific surfaces of the adsorption phases are markedly larger than in partition chromatography (cf. Table 5.2-4).

Adsorption chromatography is of particular significance for the separation of low boiling point gases such as hydrogen, nitrogen, oxygen, methane, carbon dioxide, carbon monoxide, or inert gases, as well as for light hydrocarbons.

Questions and problems

1. At a packed column with squalan as the liquid stationary phase the compounds chloroform (boiling point 61 °C) is eluted in front of carbon tetrachloride (boiling point 77 °C). If a nitrilsiloxane is used as the stationary phase the retention order is rerversed. Explain the differing elution orders based on the Herington equation.
2. Compare the phase ratio of two capillary columns with an inner diameter of 0.24 mm and 0.36 mm at the same film thickness $d_p = 0.2\,\mu m$. Wich of the columns is more suitable for the analyses of low boiling compounds and which for analysing high boiling or thermolabile compounds?
3. How is the plate height influenced by the velocity of the carrier gas?
4. Which variant of GC would you use to
    (a) analyze a multicomponent sample in a large boiling range
    (b) determine residual monomers in foils
    (c) analyze amino acids
    (d) separate isomers?
5. Why is a mass spectrometer especially suited as a spectroscopic detector in GC? Wich instrumental problems may arise with the different GC techniques?
6. Summarize methods to identify unknown components in GC!
7. What are the differences between the following open tubular columns: PLOT, WCOT and SCOT?

5.3 Liquid Chromatography

The basis of liquid chromatography (LC) is a *liquid mobile phase*. In classic liquid chromatography, as introduced by Tswett in 1906, one worked with glass columns which had an inner diameter of between 1 and 5 cm and a length of between 50 and 500 cm. In order to guarantee practical flow speeds (up to 1 mL min^{-1}) particle sizes of between 150 and 200 mm were used. However the separations often took a very long time.

Attempts to increase the flow speed by applying a vacuum or by employing pumps could not bring about any improvement, since, as we already know from chromatographic theory, an increase alone in the linear rate of the mobile phase causes one to fall out of the range of the minimum number of theoretical plates. It was recognized relatively early on that a higher performance could only be achieved by a reduction in the particle size of the carrier material. It was also clear that the required pressures could not be realized by conventional LC equipment.

The practical transformation succeeded at the end of the 1960s when particles with diameters of up to 3 or 10 µm, as well as a new generation of instrument modules were introduced. *Classical liquid chromatography* is used as ever for preparative purposes or for introductory experiments into liquid chromatography. In the following section we will consider in particular the most important elements of rapid liquid chromatography or *high-performance liquid chromatography*, known as *HPLC* for short.

5.3.1 High-performance liquid chromatography (HPLC)

Separation principles and performance

Separations using a liquid mobile phase are based on the four principles:

- *Adsorption*
- *Distribution*
- *Ion exchange*
- *Exclusion*

Historically the first principle used is adsorption in the form of *liquid-solid chromatography*. It serves to separate unpolar compounds, isomers or compound classes such as, for example, aliphatic hydrocarbons and aliphatic alcohols. The principles of partition are translated into *liquid-liquid chromatography*. *Normal phase* and *reversed-phase chromatography* belong to this. The latter method is currently one of the most frequently used analysis methods of all. Chromatography based on ion exchange is to be understood as the transfer of classic ion exchange in the batch process to column chromatography. The high performance variation of ion exchange chromatography is called *ion chromatography*. The exclusion principle is based on the molecular sieve effect; that is to say, compounds of a particular molecule size cannot penetrate the pores of the sorbent employed and

Affinity chromatography can be considered a fifth separation principle in LC. This kind of separation is based on specific interactions between a solute molecule and a molecule attached to the stationary phase. For example, an *antibody* immobilized at the stationary phase interacts specifically with a particular solute *protein*.

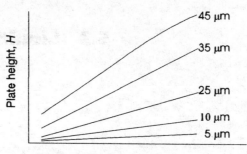

Fig. 5.3-1. Influence of the particle size of the carrier material in LC on the plate height H

are excluded compared with smaller molecules. The method associated with this is termed *gel chromatography* or *size exclusion chromatography*.

As is the rule in chromatography, the individual separation mechanisms of adsorption, distribution, ion exchange, and exclusion rarely occur in isolation, rather several principles act to a certain degree simultaneously.

Particle size of support material

It could be deduced from the dynamic theory of chromatography that the plate height depends directly on the particle size of the support material through the mass transfer coefficient, C_M (Table 5.1-3). Thus a reduction in particle size decreases the plate height and thus the efficiency of the separation column.

Figure 5.3-1 illustrates the alteration in plate height according to the $H(\bar{u})$ function depending on the particle size of the packing material. As already mentioned, no minimum for the $H(\bar{u})$ function can be observed in LC, since this would correspond to small, impractical flow rates.

Instrumentation

The components of a *classical unit* for liquid chromatography are:

- An elution medium reservoir containing the solvent for the mobile phase. At the simplest level, one can use a drip funnel.
- A separation column made of a glass tube typically with an inner diameter of 1 cm and a length of 30 cm. The packing material is kept in the column by a glass frit or glass mineral wool (cf. Fig. 5.1-1).
- A syringe or an injection valve to feed in the sample solution.
- A fraction collector with which a few millilitres of the eluate are collected manually or automatically. A flow-through cell photometer serves mainly for continuous detection in the eluent.

When using 3 or 10 μm particles, columns and connecting tubes are required to obtain flow rates of approximately 1 mL min^{-1} which can bear pressure up to 15 MPa. A classic liquid chromatographic unit cannot be used for this. The modules of a *HPLC unit* are illustrated in Fig. 5.3-2. The unit consists of a reservoir for the solvents in the mobile phase, of a pump system, of the sample feeding or injection system, possibly of a precolumn, of the separation column, and a detector. In order to minimize the peak broadening the dead volume of the unit, especially in the injection system and in the detector must be kept small.

Solvents

The solvents used as mobile phase are stored in a *reservoir* in glass or stainless steel bottles. *Dissolved gases*, which can lead to the formation of bubbles and thus to interference in the detector, as well as suspended matter, must be removed. The simplest way to clean the solvents for this is by sucking them through a millipore filter under vacuum. The dissolved gases, such as nitrogen and oxygen, can also be

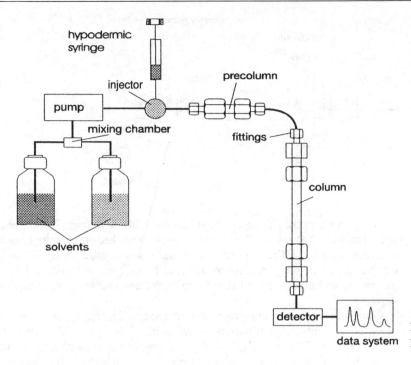

Fig. 5.3-2. Structure of a HPLC unit with precolumn

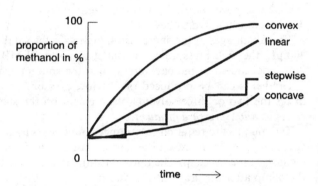

Fig. 5.3-3. Gradient shape for eluents combined binarily from the solvents methanol and water

outgased by introducing a noble gas (helium, for example) or by being processed in an ultrasound bath.

A distinction is drawn between isocratic and gradient elution. In the *isocratic* method one works with a single solvent of constant composition such as, for example, 30% volume methanol and 70% volume water. Better separation is often achieved using *gradient elution*.

There the composition of the eluents is constantly altered according to a particular programme. Solvents of varying polarity are used. For example, in a methanol/water eluent the proportion of methanol is increased from 30% (v/v) to 70% (v/v) in either a linear or another manner (cf. Fig. 5.3-3). The effect of a solvent gradient is comparable to that of a temperature gradient which we have become familiar with in GC. Shorter analysis times result and the peaks can be evaluated with similar precision across the whole chromatogram.

Pumping systems

The requirements for pumps for HPLC are manifold:

- generation of pressure up to 15 MPa
- slight residual pulsation
- chemical resistance
- constant displacement in the range between 0.1 and 10 mL min^{-1}
- reproducibility and control of the flow with a relative error of less than 0.5%

Stainless steel, teflon, or ceramics are used as materials. The high pressure strain is obtained using sapphire valves. One differentiates between reciprocating and

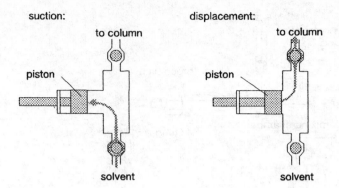

Fig. 5.3-4. Alumina (sapphire) pump head with double-ball valve

displacement pumps. *Displacement pumps* work like a syringe. A specific volume – approximately 200 mL – of the mobile phase is sucked in and discharged free of pulsation into the HPLC system. A crucial disadvantage, however, is the interruption of the delivery process in order to fill or rinse the piston. Displacement pumps are still used in micro HPLC, since the demand for mobile phases there is slight.

Reciprocating pumps are preponderant today (Fig. 5.3-4). As a rule, they are operated as double piston pumps, which work with a phase shift of 180° to suppress pulsation. They are also called oscillating (inverse) displacement pumps. To avoid direct contact of the solvent with the pump valves, the pumps are also available as *piston diaphragm pumps*. Here the piston movement is transferred to a diaphragm using hydraulics.

The advantages of the short piston pump are the small internal volumes of 40 to 400 µL, the high pressure at the outlet of up to 60 MPa, as well as the constant flow independent of the back pressure of the column and the solvent viscosity.

Gradients can be produced on the low pressure or high pressure side. If one mixes the two or three solvents of an eluent on the suction side of a pump, one refers to a *low pressure gradient*.

Two pumps are required to produce a *high pressure gradient*. One single solvent constituent, or in the case of ternary mixtures – two constituents, are presented in a constant relationship. The third constituent is admixed in on the pressure side of the pump after the gradient programme.

High pressure gradients provide more precisely composed gradients than low pressure gradients. This can be ascribed to the fact that the volume contraction in the low pressure variant can become significant when the various solvents are mixed. To avoid damaging the pump and contaminating the column particles the sample solution should be filtered prior to injection, e.g. by passing it through a 1 µm filter.

Injection system

The sample injection system must allow volumes in the range of 5 to 500 µL to be introduced. In addition, pressure should be kept in the system. Therefore for sample introduction one uses an injection system in the form of a *sample loop*. Using a microlitre syringe, the sample solution is fed in through a needle inlet into the loop of a 6-way valve (Fig. 5.3-5). At this point the eluent 2–3 conveyed from the pump flows directly into the column. By switching over the valve into the 1–2–3–4 position a definite volume flows into the system with the help of the eluent flowing through. Then the valve is put back again.

Sample injection systems which work automatically are preferred for high precision in sample introduction. They also work based on sample loops and are operated by compressed-air switching.

Column packings

The inside of columns consist preferentially of polished stainless steel. A rough surface would appreciably reduce the number of theoretical plates. In addition,

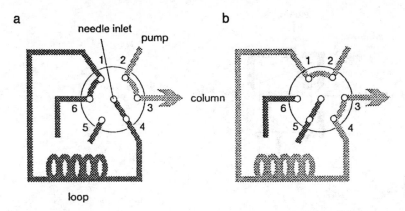

Fig. 5.3-5. Injection valve for HPLC.
(a) Filling the sample loop; (b) Injection

there are columns made of heavy-walled glass typing (duran or pyrex glass). The columns are integrated into the system in the form of cartridges (Fig. 5.3-2). The connections to the stainless steel tubes are made via fittings (conical metal sealing rings). A standard column is 250 mm long, with an inner diameter of 4.6 mm and is filled with 5 or 10 µm particles. Numbers of theoretical plates of approximately 50 000 per metre can thus be obtained.

In order to reduce the use of solvents which must be very pure for HPLC purposes, minimized columns with inner diameters of up to 1 mm and 30 to 75 mm long are increasingly being used. Numbers of theoretical plates up to 100 000 per metre can be attained using 3 µm particles (*microbore HPLC*).

Short *precolumns* are often employed to protect the separation column or for preseparation (4.5 mm inner diameter × 30 mm length). More coarse-grained packing material of particle sizes between 10 and 30 µm is used to prevent a greater pressure drop.

Packing columns with particles smaller than 20 µm is a particular problem. The high surface energy of the carrier material and the electrostatic charge connected with this no longer permit dry filling. When filling the packing material into the column with the help of a liquid, one must avoid a particle size gradient occurring in the column through sedimentation.

The packing material is suspended in a liquid to *fill up the column*. Methanol can be used for this for reversed phases. One can achieve even better packing using the so-called floating suspensions (*slurry* technique). Here, the differences in density between the solid phase and the liquid are compensated for by a specifically chosen dispersing liquid, such as dibromomethane.

Slurry means a suspension of solids in a liquid.

Detectors

Two principles are followed in HPLC for detection:

1. Detection of a general *characteristic of the mobile phase*, such as its refractive index or conductivity. There the analyte or the *solute* is indirectly indicated by the change in the characteristics of the mobile phase.
2. Detection of the *characteristics of the solute*, such as its absorption in UV, its fluorescence, or its diffusion current on an electrode.

The UV detector is most frequently used and its application accounts for over 70% of detections.

Detectors in HPLC may be classified as *bulk property detectors*, e.g. the RI-detector, and *solute property detectors*, such as UV-detectors.

Absorbance detectors

A cell in the form of a liquid flow cell is employed to measure the light absorption of an eluent at the outlet of a column. A z-shaped structure of a liquid flow cell is described in Fig. 5.3-6. In order to avoid peak broadening the cuvette volume is minimized and amounts to between 1 and 10 µL. The path-length of the cell varies between 2 and 10 mm. For measurements in the UV range the cell windows must be made of quartz.

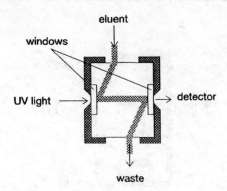

Fig. 5.3-6. Liquid flow cell for UV detection in HPLC

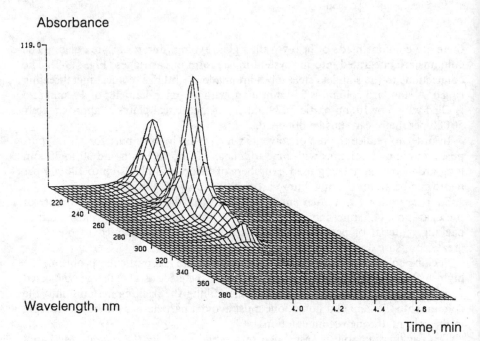

Fig. 5.3-7. 3-D diagram of a photodiode-array detector in HPLC for measuring phenanthrene

Filter photometers represent the simplest variation of a UV detector. For this, an emission line in the mercury vapour lamp is isolated and used for detection. The line at *254 nm* is typically used, as all aromatic compounds have maximum absorption at this wavelength and many other organic, as well as some inorganic substances absorb in this spectral range.

One also works at a single wavelength with a multiwavelength detector, which is fitted with a grating or prism monochromator as a *scanning photometer*.

Photodiode-array spectrometers have proved themselves for monitoring the full UV spectrum. The information is received as a 3-D or contour-line representation in which the absorbance is recorded in dependence of the retention time and the wavelength (Fig. 5.3-7). For example, from the contour-line representation one can select the most favourable wavelength for detecting individual analytes very easily.

The transmission limits of the solvents in the UV range are also given in Table 5.3-2.

In principle, one can also use photometers in the IR range as detectors in HPLC. However, at present, due to the high self-absorption of the typical eluents, such as water or methanol, they are reserved for special applications.

Fluorescence detectors

Compared with UV detectors, greater sensitivity of up to about the factor 1000 can be achieved using fluorescence detectors. In fluorescence detectors the excitation source is most frequently a mercury vapour lamp. Xenon high pressure lamps are also employed for more demanding tasks. In addition, the excitation and emission wavelengths can be selected by monochromators, or a fluorescence spectrometer is used as a detector. The intrinsic fluorescence of substances can often be exploited in the analysis of drugs, of clinically relevant compounds or of natural substances. To detect nonfluorescent compounds the substances to be determined are derivatized (see Fig. 5.3-12).

RI detector (refractometer)

The method for measuring the refractive index of the mobile phase can be employed universally as it is completely unspecific and reacts to almost all compounds. The basis is the difference in the refractive indices between the pure eluent and the eluent-containing sample constituents.

In principle, the light reflected on the prism (reflection type) or the deflected light (deflection type) can be evaluated. Figure 5.3-8 illustrates the structure of an RI detector according to the reflection principle. There the reflected portion is not directly measured, as otherwise the observation angle would have to be constantly altered. The light is rather detected after it has passed through the eluent and reflected on a steel plate. At the same time the steel plate acts as a thermostat.

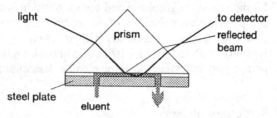

Fig. 5.3-8. Diagramatic structure of an RI detector which measures the reflected light after it has passed through the eluent

Refractometers consist of a measuring and a comparative cell (differential refractometer). The RI detector is less sensitive than, for example, the UV detector. Its high temperature sensitivity is also a disadvantage, so that the detector has to be very precisely thermostatted – if possible at $\pm 0.001\,°C$. The detector is not suitable for gradient elution. It is used for compounds which do not react in the UV range such as sugar, for instance.

Electrochemical detectors

Voltammetry, amperometry, coulometry, and conductometry can be exploited for electrochemical detection.

Conductometric detection is routinely used in ion chromatography (Sec. 5.3.2) where one works with a flow conductometer.

Coulometric detection is possible if the solute can be transformed with a coulometrically produced reagent, for example halogenide with coulometrically produced Ag^+-ions. The coulometric detector is rarely used.

To record a current-voltage characteristic using a dynamic (for example, a dropping mercury electrode) or a stationary electrode is in principle possible using a *voltammetric detector*. If one does not wish to stop the eluent, then the change of voltage on the working electrode has to be sufficiently rapid compared with the flow rate of the mobile phase in order to be able to measure the diffusion current at an even concentration. For this, potential scanning rates in the range of $1\ Vs^{-1}$ are required. And for this only measurements of stationary electrodes and the exploitation of fully reversible redox reactions can be considered. Therefore detection using a voltammogram is reserved for special investigations.

The *amperometric method* is preferred in practice. A constant potential is attached to a working electrode made of glassy carbon material, gold or platinum, and the limiting diffusion current at a given potential is measured in relation to a reference electrode. So as not to load the reference electrode, the current flows through an auxiliary electrode in the cell (Fig. 5.3-9).

An amperometric detector can be used for all substances that can be reduced or oxidized in the potential range of the working electrode employed. In order not to have to remove the oxygen in the eluent, oxidative processes in detection can be more easily handled. The selective and easily provable detection of biochemical

Elemental *speciation* considers the actual form of an element, i.e. its oxidation state and binding type.

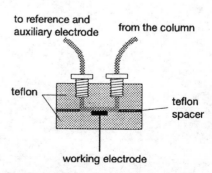

Fig. 5.3-9. Electrochemical detector with a glassy carbon working electrode

OH   CH₃

adrenaline

OH   H

noradrenaline

substances, such as the stress hormones adrenaline or noradrenaline (catechol-amine) has proved its value in the pmol range.

The disadvantage of amperometric detectors is the danger of poisoning the electrode surface in samples which contain surface-active substances, such as protein in serum or surfactants in liquid waste.

Spectroscopic detectors

We have already become familiar with one spectroscopic detector in the form of a photodiode-array detector in the UV range. Further possibilities used in practice are combinations of HPLC with *mass spectrometry* to identify organic compounds or with *atomic absorption spectrometry or atomic emission spectroscopy* to speciate elements. The characteristic features associated with this will be referred to in the section on hyphenated methods (Chap. 14).

Partition chromatography

Liquid chromatography based on distribution is the most frequently used method compared to the principles of adsorption, ion exchange, and molecular exclusion. *Polar, uncharged* compounds with *a relative molar mass smaller than 3000* are typically analyzed using distribution chromatography. Modern developments also permit polymolecular substances to be determined.

Stationary phases

In LC, *liquids immobilized* on the one hand on a carrier act as stationary phases. On the other hand, the carrier can be modified by a chemical reaction and *chemically bonded phases* are created.

Immobilized liquids

In classical LC one works with liquids which adhere to the carrier by physical adsorption. Typical carrier materials are *silica gel* with a surface area of 10 up to 500 m² g⁻¹, or *alumina* with a surface area of between 60 and 200 m² g⁻¹. Polar liquids, such as water or triethylene glycol are employed as stationary phases. The mobile phase is nonpolar and could be either, for example, hexane or diisopropyl ether.

Chromatography in which a polar stationary phase and a less polar or nonpolar mobile phase is used, is also termed *normal phase chromatography*. In contrast to this, the inverse arrangement of phases – when one is chromatographing using a relatively nonpolar stationary phase, for example hydrocarbons, and a polar mobile phase, such as water or methanol – is termed *reversed-phase chromatography*.

LC with physically immobilized liquids as the stationary phase is only of significance for classic normal phase chromatography. Due to the loose bonding of the liquid on the carrier *leaching* by the mobile phase must be avoided. Before separation, the stationary and the mobile phases are each saturated with the other solvent. Or the construction of the liquid stationary phase is undertaken using parts of the mobile phase (solvent controlled phases). Even with the safety measures mentioned *no gradient elution* can be achieved using liquids as stationary phases.

Therefore phases to which the liquid is *chemically* (covalently) *bonded* are more important in practice.

Chemically bonded phases

This type of stationary phase can be prepared for both *normal phase* and *reversed-phase chromatography*. Today chromatography on reversed phases accounts for about 75% of all applications of HPLC.

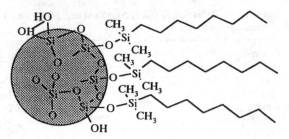

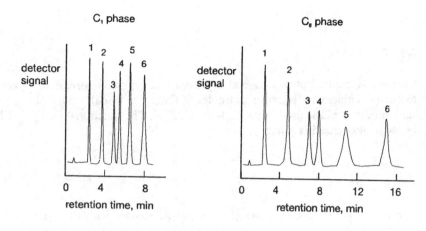

Fig. 5.3-10. Part of the surface of a C_8-(octyl)-reversed-phase. Residual silanol groups project on to the surface, and are partly bound through hydrogen bonds, as well as through oxygen atoms in the siloxane bond

Fig. 5.3-11. Comparison of retention behavior in HPLC using siloxane reversed phases with the alkyl groups methyl (C_1) and octyl (C_8). 1 – uracil, 2 – phenol, 3 – acetophenone, 4 – nitrobenzene, 5 – methyl benzoate, 6 – toluene.
Mobile phase 50/50 (v/v) methanol/water; flow rate 1 mL min^{-1}

Reversed phases

Silica gel is preponderant as carrier material in addition to alumina or resin ion-exchanger. In contrast with the classical resin ion-exchanger materials silica gel is *inelastic*; that is to say, an uptake of water does not alter the volume of the gel.

Silica gel is used as a uniform, broken, porous (pore volume approx. $1.2\,\text{mLg}^{-1}$), mechanically sturdy material in particle sizes of 3, 5 or 10 μm. Spherical material is preferred to irregular material.

The surface of fully hydrolyzed silica gel consists of silanol – (hydroxyl-) groups, as we have already seen in GC (Fig. 5.2-5). The surface constitutes about $8\,\mu\text{mol m}^{-2}$. The silica gel is allowed to react with alkyl chlorsilanes in order to become hydrophilic, and receives siloxanes (Si–O–Si grouping) as chemically bonded phases.

The alkyl group most frequently used in practice is C_{18} (*n*-octadecyl) followed by C_8 (*n*-octyl). The long chain hydrocarbon groups are aligned parallel to one another and perpendicular to the particle surface. They form a *brush-like surface* (Fig. 5.3-10).

One can regard the surface of the reversed phase as being a liquid even if up to now it is not quite clear whether the retention of the substances is more as a result of their physical adsorption on the surface or whether the surface behaves like a liquid hydrocarbon medium.

The longer the alkyl chain is, the greater the retention of the compounds becomes (Fig. 5.3-11). In addition, the maximum sample quantity usable depends on the type of hydrocarbon. The sample quantity approximately doubles when one passes from a C_4 to a C_{18} phase.

When the silica gel is silanized, the surface can at most be 50% covered, which corresponds to about 4 mmol m^{-2}. The remaining *residual silanol groups* as strongly polar groups can lead to adsorption of polar compounds (Fig. 5.3-10). This is made noticeable in a chromatogram by the *tailing* of the peaks. Therefore one endeavours to deactivate the residual silanol groups as much as possible. For this purpose they are capped by a post-reaction using trimethyl chlorsilan and methyl siloxanes ensue (cf. Sec. 5.2).

The siloxane phases are stable in the usual polar mobile phases of water, methanol or acetonitrile in the pH range of 2 to 8. They hydrolize at pH values above 8, and degradation or reorganization of the packing material occurs.

Table 5.3-1. Polar functional groups as chemically bonded phases in normal phase chromatography

Grouping	Functional group
Diol	$-(CH_2)_3OCH_2CH(OH)CH_2OH$
Cyano	$-(CH_2)_3C\equiv N$
Amino	$-(CH_2)_nNH_2$ with $n = 3$ or 4
Dimethylamino	$-(CH_2)_3N(CH_3)_2$
Diamino	$-(CH_2)_3NH(CH_2)_2NH_2$

Normal phases

Chemically bonded phases have also proved their value in *normal phase chromatography*. Table 5.3-1 provides examples of polar functional groups of the chemically modified phase used there, such as silica gel. The functional groups are listed in order of decreasing polarity.

Coating materials

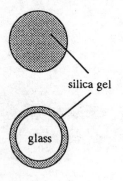

silica gel

glass

For the precolumns *nonporous glass or polymer beads* with diameters of about 30 μm are customary as support materials. The surface of this material is coated with porous layers of silica gel, alumina, or resin ion-exchanger (PLB – porous layer beads).

A liquid can be adsorbed on the layers as a stationary phase or a chemical modification attaches itself to the surface.

Mobile phases

By contrast with GC, where the mobile phase has been an *inert* gas, significant *interactions* appear between the substances and the liquid mobile phase in LC. The choice of mobile phase in LC is therefore an important factor in method development.

It is known from chromatographic theory that the retention behaviour and the appearance of constituents according to Eq. 5.1-23 depend on the number of theoretical plates N, the capacity factor k', and the selectivity factor α.

The values for k' and α can be varied by modifying the mobile phase in relation to the solvents used. Here the most important characteristic is their *polarity*.

In principle, one first selects the stationary phase, which should exhibit a similar polarity as the constituents to be separated. In accordance with this the mobile phase is chosen so that k' values occur if possible in the range between 2 and 5. If the polarity of the mobile phase lies too close to the polarity of the stationary phase, then the retention time gets too short. On the other hand, if the polarities of stationary and mobile phases are too similar, then retention times which are too long result.

Eluotropic series

Eluotropic series were established to evaluate the polarity of solvents. We use the *polarity index* after Snyder, who classified "strong" polar solvents and weak (weakly polar or nonpolar) solvents. The basis of this polarity scale is solubility measurements in dioxane, nitromethane, and ethanol. Table 5.3-2 reproduces the polarity index, P', for selected solvents in LC. It extends from the nonpolar, weak *alkanes* to the most polar of all solvents, *water*.

To evaluate the *polarity of solvent mixtures* the polarity indices of the individual solvents can be summed up. The calculation, e.g., for the polarity of a methanol/water mixture in the volume ratio 30/70 is:

Table 5.3-2. Eluotropic series for solvents in liquid chromatography in order of increasing polarity

Solvent	Polarity index, P'	Elution strength (SiO_2)	UV transmission, nm
Fluoroalkane	< -2	-0.2	200
Cyclohexane	0.04	0.03	200
n-Hexane	0.1	0.01	195
Carbon tetrachloride	1.6	0.11	265
Diisopropyl ether	2.4	0.22	220
Toluene	2.4	0.22	285
Diethyl ether	2.8	0.38	215
Methylene dichloride	3.1	0.34	230
Tetrahydrofuran	4.0	0.35	210
Chloroform	4.1	0.26	235
Ethanol	4.3	0.68	205
Acetic ether	4.4	0.38	255
Dioxane	4.8	0.49	215
Methanol	5.1	0.73	205
Acetonitrile	5.8	0.50	190
Nitromethane	6.0	0.49	380
Water	10.2	large	170

$$P_{\text{methanol/water}} = 0.3 P_{\text{methanol}} + 0.7 P_{\text{water}} = 1.53 \times 7.14 = 8.67$$

or in general for a mixture of m solvents:

$$P_{\text{mixture}} = \sum_{i=1}^{m} \Phi_i P'_i \tag{5.3-1}$$

with Φ_i signifying the volume fraction of the solvent i.

Elution strength

The elution strength of solvents is commonly characterised by the *solvent or elution strength*. Table 5.3-2 provides the elution strength for this, $\varepsilon°(SiO_2)$, which holds good for silica gel as stationary phase. From the relation with the stationary phase used, i.e., silica gel, one sees that the elution strength depends on the stationary phase contemplated. This was not the case in the relative evaluation using the polarity index.

Therefore the values for elution strength in Table 5.3-2 firstly apply specifically to the polar stationary phases in adsorption chromatography on silica gel or after division by 0.8 to alumina. If nonpolar stationary phases are used, then the eluotropic series also reverses. For example, strongly polar water possesses slight elution strength on hydrocarbon phases compared with nonpolar hexane. Polarity and elution strength should therefore be kept conceptually separate, even if they can be identical for certain phase combinations.

To conclude, let us consider elution behaviour in normal and reversed-phase chromatography. In normal phase LC where the polarity of the stationary phase is greater than that of the mobile phase, polar compounds elute last. The retention time will be all the longer, the more nonpolar the mobile phase is. By comparison, in reversed-phase chromatography the polar compounds elute first and the more polar the mobile phase has been set, the more strongly the unpolar constituents will be retained.

Choice of mobile phase

The mobile phase can be most soundly selected on the basis of a *separation mechanism* acting during separation. However, the most varied separation mechanisms often overlap one another or they cannot be investigated afresh for each problem of separation. Therefore the choice of composition of the mobile phase often occurs by *trying out* various solvent mixtures according to the experience of the operator or by a *systematic procedure* according to the principles of multivariate optimization (cf. Sec. 12.4).

The setting of the retention behaviour; that is to say, of the k' values of the compounds to be separated, does not automatically ensure its actual separation, which is determined by the selectivity factor α. A change of separating column is one means of clearly altering the selectivity factor. The composition of the mobile phase can also be further optimized to improve the selectivity factor or selectivity generally.

A mixture of the solvents methanol, acetonitrile, and tetrahydrofuran is used in *reversed-phase chromatography*, for example. Water is used to set the k' values.

By using the solvent mixtures chromatograms are developed in accordance with a particular mixture design. Subsequently the elution strength within the range of the best separations is so precisely adjusted that a large number of separation problems can be solved.

In *normal phase chromatography* a mixture of the solvents diethyl ether, methylene chloride, and chloroform is used analogously. The elution strength is modified with n-hexane.

Applications of partition chromatography

HPLC on *reversed phase bonded packings* is as a rule the first method tested if a new analytical task has to be solved. By contrast with the many stationary phases in GC, the stationary phase in HPLC is standardized as much as possible by employing C_{18}-reversed phases. The adjustment of the retention behaviour or the selectivities is achieved by varying the composition of the mobile phase.

Reversed-phase HPLC is used in practically all fields in which polar compounds have to be analyzed. Amongst these are pharmaceutical, biochemical, forensic, clinical or industrial analyses, as well as the testing of foodstuffs or of harmful substances, for example, pesticides, polycyclic aromatic hydrocarbons (PAH), or polychlorinated biphenyls (PCB).

If the selectivity of an *isocratic* elution does not suffice, then one can elute with the aid of a *solvent gradient*. In addition, the elution strength of the mobile phase can be modified by *varying the pH value* or by adding *ion pair forming reagents*. In both instances the aim is to produce neutral particles in the mobile phase which can interact with the nonpolar reversed phase. In the case of *ion-pair chromatography* the ionic solute reacts with a corresponding counter-ion. Examples of analysis in the presence of ion pairing reagents are listed in Table 5.3-3.

By selective adjustment of the *pH value* the preponderance of those *equilibrium forms* of the acidic or alkaline species can be achieved, which are of particular importance.

A further means of modifying the analytical problem is to *derivatize* the analytes. The polarity of the analytes can thereby be altered, the sensitivity raised at the detector, and the response can be more selectively set. Amino acids are, for example, be derivatized in protein hydrolysates through reaction with dansyl chloride (1-dimethylaminonaphthalene-5-sulfonyl chloride) so that fluorescent compounds are produced (Fig. 5.3-12).

Enantiomers are separated by using *chiral phases*. Coated materials are used for this, consisting of silica gel as a carrier and a polymer coating to which an optically active polymer is chemically bound. The interaction of the chiral phase can be based for example on a complex formation. This is illustrated in Fig. 5.3-13 for the analysis of optically active amino acids using a L-proline-copper (II) complex.

Table 5.3-3. Ion pairs for reversed-phase HPLC

Analytes	Counter ion	Mobile phase
Amines	ClO_4^-	0.1 M $HClO_4/H_2O$/acetonitrile
Carboxylic acids	$(C_4H_9)_4N^+$	pH 7.4
Sulfonic acids	$((C_{16}H_{33})CH_3)_3N^+$	water, propanol

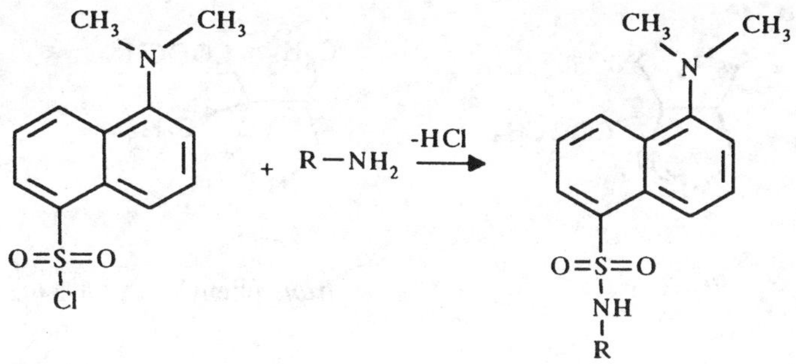

Fig. 5.3-12. Derivatization of the amino group of an amino acid or peptide residue (R-NH$_2$) using dansyl chloride to produce a fluorescent derivative

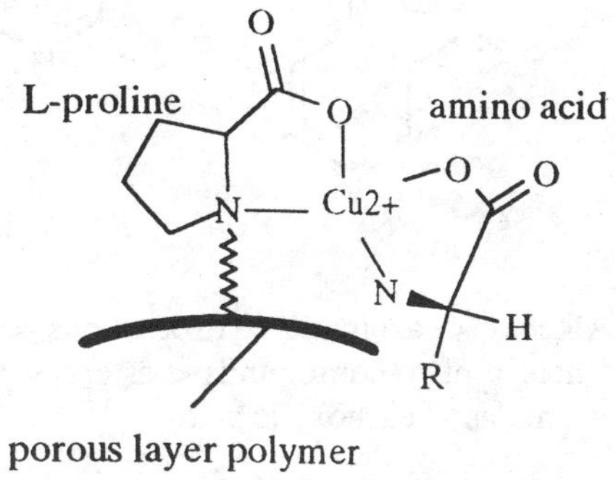

Fig. 5.3-13. Complex formation of a chiral phase taken from an optically active L-proline-copper (II) complex for separating enantiomers of amino acids

Special applications of *normal phase chromatography* are the separation of PAHs which contain alkyl chains in addition. Analysis of these is important in order to be able to analyze traces of potentially carcinogenic and mutagenic substances in air, liquid waste, and industrial processing flows. Although one can separate unsubstituted PAH compounds on reversed phases, it is not possible to separate alkylated compounds according to ring size. A diamine phase can be employed as a normal phase and heptane as the mobile phase.

Other examples of separations on normal phases are class separations of alkanes or lipids, as well as separations of steroids, sugars, or fat-soluble vitamins.

Adsorption chromatography

Adsorption or *liquid-solid* chromatography is the oldest method which was first applied by Tswett. *Silica gel* and *alumina* are used as stationary phases. Silica gel is most important, since it permits a higher sample capacity and is available in move varied qualities.

The basis for retention of the substances is the differing *adsorption processes* on the solid adsorbent, when the molecules of the mobile phase compete with those of the analyte. In contrast with the distribution between two liquids, adsorption is energetically promoted. The stronger interactions and the unimportant diffusion into the stationary phase stipulate shorter retention times than in distribution chromatography. However, the adsorption isotherms are only linear in a small concentration range, so that *overloading* of the columns can easily occur.

Adsorption is localized on *active centers* of the surface. Strongly polar molecules, such as water can deactivate the surface, so that the water content of the solvent used is an important quantity in adsorption chromatography.

Since the stationary phase is more polar than the mobile phase, which consists, for example, of methylene chloride and isooctene, adsorption chromatography is occasionally considered to be a type of normal phase chromatography.

It is better to use *elution strength ε* (Table 5.3-1) instead of the polarity index to evaluate the elution strength of the solvents. Elution strength is a measurement for the *adsorption energy* of the solvent per surface. It amounts to 0.8 times as much for silica gel compared to that of alumina. There is hardly any difference in retention behaviour between silica gel and alumina as stationary phases. The retention times increase in general in the following sequence:

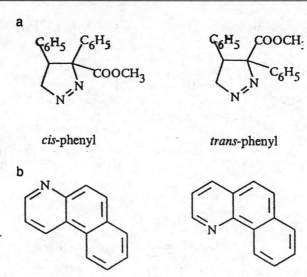

Fig. 5.3-14. Ways of separating differingly substituted molecules using adsorption chromatography. (a) Stereoisomeric forms of *cis-* and *trans-pyrazolines*. (b) *Positional isomers of azaderivatives of phenanthrene*

Alkenes < aromatic hydrocarbons < halogenated compounds and sulfides < ethers < nitro-compounds < esters ~ alcohols ~ amines < sulfones < sulfoxides < amides < carboxylic acids.

Adsorption chromatography is particularly suited to separating *unpolar substances* which are difficult to dissolve in water and thus cannot be easily chromatographed with the use of partition chromatography. As with partition chromatography it is no problem to analyze compounds with differing functional groups.

Positional isomers or stereoisomers can also be effectively separated with the aid of adsorption chromatography. Figure 5.3-14 provides two examples of this.

Just as in distribution chromatography one can proceed systematically in adsorption chromatography to optimize the composition of the mobile phase. In some cases, tests involving thin-layer chromatography are added, which one can regard as being a flat-bed variation of adsorption chromatography.

5.3.2 Ion chromatography as a classical and high-performance method

The high-performance variation of ion exchange chromatography is termed ion chromatography. Ion chromatography serves to separate or analyze ions using ion-exchangers. In particular it is a question of analyzing inorganic ions.

The *dynamic* method in ion exchange is based on the principle of elution chromatography (Fig. 5.1-1).

Classical ion exchange chromatography

Classical ion exchange chromatography is performed with porous exchange resins based on styrene-divinyl benzene copolymerisates. It was originally developed to separated the chemically very similar rare earth elements with the aid of cation exchangers. Determination of the ions collected in fractions took place using titrimetric methods.

As an example let us consider the separation of metal ions as *chloro complexes* on a strongly alkaline *anion exchanger*. The exchanger is loaded with chloride from a 12 M HCl and metal ions are introduced into the column at this hydrochloric acid concentration. At the same time nickel ions flow through the column. This is because no exchange occurs with the very weakly developed anionic chloro complexes of nickel. The other metal ions are eluted stepwise during elution with lower concentrations of hydrochloric acid. The sequence corresponds to the

Apart from ion exchange chromatography ion-exchange resins are also used in *batch procedures.*

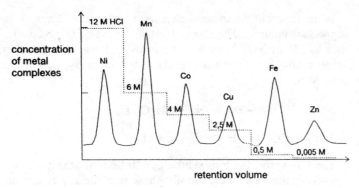

Fig. 5.3-15. Separation of metal ions as chloro complexes on a strongly alkaline anion exchanger with a stepwise gradient of the eluent HCl

stability of the chloro complexes. For copper ions, for example, complex formation is as follows:

$$Cu^{2+} + 4Cl^- \rightleftharpoons [CuCl_4]^{2-} \qquad (5.3-2)$$

Figure 5.3-15 shows the ion exchange chromatogram for the separation of Ni^{2+}, Mn^{2+}, Co^{2+}, Cu^{2+}, Fe^{3+}, and Zn^{2+}.

At the same time the *stepwise gradient* of the hydrochloric acid is plotted. The metal ions collected at the varying concentrations of hydrochloric acid can be determined individually either titrimetrically or photometrically.

Ion chromatography

Modern ion chromatography evolved in the middle of the seventies using HPLC modules. Two problems needed to be solved. For one thing the classical exchanger resins had to be developed further. Due to their compressibility, their swelling behaviour, and the slow diffusion of the analyte molecules into the pores of the resin, they were not suitable for high-performance chromatography. For another, a universal detector did not exist to detect the inorganic ions.

Packing materials

The classical ion-exchanger resins are replaced in ion chromatography by coating materials in which the synthetically produced ion-exchangers are applied to the surface of nonporous glass or polymer beads. The materials are between 30 and 40 μm in diameter (pellicular exchangers). A second variant consists of the economical coating of porous silica gel, as used in adsorption chromatography, using liquid ion-exchangers.

In both instances the *diffusion processes* are markedly speeded up compared with the classical exchangers. However the *capacity* of the new exchange materials is smaller.

Conductivity detector

Conductivity measurement is used as a universal detection principle for inorganic ions. In aqueous solutions ions exhibit a conductivity proportional to their concentration. Direct detection of the analyte in eluents as used in LC was however not possible. The concentration of electrolytes in the eluent customarily in LC is just too high to be able to recognize the very tiny signals of the eluting ions through the large basic conductivity.

Ion chromatography with suppressor column

The detection problem was solved by coupling the analytical column with a *suppressor column* and the selective choice of mobile phases.

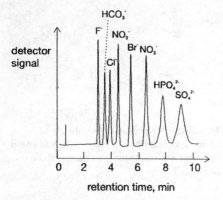

Fig. 5.3-16. Ion chromatographic analysis of anions using an eluent of 2.8 mM NaHCO₃/2.3 mM Na₂CO₃

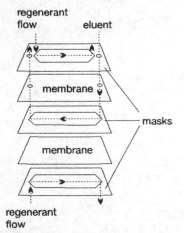

Fig. 5.3-17. Suppressor based on micromembranes. The eluent is separated from the regenerant flowing against it by ion-exchanger membranes

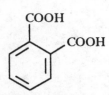

isophthalic acid

When determining *anions* one uses a *cation exchanger in the acidic form* as a suppressor column. The eluent consists of NaHCO₃/Na₂CO₃. After the anions, such as Cl⁻ or NO₃⁻ have been separated in the analytical column packed with an anion exchanger, the eluent reacts in the suppressor according to the following equations:

$$Na^+ + HCO_3^- + \overline{H}^+ \rightleftharpoons Na^+ + H_2CO_3 \tag{5.3-3}$$

$$2Na^+ + CO_3^{2-} + 2\overline{H}^+ \rightleftharpoons 2\overline{Na}^+ + H_2CO_3 \tag{5.3-4}$$

The line above the ions symbolizes those ions which are on the exchanger. By conveying the ions of the mobile phase into the barely conductive undissociated carbonic acid the conductivity of the eluent is suppressed. Since the ions to be determined, such as Cl⁻ or NO₃⁻ undergo no reaction in the suppressor their conductivity can be very sensitively detected.

Figure 5.3-16 illustrates an ion chromatogram for determining eight anions under given experimental conditions.

A cation exchanger is used as the analytical column to determine *cations*. Hydrochloric acid is suitable as a mobile phase and the suppressor column must contain an *anion exchanger in the OH form*.

The reaction of the eluent, HCl, then proceeds in accordance with the following neutralization equation:

$$H^+ + Cl^- + \overline{OH}^- \rightleftharpoons \overline{Cl}^- + H_2O \tag{5.3-5}$$

Again only the analyte ions remain in the solution as the sole conductive species, here cations such as Na⁺ or Mg²⁺.

A disadvantage of the suppressor column is the fact that it must be regenerated after a certain time (every 10 hours), as the exchanger is used up over time. In modern equipment the suppressor column is therefore replaced by a continuously operating *membrane suppressor*. Figure 5.3-17 provides a representation of the suppressor system based on membranes. The eluent flow is surrounded by two exchanger membranes, which supply the H⁺ or the OH⁻ ions depending on the type of determination. The exchange membranes are constantly renewed by the regenerant flowing in the countercurrent in the form of an acid or a base.

Instead of the layered arrangement of the flow of eluent and regenerant, as seen in Fig. 5.3-17, ring-shaped arrangements are inserted by using *hollow fibre* suppressors.

Single column ion chromatography

The suppressor column can be obviated if the conductivity of the mobile phase can be kept very low. For this, one works with *low capacity ion-exchangers* and with *low conductivity* eluents. Weak organic acids, such as, for example, isophthalic acid, benzoic acid or salycilic acid have proved their value as eluents. The pH value must be adhered to precisely in order to maintain the ion strength and thus the basic conductivity of the eluent. The existing intrinsic conductivity of the eluent is no longer suppressed using an ion-exchanger as suppressor, but is *electronically* suppressed.

Increasingly organic ions too, such as the amino acids, are being analyzed using the *single-column technique*. The analyses can be carried out with conventional HPLC equipment. However the detection capacity is smaller than in the suppressor technique.

The suppressor column is also dropped if a photometric detector can be used. We are familiar with *indirect photometric detection* in which a substance such as isophthalic acid absorbing in the UV range provides constant basic absorption. Ions eluting from the column are indirectly measured by the displacement of the isophthalic acid.

Direct photometric detection is possible using a reaction detector. For this again a membrane system serves as an efficient procedure for determining metal ions (cf. Fig. 5.3-17). A complex forming agent such as PAR (pyridil-azo-resorcin) is introduced into the eluent flow via a dialyzing membrane. The eluent remains colorless during the throughflow of the pure mobile phase. When metal ions occur in the eluent complex formation takes place on the dialyzing membrane and intensive dyeing develops above the colorless background.

5.3.3 Gel chromatography

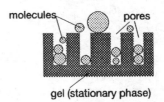

Gel chromatography is a particular form of liquid chromatography. It is based on the separation of molecules due to their differing size (Fig. 5.3-18). All molecules over a particular size are excluded on silica gel or polymer beads of a defined pore size (*size exclusion chromatography*). Molecules with molecular masses below the exclusion limit of the packing material are correspondingly retained. In contrast with all the methods of LC so far discussed, gel chromatography is not based on any chemical or physical interactions with the stationary phase.

Fig. 5.3-18. Principle of gel chromatography

Gel chromatography is significant for analyzing *high-molecular species*, such as proteins or polymers.

Two *extremes* can be differentiated in the interaction of the molecules with the porous stationary phase. Molecules with a larger diameter than the average diameter of the pores are completely *excluded* and are eluted along with the mobile phase as the first constituents. Molecules with a distinctly smaller diameter than the pores of the packing material can freely penetrate and remain the longest time in the stationary phase, and therefore are eluted last.

In gel chromatography *partitioning of molecules* between the stationary and mobile phase is based on molecuar size and to some extent on shape or polarity of the molecules.

Molecules of average diameters penetrate into the pores of the packing depending on the molecular size and partly also according to molecular shape. They appear with varying retention times between the peaks of the two extremes.

We use *retention volumes* to describe retention behaviour in gel chromatography, which as a product of the retention time and the flow volume of the mobile phase can be deduced from the chromatogram (cf. Eqs. 5.2-1 and 5.2-2).

The total volume, V_{total}, for a column packed with porous silica gel or polymer provides:

$$V_{total} = V_0 + V_p + V_{gel} \tag{5.3-6}$$

Here V_0 stands for the dead volume, and V_p for the volume of pores. In gel chromatography the dead volume derives from the void volume and corresponds to the theoretical volume for the transport of the completely excluded molecules. The volume for molecules which can move freely into the pores can be obtained from the total volumes of the dead and pore volumes; that is to say, the maximum attainable retention volumes amounts to just $V_0 + V_p$ (Fig. 5.3-19).

The *elution volume* of a molecule, V_E, which remains for a certain time in the stationary phase, depends on the dead volume and to a certain amount on K from the pore volume. This is calculated as:

$$V_E = V_0 + KV_p \tag{5.3-7}$$

The constant K corresponds to a partition coefficient which can assume values of between 0 and 1 in gel chromatography:

- In the event of a molecule being completely excluded, $K = 0$ holds good
- for molecules with free permeation into the particle pores, $K = 1$
- a partial exclusion of molecules provides K values   between 0 and 1

This simple approach only holds well for the case when no interactions, such as adsorptions, arise between the analyte molecules and the surface of the gel. If interactions occur K values larger than 1 can develop. Peaks, which correspond to the retention volume V_R customarily used in LC (Fig 5.3-19a), can then be observed outside the exclusion range.

Rearrangement of Eq. 5.3-7 provides the usual method of writing partition coefficients in accordance with Eq. 5.1-1:

$$K = \frac{V_E - V_0}{V_p} = \frac{c_s}{c_M} \tag{5.3-8}$$

The partition coefficient is a valuable quantity when comparing various packing

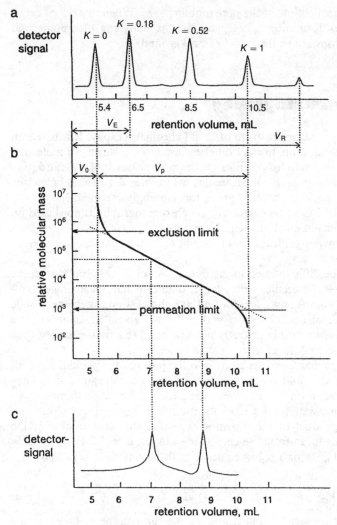

Fig. 5.3-19. Retention behavior and calibration in gel chromatography. (a) Retention of standards in the range of molecular exclusion with K values between 0 and 1 and relative molar masses between 10^6 and 10^3. The final (fifth) peak originates from a compound which has undergone a chemical interaction with the packing material. (b) Logarithmic relationship between the molar mass and the retention (elution) volume of the standard substances in A. (c) Determination of molecular mass based on the chromatogram of an unknown sample. For this elution must be carried out under the same conditions as for calibration in (b) (e.g., application volume, rate of flow)

materials. Furthermore, based on this quantity, all the theoretical fundamentals of chromatography inferred in Sec. 5.1 can be transferred to gel chromatography.

Stationary phase

Porous glasses or silica-based particles, polymers, as well as polysaccharides are used as packing materials. The particle sizes are in the range 5–10 μm.

Porous glasses and silica gels have many advantages due to their rapid equilibration in the diffusion of analyte and solvent molecules into the pores, due to the great stability of the materials even at high temperatures and the ease of manufacture of the column packings. The attainable pore sizes range from 40 to 2500 Å. Adsorption effects can prove a disadvantage in these materials. The surface is therefore often deactivated by silanization.

At the beginning, styrene-divinylbenzene copolymerisates were used exclusively as a *polymer*. The pore size of the resins is controlled by the amount of divinylbenzene and the degree of cross-linking arising from this. As these materials are hydrophobic, they can only be used in combination with nonpolar mobile phases.

Today hydrophile gels are also available which are manufactured from sulfonated divinylbenzene or polyacrylamide resins.

The pore size of typical packing materials is listed in Table 5.3-4. They are directly linked to the exclusion limits of the materials. The *exclusion limit* is that molecule size above which no retention is observed. The exclusion limit is given as relative molecular mass in Dalton.

One must note that for the information in Table 5.3-4 the calculated molecular mass is in the final analysis not decisive for chromatographic behaviour but rather

Table 5.3-4. Typical packing materials in gel chromatography

Average pore size, Å	Exclusion limit, Dalton (relative molecular mass)
Styrene divinylbenzene copolymerizate	
10^2	700
10^4	$1 \times 10^4 \cdots 20 \times 10^4$
10^6	$5 \times 10^6 \cdots 10 \times 10^6$
Silica gel	
125	$0.2 \times 10^4 \cdots 5 \times 10^4$
500	$0.05 \times 10^5 \cdots 5 \times 10^5$
1000	$5 \times 10^5 \cdots 20 \times 10^5$

the volume which a convoluted, solvated molecule occupies in the eluent. This volume is termed the *hydrodynamic volume*.

Mobile phase

The choice of mobile phase depends on the type of stationary phase. Gel chromatographic systems can be differentiated according to gel filtration and gel permeation.

In *gel filtration* one works with hydrophilic packings to separate water-soluble analytes. Therefore *aqueous eluents* are used as the mobile phase which as a rule contain buffers of pH control. *Gel permeation* exploits hydrophobic packing materials and correspondingly *nonpolar organic* solvents such as tetrahydrofuran, methylene chloride, or toluene are used. Gel permeation serves to determine sparsely water-soluble analytes.

Detectors

Detection can be undertaken in gel chromatography with the detectors proportional to concentration such as the differential refractometer or photometric detectors in the UV or the IR range.

The *flow viscosimeter* is customarily used as the special detector for high-molecular substances. There the basic viscosity of the eluent is measured. Viscosity is increased in the presence of a high-molecular analyte constituent and the signal resulting can be evaluated as a peak.

Applications

Gel chromatography serves to analyze substances with a relative molecular mass larger than 2000. We can distinguish between applications of gel filtration for separating water-soluble compounds and gel permeation for separating water-insoluble compounds.

Gel filtration is used if, for instance, high-molecular natural substances are to be separated from lower molecular substances or from salts. Thus proteins can be separated from amino acids and lower molecular peptides, if the exclusion limit of the packing material is at molecular masses of several thousand Dalton (cf. Table 5.3-4).

Gel permeation serves to separate homologs and oligomers, for example, fatty acids with relative molecular masses in the range between 100 and 350. For this a polymer with an exclusion limit at 1000 is selected.

Both types of gel chromatography can be exploited when *molecular masses* are to be determined or when the *partition of molecular masses* of natural substances or polymers has to be found.

Evaluation of retention volumes of the constituents of a sample is undertaken using standard substances which have characteristics similar to the analytes. The basis of the calibration is the logarithmic relationship between the elution volume and the relative molecule mass, M_r (cf. Fig. 5.3-19b). It can be approximated in the median permeation range by a straight line:

$$\lg M_r = b_0 - b_1 \lg V_E \tag{5.3-9}$$

with b_0 and b_1 being regression parameters and V_E representing the elution (retention) volume.

Dextrane, polyethylene glycol, sulfonated polystyrenes, or proteins are used as *water-soluble standards*. Polystyrene, poly (tetrahydrofuran), and polyisoprene are typical *water-insoluble standards*.

The principle for determining the retention volume of a sample is explained in Fig. 5.3-19c.

Advantages

The well-defined exclusion range between the dead volume, V_0, and the sum of dead and pore volume, $V_0 + V_p$, ensures *short retention times* which are determined by the partition coefficient, K. Narrow peaks capable of good evaluation are produced. Since as a rule, the sample does not enter any chemical or physical interactions, it elutes from the column *without loss*. That is to say, it can also be preparatively worked. On the other hand, the sample does not influence the column as can be the case in other liquid chromatographic processes.

Disadvantages

Due to the narrow exclusion range (K values between 0 and 1) the *peak capacity* in gel chromatography is limited (cf. Table 5.1-4). Furthermore, *analytes of similar size*, such as isomers, cannot be separated. It applies as a rule of thumb that separations based on molecular separation are only a success if the molecular masses are *at least* differentiated *by 10%*.

5.3.4   Thin-layer chromatography

In the preceding sections on LC, elution chromatography in columns was discussed exclusively. In this section, the planar methods of chromatography will be dealt with in which an *internal chromatogram* is traced (cf. Sec. 5.1).

The flat layer of material serving as a stationary phase can itself act either as a carrier, as in paper chromatography, or the stationary phase is applied as a layer on a carrier made of glass, synthetic material, or metal. The mobile liquid phase is moved by capillary forces across the layer. It can also travel across the layer with the aid of gravitation or of an electrical potential.

Paper and *thin-layer chromatography*, as well as *electrochromatography*, belong to the planar techniques.

We will limit ourselves to the most widely disseminated planar technique – *thin-layer chromatography* (TLC). TLC can be understood as being a modified form of LC. In actual fact TLC is often employed as an initial stage for column chromatographic separations, as it is easier to handle and a separation system can therefore be tried out more quickly.

TLC is the workhorse for *screening tests* in chemical, industrial, clinical, pharmaceutical, biochemical, or biological laboratories.

Stationary and mobile phases

The stationary phases used in TLC are the same materials we know from HPLC for separations based on adsorption, partition (normal and reversed-phase partition), ion exchange and size-exclusion. The materials are applied as finely dispersed particles to a flat plate with dimensions of 5×10, 10×20 or 20×20 cm. The layer thicknesses ensuing amount to 200 to 250 µm in particle diameters of 20 µm and above for *conventional plates*. One achieves about 200 separation stages during a development time of 25 mins over a distance of 12 cm.

There is a further development of plates in *HPTLC* (high-performace thin layer chromatography). Here one works with layer thicknesses of 100 μm and particle sizes of 5 μm or less. The separations are sharper and are achieved in a shorter time, about 10 min. Up to 4000 theoretical plates can be attained in 3 cm. The distinctly smaller *sample capacity* is a disadvantage of the high-performance plates.

Sample application and plate development

Application of the sample, which must contain a concentration of 0.01 to 0.1% of the analyte, is most easily undertaken using a capillary. Volumes in the range of between 0.5 to 5 mL are customary. The sample drop is applied at a distance of between 1 and 2 cm from the edge of the plate. The resultant spot should be about 5 mm in diameter for qualitative analyses. It should be even smaller for quantitative evaluation. Platinum-iridium capillaries are typical in HPTLC; with which volumes of between 100 and 200 nL are applied to the plate. In addition, there are automatic *dispenser systems* which permit the sample solution to be metered with great precision. The solvent in the sample solution is vaporized before the chromatographic process.

In order to *develop a chromatogram* the mobile phase must be moved across the stationary phase. The plate is placed in a closed container chamber for this (Fig. 5.3-20). The chamber is saturated with mobile phase – the developing solvent in the TLC. One must be careful that the sample spots on the plate do not soak into the developing solvent. The developing solvent on the plate rises upwards due to the capillary forces, however not at a constant rate. After a hyperbolic function the migration rate of the developing solvent reduces on the separation path.

Once the developing solvent has covered about two-thirds of the plate, the plate is removed and the development is thus ended. The developing solvent is allowed to dry and the plate is available for evaluating the spots.

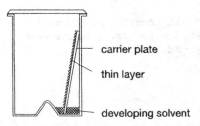

carrier plate

thin layer

developing solvent

Fig. 5.3-20. Development chamber for thin layer chromatography

Detection

In *screening tests* instrumental detection is dispensed with and one limits oneself to locating the spots and to identifying the substances in a simple manner. The following methods exist for *locating* the spots on the plate:

(a) Exploiting the available *luminescence characteristics* of the analyte. In the case of organic compounds these are as a rule fluorescence phenomena, and in inorganic compounds phosphorescence phenomena.

(b) Impregnating a *fluorescent indicator substance* in the layer bed. When being evaluated under UV light the analytes appear as pale, nonfluorescent spots. Pyrene derivatives, fluorescein, morin, or rhodamine B serve as indicator substances.

Visualization is the procedure to locate an analyte on the thin-layer plate.

(c) Spraying the layer with a *nonspedific strong oxidant*, such as HNO_3, $KMnO_4$, or H_2SO_4. Black spots develop on the plate through the oxidation of the organic compounds.

(d) Spraying on *group* or *substance specific reagent solutions*, such as ninhydrin to visualize NH_2, groupings (Fig. 5.3-21), iron (III) chloride for phenols, aniline phthalate for reducing sugars, or complex forming reagents to make metal ions visible.

A substance can be very easily *identified* by a spot if an authentic reference substance is chromatographed with it on the same plate and the *reaction or color* of the spots of the unknown and of the reference substances are compared. Identification using retention data will be discussed under the heading "applications".

Locally resolved measurement of the diffuse reflectance in the UV-VIS range using a densitometer, also called a *scanner*, is the dominating instrumental method of detection. On-line coupling of TLC can also be undertaken for IR spectroscopy, for example.

Off-line coupling of TLC is possible with all the well-known spectroscopic

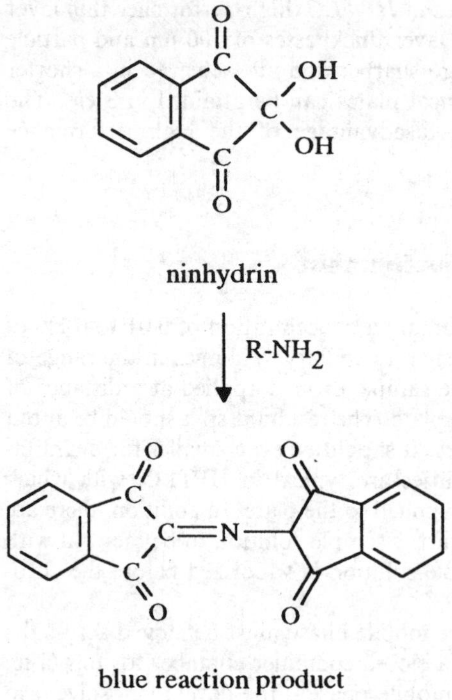

ninhydrin

R-NH$_2$

blue reaction product

Fig. 5.3-21. Formation of a blue reaction product from ninhydrin in the presence of NH$_2$ groups

methods. For this, the separated substance spot is taken from the plate and processed further for the subsequent analysis.

R_f value and capacity factor

A characteristic feature of TLC is found in the evaluation of the chromatogram which represents an *internal chromatogram*. The *retardation factor*, R_f, is used as a retention value in TLC. It is defined as the ratio of the migration distance of the analyte, z_R, to the migration distance of the mobile phase, z_M (Fig. 5.3-22):

$$R_f = \frac{z_R}{z_M} \tag{5.3-10}$$

In symmetric spots the middle of the substance spot is evaluated. In asymmetrical spots the intensity maximum is used.

The application of chromatographic fundamentals is based on the connection between the R_f value and the *capacity factor*.

The length of time the analyte remains in the mobile phase is obtained from the distance moved, z_R, divided by the linear flow rate of the mobile phase, $\bar{u}$:

$$t_M = \frac{z_R}{\bar{u}} \tag{5.3-11}$$

Within this length of time, t_M, the mobile phase travels the distance z_M. The retention time, t_R, corresponding to that of external chromatography thus emerges as:

$$t_M = \frac{z_M}{\bar{u}} \tag{5.3-12}$$

If one inserts the Eqs. 5.3-11 and 5.3-12 into the equation for the capacity factor (5.1-9), then one gets:

$$k' = \frac{z_M - z_R}{z_R} \tag{5.3-13}$$

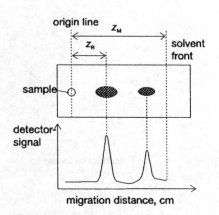

Fig. 5.3-22. Evaluation in thin-layer chromatography

or using the R_f value:

$$k' = \frac{1 - \dfrac{z_R}{z_M}}{\dfrac{z_R}{z_M}} = \frac{1 - R_f}{R_f} \qquad (5.3\text{-}14)$$

The relationship between the *partition coefficient*, K, and the R_f value can be illustrated by rearranging the relation (5.3-14) using the phase ratio, β, (Eq. 5.1-6):

$$R_f = \frac{1}{1 + k'} = \frac{1}{1 + \beta K} \qquad (5.3\text{-}15)$$

The migration distances can also be used to calculate the number of theoretical plates and the plate height. For thin-layer chromatography they are accordingly:

Number of theoretical plates: $N = 16 \left(\dfrac{z_R}{w} \right)^2 \qquad (5.3\text{-}16)$

Plate height: $H = \dfrac{z_R}{N} \qquad (5.3\text{-}17)$

Applications

We have already mentioned the innumerable applications of TLC as an easily handled *screening method*. The possibility of the *parallel development* of several samples on one plate is an important prerequisite for this.

In principle *qualitative analyses* are possible on the basis of retention data as well as of instrumental detection.

However an evaluation of the absolute R_f value is unpromising, as it is strongly influenced by the experimental conditions. Amongst these are the thickness of the layer, the moisture content of the mobile and stationary phase, the temperature, the degree of saturation of the development chamber with mobile phase, or the sample or spot size. Therefore it is better to use a *relative retention factor*, R_{rel}, which is determined for an analyte i in comparison with a standard substance *st*:

$$R_{rel} = \frac{R_{f(i)}}{R_{f(st)}} \qquad (5.3\text{-}18)$$

Generally, photometric evaluation with a scanner is used for *instrumental qualitative and quantitative analysis*. The *Kubelka Munk* function is used to transform the nonlinear signal concentration dependence. Correction of the *background signal* which is strongly determined by the quality of the layer material is particularly important for quantitative evaluation.

Compared with the elution technique in HPLC, classical TLC has some intrinsic weaknesses.

1. The developing solvent rate is *not constant*. It can only be altered through the particle size of the layer material and through the type of developing solvent. The decreasing migration rate of the developing solvent in the course of the development leads to broadening of spots. The *length of the distance* which can be selected is therefore limited and thus so are the attainable number of theoretical plates.
2. Due to the additional equilibria through the gas phase, as well as through the developing solvent mixtures on the layer, the *composition of the developing solvent* can alter during separation. This often leads to capacity factors which are difficult to reproduce and thus to difficult reproducibility of separations.

These disadvantages can be partly overcome by developments with a forced developing solvent movement, such as in OPTLC (over-pressured thin layer chromatography) or in rotational planar chromatography.

The selectivity of TLC separations can be improved by developing the plate in two dimensions using different mobile phases in each direction:

solvent B ↑

solvent A →

Questions and problems

1. List the properties of a solvent which make it suitable as a mobile phase in LC.
2. Mention typical detectors in HPLC. Which of those are unspecific and which are more selective?
3. What is the difference between adsorption of a dissolved substance on a solid and an ion-exchange process?
4. How does one separate electrolytes from nonelectrolytes?
5. Which procedures can be used to identify solutes in HPLC?
6. What makes thin-layer chromatography perfect as a screening method?
7. Exaplain why in classical ion chromatography an additional suppressor column is necessary. What possibilities exist to run ion chromatography on a single column?

In supercritical fluid chromatography a supercritical fluid is used as the mobile phase. SFC combines important advantages of gas and liquid chromatography. It is especially useful for analyzing compounds which cannot be determined using either GC or LC. This applies to all substances which for one thing are *not volatile* or cannot be vaporized without degradation, and thus cannot be analyzed with GC. For another, it is a question of compounds which cannot be directly analyzed with LC, since they contain no functional groups and therefore do not react to the usual spectroscopic and electrochemical *detectors in LC*.

It is estimated that the number of analyses of compounds which cannot be converted into the gas phase and do not possess any functional groups amounts to aproximately 25% of all separation problems. Typical problems like this are linked to the analysis of natural substances, pharmaceutical active agents, food-stuffs, pesticides, polymers, or crude oil.

We know from physical chemistry that there is a *critical temperature* for a substance above which it can no longer be converted into the liquid phase, even if higher pressures are applied. The vapour pressure belonging to the critical point is the *critical pressure*. The fluid is in a *supercritical state* close to the critical point and has characteristics which lie somewhere between those of gases and liquids.

Characteristics particularly important for chromatography on the critical point are the *density*, the *viscosity*, and the *diffusion coefficient*. Table 5.4-1 provides a comparison of these characteristics between gases, supercritical fluids, and liquids. The extraordinarily high density of the supercritical fluids conditions an extra-

Table 5.4-1. Significant characteristics of gases, supercritical fluids, and liquids for chromatography

Characteristic	Gas	Supercritical fluid	Liquid
Density, $g\,cm^{-3}$	0.6×10^{-3}–2×10^{-3}	0.2–0.5	0.6–2
Viscosity, $g\,cm^{-1}\,s^{-1}$	1×10^{-4}–3×10^{-4}	1×10^{-4}–3×10^{-4}	0.2×10^{-2}–3×10^{-2}
Diffusion coefficient, $cm^2\,s^{-1}$	1×10^{-1}–4×10^{-1}	10^{-4}–10^{-3}	0.2×10^{-5}–2×10^{-5}

Table 5.4-2. Critical quantities for mobile phases in supercritical fluid chromatography

Fluid	Temperature, T_k, °C	Pressure, p_k, Pa	Density, d_k, $g\,cm^{-3}$
CO_2	31.3	7.39	0.468
N_2O	36.5	7.27	0.457
NH_3	132.5	11.40	0.235
Methanol	239.4	8.10	0.272
n-butane	152.0	3.80	0.228
Dichlordifluoromethane	111.8	4.12	0.558
Diethylether	195.6	3.64	0.265

ordinarily *good solubility* of large nonvolatile molecules. Thus carbon dioxide in a supercritical state dissolves n-alkanes with a quantity of carbon atoms between 5 and 40 or it dissolves polycyclic aromatic hydrocarbons (PAHs).

Critical data for some substances are arranged together in Table 5.4-2. All these substances can be used as mobile phases in SFC. The range just above the critical point at temperatures of about 1.2 T_k and a pressure of 1 to 3 p_k is particularly suitable for SFC.

As one can deduce from the data in Table 5.4-2, both the critical temperatures and the critical pressures lie within the usual HPLC conditions. A chromatograph for SFC can therefore be constructed with similar equipment to that which we have become familiar with for high-performance chromatography in LC and GC.

The particular characteristics of supercritical fluids can also be used in the form of *super fluid extraction* in sample preparation. Technologically, caffeine-free coffee can be produced by the extraction of caffeine from coffee, or cigarettes with a low nicotine content can be produced in the extractive separation of nicotine from tobacco.

Instrumentation

As already mentioned, SFC equipment can be assembled from the modules as we know them from HPLC and GC. In addition, *precise temperature setting* must take place in the column. This is achieved with the help of a thermostatted column oven as is usual in GC. As a further peculiarity a device is required with which the *column pressure* is maintained on the one hand, and on the other hand the supercritical fluid at the end of the column can be released, so as to be able to *detect* it as a *gas*. For this, a capillary between 2 and 10 cm in length is used, which has an approximately 10 times smaller internal diameter (5 and 10 μm) than the separating capillary (50 to 100 μm). This is termed a back pressure device or *restrictor*. The simplest thing is to extract the separating capillary at the end of the column (Fig. 5.4-1).

The *pressure* in the system must be precisely monitored since the density of the supercritical fluid alters with the pressure and pressure differences lead directly to a change in the capacity factors. Higher pressure provides greater density. This causes the elution power of the mobile phase to increase and shorter retention times can be observed. An increase in pressure in a carbon dioxide phase from 7 MPa to 9 MPa, for example, leads to a reduction in retention time from 25 to 5 min. Thus *pressure programming* can be used in SFC as a gradient in just the way we are familiar with it for temperature in GC or the composition of the mobile phase in LC.

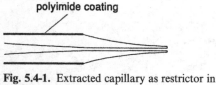

polyimide coating

Fig. 5.4-1. Extracted capillary as restrictor in capillary SFC

Stationary and mobile phases

The *stationary phase* in SFC can either be arranged as a packed column or be set up in a capillary. The *packed columns* are similar to those in HPLC separations on the basis of partition equilibria; that is to say, the separation columns have an internal diameter of between 0.5 and 4.6 mm, are up to 25 cm long, and are filled with particles between 3 and 10 μm in diameter.

In fused-silica capillaries the stationary phase is applied in the form of siloxanes as *liquids* or *chemically bonded* to the inner wall. The usual measurements of the separating capillaries are lengths of between 10 and 20 m with inner diameters of between 0.05 and 10 mm and film densities of 0.05 to 1 μm.

As explained below, the performance characteristics of separating columns in SFC are comparable to those in HPLC or capillary GC.

Carbon dioxide is most frequently used as the *mobile phase*. It is easy to handle, relatively cheap, nonpoisonous, odourless, and does not absorb in the UV range up to 190 nm. The critical CO_2 data are such that the temperature and the pressure can be varied over relatively wide ranges using equipment similar to HPLC. Sometimes an organic modifier, such as methanol or dioxane is added to the mobile phase. Other mobile phases in SFC were already listed in Table 5.4-2.

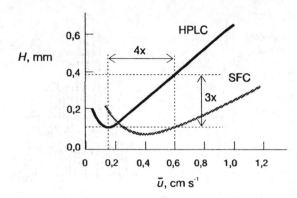

Fig. 5.4-2. Comparison of $H(\bar{u})$ function in an ODS separating mobile phase and a supercritical carbon dioxide mobile phase in SFC

Detectors

The gas present in the "restrictor" can be nonspecifically detected using the flame-ionization detector (FID) as used in GC. A prerequisite for this is low background signals such as are produced in the mobile phases in CO_2, NH_3, or N_2O.

Coupling of mass spectrometry is considerably easier to achieve by comparison with LC and is frequently used. UV-, IR-fluorescence, or flame-photometric detectors as well as TCD or ECD (cf. Sec. 5.2) can be employed as other detectors.

Performance characteristics of SFC

Two important consequences can be drawn from the comparison between the characteristics of supercritical fluids and those of gases and liquids in Table 5.4-1:

1. *Higher flow rates* can be adjusted due to the lower viscosity of supercritical fluids compared with liquids (cf. Fig. 5.4-2). The separations in SFC are therefore more rapid than those in LC and are comparable with those in GC.
2. The *diffusion rates* of supercritical liquids are somewhere between those of GC and LC. Due to this *peak broadening* is greater than in liquids but is less than in gases.

In order to explain these relationships, the $H(\bar{u})$ function on an ODS separating column for SFC in Fig. 5.4-2 is compared with one for HPLC. At a linear speed of 0.6 cm s^{-1} a plate height of 0.13 mm is achieved with SFC. The plate height in HPLC at the same linear speed is higher by a factor of 3 at 0.39 mm. According to Eq. 5.1-17 this means in practice that the peaks for SFC separation are narrower by a factor of 3 than in the HPLC chromatogram.

If in Fig. 5.4-2 one now compares the plate height of 0.13 mm which corresponds to the minimum of the $H(\bar{u})$ function in HPLC as regards the *linear velocities* attainable, then this is 0.15 cm s^{-1} for HPLC and 0.6 cm s^{-1} for SFC. Thus the linear velocity is higher for SFC by a factor of 4.

The mobile phase in SFC is not only an inert transport medium as in GC. *Interaction* of the solute with the mobile supercritical phase is very important just as in LC. It can be exploited for selected influencing of the *selectivity factor* α.

The zwitter position of SFC also has an effect when considering the partial pressures of the solute. The dissolution of the analyte in the supercritical fluid comes very close to its vaporization. However this occurs at a considerably lower temperature. Due to this, the *partial pressure* of the dissolved molecules at a particular pressure in the supercritical fluid is several orders of magnitude higher than in gases. This fact creates the basis for eluting high-molecular compounds, polymers, large biomolecules, or thermally unstable substances from the column.

Compounds with significantly higher molecular masses than in GC can be analyzed using SFC. The molar masses can be at least as big as in LC, i.e., up to 10^5. Even larger molar masses of up to 10^7 can only be separated by gel chromatography.

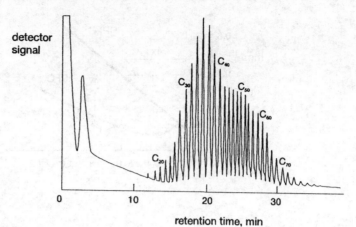

Fig. 5.4-3. SFC analysis of oligomeric polyethylenes with an average molecular mass of 740. Separating column: 10×0.01 cm packed alumina normal phase column with 5 μm particles; mobile phase: CO_2; pressure 10 MPa for 7 min, then programmed pressure increase to 36 MPa in 25 min, then at 36 MPa isobar; oven temperature 100 °C; detector: FID

Applications

The multiplicity of applications of SFC for analyzing nonvolatile compounds with relatively high molar masses was already mentioned. Natural products, drugs, foodstuffs, pesticides, surfactants, polymers, additives, crude oil, and explosives are analyzed using SFC.

The SFC chromatogram for the analysis of *oligomeric polyethylenes* is illustrated in Fig. 5.4-3. Note that a packed column has been used for this. The *pressure gradient* used is typical for SFC. One starts with a constant pressure, then one changes the pressure over a certain time in a linear manner up to a maximum pressure when one chromatographs at constant pressure to the end.

Electrophoretic procedures are based on the *migration* of ions in an electric field. For separation, one exploits the varying migration rates of the particles and suitably evaluates the selected separation effects.

Most electrophoretic procedures are not genuine chromatographic methods, since no partition of the analyte occurs between a mobile and a stationary phase. Instrumentation in classical paper electrophoresis and in modern capillary electrophoresis is, however, very similar to the corresponding planar chromatographic procedure and to capillary column chromatography, so that we can transfer the knowledge gained there to electrophoretic methods.

Besides the migration of lower molecular ions, the movement of colloidal particles, macromolecules, viruses, and complete cells can also be utilized.

Classical electrophoresis

The classical form of electrophoresis is used in a variety of ways in biochemical and molecular biological research for the separation, isolation, and analysis of proteins, polynucleotides, and other biopolymers. A distinction is drawn between carrier-free and carrier-supported electrophoresis.

Carrier-free elctrophoresis introduced by A. Tiselius (Nobel Prize 1948) is based on the migration of ions in buffer solutions which are layered above the sample solution. The buffer solution arranged in a u-shaped tube must be specifically lighter than the sample solution (Fig. 5.5-1). On application of a direct voltage using platinum electrodes the sample solution is partly separated and the ions enrich themselves on the boundary surfaces between buffer and sample solution (*boundary surface electrophoresis*). However complete separation does not occur.

Carrier electrophoresis is more frequently practiced. Here, the ions migrate on a paper carrier or a gel, such as agar, polymer, or silica gel. The carrier is saturated with a solution of a supporting electrolyte and ends in buffer solutions containing the electrodes (Fig. 5.5-1). A dc voltage of a hundred and more volts is applied to the filter paper soaked with the electrolyte solution.

The sample spots are laid down in line at one end of the carrier and migrating zones form for various types of species (*zone electrophoresis*).

The *electropherogram* can be evaluated by dyeing or photometrically. Figure 5.5-2 illustrates the classical separation of serum proteins.

Futher developments

Isoelectric focussing is used to determine *ampholytes*, such as amino acids and peptides. In this method a *pH gradient* is produced with a buffer mixture. It is known from the theory on ampholytes that substances at the pH value of the isoelectric point no longer move in an electrical field. Therefore in isoelectric focussing one achieves very special separation effects for ampholytes, since they only migrate to the isoelectric point and are concentrated there.

An *amphiprotic substance* or *ampholyte* is a species that can either provide or accept protons.

a

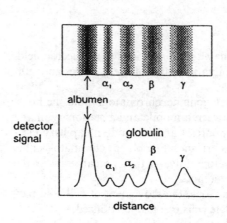

Fig. 5.5-1. Equipment for (a) carrier-free electrophoresis and (b) carrier electrophoresis on paper

Fig. 5.5-2. Electropherogram for serum proteins with photometric evaluation

6-aminocapronate
as terminator ion

morpholine ethane sulfonate
as leading ion

The method has a variety of uses in *protein analytical chemistry*. A prerequisite is a stable, nonmigrating pH gradient. This is achieved using buffer solutions with low molecular ampholytes which possess a high buffer capacity and exhibit varying isoelectrical points. They are focussed at defined positions in the electrical field and transfer their pH value to the surroundings.

An equally significant further development is *isotachophoresis* (the Greek *isos* for "equal" and *tachos* for "speed"), where the particles migrate in the electrical field at the same speed. In the methods of electrophoresis so far discussed, at a constant electric field, E, the particles move at different velocities, that is to say, the velocity of the particle i is:

$$v_i = Eu_i \tag{5.5-1}$$

Since the ion mobility, u_i, varies at the same migration speed of ions in isotachophoresis the electric field must inevitably differ in the zones:

$$v = E_1 u_1 = E_2 u_2 = \ldots E_n u_n \tag{5.5-2}$$

The less mobile a particle is, the greater will be the field intensity in the corresponding zone:

$$u_1 > u_2 > \cdots > u_n$$

$$E_1 < E_2 < \cdots < E_n$$

The electric field is uniform within the zones. Marked jumps appear at the edge of a zone. Ions moving too rapidly are slowed down in the previous zone. Ions migrating too slowly are accelerated by the higher field intensity into the subsequent zone. *Dynamic zone intensification* occurs.

How is the almost equal speed of the particles achieved in practice? In isotachophoresis a teflon capillary is used as separation distance, for example, and two electrolyte ions of differing mobility are employed. A *leading ion* is introduced at one end which is more mobile, and at the other end a *terminator ion* with less mobility is used. The sample is placed on the boundary surface of both electrolytes. The ions arrange themselves in a sequence of decreasing mobility in the course of separation and are separated in so doing. In addition, a buffering counter ion, which takes over buffering at the desired pH value, must be available for both electrolytes.

Immunoglobulins can, for example, be isotachophoretically determined if morpholine ethane sulfonate (pH 9) serves as a leading ion, and aminocapronate (pH 10.8) as a terminator ion. Isotachophoresis has become very popular in the form of the capillary technique.

In principle, by using the *capillary technique*, the important disadavantages of electrophoresis can be overcome. These disadvantages stem from the very disturbing *heat convection* of free electrophoresis, which leads to heterogeneity and thus to peak broadening. A further disadvantage of classic electrophoresis is the time-consuming *evaluation of the electropherograms*, which is in part difficult to reproduce.

Capillary electrophoresis

The transfer of further elements of capillary GC to electrophoresis has led to the comeback of this methodology. It is termed capillary electrophoresis (CE), capillary zone electrophoresis (CZE), and high-performance capillary electrophoresis (HPCE).

In principle, all electrophoretic procedures can also be transfered to the capillary technique. However, here minaturization does not serve to avoid diffusion paths, such as in GC or SFC, but serves to avoid disturbing thermal influences. The low currents flowing in CE lead to significantly less *Joule heat*. In addition, the electrically generated heat dissipates easier to the surrounding in narrow-fused silica capillaries.

Instrumentation

The block diagram of a CE system is indicated in Fig. 5.5-3. It consists of two buffer solutions, the capillary with the cooling device, the high-voltage power supply, a sample application system, the detector, and an evaluation system.

Typical *columns* of between 10 and 100 cm long and with internal diameters of between 25 and 100 μm consist of unmodified fused silica material or the capillaries are chemically or adsorptively modified. There is either a buffer solution, or a gel or polymer solution in the capillary.

The *high-voltage* applied to the Pt electrodes to form the electrical field is at 30 kV.

Sample application can be undertaken by *gravity injection*. For this, the capillary is plunged into the sample solution at its positive end and the container is raised about 10 cm above the level of the buffer solution at the negative end. Another method is *electrokinetic sampling*. Using a short impulse of 5 kV for a few seconds and, due to electroendoosmotic flow, a defined sample volume of between 5 and 50 nL reaches the capillary (see below). In automatic systems the capillary punctures a sample solution container sealed with a septum and is subsequently injected into the buffer solution.

The constituents migrate towards the opposite electrode in the capillary cooled by air or liquid mediums. Little heat is produced on account of the very high resis-

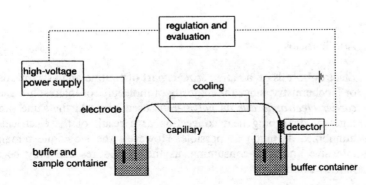

Fig. 5.5-3. Capillary electrophoresis (CE) system

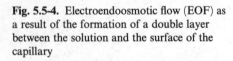

dodecyl
sulfate

tance. Moreover, the large ratio of surface-to-volume ensure good dissipation of the electrically generated heat and no peak broadening takes place. Peak broadening often only corresponds to the peak broadening theoretically expected for longitudinal diffusion. A number of theoretical plates up to 500 000 or even a million can be achieved.

UV detectors, including diode-array and fluorescence detectors are used to detect the ions directly on the column. In addition, there are minaturized electrochemical detectors.

Separation principles

As already mentioned, all the principles of electrophoretic procedures on a macro scale can be transferred to a capillary. A peculiarity of capillary electrophoresis is the *electroendoosmotic flow* (EOF) – often only called electroosmotic flow. This is produced by the formation of a *double layer* between the solution and the surface of the inner wall of the capillary. The inner wall of a fused silica capillary is negatively charged on account of dissociated silanol groups present.

This negatively charged layer attracts positive ions from the liquid, so that a positively charged liquid ring is produced. The liquid ring moves towards the negatively charged cathode (Fig. 5.5-4). The electroosmotic flow is strongly pH dependent. It can be observed at pH values above 4. The higher the pH value rises, the more clearly marked the effect is. EOF can be limited by chemical modification or coating of the inner wall of the capillary.

The profile of the electroosmotic flow is very flat (Fig. 5.5-4) compared to a hydrodynamic flow, the profile of which is parabolic. As a result of the electroosmotic flow peak broadening does not occur.

When an electrical field is applied, the positive ions normally migrate to the cathode and the negative ions migrate to the anode. Migration of *positive ions* to the cathode will not be influenced by EOF, since they move more rapidly than EOF. *Negative ions*, however, are attracted by the positive layer of the liquid ring. In so doing the movement of anions to the anode can be overcompensated by EOF. They then move slowly towards the cathode. Neutral particles migrate with EOF. As a result, one can observe consecutively first the postive, then the neutral particles, and lastly the negatively charged particles on the detector.

However, the neutral particles cannot be separated in this way. For this, micelle formers are added – *micellar electrokinetic capillary chromatography* (MECC). Micelle formers are surfactants like dodecyl sulfate. The micelle which forms is itself sufficiently hydrophobic towards its interior to be capable of enclosing the neutral molecules. Ionically charged particles emerge towards the outside where they can be electrophoretically separated like other ions (Fig. 5.5-4).

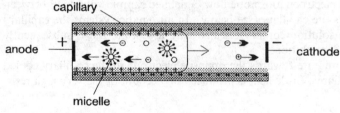

Fig. 5.5-4. Electroendoosmotic flow (EOF) as a result of the formation of a double layer between the solution and the surface of the capillary

Applications

Electrophoresis on a *macro scale* is part of routine analysis for analyses in biology or biochemistry, since the majority of molecules of interest are charged. As a rule, *carrier electrophoresis on polymers* is used here. At the same time the disadvantages of disturbing heat convection as a result of the electrical heating up are minimized as much as possible. However, the technique remains both labour-intensive and time-consuming, as the procedures can only be automated to a limited extent.

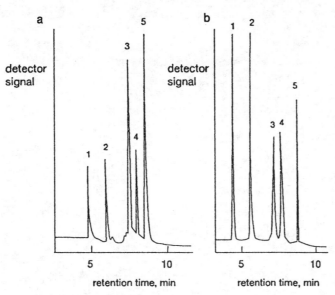

Fig. 5.5-5. Capillary electrophoresis of basic proteins on an untreated capillary (a) and a capillary laminated with polyvinylalcohol (b). 1 – cytochrome C, 2 – lysozyme, 3 – trypsin, 4 – trysinogen, 5 – chymotrypsin; capillary: actual length 57 cm, total length 70 cm, internal diameter 50 μm; buffer: 150 mM (a) or 50 mM (b) sodium phosphate at pH 3. High-voltage: 30 kV; sample application at 15 kV over 5 s; detection: 214 nm.

The most urgent problems of heat convection and of troublesome detection were overcome with the introduction of *capillary electrophoresis*. As a high performance method, CE provides the basis for the most varied analyses of *amino acids, peptides, proteins, nucleic acids*, and other *biopolymers*. Thus fragments of DNA (desoxyribonucleic acid) can be analyzed within the space of 10 minutes in a 35 cm × 50 mm capillary with the aid of a borate buffer at 8 kV high-voltage and photometric detection at 260 nm. In conventional polyamide gel electrophoresis this takes three hours.

Figure 5.5-5 shows two capillary electropherograms for analyzing *basic proteins*. If one uses untreated fused silica capillaries (Fig. 5.5-5a), then a relatively high buffer concentration must be employed in order to be able to keep the adsorption effects on the inner wall small. However if one looks closely, one can still observe clear peaktailing. If the capillary is coated with polyvinylalcohol (Fig. 5.5-5b) symmetric peaks result. A number of theoretical plates amounting to 1 000 000 can be achieved in separation on the actual length of the 57 cm long capillary.

In addition, *electrolytes*, such as cations and anions in biological liquids can be determined. In so doing *dynamic concentration ranges* emerge which can extend across three orders of magnitude.

Since *neutral molecules* have meanwhile also become susceptible to analysis with the help of micelle formation in MECC, analyses in the environmental field are also to be found, for example, to determine phenols.

Capillary electrochromatography (CEC) is a combination of CE and LC and is expected to develop into a very powerful separation technique.

Questions and problems

1. Mention the properties of a supercritical fluid that are important for it to be used as a mobile phase in chromatography.
2. What are the differences in the instruments for supercritical fluid chromatography as compared to GC and HPLC equipment?
3. Explain the importance of the pressure in SFC.
4. How is the isoelectric point exploited in electrophoresis?
5. What is the major advantage of micellar electrokinetic capillary chromatography over liquid chromatography?

5.6.1 Introduction

The concept of field-flow fractionation (FFF) was developed in the mid 1960s by J. Calvin Giddings, Professor of Chemistry at the University of Utah in Salt Lake City. Since that time the scope of the technique has been widened to span the range from small macromolecules from a few hundred mass units to colloidal species of every imaginable type, and on to particulate matters up to 100 μm diameter.

Like chromatography FFF is an elution technique, but FFF is not a strict form of chromatography. While separation in chromatography is the result of the differential partitioning of sample species between a stationary and mobile phase, separation in FFF is achieved through the partitioning into regions of different carrier velocities within the flow channel under the influence of an applied field. This applied field retains the particles more gently and with more precise control than do the intermolecular forces used for this purpose in chromatography. The FFF techniques are thus particularly useful for the study of macromolecules and colloidal particles, since these species often interact adversely or irreversibly at active interfaces or undergo shear degradation during their passage through packed chromatographic columns.

Diffusion as a basis of separation

In both techniques, chromatography and FFF, the diffusion of molecules or particles in a fluid medium plays an important role. In chromatography it is the driving force by which the molecules reach the interface, where the distribution takes place. In FFF the diffusion counteracts the field forces and it is responsible for the equilibrium layer position of each species.

Diffusion as a basis of separation methods has attracted the attention of physicists, physical chemists, and chemists for more than half a century and has found applications in all kinds of separation modes. In the 1920s W. Weaver was interested in the diffusion and settling properties of small particles in fluid systems, and in 1938 Klaus Clusius (University of Munich and later at the University of Zürich) published a large number of articles describing the separation of gas mixtures and isotopes based on the principles of thermal diffusion. He observed that a separation is achieved through diffusion in a tube along which a temperature gradient is applied. This technology became an important tool for the separation of isotopes.

Advantages of FFF techniques

A major advantage of FFF over related techniques is its theoretical simplicity. In a channel shaped like a ribbon a laminar flow is generated, and under normal operating conditions its flow profile is parabolic. Since the applied field acts uniformly in a direction perpendicular to the flow, component particles are subject to forces and displacements which are calculable (see Fig. 5.6-1). This means that particles elute with retention times which can also be calculated from, and thus predicted by first principles. The existence of this firm theoretical basis eliminates the need for complicated, time-consuming calibration procedures and considerably simplifies method development for new samples. Optimal system parameters may be calculated for any sample with known properties before the first injection, thereby saving time and materials.

5.6.2 The physical components of FFF

While each of the FFF subtechniques has concomitant requirements and restrictions in terms of technical design, the fundamental instrumental arrangement is more or less invariable from system to system. A complete apparatus consists of the FFF channel itself, the field- or gradient-generating components, some type of pumping system to drive the carrier solution, an appropriate detector, a chart recorder, a flow measuring device, and if desired, a fraction collector. The trends follow those of liquid chromatography: FFF instruments are increasingly controlled by a computer which also serves for data collection and treatment. As shown in Fig. 5.6-1, the channel is ribbonlike rather than tubular. It is cut from plastic or metal sheeting, typically 0.05 to 0.5 mm thick, to produce a long narrow spacer, typically 100 cm long and 2 to 3 cm wide.

The channel walls are constructed such that they allow transmission of the driving force (field), and thus differ from system to system.

FFF generally places no special demands on either the pumping or detection systems. Although early work dealt almost exclusively with the use of perestaltic pumps for carrier delivery, it is now realized that the unpacked channel characteristics of FFF poorly serves to dampen pulsing flows and that the use of any pulsing pump should be avoided. However, since the open channel causes relatively low resistance to flow, almost any type of pump which accommodates the desired range of flow rates is acceptable.

The selection of detector largely depends on the properties of the particles to be analyzed. Detectors suitable for HPLC are normally suitable for FFF as well.

Sample size in FFF usually ranges from 10 to 200 µL, with a concentration of under 1%. Small sample quantities lead to improved retention and better separation and should be used when detection limits permit. Moderate sample overloading is manifested in higher-than-predicted retention volumes and excessive band broadening; extreme overloading results in skewed and artefact peaks. During the course of a separation, however, the sample is diluted many times and detectability may be hindered.

A programmed reduction of the field strength could be used to enhance the speed and efficiency of the separation of a complex mixture whose molecular

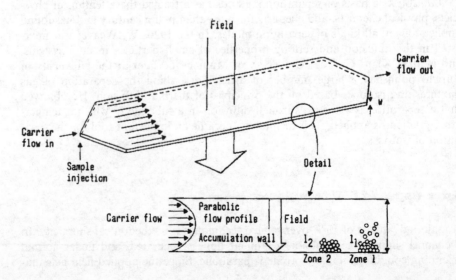

Fig. 5.6-1. Schematic view of an FFF separation channel (above) and the flow profile in such a device (below)

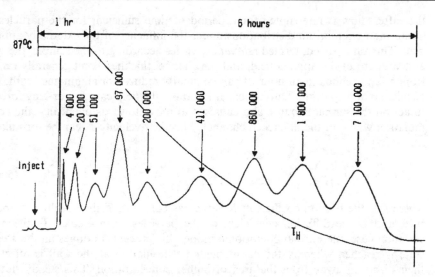

Fig. 5.6-2. Programmed thermal FFF: separation of particles with a wide range of molecular weights (with permission from J.C. Giddings et al. Anal. Chem 48 (1976) 1587)

weights vary over a wide range (see Fig. 5.6-2). In this type of programming, the force acting on a particle decreases with time, and leads to a steady increase in the particle's velocity through the channel. Today, the use of computers for the control of field strength allows this parameter to be simply and accurately varied throughout the course of a run.

The versatility of FFF is reflected in the wide diversity of samples to which it is applicable, and stems from the numerous controllable parameters affecting resolution and separation speed. The type of field, the field strength and how it is programmed, the carrier properties and velocity, and even the channel geometry, can all be varied in numerous ways to ensure optimal selectivity, speed, and accuracy.

Since the applied field separates on the basis of a physical property of the sample particles, i.e., size, density, diffusion coefficient, electrical charge or molecular weight, it is the distinguishing property of the particles that determine which of the FFF subtechniques will be most appropriate. The field must be chosen such that it effects the various particle populations contained in the sample sufficiently and differentially.

5.6.3 The different FFF subtechniques

A large number of fields or gradients can be used to bring about separation in FFF:

- the use of sedimentation and shear forces
- thermal gradients
- cross flow
- electrical fields
- magnetic fields

Further extension of the range of applicabilities occur through the combinations of these driving forces with various operating modes, which result from the different ways in which the sample components partition across the differential flow profiles in the channel. This partitioning may be diffusion controlled (normal FFF) or hydrodynamically controlled (steric FFF), and may depend upon transport coefficients. It can be regulated through the superposition of two opposing fields, resulting in a focusing of the sample on to a thin layer somewhere between the walls (hyperlayer FFF). By far the most extensively studied have been the normal and steric modes.

The elution process

A sample injected into an FFF channel begins as a plug in which the concentration of macromolecules is uniform over the entire channel width. Immediately thereafter, flow is interrupted for a period of time sufficient for the particles to attain their equilibrium distributions under the influence of the externally applied field. This time period, termed *relaxation*, varies according to the sample, the type and strength of the applied field, and the channel thickness, and generally ranges from a few seconds to an hour. It can be greatly reduced or eliminated by the use of inlet stream splitting. During relaxation time, the field causes a driving force F to act on the sample mixture and causes it to move towards one wall – the accumulation wall – of the unpacked channel. This drift velocity U can be formulated as:

$$U = \frac{F}{f} = F\left(\frac{RT}{D}\right)$$

(5.6-1)

where f is the friction coefficient, R is the gas constant, T the absolute temperature, and D the diffusion coefficient of the particles or molecules. In the most common case, the normal operational mode, this process continues until a steady state is reached whereby the components accumulation at the wall is offset by their diffusion away from the area of higher concentration. This steady state is characterized by the formation of discrete, narrow layers of particles parallel to the wall, each of which has an exponential concentration distribution with the highest concentration nearest the wall. This concentration profile exhibits a thickness of l:

$$l = \frac{D}{U}$$

(5.6-2)

which is also the mean distance of the particle from the wall. The layer thickness is most conveniently expressed as a fraction of the channel thickness, w:

$$\lambda = \frac{l}{w}$$

(5.6-3)

where λ is a dimensionless retention parameter. Combination of these three equations yields:

$$\lambda = \frac{RT}{Fw}$$

(5.6-4)

Clearly, in order for λ to be small, the applied field must be strong enough such that the energy of interaction, Fw, well exceeds the thermal energy, RT.

The laminar flow of a Newtonian fluid along any thin channel at a moderate velocity gives rise to a parabolic flow velocity profile, with the velocity approaching zero at the walls due to frictional drag. Such is the case when flow is initiated between the parallel plates of the FFF channel. The resulting differential flow amplifies the separation which has already occurred in the previous step. The discrete zones formed during the relaxation process are carried downstream towards the channel outlet at different velocities, depending upon how far each layer extends into the faster streamlines towards the center, ultimately leading to separation through difference in retention time.

As in chromatography, the retention ratio r in FFF is defined as the ratio:

$$r = \frac{V^\circ}{V}$$

(5.6-5)

where V° is the void volume of the channel and V, the retention volume of the species of interest. The open channels used in FFF enable the void volume to be determined either by calculation from the channel dimensions or by careful measurement of the retention time of a nonretained marker injected into the system.

Theoretical retention times may be calculated from the expression:

$$r = 6\lambda\left[\coth\left(\frac{1}{2\lambda}\right) - 2\lambda\right]$$

(5.6-6)

For well-retained components ($\lambda < 0.1$) this equation approaches the simple, limiting form:

$$r = 6\lambda$$

(5.6-7)

which leads to a general expression for the retention time in diffusion-controlled FFF processes of:

$$t_r = \frac{t^\circ F w}{6RT} \qquad (5.6\text{-}8)$$

Sample particles greater than about 1 μm experience a retention mechanism which deviates from that of the operational mode described above. The shift to the steric mode is characterized by a reversal in elution order; the larger the particle undergoing steric FFF, the earlier it will elute. When these large particles, whose Brownian motion is negligible, are acted upon by the field, they essentially come to rest at the accumulation wall. This tendency is opposed by the presence of hydrodynamic lift forces, which sweep the particles up and away from the wall under conditions of high velocity. Although the theory behind this retention process is not yet fully developed, it is clear that a fairly sensitive balance must be reached between the applied field and these flow-induced lift forces. If the flow rate is low compared to the field strength, particles may adsorb onto the walls and elute unpredictably or not at all. If the opposite is true and the flow rate is too high to be effectively countered by the field, the lift forces lead to extreme deteriorations in resolution. Assuming that the necessary criteria are met, initiation of flow along the channel after relaxation carries the particles downstream at rates determined by the extent to which they protrude into the flow stream: the equilibrium distance of the particle's center of gravity from the wall will be approximately equal to the particle radius. The retention equation for this operational mode may be formulated as:

$$t_r = \frac{t^\circ w}{6\gamma r}$$

where r is the particle radius and γ is a dimensionless factor approximately equal to 1.

This close proximity of the particles to the wall naturally has its drawbacks, and, in fact, it has been shown that the surface structure and properties of the accumulation wall are particularly critical to the success of separation in this mode. Nevertheless, the technique has shown enormous success for the separation of polystyrene standards.

Sedimentation FFF

Of all FFF subtechniques by far the most thoroughly studied have been thermal and sedimentation FFF. The appropriate instrumentation has been commercialized.

Particles of around 1 μm or greater, and of sufficient density, can undergo separation in sedimentation FFF even under the earth's gravitational force alone. However, the force generated by the spinning of a modified centrifuge around whose axis the channel has been curved is considerably more powerful and more easily controllable. This configuration, depicted in Fig. 5.6-3, greatly extends the range of applications. The commercial instruments are capable of maintaining fields well over 30 000 gravities, making it possible to retain particles as small as 0.01 μm. In sedimentation FFF the density of the carrier has a significant effect on the retention and should be accurately known for high precision work. Adjustment of the carrier density through the addition of density modifiers can be utilized to induce or shift retention, but technical limitations generally require the carrier to be aqueous.

When the relaxation process is complete, carrier flow is reinitiated and the particles proceed down the channel, eluting in an order determined by their size or density. The equation for the retention parameter in sedimentation FFF is:

$$\lambda = \frac{D}{sGw} \qquad (5.6\text{-}10)$$

where s is the sedimentation coefficient and G the acceleration. By carrying out runs with various carrier densities, it is possible to determine both particle size and

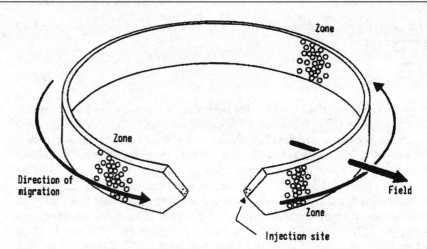

Fig. 5.6-3. Arrangement of FFF channel around the core of a centrifuge in sedimentation FFF

density simultaneously. In general, a good densitometer can be considered indispensible for any laboratory hoping to do high-precision FFF work.

Sedimentation FFF has been used for the sizing of latex samples of various polydispersities and for the characterization of aggregated polymethylmethacrylate samples. It served for the molecular weight determination of several viruses, for which it was shown that the infectivity of the virus remained essentially unaffected. It was used to fractionate polymerized serum albumin microspheres and several oil-water emulsions. From liposomes to DNA, from human and animal cells to natural water samples, from gold and silver solutions to cartilage proteoglycvans, the method has proven its ability to separate and characterize industrial, biological, and environmental samples of all kinds. However, sedimentation FFF does have practical limits dictated by particle size. Polymers, in particular, which are too small to be adequately affected by the centrifugal forces of sedimentation FFF can often be successfully characterized by thermal FFF instead.

Thermal FFF

Thermal FFF was the first of the FFF subtechniques to demonstrate success as a tool for macromolecular analysis. The walls of the channel are made of a thermally conductive material (copper), one of which contains a heating element, and another is machined to allow the circulation of coolant. By adjusting the temperature of the hot wall, it is possible to regulate the temperature drop across the thin channel dimension. Thermal diffusion is the process by which a molecule in solution undergoes migration relative to its solvent in the presence of a temperature gradient. However, very few samples seem to undergo thermal diffusion in aqueous carriers, and most experiments so far have focused on separations of synthetic polymers in organic solvents. These polymers tend to accumulate at the cold wall under the influence of the thermal gradient, with the highest molecular weight polymers driven closest to the wall and thus retained the longest. Retention in thermal FFF has been shown to be dependent both upon molecular size and composition. The layer thickness can be expressed as:

$$\lambda = \frac{D}{D_\tau w \left(\dfrac{\mathrm{d}T}{\mathrm{d}x}\right)} \tag{5.6-11}$$

where D_τ is the coefficient of thermal diffusion. Difference in polymer size are reflected in D, while differences in chemical properties are reflected in D_τ. High molecular mass polymers are driven closer to the accumulation wall by virtue of their low diffusion coefficients, and thus retained longer. Biological samples and samples requiring aqueous carriers clearly require other methods and this need is fulfilled by flow FFF.

Flow FFF

Although flow FFF is of special interest because it is, in principle, universally applicable, the selectivity range largely overlaps that of thermal FFF. The limiting factor is normally the composition of the accumulation wall, which consists of a semipermeable membrane of molecular weight cutoff low enough to ensure that the sample components do not pass through the membrane, supported by a porous frit. The active field in this case is a cross flow of carrier, which permeates the membrane and forces the sample components to be differentially distributed over the channel thickness by virtue of their diffusion coefficients alone. Since essentially all particles are affected by this type of force, the limit is only dictated by the molecular weight cutoff of the membrane. With a membrane with a 10 000 molecular weight cutoff, fulvic and humic acids down to 300 molecular weight can be characterized. Recently some of the technical problems associated with flow FFF have been resolved by an asymmetric system, where the carrier solvent is removed at the accumulation wall by a semipermeable membrane, but where there is no opposite inlet at the other wall. With such an asymmetric flow FFF system many applications for biological samples were found.

General reading

Altria, K.D., *Capillary Electrophoresis Guidebook*, Chapman and Hall, Van Nostrand Reinhold, 1995.

Foret, F., Krivankov, L., Bocek, P., *Capillary Zone Electrophoresis*, VCH, New York, 1994.

Giddings, J.C., *Unified Separation Science*, Wiley, New York, 1991.

Guichon, G., Guillemin, L., *Quantitative Gas Chromatography*, Elsevier, Amsterdam, 1988.

Haddad, P.R., Jackson, P.E., *Ion Chromatography: Principles and Applications*, Elsevier, New York, 1990.

Hamilton, R., Hamilton, S., *Thin Layer Chromatography*, Wiley, New York, 1987.

Hunt, B.J., Holding, S.R., Eds., *Size Exclusion Chromatography*, Chapman and Hall, New York, 1989.

Jennings, W., *Analytical Gas Chromatography*, Academic Press, Orlando FL, 1987.

Lindsey, S., *High-Performance Liquid Chromatography*, Wiley, New York, 1988.

Sewell, P., Clarke, B., *Chromatographic Separations*, Wiley, New York, 1988.

Small, H., *Ion Chromatography*, Plenum Press, New York, 1989.

Snyder, L.R., Glaijch, J.L., Kirkland, J.J., *Practical HPLC Method Development*, Wiley, New York, 1988.

White, C.M., *Modern Supercritical Fluid Chromatography*, Huethig Verlag, Heidelberg, 1988.

6 Kinetics and Catalysis

Learning objectives

■ To introduce students to chemical kinetics as the essential foundation for analytical methodology based on direct reaction-rate measurements and measurements relying on physical and chemical dynamic considerations, two major components of modern analytical chemistry

■ To acquaint students with measurements made on evolving systems by showing them that they provide enhanced information on chemical systems from which chemical reaction mechanisms, and the greater or lesser complexity of processes potentially accompanying them until a steady situation is reached, can be elucidated

6.1 Introduction

Thermodynamics answers such questions as: why does a reaction takes place and what total energy changes are involved in the process, by considering only the starting substances and the end products. Classical analytical methodology relies on thermodynamic constants that are determined by measuring some property of the reaction system at equilibrium. Kinetics provides answers to some questions for which thermodynamics has no plausible explanation, including what route the reactants follow for conversion into products and how long they take in the process, what properties the intermediate substances formed and consumed during the reaction have, and what energies are involved in their formation and disappearance. A host of modern analytical methods are based on kinetic parameters that are obtained by calculating the reaction rate via changes in the concentration of some reactant or product with time. In addition, kinetic constants often allow reaction mechanisms to be established in order to account for non-stoichiometric results. In any case, both thermodynamic and kinetic information is needed in order to fully understand why and how a given reaction occurs.

In short, every process, whatever its nature, takes place at a finite rate and tends to reach an equilibrium position. As a result, it encompasses two distinct regions: a kinetic (dynamic) region where the reaction system approaches equilibrium, and an equilibrium (static) region that is accessed after all the steps involved in the process have reached such equilibrium (Fig. 6.1-1). Not all reactions lend themselves to static measurements; in fact, some involve side processes, while others approach equilibrium very slowly or give rise to nonquantitative amounts of products. Under these restrictive conditions, equilibrium methods cannot compete with kinetic methods, which are thus to be preferred.

The use of catalysts for accelerating slow reactions arose from the close relationship between their effects and the kinetics of chemical reactions, which fostered development of kinetic determinations in chemical analysis. This more than justifies dealing with kinetics and catalysis jointly in this chapter.

Making analytical measurements under nonequilibrium conditions is dramatically facilitated by modern automatic data acquisition and processing techniques based on powerful instrumentation and computers, which endow kinetic measurements with precision and reliability comparable to those of equilibrium measurements.

For the above reasons, kinetics deserves a place in chemistry curricula, as it has already had in analytical chemistry handbooks and monographs devoted to kinetic methods in analytical chemistry [6.1-1], [6.1-2].

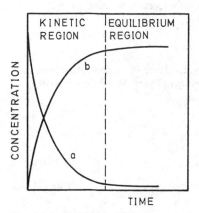

Fig. 6.1-1. Variation of the concentration of a reactant (a) or product (b) of a chemical reaction with time. The kinetic and equilibrium regions are shown. Reproduced from Pérez-Bendito, D., Silva, M., *Kinetic Methods in Analytical Chemistry*, Chichester: Ellis Horwood, 1988

6.2 The Chemical Reaction Rate

The *reaction rate* is defined as the number of moles of material consumed or formed per unit volume per unit time (curves *a* and *b* in Fig. 6.1-1) in a reaction:

$$rate = \pm \frac{dc_i}{dt} \qquad (6.2\text{-}1)$$

where c_i is the concentration of the ith reactant or product, depending on whether the negative or positive sign, respectively, is used. In the process, the concentrations of products and reactants change and gradually approach constant values as the reaction nears completion. The rate of change of such concentrations also varies (it approaches zero as the reaction develops). The major analytical implication of chemical kinetics arises from its being essentially concerned with the relationship between concentrations and the rates at which they change. Hence time always plays a prominent role in kinetic measurements, so accurate reproduction of the reaction conditions is consistently required.

The rate law or rate equation relates the rate of reaction to concentrations.

The *rate law* or *rate equation* of a reaction is an equation relating its rate at any time with the concentrations of all the substances on which it depends. The rate law for a straightforward reaction:

$$A + B \rightarrow P \qquad (I)$$

where P denotes products, is given by:

$$rate = -\frac{d[A]}{dt} = -\frac{d[B]}{dt} = \frac{d[P]}{dt} = k[A][B] \qquad (6.2\text{-}2)$$

The rate constant expresses reaction rate per unit concentration of reactant.

where the proportionality constant k is the so-called *rate constant*, which only depends on the pressure and temperature, and represents the reaction rate per unit concentration of reactant.

The reaction rate is therefore a positive quantity that is given at any time by Eq. 6.2-1, and the rate law is a differential equation given by Eq. 6.2-2, which establishes a relationship between the reaction rate and the concentration of the substances on which it depends at any time through the rate constant.

The rate of a reaction taking place in a single step:

$$B \rightarrow P \qquad (II)$$

A reaction expressed as a single step, ignoring any reverse reaction, is called an elementary reaction.

which is called an *elementary step* or *elementary reaction*, is given by:

$$rate = -\frac{d[B]}{dt} = \frac{dP}{dt} = k[B] \qquad (6.2\text{-}3)$$

where the rate in the reverse direction is assumed to be negligible.

The molecularity of an elementary reaction is the number of particles taking part in the reaction.

The *molecularity* of an *elementary reaction* (i.e., one that takes place in a single step) is equal to the number of particles that take part in it. An elementary reaction may involve a single ion or molecule of reactant or more (two or three), according to which it is called *unimolecular*, *bimolecular* or *termolecular*, its equation rate being dependent on its stoichiometry. Bimolecular elementary steps are quite commonplace since collisions between two identical or different particles are very frequent. On the other hand, termolecular reactions are much more uncommon as simultaneous collisions of three particles are comparatively rare, and elementary steps of an order higher than three are as yet unknown. Many reactions are the result of successive elementary steps of low molecularity.

The slow, rate-determining step determines the overall rate of reaction.

If one elementary step is much slower than the rest, then it is called the *rate-determining step* of the reaction, the overall rate of which is governed by the rate of such a step.

The half-time of a reaction is the time during which reactant concentration decreases to half its initial value.

One other interesting kinetic concept is the *half-life* or *half-time* of a chemical reaction, which is denoted by $t_{1/2}$ and defined as the time needed for the concentration of a reactant to fall to half its initial value. The mathematical expression for this parameter depends on the reaction order.

All of the above considerations rely on the assumption of irreversible processes and negligible reverse reactions. However, most chemical reactions are reversible, so, while the rate of the forward reaction decreases, that of the reverse reaction increases until equilibrium is reached, at which time the net reaction rate is zero. As result, the kinetic treatment involved is more complex. Further complications arise from *consecutive reactions* (e.g., $A \rightarrow B \rightarrow P$), *competitive reactions* (e.g. $A + B \rightarrow P$ and $A + C \rightarrow P'$) and *chain reactions*. *Heterogeneous reactions*, which involve two or more phases, are also rather difficult to represent in mathematical terms. All these intricate reaction types are beyond the scope of this chapter, which deals essentially with irreversible homogeneous bimolecular reactions.

6.2.1 Rate law and reaction order

The overall *order* of a reaction is the sum of the exponents of all the concentrations on the right hand side of its rate equation. Thus, the order of reaction (I), the rate law of which is given by Eq. 6.2-2, is 2, and the reaction is *second-order*. One can also refer to the reaction order with respect to a given reactant (a *partial* order), given by the exponent of its concentration on the right-hand side of the rate equation. Reaction orders as high as six and seven are known. The rate constant of a reaction of *n*th order is called an *n*th-order rate constant.

The reaction order is an empirical parameter – it may be a fraction, though not necessarily a fixed one – and as such obtained experimentally for a chemical reaction under given conditions. It should never be confused with molecularity, though. The rate of an elementary reaction is simply that given by the stoichiometry of the reaction itself; by convention, the elementary reaction:

$$a\mathrm{A} + b\mathrm{B} + \cdots \xrightarrow{k} c\mathrm{C} + d\mathrm{D} + \cdots \tag{III}$$

is assigned the following rate law:

$$-\frac{1}{a}\frac{d[\mathrm{A}]}{dt} = -\frac{1}{b}\frac{d[\mathrm{B}]}{dt} = \cdots = k[\mathrm{A}]^{a}[\mathrm{B}]^{b} \tag{6.2-4}$$

However, the partial orders in the rate equation for many reactions that cannot take place in a single step do not coincide with their stoichiometric coefficients. Such is the case with the well-known oxidation of bromide by bromate ion:

$$\mathrm{BrO_3^-} + 5\,\mathrm{Br^-} + 6\,\mathrm{H^+} \rightleftharpoons 3\,\mathrm{Br_2} + 3\,\mathrm{H_2O} \tag{IV}$$

the rate equation for which is:

$$-\frac{d[\mathrm{BrO_3^-}]}{dt} = [\mathrm{BrO_3^-}][\mathrm{Br^-}][\mathrm{H^+}]^2 \tag{6.2-5}$$

While the overall reaction is fourth-order, it is first-order with respect to bromate and bromide ion, and second order with respect to hydrogen ion. It is thus clear that the overall equation for a reaction does not allow one to predict its order or the form of its rate law, which have thus to be determined experimentally.

Provided one of the reactants involved in reaction (I) (e.g., B) is in a large excess, its concentration changes will be negligible, so [B] can be included in the rate constant k in Eq. 6.2-2 and the reaction said to be *pseudo* first-order in reactant A or *pseudo* zero-order in B, i.e.:

$$-\frac{d[\mathrm{A}]}{dt} = k'[\mathrm{A}] \tag{6.2-6}$$

$k' = k[\mathrm{B}]$ being the *pseudo* first-order rate constant with respect to A, also denoted by k_{A}. If the rate law is of the form $-d[\mathrm{A}]/dt = k'$ or $-d[\mathrm{A}]/dt = k'[\mathrm{A}]^2$, then the reaction is said to be zero-order or second-order in A, respectively.

Figure 6.2-1 shows the variation of the concentration of A with time for reactions of various orders in this reactant. As can be seen, the plots are decreasing exponential curves (the slope of which at each point is the reaction rate) except for that corresponding to the zero-order reaction in A, which is straight line because its slope $(d[\mathrm{A}]/dt)$ does not change as the reaction develops and the concentration of A decreases. The slopes of the *pseudo* first-order and second-order reactions in A approach zero as the concentration of A decreases and the reaction nears completion. The shapes of these curves are not exactly the same (see curves 2 and 3 in Fig. 6.2-1) because, even though the initial concentration is the same, the *pseudo* first-order and second-order rate constants are different and the differences are rather difficult to recognize.

Overall reaction order is the sum of the exponents of the concentrations on the right-hand side of the rate equation. Partial reaction order is the exponent of a particular reactant concentration.

Pseudo first-order reaction kinetics are observed if one of the reactants is present in large excess.

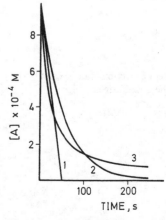

Fig. 6.2-1. Variation of the concentration of a reactant A with time for reactions of different orders with respect to A. (1) Zeroth order. (2) First order. (3) Second order

Table 6.2-1. Kinetic equations corresponding to simple irreversible reactions, $v = k[A]^a[B]^b$; $(n = a + b)$

Order			Differential form	Integral form	Half-life	Units of k
n	a	b				
0	0	0	$-\dfrac{d[A]}{dt} = k$	$[A]_0 - [A]_t = kt$	$[A]_0/2k$	$\mathrm{mol\,L^{-1}\,s^{-1}}$
1	1	0	$-\dfrac{d[A]}{dt} = k[A]$	$\ln\dfrac{[A]_0}{[A]_t} = kt^{(*)}$	$\ln 2/k$	$\mathrm{s^{-1}}$
2	1	1	$-\dfrac{d[A]}{dt} = k[A][B]$	$\dfrac{1}{[B]_0 - [A]_0}\ln\dfrac{[A]_0[B]_t}{[B]_0[A]_t} = kt$	$\dfrac{1}{k([B]_0 - [A]_0)}\ln\left(2 - \dfrac{[A]_0}{[B]_0}\right)$	$\mathrm{L\,mol^{-1}\,s^{-1}}$
2	2	0	$-\dfrac{d[A]}{dt} = k[A]^2$	$\dfrac{1}{[A]_t} - \dfrac{1}{[A]_0} = kt$	$1/k[A]_0$	$\mathrm{L\,mol^{-1}\,s^{-1}}$
-1	-1	0	$-\dfrac{d[A]}{dt} = k[A]^{-1}$	$\dfrac{1}{2}([A]_0^2 - [A]_t^2) = kt$	$3[A]_0^2/8k$	$\mathrm{mol^2\,L^{-2}\,s^{-1}}$
n	n	0	$-\dfrac{d[A]}{dt} = k[A]^n$	$\dfrac{1}{n-1}\left(\dfrac{1}{[A]_t^{n-1}} - \dfrac{1}{[A]_0^{n-1}}\right) = kt$	$\dfrac{2^{n-1} - 1}{(n-1)k[A]_0^{n-1}}$	$\mathrm{mol^{-(n-1)}\,L^{(n-1)}\,s^{-1}}$

(*) Or, in exponential form: $[A]_t = [A]_0 \exp(-kt)$
Reproduced from Pérez-Bendito, D., Silva, M., Kinetic Methods in Analytical Chemistry, Chichester: Ellis Horwood, 1988.

Alternatively, one can plot changes in the concentration of a product P, the concentration of which is related to that of A by $[A] = [A]_0 - [P]$, $[A]_0$ being the initial concentration of A. The result is an increasing exponential curve.

Table 6.2-1 lists the mathematical kinetic equations for straightforward irreversible reactions of *pseudo* zero, first, second, and nth order, together with their corresponding reaction half-lives and the units of the rate constant. All are given in both differential and integral form.

The dimensions of the rate constant depend on the reaction order. They can be determined by solving the rate equation for k and providing the units for all other quantities. Since concentrations are invariably expressed in $\mathrm{mol\,L^{-1}}$ and time in seconds, k must be expressed in $\mathrm{mol\,L^{-1}\,s^{-1}}$, $\mathrm{s^{-1}}$ and $\mathrm{L\,mol^{-1}\,s^{-1}}$ for zero-, first- and second-order reactions in A, respectively (for the dimensions of k in reactions of a different order see Table 6.2-1). The rate constant for a *pseudo* nth-order reaction has the same dimensions as that for an nth-order reaction.

First-order reactions are those of greatest analytical interest and widest use. They can readily be implemented by adding an excess of all the reactants but one so that the reaction rate is controlled by the remaining reactant, with respect to which the reaction is made *pseudo* first-order. The concentrations in Eq. 6.2-6 can be replaced with any measurable quantity provided it is directly proportional to the concentration. Thus, changes in the concentration of a reactant or product can be followed as a function of time by using a physical analytical technique.

In order to establish the rate equation for a given system, one must determine the partial reaction orders in the different variables influencing the process. This can be done in two ways depending on whether the differential or integral form of the rate equation is used. As a rule, differential methods, which entail measuring the initial reaction rate $(\tan \alpha)$, are mainly used for determining partial orders, whereas integral methods are usually employed for obtaining rate constants.

Partial orders of reactions can be calculated from the differential rate equation.

The partial order n in the differential form of a *pseudo* first-order rate equation:

$$rate = \tan\alpha = -\frac{d[A]}{dt} = k_A[A]^n \tag{6.2-7}$$

where A is the species whose partial order is to be determined and k_A the *pseudo* nth-order rate constant for such a species, can obtained by taking logarithms in Eq. 6.2-7, i.e.:

$$\log(\tan\alpha) = \log k_A + n\log[A] \tag{6.2-8}$$

which is the equation for a straight line of slope n and intercept $\log k_A$, both of which can be determined by plotting $\log(\tan\alpha)$ against $\log[A]$ at different initial concentrations of reactant A. Provided the plot is a straight line, the partial order sought will be unity, as can readily be inferred from Eq. 6.2-7.

In practice, determining the partial order in each reactant influencing the system entails changing its concentration while keeping constant and in a large excess those of the other species involved. By plotting as a function of time the values of the measured property obtained at different times and concentrations and determining the initial rates (tangents of the initial straight segments), the variations in the measured parameter are rendered exclusively due to the reactant, the partial reaction order of which can thus be determined.

Once the influence of each variable and concentration on the reaction has been determined, a rate equation of the following general form can be established:

$$v = -\frac{d[A]}{dt} = k[A]^a[B]^b[C]^c \dots \tag{6.2-9}$$

This equation is widely applicable and can include the concentrations of catalysts, hydrogen ion and solvents.

Use and interpretation of the rate law: reaction mechanism

The rate law for a reaction can be used to predict what the rate will be under given conditions and how long it will take to proceed to a certain extent. In addition, the rate law can facilitate understanding of the elementary steps by which the reaction takes place through information on the rate-determining step and any fast steps involved, which in turn can provide some insight into how and why a given reaction occurs and why it behaves as it does.

A rate law can be interpreted by envisaging a *mechanism*, which is a sequence of chemical steps that accounts for it and makes chemical sense. In fact, the mechanism of a reaction, which accounts for the overall equation of the reaction and its rate law consists of several elementary steps, each being consistent with chemical and structural principles. One pivotal concept as regards mechanisms is that of the *activated complex*, an unstable substance that can decompose to either regenerate the starting materials or form the products. The rate law for a reaction furnishes information on the step that leads to the formation of the activated complex, also called the "transition state". Every elementary reaction develops via a critical complex corresponding to the state of the potential-energy maximum along the reaction path. A distinction should be made at this point between a reaction intermediated and an activated complex. The former lies at a potential-energy minimum along the reaction path, so further activation (whether by intramolecular distortion, rearrangement or a subsequent bimolecular reaction with a different chemical component) is required to convert it into the activated complex leading to the product.

The activated complex can decompose to form either products or reactants.

A wide variety of chemical reactions have been interpreted in terms of the activated complex theory. The basic premise of this theory is that the reactants must exist in *quasi*-equilibrium with the activated complex and possess a characteristic frequency of passage over the potential-energy maximum or transition state. The ensuing scheme is reflected in reactions (V) and (VI):

$$A + B \overset{K}{\rightleftharpoons} (AB)^* \tag{V}$$

$$(AB)^* \overset{v}{\rightarrow} C + D \tag{VI}$$

where $(AB)^*$ denotes the activated complex. The equilibrium step, formulated as if it were a conventional equilibrium, is characterized by an equilibrium constant K and an associated thermodynamic function.

6.2.2 Integrated rate equations

Ordinary rate equations can be integrated in order to construct more powerful graphs than those obtained simply by plotting concentrations against time. Integration of the rate equation for a reaction of first order in A (Eq. 6.2-6) yields:

$$\ln[A]_t = \ln[A]_0 - k_A t \tag{6.2-10}$$

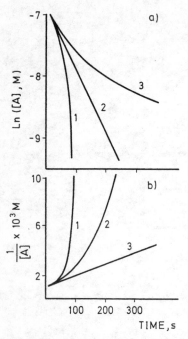

Fig. 6.2-2. Plot of (a) ln[A] and (b) 1/[A] against time for reactions of different orders in reactant A. (1) Zeroth order. (2) First order. (3) Second order

Partial orders of reaction can also be calculated from the integrated form of the rate equation.

or, in decimal form:

$$\log[A]_t = \log[A]_0 - 0.4343 k_A t \tag{6.2-11}$$

where $[A]_0$ and $[A]_t$ are the initial concentration of species A and that at reaction time t, respectively, and k_A is its rate constant.

Equation 6.2-10 can be expressed in exponential form as:

$$[A]_t = [A]_0 e^{-k_A t} \tag{6.2-12}$$

If it is a product rather than a reactant by which the reaction is followed, Eqs. 6.2-10 and 6.2-11 must be slightly modified. Thus, Eq. 6.2-10 must be transformed into Eq. 6.2-13:

$$\ln([P]_\infty - [P]_t) = \ln[P]_\infty - k_A t \tag{6.2-13}$$

where $[P]_t$ and $[P]_\infty$ are the concentrations of product formed at times t and ∞ (a long enough time for all the reactant A to be converted into product, i.e., $[A]_0 = [P]_\infty$).

Integration of the rate law for the simplest possible instance of second-order dependence on a given reactant:

$$-\frac{d[A]}{dt} = k_A [A]^2 \tag{6.2-14}$$

yields:

$$\frac{1}{[A]_t} = \frac{1}{[A]_0} + k_A t \tag{6.2-15}$$

Equations 6.2-10 or 6.2-13 and 6.2-15 are called "integrated rate equations" and allow one to determine the order in A by identifying which of the different plots they suggest gives a straight line. According to these equations, a plot of $\ln[A]_t$ or $1/[A]_t$ against time should be linear if the reaction is first-order or second-order, respectively, in A. Such plots are shown in Fig. 6.2-2. As can be seen, it is easy to tell whether the reaction is first- or second-order in A. In fact, if the curve is concave downward (Fig. 6.2-2a), the reaction is zero-order; on the other hand, if the curve is concave upward, then the reaction is of second or a higher order in A. The integrated rate equation for a reaction of zero order in A is shown in Table 6.2-1. These curve shapes are easy to distinguish. As can be seen in Fig. 6.2-1, if the first plot were one of [A] *vs* time, failure to obtain a straight line would make deciding what to try next more difficult.

Determining partial orders by using integration methods involves plotting the integral form of the rate equation for a previously assumed reaction order. Obtainment of a set of parallel straight lines corresponding to different initial concentrations of A confirms the partial order previously assumed or calculated by using the differential form of the equation. The *pseudo* nth-order rate constant can readily be calculated from the slope of the straight line obtained.

The half-time of a reaction can be calculated from the logarithmic form of the rate equation.

The half-time, $t_{1/2}$, for a first-order or *pseudo* first-order reaction is independent of the reactant concentration. This can be demonstrated by solving Eq. 6.2-10 for t and setting $[A]_t = [A]_0/2$ since:

$$t_{1/2} = \frac{\ln[A]_0 - \ln[A]_t}{k_A} = \frac{\ln\left(\dfrac{[A]_0}{[A]_0/2}\right)}{k_A} = \frac{\ln 2}{k_A} = \frac{0.693}{k_A} \tag{6.2-16}$$

Consequently, the reaction takes the same amount of time to consume half the initial concentration of reactant as from one-half to three-quarters. As can be seen from Fig. 6.2-3, half of the initial concentration or amount of reactant disappears during one half-time; half of the remaining half disappears during another half-time; half of the quarter remaining after $2t_{1/2}$ disappears during another $t_{1/2}$, so only one-eighth is left after $3t_{1/2}$ and so on. If the reaction is zero-order in A, all of A will be gone at $2t_{1/2}$; if it is second order in this reactant, then a third of the initial amount will still be left after $2t_{1/2}$, a quarter after $3t_{1/2}$, and a fifth after $4t_{1/2}$, which can readily be inferred from the corresponding integrated rate equation and the respective $t_{1/2}$ expression given in Table 6.2-1.

The equations given in Table 6.2-1 can also be used to demonstrate that the

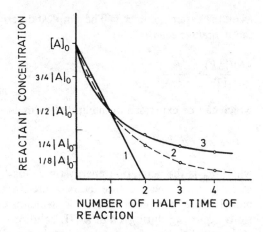

Fig. 6.2-3. Relationship between the reaction half-time and the fraction of remaining reactant for a (1) zeroth-order, (2) first-order and (3) second-order reaction with respect to reactant A.

time needed for the concentration of A to decrease to one-tenth of its initial value is equal to $1.8t_{1/2}$, $3.32t_{1/2}$, and $9t_{1/2}$ for a reaction of zeroth, first and second order in A, respectively. These values can in turn be used to determine the order in a given reactant simply by inspecting them under *pseudo nth*-order conditions. However, the above-described differential and/or integrated approaches for determining partial orders are to be preferred for this purpose.

6.2.3 Factors influencing the reaction rate

A few reactions are so rapid that every single collision between two reactant particles leads to the formation of product. Such reactions are said to be *diffusion-controlled*. Others are much slower as a result of the activated complex involved being too unstable and hence are only formed by collisions between particles with extremely high kinetic energies. The reaction between hydrogen and hydroxyl ions is at the former of these two extremes. Its rate law is given by $d[H^+]/dt = -k[H^+][OH^-]$ and its rate constant, k, is ca. 10^{11} L mol^{-1} s^{-1} (i.e., so large that only 7×10^{-12} s is needed for half the amount of hydrogen ion added to a solution containing 1 mol L^{-1} hydroxyl ion to be consumed). There are thus two broad categories of reactions in kinetic terms, viz. slow reactions (those with half-lives of 10 s or longer) and fast reactions (those reaching half-completion in less than 10 s).

The rate at which collisions between reactants occur is dependent on a number of factors, which naturally include the reactant concentrations; however, since these are explicitly comprised in the rate equation, it is more useful to focus attention on other influential factors. Such factors include the temperature, the solvent's relative permittivity and the ionic strength of electrolytes present in the reaction medium. The rate of a reaction can be altered (increased or decreased) by substances that accelerate (catalysts) or decelerate the reaction (inhibitors). Catalysis is a major phenomenon that warrants separate discussion and is dealt with at length in Sec. 6.3.

Influence of temperature

Most of the systems involved in kinetic analysis are influenced by temperature, which must be strictly controlled when making analytical measurements. Increasing the temperature also increases the kinetic energy of the reactant particles and causes them to move more rapidly and collide more frequently in solution. The increased kinetic energy furnishes the *energy of activation* required to form the activated complex.

The rate constant depends on the activation energy. In fact, an increase in the activation energy decreases the fraction of the overall number of collisions where the amount of kinetic energy required is available; the resulting decrease in the frequency with which the reaction takes place reflects in a decrease in the rate constant, which also depends on the frequency with which collisions occur, as well

The Arrhenius equation relates the activation energy of a reaction to the temperature.

as on the system geometry. The simplest description for these effects is provided by the *Arrhenius equation*:

$$\frac{d(\ln k)}{dT} = \frac{E_a}{RT^2} \tag{6.2-17}$$

which can be expressed in integrated form as:

$$k = Ae^{-E_a/RT} \tag{6.2-18}$$

where E_a is the activation energy (in $J\,mol^{-1}$), R the gas constant ($8.3143\,J\,K^{-1}\,mol^{-1}$), and T the absolute temperature (in K). A is the product of two factors, namely: (a) the *frequency factor*, Z, which is equal to the overall number of collisions occurring during 1 s in a 1-L solution containing 1 mole of each reactant, and typically – not for reactions between multivalent ions – ca. $10^{11}\,L\,mol^{-1}\,s^{-1}$; and (b) a steric factor, p, which represents the fraction of collisions where the reactants are spatially oriented in such a way that combining to form the activated complex is feasible. The steric factor depends on the structures of the reactants and varies between 0 and 1.

For a bimolecular elementary step of $p = 1$ and $E_a = 0$, the rate constant would be equal to $Z(\simeq 10^{11}\,L\,mol^{-1}\,s^{-1})$ and decreases with an increase in E_a. Thus, an activation energy of 20, 40 and $100\,kJ\,mol^{-1}$ would result in a k value of 3×10^7, 1×10^4 and $3 \times 10^{-7}\,L\,mol^{-1}\,s^{-1}$, respectively. The rate-determining step can therefore be very slow ideed if the activation energy is very high.

The effect of temperature on the rate constant can be most readily described by expressing Eq. 6.2-18 in natural logarithmic form:

$$\ln k = \ln A - \frac{E_a}{RT} \tag{6.2-19}$$

or decimal logarithmic form:

$$\log k = \log A - \frac{E_a}{2.303RT} \tag{6.2-20}$$

The activation energy of a reaction can be calculated from the integrated Arrhenius equation.

For most reactions, the steric factor and the frequency factor (both included in A), as well as the activation energy, are all nearly temperature-independent, so a plot of $\ln k$ (or any proportional quantity such as the natural logarithm of the initial reaction rate) against the reciprocal of the absolute temperature ($1/T$) is usually linear. From the slope of such a plot ($-E_a/R$), the activation energy can readily be calculated. Once known, it can be used to calculate k at any desired temperature from Eq. 6.2-19.

The rate constant for many homogeneous reactions increases by a factor of 2–3 per 10 °C increase in the temperature in the vicinity of room temperature. This holds for reactions with activation energies from 50 to $90\,kJ\,mol^{-1}$ (below and above this range, the factor is less than 2 and much higher than 3, respectively). This is a result of the rates of different reactions being affected to different extents by temperature changes, depending on their activation energies. Thus, from Eq. 6.2-21:

$$\ln k_{T_2} = \ln k_{T_1} - \frac{E_a}{R}\left(\frac{1}{T_2} - \frac{1}{T_1}\right) \tag{6.2-21}$$

The quantitative effect of temperature on the rate constant depends on the activation energy itself.

obtained by formulating Eq. (6.2-19) at two different temperatures and subtracting the two resulting expressions, it follows that heating a reaction mixture from 298 K (25 °C) to 308 K (35 °C) increases the rate constant (k_{T_2}) by only 14% if $E_a = 10\,kJ\,mol^{-1}$, but raises the factor to ca. 4 if $E_a = 100\,kJ\,mol^{-1}$ and ca. 50 if $E_a = 300\,kJ\,mol^{-1}$. Accordingly, slow reactions (those with small rate constant) will be more markedly affected by increasing temperatures than will fast reactions (those with large rate constants). In summary, the quantitative effect of temperature on the rate constant depends on the particular activation energy, and reactions that develop rapidly at room temperature can be slowed down by cooling, whereas slow reactions can be accelerated by heating.

6.2.4 Analytical use of the reaction rate

The determination of a chemical species by use of the kinetic region of a reaction (Fig. 6.2-1) entails direct or indirect measurements of its rate, which is related to the concentration as shown in Eq. 6.2-6. For this purpose, changes in the concentration of a reactant or product are measured as a function of time under the experimental conditions for a *pseudo* first-order reaction in order to construct a plot consisting of a rising or falling kinetic curve, respectively.

Determination methods based on pseudo first-order reactions

Some kinetic methods allow individual determinations of species as well as simultaneous determinations of two or more species in mixtures based on differential or integrated *pseudo* first-order reactions, even though the exact methodology used varies from application to application.

Determination of a single species

The most commonplace kinetic methods are applied to the initial portion of the curve (i.e., when the reaction has developed by only 1–3%). Such a portion is usually linear and its slope proportional to the concentration of the measured species. Under these conditions, the reaction can be assumed to be *pseudo* zero-order and Eq. 6.2-6 can be expressed in incremental form as:

$$v = -\frac{\Delta[A]}{\Delta t} = \frac{\Delta[P]}{\Delta t} = k'[A]_0 \tag{6.2-22}$$

from which the initial concentration of A, $[A]_0$, can readily be calculated. This differential method is called the *initial-rate method* and has several major assets, namely: (a) it is applicable to a variety of chemical reactions that are unaffordable by equilibrium methods because of their slowness or lack of quantitativeness; (b) it avoids problems associated with side reactions emerging as the process approaches completion; and (c) initial rate measurements made at the early stages of reaction are more precise since the slope of the kinetic curve and the signal-to-noise ratio obtained are both maximal.

The initial-rate analytical method can give precise kinetic data on reactions which are inaccessible to other methods of investigation.

Other widely used alternatives to the initial-rate method include the *fixed-time method* and the *variable-time method*. Both rely on Eq. 6.2-22 and constancy of Δt or $\Delta[A]$ (or $\Delta[P]$). Thus, application of the fixed-time method involves measuring the concentration of a reactant or product at a preset time from the start of the reaction, whereas the variable-time method entails measuring the time needed for the concentration of a reactant or product to reach a preset value. Therefore, $\Delta[A]$ (or $\Delta[P]$) and the reciprocal of time, $1/\Delta t$, will be linearly related to $[A]_0$, which allows one to construct a revealing calibration graph.

Fixed-time and variable-time methods offer alternatives to the initial-rate method.

The determination of a single species can be accomplished by using an integral method based on the integrated *pseudo* first-order equation Eq. 6.2-10 or 6.2-13; the procedure involves plotting $\ln[A]_t$ – or $\ln([P]_\infty - [P]_t)$ – against time. The initial concentration of A, $[A]_0$, can readily be calculated from the intercept of the straight line thus obtained. The fixed-time and variable-time integral methods can also be applied on the basis of Eq. 6.2-10 expressed in the following form:

$$\ln \frac{[A]_1}{[A]_2} = k_A \Delta t \tag{6.2-23}$$

by making $t_1 = 0$ (hence $[A]_1 = [A]_0$) and keeping Δt and $\Delta[A]$ constant, respectively.

Simultaneous kinetic determinations

Reaction-rate measurements are effective solutions to a perpetual analytical problem: the resolution of mixtures of closely related species. Methods for simultaneous kinetic determinations of species in mixtures rely on the different rate at which two or more species interact with a common reagent and allow the mixtures to be resolved with no prior separation. These methods are also referred to as "differential reaction-rate methods".

Kinetic multicomponent determination methods can be applied to reactions which are not pseudo first-order.

Even though a few methods exist for reactions other than those of *pseudo* first order, there are most frequently processed by using differential reaction-rate methods.

Let A and B be two substances in a mixture that react with a reagent R to yield two products P and P′ which, though different in nature, are essentially similar as regards the analytical property to be measured. Provided their respective reaction rates are sufficiently different and independent of each other (i.e., in the absence of a mutual kinetic or synergistic effect), and the process is first-order in each mixture component (i.e., $[R]_0 \gg [A]_0$ and $[R]_0 \gg [B]_0$), then the sum of the concentrations of A and B at time t will be given by:

$$C_t = [A]_t + [B]_t = [P]_\infty - [P]_t = [A]_0 \exp(-k_A t) + [B]_0 \exp(-k_B t) \qquad (6.2\text{-}24)$$

where k_A and k_B are the *pseudo* first-order rate constant of A and B, respectively.

Equation 6.2-24 is the starting point for the most popular differential kinetic methods, *viz.* the *logarithmic-extrapolation method* and the *proportional-equation method*. The applicability of these and other differential rate methods relies heavily on the relative values of the rate constants of the reactions involved and the concentration ratio between the mixture components, i.e., on k_A/k_B and $[A]_0/[B]_0$.

The *logarithmic-extrapolation method* involves taking logarithms in Eq. 6.2-24 and plotting $\ln C_t$ or $\ln([P]_\infty - [P]_t)$ as a function of time in order to obtain a curve if $k_A \neq k_B$ or a straight line otherwise. If species A disappears at a higher rate than does B (i.e., $k_A > k_B$), then $[A]_t \to 0$ and the curve eventually becomes a straight line, as shown in Fig. 6.2-4, a plot in decimal logarithmic form of Eq. 6.2-25:

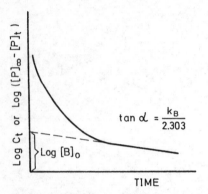

Fig. 6.2-4. Application of the logarithmic-extrapolation method to first-order reactions. Reproduced from Pérez-Bendito, D., Silva, M., *Kinetic Methods in Analytical Chemistry*, Chichester: Ellis Horwood, 1988

$$\ln C_t = \ln([P]_\infty - [P]_t) = \ln[B]_0 - k_B t \qquad (6.2\text{-}25)$$

The intercept of such a plot allows the initial concentration of the slower reacting component, $[B]_0$, to be calculated, that of the faster component being determined by difference ($[A]_0 = [P]_\infty - [B]_0$). If photometric monitoring is used, C_t, $[P]_\infty$ and $[P]_t$ can be substituted by absorbance readings.

The logarithmic extrapolation method does not require prior knowledge of rate constants.

The logarithmic-extrapolation method requires no prior knowledge of k_A or k_B – indeed, the latter can be calculated from the slope of the straight line. However, it does require at least 99% of the more reactive species to be consumed in the process if reliable results are to be expected, and if the total initial concentration of species A and B is to be accurately known.

The proportional equation method requires prior knowledge of rate constants.

Using the *proportional-equation method* for resolving a binary mixture entails measuring C_t at two reaction times and establishing two expressions similar to Eq. 6.2-24. By solving the equation system one can determine the initial concentration of each species provided their respective rate constants are known beforehand. Not only time, but indeed any other experimental variable such as the temperature or a physico-chemical property of the reactants can be used for this purpose. The equation system needed will be of the form:

$$f_1 = k_{A1}[A]_0 + k_{B1}[B]_0$$

$$f_2 = k_{A2}[A]_0 + k_{B2}[B]_0 \qquad (6.2\text{-}26)$$

where f_1 and f_2 denote a measurable parameter determined under two different sets of experimental conditions; and k_{A1}, k_{A2}, k_{B1} and k_{B2} are proportionality constants obtained separately for each component under different conditions.

For practical purposes, the rate-constant ratio, k_A/k_B, should be 3–4 or greater. However, some recent mathematical approaches such as the multipoint curve-fitting method, which takes advantage of the information provided by the entire kinetic curve (e.g., by using the Kalman filter), can be applied to mixtures with rate constant ratios of only 1.5–2.

6.2.5 Kinetic-based determinations involving uncatalyzed reactions

Application of kinetic methodology to uncatalyzed reactions such as those described above is expressly justified in two special situations where it clearly surpasses equilibrium methods, *viz.* when the process develops rather slowly or incompletely (and side reactions emerge as it approaches equilibrium) and when the available equilibrium method is rather labor-intensive and/or time-consuming. Otherwise there is little gain in sensitivity or accuracy over non-kinetic methods. In any case, the kinetic approach often provides better selectivity.

Uncatalyzed reactions are of special relevance to the analysis of mixtures of closely related compounds, for which various differential reaction-rate methods have been developed that are widely used in organic analyses and for the resolution of metal mixtures based on differences in the rate of ligand displacement from some complexes.

Kinetic methods based on uncatalyzed reactions for the determination of organic compounds, whether in isolation or in mixtures, rely on a variety of reactions including (a) oxidation of analytes; (b) bromination of substances possessing aromatic nuclei; (c) condensation or addition of amines and carbonyl compounds; and (d) hydrolysis.

Periodate ion is rather a commonplace reagent used for oxidizing organic hydroxylated compounds such as phenols, chlorophenols and vicinal glycols based on the Malaprade reaction. Phenols and their derivatives can be determined at concentrations from 50 to 500 $\mu g\,mL^{-1}$ by using a fixed-time method based on absorbance measurements at 340 nm, which is the maximum absorption wavelength of the resulting quinols and quinones [6.2-1], [6.2-2]. Some organic compounds of pharmacological interest including vitamins B_1 and C can also be determined by a redox reaction. Thiamine is oxidized by Hg(II) to thiochrome, a fluorescent compound on which its determination relies [6.2-3]. The kinetic method used for this purpose is quite sensitive (detection limit 2×10^{-8} M) and has so far been applied to the determination of thiamine in various mineral/ multivitamin formulations. Catecholamines are oxidized to *o*-benzoquinones by iridium hexachloride, and to aminochromes by periodate ion [6.2-4], which allows the determination of adrenaline and L-dopa at concentrations from 0.2 to 2.0 mM by using the stopped-flow technique. The use of periodate as oxidant enables resolution of the mixture.

Kinetic methods based on bromination reactions use bromine produced *in situ* from bromate and bromide ion in an acid medium, and Methyl Orange, the extent of decolorization of which is proportional to the concentration of analyte present. Cresols, xylenols, ethyl and phenyl phenols, paracetamol, and salicylic and acetyl-salicylic acid are some of the substances that can be determined in this way by using the variable-time method [6.1-1].

Uncatalyzed reactions are arousing steadily growing interest for the determination of individual species and resolution of mixtures as a result of the increasing availability of automated kinetic instrumentation, which has significantly expanded the number of determinations that can be routinely performed in various areas (see Table 6.4-1) – these used to be the sole patrimony of catalytic kinetic analysis.

6.3 The Phenomenon of Catalysis

6.3.1 General considerations

One of the ways of accelerating a slow reaction is by adding a minute amount of a catalyst. The chemical nature of catalysts varies widely. Thus, various metals, nonmetals and organic compounds (enzymes included) exert catalytic effects, thereby increasing the rate of some reactions in aqueous media. This phenomenon is known as *homogeneous catalysis*. If a solid phase (e.g., an electrode) or, in general, a two-phase system, is involved, the phenomenon is referred to as *heterogeneous catalysis*, which is not dealt with here.

Catalysis may be homogeneous or heterogeneous.

The recent inception of micellar media in experiments involving reaction rates has provided a novel means for increasing the rate of some catalyzed and uncatalyzed reactions. The ensuing physical phenomenon is called *micellar catalysis*, which can be considered a type of homogeneous catalysis despite the *pseudo*-phase system involved.

The current wide use of chemical catalysts in reaction-rate methods, which is the primary reason for their popularity, has been fostered by the fact that the rate of a catalyzed reaction is directly proportional to the catalyst concentration, which therefore allows development of highly sensitive methods for the determination of metals and nonmetals with catalytic properties. The high sensitivity of these methods originates from the fact that the catalyst is not consumed in the reaction, but takes part in it in a cyclic manner. Other catalysts (e.g., enzymes) are highly sensitive per se.

> A catalyst changes the rate of a reaction without modifying the overall change in the standard Gibbs energy of the reaction.

A catalyst can be defined as a substance that increases the rate of a chemical reaction without modifying the overall standard Gibbs energy change in the reaction; the process is called "catalysis" [6.3-1]. As such, a catalyst has several essential features, namely: (a) it remains chemically unchanged at the end of the reaction; (b) a small amount is often sufficient to cause the reaction to develop to a substantial extent; and (c) it does not affect the equilibrium position of a reversible reaction.

In the catalytic process, the activation energy is lowered as a result of the catalyst being continuously regenerated (catalytic cycle) in one of the reaction steps, which assures that its concentration is preserved constant throughout the process. Since the equilibrium constant for a given reaction, K, is equal to k_1/k_{-1} (k_1 and k_{-1} being the rate constants of the forward and reverse reaction, respectively) a catalyst will influence the reaction in both directions to the same extent, thereby accelerating attainment of equilibrium while not altering its position. A catalyst also increases the rate of nonspontaneous processes, so the term "negative catalysis" as applied to substances that raise the free activation energy is technically incorrect.

The reaction catalyzed by the substance to be determined is known as the "indicator reaction" or "uncatalyzed reaction"; for analytical purposes, its rate must be very low or negligible relative to the catalyzed reaction.

The rate of a chemical reaction can also be altered, whether increased or decreased, by the presence of some substances known as activators and inhibitors, respectively. The phenomenon of inhibition is usually associated with catalyzed enzymatic and nonenzymatic reactions, even though it has also been observed in other types of reaction (e.g., photochemical and electrochemical reactions). On the other hand, the phenomenon of activation is exclusively associated with catalyzed reactions. An activator can be defined as a substance that takes part in a reaction step whose activation energy is lower than that for the catalyst alone, so the activator does not exert its effect in the absence of a catalyst. This beneficial effect not only results in significantly improved sensitivity in the determination of the catalyst, but also facilitates that of the activator.

Both inhibitory and activating effects are of analytical use [6.1-1] inasmuch as they are proportional to the amount of inhibitor or activator present in solution, and both have been used in relation to enzymatic and nonenzymatic processes. In enzyme reactions, activation or inhibition arises from the presence of metal ions, whereas in nonenzyme reactions, the activating or inhibitory effect is exerted by a ligand (e.g., a hydroxylated or polyaminocarboxylic compound) or a metal ion.

6.3.2 Rate equation and reaction mechanism

The rate equation for a catalyzed reaction such as:

$$A + B \xrightarrow{C} P + Y \tag{VII}$$

where C is the catalyst, A and B the reactants, and P and Y the products, can be formulated on the basis of a mechanism involving the formation of a transient complex according to the following reaction sequence:

$$C + B \underset{k_{-1}}{\overset{k_1}{\rightleftharpoons}} CB + Y \tag{VIII}$$

$$CB + A \overset{k_2}{\rightleftharpoons} P + C \tag{IX}$$

where B, added in excess, forms the intermediate complex CB with the catalyst, and A is the monitored reactant. Two different cases can be considered according to whether reaction (VIII) or (IX) is the rate-determining step (rds). The overall rate of the process depends on the relative values of k_1, k_{-1} and k_2.

If step (IX) is the rds (i.e., if $k_2 \ll k_{-1}$), then the complex dissociation reaction will be much faster than that between the complex and reactant. In this situation, the process is said to be in "pre-equilibrium". If equilibrium of reaction (VIII), represented by the constant:

> The pre-equilibrium mechanism applies if the dissociation of the complex is much faster than the reaction between complex and reactant.

$$K_C = \frac{k_1}{k_{-1}} = \frac{[CB][Y]}{[C][B]} \tag{6.3-1}$$

is assumed to occur first, then the overall reaction rate will coincide with that of the rds, i.e., step (IX), so

$$-\frac{d[A]}{dt} = k_2[CB][A] \tag{6.3-2}$$

Substitution of [C] and [B] in Eq. 6.3-1 by $[B]_0 - [CB]$ and $[C]_0 - [CB]$, respectively ($[B]_0$ and $[C]_0$ being the initial concentration of species B and A), yields:

$$K_C = \frac{[CB][Y]}{[B]_0([C]_0 - [CB])} \tag{6.3-3}$$

Solving this equation for [CB] and substitution into Eq. 6.3-2 yields the overall rate equation:

$$-\frac{d[A]}{dt} = k_2 \frac{K_C[B]_0[C]_0}{K_C[B]_0 + [Y]} [A] \tag{6.3-4}$$

where [A] is the concentration of reactant A at time t.

If step (VIII) is the rds (i.e., $k_2 \gg k_{-1}$ and $k_2 \gg k_1$), then the reaction of the intermediate complex with reactant A will be much faster than the complex formation or dissociation. In this situation, the concentration of CB in the system will always be constant and very low, and the process said to be in a "steady state". Accordingly:

> The steady-state mechanism applies if the reaction of the intermediate complex is much faster than formation or dissociation of the complex.

$$\frac{d[CB]}{dt} = 0 = k_1[C][B] - k_{-1}[CB][Y] - k_2[CB][A] \tag{6.3-5}$$

Solving this equation for [CB] and substituting [C] and [B] for their expressions above provides the concentration of complex CB:

$$[CB] = \frac{k_1[C]_0[B]_0}{k_1([C]_0 + [B]_0 - [CB]^2) + k_{-1}[Y] + k_2[A]} \tag{6.3-6}$$

Neglecting $[CB]^2$, which is small relative to $[C]_0 + [B]_0$, in the denominator of this equation, and, taking into account that, according to the steady-state condition:

$$-\frac{d[A]}{dt} = k_1[C][B] - k_{-1}[CB][Y] = k_2[CB][A] \tag{6.3-7}$$

substitution for [CB] given in Eq. 6.3-6 into the right-hand side of Eq. 6.3-7 gives:

$$-\frac{d[A]}{dt} = \frac{k_1 k_2[C]_0[B]_0[A]}{k_1([C]_0 + [B]_0 - [CB]^2) + k_{-1}[Y] + k_2[A]} \tag{6.3-8}$$

which is the general rate equation for a process in its steady-state.

The general equations 6.3-4 and 6.3-8 can be simplified for practical purposes by replacing [A] with $[A]_0 - [P]$ ([P] being the decrease in the concentration of reactant A or increase in that of product P, measured during the reaction) since most analytical catalyzed reactions are monitored via the product. Therefore:

$$-\frac{d[A]}{dt} = \frac{d[P]}{dt} = K\alpha_c[C]_0([A]_0 - [P]) \tag{6.3-9}$$

where α_c is a function containing the remainder of the concentration terms appearing in Eq. 6.3-8 except for $[C]_0$, $[A]_0$ and $[P]$, and K is a term including all the rate or equilibrium constants involved.

On the other hand, if the uncatalyzed reaction takes place at an appreciable rate reflected in a non-negligible value for constant k_3, then the overall reaction rate will be given by:

$$\left(\frac{d[P]}{dt}\right)_{obs} = K\alpha_c[C]_0([A]_0 - [P]) + k_3[B]_0([A]_0 - [P]) \tag{6.3-10}$$

At times close to the start of the reaction, [P] can be neglected relative to $[A]_0$ and Eq. 6.3-10 be rewritten as:

$$v_0 = \left(\frac{d[P]}{dt}\right)_{t=0} = K'[C]_0 + K'_1 \tag{6.3-10}$$

where K' and K'_1 are both constant provided $[A]_0$ and $[B]_0$ are sufficiently greater than $[C]_0$, and $[B]_0 \gg [A]_0$. For this equation to hold, the reactions must be made *pseudo* first-order in the monitored reactant since the catalyst concentration, by definition, does not change during the reaction.

Chelate- and complex-formation reactions between the catalyst and a reactant are highly significant to the catalytic cycle represented by reactions (VIII) and (IX), and are the basis for a host of catalytic determinations. Many reactions, some of which involve an enzyme as the catalyst, behave according to a pre-equilibrium mechanism. The enzyme associates to a given reactant called the substrate and is released from the enzyme-substrate complex during the rate-determining step so it can catalyze the decomposition of another substrate molecule and restart the catalytic cycle.

One other way in which the catalyst may interact with one ingredient of the indicator reaction involves reacting with A to yield product P and its own activated form, C^*, which can subsequently react with B to yield the other product, Y, according to the following sequence:

$$A + C \rightarrow P + C^* \tag{X}$$

$$C^* + B \xrightarrow{\text{fast}} Y + C \tag{XI}$$

where step (X) is the rds. Many redox catalytic reactions are based on this type of mechanism, where formation of species C^* frequently involves a change in the oxidation state of the catalyst. For this mechanism to be feasible, the system in question must meet two essential requisites, namely: (a) the oxidation potential of the catalytic system, E_C, should be more positive than that of the P/A couple and more negative than that of the B/Y couple (i.e., $E_{B/Y} > E_C > E_{P/A}$); and (b) a direct interaction between A and B, though thermodynamically permitted, should be kinetically forbidden. In addition, the reaction between the catalyst and B should be very fast.

The general mechanisms discussed above are not so simple in practice; in fact, each step may involve more than one stage, one of which will be the actual rds.

Nonenzyme-catalyzed reactions

The rate equations applicable to reactions catalyzed by compounds other than enzymes are the simplified equations 6.3-9 and 6.3-10. Such reactions include ordinary catalytic reactions and Landolt-type reactions. The former encompass a variety of interactions including redox, chemiluminescence, acid–base and complex-formation reactions.

Redox reactions are by far the most commonly used indicator reactions in kinetic catalytic methods and typically involve such oxidants as hydrogen per-oxide, atmospheric oxygen, bromate, periodate, iodate and peroxydisulfate ions, Fe(III) and Ce(IV); inorganic reductants such as tin(II), iron(II), arsenic(III), iodide and thiosulfate ion; and a wide variety of organic reductants including amines, phenols and azo dyes, among others. The catalysts involved are usually multivalent metal ions that possess vacant d-orbitals and can form complexes with one of the ingredients of the indicator reaction (e.g., an organic reductant). They are generally quadrivalent (Zr, Hf, Th), quinquevalent (V, Ta, Nb), sexivalent (Mo, W), di- or trivalent (Fe, Mn, Cu, Co) metals, or metals of the platinum family (Ag, Os, Pd, Ru, Pt, Rh, Ir). Only a few inorganic anions possess catalytic activity, though. Among them, iodide, nitrite, sulfide and bromide are the most commonly used as catalysts. Some metal ions have a preferential accelerating effect on the oxidation of inorganic or organic substances by a given oxidant (e.g., manganese acts almost exclusively on reactions involving periodate ion, while vanadium ion interacts preferentially with those involving bromate).

Some of the more typical examples of redox catalyzed reactions include the iodide-catalyzed Ce(IV)–As(III) system (the Sandell–Kolthoff reaction); the iodine–sodium azide system, catalyzed by sulfur-containing compounds; the de-composition of hydrogen peroxide by Cu(II), Co(II), Mn(II) and Fe(III); the hydrogen peroxide–iodide system, catalyzed by Fe(III), Mo(VI), W(VI), Zr(IV) and Hf(IV); the Cu(II)-catalyzed hydrogen peroxide–hydroquinone system; and the Mn(II)-catalyzed periodate–Malachite Green system.

The Sandell–Kolthoff reaction is carried out in a sulfuric medium, even though the catalytic activity of iodide is reportedly 20 times higher in nitric acid. Chloride ion inhibits the catalyzed reaction at concentrations above ca. 0.34 M, but accel-erates it at low concentrations. The kinetic and mechanism for this reaction have been thoroughly studied [6.3-2]. The proposed mechanism, consistent with experi-mental facts, is as follows:

Catalyzed redox reactions are widely used in kinetic analysis.

$$I^- + Ce(IV) \xrightarrow{k_1} Ce(III) + I \tag{XII}$$

$$I + As(III) \underset{k_{-2}}{\overset{k_2}{\rightleftharpoons}} Intermediate \tag{XIII}$$

$$Intermediate + Ce(IV) \xrightarrow{k_3} As(V) + Ce(III) + I^- \tag{XIV}$$

$$I + Ce(IV) \xrightarrow{k_4} Ce(III) + I^+ \tag{XV}$$

$$I^+ + As(III) \underset{k_{-5}}{\overset{k_5}{\rightleftharpoons}} As(V) + I^- \tag{XVI}$$

It should be noted that reactions (XII), (XIII) and (XIV) on the one hand, and (XII), (XIII) and (XVI) on the other, represent two different catalytic cycles that do not conform to experimental facts individually.

The cerium–arsenite reaction is a typical catalyzed redox reaction.

The interaction between a catalyst and an organic substance (e.g., an amine or phenol) usually involves the prior formation of a charge-transfer complex between both, the stability of which is directly related to the pH of the medium; hence the marked influence of this experimental variable on the catalytic activity of a metal ion. On the other hand, inorganic substrates interact with the catalyst via electron transfers and may give rise to unstable intermediates in other steps, as in the mechanism for the Ce(IV)–As(III) system above.

Chemiluminescence (CL) reactions, which are accelerated by traces of metal ions, typically involve an oxidation process. The best known CL reaction is the oxidation of luminol (5-amino-2,3-dihydrophthalazine-1,4-dione) by hydrogen peroxide. Extremely small amounts of such metal ions as Co(II), Cu(II), Cr(III), Mn(II), Fe(III) and Ni(II) at pH 10–11 enhance the release of radiant energy, Luminol is converted into a doubly charged anion which is subsequently oxidized to an excited single state that emits radiation on decomposing to aminophthalate ion according to the following scheme [6.3-3]:

Chemiluminescence reactions often involve redox processes.

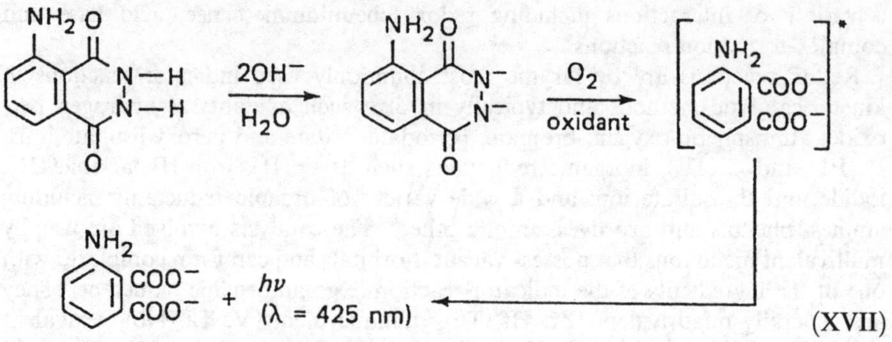

(XVII)

Alternative oxidants for this purpose include atmospheric oxygen and peroxydisulfate ion. Though much less frequently than luminol, lucigenin and lofine are also used as the substrates in this type of reaction.

Landolt reactions also involve a redox process in most cases (an acid–base or complexation reaction in others). This type of reaction is based on the Landolt effect [6.3-4], which reflects in the occurrence of an induction period when a slow reaction is coupled to a fast one via the reduction product of the first according to

$$A + B \xrightarrow[\text{slow}]{k_1} P \tag{XVIII}$$

$$P + L \xrightarrow[\text{fast}]{k_2} Y \tag{XIX}$$

Because the second reaction is faster than the first, its product (P) can only be detected after L (the "Landolt reagent") has completely disappeared by the second reaction. If reaction (XVIII) is accelerated by a catalyst (Fig. 6.3-1), the time elapsed until the product appears will be a measure of the catalyst concentration, which in turn is directly related to the length of the induction period, t_i, by an empirical equation of the form $[C]_0 = K_i/t_i$ or $[C]_0 = K_i/t_i^2$, depending on the system concerned. One typical Landolt reaction is the oxidation of iodide ion by H_2O_2, BrO_3^- or $S_2O_8^{2-}$, catalyzed by molybdenum, vanadium and copper in the presence of ascorbic acid or thiosulfate ion as the Landolt reagent.

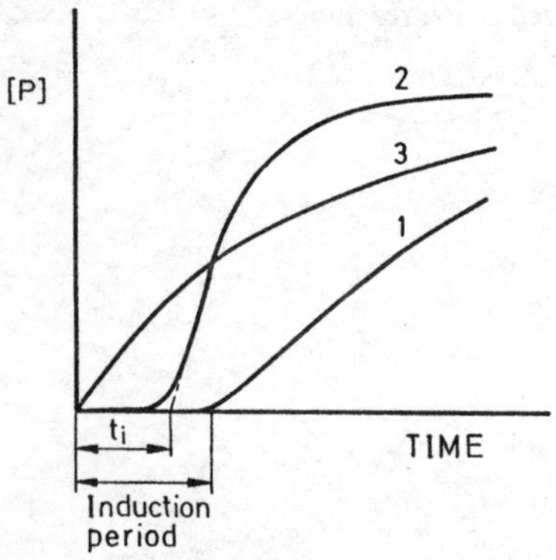

Enzyme-catalyzed reactions

Enzyme reactions are a special type of catalytic reaction. A monosubstrate enzyme reaction can be formulated in quite simple terms:

$$E + S \underset{k_2}{\overset{k_1}{\rightleftharpoons}} ES \tag{XX}$$

$$ES \xrightarrow{k_3} E + P \tag{XXI}$$

where E, S, ES and P denote the enzyme, substrate, enzyme–substrate complex and product, respectively. However, the kinetics of an enzyme-catalyzed reaction can be complicated by the occurrence of several ES complexes. On the assumptions that [E] ≪ [S] and step (XXI) is the rds (i.e., a pre-equilibrium mechanism), Michaelis and Menten formulated the reaction rate as:

$$\frac{d[P]}{dt} = v_0 = k_3[ES] \tag{6.3-11}$$

Because the previous reaction, (XX), is at equilibrium:

$$K = \frac{k_1}{k_2} = \frac{[ES]}{[E][S]} \tag{6.3-12}$$

taking into account that $[ES] = [E]_0 - [E]$ ($[E]_0$ being the overall concentration of enzyme), substituting [E] into Eq. 6.3-12 and solving it for [ES] allows subsequent

Most Landolt reactions involve redox processes; some involve acid–base or complexation processes.

Fig. 6.3-1. Kinetic curves for Landolt reactions obtained in the absence (1) and presence (2) of a catalyst. Curve 3 corresponds to the indicator reaction (A + B → P). Reproduced from Pérez-Bendito, D., Silva, M., *Kinetic Methods in Analytical Chemistry*, Chichester: Ellis Horwood, 1988

substitution into Eq. 6.3-11 in order to obtain:

$$v_0 = \frac{k_3 K[E]_0[S]}{1 + K[S]} = \frac{k_3[E]_0[S]}{K_S + [S]} \qquad (6.3\text{-}13)$$

where $K_S = 1/K$.

There are two possible extreme situations. In one, [S] is so small that $K[S] \ll 1$ and the rate is approximately given by:

$$v_0 \simeq k_3 K[E]_0[S] \qquad (6.3\text{-}14)$$

which is consistent with the observed fact that the reaction is first-order in S when its concentration, [S], is quite small (Fig. 6.3-2). On the other hand, if [S] is so large that $K[S] \gg 1$, then:

$$v_0 \simeq k_3[E]_0 \qquad (6.3\text{-}15)$$

based on the observed fact that the reaction is zero-order in S when its concentration is high enough (Fig. 6.3-2). The relative value of [S] obviously depends on that of K.

On the assumption that k_3 need not necessarily be much smaller than k_2, the steady-state approach provides a plausible explanation for the behavior of many enzyme reactions. Based on this principle, [ES] remains constant, so $d[ES]/dt = 0$ and:

$$k_1[E][S] = k_2 + k_3[ES] \qquad (6.3\text{-}16)$$

On the same grounds as for the pre-equilibrium mechanism, an expression similar to Eq. 6.3-13 can thus be formulated:

$$v_0 = \frac{k_3[E]_0[S]}{\dfrac{k_2 + k_3}{k_1} + [S]} = \frac{k_3[E]_0[S]}{K_m + [S]} \qquad (6.3\text{-}17)$$

which is called the "Michaelis–Menten equation", the term $K_m = (k_2 + k_3)/k_1$ being referred to as the "Michaelis–Menten constant". For $k_3 \ll k_2$, $K_m = K_s$ in Eq. 6.3-13, which thus becomes a special instance of Eq. 6.3-17. Therefore, K_m is only a measure of the affinity of an enzyme for the substrate when $k_3 \ll k_2$ since only under such conditions does it equal K_S. On the other hand, K_m is an empirical constant coinciding with the substrate concentration when the reaction rate is half its maximum value, v_{max} (Fig. 6.3-2), since, taking into account that $k_3[E]_0 = v_{max}$, if $K_m = [S]$ then Eq. 6.3-17 simplifies to:

$$v = \frac{v_{max}[S]}{2[S]} = \frac{v_{max}}{2} \qquad (6.3\text{-}18)$$

The enzyme concentration in an enzyme-catalyzed reaction is so small that the prior equilibrium is driven nearly to completion by excess substrate. This allows one to determine whether a reactant takes part in a prior equilibrium or the rate-determining step.

In addition to their high sensitivity, enzyme-catalyzed reactions are extremely selective. Enzymes are protein catalysts that act on a given substrate only. For example, of all the biochemical reactions that take place in the human body, the hydrolysis of urea is the only one whose rate is affected by the presence of an enzyme called urease, which catalyzes its conversion into ammonium and carbonate ions:

$$CO(NH_2)_2 + 2\,H_2O + H^+ \xrightarrow{\text{urease}} 2\,NH_4^+ + HCO_3^- \qquad (XXII)$$

Because urease is not consumed in the process, a very small amount of enzyme can effectively accelerate the hydrolysis of a very large amount of urea. This reaction is very slow [its half-time at normal body temperature (37 °C or 310 K) is ca. 10^5 years] but can be substantially expedited by using urease.

Glucose oxidase is one other highly specific enzyme. It catalyzes the oxidation of β-D-glucose to gluconic acid in the presence of many other sugars, some of which also react, but at a much lower rate.

Enzymes are quantified in international activity units. An enzyme unit is defined

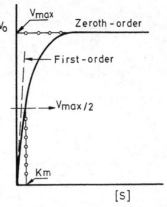

Fig. 6.3-2. Effect of the substrate concentration on the rate on an enzyme-catalyzed monosubstrate reaction

The Michaelis–Menten constant can be used to measure enzyme–substrate affinities.

as the amount of enzyme that transforms 1 μmol of its substrate at 25 °C in 1 min under optimal experimental conditions including the pH, ionic strength and substrate concentration.

Micellar catalysis

Micellar catalysis is a special type of catalysis. Micelles formed by surfactants at high enough concentrations in aqueous solutions are known to have the ability to alter the rate of chemical reactions as a result of the reactants being attracted to the micellar surface and the ensuing concentrating effect resulting in more rapid reaction. Reactants can be accommodated in various ways in micelles, viz. inside their hydrophobic core, adsorbed on the surface layer by electrostatic interactions, etc. Even though the micelle concentration does not change during the process, micelles are not strictly catalysts since they are not involved in any reaction step in a cyclic manner. The effect is thus one of physical rather than chemical catalysis.

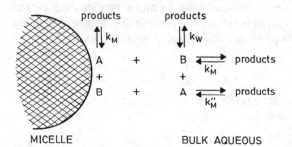

Fig. 6.3-3. Interactions in a bimolecular reaction in the presence of micelles. Reproduced from Pérez-Bendito, D., Rubio, S., *Trends Anal. Chem.*, 12, 9, 1993

The pseudo-phase kinetic model can be applied to micellar catalysis.

Figure 6.3-3 illustrates the most likely interactions in a bimolecular reaction between two species A and B in the presence of micelles [6.3-5]. k_M and k_w denote the rate constants of the reactions taking place in the micellar and aqueous phase, respectively, and k'_M and k''_M are the rate constants used when only one of the reactants associates with the micelles.

According to the *pseudo*-phase kinetic model, the second-order rate constant for a bimolecular reaction is given by [6.3-6]:

$$k_{exp} = \frac{(k_M P_A P_B + k'_M P_A + k''_M P_B)CV + k_w(1 - CV)}{(1 + k_A C)(1 + k_B C)} \tag{6.3-19}$$

where subscripts M, w, A and B denote the micellar phase, aqueous phase and reactants; P_A and P_B are the partition coefficients of the reactants (e.g., $P_A = [A]_M/[A]_w$); C is the surfactant concentration (molarity) minus the critical micelle concentration (cmc); V is the surfactant molar volume [the CV and $(1 - CV)$ factors thus represent the volume fractions of the micellar and aqueous phase, respectively]; and k_A and k_B are the binding constants of the reactants, viz. $k_A = (P_A - 1)V$ and $k_B = (P_B - 1)V$.

The efficiency of micellar catalysis is measured as the ratio between the reaction rate in the presence and absence of micelles, which depends on two factors according to the following equation:

$$\frac{k_{exp}}{k_w} = \frac{k_M}{k_w} \frac{k_A k_B}{V(\sqrt{k_A} + \sqrt{k_B})^2} \tag{6.3-20}$$

The first factor, k_M/k_w, takes account of changes in the intrinsic reactivity of the reactants in the presence of the micellar medium, whereas the second, $k_A k_B/V(\sqrt{k_A} + \sqrt{k_B})^2$, depends on how effectively the reactants are concentrated on the micellar surface, which in turn is a function of the micelle binding constants for the reactants (k_A and k_B) and the surfactant molar volume (V). Experiments involving various reactions and models have shown concentration of the reactants in a small volume of the micellar *pseudo*-phase to be responsible for the increased rates observed. This is of a great analytical potential since it can be exploited to improve the quality (sensitivity and/or selectivity) of kinetic-based determinations involving both catalyzed and uncatalyzed reactions.

6.3.3 Analytical use of catalyzed reactions

Whatever the mechanism, Eqs. 6.3-9 and 6.3-10 in Sec. 6.3.2 are equally valid for the kinetic determination of catalysts, which entails recording – usually in an automatic or semiautomatic fashion – the kinetic curve (i.e., changes in the concentration of the indicator species with time). The curve will consist of a rising or falling portion depending on whether the monitored species is a reactant or product, respectively (see Fig. 6.1-1). Concentrations can be substituted by any measurable property provided it is proportional to the concentration. A physical technique is usually employed for this purpose, viz. to facilitate continuous measurements (e.g., absorbance, potential, temperature, luminescence, conductivity) of the analytical signal.

Methods for the determination of catalysts are essentially identical to those involving noncatalyzed reactions. Those based on measurement of the induction period were described above in dealing with Landolt reactions. The initial-rate method is applied on the assumption that [P] is negligible relative to [A]. The latter can be considered a constant in Eq. 6.3-9, which can thus be simplified to Eq. 6.3-11; a plot of this latter showing the variation of [P] through the measured property as a function of times will thus be a straight segment (Fig. 6.3-4a), the slope of which is given by:

$$v_0 = \frac{d[P]}{dt} = \tan\alpha = \frac{\Delta[P]}{\Delta t} = k'[C]_0 + k'_1 \tag{6.3-21}$$

This is a variant of Eq. 6.3-11. It is directly related to the catalyst concentration and is the ultimate basis for application of this method. By plotting v_0 ($\tan\alpha$) as a function of the catalyst concentration, $[C]_0$, for several samples containing known concentrations of the catalyst (Fig. 6.3-4b), a straight line of zero (A) or nonzero intercept (B) will be obtained depending on whether the uncatalyzed reaction develops to a negligible or appreciable extent, respectively.

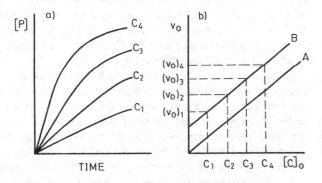

Fig. 6.3-4. Implementation of the initial-rate method. Reproduced from Pérez-Bendito, D., Silva, M., *Kinetic Methods in Analytical Chemistry*, Chichester: Ellis Horwood, 1988

The *differential fixed-time method* entails measuring the concentration of a reactant or product at a preset time from the start of the reaction and using Eq. 6.3-21 as solved for $\Delta[P]$:

$$\Delta[P] = k'[C]_0\Delta t + k'_1\Delta t \tag{6.3-22}$$

Since Δt is constant, a plot of $\Delta[P]$ as a function of $[C]_0$ will be a straight line of slope $k'\Delta t$ and intercept $k'_1\Delta t$, as shown in Fig. 6.3-5 for $k'_1 = 0$. The *differential variable-time method*, also known as the "fixed-concentration" or "constant-concentration" method, entails measuring the time needed for a present change in the reaction medium to take place. Solving Eq. 6.3-22 for $1/\Delta t$ yields:

Kinetic methods can be used for the determination of a catalyst.

$$\frac{1}{\Delta t} = \frac{k'[C]_0 + k'_1}{\Delta[P]} \tag{6.3-23}$$

Since $\Delta[P]$ is constant, a plot of $1/\Delta t$ against $[C]_0$ will be a straight line of slope $k'/\Delta[P]$ and intercept $k'_1/\Delta[P]$, as shown in Fig. 6.3-6 for $k'_1 = 0$.

If [P] is not negligible relative to $[A]_0$ in Eq. 6.3-9, the equation must be integrated over a finite, though not necessarily short, time interval. This is the basis for integrated kinetic methods (the tangent method and the fixed- and variable-time method), which are less commonplace than differential methods.

These three methods are used for the determination of metal ions and species

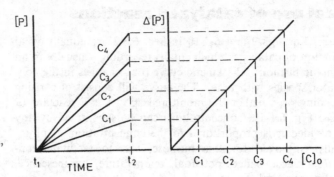

Fig. 6.3-5. Implementation of the fixed-time method. Reproduced from Pérez-Bendito, D., Silva, M., *Kinetic Methods in Analytical Chemistry*, Chichester: Ellis Horwood, 1988

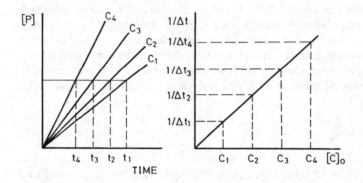

Fig. 6.3-6. Implementation of the variable-time method. Reproduced from Pérez-Bendito, D., Silva, M. *Kinetic Methods in Analytical Chemistry*, Chichester: Ellis Horwood, 1988

Kinetic methods can also be used for determination of the substrate in an enzymatic reaction.

other than enzymes (e.g., some inorganic anions). Occasionally, they are also used for determining a reactant (e.g., H_2O_2) involved in the indicator reaction. In any case, the primary application of these methods is for the determination of catalysts on account of the high sensitivity they provide (detection limits typically lie in the 10^{-9} to 10^{-7} M range but are occasionally as low as 10^{-12}–10^{-10} M). Such a high sensitivity can be further increased by using an activator, which is quite a common practice in kinetic catalytic analysis.

One other way of raising the sensitivity in kinetic catalytic analysis is by using a micellar medium. Micellar catalysis (of physical nature) and chemical catalysis are combined in a number of methods the sensitivity of which is higher by up to one order of magnitude than those of their counterparts involving a purely aqueous medium. This can be accomplished via different mechanisms, one of which involves concentration of the catalyst at the micellar surface by formation of a complex with one of the reactants of the indicator reaction [6.3-5].

Enzyme-catalyzed reactions are used for the determination of substrates in preference over that of the enzyme itself (its activity). In addition to the high sensitivity obtained in the determination of substrates (and other substances in an indirect manner), a high selectivity can also be achieved by using a suitable enzyme. Substrate determinations are based on Eq. 6.3-17, where [S] should be much less than K_m in order that the equation may be simplified to:

$$v_0 = \frac{k_3[\mathrm{E}]_0}{K_m}[\mathrm{S}] = \frac{v_{\max}}{K_m} = k'[\mathrm{S}] \tag{6.3-24}$$

which is a first-order rate equation (Fig. 6.3-2). From Eq. 6.3-24 it immediately follows that the rate of the enzyme reaction is proportional to the substrate concentration at a given, constant enzyme concentration, so the above-described initial-rate, fixed-time and variable-time methods are all applicable. For the required proportionality to hold in practice, the $[\mathrm{S}]/K_m$ ratio should be not higher than 0.2, and preferably lower than 0.05. At a constant substrate concentration, the reaction rate will be proportional to the enzyme concentration, as can be inferred from the first expression of Eq. 6.3-24.

Kinetic determinations of metal catalysts (and other substances), substrates and enzymes are so very numerous that they defy even summarization. Readers interested in the determination of metal catalysts are referred to the monographs cited in Sec. 6.1, as well as some reviews [6.3-7], [6.3-8]; the literature on enzyme reactions is overwhelmingly vast and well known.

6.4 Monitoring of the Analytical Signal in Kinetic and Catalytic Methods

The rapid, steadily growing development of methods based on reaction rates (i.e., the kinetic region in Fig. 6.1-1) is primarily the result of recent breakthroughs in analytical instrumentation based on the availability of increasingly more powerful computers for automation of analytical tasks. This has resulted in increasingly higher accuracy and precision in analytical measurements and avoidance of the typical problems posed by traditional time measurements. In fact, a growing number of commercially available instruments are currently furnished with special devices for kinetic measurements.

Rather than using the absolute value of the property to monitor the course of the analytical reaction (absorbance, fluorescence, potential), kinetic methods use its variation as a function of time, so static signals arising from the sample background cause no error. This is a clear advantage over static methods. On the other hand, kinetic methods call for strict control of time and temperature. In fact, the transduced signal to be processed should be acquired as accurately as possible as regards time. The temperature should also be strictly controlled (oscillations should never exceed ± 0.01–$0.1\,°C$) as it has a marked effect on the reaction rate (see Sec. 6.2.3).

The technique of choice to measure the rate of a reaction is dictated by its half-time. Thus, the instrumentation needed for monitoring slow reactions (those with half-times longer than $10\,s$) is simpler than that required for fast reactions (those with half-times shorter than $10\,s$) as regards both mixing of the reactants and measurement and acquisition of the analytical signal.

The analytical procedure involved in calculating the rate of a reaction consists of three steps, namely: (a) preparation, measurement, transfer and mixing of the reactants (i.e., of sample and reagents); (b) monitoring and transducing of the signal at a constant temperature, which are done by the measuring system (a photometer or potentiometer, for example), set to follow the course of the reaction through changes in a property of one reactant or product; and (c) times acquisition of the data produced for simultaneous or sequential processing by manual computation or with the aid of a computer. Completely automated systems take over all three steps, while partly automated systems usually perform steps (b) and (c) only.

General procedures used in kinetic methods involve three main steps.

The first step, mixing of sample and reagents, can be carried out in a closed or open system depending on whether or not the external experimental conditions remain constant. Closed systems are suitable for slow reactions (e.g., catalyzed reactions), whereas open systems are better suited to reactions with fast kinetics, even though they can also be advantageously applied to slow reactions in order to implement automatic routine analyses.

6.4.1 Closed and open systems

Kinetic methods entail mixing sample and reagent(s) thoroughly at a constant temperature prior to transfer to the detection system for measurement. This can be done manually (as is usually the case in closed systems) or automatically.

Manual mixing can be performed simply by homogenizing the reactants in the reaction vessel, which can be a photometric cuvette or electrode cell. The instant at which the last reactant is added (usually from a syringe) is taken as the start of the reaction ($t = 0$) and must therefore be accurately known. As a rule, the reactants are thermostated prior to mixing, as is the cell compartment, in order that the reaction mixture may reach the desired temperature in a very short time. Alternatively, the sample and reagents can be mixed in a volumetric flask and made to the mark, and a portion of the diluted mixture be transferred to the detection cell. The reactants can also be incorporated into the measuring system automatically in order to boost reproducibility. Such is the case with the computer-controlled Technicon RA 1000 analyzer and centrifugal analyzers equipped with photometric detectors, which are widely used in clinical laboratories.

Automated mixing of reactants calls for special mechanical devices when fast reactions are involved as these must be managed in open systems, which can be of the *continuous-flow* and *discrete* (*batch*) type. The stopped-flow technique is the most frequently use choice in the former type of system (for fast reactions), while stat methods and the continuous-addition-of-reagent (CAR) technique are the two most commonly employed choices in the latter. They are described in the following section.

6.4.2 Automated kinetic analysis

"Stat" methods can be used to monitor catalyzed reactions in open systems.

Stat methods are the most frequently used choices for monitoring catalyzed reactions in open systems. They involve adding a reactant to the reaction vessel at a given rate such that a characteristic property of the monitored reaction is kept constant. At the beginning, a small amount of an appropriate reaction ingredient is added until a preset value of the monitored parameter (pH, absorbance) is reached. Any deviation from this state as a result of the reaction developing further is immediately offset by automatically supplying an additional amount of the ingredient concerned, the speed of signal restoration being proportional to the catalyst concentration. This approach has been used, among others, for the kinetic determination of iodide ion by its catalytic effect on the reaction between Ce(IV) and As(III), monitored photometrically via ceric ion added as titrant at a variable rate.

The continuous-addition-of-reagent technique can be used to plot the kinetic curve.

The continuous-addition-of-reagent (CAR) technique is operationally similar to the above-described discrete approach inasmuch as it relies on the continuous addition of a reagent at a constant rate to the determinand in order to obtain kinetic curve, the slope of the straight portion of which will be proportional to the analyte concentration. The reaction rate can be monitored spectrophotometrically, fluorimetrically or even by chemiluminescence measurements. Figure 6.4-1 shows a block diagram of the instrumental setup used for this purpose [6.4-1], which consists of an autoburette (addition unit), a photometric or fluorimetric detector including an immersion probe for optical signal measurement, and a computer. This open-system batch technique is specially suited to fast uncatalyzed reactions and an advantageous alternative to the stopped-flow technique as regards instrumental and operational simplicity. The CAR technique allows mixtures of two or more analytes to be directly resolved without the need to use a differential rate method (see Sec. 6.2.4). The signal *vs* time plot obtained for a binary mixture typically shows two consecutive linear segments of different slope from which each reaction rate can be determined and subsequently related to the concentration of each component.

The stopped-flow technique is mainly used for fast reactions.

The stopped-flow mixing technique usually involves mixing the sample solution and reagent(s) by means of two driving syringes actuated automatically by a pneumatic device, in a flow-cell or mixing chamber that can also act as the observation cell. The flow is abruptly stopped by means of a third syringe and the analytical signal is then recorded as a function of time. This technique is usually required for fast reactions (with half-times of a few milliseconds to a few seconds). Its performance relies heavily on the instrument's dead time (determined by the mixing time and the transfer and stop times), which should be smaller that the reaction half-time by ca. two orders of magnitude. Figure 6.4-2 depicts a stopped-flow spectrophotometer (for spectrofluorimeter) coupled to an autosampler, under the

Fig. 6.4-1. Experimental setup used for implementation of the continuous-addition-of-reagent technique. Reproduced from Márquez, M., Silva, M., Pérez-Bendito, D., *Anal. Chim. Acta*, 237, 353, 1990

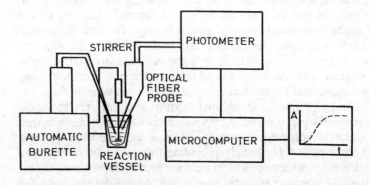

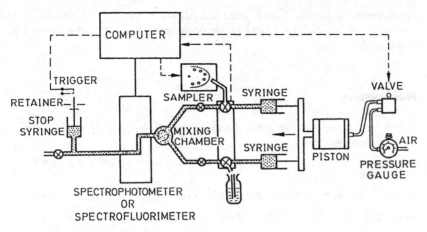

Fig. 6.4-2. Scheme of a stopped-flow spectrophotometer (or spectrofluorimeter). Reproduced from Pérez-Bendito, D., Silva, M., *Kinetic Methods in Analytical Chemistry*, Chichester: Ellis Horwood, 1988

Table 6.4-1. Some selected applications of kinetic non-enzymatic methods in environmental, clinical and pharmaceutical analysis

Sample	Species	Comments
Drinking water	Manganese	– Automatic photometric method – Standard-addition method with fluorimetric detection[a] – Segmented-flow system (30 samples/h)
	Vanadium	Prior ion-exchange separation
	Zinc, magnesium, copper and nickel[b]	Mixture resolution
	Iodide	Automatic photometric monitoring
	Phosphate, silicate	Uses miniature centrifugal analyzer
Tap water	Cobalt[c]	Removal of interfering Fe and Mg
	Nitrite	Stopped-flow method (360 samples/h)
	Hypochlorite[c]	Stopped-flow method
River water	Copper	Compared with non-flame AAS
	Iron	Also applied to sea and tap water
	Iodate, iodide	Uses Technicon I AutoAnalyzer
	Carbaryl + 1-naphthol[a]	Stopped-flow; also determined in vegetables
Sea water	Calcium + magnesium[b]	Stopped-flow mixture analysis
	Chlorinated compounds[c]	Also applied to other types of water
Waste water	Cadmium + manganese[b]	Stopped-flow method
	Cyanide	Electrolytic plating sewage
	Benzaldehyde, formaldehyde[a]	Used in manufacturing synthetic fibers
Air	Sulphide[c]	Stopped-flow/microdistillation system
	Nitrogen oxides[a]	Determination of $NO-NO_2$ mixtures
Soils	Carbofuran[a]	Stopped-flow method
Serum	Copper + iron	CAR method
	Uric acid,[a] creatinine,[a] urea[a]	Stopped-flow method
Urine	Iodide	Uses AutoAnalyzer
	Uric acid, sulfonamides[a]	Stopped-flow method
	Epinephrine + norepinephrine[a]	Stopped-flow method
	Mercury	CAR method
Pharmaceuticals	Paracetamol	Fluorimetric stopped-flow method
	Thiamine[a]	Oxidation to thiochrome
	T_3 and T_4, theophylline,[a] procaine	Stopped-flow method
	Chlorpromazine + perphenazine	Stopped-flow method
	Sulfonamides,[a] vitamin B_{12}[a]	CAR method

Reaction type: [a] Uncatalyzed, [b] Ligand-exchange, [c] Chemiluminescence

control of a computer interfaced on-line to it. This experimental setup can be used to investigate mechanisms of reactions involving fast steps.

The kinetic methods describe above have proved to be useful for the determination of a vast number of inorganic and organic species. Automated systems are especially suitable for analytical control of parameters in routine analyses involving catalyzed or uncatalyzed reactions [6.4-2]. Table 6.4-1 lists some selected applications of kinetic nonenzymatic methods to the analysis of real samples in various areas of analytical interest including environmental, clinical and pharmaceutical chemistry.

References

[6.1-1] Pérez-Bendito, D., Silva, M., *Kinetic Methods in Analytical Chemistry*, Chichester: Ellis Horwood, 1988.

[6.1-2] Mottola, H.A., *Kinetic Aspects of Analytical Chemistry*, New York: John Wiley & Sons, 1988.

[6.2-1] Buckman, N.G., Magee, R.J., Hill, J.O. *Anal. Chim. Acta* 1983, 153, 285.

[6.2-2] Sherman, L.R., Trust, V.L., Hoang, H., Talanta, 1981, 28, 408.

[6.2-3] Ryan, M.A., Ingle, J.D. Jr., *Anal. Chem* 1980, 52, 2177.

[6.2-4] Pelizzetti, E., Mentasti, E. *Anal. Chim. Acta*, 1976, 85, 161.

[6.3-1] IUPAC, *Nomenclature of Kinetic Methods of Analysis*, Pure & Appl. Chem 1993, 65, 2291.

[6.3-2] Rodríguez, P.A., Pardue, H.L., *Anal. Chem* 1969, 41, 1369.

[6.3-3] Guibault, G.G., *Practical Fluorescence: Theory, Methods and Techniques*, Chapter 9, New York: Dekker, 1973.

[6.3-4] Landolt, H., *Ber. Deut. Chem. Ges* 1986, 19, 1317.

[6.3-5] Pérez-Bendito, D., Rubio, S., *Trends Anal. Chem* 1993, 12, 9.

[6.3-6] Berezin, I.V., Martinek, K., Yatsimirskii, A.K., *Russ. Chem. Rev* 1973, 42, 787.

[6.3-7] Mottola, H.A., Pérez-Bendito, D., Mark, H.B. Jr, *Anal. Chem* 1990, 62, 411R.

[6.3-8] Mottola, H.A., Pérez-Bendito, D., *Anal. Chem* 1992, 65, 407R.

[6.4-1] Márquez, M., Silva, M., Pérez-Bendito, D., *Anal. Chim. Acta* 1990, 237, 353.

[6.4-2] Pérez-Bendito, D., Silva, M., Gómez-Hens, A., *Trends Anal. Chem* 1989, 8, 302.

Questions and problems

1. The reaction $A + B \rightleftharpoons C + D$ obeys the rate law $d[A]/dt = -k[A]$, with $k = 1.0 \times 10^3 \, s^{-1}$ at 297 K. What is the half-time for the reaction at 297 K?

   $(t_{1/2} = 7 \times 10^{-4} \, s)$

2. The rate constant for a certain reaction is equal to 8.93×10^{-7} and $8.12 \times 10^{-5} \, L \, mol^{-1} \, s^{-1}$ at 650 and 750 K, respectively. What is the activation energy for the reaction?

   $(E_a = 182.8 \, kJ \, mol^{-1})$

3. The reaction $Co(OH_2)_4Cl_2^+ + OH^- \rightleftharpoons Co(OH_2)_4ClOH^+ + Cl^-$ obeys the rate law $d[Co(OH_2)_4Cl_2^+]/dt = -k[Co(OH_2)_4Cl_2^+]$. Suggest a plausible mechanism.

   $[Co(OH_2)_4Cl_2^+ \rightleftharpoons Co(OH_2)_4Cl^{2+} + Cl^-$　Rate-determining step
   $Co(OH_2)_4Cl^{2+} + OH^- \rightleftharpoons Co(OH_2)_4ClOH^+$　Fast step]

4. The rate constant for a first-order reaction is doubled by increasing the temperature from 20 °C to 30 °C. (a) How will it be affected if the temperature is raised from 85 °C to 95 °C? (b) What is the activation energy for the reaction?

   [(a) It will increase by a factor of 1.66]
   [(b) $E_a = 52.38 \, kJ \, mol^{-1}$]

5. The decomposition of 1,1-dihydroxyethane is described by the reaction $CH_3CH(OH)_2 \rightarrow CH_3CHO + H_2O$. Its rate equation in an aqueous solution containing $0.1 \, mol \, L^{-1}$ acetic acid is $d[CH_3CH(OH)_2]/dt =$

$-k[CH_3CH(OH)_2]$ and its first-order rate constant is equal to $0.32\,s^{-1}$. What length of time is required for 90% of the 1,1-dihydroxyethane to disappear from a solution originally containing $1 \times 10^{-3}\,mol\,L^{-1}$ of this substance and $0.1\,mol\,L^{-1}$ acetic acid?

$(t = 72\,s)$

6. The rate law for a reaction $A + B \rightleftharpoons C + D$ is $d[A]/dt = -k[A][B]$ and its second-order rate constant is $2.22 \times 10^{-3}\,L\,mol^{-1}\,s^{-1}$. (a) What length of times will be required for 90% of A to disappear from a solution containing $1 \times 10^{-4}\,mol\,dm^{-3}$ A and $0.01\,mol\,dm^{-3}$ B? (b) What concentration of A would still be present after 48 h?

$[(a)\ t = 28.8\,h]$
$[(b)\ 2.16 \times 10^{-6}\,M]$

7. A second-order reaction with respect to a single component was 75% complete in 92 min when the initial reactant concentration was $0.240\,mol\,L^{-1}$. How long would be required to achieve a concentration of $0.043\,mol\,L^{-1}$ in a run with an initial concentration of $0.146\,mol\,L^{-1}$?

$(t = 121\,min)$

8. The decomposition of dinitrogen pentoxide according to $2\,N_2O_5 \rightleftharpoons 2\,N_2O_4 + O_2$ is a first-order reaction the rate law for which is $d[N_2O_5]/dt = -k[N_2O_5]$. In one experiment, a solution of N_2O_5 in CCl_4 was allowed to decompose and the concentration of reactant was found to decrease to half of its original value in $900\,s$. What was the rate constant for the reaction under the experimental conditions used?

$(k = 7.7 \times 10^{-4}\,s^{-1})$

9. The following data were obtained by monitoring the reaction

$(CH_3)_3CBr_{(aq)} + H_2O_{(l)} \rightleftharpoons (CH_3)COH_{(aq)} + H^+ + Br^-$

at 298 K:

$t(s \times 10^3)$ 5   20   60   100   150

$[(CH_3)_3CBr]$

$(mol\,L^{-1} \times 10^{-2})$ 3.80   3.08   2.33   1.76   0.502

(a) What is the reaction order with respect to $(CH_3)_3CBr$?
(b) What is the value of the nth-order rate constant at 298 K?
(c) What was the initial concentration of $(CH_3)_3CBr$?

$[(a)\ \text{First}]$
$[(b)\ -k = 1.5 \times 10^{-5}\,s^{-1}]$
$[(c)\ 0.041\,mol\,L^{-1}]$

10. What should the rate-constant ratio k_A/k_B be for a mixture of two components A and B for 99.9% of A to react during the time needed for only 0.1% of B to react?

$\left(\dfrac{k_A}{k_B} = 6900\right)$

11. The rate of oxidation of glucose (G) by oxygen, catalyzed by glucose oxidase (GO), at pH 5.6 and $0\,°C$ in the presence of an enzyme concentration of $1.17 \times 10^{-5}\,M$ is given by

$$\frac{1}{v} = \frac{k_2 + k_4}{k_2 k_4} + \frac{1}{k_1[G]} + \frac{1}{k_3[O_2]}$$

(a) What will the relationship between the reaction rate and the glucose concentration at an oxygen concentration of $1 \times 10^{-3}\,M$ be?

(b) If the reaction rate is proportional to the enzyme concentration, what will the kinetic expression including the enzyme and glucose concentrations be? Constants k_1, k_2, k_3 and k_4 are equal to $2.1 \times 10^3 \, mol^{-1} \, s^{-1}$, $650 \, s^{-1}$, $1.3 \times 10^6 \, mol^{-1} \, s^{-1}$ and $370 \, s^{-1}$, respectively.

$\{$(a) $v = 2.1 \times 10^3 [G]/(1 + 10.5[G])\}$
$\{$(b) $v = 1.8 \times 10^8 [G][GO]/(1 + 10.5[G])\}$

7 Methods of Chemical Analysis and Their Applications

This chapter deals with the application of the principles of chemical analysis to practical problems. The field extends from volumetric to gravimetric methods and from electroanalytical applications to the analysis in flowing systems and chemical sensors. Today's analytical applications are heavily influenced by the needs of analytical biotechnology and analytical biology. Therefore, essential principles of biochemical analysis and biosensors are also included.

7.1 Titrimetry (Volumetry)

Learning objectives

■ To describe titrimetric methods – quantitative methods using solution chemistry to determine the concentration of an analyte

■ To make clear that a general requirement for all volumetric methods is that the titration process is fast and that it proceeds in a definite stoichiometric ratio. To stress that the endpoint of the reaction must be easy to detect and the reaction should be specific and not influenced by other constituents of the solution, i.e., there should be no interference

■ To make the practical point that the change of the standard free energy ($\Delta G°$) of the reaction should be large, so that the equilibrium is shifted toward the products

■ To show that only a few general reactions satisfy these requirements; in homogeneous systems, successful volumetric methods are restricted to acid–base, some complexometric, and redox systems

7.1.1 Acidimetric and alkalimetric titrations

Introduction

Proton transfer reactions are usually extremely fast, although exceptions, such as the deprotonation of $CO_{2(aq)}$ are known, and therefore suitable for application in volumetric procedures. In addition, products of definite stoichiometry are formed and a number of chemical indicators and instrumental detectors are available for the endpoint detection.

Most proton transfer reactions are suitable for volumetric analysis.

The instrumental devices are usually based on pH measurement, but other techniques, such as the conductimetric methods are also in current use. The only limiting condition is the self-dissociation of the medium (water or any other solvent with acid–base properties) which may produce an acid (H_3O^+) or a base form (OH^-). This leveling effect of water on strong acids and strong bases has been discussed in Sec. 4.2. It restricts the practical range for acidimetric or alkalimetric titration to a pH range between 0 and 14. For this reason, acidimetric or alkalimetric titrations are sometimes performed in nonaqueous solvents, where a different or larger practical range may apply. In Fig. 4.2-1, some of the most popular nonaqueous solvents and their operational range are compared with water. The comparison is based on the pK values that solvents exhibit in aqueous solution.

Acidimetric and alkalimetric methods are usually based on pH measurements. With chemical indicators estimates are usually accurate to within about one pH unit and more accurate determinations are better done with electrodes which respond to the hydrogen ion concentration, or preferably the hydrogen ion activity. Several such electrodes are known and have been discussed in more detail (see Sec. 4.4).

The most popular and widely used sensor is the glass electrode, although it may exhibit some nonideal behavior in solution with high pH values or high ionic strength. A typical glass electrode shows the ideal pH response in the pH range from about 1 to 12. A constant potential difference for each pH unit that is practically identical with the theoretical (minimum) slope is then observed. The following millivolt difference is calculated per pH unit from the relationship:

$$E = E_2 - E_1 = \frac{2.3026RT}{F}(\log[H^+]_2 - \log[H^+]_1)$$

where R is the gas constant ($8.3143 \ JK^{-1} mol^{-1}$); T is the absolute temperature (K); F is the Faraday constant ($96485.0 \ C\,mol^{-1}$); and E is the potential difference observed between solution 1 and 2 with hydrogen ion concentrations $[H^+]_1$ and $[H^+]_2$, respectively.

The calculated values are:

0 °C: 54.20 mV	30 °C: 60.14 mV
10 °C: 56.18 mV	40 °C: 62.13 mV
20 °C: 58.16 mV	50 °C: 64.12 mV
25 °C: 59.16 mV	60 °C: 66.11 mV

The titration curve

The titration curve is a convenient way of presenting a volumetric procedure.

The results of a titration are best plotted in an appropriate diagram. However, before one can discuss the results of such a plot, a number of conventions must be introduced and defined. In the following discussion, an alkalimetric titration is outlined, but similar results are obtained in an acidimetric titration; the conclusions may be applied to both types of titration.

The following conventions are introduced:

1. In a theoretical treatment it is assumed that the titrant is sodium hydroxide NaOH. One is dealing with an alkalimetric titration.
2. The concentration of the titrant (NaOH) is large (>1.0 mol/L), so that no significant volume changes occur during the titration.
3. The concentration of the analyte (acid) is symbolized by C_s. Its acid form is HB and its base form is B^-. It is assumed that the acid is a neutral species (without charge) and that the base form is a univalent anion.

$$C_s = [HB] + [B^-] = [HB]_{total} = [B]_{total} \tag{7.1-1}$$

Instead of the volume of the titrant, the degree of neutralization a_s can be used in acid-base titrations (see Eq. 7.1-2).

The quantity a_s is the degree of neutralization, defined as:

$$a_s = \frac{\text{number of moles of added base}}{\text{number of moles of original acid HB}} \tag{7.1-2}$$

a_s is a normalized quantity that replaces the volume (mL) in the titration curve.

The quantity p_s is the degree of protonation, defined as:

$$p_s = \frac{[\text{protonated base}]}{[\text{total base}]} = \frac{[HB]}{C_s} \tag{7.1-3}$$

therefore:

$$[HB] = C_s p_s \quad \text{and} \quad [B^-] = C_s(1 - p_s)$$

One has to consider the following five equations which enable us to calculate the magnitude of the five unknown quantities: $[Na^+]$, $[H^+]$, $[HB]$, $[B^-]$, and $[OH^-]$:

$$[Na^+] = a_s C_s \tag{7.1-4}$$

$$C_s = [HB] + [B^-] \tag{7.1-5}$$

acid dissociation:

$$K_A = \frac{[H^+][B^-]}{[HB]} \tag{7.1-6}$$

self-dissociation of water:

$$K_W = [H^+][OH^-] = 10^{-14} \tag{7.1-7}$$

electroneutrality:

$$[Na^+] + [H^+] = [OH^-] + [B^-] \tag{7.1-8}$$

We now consider three different titrations.

Titration of a strong acid (HCl) with a strong base (NaOH)

We assume $C_s = 0.01$ mol/L. The values of the different quantities are given in Table 7.1-1.

Table 7.1-1. Hydrogen ion and hydroxide ion concentration as a function of degree of neutralization

a_S	$[H^+]$	$[OH^-]$	pH
0.0	10^{-2}	10^{-12}	2.00
0.1	9.0×10^{-3}	1.1×10^{-12}	2.05
0.5	5.0×10^{-3}	2.0×10^{-12}	2.30
0.9	10^{-3}	10^{-11}	3.00
0.99	10^{-4}	10^{-10}	4.00
0.999	10^{-5}	10^{-9}	5.00
1.0	10^{-7}	10^{-7}	7.00
1.001	10^{-9}	10^{-5}	9.00
1.01	10^{-10}	10^{-4}	10.00
1.1	10^{-11}	10^{-3}	11.00
1.5	2.0×10^{-12}	5.0×10^{-3}	11.70
1.9	1.1×10^{-12}	9.0×10^{-3}	11.95
2.0	10^{-12}	10^{-2}	12.00

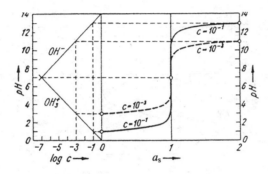

Fig. 7.1-1. Titration curve of a strong acid (HCl) with a strong base (NaOH) in aqueous solution. The left part represents the logarithmic (Hägg) diagram

The results of this calculation are plotted in Fig. 7.1-1 where the pH of the solution is plotted against the degree of neutralization. It represents the curve of any titration of a univalent strong acid with a strong base, and the abscissa may be replaced by the corresponding volume of added base.

At the point $a_s = 1.0$ the expression dpH/da_S, representing the slope of the titration curve, reaches its maximum value. It is also the endpoint of the titration. It is easily detected by the pronounced pH jump or by any acid–base indicator with a pK_i value between 4 and 10. This means that the choice of the indicator is not critical. The expression dpH/da_s is at a minimum at the beginning and at the end of the titration curve. This indicates that strong acids and strong bases themselves have a pronounced buffer capacity.

Selection of the indicator for titration of a strong acid with a strong base is not critical.

Titration of a weak acid (acetic acid) with a strong base (NaOH)

We assume $C_s = 0.01 \, mol/L$. The values of the different calculated quantities are given in Table 7.1-2. Some simplified relations have been introduced to reduce the effort of calculation. The acid dissociation constant is $K_A = 1.74 \times 10^{-5}$ or $pK_A = 4.76$.

The results are also represented in a titration curve (Fig. 7.1-2). It can be seen that at the point $a_s = 0.5$ and $a_s = 1.0$ the term dpH/da_S assumes a minimum and maximum value, respectively. At $a_s = 0.5$ the system exhibits its largest buffer capacity (since $[H^+] = [B^-]$) and therefore shows the smallest change in pH per volume of added base. The point of minimum slope can therefore be used to determine the pK_A value of an unknown weak acid (the point where $pH = pK_A$). The endpoint ($a_s = 1.0$) is again characterized by a pH jump, but it is less pronounced than in a titration of a strong acid with a strong base. Only a limited number of pH indicators can be used for the endpoint detection. The pK_i must be between 7 and 10 (see Table 4.2-2) for the choice of indicator.

Indicators for titration of a weak acid with a strong base should have pK_i values between 7 and 10.

Table 7.1-2. Titration of a weak acid (acetic acid, $pK_A = 4.76$) with a strong base (NaOH)

a_S	$[Na^+]$	$[H^+]$	$[B^-]$	$[HB]$	pH	Simplified relation
0.0	0.0	4.20×10^{-4}	4.20×10^{-4}	9.58×10^{-3}	3.38	$[H^+] = [B^-]$
0.1	1.0×10^{-3}	1.35×10^{-4}	1.14×10^{-3}	8.86×10^{-3}	3.87	$[Na^+] + [H^+] = [B^-]$
0.2	2.0×10^{-3}	6.76×10^{-5}	2.06×10^{-3}	7.94×10^{-3}	4.17	$[Na^+] + [H^+] = [B^-]$
0.3	3.0×10^{-3}	4.00×10^{-5}	3.04×10^{-3}	6.96×10^{-3}	4.40	$[Na^+] + [H^+] = [B^-]$
0.4	4.0×10^{-3}	2.58×10^{-5}	4.03×10^{-3}	5.97×10^{-3}	4.59	$[Na^+] + [H^+] = [B^-]$
0.5	5.0×10^{-3}	1.74×10^{-5}	5.02×10^{-3}	4.98×10^{-3}	4.76	$[Na^+] + [H^+] = [B^-]$
0.6	6.0×10^{-3}	1.15×10^{-5}	6.01×10^{-3}	3.99×10^{-3}	4.94	$[Na^+] + [H^+] = [B^-]$
0.7	7.0×10^{-3}	7.14×10^{-6}	7.01×10^{-3}	2.99×10^{-3}	5.13	$[Na^+] + [H^+] = [B^-]$
0.8	8.0×10^{-3}	5.40×10^{-6}	8.00×10^{-3}	2.00×10^{-3}	5.26	$[Na^+] = [B^-]$
0.9	9.0×10^{-3}	1.93×10^{-6}	9.00×10^{-3}	1.00×10^{-3}	5.72	$[Na^+] = [B^-]$
0.99	9.9×10^{-3}	1.76×10^{-7}	9.90×10^{-3}	1.00×10^{-4}	6.76	$[Na^+] = [B^-]$
1.0	1.0×10^{-2}	4.16×10^{-9}	1.00×10^{-2}	2.40×10^{-6}	8.38	$[Na^+] = [OH^-]$
1.1	1.1×10^{-2}	1.00×10^{-11}	1.00×10^{-2}	5.76×10^{-9}	11.00	$[Na^+] - [B^-] = [OH^-]$
1.5	1.5×10^{-2}	2.00×10^{-12}	1.00×10^{-2}	1.15×10^{-9}	11.70	$[Na^+] - [B^-] = [OH^-]$
1.9	1.9×10^{-2}	1.11×10^{-12}	1.00×10^{-2}	6.46×10^{-10}	11.95	$[Na^+] - [B^-] = [OH^-]$
2.0	2.0×10^{-2}	1.00×10^{-12}	1.00×10^{-2}	5.76×10^{-6}	12.00	$[Na^+] - [B^-] = [OH^-]$

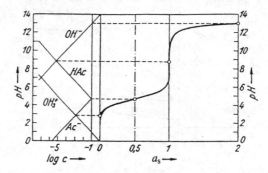

Fig. 7.1-2. Titration curve of a weak acid (CH$_3$COOH) with a strong base (NaOH) in aqueous solution

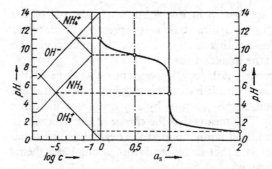

Fig. 7.1-3. Titration curve of a weak base (NH$_3$) with a strong acid (HCl) in aqueous solution

Titration of a weak base (NH₃) with a strong acid (HCl)

The results of the titration of NH$_3$, $C = 0.1\,\mathrm{mol/L}$ with HCl are represented in Fig. 7.1-3. It can be seen that at the endpoint of the titration the slope assumes a maximum value. However, the pH jump is smaller than in the titration curve of a strong acid with a strong base. It is comparable to the pH jump in the titration of a weak acid with a strong base, but it is shifted toward lower pH values.

As can be seen from the titration curve displayed in Fig. 7.1-4, mixtures of strong and weak acids in aqueous solutions can be titrated in a single sample with a single titrant (a strong base in this example) if the dissociation constants of the acids are different enough to provide for a sufficient pH jump around the equivalence points. The example shown is a mixture of 0.1 mol/L HCl and 0.1 mol/L H$_3$PO$_4$. The first equivalence point Cl marks the sum of the concentrations of HCl and H$_3$PO$_4$, because the first proton of the phosphoric acid ($pK_{A1} = 2$) and the HCl proton are indistinguishable in aqueous systems. The second dissociation of phosphoric acid ($pK_{A2} = 7$), however, marks its own equivalence point C2, in such

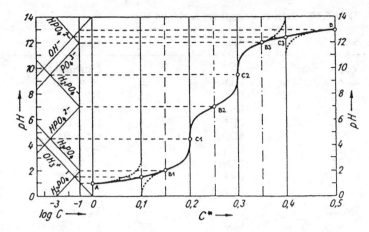

Fig. 7.1-4. Titration curve of a mixture of HCl (0.1 mol/L) and H_3PO_4 (0.1 mol/L) with NaOH in aqueous solution

a way, that the titrant volume C2 – C1 indicates the concentration of the phosphoric acid alone. From that result, the concentration of HCl can also be calculated. B₁, B₂ and B₃ are buffer points.

It is even possible to titrate extremely weak acids with a strong base if the acidity of such an acid is enhanced through a chemical reaction. Several acid enforcement techniques are discussed later in this section.

The most common acids and bases may be arranged in the following acidity order, according to the behavior in an alkalimetric or acidimetric titration (see Table 7.1-3 and Table I in Appendix 3). Typical acid-base indicators are found in Table I in Appendix 3.

Mixtures of strong and weak acids (or bases) can be titrated in one single experiment in aqueous solutions. Even for such complex systems the graphical solution via the Hägg-diagram provides for a clear insight into the system (cf. Fig. 7.1-4).

Acid enforcement techniques

Extremely weak acids cannot be titrated with bases in pure aqueous solutions. However, through certain chemical reactions weak acids may be transformed into strong acids, so that they can be titrated. These reactions involve one of the following.

Acid enforcement techniques allow titration of extremely weak acids with bases.

a) Precipitation of the anion of the weak acid. Barium ions precipitate a number of anions and they may be used for this purpose:

$$2\,HB + Ba^{2+} \rightarrow \{BaB_2\} + 2\,H^+$$

The brackets { } denote the solid state of the barium precipitate, for example:

$$2\,HPO_4{}^{2-} + 3\,BaCl_2 \rightarrow \{Ba_3(PO_4)_2\} + 2\,H^+ + 6\,Cl^-$$

b) Oxidation of the weak acid to an oxygen-richer acid, for example:

$$HNO_2 + \text{``O''} \rightarrow HNO_3$$

or:

$$H_2SO_3 + \text{``O''} \rightarrow H_2SO_4$$

c) Complexation
Examples include formation of an ester with a polyalcohol:

$$H_3BO_3 + 2\,C_3H_5(OH)_3 \rightarrow BO_4(C_3H_5OH)_2{}^- + H_3O^+ + 2\,H_2O$$

In a similar way it is possible to titrate the ammonium ion (weak acid) with a strong base after the addition of formaldehyde which converts it to hexamethylenetetramine and a free hydronium ion (urotropine reaction):

$$NH_4^+ + H_2O \rightleftharpoons NH_3 + H_3O^+ \xrightarrow{HCHO} C_6H_{12}N_3 \text{ (urotropine)} + H_3O^+$$

Table 7.1-3. Relative strengths of acids and bases

pK_A	Acid		Base	pK_B
Very strong:			**Extremely weak:**	
<-5	ClO_4H	$\rightleftharpoons$	$H^+ + ClO_4^-$	~ 23
<-5	IH	$\rightleftharpoons$	$H^+ + I^-$	~ 23
~ -3.5	BrH	$\rightleftharpoons$	$H^+ + Br^-$	~ 20
~ -2	ClH	$\rightleftharpoons$	$H^+ + Cl^-$	~ 17
~ -2	SO_4H_2	$\rightleftharpoons$	$H^+ + SO_4H^-$	~ 17
-1.74	OH_3^+	$\rightleftharpoons$	$H^+ + OH_2$	15.74
-1.32	NO_3H	$\rightleftharpoons$	$H^+ + NO_3^-$	15.32
0	ClO_3H	$\rightleftharpoons$	$H^+ + ClO_3^-$	14
0	BrO_3H	$\rightleftharpoons$	$H^+ + BrO_3^-$	14
Strong:			**Very weak:**	
1.42	$C_2O_4H_2$	$\rightleftharpoons$	$H^+ + C_2O_4H^-$	12.58
1.64	IO_3H	$\rightleftharpoons$	$H^+ + IO_3^-$	12.36
1.92	$SO_3 + OH_2$	$\rightleftharpoons$	$H^+ + SO_3H^-$	12.03
1.92	SO_4H^-	$\rightleftharpoons$	$H^+ + SO_4^{2-}$	12.08
1.96	PO_4H_3	$\rightleftharpoons$	$H^+ + PO_4H_2$	12.04
2	$S_2O_3H^-$	$\rightleftharpoons$	$H^+ + S_2O_3^{2-}$	12
2.22	$Fe(OH_2)_4^{2+}$	$\rightleftharpoons$	$H^+ + Fe(OH)(OH_2)_3^{2+}$	11.78
2.32	AsO_4H_4	$\rightleftharpoons$	$H^+ + AsO_4H_3^{3+}$	11.68
3.14	FH	$\rightleftharpoons$	$H^+ + F^+$	10.86
3.35	NO_3H	$\rightleftharpoons$	$H^+ + NO_3^-$	10.65
3.7	$HCOOH$	$\rightleftharpoons$	$H^+ + HCOO^-$	10.3
4	$NCSH$	$\rightleftharpoons$	$H^+ + NCS^-$	10
4.21	$C_2O_4H^-$	$\rightleftharpoons$	$H^+ + C_2O_4^{2-}$	9.79
Weak:			**Weak:**	
4.74	CH_3COOH	$\rightleftharpoons$	$H^+ + CH_3CO_2^-$	9.25
4.85	$Al(OH_2)_4^{2+}$	$\rightleftharpoons$	$H^+ + Al(OH)(OH_2)_3^{2+}$	9.15
6.52	$CO_2 + OH_2$	$\rightleftharpoons$	$H^+ + CO_3H^-$	7.48
6.92	SH_3	$\rightleftharpoons$	$H^+ + SH_2^-$	7.03
7	$AsO_4H_3^-$	$\rightleftharpoons$	$H^+ + AsO_4H^{2-}$	7
7	SO_3H^-	$\rightleftharpoons$	$H^+ + SO_3^{2-}$	7
7.12	$PO_4H_2^-$	$\rightleftharpoons$	$H^+ + PO_4H^{2-}$	6.38
7.25	$ClOH$	$\rightleftharpoons$	$H^+ + ClO^-$	6.75
8.69	$BrOH$	$\rightleftharpoons$	$H^+ + BrO^-$	5.31
9.24	BO_3H_2	$\rightleftharpoons$	$H^+ + BO_3H^-$	4.76
9.22	AsO_3H_3	$\rightleftharpoons$	$H^+ + AsO_3H_2^-$	4.78
9.25	NH_4^+	$\rightleftharpoons$	$H^+ + NH_3$	4.75
Very weak:			**Strong:**	
9.40	NCH	$\rightleftharpoons$	$H^+ + NC^-$	4.60
9.61	$Zn(OH_2)_6^{2+}$	$\rightleftharpoons$	$H^+ + Zn(OH)(OH_2)_5^+$	4.39
10	SiO_4H_4	$\rightleftharpoons$	$H^+ + SiO_4H_3^-$	4
10.6	IOH	$\rightleftharpoons$	$H^+ + IO^-$	3.4
10.4	CO_3H^-	$\rightleftharpoons$	$H^+ + CO_3^{2-}$	3.6
11.62	O_2H_2	$\rightleftharpoons$	$H^+ + O_2H^-$	3.38
12	$SiO_4H_2^-$	$\rightleftharpoons$	$H^+ + SiO_4H_2^{2-}$	2
12.32	PO_4H^{2-}	$\rightleftharpoons$	$H^+ + PO_4^{3-}$	1.68
13	AsO_4H^{2-}	$\rightleftharpoons$	$H^+ + AsO_4^{3-}$	1
13	SH^-	$\rightleftharpoons$	$H^+ + S^{2-}$	1
13.52	$AsO_3H_2^-$	$\rightleftharpoons$	$H^+ + AsO_3H^{3-}$	0.48
Extremely weak:			**Very strong:**	
15.74	OH_2	$\rightleftharpoons$	$H^+ + OH^-$	-1.74
~ 23	NH_3	$\rightleftharpoons$	$H^+ + NH_2^-$	~ -9
~ 24	OH^-	$\rightleftharpoons$	$H^+ + O^{2-}$	~ -10
~ 34	CH_4	$\rightleftharpoons$	$H^+ + CH_3^-$	~ -20
38.6	H_2	$\rightleftharpoons$	$H^+ + H^-$	-24.6

7.1.2 Complexometric titrations

Introduction

Complexometric titration methods are based on the stoichiometric reaction of metal ions in aqueous solutions with a class of suitable, water-soluble bidentate or polydentate ligands, called "complexones".

To be of practical use, the reaction of the metal ions and the complexone must be fast, must result in a water-soluble product, and should go nearly to completion when equivalent quantities of both substances are present. The first condition is a kinetic requirement, as in all titrations, the last implies that the stability constant of the metal complex formed is large.

The most practical ligand systems are those involving chelating polyamines and aminopolycarboxylic acids (greek *chele* means "scissors"). Most complexes with nonchelating monodentate ligands do not have sufficiently large stability constants to fulfil the stability requirement, as can be shown in the example of (monodentate) ammonia complexes with Cu^{2+} and the (tetradentate) triaminotriethylamino complex with the same metal ion.

Complexones must fulfil kinetic, solubility, and stability criteria to be suitable for titration.

a) Copper–tetrammine complexes (see also Table 7.1-4)
A series of four relatively unstable monodentate complex species is formed consecutively:

$$\begin{array}{llll}
Cu^{2+} & + NH_3 & \rightleftharpoons [Cu(NH_3)]^{2+} & \log K_1 = 4.1 \\
[Cu(NH_3)]^{2+} & + NH_3 & \rightleftharpoons [Cu(NH_3)_2]^{2+} & \log K_2 = 3.5 \\
[Cu(NH_3)_2]^{2+} & + NH_3 & \rightleftharpoons [Cu(NH_3)_3]^{2+} & \log K_3 = 2.9 \\
[Cu(NH_3)_3]^{2+} & + NH_3 & \rightleftharpoons [Cu(NH_3)_4]^{2+} & \log K_4 = 2.1 \\
\hline
Cu^{2+} & + 4NH_3 & \rightleftharpoons [Cu(NH_3)_4]^{2+} & \log K_{ST} = 12.6
\end{array}$$

b) Copper–triaminotriethylamine complex (see also Table 7.1-4)
The final complex with the tetradentate ligand is formed in a single step, resulting in a much higher stability constant of the complex and a steeper titration curve, owing to the chelate-effect. The higher stability of a chelate complex is the result of an increase of the entropy of the system. One polydentate ligand, on reaction with a hydrated metal ion (e.g., Cu^{2+} binds four water molecules) liberates several water molecules, increasing the number of unbound particles and hence the entropy (Fig. 7.1-5).

Monodentate complexes are not strong enough for complexometry as compared to polydentated complexes (see Table 7.1-4 and Fig. 7.1-5).

$$Cu^{2+} + N(CH_2CH_2NH_2)_3 \rightleftharpoons [Cu(N(CH_2CH_2NH_2)_3]^{2+} \qquad \log K_{ST} = 18.8$$

Table 7.1-4. Stability constants of some polyamine–metal complexes with copper and zinc ($\log K_{ST}$)

Complexing agent	Structure[a]	Zinc		Copper	
Ammonia	NH_3	2.3 2.3 2.4 2.1		4.1 3.5 2.9 2.1	
Ethylenediamine	$H_2NCH_2CH_2NH_2$	5.9	5.2	10.7	9.3
Diethylenetriamine	$H_2NCH_2CH_2NHCH_2CH_2NH_2$	8.9	5.5	16.0	5.3
1:2:3-Triaminopropane	$H_2NCH_2CHNH_2CH_2NH_2$	6.8	4.3	11.0	9.0
Triethylenetetramine (Triene)	$H_2NCH_2CH_2NHCH_2CH_2NH-$ $CH_2CH_2NH_2$	12.1		20.4	
β,β',β''-Triaminotriethylamine	$N(CH_2CH_2NH_2)_3$	14.7		18.8	

[a] Experimental conditions are 20 °C and ionic concentration 0.1 mol/L. Taken in part from G. Schwarzenbach, *Analyst* 80, 713 (1955)

Complexing agents

Several useful complexing agents with chelating properties, forming water-soluble chelate complexes, are commercially available. The major field of application is the quantitative determination of water hardness.

A range of complexing reagents are commercially available.

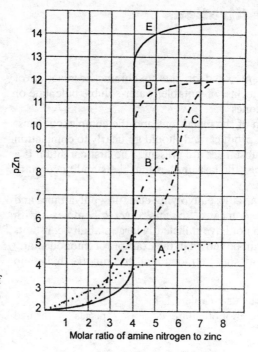

Fig. 7.1-5. Titration curves for the titration of zinc with polyamines. Titrants: (A) ammonia; (B) ethylenediamine; (C) diethylenetriamine; (D) triethylenetetramine (Triene), (E) triaminotriethylamine

a) Nitrilotriacetic acid, NTA (Titriplex I)

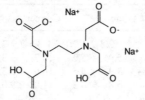

$M = 191.14\,\text{g/mol}$
tetradentate, white powder, easily soluble in alkaline solutions

b) Ethylenediaminetetraacetic acid, EDTA (Titriplex II)

$M = 292.25\,\text{g/mol}$
hexadentate, white powder, not easily soluble in water, but in alkaline solutions

Titriplex III is the mainly used complexometric agent. It is a hexadentate ligand.

c) Ethylenediaminetetraacetic acid disodium salt dihydrate (Titriplex III)

$M = 372.24\,\text{g/mol}$
hexadentate, white powder, easily soluble in water, the aqueous solution is slightly acidic

d) *trans*-1,2-Diaminocyclohexane-N,N,N',N'-tetraacetic acid monohydrate, DCYTA

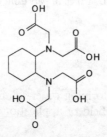

$M = 364.36\,\text{g/mol}$
hexadentate, white powder, similar to Titriplex II, but with larger K_{ST} values (see Table 7.1-5)

Table 7.1-5. Stability constants of some aminopolycarboxylic acid complexes

Metal ion	$\log K_{ST}$		
	NTA	EDTA	DCYTA
Ba^{2+}	4.8	7.8	8.0
Ca^{2+}	6.4	10.7	12.5
Fe^{3+}	15.9	25.1	29.3
La^{3+}	10.4	15.4	16.3
Mg^{2+}	5.4	8.7	10.3
Zn^{2+}	10.5	16.5	18.7

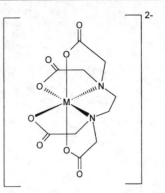

Fig. 7.1-6. Octahedral structure of a chelate complex, formed from the hexadentate EDTA and a divalent metal ion. The coordination sites are two amino-nitrogen atoms and four oxygen atoms of the carboxylate groups

EDTA is the most frequently used complexometric agent. Figure 7.1-6 shows the three-dimensional structure of the chelate, displaying the six coordination sites in the ligand, completely surrounding the central ion.

Simultaneous equilibria influencing complexometric reactions

The formation of complexes may be accompanied by simultaneous parallel reactions in complex aqueous solutions according to:

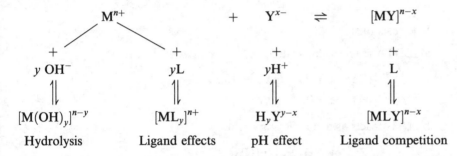

| Hydrolysis | Ligand effects | pH effect | Ligand competition |

Without such side reactions, which is the ideal case, the total amount of ligand and metal ions would be available for the main titration reaction, as characterized by the equilibrium constant K_{ST}:

$$K_{ST} = \frac{c_{[MY]}}{C_M \cdot C_Y} \qquad \text{Prerequisite: } c_M = C_M \quad \text{and} \quad c_Y = C_Y$$

Owing to each of the side reactions, alone or in combination with each other, the available equilibrium concentration of metal ions and/or ligand may be (much) smaller than the total concentrations. A factor f is introduced to correct from the ideal case (*f is between 1 and 0*) and to give the apparent or conditional stability constant K'_{ST}:

The apparent, or conditional stability constant expresses the effect of side reactions.

$$f_{Me} = \frac{c_M}{C_M} \quad \text{and} \quad f_Y = \frac{c_Y}{C_Y}$$

$$K_{ST} = \frac{c_{[MY]}}{f_M C_M f_Y C_Y}$$

The conditional stability constant K'_{ST} describes the complex stability of the real system, including all side reactions:

Complexometric reactions are part of complex simultaneous equilibria in aqueous solutions, which can significantly influence the complex stability ($\rightarrow$ conditional stability constant K'_{ST}, see Fig. 7.1-7 as an example).

$$K'_{ST} = K_{ST} f_M f_Y$$

Example: titration of Zn^{2+} with $EDTAH_2Na_2(YH_2^{2-})$ in NH_3 -solution

Overall reaction of the titration: $\quad Zn^{2+} + YH_2^{2-} \longrightarrow [ZnY]^{2-} + 2H^+$

Complex reaction: $\qquad\qquad\qquad Zn^{2+} + Y^{4-} \rightleftharpoons [ZnY]^{2-} \qquad\qquad K_{ZnY}$

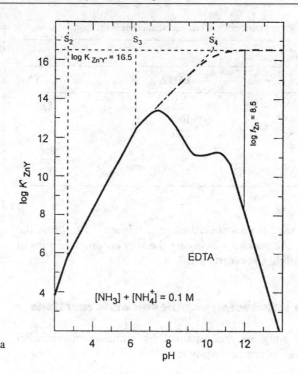

Fig. 7.1-7. Plot of the conditional stability constant K'_{ZnY} of the Zn-EDTA complex as a function of pH at $C_{NH_3}(\text{total}) = 0.1\,\text{mol/L}$

Side reactions:

a) EDTA reacts as an acid:

$$Y^{4-} + H^+ \rightleftharpoons YH^{3-} \quad pK_{S_4} = 10.3$$

$$YH^{3-} + H^+ \rightleftharpoons YH_2^{2-} \quad pK_{S_3} = 6.2$$

b) NH_3 reacts as a complexing agent:

$$Zn^{2+} + NH_3 \rightleftharpoons [Zn(NH_3)]^{2+}$$

$$[Zn(NH_3)]^{2+} + NH_3 \rightleftharpoons [Zn(NH_3)_2]^{2+}$$

$$K'_{ZnY} = K_{ZnY}f_{Zn^{2+}}f_{Y^{4-}}$$

c) OH^- reacts as a further complexing agent (pH > 10)

$$Zn^{2+} + 4OH^- \rightleftharpoons [Zn(OH_4)]^{2-}$$

Values of selected stability constants can be found in Table 7.1-5 and in the literature.

Indication in complexometry

The equivalence point of a complexometric titration is represented by the inflection point of a titration curve according to Fig. 7.1-5. It can be found by:

a) direct registration of the titration curve by electrochemical techniques (see Sec. 7.3). Major methods are amperometry, if the "free" metal ion and the metal chelate have different *half*-wave potentials, or by potentiometry (ion-selective electrodes)

b) metallochromic indicators; these are complexing agents as well, but with the special property of color changes between the free ligand and the metal complex. In addition, the stability of the indicator complex must be significantly smaller than the stability of the titrant complex with the same metal ion.

At the beginning of the titration, the indicator reacts with the metal ion, with the formation of a colored metal–indicator complex. During the titration, the stronger metal–titrant complex is formed, which causes the indicator complex to liberate its metal ions around the equivalence point, and to change into the free indicator form, also changing the color. At comparable color intensity of free and complexed form, the equivalence point is marked by the appearance of the mixed color, when 50% of the indicator has been liberated.

Metallochromic indicators are availaible for several metal ions. Here are some selected examples:

Eriochrom Black T (EBT)   (for Mg, Pb, Cd, Mn, Zn)

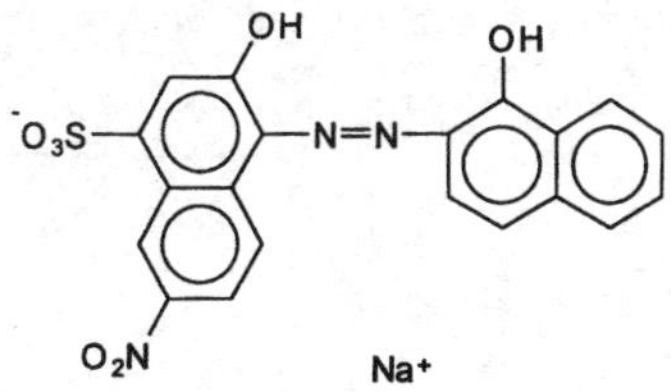

EBT is a trivalent acid, used practically as NaH_2E.

Indicator reaction: $HEBT^{2-} + Mg^{2+} \rightleftharpoons [MgEBT]^- + H^+$
                             blue   red

EBT is sensitive to oxidation, the aqueous solution is stabilized with ascorbic acid; Fe, which would block the indicator, is masked with hydroxylamine. Commercially available "indicator buffer tablets" are a mixture of an indicator based on EBT and dyes (color change red/grey/green) with a buffer for $pH = 10$.

Further useful indicators for complexometric titrations are:

Calgon carbocylic acid, for Ca, also for Cd, Mg, Mn, Zn:

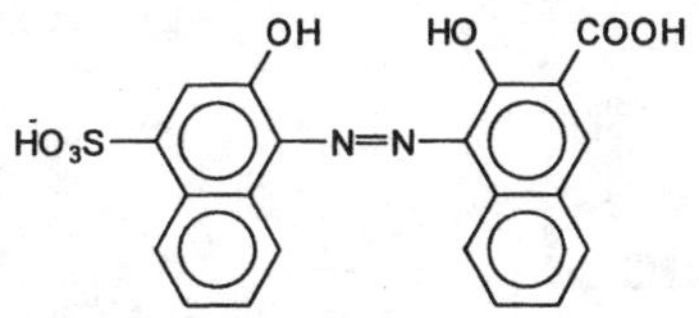

and Tiron for Fe^{3+}:

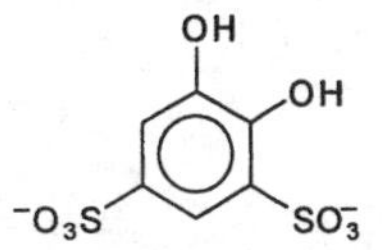

Applications of complexometric titrations

As a general rule, high selectivity for the metal ion to be determined by complexometric titrations of metal ion mixtures is required, and can be achieved by exploiting the principles of complex chemistry (see also Sec. 4.3). The following variables have to be optimized for each individual case:

- selection of ligand
- selection of indicator
- adjustment of pH
- masking of interfering cations, forming stronger chelates with the titrant
- precipitation of interfering cations

Complexometric titrations can be followed via the titration curve or by the use of indicators.

Titration methods

a) Direct titration: EDTA forms chelate complexes with many cations, but with different stability. Fe^{3+} and Ca^{2+} can be titrated with EDTA in the same solution, Fe^{3+} at $pH = 3$, and then Ca^{2+} after changing the pH to 10.

b) Back titration: for metal ions without an appropriate indicator, the sample solution is mixed with a known amount of EDTA in excess, and the excess is determined with Mg^{2+} and EBT

c) Substitution titration: the sample solution (e.g.: containing Fe^{3+} is mixed with an excess of $(MgEDTA)^{2-}$, then the freed Mg^{2+} can be determined in the usual way

d) Indirect titration: the determination of anions is possible by this technique, e.g., sulfate determination:

> precipitation of the sulfate with excess Ba^{2+}
> titration of the excess Ba^{2+} with EDTA

Examples

a) Determination of water hardness

Water hardness is represented by the sum of the concentrations of Ca^{2+} and Mg^{2+} in mmol/L, and is determined by the simultaneous titration of the two ions with EDTA (Titriplex III) and EBT at pH = 10. Masking agents for the major interferents Fe^{3+} and Mn^{2+} (hydroxylamine), Al^{3+} (triethanolamine), and Cu^{2+} (cyanide) are described in the literature.

b) Individual determination of Ca^{2+} and Mg^{2+}

The solution is first adjusted to pH = 12 with NaOH and $Mg(OH)_2$ is precipitated. Then Ca^{2+} is titrated alone with EDTA and Calgon carbonic acid. To prevent errors due to coprecipitation of Ca^{2+}, the precipitate is redissolved in HCl and $Mg(OH)_2$ precipitated a second time with NaOH. The exact Mg^{2+}-content is calculated from the difference between the total hardness and the Ca^{2+} concentration.

7.1.3   Electrochemical titrations

Electrochemical titrations are classical titrimetric methods with electrochemical endpoint detection. They are mainly potentiometric, amperometric, and conductimetric techniques, as described below and, in greater detail, in Sec. 7.3.

Potentiometric titrations

Potentiometric methods can be used to monitor the course of titration procedures, i.e., to determine the titration endpoint if an indicator electrode is available for at least one of the chemical species taking part in the stoichiometric equivalence point of the chemical reactions (either the titrant or the titrand or both). The inflexion point of the titration reaction (i.e., with the largest $\Delta E / \Delta V_R$ value) coincides with the endpoint of the reaction if the titration curve is symmetric. In any reaction where the stoichiometry is not 1:1, the titration curve is not symmetric. The titration error associated with the asymmetric titration curves can be partly overcome if the titrant concentration is determined under exactly the same experimental conditions as that of the analyte.

Typical titration curves recorded during the course of a precipitation titration are shown in Fig. 7.1-8 together with different endpoint determination approaches (direct reading, first derivative, and second derivative, respectively). It must be kept in mind that the error of potentiometric titrations is also affected by the error of the titration endpoint location.

The potentiometric method of endpoint location can also be used in different automated titrations. But, in any titration procedure, a potentiometric indicator electrode serves only as an endpoint indicator and the analytical concentration of the analyte is determined on the basis of the stoichiometric chemical reaction.

Margin notes:

Water hardness is measured as the total concentration of calcium plus magnesium.

By using electrochemical cells as detectors whole titration curves can be recorded rather than the end points alone.

Potentiometric cells with glass electrodes register acid-base titration curves, also mixtures e.g. such as in Fig. 7.1-4.

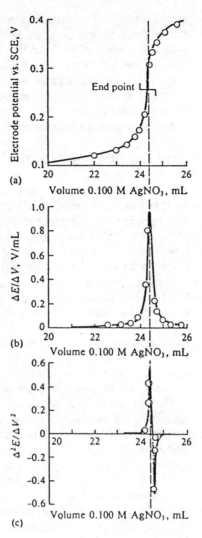

Fig. 7.1-8. Typical potentiometric titration curves for titration of chloride with silver ions

Amperometric titration

Amperometry as a detection mode is widely used in different flow analytical techniques, including flow injection technique as well as chromatographic separation methods, and also in conventional volumetric and coulometric titrations. In the latter, current measurement is used to locate the titration endpoint.

The working electrode material is more critical in continuous flow analytical techniques than in titrations, mainly because of mechanical and long-term operational stability. Different carbon electrodes, such as glassy carbon or carbon paste (a mixture of graphite powder and paraffin or silicone oil) is especially popular for high-performance chromatography and flow injection analysis. For conventional titrations, the indicator electrode is a polarizable electrode, e.g., a dropping mercury electrode (DME), a platinum, or a graphite microelectrode. The other electrode is either a reference electrode (i.e., a nonpolarizable electrode, for which the potential remains unchanged at relatively small current flow) such as a saturated calomel electrode (SCE), or another polarizable electrode. Thus, amperometric titration methods may be subdivided into:

Amperometric cells register e.g. metal ion concentrations during precipitation reactions or the rest water content of "water-free" organic solvents according to Karl Fischer.

- titrations with one polarizable electrode
- titrations with two polarizable electrodes (biamperometric titrations)

Amperometric titration with one polarizable electrode

This method involves measurement of the current during the titration in the presence of background electrolyte at a preselected working electrode potential in a stirred solution. The working electrode potential is selected so that the current is due to the electrochemical oxidation or reduction of the analyte or the titrant or both. The amperometric titration curve is constructed on the basis of the alteration of the limiting current versus titrant volume or titration percentage. The shape and characteristics of the titration curves depend on the current–voltage curve for the working electrode and the reversibility of the redox couples involved in the titration reaction.

As an example let us consider the following titration reaction:

$$Pb^{2+} + Cr_2O_7{}^{2-} \rightarrow PbCr_2O_7 \tag{7.1-9}$$

The working electrode is a dropping mercury electrode, while the reference electrode is a SCE. The working electrode potential is set and kept at a constant value with respect to the reference electrode so that the limiting current of the reducible species, Pb^{2+}, is measured as a function of volume of titrant added. If the potential is set to $0\,V$ the current remains at zero level up to the end point because Pb^{2+} is not reduced at that potential. After the equivalence point the diffusion current rises due to the added excess of the reducible chromate ions. The titration endpoint is the intersection point obtained by extrapolation of the two linear portions of the titration curve. The titration points close to the endpoint are not considered, owing to the dissociation of complexes and the solubility of the precipitates formed during the titration process.

When both titrant and analyte ions give diffusion current at the selected operating voltage a V-type titration curve is obtained. This is true for the lead-chromate-couple if the potential is set to $-1\,V$.

The rotating Pt electrode can also be used as working electrode in amperometric titrations. The increased limiting current value due to convective diffusion and the lowered residual current at an electrode of constant surface area in amperometric operation mode significantly lower the detection limit of the technique. Thus, amperometric titration can be used for quantitative analyses in the concentration range of 10^{-5} to $10^{-6}\,mol/L$. However, it must be kept in mind that the low overvoltage of hydrogen on the platinum electrode may be a serious problem when using the electrode as cathode in acidic solutions.

The relative error of amperometric titration is less than ± 2–3% RSD. The selectivity of the direct amperometric technique can be improved by appropriate selection of the titration reaction. Thus, compounds of similar $E_{1/2}$ may be determined in the presence of each other by amperometric titration with the proper selection of the titrants.

Amperometric titration with two polarizable electrodes

Amperometric titrations using two identical polarizable electrodes of equal size are also called biamperometric titrations. The current is measured while a relatively small, fixed potential (10–$100\,mV$) is applied across the electrodes as a function of the titrant volume. One electrode functions as the anode, while the other as the cathode. The shape of the titration curve is strongly dependent on the reversibility of the electrode reaction of the titrant and the analyte system. Biamperometric titrations are primarily used for the endpoint detection of redox titrations.

Let us consider, as an example, the titration of I_2 with $S_2O_3{}^{2-}$ at a small potential, ΔE applied across the two platinum electrodes. The titration reaction is:

$$I_2 + 2\,S_2O_3{}^{2-} \rightleftharpoons 2\,I^- + S_4O_6{}^{2-} \tag{7.1-10}$$

For the qualitative interpretation of the titration curves, the $I_2\,|\,I^-$ and $S_2O_3{}^{2-}\,|\,S_4O_6{}^{2-}$ redox couples must be considered. As is known, current will flow in the electrochemical cell if an electrode reaction proceeds at both electrodes. If the fixed potential applied is small, the current flows in the cell when both components of the reversible redox system are present in the solution. The magnitude of the current is determined by the component present in the solution at a lower concentration.

The $I_2 \mid I^-$ system is reversible at platinum electrodes, while the $S_2O_3^{2-} \mid S_4O_6^{2-}$ system is irreversible. Thus, at the small applied potential, the following reaction proceeds:

$$I_2 + 2e^- \rightarrow 2I^- \tag{7.1-11}$$

and the reaction:

$$S_2O_3^{2-} + 2e^- \rightarrow S_4O_6^{2-} \tag{7.1-12}$$

does not.

Accordingly at the start of the titration there is no current flow since no I^- is present in the solution to be titrated; only the residual current can be measured.

As the titration proceeds I^- is generated by the redox reaction and faradaic current flows between the Pt electrodes according to the following electrode reactions:
At the cathode:

$$I_2 + 2e^- \rightarrow 2I^- \tag{7.1-13}$$

At the anode:

$$2I^- \rightarrow I_2 + 2e^- \tag{7.1-14}$$

The current is determined by the concentration of the species of the redox couple present in the solution at lower concentration. Thus, the current has a maximum at 50% titration when the concentrations of both I_2 and I^- are equal, then it decreases owing to the consumption of the I_2 in the titration reaction. At 100% titration and beyond this point there is no I_2 present and only residual current is observed. The titration curves of this type of biamperometric titration are said to have "dead stop" end-points since the current falls essentially to zero at the end-point and remains there.

Similar titration curves can be derived by considering the current–voltage curves of the titrant and the analyte systems and the potential applied.

Karl–Fischer titration of water

The technique can be used for the determination of water in purified solvents and for water of crystallization in crystals. The Karl-Fischer reagent, containing, e.g., iodine, sulfur dioxide, pyridine in $1:3:10$ mole ratio in methanol, can be used for the direct titration of water in any solvent which does not react with sulfur dioxide and/or iodine. It must be kept in mind that solvents containing aldehydes and ketones cannot be titrated since they bind sulfur dioxide.

The Karl–Fischer method is used for the determination of water.

The basis of the method is that sulfur dioxide reacts with iodine in the presence of water according to the following reaction leading to equilibrium:

$$SO_2 + I_2 + 2H_2O \rightleftharpoons SO_4^{2-} + 2I^- + 4H^+ \tag{7.1-15}$$

The equilibrium can be shifted to the right-hand side owing to the presence of pyridine:

$$\tag{7.1-16}$$

The water equivalent of the Karl-Fischer standard is determined with a methanol solution containing a known amount of water, e.g., $CH_3COOH \cdot 3\,H_2O$.

In the course of the titration, the variation of the current as a function of a known volume of standardized Karl-Fischer reagent, i.e., $i\text{--}V_{Reagent}$ (mL) is recorded. The titration endpoint is determined by dead stop endpoint detection.

Conductometric titrations

Conductometric methods often involve the use of selective reagents.

The specificity of conductometry is enhanced by the addition of a selective reagent. This fact is used in conductimetric titrations in which the variation of the electrical conductivity of a solution as a function of the reagent addition V_R is monitored.

The quantitative determination is based on the determination of the reagent consumption until chemical equivalence is reached. The titration endpoint is obtained by the intersection of two straight lines fitted to the linear parts of the titration curves (i.e., conductance–titration volume data pairs) recorded, during the course of titration.

The main advantages of conductance measurements over classical and potentiometric endpoint detection are:

- they can be used for the titration of colored or turbid solutions
- the titration curve can be constructed on the basis of a few data points, far from the chemical equivalence point

The latter allows one to overcome problems such as the slow potential response in the neighborhood of the chemical equivalence point in potentiometric titrations. Thus the speed of titrations can be significantly enhanced.

Conductivity endpoint detection is used especially in acid–base, and precipitation titrations. As an illustration, let us consider the following acid–base titration:

$$\underline{H^+} + \underline{A^-} + M^+ + OH^- = H_2O + \underline{M^+} + \underline{A^-} \tag{7.1-17}$$

The underlined ions are present in the solution before the titration endpoint.

Conductometric cells register acid-base titrations also for mixtures of weak acids in aqueous, eventually colored or turbed solutions.

Thus with the addition of titrant (MOH), H^+ ions are replaced by metal ions M^+, while the anion concentration is unchanged. Until the chemical equivalence point is reached, the number of ionic species is unchanged; only the hydrogen ions are exchanged for metal ions, and since the mobility of the hydrogen ions are much greater than that of metal cations, the conductance of the solutions decreases until the equivalence point is reached. After the equivalence point the conductance corresponds to the excess of hydroxyl ions added, compared with that corresponding to the equivalence point. Thus the conductance increases after the equivalence point as the base concentration increases, and the corresponding titration curve (strong acid with strong base) is V-shaped.

In contrast, in the titration of a weak acid with a weak base, the conductivity of the starting solution is low, owing to limited dissociation. The salt produced in the course of the titration dissociates completely resulting in a higher conductivity. However, after the equivalence point is reached, the concentration of the strong electrolyte remains constant, and only that of the weak base increases.

In general a well-defined titration point is only obtained if a strong acid is titrated with strong base or a weak acid with weak base.

Hydrolysis, dissociation of the reaction product, or solubility phenomena give rise to curvature of the titration curve in the vicinity of the endpoint. At a distance away from the end point i.e., from 0–50%, and between 150–200% of the equivalent volume of titrant, these effects are suppressed. By extrapolating these linear portions, the position of the endpoint can be determined precisely.

As in all titration techniques, the titrant must be 10 times as concentrated as the solution being titrated in order to keep the volume change small. If necessary a correction may be applied; all conductance reagents are multiplied by the terms $(V + v/V)$, where V is the initial solution volume, while v is the titrant volume added.

Questions and problems

1. What are the common principles of all titrimetric methods of analytical chemistry?
2. What different types of volumetric methods do you know?
3. Draw the titration curve for the acidimetric titration of 0.01 M NaOH with 0.1 M HCl and describe the different parts of the curve as well as the pH of the end point using the Hägg-diagram of the system.
4. Using the appropriate Hägg-diagram, draw the titration curve for the acidimetric titration of 0.01 M NH_4OH with 0.1 M HCl and describe the different parts of the curve as well as the pH of the buffer point and the end point!
5. Upon titration of H_3PO_4 with NaOH in aqueous solutions, according to Fig. 7.1-4, only 2 equivalence points (C_1 and C_2) and 1 buffer point (B_2) can be visually observed. Explain, why the other points are hidden in the curve!
6. How can weak acids (bases) be enforced before titration?
7. What type of ligands can be successfully used for complexometry and why?
8. Describe the function of complexometric indicators.
9. Under what experimental conditions can one register titration curves?

7.2 Gravimetry

Learning objectives

■ To set out the advantages and disadvantages of gravimetry
■ To show, how a reproducible precipitation is obtained, and to illustrate the critical issues of gravimetry
■ To introduce into the principles and applications of electro-gravimetry

7.2.1 Introduction

In gravimetry, an analyte (in most cases ionic) is precipitated in the form of an insoluble compound of defined stoichiometry. After collection and drying, the product is weighed on an analytical balance and, from the mass and known stoichiometry, the original analyte is quantitatively determined.

Gravimetric methods are widely used in standardization processes, although volumetric and instrumental techniques have superseded gravimetric methods in most routine and general analytical investigations. Generally speaking, gravimetric methods are extremely accurate, owing to the fact that it is possible to weigh substances to great accuracy with analytical balances. It is common practice to determine a weight to 5 digits; however, even the most sensitive volumetric and instrumental techniques are rarely more precise than about 1–0.1%.

> Gravimetric methods offer extremely high accuracy.

The accuracy (trueness and precision) of a gravimetric method depends on the precipitation technique and also on the properties of the precipitate. If high accuracy is demanded, the following requirements must be met:

- definite and reproducible stoichiometric composition of the precipitate
- low solubility in the mother liquor and in the wash solvent (the analyzed substance must be quantitatively incorporated into the final precipitate)
- minimal interference from other elements and components of the system
- low surface area of the precipitate (crystalline) so that adsorption of impurities is minimal
- properties that enable convenient separation of the solid from the mother liquid and efficient washing with a suitable solvent
- thermal stability, so that the precipitate may be dried conveniently without changing its composition
- stability of the dried product (hygroscopic properties of the substance are troublesome)

7.2.2 Gravimetric techniques

It is of the utmost importance that the precipitation leads to a filtrable product of highest purity. Therefore some skill and experience is required when gravimetric methods are to be applied.

Some of the above-mentioned properties can be influenced by the precipitation technique. Crystalline substances generally have a definite stoichiometric composition, they are easy to separate from the original solution, either by filtration or centrifugation, and they may be efficiently washed with a minimum volume of the rinsing solvent. In addition, crystalline substances are characterized by a relatively small specific surface area. Therefore, the surface activity is relatively small (par-

> Control of the precipitation technique is important for accurate gravimetric determination.

ticularly when compared with amorphous or colloidal precipitates) and the amount of adsorbed or occluded material is minimal.

Crystal habit and size can be controlled to a certain extent by the precipitation technique. Rapid mixing of two concentrated reagent solutions results in the formation of undesirable microcrystalline products. If the solubility of a substance is extremely temperature dependent, small crystals are obtained by sudden temperature changes, resulting in a lower solubility. If a hot solution is cooled very slowly, the crystal seeds have an opportunity to grow before new crystallization centers are formed. Supersaturated solutions are subject to sudden crystallization processes that produce a heavily occluded material; mother liquor with all its constituents is occluded in the precipitate. These impurities cannot be removed by washing.

As a general rule, hot solutions should be used to form precipitates with the appropriate features necessary for the filtration. Add the reagents dropwise and choose as small a concentration as possible. Stir during the addition of any reagents so that you achieve a homogeneous distribution of all solutes.

A number of volatile solutes and electrolytes are known. Solutions of such chemicals should not be heated, otherwise substantial losses may occur. For this reason, the heating of a solution containing a volatile acid (H_2S, HCl, etc.) or base (NH_3, etc.) is not recommended. For instance, in the gravimetric determination of hydrochloric acid with silver nitrate, the acid solution should not be heated before the addition of $AgNO_3$. However, after addition of $AgNO_3$, a temperature close to the boiling point (80 °C) is recommended in order to obtain a precipitate which can be conveniently filtered. Therefore, chloride solutions used in gravimetric determinations should not be heated before the addition of $AgNO_3$, since the precipitation is carried out in an acidic medium (addition of HNO_3) and therefore contains volatile HCl.

Before filtration, the precipitation should have settled to the bottom of the vessel, so that the most of the liquid can be removed by decantation.

In the early days the precipitation was collected on a filter paper, which was then burned completely in a platinum crucibles to leave the final product only. Later, special ceramic crucibles were made available as filtering devices. In both applications the washing process is essential for the accuracy of final results. To obtain a dry product, the crucible containing the precipitation is dried in an oven. From the weight difference between the empty crucible and crucible plus precipitate, the weight of the precipitate is determined, and from that the weight of the analyte is calculated.

7.2.3 Evaluation of gravimetric analyses

The aim of a gravimetric analysis is the determination of the concentration w_A of an analyte in a given sample using the laws of stoichiometry. The result is expressed as follows and is based on two precise weighings:

$$w_A = \frac{m_A}{e} 100 = \frac{af}{e} 100$$

Provided we have a pure precipitate, the concentration ϖ_A of the component to be determined can be calculated – based on two weighings – using the laws of stoichiometry.

where w_A is the content of the component sought (%); m_A is the mass of the component sought (mg); e is the mass of the sample before gravimetric treatment (mg); a is the mass of the substance weighed after precipitation, drying, etc. (mg); and f is the stoichiometric factor.

The stoichiometric factor f is the ratio of the molar masses of component A and its weighing form:

$$f = \frac{\text{formula weight of the substance sought (g.mol}^{-1})}{\text{formula weight of the substance weighed (g.mol}^{-1})}$$

f 100 is the percentage of the substance sought in the substance weighed.

The following table gives some examples of the calculation of stoichiometric factors.

Species sought	Species weighed	Stoichiometric factor	
MgO	$Mg_2P_2O_7$	$\dfrac{2 \cdot M(MgO)}{M(Mg_2P_2O_7)}$	0.3622
Fe_3O_4	Fe_2O_3	$\dfrac{2 \cdot M(Fe_3O_4)}{3 \cdot M(Fe_2O_3)}$	0.9666

The stoichiometric factor f should be as small as possible, in order to provide for a high sensitivity of the method, and to keep the weighing error small. That can be obtained by using organic precipitation reagents with high molar mass.

Species sought	Precipitation reagent	Species precipitated	Species weighed	Stoichiometric factor
Fe^{3+}	OH^-	$Fe(OH)_3$	Fe_2O_3	0.6994
Al^{3+}	8-Hydroxyquinoline	$Al(HQ)_3$	Al_2O_3	0.5293
Al^{3+}	8-Hydroxyquinoline	$Al(HQ)_3$	$Al(HQ)_3$	0.05873
Ni^{2+}	Dimethylglyoxime (sodium salt)	$Ni(DMG)_2$	$Ni(DMG)_2$	0.2032

A practical example is given for the gravimetric determination of Ba^{2+} after precipitation with H_2SO_4 as $BaSO_4$, filtration, and thorough drying in a quartz crucible:

Sample weight: 0.6537 g
Weight of $BaSO_4$: 0.4288 g

$$f = \frac{M(Ba)}{M(BaSO_4)} = \frac{137.34}{233.40} = 0.5884 \qquad w_{Ba} = \frac{af}{e} \times 100 = 38.6\%$$

7.2.4 Electrogravimetry

In contrast to voltammetric experiments (see Sec. 7.3), electrolysis can be performed under conditions where the composition of the solution is altered, owing to the quantitative electro-oxidation or electro-reduction of solution components. These bulk electrolysis techniques are characterized by a large ratio of working electrode area A to solution volume V and convective diffusion-controlled mass transport conditions. The use of large electrodes permits the flow of relatively high current and makes possible the fast conversion of material into product in electro-oxidation or electro-reduction. The basic principles governing electrode reactions in bulk electrolysis are the same as discussed in connection with microelectrolysis (voltammetric) techniques.

Quantitative oxidation and reduction reactions are the basis of electrogravimetry.

Bulk electrolysis techniques can be classified according to the quantities measured. In electrogravimetric experiments, the weight of the material deposited at one of the electrodes (the working electrode) in a pure form of known stoichiometry is determined. Electrogravimetry belongs to the group of microdeterminations gravimetry using the current as reagent.

Electrogravimetry is among the oldest electroanalytical techniques and had become a useful analytical tool in the analysis of metal ions. It is now used primarily for the electroseparation of metals from complex mixtures.

In coulometric methods, however, the total amount of electricity Q consumed in the electrolysis serves as the basis for the quantitative determination of the amount of analyte present or reagent produced (see Sec. 7.3).

Both electrogravimetric and coulometric methods are highly accurate (true and precise) analytical methods. The evaluation of the analyte concentration requires no preliminary calibration with standard solutions.

Principle of electrolysis with controlled current

In the electrolytic process a chemical reaction is driven in the nonspontaneous direction by an external source of electricity, by applying a dc voltage to the cell consisting of two electrodes in contact with the solution. The total voltage applied across the cell is distributed in the following manner if $E_{appl} < E_{equ}$:

$$E_{appl} = E_a - E_c + iR \tag{7.2-1}$$

where E_a is the anode potential; E_c is the cathode potential; i is the current; and R is the resistance.

Each electrode potential is considered as the sum of a reversible potential and the overpotential. The reversible potential depends on the standard potential of the electrochemical half-reaction taking place at the electrode and the activities of the relevant substances in the vicinity of the electrode surface.

Let us consider the following overall reversible reaction:

$$O + ne^- \rightleftharpoons R \tag{7.2-2}$$

where both O and R are soluble.

The rate of the electrode reaction depends on the applied potential and the concentrations of the substances O and R are expressed by the Nernst equation as:

$$E_{appl} = E^0 + \frac{RT}{nF} \ln \frac{c_O}{c_R} \tag{7.2-3}$$

As the electrolysis proceeds, i.e., as the metal deposition proceeds, the electrode potential decreases with decrease of the metal cation concentration, c_O in the solution.

The degree of completion of the bulk electrolysis process can be predicted from the applied potential and the Nernst equation. Supposing that in the beginning, the concentration of O is $c_{O,i}$, E^0 is the standard electrode potential, V_s is the volume of solution and x is the fraction of O reduced to R at the electrode potential, E then moles of O at equilibrium $= V_s c_i (1 - x)$, while moles of R at equilibrium $= V_s c_i x$; and

$$E_{appl} = E^0 + \frac{RT}{nF} \ln \frac{V_s c_i (1 - x)}{V_s c_i x} = E^0 + \frac{RT}{nF} \ln \frac{(1 - x)}{x} \tag{7.2-4}$$

For example at 99.99% completeness of reaction of O to R (i.e., $x = 0.9999$) the working electrode potential is:

$$E_{appl} = E^0 + \frac{0.059}{n} \log \left(\frac{0.0001}{0.9999} \right) \cong E^0 + \frac{(0.059)}{n} \cdot \log 10^{-4} \tag{7.2-5}$$

or $236/n$ mV more negative than E^0 at 25 °C.

Similarly, for electrogravimetric conditions, when metal ions are deposited, e.g., on an inert electrode such as platinum:

$$O + ne^- \rightleftharpoons R \text{ (solid)} \tag{7.2-6}$$

The Nernst equation yields:

$$E_{appl} = E^0 + \frac{RT}{nF} \ln[c_i(1 - x)] \tag{7.2-7}$$

where c_i is the initial concentration of oxidant; and x is the fraction of the oxidized form reduced at electrode potential E.

Considering the extent of electro-reduction, the potential shift due to the given degree of completion of electrolysis can be calculated (Eq. 7.2-4).

From this it follows that an alteration for the electrode potential due to electrolysis must be kept in mind when designing a bulk electrolysis experiment for the selective determination of metals.

The change in electrode potential as an electrolytic reaction proceeds must be accounted for in bulk electrolysis determinations.

Let c_A and c_B be the initial concentration of components A and B, respectively, and E_A^0 and E_B^0 the redox potential values of the relevant half-reactions. If $E_A^0 > E_B^0$ then the condition for the selective separation at $x = 0.9999$ is:

$$E_A^0 + \frac{0.0592 \times 4}{n} \; E_B^0 + \frac{0.059}{n} \log c_B \qquad (7.2\text{-}8)$$

$$\Delta E = \frac{0.059}{n}(4 + \log c_B) \qquad (7.2\text{-}9)$$

In the foregoing it was assumed that the electrochemical reduction of both metal cations is reversible.

For the case shown, the extent of the electrode reaction will be governed by equilibrium conditions, but the rate of electrolysis will be small at potentials predicted by the Nernst equation. The rate of electrolysis is naturally increased with increase of the electrode potential applied, and reaches a limiting value at a potential corresponding to the plateau region of the current–voltage curves. As a consequence the actual potential for the electrolysis is usually selected on the basis of experimental current–voltage curves of microelectrodes (see Sec. 7.3) recorded under conditions close to those of the intended bulk electrolysis.

The selectivity of electrolysis techniques (electroseparation) can also be affected by the composition of the supporting electrolyte chosen for electrolysis. For the separation of two oxidation | reduction systems with similar E^0 values (e.g., $\Delta E^0 < 236$ mV at $25\,^\circ$C), the addition of a complexing agent, which complexes one of the metals may improve the separation. The addition of an agent that forms complex with a metal ion usually shifts the E^0 value of the metal–ion | metal redox system to more negative values. This approach can be employed for the separation of copper(II) from bismuth(III) ($E_{Cu}^0 = 0.345$, $E_{Bi}^0 = 0.320$). When the deposition is carried out in a tartrate-ion-containing buffer of pH 6 the quantitative electroseparation of the two metal ions can be accomplished. The metal deposition is often disturbed by H_2 formation, which can be avoided simply by making the solution more alkaline.

On the basis of the foregoing calculation, one can determine the standard potential difference theoretically needed for determining one ion without interference from another. This potential difference is 240 mV/25 °C for singly charged ions, 120 mV/25 °C for doubly charged ones.

The working electrode potential can be altered only by variation of the potential applied to the electrolysis cell. However, as Eq. 7.2-4 predicts, E_{appl} affects all potential values developed in the cell, all of which vary as the electrolysis proceeds. Thus, the only practical way of keeping the electrode potential at a preselected value, is to measure the cathode potential continuously with respect to a reference electrode. The cell potential can then be adjusted within the desired range. This type of analysis technique is called *controlled cathode potential electrolysis* or *controlled anode potential electrolysis*.

The working electrode potential with respect to a third reference electrode is controlled automatically with a potentiostat.

Current efficiency

The current efficiency for an electrode reaction (*i*th reaction) is equal to the fraction of total current rendered for the *i*th reaction:

$$\text{current efficiency for the } i\text{th reaction} = \frac{i_i}{i_{total}} v \qquad (7.2\text{-}10)$$

where i_{total} is the total current passing through the electrolysis cell.

100% current efficiency means that no side reaction takes place at an electrode, i.e., there is only one working electrode process.

It is beneficial for bulk electrolysis to be carried out with high current efficiency in order to reduce analysis time. In electrogravimetric experiments, 100% current efficiency is not required unless the side reactions produce insoluble reaction

High current efficiency reduces the analysis time.

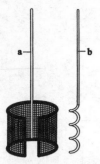

Fig. 7.2-1. Electrode designs for controlled potential bulk electrolysis

products. However, for coulometric titrations, 100% current efficiency is a prerequisite.

Electrolysis cells

Electrode designs used for controlled potential bulk electrolysis experiments are shown in Fig. 7.2-1. As cathode, Pt gauze or foil cylinders of diameter 2–3 cm are commonly used. Platinum electrodes have the advantage of being relatively non-reactive, and can be used for the deposition of a number of different metal ions. However, a protective coating is required before the electrolysis of certain metal ions, notably tin and zinc, to avoid alloy formation.

Amalgam-forming metals can be deposited on Hg cathodes. This procedure is especially used for removing easily reducible elements as a preliminary to the analysis.

The anode can also be made of Pt. Proper positioning of the auxiliary electrode is essential in order to provide uniform current density across the working electrode surface. The location of the reference electrode is also important for the long-term stability of the reference electrode and the iR drop.

The working and the auxiliary electrodes are often placed in separate compartments in order to avoid mixing of the electrolytes. The two compartments are usually separated by a sintered glass disk or an ion-exchange membrane of relatively low resistance, that does not contribute appreciably to the overall cell resistance. The use of separators can be avoided by the judicious choice of the auxiliary reaction, in which solid products, such as AgCl or inert gaseous products (e.g., N_2) are formed.

The cell resistance and the proper positioning of the reference electrode in the cell are important when the electrolysis is carried out in nonaqueous solvents with lower dielectric constant (e.g., acetonitrile, dimethylformamide, ammonia).

Electrogravimetric methods

Electrogravimetric methods can be used for quantitative separation of metals.

Electrogravimetric determinations require adherent, smooth deposits. The quality of the deposit depends, among other things, on the form of the metal ion in the solution. In general, depositions from solutions of complexed ions are smoother and more adherent than those obtained from solution containing the aquo form of the ion.

The time required for electrogravimetric analysis depends on the current density. However, too high current values may result in deposits of poorer quality. Hydrogen formation during deposition also results in rougher deposit.

The sensitivity of an electrogravimetric method is limited by the difficulty in determining the small difference in weight between the electrode and electrode plus deposit. In general, the deposit of 0.2–5 mmol material is especially advantageous. Electrogravimetric determinations have been supplemented by coulometric methods, except if the 100% current efficiency requirement cannot be fulfilled.

Electroseparation

In the previous section, the conditions for selective quantitative deposition of one metal M_1 on a solid electrode were discussed. The selectivity in this case means that no appreciable deposition of the other metal M_2 occurs under the experimental conditions selected. Thus, for complete deposition of M_1 ($\geq 99.9\%$) as Hg amalgam, the electrode potential E must be selected as:

$$E \leq E_1^0 - \frac{0.18}{n_1}\text{V at 25°C} \tag{7.2-11}$$

(where E_1^0 is the standard potential, n_1 is the number of electrons taking part in the electrode reaction), while for 0.1% deposition of M_2:

$$E > E_2^0 + \frac{0.18}{n_2} \text{V at } 25\,^{\circ}\text{C} \tag{7.2-12}$$

Thus for 99.9% separation of M_1;

$$\Delta E = E_1^0 - E_2^0 = 0.18\,(n_1 + n_2) \tag{7.2-13}$$

that is for ions of valency one, it is 0.36 V.

For electroseparation the Hg pool cathode is the most frequently used.

Questions and problems

1. What are the experimental conditions to obtain the high accuracy for gravimetric analyses in practice?
2. How does one enhance the chances to obtain a pure precipitate?
3. Compare the evaluation of gravimetric analyses and of titrimetric analyses.
4. Describe the principles of electrogravimetry and electroseparation.
5. How can one determine the standard potential difference needed for determining one metal ion without interference from another?
6. What types of electrodes are used in electrogravimetry?

7.3 Electroanalysis

Learning objectives

■ To discuss the operational principles of the basic electro-analytical techniques

■ To describe the type and the properties of different potentiometric and voltammetric electrodes in detail

■ To outline the potential use and capability of individual electroanalytical techniques

Electroanalytical methods are among the most powerful and popular techniques used in analytical chemistry. They are the product of intensive research in electrochemistry, a discipline that was developed almost one hundred years ago. Electroanalytical methods are some of the earliest instrumental analysis techniques, that revolutionized analytical chemistry in the twentieth century.

The basic electroanalytical techniques are *potentiometry*, *voltammetry* or *polarography*, *coulometry*, and *conductimetry*. All these techniques have found applications in volumetric techniques, which have been covered in Sec. 7.1.3. However, electroanalytical techniques also play a significant role in the development of detectors for chromatographic and related methods (see Chap. 5) and are an important branch of chemo- and biosensor development (Chaps. 7.7 and 7.8, respectively).

7.3.1 Potentiometry

Potentiometry is one of the most simple electroanalytical techniques, being primarily used for pH measurements in different samples and for the determination of ionic constituents in biological fluids (e.g., blood and urine), but also as a transduction mode in monitoring selective interactions in molecular sensor devices or in the course of chemical reactions.

The principle of potentiometric measuring techniques is the measurement of the cell potential, i.e., the potential difference between two electrodes (the indicator and the reference electrodes) in an electrochemical cell under zero current conditions with the aim of obtaining analytical information about the chemical composition of a solution. In potentiometry, the chemical sensor is the indicator electrode and its potential and the scope of the application of the technique depends on the nature and the selectivity of the indicator electrode. There are two major types of analysis techniques in potentiometry: namely when the cell potential is determined and correlated to the activity or the concentration of the individual chemical species (direct potentiometry) or when the variation of the cell potential is monitored as function of the reagent addition to the sample (potentiometric titration).

Direct potentiometry vs. potentiometric titration

The function of the measuring instrument is primarily to ensure that no significant current is drawn from the electrochemical cell. For this purpose either a potentiometer or a voltage follower can be used, although the latter has become dominant for cell voltage measurements.

Potentiometric cell

In potentiometry, the solution containing a species (analyte) whose activity (concentration) is to be determined and the indicator electrode constitute an electrochemical half-cell, while the second half-cell is the reference electrode assembly (Fig. 7.3-1). The indicator electrode is the chemical sensor and its potential E_{ind} is

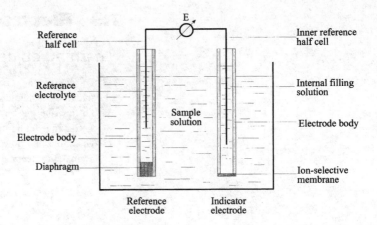

Fig. 7.3-1. Schematic representation of a potentiometric measuring cell

proportional to the logarithm of the analyte activity (concentration), while the reference electrode in a solution of constant composition has a fixed potential E_{ref}, which is assumed to be independent of the analyte containing solution composition in the cell. Both electrodes are ideally nonpolarizable electrodes.

The potential of the cell E can be expressed by the following equation:

$$E = (E_{ind} - E_{ref}) + E_{jnct} \qquad (7.3\text{-}1)$$
$$\text{right} \quad \text{left}$$

where E_{ind} is the indicator electrode potential, E_{ref} is the reference electrode potential, and E_{jnct} is the liquid junction potential.

Accordingly, the cell potential is defined as the potential of the electrode on the right of the potentiometric cell with respect to that on the left while the corresponding half-reactions are expressed in terms of their reduction potentials.

Cell potential and cell voltage

For example

$$Ag \,|\, AgCl, KCl \,\|\, Ce^{3+}, Ce^{4+} \,|\, Pt \qquad (7.3\text{-}2)$$

Left-hand (reference) electrode:

$$AgCl(s) + e^- \rightleftharpoons Ag(s) + Cl^- \qquad E^0 = 0.222\,V \qquad (7.3\text{-}3)$$

Right-hand (indicator) electrode, e.g. Pt:

$$Ce^{4+} + e^- \rightleftharpoons Ce^{3+} \qquad E^0 = 1.44\,V \qquad (7.3\text{-}4)$$

The corresponding electrode potentials are:

$$E_{left} = 0.222 - 0.05916 \log[Cl^-] \qquad (7.3\text{-}5)$$

$$E_{right} = 1.44 - 0.05916 \log \frac{[Ce^{3+}]}{[Ce^{4+}]} \qquad (7.3\text{-}6)$$

The cell potential at 25 °C:

$$E = E_{right} - E_{left} = \left(1.44 - 0.05916 \log \frac{[Ce^{3+}]}{[Ce^{4+}]}\right) - (0.222 - 0.05916 \log[Cl^-])$$

$$= 1.218 - 0.05916 \log \frac{[Ce^{3+}]}{[Ce^{4+}][Cl^-]} \qquad (7.3\text{-}7)$$

Since the chloride ion concentration is fixed in the reference half cell, the cell potential reflects the changes in the quotient $[Ce^{3+}]/[Ce^{4+}]$.

In a properly designed potentiometric cell, E_{ref} is constant, and E_{jnct} is either constant or negligible.

Types of potentiometric electrodes

Conventional electrodes for potentiometry are described as types or classes:

(a) Classical electrode types

Class 0. Inert metals (e.g. Pt, Au) in contact with a solution of redox couple, e.g., Pt | Ce^{4+}, Ce^{3+} system. Ideal inert materials exchange electrons reversibly with the electrolyte components and are not subject to oxidation or corrosion themselves;

Class 1. Reversible metal/metal ion (ion exchanging metals bathed in electrolytes containing their own ions), e.g. Ag/Ag^+;

Class 2. Reversible metal/metal ion saturated salt of the metal ion and excess anion X^-, e.g. $Ag/AgX/X^-$;

Class 3. Reversible metal/metal salt or soluble complex/second metal salt or complex and excess second cation, example Pb/Pb oxalate/Ca oxalate/Ca^{2+}.

(b) Membrane-type devices (ion-selective electrodes)

Reference electrodes

A reference electrode made up of phases having constant chemical composition exhibits a known and constant (analyte concentration independent) potential. The important requirements for a satisfactory reference electrode system are reversibility, reproducibility, and stability in time. Reversibility means that the direction of the electrode reaction can be changed simply by changing the polarity of the electrode. The reproducibility is expressed as the standard deviation of the cell potential at consecutive measurements in a solution of given concentration, while the stability of the electrode is disscussed in terms of drift in flowing solutions and the residual standard deviation of the response in a given solution. The reference electrodes used in analytical practice are secondary standards and most commonly electrodes of the second type. The potentials of the reference electrodes are determined with respect to the standard hydrogen electrode (SHE, primary standard).

The calomel electrode is by far the most widely used reference electrode

Silver/silver chloride electrode (Ag|AgCl, KCl)

The silver/silver chloride electrode is prepared, e.g., by coating a silver wire anode with silver chloride electrolytically in 0.1 mol/L chloride solution. The silver-chloride-coated wire is immersed in KCl solution of known concentration, usually saturated, i.e., 3.8, 1.0, or 0.1 mol/L (Fig. 7.3-2a).

The response of the silver/silver chloride electrode is based on the following reaction:

$$AgCl(s) + e^- \rightleftharpoons Ag(s) + Cl^- \qquad (7.3-8)$$

The silver/silver chloride electrode is the most reproducible reference half-cell, with good electrical and chemical stability at 25 °C. The potential of the electrode is affected by all solution constituents that alter the silver ion concentration. Thus, the electrode cannot be used directly (i.e., without an additional salt bridge) in solutions that contain proteins, bromide, iodide or sulfide ions that form insoluble compounds (precipitates) with the silver ions, in the presence of AgCl-complexing agents such as CN^-, SCN^-, or in strong oxidizing or reducing media.

The temperature hysteresis effect is very small with the silver/silver chloride reference electrode. For this reason it is recommended for applications in which the temperature cannot always be held constant or is above 80 °C.

Calomel electrode (Hg|Hg$_2$Cl$_2$, KCl)

Calomel electrodes include a platinum wire immersed in a paste of mercury and mercury(I)chloride placed in a glass tube. The internal filling is potassium chloride of known concentration, saturated with mercury (I) chloride (Fig. 7.3-2c). The potential of the calomel electrode is determined by the following reaction:

$$Hg_2Cl_2(s) + 2e^- \rightleftharpoons 2Hg(l) + 2Cl^- \qquad (7.3-9)$$

and expressed by the Nernst equation as:

$$E = E^{0\prime} - \frac{0.05916}{2} \log[Cl^-]^2 = E^{0\prime} - 0.05916 \log[Cl^-] \qquad (7.3-10)$$

The Nernst equation

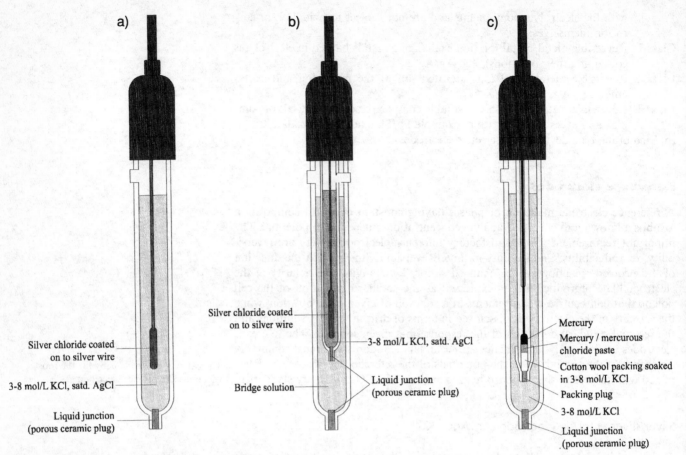

Fig. 7.3-2. Reference electrode designs. (a) Silver/silver chloride electrode. (b) Double junction silver/silver chloride electrode. (c) Calomel electrode

Most commonly a calomel electrode with saturated (3.8 mol/L at 25 °C) KCl (SCE) is used because it is easy to prepare, the presence of KCl crystals guarantees its constant solution concentration; and it has a potential +0.241V at 25 °C. The calomel electrode is not suggested for use as a reference electrode at temperatures above 75 °C.

Conversion of different potential scales is illustrated in Fig. 7.3-3.

Liquid junction potential

The liquid junction is a solution contact zone between two different electrolyte solutions, usually the reference electrode assembly and the sample. It is a nonselective interface, the site of the troublesome liquid junction (or diffusion) potential, E_{jnct}.

In a potentiometric cell, the reference electrode is connected to the sample solution via a salt bridge electrolyte. The salt bridge ensures electrical contact between the two half-cells of the electrochemical cell. Mixing of the two nonidentical phases (of dissimilar composition) can be minimized if the contact between the two phases is made small, by ensuring capillary contact, sintered glass diaphragms, or agar-agar plugs. The nature of the liquid junction (sample solution | salt bridge contact region) has an influence on the reproducibility of the cell voltage measurements. The value of the liquid junction potential is, in the majority of cases, unknown, and sets a limit on the accuracy of direct potentiometric measurements. In the cell diagram, the liquid junction is represented by a double line, $\|$.

Various types of liquid junctions are demonstrated in Fig. 7.3-4. In the junction zone (e.g., a diaphragm) a concentration gradient develops as the driving force for diffusion. Owing to the differences in mobilities of the ions at the interface (junc-

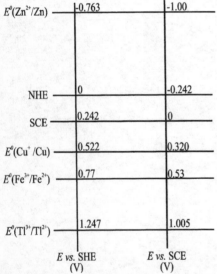

Fig. 7.3-3. Voltage conversion between SHE and SCE scales

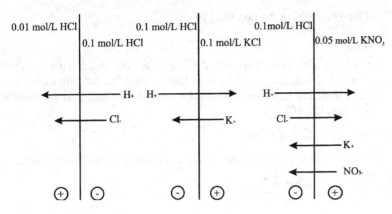

Fig. 7.3-4. Schematic presentation of the structure of different types of liquid junction. (From Bard, A.J. and Faulkner, L.R. (1980), *Electrochemical Methods*, New York: Wiley, p. 63, with permission)

tion) charge separation, i.e., face boundary potential will develop, resulting in the movement of ions at the interface at equal rates. The steady state potential thus developed is called the liquid junction or diffusion potential.

A useful approach for minimizing liquid junction potential is to introduce an intermediate salt bridge solution, where the solution in the bridge has ions of equal mobility as KCl:

$$Hg \,|\, Hg_2Cl_2 \,|\, HCl \,(0.1 \,mol/L) \,\|\, KCl \,(c) \,\|\, KCl \,(0.1 mol/L) \,\|\, Hg_2Cl_2 \,|\, Hg$$

$$(7.3\text{-}11)$$

As the concentration c of KCl increases, the value of the junction potential decreases, because the series junction potentials become similar in magnitude and have opposite polarities, thus they tend to cancel.

Double-junction reference electrodes are commercially available for precise cell potential measurements (Fig. 7.3-2b). By the use of double-junction salt bridges, in addition to the decrease of the junction potential, contamination of the sample solution with the internal filling solution of the reference electrode can be overcome. Double-junction reference electrodes are especially used for precise ion-selective electrode measurements.

In direct potentiometry it is essential that the liquid junction be the same for the calibration solutions and the sample solutions. In other hand, in analytical practice the liquid junction potential is approximated by the Henderson formalism [7.3-1].

Potentiometric indicator electrodes

The indicator electrode is the chemical sensor in the potentiometric cell, generating a potential, which is proportional to the logarithm of the activity (concentration) of the chemical species it responds to. The indicator electrode potential can be described by the Nernst equation or an equation derived from it.

Classical potentiometric electrodes

Metallic indicator electrodes

At *electrodes* of the *first* kind (e.g., Ag|Ag$^+$) the potential determining interfacial equilibrium is:

$$M^{n+} + ne^- \rightleftharpoons M \qquad (7.3\text{-}12)$$

At equilibrium the electrochemical potentials of the species distributing between the two contacting immiscible phases are equal:

$$\bar{\mu}_{M^{n+}} + n\bar{\mu}_e = \bar{\mu}_M \qquad (7.3\text{-}13)$$

where $\bar{\mu}_M$ is the electrochemical potential of metal M in the metallic phase, $\bar{\mu}_{M^{n+}}$ is the electrochemical potential of metal ion M^{n+} in the solution phase, $\bar{\mu}_e$ is the electrochemical potential of electrons in the metallic phase, and n is the number of electrons exchanged per mole.

The electrochemical potential is:

$$\bar{\mu} = \mu + nF\phi \tag{7.3-14}$$

By solving the equations for $\Delta\phi$, i.e., for the equilibrium Galvani potential ($\Delta\phi = \phi_{metal} - \phi_{solution}$) and by considering that the activities of the electrons and metal atoms are equal to unity in the metal phase and that the chemical potential μ of a species, M^{n+} is

$$\bar{\mu}_{M^{n+}} = \bar{\mu}^{\circ}_{M^{n+}} + RT \ln a_{M^{n+}} \tag{7.3-15}$$

the classical Nernst equation for the electrode potential is obtained:

$$E = E^0 - \frac{2.303\,RT}{nF} \log \frac{1}{a_{M^{n+}}} \tag{7.3-16}$$

where E^0 is the standard electrode potential, $2.303RT/nF$ is the so-called Nernst factor, and $a_{M^{n+}}$ is the activity of the metal ion.

As the potential of this type of electrodes is the result of an electron exchange equilibrium, its potential is affected by other redox couples in the solution.

Electrodes of second kind

At *an electrode of the second kind* (e.g. Ag|AgCl, (x mol/L) KCl) the potential generating electrochemical process is also a redox process:

$$Ag^+ + e^- \rightleftharpoons Ag \tag{7.3-17}$$

The silver ion concentration, however is determined by the following solubility equilibrium:

$$AgCl(s) \rightleftharpoons Ag^+ + Cl^- \quad K_{sp} = 1.8 \times 10^{-10} \quad (25\,°C) \tag{7.3-18}$$

By combination of the two equations, the electrode potential can be expressed as:

$$E = \underbrace{E^0_{Ag^+} + 0.05916 \log K_{sp}}_{E^{0\prime}_{Ag,AgCl}} - 0.05916 \log[Cl^-] \tag{7.3-19}$$

$$E = E^{0\prime}_{Ag,AgCl} - 0.05916 \log[Cl^-]$$

Thus, the electrode responds reversibly to the chloride ion activity, ideally the potential variation is 59.16 mV (at 25 °C) for each tenfold change in $[Cl^-]$.

The electrode potential is influenced by any species that affects the activity of the silver ion, e.g., the presence of AgCl-complexing agents such as CN^-, SCN^-; the presence of species which form less-soluble compounds with Ag^+, such as S^{2-}, Br^-, I^-, and strong oxidizing and reducing media. Under this condition the Ag|AgCl electrode cannot be immersed directly into the test solution but should be connected by a salt bridge.

The electrode has a fixed potential only if the chloride ion concentration in the electrode assembly is kept constant.

Redox electrodes

Pt, C, and Au electrodes are relatively inert and are used to transmit electrons to or from a reducible or oxidizable species in solution. The value of the Pt indicator electrode potential (half-cell potential) is determined by the activities of the oxidant (ox) and reductant (red) of the solution according to the Nernst equation:

$$E = E^0 - \frac{0.0592}{n} \log \frac{a_{red}}{a_{ox}} = E^0 - \frac{0.0592}{n} \log \frac{\gamma_{red}}{\gamma_{ox}} - \frac{0.0592}{n} \log \frac{c_{red}}{c_{ox}}$$

$$= E^{0\prime} - \frac{0.0592}{n} \log \frac{c_{red}}{c_{ox}} \tag{7.3-20}$$

where $E^{0\prime}$ is the formal potential.

The formal potential is defined as the potential of the half-cell at unit concentrations of both oxidant and reductant at the specified level of an electrochemically inert electrolyte.

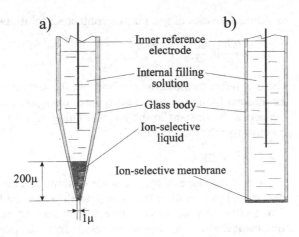

Fig. 7.3-5. Ion-selective electrodes. (a) Micro-electrode design. (b) Membrane type of macro-electrode arrangement

Ion-selective electrodes

Ion-selective electrodes (ISE) are electrochemical half-cells, consisting of an ion-selective membrane (i.e., a selective interphase), an internal filling solution, and an internal reference electrode (conventional construction) (Fig. 7.3-5) or of an ion-selective membrane and a solid contact ("all-solid-state" configuration). Such an electrode allow selective determination of the activity of certain ions in the presence of other ions; the sample under test is usually an aqueous solution. These devices are distinct from systems that involve redox reactions (electrodes of zeroth, first, second, and third kinds, although they often contain a second kind electrode as an "inner" or "internal" reference electrode.

In ion-selective electrode potentiometry, the other half cell is composed of an external reference electrode, and the contact between the two half-cells is preferably maintained by an intermediate salt bridge, which can be placed adjacent to the reference electrode compartment (double-junction reference electrode) (Fig. 7.3-2). The schematic (shorthand) notation of an electrochemical cell incorporating an ion-selective electrode as indicator electrode is:

$$\Delta\phi_1 \qquad\qquad \Delta\phi_2 \qquad \Delta\phi_{jnct} \qquad\qquad E_m \qquad\qquad\qquad\qquad \Delta\phi_3$$

$$Hg\,|\,Hg_2Cl_2, KCl(std)\,\|\,salt\ bridge\,\|\,sample\,|\,membrane\,|\,internal\ solution, AgCl\,|\,Ag$$

$$\qquad\qquad Reference\ electrode \qquad\qquad\qquad\qquad\qquad ISE$$

The total electrical potential difference measured between the two reference electrodes is composed of a number of local potential differences, constituting the potential response of the potentiometric cell:

$$E = (\Delta\phi_1 + \Delta\phi_2 + \Delta\phi_3) + \Delta\phi_{jnc} + \Delta\phi_m = \Delta\phi_0 + \Delta\phi_{jnc} + E_m \qquad (7.3\text{-}21)$$

where $\Delta\phi_0$ is the reference cell potential which is independent of the analyte concentration, $\Delta\phi_{jnc}$ is the liquid junction potential, and $\Delta\phi_m$ is the membrane potential.

The membrane potential, E_m, describes the performance of the ion-selective membrane electrode (Fig. 7.3-6). For a membrane which is ideally selective for ion A, the zero-current membrane potential is a direct measure of the corre-

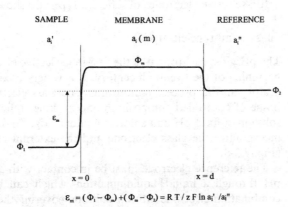

Fig. 7.3-6. Potential distribution at the ion-selective membrane | solution interface and in the bulk membrane

sponding activities of ion i in the contacting solutions:

$$E_m = \frac{RT}{z_i F} \ln \frac{a_i'}{a_i''} \qquad (7.3\text{-}22)$$

where a_i' refers to the activity of ion i in the sample solution, a_i'' to that of ion i in the internal filling solution, and z_i is the charge of ion i.

Since a_i'' is constant in the internal filling solution, the cell potential of an ion-selective membrane electrode is

$$E = E_i^{0'} + S \log a_i' \qquad (7.3\text{-}23)$$

where S is the slope of the electrode calibration graph ($\Delta E / \Delta \log a_i'$) which in the ideal case, is identical to the Nernstian slope; i.e. $59.16 \, \mathrm{mV}/z_i$ ($25\,^\circ\mathrm{C}$).

In analytical practice, we have to consider an additional contribution to the total measured ion activity which results from the presence of interfering species B in the sample solution. The response of real membrane electrode systems can successfully be described by the Nikolsky–Eisenmann equation:

$$E = E_i^0 + S \log \left[a_i' + \sum_{i \neq k} K_{ik}^{\mathrm{pot}} (a_k')^{z_i/z_k} \right] \qquad (7.3\text{-}24)$$

where K_{ik}^{pot} is called the selectivity coefficient, z_k is the charge number of interfering ion, k; a_k is the activity of the interfering ion, k.

The potentiometric selectivity coefficient (K_{ik}^{pot}) expresses the selectivity of the ion-selective electrode toward ion, k with respect to the principle ion, i. In other words, the selectively coefficient expresses the extent to which the sensor is more selective to ion i than ion k. For example $K_{ik}^{\mathrm{pot}} = 10^{-3}$ means that the activity of ion k can be 1000 times higher than that of ion i in order to contribute to the cell potential to the same extent as ion, i. The selectivity coefficients of individual electrodes are listed in textbooks.

Ion-selective membrane electrodes may be classified according to the nature of the basic membrane materials, into the following categories:

Primary ion-selective electrodes
- glass membrane electrodes
- crystalline (or solid-state) membrane electrodes
- liquid membrane electrodes
  - liquid ion-exchanger membrane electrodes
  - neutral carrier-based liquid membrane electrodes

Compound or multiple membrane ion-selective electrodes
- molecular sensing devices such as gas-sensitive and enzyme electrodes, the potentiometric detection unit of which is based on conventional potentiometric electrodes of the type listed above (see Sec. 7.7).

Ion-selective field effect transistors (ISFETs),

These kinds of electrodes are hybrids of ion-selective electrodes and metal-oxide field effect transistors (MOSFETs). In the ISFETs the metal gate of a MOSFET has been directly replaced by or contacted with a solid or a liquid ion-sensitive membrane. The response of such miniaturized sensors is linked to a current (Sec. 7.7).

Ion-selective electrodes of different types are shown in (Fig. 7.3-7).

Glass membrane electrodes

The pH glass electrode was the first ion-selective electrode to be discovered, at the beginning of the twentieth century. It is widely used for measuring pH in the laboratory and in process monitoring. The glass electrode consists of a glass membrane of controlled composition and an inner reference electrode immersed in a solution of fixed pH and chloride content (Fig. 7.3-7a). The combination electrode incorporates the glass electrode and the external reference electrode in one unit (Fig. 7.3-8).

The reference electrode must be in contact with a sample solution of unknown pH through a liquid | liquid junction, which can be ensured by immersing the combination electrode into the solution covering the ceramic plug.

Types of ion selective electrodes

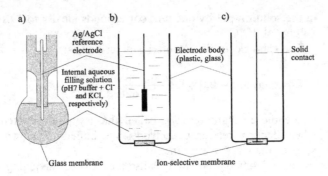

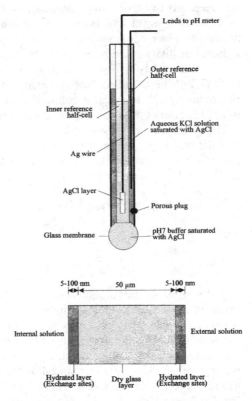

Fig. 7.3-7. Design of ion-selective electrodes of different nature. (a) Glass membrane electrode. (b) Crystalline membrane electrode; membrane type. (c) Crystalline membrane electrode; with solid contact

Fig. 7.3-8. Structure of a combination glass electrode; as well as that of the glass membrane

Stable potential values are obtained with all glass electrodes after a definite "soaking time" in dilute buffer solution. During the soaking process the glass surface adsorbs water and a swollen hydrated layer is formed. Parallel to this, silanol groups are formed, and there is a tendency for exchange of sodium ions in the swollen hydrated layer for hydrogen ions in the sample. The cross section of the glass membranes thus formed is shown in Fig. 7.3-8.

The membrane potential develops because the silica network due to the strong basic nature of the silica groups has a strong affinity for certain cations, particularly H^+, to adsorb at the anionic fixed sites:

$$\equiv SiOH_{surface} \rightleftharpoons \equiv SiO^-_{surface} + H^+_{solution} \tag{7.3-25}$$

The specific adsorption of hydrogen ions creates a charge separation that alters the interfacial potential difference at each surface, while the sodium ions transport the charge across the surface. The mechanism of the glass electrode response is characterized by the so-called fixed charge membrane model.

The potential difference measured between the two reference electrodes of a combination glass electrode depends on the pH of the solution outside the glass membrane (in the test solution), since all the other interfacial potentials are kept constant by keeping the composition of the contacting phases constant. A change

The glass electrode potential

in the solution pH by one unit corresponds ideally to 59.16 mV change in the cell voltage at 25 °C.

The glass electrode potential is described as:

$$E = \text{constant} - 0.059 \log \frac{a_{H^+}(\text{inside})}{a_{H^+}(\text{outside})} \qquad (7.3\text{-}26)$$

where the constant term, is called the asymmetry potential which is measured when identical reference electrodes and buffers of known pH are placed on both sides of the glass membrane.

The asymmetry potential is ascribed to the inequality of the two sides of the glass membrane, and can be corrected for by calibration in buffers of known pH.

As early as 1923 it was shown that, especially at higher pH, the glass electrode also responds to other monovalent cations such as silver, sodium, and potassium ions [7.3-2] Lengyel and Blum [7.3-3] systematically studied the correlation between the glass composition and electrode function, and concluded that the sodium sensitivity could be enhanced by the introduction of certain trivalent metal oxides, mainly Al_2O_3, B_2O_3 etc. It was also found that, by replacing Na_2O with Li_2O [7.3-4] the so called "sodium error" could be reduced and such membrane electrodes with this new glass composition could be used for pH measurement up to pH 13, even at high sodium ion concentration.

The systematic investigation of the ion-selectivity of the glass electrode proved that the surface structure of the glass membrane material at the membrane solution interface is the primary influence on the selective response. The selectivity coefficient $K_{H,Na}^{Pot}$ is the product of the surface ion-exchange equilibrium constant:

$$Na^+{}_{surf} + H^+{}_{sol} \rightleftharpoons H^+{}_{surf} + Na^+{}_{sol} \qquad (7.3\text{-}27)$$

while

$$K_{H,Na} = \frac{a_{H^+,surf} \cdot a_{Na^+,sol}}{a_{H^+,sol} \cdot a_{Na^+,surf}} \qquad (7.3\text{-}28)$$

and the relevant ion mobility ratios u_{Na^+}/u_{H^+} of the exchanging ions in the hydrated glass layers. Thus by judicious alteration of the glass membrane composition, the selectivity coeffients can also be altered.

Owing to the sodium sensitivity of the pH-sensitive glass membrane at high pH (i.e., low a_{H^+}) and high a_{Na^+} values, the measured pH values are lower than the true pH values. The error introduced by the presence of the sodium ions in the pH determination is called the "alkaline" or "sodium" error. The reasons for the "acidity" error, i.e., for the deviation of the ΔpH vs. pH graph in the acidic region, are not clear. How the acidity and sodium error vary with glass composition is demonstrated in Fig. 7.3-9.

Glass membranes with optimized selectivities for Na^+ are:

$$Na_2O, Al_2O_3, SiO_2 \qquad [K_{Na,K}^{Pot} \approx 10^{-3}] \qquad (7.3\text{-}29)$$

   11%   18%   71%

while for K^+

$$Na_2O, Al_2O_3, SiO_2 \qquad [K_{K,Na}^{Pot} \approx 10^{-1}] \qquad (7.3\text{-}30)$$

   27%   5%   68%

Before the first use, the glass electrode must be kept in dilute buffers in order to establish the hydrated surface layer. Between measurements, the electrodes are stored in water to retain its hydrated surface layer.

Crystalline (solid state) electrodes

The sensing element (the membrane material) of a crystalline (solid state) ion-selective electrode consists of an ionically conducting, sparingly soluble salt in ionogenic solvent (water) such as AgCl, Ag_2S, LaF_3, or $Ag_2S + CuS$, $Ag_2S + PbS$, and $Ag_2S + CdS$, constructed either in homogeneous or heterogeneous form. Homogeneous membranes are made up entirely of the salt, either as a slide of

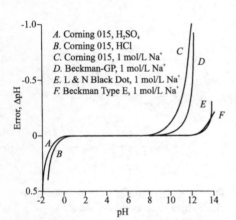

Fig. 7.3-9. Acid and alkaline error for selected glass electrode at 25 °C. (From Bates, R.G. (1973), *Determination of pH*, 2nd ed., New York: Wiley, p. 365, with permission)

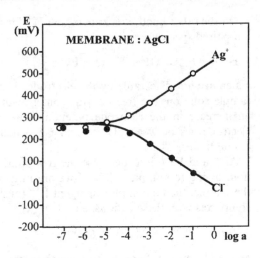

Fig. 7.3-10. The response of a silver chloride based ion-selective electrode to silver and chloride ions

crystal or a pressed pellet of appropriate shape. In heterogeneous membranes, the active salt is incorporated in an inert supporting material, such as silicone rubber.

The silver-halide-based crystalline electrodes are designed primarily to detect anions. In practice, however, these sensors can be used either to detect silver or its counter anion (X^-) (Fig. 7.3-10).

The potential response functions for Ag^+ and X^- ions are, respectively:

$$E = E^0_{Ag^+} + \frac{RT}{F} \ln a_{Ag^+} = E^0_{X^-} - \frac{RT}{F} \ln a_{X^-} \qquad (7.3\text{-}31)$$

$$E^0_{X^-} = E^0_{Ag^+} + \frac{RT}{z_i F} \ln K_{sp,AgX} \qquad (7.3\text{-}32)$$

where $K_{sp,AgX}$ is the solubility product of the relevant silver halide salt

The *detection limit* of silver halide based ion-selective electrodes as determined by the membrane material, is related to the minimal silver ion activity, $a_{Ag^+,min}$ set by the dissolution process at the membrane solution interface. For this, primarily two interfacial processes must be considered:

Detection limit of AgX based ion selective electrodes

1) Dissolution is governed by the solubility product of the electrode membrane material (e.g., AgCl-based membranes).
2) Leaching out of silver ions from the membrane where it is accumulated due to coprecipitation, adsorption, etc.

Accordingly, electrodes based on insoluble salts cannot be considered as inert, and owing to the above type of surface reactions, the detection limit of such electrodes lies in the range of 10^{-5}–10^{-6} mol/L.

One of the most valuable electrodes of the crystalline membrane type is the fluoride electrode, since there are only a few analytical methods that allow simple selective determination of fluoride. The electrode membrane consists of a disk of LaF_3 single crystal doped with Eu(II) in order to increase membrane conductivity (Doping here means that, by the addition of Eu(II), a small amount of La(III) is replaced). Its function is based on the selective adsorption of fluoride ions at the surface of the electrode resulting in charge separation, while in the membrane the electrical conductivity is due to the mobility of fluoride ions alone. The internal filling solution contains 0.1 mol/L NaF and 0.1 mol/L NaCl in contact with a Ag|AgCl electrode.

The electrode has a working range of 10^0–10^{-6} mol/L for fluoride ions. The only major interfering anion is OH^- as it can participate in a surface ion exchange reaction:

$$LaF_3 + 3\,OH^- \rightleftharpoons La(OH)_3 + 3\,F^- \qquad (7.3\text{-}33)$$

Thus, at higher pH values, a layer of $La(OH)_3$ is built up since the solubility of $La(OH)_3$ corresponds roughly to that of LaF_3. The relevant potentiometric selectivity coefficient is $K^{pot}_{F,OH} = 0.1$, while the selectivity of the electrode for fluoride over other anions is several orders of magnitude higher.

On the other hand, it is known that H^+ ions combine with fluoride ions to form ion-paired species:

$$H^+ + 2\,F^- \rightleftharpoons HF + F^- \rightleftharpoons HF_2^- \tag{7.3-34}$$

which are not detectable by the electrode. Therefore, by decreasing the pH in the sample solution, the free fluoride content will also be decreased, which results in an increase in the electrode potential. The working pH range for the fluoride electrode lies between pH 5.5 and 6.5, which can be maintained by acetate or citrate buffer.

Al^{3+} and Fe^{3+} ions decrease the free fluoride content in a F^--containing solution, owing to complexation. If we are interested in the total fluoride content of the sample, then a complexing agent L^{3-} must be added which forms more stable complexes with these cations, e.g.:

$$AlF_6^{3-} + L^{3-} \rightleftharpoons Al-L + 6\,F^- \tag{7.3-35}$$

and will set the fluoride free. Such complexing agents are citrate or ethylenediaminotetraacetic acid (EDTA). Accordingly, by the proper selection of measuring conditions, the effect of pH and other complexing agents such as Al^{3+} and Fe^{3+} can be eliminated.

Liquid membrane electrodes

Liquid membranes are three (or four) component membranes with electrically (positively or negatively) charged or electrically neutral mobile ion carriers. The mobile carriers are compounds with ion-binding properties (sites), which are incorporated most commonly into plasticized poly(vinyl chloride) matrices.

Among the charged carriers, the liquid ion-exchanger-based electrodes for monitoring nitrate, and calcium ion activities are of the utmost interest. The active membrane ingredient for a nitrate-sensitive electrode is, e.g., tris (4,7-diphenyl-1,10-phenanthroline)Ni^{2+}-nitrate in *p*-nitrocymene, while for a calcium-ion-selective electrode the ion-exchanger is, e.g., calcium dodecylphosphate dissolved in dioctylphenylphosphonate. The potential determining reaction, e.g., for a calcium ion response at each membrane | solution interface is:

$$[(RO)_2PO_2^-]_2Ca \rightleftharpoons 2(RO)_2PO_{2,surf}^- + Ca_{soln}^{2+} \tag{7.3-36}$$

The ion exchanger facilitates the transport of Ca^{2+} across the membrane. The calcium electrode has a working range of 10^{-1}–10^{-6} mol/L and the major interfering ions are H^+, Zn^{2+}, Fe^{2+}, Mg^{2+}, etc.

Following up the pioneering idea of Stefanac and Simon [7.3-5] to use antibiotics as neutral carriers in liquid membrane electrodes, a great number of electrically neutral ligands with different structures, called ionophores, were designed for ion-selective potentiometry. The neutral carriers are antibiotics, synthetic acyclic compounds, cyclic polyethers (crown compounds) (Fig. 7.3-11) or more recently calix(4)-arene derivatives. A common feature of these ionophores is that they have a stable conformation with polar binding sites (cavity) for the uptake of a cation, surrounded by a lipophilic shell (Fig. 7.3-12). Nevertheless, the ionophores should have sufficient flexibility to allow a fast ion exchange.

The ionophore is incorporated most commonly into a plasticized poly(vinyl chloride) membrane, typically consisting of 1 wt% ionophore, 66 wt% nonpolar solvent (plasticizer), and 33 wt% PVC. Lipophilic salt additives are added to the membrane with the aim of decreasing membrane resistance and improving ion selectivity. The most frequently used membrane constituents are summarised in Table 7.3-1.

The potential-generating membrane solution interfacial process is:

$$M_{soln}^{z+} + nS_{membr} \xrightleftharpoons{K_{m,n}} MS_{n,membrane}^{z+} \tag{7.3-37}$$

where $K_{m,n}$ is the distribution parameter and S stands for the ionophore.

The potentiometric ion selectivity of the membrane is a function of the stability constants and of the distribution coefficients of complexes formed with the analyte ion, A (primary ion) and with the interfering ion, B respectively in a surface ion-

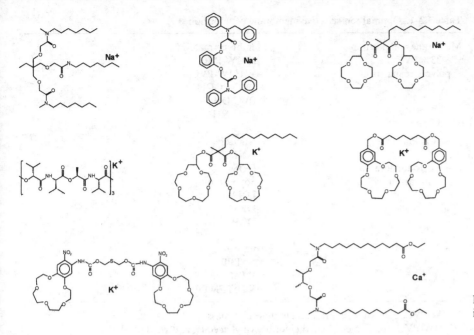

Fig. 7.3-11. Chemical structure of different ionophores for liquid membrane electrodes

DIAMETER IN A OF CROWN ETHERS AND SOME CATIONS IN CRYSTALS					
CATION DIAMETER	Li^+ 1.2	Na^+ 1.9	K^+ 2.66	Cs^+ 2.96	NH_4^+ 3.34
CROWN ETHER DIAMETER	14-Cr-4 1.3	15-Cr-5 2.0±0.2	18-Cr-6 2.9±0.2		21-Cr-7 3.85±0.45

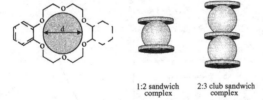

1:2 sandwich complex   2:3 club sandwich complex

Fig. 7.3-12. Structure of biscrown-ether-alkaline ion complexes

exchange process:

$$B^+_{soln} + AS^+_{membr} \rightleftharpoons BS^+_{membr} + A^+_{soln} \qquad (7.3\text{-}38)$$

For liquid membranes incorporating ionophores that form $1:1$ complexes with cations, the selectivity behavior can be approximated as:

$$K^{pot}_{AB} \approx \frac{\beta_{BS}}{\beta_{AS}} \qquad (7.3\text{-}39)$$

since the distribution coefficient of ionophore complexes are independent of the nature of the central ion. (β_{BS} and β_{AS} are the relevant stability constants.)

The tailored design of ionophores for coordination of different cations in liquid membrane electrodes proved to be highly selective, especially to alkaline and alkaline earth cations, such as Ca^{2+}, Na^+ and K^+, which are of biological importance. The analytically important parameters of a significant, biscrown-ether derivative based potassium sensor are shown in Fig. 7.3-13.

The interesting feature of liquid membrane electrodes is that they can be fabricated in different shapes and sizes. Micropipet-type potassium-selective sensors with a few μm tip size are of great value for physiologists and life scientists for the in vivo monitoring of potassium ion activity, e.g, in extracellular fluids. Potassium liquid membrane electrodes are about 10^4 times as sensitive (selective) to potassium as to sodium. In respect of the lifetime of liquid membrane electrodes, the

Liquid membrane electrodes for clinical analysis

Table 7.3-1. Optimal ion-selective membrane composition

Membrane composition	Membrane materials
~33% PVC polymeric matrix	HMW-PVC VHMW-PVC PVC-COOH PVC-OH PVC-NH$_2$
~66% Plasticizer	Plasticizers o-NPOE
~1% Ionophore (ion carrier)	DOS Soybean oil DNA DBS DOP
~0.5% Additives	Additives NaTPB KTpClPB KTbTFMPB

HMW-PVC, poly(vinyl chloride) high molecular mass
VHMW-PVC, copolymer of vinyl chloride, vinyl acetate, and vinyl alcohol
o-NPOE, 2-Nitrophenyl octyl ether
DOS, bis(2-ethylhexyl) sebacate
DNA, bis(1-butylpentyl) adipate
DBS, dibutyl sebacate
DOP, bis(2-ethylhexyl) phthalate
NaTPB, sodium tetraphenylborate
KTpClPB, potassium tetrakis(4-chlorophenyl)borate
KTbTFMPB, potassium tetrakis[3,5-bis(trifluoromethyl)phenyl]borate

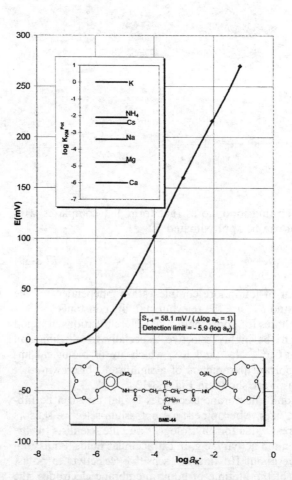

Fig. 7.3-13. Analytically important parameters of a biscrown-ether based potassium sensor; S is the slope value

$$\frac{\Delta E}{\Delta \log a_{K^+} = 1}$$

$\log K_{KM}^{pot}$ is the selectivity coefficient determined at 0.1 mol/L activity level with the so called separate solution method (Guilbault, G.G., et al. (1976) *Pure Appl. Chem. 48*, 127).

lipophilicity of the ionophores, and also those of all the membrane components (e.g., plasticizer, salt additives, etc.) are of special importance.

Compound (multiple-layer) ion-selective electrodes

Sensitised potentiometric devices (electrodes) consist of a base sensing element (an ion-selective electrode) and a modifying membrane layer which acts as an additional selective interface. The nature of the modifying layer (chemically active or not) induces a further selectivity to the signal. Thus, by this construction the sensitivity of the base element is extended toward the detection of different inorganic gaseous and organic molecules. These devices are considered in Sec. 7.7 on chemical sensors.

Direct potentiometric measurement with calibration

Direct potentiometry can be used for the determination of concentration or activity of a chemical species for which only a selective indicator electrode is available. The technique is based on comparison of the indicator electrode potential (or the cell potential if all potential terms other than that of the indicator electrode are kept constant) measured in the analyte solution with that determined in two or more standard solutions of known analyte concentration. The ion-selective electrodes, including the pH sensitive glass electrode function, cannot be accurately described with the Nernst equation (E^0, S).

Measurement of pH

Due to the difficulties of thermodynamically determining single ion activities as well as single electrode potentials, the experimental method established for the determination of pH does not allow the determination of pH values defined on the basis of hydrogen concentration or activity.

The term "operational pH" is based on assigned pH values of reference standard solutions, namely:

$$pH = pH(s) + \frac{E - E_s}{2 \cdot 302RT/F} \tag{7.3-40}$$

where pH(s) is the assigned pH value of the standard reference solution, E_s is the corresponding cell potential, and E is the cell potential measured in solution of unknown pH.

The operational pH scale was established by conventional assignment of pH values to standard solution(s) of known composition based on the measurements in the following potentiometric cell without liquid junction, called a Harned-type cell:

$$Pt \,|\, H_2 \; (1 \text{ atm}) \; H^+ \, (a_{H+} = 1), Cl^- \,|\, AgCl \,|\, Ag \tag{7.3-41}$$

The "operational pH" definition had been accepted by NIST (National Institute of Standard and Technology, USA), similar organizations in other countries and IUPAC (International Union of Pure and Applied Chemistry).

For calibration of the glass electrode, which is predominantly used for pH measurements, two-point calibration has been suggested and accepted. The pH of the calibration standards should bracket the pH of the unknowns and should not differ from each other by more than $\Delta pH = 5$, since the slope S of the electrode calibration graph, E vs pH, is not the same over the entire pH range.

Acceptance of the term operational pH means the realization of the fact that the electrochemically determined pH is not exactly the same as that defined on the basis of the hydrogen ion activity or of the hydrogen ion concentration, but rather the result of conventional operational procedure.

pIon (pX) determination

Direct potentiometry offers a way to determine the concentration or the activity of the chemical species for which ion-selective electrodes are available with the help of a calibration graph. The electrode calibration should be carried out for two reasons; (a) to check the properties of the potentiometric cell; and (b) to serve as a basis for activity or concentration evaluation. As known in practice, the response of ion-selective electrodes deviates from the ideal Nernstian behavior, and can be approximated by the following equation:

$$E = E' \pm S \log a_I \qquad\qquad (7.3\text{-}42)$$

where E is the cell voltage, E' is the constant term, whose value corresponds to a solution where $\log a_I = 0$, S is the slope of the electrode response, theoretically $= 59.16\,mV/pIon$ unit for a monovalent ion at $25\,°C$ (positive for cations, negative for anions) and a_I is the activity of the measured ion, I.

According to Eq. 7.3-42 E vs $\log a_I$ provides, over a wide activity range, a straight line calibration graph which can be constructed on the basis of two-point calibration, using at least two standard solutions.

Ion activity determination using an activity calibration graph

The calibration standards are prepared by serial dilution of a stock solution (e.g., $0.1\,mol/L$) prepared by accurately weighing a certain amount of the salt of the corresponding analyte (measured) ion. The relevant activity values are calculated on the basis of single ion activity coefficients, determined by an independent method, or calculated by the extended Debye–Hückel equation (see Chap. 4.1), bearing in mind that all methods for calculation of single ion activity coefficients are only approximations.

The problems associated with pIon determination on the activity scale are the same as those in very precise pH measurements.

In respect to the analytical accuracy of the pIon determination on an activity scale, one must keep in mind that the accuracy of the analytical measurements cannot exceed that of the calibration standards. An error of $1\,mV$ in the potential determination correponds to 4% relative error for a monovalent analyte, while it is 8% for divalent analyte.

Determination of concentration using concentration standards

The primary requirement for calibration standards is that the activity coefficient of the measured ion be the same in the standard and in the sample solutions. However, this can only be fulfilled if the ionic strength:

$$\left[I = \frac{1}{2} \sum_i z_i^2 c_i \right]$$

is the same in the standard and in the sample solutions (see Sec. 4.1).

There are two ways to prepare solutions as similar in nature as possible:

1) Standards are prepared with the same matrix, which is comparable to the sample. In blood analysis this is usually the case for serum samples, for which the concentration of the ionic constituents are known from independent measurements.
2) The other method is the addition of inert electrolyte (not detectable by the ion-selective electrode) in excess to both sample and standard solutions to maintain the ionic strength so high that its variation among the various solutions can be neglected (e.g., $1:10$ dilution of the solutions with $5\,mol/L$ $NaNO_3$).

For the analysis of specific ions, e.g, S^{2-}, CN^-, F^-, multipurpose conditioning solutions are available. For the analysis of fluoride in water samples total ionic strength adjusting buffers of differing compositions have been elaborated. (For example TISAB I consists of $1\,mol/L$ NaCl, $0.25\,mol/L$ acetic acid, $0.75\,mol/L$ sodium acetate, and $0.001\,mol/L$ sodium citrate.) The fluoride ion standards and water sample are diluted $1:1$ with TISAB in order to maintain the total ionic strength to a constant value and to adjust pH ≈ 5.5. The role of citrate is to release the fluoride ions from Al^{3+}, Fe^{3+} complexes.

Standard addition methods

The method of standard addition is suggested for analyte concentration evaluation if the sample matrix is complex or unknown. This technique is based on the observation of the change in cell potential on addition of a precisely measured volume of standard solution containing the measuring ion to a known volume of sample. The volume of the standard solution added must be small, so as not to alter significantly the ionic strength, the fraction of the free measured ion in the sample, and the liquid junction potential.

The cell potential measured in a sample of unknown concentration before the addition of the standard is:

$$E_1 = E' + S \log(c_x f_x k_x) \tag{7.3-43}$$

After the addition of a certain amount of a standard, it is:

$$E_2 = E' + S \log[(c_x + \Delta c) f_x k_x] \tag{7.3-44}$$

where c_x is the unknown concentration, f_x is the activity coefficient of the analyte ion in the sample, k_x is the fraction of the free measured ion in the presence of complexing agent, and Δc is the concentration change due to the addition of the standard.

By combing the two equations:

$$\Delta E = E_2 - E_1 = S \log \frac{c_x + \Delta c}{c_x}$$

and

$$\tag{7.3-45}$$

$$c_x = \Delta c (10^{\Delta E/S} - 1)^{-1}$$

The amount of added analyte is suggested to be about 100% of the original analyte. With multiple additions, not only the analyte concentration, but also the slope S of the electrode response can be determined.

7.3.2 Voltammetry

Voltammetry comprises microelectrolysis techniques in which the working electrode potential is forced by external instrumentation to follow a known potential–time function and the resultant current–potential and current–time curves are analyzed to obtain information about the solution composition. Depending on the shape of the potential–time perturbing signal and on the mode of the analyte transport, voltammetric techniques can be subdivided as *linear potential sweep (dc) voltammetry*, *potential step methods*, *hydrodynamic methods*, and *stripping voltammetry*.

Voltammetric techniques are distinct analytical tools for the determination of many organic and inorganic substances (primarily metal ions), which can be oxidized or reduced electrochemically (electroactive species) at trace level. The simultaneous determination of several analytes may also be possible with a voltammetric technique. Due to the analysis circumstances where the working electrode is a microelectrode, the sample solution volume is relatively large and the analysis time is short, the bulk concentration of the analyte is not changed significantly during the analysis. Thus, repeated measurements can be carried out in the same solutions. The selectivity of the techniques is moderate, but can be largely enhanced, e.g., by the combination of liquid chromatography with electrochemical detection.

The development of voltammetric techniques dates back to the pioneering work of Jaroslaw Heyrovsky (dc voltammetry with a dropping mercury electrode) in 1922 which was acknowledged by the Nobel Prize in 1959.

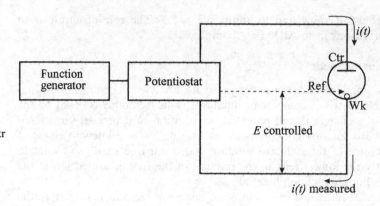

Fig. 7.3-14. Schematic experimental arrangement for controlled potential experiments; Wk is the working electrode, Ctr is the counter electrode, Ref is the reference electrode. (From Bard, A.J. and Faulkner, L.R. (1980), *Electrochemical Methods*, New York: Wiley, p. 137, with permission)

The electrochemical cell for voltammetry is the electrolytical cell with 1 polarizable electrode (mercury drop electrode or Pt-tip electrode)

The voltammetric cell incorporates the sample solution and a pair of electrodes of unequal size, as well as a reference electrode. The microelectrode is the working electrode, while the other current-conducting electrode is the auxiliary or counter-electrode. A *potentiostat* controls the voltage between the working electrode and the auxiliary electrode in order to maintain the potential difference between the working and the reference electrodes according to a preselected voltage–time program supplied by the function generator (or a microcomputer). The potential difference between the working and the reference electrodes is measured by a high-input feedback loop (without current flow). The potentiostat is considered as "active instrumention," which has an external control on the working electrode potential. Fig. 7.3-14 shows the schematic design of this type of controlled potential experiment. However, for voltammetric analysis in aqueous solutions, a simpler two-electrode instrumentation (working and reference electrode incorporated) can also be used.

The working electrode is an *ideally polarizable electrode*, i.e., the electrode shows a large change in potential when an infinitesimally small current is passing through. The polarizability of the electrode is characterized by the horizontal region of the *i–E* curve of the electrode defining a potential window for analytical purposes, within which the electrochemical oxidation/reduction measurements can be made. On the contrary, an *ideally nonpolarizable electrode* is an electrode of fixed potential, whose potential remains constant for the passage of a relatively small current, as is the case in microelectrolysis experiments. Nonpolarizable electrodes, such as electrodes of the second kind, or the large-surface mercury pool electrode, are used as reference electrodes in voltammetry. The auxiliary or counter-electrode is a current-conducting electrode (e.g., Pt wire).

The voltammetric experiments are usually performed in quiescent solution in a presence of a large excess of inert salt, called supporting electrolyte.

Working electrodes

In voltammetry, different *working electrodes* can be used; among the most commonly employed ones are the following.

The dropping mercury electrode (DME)

In classical polarography the dropping mercury microelectrode is used as a micro working electrode. The mercury drop is formed at the end of a glass capillary (length 10–20 cm, inside diameter 0.05 mm) as the capillary is connected by flexible tube to a reservoir about 50 cm in height. The mercury droplets are of highly reproducible diameter and of a life time ranging from 2 to 6 s. The drop-time is controlled by the height of the Hg reservoir, i.e., the hydrostatic pressure of Hg. The device is sometimes connected to a "knocker" to control the lifetime of drops. The advantages of the dropping mercury electrode are the following: 1) the constant renewal of the electrode surface eliminates the contamination of the electrode surface, which results in reproducible current–potential data; 2) the charge transfer overvoltage of hydrogen ions present in the aqueous solvent is high on Hg, thus its reduction does not disturb the study of the reduction processes of the electroactive species having more negative potential than the reversible potential of the proton discharge. In acidic solution, e.g., in 0.1 mol/L HCl solution at a potential more negative than $-1.2\,\text{V}$, hydrogen gas evaluation is observed; 3) Hg forms amalgams with many metals thus lowering their reduction potential.

The polarization range of the mercury electrode vs. SCE in aqueous solution in the absence of O_2 is between $+0.3$ and about $-2.7\,\text{V}$. The cathodic potential limit is determined by the reduction of the cations of the so-called supporting electrolyte (a supporting electrolyte is added in order to ensure diffusion-controlled mass transport conditions). The most negative polarization limit can be achieved in the presence of tetraalkylammonium cations, while the oxidation of Hg:

$$2\,\text{Hg} \rightleftharpoons \text{Hg}_2^{2+} + 2\,\text{e}^- \tag{7.3-46}$$

sets a limit of polarization in the positive potential range. From this it can be concluded that reducible electroactive substances can be primarily analyzed with the Hg drop microelectrode.

The static mercury drop electrode (SMDE)

This electrode construction features the same elements of the dropping mercury electrode, e.g., Hg reservoir and glass capillary, but the reproducibility of drop size and the drop time are controlled more precisely (Fig. 7.3-15). This is ensured by incorporating a solenoid-activated valve between the Hg reservoir and the capillary, and a mechanical knocker attached to the capillary. Thus, by appropriate setting of the opening and closing times of the valve, the time of dislodging of the drop can be set, and the device can be operated as a hanging mercury drop or in a dropping operation mode with controlled drop time.

The SMDE has all the characteristics of the dropping mercury electrode that make it specially advantageous for routine analytical work. An additional feature is due to the constant surface area of the drop, which will be discussed in connection with highly sensitive, current-sampled pulsed techniques (see section *Pulsed polarography*).

Solid electrodes

For the determination of oxidizable components, solid electrodes are generally used. Among the different electrodes (e.g., Pt, Au, C) used in electrolysis experiments, the different types of graphite electrodes have gained wide application in analytical voltammetry. This is attributed to features such as a large useful anodic potential range, a low electrical resistance, and an easily renewable electrode surface. Carbon paste [7.3-6] and glassy carbon electrodes [7.3-7] are used to monitor electrooxidation processes both in solution at rest, and in flowing electrolyte, e.g., in high-performance liquid chromatography. With the latter technique the analysis of electrochemically oxidizable substances, e.g., biogenic amines at the nanogram and picogram levels, can be accomplished in a low-volume voltammetric detector cell (Fig. 7.3-16).

The limit of useful anodic potential range is $+1.5\,\text{V}$ given by the evolution of O_2.

Linear sweep (dc) voltammetry (LSV)

A linear (dc)potential sweep (i.e., E vs. t is linear) is applied between the working and the auxiliary electrodes of the electrochemical cell. The working electrode is a polarizable microelectrode, e.g., the dropping mercury electrode, the static mercury electrode, or any of the solid electrodes, while the auxiliary and the reference electrodes are electrodes of large surface area and relatively nonpolarizable.

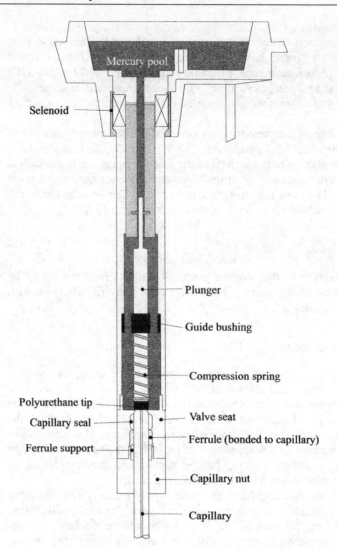

Fig. 7.3-15. The schematic representation of the commercial static mercury drop electrode (the mechanical knocker is not shown). (Courtesy of EG&G Princeton Applied Research, Princeton N.Y.)

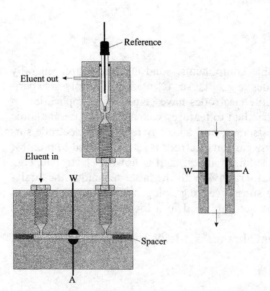

Fig. 7.3-16. Thin-layer cell incorporating a graphite (e.g., glassy carbon) working electrode where W is the working electrode; A is the auxilary electrode. (Courtesy of Bioanalytical Systems Inc. West Lafayette, IN)

Linear sweep voltammetry with DME (polarography)

In polarography, the dropping mercury electrode is used as a working electrode, while the auxiliary electrode is a mercury pool electrode. Most commonly, in the analysis of aqueous samples, the mercury pool electrode serves as a reference electrode, also. The applied potential is:

$$E_{appl} = E + iR = E_{eq} + \eta + iR \qquad (7.3\text{-}47)$$

where iR is the ohmic potential drop (E_{ohm}) required to force the current to flow through the cell. Its value is appreciable only at the μA level of current flow i if the value of the solution resistance R is high. Its value is lowered by the addition of an inert electrolyte, the supporting electrolyte; E_{eq} is the "open circuit" cell potential, i.e., equilibrium potential; and η is the overpotential, which acts to increase the applied potential required for electrolysis (charge transfer reaction).

The overpotential, i.e., the charge transfer overpotential, is a measure of the activation energy of a heterogeneous chemical reaction (charge transfer reaction), and it is needed to overcome the activation barrier of a reaction. The value of the overpotential of an electrode reaction is different for different electrode materials.

The voltage program ($E_t = E_i \pm vt$, where E_i is the starting potential and v is the rate of polarization) is shown in Fig. 7.3-17. The current will flow due to electrolysis. The resultant current vs. potential recordings (polarograms) are used for analysis and the current is measured continuously during the life of a growing Hg drop (Fig. 7.3-18 curve A).

The transport of solute to the electrode depends on three factors: diffusion; stirring (convection); and electrostatic attraction. In polarography it is important to minimize the latter two mechanisms and to ensure that analyte transport to the electrode surface occurs only by diffusion. The effect of electrostatic attraction (or repulsion) of analyte ions by the electrode is reduced to a negligible level by the presence of a high concentration of supporting electrolyte, e.g., 1 mol/L HCl. The concentration of the supporting electrolyte is 50–100 times higher than that of the analyte.

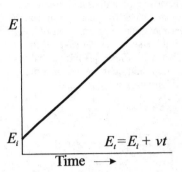

Fig. 7.3-17. Potential vs. time programs: Linear scan;

Polarograms (i vs. E relationships)

The current–potential recording under diffusion-controlled polarographic conditions is called a polarogram. In Fig. 7.3-18 typical polarograms recorded (a) in a quiescent solution containing 0.1 mol/L HCl and 5×10^{-3} mol/L Cd^{2+} (curve A) and (b) in the supporting electrolyte (0.1 mol/L HCl) alone (curve B) are exhibited.

The reduction of Cd^{2+} at the dropping Hg electrode:

The current-potential recording under diffusion controlled polarographic conditions is called *polarogram*.

$$Cd^{2+} + 2e^- \rightleftharpoons Cd(Hg) \qquad O + ne^- \underset{k_b}{\overset{k_f}{\rightleftharpoons}} R \qquad (7.3\text{-}48)$$

results in a step-shaped increase in the current.

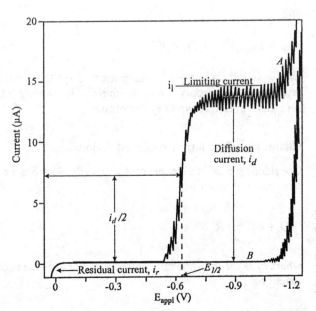

Fig. 7.3-18. Polarograms recorded (A) in a solution containing 5×10^{-4} mol/L Cd^{2+} and 1 mol/L HCl and (B) in a 1 mol/L HCl solution. (Adapted from Sawyer, D.T. and Roberts Jr., J.L. (1980), *Experimental Electrochemistry for Chemists*, New York: Wiley, with permission)

The classical polarogram has three current regions, a *residual current region* at small negative electrode potentials and a *limiting current region*, where the polarogram deviates from *Ohm's* law due to polarization of the cathode and a *charge transfer controlled region* in between, where qualitative (halve wave potential) and quantitative (diffusion current) information are contained

The polarogram has three current regions:

a) *residual current* region at small negative electrode potentials

b) *charge transfer controlled* region at electrode potential values corresponding to or higher than the reduction potential of Cd^{2+}. In this potential range the current, and the charge transfer rate increase exponentionally with the applied voltage (the rate of the reaction is kinetically controlled and the steady state current is diffusion supplied)

c) *limiting current* region the current levels off as the electrode reaction is diffusion controlled

The corresponding alteration of the Cd^{2+} concentration at the electrode surface ($c_{Cd(X=0)}$) as a function of applied potential is shown in Fig. 4.4-7a. The current increases at about $-1.2\,V$ due to the reduction of H^+. The current oscillation is ascribed to the fact that the polarographic analysis is carried out for drops of mercury of varying surface area. When a drop falls off the current approaches zero.

The residual current (i_r)

The residual current has two components, the faradaic current, and the charging (condensor) current. The faradaic component of the residual current is ascribed to the electrochemical reduction or oxidation of impurities of the supporting electrolyte. The supporting electrolyte concentration is relatively high compared with that of the analyte; the concentration and consequently the contribution of the impurities to the residual current can be significant.

The condenser current is the current required to charge or discharge the electrical double layer formed at the working electrode | electrolyte interface which acts as a capacitor. The variation of the charging current recorded in $0.1\,mol/L$ HCl as a function of potential is shown in Fig. 7.3-18 curve B. (The zero current value is observed at $-0.4\,V$, which coincides with the electrocapillary maximum at which the mercury drop has no charge.) Each drop needs to be charged or discharged; thus the value of the charging current does not die away in the course of the polarographic experiments. The charging current varies with potential as well as the electrode surface area:

$$i_c = kC_i \frac{dA}{dt} (E_{appl} - E_z) \tag{7.3-49}$$

$$A = 0.85m^{2/3}t^{2/3}$$

$$\frac{dA}{dt} = 0.85m^{2/3}t^{-1/3} \tag{7.3-50}$$

and

$$i_c = kC_i m^{2/3} t^{-1/3}(E - E_z) \tag{7.3-51}$$

where A is the electrode surface area, C_i is the integral capacity of the double layer, m is the mercury flow rate (mg/s), t is the time (s), and E_z is the potential of zero charge, electrocapillary maximum.

Charge transfer controlled current region

Considering a reduction process at which electrons are transferred between oxidant and reductant as:

$$O + ne^- \overset{k_f}{\underset{k_b}{\rightleftharpoons}} R \tag{7.3-52}$$

where k_f is the forward rate constant, while k_b is the rate constant of the backward reaction.

When $E_{appl} < E_{eq}$, which means that it is more negative than the "open circuit" cell potential, measured at $i = 0$, then a steady state current will flow across the cell, the rate of which is governed by processes such as an electron transfer at the electrode and mass transfer of the oxidant from the bulk solution to the electrode surface, and a new equilibrium will be established:

$$E_{appl} = E^0 - \frac{0.05916}{n} \log \frac{c_{R(x=0)}}{c_{O(x=0)}} \tag{7.3-53}$$

where $c_{R(x=0)}$ and $c_{O(x=0)}$ are the concentration of the oxidant and reductant at the electrode surface ($x = 0$).

The magnitude of the net current:

$$i = nFA(k_f c_O - k_b c_R) \tag{7.3-54}$$

where A is the electrode surface area.

The rate constant of the charge transfer reaction depends on E_{appl} as:

$$k_f = k_f^0 \exp\left[-\frac{\alpha n F E_{appl}}{RT}\right] \tag{7.3-55}$$

and

$$k_b = k_b^0 \exp\left[\frac{(1 - \alpha)n F E_{appl}}{RT}\right] \tag{7.3-56}$$

where k^0 is the rate constant measured at $E_{appl} = 0$, i.e., the rate constant of the relevant chemical reaction, α is a factor expressing the fraction of the electrode potential (interfacial potential) contributing to the rate constant of the relevant chemical reaction.

The above expressions indicate that the electrode potential affects the rate constants of the relevant chemical reactions via the alteration of activation energies. For example, when the potential of a cathode is made more negative, it facilitates the transfer of the reactant as it raises its potential energy by an amount $\alpha n F E$, which results in a decrease of the activation energy. On the contrary, a more negative electrode potential decreases the rate of the charge transfer of the anode reaction.

Accordingly:

$$i = nFA\{c_O k_f^0 \exp[-\alpha n F E_{appl}/RT] - c_R k_b^0 \exp[(1 - \alpha)n F E_{appl}/RT]\} \tag{7.3-57}$$

As predicted by Eq. 7.3-57 the current in this region varies exponentially with E_{appl}. This is due to the fact that the rate of the electrode reaction is slower than the rate of diffusion of the analyte.

Limiting current

The rate of the electrode reaction depends on E_{appl} as described by the Butler–Volmer equation (Eq. 7.3-57) discussed above. However, when the kinetics of the charge transfer and coupled chemical reactions are very rapid, the rate of the electrode reaction v is controlled by the mass transfer rate:

$$v = v_{mt} = \frac{i}{nFA} \tag{7.3-58}$$

In polarography, the polarograms are recorded in unstirred (quiescent) solutions, in the presence of a supporting electrolyte whose concentration is 50–100 times higher than the analyte; thus the concentration gradient created by the electrolysis is the driving force for the diffusion of analyte species:

$$\frac{i}{nFA} = m_O[c_O - c_{O(x=0)}] \tag{7.3-59}$$

where c_O is the concentration of the oxidant, O in the bulk solution, $c_{O(x=0)}$ is the concentration at the electrode surface, and m_O is the mass transfer coefficient.

The value of $c_{O(x=0)}$ is a function of the electrode potential. The largest rate of mass transfer of oxidant, O occurs when $c_{x=0} = 0$. The highest current value, recorded when the rate of diffusion of oxidant has become constant, is called the *limiting current* (Fig. 7.3-18, curve A).

The limiting current has been derived on the basis of diffusion kinetics by considering Fick's first and second laws. The faradaic current at time t is

$$i_t = nFAf_{(x=0,t)} \tag{7.3-60}$$

where $f_{(x=0,t)}$ is the material flux at $x = 0$ as a function of time.

Fick's first law describes the material flux and its concentration as a function of location and time. Thus, the flux of the substance at the electrode surface ($x = 0$) is:

$$f_{(x=0,t)} = \left(\frac{dN}{dt}\right)_{x=0,t} = D_O\left(\frac{\partial c_O}{\partial x}\right)_{x=0,t} \tag{7.3-61}$$

while Fick's second law expresses the concentration variation of oxidant c_O, as a function of time:

$$\left(\frac{\partial c_O}{\partial t}\right)_{x=0} = D_O\left(\frac{\partial c_O}{\partial x}\right)_t \tag{7.3-62}$$

To obtain the diffusion-limited current i_d and the concentration profile $c_O(x, t)$ the linear diffusion equations have been solved under the boundary conditions:

$$c_{O(x,0)} = c_O$$

$$\lim_{x \to 0} c_{O(x=0)} = c_O$$

$$c_{O(0,t)} = 0 \qquad (\text{for } t > 0)$$

by considering the diffusion profile affected by the electrode geometry.

By considering a semi-infinite diffusion to an expanding plane, Ilkovič derived the polarographic limiting current as:

$$i_d = knD^{1/2}c_O m^{2/3} t^{1/6} \tag{7.3-63}$$

where k is the Ilkovič constant ($k = 703$); D is the diffusion coefficient (cm^2/s), c_O is the concentration of the substance in the bulk of the solution (mol/L), m is the mercury flow rate (mg/s), and t is the drop time (s).

Among the factors determining the diffusion current, the most important is the analyte concentration; the diffusion current is proportional to bulk analyte concentration. This well-defined correlation between i_d and the concentration of the electroactive species in the solution bulk, as described by the Ilkovič equation, is the basis for the application of polarography in quantitative analytical chemistry.

With the determination of i_d, it must be kept in mind that: (a) $i_d = i_1 - i_r$, meaning that the value of the residual current i_r must always be subtracted from the limiting current value i_1; b) the method of evolution of i_d must be the same within one set of experiments. The correction for i_r can be done either by extrapolating its value to the potential at which i_d is determined graphically, or by determination in the supporting electrolyte alone.

The quantities m and t depend on the dimensions of the capillary of the dropping mercury electrode. The product $m^{2/3}t^{1/6}$ is called the *capillary constant*, which reflects the effect of the electrode characteristics on the diffusion current. The capillary constant and thus i_d varies with the square root of the height of the Hg column h. Consequently the measurement of i_d in function of h provides a way to proof that the current is, indeed diffusion controlled.

The diffusion current shows a temperature dependence of about 2% per °C as the diffusion coefficient D varies with temperature. Thus, the variation of temperature can affect the quantitative determination. By studying the effect of temperature variation on the diffusion current, it was concluded that if the determination of the polarographic current i_d is to be measured to an accuracy less than 1% then the electrochemical cell needs to be thermostated within $\pm 0.5\,\text{C}^{\text{O}}$.

Half-wave potential

For a *reversible redox electrode reaction*, in the potential region corresponding to $i > 0$ and $i < i_d$ the correlation between the electrode potential and the concentrations of the oxidant ($c_{O(x=0)}$) and reductant ($c_{R(x=0)}$), can be expressed by the Nernst equation:

$$E_{\text{appl}} = E_0 - \frac{RT}{nF} \ln \frac{c_{R(x=0)}}{c_{O(x=0)}} \tag{7.3-64}$$

The electrode surface concentrations caused by the faradaic current flow creates a concentration gradient and the concomitant diffusion provides the supply of the electroactive species.

Let us consider the redox process:

$$O + ne^- \rightleftharpoons R \tag{7.3-65}$$

where the R is produced by the electrode reaction alone, and remains soluble.

The reduction current in the rising part of the polarogram:

$$i = kD_O^{1/2}(c_O - c_{O(x=0)}) \tag{7.3-66}$$

and

$$i = k'D_R^{1/2} c_{R(x=0)} \tag{7.3-67}$$

If the reactions are reversible, the equation correlating the current and the potential in the polarogram is:

$$E = E_0 - \frac{0.05916}{n} \log \frac{i}{i_d - i} + \frac{0.05916}{n} \log \frac{D_R^{1/2}}{D_O^{1/2}} \tag{7.3-68}$$

which describes the sigmoid part of the current–voltage curve.

At $i = id/2$, if $k \approx k'$:

$$E = E_{1/2} \approx E_0 + \frac{0.05916}{n} \log \frac{D_R^{1/2}}{D_O^{1/2}} \tag{7.3-69}$$

and since $D_O = D_R$:

$$E_{1/2} = E_0 \tag{7.3-70}$$

The half-wave potential is independent of the concentration and is practically the same as the standard electrode potential for the relevant half-cell reaction. However, it must be pointed out that $E_{1/2} \neq E_0$ for electrode reactions, where the reaction products form an amalgam with Hg.

As the half-wave potential is independent of solution analyte concentration and dependant primarily on the nature of the analyte, it can be used for the identification of analyte species taking part in the electrode reaction. However, this can be done only with caution because several factors, e.g., the pH of the solution, the equilibrium constant of homogeneous cell reaction can affect the value of $E_{1/2}$. The effect of pH is pronounced at the electrochemical oxidation of organic compounds, which involves proton consumption or production of hydrogen ion:

$$O + ne^- + 2\,H^+ \rightleftharpoons RH_2 \tag{7.3-71}$$

then $E_{1/2} = E_0 - \text{pH}$. To avoid the effect of pH on $E_{1/2}$ good pH buffering of the solution is required.

Factors affecting the shape of the polarogram (the evaluation of the diffusion current)

The evaluation of the diffusion current i_d is influenced by:

- the residual current
- current maxima
- the presence of oxygen

The *current maxima* of different shapes can occur in the diffusion current region, at the beginning of the plateau section of the polarogram. They are common in polarography when reduction or oxidation of cationic, anionic species or neutral molecules, occurs.

The interpretation of current maxima, i.e., the reasons for current maxima have not yet been fully elaborated. In general, they are ascribed to the convection of the solution layer in the vicinity of the working electrode which is induced by the inhomogeneous charge distribution in the dropping Hg electrode. The current maxima can be suppressed by the addition of traces of polarographically inactive surfactants, such as gelatin, methyl red or other dyes, or Triton X-100 (a commercially available surfactant) to obtain the desired effect. Commonly, 0.1–0.2 ml 0.5% gelatin is added to 10 ml sample solution. Above a certain concentration level ($>10^{-4}\%$), the diffusion current will show variation with the supressor concentration, too. However, this effect is not specific for a particular type of supressor molecule, since it is due to the viscosity change in the test solution upon supressor addition. *It is worth noting that current suppressors are required only in dc polarography with the dropping mercury electrode.*

The oxygen dissolved in solutions in contact with air can be reduced at the dropping Hg electrode, resulting in two polarographic waves. The first wave is due to

$$O_2 + 2\,H^+ + 2\,e^- \rightarrow H_2O_2 \qquad E_{1/2} \sim 0.1\,V \text{ (vs. SCE)} \qquad (7.3\text{-}72)$$

while the second one

$$H_2O_2 + 2\,H^+ + 2\,e^- \rightarrow 2\,H_2O \qquad E_{1/2} \sim 0.9\,V \text{ (vs. SCE)} \qquad (7.3\text{-}73)$$

Both waves are of equal height. Since the oxygen reduction waves appear in the 0–1 V potential range, the presence of O_2 interferes with many other species, which can be reduced in this potential range. This interference, however, can be eliminated by the removal of O_2. Most commonly the O_2 is removed from solutions of all pH by nitrogen or argon purging for about 10 min, and by maintaining a nitrogen blanket over the solution or in alkaline media by Na_2SO_3 solution (e.g., 0.2 ml saturated Na_2SO_3 is added to 10 ml solution).

Polarography as an analysis technique

Mixture analysis is possible (see fig. 7.3-19)

Many inorganic and organic compounds exhibit polarographic activity, i.e., undergo electrochemical reactions in the "useful" potential range of the dropping mercury electrode. The positive potential limit is +0.3 V (vs. SCE), at which the electrode material (Hg) is oxidized, while the negative potential limit at which the solvent or the supporting electrolyte is reduced lies between −1.8 V and −2.2 V (vs. SCE). Typical polarograms for different inorganic species are shown in Fig. 7.3-19. The types of organic bonds that can be reduced at the dropping mercury electrode are summarized in Table 7.3-2. One or more reducible groups being present in an organic molecule provide its polarographic activity.

The general principles of preparing solutions for polarographic measurements are:

- the solution should contain polarographically inactive supporting electrolytes, such as salts, acids or buffers
- a certain amount of current maxima supressor should be present in the solution
- oxygen should be removed prior to the polarographic measurement

For the analysis of organic molecules, nonaqueous solvents are often used. As alcohols, a mixture of alcohol and water, acetonitrile, dimethyl formamide, and as

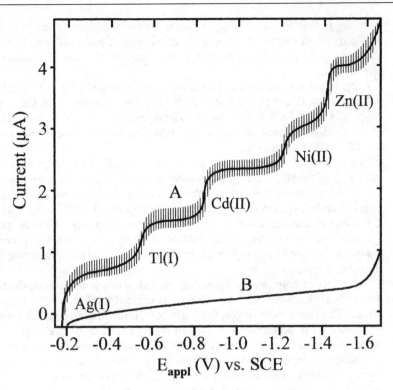

Fig. 7.3-19. Polarograms of (A) approximately 0.1 mmol/L each of silver(I), thallium(I), cadmium(II), nickel(II), and zinc(II), listed in the order in which their waves appear, in 1 mol/L ammonia/1 mol/L ammonium chloride containing 0.002% Triton X-100 and (B) the supporting electrolyte alone. (From Meites, L. (1967), *Polarographic Techniques*, 2nd ed., New York: Wiley, p. 164, with permission)

Table 7.3-2. Some polarographically active functional groups

$\phi\ \overset{\displaystyle \vert}{C}{=}O$	$\!\!>\!CHO$	CX_n	$-NO_2$
$\phi\ \overset{\displaystyle \vert}{C}{=}C\!\!<$	$\!\!>\!C{=}N{-}$		$-NO$
$C{\equiv}C{-}$	$-C{\equiv}N$	$\overset{\displaystyle \vert}{\underset{\displaystyle \vert}{C}}X$	$-NHOH$
$\!\!>\!C{=}\overset{\displaystyle \vert}{C}{-}\overset{\displaystyle \vert}{C}{=}C\!\!<$	$-N{=}N{-}$		$-ONO$
$\!\!>\!C{=}\overset{\displaystyle \vert}{C}{-}\overset{\displaystyle \vert}{C}{=}O$	$-O{-}O{-}$	$O{=}\overset{\displaystyle \vert}{C}{-}\overset{\displaystyle \vert}{C}X$	$-ONO_2$
$O{=}\overset{\displaystyle \vert}{C}{-}\overset{\displaystyle \vert}{C}{=}O$	$-S{-}S{-}$		$-NO{=}N{-}$

X HALOGEN ATOM

supporting electrolytes tetraalkylammonium perchlorate, and lithium perchlorate are utilized most frequently. A three-electrode potentiostat is needed to overcome problems associated with iR drop.

The polarograms have two distinct characteristics, namely $E_{1/2}$ and i_d, which serve as a basis of using polarography for analytical purposes. The value of $E_{1/2}$ is characteristic of the electroactive species and the medium in which it is determined. Thus the experimentally determined $E_{1/2}$ can be used (with caution) on the basis of tabulated $E_{1/2}$ values for the qualitative identification of an unknown substance.

Polarography is used mainly for quantitative purposes. The linear relationship between i_d and analyte bulk concentration gives a basis for this. For the concentration evaluation, either the calibration graph or the standard addition method can be used.

If the diffusion-controlled conditions are maintained then the so-called classical polarography is suitable for the determination of analytes in the concentration range of $5 \times 10^{-5} - 5 \times 10^{-3}$ mol/L. The detection limit is determined by the value of the residual current, which sets the detection limit in the concentration range of $1 \times 10^{-5} - 5 \times 10^{-5}$ mol/L.

The resolution of the method depends on the difference of the half-wave potentials, $E_{1/2}$ of the electroactive substances A and B, and on their concentration ratio. Two polarographic waves are separated when $\Delta E_{1/2}$ of the relevant analytes is larger than 150–200 mV, and their concentrations are of the same magnitude. But if the concentrations differ from each other appreciably, then the possibility of the selective determination depends on whether the half-wave potential of the analyte being present in higher concentration is less or more negative than that of the other substance.

The effect of complexing agents on the half-wave potentials of the reduction of metal ions is known. This phenomenon is utilized in the resolution of overlapping waves. The classical example, for the separation of polarographic waves by complexation is the determination of lead in the presence of thallium at a dropping mercury electrode. Both metals reduce practically at the same potential in acidic and neutral solution; the two polarographic waves of the two ions overlap. However, in alkaline solution the lead forms a hydroxo complex while the thallium remains as a free ion. Thus a difference of 300 mV exists between the half-wave potentials which is large enough to allow the quantitative evaluation of the two ions in about the same concentration.

Evaluation of diffusion currents (i_d)

When diffusion currents are determined it should be kept in mind that:

1) The residual current i_r exists in the limiting current region; therefore its value should be subtracted from the plateau current i_l, i.e., $i_d = i_l - i_r$.
2) The current evaluation method should be the same within one series of analyses.

Different graphical methods have been suggested for the determination of i_d. All are based on the assumption that the residual current increases linearly with potential, thus its value at any potential can be determined by extrapolation.

Accuracy of polarographic determination

The reproducibility of i_d and the error of the graphical method used to evaluate the diffusion current determine primarily the accuracy of the polarographic concentration determination. The former is good, owing to the renewable surface of the electrode. Under controlled experimental conditions, these account for an error in the determined concentration of 1–5%.

Linear sweep voltammetry (LSV) with stationary electrodes

A great variety of solid electrodes have been employed in different voltammetric techniques over the years. Their increasing popularity can be attributed to the fact that the oxidation of many organic molecules cannot be studied by the mercury electrode because of its limited anodic potential range. As solid electrodes noble metals, especially platinum and gold, and carbon (graphite) electrodes have gained the widest acceptance. Platinum and gold are used basically in nonaqueous media, while the carbon electrodes are employed in aqueous solutions.

The potential of the working electrode (i.e., an electrode of constant surface area) is swept linearly. The sweep voltage can be expressed as:

$$E_t = E_i \pm vt \tag{7.3-74}$$

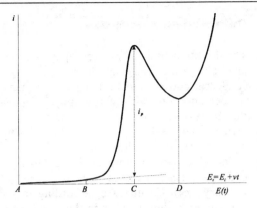

Fig. 7.3-20. Typical current-potential curve (voltammogram) for linear sweep voltammetry with electrode of constant surface area; E_i is the initial potential, v is the rate of polarisation

where E_i is the starting potential, v is the rate of polarization; typically 0.001–0.1 V/s, $\pm$ indicates the direction of the potential change, and t is the time interval of electrolysis.

Let us consider the reduction of substance O in a solution containing an excess of background electrolyte. The solution is unstirred, and linear diffusion is maintained. E_i is set at a value where no reduction takes place.

Typical current–voltage (LSV) response curve is shown in Fig. 7.3-20, which is interpreted on the basis of diffusion mass transfer kinetics.

The response curve has three distict regions. In the potential range which is more positive than E^O for the reduction of O, only residual current flows $(A - B)$. When the electrode potential approaches E^O the reduction begins and faradaic current starts to flow $(B - C)$. In this region, the charge transfer reaction is fast, and the current increases exponentially with the electrode potential. Owing to the current flow, the surface concentration of O decreases and the concentration gradient, i.e., the flux to the surface will increase. As the E exeeds E^O the surface concentration of O will approach zero, and the mass transfer rate to the electrode surface reaches a maximum value. With the further increase of potential it decreases, owing to the increases of the deplation layer thickness $(C - D)$.

The peak current i_p measured under diffusion-controlled conditions is of analytical significance. For the peak current, Randles and Ševčik derived the following equation:

$$i_p = kD^{1/2}Av^{1/2}c \tag{7.3-75}$$

where k is a constant, called the Randles–Ševčik constant (its value is 2.7×10^{-2}), D is the diffusion coefficient, n is the number of electrons taking part in the electrode reaction, v is the rate of potential change (rate of polarization, scan rate), and c is the concentration of the electroactive component.

The equation shows that i_p shows a linear dependence on concentration if A, D and v are constant in the LSV experiment. ($i_p = i - i_r$ at E_p, where E_p is the peak potential.)

Application of electrodes of constant surface area involves the advantage that the value of the residual current is one order of magnitude lower than that at a dropping mercury electrode. This is an important feature of electrodes of constant surface area as far as the analytical application is concerned, since it results in a detection limit which is one order of magnitude smaller. The disadvantage of the use of an electrode of constant surface area is that, in certain cases, its surface can be contaminated as the product of the electrode reaction used for analytical purpose remains at the surface of the electrode, forming an insoluble layer. The renewal of the electrode surface can be carried out by polishing the electrode; however there may be some differences in the renewed and original surface area (about 5–10%). This requires recalibration! The contaminated surface effect, the so-called "filming effect", can be decreased with the decrease of the concentration of the analyte.

Linear sweep voltammetry with a stationary electrode, where the excitation signal is a linear potential scan with an isoscale triangular waveform is called *cyclic voltammetry*. This technique allows one to study electrode reaction mechanisms.

Potential step methods

To lower the detection limit, the residual current must be suppressed by special techniques

The detection limit of polarography is determined by the residual current, more precisely by the charging current. Utilizing the different time dependence of the condenser current and Faraday current, new techniques were developed to lower the detection limit of polarography [7.3-8, 7.3-9]. These techniques are based on the measurement of current as a function of time (chronoamperometry), after applying a potential pulse and the current is not measured continuously, but rather within a certain time interval.

So-called current-sampled d.c. polarography (its early name was Tast polarography) takes advantage of the fact that the diffusion current, i_d, at the dropping mercury electrode increases according to $t^{1/6}$ while the condenser current is decreasing according to $t^{-1/3}$ in the course of the life time of the mercury drop. By optimizing the current sampling time, i.e., when it is measured in a narrow interval of the last period of the life time of the mercury drop, the detection limit of the method can be decreased by one order of magnitude owing to the improved i_d/i_r ratio.

The waveform applied in a basic potential step experiment is shown in Fig. 7.3-21. Let us consider its effect on the interface between an electrode of constant surface area and an unstirred solution containing a supporting electrolyte and an electroactive species (diffusion-controlled conditions!). E_1 is selected in the potential region where no faradaic process occurs, and E_2 is the mass-transport-limited region of, e.g., the reduction of O. The variation of the diffusion-limited current i_d as a function of time at an electrode of constant surface area had been derived by Cottrell on the basis of linear diffusion equation:

$$i_d(t) = \frac{\eta FAD_O^{1/2} C_O}{\pi t^{1/2}} \tag{7.3-76}$$

a)
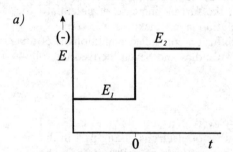

b)
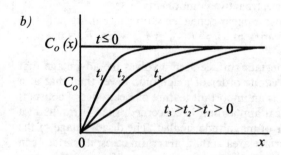

c)
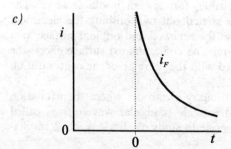

Fig. 7.3-21. Basic potential step experiment. (a) Wave form applied. (b) Concentration profiles for an electroactive species at E_2. (c) i-t profile following the step change in potential

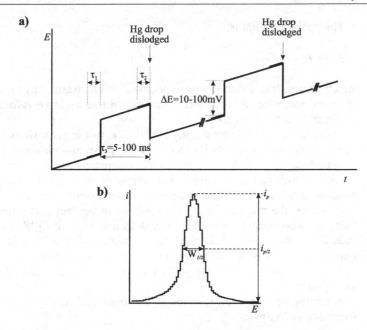

Fig. 7.3-22. (a) Potential program applied in DPP experiment where τ_1 and τ_2 are the current sampling time. (b) Current-voltage curve recorded in differential pulsed mode (DPP curve)

As the Cottrell equation shows, the faradaic current is decreasing according to $t^{-1/2}$. Thus, by the proper selection of the sampling time in pulsed techniques, the ratio of the faradaic and the charging current can be favorably affected.

Differential pulse polarography (DPP)

The potential–time perturbing signal applied to the electrochemical cell in differential pulse polarography is shown in Fig. 7.3-22. The potential pulses of fixed, but small amplitudes (10–100 mV) ΔE, are superimposed periodically on a dc potential that increases linearly with time. The rate of potential increase, i.e., the rate of polarization is in the range 0.1–0.2 V/min as in classical dc polarography. The duration of pulses is about 5–100 ms.

The working electrode is most commonly a dropping mercury electrode with controlled drop time (static mercury drop electrode, SMDE) and one pulse is applied at each drop. This requires the synchronization of the drop's life time with the pulses.

The current intensities are measured twice during each drop's life time, namely just before the application of the pulse and in the last period of the pulse duration. The instrument records the current difference, Δi at each pulse as a function of the linearly increasing dc voltage. As a consequence the current–voltage curve, i.e., the DPP curve is a peak-shaped derivative voltammogram (Fig. 7.3-22). The steps on the voltammogram correspond to the life time of the drops.

The current intensity at the peak height (at the current maximum) can be described as:

$$i_P = \frac{(\pi F)^2}{4RT} A_{\tau 3} c_O \Delta E \sqrt{\frac{D_O}{\pi t}} \tag{7.3-77}$$

where ΔE is the amplitude of the pulses, τ_3 is the duration of the pulse, A is the average surface area during the duration of the pulse, and c_O is the analyte concentration, while the other symbols have their usual meaning.

Equation 7.3-77 demonstrates that the peak height is directly proportional to the analyte bulk concentration.

The half-peak width is:

$$w_{1/2} = 3.52 \frac{RT}{nF} \tag{7.3-78}$$

Thus, by substituting the values of the constant terms, one obtains a value of $w_{1/2}$ equal to 90.4 mV for components taking part in the electrode reaction with $n = 1$, while it is 45.2 for components with $n = 2$.

The peak potential is:

$$E_p = E_{1/2} - \frac{\Delta E}{2} \qquad (7.3\text{-}79)$$

which demonstrates that the peak potential of an electroactive species taking part in an electrochemical reduction shifts toward less negative values by the increase of the pulse amplitude ΔE.

Differential pulse polarography (DPP) due to the peak-shaped current-voltage signal has a higher selectivity than normal pulse and classical dc polarography. Components having $\Delta E_{1/2} \sim 90\,\text{mV}$ show up separately in the DPP recordings compared with dc polarography. The resolution of the peaks is independent of the concentration ratios of the corresponding electroactive species.

However, the main advantage of DPP lies in the fact that it increases the sensitivity of polarographic methods. The detection limit of DPP is orders of magnitude lower than that of classical dc polarography. This can be ascribed to two reasons, the first is an enhancement of the faradaic current due to the potential pulse (Eq. 7.3-77), while the second is an appropriate selection of the current sampling time.

Summing up, the main features of the analytical capability of the pulsed technique are as follows:

- improved selectivity, resolution of the DPP peaks
- low detection limit, e.g., $10^{-8}\,\text{mol/L}$
- the normal working range of the technique allows the use of supporting electrolyte with concentrations as low as $10^{-3}\,\text{mol/L}$ reducing the contribution of the contaminants in the supporting electrolyte to the residual current

Stripping analysis methods

Electrochemical stripping analysis methods incorporate a variety of electrochemical procedures which are based on

Electrolytic enrichment is used for lowering the detection limit

1) The electrochemical deposition or accumulation of the analyte at a voltammetric electrode of constant surface area in controlled potential electrolysis.
2) The stripping or dissolution of the analyte from the electrode by a voltammetric technique or chemical reaction.

As a common feature all stripping methods include a preconcentration step (deposition), but they may differ in the techniques used for the analysis step. When a voltammetric technique is used for the dissolution step, the technique is called voltammetric stripping analysis (VSA) (Fig. 7.3-23), while in potentiometric stripping analysis (PSA) a chemical reaction is the basis of the stripping step.

As stripping analysis is based on the eletrochemical preconcentration of the analyte, it permits the analysis in the concentration range of 10^{-6}–$10^{-8}\,\text{mol/L}$. Thus it is a useful technique in trace analysis of metals.

Anodic stripping voltammetry is used primarily for the determination of amalgam-forming metals, while *cathodic stripping voltammetry* is employed for determining species that form insoluble salts with mercury on the electrode surface. According to the latter technique in the anodic preconcentration step, Hg(I) ions are generated, and analytes forming insoluble salts with Hg(I) ions precipitate on the surface of the electrode. The cathodic scan of the potential results in the reduction of the insoluble salt to Hg(I) ion and the relevant anions give rise to a cathodic peak. Similar to Hg, Ag electrode can also be used in cathodic stripping voltammetry. With cathodic stripping voltammetry, especially halide ions, sulfate, oxalate, and mercaptans can be analyzed.

7.3.3 Amperometry

Amperometry is a special case of dc voltammetry, where the potential is kept constant at a certain fixed value within the limiting current region

Amperometry is a voltammetric method, which is based on the measurement of current at a fixed operating potential in stirred (or flowing) solutions or at a rotating working electrode. The current is the result of electrochemical oxidation

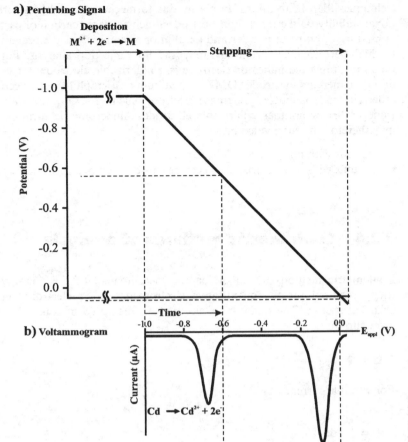

a) Perturbing Signal

Fig. 7.3-23. Voltammetric stripping analysis. (a) Perturbing signal for stripping determination of metal ions. (b) Linear sweep voltammogram

or reduction of the electroactive compound after applying the potential pulse across the working and the auxiliary electrodes. Hydrodynamic voltammetry, i.e., linear sweep dc voltammetry performed in flowing (or stirred) solutions is used for selecting the operating potential, which corresponds to the limiting current plateau of the current–voltage curve obtained for an electrode of constant surface area in a solution containing the electroactive compound and the supporting electrolyte.

The limiting current measured at the electrochemical reduction of analyte, O, under hydrodynamically controlled conditions, i.e., under convective diffusion conditions can be expressed as:

$$i_l = nFAm_O c_O \tag{7.3-80}$$

where m_O is the mass transport coefficient, which is flow rate dependent, and c_O is the analyte concentration.

If the mass transport coefficient is constant, then the concentration variation of the analyte can be monitored by the current measurement. The constancy of m_O can be maintained by ensuring a constant diffusion layer thickness, by stirring or flowing the solution, or rotating the working electrode by a constant rate, as:

$$m_O = D_O/\delta_O \tag{7.3-81}$$

where D_O is the diffusion coefficient, and δ_O is the diffusion layer thickness.

Within the stagnant solution layer (Prandtl layer) linear diffusion conditions are assumed. In amperometry, as the solution is moving, the diffusion layer thickness is constant, thus a steady-state current is monitored, which is independent of time.

Amperometry as a detection mode is widely used in a number of flow analytical techniques, including flow injection technique as well as chromatographic separation methods, and also in conventional volumetric and coulometric titrations. In the latter the current measurement is used to locate the titration end point. Amperometry has achieved an important application as a transduction mode in biosensors.

The working electrode material is more critical in continuous flow analytical

Amperometry can be used for the end point detection in titrations or in a direct way (e.g. amperometric sensor for O_2-determination)

techniques than in titrations, primarily due to mechanical and long-term operational stability. Different carbon electrodes, such as glassy carbon or carbon paste (a mixture of graphite powder and paraffin or silicone oil), are especially popular for high-performance chromatography and flow injection analysis. For conventional titrations the indicator electrode is a polarizable electrode, for example, a dropping mercury electrode (DME), a platinum or a graphite microelectrode. The other electrode is either a reference electrode, such as a saturated calomel electrode (SCE) or another polarizable electrode. Amperometric titration methods may therefore be subdivided into:

- titrations with one polarizable electrode
- titrations with two polarizable electrodes (also called biamperometric titrations)

In coulometry, the reagent used in titrations is produced *in situ* by electrochemical reactions under 100% stoichiometry and current efficiency

7.3.4   Coulometric methods of analysis

Coulometric methods are based on the measurement of the quantity of electricity required for the quantitative electrooxidation or electroreduction of a chemical species. The quantity of electricity is measured in coulombs. For constant current:

$$Q = it \tag{7.3-82}$$

For variable current:

$$Q = \int_0^t i \, dt \tag{7.3-83}$$

The quantity of electricity can be expressed in Faraday, one Faraday, i.e., $96\,485\,C$ produces an equivalent weight of element. On this basis, one can correlate the mass of a chemical species that is formed at an electrode to the charge passed through the electrolysis cell:

$$m_k = Q_k K = Q_k \frac{M_k}{nF} = \frac{M_k}{nF} \int_0^t i_k \, dt \tag{7.3-84}$$

where m_k is the the amount of analyte k formed in the electrolysis, M is the formula mass of analyte k, n is the number of electrons taking part in the electrode reaction, F is the Faraday constant, and K is a constant, characteristic of the quality of the analyte taking part in the electrode process.

In contrast to electrogravimetry, it is essential that the current passing through the electrolysis cell is consumed only by the electrochemical reduction or oxidation of chemical species, since the amount of analyte is calculated from the quantity of charge passed at 100% current efficiency. In general, one can state that 100% current efficiency conditions are established if there are no side reactions. Side reactions may be caused by the electrochemical reactions of the solvent, the electrode material, the components of the supporting electrolyte, or of the products of the electrolysis undergoing secondary electrode reaction.

For investigating the role of the solvent, one must consider that in electroanalysis, solvents with autodissociation are used; thus the solvent at a given electrode potential will also undergo electrochemical oxidation/reduction. Accordingly, there exists a potential window for every solvent within which electrochemical oxidation/reduction can be used for analytical purposes. In aqueous solvents, this range is given by the reversible potentials of the H_2 and O_2 electrodes, which at pH 0 are in the range of 0 and 1.23 V, respectively. In practice however, this potential window is always larger, owing to kinetic limitations, and one can measure potentials at values more negative than 1.23 V and more positive than 0, depending on the activation energies of the different electrode materials.

Coulometric methods can be divided into two groups:

1) *Direct coulometric methods* which are based on the measurement of the quantity of electricity required for the quantitative electrooxidation or electroreduction of the analyte.
2) *Indirect or reagent-generating coulometry* which is based on the electrolytic generation of a reagent, for titrimetric analytical purposes. For this reason, the latter method is also called coulometric titration.

The coulometric measurements are performed in order to achieve 100% current efficiency, either at controlled electrode potential (controlled potential coulometry, i.e., *potentiostatic coulometry*) or at constant current, i.e., *constant current coulometry*.

Potentiostatic coulometry, as in controlled potential electrolysis, is performed at a preselected, constant working electrode potential. At the potential selected, only a single reaction is responsible for the current flow in the electrolysis cell. The current signal is recorded during the experiments as a function of time, and electronic integration of the current–time curve, provides the analytical signal, i.e., the quantity of electricity. Another possibility is to measure the quantity of electricity used up in the electrolysis of the analyte with chemical coulometers, connected in series to the analytical cell.

In controlled potential coulometry the analyte concentration is determined from the total number of coulombs used in the electrolysis. This can only be done if:

1) The stoichiometry of the electrode reaction is known
2) Only a single electrode reaction is involved, i.e., no side reaction of different stoichiometry takes place
3) The electrode reaction proceeds close to 100% current efficiency.

These requirements are rather similar to those set for the titration reactions in normal titrimetric methods.

The current in a potentiostatic coulometric experiment is monitored during the electrolysis, usually with a strip-chart recorder, so that the background current can be determined and the completion of electrolysis is observed.

The current value at any time for solution bulk electrolysis is:

$$i_t = nFAD\frac{c_O}{\delta}\exp\left(-\frac{DA}{V\delta}\,t\right) \tag{7.3-85}$$

where n is the number of electrons taking part in the electrode reaction, c_O is the starting concentration of species O, V is the solution volume, and δ is the diffusion layer thickness.

According to Eq. (7.3-85) the current decreases exponentially as a function of time; and asymptotically approaches the background level. Thus, an electrolysis takes about 30–100 min. The shape of a characteristic i–t curve is shown in Fig. 7.3-24.

In solid-phase electrolysis, e.g., at the anodic dissolution of metals, the current at any time is:

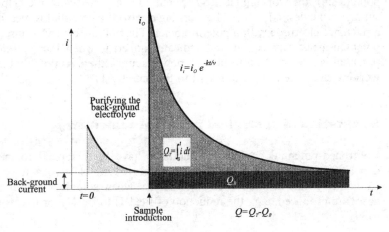

Fig. 7.3-24. Current-time behavior in controlled potential coulometry; where Q_T is the total number of coulombs, Q_B is the charge used in the electrolysis of impurities of the background electrolyte, while Q_S is the charge consumed in the electrolysis of the sample solution

$$i_t = nFAk_a \exp\left[-\frac{AD}{V\delta\left(\dfrac{D}{k_c f_M \delta} + 1\right)} t \right]$$ (7.3-86)

where n, F, A, D, δ and t have the usual meaning, k_c and k_a are the heterogeneous rate constants for the cathodic and anodic reactions, respectively, and f_M is the activity of metal ions generated.

As the current varies in the course of the electrolysis, i.e., as a function of time, the quantity of charge is determined with the help of a current integrator. According to MacNevin and Baker [7.3-10] the time of coulometric determination carried out at controlled potential can be decreased appreciably if the i–t curve is used for the determination of Q. Thus:

$$Q = \int_0^t i_t \, dt$$ (7.3-87)

From Eq. (7.3-85), if $t = 0$,

$$i_o = nFAD \frac{c_o}{\delta}$$

and introducing β as $= DA/V\delta$, one obtains:

$$Q = \int_0^t i_o \exp(-\beta t) = i_o \beta$$ (7.3-88)

The slope $(-\beta/2303)$ of the $\log i$–t function provides the value of β, while the intercept provides i_o. For the determination of the Q value it is sufficient to measure 3–4 points, on the basis of which one can construct the relevant $\log i$–t function and the values of i_o and β. This approach cannot be used for the determination of Q for electrolysis with Hg electrodes, since the Hg surface is not constant.

The determination of the end point of the measurement does not require a separate indicator system; the method is accurate, the selectivity is excellent.

The quantity of electricity can be determined by chemical coulometers, e.g., with a H_2–N_2 coulometer. In this case a hydrazine-sulfate-containing solution is electrolyzed:

$$N_2H_5{}^+ \rightarrow N_2 + 2\,H_2 + H^+$$ (7.3-89)

The process can be directly related to a chemical standard (e.g., silver coulometer) which is characterized by high accuracy and precision. However, it is inconvenient and time consuming to use.

Electrochemical, and more recently, operational amplifier integrators give direct read-out in coulombs and can be employed to record Q–t curves during electroanalysis.

For controlled potential coulometry, an automatic potentiostat is required. The potentiostat is generally equiped with a current recorder, which provides the current–time plot subsequently evaluated with an electronic integrator. In a potentiostat, the working electrode potential is adjusted with the help of a power supply to the desired value. The working electrode's potential is measured against a reference electrode with a potentiometer. The null instrument is inserted into the potential-measuring circuit and indicates an error when the working electrode potential is altered. As a result, the power supply alters its potential and thus the working electrode potential is kept to the preset value.

Application of controlled-potential coulometry

Coulometry allows the use of chemically unstable reagents, such as Br_2 in titrations with high accuracy.

Controlled-potential coulometric methods have been applied to many electrodeposition reactions which are also employed in electrogravimetric analysis. Coulometry can also be used for analysis when soluble reaction products or gases are produced, e.g., the reduction of $Fe(III)$ to $Fe(II)$, or oxidation of N_2H_4 to N_2.

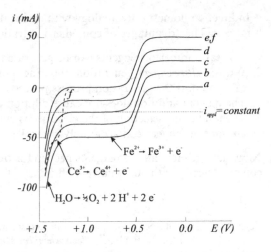

Fig. 7.3-25. Schematic current-potential curves for the oxidation of Fe^{2+} in 1 mol/L H_2SO_4 at different degrees of the electrolysis process; curve a, 0; curve b, 0.25; curve c, 0.5; curve d, 0.75; curve e, l; curve f, of excess Ce^{3+}

Controlled current methods

The quantity of electricity can be determined without special instruments if the electrolysis is performed at constant current. Since the current intensity is known, only the time of electrolysis needs to be measured, for the determination of the quantity of electricity:

$$Q = it \tag{7.3-90}$$

Bulk electrolysis processes under controlled current conditions can be studied on the basis of current–voltage curves as in Fig. 7.3-25.

Figure 7.3-25 shows the anodic oxidation of Fe(II) ions in sulfuric acid.

If the value of constant current is 25 mA less than the limiting current, the oxidation of Fe^{2+} at the anode proceeds. As the time of electrolysis increases the Fe(II) concentrations decrease in the solution. Curves b, c, d correspond to 25, 50, 75, and 100% completion of electroanalysis, respectively. When the Fe(II) concentration is less than that required for maintaining the constant current, the current can only remain at the value applied if the anode potential shifts to more positive values, at which O_2 is produced:

$$2\,H_2O \rightarrow 4\,H^+ + O_2 + 4\,e^- \tag{7.3-91}$$

(curves c, d, e)
With the decrease of the Fe(II) concentration, more and more charge is consumed for the oxidation of water, which means that the 100% current efficiency condition for Fe(II) is not fulfilled.

Depending on the value of the current applied, O_2 formation starts earlier or later in the course of electrolysis. Therefore, if the analyte is oxidized or reduced directly in the electrolysis, in the majority of cases 100% current efficiency cannot be obtained.

This difficulty can be overcome if a so-called depolarizer is added to the water. The depolarizer can be oxidized at a less positive potential than the analyte. If Ce(III) ions are added to the solution Ce(III) ions are oxidized to Ce(IV) at 1.15 V, which oxidizes Fe(II)→Fe(III). This way the oxidation of water is eliminated. The final result is the same, as Fe(II) would be oxidized directly with 100% current efficiency.

For attaining 100% current efficiency, the depolarization is used in 50–10 000-fold excess.

In indirect coulometry, the analyte does not take part in the electrochemical reaction, but rather constant current coulometry is used for the electrochemical generation of a titration reagent, either from an appropriately selected reagent or the working electrode material itself. The reagent undergoes a fast and quantitative chemical reaction with the analyte. The completeness of the chemical reaction needs to be indicated. For the end point detection of coulometric titration potentiometry and amperometry are the most widely used sensitive methods. The electric signal of the indicator electrode can also be used for automation of coulometric titrations.

Indirect coulometry, by analogy with classical titrations, are called coulometric titrations. The advantages of coulometric titrations are:

1. A great number of reagents can be generated coulometrically.
2. With *in situ reagent* generation, the exact concentration of the reagent can be calculated, if the generation process proceeds with 100% current efficiency.
3. The reagent addition does not cause dilution of the analyte containing solution.
4. With judicious selection of the electrolysis current and the electrolysis time, the amount of reagent can accurately controlled.

Nowadays, almost all the reagents needed for titration processes can be generated coulometrically with 100% current efficiency (Table 7.3-3).

Table 7.3-3. Typical electrogenerated titrants and substances determined by coulometric titration [A.J. Bard, L.R. Faulkner, *Electrochemical Methods*, p. 384. New York: Wiley, 1980.]

Electrogenerated titrant	Generating electrode and solution	Typical substances determined
Oxidants		
Bromine	$Pt/NaBr$	$As(III)$, $U(IV)$, NH_3, olefins, phenols, SO_2, H_2S, $Fe(II)$
Iodine	Pt/KI	H_2S, SO_2, $As(III)$, water (Karl Fischer), $Sb(III)$
Chlorine	$Pt/NaCl$	$As(III)$, $Fe(II)$, various organics
Cerium(IV)	$Pt/Ce_2(SO_4)_3$	$U(IV)$, $Fe(II)$, $Ti(III)$, I^-
Manganese(III)	$Pt/MnSO_4$	$Fe(II)$, H_2O_2, $Sb(III)$
Silver(II)	$Pt/AgNO_3$	$Ce(III)$, $V(IV)$, $H_2C_2O_4$
Reductants		
Iron(II)	$Pt/Fe_2(SO_4)_3$	$Mn(III)$, $Cr(VI)$, $V(V)$, $Ce(IV)$, $U(VI)$, $Mo(VI)$
Titanium(III)	$Pt/TiCl_4$	$Fe(III)$, $V(V,VI)$, $U(VI)$, $Re(VIII)$, $Ru(IV)$, $Mo(VI)$
Tin(II)	$Au/SnBr_4(NaBr)$	I_2, Br_2, $Pt(IV)$, $Se(IV)$
Copper(I)	$Pt/Cu(II)(HCl)$	$Fe(III)$, $Ir(IV)$, $Au(III)$, $Cr(VI)$, IO_3^-
Uranium(V), (IV)	Pt/UO_2SO_4	$Cr(VI)$, $Fe(III)$
Chromium(II)	$Hg/CrCl_3(CaCl_2)$	$O_2Cu(II)$
Precipitation and complexation agents		
Silver(I)	$Ag/HClO_4$	Halide ions, S^{2-} mercaptans
Mercury(I)	$Hg/NaClO_4$	Halide ions, xanthate
EDTA	$Hg/HgNH_3Y^{2-}$*	Metal ions
Cyanide	$Pt/Ag(CN)_2^-$	$Ni(II)$, $Au(III,I)$, $Ag(I)$
Acids and bases		
Hydroxide ion	$Pt(-)/Na_2SO_4$	Acids, CO_2
Hydrogen ion	$Pt(+)/Na_2SO_4$	Bases, CO_3^{2-}, NH_3

*Y^{2-} is ethylenediamine tetra acetate anion

References

[7.3-1] Henderson, P., *Z. Phys. Chem.*, 59, 118 (1908); 63, 325 (1907).

[7.3-2] Horovitz, K., *Z. Phys. Chem.*, 15, 369 (1923).

[7.3-3] Lengyel, B., Blum, E., *Trans. Faraday Soc.*, 30, 461 (1934).

[7.3-4] Perley, G.A., *Anal. Chem.*, 21, 391 (1949).

[7.3-5] Stefanac, Z., Simon, W., *Chimia*, 20, 463 (1966); *Microchem. J.*, 12, 125 (1967).

[7.3-6] Adams, R.N., *Anal. Chem.*, 30, 1576 (1958).

[7.3-7] Yamada, S., Sato, H., *Nature*, 193, 261 (1962).

[7.3-8] Osteryoung, J.G., Schreiner, M.M., *CRC Critical Reviews in Analytical Chemistry*, 19, 1 (1988).

[7.3-9] Borman, S.A., *Anal. Chem.*, 54, 6 (1982).

[7.3-10] MacNavin, W.M., Baker, B.B., *Anal. Chem.*, 24, 986 (1952).

General reading

Bard, A.J., Faulkner, L.R., *Electrochemical Methods; Fundamentals and Applications*, John Wiley and Sons, New York, Chichester, Brisbane, Toronto, 1980.

Bond, A.M., *Modern Polarographic Methods in Analytical Chemistry*, Marcel Dekker, Inc. New York and Basel, 1980.

Borman, S.A., *New Electroanalytical Pulse Techniques, Anal. Chem.*, 54, 698 (1982).

Freiser, H. (Ed.), *Ion-Selective Electrodes in Analytical Chemistry*, Vol. 1 and 2, Plenum Press, New York and London

Hobart, H.W., Lyune, L.M., Jr., Dean, J.A., Settle, F.A., Jr., *Instrumental Methods of Analysis*, 2nd ed., Wadsworth Publishing Company, Belmont, California (Division of Wadsworth, Inc.), 1988.

Kissinger, P.K., Heineman, W.R. (Ed.), *Laboratory Techniques in Electroanalytical Chemistry*, Marcel Dekker Inc., New York and Basel, 1984.

Kolthoff, I.M., Liugare, J.J., *Polarography*, Vol. I., 2nd ed., Interscience Publishers, New York, London 1952.

Meier, P.C., Ammann, D., Osswald, H.F., Simon, W., *Ion-Selective Electrodes in Clinical Chemistry*, Medical Progress through Technology, 5,1–12 (1977).

Osteryoung, J.G., Schreiner, M.M., *Recent Advances in Pulse Voltammetry*, CRC Critical Reviews in Analytical Chemistry, 19, 1 (1988).

Pungor, E., *Oscillometry and Conductometry*, Pergamon Press, Oxford, London, Edinburgh, New York, Paris, Frankfurt, 1965.

Wang, J., *Analytical Electrochemistry*, VCH, New York, 1994.

Willard, H.H., Merrit Jr., L.L., Dean, J.A., Settle Jr., F.A., *Instrumental Methods of Analysis*, 2nd ed., Wadsworth Publishing Company, Belmont, California (Division of Wadsworth, Inc.), 1988.

Questions and problems

1. Describe and compare the principles of the 4 major electroanalytical methods!
2. In potentiometry, why do we have to measure under "currentless" conditions?
3. Describe the different types of potentiometric electrodes.
4. Write down the Nernst equation for
   (a) $Fe^{2+} \rightarrow Fe^{3+} + e^-$ and for
   (b) $MNO_4^- + 8\,H^+ + 5\,e^- \rightarrow Mn^{2+} + 4\,H_2O$
   also in graphical form and describe the effect of the pH to the potentials of reactions a) and b) using the data in the Appendix.
5. Describe the mechanism of the pH-glass electrode.
6. Describe the mechanism for the liquid membrane electrode for pK.
7. What is a "polarizable electrode"?
8. Draw a classical polarogram and discuss the three current regions to be seen as well as the analytical information to be extracted from that curve.
9. Under what conditions is mixture analysis possible by classical dc-polarography?
10. Describe the principles of differential *pulse polarography* and *anodic stripping voltammetry* and compare both with the classical method.
11. Describe the principles of *amperometry* and their major analytical application.
12. Why is 100% current efficiency important for the application of *coulometry*?

7.4   Flow Injection Analysis

Learning objectives

■ To provide an introduction to automated assays
■ To describe the basic principles of FIA
■ To demonstrate the capabilities of FIA in relation to batch assays and conventional continuous flow systems
■ To show that FIA allows one to augment existing analytical techniques
■ To show how FIA offers novel analytical procedures which are not feasible by conventional means
■ To highlight the potential of FIA in selected practical assays

7.4.1   Batch and continuous flow analysis

Traditionally, it has been taken for granted that the only sensible way to perform chemical analysis is to mix the analyte species and the reagent(s) homogeneously and wait for chemical equilibrium to the established. Generations of analytical chemists have been taught this approach, irrespective of whether one was dealing with batch assays or, as they became available in the late 1950s, with continuous-flow procedures.

In *batch* assays, the analyte solution remains within a suitable container (say, a test tube or a flask; Fig. 7.4-1a) to which appropriate reagents are added, and after thorough mixing, possible heat transfer, followed by waiting for chemical equilibrium to be attained, the contents of the container are then transferred to a measuring cell of a suitable detector, such as the cuvette of a spectrophotometer. The entire procedure may be mechanized or automated, as evidenced in the large number of different types of (predominantly clinical chemical) apparatus which have become commercially available. Apart from the centrifugal analyzers, they are basically all founded on the conveyor belt principle (Fig. 7.4-1b), i.e., the container progresses through individual stations where various unit operations are effected, e.g., addition of analyte sample, reagents, mixing and heating, incorporation of a delay time to ensure complete reaction, and finally detection. Regardless of how the analyzers are designed, they share the common feature that the solution is stationary while the container is moving, precisely mimicking what is done in a manual batch approach.

In *continuous-flow* analysis, the opposite approach is employed: the system is *stationary* while the solution moves in a set of conduits (tubes). Yet, as demonstrated with the introduction of the first commercially available continuous-flow system, the AutoAnalyzer, marketed by Technicon (Fig. 7.4-1c), the conceptual approach to dealing with chemical assays in this manner was by no means new; homogenization and attainment of chemical equilibrium, resulting in so-called steady-state conditions. In order to secure the latter, and to ensure the identity of individual samples, air bubbles were introduced into the various streams. Although ingenious, this approach also required that steady-state conditions were attained, because, with a mixture of solution and air bubbles, the timing in the system is poorly defined. This idea of performing continuous-flow analysis was challenged in the mid-1970s when flow injection analysis (FIA) was conceived and described in Denmark [7.4-1], and this analytical approach has revolutionized the way of performing chemical assays. While previously it had been necessary to operate via steady-state conditions, FIA proved that this was not essential, allowing the mixing of sample and reagents to be performed in a highly controlled manner and under very precise and reproducible conditions.

In *batch* assays, the solution is stationary while the container is moving.

In *continuous-flow analysis*, the system is stationary while the solution moves.

Steady-state conditions are *not* required in FIA.

7.4.2   Principles

FIA has several important characteristics compared with traditional continuous-flow measurements: higher sampling rates (typically 100–300 samples/h); enhanced

Automated measurements with FIA are economical, even for small sample series.

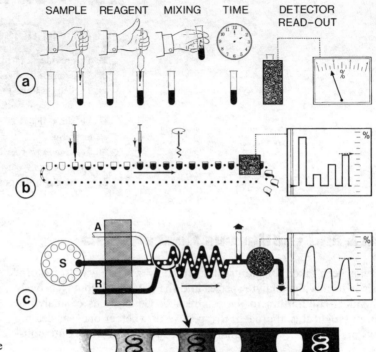

Fig. 7.4-1. The parallel between manual and automated operations that have to be performed in the course of a typical photometric assay: a) Manual handling; b) Discrete belt-type analyzer; c) Continuous-flow air-segmented analyzer, with a detail of the air and liquid segments, showing the mixing pattern that leads to homogenization of individual liquid segments. All assays aim to perform the measurements at a steady-state (stable readout, i.e., "flat top")
From [7.4-3] courtesy John Wiley & Sons

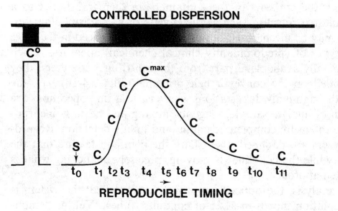

Fig. 7.4-2. Dispersed sample zone, of original concentration C^0 injected at position S, and the corresponding recorder output. Corresponding to each individual concentration C along the gradient there is a specific value of the dispersion coefficient D, each of which can be related to a fixed delay time t_i elapsed from the moment of injection t_0
From [7.4-3] courtesy John Wiley & Sons

FIA is based on *sample injection, controlled dispersion*, and *reproducible timing*.

Any point of the concentration gradient can be used for the analytical readout. Often the maximum peak height is used.

response times (often less than 1 min between sample injection and detector response); much more rapid start-up and shut-down times (merely a few minutes for each); and, except for the injection system, simpler and more flexible equipment. The last two advantages are of particular importance, because they make it feasible and economic to apply automated measurements to a relatively few samples of a nonroutine kind. No longer are continuous-flow methods restricted to situations where the number of samples is large and the analytical method highly routine.

FIA has undergone certain changes, e.g., in recent years it has been supplemented by SIA (sequential injection analysis [7.4-2]) but its fundamentals remain the same as originally defined [7.4-1, 7.4-3] (Fig. 7.4-2): injection of a well-defined volume of sample; reproducible and precise timing of the sample and of the manipulations it is subjected to in the system, from the point of injection to the point of detection (so-called controlled, or rather controllable, dispersion); and the creation of a concentration gradient of the injected sample, providing a transient, but strictly reproducible readout of the recorded signal. The combination of these features, as recorded by the detector, which may continuously observe an absorbance, an electrode potential, or any other physical parameter as it changes on passage of the sample material through the flow-cell, make it unnecessary to achieve chemical equilibrium (steady-state conditions). Any point on the path toward the steady-

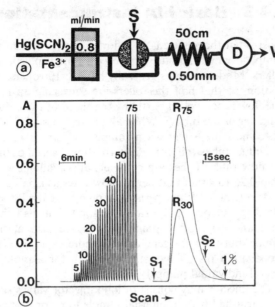

Fig. 7.4-3. a) Flow diagram for the spectrophotometric determination of chloride: S is the point of injection, D is the detector, and W is the waste; b) Analog output showing chloride analysis in the range of 5–75 ppm Cl for the system shown in (a). To demonstrate the repeatability of the measurements, each sample was injected in quadruplicate. The injected volume was 30 µL, sampling rate was approximately 120 samples/h. The fast scan of the 30-ppm sample (R_{30}) and the 75-ppm sample (R_{75}) on the right show the extent of carry-over (less than 1%) if samples are injected in a span of 38 s (difference between S_1 and S_2)
From [7.4-3] courtesy John Wiley & Sons

state signal (Fig. 7.4-1c) is as good a measure as the steady-state itself, provided that this point can be reproduced repeatedly, and this is certainly feasible in FIA with its inherently exact timing. The analytical readout in FIA can be made at any point along the concentration gradient created, but in most cases the peak height corresponding to the concentration C^{max} in Fig. 7.4-2 is used, because it is readily identified. Yet, as detailed in Section 7.4.10, the exploitation of all the concentrations along the gradient has formed the basis for a series of entirely novel analytical approaches. For the time being, we will, however, use the peak maximum as the analytical response.

The characteristics of FIA are probably best illustrated by a practical example, such as the one shown in Fig. 7.4-3, which depicts the spectrophotometric determination of chloride in a single-channel system, based on the following sequence of reactions:

Only a few microliters of sample volume are required.

FIA measurements are very rapid.

$$Hg(SCN)_2 + 2Cl^- \rightleftharpoons HgCl_2 + 2SCN^-$$

$$Fe^{3+} + SCN^- \rightleftharpoons Fe(SCN)^{2+}$$

The analytical sequence is founded on the reaction of chloride with mercury(II) thiocyanate with the ensuing release of thiocyanate ions which subsequently react with iron(III) to form the intensely red iron(III) thiocyanate complex, the absorbance of which is measured. The samples, with chloride content 5–75 ppm, are injected (S) through a 30 µL valve into the carrier solution containing the mixed reagent, pumped at a rate of 0.8 mL/min. The iron(III) thiocyanate is formed and forwarded to the detector (D) via a mixing coil (0.5 m long, 0.5 mm ID), as the injected sample zone disperses within the reagent carrier stream. The mixing coil minimizes band broadening (of the sample zone) owing to centrifugal forces, resulting in sharper recorded peaks. The absorbance A of the carrier stream is continuously monitored at 480 nm in a micro-flow-through cell (volume 10 µL) and recorded (Fig. 7.4-3b). To demonstrate the reproducibility of the analytical readout, each sample in this experiment was injected in quadruplicate, so that 28 samples were analyzed at seven different chloride concentrations. As this took 14 min, the average sampling rate was 120 samples/h. The fast scan of the 75- and 30-ppm sample peaks (shown on the right in Fig. 7.4-3b) confirms that there was less than 1% of the solution left in the flow-cell at the time when the next sample (injected at S_2) would reach it, and there was no carry-over when the samples were injected at 30-s intervals. These experiments clearly reveal one of the key features of FIA; all samples are sequentially processed in exactly the same way during passage through the analytical channel, or, in other words, what happens to one sample happens in exactly the same way to any other sample.

7.4.3 Basic FIA instrumentation

Peristaltic pumps are mostly used for
propelling liquids.

Most often the solutions in a flow injection system are propelled by a peristaltic pump, in which liquid is squeezed through plastic or rubber tubing by moving rollers. Modern pumps generally have 8–10 rollers, arranged in a circular configuration, so that half the rollers are squeezing an individual tube at any instant. The flow is controlled partly by the speed of the motor, and partly by the internal diameter of the tubing. While all pump tubes have identical wall size, so that they fill to the same extent when completely compressed, the internal diameters of the individual tubes will, for a fixed rotational speed of the peristaltic pump, determine the flow rates. Tubes are commercially available with internal diameter ranging from 0.25 to 4 mm that permit flow rates as small as 0.0005 mL/min and as great as 40 mL/min. While peristaltic pumps generally suffice for most applications, allowing not only pulse-free operation, but also the operation of several tubes simultaneously, other pump devices are also applied in FIA, such as piston pumps, but they are generally far more expensive (besides, they only permit the propagation of a single stream, which for a multiline manifold would call for several individual pumps).

FIA is economical on sample and reagents and generates minute amounts of waste.

The injector may consist of loop injector valves similar to those used in HPLC, i.e., furnished with in internal sample loop, or more commonly, a dedicated FIA-valve comprising a rotor and a stator furnished with four, six, or more individually accessible ports, where the injected sample volume, normally $1–200\,\mu L$ (typically $25\,\mu L$) is metered via an external loop of appropriate length and internal diameter. Because the injected sample volume is so small, it does not require much reagent per sample cycle, which not only makes FIA a simple, microchemical technique, capable of providing a high sampling rate at the expense of minimum sample and reagent consumption, but also a system which generates minute amounts of waste. This is important because the disposal of chemical wastes in many instances is even more expensive than the chemicals themselves.

The conduits used in FIA manifolds mostly consist of narrow-bore plastic tubes (of materials such as PVC or PTFC), typically of inner diameter 0.5–0.8 mm. The tubes are normally coiled or knotted in order to minimize the dispersion (promote secondary flow pattern), but as a rule the tube lengths of the FIA manifold should be made as short as possible in order to avoid adverse dilution of the injected sample solution.

Sequential injection analysis (SIA), which in essence is a subclass of FIA, depends on time-based, sequential aspiration of well-defined sample and reagent zones into a holding coil by means of a directional (selector) valve. The flow is then reversed and the stacked zones are forwarded via a single-line reactor coil to the detector. During these steps, the sample and reagent zones penetrate into each other, and in the region of mixing a product is formed. Except for the selector valve, the components for SIA are identical to those employed in FIA. Although this variant has certain advantages for given applications (e.g., minute consumption of sample and reagent solutions), it also has severe practical limitations (mainly that it merely permits a single-line manifold), and for these reasons SIA will not be discussed further in this section.

FIA is compatible with virtually any type of detection device.

FIA (or SIA) is virtually compatible with any type of detector. One of the reasons that FIA has achieved so much success is that not only is it a general solution handling technique, applicable to a variety of tasks, but it fits with virtually any type of detection device. This will be demonstrated in the following sections, but first emphasis will be placed on the controlled dispersion, how it can be manipulated, and what purpose it serves.

7.4.4 Dispersion in FIA

The dispersion is quantified by the dispersion coefficient D.

The degree of dispersion, or dilution, in a FIA system is characterized by the dispersion coefficient D. Let us consider a simple dispersion experiment. A sample solution, contained within the valve cavity prior to injection, is homogeneous and has the original concentration C^0 that, if it could be scanned by a detector, would yield a square signal with height proportional to the sample concentration (Fig.

7.4-2). When the sample zone is injected, it follows the movement of the carrier stream, forming a dispersed zone the shape of which depends on the geometry of the channel and the flow velocity. Therefore, the response curve has the shape of a peak, reflecting a continuum of concentrations (Fig. 7.4-2, bottom), and forming a concentration gradient, within which no single element of fluid has the same concentration of sample material as its neighbors. It is useful, however, to view this continuum of concentrations as being composed of individual fluid elements, each having a certain concentration of sample material C, since each of these elements is a potential source of readout (see also Sec. 7.4.10).

In order to design a FIA system rationally, it is important to know how much the original sample solution is diluted on its way to the detector, and how much time has elapsed between sample injection and readout. For this purpose, the dispersion coefficient D has been defined as the ratio of concentrations of sample material before and after dispersion has taken place in that element of fluid that yields the analytical readout, that is:

$$D = C^0/C$$

which, for $C = C^{max}$, yields

$$D = C^0/C^{max} \ (0 < D < \infty)$$

Hence, if the analytical readout is based on maximum peak height measurement, the concentration within that imaginary fluid element which corresponds to the maximum of the recorded curve C^{max} has to be considered. Thus, with knowledge of D, the sample (and reagent) concentrations may be estimated. The determination of D of a given FIA manifold is readily performed. The simplest approach is to inject a well-defined volume of a dye solution into a colorless stream and to monitor the absorbance of the dispersed dye zone continuously by a spectrophotometer. To obtain the D^{max} value, the height (i.e., absorbance) of the recorded peak is measured and then compared with the distance between the baseline and the signal obtained when the cell is filled with undiluted dye. Provided that the Lambert–Beer law (see Sec. 9.1) is obeyed, the ratio of respective absorbances yields a D^{max} value that describes the FIA manifold, detector, and method of detection. Note that this definition of the dispersion coefficient considers only the physical process of dispersion and not the ensuing chemical reactions, since D refers to the concentration of sample material prior to and after the dispersion process alone has taken place. In this context it should be emphasized that any FIA peak is a result of two kinetic processes, which occur simultaneously: the *physical* process of zone dispersion, and the *chemical* processes resulting from reactions between sample and reagent species. The underlying physical process is well reproduced for each individual injection cycle; yet it is not a homogeneous mixing, but a dispersion, the result of which is a concentration gradient of sample within the carrier solution.

The definition of D implies that, when $D = 2$, for example, the sample solution has been diluted 1:1 with the carrier stream. The parameters governing the dispersion coefficient have been the subject of detailed studies. In short, in can be stated that the most powerful means of manipulating D are the injected sample volume, the physical dimensions of the FIA system (lengths and internal diameters of the tubes), the residence time, and the flow velocity. Additional factors are the possibility of using a single line rather than a confluence manifold with several conduits, and of selecting the element of measurement on any part of the gradient of the dispersed sample zone other than that corresponding to the peak maximum [7.4-3].

For convenience, sample dispersion has been defined as *limited* $(D = 1–2)$, *medium* $(D = 2–10)$, *large* $(D > 10)$, and *reduced* $(D < 1)$; the FIA systems designed accordingly have been used for a variety of analytical tasks. Limited dispersion is used when the injected sample is to be carried to a detector in undiluted form, i.e., the FIA system serves as a means of rigorous and precise transport and sample presentation to the detection device (such as an ion-selective electrode or an atomic absorption spectrophotometer). Medium dispersion is employed when the analyte must mix and react with the carrier stream to form a product to be detected. Large dispersion is used only when the sample must be diluted to bring it within the

The dispersion coefficient D of the sample is defined as the ratio of concentrations of sample material before and after dispersion has taken place in that element of fluid that yields the analytical readout.

The dispersion coefficient for a given FIA system is readily determined.

The dispersion coefficient is a measure of the extent of dilution of the injected sample at readout.

Any FIA peak is the result of of two kinetic processes, which occur simultaneously: the *physical* process of dispersion and the superimposed *chemical* process due to reaction between sample and reagents.

measurement range. Reduced dispersion implies that the concentration of the sample detected is higher than the concentration of the injected sample, i.e., on-line preconcentration is effected (e.g., by incorporation of an ion-exchange column or via coprecipitation, Sec. 7.4.9).

FIA is a generally applicable solution handling technique.

FIA provides for analytical procedures which are difficult or impossible to execute by conventional schemes.

The following sections give examples covering the various types of dispersion patterns. Considering the extensive literature on FIA (by early 1996 the count of FIA-papers had passed 7000 to which should be added more than 25 monographs), it is of course impossible to cover more than a few aspects, and the reader is encouraged to consult the literature for further information [7.4-4], which shows that FIA is established as a powerful method in modern analytical chemistry. Originally devised as a means of providing serial analysis, FIA soon became much more than this, i.e., a generally applicable solution handling technique, in the analytical laboratory as well as in industrial fields such as process control. Additionally, it was shown that it offered unique analytical possibilities, providing means for devising analytical procedures which are otherwise difficult or impossible to implement by conventional schemes, such as assays based on the generation and measurement of metastable constituents, allowing kinetic discrimination schemes, and exploring detection principles relying on bio- and chemiluminescence. For the same reason, FIA in many contexts has been replaced by FI, emphasizing that the flow injection system is used for solution handling rather than merely for analysis.

7.4.5 FIA for reproducible and precise sample presentation (limited dispersion)

Because of its inherently strict timing, an obvious application of FIA is to use it for precise and reproducible presentation of a given sample to the detector, thereby ensuring that all conditions during each measuring cycle are rigorously maintained. This might for instance be advantageous if the detector is an electrode or a sensor, the response of which is diffusion controlled, or if the behavior of the detection device is affected by the time that the sample is exposed to it. Examples of the former group are ion-selective electrodes and biosensors, and of the latter group the combination of FIA with atomic absorption spectrometry (AAS).

ISEs are advantageously operated in the dynamic mode of FIA.

Interfering ions can be kinetically discriminated.

Thus, in potentiometry it is observed that many ion-selective electrodes (ISE) operated in the dynamic mode facilitate fast and reproducible readout. ISEs are generally characterized by fairly long response times to reach steady-state conditions, and therefore it can be difficult to ascertain exactly when to make the readout. In FIA this decision is left entirely to the system, because the sample reaches the detector after a time governed exclusively by the manifold selected. But ISEs also often exhibit kinetic discrimination toward the ion under investigation and interfering species (i.e., during the short duration of sample exposure the response of the sensor to the different species can vary considerably, which in turn can be exploited to increase the selectivity and the detection limit of the sensor). This is illustrated in Fig. 7.4-4, where it can be observed that if the readout is taken at

Fig. 7.4-4. Typical potential–time response profiles of an ion-selective electrode for the primary ion (A) and an interfering ion (B). If the readout is taken at the time marked by the vertical dotted line, the contribution of B is minimized

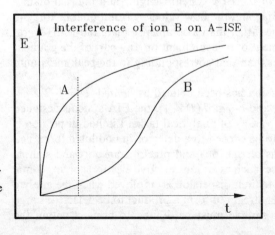

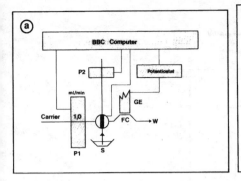
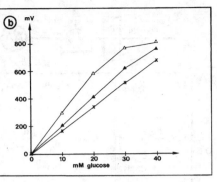

Fig. 7.4-5. a) FIA manifold for the detection of glucose with an amperometric sensor to detect enzymatically generated hydrogen peroxide; b) Calibration graphs for glucose in the concentration range 0–40 mmol/L at three different flow rates: (△) 0.50, (▲) 0.75, and (×) 1.00 mL/min
From [7.4-5] courtesy Elsevier Science Publishers

steady-state (at time *t* on the far right), the interfering ion B exerts considerable interference, yet if the signal is recorded at the time marked by the dotted vertical line, the interference is (at the expense of a slightly lower recorded signal for the primary ion A) virtually eliminated.

The same concept of manipulating the sample exposure time can be extended to more complex sensors such as enzyme sensors, in which a membrane containing one or more immobilized enzymes is placed in front of the active surface of a detection device. The analyte is transported by diffusion into the membrane and here degraded enzymatically, forming a product which can then be sensed optically or electrochemically. A condition for obtaining a linear relationship between analyte and signal is, however, that pseudo-first-order reaction kinetics are fulfilled [7.4-5], i.e., the concentration of converted analyte reaching the detector surface must be much smaller than the Michaelis–Menten constant. Since this constant for most enzyme systems is of the order of 1 mmol L^{-1}, and many sample matrices (e.g., human sera or blood) contain much higher concentrations, the use of enzyme sensors in static (batch) systems often calls for complicated use of additional membrane layers aimed at restricting diffusion of analyte to the underlying enzyme layer, or in the case of electrochemical detectors, for chemically modified electrodes where the electron transfer is governed by appropriate mediators.

However, when operated in FIA, the degree of conversion can be simply adjusted by the *time* the analyte is exposed to the enzyme layer, and therefore the amount of converted analyte can be regulated directly by adjusting the flow rate of the FIA system. This approach is demonstrated in Fig. 7.4-5a which shows a system for the determination of glucose by means of an amperometric sensor (GE) incorporating glucose oxidase, while Fig. 7.4-5b shows the calibration plots for three different flow rates. By increasing the flow rate from 0.5 to 1.0 mL/min, the linear measuring range can readily be expanded from 20 to 40 mmol/L glucose, ensuring that the system can be used for physiological measurements. Additionally, the exposure time can be exploited for kinetic discrimination, taking advantage of differences in diffusion rates of analyte and interfering species within the membrane layer of the electrode. Thus, with the system in Fig. 7.4-5 it was shown that total spatial resolution of the signals due to glucose and paracetamol were achieved, even when paracetamol was added at excessive levels. Besides, attention should be drawn to the fact that the operation of a (bio)sensor in the FIA mode ensures constant monitoring of the sensor itself, i.e., while the recorded signal is a measure of the concentration measured, the baseline signal indicates the stability of the sensor.

AAS is one of several analytical instruments where the performance can benefit, and in some cases be significantly enhanced, when combined with FIA. Thus, by using a system consisting merely of a connecting line between the pump/injection device and a flame-AAS instrument (Fig. 7.4-6), several advantages can be gained by the possibility of repetitive and reproducible sample presentation. As seen in Fig. 7.4-6, the sample frequency can be improved compared with the traditional aspiration of sample solution (Fig. 7.4-6a), but what is more important is that during that time normally required to aspirate one sample, it is possible to inject two individual sample aliquots, which means that the FIA approach entails not only improved precision, but also improved accuracy (Fig. 7.4-6b). Furthermore, one can take advantage of the fact that the sample is exposed to the detector for

A linear relationship between analyte concentration and signal response of a biosensor requires that pseudo-first-order kinetics are fulfilled. Such conditions are readily achieved in a FIA system.

Conditions for pseudo-first-order kinetics can be obtained by adjusting the *time* that the sample is exposed to the sensor, i.e., by adjusting the *flow rate* of the FIA system.

The combination of AAS and FIA enhances the performance of the detection device.

The short sample exposure time eliminates or greatly reduces the risks of matrix interference/salt contamination.

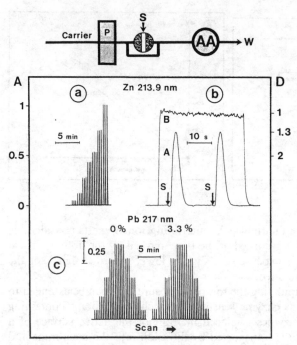

Fig. 7.4-6. Single-line FIA manifold for determination of metal ions by flame atomic absorption spectrometry (AA). Recordings obtained at a flow rate of 4.9 mL/min and an injected sample volume of 150 µL.
a) Calibration run for zinc as obtained by injection of standards in the range 0.10–2.0 ppm; b) Recorder response for the 1.5 ppm standard as obtained by (A) injection via the FIA system and (B) continuous aspiration in the conventional mode (also at 4.9 mL/min). *D* represents the dispersion coefficient value, which in (B) is equal to 1; c) Calibration runs for a series of lead standards (2–20 ppm) recorded without (0%) and with (3.3%) sodium chloride added to the standards After [7.4-3] courtesy John Wiley & Sons

only a very short period of time. The rest of the time the detector is cleansed by the carrier solution, which means that the wash-to-sample ratio is high, and therefore the chance of clogging the burner, due for instance to a high salt content, is vastly reduced or eliminated. This can be seen in Fig. 7.4-6c by comparing the Pb calibration runs for pure aqueous Pb standards and standards prepared in a matrix simulating seawater (3.3% NaCl), where it is obvious that the two calibration runs are identical. This feature was even more dramatically demonstrated by Schrader et al. [7.4-6] who, in an experiment comprising 160 repetitive injections of a 1 ppm Cu standard prepared in a 30% NaCl solution, found that hardly any deterioration of the recorded signal was observed.

We now turn our attention to procedures where the dynamics of the dispersion process are superimposed by chemical reactions, and where FIA allows us to exploit the kinetic behavior of the chemistry taking place, i.e., where a kinetic effect is being used either to increase the selectivity of an analytical procedure, or indeed to obtain information which is not at all accessible by conventional batchwise operation. This is well illustrated in the so-called FIA conversion techniques.

7.4.6 FIA conversion techniques (medium dispersion applications)

FIA conversion techniques are procedures by which a nondetectable species is converted into a detectable component through a kinetically controlled chemical reaction.

Conversion techniques offer the advantages of *kinetic discrimination* and *kinetic enhancement*.

Kinetic discrimination relies on exploiting differences in reaction rates of the primary reaction from those of side reactions.

The system shown in Fig. 7.4-3, where the detection of chloride obviously calls for a chemical reaction in order to sense this species, is a good example of the use of medium dispersion. In designing such systems it is important to bear in mind that the dispersion should be sufficient to allow partial mixing of sample and reagent, yet it should not be so excessive as to dilute the analyte unnecessarily, which would lower the detection limit. Most FIA procedures are based on medium dispersion, because the analyte must be subjected to some form of intelligent "conversion". In their widest sense, FIA conversion techniques can be defined as procedures by which a nondetectable species is converted into a detectable component through a kinetically controlled chemical reaction, with the aim of performing either appropriate sample pretreatment, reagent generation, or matrix modification. In performing these tasks, *kinetic discrimination* and *kinetic enhancement* offer further advantages. In kinetic discrimination, the differences in the rates of reactions of the reagent with the analyte and the interferents are exploited. In kinetic enhancement, the chemical reactions involved are judiciously driven in the direction appropriate to the analyte of interest. While, in batch chemistry, the processes are forced to equilibrium, so that subtle differences between reaction rates cannot be exploited, in the FIA mode, small differences in reaction rates with the same reagent result in different sensitivities of measurement.

A simple demonstration of a homogenous conversion procedure exploiting kinetic discrimination is given by the following reaction sequence, aimed at analyzing chlorate in process liquor:

$$2\,ClO_3^- + 10\,Ti^{3+} + 12\,H^+ \rightarrow 10\,Ti^{4+} + Cl_2 + 6\,H_2O \qquad \text{(fast)}$$

$$Cl_2 + LMB \rightarrow MB \qquad \text{(fast)}$$

$$MB + Ti^{3+} \rightarrow LMB + Ti^{4+} \qquad \text{(slow)}$$

The assay is performed by injecting a sample of chlorate into an acidic carrier steam of titanium(III), which is subsequently merged with a second stream of leucomethylene blue (LMB). While the first two of these reactions are very fast, the reduction of the blue species MB by the third reaction is slow. Thus, the chlorate concentration can readily be quantified via the absorbance of the MB species generated by the second reaction, while re-formation of LMB takes place after the sample plug has passed the detector.

Another procedure illustrating this approach is the assay of thiocyanate, a species of interest in clinical chemistry, because it is not naturally present in humans to any significant extent, except in tobacco smokers. The half-life of thiocyanate in the body is approximately 14 days, and therefore it is easy, via analysis of body fluids (saliva, blood, or urine) to distinguish smokers from nonsmokers. A fast, simple, and convenient FIA method for the determination of thiocyanate relies on the reaction of this compound with 2-(5-bromo-2-pyridylazo)-5-diethylaminophenol (5-Br-PADAP) in acidic media in the presence of dichromate as oxidizing agent (Fig. 7.4-7a), resulting in the generation of an intensely colored (reddish), albeit transient, product of high molar absorptivity. The latter feature makes the method ideal for quantifying low levels of thiocyanate, while the transient nature of the product implies that it is generated rapidly, but then fades away, having a life time of ca. 10 s. Hence, it is important that the readout is taken at the point where the color development is at its maximum. The FIA system used is shown Fig. 7.4-7b. The sample (50 µL of saliva) is injected into a carrier stream of water, which is subsequently merged with the reagent (5-Br-PADAP) and ultimately with the oxidizing reagent, the acid concentration in the final mixture being ca. 2 mol L^{-1}. Because of the reproducible timing of the FIA system, it is possible to detect and quantify the metastable, colored constituent. There is, however, an additional problem. Even if no thiocyanate is present, the 5-Br-PADAP and the dichromate will gradually react with each other, yielding a component which absorbs at the wavelength used (570 nm), i.e, a passive background signal. To make things even more complicated, the signal due to the background increases with reaction time. In other words, one is faced with a situation such as that shown in Fig. 7.4-8, where the transient signal of the analyte as a function of time first increases and then decreases, while the background signal steadily increases.

Transient or metastable reaction products of interesting analytical characteristics, such as high molar absorptivity, can be exploited in FIA.

The reproducible timing of FIA is the basis for measuring metastable constituents.

FIA allows one to discriminate against a varying background signal.

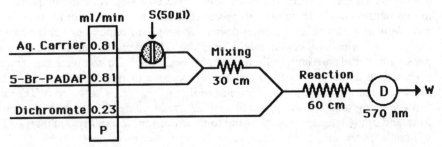

Fig. 7.4-7. Reaction scheme for thiocyanate with 5-Br-PADAP which in the presence of an oxidant (e.g., dichromate) in acidic media results in the generation of a metastable product, which has analytically interesting characteristics. FIA manifold for determination of the metastable reaction product
From [7.4-7] courtesy Royal Society of Chemistry

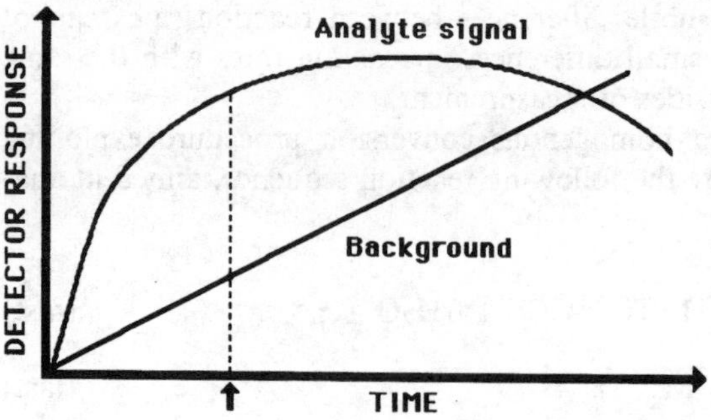

Fig. 7.4-8. Exploitation of FIA for quantification of a metastable analyte species in a system which additionally entails a gradually increasing background signal response. Because of the precise and reproducible timing of FIA, it is possible to take the readout at that time which corresponds to the largest difference between the analyte and background signals (marked by the arrow and vertical dotted line)

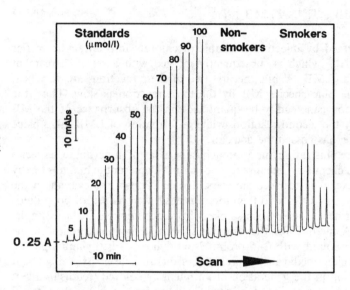

Fig. 7.4-9. Readouts for thiocyanate as obtained with the FIA manifold shown in Fig. 7.4-7. To the left is shown a calibration run consisting of a series of thiocyanate standards, each sample injected in duplicate. To the right are given the output signals for ten saliva samples, also injected in duplicate, where the first five are from nonsmokers and the remaining five samples are from smokers. Note that the baseline corresponds to a very high background signal, yet the readouts are all very reproducible
From [7.4-7] courtesy Royal Society of Chemistry

Selectivity enhancement can be achieved by using packed reactors, incorporating for instance ion-exchange materials.

FIA allows the exploitation of reagents generated *in statu nascendi*.

By using FIA it is possible, by appropriate design of the analytical system, to adjust the sample residence time so that the detection can be effected at precisely that time where the difference between the two signals is at its maximum (indicated by the arrow and dotted vertical line). Figure 7.4-9 shows a number of actually registered signals, as obtained with the FIA system shown in Fig. 7.4-7. On the left are shown the signals for a series of aqueous standards (in the range 5–100 µmol L^{-1}, each injected in duplicate), and on the right are shown five saliva samples from smokers and then five from nonsmokers (both series again injected in duplicate). As can be seen, the smokers and nonsmokers fall into two distinct groups, but what is more significant from an analytical point of view is to observe that, although the background signal is relatively high (ca. 0.25 AU) and the individual signals are in the mAU range, the repeatability of all the duplicates is very satisfactory.

The possibilities of achieving selectivity enhancement in FIA become even greater in *heterogeneous* conversion techniques, where it is possible to incorporate into the manifolds steps such as gas diffusion, dialysis, solvent extraction, or the use of packed reactors. Thus, ion-exchangers have been employed as column materials in order to transform a sample constituent into a detectable species, an example being the determination of various anions by atomic absorption spectrophotometry: For instance, cyanide has been assayed via interaction with a column containing CuS, resulting in the formation of soluble tetracyanocuprate, allowing the cyanide to be quantified indirectly by AAS by means of the stoichiometric amount of copper released from the solid surface [7.4-3]. As mentioned earlier, ion-exchangers can also be used to preconcentrate an analyte in order to accommodate it to a particular detection device, or simply to remove unwanted matrix components which otherwise might interfere. However, the most widely used packing materials are immobilized enzymes (Sec. 7.4-7). To complete the picture, it should be noted that reactors containing oxidants or reductants have been devised to generate reagents in *statu nascendi* (e.g., Ag(II), Cr(II), or V(II)), advantage being taken of the protective environment offered by the FIA system to form and apply reagents which, owing to their inherent instability, are impractical to handle under normal analytical conditions.

7.4.7 FIA systems with enzymes

When applied in the FIA mode, the use of immobilized enzymes, packed into small column reactors, offers not only the selectivity, economy, and stability gained by immobilization, but also ensures that strict repetition, and hence a fixed degree of turnover from cycle to cycle, are maintained. In addition, by obtaining a high concentration of enzyme immobilized within a small volume, the ensuing high activity facilitates an extensive and rapid conversion of substrate at minimum dilution of the sample, the small dispersion coefficient in turn promoting a lower detection limit. The enzymes needed for a particular assay can be incorporated into a single packed reactor, or into sequentially connected reactors in order to process the required sequence of events and also to afford optimal operational conditions for the reactions occurring in the individual reactors. Both optical and electrochemical detectors can be used to monitor the reaction. Thus, for oxidases, which generate hydrogen peroxide, detection via chemiluminescence can be employed, based on the reaction with luminol.

Chemi- and bioluminescence are particularly fascinating and attractive detection approaches, primarily because of the potentially high sensitivity and wide dynamic range of luminescent procedures, but also because the required instrumentation is fairly simple. Furthermore, luminescence has an added advantage over most optical procedures: as light is produced and measured only when sample is present, there is generally no problem with blanking. However, luminescent reactions usually generate transient emissions, because the intensity of the light emitted is proportional to the reaction rate rather than to the concentrations of the species involved (Fig. 7.4-10). Hence, the radiation is most often emitted as a flash which rapidly decreases, and for this reason the conventional approach of quantification has been to integrate the intensity over a fixed period of time and relate this to the amount of analyte. It is obvious, however, that if the measurement of the light intensity dE/dt can be made under precisely defined and reproducibly maintained conditions so that all samples are treated, physically and chemically, in exactly the same manner (i.e., the measurements can be taken repetitively at identical delay times t_i, Fig. 7.4-10), it is possible directly to relate any dE/dt value (and preferably the one corresponding to the maximum emission, Δt) to the analyte concentration. This is feasible by means of FIA, and therefore the combination of luminescence and FIA has revolutionized the application of bio- and chemiluminescence as analytical chemical detection procedures. An example of the use of chemiluminiscence is given in Fig. 7.4-11, which shows a system for determination of glucose by means of immobilized glucose oxidase. The sample is injected via the valve into a carrier stream of buffer (in order to ensure constant pH) and then guided to the enzyme reactor. Here the glucose is degraded to form hydrogen peroxide which is subsequently mixed with luminol and hexacyanoferrate(III), leading to the generation of chemiluminescence which is monitored by a set of photodiodes. Numerous applications of enzyme assays in FIA have been reported, advantage being taken of the fact that these components constitute the selective

Enzymes immobilized in packed reactors are advantageously used in FIA.

Chemi-/bioluminescence and FIA form an ideal combination, because luminescent reactions generate transient emissions.

Enzymes are exploited to provide the selective link in the analytical chain, while FIA is a powerful tool in facilitating the quantitative evaluation.

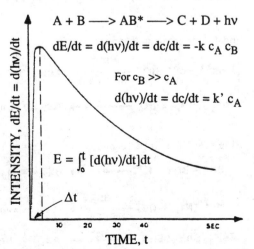

$$A + B \longrightarrow AB^* \longrightarrow C + D + h\nu$$

$$dE/dt = d(h\nu)/dt = dc/dt = -k\, c_A\, c_B$$

For $c_B \gg c_A$

$$d(h\nu)/dt = dc/dt = k'\, c_A$$

$$E = \int_0^t [d(h\nu)/dt]dt$$

INTENSITY, $dE/dt = d(h\nu)/dt$

TIME, t

Fig. 7.4-10. Typical course of light generation ($h\nu$) in a bio- or chemiluminescent reaction as a function of time. Quantification can be accomplished either by relating the amount of analyte to the energy released (i.e., integrating the area under the curve), or by using FIA and determining the intensity dE/dt after a fixed period of time Δt, which, if pseudo-first-order reaction kinetics are fulfilled ($c_B \gg c_A$), is directly proportional to the concentration of the analyte

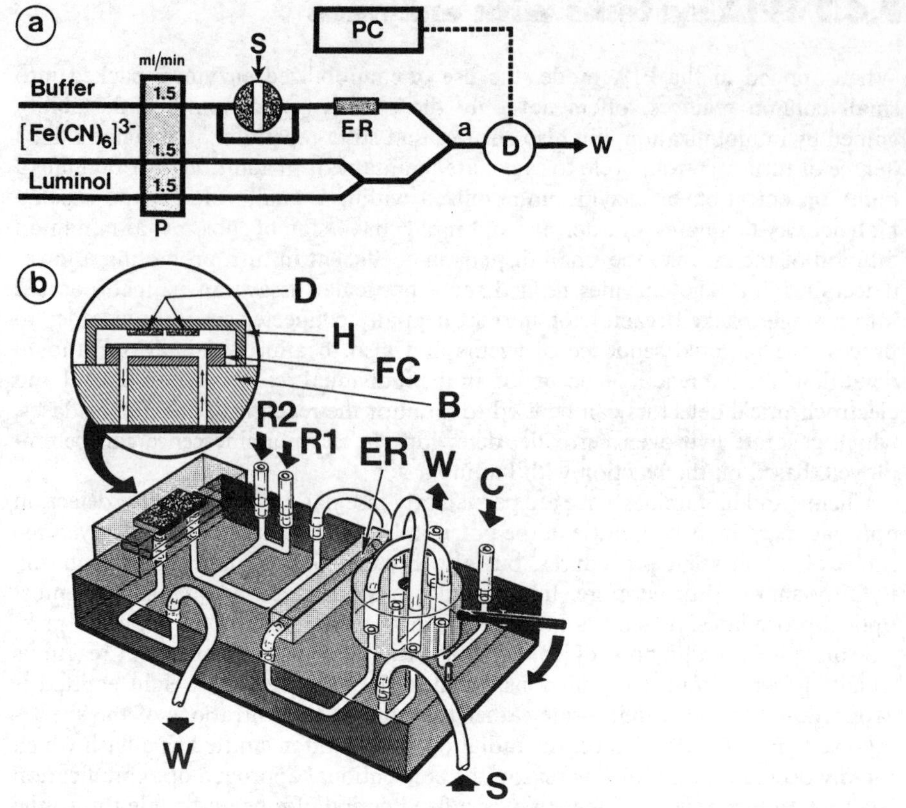

Fig. 7.4-11. a) Manifold for the determination of glucose with detection by chemiluminescence. The sample S is injected into a carrier stream C of buffer and propelled to a reactor ER containing immobilized glucose oxidase, in which the glucose of the sample is degraded, leading to the formation of hydrogen peroxide. After confluence with luminol and hexacyanoferrate(III), the sample zone is finally guided through a short channel a (2 cm, corresponding to 16 μL) into the light detector D, the output from which is led to a computer; b) Integrated FIA microconduit accommodating the manifold components shown for the system in (a). C represents the carrier stream of buffer, Rl is luminol, and R2 is the hexacyanoferrate(III). The flow cell FC comprises two photo diodes D contained in a housing H mounted on the microconduit base plate B
From [7.3-4] courtesy John Wiley & Sons

link in the analytical chain, which therefore becomes selective overall. For the same reasons, FIA is not only complementary to (bio)sensors, but in many cases it can be an attractive alternative [7.4-8].

7.4.8 Flow injection–hydride generation schemes

The problems associated with hydride generation schemes are generally overcome when executed under FIA conditions.

Interference due to the presence of transition metal ions such as Cu, Co, and Ni are eliminated or kinetically discriminated against, to the benefit of the primary hydride-forming reaction.

Several elements (such as As, Sb, Bi, Se, Te, and Ge) can, by reaction with a strong reducing agent, such as sodium tetrahydroborate, become chemically converted to their hydrides. Gaseous hydrides can be readily separated from the sample matrix and guided to the heated quartz flow-through cell of an AAS instrument, where they are atomized by heating and excited by radiation, so that the elements of interest can be selectively quantified. Originally, the hydride generation technique was introduced as a batch procedure, but this involved several problems, as illustrated in Fig. 7.4-12. The conversion of the analyte itself must

Hydride generation/atomization:

$$As^{3+}, Sb^{3+}, Sn^{4+} \xrightarrow[\text{Acid (HX)}]{BH_4^-} AsH_3, SbH_3, SnH_4 \uparrow$$

$$AsH_3, SbH_3, SnH_4 \xrightarrow{\Delta} As, Sb, Sn + nH_2$$

Side reactions/interferences:

$$BH_4^- + 3HX + H^+ \longrightarrow BX_3 + 4H_2$$

$$Me^{2+}(Ni, Cu, Co) \xrightarrow[\text{Acid (HX)}]{BH_4^-} Me^0 \text{ (slower)}$$

$$AsH_3, SbH_3, SnH_4 \xrightarrow{Me^0} As, Sb, Sn + nH_2$$

Fig. 7.4-12. Reactions taking place in the generation of gaseous hydrides (as exemplified for As, Sb, and Se), and possibly concurrent side reactions. The latter can to a large extent be suppressed, or even eliminated, by using FIA

necessarily take place in an acidic medium. As shown in the figure, there are, however, possibilities for side reactions and interferences. The tetrahydroborate itself can react with acid and form hydrogen, whereby the reagent is wasted for the hydride formation. Therefore, the tetrahydroborate must be prepared in a weakly alkaline medium and mixed with the sample and the acid precisely when it is required, and under very controlled conditions. A serious possibility for interference is the formation of free metals or metal boride precipitates, particularly of Ni, Cu, and Co. If ionic species of these metal constituents are present in the sample, they become reduced by the tetrahydroborate, giving rise to the formation of colloidal free metals or metal borides, which have been shown to act as superb catalysts for degrading the hydrides before they can reach the measuring cell. However, because of the dynamic conditions prevailing in FIA, and because of the inherently short residence time of the sample in the system, these side reactions can to a large extent be eliminated or kinetically discriminated against at the expense of the main reaction. If side reactions occur, the precise timing of the FIA system ensures that they take place at exactly the same extent for all samples introduced [7.4-3, 7.4-9].

7.4.9   On-line sample conditioning and preconcentration

One of the characteristics of FIA is that it allows on-line sample conditioning or on-line preconcentration to be effected judiciously. These tasks can be accomplished either by incorporation of suitable column reactors (e.g., ion-exchangers to retain the ionic species, injected in a large volume of sample solution, which later can be eluted by a small volume of injected eluent so that the recorded signal fall within the working range of the instrument [7.4-3, 7.4-9]), or by intelligent design of the FIA system per se. A good example is the determination of low levels of selenium(IV) and arsenic(V/III).

Assay of trace levels of Se(IV) (or As) is of particular interest in aqueous samples, such as drinking water. Although the sophisticated combination of inductively coupled atomic emission spectrometry (Sec. 8.1) and mass spectrometry (ICP-MS, Sec. 8.5) has the potential for measuring minute concentrations, it suffers in this context from the inherent problem of $Ar-Ar^+$ dimer interference at m/z values identical to that of selenium, and $Ar-Cl^+$ identical to As. Therefore, an alternative approach has to be selected which must include a preconcentration step. In solving this problem, much inspiration was found in work by Fang [7.4-9, 7.4-10] who described an ingenious and simple FIA approach to assaying trace metal elements, comprising preconcentration by coprecipitation with an appropriate reagent, followed by dissolution. Developing an idea originally proposed for batch assays, Fang managed to incorporate the procedure on-line into FIA, not by using a filter to collect the precipitate, but simply by using a knotted reactor made of Microline tubing. While such a device had previously been employed for promoting radial dispersion at the expense of axial dispersion [7.4-3], it proved superbly suitable as a collector for the coprecipitate which simply adhered to the walls of the knotted tube. In his experiments, Fang formed an "organic" coprecipitate which was dissolved in an organic liquid. However, in the present case, where it was necessary not only to use the coprecipitation approach for preconcentration, but also to take advantage of hydride generation of the Se(IV) in order to separate this species from the matrix constituents (Sec. 7.4.8), it was imperative to find an "inorganic" precipitate which ultimately could be dissolved in an inorganic solvent. This was accomplished by using La(III) which at around pH 9.5, precipitates as $La(OH)_3$ and coprecipitates Se(IV) and As(III). The FIA system used is shown in Fig. 7.4-13a, in the "fill" position, where the sample, via time-based injection, is precipitated in the knotted reactor (*KR*, 100 cm Microline tubing 0.5 mm ID) by simultaneous addition of an ammonia buffer and an La(III) solution. After a sufficient amount has been collected, the injection valve is switched and the precipitate is dissolved (Fig. 7.4-13b) by a carrier of $1.0\,mol\,L^{-1}$ acid, and subsequently mixed with the tetrahydroborate solution, the selenium hydride generated ultimately being guided by means of an auxiliary Ar stream to the flow-cell of the AAS instrument.

A knotted reactor is an effective means for preconcentration via coprecipitation.

The use of hydride generation requires that the coprecipitate is soluble in an inorganic eluent.

FILL-POSITION

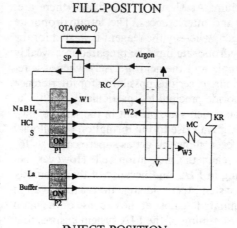

INJECT-POSITION

Fig. 7.4-13. Schematic diagram of the flow injection-hydride generation-atomic absorption spectrometry (FI-HG-AAS) system for on-line coprecipitation-dissolution of selenium or arsenic. a) The system is shown in the "fill" position, where sample is aspirated by pump P1 and mixed on-line with buffer and coprecipitating agent, La(III). The coprecipitate generated is entrapped in the knotted reactor, KR; b) The system is shown in the "inject" position, where the valve V has been switched over permitting the eluent, HCl, to pass through the knotted reactor to dissolve and elute the precipitate and forward it to mixing with the reductant, NaBH$_4$, leading to the formation of the hydride. The hydride is separated from the liquid matrix in the gas–liquid separator SP and subsequently by means of an auxiliary stream of argon gas guided to the heated quartz cell QTA of the AAS instrument
After [7.4-11] courtesy Royal Society of Chemistry

Using this approach, a lower limit of detection for Se(IV) of $0.006\,\mu g\,L^{-1}$ was achieved, and the RSD at the $0.1\,\mu g\,L^{-1}$ level was as low as 3%, while the detection limit for As(V/III) was $0.003\,\mu g\,L^{-1}$. Furthermore, 400 times excess of Ni(II) and Cu(II) could be tolerated without any significant adverse effects on the readout.

7.4.10 Exploiting the physical dispersion process: FIA gradient techniques

FIA gradient techniques rely on using one or several of the concentration elements along the concentration gradient created.

FIA gradient techniques have yielded a series of novel analytical applications.

Up to this point we have by and large confined ourselves to using the peak maximum as our analytical readout. However, any FIA peak that we create by injecting a sample necessarily contains a continuum of concentrations representing all values between zero and C^{max} (Fig. 7.4-2), irrespective of whether we intend to use only a single element of it, such as the peak maximum, to obtain our analytical result. However, because the carrier stream is noncompressible, and its movement can therefore be strictly controlled and reproduced from one measuring cycle to the next with high precision, each and every element of this concentration gradient is inherently characterized by two parameters: a fixed delay time t_i (Fig. 7.4-2), elapsed from the moment of injection t_0; and a fixed dispersion value, i.e., the dispersion coefficient of the injected sample is $D_S = C_S^0/C_S$, while that of the reagent is $D_R = C_R^0/C_R$. Therefore, it is within our power reproducibly to select and exploit any element, or elements, simply by identifying the time t_i associated with the particular element, its concentration being given by $C = C^0/D(t_i)$. This feature has formed the basis for the development of a number of FIA gradient techniques (Fig. 7.4-14), opening up a series of novel analytical applications. A compilation of various gradient techniques is presented in Table 7.4-1 [7.4-3, 7.4-12].

The stopped-flow method relies on arresting the flow of the sample–reagent mixture.

One application that deserves special emphasis is the stopped-flow method. As mentioned in Table 7.4-1, this approach is used either to stop the flow of the sample/reagent mixture in order to obtain increased reaction time without ex-

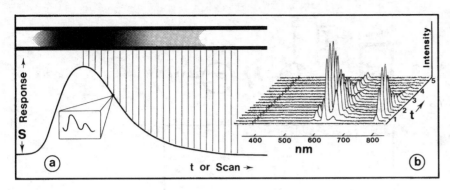

Fig. 7.4-14. Gradient scanning based on selection of readouts via delay times (each corresponding to a different sample/reagent ratio) during which a detector rapidly scans a range of wavelengths (a), thus creating an additional dimension on the time–concentration matrix (b), showing a series of successive emission spectra recorded on the ascending and descending part of a dispersed zone, containing Na, K, and Ca injected into an atomic emission spectrometer furnished with a fast scanning monochromator
From [7.4-3] courtesy John Wiley & Sons

Table 7.4-1. Examples of FIA gradient techniques

Gradient dilution
 Selecting and using for the analytical readout specific fluid elements along the concentration gradient, the concentration being $C = C^0/D(t_i)$. To be used, for instance, to accommodate the concentration of sample to the dynamic range of a detector

Gradient calibration
 Identifying and exploiting a number of elements along the gradient, the concentrations of which are given through the dispersion coefficient of the individual elements. A multipoint calibration curve can be obtained from a single injection of a concentrated sample

Stopped-flow
 Increase of sensitivity of measurement by increasing residence time, or quantifying sample concentration by measuring a reaction rate under pseudo-zero-order kinetics see Fig. 7.4-15

Gradient scanning
 Combining the use of gradient dilution with the use if a dynamic detector which, for each concentration level, is able continuously to scan a physical parameter, such as wavelength or potential (see Fig. 7.4-14)

Titration
 Identifying fluid elements on the ascending and descending parts of the concentration gradient where equivalence between titrand and titrant is obtained, and relating the time difference between these elements to the concentration of injected analyte

Penetrating zones
 Exploitation of the response curves from the concentration gradients formed when two or more zones are injected simultaneously. In addition to acting as an economical way of introducing sample and reagent solutions, it can be used for measuring selectivity coefficients, and to make standard additions over a wide, controllable range of standard/analyte concentration ratios

cessive or undue dilution (as would be the case if the increased reaction time were to be obtained by pumping through a long reaction coil), or, if the stop sequence is effected within the detector itself, to monitor the reaction as it progresses, allowing us to see it as it happens. If the conditions are manipulated to simulate pseudo-zero-order kinetics (i.e., by adding excess of reagent and assuming that the sample concentration does not change during the observation period), the reaction rate, i.e., the slope of the recorded stopped-flow curve, becomes directly proportional to the concentration of analyte (sample), as shown in Fig. 7.4-15. The strength of this approach stems from the high reproducibility with which various segments of the concentration gradient formed by the injected sample can be selected and arrested witin the observation field of the detector, and the fact that we can change the reaction conditions to meet the pseudo-zero-order reaction criteria simply by selecting the point (or points) on the dispersion profile which fulfill this condition, i.e., the time we effect the actual stop of the zone. Reaction rate measurements, where the rate of formation (or consumption) of a certain species is measured over a larger number of data points, not only improve the reproducibility of the assay, but also ensure its reliability. This benefit occurs because interfering phenomena such as blank values, existence of a lag phase, and nonlinear rate curves, may be readily identified and eliminated. Because of the inherent automatic blank control, the stopped-flow approach is an attractive option in applications such as clinical chemistry, biotechnology, and process control, where sample matrices of widely different blank values are often encountered.

Stopped-flow can either be used to obtain increased reaction time without excessive dilution, or for monitoring the chemical reaction in order to make reaction rate measurements.

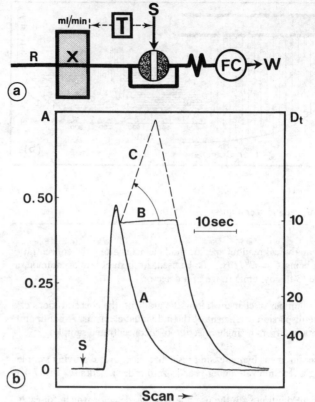

Fig. 7.4-15. a) Simple stopped-flow FIA manifold. When the sample S is injected, the electronic timer T is activated by a microswitch on the injection valve. The time from injection to stopping the pumping (delay time) and the length of the stop period can both be preset electronically; b) The principle of the stopped-flow FIA method as demonstrated by injecting a dyed sample zone into a colorless carrier stream and recording the absorbance by means of a flow-through cell: (A) Continuous pumping; (B) 9 s pumping, 14 s stop period, and continuous pumping again: (C) the dashed line indicates the curve that would have been registered if a zero- or pesudo-zero-order chemical reaction had taken place within the flow-cell during the 14 s stop interval, i.e., the slope would then be directly proportional to the concentration of analyte

From [7.4-3] courtesy John Wiley & Sons

FIA provides for novel and unique analytical procedures.

7.4.11 Concluding remarks

Although a number of FIA applications have been detailed in order to demonstrate the versatility and applicability of FIA, they are merely a fraction of those which have appeared in the literature. However, it is hoped that they may serve as inspiration for the reader. In an earlier paper [7.4-13], this author wrote that "FIA is still far from fully exploited, *because the ultimate test for an analytical approach is not that it can do better what can be cone by other means, but that it allows us to do something that we cannot do in any other way*". And FIA does allow us to make unique applications. The only limitation is simply our own ingenuity.

References

[7.4-1]　Ruzicka, J., Hansen, E.H., *Anal. Chim. Acta* 1975, 78, 145–157.

[7.4-2]　Ruzicka, J., Marshall, G.D., *Anal. Chim. Acta* 1990, 237, 329–343.

[7.4-3]　Ruzicka, J., Hansen, E.H., *Flow Injection Analysis*. 2nd ed. New York: John Wiley & Sons, 1988.

[7.4-4]　For monographs on FIA, see ref.3; Valcarcel, M., Luque de Castro, M.D., *Flow Injection Analysis: Principles and Applications*. Chichester: Ellis Horwood, 1987; Karlberg, B., Pacey, G.E., *Flow Injection Analysis. A Practical Guide*. New York: Elsevier, 1989.

[7.4-5]　Petersson, B.A. *Anal. Chim. Acta* 1988, 209, 231–237.

[7.4-6]　Schrader, W., Portala, F., Weber, D., Fang, Z.L., in *5. Colloquium Atomspektrom. Spurenanalytik*, Welz, B. (Ed.), Perkin-Elmer, 1989; pp. 375–383.

[7.4-7]　Bendtsen, A.B., Hansen, E.H., *Analyst* 1991, 116, 647–651.

[7.4-8]　Hansen, E.H., *Talanta* 1994, 41, 939–948.

[7.4-9]　Fang, Z.L., *Flow-Injection Separation and Preconcentration*. Weinheim: VCH, 1993.

[7.4-10]　Fang, Z.L., *Flow Injection Atomic Spectrometry*. New York: John Wiley & Sons, 1995.

[7.4-11]　Nielsen, S., Sloth, J.J., Hansen, E.H., *Analyst* 1996, 121, 31–35.

[7.4-12] Hansen, E.H., *Fresenius' Z. Anal. Chem.* 1988, 329, 656–659.

[7.4-13] Hansen, E.H., *Quim. Anal.* 1989, 8, 139–150.

Questions and problems

1. What is the distinctive difference between batch and continuous-flow assays?
2. What are the characteristic differences between FIA and the conventional continuous-flow approach (AutoAnalyzer)?
3. How is the dispersion coefficient D in FIA defined, and how is it determined for a given FIA system?
4. Does the dispersion coefficient yield any information on the chemical reactions involved in a given assay?
5. Why is it advantageous to operate ion-selective electrodes and biosensors in the FIA mode?
6. Many biosensors are based on the use of enzymes, because of the high selectivity offered by thè enzymes. If a linear relationship is to be obtained between the concentration of analyte and the signal response, it is a prerequisite that pseudo-first-order reaction kinetics are achieved. What does that mean? And how can such conditions be accomplished in a FIA system?
7. The combination of FIA and AAS is known to offer specific advantages. What are they?
8. What does kinetic discrimination mean, and how can it be used in FIA? Give an example.
9. What does reagents *in statu nascendi* mean, and why is FIA applicable for their use?
10. Why is the use of FIA attractive in combination with detection relying on bio- or chemiluminescence?
11. What does hydride generation imply? And why is the use of FIA in this context beneficial?
12. On-line sample preconcentration is used in FIA in order to reach lower limits of detection. Give two examples of how preconcentration can be achieved.
13. In hydride generation procedures, it is known that certain transition metals can interfere. Explain how they act in this respect.
14. What is meant by FIA gradient techniques, and what do they rely on?
15. Stopped-flow meaurements are used in FIA. What purposes do they serve?

7.5 Thermal Analysis

Learning objectives

■ To describe the basic principles and instruments for the major thermoanalytical techniques: TG, DTA, DSC, EGA, TMA, and some of their combinations.

■ To discuss the choice of parameters, limitations of the method, and sources of error for TG and DTA/DSC.

■ To guide the student in the use of thermoanalytical techniques for analytical problem solving by presenting some selected examples from materials research.

7.5.1 Introduction

Thermal analysis is defined by ICTAC (international confederation for thermal analysis and calorimetry) as "a group of techniques in which a property of the sample is monitored against time or temperature while the temperature of the sample, in a specified atmosphere, is programmed. The programme may involve heating or cooling at a fixed rate of temperature change, or holding the temperature constant, or any sequence of these." The key words here are *sample, programmed heating or cooling*, and *temperature*, but it is obvious that this broad definition encompasses methods which go beyond the conventional analytical chemistry. Nevertheless, the main thermoanalytical techniques, for instance thermogravimetry and differential thermal analysis, are commonly used to solve analytical problems since they often provide an elegant and time-saving way of doing this.

Table 7.5-1 lists the main techniques, giving their ICTAC approved names and acronyms together with the properties monitored against time or temperature. A few of these methods have already existed for some hundred years (TG, DTA, and dilatometry, for instance) while the others were more recently developed to solve specific problems in materials research. For problem solving, the thermoanalytical methods are versatile because a variety of methods are available and, furthermore, because the methods can be combined with each other to provide more information from the same sample during a single measurement (so-called simultaneous methods, where the most common combination is TG + DTA). There is also a well-developed instrument industry providing the user with a choice of modern computer-controlled thermoanalytical instruments with data processing and automated sample handling capabilities. Thermoanalytical meth-

A large variety of thermoanalytical techniques is available to meet the needs of materials characterization and analysis.

Table 7.5-1. Main thermoanalytical techniques

Property measured	Technique	Accepted acronym
Mass	Thermogravimetry or	TG
	Thermogravimetric Analysis	TGA
Volatiles	Evolved Gas Analysis	EGA
Temperature	Differential Thermal Analysis	DTA
Heat or heat flux	Differential Scanning Calorimetry	DSC
Mechanical properties	Thermomechanical Analysis	TMA
	Dynamic Mechanical Analysis	DMA
Dimensions	Thermodilatometry	
Acoustical properties	Thermosonimetry	
	Thermoacoustimetry	
Electrical properties	Thermoelectrometry	
Magnetic properties	Thermomagnetometry	
Optical properties	Thermooptometry	
Radioactive decay	Emanation Thermal Analysis	ETA

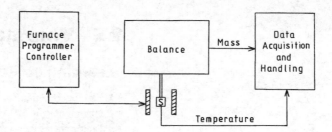

Fig. 7.5-1. Schematic representation of a thermobalance

ods nowadays find widespread use in materials' characterization and analysis as well as in quality and process control. The materials studied include polymers, pharmaceuticals, ceramics, metals, and alloys. They can also be used to study the precursors for synthesis of new materials and to help establish the optimum conditions for a synthetic process.

In the following sections, the main thermoanalytical methods including their combinations will be surveyed focusing on the techniques and problems related to analytical chemistry.

7.5.2 Thermogravimetry (TG)

Thermogravimetry (TG) is probably the most commonly employed single thermoanalytical technique.

Thermogravimetry (TG) or thermogravimetric analysis (TGA) is one of the basic methods in thermal analysis. The instrument is built around a furnace where the sample is mechanically connected to an analytical balance; hence the name *thermobalance* for a TG instrument. The thermobalance was originally developed by K. Honda in 1915 but since then the instrument has been greatly improved as regards to sensitivity, automatic recording of the Δm *vs.* T curve, and the controlling of the instrument including heating rate, atmosphere, etc.

The thermobalance

The three essential parts of a modern TG instrument are the balance, the furnace, and the instrument control/data handling system. Figure 7.5-1 gives a schematic representation of a typical thermobalance.

The balance

A sensitive and reliable analytical balance is a central part of a TG instrument and it is therefore understandable that many of the TG instrument companies are former or present balance manufacturers. Sensitivities typically in the order of 1 µg and maximum loads of 1 g are required for the balance. In most cases the actual samples in a TG experiment weigh from 10 to 50 mg. Several types of balance mechanism are possible including beam spring, cantilever, and torsion balances, but a null-point weighing mechanism is favored because then the sample always remains in the same heating zone of the furnace.

A very popular balance concept introduced in the early 60s is depicted in Fig. 7.5-2. The electromagnetic balances are relatively insensitive for vibrations and they have high sensitivity (up to 0.1 µg) and thermal stability. When a sample is added to the left the beam is displaced from the equilibrium. This is detected by a photocell system which starts an electric torque motor to restore the beam's original position. The restoring force is proportional to the change in weight and to the current consumed by the motor.

The furnace

The temperature range of a furnace used in thermobalances depends mainly on the materials used for its construction. If the range extends up to 1000–1100 °C

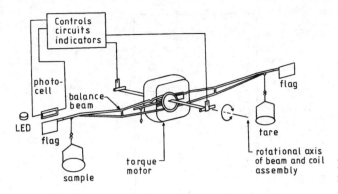

Fig. 7.5-2. Principle of an electronic microbalance with a null-point mechanism

fused quartz tubes together with Kanthal-type heating-element materials may be used, but temperatures up to 1500–1700 °C require other ceramic refractories such as alumina or mullite. Most thermobalance manufacturers offer instruments which can reach 1500 °C but only a few make instruments which can be used above these temperatures because of the material problems involving heating elements, furnace construction, as well as the thermocouples used for temperature measurement.

There are basically three ways to place the sample relative to the balance and furnace (Fig. 7.5-3), each of which have their own advantages and disadvantages. In all cases it is important that the sample is within the uniform temperature zone of the furnace and that the balance mechanism is protected from radiant heat and corrosive gases either evolving from the sample or used as reactive atmosphere. While all commercial thermobalances offer the use of an inert (nitrogen or argon) or oxidative (air or oxygen) atmosphere, only very few are designed to be used with corrosive and reactive atmospheres, e.g., chlorine and sulfur dioxide.

Instrument control and data handling

Recently, the PC has been incorporated into most commercially available instruments to take care of the heating and cooling cycles as well as data storage and handling. It can also calculate the 1st derivative of the Δm *vs.* T curve (TG); this is called the derivative thermogravimetric (DTG) curve. The DTG curve can significantly aid the interpretation of TG curves by resolving overlapping thermal reactions. Another way to resolve the reactions and approach thermodynamic equilibrium is to use isothermal heating or a very slow heating rate. In the quasi-isothermal TG (also called high-resolution or controlled-rate TG) the heating is slowed down when a weight change begins. This gives an enhanced resolution but, on the other hand, requires more time for a TG run. The loss of time can be partly compensated by setting a relatively fast rate for those regions where no changes occur.

Sources of error

Several factors influence the correctness of the measured mass and temperature. These factors arise either from the TG instrument and parameters chosen for experiment or from the sample and the atmosphere. Fortunately, most of the factors can be controlled or the data corrected for their influence.

Table 7.5-2 gives the major factors affecting a thermogravimetric curve. Most of them influence the correctness of temperature recording, and in some cases the effect can be dramatic. Figure 7.5-4 gives an example how much the combined effect of heating rate and sample size can effect the reaction temperatures. Both smaller sample size and slower heating rate lower the temperatures where a thermal reaction occurs, the combined effect can be well over 100 degrees. Temperature calibration of a TG instrument can conveniently be carried out with standards based on magnetic transition. If, for instance, a Ni-metal standard is weighed under an external magnetic field in a thermobalance it looses ferromag-

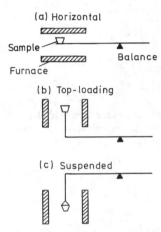

Fig. 7.5-3. The three main possibilities to place the sample relative to the balance and furnace

Table 7.5-2. Major factors affecting the recorded mass (m) and temperature (T) in thermogravimetry

---

Buoyancy (m)
Condensation and reaction (m)
Electrostatic effects (m)
Heating rate (T)
Gas flow (T)
Sample holder (T)
Reaction enthalpy (T)
Sample size and packing (T)

Several factors influence the TG curves and may cause errors.

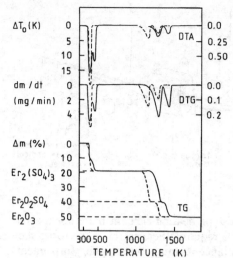

Fig. 7.5-4. TG, DTG and DTA curves for the decomposition of $Er_2(SO_4)_3 \cdot 8\,H_2O$ in air under two different experimental conditions: (I) 200 mg, 10 °C/min (solid line) and (II) 20 mg, 2 °C/min (broken line)

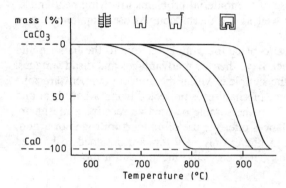

Fig. 7.5-5. Influence of the sample holder geometry on the decomposition of calcium carbonate ($CaCO_3$)

netism at its Curie point, $T_c = 353$ °C. This can clearly be seen as a jump in the Δm vs. T curve.

The sample holder may also have a tremendeous influence especially when the surrounding atmosphere is in a chemical equilibrium with the sample. A well-known example is the decomposition of calcium carbonate (Fig 7.5-5) where an open sample holder allows the generated CO_2 to be swept effectively away by the flowing gas. On the other hand, the labyrinth crucible on the right prevents the CO_2 from escaping before its partial pressure exceeds the ambient pressure (1 atm) and thus the decomposition starts first at 900 °C. The two sample holders in the middle are more open than the labyrinth crucible, but also here the effect of self-generated atmosphere on the decomposition temperatures is obvious compared to the very open structure of the first tray-type holder.

Applications of TG

One of the early applications of TG was to accurately define the conditions for the drying or ignition of analytical precipitates. Although this analytical application has lost its significance, there are still several problems for which TG can provide an answer. For instance, it can give the water content of a sample or even differentiate between the adsorbed water and constitutional water because they are usually expelled at different temperatures.

Another practical example is the proximate analysis of coal and other similar fuels (Fig. 7.5-6). If the heating is first carried out in an inert atmosphere (N_2) the amounts of moisture and volatiles can be read from the thermogram. Then at fixed temperature the thermobalance automatically switches the atmosphere into an oxidizing one whereupon the carbon is burnt and its content as well as that of ash can be read from the TG curve. The accuracy of results obtained with a TG instrument are comparable with those of the standard batch method requiring much more manual work.

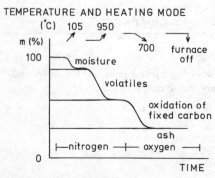

Fig. 7.5-6. Proximate analysis of coal with TG

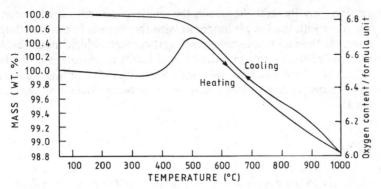

Fig. 7.5-7. TG data for $YBa_2Cu_3O_{6.5}$ showing the effect of heating (loss of oxygen) and cooling (regaining the oxygen up to $x = 6.9$)

A second example concerns the new oxide superconductors. Since their discovery in 1986–87 most of the application-oriented research has been focused on the so-called 1-2-3 compound or $YBa_2Cu_3O_{7-x}$. The oxygen content defined by x is of crucial importance for the superconduction properties. For a high critical temperature (90 K) x should be small, i.e., the oxygen content should be close to 7. The oxygen content is controlled by slow cooling in air and this can be monitored and the oxygen content determined by thermogravimetry (Fig. 7.5-7).

7.5.3 Differential thermal analysis (DTA) and differential scanning calorimetry (DSC)

While thermogravimetry is measuring the mass change of the sample upon heating or cooling, the techniques differential thermal analysis (DTA) and differential scanning calorimetry (DSC) are concerned with the measurement of *energy changes*. Both methods are closely linked with each other yielding the same kind of information. From a practical point of view the distinction lies in the operating and construction principle of the instruments: in DTA the *temperature difference* between sample and reference is measured, while in DSC the temperatures of the sample and the reference are kept identical and the difference in heating power needed for this is monitored. The classical DTA is the oldest method of thermal analysis because Le Chatelier already introduced it in 1887. Today DTA and DSC are probably the most widely applied thermoanalytical techniques.

DTA and DSC are capable for detecting energy changes when a sample is heated or cooled. The phenomena detected may be physical or chemical in nature.

Basic principles and apparatus for DTA and DSC

DTA

When a sample (S) and a reference material (R) are uniformly heated in a furnace and an endothermic effect takes place in the sample, its temperature T_S will lag behind the temperature of the reference T_R. The temperature difference $\Delta T = T_S - T_R$ is recorded against the temperature T_R which in practice equals the furnace temperature and a DTA curve is obtained (Fig. 7.5-8a). Similarly, an exothermic reaction gives rise to a peak but in the opposite direction.

The reference material should have the following properties. First, it should not

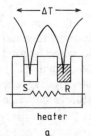

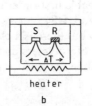

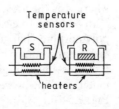

Fig. 7.5-8. Schematic diagram of the three main differential thermoanalytical techniques: (a) classical DTA, (b) Boersma-type DTA, and (c) DSC

undergo any thermal changes in the temperature range used. Secondly, it should not react with the sample holder or with the thermocouple. The third requirement concerns thermal conductivity and heat capacity which should be closely similar to those of the sample in order to avoid drift or curvature in the baseline of the DTA curve. For inorganic samples, alumina (Al_2O_3) or SiC are usually employed as reference materials, while for organic polymers silicon oil can, for instance, be used.

DSC

Contrary to the operating mode of DTA, the temperature difference between the sample and the reference is kept at zero in DSC, i.e., $\Delta T = T_S - T_R = 0$. This is achieved by independent heaters and the method is called power-compensating DSC (Fig. 7.5-8c).

In addition to the classical DTA and the power-compensating DSC, there is a third variant of DTA and DSC instruments, namely the calorimetric or Boersma DTA which is sometimes also called heat-flux DSC (Fig. 7.5-8b). In this version, as in classical DTA, temperature difference is monitored. The sample and reference are positioned on a heat-flux plate which generates a very controlled heatflow from the furnace wall to the sample and reference. Thus in Boersma DTA/heat-flux DSC the construction yields a ΔT proportional to heat-flux difference between the sample and the reference. All versions of DTA and DSC yield similar information but usually the operating temperature range of power-compensating DSC instruments is more limited (typically up to 700 °C) than that of the DTA instruments where the high-temperature versions may reach 1500 °C or higher temperatures. On the other hand, the sensitivity of instruments operating below 700 °C is higher than that of high-temperature versions and consequently the sample may be much smaller (in the order of a few milligrams); as reference, only an empty sample holder is sufficient.

Calibration and interpretation of the DTA and DSC curves

For quantitative measurements a DSC or DTA instrument has to be calibrated by standard reference materials.

A rather comprehensive set of standard reference materials (SRM) for DTA and DSC are available. The SRMs have been issued by NIST (National Institute for Standards and Technology) in cooperation with ICTAC and they cover for temperature calibration the range of −32 °C to 925 °C. The calibration is based on sharp transitions which these materials undergo at certain temperature, for instance the polymorphic transition of α-quartz to β-quartz at 573 °C. Enthalpy (and simultaneous temperature) calibration is based on melting enthalpies of high-purity metals. The metals include indium, tin, lead, zinc and aluminium covering the temperature range of 156 to 660 °C or the typical operating range of an DSC instrument. For higher temperatures, silver (m.p. 961 °C) or gold (m.p. 1064 °C) may be used. DSC can also be used to determine the heat capacity C_p of a sample but a standard, usually sapphire (single crystal Al_2O_3), is also required.

In quantitative determinations a problem arises from the integration of the peak area which depends on the baseline definition. Usually the computer programs of a DSC and DTA instrument include several possibilities for baseline extrapolation (Fig. 7.5-9), but the user must make an appropriate choice. The measured enthalpy change $\Delta H(\text{J/g})$ is directly proportional to the peak area

$$\Delta H = Ak/m \qquad (7.5-1)$$

where A is the area, k is the calibration factor, and m is the mass of the sample.

In addition to clearly distinguishable first-order transitions, such as melting and polymorphic changes, the DTA and DSC curves can be used to detect second order transitions where ΔH is zero but there is a change in the heat capacity (C_p). Examples include glass transition of polymers and Curie point of ferromagnetic materials. The procedure to determine the glass transition temperature is shown in Fig. 7.5-10.

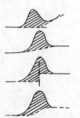

Fig. 7.5-9. Some possible alternatives to define the baseline and peak area in DTA/DSC

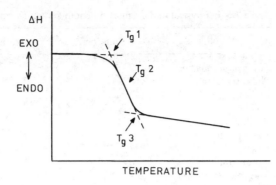

Fig. 7.5-10. The glass transition temperature (T_g2) can be determined as midpoint temperature between the onset (T_g1) and endset (T_g3) temperatures

Table 7.5-3. The influence of sample and operational parameters on resolution and sensitivity

Parameter	Maximum resolution	Maximum sensitivity
Sample size	small	large
Particle size	small	large
Sample packing	dense	loose
Heating rate	slow	fast
Atmosphere	high conductivity	low conductivity

As in thermogravimetry, the operational parameters influence the DTA and DSC curves. For a proper interpretation of the thermal effects, the key factors to be considered are the resolution of the peaks and sensitivity. Table 7.5-3 gives a simplified picture of how the various parameters influence the curves. In addition, the instrument design has a distinct influence. Maximum resolution is achieved when the sample and reference holders are linked together in a block-type arrangement while isolated holders lead to higher sensitivity. Sample cup (open, closed, or closed with a pinhole) has an effect too.

Applications of DTA and DSC

DTA and DSC have a very wide range of applications with respect to the type of materials as well as physical and chemical phenomena which can be studied. Table 7.5-4 lists the main application areas and in the following a few selected examples will be discussed. These include problems from both qualitative and quantitative analysis of materials.

Purity analysis

One of the most frequent uses of DSC in organic and pharmaceutical industries is the purity analysis. This is based on the van't Hoff equation:

$$T_0 - T_m = \frac{RT_0^2}{\Delta H} x \qquad (7.5\text{-}2)$$

where T_0 and T_m are the melting points of the pure and unpure substance, respectively; ΔH is the melting enthalpy of the pure substance while x is the mole fraction of the impurity. The van't Hoff equation is based on the assumption of an ideal entectic system; solid solution or compound formation between major material and impurity should be excluded.

Based on a single standardized DSC run the computer program of the instrument can calculate the degree of purity with a satisfactory accuracy. Qualitatively, the degree of purity using the van't Hoff equation is visible in the shape of the melting curve (Fig. 7.5-11).

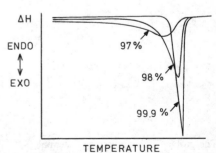

Fig. 7.5-11. The purity of a sample is reflected in the shape and position of its melting curve. DSC melting curves of samples with 97, 98.5 and 99.9% purity (from left to right)

Table 7.5-4. Physical and chemical phenomena which can be detected by DTA/DSC

	Endothermic		Exothermic	Not detectable with TG
Physical phenomena				
Phase transitions	X		X	X
Melting	X			X
Boiling	X			
Sublimation	X			
Adsorption			X	
Desorption	X			
Absorption			X	
Magnetic transition[a]				X
Glassy transition[a]				X
Heat capacity change[a]				X
Chemical phenomena				
Chemisorption			X	
Decomposition	X	or	X	
Oxidation	X	or	X	
Reduction	X	or	X	
Burning			X	
Polymerization			X	X
Polycondensation			X	
Solid state reactions	X	or	X	X
Catalytic reactions			X	X

[a] 2nd order transition, only the heat capacity changes

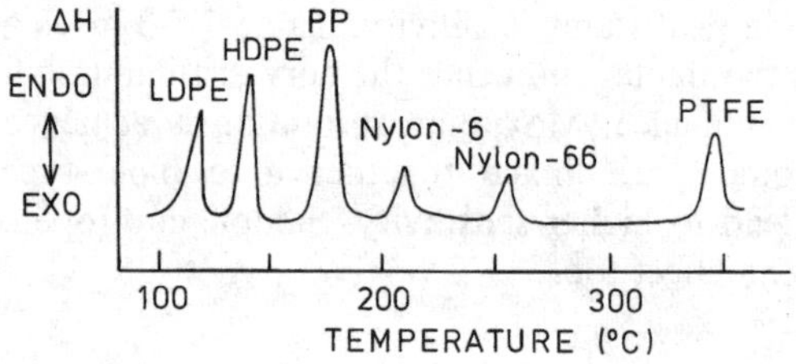

Fig. 7.5-12. DSC curve of plastic waste. LDPE and HPDE: low and high density polyethylene, respectively, PP: polypropylene, PTFE: polytetrafluoroethylene (teflon)

Other applications

DTA/DSC can be used as a fingerprint technique for qualitative analysis. Fig. 7.5-12 shows a DSC analysis of plastic waste where six different polymers can be recognized on the basis of their melting or transition temperatures.

Figure 7.5-13 shows an example of quantitative analysis. Blast furnace slag is obtained in large quantities as a side product during iron and steel production. For some of its applications it is necessary to know its degree of crystallinity. This can be determined by DTA by measuring the area of the exothermic peak above 700 °C which is due to crystallization of the glassy part of the slag (Fig. 7.5-13a). The relationship between the peak area and the degree of noncrystallinity (glass) as determined by a reference method (optical microscopy) is linear (Fig. 7.5-13b).

7.5.4 Combined techniques

Simultaneous techniques

The problem solving power of thermal analysis is often enhanced by using simultaneous methods rather than a single technique.

Quite often it is found out that information provided by a single TA technique (for instance, TG) is not sufficient to solve a particular problem. Complementary information must then be sought after by carrying out an experiment in another instrument (for instance, DTA). This is time-consuming and differences in instrument constructions make it difficult to compare the two sets of data with each other and to draw firm conclusions.

(a)

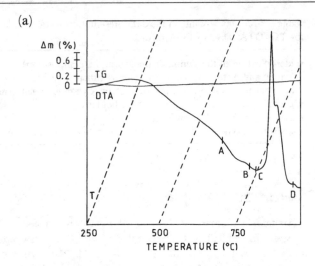

(b)

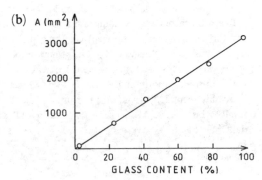

Fig. 7.5-13. (a) TG and DTA curves of blast furnace slag showing only an exothermic transition due to crystallization above 700 °C and (b) the dependence of the DTA peak area on the degree of noncrystallinity (glass content)

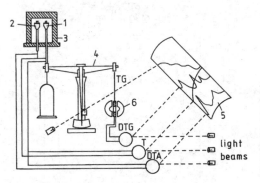

Fig. 7.5-14. The principle of a simultaneous TG-DTA (and DTG) instrument from the late 1950s. *1, 2* sample and reference holders, respectively, *3* furnace, *4* balance, *5* photographic recording. Note the production of DTG signal by a coil moving inside a permanent magnet (*6*)

Because TG and DTA (or DSC) together provide complementary information the advantages of carrying out simultaneous measurements on the same sample are obvious. The technical problem of combining TG and DTA was first solved in 1955 by F. and J. Paulik together with L. Erdey in a rather elegant way, considering then available electronic components (Fig. 7.5-14). This pioneering construction was soon followed by several other designs and now a number of companies are producing simultaneous TG-DTA instruments. Typically the instruments have a temperature range up to 1500 °C and a capability for deriving and plotting the DTG curve from TG data.

The DTA part in a combined TG-DTA instrument is a compromise, however, in comparison to a dedicated DTA/DSC instrument and therefore a combined instrument is seldom used for true quantitative measurements. Nevertheless, it provides additional information which is often crucial for the correct interpretation of the thermal degradation processes. For a combined TG-DTA instrument the same factors which influence the separate TG and DTA curves should be taken into account and therefore it is important to know and report the experimental conditions properly (Table 7.5-5).

Table 7.5-5. Major points in the data presentation recommendations by ICTAC and IUPAC for TG, DTA, DSC or EGA record

- identification of the sample (name, formula, composition)
- source and history of the sample (pretreatments, etc.)
- atmosphere (composition, pressure; static, dynamic or self-generated)
- geometry and material of the sample holder
- sample weight and its packing
- instrument type, heating rate, and temperature programme used.

Worked example:

An unknown compound (X) was analyzed by simultaneous TG/DTA in nitrogen. The final residue corresponding to a weight loss of 72.5%, as measured from the TG curve, was identified as CoO by X-ray diffraction using the JCPDS-reference file. Additional information obtained by EGA-MS revealed that in the first two steps ($1 \Rightarrow 2$ and $2 \Rightarrow 3$) only water was expelled while in the step $3 \Rightarrow 4$ SO_2 and SO_3 were evolved in a molar ratio $1:2$.

Based on these data the composition of the unknown compound can now be calculated and the reaction mechanism resolved. First, the composition of X can be calculated based on the total weight loss and EGA information (X must contain water, sulfur and oxygen). A reasonably close fit between the calculated (73.3%) and observed (72.5%) weight losses indicates the composition $X = CoSO_4 \cdot 7 H_2O$ (see the table below). Steps $1 \Rightarrow 3$ can now be readily identified as loss of water in two stages. The formation of Co_3O_4 in step $3 \Rightarrow 4$ results from the EGA data ($SO_2:SO_3$ ratio $1:2$) and is corroborated by the weight loss data.

Reaction	Total weight loss (%)	
	observed	calculated
$1 \to 3$ $CoSO_4 \cdot 7 H_2O \to CoSO_4 + 7 H_2O$	45.0	44.8
$3 \to 4$ $CoSO_4 \to \frac{1}{3} Co_3O_4 + \frac{1}{3} SO_4 + \frac{2}{3} SO_3$	70.0	70.6
$4 \to 5$ $\frac{1}{3} Co_3O_4 \to CoO + \frac{2}{3} O_2$	72.5	73.3

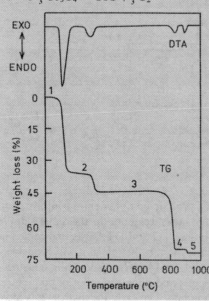

Another simultaneous measurement often connected to TG or DTA is the evolved gas analysis (EGA) to be discussed in the next chapter. DTA has also been combined to hot-stage microscopy in a commercially available instrument.

Figure 7.5-15 shows an example of the use of a simultaneous TG-DTG-DTA measurement to find out the optimum reaction conditions for the preparation of europium-activated yttrium oxosulfide $Eu:Y_2O_2S$ which is the red phosphor in color TV. The synthesis is started from coprecipitated Eu and Y sulfites in a molar ratio $6:100$ which are then reduced by carbon monoxide. The TG curve shows the

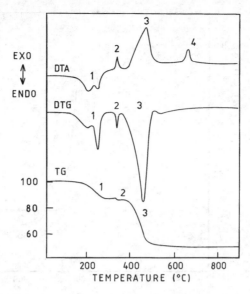

Fig. 7.5-15. Simultaneously recorded TG, DTG and DTA curves for the reduction of coprecipitated sulfites $(Ln, Ln')_2(SO_3)_3 \cdot 3\,H_2O$, Ln = Y, Ln' = Eu (6%)

dehydration (1) and the reduction of Eu- (2) and Y-sulfites (3). Both DTG and DTA curves indicate that the dehydration takes place in two steps but only the DTA curve is able to show a phase transition (crystallization) of the oxosulfide (4) at 650 °C. The overall weight loss (46.6%) is smaller than the theoretical value for the reduction (47.9%) indicating that minor side reactions occur.

Ex situ combinations

In actual problem solving with TA, it is often necessary to identify the intermediates and thermally-induced processes by ex situ measurements of samples taken in the middle or at the end of the processes. All forms of spectroscopic and diffraction techniques may be used for this purpose but the most common and powerful methods are probably X-ray powder diffraction (see 7.5-6) and Fourier Transform Infrared Spectroscopy (FTIR).

An example of the use of Mössbauer spectroscopy to identify reaction intermediates and end products is given in Fig. 7.5-16. Ferrous sulfate is produced as a by-product in excess of demand in several industrial processes. Thermal analysis has been used to study its conversion to more valuable products. In a reducing carbon monoxide atmosphere the reaction mechanism is complicated and the products can not be identified on the basis of TG curve only. Fig. 7.5-16a shows the TG curves recorded at two heating rates and the points where samples $(a-d)$ were taken for a Mössbauer analysis (Fig. 7.5-16b). The positions of the Mössbauer peaks and their splittings can then be used to unequivocally identify the various iron-containing phases. From the TG curve it is already obvious that sample (a) is anhydrous $FeSO_4$ which is also the dominant phase in sample (b). In the next step (c) two ferromagnetic phases (FeS and Fe) are evident from the Mössbauer spectra. The end product (d) contains three phases: Fe, FeS and Fe_3C. The formation of cementite (Fe_3C) from the CO atmosphere $(2\,CO \rightleftharpoons CO_2 + C)$ explains the slight weight gain at higher temperatures.

7.5.5 Evolved gas analysis (EGA)

Most thermally induced reactions at higher temperatures involve volatile species evolving from the sample. Because TG can only give the overall mass loss and DTA/DSC the enthalpy change, it is often difficult to interpret the reaction mechanisms. This is the case especially when overlapping reactions occur or when there are several gaseous species with nearly the same molecular weight, for instance H_2O and NH_3.

Fig. 7.5-16. (a) Thermal decomposition of $FeSO_4 \cdot H_2O$ in carbon monoxide. The calculated (theoretical) weight levels are given for the possible products. The (a–d) letters refer to samples taken for Mössbauer experiments shown in Fig. 7.5-16b

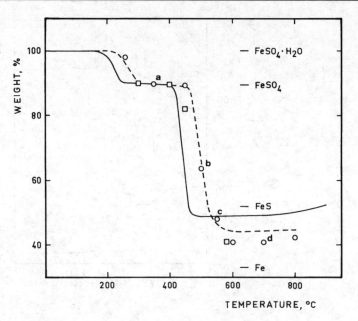

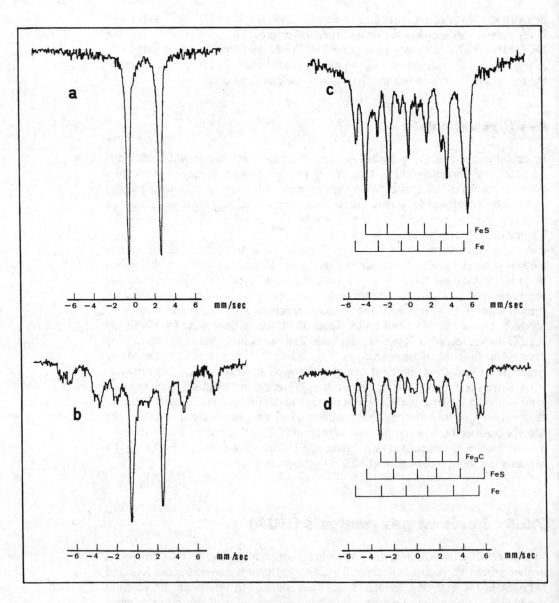

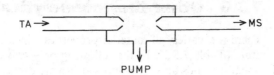

Fig. 7.5-17. Principle of a jet separator

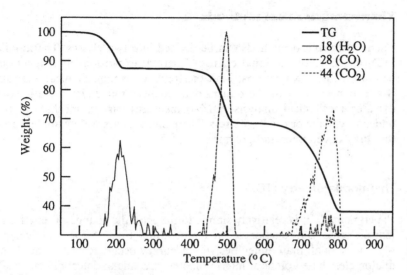

Fig. 7.5-18. A simultaneous TG-MS-EGA for $CaC_2O_4 \cdot H_2O$

Evolved gas analysis (EGA) is a gas-analytical technique which can be used in connection with a TA measurement (most commonly with TG). Several methods can be used for EGA ranging from simple gas detection methods (evolved gas detection, EGD) to the more elaborate ones based on gas chromatography (GC), Fourier transform infrared (FTIR) or mass spectrometry (MS). EGA methods based on GC were already described in the 1960s but due to its inherent slowness gas chromatography is nowadays less widely used than FTIR or MS.

Mass spectrometry offers several advantages over EGA including, besides high sensitivity, the possibility for unequivocal, simultaneous and fast detection of several gaseous species within the available mass range of the instrument. In addition to the problem of representative sampling, the crucial problem is on how to construct the interface between the MS and the TA instrument, in particular a TG analyzer.

Because the quadrupole and time-of-flight MS instruments operate under high vacuum (10^{-8}–10^{-10} atm) only two choices are left for the TG-MS interface: (a) operating the thermobalance only under vacuum and coupling it directly to MS, or (b) letting the thermobalance to work at atmospheric pressure and reducing the pressure through the interface. Obviously the second alternative is the desired one for TA experiments and several interface constructions have been reported for an atmospheric pressure TG-MS coupling. These include heated capillary, double orifice and jet separator systems (Fig. 7.5-17). Each system has its advantages and disadvantages; for instance, the heated capillary is probably the simplest system to construct but the capillary easily gets clogged when the evolved species condensate. Commercially available instruments are based either on heated capillary or double orifice interface. As regards the mass spectrometer, its limitations are mainly in the mass range if a quadrupole type instrument is used. An inexpensive quadrupole MS instrument usually covers the mass range up to 200–400 which is sufficient for inorganic species (Fig. 7.5-18) but not for organic high molecular weight fragments.

FTIR, on the other hand, does not require an interface to reduce the pressure, but a simple heated tubing is sufficient to connect the FTIR gas cell with the TA instrument. Difficulties with FTIR are connected with the fact that it cannot detect homopolar molecules such as O_2 and N_2 nor differentiate between structurally very closely related molecules such as hydrocarbon homologs.

Evolved gas analysis (EGA) by MS or FTIR is necessary to resolve complex reaction mechanisms involving gaseous species.

7.5.6 Other thermoanalytical techniques

There is a variety of material properties which can be monitored against temperature (Table 7.5-1). In the following only two of the most frequently used additional methods and their applications are discussed.

Thermomechanical methods

Thermomechanical methods are applied for the study of polymers and ceramics.

Thermomechanical methods can be divided into two classes. Thermodilatometry (TD) measures dimensional changes (thermal expansion) as a function of temperature without an external load or stress. Thermomechanical analysis (TMA) is also concerned with the dimensional changes but under a static stress, while dynamic mechanical analysis (DMA) measures various mechanical parameters under a dynamic or oscillary stress. These methods are widely applied especially in the study of ceramics and polymers.

Thermodilatometry (TD)

Experimentally it is relatively simple to measure the change of length in a particular direction as a function of the temperature (Fig. 7.5-19). Only when the sample is very small may difficulties with accuracy occur. However, modern thermodilatometers based on laser interferometers are able to detect dimensional changes well below 1 µm. The determination of thermal expansion in various directions gives the extent of anisotropy and in this respect it is more informative than the determination of the change in volume only.

Thermal expansion data are useful, for instance, for determining whether or not a substrate and thin film on it are compatible. If the difference as a function of temperature is too large, then the film will crack or peel off. In addition, TD can be used to detect phase transformations.

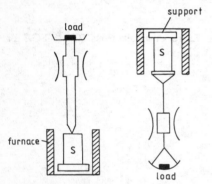

Fig. 7.5-19. Schematic representation of two TMA measurements with the (a) penetration and (b) extension mode

Thermomechanical analysis (TMA) and dynamic mechanical analysis (DMA)

Unlike TD measurements, in TMA and DMA a significant stress is applied on the sample. TMA is actually an extension of the TD technique where a load can be used to compress or extend the sample, for instance. DMA requires a more complicated experimental setup because the varying stress is applied periodically (sinusoidally).

Both TMA and DMA can be applied to study solid polymers, films and fibres as well as viscous fluids and gels. Applicational examples include the determination of glass transition temperature of a polymer and dynamic modulus of a fibre (DMA).

High-temperature X-ray diffraction

X-ray diffraction (XRD) is a powerful technique for the identification of crystalline phases and their mixtures. Due to the availability of computer-accessible files of XRD data the identification of phases is often straightforward provided that the phases are crystalline and their concentration is above the detection limit of the method (1–2%).

XRD can be used as an *ex situ* technique for the identification of intermediates and end products in a thermal process. This is a popular approach because it does not require any special instrumentation; a normal powder diffractometer will do. For the identification of the crystalline phases a powder diffraction file either on cards or as a computer file is necessary. Naturally a reference file can be built up by the user himself but in most cases it is more convenient to consult the JCPDS file (joint committee for powder diffraction standards).

X-ray diffraction measurements can be carried out either *ex situ* or *in situ*.

In situ high-temperature XRD is a true thermoanalytical method because the changes in the crystal structure of the sample can be monitored while the sample

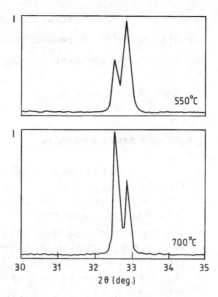

Fig. 7.5-20. A high-temperature XRD record showing how the orthorhombic (above) $YBa_2Cu_3O_{7-x}$ changes into the tetragonal phase when the temperature exceeds 600 °C

holder is heated using a temperature programme. Several high-temperature furnace constructions are commercially available. The highest achievable temperature may be as high as 2500 °C but then the instrument must operate in vacuum and only refractory metals such as molybdenum (m.p. 2625 °C) or tungsten (m.p. 3410 °C) can be used as materials. Normally the temperature range extends to 1400–1500 °C allowing the use of platinum (m.p. 1770 °C) even in an oxidizing atmosphere.

The applications of high-temperature XRD include the detecting of phase transitions, monitoring solid state reactions and measuring thermal expansion by determining the lattice constants as a function of temperature.

While a normal detector in a powder diffractometer scans over the measured 2Θ-range a position-sensitive detector (PSD) can measure simultaneously a broader range (5–120 degrees in 2Θ). The PSDs have become very popular in high-temperature work because fast changes can be followed as a function of temperature. A good example is the orthorhombic-to-tetragonal transition in the superconducting $YBa_2Cu_3O_{7-x}$ (so-called 123 compound) which occurs around 600 °C (Fig. 7.5-20). The orthorhombic phase is superconducting with a critical temperature $T_c = 90$ K.

High-temperature XRD has been combined with DSC and this combination is even more powerful for detecting and identifying phase changes than either one of these techniques alone.

General reading

Brown, M.E., *Introduction to Thermal Analysis, Techniques and Applications*, London: Chapman and Hall, 1988.

Charsley, E.L., Warrington, S.B., (Eds.), *Thermal Analysis – Techniques and Applications*. London: Royal Society of Chemistry, 1992.

Chung, D.D.L., De Haven, P.W., Arnold, H., Ghosh, D., *X-ray Diffraction at Elevated Temperatures. A Method for in situ Process Analysis*. New York: VCH Publishers, Inc., 1993.

Gallagher, P.K., *Thermoanalytical Methods*, Ch. 7 in Cahn, R.W., Haasen, P., Krämer E.J., (Eds)., *Materials Science and Technology – A Comprehensive Treatment*, Vol 2A, Weinheim: VCH Publishers Inc., 1992.

Hatakeyama, T., Quinn, F.X., *Thermal Analysis. Fundamentals and Applications to Polymer Science*, Chichester: John Wiley & Sons, 1994.

Hemminger, W.F., Cammenga, H.K., *Methoden der Thermischen Analyse*, Berlin: Springer-Verlag, 1989.

Hill, J.O., (Ed.), *For Better Thermal Analysis*. 3rd ed. London: ICTA, 1991.

Keatch, C.J., Dollimore, D., *An Introduction into Thermogravimetry*. 2nd ed. London: Heyden and Sons, 1975.

Van der Plaats, G., *The Practice of Thermal Analysis*, Switzerland: Mettler, 1991.

Pope, M.I., Judd, M.D., *Differential Thermal Analysis*, London: Heyden and Sons, 1977.

Turi, E.A., *Thermal Characterization of Polymeric Materials*, New York: Academic Press, 1981.

Wendlandt, W.W., *Thermal Analysis*. 3rd ed. New York: John Wiley and Sons, 1986.

Questions and problems

1. Which parameters effect the TG curve and how?
2. Describe the ways used to enhance the resolution of TG data.
3. What are the physical and chemical phenomena which can be detected by DTA/DSC but not by TG?
4. Which choice of parameters leads to maximum resolution in DTA/DSC?
5. What is the recommended information which should accompany the TG or DTA/DSC curve in order to facilitate its interpretation?
6. How is the temperature scale calibrated in (a) TG and (b) DTA/DSC runs?
7. What is the difference between first and second order transition from the point of view of DSC/DTA? Give an example of a second order transition detectable by DTA/DSC?
8. What are the usual simultaneous combinations of TA techniques and what advantages do they offer as compared to separate techniques?
9. How can the EGA analysis be carried out? What are the difficulties and advantages of the various approaches?
10. Write the chemical equations for the thermal decomposition of calcium oxalate using the MS-EGA data depicted in Fig. 7.5-18. Calculate also the theoretical weight losses for each step and compare them with the observed weight losses.
11. How do the TD, TMA and DMA techniques differ in principle from each other?
12. How could you additionally characterize (*ex situ*) the final residue which you have obtained from a TG run if the original sample is (a) an inorganic salt (b) an organic polymer?
13. List some application areas of (a) TG and (b) DTA/DSC where these thermoanalytical methods are competitive or superior over other analytical techniques. Try also to explain why this is so.

7.6.1 Introduction

The aim of elemental organic analysis in general is the determination of the elemental composition of organic compounds. Major constituents of organic substances are C, H, N, S and O besides halogens, P and metals. A complete knowledge of the percentage of the elements in a compound in addition to its molecular weight makes the calculation of the molecular sum formula possible. Therefore techniques for the determination of the percentage of the most frequent constituents of organic compounds, C, H, N and O in milligram amounts of organic samples have been developed already early in the 20th century. It is remarkable, that for that purpose new microbalances had to be developed first, in order to meet the high sensitivity requirements for weighing in the samples for that purpose, namely 0.1 to 3 mg samples.

The principle of the CHN-technique, as developed by Fritz Pregl in Innsbruck and Graz (Nobel Prize 1923) is also still valid in today's modern instrumental techniques of Elemental Organic Analysis. It consists of disintegration of a precisely weighed amount of a sample by rapid combustion (oxidation) in an oxygen stream at high temperatures (1000 °C to 1800 °C), of separation of the reaction products (what are they, if the sample consists of C, H, N only?) and of subsequent detection of the minute amounts of the gases produced by the combustion as well as a final calculation step.

In order to determine C, H, N and O the sample has to be disintegrated first.

At the given reaction conditions for combustion (see below), organic compounds disintegrate to form CO_2, H_2O, N_2 and NO_2. Oxygen has to be determined in an additional sample after a combustion step in Helium, resulting in CO, which can be converted into CO_2 and determined separately.

7.6.2 Methods based on selective adsorption for separation

CH-Analyzer, historic system F. Pregl (see Fig. 7.6-1):

The catalytically supported combustion (sample in a Pt-vial) took place at 750 °C in a constant oxygen stream. As oxidation support Pregl proposed CuO/PbCrO$_4$ at 750 °C and PbO_2 at 190 °C (why?). The resulting combustion products H_2O and CO_2 were then selectively absorbed in preweighed tubes containing $Mg(ClO_4)_2$ and NaOH on asbestos (!) respectively. The measurement of the C- and H-content of the sample was based on the tedious gravimetric technique. The also produced gases NO_x and N_2 were not detected at that time (why?). Because of the experimental constraints, the technique has been replaced by modern schemes allowing automized determinations.

CHN-Analyzer, system W. Simon (see Fig. 7.6-2):

In this first commercial version of an automatic elemental organic analyzer (1963), Pregl's combustion and adsorption principles were used in a refined

Disintegration in O_2-atmosphere for C, H, N analysis.

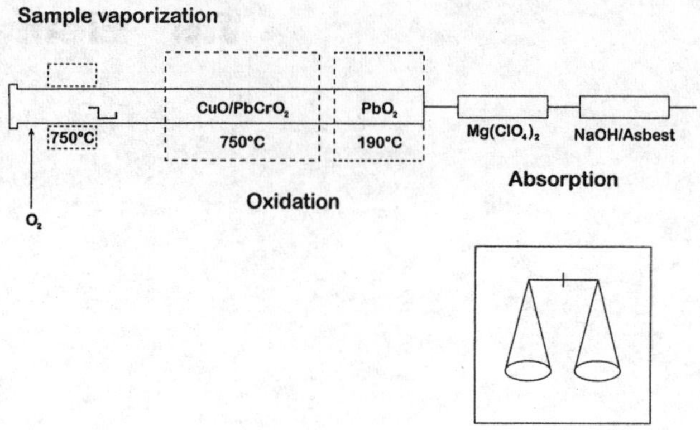

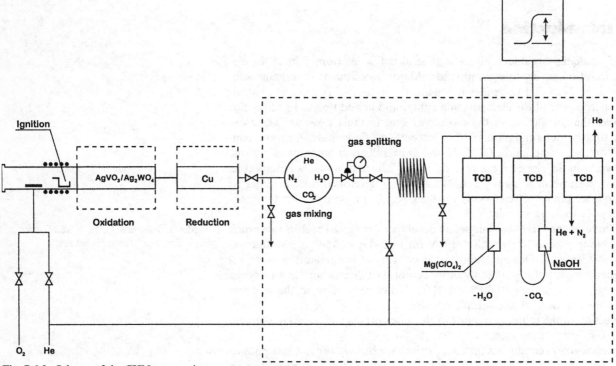

Fig. 7.6-1. Scheme of the first apparatus for simultaneous C- and H-determination according to F. Pregl

Fig. 7.6-2. Scheme of the CHN-automatic analyzer according to W. Simon

version, but the gravimetrical determination was replaced by differential measurement of the thermal conductivity of the reaction gases.

Combustion: At ca. 1000 °C in oxygen (1–3 mg of organic samples in a Pt-vial), with $AgVO_3/AgWO_4$ as combustion aid), the eventually formed NO_x is reduced in the subsequent reduction unit (Cu at 650 °C) to N_2 and excess of oxygen is bound as CuO. After the combustion is complete (several seconds) the O_2 is turned off and replaced by a He-carrier stream. Halogens can be removed by adsorption on Ag-wool, SO_2 by oxidation with MnO_2.

Adsorption and detection of reaction gases: As can be seen in Fig. 7.6.2, the gas mixture consists after the combustion/reduction unit of H_2O, CO_2, N_2 and He. By subsequently removing H_2O and then CO_2 using Pregl's method the amount of the reaction gases can be determined by 3 dedicated *TCD*'s (*Thermal Conductivity Detector*). The total procedure takes 14 minutes.

CHNS + O-Analyzer, System E. Pella (see Fig. 7.6-3):

This modern analyzer works with a modified combustion system and a completely different separation unit for the reaction gases based on gas chromatography.

The detection is performed by TCD's as in the Simon version above. The Pella-system allows for the simultaneous determination of C, H, N and S in one single microsample and of O in a different sample, under heating in He, but using the same GC separation principle.

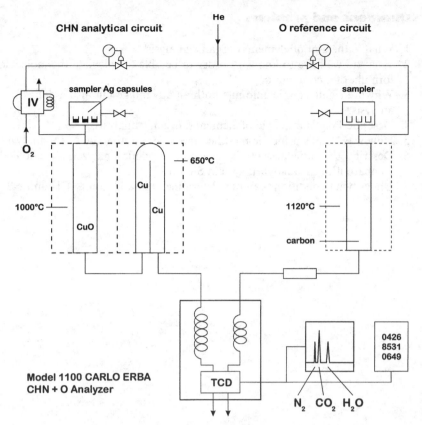

Fig. 7.6-3. Scheme of the CHNS + O automatic analyzer according to E. Pella

Combustion: The sample (typically 1 mg) is introduced into the combustion unit in a Sn-capsule. A local combustion temperature of 1800 °C is obtained by strong exothermic effects, when burning the sample in the Sn-capsule in oxygen at an oven temperature of 1000 °C. This *dynamic flash combustion* guarantees a complete disintegration also of halogenated organic samples and other, highly heat resistive substances, such as organometallic or even inorganic compounds. CuO or WO_3 are used in the same unit as combustion aid. The reduction of NO_x to N_2 and of SO_3 to SO_2 are performed in the reduction tube by Cu at 650 °C.

Separation and detection: The reaction gas mixture is then separated by gas chromatography, using a specially developed solid phase, Poropak QS in a packed column. The reaction gases are eluted in the sequence N_2, CO_2, H_2O and SO_2 and quantitatively determined by a *TCD*. CHNS can be determined simultaneously in one sample of ca. 1 mg in approximately 10 minutes.

Oxygen determination

The oxygen content of organic samples is important information, especially in today's modern fuels, where oxygenated compounds are used to replace lead as anti-knocking agent. Automatic liquid injection devices have been developed for such applications and allow liquids to be injected directly into the pyrolysis chamber without weighing. Oxygen analysis is performed by heating organic samples in He at 1120 °C, hereby producing CO, N_2, H_2, CH_4 and acid gases (see Fig. 7.6-3, right part). The gas mixture is subsequently passed through nickelized carbon as efficient reduction catalyst (why is efficiency so important here?) and an alkaline scrubber, to remove the acidic gases. The remaining gases are injected into a GC-column packed with a molecular sieve as stationary phase. As detector a TCD is used as before. From the CO-measurement and the known sample weight, the quantitation of the oxygen content of the sample becomes possible in the usual way. A complete oxygen analysis takes 5 minutes.

Disintegration in He-atmosphere for O-analysis.

Questions and problems

1. What is the aim of elemental organic analysis?
2. What do we have to know, in order to be able to calculate the molecular sum formula of a compound?
3. Why are small sample amounts both an advantage but also a difficulty for the analysis?
4. Describe Pregl's principle of elemental organic analysis!
5. How was Pregl's principle modified in modern automatic CHN-analyzers?
6. Describe the principles of the elemental organic analyzer, system Pella and compare it to the analyzer, system Simon!
7. Under what conditions can one determine O in addition to CH and N?

7.7.1 Introduction

The development of chemical sensors is important with respect to several objectives. A first aim is the substitution of classical analytical methods by sensor measurements. A typical example here is the systematic replacement of flame photometers for determination of blood electrolytes, such as lithium, sodium, potassium, and calcium, by ion-selective electrodes.

A second objective results from the increasing requirements for automation of industrial processes. Measurement of chemical parameters is needed for monitoring and controlling a process. Thus, in process analytical chemistry (Chap. 16) the methods for on-line analysis are based on the availability of chemical sensors. Special requirements in industry are robustness, long-term stability, or temperature insensitivity of the sensors. Additional needs arise in connection with biotechnological processes where multifunctional, biocompatible sensors are required.

Monitoring chemical quantities is also important in the environment, in industrial plants, in safety techniques and in medicine. Monitoring of pH and turbidity of *water* is a relatively easy task. pH can be measured by a glass electrode, and turbidity by means of an optical sensor. Conceivably, the pollution of water by heavy metals and organic compounds may be monitored continuously by sensors rather than by off-line methods in the laboratory. Chemical sensors will be needed not only for individual analytes but also for measurements of sum parameters.

The control of the emission of flue gases or the monitoring of air contaminants and emissions of SO_2 are important requirements in the *environmental field*. Gas sensors based on optical or microelectronic principles are already available. Continuous monitoring of contaminants in soils or solid waste is more difficult. For this purpose, sensors are needed that provide chemical information by remote measurements.

For monitoring of exposure to hazardous substances, inexpensive sensors are needed that can be mass-produced. This kind of sensor is expected to detect long-term exposition or maximum loads. In the mining industry, explosive or hazardous gases need to be determined, while for air-conditioning, the humidity and CO are important.

Another area is *in-situ* monitoring. For example, during a heart operation, the ratio of potassium to sodium needs to be measured since it is an important indicator for the patient's condition. The time-dependent monitoring of drugs, anesthetics, and metabolites in the body during the course of an operation is another active field of sensor application in medicine. There is a growing demand for sensors that can be used as noninvasive devices, e.g., for determination of glucose in blood by recording the NIR spectrum through the finger nail or skin.

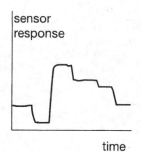

sensor response

time

7.7.2 Characteristics and basic principles

Chemical sensors are defined as follows:

Chemical sensors are (small) devices capable of continuously recognizing concentrations of chemical constituents in liquids or gases and converting this information in real-time to an electrical or optical signal.

Table 7.7-1. Typical parameters for chemical sensors

Requirement
detection limit
working range
selectivity/specificity
drift
reproducibility
miniaturization
mechanical stability
response time
long-term stability
compatability to pressure, temperature, explosiveness, radioactivity, biological conditions, sterilization

Table 7.7-2. Overview of sensing schemes and types of sensors

Sensing scheme	Sensor type
Conductivity change	Metal oxide semiconductors
	Organic semiconductors
Potential change	Ion-selective electrodes
	Solid-state gas sensors
	Field-effect transistors
Current change	Amperometric sensors (oxygen and enzyme electrodes, immunosensors, solid state gas probes)
Change of resonance frequency	Piezoelectric crystal resonators
	Surface acoustic wave sensors
Change of optical properties	Transmission/absorption, scatter, reflectance, fluorescence, refraction, decay time, polarization
Thermal effects	Thermal/calorimetric sensors
	Pellistors

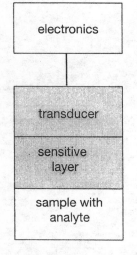

Schematic sensor

In principle, a sensor consists of a chemically sensitive layer – the recognition system – a transducer for converting the chemical information into an electrical or optical signal, and data evaluation electronics, usually integrated into the sensor. As an example, for an ion-selective electrode (Sec. 7.3.1) the chemical-sensitive layer is a solid-state or liquid membrane and the transducer is based on the electrical potential, evaluated by means of a voltmeter.

The performance characteristics of a sensor are derived from general analytical criteria as well as from specific requirements, such as long-term stability. Table 7.7-1 lists typical parameters that characterize a sensor.

No universal basic principle exists for sensor development. Every practical task requires a specific adaption of the measuring principle, so that sensors are known which are based on electrical, optical, gravimetric, or thermal principles.

An overview of basic principles for developing chemical sensors is given in Table 7.7-2. Many of these principles are introduced elsewhere in this volume. This section concentrates on specific aspects of sensor construction, and new principles, such as microelectronic or fiber optical sensors are explained in more detail.

7.7.3 Electrochemical and microelectronic sensors

Potentiometric sensors

Sensors based on potentiometry are best represented by the pH glass electrode and ion-selective electrodes (ISEs) as discussed in Sec. 7.3.1. Potentiometric measurements can also be used to develop sensors for determination of gases, e.g., carbon monoxide or ammonia, and of organic compounds, such as urea.

Solid state gas sensors

Solids show conductivity at elevated temperatures due to the presence of ions. This effect can be used to construct a gas sensor. Solid state electrodes with oxide ion conductivity are especially important. These electrodes are redox electrodes. A typical material is ZrO_2 doped with CaO or Yb_2O_3, giving a crystalline structure with cationic vacancies in the lattice which account for ion conductivity. Solid state electrolyte sensors based on ZrO_2 are suitable for determination of oxygen in combustion gases or for monitoring metallurgical processes, where oxygen is to be determined in molten iron at temperatures of about 1000 °C. The redox equilibrium is:

$$O_2 + 4e^- \rightleftharpoons 2O^{2-} \tag{7.7-1}$$

The O^{2-} ions tend to migrate into the vacancies causing the vacancy to be displaced. For measurements in molten iron, the solid state electrolyte is contacted by a platinum electrode and the resulting potential is measured with a second metal electrode in the molten iron. When measuring O_2 in gases, the solid state electrolyte is implanted in a porous platinum electrode and the electrode potential is measured relative to a reference signal. Usually, air is used as the reference medium.

According to the Nernst equation the relationship between the electrode potential E and the partial oxygen pressure in the reference $p_{reference}$ and sample chamber p_{sample} is:

$$E = \frac{RT}{4F} \ln \frac{p_{reference}}{p_{sample}} \tag{7.7-2}$$

The so-called λ-probe is commonly used for measuring automobile exhaust gases. This probe is based on the fact that for a value of $\lambda = 1$, at which the exact stoichiometric ratio of fuel to air is valid, a large jump in the oxygen partial pressure occurs. This change can be used to control the fuel to air mixture.

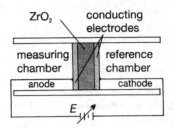

Solid state oxygen sensor for hot gases

Gas-permeable membrane sensors

Gas-sensitive potentiometric sensors incorporate an electrochemical cell with an ion-selective electrode and a reference electrode. Both are immersed in an internal electrolyte solution. The internal electrolyte is separated from the measuring solution by a gas permeable membrane (cf. Fig. 7.7-1). The microporous or homogeneous membrane is typically 0.1 mm thick. Microporous membranes are manufactured from hydrophobic polymers, e.g., polytetrafluoroethylene (PTFE) or polypropylene. Such membranes consist of 70% of pores with pore diameters of less than 1 µm, so that gases can permeate by effusion, while water or ions are rejected by the hydrophobic membrane.

Homogeneous membranes are prepared from silicon rubber. The gas dissolves in the membrane and diffuses through it. In order to warrant fast gas transfer, the membranes are often much thinner than porous membranes, i.e., about 0.02 mm thick.

Let us consider the mode of operation of a potentiometric gas sensor for CO_2. The gas penetrates the membrane and dissolves in the internal electrolyte that consists of a $NaHCO_3/NaCl$ solution. Protolysis of CO_2 occurs according to:

$$CO_2 + H_2O \rightleftharpoons HCO_3^- + H^+ \tag{7.7-3}$$

A relatively high concentration of HCO_3^- ions in the internal solution guarantees a direct relationship between the CO_2 concentration in the external solution and the hydrogen ion activity in the internal solution. The conventional glass electrode is used as hydrogen ion-selective electrode and calibration is based on the Nernst equation. It is also possible to use pH measurements to determine other gases, such as SO_2 or NO_2. The pertinent chemical reactions are given in Table 7.7-3.

For membrane electrodes interference is restricted to gases that react with the internal solution in a way similar to the analyte gases. This is observed, e.g., when determining NO_2 in the presence of CO_2 or SO_2. In such cases, the pH glass electrode is replaced by an ion-selective electrode. For the determination of NO_2,

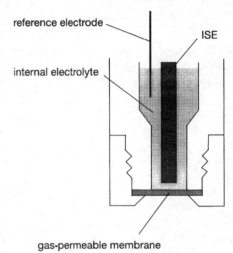

Fig. 7.7-1. Scheme of a gas-sensitive probe based on potentiometric measurements

Table 7.7-3. Types of commercial potentiometric gas sensors

Analyte	Reaction in the internal solution	Sensitive electrode
CO_2	$CO_2 + 2H_2O \rightleftharpoons HCO_3^- + H_3O^+$	pH glass
NO_2	$2NO_2 + 3H_2O \rightleftharpoons NO_2^- + NO_3^- + 2H_3O^+$	pH glass or NO_3^- ISE
SO_2	$SO_2 + 2H_2O \rightleftharpoons HSO_3^- + H_3O^+$	pH glass
NH_3	$NH_3 + H_2O \rightleftharpoons NH_4^+ + OH^-$	pH glass
H_2S	$H_2S + 2H_2O \rightleftharpoons S^{2-} + 2H_3O^+$	Ag_2S ISE
HCN	$HCN + H_2O \rightleftharpoons CN^- + H_3O^+$	Ag_2S ISE
HF	$HF + H_2O \rightleftharpoons F^- + H_3O^+$	LaF_3 ISE

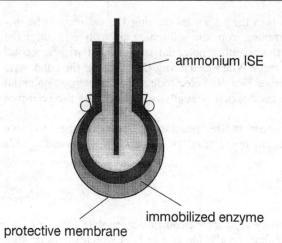

Fig. 7.7-2. A biocatalytic membrane sensor based on an ISE for ammonium ions and the enzyme urease

ammonium ISE

immobilized enzyme

protective membrane

a nitrate ISE can be used, as can be understood from the reaction equation in Table 7.7-3. This Table gives some additional examples for using ISEs as sensitive electrode. The slow response times and low sensitivities of this kind of gas sensor are disadvantageous.

Biocatalytic membrane sensors

ISEs may be covered with an enzyme that catalyzes a biochemical reaction. For example, the enzyme urease catalyzes the hydrolysis of urea to give ammonia and carbon dioxide. The reaction products can be monitored by means of the above gas-sensitive ISEs. The combination of appropriate enzymes with electrodes allows, numerous organic species to be detected. These represent one possibility for the construction of *biosensors* (cf. Sec. 7.8). The application of enzymes as biocatalysts leads to reactions that occur under mild conditions of pH and temperature, that are selective, and consume minimal amounts of substrate.

In order to minimize the consumption of the relatively expensive enzymes, devices are preferred where the enzyme can be immobilized. Immobilization is performed either directly on the surface of the electrode, or the enzyme is immobilized in a solid state reactor and the sample is passed through the reactor and the products are detected at its end.

An example of a substrate-specific membrane electrode is the determination of urea by means of an ammonium-sensitive glass electrode as the ISE. The enzymatic reaction is based on the hydrolysis of urea in the presence of the enzyme urease:

$$NH_2CONH_2 + 2\,H_2O + H^+ \xrightarrow{\text{Urease}} 2\,NH_4^+ + HCO_3^- \qquad (7.7\text{-}4)$$

Urease is immobilized in a polyacrylamide layer mounted on the surface of the NH_4^+ glass electrode as the ISE. Figure 7.7-2 shows a schematic arrangement of this biosensor. Problems with this device may arise from the limited selectivity of this type of ammonium ISE. In most biological matrices, interferences occur with respect to sodium and potassium ions. As an alternative the detection may be replaced by a gas-sensitive ammonia electrode (Table 7.7-3). In this case, however, the sensitivity is restricted because the pH value necessary for the detection and the enzymatic reaction are different. Maximum sensitivity for the gas-sensitive ammonia sensor is obtained at pH between 8 and 9, where most of the NH_4^+ is converted to ammonia. The enzyme reaction, however, is optimal at about pH 7.

Commercial systems for routine analysis of urea work on the basis of a bead reactor with the immobilized enzyme. The sample solution passes through the reactor. After alkalinization of the solution, the ammonia is detected by a gas-sensitive ammonia ISE.

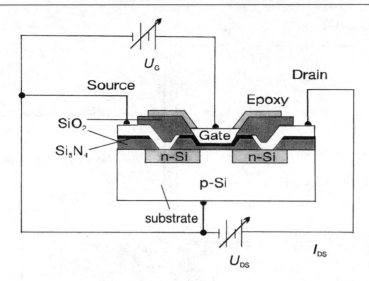

Fig. 7.7-3. Schematic of a metal oxide field-effect transistor (MOSFET)

Field-effect transistors

The potentiometric sensors considered up to now consist in general of a sensitive membrane connected to the measuring instrument by a solid contact or an ion-conducting salt bridge. Electrical amplification and digitization occur separately in the voltage recording device.

Microelectronic integration of the sensor and signal processing is possible, if the potentiometric sensor is based on a *field-effect transistor*. This chip may be chemically sensitive (CHEMFET) or specifically ion selective (ISFET). Bergveld developed FET sensors on the basis of metal oxide silicon field effect transistors (MOSFET).

The structure of such a transistor is shown in Fig. 7.7-3. The transistor consists of a substrate of p-type Si, with two n-type Si-enriched regions formed on its surface. The n-type Si region electrodes, source and drain, are connected to the surroundings via metal contacts such as vapor-deposited aluminum electrodes. By means of an insulator, SiO_2, the Si regions are almost completely covered so that connection feasible only through the drain and source electrodes. The silicon nitride layer, Si_3N_4, can be recognized as an additional passivation layer (Fig. 7.7-3).

If a voltage U_{DS} is applied between the source and drain, no current flows along the connection channel since the pn-diode channel is shut, either between the source electrode and the substrate or between the drain electrode and the substrate. In the chip, however, there is a further contact, consisting of a metal oxide layer (the "gate"). This layer is electrically separated from the substrate by the SiO_2 insulator.

If an additional voltage U_G is applied between gate and substrate an electric field is produced in the n-channel between the n-Si regions, so that a current I_{DS} flows between drain and source. The size of the current is determined by the voltage. Of course, a minimum voltage is necessary to produce a current. Because of the high resistance between gate and substrate, the input current becomes negligible. The high resistances of field-effect transistors make them very suitable as inputs for pH or ion meters and for signal amplification in conventional voltmeters. The influence of the drain source voltage is due to changes in the electrical characteristic of the transistor (Fig. 7.7-5).

For construction of an ion-selective FET, the original gate has to be removed or replaced by another material. If removed, the lower Si_3N_4-layer acts as conductive layer. If replaced by a new ion-selective layer, e.g., a membrane of a suitable ionophore (Fig. 7.7-4) it acts as an ISE. Electrical contact is obtained by a reference electrode, as commonly used for potentiometric measurements. The potentials resulting from the applied gate voltage U_G and from the ion-selective layer, add together, and the measured current I_{DS} depends on the activity of the potential-forming ions. As a result, the electrical characteristic of the FET changes. In the optimum case, the change of the characteristic is 59 mV for a one-electron step for concentration changes of one order of magnitude (Fig. 7.7-5).

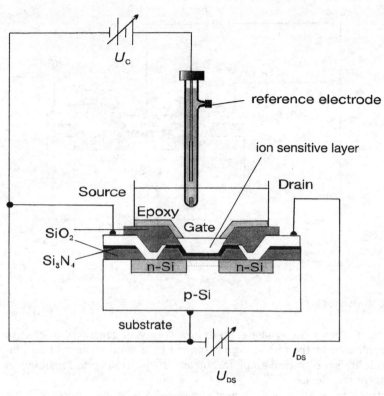

Fig. 7.7-4. Sensor based on an ion selective field-effect transistor (ISFET)

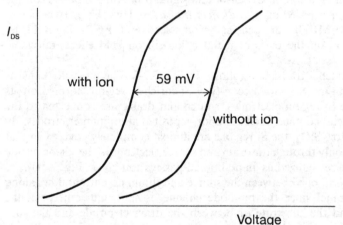

Fig. 7.7-5. Electrical characteristics of an ISFET: the drain current I_{DS} depends on the sum of the applied gate voltage U_G and the potential of the ion-sensitive layer

For practical measurements with an ISFET, the drain source current is kept at a constant working level. The change of the gate voltage with ion concentrations is compensated by an additional control voltage. This feedback voltage is proportional to the activity of dissolved ions as for the case of conventional measurements according to the Nernst equation. The first application of an ISFET was for pH measurement, and the best Nernstian response is obtained if the gate is made of Ta_2O_5.

Further applications of FETs for the determination of other ions, but also of gases, enzymes, and antibodies, are given in Table 7.7-4. In principle, all layers applied to ISEs can also be transferred to microelectronic sensor devices.

The performance of an enzyme-coated macrosensor is discussed in detail in Section 7.8.3. Immunosensors, in turn rely on the basic principles of radio-immunoassay (Sec. 7.8.3).

The many methodological variations for developing sensors based on micro-lithographics are hindered by severe problems. In microelectronic circuits, the sites used for chemical sensing are usually hermetically sealed in metal cans or encapsulated in plastic or ceramic blocks. This is because of the high sensitivity of these devices to changes in the ambient environment, such as alterations in humidity, irradiation, or temperature. These interactions are based on forces that can usually be neglected in the macro range, e.g., surface tension, diffusion effects, van der

Table 7.7-4. Chemically selective field-effect transistors

Type	Gate	Analyte
Gas-sensitive FET (GASFET)	Pd	H_2, NH_3, CO
Ion-sensitive FET (ISFET)	Ion-selective layers: Ta_2O_5, Al_2O_3, BN	H^+
	Gas-permeable membrane + Al_2O_3	NH_3, CO_2
	Valinomycin	K^+
Enzyme-coated FET (ENFET)	Gel- or polymer layer with immobilized enzyme	Urea, glucose, penicillin, H_2, acetylcholine
Immuno FET (IMFET)	Antigen or antibody	Albumin

Table 7.7-5. Examples of amperometric gas sensors

Analyte	Reaction	Working electrode
CO	$CO + 3H_2O \rightleftharpoons CO_2 + 2H_3O^+ + 2e^-$	Pt
NO_2	$NO_2 + 3H_2O \rightleftharpoons NO_3^- + 2H_3O^+ + e^-$	Au
SO_2	$SO_2 + 6H_2O \rightleftharpoons SO_4^{2-} + 4H_3O^+ + 2e^-$	Au
Cl_2	$Cl + 2e^- \rightleftharpoons 2Cl^-$	Pt
H_2S	$H_2S + 2Ag + 2H_2O \rightleftharpoons Ag_2S + 2H_3O^+ + 2e^-$	Ag
HCN	$Ag + CN^- + H_2O \rightleftharpoons AgCN + H_3O^+ + e^-$	Ag

Waals forces, or quantum tunneling effects. In the field of microelectronic chemical sensors, these interactions become very important.

Amperometric sensors

The use of voltammetric measurements as a sensor principle is well known for the amperometric Clark sensor for oxygen measurements (Sec. 7.3). The working electrode of the Clark sensor is a platinum electrode linked to a silver anode. The sensor can be modified to circumvent the need for regeneration of the silver electrode. For example, silver can be used in the working electrode and lead in the auxiliary electrode. Potassium hydroxide serves as the electrolyte. In contrast to the reduction of oxygen at the silver working electrode, oxidation of lead represents the electron-generating anodic reaction:

$$Pb + 6OH^- = [Pb(OH)_6]^{2-} + 4e^- \qquad (7.7-5)$$

Since lead hydroxide is soluble, there is no increase in the resistance which is in contrast to the Clark electrode, where silver hydroxide precipitates at the silver anode.

Amperometric sensors for other gases are also known (Table 7.7-5). Sensors without a membrane are used for monitoring chloride in drinking water. Sensors for carbon monoxide are common in the coal mining industry. This kind of semi-quantitative sensor responds rapidly in emergencies, but cannot compare – in terms of resolution of the signal – to membrane sensors, such as the Clark electrode.

A combination of the Clark sensor with an enzyme layer is discussed in Section 7.8.3 for the determination of glucose by glucose oxidase. This scheme can be transferred to other oxidative enzymatic reactions, e.g., for the determination of galactose based on galactose oxidase, or of uric acid by means of uricase.

Conductivity sensors

Electrical conductivity is another principle that can be used for gas sensing purposes. The respective sensors consist of metal oxides, such as SnO_2, ZnO, TiO_2 or Fe_2O_3 with *n*-type conductivity. The scheme of a SnO_2 thin-layer sensor is shown

Combination of semiconducting sensors in an array is sometimes called an *electronic nose*. Based on pattern recognition methods, the responses can be used to analyze an unknown gas, both qualitatively and quantitatively.

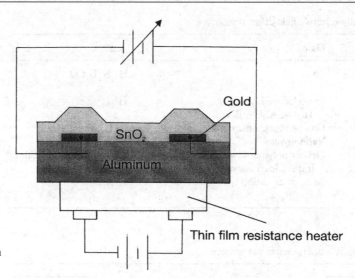
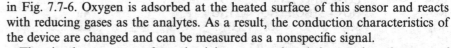

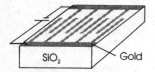

Fig. 7.7-6. Gas sensor based on an aluminum substrate covered with a semiconductor

Interdigitized electrodes

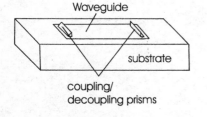

Planar waveguide

in Fig. 7.7-6. Oxygen is adsorbed at the heated surface of this sensor and reacts with reducing gases as the analytes. As a result, the conduction characteristics of the device are changed and can be measured as a nonspecific signal.

The simple structure of conductivity sensors has led to a broad range of industrial and domestic applications for detection of H_2, PH_3, NH_3, SO_2, CO, CH_4, O_2, and other gases.

Conductivity is also exploited in so-called *chemiresistors*. This kind of sensor is constructed from a thin film of an organic semiconductor laid down over inter-digitized electrodes. Typical film materials are phthalocyanines, which have a chemical structure similar to hemin and chlorophyll. Depending on the central atom of the complexing agent, sensors for CO (Zn-phthalocyanine) or NO_2 (Pb-phthalocyanine) can be prepared.

7.7.4 Optical sensors

The development of optical sensors became feasible with the availability of optical fibers in the visible range. Recent developments are concerned with the extension of the visible spectral range to include fiber-optic sensors in the UV, near- and mid-IR range. Apart from fiber-optical sensors, there is also an interest in sensors based on planar optics.

In general, fiber optical sensors generate an optical signal in proportion to the analyte concentration.

Measuring inherent optical properties

The simplest type of optical fiber sensor involves coupling an optical fiber to a spectrophotometer. With this device, the color or fluorescence of solutions or of biological matter can be measured. In order to illustrate the applicability of fiber-optical sensors some fundamentals of *waveguides* are considered.

Fiber cables consist of a core (the waveguide) and a cladding both made from glass, quartz, or plastic. The cable diameters range from 0.05 μm to 0.6 cm. The light propagates down a single fiber or a fiber bundle. The bundles can be arranged randomly or in a fixed position at the input and output of the fiber. In the latter case, a whole picture can be transmitted by the fiber cable. These fibers are important in medical diagnosis (endoscopy) for inspecting organs, such as the stomach.

Guiding light through an optical fiber is demonstrated in Fig. 7.7-7. When light reaches the waveguide, part of it is transmitted and part is totally reflected. For the total reflection of light to occur, a critical angle θ is necessary, and the refractive index of the *core* n_1 has to be higher than the refractive index of the *cladding*

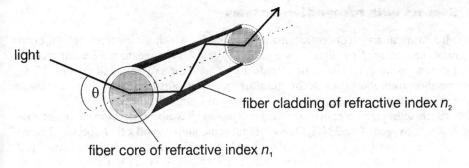

light

θ

fiber cladding of refractive index n_2

fiber core of refractive index n_1

Fig. 7.7-7. Reflection of light in an optical fiber

monomode fiber

multimode fiber

n_2. The refractive index of a glass core is ~ 1.6 and that of glass cladding ~ 1.5. For measurements in the mid-IR range other materials are necessary, such as chalcogenide glasses (AsSeTe) or polycrystalline silver halide fibers, that are transparent between 2 and 20 μm. The refractive index of a silver halide fiber consisting of 75% AgBr and 25% AgCl is 2.21. This core material is cladded by a plastic cover of refractive index 1.5.

The critical angle θ is determined by the *numerical aperture (NA)* of the waveguide. *NA* depends on the refractive index of the core n_1 and of the cladding n_2:

$$NA = \sin \theta = \sqrt{n_1^2 - n_2^2} \qquad (7.7\text{-}6)$$

The larger the numerical aperture, the more light the waveguide can capture and transmit.

Commonly, *multimode waveguides* are used. In these fibers the light propagates in many modes down the waveguide. Attenuation of light does not occur in the same proportion for the different modes. The refractive index is uniform and changes abruptly at the cladding interface. *Monomode waveguides* possess a very small core diameter, typically 2–10 μm, so that the light propagates almost linearly down the fiber. The thickness of the optical cladding should be at least ten times that of the core. The coupling of light into a monomode fiber is difficult, since a narrow angle of incidence has to be achieved. Therefore, highly focused light sources, such as lasers or laser diodes are needed.

Waveguides are coupled to a spectrophotometer as bifurcated fibers (*Y-cables*). Figure 7.7-8 shows a typical arrangement of a plain fiber "sensor" for measuring the absorption in liquid samples. Light propagates down the fiber into the sample solution and is redirected at a reflector into a second fiber to the spectrophotometer. The path-length of the light is twice the distance between the end of the fiber and the reflector.

The waveguide merely serves to transfer the radiation. Instead of a photometer the fibers may be linked by means of an *optoelectronic circuit*; in this case, the light source is a light emitting diode (LED) and the detector a photodiode.

Applications of plain optic fiber sensors are given in Table 7.7-6. Whereas a photometric titration based on an optoelectronic sensor is easily achieved, monitoring of chemical processes or ground-water is much more complicated. For example, the direct detection of *organic compounds* in ground-water is feasible by fluorescence measurements. Although individual chemicals cannot be determined, the quality of water can be monitored by a combination of fiber optics, laser enhancement, and quantitative Raman spectroscopy. With such a system, contaminants in the ppb-range can be monitored at distances up to 1000 m, assuming that at least some of the contaminants are fluorescent.

Table 7.7-6. Applications of plain fiber optic sensors

Analyte	Measured optical property	Application range
Copper ions	Absorption at 930 nm	Copper electroplating
Organic compounds	Fluorescence	Ground-water monitoring
Hemoglobin	Diffuse reflectance at 600–750 nm	Blood oximetry
Halothane	Absorption in the NIR	Anesthetics

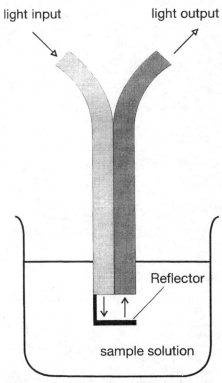

light input

light output

Reflector

sample solution

Fig. 7.7-8. Fiber optic sensor for measuring light absorption in a sample solution. The fiber cables consist either of single fibers or of fiber bundles

Sensors with recognition systems

Immobilization can be mechanical, electrostatic, or covalent.

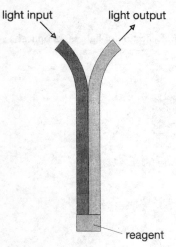

light input   light output

reagent

Fig. 7.7-9. Optical sensor using an immobilized reagent

Measuring analyte concentrations in samples for which no inherent optical properties can be used requires a recognition system with a measurable color change. For this purpose, an indicator is frequently bound directly to the fiber (Fig. 7.7-9); common immobilization techniques are based on adsorption on (ion) exchange resins, inclusion into polymers, or by covalent immobilization.

Such sensors, also called *optrodes* or *optodes*, have been developed for measuring pH, oxygen, CO_2, NH_3, heavy metal ions, and several other species. The respective indicator is immobilized at the fiber tip and its color change is measured.

In order to obtain a reversible sensor, the chemical equilibria between the immobilized reagent and the analyte in solution (or in the gas phase) have to be taken into account. Consider the simplest case of a reaction of analyte A with the immobilized reagent $\overline{R}$:

$$A + \overline{R} \rightleftharpoons \overline{AR} \tag{7.7-7}$$

(To distinguish species in solution, such as A, from those at the resin the bar is used for characterizing immobilized species.)

For equilibrium formation between the complex at the fiber, $\overline{AR}$, and the immobilized reagent $\overline{R}$ it is valid:

$$K = \frac{[\overline{AR}]}{[A][\overline{R}]} \tag{7.7-8}$$

In the case of measurement of the optical properties of $\overline{AR}$ we obtain for the signal to concentration relationship:

$$[\overline{AR}] = K[A][\overline{R}] \tag{7.7-9}$$

Under the assumption that the equilibrium concentration of A is almost identical to the total concentration of the analyte in solution, $[A] = c_A$ and that the concentration of the immobilized reagent equals the difference between the total reagent concentration c_R and the product concentration, i.e., $\overline{R} = c_R - \overline{AR}$, Eq. (7.7-9) can be rewritten in terms of the constant quantities K and c_R as follows:

$$[\overline{AR}] = \frac{Kc_A c_R}{1 + Kc_A} \tag{7.7-10}$$

Figure 7.7-10 shows a typical dependence of complex concentration in the immobilized layer on analyte concentration.

The calibration curves are linear only at low analyte concentrations, i.e., for $Kc_A \ll 1$ or $c_A \ll 1/K$. At high analyte concentrations, the curve approaches a limiting value because of saturation of the reagent by the analyte.

The nonlinearity of the calibration curves is not a significant problem if a microprocessor is used for data evaluation. The decrease in sensitivity with increasing concentration remains a disadvantage. Often, however, the sensors are needed only for narrow concentration ranges where linearity can be achieved.

Fig. 7.7-10. Relationship between concentration of complexed reagent [$\overline{AR}$] and analyte concentration C_A. The optical signal is usually directly related to [$\overline{AR}$]. Equilibrium constant $K = 1660\,\text{L/mol}$, $\overline{R} = 10^{-4}\,\text{mol/g}$

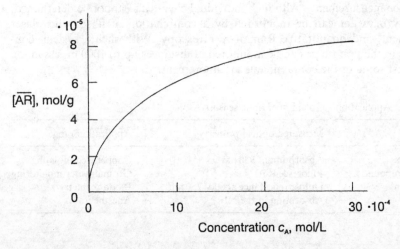

Table 7.7-7. Chemical immobilization techniques for optical sensors and methods for surface modification

Polymer	Reactive group	Modified by reaction with	Partner
Cellulose	Aminoethyl	Bromocyan, ethylenediamine	Carboxylate, sulfonic acids
	Carboxyethyl	Chloroacetic acid	Amines
Glass, silica gel	Aminopropyl	γ-aminopropyl-triethoxysilane	Carboxylic acids, aldehyde
	Vinyl	Triacetoxyvinylsilane	Nucleophiles
Polyacrylamide	Carboxyethyl	Strong alkalies and acids	Amines, proteins

Table 7.7-8. Fiber optic sensors with immobilized reagents (optodes)

Analyte	Reagent/immobilization	Measuring principle
pH (2–5)	Congo red/cellulose acetate	Reflectance
pH (4–7)	Fluoresceinamine/glass	Fluorescence
Al^{3+}	Morin/cellulose	Fluorescence
K^+	Valinomycin + MEDPIN/PVC	Reflectance
Cl^-	Fluorescein/silver colloid	Fluorescence
Humidity	$CoCl_2/Co(H_2O)_nCl_2$/gelatine	Reflectance
O_2	Ru-trisbipyridyl/silicone	Phosphorescence
NH_3	Bromothymol blue (silicone)	Evanescent waves
Albumin	Bromocresol green/cellophane	Reflectance

Selection of the indicator reagent must take into consideration that reagents are to be applied which form less stable complexes. This guarantees that the reagent concentration can be adjusted to the range of analytical interest. This is in contrast to classical analytical applications of equilibrium reactions, where high equilibrium constants and reagent excess are the rule.

Chemical immobilization is based on covalent binding of the indicator to the fiber matrix. This kind of immobilization provides a rather strong coupling of the indicator and minimizes *leaching*. A prerequisite for chemical immobilization is the availability of reactive groups at the indicator or the support matrix. As a rule, these groups are generated by preliminary activation steps at the matrix. The most frequently used support materials are cellulose, dextrane, or agarose. In addition, artificial polymers are used, e.g., polyacrylamide. Examples of chemical immobilization techniques are given in Table 7.7-7. The techniques can be taken over from the research experience in polymer and surface chemistry as well as from modification of stationary phases in chromatography.

Fiber-optic sensors are not restricted to the determination of pH. Optodes have been developed for cations, anions, gases, organic compounds, and for measuring ionic strength (Table 7.7-8). Absorption and reflectance based sensors are less common than fluorescence sensors because of the higher sensitivity of the latter.

Transfer of the ion-selective electrode principle to optical sensors has led to the development of *ion-selective optodes*. One possibility consists in the application of ion exchange equilibria between the solution and the PVC membrane of the optode. Determination of potassium ions is feasible on the basis of the equilibrium:

$$K^+ + \overline{HC^+} + \bar{I} \rightleftharpoons \overline{KI^+} + H^+ + \overline{C} \tag{7.7-11}$$

where $\overline{HC^+}$, $\overline{C}$ are the protonated and deprotonated form, respectively, of the chromophore immobilized in PVC, and $\bar{I}$ and $\overline{KI^+}$, respectively, are the ionophore and its complex with potassium ions in PVC. The equilibrium is driven by the electroneutrality condition.

A versatile ionophore is valinomycin, familiar from ion-selective electrode applications. The chromophore is a hydrophobic compound, e.g., MEDPIN (Miles (Bayer)):

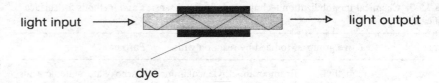

light input ⟶   ⟵ light output

Fig. 7.7-11. Optical sensor for measurement of evanescent waves

dye

The advantage of ion-selective optodes is their large working range, similar to that of ISEs. In addition, the many ionophores developed for ISEs can also be used in optodes.

Optodes based on evanescent waves

The measurement of *evanescent waves* is feasible if the indicator is immobilized on the core of a waveguide (Fig. 7.7-8). For determination of ammonia, a solution of bromothymol blue in silicone is deposited on the waveguide (Fig. 7.7-11). At each internal reflection in the waveguide, a fraction of the beam extends into the coating and interacts with the indicator. The same phenomenon is known in IR spectroscopy as the ATR technique (Sec. 9.2).

In the determination of ammonia the evanescent wave is affected by the actual color of the dye in proportion to the ammonia concentration. The differences can be quantified, e.g., by measuring absorbance.

Table 7.7-8 gives some further examples for developing fiber-optical sensors.

Advantages of optical sensors over electrochemical sensors are the following:

- All the spectroscopic information is available. If necessary, this information can be monitored at several measuring sites (multiplex advantage).
- There is no electrical interference, so that measurements are possible in high electric fields, e.g., during electrolysis or in transformers.
- No reference electrode is necessary.
- The reagent phase can be made low-cost, so that one-way sensors can be designed.

Disadvantages of optical sensors are:

- Ambient light influences the measurements. This interference has to be reduced, e.g., by pulsing the light source.
- The reversibility of optodes in liquid samples is often poor and repeated regeneration of the sensor becomes obligatory.
- The long-term stability is frequently limited, owing to leaching of the indicator.
- The establishment of an equilibrium between the analyte in solution and the immobilized reagent (Eq. 7.7-10) results in relatively narrow working ranges, apart from ion-selective optodes with a logarithmic signal–concentration relationship.

catalytic layer

wire

inert oxide layer
(ThO_2, Al_2O_3)

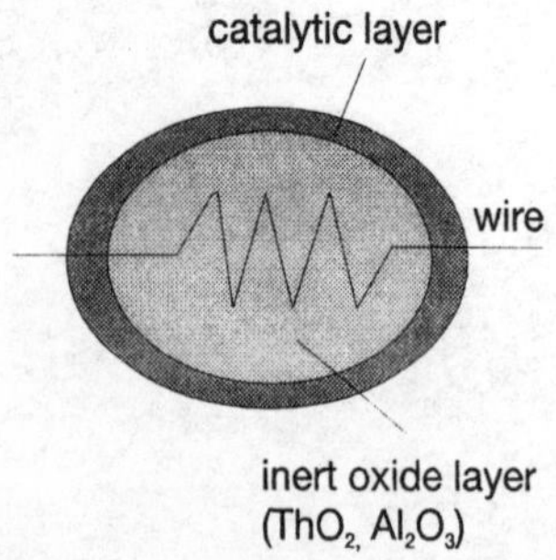

Fig. 7.7-12. Pellistor consisting of a catalytically active Pt or Pd layer and an inactive sinter pearl made of an oxide

7.7.5 Thermal sensors

The measurement of reaction enthalpy is utilized in gas sensors based on a catalytic reaction (*pellistors*). Such catalytic sensors consist of a heating wire embedded in a sinter pearl and a catalytically active layer doped with Pt or Pd metal (Fig. 7.7-12). The wire coil is heated to about 550 °C. Reducing gases, such as CO or CH_4, are oxidized by adsorbed oxygen, and the heat of reaction can be measured by the increasing resistance of the coil. The oxidation rate at the surface of the sensor is proportional to the concentration of the analyte gas. The precision of this unspecific resistance measurement can be enhanced by reference measurements with respect to an inactive sinter pearl.

Calorimetric sensors are based on miniaturized calorimeters and are applied to solutions (cf. Sec. 7.5). In the simplest case the sample passes through a reactor at the end of which the heat of reaction is measured by a thermistor. For determination of substrates (cf. Sec. 7.8) an enzyme is immobilized in the reactor. Such devices are not sensors in the strict sense of our definition since monitoring is discontinuous. Conventionally they are termed sensors. The detection of reaction heat becomes practicable by utilizing a particular enzyme reaction. Determination of urea, penicillin, glucose, sucrose, cholesterol, or lactate is feasible.

7.7.6 Mass-sensitive sensors

The change of mass and the subsequent change in resonance frequency in the presence of an analyte gas is used in mechano-acoustic sensors. The most important are the piezoelectric quartz crystal resonator and surface acoustic wave (SAW) sensors.

The principle of a quartz crystal resonator is explained in Fig. 7.7-13. A quartz crystal resonator that has been freed from its encapsulation is covered by an organic, gas-absorbing layer. The quartz is arranged as a frequency-determining element in an oscillator circuit. By deposition of gas onto the planar crystal surface the mass is changed. As a result the resonant frequency changes as well. Operation is performed in the thickness shear mode. In this mode the surfaces of the crystal are moved against each other. The following relationship holds between mass change Δm and the change in frequency Δf:

$$\Delta f = -2.3 \times 10^6 f_0^2 \left(\frac{\Delta m}{A}\right) \tag{7.7-12}$$

where A is the quartz surface area, and f_0 is the resonant frequency of the uncovered quartz.

Frequencies of 10–15 MHz can be resolved with resolution of up to 0.01 Hz. Thus masses as low as 10 pg can be determined. Different coatings are used as absorbing layers. For example, organophosphorous compounds can be determined by means of a copper diamine substrate linked to a polymer resin. With this material, a reversible sensor is formed that responds to diisopropyl methylphosphonate at 20 ppb concentrations. Mass-sensitive sensors are also used for monitoring gas pollution or for determination of dust in air.

More sensitive devices can be obtained by surface acoustic wave (SAW) sensors. This is because of the higher frequencies involved. According to Eq. 7.7-12, resonant frequencies f_0 up to 1 GHz are common, yielding detectable masses down to the femtogram level. The operation of an SAW sensor is explained in Fig. 7.7-14.

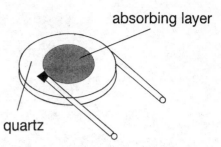

Fig. 7.7-13. Piezoelectric crystal resonator sensor

The piezoelectric quartz crystal resonator is also termed a *quartz microbalance*.

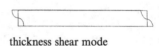

thickness shear mode

The piezoelectric quartz crystal resonator generates *bulk acoustic waves* in constrast to *surface acoustic waves* observed with an SAW device.

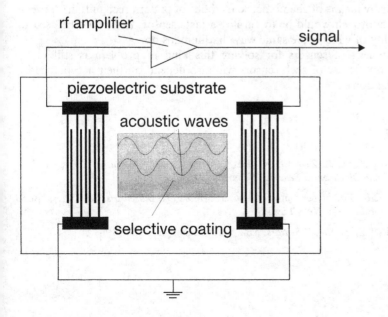

Fig. 7.7-14. Surface acoustic wave (SAW) gas sensor, e.g., for determination of styrene vapor at 5 ppm with PtCl$_2$(ethylene)(pyridine) as the selective coating

On the surface of a piezoelectric substrate a set of interdigitized electrodes is laid down by integrated circuit technology. After applying an rf signal to this transmitter a seismic Rayleigh wave passes along the surface. If the substrate is covered with a gas-sensitive coating, changes in the ambient gas can be detected by frequency measurements. The selectivity of the device is determined by the chemically sensitive coating.

7.7.7 Sensor arrays

Most sensors are not specific, but respond simultaneously to several species. In order to compensate for the lack of selectivity, measurements over several sensor channels are necessary, i.e., at several wavelengths, potentials, currents, or resonant frequencies.

Multiple channel sensors or *sensor arrays* can be formed by aggregation of several single sensors. For example, piezoelectric quartz crystals may be coupled in an array to form a novel device and operated simultaneously. Sensor arrays based on field-effect transistors consist of a single circuit, where individual sensors are formed of different coatings (Fig. 7.7-15). Optical sensors can also be run in the multiple channel mode, if a spectral range is evaluated, rather than a single

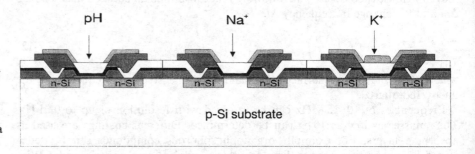

Fig. 7.7-15. Array of ISFETs for analysis of blood electrolytes. The pH-sensitive layer is based on a Si_3N_4 gate. For measurement of sodium, the Si_3N_4/SiO_2 insulator is doped with Al and Na; potassium is measured via a sensitive layer containing valinomycin

wavelength, e.g., by means of a diode array spectrophotometer. The channel is then the recorded wavelength.

Data evaluation for multiple channel sensors is carried out by the principles of simultaneous multicomponent analysis, as explained in Sec. 12.5.4.

Up to now sensor development has been concentrated on the construction of sensors for the analysis of individual chemical species. Thus, the general approach adopted in off-line analysis has merely been transferred to the development of sensors. For example, assessment of wine quality is performed by determination of individual components, followed by data evaluation on the basis of some trace components or by means of chemometric methods of pattern recognition. A much more natural approach would be to employ a taste sensor, that could be used to judge the quality of wine in the same way that humans do.

The development of sensors for solving this kind of problem is still in its infancy. The routine use of such sensors will also depend on their acceptance by official organizations.

References

[7.8-1] Göpel, W., Hesse, J., Zemel, J.N. (Eds.), *Sensors: A Comprehensive Survey, Vol. 2, Chemical and Biochemical Sensors*. Weinheim: VCH, 1991.

[7.8-2] Wolfbeis O.S. (Ed.), *Fiber Optical Chemical Sensors and Biosensors*, CRC Press, Boca Raton, Florida, 1991, Vols. I and II.

[7.8-3] Edmonds, T.E., *Chemical Sensors*. New York: Chapman & Hall, 1988.

Questions and problems

1. Electrochemical measurements of ions in solutions are strongly influenced by the ionic strength. What effect of ionic strength is to be expected if optodes are used?
2. Votammetric sensors are usually based on amperometry. Does a polarographic sensor make sense?
3. Optodes exploit the color change of the immobilized indicator. What does the shape of the calibration curve look like?
4. How can one correct for a lack of specificity and/or selectivity of a sensor?
5. What are the differences and what are the similarities between diode arrays and sensor arrays?

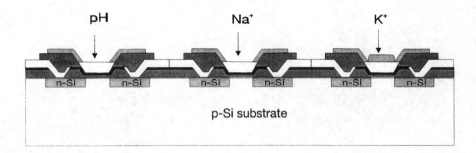

7.8 Biosensors

Learning objectives

■ To understand how to bring together the reagents and detection
method of a bioassay, and form an analytical *device*.
■ To make decisions about the choice of measurement parameter
and thus the type of transducer.
■ To consider how the reagents can be immobilised in the vicinity
of the transducer.

7.8.1 Introduction

Biosensors ... what does this word imply? This volume has been largely concerned
with the detection of a reaction in solution between a sample and added reagents
by an analytical method (chemical or physical); sometimes a sample preparation
or a separation step is involved in addition to the analytical step. A biosensor, on
the other hand, is a 'reagentless' system. Strictly speaking, it is more correct to
say that the reagents are integral and therefore are not added by the user – which
is an advantage. The other desirable feature of the biosensor is that no special
sample preparation, is required. Consequently, biosensors are most likely to be
concerned with immobilized reagents – that is to say, with reactions initiated
within immobilized layers and their interrogation. However, these are features
already encountered in the chapter on Chemical Sensors. So what makes biosensors different?

*Biosensors belong to the Molecular Sensor family, and so comprise an analyte-
selective interface in close proximity to, or integrated with, a transducer (Fig. 7.8-1),
whose function it is to relay the interaction between the surface and analyte either
directly or through a chemical mediator. In biosensors the analyte-specific surface
employs biomolecules, biorecognition sites or analogues thereof.*

The transducer is the detector. It monitors the chemical or biochemical reaction
initiated by the addition of the sample. The nature of the transduced parameter
depends on the type of bioanalytical event occurring upon detection of the analyte.
However, since several parameters can often alter along an analytical reaction
pathway, the choice of device is not necessarily restricted to a single transducer.
Some of the many analytical methods presented could be adapted into a sensor
format. Look back through the examples given in other chapters and consider
whether it would be posible to make the conversion.

This chapter considers some of the fundamental principles and the necessary
elements in building a biosensor device. It must be remembered that not all sensors
have roots in traditional analytical methods. Some biosensors have no analytical
analogue with which they can be compared; they have been developed purely as
sensors.

Probably any measurable physicochemical parameter could be employed in a
biosensor, although it would not always be practical to do so! The most widely
encountered devices use the parameters summarized in Fig. 7.8-2. In this chapter
some examples using optical and electrochemical techniques will be explored further. Exploration into other biosensors can be found in the specialist volumes on
biosensors [7.8-1]–[7.8-4].

Biosensors are analytical devices.

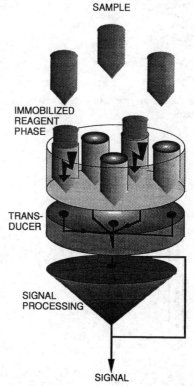

Fig. 7.8-1. Schematic diagram of a biosensor

The biorecognition element and the transducer

The biorecognition element of a biosensor is a protein, macromolecule or complex
with specific surface or interior recognition sites, essential to the recognition of

The biorecognition element imparts analyte
selectivity and causes an analytical signal.

Fig. 7.8-2. Transduction parameters and device type

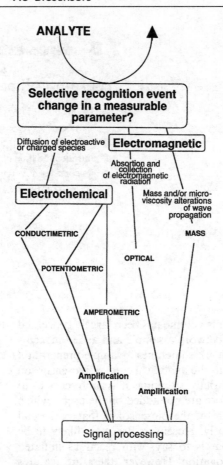

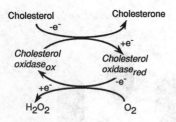

Scheme 7.8-1. Enzyme catalysed oxidation of cholesterol

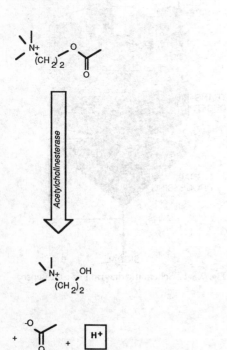

Scheme 7.8-2. Enzyme catalysed conversion of acetylcholine to acetate and choline, causing an increase in [H⁺]

analyte. The element imparts the required analyte selectivity to the transducer. The type of reaction catalyzed by the enzyme will decide the transducer. The analyte, and thus the transduction methods available determines the nature of the biorecognition element. Consider two examples in which an enzyme is employed in devising a sensor for that enzyme's substrate. In Scheme 7.8-1, the enzyme reaction involves an *electron transfer*, which means that that cholesterol assay could be transduced by an *amperometric* electrochemical sensor. Scheme 7.8-2 involves a change in [H⁺] which means that this acetylcholine conversion could be monitored by a pH electrode or a pH-sensitive dye in an optical device. Other enzymes may be utilized in the monitoring of reactions involving hydrolysis, esterification, cleavage, etc.; the analyte is usually the enzyme substrate. Consider how the analysis could be achieved if you could not find a suitable enzyme reaction for your analyte *but* you found that it was an *inhibitor* of an enzyme reaction?

Alternatively, the required analyte might be antigenic, and so could be formatted into an immunoassay (cf. Sec. 7.9). A label would usually be employed to follow this assay, and the transducer would depend upon the nature of the label (Fig. 7.8-3). However, as will be seen later in this chapter, attempts have also been made to follow the complex formation between antigen and antibody without a label.

Biorecognition elements are not confined to these macromolecules. Whole cells or cellular components may offer the required analyte-sensitivity, and the transduced signal varies according to the cellular process involved. The action of a signal analyte might be monitored due to its effect on gross cell respiration (which can be followed by an oxygen electrode), or membrane transport characteristics, or enhanced or inhibited charge transfer. Whatever the parameter, the cells must be immobilized at the transducer.

Nonbiological materials are also used if a synthetic analogue can be developed to show all the natural characteristics of selectivity with enhanced tranduction. Such possibilities look very attractive for biosensor deployment. Nevertheless, efforts at designing analyte-selective sites in a 'biosensor-friendly' format require a little more time and perfection if they are to be as good as the natural macromolecules which are to be emulated.

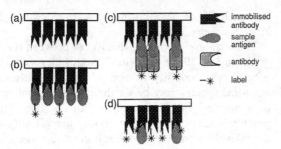

Fig. 7.8-3. Formats for immunosensors.
* is the label.
(a) Antibody modified surface of a transducer;
(b) Competitive immununoassay, [*] $\propto$ 1/[Ag];
(c) Sandwich immunoassay, [*] $\propto$ [Ag];
(d) Homogeneous immunoassay, [*] $\propto$ 1/[Ag]

7.8.2 Producing the biological surface

Irrespective of the method described in Fig. 7.8-2, or the type of biorecognition event which is used in the analysis to give selectivity for the analyte, there is a common feature of all the types of biosensor. *The biorecognition element is immobilized at a surface*, usually the surface of the transducer, where the signal is detected.

Methods of immobilization fall into different categories, and choosing the best method is not always straightforward. Sites within a macromolecule must be uncompromised by immobilization. However, it is not easy to predict orientation of reaction during an immobilization, since there are always several reactive groups or centres which could be involved. In essence there are two types of immobilization: entrapment and binding.

The latter can involve binding energies from *10 to 500 kJ mol⁻¹*, so it is not surprising that the results can be very different. Entrapment is often more straightforward, but is subject to serious leaching of the reagent. This leaching is particularly acute when the reagent molecules are smaller than the pores of the entrapment matrix, unless some binding to the matrix is also involved; in which case, the immobilization has binding character as well as entrapment.

The biorecognition element and other reagents must be immobilized.

Theoretical considerations in designing an immobilization

Reaction at an interface

Apart from the fact that one reagent is stationary and one is under diffusion control, a basic difference exists between the immobilization binding process (chemisorption) and a bimolecular reaction in the homogeneous phase. In the latter case, the reactivity of the reactants remains constant regardless of the degree to which the reaction has proceeded, so long as neither reagent becomes limiting. The reaction thus follows the same kinetics at $t = 0$ and $t > 0$ (Fig. 7.8-4a). During immobilization, on the other hand, individual adsorption sites cannot be considered to be independent of one another; adsorption at one site may affect

Reactions on a surface are heterogeneous.

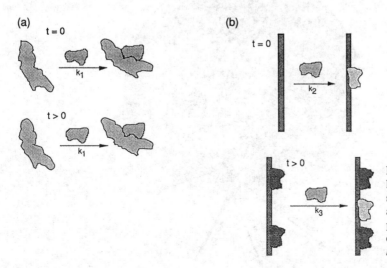

Fig. 7.8-4. Reactivity of reaction in (a) homogeneous phase: reaction remains the same at $t = 0$ & $t > 0$. Reaction constant $= k_1$ at $t = 0$ and $t > 0$ and (b) heterogeneous phase: reaction changes according to surface coverage. Reaction constant $= k_2$ at $t = 0$ and k_3 at $t > 0$

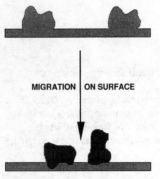

Fig. 7.8-5. Migration of adsorbate

the reactivity of neighboring sites considerably (Fig.7.8-4b), and so the reaction kinetics change with time.

The attractive and repulsive forces between the adsorbate and the adsorbent can extend up to 50 nm and their magnitudes vary. Following the initial adsorption process, the mobility of the adsorbate is also a point for consideration. The adsorbed species may be localized at its initial adsorption site for the entire time, or it may migrate from one site to another (Fig. 7.8-5). In the latter instance, potential energy barriers separating neighboring sites have to be overcome. Since the activation energy for migration is frequently less than the bonding energy this can be quite easy.

Every surface has roughness (heterogeneity). If the roughness is of a comparable size to the adsorbate, then the adsorbate will be able to interact with different planes of the surface; each plane of a given orientation will have a different reactivity. It is to be expected that these planes will react in order of decreasing reactivity. Even if the surface is homogeneous and all the adsorption sites identical, there is still a decrease in the heat of adsorption with increasing coverage, due to dipole-dipole interactions between adsorbed molecules and incoming molecules. For an ideal, non-mobile, simple adsorbed layer, the decrease in the heat of adsorption is linear with coverage. In a mobile layer the heat of adsorption remains nearly constant initially and then shows a step inflection at 50% coverage. This indicates that the result of high mobility is to minimize lateral interactions.

Range of forces

During attachment of a biorecognition molecule to a surface, the collision of the molecule with the surface must be followed to which there is a limited sequence of possible outcomes (Fig. 7.8-6):

1. Reflection back into the bulk of the adsorbate, with (inelastic) or without (elastic) transfer of energy.
2. Transfer of energy (inelastic collision) such that the molecule is unable to 'climb' out of the potential well at the surface and is in an excited physisorbed state, which is associated with comparatively weak forces, e.g., van der Waals, and a low enthalpy of adsorption of $\sim -40 \, \mathrm{kJmol^{-1}}$.
3. Subsequent possible processes:
   a) *Further loss of energy to the surface at the same site.*
   b) *Migration over the surface with loss of energy at other sites.*
   c) *Desorption with a gain in energy from the adsorbent.*
   d) *Transfer to the chemisorbed state, either at the initial site or after migration.*

Chemical adsorption is associated with the formation of a chemical bond with a typical enthalpy change of $\sim -40 \, \mathrm{kJmol^{-1}}$.

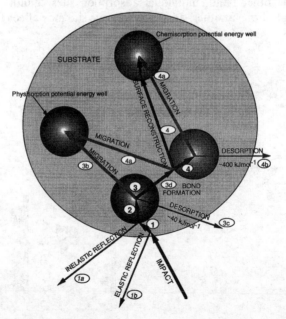

Fig. 7.8-6. The sequence of events at a surface leading to chemisorption

4. Once chemisorption has occurred, further possibilities exist.
   a) *Migration of the chemisorbed species.*
   b) *Desorption from the chemisorbed state.*
   c) *Further chemisorption giving multiple attachment.*
5. Surface reaction may take place between the incoming molecule and another species already adsorbed, but not directly involve the substrate.

Chemisorption has a large enthalpy of adsorption, greater specificity of orientation and more stable layers than physisorption. Nonetheless, on a surface containing many sites capable of physical and chemical adsorption, a mixed population of adsorbed species is still likely to accumulate, whose nature varies with temperature and rate of deposition.

Once the biorecognition surface has been created, its reaction with analyte will not be homogeneous, but will also involve chemisorption.

Methods of immobilization

The immobilization of the biorecognition element can happen in one of two ways. The biorecognition molecules can be attached directly to the transducer surface, or else a'carrier substrate' can be modified with the biorecognition molecule, which is subsequently, or sometimes simultaneously, deposited on the transducer.

Physical adsorption

Physical adsorption of the biorecognition molecule onto the surface of the transducer is the simplest of the immobilization processes. Carbon has been one of the most popular adsorbent materials for enzyme electrodes [7.8-5]–[7.8-8]. Polystyrene has provided a particularly useful surface for optical transduction in solution assay and thus seems to have potential in creating a sensor format. However, this kind of adsorption is very often too weak to prevent a fast leaching rate of the reagents, and thus the sensor may need to be covered with an analyte-permeable membrane able to retain the biorecognition molecule on the transducer.

Physical adsorption involves low energy forces.

Although physical adsorption is simple, it can also be the least satisfactory since little control over the orientation or site of attachment is possible without altering the surface itself or influencing the activity of the adsorbed species! This has been illustrated by Duschl and Hall [7.8-9], who compared human immunoglobulin G (hIgG) adsorbed on SiO_2 and on an aminosilane treated SiO_2 surface. The two had apparently similar hIgG monolayers, which implied that they would be good biosensing surfaces. However, the reaction of the antibody, anti-hIgG, with these two layers was quite different; the former bound only about three antibody molecules per molecule of adsorbed hIgG, while the latter could support five or six. This indicates different interactions between the adsorbed protein and the surface, giving nonidentical recognition surfaces in the two models.

In all methods, the attachment of protein to the transducer surface for immunoassay is difficult since it must be achieved without compromising the active site. If the modified surface is to be used directly without a label or mediator, all nonspecific interactions must be prevented, since 'direct', labelless measurements which usually involve monitoring a change in mass or refractive index at the surface, cannot distinguish the specific and non-specific interactions. A 'close-packed' surface immobilization may not suffer from non-specific binding, but nevertheless, is often still not effective for antibody-antigen complex formation because of steric hindrance. Conversely, a widely-spaced packing usually allows non-specific interactions with the underlying surface to occur, and so large, false-positive signals may be recorded.

Physical retention in polymer matrices

A favorite method for the entrapment of enzymes and whole cells is by cross-linking to give a gel matrix where the receptor species is physically retained. The

Cross-linking reagents can be used to create a polymer matrix in which the reagent is trapped.

Table 7.8-1. Bifunctional reagents

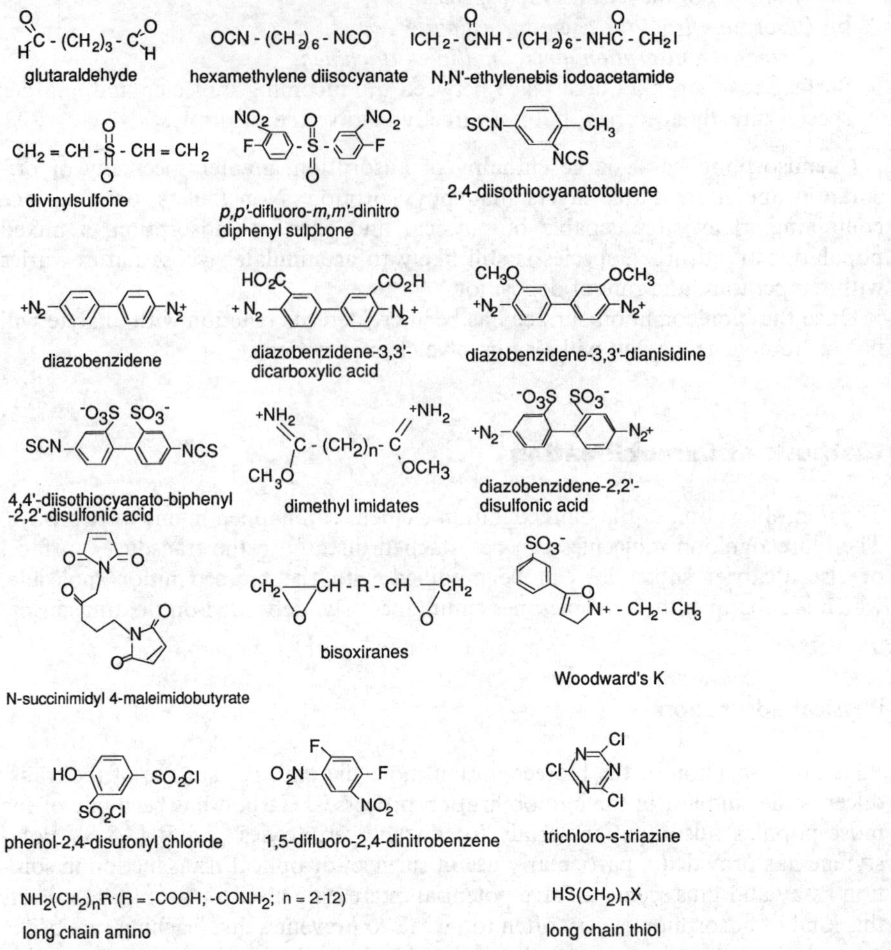

glutaraldehyde

hexamethylene diisocyanate

N,N'-ethylenebis iodoacetamide

divinylsulfone

p,p'-difluoro-m,m'-dinitro diphenyl sulphone

2,4-diisothiocyanatotoluene

diazobenzidene

diazobenzidene-3,3'-dicarboxylic acid

diazobenzidene-3,3'-dianisidine

4,4'-diisothiocyanato-biphenyl -2,2'-disulfonic acid

dimethyl imidates

diazobenzidene-2,2'-disulfonic acid

N-succinimidyl 4-maleimidobutyrate

bisoxiranes

Woodward's K

phenol-2,4-disufonyl chloride

1,5-difluoro-2,4-dinitrobenzene

trichloro-s-triazine

$NH_2(CH_2)_nR$ (R = -COOH; -CONH$_2$; n = 2-12)

long chain amino

$HS(CH_2)_nX$

long chain thiol

methods of obtaining this 'gel' are numerous. Glutaraldehyde cross-linking is probably the most widely tried method of immobilization and is still generally favoured for reagent phase immobilization in flow injection analysis, despite the problems with its bulk fabrication. In any case, many bifunctional agents are employed; Table 7.8-1 lists some examples of those most frequently encountered, which are principally reactive towards $-NH_2$, $-OH$ and $-SH$ groups.

All kinds of polymers (e.g., gelatine, agar, cyclodextrins, polypyrrole, polyaniline, polyvinylchloride, polyvinylalcohol, polyvinylpyridene, polyacrylamide, and many other polymers and gels [7.8-10]–[7.8-12]) have also been thoroughly investigated as matrices for entrapment, but none are ideal. Some are clear and permeable so they have obvious potential in different types of sensors but an immobilization matrix must also show adequate 'storage characteristics, particularly in its stabilisation of the immobilized biorecognition molecule. For aqueous or water-containing applications the hydration properties are also important. A newer approach has been to use film-forming emulsion polymers where pH, isotonicity and hydration properties can be tailored to the biorecognition system and application [7.8-13]. This means that a similar immobilization protocol can be employed, irrespective of whether it is an enzyme, antibody or whole cell. The biosystem-containing emulsion can be prepared in aqueous media even though the film becomes insoluble after formation.

Low temperature cured sol-gel silicate glasses can be made [7.8-14]–[7.8-16], but those which take weeks to 'gel' may not be ideal for the less stable biological molecules. Nevertheless, many other low temperature glasses cure faster. For smaller proteins and molecules, physical retention alone, *irrespective of the method*, is insufficient to achieve long-term retention in many cases.

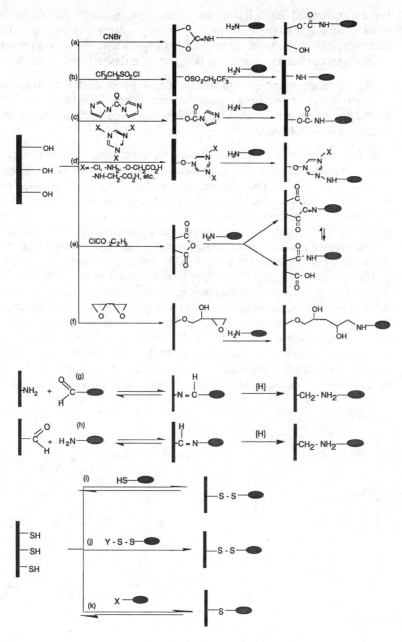

Fig. 7.8-7. Coupling chemistries

Surface modification

In this approach, the substrate may be the transducer itself or else a polymeric material that will be deposited on the transducer in a subsequent operation. Not all transducers have suitable coupling groups on their surface and so it is often better to make the linkage to a material which can be subsequently deposited onto the transducer. The majority of these materials are polymers, e.g., maleic anhydride copolymers, methacrylic acid/anhydride copolymers and their derivatives [7.8-17]–[7.8-20], or other surfaces with free $-NH_2$, $-SH$, $-COOH$ groups etc. Such supports are deposited over the transducer in a thin layer. The pendant reactive groups through which the biorecognition molecule will be attached, can be added to the polymer, by derivatisation of the monomer and copolymerisation, or by direct reaction with the preformed polymer. Examples of some of the coupling chemistries for coupling to different reactive groups on the biorecognition molecule are shown in Fig.7.8-7(g–k).

Many other surface groups on the immobilization substrate must be functionalized prior to covalent linkage. Silane coupling agents of the type X_3SiL or $XSiR_2L$, where X is a hydrolyzable group and L is a ligand (or one capable of

The reagent may be attached by a covalent bond, either directly to the transducer or to a layer deposited on the transducer.

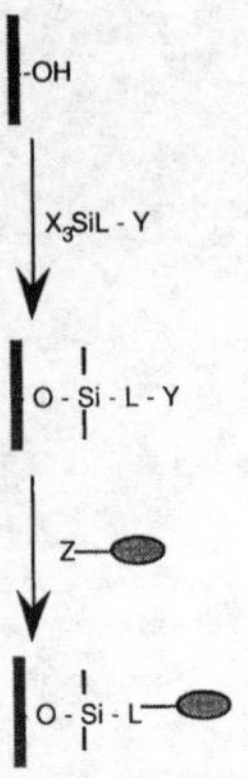

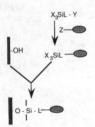

Scheme 7.8-3. Immobilization onto a silane derivatized substrate

Scheme 7.8-4. Immobilization of a silane derivatized molecule

being transformed into a ligand) are attractive reagents for functionalising surface –OH groups and have been employed widely – especially on silica. The thermal and hydrolytic stability of the resulting linkage is high on such supports.

Two approaches are possible in designing the desired surface modification:

a) the coupling agent is attached to the support and then reacted with the macromolecule requiring immobilization (Scheme 7.8-3);
b) the macromolecule-silane complex is formed and then condensed directly with the surface (Scheme 7.8-4).

Both methods show advantages and disadvantages. If there is more than one surface ligand, then the first approach could lead to an inhomogeneous surface product. On the other hand, if there is more than one reactive group on the macromolecule, then reaction with the silane would lead to a mixture of silane macromolecular complexes which in turn would give a mixture of immobilization products. The former approach is generally preferred since the range of ligands can be readily extended after attachment, by standard organic reactions that are not readily applicable to the usual X_3SIR coupling reagents themselves.

The surface –OH group has been popular in other coupling chemistries (Fig. 7.8-7a–f), but other groups may also be employed, e.g. amido, –COOH etc. (Fig. 7.8-8). On metal surfaces such as silver and gold, a particularly effective means of creating a useful functional group for immobilization, is to use self-assembling layers [7.8-21] such as long chain thiols, with a terminal group for coupling to the recognition molecule. In the case of DNA immobilization an –NH$_2$ terminal group allows coupling through the thymine ring (Fig. 7.8-8f) [7.8-22]. Many other compounds of the structure $Y(CH_2)_nX$ having two terminal functional groups

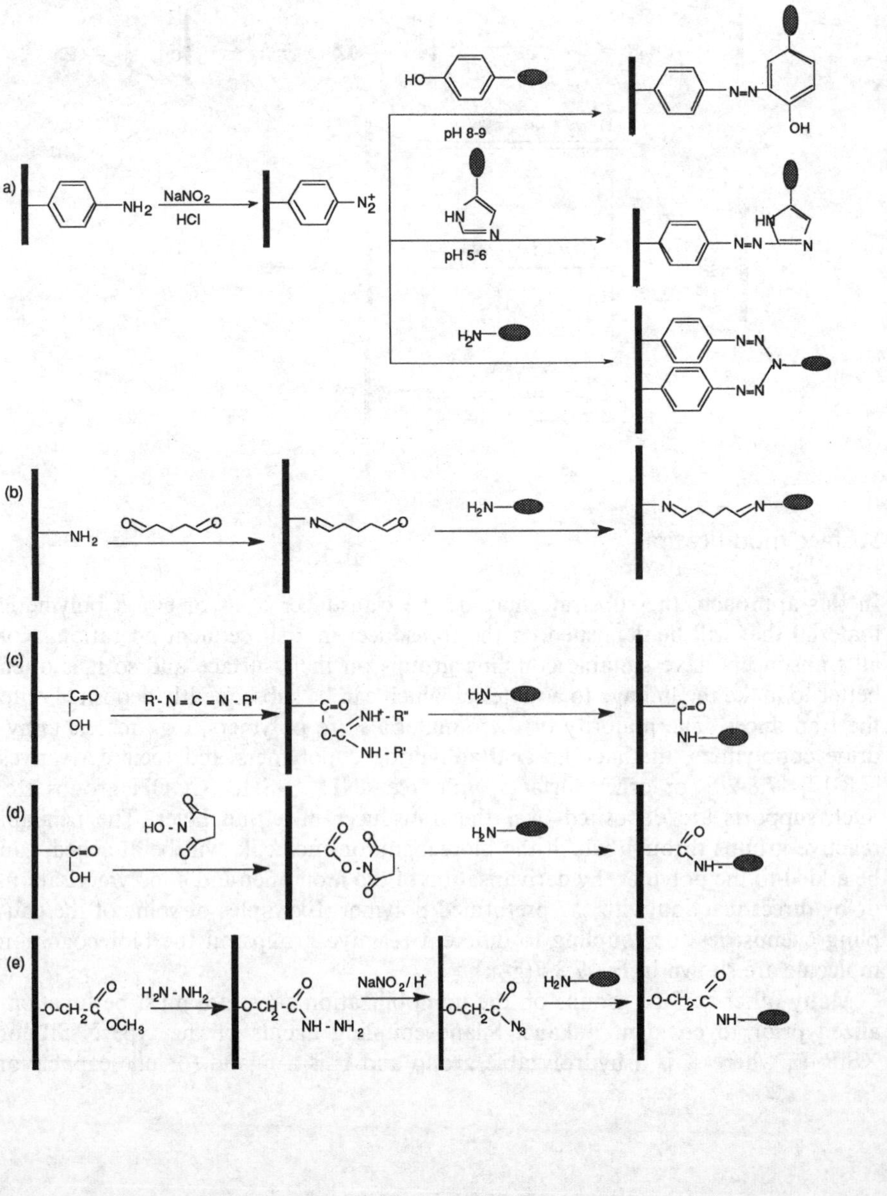

Fig. 7.8-8. Immobilization chemistries

separated by a hydrophobic chain (spacer) will organize themselves at a surface according to the reactivity of the functional groups. If Y is chosen to react with the surface then an ordered layer, including the spacer, can be formed in one step. Different functional groups can be introduced provided that they will not compete with the Y head group for coordination with substrate and they will not be so large that they prevent close packing of the hydrocarbon chains.

In designing the immobilization of the biorecognition system it is frequently necessary for a spacer between the surface and the biomolecule. In principle it could also be attached as an integral part of the functionalizing reagent (as seen above), or as part of the bioactive molecule. An obvious choice for a more hydrophilic and more rigid spacer is that of an oligopeptide. The terminal α-amino group of the peptide for example, can be coupled to a cyanobromide activated support and free the –COOH end of the coupled peptide used for linkage to the bioactive molecule. The potential disadvantage of such a spacer is the danger of cleavage by proteases present in biological samples, but many of the other chemistries mentioned here could also be susceptible to this cleavage.

Deposition of the immobilization layer

Even having achieved the successful immobilization of the biorecognition molecule, the task of applying the layer to the sensor surface may not be complete. In the examples given above, either the immobilization reagents or the polymer matrix has to be deposited accurately and reproducibly onto the surface of the sensor. The thickness and diffusion characteristics of this layer will determine the character of the final response signal.

In the field of electronics, deposition and definition are performed to precise limits of accuracy. Thick and thin film technologies have become tools in the fabrication of miniaturized electronic circuits. Several different techniques are available, all of them suitable for adaptation, for the deposition of layers containing the biorecognition element onto a transducer.

Materials which have been modified with reagents must be accurately deposited on the transducer.

Screen printing is a thick film method which can be used to deposit a range of electrode materials (including carbon, Ag/AgCl and noble metals) as well as the biorecognition layer, mediators, or protective overlayers. Typically, the accuracy of the deposition is controlled to within 500 μm, although with certain materials, 100 μm has been reported.

Ink-jet printing is able to dispense single drops or up to 50 000 drops per second, of a conducting solution, such as a buffered protein solution. The droplet size can be selected between 0.3 and 1.5 nl, resulting in a 1–5 μm definition depending upon the surface tension of the liquid and the hydrophobicity and absorbancy of the substrate on which is it printed.

Conventional photoresists can be patterned with a resolution approaching 250 nm using a 350–450 nm light source, or less than 100 nm using an electron beam technique. With protein/polymer layers, however, the practical limitations of photolithography are estimated to be greater than 150 μm because of the light scattering induced by the protein.

7.8.3 Achievement of biotransduction

Amperometric

Interpretation of the signal

The steady state current at an amperometric electrode under diffusion control is (from Sec. 7.3):

$$I_d = \frac{nFD[S]}{d} \qquad (7.8\text{-}1)$$

where d is the diffusion layer thickness, D the diffusion coefficient of the measured species in the layer and $[S]$ the substrate concentration. However, the rate of an

The current measured at an electrode under potentiostatic control can be related to the kinetics of the enzyme reaction.

enzyme catalyzed reaction is given by:

$$\frac{d[S]}{dt} = \frac{k_2[E_0][S]}{K_M + [S]} \tag{7.8-2}$$

where K_M is the Michaelis Menten constant and E_0 the total enzyme concentration. Assuming that in Scheme 7.8-5 the reoxidation of the enzyme is the fast step and does not limit the reaction, then, at the simplest level, the current at an electrode under enzyme kinetic control could approximate to:

$$I_k = \frac{nFdk_2[E_0][S]}{K_M + [S]} \tag{7.8-3}$$

This gives a maximum current response I_{max} when $[S] \gg K_M$.

$$I_{max} = nFdk_2[E_0] \tag{7.8-4}$$

so that the diffusion limited current could be described by:

$$I_d = \frac{I_{max}D[S]}{d^2k_2[E_0]} \tag{7.8-5}$$

and the kinetic current:

$$I_k = \frac{I_{max}[S]}{K_M + [S]} \tag{7.8-6}$$

Mediators can be used as "charge transfer messengers" between electrode and enzyme.

which suggests that, for a given enzyme concentration, I_d decreases with increasing d and increases with increasing D, whereas I_k increases with increasing d and is independent of D. This presumes that the concentration of all species throughout the layer is the same and that consumption of any substrate at the outer surface does not lead to a depletion near the electrode (where the measured reaction is taking place). In practice, a non-linear enzyme reaction is more probable and an accurate calibration of the electrode must take this into consideration. Several models have been devised that are able to predict the concentration gradients through the immobilized layer [7.8-23]–[7.8-24]. The need to consider concentration gradients and cosubstrate limitations is important in ensuring that the biosensor performs in the way in which it was intended.

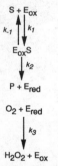

Scheme 7.8-5. Kinetic steps of oxidase enzyme catalysis

Mediators

In Scheme 7.8-5, it seems as if the electron transfer might occur from the reduced enzyme directly to the electrode. In practice this rarely occurs and, even for the enzymes where it has been observed, it cannot be achieved with repeated efficiency and so is not ideal for an analytical signal. The electrode signal must therefore depend upon an intermediary. Such a cosubstrate intermediary, which can reoxidize the enzyme and itself be reoxidised at the electrode, is called a *mediator*.

Scheme 7.8-6 shows the three classes of enzyme most frequently encountered in amperometric enzyme electrodes. Many electron acceptor molecules and complexes have been considered for the role of mediator; it can be seen in Scheme 7.8-6a that the oxidases can use their natural mediator oxygen, because the product of its interaction with the enzyme is hydrogen peroxide, which is electrochemically active. In other cases, synthetic mediators such as ferrocene (I) (η^5-bis-cyclopentadienyl-iron) derivatives and many others [7.8-25]–[7.8-27] are required.

In most cases, a mediator is sought which will complete the enzyme redox cycle and return to its initial oxidized form. In this situation, it is always recycled, and should not cause the reaction to become limited by its depletion. Unfortunately, this is not always the case, and cosubstrate limitation may exist. The two cases can be identified by their different characteristic responses.

As discussed in the previous section, the biorecognition element of the biosensor exists in a layer next to the transducer. For an amperometric enzyme-linked sen-

(I)

ferrocene

sor, the magnitude of the current signal is optimised according to the Thiele moduli, Φ_m^2 and Φ_s^2:

$$\Phi_m^2 = \frac{d^2 k_2 [E_0]}{D_m [Med_b]} \qquad (7.8\text{-}7)$$

$$\Phi_s^2 = \frac{d^2 k_2 [E_0]}{D_s [S_b]} \qquad (7.8\text{-}8)$$

where $[Med_b]$ refers to the bulk mediator concentration. These are the modelling parameters [7.8-28], which enable the change in the concentration gradient through the enzyme layer to be predicted. From the concentration gradient the current can be predicted (compare with Eq. 7.8-5). The important information that we can extract from these parameters is the d^2 relationship (Eqs. 7.8-7 and 8) which shows how sensitive the response will be with varying thickness of the enzyme layer, and also the importance of controlling the concentration of enzyme $[E_0]$ for the biosensor to be reproducible. It is also possible to predict what would happen to the electrode response if one of the parameters in the Thiele modulus were varied. For example, Fig. 7.8-9 shows the responses for an enzyme electrode with different layer thicknesses, where (a) the mediator is recycled, and (b) the mediator is consumed. In (a) the thicker layer leads to a higher signal, but does not alter the linear range of the device; whereas in (b), the effect of cosubstrate limitation when the mediator is not recycled causes the signal to increase with thickness, *but* with a smaller dynamic range.

Oxidase electrodes

The first report of a biosensor has been attributed to Clark and Lyons [7.8-29] who described an 'enzyme electrode' that used glucose oxidase as a selective bio-recognition molecule for glucose. The enzyme was held next to a platinum electrode in a membrane sandwich. The Pt electrode, polarized at $+0.6\,V$ vs SCE, responded to hydrogen peroxide produced by the enzyme reaction (Scheme 7.8-7). This report lead to the development of the glucose analyser for the measurement of glucose in whole blood (Yellow Springs Instrument 23 YSI, 1974). It would be difficult to find a better matched transduction of a biological system than that available in such amperometric enzyme biosensors – i.e., the *electron transfer* concerned with a *redox enzyme*. Nevertheless, the parameter measured is not the direct electron transfer between enzyme and electrode, but the oxidation of a product of the transformation – hydrogen peroxide. The transduced electrode signal is therefore an indirect one.

Although this is indirect, it appears most satisfactory. Unfortunately, in some situations it is unsuitable to base the measurement on hydrogen peroxide. Such examples are low pO$_2$ where the reaction will become limited or in samples where measurement of peroxide on an electrode suffers from considerable interference from other electroactive species (such as Vitamin C, paracetamol, etc.). In these situations a mediator is required, whose oxidising power can compete with that of

The amperometric signal results from the electroactive species which diffuses to the electrode and the size of the signal is modulated by the properties of the diffusion layer.

Oxidase enzymes which produce hydrogen peroxide can be linked to the amperometric determination of peroxide.

(a) FAD-Oxidase Enzymes

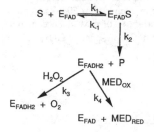

(b) NAD-Dehydrogenase Enzymes

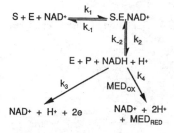

(c) PQQ-Dehydrogenase Enzymes

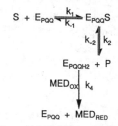

Scheme 7.8-6. Common assay schemes involving redox enzymes (a) FAD oxidases, (b) NAD dehydrogenases, (c) PQQ dehydrogenases

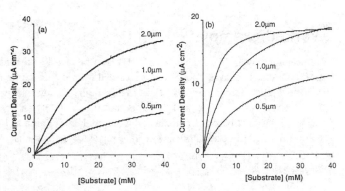

Fig. 7.8-9. Effect of layer thickness on the normalised response for a mediated redox electrode, where (a) the mediator is recycled (b) the mediator is consumed

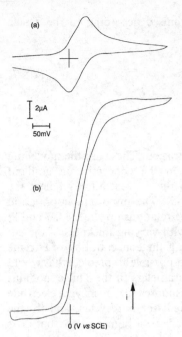

Fig. 7.8-10. (a) Cyclic voltammogram of Diaminodurene, recorded at $50\,mVs^{-1}$ in solution containing 0.05 mM glucose, at pH7. (b) as (a), but in the presence of GOD ($24\,\mu M$)

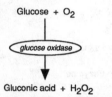

Scheme 7.8-7. Oxidation of glucose catalysed by glucose oxidase

Scheme 7.8-8. Reversible transfer of electrons to and from a mediator

Scheme 7.8-9. Interaction of a redox mediator with a redox enzyme

the natural re-oxidant of the enzyme. The kinetics of this redox catalysis are therefore of prime importance. As described in Sec. 7.2, a current/voltage plot of a reversible redox couple (Scheme 7.8-8) in solution obeys the *Randles–Sevcik* equation:

$$I_P = -0.4463nF\left(\frac{nF}{RT}\right)^{1/2}[S]D^{1/2}v^{1/2} \tag{7.8-9}$$

In the case of a coupled reaction (e.g. Scheme 7.8-9) with a fast scan speed (v) and small k_4, the current/voltage plot may show no change. As v becomes comparable with k_4, Med_{ox} will be removed from the system by the enzyme reaction and the 'apparent' concentration of Med_{red} will increase. At very slow scan speeds the reverse peak may disappear altogether and the cathodic current reach a limiting plateau (Fig. 7.8-10) that is independent of sweep speed, given by:

$$I = nF[Med_{ox}](Dk_5[E_{red}])^{1/2} \tag{7.8-10}$$

Identification of the catalytic mechanism is easily made through analysis of $i/v^{1/2}$ which increases with v. A working curve published by Nicholson and Shain in 1964 [7.8-30] allows k_4 to be estimated from the ratio of the catalysed current (I_k) to the diffusion limited current (I_d). When the enzyme is saturated with substrate ($[S] \gg K_M$), the model can be approximated to a first order reaction, thereby allowing a pseudo first order rate constant, k, to be obtained. If k is calculated for different concentrations of $[E]$ therefore, a second order rate constant, "k", is described.

Worked example:

Ferrocene monocarboxylic acid (FcCOOH) is to be tested as a mediator for glucose oxidase (GOD). The data in Table 7.8-2 is collected from the oxidation peak current obtained from the cyclic voltammogram of FcCOOH in buffer solution; (i) in the absence of GOD ($I = I_d$ given by Eq. 7.8-7), and (ii) in the presence of GOD and a saturating amount of glucose (0.05 M) (I becomes dominated by the catalytic reaction as v decreases).

For a fixed concentration of enzyme, I_k is related to I_d by the factor:

$$\sqrt{\frac{kRT}{nFv}} = \sqrt{\frac{k}{a}} \tag{7.8-11}$$

Table 7.8-2. Data from the oxidation current peak for ferrocene monocarboxylic acid in the presence of glucose with (I_k) and without (I_d) glucose oxidase.

SCAN SPEED v (mVs^{-1})	(i) I_d (mA)	5.7 µM [GOD]	12.6 µM [GOD]	(ii) I_k 16.4 µM [GOD]	23.7 µM [GOD]	32.4 µM [GOD]
100	17.0	19.4	23.4	24.6	27.8	33.4
75	14.6	17.4	21.8	23.8	27.2	32.6
50	11.8	15.4	20.4	23.9	26.6	32.0
25	8.2	13.4	19.4	22.2	25.8	31.0
10	5.0	12.8	18.8	21.6	25.0	29.8
5	3.6	12.7	18.4	21.2	24.6	28.8
2.5	2.4	12.6	18.0	20.8	23.8	28.0

For a measured I_k/I_d, $\sqrt{(k/a)}$ can be obtained from the working curve [7.8-30]. When $k > a$ (low v), the relationship is linear and dominated by the catalytic process. In this region, a plot of (k/a) versus $1/v$ gives a straight line of slope kRT/nF. Figure. 7.8-11 plots k/a for different GOD concentrations, yielding k.

$$k = k''[\text{GOD}] \tag{7.8-12}$$

where k is the pseudo first order rate constant, and k'' the second order rate constant. Figure 7.8-12 takes k, extracted from Figure 7.8-11 and plots it against [GOD] to obtain k''. A second order rate constant of 1.9×10^5 l mol^{-1} s^{-1} is obtained. This compares with 15×10^5 l mol^{-1} s^{-1} quoted for the oxidation of GOD by oxygen and so that this ferrocene derivative is shown to be a good mediator.

For an oxidase electrode, designed to monitor the oxidation current of Med_{red}, and which can employ either oxygen or the mediator in the reoxidation step, the influence of oxygen on the accuracy of the signal becomes important. In this situation, k_3 and k_4 (Schemes 7.8-5 and 7.8-9, Scheme 7.8-6a) are competing. Such a mediated enzyme electrode can under-read if the mediation kinetics (k_4) do not out-compete the oxidation by molecular oxygen (k_3).

A mediator can be used instead of oxygen to reoxidize the oxidase, but it must compete with oxygen.

Optimization of one enzyme-linked assay can lead to adjacent schemes and have implications beyond the analyte of original interest. For example, Scheme 7.8-10a/b shows how the glucose oxidase system can be applied in a 'competition' reaction for glucose with other enzymes using glucose as a substrate, and thus lead, for example, to an assay for creatine, hexokinase or creatine kinase [7.8-31]. Alternatively, when the mediator is used as a label for an antigen, the antigen-

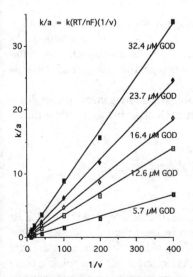

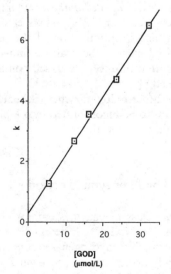

Fig. 7.8-11. k/a versus [GOD] yielding a set of k values

Fig. 7.8-12. k versus [GOD] yielding k''

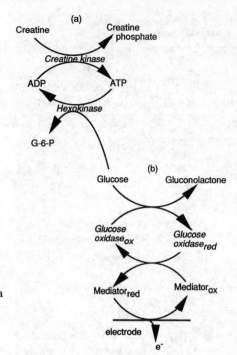

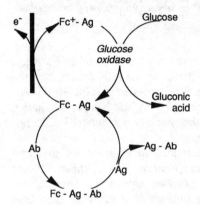

Scheme 7.8-10. Reaction pathways utilising a redox enzyme linked amperometric measurement via a redox mediator. (b) assay for glucose; (a) + (b) assay for creatine, creatine kinase or hexokinase

Scheme 7.8-11. Reaction pathway utilising a redox enzyme linked amperometric measurement via a redox mediator for competitive immunoassay using Fc-labelled antigen

mediator complex could be used in an enzyme-linked immunoassay (Scheme 7.8-11). This latter scheme [7.8-32] depends on the *ferrocene (Fc)-labelled antigen-antibody complex* being *deactivated* for electron transfer to the enzyme, and the *uncomplexed labelled antigen* remaining *active*.

Such mediator-linked assays can seem like elegant solutions for a biosensor. Nevertheless, optimization of the reagents alone is not a complete solution. As discussed earlier, the reagents must be immobilized to interact with both analyte and transducer as a self-contained system before the 'biosensor' label is attached.

The other redox enzymes (Scheme 7.8-6) involve the same general principles if they are to be incorporated into a biosensor, but each class involves subtle differences in consideration.

NAD-linked enzyme electrodes

Reduction of NAD^+ at an electrode leads to dimers and is not a good analytical signal.

A large class of enzymes employ the nicotinamide nucleotide, NAD, or else its phosphate, NAD(P), instead of molecular oxygen in their redox processes. These are normally dehydrogenases concerned with hydrogen transfer reactions, and are typically as described in Scheme 7.8-6b. The NAD cosubstrate is a soluble component which is associated with the enzyme during reaction but is able to leave its binding site in the enzyme and move into bulk solution.

Many NAD-linked hydrogenases also exist where the enzyme reaction is tuned to operate in the reverse direction – thus leading to NAD^+ production. In a biosensor format, a signal would thus be derived from the reduction of NAD^+. Reduction of NAD^+ at an electrode tends to lead to a dimer via a radical polymerization rather than to the reduced co-factor, NADH. These dimers become adsorbed on to the electrode surface thus making NAD^+ unsuitable for use as a biosensor component.

Oxidation of NADH to NAD^+ can be carried out on a graphite electrode, but usually at a high overpotential or accompanied by electrode fouling [7.8-33]; it is therefore more desirable to perform an indirect oxidation via a mediator or link it to another enzyme reaction in a cascade, as shown before in Scheme 7.8-10. These redox agents require the same properties as described in the preceding section and, although the kinetics of the mediation must not limit the reaction, here there is no competition with molecular oxygen.

> Oxidation of NADH at an electrode usually needs a high overpotential.

Pyrroloquinoline quinone enzymes

Quinoproteins utilize quinone species as cofactors and prosthetic groups. In most bacterial quinoproteins, pyrroloqinoline quinone (PQQ) is tightly but noncovalently bound. Most of the quinoproteins are dehydrogenases, and often do not require any further soluble cofactors: the natural electron acceptor being a copper protein or membrane bound ubiquinone. These enzymes can be used in the same way as oxidases. Routine assays employ artificial electron acceptors (Scheme 7.8-6c), and, like the NAD-linked systems discussed above, this artificial mediation is not susceptible to oxygen interference in the same way as for oxidases because O_2 is not a cosubstrate for the enzyme [7.8-34], [7.8-35].

Cells and cellular components

This discussion has so far concentrated on the enzyme as the biological recognition component. This component need not always be an isolated enzyme, but may instead be a whole cell [7.8-36] or cellular component. Microbial biosensors have been described with some specificity for analytes ranging from CO_2 to Vitamin B. Tissue preparations have also been described which utilise sources whose variety appears limited only by the imagination: banana pulp or apple powder, for example, contain a high concentration of polyphenol oxidase which can be used in the estimation of dopamine and catechols [7.8-37], [7.8-38]. In the majority of instances these cell based devices have been related to a change in the O_2 tension which has been monitored at an amperometric electrode mounted behind the cell layer. This relates the cell 'respiration' to the presence of a given analyte and, like the measurements already described, it is an indirect signal.

> Whole cells can be used as the recognition element.

Unlike the free enzyme systems discussed above, transport of both the analyte and the 'messenger species' between the electrode and the biorecognition element within the cell, may be restricted by the cell walls and membranes, and the chemistry of the redox messenger molecule (the mediator) must be chosen to have the right membrane transport properties, as well as the right electrochemistry. Factors such as lipophilicity and surface interactions play a more important role than in enzyme preparations.

An example of cell mediation can be found in the monitoring of herbicide residues via the photosynthetic electron transport (PET) pathway by utilizing cyanobacteria or thylakoid membranes [7.8-39]. For many herbicides the mode of action is as inhibitors of PET, often acting between the two photosystems (PSI and PSII) as indicated in Fig. 7.8-13, and the result is a decrease in the photocurrent.

> Some herbicides can be assayed by looking at their inhibition of the photosynthetic pathway.

In principle, it would be possible to have one or several redox cycles between the redox protein of interest and the electrode. For example, in the PET system described above, there is an NADPH-dependent dehydrogenase, which is located in the plasma membrane, which is connected with the PET output (top of Fig. 7.8-13). This can be measured via mediation (e.g., with ferricyanide) and avoids such problems as the need to traverse membranes by the mediator; however,

NADP is a longway 'downstream' of the original inhibition site and so such an indirect route is more likely to be susceptible to interference by other species of 'non-analyte' origin.

p-Phenylenediamines are good electron acceptors for PSII. In general C-substituted p-phenylenediamines are able to enter the cell (Fig. 7.8-14), are reduced at PSII and exit from the cell. N-substituted derivatives however, remain in the lumen after reduction at PSII, and therefore do not complete the mediation by carrying the message to the electrode [7.8-40].

Potentiometric

Potentiometric measurements are related to the Nernst equation.

Not all analytes can be readily assayed via a redox system and parameters other than electron transfer must be probed. Indeed, even where a redox enzyme is available for the analyte in question, it may not always be desirable to deploy an amperometric technique. An alternative is to consider a potentiometric measurement.

Interpretation of the signal

The measured potential (E) is given by the Nernst equation (Sec. 7.3), and for a change in activity of an ion i:

$$E = E^0 \pm \frac{RT}{nF} \ln a_i \tag{7.8-13}$$

so that the measurement parameter is logarithmically related to the concentration of the measurand. As has already been seen (Sec. 7.3), the trick with a potentiometric measurement is to make it respond selectively to just one ion; the variation in the case of a biosensor is that there should be a change in an ion transport as a result of the biorecognition event. [H^+] is the primary variable, although the change in pH may come about directly, or through equilibria with CO_2 or NH_3. In the latter case, an ammonium ionophore-containing membrane or pH-sensitive surface could be employed.

In an *enzyme*-linked pH change, the final response of the pH sensor depends on the balance of all the equilibria involving H^+: protonation and deprotonation reactions of the enzyme reaction products, buffering capacity etc. Considering just the direct 'output' from the enzyme reaction, without the equilibria associated with buffering capacity etc., and assuming that [H^+] is directly related to substrate (S) consumption, then the measured potential will be related to [S]. As seen in the previous discussion concerning an amperometric steady-state model, the behaviour is predicted by the superimposition of the diffusional considerations in the immobilized enzyme layer on the enzyme kinetics. Here there is no mediator involvement or competition with molecular oxygen but instead a direct measurement of a product of the enzyme reaction. Using a similar development as described before, and starting from:

$$\frac{d[S]}{dt} = \frac{D_s d^2[S]}{dy^2} = \frac{k_2[E_0][S]}{K_M + [S]} \tag{7.8-14}$$

where y is the distance across the membrane, produces the same Theile modulus, giving information about the source of change in signal magnitude.

$$\Phi^2 = \frac{d^2 k_2[E_0]}{D_s[S_b]} \tag{7.8-15}$$

Applying suitable boundary conditions gives substrate and/or product concentrations across the enzyme layer. The solution shows that where $K_M \gg [S_b]$, the surface concentration at the underlying potentiometric membrane is related to [S_b]. However, when $K_M \ll [S_b]$ the response will depend on the relative rates of the enzyme-limited reaction and diffusional mass transport, and is independent of [S_b].

As mentioned before, the balance between enzyme-related and buffering equilibria contributes to the measured signal. A theoretical kinetic model considering

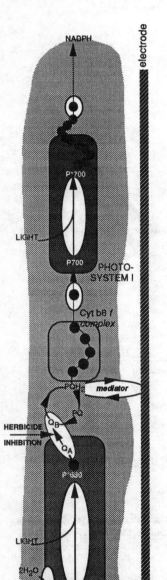

Fig. 7.8-13. Schematic of photosynthetic electron transport chain (PET). Input H_2O and Light; output NADP and O_2. Proteins of the PET represented by their shorthand abbreviation, or the symbol, ●

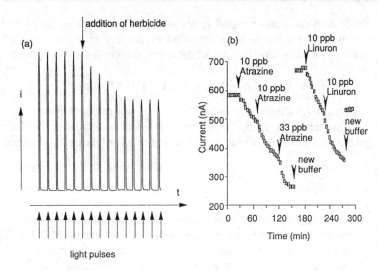

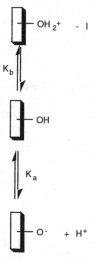

Fig. 7.8-14. Diaminodurene (C-substituted p-phenylenediamine) mediated Photocurrent in cells of Synechococcus. (a) Electrode current in response to light pulse. Charge transport in PSII is activated and intercepted by mediator at the Q protein (fig 13). (b) Detection of 2 PET active herbicides. The peak photocurrent is plotted vs time

all these associated phenomena predicts the steady-state response of an enzyme-based pH sensor [7.8-41]; the existence of a 'Nernstian' slope is not predicted and does not have the same significance as found in an ion-selective electrode making a direct ion-transport measurement.

An attempt has been made to rationalize the effect due to a protein at an interface [7.8-42], and to show how an oxide FET interface can be employed to monitor the dynamic response of proteins. At the amphoteric interface of an inorganic oxide, surface hydroxyl groups exist in an equilibrium (Scheme 7.8-12) and thus the surface potential, Ψ, will be influenced by the bulk pH ([H_b^+]), depending on the buffering capacity of the surface. Surfaces having large values of $-OH^+$ and $-O^-$ keep the value of [H_s^+] (the 'solution' H^+ in close proximity to the surface) constant over a large pH range and ψ will then respond in a Nernstian way to [H_b^+].

When a protein is also on the surface the equilibria become more complex. A protein is made up of many peptides containing ionizable groups each with its own pH equilibrium (Scheme 7.8-13). The ratio of the activities of the internal (i_{in}) and external (i_{ex}) mobile ions within this 'system', gives rise to the Donnan potential:

$$\psi_D = \frac{RT}{nF} \ln \frac{a_{i_{ex}}}{a_{i_{in}}} \qquad *.14$$

At pH's displaced from the isoelectric point, pI, this potential will give the difference between the 'internal pH' of the protein and the solution. Any changes in $a_{i_{ex}}$ will cause a change in ψ_D and induce proton release or take-up by the protein to restore equilibrium. During this restoration, the immediate environment of the protein will experience a transient surge in the pH due to the influx/eflux of protons. If the protein is immobilised in the vicinity of an ion selective field effect transistor (ISFET), this ion concentration stimulus can be measured, because the ISFET reacts very fast to changes in surface charge.

Enzyme linked assays

Potentiometric measurements have been most frequently developed around pH sensitive electrodes and the same analytical reagents have been used in pH-FETs. Any of the enzyme pathways which result in a change in H^+ can be applicable (e.g. schemes 7.8-2 and 7.8-14).

The use of biorecognition molecules. linked to an assay, induce a specificity to the response, which biosensors use as the basis of their operation. However, the specificity is only as good as the specificity of the biorecognition system itself. A less specific recognition system may be a disadvantage, but in some instances a specificity for other substrates can be used to advantage. For example, the catalytic cleavage of carbon-fluorine bonds by horseradish peroxidase, provides a useful means of detection of biorecognition system itself. A less specific recognition system may be a disadvantage but in some instances a cross-specificity for

Scheme 7.8-12. Amphoteric equilibrium at interface of inorganic oxide

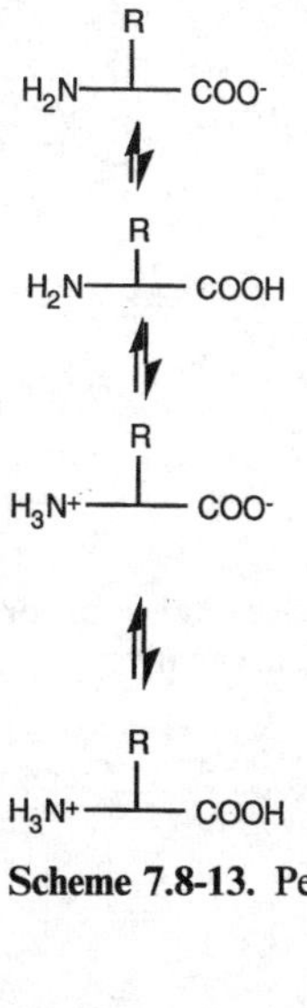

Scheme 7.8-13. Peptide equilibria

A potentiometric signal can often be induced indirectly by causing a change in surface charge or in the selective transport of a species.

other substrates can be used to advantage. For example, the catalytic cleavage of carbon-fluorine bonds by horseradish peroxidase in the presence of a fluoride ion sensor [7.8-43] provides a useful means of detection of organofluoride compounds.

$$H_2O_2 + X-F_n \xrightarrow{peroxidase} X-F_{n-1} + F^- + H_2O \qquad (7.8\text{-}16)$$

Section 7.3 shows other potentiometric measurement systems which might also be suitable for coupling to an enzyme reaction. Consider the different classes of enzyme and identify which would be suitable for linking to a potentiometric assay.

Immunoassay

Another application of potentiometric sensors and FET devices is in immuno-assay involving antibody-antigen complex formation [17.8-44]. The theoretical basis for these measurements is not always clear. For example, Aizawa [7.8-45] showed that a membrane specific for the Wassermann antibody, prepared by casting a lipid antigen of cardiolipin, phosphatidylcholine and cholesterol in triacetylcellulose, and mounted in a potentiometric cell, would respond to the antibody. It is suggested that this response results from changed ion-exchange properties in the membrane when the antibody is bound. Schasfoort *et al*, [7.8-46] have reported a means of applying a series of step changes in electrolyte concentration and monitoring the transient membrane potential using a fast responding H^+ sensitive FET (see Sec. 7.8-3) as a method of immunosassay. This technique would be appropriate for the measurement of any event at the surface of a FET, which involved a change in surface charge.

Optical

Optical biosensors operate in an extrinsic or intrinsic mode.

In the previous sections on electrochemical biosensors, several different bioassays have been manipulated to create a biosensor. In each example it was shown how the biorecognition mechanism could be linked into a physical detection method, either directly, or via mediators. Using *the same approach*, biosensors can be devised using different transduction methods. This section will not, therefore, repeat the philosophy of this approach but will focus more on the 'peculiarities' of optical transduction.

Optical biosensors have been considered in two modes which differ according to optical configuration. In the intrinsic mode the incident wave is not directed through the bulk sample, but propagates in a wave guide and interacts with the sample at a surface within the evanescent field. In the extrinsic mode, the incident light interacts directly with the sample, either passing through, or reflected from the sample phase. In this case the light source does not have to be carried within a wave guide although it is sometimes convenient to capitalize on the ability to carry the light source to a remote sensing site along an optical fiber.

Some assays of biological media can be performed without the intervention of the biorecognition surface, which is a characteristic of a biosensor, since the analyte possesses an intrinsic optical property. For example, *in situ* oxygen levels have been measured through the shift in absorbtion between hemoglobin and the oxy-hemoglobin couple. By contrast, in an optical biosensor, the assay depends on a biorecognition event in an immobilized transduction layer; and in the same way that indirect measurements and mediators may be used in the electrochemical methods, so too may labels be employed for optical systems. In optical assays, however, such labels may be 'passive' or may themselves react with the analyte or a product from the bioreaction. Many of these assay possibilities have been discussed in earlier chapters.

Optical labels may be "passive" indicators or be part of the reaction with the analyte.

In general, assays relying on the human eye for detection have tended to use chromogenic, colored reagent labels, whereas the most widely employed dyes for immunoassay have been fluorescein derivatives. Fluorescein derivatives are in fact less than ideal for sensor applications since they tend to be pH sensitive and the short excitation wavelength (λ_{ex}) limits the optical materials that can be employed. Fluorophores with longer λ_{ex} and an adequate Stokes shift are therefore favoured (e.g. Texas Red). The measurement may be of fluorescence directly or, if a quenching reaction is involved (e.g., O_2 quenching of fluorophores), based on the fluorescence life-time. In this case, the Stern-Volmer relationship becomes

$$\frac{\tau_0}{\tau} = 1 + K[S] \tag{7.8-17}$$

where K is the quenching constant and τ the fluorescent life-time [7.8-47], [7.8-48]. It must be remembered that in a biorecognition-linked assay, a quenching dependency can be prone to severe interference and so considerable thought about the construction of layers has to be given so as to aviod this.

On the whole, it is desirable to use labels with long wavelength absortion and emission maxima. This is a preference both where waveguides are to be employed, since most commonly available fibres have poor transmittance below a wavelength of 420 nm and in all formats since the simple, cheaper light sources (light-emitting diodes and diode lasers) are mostly at wavelength above 450 nm. The longer wavelength can also have additional advantages of a lower fluorescent background, and the higher sensitivity of available photodetectors. It should be remembered that it is not always straightforward to work within these constraints and that some of the most sensitive labelling reagents operate at lower wavelengths.

Interpretation of the signal

Let us consider an example where the label reacts with the anlayte. If the interaction between a label and analyte can be described very simply, according to the equilibrium:

$$A + L \leftrightarrow A - L \tag{7.8-18}$$

so that:

$$K = \frac{[A - L]}{[A][L]} \tag{7.8-19}$$

and the total label concentration in the immobilized phase is:

$$[L_T] = [A - L] + [L] \tag{7.8-20}$$

then

$$[L] = \frac{[L_T]}{1 + K[A]} \quad \text{and} \quad [A - L] = \frac{K[A][L_T]}{1 + K[A]} \tag{7.8-21}$$

and

$$[A] = \frac{1}{K} \left\{ \frac{[L_T] - [L]}{[L]} \right\} \quad \text{or} \quad [A] = \frac{1}{K} \left\{ \frac{[A - L]}{[L_T] - [A - L]} \right\} \tag{7.8-22}$$

> If the analytical reaction can be described as a reversible equilibrium, a linear analytical signal will not result.

These equations imply that the response to analyte may not appear linear across the entire concentration range of interest. In a bulk solution assay this problem is often overcome by ensuring that the amount of label employed is adjusted to cause its reaction with analyte to fall into a region where the signal is proportional to the concentration of L or $A - L$, depending on which causes the optical signal (e.g., for measurement of $[A - L]$, this would be when $[L_t] \gg [A - L]$). By contrast, in the immobilized phase it is usually more difficult to manipulate the concentrations to attain this range. Instead, a two wavelength reference can be proposed where $[A - L]$ and $[L]$ have different absorptions giving an $[L_T]$ independent ratio:

$$\frac{[A - L]}{[L]} = K[A] \tag{7.8-23}$$

The second way to use labels is as 'passive' indicators. There are several formats in which a label could be thus employed. For example, in a simple competitive binding assay in solution the response can be modelled by the following two equilibria: firstly, between an antibody or binding protein (Bp) and the analyte (A), and secondly, between the antibody or binding protein and a labelled analyte analogue (a*) – see Eqs. 7.8-24 and 7.8-25.

$$A + Bp \leftrightarrow A : Bp \qquad K_A = \frac{[A : Bp]}{[A][Bp]} \tag{7.8-24}$$

$$a^* + Bp \leftrightarrow a^* : Bp \qquad K_a^* = \frac{[a^* : Bp]}{[a^*][Bp]} \tag{7.8-25}$$

The assay must be able to distinguish between the reagents on the left and right-hand sides of the equilibria.

These equilibria would be expected to hold when Bp is immobilized if the affinity constants are independent of concentration and limitations imposed on reaction with Bp by the immobilisation. The equilibria will be independent of concentration if the binding at one Bp does not effect the binding on an adjacent Bp. If this is true and

$$[a_T^*] = [a_T^*] + [a^* : Bp] \tag{7.8-26}$$

$$[A_T] = [A] + [A : Bp] \tag{7.8-27}$$

$$[Bp_T] = [Bp] + [a^* : Bp] + [A : Bp] \tag{7.8-28}$$

then [a*] or [a* : Bp] will be measured, depending on the way in which the sensor is formatted, and their concentrations influenced by the relative affinity ratios K_A and K_{a^*}. By substitution in the equations above:

$$[Bp_T] = \left(\frac{1 + K_A[A]}{K_{a^*}}\right)\left(\frac{[a_T^*] - [a^*]}{[a^*]}\right) - [a^*] + [a_T^*] \tag{7.8-29}$$

so that the normalised signal due to $[a^*]/[a_T^*]$ is given by:

$$\frac{[a^*]}{[a^*T]} = 1 + \left(\frac{1 + K_A[A]}{K_{a^*}[a^*]}\right)\left(\frac{[a_T^*] - [a^*]}{[a_T^*]}\right) - \frac{[Bp_T]}{[a_T^*]} \tag{7.8-30}$$

and

$$[A_T] = A\left[\left\{\frac{[a^*T] - [a^*]}{K_{a^*}[a^*]}\right\}K_A + 1\right] \tag{7.8-31}$$

The ratio of $[Bp]_t : [a_T^*]$ as well ad $K_A : K_{a^*}$ are critical in tuning the range of the response and can be designed for optimal performance.

As can be deduced from this discussion, one of the main tasks required of an optical biosensor, irrespective of whether it uses a 'passive' or 'active' label, is to be able to distinguish 'free' or unlabelled analyte from the 'bound' or labelled analyte *without* performing a separation step in the assay, as one might do in classical solution-based bioassay!

Extrinsic labelled assays

Enzyme-linked extrinsic optical sensors

Optical measurements can be linked to peroxide or other products of enzyme reactions in just the same way as electrochemical sensors.

As we have seen in an earlier chapter, many of the traditional bioassays are based on optical measurements, but these are solution assays without immobilized reagents and sometimes with separation steps. Nevertheless, they already have many of the necessary analytical components to create a biosensor, so how can the transition be achieved? Optical biosensors are concerned with the interrogation of reactions at surfaces; the primary need is therefore to move the reaction on to a surface. Many of the 'home test' kits have undergone this transition and now give optically detectable measurements. Consider, for example, designing a glucose assay kit for use by a diabetic. We have seen in a previous section that glucose can be assayed in an enzyme-linked reaction involving glucose oxidase (Scheme 7.8-7). In the earlier section "Oxidase electrodes", an amperometric glucose assay was discussed that was based either on hydrogen peroxide or else a synthetic mediator.

Earlier in this chapter the bioassay was also related to hydrogen peroxide, but in this instance, the peroxide was detected by linking into another reaction culminating in the production of a colored complex in the presence of peroxide. The optical biosensor 'test strip' must include all the reagents of this assay in an immobilized form. The sample is placed on the strip and color development is monitored by a 'reader' containing a light source and detector. In this instance the chromogenic complex formed is the 'label'; it is an 'active' label since it is involved in the reaction with peroxide but, unlike the steps above (Eqs. 7.8-18 to 7.8-23), its formation is a *fast* irreversible step occurring alongside the main enzyme mechanistic pathway. There is a linear dependency between the signal and the peroxide concentration: the concentration of the latter depends on the kinetics of the enzyme steps and not on the kinetics of the development of the color.

A different example employs an enzyme-linked reaction in which a change in H^+ is involved (e.g., see Schemes 7.8-2 and 7.8-14). Here the 'label' would be one whose optical properties change with pH. In this instance the labelling step is an equilibrium (Eq. 7.8-17), which means that a linear dependency between analyte and signal only exists under certain conditions. The choice of pH indicator is influenced by the pH range of interest and the pK_a of the immobilized chromophore/fluorophore (which may be different to that in solution). A large number of such compounds exist, with different pH ranges and different wavelength characteristics, so the feasibility of linking a reaction seems high. However, as mentioned in the earlier section on potentiometric measurement, the problem associated with monitoring a reaction via a pH change is the characterization of the instrinsic pH of the sample, its buffering capacity and the sensitivity of the equilibrium to ionic strength, all of which can have a considerable influence on the assay and must be taken into account when interpreting the results. These complications can also make the route undesirable.

Nonetheless, in the same way that the electrochemical sensors utilize a response from a cosubstrate *or* product of the enzyme reaction, so too can a corresponding optical route be devised. Many of the assays discussed in preceding chapters would be suitable for reformatting in this way. For example, xanthine oxidase immobilized to a suitable substrate can be used to determine hypoxanthine and xanthine [7.8-49]. The hydrogen peroxide produced in the oxidase reaction (see Scheme 7.8-6a) can be detected by a chemiluminescent mechanism, based on luminol (Scheme 7.8-15). As can be inferred from the scheme, the sensor requires the co-immobilization of peroxidase and a supply of luminol – which makes it especially popular in flow injection analysis systems since luminol is water soluble. Unfortunately, there is often a pH 'mismatch' in the system because a basic pH of around 10–12 is generally required for the chemiluminescence whereas many enzymes have a lower pH optimum.

An attractive variation is to use a bioluminescent assay, since here too no excitation source is required. Bioluminescence has been primarily associated with the firefly; the process of interest in an assay linked to the firefly's luminescence, is that associated with the *luciferase* enzyme-catalysed oxidation of the luciferin heterocycles (Scheme 7.8-16) or the bacterial *luciferase* reaction with reduced flavin cofactor (Scheme 7.8-17). The reaction alone does not offer a general assay method, but in conjunction with cofactors such as ATP, FMN or NADH, its links with other metabolic path ways is ensured. Detection limits as low as 10^{-15} mol NADH have been recorded with linearity over five orders of magnitude.

Extrinsic labelled immunoassay

In an immunoassay the measurand is not linked to a product of a catalyzed reaction, but is a label, marking the presence of a complexed or free antigen or antibody. The immunoassay principle is discussed in Sec. 7.9. As well as being used in the enzyme-linked test kits (e.g., glucose), immunoassay is also widely found self-test kits; for example, in a pregnancy test kit – although here the result usually relies on the user to see a color development and it is not monitored automatically (i.e., the detector is the human eye!). In such a test strip, separation of labelled and unlabelled complexes is usally performed by a phase or chroma-

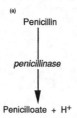

(a)
Penicillin

penicillinase

Penicilloate + H^+

(b)
Urea

urease

NH_4^+ + HCO_3^-

Scheme 7.8-14. pH-linked enzyme catalysed reactions

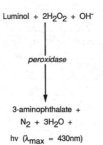

Luminol + $2H_2O_2$ + OH^-

peroxidase

3-aminophthalate +
N_2 + $3H_2O$ +
hv (λ_{max} = 430nm)

Scheme 7.8-15. Luminol

Bioluminescence has been shown to achieve a very low detection level.

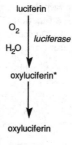

luciferin

O_2 ⎫ *luciferase*
H_2O ⎭

oxyluciferin*

oxyluciferin
+hv (λ=562nm)

Scheme 7.8-16. Luciferase catalysed light generation

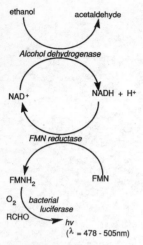

Scheme 7.8-17. Utilizing bacterial luciferases, to form an excited state complex with reduced flavin cofactor ($FMNH_2$), in the assay of ethanol

Measurement integrity depends on ability depends on ability to separate bound and free label.

tographic separation. The assay then follows the binding protein-analyte model (Eqs. 7.8-24 to 7.8-31) in which the label is most often an enzyme and the color is developed in subsequent reaction steps.

In principle, it should be possible to extend these general systems to give a more quantitative read-out in cases where the assay demands more than just a *positive/ negative* output. Several configurations for the sensor are possible depending on whether competitive or sandwich assay format is preferred. In theory, an especially viable alternative would seem to be the competitive displacement of a fluorescent label. As this is an equilibrium, fouling or contamination of the surface should not alter the absolute result, although it might obviously influence the signal-to-noise ratio [7.8-50]. A possible drawback is that the success of such displacement assays is very much dependent on the relative binding kinetics of the label-conjugate and the sample. These need to be significantly different to achieve a quantiative result.

A common factor in all these assays is that the measurement integrity depends on a separation step: the two forms of the label need to be physically separated from one another to allow the assay to distinguish them. One way to achieve this separation is to cause the reaction to take place in the 'blind volume' of the sensor while performing the measurement within a defined detection window. Such a set-up can be established using a two compartment cell: the biorecognition reagent (e.g., a binding protein) is immobilized in one compartment, i.e., the blind volume, while label and analyte move between the two compartments. The assay depends then on the detection area being situated in the correct region.

A natural detection zone is created by the acceptance cone at the end of a fiber (Fig. 7.8-15a). A typical sensor might thus use two fibers to guide the light to and from the distal end of the fiber [7.8-51], where the sensing chemistry is localized. Optical fibers have an effective field of view which is described by the numerical aperture (NA) of the fiber and, as can be seen from Fig. 7.8-15a, the higher the NA, the greater the field of view. In the two-fiber configuration described above, a poor choice of fiber can result in a large 'blind volume' (Fig. 7.8-15b), which will obviously reduce the measured signal. Nonetheless, the focussing of the field to give visible and blind volumes can be used to advantage in a competitive assay format (Fig. 7.8-16). In this configuration a binding protein (Bp) is immobilized in the blind volume and the sample (A) and labelled analogue (a*) compete for binding sites, leaving unbound a* diffusing in the illuminated volume, where it is measured.

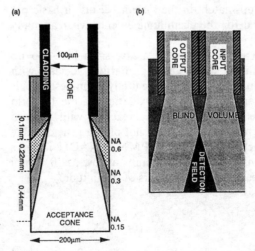

Fig. 7.8-15. (a) Field of view in sample medium at the end of a fiber, as a function of the numberical aperture (NA). (b) Detection field in a sample, using two adjacent fibers for excitation and detection

Worked example:

Concanavalin A (ConA) (Bp in the model above) was bound to hollow dialysis tubing mounted at the end of a fiber optic in the blind volume (Fig. 7.8-16) [7.8-52]. [Bp_T] was estimated to be 10 mM, based on the loading per unit volume of tubing. The fiber had a diameter of 100 μm and numerical aperture of 0.3. For a dialysis tubing of internal diameter

200 μm, the blind volume length was ~0.2 mm (Fig. 7.8-16a). A solution of FITC-labelled dextran (a*) was allowed to infuse the fiber to a total concentration [a*$_T$] of 1.5 mM. The sensor was thus 'charged' for use. Glucose (A) was introduced, diffused into the dialysis tube and displaced dextran (a*) from ConA competitively ($K_A = 3.2 \times 10^2$ M^{-1}; $K_{a^*} = 7.5 \times 10^4$ M^{-1}). Table 7.8-3 gives the fluorescence output due to a*, measured in arbitary units by a photodiode detector. From experimental calibration, the maximum signal (for [a*] = [a*$_T$] = 1.5 mM) can be estimated, and thus [a*]/[a*$_T$] obtained.

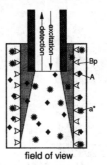

Fig. 7.8-16. 'Blind volume' immobilization of binding protein for competitive assay

Table 7.8-3. Date from a 'blind volume' competitive assay of glucose

GLUCOSE (A) (mM)	Free DEXTRAN (a*) arb units	Total DEXTRAN (a*$_T$) arb units	[a*]/[a*$_T$]	Free DEXTRAN (Calc) mM
2.5	2.4	13	0.185	0.28
5.0	2.9	13	0.223	0.34
7.5	3.5	13	0.269	0.40
10.0	4.2	13	0.323	0.48
12.5	5.1	13	0.392	0.59
15.0	6.1	13	0.469	0.70
17.5	6.7	13	0.515	0.77
20.0	7.2	13	0.554	0.83
22.5	7.5	13	0.577	0.86
25.0	8.1	13	0.623	0.93
27.5	8.3	13	0.638	0.96
30.0	8.7	13	0.669	1.00
32.5	8.9	13	0.685	1.03
35.0	9.0	13	0.692	1.04
37.5	9.3	13	0.715	1.07
40.0	9.4	13	0.723	1.08
42.5	9.5	13	0.731	1.10
45.0	9.6	13	0.738	1.11

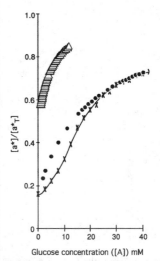

Fig. 7.8-17. Experimental response from a blind volume competitive binding assay, compared with modelled data points, assuming simple competitive binding kinetics. x – experimental; ● – calculated; Δ – predicted

Plotting fluorescence output versus glucose concentration (Fig. 7.8-17a) reveals that for [A$_T$] > 20 mM, [a*] becomes insensitive to changes in [A$_T$]. Figure 7.8-17c predicts the output, based on Eqs. 7.8-30 and 7.8-31 above and taking the values for [a*$_T$], K_{a^*}, K_A and [Bp$_T$] as given. It can be seen that the predicted range for the sensor is ~0.5 mM glucose. Several factors may influence this considerable deviation. The model is based on a homogeneous reaction. It has already been noted in the earlier section "Theoretical Considerations in Designing an Immobilisation", that heterogeneous reactions involving interactions at a surface, have kinetics that can be a function of surface coverage and that K_{a^*} and K_A are probably influenced by total sugar concentration ([a*$_T$] + [A$_T$]). The model should also consider the concentration of Bp within the diffusion layer thickness of the immobilized layer (not the averaged concentration based on the total volume). In fact a better 'fit' to the experimental data is obtained if an immobilized [Bp$_T$] of 70 mM is assumed (Fig. 7.8-17b), but real characterization of the response must also take into account diffusion of a* from the reaction layer in the 'blind volume', across the field of view. The signal will only be independent of diffusion for concentrations where equilibration time is less than the diffusion time.

Intrinsic labelled assays

In this mode, the reagents are immobilized on a surface of a waveguide which is in contact with the sample phase. Light propagating along the guide is able to interact with the sample phase (refractive index n_2), which is within about 0.5–1.5 wavelengths of the interface; this region is the evanescent field. Considering the ray model for light propagation in a wave guide (refractive index n_1) (Fig. 7.8-18) it can be seen that the propagation follows a series of total internal reflections. This can be seen when $n_1 > n_2$ from Snell's law,

Waveguides provide an intrinsic mode optical sensor.

$$\frac{\sin \theta}{\sin \phi} = \frac{n_2}{n_1} \qquad (7.8\text{-}32)$$

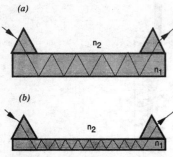

(a)

(b)

Fig. 7.8-18. Ray diagram model of light propagating through (a) 'thick' film and (b) 'thin' film waveguides

where θ is the incident angle and ϕ is the angle of transmission. When $\sin \theta > n_2/n_1$ there is no real solution for ϕ, the angle of transmission is *complex* and these are the conditions of total internal reflection. The transmitted wave is an evanescent wave normal (z-direction) to the interface (x-direction) with the electric field, E, which is given by the equation

$$E = E_0 \exp \frac{\left\{ 2\pi n_1 \left(\sin^2 \theta - \left(\frac{n_2}{n_1} \right)^2 \right)^{1/2} z \right\}}{\lambda}$$

where only the z component remains (under other coditions the electric field will include an x component) so that the field is decaying in the z direction with a penetration depth d_p, given by the equation

$$d_p = -\frac{\lambda}{2\pi n_1 (\sin^2 \theta - (n_2/n_1)^2)^{1/2}} \qquad (7.8\text{-}33)$$

which indicates that the attenuation of the signal will be greatest at the interface and zero in the bulk.

This effectively means that there is a 'detection zone' at the interface, and a 'blind zone' in the bulk. Although this configuration has the advantage that the light does not pass through the bulk sample, it has the disadvantage that the interaction between the light and surface immobilized reagents is less in normal multimode fibers than with reagents in a bulk solution. Considering the ray model for the propagation of light through a guide, it can be seen that interaction with the surface only occurs at N 'points' of total internal reflection within the wave guide (see Fig. 7.8-20). N is given by the equation,

The "detection zone" is at the surface of the waveguide.

$$N = \frac{L}{2a} \cot \theta \quad \text{where } L \text{ is length and } 2a \text{ the thickness.} \qquad (7.8\text{-}34)$$

Thin film wave guides increase the interaction because 'a' is smaller [7.8-52], but their fabrication is far from straightforward. Ion exchange at the surface of a glass slide, giving a layer of higher refractive index, is one solution. Another is the utilization of low temerature cured phosphate glasses. Thick film waveguides are comparatively easy to make and offer a low cost sensor material. Nevertheless, using suitable thin film waveguides would be preferable in many instances so as to allow many of the extrinsic mode assays to be converted into intrinsic optical devices with adequate sensitivity.

Optical interrogation without labels

At every interface light is refracted and/or reflected giving information about the materials each side of the interface.

From Eq. 7.8-33 it can be seen that at a given wavelength the electric field for the evanescent wave will change according to the refractive index ratio of the two phases (waveguide, n_1 and surface medium, n_2). In a biosensor where the surface layer is the assay reaction zone, the reaction can be followed directly if it causes a change in the optical characteristics of the layer (like refractive index or thickness). There are several ways in which these changes can be amplified so that they can be used as an analytical signal.

Interference techniques

When two waves combine they interfere with each other, either constructively or destructively. The utilization of this phenomenon is called *interferometry*. When used in a biosensor, it usually involves the manipulation of reflection in the intrinsic mode. Two examples will be discussed here.

The model for reflection described above only considers reflection at a single interface. However, if several layers are placed together, then at each interface both reflection and transmission occur (Fig. 7.8-19). This ultimately leads to a 'compound' reflection, which is the combination of the reflections from each interface. The intensity of the recorded reflectance is a function of the thickness of each layer and the complex refractive index. It is very sensitive to small changes in

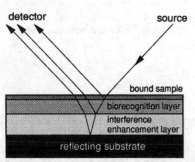

Fig. 7.8-19. Reflectance from multilayers

either parameter, and thus cn be used to detect interactions at the surface of the multilayer. For maximum sensitivity, the dielectric nature of the materials have to be carefully chosen (the dielectric constant, ε is related to the refractive index $n = \varepsilon^{1/2}$). A nonabsorbing dielectric (e.g., SiO_2) with dielectric constant $\varepsilon_1 = \varepsilon_{1,r}$ (i.e., the dielectric constant only has a real term) is deposited to a thickness of the order of $\lambda/4$ on a highly reflecting substrate with dielectric constant $\varepsilon_0 = \varepsilon_{0,r} - i\varepsilon_{0,i}$ (e.g., silicon with a complex dielectric constant). Changes in the biorecognition layer deposited on top of the SiO_2 layer can then be followed by measuring and analysing the reflectance, since the reflected waves have been modulated by the layers. This multilayer system has been demonstrated as a transduction system for antibody-antigen binding [7.8-9]. Nonetheless, in common with all such 'unlabelled' immunoassay methods, non-sepcific binding must be eliminated or fully characterised if this is to have potential as a biosensor.

Such considerations demonstrate the problem of obtaining a reference point in these biosensors. Two types of calibration are necessary. The first concerns the calibration of the analyte; the second requires information on the background signal. A background can be obtained with reference to a blank, the question is: how can this be incorporated into a biosensor? In the case of waveguides interacting with the sample via the evanescent field, this requires two distinguishable surfaces: one for the specific reactions and one for the non-specific reactions. One way is in a Mach-Zehnder-like configuration (Fig. 7.8-20). First the incident light is split; it then travels through the reference or sample waveguide and subsequently recombines. The intensity of the received, multiply-reflected beam is influenced by the optica pathlength and the reflectivity at the interface. The reflectivity will of course be different at the surfaces modified by biorecognition element and reference interface. When non-specific interactions have been eliminated, this configuration has achieved measurements of antibody-antigen binding.

Grating couplers

A grating coupler allows light to be coupled into a waveguide. The grating is a periodic structure (e.g., triangular, saw-tooth, sinusoidal) (Fig. 7.8-21) which is embossed on the surface of the guide. As the light is incident on the grating it excites a guided mode within the waveguide, dependent on the grating period. A way of enhancing the effect of a refractive index change at an interface is to utilize changes occurring at the surface of a *grating coupler* on a waveguide. For example, in Fig. 7.8-21, light incident in the vicinity of the coupler, will excite guided modes in the waveguide depending on the refractive index of the adjacent medium. In this format the surface has been sensitized to precise changes in refractive index. This format is, therefore, suitable for conducting unlabelled immunoassays since the binding between immobilised antibody and sample antibody gives a greater density of these molecules within a wavelength of the surface (note that the penetration depth is of the order of a wavelength, Eq. 7.8-34) and thus will result in a change of refractive index. For example, a competitive immunoassay can be demonstrated for pesticide detection, where a triazine hapten is covalently immobilized to an aminosilane modified coupler surface, and the binding of antibody observed by measuring the incoupling angle [7.8-53, 7.8-54]. In order to detect the presence of the triazine pesticides, the sample had to be pre-incubated with the antibody, and then introduced to the immobilized hapten. Obviously, to ensure that the user need have no intervention in this assay procedure, this incubation step must be integrated into the design.

Surface plasmon resonance (SPR)

In this technique, the waveguide is covered by a thin metal (plasmon) film, and changes within the field propagating along the surface of the metal film are probed. This is achieved since the action of an exterior electric field (in the waveguide) of the right strength (see Eq. 7.8-33 for the electric field of an evanescent wave) transfers energy from the waveguide and oscillating surface charges are produced that propagate along the metal surface. The oscillations must couple with E_z, the field extending in the z direction, so the metal surface must be the correct distance from the surface of the waveguide within the penetration depth, d_p, of the evanescent field. For a surface plasmon wave existing between two

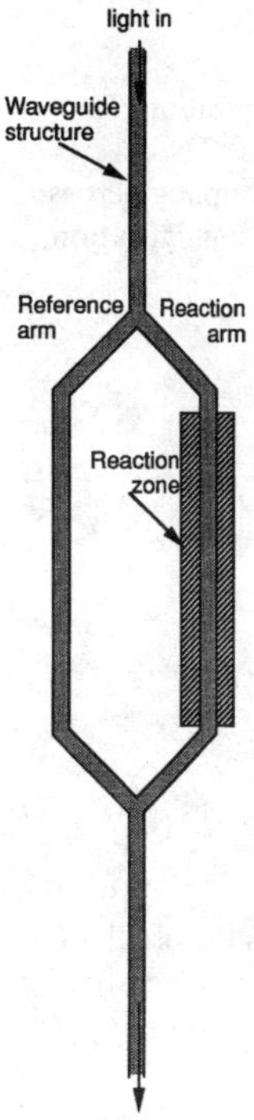

Fig. 7.8-20. Mach-Zehnder-like interferometer

Different optical configurations can be devised to enhance the measurement of changes in refractive index.

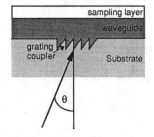

Fig. 7.8-21. Grating coupler

Fig. 7.8-22. Excitation configurations for surface plasmon resonance. (a) The Kretschman configuration, coupling across a thin metal film. (b) The Otto configuration, coupling across an air gap

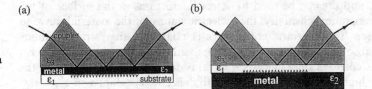

media of dielectric constants ε_1 and ε_2, the wave is confined at the interface with wave vector $\mathbf{k}_x$ in the propagation direction and $\mathbf{k}_{zl}$ and $\mathbf{k}_{z2}$ in the +z and −z directions. Continuity across the interface requires that:

$$\frac{\mathbf{k}_{z1}^2}{\varepsilon_1^2} = \frac{\mathbf{k}_{z2}^2}{\varepsilon_2^2} \tag{7.8-35}$$

An expression for $\mathbf{k}_x^2$ can therefore be obtained

$$\mathbf{k}_x^2 = \left(\frac{\varepsilon_1 \varepsilon_2}{\varepsilon_1 + \varepsilon_2}\right)\left(\frac{\omega}{c}\right)^2 \tag{7.8-36}$$

Surface plasmon resonance is the oscillation of surface charges (electrons in a metal) propagating at the surface.

The surface plasmon wave can be excited by incident light of the same frequency and $\mathbf{k}_x$ component. For light incident on a metal of dielectric constant (ε_2) from a dielectric material (ε_3) the $\mathbf{k}_x$ component will be:

$$\mathbf{k}_x^2 = \varepsilon_3 \left(\frac{\omega}{c}\right)^2 \sin^2\theta \quad \text{where } \theta \text{ is the angle of incidence} \tag{7.8-37}$$

The plasmon resonance at the $\varepsilon_1/\varepsilon_2$ interface of the metal can therefore be excited when the $\mathbf{k}_x$ components of the incident and plasmon waves match:

$$\mathbf{k}_x^2 = \left(\frac{\varepsilon_1 \varepsilon_2}{\varepsilon_1 + \varepsilon_2}\right)\left(\frac{\omega}{c}\right)^2 = \varepsilon_3 \left(\frac{\omega}{c}\right)^2 \sin^2\theta \tag{7.8-38}$$

For this to occur $\varepsilon_3 > \varepsilon_1$.

Experimentally, resonance is observed as a minimum in the reflectance at an angle of incidence θ. As can be deduced from Eq. 7.8-8 changes in ε_1 will require a change in θ to maintain the resonance condition, so monitoring the position of the reflectance minimum is very sensitive to reaction on the metal surface (in ε_1). This is the source of signal measured in an SPR sensor, since the surface of the metal can be modified with a biorecognition molecule which will respond to analyte, changing ε_1 at the interface.

The earlier reports of SPR in the detection of protein binding dealt with investigations of unlabelled interactions having, for example, an antibody immobilized on the metal film [7.8-55], [7.8-56]. It soon became clear that although bound antigen could be detected, in some analytical environments there was often no means of distinguishing this signal from any other non-specific interaction with the surface [7.8-57]. Refractive index labels and other optical labels can circumvent this problem and, since SPR is so sensitive, can offer low detection limits.

Most of the applications of this technique have been in immunoassay or other protein binding assays rather than enzyme-linked assays. To find whether there are other applications, it is necessary to work out whether the parameter measured can be used. In this case the assay must involve a change in surface ε, to be able to employ this technique. Many of the assays which rely on binding tetween an immobilized biorecognition molecule and an analyte might be suitable if the difference between the unbound recognition molecule and analyte-bound recognition molecule is sufficient – and if the reagents can be successfully immobilised with retention of activity. For example, a DNA probe sequence was immobilized on a thiol-modified silver film deposited on a waveguide in a SPR format [7.8-22]. The probe showed a characteristic change in the SPR excitation on binding complementary DNA (Fig. 7.8-23). However, when the same DNA probe sequence was bound to the surface via a silane, the probe could not couple the complementary DNA and thus failed as a biosensor. This observation brings us full circle to consider the method of immobilization of the biorecognition element as paramount in the construction of a biosensor.

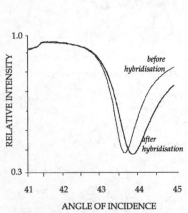

Fig. 7.8-23. Immobilized DNA probe monitored by surface plasmon resonance at a silver film

7.8.4 Conclusions

This chapter has looked briefly at the components of biosensors. It has concentrated on two main classes of biosensor, namely – electrochemical and optical – but the reader should not be misled into believing that the story begins and ends with these examples. As discussed at the start, the possibilities are as wide as your imagination. The criteria for success are the choice of a well-matched measurement parameter, a good method of immobilization, which can be readily developed for bulk fabrication, and the relevance of your measurement to the chosen application!

References

[7.8-1]   Hall, E.A.H., *Biosensors*. Milton Keynes: Open University Press, 1990.

[7.8-2]   Cass, A.E.G. (Ed.), *Biosensors: A practical approach*. Oxford: IRL Press, 1990.

[7.8-3]   Göpel, W., Hesse, J., Zemel, J.N. (Eds), *Sensors: A Comprehensive Survey*. Weinheim: VCH, 1991, Vols. 2 and 3.

[7.8-4]   Wolfbeis, O.S. (Ed.), *Fiber Optic Chemical Sensors and Biosensors*. Florida: CRC press, 1991, Vols. I and II.

[7.8-5]   Ikeda, T., Hamada, H., Miki, K., Senda, M., *Agric. Biol. Chem. (Japan)* 1984, 49, 541–543.

[7.8-6]   Randriamahazaka, H.N., Nigretto, J.M., *Electroanalysis* 1993, 5, 231–241.

[7.8-7]   Amine, A., Kaufman, J.M., Patriarche, G.J., *Talanta* 1991, 38, 107.

[7.8-8]   Wang, J., Wu, L.H., Lu, Z.L., Li, R.L., Sanches, J., *Anal. Chim. Acta* 1990, 228, 251

[7.8-9]   Duschl, C., Hall, E.A.H., *J. Coll. Int. Sci.* 1991, 114, 368–380.

[7.8-10] Kutner, W., Storck, W., Doblhofer, K., *J. Incl. Phenom.* 1992, 13, 257.

[7.8-11] Hale, P.D., Lan, H.L., Boguslavsky, L.I., Kagan, H.I., Okamoto, Y., Skotheim, T.A., *Anal. Chim. Acta* 1991, 251, 121.

[7.8-12] Yon Hin, B.F.Y., Smolander, M., Compton, T., Lowe, C.R., *Anal. Chem.* 1993, 65, 2067.

[7.8-13] Martens, N., Hall, E.A.H., *Anal. Chim. Acta* 1994, 292, 49–63.

[7.8-14] Hench, L.L., West, J.K., *Chem. Rev.* 1990, 90, 33–72.

[7.8-15] Braun, S., Rappoport, S., Zusman, R., Avnir, D., Ottolenghi, M., *Materials Lett.*, 1990, 10(1,2), 1–5.

[7.8-16] Narang, U., Psasad, P.N., Bright, F.V., Ramanathan, K., Kumar, N.D., Malhotra, B.D., Kamalasanan, M.N., Chandra, S. *Anal. Chem.* 1994, *66*, 3139.

[7.8-17] Hall, C.E., Hall, E.A.H., *Anal. Chim. Acta* 1993, 281, 645–653.

[7.8-18] Urban, G., Jobst, G., Kepplinger, F., Aschauer, E., Tilado, O., Fasching, R., Kohl, F. *Biosensors Bioelectronics* 1992, 7, 733.

[7.8-19] Urban, G., Jobst, G., Aschauer, E., Tilado, O., Svasek, P., Varahram, M., Ritter, Ch., Riegebaure, J., *Sensors and Actuators* 1994, 18–19, 592.

[7.8-20] Hall, E.A.H., Hall, C.E., Martens, N., Mustan, N., Datta, D, "Uses of Immobilized Biological Compounds," in: *ASI Series 252*: Guilbault, G.G., Mascini M., (Eds.) Kluwer Academic Publishers, 1991.

[7.8-21] Willner, I., Riklin, A., *Anal. Chem.* 1994, 66, 1535.

[7.8-22] Liley, M.J., *"Surface Plasmon Resonance for the detection of DNA Hybridisation" PhD Thesis*, University of Cambridge, 1990.

[7.8-23] Martens, N., Hindle, A., Hall, E.A.H., *Biosensors Bioelectronics* 1995, 10?

[7.8-24] Leypoldt, J.K., Gough, D.A., *Anal. Chem.* 1984, 56, 2896.

[7.8-25] Cass, A.E.G., Francis, G., Hill, H.A.O., Higgins, I.J., Aston, W.J., Plotkin, E.V., Scott, L.D.L., Turner, A.P. F., *Anal. Chem.* 1984, 56, 667.

[7.8-26] Degini, Y., Heller, A., *J. Am. Chem. Soc* 1988, 110, 2615.

[7.8-27] Schlapfer, P., Mindt, W., Racine, P., *Clin. Chim. Acta* 1974, *57*, 283.

[7.8-28] Martens, N., Hall, E.A.H., *Anal. Chem.* 1994, 66, 2763–2770.

[7.8-29] Clark, L.C. Jr., Lyons, C., *Ann. N. Y. Acad. Sci.* 1962, 102, 29.

[7.8-30] Nicholson, R.S, Shain, I., *Anal. Chem.* 1964, 36, 706.

[7.8-31] Davies, P., Green, M.J., Hill, H.A.O., *Enzyme Microb. Technol.* 1986, 8, 349–352.

[7.8-32] diGleria, K., Hill, H.A.O., McNiel, C.J., Green, M.J., *Anal. Chem.* 1986, 58, 1203.

[7.8-33] Wang, J., Chen, Q., *Elecroanalysis* 1994, 6, 850.

[7.8-34] Khan, G.F., Kobatake, E., Ikariyama, Y., Aizawa, M., *Anal. Chim. Acta* 1993, 281, 527–533.

[7.8-35] D'Costa, E.J., Higgins, I.J., Turner, A.P.F., *Biosensors* 1986, 2, 71–87.

[7.8-36] Wang, J., Naser, N., *Anal. Chim. Acta* 1991, 242, 259.

[7.8-37] Deshpande, M.K., Hall, E.A.H., *Biosensors Bioelectronics* 1990, 5, 431–448.

[7.8-38] Chen, Y., Tan, T.C., *Biosensors Bioelectronics* 1994, 9, 401–410.

[7.8-39] Carpentier, R., Lemieux, S., Mimeault, M., Purcell, M., Goetze, D.C., *Bioelectrochemistry and Bioenergetics* 1989, 22, 391–401.

[7.8-40] Martens, N., Hall, E.A.H., *Photochem and Photobiol* 1994, 59, 91–98.

[7.8-41] Glab, S., Koncki, R., Hulanicki, A., *Electroanalysis* 1991, 3, 361–364.

[7.8-42] Bergveld, P., "Stimulus Response Measurements on Protein Containing Membranes, Deposited on an ISFET Surface", in: *Uses of Immobilized Biological Compounds*: Guillbault, Mascini (Eds.), Netherlands: Kluwer, 1993, pp. 289–308.

[7.8-43] Cowell, D.C., Dowman, A.A., Lewis, R.J., Pirzad, R., Watkins, S.D., *Biosensors Bioelectronics* 1994, 9, 131–138.

[7.8-44] Schasfoort, R.B.M., Bergveld, P., Bomer, J., Kooyman, R.P.H., Greve, *J., Sensors and Actuators* 1989, 17, 531.

[7.8-45] Aizawa, M., Kato, S., Suzuki, S., *J. Membrane Sci.*, 1977, 2, 125–130.

[7.8-46] Schasfoort, R.B.M., Bergveld, P., Kooyman, R.P.H., Greve, J. *Biosensors Bioelectronics* 1991, 6, 477–489.

[7.8-47] Lippitsch, M.E., Pusterhofer, J., Leiner, M.J.P., Wolfbeis, O.S., *Anal. Chim. Acta* 1988, 205, 1.

[7.8-48] Bannwarth, W., Schmidt, D., Stallard, R.L., Hornung, C., Knorr, R., Mueller, F., *Helv. Chim. Acta.* 1988, 71, 2085.

[7.8-49] Hlavay, J., Haemmerli, S.D., Guilbault, G.G., *Biosensors Bioelectronics* 1994, 9, 189–195.

[7.8-50] Krull 1989

[7.8-51] Offenbacher, H., Wolfbeis, O.S., Fürlinger, E., *Sensors and Actuators* 1986, 9, 73–79.

[7.8-52] Mansouri, S., Schultz, J.S., *Bio/Technology* 1984, 2, 385.

[7.8-53] Nellen, P.M., Lukosz, W., *Biosensors Bioelectronics* 1991, 6, 517–525.

[7.8-54] Bier, F.F., Schmid, R.D., *Biosensors Bioelectronics* 1994, 9, 125–130.

[7.8-55] Cullen, D.C., Brown, R.G.W., Lowe, C.R., *Biosensors* 1987/88, 3, 211–225.

[7.8-56] Liedberg, B., Nylander, C., Lundström, I., *Sensors and Actuators* 1983, 299–306.

[7.8-57] Cullen, D.C., Lowe, C.R., *Sensors and Actuators* 1990, B1, 576–579.

Questions and problems

1. What changes need to be made to convert suitable solution-based bioassays to biosensors? Which bioassays are ideal for this conversion? Which bioassays would have problems in this conversion?
2. What types of biological recognition systems are normally used in biosensors? What types of assay is each group suitable for? What ideas do you have for other biorecognition systems which could be used?
3. Redox enzymes are often used in amperometric biosensors, to measure the enzyme's substrate. How do they work? What are the different classes of

redox enzyme? What additional reagents are required in the amperometric biosensor?

4. What analyte was measured in the first biosensor? How was the measurement achieved?

5. There are different approaches to producing an immobilized reagent layer, what are they? When a molecule interacts with a surface, what are the sequence of events which may occur?

6. Give some examples of biorecognition events that result in a change of pH. How can this change be measured in a biosensor format?

7. How would you design an optical biosensor to measure glucose and cholesterol?

8. Bioluminescence is a very sensitive assay system. Describe the steps of this assay, and describe how it can be linked into other systems to measure different analytes.

9. An optical biosensor can be devised with the reagents on the surface of a waveguide. Light which propagates along the waveguide is able to interact with the reagent on the surface. The interaction can be monitored by recording changes in the propagated light. How does this happen? If the reagent was placed in the bulk solution, instead of on the surface of the waveguide, what effect would this have on changes in the propagated light?

10. Why is it not possible to excite the surface plasmon by shining light on a flat metal surface? What are the different configurations used for surface plasmon resonance?

11. What factors might effect the accuracy of a biosensor? Give examples with reference to particular types of biosensor.

7.9 Immunoassay

Learning objectives

■ To learn about antibody-antigen equilibria and how to use them in analysis
■ To understand different immunoassay formats
■ To discover how immunoassay can be followed using labels

7.9.1 Introduction

Immunoassay exploits the unique specificity of an *antibody* binding an *antigen* in order to *selectively* recognize and determine analytes that are either antibodies or antigens. High selectivity can be obtained, because other interfering compounds in the sample may not be recognised in an immunoassay. Antibodies represent one of the major classes of protein, collectively known as immunoglobulins (Ig). Within this family, the most abundant antibodies are IgG ($\sim 70\%$), a "Y"-shaped protein (Fig. 7.9-1) consisting of two identical heavy – light chain heterodimers, linked by disulfide bridges. There are two binding sites, formed by the variable globular domains of the heavy and light chains in the Fab region, and the specificity of these sites for binding an antigen is determined by the amino acids of the complementarity – determining regions. There are a maximum of 17 amino acids in the binding cavity, so the binding site on the antigen must be of this order of size. An antigen need not be a protein; it can be any macromolecule which is capable of inducing an immune response, and thus causing the formation of an antibody against it. Low molecular mass molecules (such as hormones or drugs that may be of interest in analysis) are not antigenic, but antibodies with a binding site specific to them, can be created by coupling the low molecular mass molecule to a protein carrier. These low molecular mass compounds are known as haptens.

Antibodies are produced by the B-lymphocytes, each B-cell displaying one specificity on its surface, so that the foreign antigen only becomes bound to those cells with "matching" binding sites. This binding stimulates these cells to divide, and to produce large amounts of the IgG with the same specificity. On the antigen, the recognition site may only be a small area of the total molecule, and thus there may be several potential recognition sites on the same antigen. The result of this process is therefore the production of *antisera* which contain a heterogeneous mixture of antibody species, with varying affinity. Such antisera is known as polyclonal, and unlike monoclonal antibodies, they exhibit recognition kinetics contributed by each of the activated recognition sites or *epitopes*.

As antibody (Ab) and antigen (Ag) approach, the primary binding force is ionic (long-range interaction), operating over distances of 10 nm. Slow exclusion of hydration water, allows the formation of hydrogen bonds and at a distance of 0.5–0.15 nm, the van der Waals forces (short range interaction) between dipoles on adjacent atoms, become more important, and the bond strengthens. The binding can be described by the equilibrium:

$$K_{eq} = \frac{[\text{Ab-Ag}]}{[\text{Ab}][\text{Ag}]} \qquad (7.9\text{-}1)$$

where K_{eq} is the equilibrium constant and Ab-Ag the antibody-antigen complex. Values of K_{eq} typically range from $10^6\,\text{Lmol}^{-1}$ to $10^{12}\,\text{Lmol}^{-1}$ but values less than $10^8\,\text{Lmol}^{-1}$ are generally of no use in immunoassay.

Defining the total antibody concentration, $[\text{Ab}]_t$, as the sum of the concentration of bound, $[\text{Ab-Ag}]$, and free antibody $[\text{Ab}]$:

$$[\text{Ab}_t] = [\text{Ab}] + [\text{Ab-Ag}] \qquad (7.9\text{-}2)$$

An antibody is a protein with a specific binding domain for antigens.

Antigens are molecules which are capable of inducing an immune response.

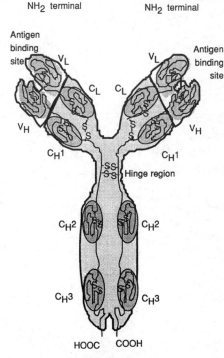

NH₂ terminal   NH₂ terminal

Fig. 7.9-1. IgG structure heavy (subscript H) and light (subscript L) chains, showing antigen binding site

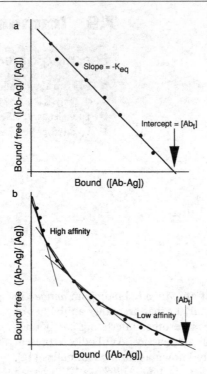

Fig. 7.9-2. Scatchard plot for (a) monoclonal antibody; (b) polyclonal antibody

we can rearrange Eq. 7.9-1

$$\frac{[\text{Ab-Ag}]}{[\text{Ag}]} = K_{eq}[\text{Ab}] = K_{eq}[\text{Ab}_t] - K_{eq}[\text{Ab-Ag}] \tag{7.9-3}$$

In a Scratchard plot, the ratio [Ab-Ag]/[Ab] is plotted against [Ab-Ag]. In such a plot-K_{eq} is the slope, and the intercept with the x-axis is [Ab]$_t$.

Antibodies may be monoclonal or polyclonal.

A serum with single epitope specificity represents a monoclonal system. In the corresponding Scratchard plot, a straight line is observed (Fig. 7.9-2a). In a polyclonal system, the Scatchard plot is more complex. It is characterized by a multiple set of straight lines (Fig. 7.9-2b)

In designing an immunoassay, the objective is to follow the equilibrium between antibody and antigen. The assay investigates the binding between antigen and antibody and distinguishes between bound and unbound antigen. Since this is an equilibrium (Eq. 7.9-1), it is not linear with antigen concentration. Therefore, it is important to know the characteristic binding response, as a function of antigen concentration.

From Eq. 7.9-1 and:

$$[\text{Ag}_t] = [\text{Ag}] + [\text{Ab-Ag}] \tag{7.9-4}$$

$$\frac{[\text{Ab-Ag}]}{[\text{Ag}]_t} = \frac{1}{1 + 1/(K_{eq}[\text{Ab}])} \tag{7.9-5}$$

$$\frac{[\text{Ab-Ag}]}{[\text{Ag}]_t} = \frac{([\text{Ab-Ag}]/[\text{Ag}])}{[\text{Ab-Ag}]/[\text{Ag}] + 1} \tag{7.9-6}$$

The curves resulting from this relationship, at different antibody concentrations, are shown in Fig. 7.9-3a. It can be seen that, for a given concentration of antibody, the binding fraction varies only as a function of antigen concentration in a critical range. Below this is a region of constant proportional binding, where changes in antigen concentration give imperceptible changes in the binding fraction. It is apparent from Fig. 7.9-3b that increasing the concentrations of antibody causes the region of maximum change (i.e., the potential dynamic measurement range) to shift to higher antigen concentrations.

On the other hand, Fig. 7.9-3c shows that, for sufficiently low concentrations of antibody ($<0.01/K_{eq}$), the fraction of occupied antibody sites reflects the analyte

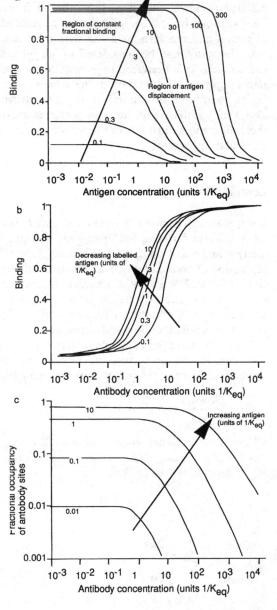

Fig. 7.9-3. (a) Binding fraction vs antigen concentrations for different antibody concentrations (b). Binding fraction vs antibody concentrations for different antigen concentrations. (c) Fractional occupancy of antibody sites vs antibody concentration, for different antigen concentrations

concentration *and* becomes independent of total antibody concentration. However, this has not been widely exploited, although it implies that the concentrations involved are so small that the depletion of antigen from the sample due to binding is negligible, with the advantage of being independent of sample volume.

In order to follow this binding, irrespective of the concentration levels involved, a label is normally introduced, which can be monitored. The means of achieving this (see below) can differ, but the general philosophy of design must consider the constraint imposed by the binding equilibrium and the necessary dynamic range.

Assay formats

All immunoassays depend on measurement of the "fractional occupancy" of the recognition sites. This implies that the assay principle relies on the assessment of occupied sites or, indirectly, on measuring unoccupied sites. As will be shown in this section, considering the fractional occupancy of antibody sites leads to the conclusion that, if the residual unoccupied sites are measured, maximum sensitivity is achieved when antibody concentrations tend to zero; the formats which satisfy

Immunoassay follows the "fractional occupancy" of recognition sites.

this condition, are usually termed "competitive". Conversely, techniques which measure occupied sites directly require high levels of antibody, and are termed "non-competitive". It should however be realized that these terms are not entirely objective, and the paradox arises that, particularly in the "non-competitive" format, the assay may be considered to fall into either category, depending on the chosen step. It is therefore important to remember that these rather ambiguous terms apply to consideration of the *primary recognition* event. Because of the different conditions for optimization, competitive and non-competitive immunoassays may differ significantly in performance over a given concentration range; these differences must be considered when deciding on the optimum format for a particular application.

Competitive assay

Competitive immunoassay involves the competition between the sample analyte and labelled analyte for a limited number of recognition sites.

Competitive immunoassay gives a measure of sites not occupied by sample; it may have different formats, but in essence, the sample analyte is mixed with labelled analyte and these compete for binding sites (Fig. 7.9-4). The method requires separation of bound and free antigen or antibody, in order to determine their relative amounts. For an antigen analyte, the presence of labelled antigen provides a means of assessing the relative partitioning. From the curves in Fig. 7.9-3, it can be deduced that optimum sensitivity can be obtained by decreasing the concentration of labelled antigen and antibody.

Let us examine the component signals in a competitive assay: considering first the condition where there is no sample and the signal response due to bound antibody-antigen in the presence of labelled antigen only is given by:

$$S_0 = \frac{[\text{Ab-Ag}]}{[\text{Ag}]_t} \times S_t \tag{7.9-7}$$

where S_t is the signal which would result from direct measurement of $[\text{Ag}]_t$. The non-specific binding response S_{ns} is a function of the fractional non-specific binding of remaining free antigen.

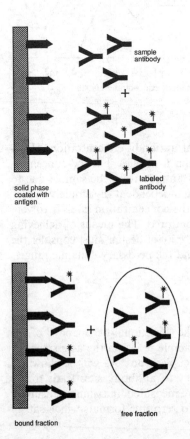

Fig. 7.9-4. Competitive assay for antibody testing

$$S_{ns} = \left\{ 1 - \frac{[Ab - Ag]}{[Ag]_t} \right\} \times S_t . B_{ns} \qquad (7.9\text{-}8)$$

where B_{ns} is the fraction of non-specifically bound antigen. Thus, as the concentrations of labelled antigen and antibody are reduced, in order to move the dynamic measurement region to lower concentrations, the errors due to non-specific binding begin to have a greater impact on the precision of measurement. Every combination of specific and non-specific binding will have its own optimal combination of antibody and labelled antigen combinations. Total elimination of non-specific binding or separation of bound and free phases is not achieved with 100% efficiency, so that the concentration of antibody generally has to be raised to increase the specific binding contribution.

In the presence of labelled antigen only, S_O is given by Eq. 7.9-7 so that the signal due to bound antigen, normalised to $[Ag]_t$ is given by the binding fraction. However, in the presence of sample (unlabelled antigen):

$$[Ag]_t = [Ag^\bullet]_t + [Ag^*]_t \qquad (7.9\text{-}9)$$

$$[Ag\text{–}Ag] = [Ab\text{-}Ag^\bullet] + [Ab\text{-}Ag^*] \qquad (7.9\text{-}10)$$

$$[Ag] = [Ag^\bullet] + [Ag^*] \qquad (7.9\text{-}11)$$

where $Ag^\bullet$ is the sample antigen and Ag^* is the labelled antigen, the latter giving

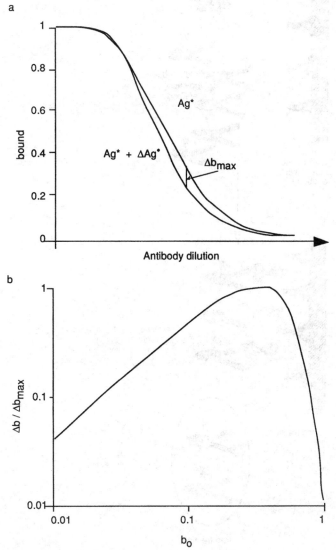

Fig. 7.9-5. (a) Antibody dilution curve for labeled antigen Ag*, and addition of a small increment of unlabeled antigen Ag*. (b) Change in binding Δb, following addition of ΔAg*, at different original binding fractions

shown for a given concentration of labelled antigen) that the signal due to bound Ag^* decreases following a small addition of sample antigen (unlabelled) $[\Delta Ag^*]$.

It is evident from these curves that the change in signal as a result of $[\Delta Ag^*]$ varies as a function of the antibody concentration used in the assay (Fig. 7.9-5b), and that the antibody level yielding maximum sensitivity depends on the labelled and target antigen concentrations. Traditionally, assay systems have tended to err on the high side for reagent concentrations, with binding fractions of 0.4–0.5. Indeed, from Fig. 7.9-3a and b, this region appears too high to lead to a large change in the binding fraction with concentration for a low target concentration range of antigen. However, the signal due to non-specific binding cannot be ignored, and sensistivity is traded against the lower interference due to non-specific binding at higher antibody concentrations.

Although the competitive immunoassay format is popular, it has practical disadvantages as well as advantages. An obvious disadvantage is that the labelling of the analyte, Ag^* may change or totally remove the binding capacity with Ab, the latter particularly if the labelling has involved a critical epitope in the binding. If the binding kinetics are different for the labelled and unlabelled antigen, this must also be included in the calculations. Occasionally, particularly where labelled haptens are involved, the label can enhance recognition. This has been seen when the labelled conjugate involves the same position on the hapten as that previously used in the protein carrier, required in antibody production.

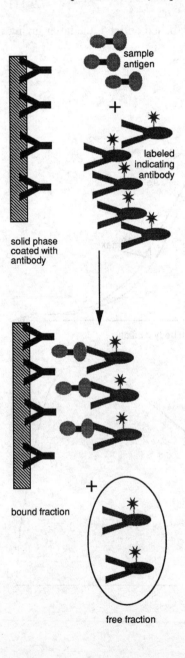

solid phase coated with antibody

sample antigen

labeled indicating antibody

bound fraction

free fraction

Fig. 7.9-6. Immunometric (sandwich) immunoassay of antigen analyte

Sandwich assay

As mentioned above, the recognition site on an antigen is only a small area, and thus there may be several different epitopes, which are spatially separated on the antigen. This feature can be exploited in a "sandwich" assay; the analyte is captured by one antibody Ab1, allowing it to be separated from the rest of the sample. The captured antigen is then incubated with excess of a second, labelled antibody Ab2, which binds only to the existing antibody-antigen complex (Fig. 7.9-6).

The labelled antibody Ab2 must be able to bind specifically to an exposed surface of the bound antigen. In the perfect sandwich assay, there would be no signal in the absence of antigen, because there would be no binding site available for the labelled antibody. In practice, there is always some signal arising from non-specific binding of Ab2 to Ab1 so that, even in the absence of total signal:

$$S_O = B_{ns}S_t \tag{7.9-12}$$

where S_t is the signal which would result from $[Ab2]_t$.

In the presence of the Ab1 layer, the binding of Ag is described by K_{eq1} according to Eqs. 7.9-1–7.9-6 (where Ab1 = Ab and $K_{eq1} = K_{eq}$), but this is then monitored by the binding of Ab2 to the bound antigen Ag$_b$, described by K_{eq2},

$$[Ag_b] = \frac{[Ab2_b]}{[Ab2_f]K_{eq2}} + [Ab2_b] \tag{7.9-13}$$

The monitored signal can distinguish between bound (Ab2$_b$) and free (Ab2$_f$) labelled antibody. Usually the bound antibody is measured so Eq. 7.9-13 suggests that it would be advantageous to use excess of Ab2 such that:

$$[Ab2_f] \gg [Ab2_b] \tag{7.9-14}$$

and Eq. 7.9-13 becomes

$$[Ag_b] \approx [Ab2_b] \tag{7.9-15}$$

However, there is an optimal concentration of labelled antibody. Higher concentrations increase non-specific binding to the capture antibody Ab1, and lower concentrations become more susceptible to measurement errors. On the other hand, the concentration of Ab1 does not influence sensitivity, so long as it in adequate supply to capture the sample antigen (normally at concentrations greater than $\sim 10^{-9}\,\mathrm{mol\,L^{-1}}$ equivalent, but $\sim 10^{-6}\,\mathrm{mol\,L^{-1}}$ equivalent is usually chosen to achieve a region of constant binding fraction >90%).

7.9.2 Design format

General considerations

The last section has outlined the basic theoretical principles, but it is possible to design an almost unlimited number of variations of recognition bindings in sequences and sandwiches, which ultimately achieve the same purpose, namely the estimation of analyte concentration! The common features of classical immunoassay techniques are that they do not measure the analtye directly, but via a labelled analyte analog or labelled analyte-recognition system. They also require separation of signal due to free and bound labelled marker; this latter requirement is a primary consideration in the assay.

Separation of bound and free fractions can be approached in different ways. In a liquid phase solution assay, precipitation, centrifugation and decantation are used, but they give rather variable results, and so are not widely favored. The first major advance on this design was made by Miles and Hales (1968), who proposed the removal of unbound labelled antibody by the introduction of a large excess of solid phase antigen (Fig. 7.9-7).

This development initiated the development of solid-phase immunoassay. However, this introduces additional complexity, particularly in terms of the thermodynamics of the interactions. Solid-phase assays differ from liquid phase assays,

In sandwich immunoassay, the sample analyte is captured by an excess of one antibody and the amount of captured analyte estimated by a second antibody, thus forming a sandwich.

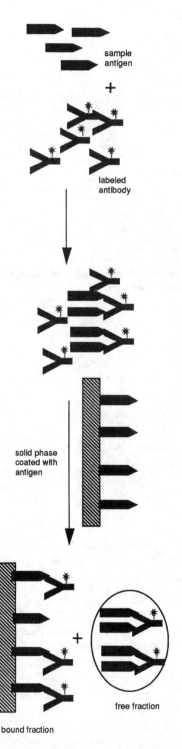

Fig. 7.9-7. Separation of bound and free fractions using an excess of solid phase antigen

both in terms of the reactivity of the immobilized antibody or antigen, and in the differences between the reaction kinetics in solution and those involving a solid phase (see also Sec. 7.8).

Surface immobilization effects

Surface immobilization may cause a reconfiguration of the protein, changing the kinetics of the immunoassay.

In a solid phase immunoassay, the solid phase is a "carrier substrate". It cannot however be considered as an inert passive component. Both chemical and physical adsorption of antibody or antigen can be employed, although the majority of immunoassays rely on the latter, non-covalent adsorption. As discussed in Sec. 7.8, even physical adsorption is not a fully characterized process. Coupling of proteins is probably hydrophobic in nature, and yet the native configuration of proteins in bulk aqueous solution tends toward surface hydrophilic groups, burying the hydrophobic groups within the polymer. It therefore seems inevitable that the binding to hydrophobic surfaces induces a confirmational change in the protein, and this may cause the loss of critical epitopes, either because they become buried, or else because the epitope itself is hydrophobic and becomes involved in the binding. It is therefore not unusual to find antibodies that work well in liquid phase assay, but fail in solid phase systems.

Developing a theoretical model for a solid-phase immunoassay can also be difficult, since several basic assumptions and concepts do not fully apply. In most solid-phase assays, the forward rate constant between the solution species and the immobilized recognition phase, is limited by diffusion, rather than the affinity constant for the recognition event. This is primarily influenced by the viscosity of the assay solution and the relative distribution of binding sites, but is also modulated by the degree of the alignment of the solution species approaches the surface.

"Concentration" of an immobilized species is an inappropriate term.

Since the biorecognition surface is a solid phase, the structure of the boundary layer at its surface, influences the reaction and the attainment of a steady state. "Concentration" is largely inappropriate to the discussion of an immobilized reagent phase in a bulk solution, however, in the boundary layer the local "concentration" of immobilized reagent is very high. Consider for example, an antibody immobilized at a surface, in a solution containing antigen (Fig. 7.9-8). The antigen binds to the immobilized antibody, causing an antigen depletion layer next to the surface, which is replaced by antigen from the bulk solution, limited by diffusion.

The surface of the carrier phase is generally considered to have a homogeneous distribution of immobilized antibody, but this is an oversimplification. Evidence suggests that the surface has a fractal geometry, with adsorbed antibodies collecting in islands of high density, separated by sparesely populated regions (Fig. 7.9-8b). The normal association/dissociation equilibrium between antibody and antigen also tends to become distorted, because the extremely high "local" surface concentration of antibody in the island clusters discourages dissociation of antigen, and thus once the antigen is bound, the process is essentially irreversible.

Physical methods of achieving separation of bound and free label

Separation of bound and free label should not distort the ratio between them.

An ideal separation method should not affect the bound: free label ratio, and should give complete separation of the two forms. However, as has been stated earlier, this is never achieved. In any case, the receptor-ligand binding is reversible, so the separation procedure must not change the reaction conditions, so that dissociation occurs. The great variety of separation methods reflects the diversity of properties of the receptor-ligand complex, and the demands made on the assay method. For example, the primary need may be speed and ease of use, rather than sensitivity or reproducibility. The result may be required as a quantitative measure, or as a positive/negative indication. All these criteria alter the specification of the assay, and the method of separation of the bound and free label in the assay must reflect these needs.

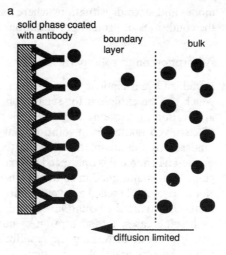

a

solid phase coated
with antibody   boundary
                       layer   bulk

diffusion limited

b

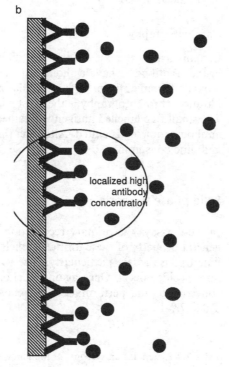

localized high
antibody
concentration

Fig. 7.9-8. (a) Boundary layer formation.
(b) Fractal geometry effects

Liquid phase

Decantation

This is conceptually the simplest method of separation; any of the associated
methods are suitable (racking, aspiration, etc.), but they can be awkward in practice, each analysis requiring a modification of the system and technique to obtain
adequate results.

Precipitation

Precipitation of immunoglobulins is readily achieved by "salting out" at neutral
pH, where they are not particularly soluble, so that in the case of a labelled antigen, the bound label can be precipitated from solution together with free residual
antibody, thus leaving free label in the supernatant. Ammonium sulfate is usually
the salt preferred, but the method suffers from a high level of non-specific binding.

A better precipitation agent is polyethylene glycol (PEG). This acts in a similar
way to a salt, removing water from the IgG structure and causing it to denature,
thus precipitating it from solution, but it provides a more controlled system. The
method has been widely used in assays for a number of different types of hor-

mones and steroids, situations where the hapten antigen remains in solution under the conditions of protein precipitation.

Adsorption on to solid particles

Proteins have a tendency to adsorb on a variety of differing materials, a property which can be exploited for separation. Cellulose, glass and silica have all found application for proteins. Powdered charcoal is the classical method for the removal of dissolved matter from solution, but in the case of IgG proteins, adsorption is hindered by the mis-match between the large protein volume and small charcoal pores. The charcoal is optimized by coating with dextran and IgG, and IgG-antigen complex is maintained in solution, whereas antigen is adsorbed onto the charcoal. In the case of a labelled antigen, the free label is removed from the solution by this method, leaving the bound label for determination in solution.

In some cases, the adsorption of antigen by the charcoal can become a competing process, disturbing the equilibrium. In any case it can be employed only where the antigen-label conjugate is sufficiently small to ensure efficient separation.

Chromatography

In many assays there is a significant difference in the sizes of the receptor and the ligand. Antibodies have a molecular mass of about 160 000, and can be readily separated from antigens with molecular masses less than 80 000. Size-exclusion gel filtration chromatography is the most widely employed chromatographic method. The small free labelled antigen is retained on the column, while the large antigen-antibody complex is eluted. Although this gives a good separation, it is expensive and time consuming, so it is unsuitable for routine, repetitive use.

Solid phase supports

In these assays the primary recognition system is attached to a solid support. A general property of these formats is their low non-specific binding, compared with liquid phase assay, particularly since washing steps can be introduced to reduce non-specific effects. Once again the choice of method is dependent on factors of convenience and performance. The choices and compromises are summarized in Table 7.9-1.

Table 7.9-1. Separation systems for immunoassay

Support	Example	Advantage	Disadvantage
Particle <20 µm	Porous glass	Easy dispensing	Separation slow
	Latex	High binding capacity	(centrifugation or other
	Magnetic particles	Fast reaction	method required)
	Microcrystalline cellulose		Magnetic precipitation slow
Particle <1 mm	Magnetic particles	Medium binding capacity	Agitation required to keep
	Sepharose beads	Magnetic precipitation fast	particle in suspension during reaction
	Sephadex beads	No centrifugation required	Slower kinetics for larger particle
Particle >1 mm	Polystyrene	No centrifugation required	Low binding capacity
			Slow reaction kinetics
			Variability in dispensing and coupling
Membrane	Nylon	No centrifugation required	Low binding capacity (reaction pathway short)
		Easy to use	Variability in binding
			Slow kinetics
Solid surface	Coated tubes	No centrifugation required	High non-specific binding
	Dipsticks		Variability in binding
	Microtiter plates	Easy to use	Slow kinetics

Beads and particles

Glass beads were among the first candidates for solid phase immobilization. IgG adsorbs onto either glass or plastic surfaces, so that this is a passive immobilization method, which is optimal at neutral pH. Higher protein loadings are obtained with particulate dimensions where 10–15% protein can become adsorbed from an incubation solution. Covalent attachment is also employed, but this usually requires greater preparation of the immobilized layer (coupling chemistry similar to those in Figs. 7.8-7 and 7.8-8). The need for this depends on the application and further treatment of the immobilized IgG.

Particle diameter is not necessarily directly related to binding capacity, since the latter also depends on surface area and surface character. However, Sephadex and celluloses are found generally to have higher binding capacities than nylon, polystyrene, or glass, although the latter two examples can be easily produced in smaller particles.

Of the particulate materials, latex (diameter $<20\,\mu m$) is popular due to its spherical geometry and high surface area. It is also of similar density to water, so it stays in suspension during the assay. Separation is then achieved by centrifugation, decantation, microfiltration, etc. However, the success of this material for immobilisation has also led to the invention of other ingenious methods of separation. For example, the inclusion of iron oxide (Robinson *et al.*, 1973) or chromium oxide causes the material to become paramagnetic, and the particles can be manipulated in a magnetic field.

For larger particles or beads, which do not normally stay in suspension, this paramagnetic doping and application of magnetic field, can be used in the converse manner, to maintain the beads in motion during the incubation step.

Membrane filtration

In this method the sample is "filtered" through a membrane in which the capture molecule (antibody or antigen) is immobilized. The unbound material is then assayed in the filtrate. The use of fibrous membranes maximises the surface area and thus binding capacity; glass fiber and cellulose fiber have been used, but a more common membrane material is nylon. An improved interaction pathway can be achieved by immobilizing the capture molecule onto polymer particles (as above), which are then entrapped in the membrane.

Immunochromatography

A further extension of these ideas is to incorporate the capture molecule in a band in a chromatographic system. In most systems using this format, the capture molecule (e.g., antibody) is immobilized on the chromatographic support material (paper, silica, polymer, or gel etc.), and fixed in a band, so that the sample (antigen) mixed with labelled antigen reacts competitively for the capture recognition sites, as it progresses past the band. The unbound material can then be assayed as it migrates beyond the capture band. Alternatively, the amount of labelled antigen captured in the band may be estimated.

Variations on this theme have also been developed: for example the capture band can be extended to the full length of the chromatographic system, so that measurement is based on the determination of the length over which the labelled antibody is captured. This is related to the *total* antigen concentration, and since the concentration of labelled antigen, added to the sample, is always the same, then for a given surface density of immobilized capture antibodies, this length depends on the sample concentration of antigen.

Plates, wells, and tubes

Probably the most convenient immobilisation surfaces for routine immunoassay are polystyrene microtitration plates, tubes, and wells. However, their effective use is critically dependent on the reproducibility of coupling. Irradiation and chemical pretreatments of the surfaces have been developed to enhance binding capacity and reproducibility, and arrays of mass produced microtitration wells improve the reproducibility of "identical" surfaces for comparative immobilization, so that the

Solid phase supports include particles, beads and surfaces; reactive surface area influences binding capacity.

assay can be readily calibrated and at least partially automated. Very many competitive and sandwich assays have been developed using these antibody- or antigen-coated surfaces.

The main disadvantage, particularly with microtitration plates, is the small surface area for reaction. The plates have about 10% of the capacity of the tubes, so that the signal that is measured in the assay must be strong enough to deal with these small quantities. In spite of these limitations, assays based on both tubes and plates have been extremely successful, probably mainly because of their ease of use rather than ultimate signal performance.

The most widely employed material is polystyrene, and in some cases, the capacity is enhanced by drying the antibody onto it; one of the main advantages is the high level of adsorption, even though the surface area is small compared with the particles discussed above. However, this absorption advantage also promotes non-specific binding, so that the assays always include a "blocking" step, utilizing a protein such as casein, to bind with any unpopulated areas of the polystyrene surface after the primary recognition molecule has been attached.

7.9.3 Labels

Radioactive labels

The usual choice of label is ^{125}I, a γ emitter, or ^3H (tritium), a β emitter, although ^{57}Co (for vitamin B_{12}) and ^{14}C have also been used. Tritium is less sensitive than ^{125}I, since the β particles emitted are weak ($E_{max} = 18.6\,keV$; $E_{av} = 5.5\,keV$), and are largely absorbed by the walls of the vessel. On the other hand, the γ-rays emitted by ^{125}I ($E = 28-35\,keV$) are not absorbed by the vessel, and the counting efficiency, using a scintillation counter, is typically of the order of 80%.

^{125}I and ^3H are the most common radioactive labels for immunoassay.

In some cases the label can be attached by simple isotope replacement, but although the hydrogens can be substituted in most potential analytes by tritium, the frequency of naturally occurring iodine-containing analytes is much rarer (e.g., thyroxine T_4). In order to use ^{125}I as a label therefore, it is usually necessary to attach it as a "tag".

Proteins

The simplest method for protein labelling is by direct substitution into the aromatic ring of tyrosine or histidine. However, the decay of the radioactive nuclide releases energy into the molecule which exceeds that involved in the chemical bonds, and therefore there is a danger that adjacent bonds will be ruptured and the molecular structure lost. Thus, although high substitution levels offer greater sensitivity, the attempted substitution level is about one ^{125}I per protein molecule, since higher levels tend to cause radiolysis, deactivating the protein.

Several methods for achieving this iodination have been proposed, but the major advances in radioimmunoassay (RIA), followed the development of the chloramine-T oxidation method of Geenwood and Hunter (1961; Scheme 7.9-1a). However, this non-specific oxidation agent is not suitable in every application, and an enzyme method, employing lactoperoxide with hydrogen peroxide has also gained popularity (Marchalonis, 1969; Scheme 7.9-1b).

Haptens and polypeptides

Labelling of peptides and proteins usually involves the aromatic ring of tyrosine or histidine.

When the peptide or hapten does not involve an aromatic ring suitable for iodine substitution, e.g., tyrosine, the usual approach is to produce a conjugate with a labelled molecule. An obvious choice for such a labeling system is a tyrosine-like residue, and the derivatives used for this purpose can be tyrosine methyl ester, tyramine or histamine, etc. (Scheme 7.9-1c). However, some reiodinated compounds are commercially available (e.g., the N-hydroxy-succinimide ester of

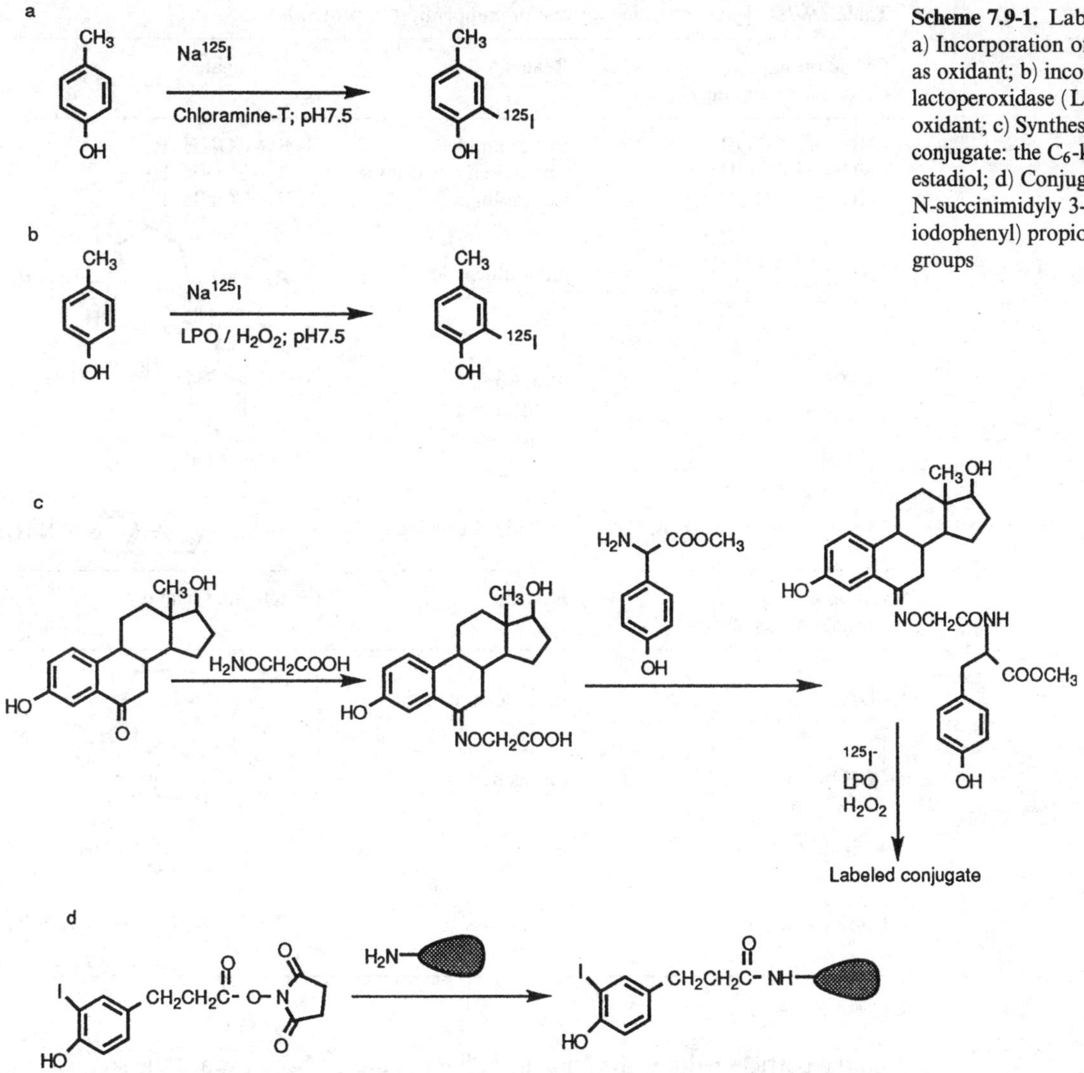

Scheme 7.9-1. Labeling methods for [125]I. a) Incorporation of iodine withchloramine-T as oxidant; b) incorporation of iodine with lactoperoxidase (LPO) | hydrogen peroxide as oxidant; c) Synthesis of a tyrosine methyl ester conjugate: the C_6-keto group of 6 keto-17β-estadiol; d) Conjugation labeling in which N-succinimidyly 3-(4-hydroxy 5-{[125]I} iodophenyl) propionate reacts with free amino groups

3-(p-hydroxyphenyl)propionic acid; Scheme 7.9-1d) so that direct reaction of this reagent with free amino groups results in an amide bond coupling. In general, the reagents chosen for the coupling depend on the availability of reactive groups in the peptide or hapten (Table 7.9-2).

Light scattering particle labels

The broad concept of agglutination in microbiological assays was explored in the 1920s, and this principle can be used to follow complex formation between antibody and antigen, where multivalent binding is involved. A polyvalent protein will react with its antibody to form a precipitated complex, but the reaction is highly dependent on factors such as ionic species and strength, and is complicated by variations in reactivity between the different antigenic sites on the antibody. In principle, the complex-formation can be followed by the change in the light scattering properties of the medium, as the reaction proceeds, but ideally the complex should remain in suspension.

A more controlled development in this direction involves the use of particle labels to enhance the light scattering properties of a labelled conjugate. When monochromatic light illuminates a particle, the incident field induces a fluctuating dipole and light is scattered perpendicular to the dipole axis, in directions which give information about the particle size. The pattern of scatter depends on the relationship between the particle volume and the wavelength. When the latter is much larger than the particle, light is scattered equally in all directions, but as the particle size increases, interference between the scattered light at different points

Light scattering from particles gives information about a particle or aggregate size, which can be used to follow immunoassay.

Table 7.9-2. Typical coupling agents for conjugation to proteins

Group on hapten/ peptide for coupling	Reagent	Conjugate
$-NH_2$ or $-COOH$	carbodiimide	$R_1-CONH-R_2$
$-NH_2$ or $-COOH$	isobutyl chloroformate	$R_1-CONH-R_2$
$-NH_2$ or $-COOH$	isoxazolium	$R_1-CONH-R_2$
$-NH_2$	glutaraldehyde	$R_1-NH-\ \ -NH-R_2$ (HO　OH)
$-NH_2$	diisocyanates (toluene-2-4-diisocyanate)	NHCONH-R_1 / NHCONH-R_2
Tyrosine, histine or lysine	diazotized benzidene	$R_1-N{:}N-\bigcirc-\bigcirc-N{:}N{\cdot}R_2$

Group on hapten/ peptide for activation	Reagent	activated ligand
$-OH$	succinic anhydride	(O=C–R–C(=O)–OH)
$-OH$	phosgene	$R-\overset{O}{C}-Cl$
Vicinal $-OH$	BrCN	$-O\ \ C{=}NH$ / $-O$
keto or aldehyde	O-(carboxy methyl) hydroxylamine	$R-CH{=}N-O-CH_2COOH$

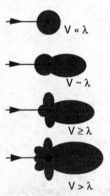

Fig. 7.9-9. Light scattering perpendicular to the dipole axis, as a function of particle volume V

on the particle reduces the amount of light scattered back toward the source, and produces "lobes" of high intensity (Fig. 7.9-9).

For measurements where the particles are moving under Brownian motion, the intensity of light collected at a fixed angle fluctuates according to that motion, since it depends on the time taken for the particle to diffuse a distance comparable with the wavelength. The scattered light at a fixed angle has a phase shift and a scattering vector K, given by:

$$K = \frac{4\pi n}{\lambda} \sin\left(\frac{\theta}{2}\right) \tag{7.9-16}$$

where n is the refractive index of the suspension phase and θ is the scattering angle. The intensity of detected light fluctuates on a time scale comparable with the time for a particle to diffuse a distance sufficient to induce a phase shift of π this being inversely proportional to K.

Gold is the most popular metal "tag" for this technique, owing to its high refractive index and relative ease of preparation. Particles can be produced to uniform sizes, ranging from about 5 to 60 nm. However, latex particles also offer ease of synthesis, plus a greater versatility to manipulate their precise properties. The best performance is obtained with a narrow size distribution, so that suspension is maintained during the assay, without excess agitation. Synthesis is usually performed in one of three main ways:

• suspension polymerization
• emulsion polymerization.
• swollen emulsion polymerization.

The processes involves the generation of a detergent micelle, which takes up monomer and polymerization initiator. The average resultant polymer particle size is controlled by the reagent concentrations, temperature and mixing rate. Monomers with reactive side groups (e.g., acrylic acid, –COOH; glycidyl methacrylate, epoxide group; acrylamide, –NH₂) allow subsequent coupling to the antigen or antibody. The coupling of antibody to the particles shows that there is an optimal loading, and that beyond this concentration, steric hindrance or depletion of antigen at the surface (see the discussion of surface immobilization techniques in Sec. 7.8.2) during binding, reduces the effectiveness. Similar problems also arise for antigen coupling to the particles.

Non-specific binding effects can also be a source of problems. The surface change of the particle influences particle aggregation and adsorption of reagent, so that conditions have to be optimized to minimize van der Waals attraction between particles, by ensuring sufficient coulombic repulsion, while not inhibiting the antibody antigen reaction.

Fluorescent labels usually give better sensitivity than colorimetric labels, because the excitation and detection wavelengths are different.

Fluorescent and chemiluminescent labels

The principle with these optical labels is much the same as the radioactive labels, except that the method of detection involves spectrophotometric measurement. However, most straightforward colorimetric label assays, where the wavelength of excitation and detection are the same, rarely achieve the desired sensitivity for an immunoassay. However, an order of magnitude improvement is gained by separating these wavelengths. There are two main classes of labels which are suitable here; the fluorophores and the chemiluminescent agents.

The fluorophores are promoted to a higher energy level by absorbing light in one wavelength region; in the excited state there is a rapid radiationless energy transfer to the lowest excited state and then a return to the ground state by emission of a photon at a longer wavelength (Fig. 7.9-10a, b); the excited state has a typical life time of $<10^{-8}$ s. The difference between the two wavelengths is known as the Stokes shift, and the larger the shift, the easier it becomes to separate the measurement of the emitted wavelength from the background incident wavelength intensity. When the quantum efficiency (the proportion of reemitted energy) is high (i.e., approaches 1), the label will be most sensitive.

During the life-time of the excited state, the molecule may also be able to participate in a chemical reaction, thus transferring energy to another species and returning to the ground state. Such a *quenching* process can cause considerable loss of signal. The kinetics of fluorescence quenching can be described by competition between two parallel processes:

Fluorescence: $M^* \xrightarrow{k_f} M + h\nu$

Quenching : $M^* + Q \xrightarrow{k_q} M + Q^*$

which gives a loss of fluorescence according to:

$$-\frac{d[L^*]}{dt} = k_f[L^*] + k_q[L^*][Q] \tag{7.9-17}$$

thus reducing the fluroscent yield according to:

$$\frac{I_f}{I_O} = \frac{k_f[L^*]}{k_f[L^*] + k_q[L^*][Q]} = \frac{1}{1 + k_q/k_f[Q]} \tag{7.9-18}$$

where I_O is the initial intensity of fluoroescence and I_f the final intensity following quenching.

Following the onset of excitation, the fluorescence intensity depends on the concentration of fluorophore in the excited state and is a function of time:

$$I(t) = \Phi \frac{d[L^*]}{dt} \tag{7.9-19}$$

where Φ is the quantum yield. If excitation involves a short pulse, the intensity profile will be of the form seen in Fig. 7.9-11, where the initial increase in intensity following illumination is followed by a decay, as the excitation is termi-

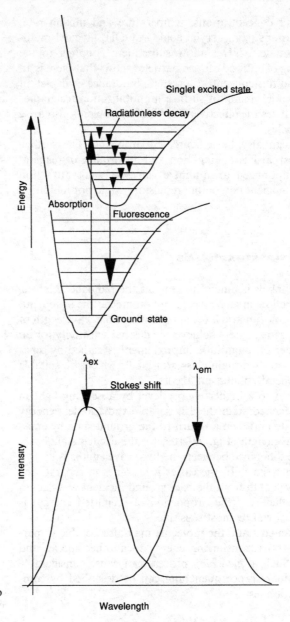

Fig. 7.9-10. The sequence of steps leading to fluorescence

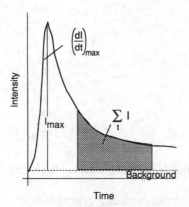

Fig. 7.9-11. Luminescent intensity as a function of time, illustrating different signal processing "windows"

"Ideal" fluorescent labels would have a large Stokes shift and be excitable with a low cost light source.

nated. This profile reveals several different possibilities for measurement. Plotted on the same time scale, each fluorescent molecule shows different results for the parameters labelled, since decay times and quantum yields are specific to a particular fluorophore.

The most common fluorophores used directly as labels have traditionally been fluorescein or rhodamine derivatives, in the former case coupling of the label often

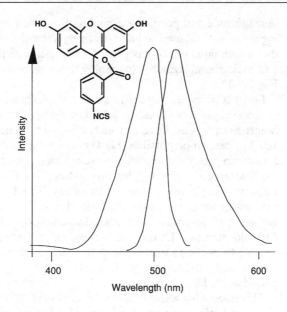

Fig. 7.9-12. Fluorescence excitation and emission spectra of fluorescein isothiocyanate labeled protein

Table 7.9-3. Fluorescence properties of some fluorescent labels used in immunoassay

Flurophore	λ_{ex} (nm)	λ_{ex} (nm)	Decay time (ns)	Molar Absorption (L/mol)	Quantum yield
Fluorescein	492	520	4.5	72000	0.85
Rhodamine B-isothiocyanate	550	585	3.0	103000	0.7
Lissamine-rhodamine B-sulphonyl Cl	530; 565	595	1.0		
Anilino-naphthalene sulphonic acid	385	471	16		0.8
2-methoxy 2,4 diphenyl-3(2H) furanone	390	480			0.1
Methyl-umbelliferone	323	386		16000	
Dansyl Cl	340	480–520	14		0.3
Lucifer yellow	430	540	3.3	13000	0.2
Erythrosin	492	517	108	101000	0.01
Ru-(bpy)$_3$	450	625	250–500	14000	0.028
pd-Coprophorphyrin	396	618	700000	200000	0.17
Pyrenes	346	395		16000	
Texas Red	596	615		80000	0.3
Eosin	520	545		101000	
Fluorescamine	394	475	7.0		0.1
Chlorophyll	430–453	648–669			

uses the isothiocyanate derivative as the active intermediate. The excitation/emission wavelengths of these labels are well suited for most fluorimeters (Fig. 7.9-12), although their position and the small Stokes shift, make it difficult to find low cost diode excitation sources, which could be used to produce a small dedicated instrument.

However, new fluorescent probes are continually being developed, to cover an ever wider spectrum and tailor to specific wavelength needs. Table 7.9-3 summarizes some of the candidates; the hydroxy derivatives of coumarin (the umbelliferones) are quite sensitive fluorophores, which have been extensively manipulated to show different fluorescent properties. Many of the derivatives of this aromatic phenol, such as the phosphate esters, glycosides etc., do not fluoresce, but the fluorescent species can be produced on hydrolysis, a property which can also be exploited in labelled immunoassay as discussed previously.

The detection limits for these fluorescent reagents are determined by background interference and quantum efficiency. In the former case, the inherent fluorescence of the sample, particularly if it contains e.g., serum proteins, NADH, or bilirubin, can be very high. Free NADH, for example, shows an absorption maximum at 340 nm, with a strong fluorescence at 400 nm, both of which change

Background fluorescence and quantum
efficiency influence detection limits.

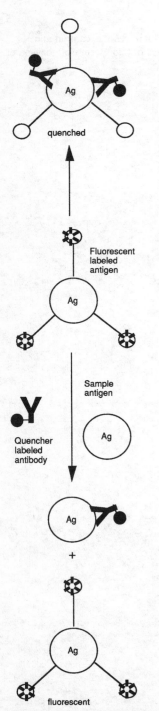

Fig. 7.9-13. Fluorescence excitation transfer
immunoassay

in magnitude and position on binding to protein and/or substrate. These interfering compounds all tend to have relatively short decay times following excitation, so that a technique which has proved a successful method for separation of the label and background signals is to employ time resolved flurorescence measurement (see Fig. 7.9-11).

For this purpose, some lanthanide complexes with certain organic ligands, such as the europium chelates, with Stokes shifts in excess of 200 nm and decay times longer than 500 ns, have proven useful, particularly since quantum yields of 30–100% appear to be possible. These chelates have been employed in different ways; for example, non-fluorescent Eu-complexing ligands can be used as the label, and the fluorescence initiated by the addition of Eu, following completion of the immuno-complexation stage of the assay. The Eu^{3+} ion is hydrated by eight or nine water molecules in solution. In this state, it is only weakly fluorescent, but exclusion of these water molecules, by complexing the organic ligands enhances the Eu^{3+} fluorescence. During the life-time of the excited state, the molecule may also be able to participate in a chemical reaction, thus transferring energy to another species and returning to its ground state. Such a quenching process can cause considerable loss of signal.

This quenching effect can also be exploited to advantage in fluorescence excitation transfer immunoassay. In principle, no separation of bound and free phases is required. The quencher is an acceptor molecule on, say, the antibody (Fig. 7.9-13), with an excitation maximum close to the emission wavelength of the original 'donor' fluorophore which is on the antigen. The rate of transfer is inversely proportional to the sixth power of the distance between the molecules, and also depends on the quantum yields and the "match" between donor and acceptor wavelengths.

$$E_{\text{ff}} = \frac{\Phi_q - \Phi_O}{\Phi_O} = \frac{r^{-6}}{r^{-6} + R_O^{-6}} \tag{7.9-20}$$

where Φ_O and Φ_O are the quantum yields of the donor in the presence and absence of acceptor, and R_O is the distance at which transfer is 50% efficient.

Fluorecein and rhodamine are a regularly employed acceptor-donor couple, and achieve 50% reduction of donor fluorescence at ∼50 Å, compared with a typical intramolecular distance between receptor sites in IgG of the order of 100 Å. In order to achieve the closest proximity between donor and acceptor, labeling of the antibody should be close to the binding site; however, such site-directed modification of the antibody with the label is difficult, owing to the lack of unique groups on the protein structure. Consequently, several flurophores are introduced in a more random fashion, but too high a level of labeling can lead to insolubility, so that 10–12 fluorophore per IgG is typical. Similar considerations apply to the donor, but in the case of a hapten this may be limited to a 1:1 conjugate, and for a protein, high loadings of fluorophore may result in reduced quantum efficiency, owing to self-quenching. In principle recording the emission of the acceptor should give the best signal, but because some direct excitation of the acceptor can occur at the donor excitation wavelength, this gives a high background. The chosen approach is therefore to monitor the quenching of the donor.

Other molecular properties can also be exploited to distinguish the signal from the background, and in some cases to separate the signals due to bound and unbound labelled antigen. For example, where the antigen is small, the fluorescein-labelled analog is able to rotate at high speed. On the other hand, the complex with antibody has a much slower rotation rate. If a population of fluorescent molecules is excited by linearly polarized light, then the probability of exciting any particular molecule is determined by its orientation. If the molecules have been "frozen" in position, so that no rotation occurs, then the result of is seen in the fluorescence observed in the transverse electric (TE, $\perp$) and transverse magnetic (TM, $\parallel$) planes separately, and the maximum value of the polarisation P would be 0.5:

$$P = \frac{I_\parallel - I_\perp}{I_\parallel + I_\perp} \tag{7.9-21}$$

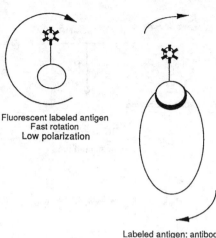

Fluorescent labeled antigen
Fast rotation
Low polarization

Labeled antigen: antibod
Slow rotation
High polarization

Fig. 7.9-14. Principle of fluorescence polarization immunoassay. Coupling of antibody to fluorophore-labeled antigen increases the polarisation of the fluorescence

Table 7.9-4. Detection limits proposed for some hapten analytes in different formats

Format	Analyte	Detection limit (mol L^{-1})
Substrate label	Theophylline	3×10^{-8}
Quenching	Gentamicin	2×10^{-6}
Excitation transfer	Codeine	1.5×10^{-9}
	Morphine	1.5×10^{-10}
Polarization	Gentamicin	4.5×10^{-6}
	Phenytoin	3.5×10^{-6}

Increasing Brownian motion tends to decrease the polarization. This principle is employed in fluorescence polarization, since the rotation of the fluorescent labelled antigen diminishes on binding to antibody (Fig. 7.9-14). Thus at low sample antigen concentration, the fluorescence polarisation is greatest, since the greatest amount of labelled antigen is bound.

For a given fluorophore, the change in fluorescence polarization depends on the molecular weight of the antigen and the nature of the complex with antibody; there must be an adequate change in molecular mass, and this may not be achieved for large antigens. It is generally accepted that fluorescence polarization immunoassay is not applicable to antigens above molecular mass 20000.

A major problem with homogeneous assays as described above, is the influence of interferents, so that the theoretical sensitivity cannot be realized. Table 7.9-4 compares some detection limits claimed for some different fluorescent label formats.

Unlike flurophores, chemiluminescent agents do not require incident light for excitation, but emit light as a result of a chemical reaction. For example, aryl acridinium esters, attached to the receptor molecule via the ester moiety, can be cleaved at the ester linkage, by alkaline hydrolysis in the assay step, to produce an unstable luminescent *N*-methylacridone, which decomposes with the emission of light (Scheme 7.9-2). The maximum intensity from these labels occurs typically within 0.4 s, with a decay half-life of 0.9 s.

This compares with the very slow emission (~ 25 s after initiation) from iso-luminol derivatives. Luminol and its derivatives belong to the family of cyclic acyl hydrazides, and emit blue light (425 nm) when oxidised. Isoluminol has a lower quanatum yield than luminol, but unlike luminol, this yield is not seriously reduced on coupling to proteins via the amino group. Isoluminol derivatives with various bridging groups for attachment to haptens have been synthesized, many of these are directed toward a peptide linkage, achieved for example, by a carbodiimide- or succinimide-activated reaction mechanism (Scheme 7.9-3). As shown in the next section, chemiluminescent monitoring has been significantly improved by the identification of a series of compounds which increase light emission when added to the reaction.

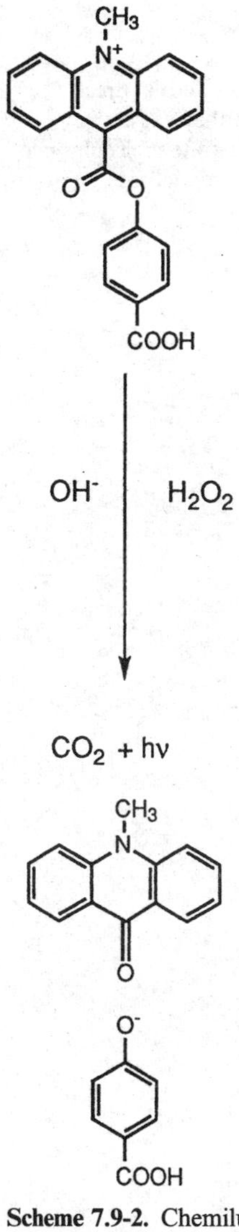

OH^- | H_2O_2

$CO_2 + h\nu$

Scheme 7.9-2. Chemiluminescent reaction with an aryl acridinium ester

Chemiluminescent labels produce light as a result of an unstable luminescent product of a reaction.

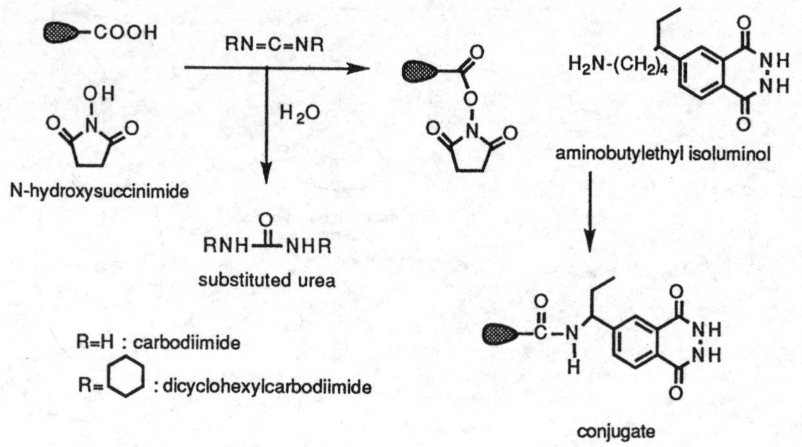

Scheme 7.9-3. Modified carbodiimide reaction for isoluminol conjugate, via a free carboxyl group

Enzyme labels

General principle

Enzyme labels do not produce a direct measurement, but since they cause the conversion of many substrate molecules per enzyme molecule, the potential for enhancement is high.

Enzymes are probably the most widely used label in immunoassay, and yet they do not produce a direct measurement, in the same way that isotopes of fluorescent labels are able to. Instead, the measurement may involve the detection of the consumption of the enzyme substrate, or the accumulation of product, thus requiring another reagent in the assay, and seeming to introduce additional complexity to the format. Nevertheless, since a single enzyme molecule can cause the conversion of many substrate molecules, the potential for signal enhancement is very high, and this advantage usually exceeds the frequently cited disadvantage of the incompatible optimization conditions of the final enzyme assay stage and the antibody-antigen binding.

The two most commonly employed enzymes used as labels are alkaline phosphatase (AP) and horseradish peroxidase (HRP). The choice depends on the ability to link these enzymes' substrate conversions to optically detectable pathways, and the ready conjugation of the enzyme without loss of activity. Alkaline phosphatase is a dimeric glycoprotein of molecular mass 140 000, containing many free amino acids for conjugation. Similarly, horseradish peroxidase has four lysine residues for conjugation, and has molecular mass of 44 000. In some cases the carbohydrate residues can also be used for conjugation.

The methods used for conjugation of the enzyme to the antibody or antigen have been developed according to the final format and special requirements of the immunoassay. There are two general approaches: homobifunctional agents; and heterobifunctional agents. Many of the reagents involved are the same as those already highlighted for the production of radioactive labelled conjugates (Table 7.9-2), and similar considerations also apply here, except that the conjugate links to a protein-label in this instance. It may be necessary to protect the active site of the enzyme during the coupling by including the enzyme substrate in the preparation; this is particularly important when the coupling is between an enzyme and hapten, since the hapten may bind at or close to the active site, masking recognition sites on both hapten and enzyme.

The coupling to the hapten should generally involve the same group as that employed to generate the immunogen in the first instance so that the recognition site is not masked by the ligand. For example, most assays for cortisol employ an antiserum raised to a cortisol-3-(*O*-carboxymethyl) oxime conjugate, since this leaves the unique D-ring-17 position available for recognition, thus creating specificity for the corticosteroid structure (Fig. 7.9-15). A hapten conjugate, labelled in the C-21 position, then shows no affinity with the antibody, so that this assay would fail.

Attachment of the enzyme label must not block the recognition site!

In the same way in which a spacer molecule may be required between the hapten and protein to generate the immunogen, binding reactivity with the antibody may be enhanced for the labelled conjugate, if a suitable spacer is included between hapten and enzyme. The most popular homobifunctional agent for enzyme-coupling is glutaraldehyde, but it is prone to produce very heterogeneous complexes, with significant enzyme-enzyme cross-linking, as well as the desired enzyme-conjugate. Heterobifunctional agents avoid this problem, with different degrees of success, depending on the relative reactivities of the different groups. *N*-hydroxy succinimide esters of dimalemide and carbodiimides have been widely employed but many different coupling agents are available (see also Sec. 7.8.2).

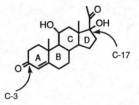

Fig. 7.9-15. Cortisol, showing the general steroid structure, with the unique "recognition" center for cortisol at C-17. Conjugation through C-3, does not mask this center

Homogeneous enzyme immunoassay

An interesting extension of the idea of using enzyme labels is to monitor the bound:free ratio by the modulation of enzyme activity due to the antibody binding in the conjugate, or due to a second binding of anti-enzyme antibodies. These assays circumvent the need for a separation step, and thus can simplify the assay protocol. Several homogeneous enzyme immunoassays (EIA) have been devised.

Enzyme multiplied immunoassay (EMIT®)

In a competitive assay, an enzyme-labelled hapten can compete with sample hapten, for a limited supply of anti-hapten antibody. If the binding of antibody to the hapten conjugate inhibits the enzyme activity, then the higher the sample hapten concentration, the lower the binding of antibody to conjugate, and therefore the lower the inhibition (Fig. 7.9-16). The first EMIT was developed for morphine, using lysozyme as the inhibition sensitive enzyme, but glucose-6-phosphate dehydrogenase has been found to be the most effective in modulation of enzyme activity for this assay. Applications of EMIT have also been targeted toward other low molecular mass analytes, in particular within the same family of drugs of abuse and therapeutic agents. The main disadvantage with this format is that it equilibrates very slowly with the solution hapten, so that it is necessary to preincubate sample and antibody before adding enzyme conjugate. Modifications of the general principle can also be applied for higher molecular mass analytes.

Enzyme cofactor label

An advantage of using an enzyme *co-factor* as the label, rather than the enzyme, is that it can easily be included in an amplification scheme (Fig. 7.9-17) and can be used in a homogeneous assay system. Hapten-cofactor conjugates are employed (e.g., NAD-hapten) and, in the presence of the NAD-linked enzyme and substrate, the antibody-bound conjugate is not active, whereas the free conjugate is able to participate in the enzyme cycling. Each time the hapten-cofactor conjugate undergoes the amplification cycle, a stoichiometric amount of product is produced, and the amount of product accumulates as the assay proceeds.

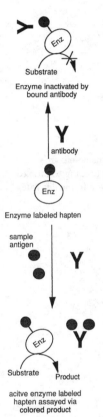

Enzyme inactivated by bound antibody

antibody

Enzyme labeled hapten

sample antigen

acitve enzyme labeled hapten assayed via colored product

Fig. 7.9-16. The EMIT assay principle. Antibody binding to hapten-labeled enzyme alters the activity of the enzyme

Signal generation

Colorimetric signal generation

The principle idea here is that the enzyme substrate should be converted to a colored product. In the case of alkaline phosphatase, *p*-nitophenyl phosphate or a similar aromatic phosphate is normally employed:

For enzyme labelled colorimetric immunoassay, the enzyme reaction must result in a colored product.

p-nitrophenyloxide
(λ_{max} 440nm)

but horseradish peroxidase has been shown to be capable of greater sensitivity. In the latter case, the label acts as cosubstrate with peroxide and a hydrogen donor. The final product is the oxidized hydrogen donor, which is chosen according to its

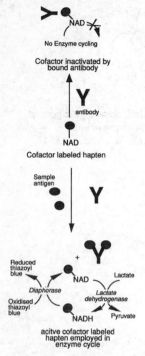

Fig. 7.9-17. Co-factor labeled enzyme immunoassay. Antibody binding to labeled hapten does not participate in the enzyme reaction

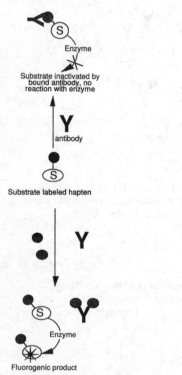

Fig. 7.9-18. Substrate labeled fluorescence immunoassay. Antibody bound to labeled hapten inhibits it as a substrate for the enzyme

Bioluminescence offers a very sensitive pathway for determination of an enzyme label.

Table 7.9-5. Peroxidase cosubstrates for colorimetric detection

Cosubstrate	λ_{max} (oxidized form) (nm)
ABTS (2,2'-azino-bis (ethyl-benzothiazoline-6-sulfonate))	415
OPD (O-phenylenediamine)	492 (acidified)
TMB (3.3'.5,5'-tetramethylbenzidine)	450 (acidified)

absorbance properties in the visible spectrum. Common peroxidase cosubstrates, with suitable absorbance characteristics in their oxidized form, are shown in Table 7.9-5. 3.3'.5,5'-tetramethylbenzidine (TMB) gives the highest absorbance values and lowest background, compared with many other substrates which have been explored, such as 2,2'-azino-bis(ethylbenzothiazoline-6-sulfonate) (ABTS) or O-phenylenediamine.

Fluorimetric signal generation

Alkaline phosphatase is more suitable than HRP for fluorescence measurement. Here the most common substrate is 4-methylumbelliferyl phosphate (or related substrates), which is dephosphorylated to produce 4-methylumbelliferone, which emits at 448 nm, when it is excited at 365 nm. Consequently, the advantage of AP is that fluorescent compounds can be devised which can be phosporylated, thus producing a fluorophore generating substrate for the rather non-specific AP enzyme. In principle therefore, an extensive family of flurophores with different wavelength and decay properties could be synthesised as suitable substrates for the enzyme label. This enzyme-linked approach to a fluorescent measurement has the obvious advantage of a much enhanced signal, compared with the direct fluorophore label, since each enzyme label initiates many fluorescent molecules.

However, this advantage is not true of all formats; for example, assays based on the EMIT principle, where a hapten is substrate labelled, and produces a *fluorogenic product* in the enzyme catalysis, only when there is no complex formed with the antibody (Fig. 7.9-18). In this instance, the β-galactosyl umbelliferone-hapten conjugate can act as the labelled analyte analog in a competitive assay with limited antibody. The antibody-hapten conjugate complex is inhibited as a substrate for the enzyme β-galactosidase enzyme, whereas the remaining free conjugate is hydrolyzed by the enzyme releasing a fluorogenic product.

Chemiluminescent signal generation

Alkaline phosphatase is also particularly useful in designing the generation of a luminescent signal, since many aromatic phosphates produce an unstable anion, on cleavage of the phosphate group, which decomposes with the emission of light (for example, adamantyl 1,2-dioxetane arylphosphate). However, the majority of luminescent measurements which are performed are more likely to be peroxide-linked. Luminol as cosubstrate with peroxide is oxidized by HRP to produce an unstable intermediate, which decays to 3-amino phthalate, with the emission of light (Scheme 7.9-4). Although chemiluminescent reactions can yield a very sensitive measurement, they nevertheless have quantum yields typically less than 20%.

Enhancement of several orders of magnitude of the output from the reaction in Scheme 7.9-4 can be achieved by the addition of an "enhancer" (Fig. 7.9-19). These enhancers are typically phenols and naphthols, although other molecules have also been shown to exhibit the effect (Table 7.9-6). The precise mechanism of this enhancement is not fully elucidated. However, it is probable that the enhancer acts as a radical transfer mediator between the enzyme reaction with peroxide and luminol, thus improving the "slow step" of the formation of luminol radicals.

Bioluminescence is also a chemiluminescent pathway, primarily associated with the mechanism responsible for the production of light by fireflies. It involves the enzyme *luciferase*, which catalyzes the oxidation of a heterocyclic organic molecule (a *luciferin*). The luciferins vary in structure, but many of the luciferase cata-

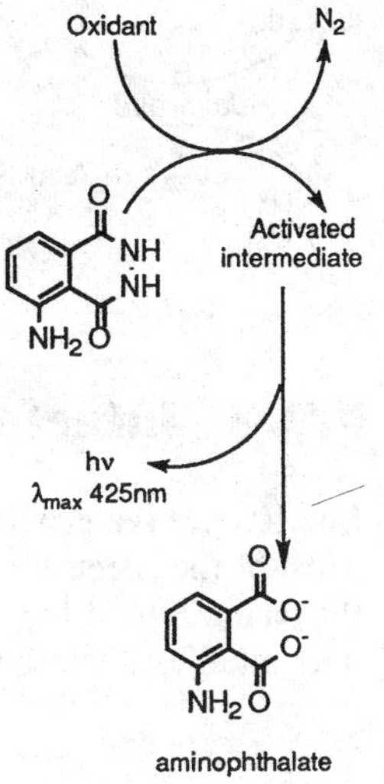

Oxidant   N₂

Activated
intermediate

hv
λ_{max} 425nm

aminophthalate

Scheme 7.9-4. Chemiluminescent oxidation of luminol by horseradish peroxidase

Table 7.9-6. Examples of chemiluminescent enhancers in horseradish peroxidase catalyzed oxidation of luminol

Phenol and Naphthol Derivatives		
luciferin	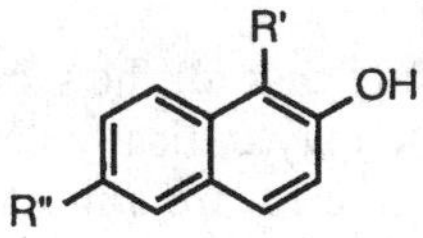 R' = R" =	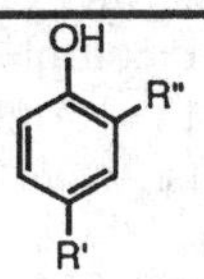 R' = R" =
R =	2-Naphthol	- I - H *p*-Indophenol
- H 6-hydroxy benzothiazole	- H - H	- Cl - Cl 2,4 Dichlorophenol
- CN 2-cyano-6-hydroxy benzothiazole	- Br - H 1-bromo-2-naphthol	*p*-hydroxycinnamic acid
	- Br - Br 1,6 dibromo-2-naphthol	 *p*-phenylphenol
		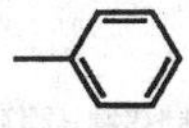 2-chloro-4-phenylphenol
Aromatic amine derivatives		
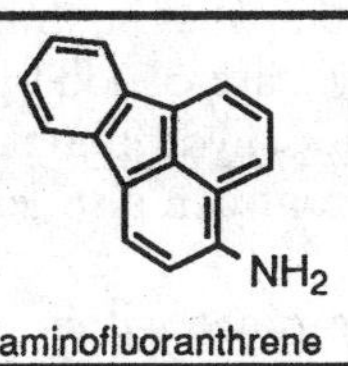 3-aminofluoranthrene	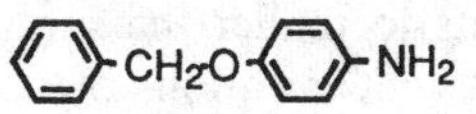 4-benzoxyaniline	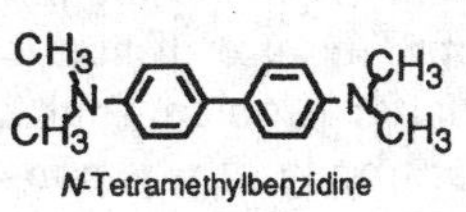 *N*-Tetramethylbenzidine

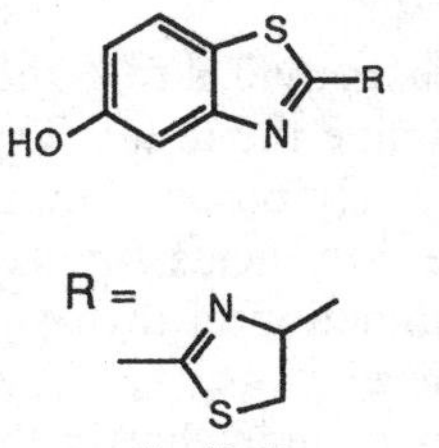

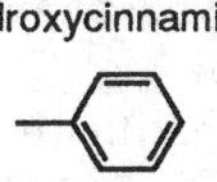

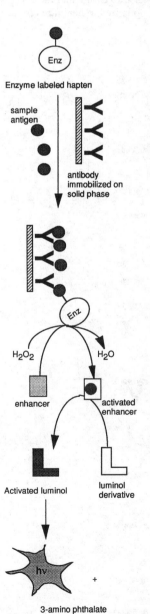

Enzyme labeled hapten

sample antigen

antibody immobilized on solid phase

H₂O₂   H₂O

enhancer   activated enhancer

Activated luminol   luminol derivative

hv   +

3-amino phthalate

Fig. 7.9-19. Enzyme immunoassay with enhanced luminescence end point

lyzed reactions involve cofactors such as ATP, FMN, or NADH, so that the use of these cofactor labels (see Fig. 7.9-17) introduces a route into the luciferase catalyzed pathway (Scheme 7.9-5). On the other hand, an enzyme amplification is achieved by an NAD-linked enzyme label. For example, Fig. 7.9-20 shows the NAD⁺-initiated enzyme cycling, which results in a luminescent signal. Such NADH-linked assays can show a detection limit as low as 10^{-15} mol NADH, and linearity over five orders of magnitude.

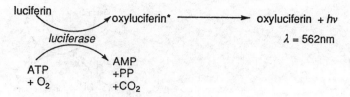

Scheme 7.9-5. ATP in firefly luciferase light generating cycle

7.9.4 Interferences

Interference is a general term for any factor which causes a bias in the assay result. Most of the effects seen are classified as "matrix" effects, but this does not identify the real nature of the problem. There may be several sources of interference, which may manifest themselves in different ways.

Effective analyte concentration

Care must be taken to ensure that the assay conditions are optimized to minimize errors.

In immunoassay, the analyte is required to undergo a binding equilibrium with a recognition molecule. If this binding is disrupted, or if the analyte is diverted toward binding with other molecules, the effective analyte concentration appears reduced. In many cases, the analyte of interest exists in the sample in both free and complexed forms; for example, many hormones may be in equilibrium with a hormone-globulin complex, or hydrophobic steroids may be partitioned into lipid vesicles. In the former case, the effect is usually overcome by the addition of blocking agents, while lipid capture is disrupted by the inclusion of selective detergents.

However, if care is not taken to ensure that all the sample analyte is free, then pretreatment (or storage) of the sample may critically determine the result. Thyroxine (T_4) and many other small molecules, are almost entirely protein bound, but are displaced by non-esterified fatty acids. Since these fatty acids form on storage of the sample, owing to the action of lipases, determination of thyroxine can become an estimate of sample age, rather than analyte level.

Complexation may also involve ions rather than proteins or lipid, but in this case the absence of the complex may be the source of the problem, since many protein analytes contain divalent cation binding sites, and the antibodies raised to these proteins in vivo can recognize the configuration held by the cation complex (principly Ca^{2+} or Mg^{2+}); thus, in the absence of these cations, the conformation may change, so that the recognition event with antibody becomes inefficient, and the assay fails.

Antibody binding efficiency

For a given antibody, the binding of antigen is influenced by the availability of the binding site and its conformational integrity. Higher than anticipated binding can also result from non-specific interaction of the analyte. In the case of solid-phase assay systems, the use of surfactants to render the surface less hydrophobic can reduce the latter problem, but high concentrations of surfactant can also lead to loss of the antibody, if it is non-covalently bound.

Like all assay techniques, there is always a usable working range within which the assay is valid. For sandwich assays, very high analyte concentrations will saturate both capture and monitoring antibodies, so that the apparent measured concentration of capture antibody/analyte/monitoring antibody is lower than expected (Fig. 7.9-21). This phenomenon is called the high-dose effect.

A low-dose effect is seen in competitive assays where the binding of labelled antigen, at low sample antigen concentrations, is greater than for zero analyte, owing to positive cooperation.

Less predicable are the errors due to endogenous antibody, able to bind with one of the assay reagents. This problem is similar to those encountered with endogenous binding of antigen, discussed above, but here the remedies are less

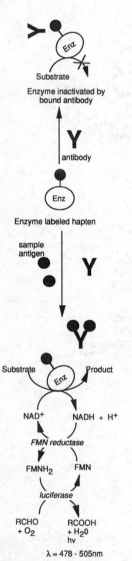

Fig. 7.9-20. Linking of the enzyme label into a bioluminescent pathway

obvious. Sandwich assays are most prone to interference by heterophilic IgG and IgM antibodies in the sample, able to bind to the capture antibody and present a recognition site for the monitoring antibody. The existence of this problem is not surprising, since antibodies to animal IgG in the sample may cross react with the reagent antibodies, but the inclusion of serum or immunoglobulins from the same species as the reagents is usually effective in preventing interference of this type. Alternatively, the use of just the Fab fragment, containing the specific recognition site as the monitoring antibody, and not the whole IgG molecule, can avoid the false signal generation.

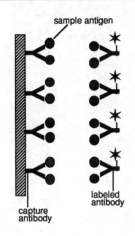

Signal generation error

The main sources of error in signal generation arise from the inhibition of the label in some manner. For example, a flurophore may become quenched by endogeneous sample species, or enzyme label may be affected by the presence of inhibitors or enhancers. As has been shown in the previous discussion, these effects can be used as an important feature of the assay, but where this is not the case, they can introduce errors.

Unfortunately, as with many interfering species, it is not always easy to predict or eliminate the problem, so that extensive testing of the assay in the environment in which it is to be employed, is obviously essential to the integrity of the method.

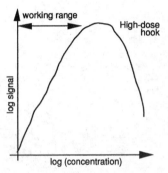

Fig. 7.9-21. High dose "hook" effect

General reading

Bangs, L.B. *Uniform latex particles*. Seradyn Diagnostics Inc. Particle Technology Division 1987.

Diamandis, E.P., *Clin. Biochem.* 1988, **21**, 139.

Hunter, W.M., Corrie, J.E.T., *Immunoassay for Clinical Chemistry*. Churchill Livingstone, Edinburgh 1983.

Langone, J.J., van Vunakis, H., *Imunochemical Techniques, Part A. Methods in Enzymology*. Vol. 70, Academic Press, New York 1988.

Langone, J.J., van Vunakis, H., *Immunochemical Techniques, Part B. Methods in Enzymology*. Vol. 73. Academic Press, New York 1988.

Langone, J.J., van Vunakis, H., *Immunochemical Techniques. Part C. Methods in Enzymology*. Vol. 74. Academic Press, New York 1988.

Langone, J.J., van Vunakis, H., *Immunochemical Techniques, Part D. Methods in Enzymology*. Vol. 84. Academic Press, New York 1988.

Miles, L.E.M., Hales, C.N., *Nature* 1968, **219**, 186–189.

Price, C.P., Newman, D.J. (Eds.) *Principles and Practice of Immunoassay*. New York, Stockton Press 1991.

Robinson, P.J., Dunhill, P., Lilly, M.D., *Biotech. Bioeng.* 1973, **15**, 603–6.

Rubenstein, K.E., Schneider, R.S., Ullman, E.F., *Biochem. Biophys. Res Commun.* 1972, **47**, 846–851.

Thorell, J.I., Larson, S.M., *Radioimmunoassay and related techniques*, Mosby, St Louis 1978.

Part III
Physical Analysis

8 Elemental Analysis

Learning objectives

- To introduce the main methods based on physical principles for elemental analysis: atomic emission spectrometry, atomic absorption spectrometry, X-ray fluorescence spectrometry, activation analysis, and inorganic mass spectrometry
- To describe the physical principles as well as the instrumentation
- To describe the analytical figures of merit and the various fields of applications

Introduction

The nature of elemental analytical chemistry has been drastically modified over the past three decades because of the development of instrumental analysis. Many techniques have emerged and have gained wide acceptance for routine analysis through commercially available systems. Some examples are inductively coupled plasma emission, inductively coupled plasma mass spectrometry, and graphite furnace atomic absorption spectrometry. Some other techniques are still at a research stage such as resonance ionization mass spectrometry, but their analytical capabilities are so promising that their introduction should occur in the near future.

The various instruments are becoming more and more sophisticated as is the associated software. However, except for neutron activation analysis which requires the use of a nuclear reactor, each instrument used for elemental analysis can be described as a series of modules which can be summarized as follows:

a) *Generation of the physical information* which includes the absorption of either an external energy or an internal energy (including combination of several types of energy), followed by the production of the physical information in the form of emission or absorption of particles or waves (Table 8-1)

b) *Isolation or extraction of the physical information* in order to select the useful signal from the whole information using, for instance, a wavelength-dispersive system, an energy-dispersive system, a mass spectrometer, etc.

c) *Conversion of the physical information into electrical information* (current or voltage) using a detector such as a detector of photons (photomultiplier tube, diode, thermal detector, etc.) or of ions (Faraday cup, electron multiplier, etc.)

d) *Processing of the electrical information* by the electronics to obtain an effective signal-to-noise ratio and a signal which can be processed by the software

e) *Conversion by the software of the information to analytical information* (qualitative analysis, use of statistics to provide a calibration curve, limits of detection, precision, uncertainty, etc.)

f) *Editing of the analytical information* through the use of a computer: report, storing, evaluation, quality control, traceability, etc.

The software is used not only to process data, but also to drive the various components of the system in order to obtain partial or full automation. Its role is, therefore, becoming more and more important.

In elemental analysis, an instrument must fulfil several specifications in order to obtain appropriate analytical performance and instrument characteristics. Analytical performance includes the quality of the results and the quality of the system. The analytical quality of the results is related to the *accuracy*, i.e., both to the *precision* and *trueness* (see Sec. 3.2.1).

The analytical quality of the system includes:

- the number of elements which can be determined by the method
- long-term stability
- selectivity
- robustness, i.e.; the absence of matrix and interelement effects

- sensitivity and low limits of detection
- linearity and dynamic range

A summary of the analytical performance of the most commonly-used analytical instruments is given in Table 8-2.

Ideal instrument characteristics are related to instrument operation and economic aspects. The instrument operation characteristics may be summarized as follows:

- ease of operation
- ease of maintenance
- full automation
- use of any form of the sample, solid, liquid, or gas
- low sample consumption, if any
- small size of the system

while the economic aspects include:

- high sample throughput
- reliability
- safety
- low capital investment
- low running cost

Table 8-1. Examples of energy supply and production of physical information

Energy	Physical information	Analytical method
X-ray photons	X-ray photons	X-ray fluorescence spectrometry (XRF)
X-ray photons	electrons	photoelectronic spectrometry (XPS)
Electrons	electrons	Auger electron spectrometry (AES)
Ions	ions	secondary ion mass spectrometry (SIMS)
Photons	photons	laser induced fluorescence spectrometry (LIFS)
Thermal energy	photons	atomic emission spectrometry (AES)
Thermal energy	ions	thermo-ionization spectrometry (TIMS)
Electrical field	ions	glow discharge mass spectrometry (GDMS)
Electrical field	photons	glow discharge atomic emission spectrometry (GDAES)

Table 8-2. Summary of the analytical performance and the characteristics of the most important instruments used in elemental analysis. The analytical performance involves the limits of detection (LOD) either in solution (ng ml^{-1}) or in solid (ppm), the robustness (absence of matrix effects), the selectivity (absence of spectral interferences), and the precision. The instrument characteristics involve the ideal form of the sample, liquid or solid, the minimum sample consumption, and the maximum salt concentration in the case of solutions. AES: atomic emission spectrometry, AAS: atomic absorption spectrometry, MS: mass spectrometry, ICP: inductively coupled plasma, GDL: glow discharge lamp, GF: graphite furnace, TI: thermo-ionization, SS: spark source, LIFS, laser-induced fluorescence spectrometry, WD-XRF: wavelength dispersive X-ray fluorescence spectrometry

System	Liquid sample	Solid sample	Sample vol. (mL)	Max. matrix conc. (g L^{-1})	LOD ng mL^{-1}	LOD ppm	Sequential multielement	Simultaneous multielement	Matrix effects	Spectral interferences	Precision %RSD
Spark-AES	a)	ideal	a)	a)	a)	1–10	yes	yes	large	significant	1
Arc-AES	possible	ideal	a)	c)	a)	0.1–1	yes	yes	large	significant	5–10
Flame-AES	ideal	a)	5–10	30	1–100		yes	yes	large	significant	0.5–1
ICP-AES	ideal	possible	1–10	10–100	0.1–10		yes	yes	small	large	0.5–1
GDL-AES	possible	ideal	a)	a)	a)		yes	yes	small	significant	a)
Flame-AAS	ideal	a)	5–10	30	1–10^3		possible	no	large	few	0.5–1
GF-AAS	ideal	possible	0.01–0.1	200	10^{-2}–0.1		possible	yes	moderate	few	3–10
TIMS	ideal	a)	0.002	1	d)		yes	yes	c)	few	0.05–0.5
ICP-MS	ideal	possible	1–10	0.1–0.5	10^{-3}–10^{-2}		yes	yes	moderate	significant	1–3
SSMS	a)	ideal	a)	a)	d)	10^{-3}–10^{-2}	no	yes	large	moderate	a)
GDMS	a)	ideal	a)	a)	a)	10^{-3}–10^{-2}	yes	yes	small	significant	a)
Furnace-RIMS	ideal	possible	0.001–1	b)	b)		no	no	b)	negligible	b)
GF-LIFS	ideal	possible	0.01–0.1	10	10^{-3}–10^{-2}		no	no	moderate	negligible	5
WD-XRF	possible	ideal	c)	c)	c)	0.1–10^4	yes	yes	large		1

a) not applicable; b) no accepted values because the technique is still at a research stage; c) depends on the analytical problem or the sample preparation; d) not used for this purpose.

The most adequate form of the sample is given in Table 8-2.

In this chapter five different types of generation of the physical information will be described:

1. Emission of photons (atomic emission spectrometry)
2. Absorption of photons (atomic absorption spectrometry)
3. Absorption of X-ray photons (X-ray fluorescence spectrometry)
4. Emission of charged particles (activation analysis)
5. Emission of ions (inorganic mass spectrometry)

They are mainly based on the use of samples in the liquid or solid forms. Depending on the initial form of the sample, sample pretreatment or preparation may be necessary for an appropriate sample presentation or introduction.

8.1 Atomic Emission Spectrometry

8.1.1 Introduction

Atomic emission spectrometry (AES) is probably one of the oldest methods – used for elemental analysis [8.1-1, 8.1-2]. The first observations of emission using an alcohol flame were made as early as the beginning of the 19th century by Brewster, Herschel, Talbot, and Foucault. Talbot even suggested that flame emission spectrometry could replace "laborious chemical analysis methods". Their results were the basis of the work of Bunsen and Kirchhoff, and can be considered as the real start of emission spectroscopy.

Bunsen and Kirchhoff were the pioneers in emission spectroscopy.

During the second part of the last century, Crookes, Reich and Richter, Janssen, Champion, Pellet, and Grenier gave further evidence of the interest in flame spectroscopy. In 1877, Gouy designed a pneumatic nebulizer to improve the control of the amount of sample introduced into the flame, and he demonstrated that the intensity of the radiation was proportional to the amount of sample. It can be considered that modern spectroscopy actually started with the work of Lundegardh in 1928. He made use of an air/acetylene flame and a pneumatic nebulizer and was able to construct calibration graphs to perform quantitative analysis. The first commercially available flame emission spectrometer was introduced by Siemens and Zeiss in the mid-1930s. The first monography, entitled "Flame Photometry" was written by Ramirez Muñoz and appeared in 1955. Currently, flame emission spectrometry is still in use, although since the beginning of the 60s new radiation sources such as plasmas have largely replaced flames.

Early studies of the spark and arc were performed by Wheatstone in 1834. Near 1850, sparks were produced by using the Rhumkorff induction coil. Arc and spark discharges have been used for emission spectroscopy since the 1920s; they were capable of analyzing most of the elements in the periodic table in solid form, which overcame one of the limitations of flame spectroscopy. Detection was achieved by using photographic plates. These plates were later replaced by the photomultiplier tube which is still in use today. Commercially available systems were introduced at the end of the second world war, and the first modern direct-reader spectrometer was introduced at the end of the 1940s. It should be noted that, although significant improvements were made to the various systems, the basic principle of the direct-reader remained the same until the recent introduction of multichannel detection.

More recently, at the beginning of the 1960s, different types of plasmas were described for analytical applications and the first commercially available emission spectrometers making use of plasma radiation sources were introduced in the 1970s. Plasmas are currently the most commonly-used radiation sources and, as evidence of their popularity, one may simply consider the large number of manufacturers involved in the production of commercially available plasma emission systems.

The success of AES may be explained by its universatility and its multielement

Both qualitative and quantitative analysis can be performed by emission spectrometry.

capability. AES is capable of both qualitative and quantitative analysis over a wide range of concentration. A large number of radiation sources has been studied and are currently available for both the analysis of solids and liquids. In this section, we will deal mainly with flame and plasma, preferentially for liquid samples, and arc and spark for solids.

8.1.2 Principle

An emission spectrum consists of lines produced by radiative de-excitation from excited levels.

AES is based on the production and the detection of line spectra emitted during the radiative de-excitation process of electrons that undergo a transition between upper excited levels and lower and ground levels. These electrons belong to the outer shells of the atoms and are called *optical electrons*. Line spectra is specific for an element, and the adequate selection of a given line and its isolation by means of a dispersive system allow the analyst to verify the presence of this element and to determine its concentration. An atomic emission spectrometer will therefore consist of a radiation source, a sample introduction or presentation system, an optical dispersive system, a detector, and electronics for data acquisition, processing, and editing.

Each element of the periodic table has a given number of electrons equal to its atomic number. The electrons have a probability of being located in shells and subshells around the nucleus according to quantum theory. The quantum theory was first introduced by Planck who assumed that electromagnetic energy is emitted or absorbed in discrete values, which means that energy is discontinuous. In the case of the electrons in a free atom, four quantum numbers are used to identify the energy state of an electron:

The energy state of an electron is identified by four quantum numbers.

1. The principal quantum number n ($n = 1$ to 7 for atoms in the ground states), which corresponds to shells named K, L, M, ... Q.
2. The orbital angular momentum quantum number l ($l < n, l = 0, 1, 2, ...$), which corresponds to subshells s, p, d, f (named after sharp, principal, diffuse, and fundamental).
3. The orbital magnetic quantum number m_l (any integer such as $-l < m_l < +l$).
4. The spin quantum number m_s ($m_s = \pm 1/2$).

The maximum number of electrons in the subshell is given by the Pauli's exclusion principle according to which no two electrons can have the same four quantum numbers. This means that the maximum number of electrons per subshell and shell is $2(2l + 1)$ and $2n^2$ respectively, i.e., for the subshells s, p, d, and f, the number of electron is 2, 6, 10, and 14, respectively.

Let us consider some examples: for Al, Mn, and Mo, the number of electrons is 13, 25, and 42, respectively. The location of the electrons on shells and subshells, i.e., the electronic configuration, is given in Table 8.1-1. The electrons that will be of concern in atomic emission spectrometry (optical electrons) are those located in the outer subshells, such as electrons $n = 3$ and $l = 1$ for Al, $n = 4$ and $l = 0$ for Mn, and $n = 5$ and $l = 0$ for Mo.

The electronic configuration of Al can be summarized by writing $1s^2\, 2s^2\, 2p^6\, 3s^2\, 3p^1$. When a free atom does not absorb any energy, its optical electrons are located on a subshell as mentioned above, $l = 1$ for Al. This is the state of lowest energy or ground state. By convention, the energy of the ground state is referred to as zero ($E = 0$). When a free atom absorbs energy, either by internal energy

Table 8.1-1. Electronic configuration of Al, Mn, and Mo

n quantum number	1	2		3			4				5
Shell	K	L		M			N				O
l quantum number	0	0	1	0	1	2	0	1	2	3	0
Subshell	s	s	p	s	p	d	s	p	d	f	s
Al	2	2	6	2	1						
Mn	2	2	6	2	6	5	2				
Mo	2	2	6	2	6	10	2	6	5		1

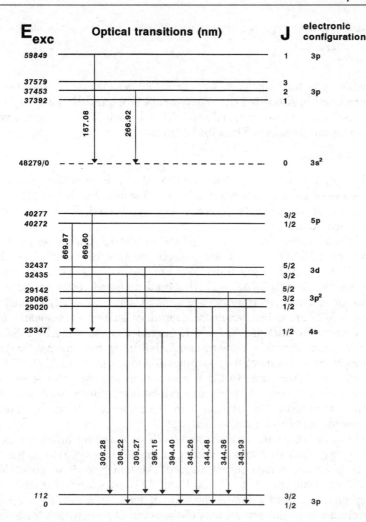

Fig. 8.1-1. Examples of ground and excited state levels of the Al atom and Al ion; excitation energies, E_{exc}, in cm^{-1} from the ground state of either the atom or the ion; possible optical transitions (the wavelength of the optical transitions is given in nm); the quantum number, J; and the electronic configuration of Al

(collisions) or by external energy (radiations), the electrons will move to outer shells and subshells, i.e., to higher quantized energy levels ($E_j, E_k, E_l, ...$). These are called excited states. An example of some excited states is given for Al in Fig. 8.1-1. Some excited states are indicated, corresponding to subshells with higher energy such as 4s, 4p, 3d, and 5p.

For a given electronic configuration, the atom as a whole can be characterized by a quantum number J. The value of J is usually obtained by using the Russell-Saunders coupling scheme. A vector L is obtained by summing the l vectors of the different electrons and a vector S is similarly obtained by summing the s vectors. We have:

$$J = L + S \tag{8.1-1}$$

J is called the total electronic angular momentum. The total electronic angular momentum quantum number J is either integral or half-odd integral.

To each level will correspond an energy (usually expressed either in cm^{-1} or eV, with $1\,\text{eV} = 8065.54\,\text{cm}^{-1} = 96.4853\,\text{kJ}\,\text{mol}^{-1}$) and a quantum number J. When the energy that is absorbed is too high, e.g., $48\,279\,\text{cm}^{-1}$ for Al (Fig. 8.1-1), the electron no longer belongs to the atom, and the atom is ionized. There is a ground state for the ion, 3s^2, and higher excited levels, such as 3p.

When the electron returns to a lower state or to the ground state, a portion of the de-excitation can occur either via a radiative process or by a collisional process. In the case of radiative de-excitation, there is an emission of electromagnetic radiation. If the transition occurs between an upper level, E_m and a lower level E_k, radiation is emitted whose frequency v is given by:

$$h v = E_m - E_k \tag{8.1-2}$$

where h is the Plank's constant ($h = 6.626 \times 10^{-34}\,\text{J}\,\text{s} = 3.336 \times 10^{-11}\,\text{cm}^{-1}\,\text{s}$). In

The relation $hc/\lambda = E_m - E_n$ is the basic equation in emission spectrometry and is evidence of the relation between energy and wavelength.

AES, wavelength λ is commonly used instead of frequency with:

$$\lambda = \frac{c}{\nu} \tag{8.1-3}$$

where c is the velocity of light ($c = 299\,792\,458\,\text{m s}^{-1}$). Wavelength used to be expressed in Ångström (Å). They should be expressed in $\text{nm}(10^{-9}\,\text{m})$.

Not every transition is allowed between each of the possible levels. Some transitions are *forbidden*. The allowed transition must follow selection rules such as:

$$\Delta J = 0, \pm 1$$

It should be noted that the transition $0 \to 0$ is forbidden. A second rule is the change of parity. Symbol I will be used for lines emitted by atoms, symbol II, for lines emitted by singly-ionized atoms, symbol III, for lines emitted by doubly-ionized atoms, etc.

A line whose transition returns to the ground state is called a resonance line.

For instance, when considering the 3d and 3p Al atom levels (Fig. 8.1-1), transitions with $\Delta J = 0, \pm 1$ are allowed because of a change in the parity. The $3/2 \to 3/2$, $3/2 \to 1/2$ transitions are allowed ($\Delta J = 0, 1$, respectively), in contrast to $5/2 \to 1/2$ ($\Delta J = 2$). It should be noted that transitions corresponding to an electron returning to the ground state ($J = 1/2$) or quasi-ground state ($J = 3/2$) are called *resonance transitions* leading to *resonance lines*. Examples of resonance lines are Al I 396.15 nm, Al I 394.40 nm, Al I 308.22 nm, and Al I 309.27 nm. Similarly, allowed and forbidden transitions may be described for the singly-ionized atom of Al. Examples of allowed lines are Al II 167.08 nm and Al II 266.92 nm. According to the selection rules and the possible excited levels, each element of the periodic table may exhibit a set of lines (spectrum) which is specific for this element. This explains why the presence of a combination of lines of an element permits qualitative analysis.

For Al, there are 46 electronic levels of the atom below the ionization limits, corresponding to about 118 lines in the 176 nm–1000 nm range. There are 226 levels for the singly-ionized atom of Al, which leads to about 318 lines in the 160 nm–1000 nm range. The Al I and Al II are species that emit relatively simple spectra, i.e., with a limited number of lines. In a similar range of wavelength, uranium can emit several tens of thousand of lines, which leads to probably one of the most complex spectra observed. However, if the resonance lines can be observed in any radiation source, then the lines originating from highly excited states can be observed only with high temperature radiation sources or under specific excitation conditions.

It should be noted that the *background radiation* is the radiation emitted by the sample when each component is present except the analyte. Background emission consists of the lines emitted by the other elements (concomitants) and the continuum arising from nonquantized transitions.

Quantitative analysis is possible if the intensity of the line can be related to the concentration of the emitting species. The intensity of a line is proportional to:

1. The difference in energy between the upper level, E_m, and the lower level, E_k, of the transition.
2. The population of electrons, n_m, in the upper level, E_m.
3. The number of possible transitions between E_m and E_k per unit time. This value is expressed by the transition probability A, and has been defined by Einstein.

Therefore, the intensity l is proportional to:

$$l \sim (E_m - E_k) \cdot A \cdot n_m \tag{8.1-4}$$

The relation between the population of the various levels has been described by Boltzmann. If we consider the population n_m and n_k of the level E_m and E_k, respectively, their ratio can be given by the Boltzmann law:

$$\frac{n_m}{n_k} = \frac{g_m \cdot \exp(-E_m/kT)}{g_n \cdot \exp(-E_k/kT)} \tag{8.1-5}$$

where k is the Boltzmann constant ($k = 1.380 \times 10^{-23}\,\text{J K}^{-1} = 0.695\,\text{cm}^{-1}\,\text{K}^{-1} = 0.8617 \times 10^{-4}\,\text{eV K}^{-1}$), T is the temperature of the radiation source and g is the statistical weight $(2J + 1)$, J being the total electronic angular momentum quantum number. As the population of excited levels is proportional to the exponential of $(-E)$, the population decreases very rapidly when E increases. A possible way to overcome this limitation is the use of high temperature radiation sources such as plasmas. In the case of the ground state, $E = 0$ and:

$$\frac{n_m}{n_0} = \frac{g_m \cdot \exp(-E_m/kT)}{g_0} \tag{8.1-6}$$

In order to relate n_m to the total population of the levels of the atom (or the ion), N,

$$N = n_0 + n_1 + \cdots + n_m + \cdots$$

it is possible to sum the terms such as $g_m \cdot \exp(-E_m/kT)$ for all possible levels and to define the partition function Z as follows:

$$Z = g_0 + g_1 \cdot \exp(-E_1/kT) + \cdots + g_m \cdot \exp(-E_m/kT) + \cdots \tag{8.1-7}$$

The Boltzmann law is modified to:

$$\frac{n_m}{N} = \frac{g_m \cdot \exp(-E_m/kT)}{Z} \tag{8.1-8}$$

The partition function is, therefore, a function of T. However, in the range of temperature of most analytical radiation sources used for analytical applications, i.e., 2000–7000 K, this variation may be small or negligible.

The intensity of a line can therefore be written as:

$$l = \Phi\left(\frac{h \cdot c \cdot g_m \cdot A \cdot N}{4\pi\lambda Z}\right) \exp(-E_m/kT) \tag{8.1-9}$$

where Φ is a coefficient to account for the mission being isotropic over a solid angle of 4π steradian.

When a radiation source is stable enough to exhibit a constant temperature, the function of partition, Z, will remain constant, and the number of atoms (or ions) N will be proportional to the concentration, c. For a given line of the element to be analyzed, g_m, A, λ, and E_m are constant. Therefore, l is proportional to c, which makes it possible to conduct quantitative analysis. In *relative* quantitative analysis, a series of standards is used to construct a calibration curve, i.e., the intensity as a function of the concentration of the analyte(s). The intensity from an analyte in an unknown sample is used with the calibration curve to deduce its concentration. Theoretically, an *absolute* quantitative analysis could also be carried out, i.e., without the use of a calibration procedure. However, an absolute quantitative analysis would require knowledge of the temperature, the solid angle of emission, etc. These measurements cannot be easily performed for routine analysis.

It should be noted that, in the case of a constant concentration of the analyte, any small variation in the characteristics of the radiation source can lead to a variation in the temperature, and a subsequent variation in the line intensity, because of the change in the population of the excited level. When considering the resonance Al I 396.15 nm line $(E_m = 25\,347\,\mathrm{cm}^{-1})$, an increase of 100 K in the temperature of the radiation source will correspond to an increase in the exponential term $(-E_m/kT)$ of about 50% and 5% at 3000 K and 6000 K, respectively. This explains why good stability of the source characteristics is required so as to obtain good repeatability and reproducibility and to avoid any drift of the analytical signal.

In atomic emission spectrometry, a source will actually have two roles: the first step consists of the atomization of the sample to be analyzed so as to obtain free atoms, usually in the ground state; the second consists of the excitation of the atoms to higher energy states. An ideal source used for emission spectrometry should exhibit excellent analytical performance as well as efficient instrumental characteristics. The analytical performance includes the number of elements that can be determined, the precision and accuracy, the selectivity, the absence of physical and chemical interferences, the long-term stability, the concentration dynamic

The intensity of the line is proportional to the concentration of the atoms.

Only relative quantitative analysis is performed in emission spectrometry, using standards with known concentrations of the element(s) to be analyzed.

range, and the limits of detection. Moreover, an emission system should be able to handle any type of samples, irrespective of their form, liquid, solid or gas, with the possibility of using a limited amount of sample. The instrument characteristics of interest include ease of operation and maintenance, automation, sample throughput, reliability, and size of the system. Some considerations should also be given to the capital investment and the running cost.

8.1.3 Radiation sources

Radiation sources currently used in AES can be conveniently classified into two categories:

1. Sources mostly suited to the analysis of solutions. Flames and plasmas are the most commonly-used sources.
2. Sources mostly suited to the direct analysis of solids. The arc and, to a larger extent, the spark are mostly used.

Flames

A flame is obtained by the chemical reaction between a fuel gas and an oxidant gas.

As stated earlier, the flame is the oldest radiation source used for AES. A flame is an exoergic-controlled chemical reaction between at least two elements or compounds in gaseous form, one being the fuel (acetylene, propane), the other the oxidant (air, oxygen, nitrous oxide) [8.1-3–8.1-8]. The energy results from the heat of combustion of the fuel. Flames are usually created at atmospheric pressure. A typical reaction is:

$$C_3H_8 + 5O_2 \rightarrow 3CO_2 + 4H_2O$$

Several gas mixtures can be used to create a flame for analytical applications. Commonly-used flames are the propane-air flame, the acetylene-air flame, and the acetylene-oxygen flame, which exhibit temperatures of 2200 K, 2500 K and 3300 K, respectively. The increase in temperature for the acetylene-oxygen flame compared with the acetylene-air flame is due to the absence of nitrogen which absorbs energy. Flames can be used under stoichiometric conditions or can be fuel-rich, i.e., with an excess of fuel to minimize the formation of analyte oxides. One interesting feature of a flame is that it is a self-sustaining process as long as fuel and oxidant are provided. In other words, it is not necessary to provide external energy. The sample in a liquid form can be introduced into the flame so that it will be desolvated, vaporized, dissociated, and then atomized before being excited.

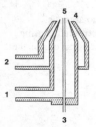

Fig. 8.1-2. Schematic drawing of the direct injection burner. 1: oxidant inlet, 2: fuel inlet, 3: capillary injector tube, 4: burner tip, 5: capillary tip

A flame is produced by means of a burner which supplies the two gases and the sample to be analyzed. In a direct injection burner (or total consumption burner), the sample, in the form of a solution, is aspirated via a capillary and injected directly into the flame by means of an aspirating gas, usually the oxidant. The fuel is mixed with the oxidant and the sample at the orifices of the burner (Fig. 8.1-2). Such a flame is usually turbulent. Because the fuel and the oxidant are mixed above the burner, there is no risk of explosion, even with gas mixtures having a high burning velocity such as acetylene-oxygen (11 m s^{-1}).

In a premix burner, the solution is aspirated and nebulized by means of the oxidant through a mixing chamber. The resulting aerosol-oxidant mixture is then mixed with the fuel before entering the burner itself. In contrast to the previous design, selection of the smaller droplets will occur in the chamber, which results in a fine aerosol reaching the flame. This ensures a more complete evaporation of the droplet and atomization of the species. However, the efficiency of nebulization is typically of the order of 5%. Such flames are laminar in structure. In a premix burner, it is essential that the exit velocity of the fuel-oxidant mixture is higher than the flame propagation velocity so to avoid any flashback condition and explosion.

Flame spectrometry is mainly used as a low-cost system for the determination of alkali and alkaline earth elements.

Flame emission spectrometry has been largely replaced by flame atomic absorption spectrometry. However, some low cost systems are still commercially available for the determination of alkali and alkaline earth elements.

Plasmas

A *plasma* is an ionized gas that is macroscopically neutral, i.e., with the same number of positive particles (ions) and negative particles (electrons). If a mono-atomic gas, X, is used, a plasma can be described by the following equilibrium:

$$X = \sum_{n=1}^{q} X^{n+} + \sum_{n=1}^{q} n.e \tag{8.1-10}$$

where X^{n+} is an ion with n charges and e represents an electron. Some properties of ideal gases such as the pressure and the volume still apply in contrast to other properties such as the viscosity and the thermal conductivity that significantly differ from those of ideal gases because of the presence of charged particles.

In contrast to a flame, it is necessary to supply an external energy in the form of an electrical field in order to ionize the gas and to sustain the plasma, which, in turn, will transmit part of this energy to the sample to atomize and excite it. Plasmas may be classified according to the kind of electrical field which is used to create and sustain the plasma:

A plasma usually consists of an ionized rare gas.

- *direct current plasma* (DCP) is obtained when a direct current field is established across electrodes
- *inductively coupled plasma* (ICP) is obtained when a high-frequency field is applied through a coil
- *microwave induced plasma* (MIP) is obtained when a microwave field is applied to a cavity

Historically, the DCP was the first described and commercialized plasma. However, the ICP is currently the most commonly used plasma because of some unique properties [8.1-9] – [8.1-12].

In the case of the ICP, a high-frequency generator (usually operating at 27 MHz or 40 MHz) is used to produce a hf field through an induction coil. There are, therefore, no contaminating electrodes. Several types of oscillators can be used to generate the hf field: the crystal-controlled oscillator and the free-running oscillator are commonly used with variations in design. The power is of the order of 1–2 kW, and the power stability is a crucial parameter for avoiding any drift in the properties of the plasma. The generator must exhibit enough flexibility to compensate for any variation of the plasma impedance due to change in its load, i.e., during injection of different types of solutions (aqueous or organic).

The inductively coupled plasma is the most commonly used plasma.

The gas that is used to generate the plasma (plasma gas) is argon. Like any noble gas, argon is a monoatomic element with a high ionization energy (15.76 eV), and is chemically inert. Consequently:

- a simple spectrum is emitted by argon in contrast to a flame where primarily molecular spectra are observed
- argon has the capability to excite and ionize most of the elements of the periodic table
- no stable compounds are formed between argon and the analytes.

It should be noted, however, that some unstable molecular excited or ionized species can be formed within the plasma, e.g., ArH. They usually dissociate after their de-excitation (see Sec. 8.5). Argon is also the cheapest noble gas since its concentration in the air is 1%. The only limitation in using argon is the poor thermal conductivity of this gas compared with molecular gases such as nitrogen and hydrogen.

The plasma is created in a torch. This torch plays the following roles: provides electrical insulation between the plasma and the coil, and confines and stabilizes the plasma for introduction of the sample. Because of the nature of the hf field and the resultant skin effect, the energy from the generator is mainly deposited in the external part of the plasma. There is consequently a zone along the axis of the plasma where the viscosity is lower. This results in the creation of a central channel that facilitates the injection and the confinement of the sample. Current torch design makes use of three concentric tubes: an outer tube to confine and insulate

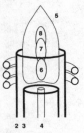

Fig. 8.1-3. Schematic drawing of an inductively coupled plasma torch. 1: induction coil, 2: outer tube, 3: intermediate tube, 4: sample injector tube, 5: plasma, 6: atomization zone, 7: atomic line emission zone, 8: ionic line emission zone. The plasma (outer) gas is introduced between the outer and the intermediate tube, the auxiliary (intermediate) gas is introduced between the intermediate tube and the injector tube, and the carrier gas is introduced via the injector tube

Samples are usually introduced into the ICP in the form of liquid aerosols produced by pneumatic nebulization of solutions.

the plasma; an intermediate tube which serves to accelerate the plasma gas which is introduced between the outer and intermediate tubes, and the injector tube for sample introduction (Fig. 8.1-3). The external tube is made of silica because of its refractivity and transparency to radiation. The observation of the plasma can be carried out perpendicularly to the axis of the plasma (lateral viewing) or along the axis (axial viewing).

Total argon consumption is in the $10-15\,L\,min^{-1}$ range. Solutions are introduced into the plasma in the form of an aerosol. This aerosol is produced by means of a pneumatic nebulizer [8.1-13]. As the average droplet diameter, $20\,\mu m$, is too large to ensure complete volatilization in the plasma, a spray chamber (of the double-pass, or cyclone type) is added to retain the large droplets. Only particles of the order of a few micrometers reach the plasma. The overall sample introduction efficiency is a few %.

Three types of nebulizers are commonly used: the *concentric* nebulizer, the *cross-flow* nebulizer, and the *V-type* nebulizer. The V-type nebulizer has been designed to be less subject to blocking due to particles in suspension in the solution. Liquid consumption is typically in the $1-2\,mL\,min^{-1}$ range. Almost in any instrument, a peristaltic pump is used to feed the nebulizer to ensure a constant uptake rate regardless of the liquid viscosity. Although the standard means of introducing the sample is through nebulization of liquid solutions, the ICP can also be used to analyze solids.

Solid samples can be in the form of fine suspensions in a liquid (slurries). The slurries are introduced into the plasma using, for instance, a V-type nebulizer. Alternatively, the fine particles can be produced by the ablation of a solid with the use of either a spark system for conductive samples or a laser for any type of samples. The particles are taken up by a stream of argon to the plasma [8.1-14]. The sample can also be packed into a graphite crucible which is introduced at the bottom of the plasma. The introduction of the sample is then obtained by volatilization (solid sample insertion) [8.1-15].

Arc

An arc is a stable electrical discharge of high current density and low burning voltage between two or more electrodes [8.1-16–8.1-18]. The voltage between the electrode gap is up to 50 V whereas the current is in the 2 A–30 A range (medium current arc). The discharge can be initiated by separating the two electrodes first brought into contact. An alternative is the use of an external high voltage spark ignition. The shape of the plasma formed in this discharge is related to the electrode gap (up to 20 mm) and the shape, the power, and the composition of the sample. Among the possible configurations, the free-burning arc is the most widely used. In this configuration, the arc is formed from both the vapour of the sample and the surrounding gas, and burns freely in space. This is in contrast to a gas-stabilized arc, where the arc is stabilized by a gas stream flowing around the arc. Free-burning of the arc leads to wandering of the discharge, and consequently, high fluctuations of the signal. This is why this type of arc is mostly used for qualitative analysis. Both dc voltage and ac voltage can be used to sustain the arc. Wandering of the arc can be reduced by applying an ac voltage to the electrodes. The arc is thus continually interrupted and reformed.

The most commonly-used material for electrodes is graphite. Graphite exhibits some interesting features: there is no contamination by elements other than carbon, it has some excellent conductivity and refractive properties, and its cost is low. One of the electrodes is used to present the sample, which is usually in the form of a powder, to the discharge. The discharge is created between the surface of the sample and the other electrode (counter electrode) (Fig. 8.1-4). This results in the consumption of the sample and in the formation of a pit. Selective and erratic volatilization can occur. The sample can also be placed into a conical hole in one of the graphite electrodes. This configuration is used to determine easily vaporized elements in the presence of a stable matrix. The metal is melted during the discharge and forms a globule. The arc is, therefore, called a globular arc (Fig. 8.1-4). Because of the time-dependent volatilization, peak intensities appear at different times.

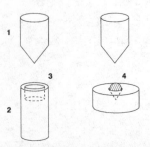

Fig. 8.1-4. Principle of an arc. Left: standard arc, 1: graphite counter-electrode, 2: carrier electrode, 3: cup for sample. Right: globular arc, 4: sample globule

Spark

Sparks are intermittent, oscillating electrical discharges of high voltage and relatively low average current between at least two electrodes [8.1-16]–[8.1-18]. One electrode consists of the sample to be analyzed whereas the other is usually made of tungsten (Fig. 8.1-5). Sparks differ from an ac arc. A spark duration is typically of the order of a few μs. The space between the electrodes is called the analytical gap and is in the 3–6 mm range. There is a wide variety of spark type, depending on the spark generator principle and characteristics. Sparks can be classified according to the applied voltage: high tension sparks (10–20 kV), medium tension sparks (500–1500 V) and low tension sparks (300–500 V). High tension sparks can be self-ignited whereas medium and low tension sparks are externally ignited by means of a high-tension pulse synchronized with the spark frequency. Precision is improved at the detriment of limits of detection when increasing the voltage. This is why low tension sparks seem to be a good compromise.

The spark frequency used to be synchronized with the power line frequency. Currently, it is synchronized with a built-in oscillator. The frequency is in the 100–500 Hz range for commercially available spark systems. Most systems make use of solid-stage generator technology. It is also possible to control the form of the spark wave. In particular, the duration of the pulse can be extended up to 700 μs to obtain a discharge with characteristics close to an arc, so as to improve the limits of detection and the determination of trace elements. Unidirectional discharge is used to protect the electrode and, therefore, to increase its lifetime. In any case, a high energy spark is applied during a prespark period to condition the surface of the sample and to minimize the interference effects. A special application is the use of a rotating electrode (rotrode) for the analysis of wear metals (i.e. metals originating from the wear of an engine) in oils. This system overcomes the difficulty of analyzing liquids with the spark. A thin film of oil is deposited onto the disk during its rotation, and the spark occurs in the analytical gap between the disk and another high voltage electrode.

Argon is often used instead of air to flush the spark stand. Argon is transparent to the UV radiation and does not react with the electrode.

For each spark, a new spot of the sample is struck. Over a series of sparks, an averaging is consequently obtained, which results in a high precision of the analyte signal.

Other radiation sources

Low pressure discharges are radiation sources in which the emission of light is obtained by electrical discharges between two electrodes at pressures <100 kPa. The sample to be analyzed usually forms the cathode. The material is removed during the discharge by atomic and ionic bombardment. The phenomenon is called *cathode sputtering*. A glow discharge is obtained near the cathode. Its size and intensity depend on the current intensity. Several types of discharges have been used as radiation sources including arc discharges, Geissler lamps, and hollow-cathode lamps. At the end of the Sixties, Grimm designed a type of glow discharge where a flat sample acted as the cathode (Fig. 8.1-6). The sample could, therefore, be readily inserted in the lamp [8.1-19–8.1-20].

In the glow discharge lamp (GDL), the current is usually less than 100 mA and the electrical field can reach several kV mm^{-1}. High tension makes it possible to obtain efficient sputtering. Various processes are involved in the emission of the glow discharge, such as electron impact for excitation and ionization, and Penning ionization. GDLs can be used for both bulk and depth-profiling analysis. The use of a dc discharge permits the analysis of conductive materials only, whereas the more recent use of rf discharges (e.g., 13 MHz) makes it possible to analyze nonconductive materials. GDLs have gained acceptance for the analysis of solids and are commercially available.

Laser-produced plasmas have been used since the beginning of the 1960s as radiation sources in atomic spectrometry [8.1-21, 8.1-22]. Pulsed lasers are generally used to create a transient plasma at the surface of the target, which implies the use of a time-resolved detection. Different types of laser can be used, including

A spark discharge is obtained by applying a high voltage between two electrodes.

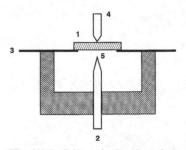

Fig. 8.1-5. Schematic drawing of a spark system. 1: conductive sample acting as an electrode, 2: tungsten counter-electrode, 3: sample holder made of insulating material, 4: electrical connection, 5: analytical gap

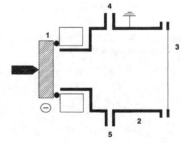

Fig. 8.1-6. Schematic drawing of a glow discharge lamp. 1: sample, 2: discharge cell, 3: silica window, 4: gas inlet, 5: vacuum

In contrast to flame, plasma, and spark, GDL is produced at reduced pressure.

the excimer lasers (194 nm, 308 nm), the Nd:YAG laser (1064 nm, 536 nm, 355 nm, 266 nm), and the CO_2 laser (10.6 µm). The current trend is to use UV lasers. More details concerning the laser-produced plasmas are given in Section 8.5.

8.1.4 Spectrometers

A spectrometer used for atomic emission is a spectral instrument which disperses, in space, light emitted by the radiation source, isolates specific spectral bands containing analyte lines or selected background regions, and measures line or background intensities by means of one or several detectors. A spectrograph differs from a spectrometer in that the whole range of the spectrum allowed by the system is recorded on a photographic plate. With spectrographs, qualitative analysis is carried out by verifying the presence of several lines of the analytes concerned, and quantitative analysis is performed by measuring the line intensities by means of a densitometer.

> Two types of dispersive systems are used in emission spectrometry: the monochromator and the polychromator.

Although they are still in use, spectrographs have been commercially replaced by spectrometers. Historically, spectroscopes were used with visual observation of the spectra. Conversely, spectrometers use photoelectric detectors and they used to be classified in two categories: monochromators and polychromators. A *monochromator* is a spectrometer that isolates a specific and single spectral band in time. This spectral bandpass can be fixed. Alternatively, the monochromator can scan continuously a given range of wavelength or move sequentially from one specific spectral band to another. In the latter case, the system is called a sequential spectrometer. A *polychromator* is a spectrometer that isolates several specific spectral bands at the same time. A polychromator is also called a simultaneous system. The term direct-reader is also used to differentiate the use of detectors from that of a photographic emulsion. Before the introduction of multichannel detection, the selection of these spectral bands was fixed. The recent introduction of multichannel detectors such as charge transfer device (CTD) detector has modified the spectrometer classification. A CTD detector can replace a photographic plate and is setup at the same location to provide a combination of a spectrograph (whole acquisition of the spectrum) and a spectrometer (measurement of the intensities of any part of the spectrum).

It should be noted that background measurement at the location of the analyte wavelength or in its vicinity is necessary for obtaining the net line intensity, when the background intensity contributes significantly to the gross line intensity. This is typically the case for plasmas.

> The grating has replaced the prism as the dispersive component in a spectrometer.

Two different types of dispersive components have been used: the *prism* and the *diffraction grating*. Most modern equipment make use of the diffraction grating because of its better dispersion characteristics. A diffraction grating consists of periodic, parallel grooves or lines either on a flat surface or on a concave surface, which impose a periodic variation in amplitude and phase of an incident wave. The first diffraction grating was designed by Fraunhofer and the first concave grating was ruled by Rowland in 1890. Only reflection gratings are currently used.

The characteristics of a diffraction grating are:

- the distance d between two successive grooves
- the groove density (i.e., per unit of length), n
- the width, W, of the grating
- the total number of grooves, $N = nW$
- the angle, θ, of the normal to the groove surface with the grating normal, in the case of saw-tooth shaped grooves (Fig. 8.1-7)

Typical values of n are in the 1000–4800 groove mm^{-1} range. The width of the grating can be larger than 100 mm. Both standard gratings (groove density >600 groove mm^{-1}, use of the large side of the triangular profile) and echelle gratings (low groove density <100 groove mm^{-1}, use of the small side of the profile, large incidence angle) are used in AES.

When a parallel beam is incident to a plane grating with a given angle, α, to the grating normal, the grating equation is used to calculate the diffracted beam angle, β, for a given monochromatic wavelength, λ. Each groove acts as a narrow mirror

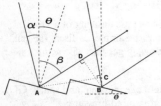

Fig. 8.1-7. Principle of a reflection plane grating. The grooves are illuminated by a parallel incident beam with an angle, α, with the normal to the grating. The light is diffracted in the direction β. The net path difference between the two waves is BC − AD. The distance between two successive grooves is AB = a. This distance is related to the groove density, n, with $n = 1/a$. The normal to the groove surface makes an angle θ, with the normal to the grating

to reemit light. The various reflected beams of light will recombine to interfere and to produce diffraction phenomena. All rays will be reinforced when the net path difference results in constructive interference, i.e., when the net path difference is a multiple of the wavelength, λ.

From Fig. 8.1-7, it can be deduced that the net path difference is equal to:

$$BC - AD = k\lambda$$

where k is a positive or negative integer called the order. The distance AB corresponds to the interval, a, between two successive grooves, with $a = 1/n$.

As $\sin\alpha = BC/a$ and $\sin\beta = AD/a$, the equation can thus be rewritten as:

$$\sin\alpha - \sin\beta = \frac{k\lambda}{a} \tag{8.1-12}$$

When using the groove density, the grating equation becomes:

$$\sin\alpha - \sin\beta = kn\lambda \tag{8.1-13}$$

When incident and diffracted beams lie on the same side of the grating normal, because $\sin(-\beta) = -\sin\beta$, the grating equation is given by:

$$\sin\alpha + \sin\beta = kn\lambda \tag{8.1-14}$$

The zero order, $k = 0$, corresponds to specular reflection ($\alpha = -\beta$).

The highest wavelength λ_{max} that can be obtained from a given grating is for the maximum value obtained for the first side of the grating equation, i.e., for the maximum value of $\sin\alpha + \sin\beta$. The highest value that can be reached by α and β is $90° - \theta$. Therefore, both $\sin\alpha$ and $\sin\beta$ are very close to 1. The value of λ_{max} is:

> The line number of a grating not only influences the resolution, but also the highest wavelength which can be diffracted.

$$\lambda_{max} < \frac{2}{kn}$$

Therefore, the highest wavelength that can be reached for a grating with 2400 and 3600 grooves mm^{-1} is 830 nm and 550 nm, respectively. Similarly, a grating with 3600 grooves mm^{-1} that operates in the 2nd order leads to a maximum wavelength of 275 nm. Incident angles near $90° - \theta$ are practically never used. The actual maximum wavelength that is used is usually lower than the theoretical one. The selection of the grating is, therefore, related to the wavelength of the analytical lines of the elements to be determined. The ideal range would be 120–770 nm. The 580–770 nm range is mostly used for the determination of the alkali elements. The wavelength region below 190 nm can be observed only in the absence of oxygen whose molecular bands absorb such radiation. It is then necessary to either purge the optical system with nitrogen or to work under vacuum conditions. The region between 160 and 190 nm is used for elements such as Al, P, and S, whereas the region below 160 nm is used for elements such as O, Cl and N.

Since the order, k, can have various integer values, this means that a given wavelength, λ, may be observed for various diffraction angles, β. An example is given in Fig. 8.1-8 for a 1200 grooves mm^{-1} grating and an incident angle, $\alpha = 20°$. A 200–500 nm wavelength range has been selected. It can be seen that the 200 nm wavelength can be observed at seven different locations ($k = 1$–$5, -1, -2$) whereas the 500 nm can be observed only at three locations ($k = 1, 2, -1$). It can be easily deduced that the number of orders that can be observed for a given wavelength and incident angle is given by:

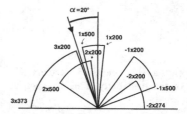

Fig. 8.1-8. Example of formation of the orders for a beam with an incident angle $\alpha = 20°$. The groove density is 1200 mm^{-1}. The first number indicates the order while the second number gives the wavelength in nm, e.g.: 2×500. Calculation was performed for the 200–500 nm range

$$\frac{(\sin\alpha - \sin(\beta - \theta))}{n\lambda} < k < \frac{(\sin\alpha + \sin(\beta - \theta))}{n\lambda} \tag{8.1-15}$$

which is very close to:

$$\frac{(\sin\alpha - 1)}{n\lambda} < k < \frac{(\sin\alpha + 1)}{n\lambda}$$

where k should take only integer values. From the example given in Fig. 8.1-8 one obtains:

$$-1.09 < k < 2.2 \text{ for } \lambda = 500 \text{ nm}$$

that is to say $k = -1, 1, 2$ as previously seen.

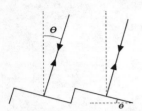

Fig. 8.1-9. Littrow mounting where both the angle of incidence α and the angle of diffraction β are equal to the angle of the normal to the groove facet and the normal to the grating θ (blaze angle), to estimate the blaze wavelength λ, through the relation $2\sin\theta = kn\lambda$

The interferometric grating is the most commonly used type of grating in emission spectrometry.

The current success of emission spectrometry is based on the availability of highly effective gratings.

If a given wavelength can be observed at different locations, for a given diffraction angle β, several wavelengths of difference orders can be observed so that the $k\lambda$ product remains constant. For the example given in Fig. 8.1-8, the use of a diffraction angle, $\beta = 14.95°$ will result in the observation of 500 nm for $k = 1$ and 250 nm for $k = 2$. It will then be necessary to use either filters to reject one of the two wavelengths, or to use a detector whose response is limited to a given wavelength range. The situation is far more complex with the echelle grating. Usually, the groove density is very low, $n < 100$ groove mm^{-1}, which implies the use of high orders (up to $k = 120$). In that case, the use of filters is not sufficient, and it is necessary to use a pre-disperser or a post-disperser.

The efficiency of the grating is the ratio of the diffracted to the incident spectral radiant power of a given wavelength range and order. If the angle, θ, is equal to zero, the efficiency is maximum for the zero order, i.e., the grating acts as a mirror and the energy of the diffracted light is reduced to a minimum. This is why an angle, θ, is given to the surface of the grooves with respect to the plane of the grating (Fig. 8.1-9). The efficiency is usually evaluated using the Littrow optical mounting where both the incident angle and the diffracted angle are equal to θ (Fig. 8.1-9). The grating equation becomes:

$$2\sin\theta = kn\lambda \tag{8.1-16}$$

Under these conditions, λ is the wavelength for which a high efficiency is observed for a given order. This is why θ is also called the *blaze angle*. For a 2400 groove mm^{-1} grating and an angle, $\theta = 27.5°$, the optimum wavelength is 385 nm in the first order and 192 nm in the second order. Selection of an appropriate angle allows the grating manufacturer to optimize the grating for a given wavelength range, depending on the application.

Two types of gratings are currently in use: the *ruled grating* and the *interferometric grating*. The first type is obtained by a ruling machine by burnishing grooves with a diamond stylus. This is an expensive and labor-intensive process. Practically, only replicates are commercially available. The advantage of such a grating is the possibility of obtaining a given blaze angle. The second type of grating is obtained using interference fringes formed at the intersection of two beams of laser light. A series of fringes produces a periodic pattern in a photosentive layer. A chemical treatment is then used to dissolve the photosensitive material. Peaks and valleys are thus obtained from the reflecting grooves. A vacuum aluminum coating is then processed. Interferometric gratings are also called holographic gratings. Standard interferometric gratings exhibit a sinusoidal groove profile. In order to obtain a blaze effect, i.e., a triangular profile, ion etching is used.

Original interferometric gratings are commercially available. Interferometric gratings are preferred in the case of concave shape, whenever high groove density is required, and when working in the UV. Both technologies are used for standard gratings whereas echelle gratings can be obtained only by ruling. Improvement in the production of gratings has certainly been a major advance in instrumentation for AES.

Although a large number of optical mountings have been described when standard gratings are used, a limited number of designs are currently used. Polychromators make use of a concave grating and are mainly based on the *Paschen-Runge* configuration. Concave gratings are not only used for diffraction but also for collimation.

The Paschen-Runge mounting is based on the use of the *Rowland circle*. A spherical concave grating with a radius, R, is set up on the perimeter of a circle with a diameter equal to R (Fig. 8.1-10). The grooves of the grating are normal to the place of the circle. One advantage of this mounting is that if an entrance slit is mounted on the circle, the optical system produces diffracted beams which are also focused on the circle. This is an advantage from a mechanical point of view, since a circle, or part of a circle, is easy to manufacture.

A series of exit slits and detectors can be set up according to the element and line selection (Fig. 8.1-10). Up to 40–50 exit slits can be mounted. It should be noted that the presence of several orders is not necessarily a drawback for polychromators. When two lines have very close wavelengths, it would be impossible

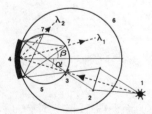

Fig. 8.1-10. Schematic drawing of a polychromator based on the use of a concave grating. 1: radiation source, 2: focusing system, 3: entrance slit, 4: concave grating, 5: Rowland circle, 6: grating circle, 7: exit slits

to position two exit slits at almost the same location. Therefore, different orders can be used to position the two exit slits at different locations and thus obtain both wavelengths separated in space.

This optical mounting can also be used as a monochromator. Line selection is then obtained by moving the exit slit and the detector. In the grating equation, where α has a constant value, the value of λ is obtained by changing that of the angle β (Fig. 8.1-11). An alternative is the displacement of the detector behind fixed exit slits, the final adjustment being obtained by a small displacement of the entrance slit. Therefore, the angle β has only discrete values and the exact value of λ is obtained by a small adjustment, $\Delta\alpha$, of the incidence angle (Fig. 8.1-11). In order to reduce the displacement of the exit slit/detector along the Rowland circle, several entrance slits can be used to obtain several angles of incidence. Optical fibers are used to conduct the light onto the entrance slits.

Plane gratings are used only for monochromators. Mirrors are necessary for the entrance and exit collimations. When a single concave mirror is used for this purpose, the optical mounting is called the *Ebert mounting* whereas when two mirrors are used, this is the *Czerny-Turner mounting* (Fig. 8.1-12). The focal length, f, of the mirror(s) defines the focal length of the dispersive system. Line selection is obtained by rotation of the grating. In this case, the grating equation can be used for calculation of the diffracted wavelength with $\alpha - \beta$ equal to a constant value.

Because of the high number of orders observed with the use of an echelle grating, most of the commercially available dispersive systems make use of a post-disperser. This disperser is a dispersive component, such as a prism or a grating, which is set up at right angles to the plane of the diffracted beam so to obtain a cross dispersion. This mounting results in a 2-dimensional spectrum. This is why the echelle-grating-based dispersive system can be used either as a monochromator or as a polychromator.

In the monochromator configuration, wavelength selection is obtained either by rotating the grating and the prism, or by moving an exit slit and the associated detector in the focal plane. In the polychromator configuration, a series of exit slits with their corresponding detectors is mounted in the focal plane. An alternative is the use of a 2-dimensional detector such as a CTD, which replaces both the slits and the photomultiplier tubes.

The performance of an optical dispersive system for spectral isolation is characterized by the reciprocal linear dispersion and the practical resolution. The reciprocal linear dispersion, $d\lambda/dx$, is the capability of the dispersive system to spread the spectrum in the focal field and is expressed in nm mm^{-1}. The better the reciprocal linear dispersion, the lower its value. Derivation of the grating equation in terms of the diffracted angle, β, leads to:

$$\frac{d\lambda}{dx} = \frac{\cos\beta}{knf} \qquad (8.1\text{-}17)$$

where f is the focal length of the dispersive system. The reciprocal linear dispersion is slightly wavelength-dependent since the value of β, and, therefore, $\cos\beta$, varies as a function of λ. This is illustrated in Table 8.1-2.

To improve the reciprocal linear dispersion, i.e., to decrease its value, it is necessary to increase at least one of the three factors, k, n, or f. For size reasons and thermal stability, most commercially available dispersive systems have a focal length in the 0.5–1 m range. Therefore, it is necessary to increase the groove density and to use orders higher than one. The influence of these two parameters is illustrated in Table 8.1-3.

It should be noted that the dispersion is not only dependent on the order and the groove density but is slightly dependent on $\cos\beta$. A good reciprocal linear dispersion is therefore obtained by using a high groove number and high orders, but at the detriment of the wavelength coverage, since the maximum wavelength will decrease. Dispersion is typically in the 0.1–1 nm mm^{-1} range for commercially available dispersive systems.

The practical resolution of a dispersive system is its capability to separate two adjacent spectral lines. *Resolution* is usually expressed in wavelength units (pm), whereas the *resolving power* is the ratio of the wavelength, where the measurement was carried out, to the resolution. The resolving power has, therefore, no units.

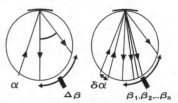

Fig. 8.1-11. Schematic drawing of different types of monochromators based on the use of a concave grating. Left: continuous displacement of the exit slit and the detector (change in the angle of diffraction β). Right: discrete displacement of the detector behind a series of exit slits and small continuous displacement of the entrance slit for the final tuning of the wavelength (both changes in the angle of incidence α and the angle of diffraction β)

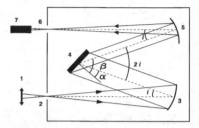

Fig. 8.1-12. Schematic drawing of a Czerny-Turner monochromator based on $\beta - \alpha = 2i$. 1: focusing lens, 2: entrance slit, 3: collimating concave mirror, 4: rotating plane grating, 5: camera concave mirror, 6: exit slit, 7: detector

The reciprocal linear dispersion is improved when the focal length and the groove number of the grating are increased.

Table 8.1-2. Variation of the reciprocal linear dispersion as a function of the wavelength. A 3600 groove mm^{-1} plane grating was used with a focal length of 1 m and a value of $\alpha - \beta = 14.74°$

Wavelength (nm)	β (°)	nm mm^{-1}
170	24.21	0.253
200	27.51	0.246
300	39.18	0.215
400	52.69	0.168
500	71.16	0.090

Table 8.1-3. Influence of the groove density (plane grating) and the order on the reciprocal linear dispersion (at 230 nm) for a 1-m monochromator ($\alpha - \beta = 14.74°$)

Groove density (mm^{-1})	2400	3600	2400	3600
Order	1	1	2	2
Reciprocal liner dispersion (nm mm^{-1})	0.382	0.235	0.157	0.061

Table 8.1-4. Examples of theoretical resolution (Rayleigh criterion) and resolving power (at 230 nm) for several gratings

Groove density (mm^{-1})	Order	Width (mm)	Resolution (pm)	Resolving power
1080	1	60	3.5	65 000
1080	2	60	1.7	130 000
1800	1	84	1.5	150 000
2400	1	110	0.9	260 000
4200	1	84	0.7	350 000
2400	2	140	0.3	670 000
3600	2	140	0.2	1 000 000

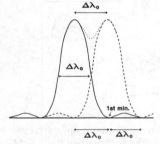

Fig. 8.1-13. Separation of two diffraction patterns formed by two adjacent lines. According to the Rayleigh criterion, two lines are separated when the maximum of the first pattern (——) corresponds to the first minimum of the second pattern (– – –). The resolution $\Delta\lambda_0$ is then the distance between the central wavelengths of the two lines

The practical resolution depends on three majors parameters:

– the theoretical resolution limited by the diffraction of the grating
– the spectral bandpass
– optical aberrations.

Because of the diffraction phenomena produced by the grooves of the grating, even an incident monochromatic light will result in a diffraction pattern. This diffraction pattern (Fig. 8.1-13) exhibits a principal maximum and a series of minima and maxima. Two adjacent monochromatic lines of equal intensity will produce two diffraction patterns. Theoretical resolution is based on the *Rayleigh criterion*: the two diffraction patterns are considered as separate, and therefore, so too are the two lines when the first maximum intensity of the first pattern corresponds to the first minimum of the second pattern (Fig. 8.1-13).

Actually, this definition of the theoretical resolution is rather optimistic, as the convolution of the two diffraction patterns does not result in a full separation of the two patterns. The minimum intensity located between the two maximum intensities (Fig. 8.1-13) is 80% of these maxima. It would have been better to have selected the second minimum. It can be easily demonstrated that the resolution, $\Delta\lambda_0$, defined according to the Rayleigh criterion, is related to the grating characteristics only, since the resolving power is equal to:

$$\frac{\lambda}{\Delta\lambda_0} = knW = kN$$

The availability of high groove-number gratings with a large size permits the spectroscopist to obtain high theoretical resolution (Table 8.1-4).

In the UV region of the spectrum, the theoretical resolution is always lower than 2 pm. Most modern gratings provide a theoretical resolution better than 1 pm. The second factor influencing the practical resolution is the *spectral bandpass* of the dispersive system. The spectral bandpass is the result of the product of the reciprocal linear dispersion by the resultant spectral slit width. This slit width is either the width of the exit slit or the image of the entrance slit, whichever is greater. Slit widths are usually in the 10–100 μm range. Some examples of spectral bandpass are given in Table 8.1-5.

It can be seen that the spectral bandpass covers a wide range of values, but in all instances, it is difficult to reach values lower than 1 pm. Practically, because of the high quality of the modern gratings (interferometric production, high groove number, large width), the theoretical resolution is no longer a limitation to the practical resolution. This limitation is mostly due to the spectral bandpass and optical aberrations. The practical resolution is usually directly determined by using a narrow emission line in the wavelength region of interest. Its value is

The practical resolution in the UV is mainly spectral bandpass limited.

Table 8.1-5. Examples of spectral bandpass for various gratings, orders, focal lengths, and slit widths

Groove density (mm^{-1})	Order	Focal length (cm)	Slit width (µm)	Spectral bandpass (pm)
2880	1	45	30	25
1800	1	100	20	10.7
2400	1	100	20	7.6
2400	2	100	10	1.6

obtained by measuring the width of the line profile at half peak-maximum intensity. Current practical resolution in the UV is in the 5–30 pm range, which is significantly higher than the theoretical resolution.

For sources exhibiting a high background radiation, such as the ICP, it may be necessary to subtract the value of the background intensity from that of the gross line intensity to obtain the net line intensity.

8.1.5 Detection

The use of the photographic plate for light detection has been declining over the past two decades. It has been replaced by the *photomultiplier tube* (PMT). The PMT consists of two parts:

- the photocathode which converts the incident photons into electrons
- the dynodes which serve to amplify the number of electrons

The advantages of the PMT are numerous: large wavelength coverage (160–900 nm is currently used), large dynamic range, high amplification gain, and low noise. However, the PMT is a single detector. This is why there is current trend to replace the PMTs in polychromators by a multichannel detector. This can be a photodiode array or a charge transfer device (CTD) such as a charge-coupled device (CCD) or a charge-injection device (CID).

The photomultiplier tube is the most commonly used detector, but there is a trend to use multichannel detection as a substitute.

Although the principle of the optical systems currently used in AES has been described several decades ago, improvement in the manufacturing processes (grating, correction of aberrations, precision in machining and optical alignment, precision and speed in grating rotation or exit slit displacement) along with the availability of new radiation sources explains why AES techniques are so well-established and have been commercially successful.

8.1.6 Analytical performance

Among the main analytical performance characteristics of an analytical method for elemental quantitative analysis the precision, the accuracy, the limits of detection, and the absence of interferences are of particular relevance.

Accuracy, precision, limits of detection, and absence of matrix effects are the most important characteristics of emission spectrometry as an analytical method.

Precision is related to the various noise sources in the AES system: shot noise due to the random emission and arrival of photons at the detector, flicker noise due to some possible instabilities of the apparatus, and detector noise. Precision can also be limited by the heterogeneity of the sample when solid samples are analyzed directly. High tension spark, flame, and plasma systems exhibit good precision with %RSD as low as 1% or even less. These values can be obtained when working at least 20–50 times above the limits of detection. In contrast, the precision obtained with an arc system is relatively poor, in the 5–10%RSD. This is why the arc is mostly used for qualitative or semiquantitative analysis.

The accuracy is mainly related to the availability of reference materials. For the analysis of metals and alloys by spark, numerous certified reference materials (see Sec. 3.3.4) are available. In contrast, because the ICP can analyze any sample that can be put in solution, the availability of reference materials can be a problem for some new materials, e.g., ceramics and polymers.

The limits of detection are expressed in $mg\,kg^{-1}$ (ppm) for solid samples and $ng\,mL^{-1}$ (or $\mu g\,L^{-1}$) for liquid samples. The arc can determine 60–70 elements with the limits of detection in the 0.1–1 ppm range. The limits are usually better than those obtained with spark emission (1–10 ppm). However, these limits of detection depend strongly on the matrix and, in the case of a spark, on the tension. For instance, the limits of detection with the use of spark emission are better in an aluminum matrix than in a steel matrix.

The use of glow discharge lamps leads to limits of detection of the order ppm. Limits of detection are in the $1–1000\,ng\,mL^{-1}$ range in flame AES, where only atomic lines are used. The use of flame makes it possible to determine 40–45 elements. The best results are obtained for alkali and alkaline earth elements. Results are not as good for elements such as As, B. Be, Cd, Sb, Se, Si, and Zn because their resonance lines are below 270 nm and the temperature of the flame is too low to efficiently populate the first excited states of these elements. For elements such as Ce, La, Th, and U, even a fuel-rich $C_2H_2–N_2O$ flame has a low temperature. Currently, flame-AES is mostly used as a low cost system for the determination of alkali elements (Na, K) at the $\mu g\,mL^{-1}$ level.

It is possible to determine most of the elements in the periodic table with an ICP. Limits of detection are in the $0.1–50\,ng\,ml^{-1}$ range. If the sample was originally a solid, this corresponds to limits of detection lower than 1 ppm in the solid since the salt concentration can be easily up to $10\,g\,L^{-1}$. Both ionic and atomic lines are used for the analysis, the ionic lines usually providing the best sensitivity. Best results are obtained for elements such as Ba, Be, Ca, Mn, Mg, and Ti. It should be noted that alkali elements exhibit a relatively poor limit of detection because the temperature of the plasma is too high for these elements.

Chemical interferences in flames are mainly due to the formation of compounds in the reaction zone which hamper the determination of an element. The temperature of the flame is too low to fully dissociate these compounds. For instance, the determination of Ca is interfered by the presence of P. This can be partly overcome by the use of releasing agents.

Chemical interferences in plasmas are minimized because of the comparatively high temperature of the ICP. However, the high temperature reached by the plasma results in the emission of a line-rich spectrum, particularly when elements such as U, Fe, and Co are present in the sample. This problem of spectral interferences is a possible limitation of the ICP.

8.1.7 Applications

The arc, associated with the photographic plate, is still in use in industry for qualitative analysis. Some systems are commercially available with PMTs such as the globular arc, for instance, for the determination of trace elements in pure copper. Sensitivity is better than X-ray fluorescence spectrometry, and there is little sample preparation since the system can directly analyze wires and drillings.

Spark AES is widely used for the direct analysis of metal and alloys such as steel, stainless steel, nickel, and nickel-based alloys, aluminium, and aluminium-based alloys, copper, and copper-based alloys, etc. In the steel industry, this technique is unsurpassable because of the speed and the precision of analysis. Spark AES can also be mounted in a pistol and connected to a mobile system for on-site control and identification of unknown samples with laboratory-like accuracy. The major limitation of spark AES is the need to construct a calibration graph for each type of sample due to the matrix effects on the analyte intensity. For instance, it is necessary to store different calibration procedures for steel and aluminium alloys.

ICP-AES can be used for analyzing any sample that can be put in solution. This covers a large field of applications including metals and alloys, geological materials, environmental samples, biological and clinical samples, agricultural and food samples, material for electronics, wear metals in oils, high-purity chemical reagents, etc. The major limitation of this technique is the need to use solutions, which can be time-consuming and tedious when solids have to be analyzed. This explains the current trend to carry out the direct analysis of solids by means of

Spark emission spectrometry in widely used for the analysis of solid samples, and ICP emission spectrometry is the most commonly used method for the analysis of samples in a form of solutions.

spark or laser ablation, introduction of slurries or direct insertion of solids into the plasma.

References

[8.1-1] Schrenk, W.G., *Appl. Spectrosc.* 1986, 40, 19.

[8.1-2] Mavrodineanu, R., Boiteux, H., *Flame Spectroscopy*. New York: John Wiley, 1965.

[8.1-3] Gaydon, A.G., *The Spectroscopy of Flames*. 2nd edition. New York: Halsted Press, 1974.

[8.1-4] Dean, J.A., Rains, T.C., *Flame Emission and Atomic Absorption Spectrometry*. New York: Marcel Dekker, 1975.

[8.1-5] Thompson, K.C., Reynolds, R.J., *Atomic Absorption, Fluorescence and Flame Emission Spectroscopy. A Practical Approach*. 2nd Edition. London: Charles Griffin Company, 1978.

[8.1-6] Alkemade, C. Th. J., Hermann, R., *Fundamentals of Flame Spectroscopy*. New York: Halsted Press, 1979.

[8.1-7] Alkemade, C. Th. J., Hollander, Tj., Snelleman, W., Zeegers, P. J. Th., *Metal Vapors in Flames*. Elmsford: Pergamon Press, 1982.

[8.1-8] Cresser, M.S., *Flame Spectrometry in Environmental Chemical Analysis: A Practical Guide*. Cambridge: The Royal Society of Chemistry, 1994.

[8.1-9] Boumans, P.W.J.M. (Ed.), *Inductively Coupled Plasma Emission Spectrometry; Part I: Methodology, Instrumentation and Performance; Part II: Applications and Fundamentals*. New York: John Wiley, 1987.

[8.1-10] Montaser, A., Golightly, D.W. (Eds.), *Inductively Coupled Plasmas in Analytical Atomic Spectrometry*. New York: VCH Publishers, 1987.

[8.1-11] Moore, G.L., *Introduction to Inductively Coupled Plasma Atomic Emission Spectrometry*. Amsterdam: Elsevier, 1989.

[8.1-12] Thompson, M., Walsh, J.N., *Handbook of Inductively Coupled Plasma Spectrometry*. Glasgow: Blackie, 1989.

[8.1-13] Sneddon, J., (Ed.), *Sample Introduction in Atomic Spectroscopy*. Amsterdam: Elsevier, 1990.

[8.1-14] Moenke-Blankenburg, L., *Spectrochim. Acta Rev.* 1993, 15, 1.

[8.1-15] Karanassios, V., Horlick, G., *Spectrochim. Acta Rev.* 1990, 13, 89.

[8.1-16] Boumans, P.W.J.M., *Theory of Spectrochemical Information*. Bristol: Adam Hilger, 1966.

[8.1-17] Ingle, J.D. Jr. Crouch, S.R., *Spectrochemical Analysis*. Englewood Cliffs: Prentice Hall, 1988.

[8.1-18] Slickers, K., *Automatic Atomic-Emission Spectroscopy*. Giessen: Brühlsche Universtätsdruckerei, 1993.

[8.1-19] Broekaert, J.A.C., *Applied Spectrosc.* 1995, 49, 12A.

[8.1-20] Marcus, K., (Ed.), *Glow Discharge Spectroscopies*. New York: Plenum Press, 1993.

[8.1-21] Radziemski, L.J., Cremers, D.A., (Eds), *Laser-induced Plasmas and Applications*. New York: Marcel Dekker, 1989.

[8.1-22] Moenke-Blankenburg, L., *Laser Microanalysis*. New York: Wiley, 1989.

Questions and problems

1. What is an optical electron?
2. What are the four quantum numbers used to identify the energy state of an electron?
3. What would be the electronic configuration for As whose the number of electrons is 33?
4. What is the wavelength (in nm) corresponding to a transition between an upper energy state and a lower energy state of 40 000 and 15 000 cm^{-1}, respec-

tively and between an upper energy state and a lower energy state of 6 and 3 eV, respectively?

5. What is a resonance line?

6. Is a transition $J = 0$ to $J = 0$ allowed or forbidden?

7. Why does the intensity of an emission line of a given element make it possible to measure the concentration of this element?

8. Why it is important that the temperature of a radiation source remains constant when this source is used for quantitative measurements?

9. The two Cd 326 nm and Cd 228 nm lines have an excitation energy of 3.80 and 5.42 eV, and a gA value of $0.009 \, 10^{-8}$ and $12 \, 10^{-8}$, respectively. What is the most intense line for $T = 2000 \, \text{K}$, and $T = 5000 \, \text{K}$. For which temperature the two lines have the same intensity?

10. Why can flames not be used for the analysis of refractory elements?

11. What is the difference between a turbulent flame and a laminar flame?

12. Could you define a plasma?

13. What is the difference between an inductively coupled plasma and a direct current plasma?

14. Why is argon widely used for producing plasmas?

15. Why is the efficiency of a pneumatic nebulizer poor?

16. What are the major differences between an arc and a spark?

17. Why has the grating replaced the prism for dispersing the light in modern emission spectrometers?

18. What is the difference between a spectroscope, a spectrograph, and a spectrometer?

19. What is the difference between a monochromator and a polychromator?

20. Is a plane grating used as a dispersive component for polychromators?

21. What are the advantages and the limitations of the photomultiplier tube use a detector in emission spectrometry?

22. Why can a grating with a groove density of 3600 grooves mm^{-1} not be used for dispersing light near 700 nm?

23. What is the groove density of a grating for which the incident angle is 60°, the diffraction angle is 21° and the wavelength is 340 nm in the first order?

24. What is a blaze angle and what it is used for? What is the optimum wavelength corresponding to a blaze angle of 30° and 3600 grooves mm^{-1}?

25. How is an interferometric grating produced?

26. What is an echelle grating? Why is post-dispersion often used?

27. What is the influence of the groove density of the grating on the reciprocal linear dispersion?

28. What is the difference between resolution and resolving power?

29. What is the analytical method based on emission spectrometry providing the best limits of detection?

30. Is the ICP very sensitive to matrix effects?

8.2 Atomic Absorption Spectrometry

8.2.1 Introduction

In contrast to atomic emission spectrometry (AES), atomic absorption spectrometry (AAS) in terms of analytical method is a recent technique, having been described by Walsh in 1955 [8.2-1]. The first atomic absorption observations were of the Fraunhofer lines in the solar spectrum and were made in 1802 by Wollaston. The flame was first used as a source of atomization and was available in the early 1960s, whereas the graphite furnace was described in the 1960s by L'vov [8.2-2] and Massman [8.2-3] and made commercially available in 1970.

8.2.2 Principle

AAS is based upon the absorption of radiation by free atoms, usually in the ground state [8.2-4–8.2-14]. By selecting a wavelength for a given element that corresponds to an optical transition between atoms in the ground state and atoms in an excited level, the absorption of the radiation leads to a depopulation of the ground state. The value of the absorption is related to the concentration of the atoms in the ground state, and therefore, to the concentration of the element. By measuring the amount of radiation absorbed, a quantitative determination of the amount of analyte can be made.

An atomic absorption spectrometer will therefore consist of a primary radiation source which produces the radiation to be absorbed, a source of free atoms with an associated sample introduction system, an optical dispersive system, a detector, and electronics for data acquisition, processing, and editing (Fig. 8.2-1). The presence of free atoms must be obtained in the path between the primary radiation source and the detector.

Electronic configuration and possible optical transitions have been described in Section 8.1.2. Optical transitions used for AAS are generally between the ground state, or near the ground state, and the first excited levels (resonance levels). The population of the various levels is described by the Boltzmann law, which indicates that this population is proportional to the exponential of $(-E)$, where E is the excitation energy of the level of concern (Sec. 8.1). A consequence is that the population decreases very rapidly when E increases. In other words, when the temperature of the source of free atoms remains below 5000 K, most atoms are in the ground state. Since the absorption of the radiation is proportional to the number of atoms in the ground state, this explains why AAS is an efficient method. Moreover, absorption spectra are far more simple than emission spectra. In contrast to atomic emission spectrometry, there is little possibility of spectral interferences due to line coincidence.

In the case of Al, the most commonly-used lines are Al I 309.28 nm, Al I 308.22 nm, i.e., a transition between the 3p and 3d subshells; and Al I 396.15 nm and Al I 394.40 nm, i.e., a transition between the 3p and 4s subshells (Fig. 8.2-2).

Atomic absorption spectrometry is a highly selective analytical method based on the absorption of radiation emitted by a primary radiation source by the atoms in the ground state, the absorption being related to the concentration of the element.

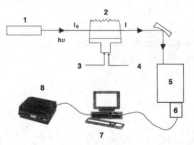

Fig. 8.2-1. Principle of atomic absorption spectrometry. 1: primary radiation source, 2: atomizer, 3: sample, 4: combustion gases, 5: optical dispersive system, 6: detector, 7: data acquisition and processing, 8: data editing

Even at high temperatures, most atoms remain in the ground state, due to the exponential population of the excited levels.

Atoms excited by the absorption of the radiation will re-emit emission lines not only at the same wavelength as the absorbing radiation but also at different wavelengths. This radiative de-excitation is called atomic fluorescence. When the same wavelength is observed, the process is called *resonance fluorescence*. The fluorescence process is actually the combination of absorption and emission processes. Fluorescence spectra are usually very simple and are superimposed onto the more complex emission spectra, the fluorescence lines being more intense than the equivalent emission lines.

When an intense primary radiation source is used (e.g., a laser), atomic fluorescence spectrometry (AFS) can be used as an analytical method. In this case, the

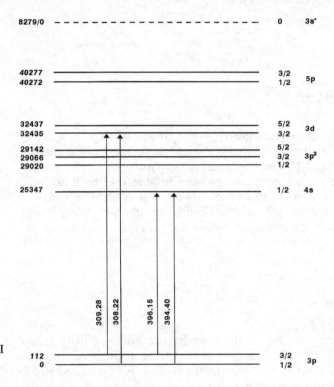

Fig. 8.2-2. Ground and excited states of Al I and optical transitions used in AAS

primary radiation source is usually out of line with the rest of the optical system, so that the detector will receive only the fluorescence signal. Actually, laser-induced atomic fluorescence spectrometry is probably one of the most sensitive analytical methods. However, laser-induced AFS is not commercially available because of the difficulty of using a laser in the UV region.

Quantitative analysis will be possible in AAS if the absorption of the radiation can be related to the concentration of the element to be analyzed. When the primary radiation source emits a line with a wavelength similar to that of the analyte line and a width narrower than that of the analyte line, the absorption of the incident light can be calculated. The transmission factor, T, is the ratio of the transmitted intensity, I, to the incident intensity, I_0. An absorption factor, α, is also defined as:

Atomic absorption spectrometry is used only for quantitative analysis.

$$\alpha = 1 - T \tag{8.2-1}$$

From Beer's law, it can be written:

$$\alpha = 1 - e^{k(\lambda).l} \tag{8.2-2}$$

where $k(\lambda)$ is the absorption coefficient and l the length of absorption. In the case of an absorption process from the ground state, 0, to a higher state, m, the coefficient is:

$$k(\lambda) = A \cdot \frac{g_m}{g_0} \cdot \frac{\lambda^4}{8\pi c} \cdot \frac{1}{\Delta\lambda_{eff}} \cdot n_0 \tag{8.2-3}$$

where g_m and g_0 are the statistical weights of the excited state and the ground state, respectively, and $\Delta\lambda_{eff}$ is an effective width of the analyte line, i.e., the width corrected by a coefficient. The population of the ground state, n_0, is practically equivalent to the total population.

The absorbance, A_{abs}, is defined as:

$$A_{abs} = -\log T = \log\left(\frac{I_0}{I}\right) = -\log(1 - \alpha) \tag{8.2-4}$$

The value of the absorbance is 0.0044, 0.301, 1, and 2 for 1%, 50%, 90%, and 99% absorption, respectively.

It can be easily deduced that the absorbance is:

$$A_{abs} = 0.434 \cdot A \cdot \frac{g_m}{g_0} \cdot \frac{\lambda^4}{8\pi c} \cdot \frac{1}{\Delta\lambda_{eff}} \cdot n_0 \cdot l \tag{8.2-5}$$

By using the logarithm of the transmission factor makes it possible to obtain a linear relationship with the analyte concentration. This relation is practically linear over a limited range of concentrations because of change in the length of absorption, variation in the line width, atom distribution in the source, etc. Linearization can be obtained by means of the software.

The oscillator strength was first used instead of the transition probability (see Sec. 8.1). The oscillator strength is dimensionless and varies in the 0–1 range. Its value can be related to the transition probability, A:

$$A = 6.67 \times 10^{13} \frac{g_0}{g_m} \frac{f}{\lambda^2} \tag{8.2-6}$$

where λ is expressed in nm.

The absorbance is proportional to the concentration of the element.

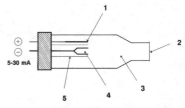

Fig. 8.2-3. Principle of a hollow cathode lamp. 1: anode, 2: silica window, 3: gas (Ar or Ne), 4: hollow cathode, 5: glass envelope

8.2.3 Primary radiation sources

The most commonly-used primary radiation sources are the hollow-cathode lamp (HCL) and the electrodeless discharge lamp (EDL). They both belong to low pressure discharges (see Sec. 8.1). The hollow-cathode lamp consists of a hollow cathode made of a highly pure metal whose spectrum is to be produced (Fig. 8.2-3) with an inner diameter in the 2–5 mm range. In some cases, the cathode can be made of several metals to obtain a multielement HCL. However, such a lamp is not widely used because of compromise conditions which result in a loss in sensitivity. The cathode and anode are set up in a glass cylinder. A high voltage and a current of up to 30 mA are used to produce a discharge which takes place entirely in the hollow cathode. The value of the current is a compromise between line intensity and line broadening due to self absorption. The fill gas is either Ar or Ne with a pressure in the 1–5 torr region. Ne is preferred for elements with high ionization potential. Narrow, intense lines are emitted following sputtering and excitation processes by the ions of the fill gas (Fig. 8.2-4). A transparent silica window is used for light transmission. HCLs are usually set up on a manual or automatic turret, and are warmed up before use. For simultaneous multi-element AAS systems, a beam combiner is used to combine the incoming beams from several HCLs. The success of the AAS method is clearly related to the availability of the HCLs.

In the case of volatile elements such as As or Se, the energy emitted by HCL can be relatively low. An alternative is the use of EDL which consists of a sealed silica tube containing the element or salt of interest [8.2-15] with Ar as a fill gas (Fig. 8.2-5). A discharge is sustained by means of a radiofrequency field through an antenna or a coil. The *rf* energy serves both for the element vaporization and excitation. The emission arising from the EDL is usually more intense than that emitted from a HCL. EDLs are generally used for As or Se determinations, but sometimes also for Cd, Hg, Pb, or Sb.

For a similar purpose another intense primary radiation source can be used, the boosted discharge HCL. This lamp uses the existing instrument lamp current supply but add a secondary boost discharge both to increase the excitation of atoms sputtered by the cathode and to minimize self-absorption. Emitted in-

The hollow cathode lamp emits intense, narrow lines of the element which constitutes the cathode.

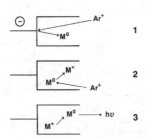

Fig. 8.2-4. Mechanisms in the hollow cathode lamp. 1: sputtering of the atoms, 2: excitation of the atoms by the ions of the filing gas, 3: radiative de-excitation of the excited atoms

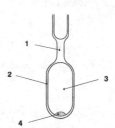

Fig. 8.2-5. Principle of the electrodeless discharge lamp. 1: welded tube, 2: silica cell, 3: He or Ar gas, 4: compound

Modulation of the beam emitted by the primary radiation source is performed in order to discriminate the light emission of the element present in the atomizer.

tensities are 5–15 times higher than for a standard HCL, resulting in a significant improvement in the signal-to-noise ratio.

In order to discriminate the light emitted by the primary radiation source from that emitted by the element present in the atomizer, it is necessary to perform source modulation. This can be achieved mechanically by means of a rotating chopper or electronically by modulating the HCL current using a pulsed power supply. Pulsing the current modulation is the most commonly-used way of modulation in commercially available systems. A synchronous detection is then used to eliminate the unmodulated dc signal emitted by the atomizer from the modulated ac signal of the lamp.

Some other primary radiation sources have been described such as the diode lasers or the continuum source. The continuum source consists of a high-pressure Xe arc lamp emitting a strong continuum intensity, i.e., without line emission. This results in a large flexibility in line selection for the primary radiation. Continuum sources have been mostly used for multielement AAS [8.2-16]. Diode lasers would be probably the ideal sources for AAS as they emit highly intense and narrow lines. However, they are currently limited to wavelengths higher than 620 nm, which hampers their wide use in AAS. A possibility is to double the frequency in order to work down to 310 nm [8.2-17].

8.2.4 Source of free atoms

The role of an *atomizer* is to efficiently convert the sample to free atoms mostly in the ground state. The free atoms must be located in the light path between the primary radiation source and the dispersive system so that a long absorption length is obtained. An ideal atomizer would result in a complete atomization of the sample. Two types of atomizer are commonly used, the *flame* and the *electrothermal atomizer* (furnace).

In atomic absorption spectrometry, the flame is formed in a burner with a long slot in order to increase the absorption pathlength.

Among the possible flames using premix burners (see Sec. 8.1) the air acetylene and the nitrous oxide (N_2O)-acetylene flames are the most common. The major difference in comparison with atomic emission spectrometry is the shape of the burner. The burner has a 5–10 cm slot in order to increase the absorption path length. A laminar flame is then obtained. The slot is designed to avoid blockage from high salt concentrations. The air-acetylene flame is commonly used for the determination of elements whose oxides are nonrefractory such as Ca, Cr, Fe, Co, Ni, Mg, Mo, Sr, and others; whereas the nitrous oxide-acetylene flame is used for elements such as Al, Si, Ta, Ti, V, Zr, etc. because of a higher temperature (>2600 °C). However, for the N_2O–C_2H_2 flame, a careful control of the various flows is necessary to avoid a flashback due to the high burning velocity. The flame is first ignited with an air-C_2H_2 mixture, then the N_2O flow is turned on and replaces the air flow. A similar procedure is carried out before extinguishing the flame. These procedures are usually fully automatized. It is also possible to adjust the fuel to oxidant ratio to obtain a fuel-rich (excess of fuel) flame for a better atomization efficiency. Usually, the viewing height can be adjusted to suit the element and the type of flame.

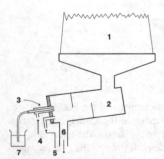

Fig. 8.2-6. Principle of flame-AAS. 1: flame, 2: burner, 3: nebulizer, 4: oxidant gas, 5: fuel gas, 6: drain, 7: liquid sample

As with atomic emission spectrometry (see Sec. 8.1) a nebulizer is used to produce an aerosol by means of the oxidant flow. A selection of fine droplets is obtained in a spray chamber equipped with paddles or impact bead (i.e., a bead where the largest droplets will impact on) where the aerosol is mixed with the fuel gas (Fig. 8.2-6) and finally reaches the burner head. A steady absorption process is therefore obtained. However, the residence time of the free atoms in the absorption pathlength is short. This is one of the reasons for the relatively poor sensitivity of flame-AAS along with the low efficiency of the nebulizer-burner system and the dilution of free atoms by the combustion gases. A typical sample consumption is of the order of 3–5 mL min^{-1}. It should be noted that the use of organic solvents increases not only the efficiency of nebulization but also the temperature of the flame.

An alternative to the injection of wet aerosols with flame-AAS is the introduction of volatile species. Hydride generation is the most commonly-used system

[8.2-18] for elements such as As, Bi, Ge, Pb, Sb, Se, and Sn. Hydrides are usually obtained by adding sodium borohydride to acidified solutions.

The alternative to the flame is the electrothermal atomizer (ETA). This atomizer makes use of an electrically heated refractory material on which the sample is deposited. A transient formation of free atoms is then obtained. One significant advantage of the ETA is the increase in the residence time of the free atoms compared with the flame. Although various forms have been described for the atomizers, commercially available ETAs make use of a cylindrical tube also called furnace whose length is in the 18–28 mm range (Fig. 8.2-7). Specifications include a complete atomization of the sample, fast formation of the free atoms, confinement of the atoms, complete removal of the sample after the atomization, fast heating time, ease of use, and low cost. The material must have a high electrical conductivity, must be refractory, and must exhibit high thermal resistance and a long durability. Refractory materials such as W or Ta seem to be suited for this purpose since their melting temperature is higher than 2600 °C. However, they exhibit several drawbacks such as their extreme brightness at high temperature. This is why the graphite furnace (GF) is widely used. A further advantage is that the presence of carbon allows the possible carbo-reductions of several analyte oxides to free atoms. Nowadays, the most commonly-used furnaces are made in electrographite coated by a layer of pyrolitic graphite which minimizes losses by diffusion of the atoms into the porous substrate, and improves the atomization process of many elements. The usual graphite furnaces (coated or not) may be heated up to 3000 °C without any mechanical problems.

A major improvement has been the introduction of the platform [8.2-19, 8.2-20] which consists of a thin graphite plate onto which the sample is deposited (Fig. 8.2-7). The platform is mainly heated by the radiation from the wall of the tube so that the increase in the temperature of the sample is delayed with that of the wall and the gaseous phase. Therefore, the atomization of the sample occurs after the wall and the gaseous phase have reached a temperature plateau. An enhancement of sample dissociation is obtained, in particular for volatile elements. Moreover, interference effects are minimized.

The electrothermal program consists of several progressive heating stages (generally three: drying, ashing, and atomization) (Fig. 8.2-8). The first stage (drying) ensures the desolvation of the sample in order to remove the solvent by evaporation. For aqueous solutions, a temperature slightly above 100 °C is used. The length of the evaporation stage is related to the amount of the sample ($0.5 \, \mu L \, s^{-1}$). The second stage is the ashing (pyrolysis, charring) of the solid residue remaining after the drying step. The ashing step ensures the removal or simplification of organic or inorganic matrix and consequently represents the most important step of the electrothermal program. The matrix is thermally decomposed at an intermediate temperature (300–1500 °C as a function of the analyte). The objective is to remove the maximum of the matrix while keeping the analyte entirely within the atomizer in a stable form so that atomization can proceed with a minimum interference from the matrix components. During the ashing step, an alternative gas (generally air or oxygen) or a chemical modifier can be used to achieve a better matrix decomposition. When the ashing step is complete, the residue should consist only of the analyte in its appropriate molecular form, and a minimum of matrix inorganic salts which are thermally stable at the ashing temperature used. The third stage is the atomization step in which the dissociation of the analyte molecular species occurs at high temperature (1200–2700 °C) and a transient formation of free analyte atoms is then obtained. A fast rise in temperature must be achieved ($2000 \, °C \, s^{-1}$).

The first two stages are usually performed with a flow of argon to avoid any oxidation. The argon flow is stopped during the atomization stage to extend the residence time of atoms in the observation volume and, consequently, to enhance sensitivity of the atomic absorption measurement performed during this step. Then a cooling period is observed. Cleaning of the furnace can be obtained by using a high current to eliminate any remaining material. In some cases, a cool-down step is used after the ashing step, generally in order to improve the atomization efficiency of refractory elements.

Preselected temperatures are measured by means of either an optical sensor or

Electrothermal atomization by means of graphite furnace allows a significant increase in the residence time of the free atoms compared with flames.

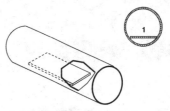

Fig. 8.2-7. Principle of the furnace equipped with a platform. 1: platform

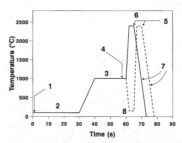

Fig. 8.2-8. Principle of the electrothermal program in GF-AAS. 1: Ar flow is on, 2: drying step, 3: ashing step, 4: Ar flow is off, 5: Ar flow is on, 6: atomization step, 7: cooling period, 8: cool-down procedure

Several progressive heating stages are carried out with the graphite furnace to obtain an efficient atomization of the sample.

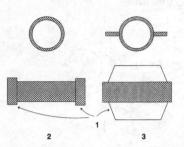

Fig. 8.2-9. Schematic end-on (top) and side-on (bottom) view of the longitudinal heating (2) and transverse heating (3) of a graphite furnace. 1: electrical connections.

the power consumption using a feedback electronic system. Both temperature and duration of each step can be optimized for a given element and matrix. Although "cookbooks" are available to describe these parameters for the most common matrices, optimization remains a crucial part of the AAS technique and requires some expertise. The residence time of the atoms is in the 0.1–1 s range, i.e., significantly longer than for a flame.

Two modes of furnace heating are currently used (Fig. 8.2-9): the end-heated (longitudinal-heated) mode and the side-heated (transverse-heated) mode [8.2-21]. The more recent side-heated mode has been developed to reduce the temperature gradient observed between the tube ends and tube center with the end-heated mode. This temperature gradient is responsible for undesirable effects during the atomization step, e.g., recondensation phenomena and corresponding memory effects, observed mainly during the determination of refractory elements. The side-heated tubes are shorter than end-sided ones: a decrease in the sensitivity is then observed in the case of side-heated tubes, due to the more pronounced loss in atomic vapour. To restore the sensitivity, end caps may be used. In any case, the total cycling time is of the order of 1–3 min. The sample consumption is of the order of 5–100 µL per cycle and the lifetime of the tube is in the 200–1000 cycles range, depending on the applied temperature and composition of the samples.

8.2.5 Optical dispersive systems

The design of the dispersive systems is similar to that described for atomic emission spectrometry (Sec. 8.1). Most commercially available AAS systems are based on a single-element determination over time. A monochromator is therefore used for line selection based on the use of a rotating plane grating. A simultaneous multielement system has been recently introduced making use of a 79 lines mm^{-1} echelle-grating-based polychromator equipped with a cross dispersion to obtain two-dimensional spectra.

The major difference with atomic emission spectrometry is the fact that the practical resolution is less crucial in AAS. Short focal lengths (0.25–0.5 m), 1200–1800 lines mm^{-1} grating, and slit widths in the 0.2–1 mm range are commonly used. The useful wavelength coverage is in the 180–860 nm range while the reciprocal linear dispersion is in the 0.1 to 2 nm mm^{-1} range.

A double-beam design is used to compensate for a possible drift of the primary radiation source.

Both single- and double-beam systems are used. In a double-beam system the beam arising from the primary radiation source is split so that one part of the beam is going through the atomizer while the second part is going directly to the entrance of the dispersive system (Fig. 8.2-10). The beams are recombined before entering the dispersive system. A double-beam design is aimed at compensating for any drift that originates at the primary radiation source level since stability is required between the measurement of the blank and the unknown solutions. However, it does not compensate for drift in the atomizer or for change in the profile of the line emitted by the primary radiation source. Since beam splitting is performed with a double-beam system, the signal to noise ratio is degraded compared with a single-beam system.

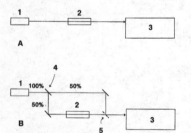

Fig. 8.2-10. Principle of a single-beam (A) optical system and a double-beam (B) optical system. 1: primary radiation source, 2: atomizer, 3: dispersive system, 4: semi-transparent mirror, 5: rotating disk

8.2.6 Detectors

Monochromators make use of the photomultiplier tube whereas the echelle-grating-based polychromator utilizes a new solid-state detector which incorporates a set of photodiodes. About 60 photodiodes are used for both primary and secondary lines of the most important elements.

8.2.7 Signal measurements

In flame AAS, the steady signal is integrated over a period of time of several second. The GF-AAS results in a transient, peak-shaped signal. Both peak height and peak area are used to measure the absorbance. Peak height measurements corresponding to the maximum absorbance value lead generally to a better sensitivity, but are mainly governed by the kinetics of the atomization process (changes in peak shapes with different matrices) and then are much more sensitive to chemical interferences. Peak area measurements are more tolerant to matrix effects as the measurements take into account the total number of absorbing atoms and are, therefore, more reliable.

8.2.8 Sensitivity

The sensitivity is the variation in the signal for a given variation in the concentration of the analyte. In GF-AAS, the sensitivity is described as the characteristic mass, i.e.; the concentration expressed in pg which leads to an absorbance, A, equal to 0.0044, which is equivalent to 1% absorption. The characteristic mass is used to verify whether a system is operating properly and to compare AAS systems. Characteristic mass depends not only on the type of element but also on the way the signal is measured (peak height or peak area). Characteristic mass values range between 0.3 pg (Cd, Mg, Zn) to 2000 pg (P).

The characteristics mass, i.e., the concentration for a 0.0044 absorbance, is used to evaluate the sensitivity of graphite furnace atomic absorption spectrometry.

8.2.9 Chemical interferences *or ionization.*

Both flame and furnace AAS can be sensitive to matrix interferences, which correspond to a change in the formation of free atoms. In the case of flame, the limited temperature does not ensure a full dissociation and atomization of thermally stable compounds in the gaseous phase. A well-known example is the interference of phosphate on Ca which results in the formation of stable Ca phosphates.

When using an air-acetylene flame, the absorbance of Ca decreases in the presence of phosphate. It is then necessary to add a buffer such as La which will form a thermally stable compound with phosphate so that Ca will be released to form free atoms. An alternative is the use of a hotter flame such as the N_2O-acetylene flame. The presence of easily ionized elements such as the alkali elements modifies the equilibrium between ions and neutral atoms. This ionization interference can be overcome by adding an excess of Cs to produce a large number of electrons so that the presence of an element such as Na with a lower amount than Cs will not significantly change the total number of electrons.

The ashing step is a crucial stage in GF-AAS with the possibility of loss of volatile species. The main use of chemical modifiers in GF-AAS [8.2-22] is to increase the difference in volatilization between the analyte and the matrix to obtain for instance a less volatile analyte (analyte modifier) or a more volatile matrix (matrix modifier). Chemical modifiers can also be used to reduce reactions between the analyte (or the matrix) and the graphite surface. Modifiers are usually in the form of liquid inorganic salts such as $Pd(NO_3)_2$, $Ni(NO_3)_2$, $Mg(NO_3)_2$, NH_4NO_3, $NH_4H_2PO_4$, etc., although organic additives such as ascorbic acid, thiourea, EDTA, Triton X-100, etc., have also been described. Inorganic salts can also be combined to obtain mixed modifiers such as $Mg(NO_3)_2 + Pd(NO_3)_2$ in order to obtain a fairly universal chemical modifier [8.2-23].

An example of a matrix modifier is NH_4NO_3 which will be used to remove chloride salts:

$$NaCl + NH_4NO_3 \rightarrow NH_4Cl + NaNO_3$$

In this case, NH_4Cl is removed during the ashing stage and the residual $NaNO_3$ does not interfere during the atomization.

Chemical modifiers are used to minimize matrix effects in graphite furnace atomic absorption spectrometry.

Many volatile species such as As, Se, Zn, and Cd can be stabilized by using analyte modifiers such as $Pd(NO_3)_2$ or $Ni(NO_3)_2$. Although the mechanisms involved in the stabilization of the analyte have not been clearly elucidated, intermediate species such as Pd_5As_2 or $NiSe$ have been reported [8.2-24]. Using efficient analyte modifiers such as Pd, highly volatile elements such as As or Se can be thermally stabilized to temperatures higher than 1300 °C without significant loss.

8.2.10 Spectral interferences

In AAS, the possibility of spectral interferences arising from a spectral line of another element within the spectral band pass of the dispersive system is rather small. Moreover, these spectral interferences are well-characterized. Examples are Cd 228.802 nm and As 228.812 nm, Al 308.215 nm and V 308.211 nm, and Sb 217.023 nm and Pb 216.999 nm. The selection of another sensitive line (in the case of deuterium lamp correction) or the use of Zeeman effect background correction make it possible to solve the problem.

Far more complex is the presence of nonspecific absorptions arising from the components of the matrix. They are either due to light scattering of solid or liquid particles still present in the atomizer or due to molecular bands. This can result in a significant increase in the background and, therefore, an increase in the absorbance of the signal. Diffusion from particles can arise in the flame when high salt concentration solutions are nebulized. It is more often observed in GF-AAS. The light diffusion follows the Rayleigh's law, which means that the diffusion coefficient increases when the wavelength decreases. A well-known example of diffusion is given by the NaCl solid or liquid particles.

Molecular bands are emitted by species, molecules, and radicals, either present in the matrix or formed during the atomization processes. Their structure depends on the spectral bandpass of the dispersive system. Most of the time, they look like nonstructured spectra. Classical examples of molecular bands are produced by halides and are observed in the 200–400 nm range.

Background subtraction can be necessary to compensate for an increase in the background due to nonspecific absorptions arising from the matrix. The deuterium lamp method and the Zeeman effect method are the most commonly used methods.

Because of the excess in absorbance, it is necessary to carry out background subtraction in order to obtain the accurate absorbance of the analyte. Besides the possibility of correcting with a true blank solution, which would be the ideal solution for eliminating the compounds responsible for the excess in absorbance, and for optimizing the various heating steps, three methods are currently commercially available for background subtraction: the deuterium lamp method, the Zeeman effect method, and the Smith-Hieftje method. The principle is to measure sequentially the absorbance of the background and that of the analyte and background, the absorbance of the analyte being obtained by subtraction. As previously mentioned, flame-AAS exhibits a steady signal, whereas GF-AAS produces a transient signal. In the former case, background subtraction can be performed before or after the analyte signal measurement (Fig. 8.2-11), while for the latter case, a bracketing method must be conducted, i.e., subtraction before and after followed by an interpolation [8.2-25].

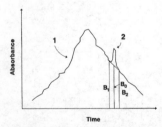

Fig. 8.2-11. Example of GF-AAS low absorbance analyte signal (2) in the presence of a large background (1). B_0: true value of the background absorption during the analyte absorption, B_1 and B_2: measured background absorption before and after the analyte absorption, respectively

The deuterium lamp consists of an arc in a deuterium atmosphere. A continuum emission is observed in the 200–380 nm range, with a maximum near 250 nm. The dispersive system receives alternatively (up to 150 Hz) the emission of the HCL and that of the deuterium lamp, the intensities of the two radiation sources being adjusted to be equal (Fig. 8.2-12). Therefore, measurement of the total absorbance of the HCL (or EDL), i.e., analyte and background, and the measurement of the absorbance of the deuterium lamp, are performed. Since the spectral bandpass of the dispersive system is usually large compared to the analyte line width, the absorbance of the analyte is negligible compared to that of the background when the absorbance of the deuterium lamp is measured. The decrease in the emission of the deuterium lamp corresponds, therefore, to the absorbance from the background, and can be subtracted to the absorbance observed with the HCL. This correction method is simple, but works efficiently mainly with nonstructured backgrounds and below 360 nm. The deuterium lamp method can be applied to both flame-AAS or GF-AAS up to absorbance values of 1 and does not degrade the sensitivity.

The second method is more complex and is based on the Zeeman effect [8.2-26], called after the Dutch physicist. This effect arises when a strong magnetic field is applied to the atomic absorbance signal. The spectral line can split into three or more components. This splitting involves the magnetic quantum number, M_j, which can take $g = 2J + 1$ values, where g is the Landé factor and J is the total electronic angular momentum quantum number (see Sec. 8.1.2). A state with an energy, E_0, can split into several states of energy, E, so that:

$$E = E_0 + \mu_B B M_j g \tag{8.2-7}$$

where μ_B is the Bohr magnetron and B the magnetic field expressed in Gauss unit.

The number of states depends on the value of M_j and their separation on the magnetic field. Transitions between the new degenerate lower and upper states are allowed if:

$$\Delta M_j = 0, \pm 1$$

The $\Delta M_j = 0$ components are called the π components, whereas the $\Delta M_j = \pm 1$ are called the σ components. The π components are linearly polarized parallel to the magnetic field, while the σ components are linearly polarized perpendicular to the magnetic field. The normal Zeeman effect (called normal because it can be explained by classical physics) arises when both states are split by an equal amount of energy. A triplet is then obtained (Fig. 8.2-13) with a central π component and two σ^- and σ^+ components so that the sum of the intensity of the three components is equal to the nonsplitted line intensity (Fig. 8.2-14). In this case, the component separation is directly proportional to the magnetic field. The anomalous Zeeman effect is more complex and leads to the formation of a multiplet whose separation is a more complicated function of the magnetic field (Fig. 8.2-15).

Background correction using the Zeeman effect is based on the fact that the nonspecific absorbance is unaffected by the presence of the magnetic field. The π component is absorbed by both the analyte and the background absorbing species, while the σ components are only absorbed by the species. Through the use of polarizers and/or modulation of the magnetic field, background subtraction can be conducted in a sequential manner.

Although the magnetic field could be applied to the primary radiation source (direct Zeeman effect), most commercially available AAS systems apply the magnetic field to the atomizer (inverse Zeeman effect), which is mostly the graphite furnace. The magnetic field is characterized by its mode, transverse or longitudinal, and by its frequency, dc (permanent magnet) or ac (up to 120 Hz). The current trend is to use an ac magnetic field up to 10 kG [8.2-27]. In the transverse mode, the magnetic field is applied perpendicularly to the beam. When the field is off, the total absorbance is measured (analyte and background) while when the field is on, only the background is measured. In the longitudinal mode, the magnetic field is applied parallel to the beam, which means that the π components is no longer observed. There is, therefore, no need for polarization.

Due to the background measurement performed exactly on the analytical wavelength (while the background is measured within the whole bandpass with deuterium lamp correction), the Zeeman effect correction is very efficient for structured background. Zeeman effect correction is also effective at any wavelength. There are, however, some limitations. A slight decrease in the sensitivity can be observed (down to 20%) and a "rollover" (bending back) of the calibration curve can be obtained.

The Smith-Hieftje method makes use of a pulsed HCL [8.2-28]. When a high current is used (up to 500 mA), self-absorption of the analyte line occurs so that a significant broadening and self-reverse line profile are observed. Only the absorption of the background is then measured. With the standard low current, both the analyte and the background absorbances are measured. This method is simple, does not require additional components, and can be used with any type of atomizer. The main limitation is a degradation in the sensitivity. In practical, modified HCLs must be used.

Although it is claimed that background correction can be performed up to high values of the absorbance, the error can be significantly high for absorbance values above 2.

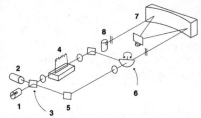

Fig. 8.2-12. Principle of a double-beam AAS system equipped with a deuterium lamp background correction device. 1: primary radiation source, 2: deuterium lamp, 3: semi-transparent mirror, 4: atomizer, 5: mirror, 6: semi-rotating mirror, 7: dispersive system, 8: detector

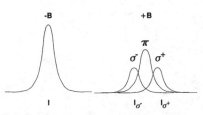

Fig. 8.2-13. Zeeman effect on the Cd I 228.8 nm line with $\Delta Mj = 0$ (π component), $\Delta Mj = +1$ (σ^+ component) and $\Delta Mj = -1$ (σ^- component)

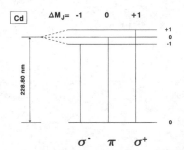

Fig. 8.2-14. Splitting of the intensity during the normal Zeeman effect. $I = \Sigma I_\sigma + \Sigma I_\pi$ and $\Sigma I_\sigma = \Sigma I_\pi$

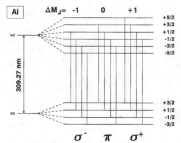

Fig. 8.2-15. Zeeman effect on the Al I 309.27 nm line with $\Delta Mj = 0$ (π component), $\Delta Mj = +1$ (σ^+ component) and $\Delta Mj = -1$ (σ^- component)

8.2.11 Recent developments in AAS

Graphite furnace atomic absorption spectrometry is one of the most sensitive method for elemental analysis.

Both flame-AAS and GF-AAS are mature techniques [8.2-29]. Flame-AAS is mostly used because of its ease of use, low cost, and reliability. The technique is well-documented, presents few interferences and ensures relatively good limits of detection. For the lowest analyte concentrations, GF-AAS must be used. AAS is certainly the most commonly-used method for elemental analysis. Evidence is given by the number of instrument companies present in this field and the number of units sold per year (about 4000 units).

There is, however, a trend to develop simultaneous multielement measurements through the use of multichannel detection. Such systems have been made commercially available, allowing the simultaneous determination of up to 6 elements. Since a graphite furnace is used, compromise conditions for the ashing and atomization steps must be selected in order not to deteriorate the sensitivity. Use of so-called universal modifiers is most helpful.

A further trend is to try to reduce the GF electrothermal program duration down to a minute. For the analysis of some matrices, so-called fast programs have been described without the ashing stage and without the addition of a modifier.

Direct GF-AAS analysis of solids has also been developed [8.2-30]. Due to several reasons such as the lack of availability of appropriate standard material, this method has been progressively abandoned. An interesting alternative is the use of solid suspensions (slurries). The sample is introduced into the GF as slurries. Electrothermal programs and modifiers are usually similar to those used for the analysis of solutions. A calibration procedure can be performed using simple aqueous standards with the same sample throughput as for the analysis of solutions. Because very small sample amounts are analyzed, the greatest problems originates from the initial sample homogeneity. Nevertheless, because there is no need for sample dissolution, analysis of solids has been applied to a large variety of samples such as metals and alloys, biological materials, polymers, coal, hair, environmental samples, and many more.

References

[8.2-1] Walsh, A., *Spectrochim. Acta* 1955, 7, 108.

[8.2-2] L'vov, B.V., *Spectrochim. Acta* 1961, 17, 761.

[8.2-3] Massman, H., *Spectrochim. Acta* 1968, 23B, 215.

[8.2-4] Dean, J.A., Rains, T.C., *Flame Emission and Atomic Absorption Spectrometry*. New York: Marcel Dekker, 1975.

[8.2-5] Robinson, J.W., *Atomic Absorption Spectrometry*. 2nd ed. New York: Marcel Dekker, 1975.

[8.2-6] Beaty, R.D., *Concepts, Instrumentation and Techniques in Atomic Absorption Spectrophotometry*, Perkin-Elmer, 1978.

[8.2-7] Thompson, K.C., Reynolds, R.J., *Atomic Absorption, Fluorescence and Flame Emission Spectroscopy. A Practical Approach*. 2nd ed. London: Charles Griffin Company, 1978.

[8.2-8] Price, W.J., *Spectrochemical Analysis by Atomic Absorption*. Chichester: John Wiley, 1979.

[8.2-9] Welz, B., *Atomic Absorption Spectrometry*. Weinheim: VCH, 1985.

[8.2-10] Ingle, J.D. Jr, Crouch, S.R., *Spectrochemical Analysis*. Englewood Cliffs: Prentice Hall, 1988.

[8.2-11] Hoenig, M., de Kersabiec, A.M., *L'atomisation Electrothermique en Spectrométrie d'Absorption Atomique*. Paris: Masson, 1989.

[8.2-12] Haswell, S., *Atomic Absorption Spectrometry: Theory, Design and Applications*. Amsterdam: Elsevier, 1991.

[8.2-13] Lajunen, L.H.J., *Spectrochemical Analysis by Atomic Absorption and Emission*. Cambridge: The Royal Society of Chemistry, 1992.

[8.2-14] Cresser, M.S., *Flame Spectrometry in Environmental Chemical Analysis: a Practical Guide*. Cambridge: The Royal Society of Chemistry, 1994.

[8.2-15] Sneddon, J., Browner, R.F., Keliher, P.N., Winefordner, J.D., Butcher, D.J., Michel, R.G., *Prog. Anal. Spectrosc.* 1989, 12, 369.

[8.2-16] O'Haver, T.C., Messman, J.D., *Prog. Analyt. Spectrosc.* 1986, 9, 483.

[8.2-17] Niemax, K., Groll, H., Schnürer-Patschan, C., *Spectrochim. Acta Rev.* 1993, 15, 349.

[8.2-18] Dedina, J., *Prog. Anal. Spectrosc.* 1988, 11, 251.

[8.2-19] L'vov, B., Pelieva, L.A., Sharnopolsky, A.I., *Zh. Prikl. Spektrosk.* 1977, 27, 395.

[8.2-20] L'Vov, B.V., *Spectrochim. Acta* 1978, 33B, 153.

[8.2-21] Frech, W., Johnsson, S. *Spectrochim. Acta* 1982, 37B, 1021.

[8.2-22] Ediger, R.D., *At. Absorpt. Newsl.* 1975, 14, 127.

[8.2-23] Shan Xiao-Quan, Wen Bei, *J. Anal. Atom. Spectrom.* 1995, 10, 791.

[8.2-24] Tsalev, D.L., Slaveykova, V.I., Mandjukov, P.B., *Spectrochim. Acta Rev.* 1990, 13, 225.

[8.2-25] Holcombe, J.A., Harnly, J.M., *Anal. Chem.* 1986, 58, 2607.

[8.2-26] Brown, S.D., *Anal. Chem.*, 1977, 49, 1269A.

[8.2-27] De Loos-Vollebregt, M.T.C., De Galan L., *Spectrochim. Acta* 1978, 33B, 495.

[8.2-28] Smith, S.B., Hieftje, G.M., *Appl. Spectrosc.* 1983, 37, 419.

[8.2-29] Wach, F., *Anal. Chem.* 1995, 67, 51A.

[8.2-30] Bendicho, C., De Loos-Vollebregt, M.T.C., *J. Anal. Atom. Spectrom.* 1991, 6, 353.

Questions and problems

1. What is the principle of atomic absorption spectrometry?
2. Why do most atoms remain in the ground state in a radiation source?
3. Why is the concept of absorbance used in atomic absorption spectrometry?
4. What is the principle of the hollow cathode lamp and what are its main advantages?
5. Why are burners with a long slot used in atomic absorption spectrometry?
6. Could you justify the use of graphite as a materials for furnace in atomic absorption spectrometry?
7. Could you describe the various steps during an electrothermal program?
8. Why is a high resolution of the dispersive system not necessary in atomic absorption spectrometry?
9. What is the difference between peak height and peak area measurements in graphite furnace atomic absorption spectrometry? What type of measurement should be used near the limit of detection?
10. What is the percentage of transmitted light for an absorbance of 0.5 and 0.01?
11. Could you define the characteristic mass? What is the characteristic mass in pg if a mass of 7.2 pg leads to an absorbance of 0.06?
12. What are an analyte modifier and a matrix modifier? Give examples.
13. Can an ashing temperature of 1600 °C be used for the determination of Cd?
14. Why may background subtraction be required in atomic absorption spectrometry?
15. What are the advantages and the limitations of the deuterium lamp method and the Zeeman effect method for background correction?
16. Can a deuterium lamp be used for background correction of the Na line at 589 nm?

X-ray fluorescence (XRF) is an instrumental analytical technique for the elemental analysis of solids and liquids with minimal sample treatment. The sample is irradiated with X-rays. The atoms in the sample are excited and will emit characteristic X-rays. The energy (or wavelength) of these characteristic X-rays is different for each element. This forms the basis for qualitative analysis. The number of characteristic X-rays of a certain element is proportional to its concentration, providing the basis for quantitative analysis. In principle all elements from boron to uranium can be determined. Trace element analysis (ppm range) and the determination of minor and major element concentrations (% range) can be performed on the same sample. Depending on how the characteristic X-rays are measured, one can distinguish between wavelength-dispersive and energy-dispersive X-ray fluorescence spectrometry (WD-XRF and ED-XRF).

> X-ray fluorescence allows the direct elemental analysis of solid matter.

In 1913, Coolidge demonstrated the possibility of WD-XRF using a high-vacuum X-ray tube. In 1948, Friedmann and Birks developed the first prototype XRF spectrometer. Energy-dispersive systems were introduced after the development of Si(Li) detectors around 1965 at Lawrence Berkeley Laboratories.

Today, XRF is widely used in laboratories dealing with multielement analysis. It is used for industrial process control (metallurgy, cement, glass, and ceramic industry) and in such diverse areas as mining, materials research, and environmental analysis. Some 14000 XRF instruments are in operation worldwide, the majority (~80%) are wavelength-dispersive systems.

> The steel, cement, and glass industries uses WD-XRF for production control.

Details on various aspects of XRF and its applications can be found in a number of monographs [8.3-1]–[8.3-5].

8.3.1 Principles

Wilhelm Conrad Roentgen discovered X-rays in 1895 while experimenting with discharge tubes at the University of Würzburg. Since the properties of the radiation did not match those of any known type, the name X (for unknown) radiation was given.

X-rays are part of the electromagnetic spectrum situated between ultraviolet radiation and gamma rays (Fig. 8.3-1). Diffraction of X-rays by matter is best described by considering X-rays as electromagnetic waves characterized by their wavelength λ. Properties such as absorption and scattering can be understood by considering X-rays as photons with a certain energy E. Equation 8.3-1 gives the relation between energy and wavelength:

> X-rays are photons (electromagnetic radiation) with energies situated between UV radiation and gamma-rays.

$$E = \frac{hc}{\lambda} \qquad (8.3-1)$$

h is Planck's constant (6.6254×10^{-34} J s) and c is the velocity of the propagation of the wave (3.00×10^8 m/s in vacuum), λ is in meters and E in joules. In X-ray spectrometry the wavelength is expressed in Ångstrom units ($1\,\text{Å} = 0.1\,\text{nm} = 10^{-10}$ m) and the energy is in kiloelectronvolts (keV). One eV is equal to the

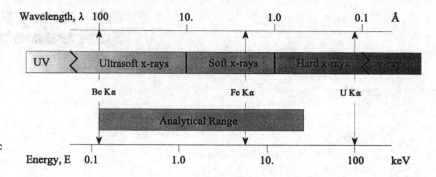

Fig. 8.3-1. X-ray part of the electromagnetic spectrum

amount of energy that an electron acquires when it is accelerated by a potential of one volt. Since $1\,J = 6.24 \times 10^{15}\,keV$, Eq. 8.3-1 becomes:

$$E[\text{keV}] = \frac{12.4}{\lambda[\text{Å}]} \tag{8.3-2}$$

An X-ray with an energy 6.40 keV (Fe $K\text{-}L_{3,2}$) has a wavelength of 1.94 Å. X-ray fluorescence analysis normally deals with X-rays between 100 and 0.5 Å (0.100 and 25 keV). X-rays with wavelengths longer than 1 Å are termed "soft" X-rays; shorter wavelength X-rays are called "hard" X-rays.

Interaction of X-rays with matter

The Bohr approximation of the atom can be used to describe the interaction of X-rays with matter. In this model, the Z electrons, defining the atomic number of the element, are grouped in shells designated K, L, M, N, O, and P, corresponding to the principal quantum number $n = 1, 2, 3, 4, 5$, and 6. A shell can have at maximum $2n^2$ electrons. In each shell the electrons are further distinguished by their azimuthal ($l = 0, \ldots, n - 1$), magnetic ($m = -l, \ldots, 0, \ldots, l$) and spin ($s = -1/2, +1/2$) quantum numbers. Two electrons in an atom cannot have the same set of quantum numbers (Pauli's exclusion principle). The electrons in an atom occupy discrete energy levels. K-shell electrons are more tightly bound than the L-shell electrons. The number of energy levels (or subshells) in each shell is equal to the number of allowed values of j (spin-orbit coupling), $j = |l \pm 1/2|$. The K-shell has one energy level, the L-shell has three, denoted L_1, L_2, and L_3, and the M-shell has five as shown in Table 8.3-1.

Because of their high energy, X-rays can eject electrons from inner atomic orbitals.

Table 8.3-2 lists the energy levels of the K- and L-electrons in a copper atom. The binding energy of inner electrons in the atom is of the same order of magnitude

Table 8.3-1. Energy levels of the K-, L- and M-shells (n principal quantum number, l azimuthal quantum number, j spin-orbit coupling)

Shell	n	$l = 0, \ldots, n-1$	$j = \|l \pm \frac{1}{2}\|$	Max. number of electrons $2j + 1$
K	1	0	1/2	2
L_1	2	0	1/2	2
L_2	2	1	1/2	2
L_3	2	1	3/2	4
M_1	3	0	1/2	2
M_2	3	1	1/2	2
M_3	3	1	3/2	4
M_4	3	2	3/2	4
M_5	3	2	5/2	6

Table 8.3.2. Energy levels in copper

Subshell	K	L₁	L₂	L₃
Binding energy (keV)	8.981	1.102	0.953	0.933

as the energy of the X-ray photons. This is the reason why X-rays can interact with the inner shell electrons.

The interaction of X-rays with atomic electrons results in either absorption or scattering of the photon. The photoelectric absorption is the dominant interaction. It causes the generation of the characteristic X-rays in the sample. Scattering is responsible for most of the continuum observed in XRF spectra because part of the exciting radiation is scattered by the sample and enters the detector system.

Photoelectric absorption

In the photoelectric absorption process (Fig. 8.3-2) a photon is completely absorbed by the atom and an (inner shell) electron is ejected. Part of the photon energy is used to overcome the binding energy of the electron and the rest is transferred to the electron in the form of kinetic energy. After the interaction, the atom (actually ion) is left in a highly excited state. A vacancy has been created in one of the inner shells. The atom will almost immediately return to a more stable electron configuration by emitting an Auger electron or a characteristic X-ray photon. The latter process is called X-ray fluorescence and will be discussed in detail later on.

Photoelectric absorption can only occur if the energy of the photon is equal or higher than the binding energy of the electron. An X-ray photon with an energy of 20 keV can eject a K-electron or a L-electron from Cu; however, an X-ray of 5 keV can only eject a L-electron from this atom (see Table 8.3-2).

Figure 8.3-3 shows how the photoelectric cross section of copper varies with the energy of the interacting photon. At high energies, e.g., 50 keV, the probability of ejecting a K-electron is rather low and the probability of ejecting an L-electron is even lower. As the energy of the X-ray photon decreases, the cross section increases, i.e., more vacancies will be produced. At 8.98 keV there is an abrupt decrease in the cross section because X-rays with lower energy can only interact with the L- and M-electrons. Then the cross section increases again until discontinuities corresponding to the binding energy of the L₃-, L₂- and L₁-electrons are observed. These discontinuities in the photoelectric absorption cross section are called *absorption edges*. The ratio of the cross section just above and just below the absorption edge is called the jump ratio, *r*. As fluorescence is the result of selective absorption of radiation, followed by spontaneous emission, an efficient absorption process is required. In other words, in XRF analysis an element can be determined with high sensitivity when the exciting radiation has its maximum intensity at an energy just above the K-edge of that element. Table 8.3-3 lists the K, L₃, and M₅ absorption edge energies and wavelengths for some elements.

The vacancy in the inner shell is filled with an electron from a higher shell and this can result in the emission of characteristic X-rays.

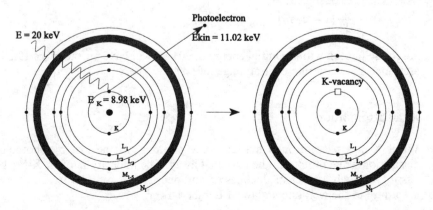

Fig. 8.3-2. Photoelectric absorption of a 20 keV x-ray photon by a K-electron of copper causes the ejection of a photoelectron and the creation of a K-vacancy

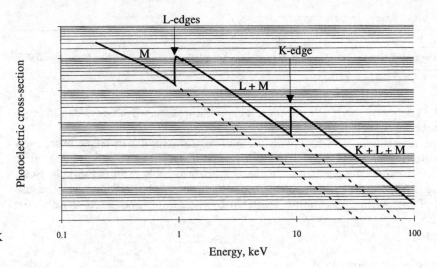

Fig. 8.3-3. Cross section of copper for photoelectric interaction as function of the energy of the interacting x-ray photon. The K absorption edge is at 8.98 keV

Table 8.3-3. Absorption edge energies and wavelengths for some elements

Element		Absorption edge energy (keV)			Absorption edge wavelength (Å)		
		K	L_3	M_5	K	L_3	M_5
C	6	0.283			43.767		
Si	14	1.837	0.098		6.745	127	
Ca	20	4.037	0.349		3.070	35.5	
Cr	24	5.987	0.574		2.070	21.6	
Fe	26	7.109	0.708		1.743	17.5	
Ni	28	8.329	0.853		1.488	14.5	
Sr	38	16.101	1.940		0.770	6.387	
Rh	45	23.217	3.001		0.534	4.130	
Ba	56	37.399	5.245	0.780	0.331	2.363	15.89
W	74	69.479	10.196	1.814	0.178	1.216	6.83
Pb	82	88.037	13.041	2.502	0.141	0.950	4.955
U	92	115.610	17.160	3.545	0.108	0.722	3.497

Elastic and inelastic scattering

Scattering is the interaction between radiation and matter which causes the photon to change direction. If the energy of the photon is the same before and after scattering, the process is called *elastic* or *Rayleigh scattering*. Elastic scattering occurs on bound electrons and forms the basis of X-ray diffraction. If the photon loses some of its energy the process is called *inelastic* or *Compton scattering*. This type of scattering occurs when X-ray photons interact with weakly bound electrons. After inelastic scattering over an angle θ, a photon, with initial energy E, will have a lower energy E' given by the Compton equation:

$$E' = \frac{E}{1 + \dfrac{E}{511}(1 - \cos\theta)} \tag{8.3-3}$$

An X-ray photon with an initial energy of 20 keV will have an energy of 19.25 keV after inelastic scattering over an angle of 90°.

X-ray attenuation

When X-rays pass through matter, some photons will be lost by photoelectric absorption while others will be scattered away. The intensity I_0 of an X-ray beam passing through a layer of thickness d and density ρ is reduced to an intensity I according to the well-known law of Lambert-Beer:

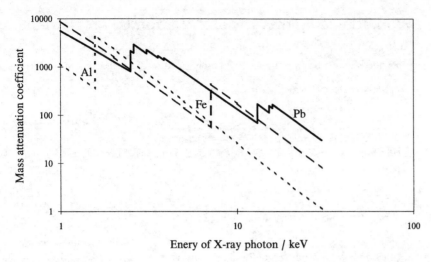

Fig. 8.3-4. Mass attenuation coefficient in cm²/g of Al, Fe, and Pb as function of the X-ray energy. — Pb, - - - Fe, and ... Al

$$I = I_0 e^{-\mu\rho d} \tag{8.3-4}$$

The number of photons (the intensity) is reduced but their energy is unchanged. The term μ is called the *mass attenuation coefficient* and has the dimensions of cm^2/g. The mass attenuation coefficient is obviously related to the cross section for photoelectric absorption, elastic and inelastic scattering, and depends on the type of atoms constituting the matter, and on the energy of the interacting photons.

Figure 8.3-4 shows the log-log plot of the mass attenuation coefficient of aluminium, iron, and lead for X-rays with energies between 1 and 50 keV. The absorption edge discontinuities due to photoelectric absorption are clearly visible. Low Z materials (Al) attenuate X-rays less than high Z materials (Pb). High energy (hard) X-rays are attenuated less than low energy (soft) X-rays.

Low-energy X-rays are more strongly absorbed by matter than high-energy X-rays. High-Z materials (e.g., lead) attenuate X-rays more strongly than low-Z (e.g., carbon).

The mass attenuation coefficient of a complex matrix, e.g., a compound, a mixture or an alloy, can be calculated from the mass attenuation coefficient of the n constituting elements

$$\mu_M = \sum_{i=1}^{n} \mu_i w_i \tag{8.3-5}$$

where μ_i is the mass attenuation coefficient, and w_i the weight fraction of element i in the sample considered.

The mass attenuation coefficient μ plays a very important role in quantitative analysis. Both the exciting primary radiation and the fluorescence radiation are attenuated in the sample. To relate the observed fluorescence intensity to the concentration, this attenuation must be taken into account. Since the mass attenuation coefficient is very high for low energy X-rays (long wavelength), accurate light element analysis is more difficult with XRF. Table 8.3-4 contains mass attenuation coefficients for some elements at various X-ray energies corresponding to characteristic X-ray lines. The notation K-L$_{3,2}$, used in Table 8.3-4, refers to the Kα line (see next paragraph).

The attenuation of X-rays in the sample complicates the quantitative analysis.

Table 8.3-4. Mass attenuation coefficients in cm^2/g of some elements (absorber) for X-ray energies corresponding to the K-L$_{3,2}$ (Kα) emission of O, Si, Fe, and Ni

Line	Radiation keV	Å	Mass attenuation coefficient μ in cm^2/g of absorber O	Si	Fe	Pb
O K-L$_{3,2}$	0.525	23.6	1270	7930	4000	12 300
Si K-L$_{3,2}$	1.74	7.13	1050	360	2490	1970
Fe K-L$_{3,2}$	6.40	1.94	23.5	124	75.5	426
Ni K-L$_{3,2}$	7.47	1.66	14.6	78.7	389	286

Worked example: X-ray attenuation by an element

Consider an iron foil with a thickness of 1 μm, the density of iron, ρ_{Fe}, is 7.9 g/cm³. How big is the intensity of a beam of Fe K-L$_{3,2}$ photons (6.40 keV) after passing through this foil?

$$\mu_{Fe}(6.40\,keV) = 75.5\,cm^2/g$$

$$I/I_0 = \exp(-75.5 \times 7.9 \times 0.0001) = 0.942$$

94.2% of this photons pass the foil (5.8% get absorbed).

How big is the intensity of a beam of Ni K-L$_{3,2}$ photons, having an energy of 7.47 keV, after passing the same foil?

$$\mu_{Fe}(7.47\,keV) = 389\,cm^2/g$$

$$I/I_0 = \exp(-389 \times 7.9 \times 0.0001) = 0.735$$

In this case, only 73.5% of the photons pass the foil or 26.5% get absorbed! The mass attenuation coefficient of iron at 7.47 keV is considerably higher that at 6.40 keV because the absorption edge of iron is at 7.109 keV.

How would be the situation for oxygen K X-rays (0.525 keV)?

$$I/I_0 = \exp(-4000 \times 7.9 \times 0.0001) = 0.042$$

Only ~4% of the O K X-rays will pass through 1 μm of iron. When determining oxygen in steel, the signal will come only from a micron thick layer!

Worked example: X-ray attenuation by a compound

The mass attenuation coefficient of silicon dioxide ($\rho = 2.65$ g/cm³) for Fe K-L$_{3,2}$ X-rays is:

$$\mu_{SiO_2} = W_{Si}\mu_{Si} + W_O\mu_O$$

$$\mu_{SiO_2} = 0.47 \times 124 + 0.53 \times 23.5 = 70.7\,cm^2/g$$

A SiO$_2$ layer of 37 μm will reduce the Fe X-ray intensity to 50%:

$$I/I_0 = \exp(-\mu\rho d) = 0.5$$

$$-\mu\rho d = \ln(0.5) \text{ or } d = \frac{-\ln 0.5}{\mu\rho} = 0.0037\,cm$$

Characteristic X-ray emission

The energy (or wavelength) of the characteristic X-rays is used to identify the elements in the sample; the number of these characteristic X-rays is a measure of the concentration.

Characteristic X-ray spectra are relatively simple and originate from transitions between K, L, M and N shells.

After photoelectric absorption the atom is in a highly excited state. The vacancy created by the photoelectric absorption will be filled by an electron from a higher shell. The energy difference between those two states, e.g., vacancy in the K-shell and vacancy in the L$_3$-shell, can be emitted as an X-ray photon. These X-rays are called "characteristic" because their energy (or wavelength) is different for each element, as every element has its own energy level.

The emission of characteristic X-rays is governed by quantum mechanical selection rules. The transition of an electron from one shell to another must obey $\Delta n > 0$, $\Delta l = \pm 1$, and $\Delta j = 0$ or ± 1. Lines that result from those transitions are called *allowed* or *diagram lines*. Lines resulting from forbidden transitions have very low intensity. The so-called "satellite" lines are due to transitions in atoms with two or more vacancies.

K-lines

With K-lines we denote all characteristic X-ray lines that originate from a vacancy in the K-shell. Figure 8.3-5 shows the transition diagram of iron. The K-vacancy can be filled by electrons from the L$_2$, L$_3$, or M$_{3,2}$ level. The L$_3 \to$ K transition results in an X-ray with an energy of 6.404 keV (1.9360 Å). This X-ray line is denoted Fe K-L$_3$ or Fe Kα_1. The other Kα line, Kα_2, is due to the L$_2 \to$ K transition. Both lines are diagram lines because the selection rules are obeyed. Since

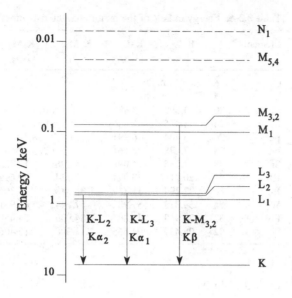

Fig. 8.3-5. Transition diagram of iron showing the origin of Kα and Kβ lines

Table 8.3-5. Principal X-ray diagram lines, IUPAC, and Siegbahn notation, and their intensity relative to the major line in each subshell

	Notation		Relative intensity
	IUPAC	Siegbahn	
K-lines	$K-L_3$	$K\alpha_1$	100
	$K-L_2$	$K\alpha_2$	~50
	$K-M_3$	$K\beta_1$	~17
	$K-M_2$	$K\beta_3$	~8
L_3-lines	L_3-M_5	$L\alpha_1$	100
	L_3-M_4	$L\alpha_2$	~10
	$L_3-N_{5,4}$	$L\beta_{2,15}$	~25
	L_3-M_1	Ll	~5
	L_3-N_1	$L\beta_6$	~1
L_2-lines	L_2-M_4	$L\beta_1$	100
	L_2-N_4	$L\gamma_1$	~20
	L_2-M_1	$L\eta$	3
	L_2-O_1	$L\gamma_6$	3
L_1-lines	L_1-M_3	$L\beta_3$	100
	L_1-M_2	$L\beta_4$	~70
	L_1-N_3	$L\gamma_3$	~30
	L_1-N_2	$L\gamma_2$	~30
M-lines	M_5-N_7	$M\alpha_1$	
	M_5-N_6	$M\alpha_2$	
	M_4-N_6	$M\beta$	

the energy difference between the two lines is rather small and not always resolved by the spectrometer, one uses the notation $K-L_{3,2}$ (or Kα) to refer to this doublet. Lines involving transitions from the M- or N-shell are called Kβ lines. Kβ lines have a higher energy (shorter wavelength) than Kα lines. The original notation of X-ray lines (Kα, Kβ ...) was introduced by Siegbahn in 1920. The more systematic IUPAC notation, involving the initial and the final state of the atom, ($K-L_3$, $K-M_{3,2}$) is now preferred (Table 8.3-5).

Not all transitions are equally probable, resulting in different intensities for the various K-lines (Table 8.3-5). The $K-L_3$ ($K\alpha_1$) line is the most intense one. The ratio of the $K-L_2$ to the $K-L_3$ line is approximately 0.5 because the L_3 level has four electrons compared to two for the L_2 level. The further away the electrons, the less likely the transition, therefore the Kβ lines represent only ~13% of the total intensity of the K-lines. The $K-L_{3,2}$ (Kα) line is most often used in XRF analysis because it is the K-line with the highest intensity. The $K-M_{3,2}$ (Kβ) line is only used when there is a strong interference (overlap) between the $K-L_{3,2}$ line and a line from another element.

Table 8.3-6. Energy in keV of the major characteristic lines of some elements

Element		K-L$_3$ (Kα_1)	K-L$_2$ (Kα_2)	K-M$_3$ (Kβ_1)	K-M$_2$ (Kβ_3)	L$_3$-M$_5$ (Lα_1)	L$_2$-M$_4$ (Lβ_1)	M$_5$-N$_{5,7}$ (Mα)
C	6	0.525						
Si	14	1.740						
Ca	20	3.692	3.688		4.013	0.341		
Cr	24	5.415	5.406		5.947	0.573	0.583	
Fe	26	6.404	6.391		7.058	0.705	0.719	
Ni	28	7.478	7.461		8.265	0.852	0.869	
Sr	38	14.165	14.098	15.836	15.825	1.807	1.872	
Rh	45	20.216	20.074	22.724	22.699	2.697	2.834	
Ba	56	32.194	31.817	36.378	36.304	4.451	4.828	
W	74	59.318	57.982	67.244	66.951	8.398	9.672	1.775
Pb	82	74.969	72.804	84.936	84.450	10.522	12.614	2.345
U	92	98.439	94.665	111.300	110.406	13.615	17.220	3.171

L- and M-lines

L-lines result from transitions of outer electrons filling the vacancies in the L-shell. Since the vacancy can be in any of the three subcells, L$_1$, L$_2$ or L$_3$, the L-line spectrum is more complex than the K-line spectrum. The most important lines are given in Table 8.3-5.

L-lines are used to determine elements with an atomic number greater than 45 (Rh). The K-lines of these elements are difficult to excite because of the high K absorption edge energy. Moreover, characteristic X-rays with energy above 25 keV are difficult to measure with standard X-ray spectrometers. The L$_3$-M$_{5,4}$ (Lα) line is the most suitable analytical line. In case of interference, the L$_2$-M$_4$ (Lβ) line can be used.

M-lines, due to a vacancy in one of the five M subcells, are seldom used in XRF. The M-lines of the heavy elements (Pb) can interfere with K- and L-lines from lower Z elements. Characteristic X-ray energies for some elements are given in Table 8.3-6, the corresponding characteristic wavelengths can be obtained via Eq. 8.3-2.

Fluorescence yield and Auger electron emission

The energy released when an L-electron drops into a K-vacancy can be transferred to another L- (or M-) electron rather than be emitted as a photon. The electron receives sufficient energy to leave the atom and is called an Auger electron. It is clear that characteristic X-ray and Auger electron emission are two competing processes. An atom with a vacancy in one of its shells will either emit a characteristic X-ray photon or an Auger electron. The probability that the vacancy will result in X-ray emission is called the fluorescence yield, e.g., for the K-shell:

$$\omega_K = \frac{\text{Number of K X-rays emitted}}{\text{Number of K vacancies created}} \qquad (8.3\text{-}6)$$

ω_K is very small ($\sim$0.01) for elements below sodium (Z = 11). For high Z elements ω_K tends toward one (see Fig. 8.3-6). Because of this, XRF is less sensitive for light elements. Although many vacancies are created, only few characteristic X-rays are emitted.

Complete tables with X-ray data (absorption edges, characteristic lines, mass attenuation coefficients) can be found in handbooks [8.3-4–8.3-6].

8.3.2 Instrumentation

Most XRF instruments use an X-ray tube to excite the sample.

XRF instruments are built around an X-ray source, a sample holder, and a spectrometer. The primary X-rays from the source are used to excite the atoms in the

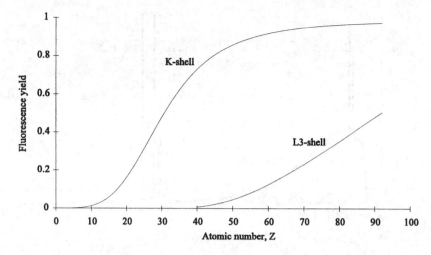

Fig. 8.3-6. Fluorescence yield of the K and the L_3 shell as function of the atomic number of the elements

sample. The spectrometer measures the wavelength (or energy) and the intensity of the (fluorescence) radiation emitted by the sample. Since the principles of the wavelength-dispersive and energy-dispersive spectrometers are quite different they will be discussed separately. X-ray tubes are the most commonly used source of primary X-rays in XRF. Radioisotope sources can be utilized in energy-dispersive instruments.

X-ray tubes

A schematic view of a typical Coolidge X-ray tube is given in Fig. 8.3-7. The tungsten filament (the cathode) and anode are mounted in an highly evacuated glass tube. The anode is made from a very pure metal such as chromium, molybdenum, rhodium, silver, or tungsten. The filament is heated by the current from a low voltage power supply. This causes thermionic emission of electrons from the W-wire. When the negative high-voltage (e.g., $-30\,\mathrm{kV}$) is applied to the filament, the electrons will be accelerated to the anode, which is at ground potential, and bombarded with high energy (30 keV for an high-voltage of 30 kV). The generated X-rays escape from the tube via a beryllium window.

The interaction of the electrons with the atoms of the anode causes the production of continuous and characteristic X-rays. The continuous X-rays are the result of decelerating collisions between the energetic electrons and the target atoms. At each collision the electron is decelerated, and the kinetic energy lost is emitted as an X-ray photon. In just one collision the electron can lose anything

X-ray tubes emit continuum (Bremsstrahlung) X-radiation and X-ray lines characteristic of the anode.

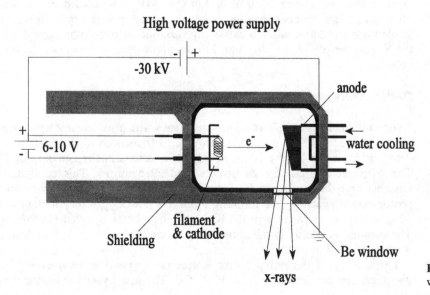

Fig. 8.3-7. Schematic diagram of a side window x-ray tube

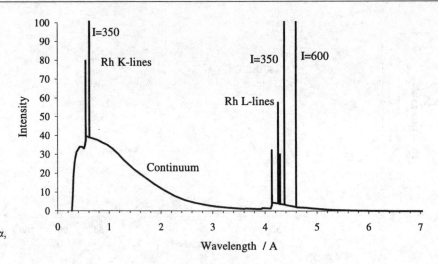

Fig. 8.3-8. X-ray spectrum of a Rh-anode tube operated at 45 kV, the intensity of Rh Kα, Lα, and Lβ₁ lines are indicated

between zero and all of its energy, resulting in a continuous spectrum up to an energy corresponding to the accelerating voltage; e.g., if the tube is operated at an accelerating voltage V of 45 kV then $E_{max} = 45$ keV or $\lambda_{min} = 12.4/V = 0.28$ Å. This relation between the short wavelength limit and the applied tube potential is called the Duane-Hunt law. The continuous or white radiation is also called "Bremsstrahlung" (German, literally break-radiation). The impinging electron can also, via a process similar to the photoelectric effect, create vacancies in the atoms. To do that, the electrons must have an energy higher than the binding energy of the orbital electrons. The de-excitation of these vacancies results in the emission of characteristic X-rays from the anode, as explained in the previous paragraph on characteristic X-ray emission.

The shape of the continuum depends mainly on the applied high-voltage. Figure 8.3-8 shows the spectrum from a Rh-anode X-ray tube operated at 45 kV. The Bremsstrahlung continuum reaches a maximum at $\sim 1.5\lambda_{min}$ (or at $2/3 E_{max}$). The total intensity of the continuum increases with the atomic number of the target and varies linearly with the tube current. The thickness of the beryllium window affects the low energy part of the spectrum. The correct choice of anode material and operating voltage allows the optimum excitation of a certain range of elements by the continuum as well as by the characteristic lines of the tube. To excite high Z elements in the sample a high accelerating voltage must be used.

The efficiency of an X-ray tube is rather low; only 1% of the electric power is concerted to X-rays, the rest is dissipated as heat. Wavelength-dispersive XRF uses tubes with an input power of 3 kW (e.g., 100 mA at 30 kV). This high power requires cooling of the anode with water to avoid melting. Energy-dispersive XRF systems have a better geometrical efficiency and can only be operated at low count rates so that low power (~ 30 W or 1 mA at 30 kV) air cooled X-ray tubes are often used. The filament heating and high voltage power supply must be very stable for quantitative work because any variation in voltage or current will alter the X-ray intensity of the tube and thus the fluorescence intensity of the sample.

Radioactive sources

Radionuclide X-ray sources are used in portable XRF instruments.

Some radionuclides emit X-rays or gamma rays with a sufficiently low energy that can be used in XRF. Gamma rays are due to transitions in the nucleus, but the decay of some radioactive isotopes, e.g. ^{55}Fe, results in the emission of X-rays. The ^{55}Fe nucleus has $Z = 26$ protons and 29 neutrons. This configuration is unstable and the nucleus will capture an (K) orbital electron, transforming a proton into a neutron. The resulting atom has now 25 protons (manganese) and 30 neutrons and a vacancy in the K-shell. This process is called *electron capture*. The vacancy will decay in the normal way by emitting a Mn K-L₃,₂ or Mn K-M₃,₂ X-ray.

The activity of the radioisotopic source is expressed in becquerels (1 Bq = 1 disintegration per second = 2.7×10^{-11} Ci). The activity of the source decreases

Table 8.3-7. Some sources commonly used in radioisotope excited XRF

Radioisotope	Half-life, $t_{1/2}$ (years)	Radiation	Energy, keV
^{55}Fe	2.7	Mn-K X-rays	5.89, 6.49
^{109}Cd	1.27	Ag-K X-rays	22.10, 24.99
		γ-ray	88
^{241}Am	433	Np-L X-rays	13.95, 17.74
		γ-ray	60

with time. After a time equal to the half-life, $t_{1/2}$, the intensity of the source is reduced to 50% of its initial value. Table 8.3-7 lists some radioactive sources commonly used in XRF. The Ag-K X-rays from ^{109}Cd are suited to determine the elements from calcium to zirconium via their K-lines. After about 3 years the source needs to be replaced.

Safety aspects prevent the use of very active sources. Sources used in XRF instruments typically have an activity between 100 and 300 MBq ($\sim$3 to 10 mCi). Because of the low geometrical efficiency of the wavelength dispersive spectrometer these sources can only be used in combination with an energy dispersive spectrometer. The use of these radioactive sources allows the construction of portable ED-XRF systems that can be operated in the field.

Wavelength-dispersive instruments

Crystal spectrometer

Wavelength-dispersive (WD) X-ray spectrometers [8.3-7] are based on the principle of Bragg diffraction. Figure 8.3-9 shows a parallel beam of X-rays with wavelength λ incident on a crystal at an angle θ. The X-ray wave is elastically scattered by the atoms in the first crystal plane (a). The resulting diffracted wave has the same wavelength and is 'reflected' under an angle θ. The same happens on the next and subsequent crystal planes. However, the X-rays reflected from the second plane (b) will travel a distance xyz = $2d \sin \theta$ further than those interacting with the first plane. X-ray waves which were in phase before the reflection will thus in general not be in phase after reflection and the intensity of the diffracted beam will be nearly zero. The intensity of the diffracted beam will be different from zero only if the difference in distance travelled equals an integer number of wavelengths (constructive interferences, as is the case in Fig. 8.3-9). This leads to the condition for Bragg diffraction:

$$2d \sin \theta = n\lambda \quad (n = 1, 2, \ldots)$$ (8.3-7)

$n = 1$ corresponds to the most intense first order diffraction.

The layout of an WD-spectrometer is given in Fig. 8.3-10. The slit collimator ensures that a parallel beam of fluorescence X-rays, emerging from the sample, falls onto the crystal at an angle θ. The detector is placed at an angle 2θ with respect to the incident beam so that the X-rays diffracted from the crystal at an

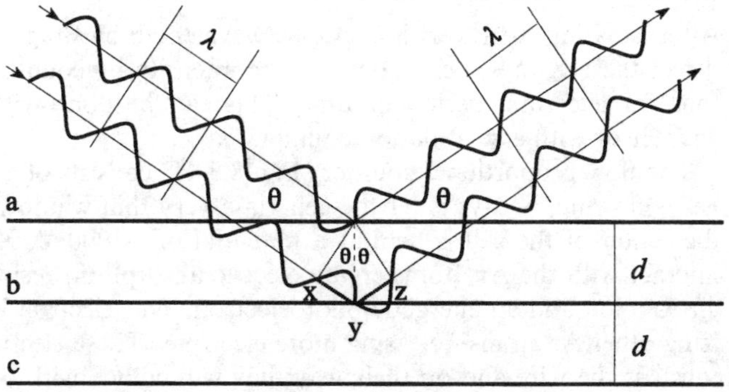

Wavelength-dispersive X-ray spectrometers are based on the principle of Bragg diffraction.

Fig. 8.3-9. Principle of Bragg diffraction used in a wavelength-dispersive spectrometer

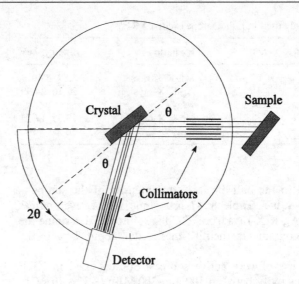

Fig. 8.3-10. Layout of a wavelength-dispersive spectrometer

Table 8.3-8. Some crystals used in wavelength-dispersive spectrometers

Crystal and plane	$2d$ (nm)	Theoretical wavelength interval (Å)	Typical element range
LiF (200) Lithium fluoride	0.402	3.88–0.52	K–Cd (K-lines) Sn–U (L-lines)
PET (002) Pentaerythritol	0.874	8.44–1.14	Al–Cl
PX-1 Multilayer	5.1	49.3–6.66	O–Mg

angle θ are measured. A collimator is also placed in front of the detector. The crystal and the detector are placed on a goniometer so that the rotation of the crystal over an angle θ results in the displacement of the detector along the circle over an angle 2θ. The minimum and maximum wavelength that can be measured depends on the 2θ range and the spacing d of the crystal. Typically, 2θ can vary between 15° and 150°. Using a LiF(200) crystal with $d = 0.201$ nm (Table 8.3-8), the minimum and maximum wavelength are then respectively 0.52 and 3.88 Å (first order reflection). This corresponds to the Kα fluorescence radiation from potassium (3.744 Å) to cadmium (0.536 Å). Therefore, one uses different crystals for different wavelength ranges. Table 8.3-8 lists some commonly used crystals.

Pentaerythritol is an organic crystal. For very large spacings (not very much dispersion required) one uses synthetic multilayers rather than crystals (e.g., PX-1 is the commercial name of one of these multilayers). The useful element range is often smaller than the theoretical wavelength range indicates, because the reflectivity of the crystal might become very low.

Flow proportional and scintillation counter

After isolating X-rays with a specific wavelength by Bragg diffraction, we need to detect these X-rays, i.e., determine the intensity by counting the number of photons for a certain period of time. This can be done with a flow proportional counter or with a scintillation counter.

The flow proportional counter (Fig. 8.3-11) consists of a cell through which Ar gas is flowing. X-rays enter the cell via a very thin window. The tungsten wire at the center of the cell is held at a potential of +1000 V. X-rays entering the cell interact with the Ar atoms (photoelectric absorption) resulting in the creation of an Ar$^+$ ion and an energetic photoelectron. This electron loses its energy by ionizing other Ar atoms, releasing more electrons. These electrons will be accelerated towards the wire and on their way they will collide with other Ar atoms causing

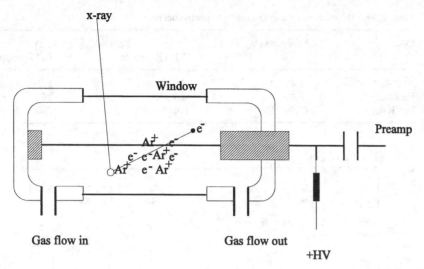

Fig. 8.3-11. Schematic diagram of a flow proportional counter used for the detection of low energetic X-rays

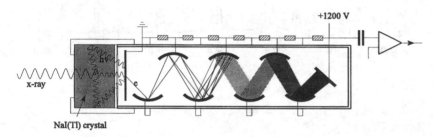

Fig. 8.3-12. Schematic diagram of a scintillation counter (NaI(Tl)-crystal + electron multiplier) used for the detection of higher energetic X-rays

further ionization and release of electrons. The total number of electrons produced in this way becomes very large but remains proportional to the initial number of electrons and thus proportional to the energy of the X-ray (proportional counter). Finally all electrons will arrive at the wire causing momentary charging of a capacitor. The preamplifier connected to this capacitor converts the charge pulse into a voltage pulse of a few hundred mV. Thus for each X-ray entering the proportional counter one voltage pulse is created with a height roughly proportional to the energy of the X-ray.

For short wavelength X-rays the efficiency of the proportional counter becomes very low. High energy photons pass through the gas without being absorbed. Therefore, below 2 Å one uses a "scintillation counter" (Fig. 8.3-12). A thallium activated sodium iodide, NaI(Tl), single crystal is used as a scintillator. The absorption of an X-ray by the crystal results in the emission of light photons with a wavelength 410 nm. These photons fall onto the photocathode of a photo-multiplier where again electrons are produced. These electrons are accelerated to the first dynode of the electron multiplier. At impact two or more secondary electrons are produced and these are accelerated to the second dynode where more electrons are produced. At the last dynode the charge is sufficiently high so that the preamplifier can produce a voltage pulse. The scintillation counter thus also produces one pulse for each X-ray that enters the detector and the height of the pulse is again proportional to the energy of the X-ray.

The preamplifier output pulse is further processed by a linear amplifier and a discriminator. Finally the pulses are counted for a preset amount of time. The pulses can also be sent to a ratemeter, which displays on an analogue scale the X-ray count rate in counts per second (cps) and further to a strip chart recorder producing a trace of the X-ray intensity as a function of the 2θ value. Today most WD spectrometers are controlled by a computer and the counts are stored in memory for display and further processing.

The processing of an X-ray event by detector and electronics takes a finite amount of time. After the arrival of one X-ray the detection system is said to be "dead" for some time because any X-ray arriving within this "dead time" will not be counted. The dead time is in the order of 200 to 300 ns, so that count rates of up to 10^6 counts per second can be handled.

Table 8.3-9. Resolution of a WD and ED spectrometer

Line	Energy (keV)	Peak width ΔE in eV	
		WD spectrometer	ED spectrometer
Al K-L$_{3,2}$	1.49	15	115
Fe K-L$_{3,2}$	6.40	60	160
Mo K-L$_{3,2}$	17.44	170	230

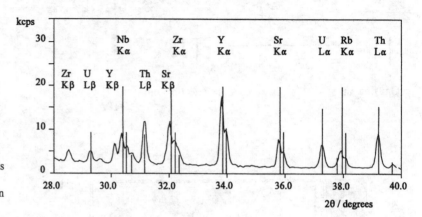

Fig. 8.3-13. Wavelength dispersive spectrum of a geological sample taken by scanning a LiF(220) crystal from 28 to 40°. The elements are in the concentration range of 100 to 1000 ppm. (Adopted from Philips Application Study 92002)

Wavelength-dispersive spectrometers have a better spectral resolution than energy-dispersive spectrometers, especially at lower energies.

Wavelength-dispersive spectrometers need a sophisticated and very precise mechanical design.

Resolution

The resolution of the spectrometer, $\Delta E/E$ depends on the crystal and collimators used. Finer collimators will give a better resolution. ΔE is measured as the full width of the peak at half its height. Typically $\Delta E/E$ has a value around 0.01. At long wavelengths the peaks are narrower than at short wavelengths (Table 8.3-9). The resolution of the WD spectrometer is sufficient to separate the lines of adjacent elements so that peak overlap is only a minor problem in WD spectrometry.

Figure 8.3-13 shows the WD spectrum of a geological material between 0.69 and 0.98 Å (2θ scan from 28 to 40°, LiF(220), $2d = 0.285$ nm). Some peak overlap between K- and L-lines is observed at this short wavelength region. Notice that the vertical scale is in kilocounts per second (kcps).

WD-XRF configurations

Modern WD-XRF instruments are sophisticated instruments with a very complex design. Mechanical functions include the rotation of the goniometer, the selection of one of the available diffraction crystasl, detector, collimators, and filter between tube and sample. All these functions as well as the high-voltage generator and the X-ray counting system are computer controlled. Programs for qualitative and quantitative analysis are also running on this computer.

One distinguishes between *sequential* and *simultaneous* WD-XRF instruments. Sequential (or single channel) instruments have one goniometer. The concentration of different elements is determined by moving the goniometer to the correct 2θ angle and measuring the fluorescence intensity for 1 to 100 s "integration time". Therefore, the entire measurement can take up to 30 minutes for a complex multielement sample. In simultaneous (or multi-channel) instruments this disadvantage is overcome by placing a number of crystal-detector combinations (similarly to polychromators in UV-VIS atomic emission) with fixed 2θ angles around the sample. Some instruments can have up to 30 channels. Multielement analysis of a fixed set of elements can be performed in a few seconds to a few minutes. These type of instruments are ideally suited for process control, e.g., in the steel industry. Combined instruments with one sequential and a limited number of fixed spectrometers also exist. An artist's impression of such a configuration is given in Fig. 8.3-14.

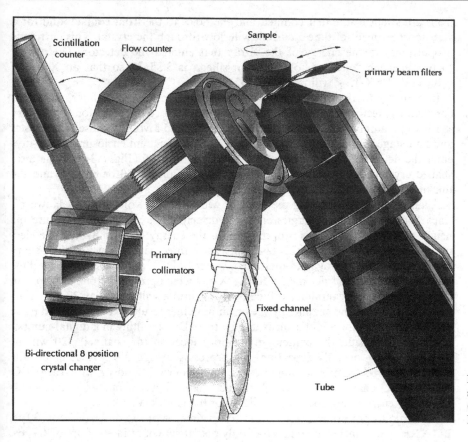

Fig. 8.3-14. Internal view of a combined sequential and fixed channel WD-XRF. (Courtesy of Philips Analytical X-ray, PW2400, adapted)

Energy-dispersive instruments

Semiconductor detector

In the energy-dispersive spectrometer the 'dispersion' (the selection of the particular energy) and the counting of the number of X-rays (of that particular energy) is done in one step. Energy-dispersive spectrometers are built around a liquid nitrogen cooled semiconductor crystal. Either a lithium doped silicon, Si(Li), or a hyperpure germanium, HPGe, single crystal is used. In these crystals, the energy difference between the valence and the conduction band is around 4 eV. At room temperature, a number of electrons are in the conduction band so that the crystal is a (semi) conductor. By cooling the crystal to liquid nitrogen temperature (−196 °C) almost all electrons will remain in the valence band and no current can flow when a voltage is applied over the crystal. Lithium is drifted into the silicon crystal to compensate for charge carriers due to impurities.

The structure of a Si(Li) detector is shown in Fig. 8.3-15. A negative voltage of ∼500 V is applied to the front contact of the crystal. When an X-ray enters the detector, its energy is absorbed by the crystal. This results in the formation of so-called "electron-hole" pairs. Electrons are promoted from the valence to the conduction band, leaving positive "holes" in the valence band. Thus the crystal becomes temporarily conducting. Because of the applied bias voltage, the elec-

At the heart of an energy-dispersive XRF instrument is a single crystal (Si or Ge), semiconductor detector.

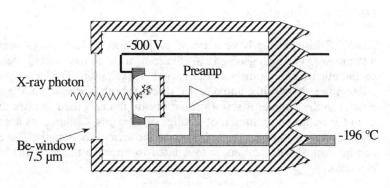

Fig. 8.3-15. Structure of a Si(Li) detector, the detector head is connected to a liquid nitrogen dewar

trons are swept to the rear contact, and the holes to the front contact, and for a very short moment of time a current will flow through the crystal. This current is proportional to the energy of the X-ray that entered the detector. The energy to produce one "electron-hole" pair in silicon is 3.85 eV, so that an X-ray of 6.400 keV (Fe K-L$_{3,2}$) will produce 1662 electrons.

In contrast to the flow-proportional and the scintillation counter, the semiconductor detector has no internal amplification. To compensate for this, a very sensitive preamplifier is used to convert the charge to a voltage pulse. The crystal and first stage of the preamplifier are mounted in a vacuum enclosure and X-rays enter the detector through a very thin beryllium window (Fig. 8.3-15). The disc-shaped crystal itself is rather small. The area is typically 30 or 80 mm^2 and the thickness 3 or 5 mm.

Energy-dispersive spectrometers measure all X-rays emerging from the sample (quasi) simultaneously; in a wavelength-dispersive spectrometer the recording is done sequentially.

Again, each X-ray entering the detector will cause one voltage pulse. However, since X-rays of various energies enter the detector, one also has to measure the height of the pulse which is proportional to the energy of each X-ray. The electronic circuit that performs this task consists of three parts: a linear (pulse shaping) amplifier, an analogue-to-digital converter (ADC), and a memory. To illustrate its function, consider the Fe K-L$_{3,2}$ X-ray which produced 1662 electrons in the detector. The preamplifier will convert this into a voltage of say 32 mV. Further amplification and shaping by the linear amplifier results in a bell-shaped pulse of 3.20 V. This pulse height is measured by the ADC, resulting in a digital number of 320. As a result, the content of memory location (or channel) 320 will be incremented by one. By repeating this process for each X-ray that enters the detector the spectrum is accumulated in the memory. Memories with 1024 (1K) or 2048 (2K) channels are used. With each channel corresponding to 20 eV this covers the energy range between 0 and 20 or 0 and 40 keV.

The time to process an X-ray event (dead time) is of the order of 10 to 30 µs. ED-XRF spectrometers can therefore only operate at count rates of up to 40 kcps. The efficiency of the Si(Li) detector drops at low energies (<2 keV) because the X-rays are absorbed in the Be-window.

Resolution

X-rays with different energies will be counted in different channels. However, not all X-rays with the same energy will be accumulated in the same channel. Physical processes in the detector cause fluctuations in the number of electrons produced for a given X-ray energy, and electronic noise in the amplifiers cause a further fluctuation of the pulse height. As a result, counts will be accumulated below and above the expected channel and a near-Gaussian peak, rather than a sharp line, will be observed in the spectrum. The peak width is expressed as its full-width-at-half-maximum (FWHM) in eV. The resolution of the spectrometer is specified as the FWHM of the Mn K-L$_{3,2}$ peak (5.89 keV) and is typically around 150 eV. In Table 8.3-9 the peak width of the K-L$_{3,2}$ of aluminium, iron, and molybdenum are given. The peak width increases with energy. The resolution of the ED spectrometer is too low to separate the K-L$_{3,2}$ peak from the K-M$_{3,2}$ line of the preceding element. This peak overlap is the major drawback of the ED spectrometer. The WD spectrometer has a much better resolution at low energies. At higher energies the difference in resolution between the two spectrometers is insignificant.

ED-XRF configurations

Energy-dispersive spectrometers have a simple mechanical design, but use sophisticated electronics to process the very weak signals from the detector.

ED-XRF instruments have a much simpler mechanical design than WD-XRF instruments. The high geometrical efficiency of the semiconductor detector allows, on the other hand, a much greater variety in excitation. The basic system involves a low power X-ray tube and a Si(Li) detector both placed at an angle of 45° with respect to the sample. A set of primary beam filters is used to alter the tube spectrum for optimum excitation of a range of elements. Collimators are used to confine the excited and the detected beam to the area of the sample. The entire system can be flushed with helium or evacuated to improve the measurement of low Z elements.

a) Secondary target ED-XRF

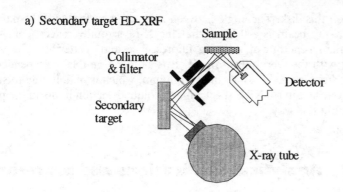

b) Total reflection XRF

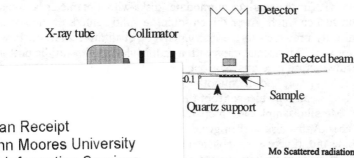

Fig. 8.3-16. Two typical energy-dispersive XRF setups. (a) Secondary target system, (b) total reflection system (TXRF)

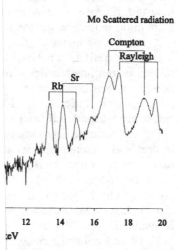

Fig. 8.3-17. ED-XRF spectrum of a geological standard (JG1) measured for 3000 s with a Mo secondary target

Energy-dispersive XRF instruments can have various excitation – detection geometries: direct tube, secondary target, radioisotopic, total reflection, etc.

In ... econdary target (Fig. 8.3-16a).
Fl ... ic disk (the secondary target).
ta; ... xcited the sample. The advan-
Th ... X-ray tube is largely removed.
ba ... iation results in a much lower
in ... letection limit. The excitation
di ... c. By interchanging the target,
... The low efficiency of the fluo-
rescence in the secondary target means that the use of X-ray tubes with higher power is required. An energy-dispersive X-ray spectrum of a geological standard measured with such a system is shown in Fig. 8.3-17. Molybdenum was used as a secondary target and the spectrum was acquired for 3000 s. Notice the considerable overlap between the peaks. The peaks above 16 keV are due to elastic and inelastic scattering of the Mo K-radiation on the sample.

A radioisotope XRF system often simply consists of an angular source placed around the head of the Si(Li) detector. The sample is placed above the detector and the source [8.3-8].

Figure 8.3-16b is a schematic diagram of a total-reflection X-ray fluorescence (TXRF) setup [8.3-9, 8.3-10]. The sample support is a quartz disk with an optically flat surface. The sample, e.g., the residue of an evaporated drop of water, is

Total reflection XRF is a very interesting technique for analysis of liquids; detection limits in the ppb range can be obtained.

placed on this disk. The angle between the quartz surface and the narrowly collimated X-ray beam is $\sim0.1°$ so that the X-rays totally reflect from the support
rather than penetrate it, causing fluorescence and scattering. The X-rays only
interact with the (very thin) sample. It is as if the sample is suspended in the air
without support. As a result, the continuum, which is mainly due to scattering, is
virtually absent in a TXRF spectrum. Absolute detection limits in the pg range are
obtained in this way.

8.3.3 Analytical applications and procedures

Sample preparation

Samples for XRF analysis can be in the form
of polished metal disks, pellets of pressed
powders, liquids, thin films, and particulate
material on a filter.

In principle, XRF requires little sample preparation. A standard sample holder of
an XRF instrument accommodates disks with a diameter between 1 and 3 cm and
up to 1 cm high. After dimensioning a solid sample to the correct size it can be
directly presented to the instrument for qualitative analysis. Powders and liquids
are poured in special cups with a thin Mylar window at the bottom. However, for
truly quantitative analysis, one needs to pay special attention to the homogeneity
of the sample and the smoothness of the sample surface, because the characteristic
X-rays can only escape from the first few micrometers of the sample surface.

To obtain good quantitative results, the
sample must be homogeneous and the
irradiated surface must be very smooth.

Metallic samples are prepared by machining them into a disk. One of the surfaces of this disk is then grinded with an abrasive such as alumina or silicon carbide and, if necessary, polished with diamond paste (e.g., the analysis of copper in
aluminium requires a surface roughness below 10 μm).

A number of materials (e.g., ores, minerals, sediments, and freeze-dried biological material) are analyzed by XRF by first pulverizing the material and then
milling it to a fine powder with particles typically less than 50 μm diameter. The
powder is then pressed into a pellet. If necessary, 5 to 10% binder material (cellulose, polyvinyl acetate) is added to make the pellet mechanically stable. Pressed
pellets are adequate for trace and minor elements analysis.

For the accurate determination of major elements such as Al, Si, K, Ca, and Fe
in ores and minerals a different sample preparation technique is used: the powdered sample is mixed with a flux material such as sodium tetraborate ($Na_2B_4O_7$)
in a ratio of 1 to 10 [8.3-11]. This mixture is heated at about 900 °C in a platinum
crucible until the sample is completely dissolved. The melt is then poured into a
mould. After solidification, a homogeneous glass bead with a very smooth surface
is obtained.

Trace and minor elements in liquids can be measured directly without sample
preparation. A typical example is the determination of sulphur in the concentration range between 1 and 100 ppm in petroleum products. For lower concentrations, preconcentration techniques are required. Transition metal ions in water
can be collected on an ion-exchange resin such as Kelex-100. The loaded resin in
then pressed into a pellet and analyzed in the conventional way. Total-refection
XRF allows the direct analysis of water with detection limits in the ppb range by
simply spotting a drop onto the "reflector" (Fig. 8.3-16b).

To study inorganic air pollution, aerosol
particles are collected on a filter and this filter
is analyzed directly with XRF.

Particulate material such as atmospheric particles and particles suspended in
liquid can be analyzed after filtration over a suitable filter. After pumping several
m^3 of air through a Nuclepore filter with a pore size of 0.4 μm, the aerosol particles form a thin layer on the filter. This thin film can be analyzed directly with
XRF with detection limits in the order of ng per cubic meter of air for the transition metals and the heavy elements. XRF is therefore very suitable for air pollution monitoring.

Qualitative analysis

Elements in the sample are identified
(qualitative analysis) by the energy (or
wavelength) of the characteristic K, L, or M
lines.

Qualitative analysis is in principle very simple with XRF and is based on the
accurate measurement of the energy, or wavelength, of the fluorescence lines
observed. Since WD-XRF spectrometers operate sequentially, a 2θ scan needs to
be made. The identification of trace elements can be complicated by the presence
of higher order reflections or "satellite" lines from major elements. With energy-

dispersive XRF the entire X-ray spectrum is acquired simultaneously. The identification of peaks is, however, hampered by the lower resolution of the ED spectrometer. Qualitative analysis software assists the spectroscopist by showing KLM markers on the spectrum display. KLM markers indicate the theoretical position of the K, L, and M lines of an element as vertical lines. When these lines coincide with the observed peak maxima in the spectrum an element is positively identified (as it is customary in atomic emission).

Quantitative analysis

Quantitative analysis is based on the relation between the intensity of the characteristic line of an element and the concentration of that element in the sample. This relation is very well-known from the fundamental principles involving the interaction of X-rays with matter, but is unfortunately not very simple. Quantitative analysis is further complicated by the fact that one needs to use the net, i.e., continuum corrected and interference free intensity of the characteristic lines.

> Quantitative analysis is based on the (not so simple) relation between the net X-ray intensity of the characteristic lines and the concentration.

Quantitative analysis with WD-XRF starts with counting the X-rays at the peak maximum and at the adjacent background during 1 to 100 s. The net count rate (counts per second) of an elements is then obtained by subtracting the continuum (background) count rate from the peak count rate. Several elements are measured sequentially unless a simultaneous instrument is used. In ED-XRF, one accumulates the entire spectrum during a hundred to several thousands of seconds. The determination of the net peak intensity is more complicated in the case of ED-XRF because of the lower peak-to-background ratio and the overlap between lines of different elements (poorer SNR and selectivity than WD-XRF). Sophisticated mathematical procedures relying on least squares fitting of the spectral data are used to determine the net peak area of the characteristic lines [8.3-12].

The remaining part of the quantitative analysis procedure is basically the same for ED- and WD-XRF. The conversion of net intensity to concentration is hampered by the so-called "matrix effects". Indeed the primary X-rays, responsible for creating the vacancies in the analyte atom, are attenuated as they enter the sample. In the same way, the fluorescence X-rays produced at a certain depth in the sample will be attenuated on their way out to the detector. In first approximation, the concentration w_i of an element i is proportional to the count rate I_i and inversely proportional to the mass attenuation coefficient μ_M of the matrix for that line:

> The net X-ray intensity depends not only on the concentration of that element, but also in the concentration of all other elements in the sample (matrix effect).

$$W_i \propto \frac{I_i}{\mu_M} \tag{8.3-8}$$

This shows the importance of the mass attenuation coefficient in quantitative analysis. The net intensity of the Fe K-$L_{3,2}$ line resulting from a concentration of 1% iron will therefore strongly depend on the composition of the matrix as illustrated in Table 8.3-10. The relationship between the intensity and the concentration is further complicated by the "enhancement" effect. Vacancies in the analyte atoms are not only produced by the primary X-rays, but might also result from the photoelectric absorption of characteristic X-rays with sufficient high energy generated in the sample itself. The process of generating characteristic X-rays in the sample by the primary radiation is called primary fluorescence. The production of characteristic X-rays by other characteristic X-rays generated in the sample is called secondary fluorescence. If secondary fluorescence occurs, the observed X-ray intensity will be higher than expected. The enhancement effect is particularly strong in Cr–Fe–Ni alloys where the Fe K-$L_{3,2}$ X-rays (6.400 keV) efficiently excite the chromium atoms ($E_K = 5.989$ keV), and the Ni K-$L_{3,2}$ X-rays can excite both iron and chromium. The combination of matrix absorption and enhancement

Table 8.3-10. Intensity in cps of the Fe K-$L_{3,2}$ line for an iron concentration of 1% in various matrices

Matrix	C	Al	Cr	Ni	Pb
Int. Fe K-$L_{3,2}$	1200	108	22	79	20

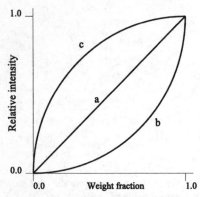

Fig. 8.3-18. Possible relations between the intensity of a characteristic line and the concentration of an element, (a) no absorption and no enhancement, (b) absorption, and (c) enhancement

effects cause in general the calibration curves in XRF to be nonlinear as illustrated in Fig. 8.3-18. Different quantitative analysis schemes are used depending on the type of matrix, the concentration range, and the availability of standards.

Quantitative analysis of thin films, such as filters loaded with aerosol particles, can be done by simply comparing the count rate for a particular element in the sample with the count rate observed in a thin film standard, because matrix effects are virtually absent.

Worked example: Analysis of air particulate matter

Air particulate matter (aerosol particles) are collected by passing $10\,m^3$ of air through a Nuclepore filter with an effective sampling area of $13\,cm^3$. An energy-dispersive XRF instrument with a Mo-anode X-ray tube operating at 40 kV was used to perform the elemental analysis of the particulate matter collected on the filter. The instrument was calibrated with thin film metal standards. X-ray attenuation in these standards and in the aerosol filter are negligible for the elements considered. From the measurement of the standards, we obtain the sensitivity for the $K\alpha$-line expresses as counts per second per $\mu g/cm^2$ for each element:

	Ca	V	Fe	Cu	Zn
Sensitivity, m $Count.s^{-1}/(\mu\,gcm^{-2})$	0.42	1.31	2.98	5.64	6.76

The measurement uncertainties in these sensitivities are mainly due to the uncertainties in the area density of these standards (μgcm^{-2}) and are estimated to be 5% (relative standard deviation).

Next the aerosol filter was measure under the same condition during 1000 seconds. The net peak areas of the $K\alpha$-lines in the measured spectrum are:

	Ca	V	Fe	Cu	Zn
Intensity, I Counts/1000 s	1260 ± 56	110 ± 60	7440 ± 136	282 ± 37	3549 ± 80

Here the uncertainties are due to counting statistics and are obtained as part of the spectrum evaluation process. The counts for vanadium are not significant because the estimated peak area (110 counts) is less than 3 times the standard deviation. For vanadium we can only give a upper limit.

By dividing the net counts by the measuring time and the sensitivity (Eq. 8.3-10), we obtain the concentration on the filter in $\mu g/cm^2$. Using the area of the filter ($13\,cm^2$) and the volume of air passed through it, we can calculate the concentration of the elements in the air in $\mu g/m^3$.

	Ca	V	Fe	Cu	Zn
Conc., $\mu g/cm^2$ on the filter	3.00 ± 0.20	<0.14	2.50 ± 0.13	0.050 ± 0.007	0.525 ± 0.029
Conc., $\mu g/m^3$ in the air	3.90 ± 0.17	<0.18	3.25 ± 0.17	0.065 ± 0.009	0.683 ± 0.037

The uncertainties in the concentrations are obtained by combining the uncertainty in the sensitivity with the uncertainty due to counting statistics.

$$\frac{s_C}{C} = \sqrt{\left(\frac{s_I}{I}\right)^2 + \left(\frac{s_m}{m}\right)^2}$$

These uncertainties reflect the measurement error (repeatability). Uncertainty due to sampling (which may be large when sampling aerosols) is not taken into account.

Linear calibration curves can be used to determine trace and minor element concentrations in alloys, mineral pellets or liquids provided that the major element concentrations of standards and unknowns are very similar. In this case the matrix effect remains the same.

Worked example: Analysis of carbon

A wavelength dispersive instrument is used to determine carbon in cast iron in the concentration range between 2 and 4%. Although the attenuation of the C-Kα X-rays is very considerable in iron, a linear calibration curve can be used since the matrix (cast iron) has nearly the same composition for all samples that need to be analysed. Seven standards were used to set up the calibration curve. The carbon content in these standards was determined very carefully using a classical reference method. The net C Kα X-ray intensities and the carbon concentration for these 7 standards are given in the table:

C concentration in wt%	C Kα X-ray intensity in counts/s
2.32	158
2.93	209
3.45	243
3.89	274
2.87	204
3.80	262
3.46	237

The relation between X-ray intensity and concentration was established by linear (straight line) regression as:

$$\text{Int C K}\alpha(\text{cps}) = (-0.05 \pm 10) + (69.9 \pm 3.1) \times \text{Conc}$$

In the figure this calibration curve is shown.

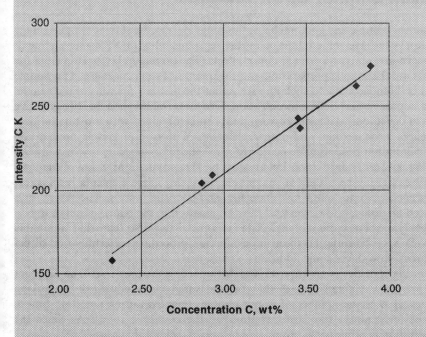

Next an unknown cast iron sample was measured, yielding an C Kα x-ray intensity of 233 counts/s, so that the C concentration was estimated at 3.28 %wt. Chemical analysis of this sample gave a concentration estimate of 3.34 %wt or a relative difference of 1.8%.

To determine major and minor elements in complex samples more elaborated matrix correction procedures need to be applied. They can be roughly divided into two categories: the influence coefficient model and the fundamental parameter model. The Lachance-Traill model [8.3-13] for a sample containing three elements a, b, and c uses the following equation:

$$w_a = R_a(1 + \alpha_{ab}w_b + \alpha_{ac}w_c)$$

$$w_b = R_b(1 + \alpha_{ba}w_a + \alpha_{bc}w_c) \tag{8.3-9}$$

$$w_c = R_c(1 + \alpha_{ca}w_a + \alpha_{cb}w_b)$$

R_a is the relative intensity, i.e., the intensity of element a in the sample divided by the intensity of the pure element, w_a is the weight fraction of element a in the sample. If no matrix effects are present, then R_a is the correct value of the concentration of element a in the sample ($w_a = R_a$). The terms α_{ab} and α_{ac} correct for the influence that the presence of elements b and c have on the relative intensity of a. They are called *influence coefficients*. These influence coefficients are determined by measuring standards or are calculated from theory. Once the coefficients are known, the above set of equations can be solved iteratively to obtain the concentration of the elements in the unknown samples. Under favourable conditions very accurate quantitative results (relative standard deviation of 0.3 to 0.1%) can be obtained by this and other similar correction models [8.3-14].

Fundamental parameter quantitative analysis uses physical principles to relate the concentration of the analyte to the observed net X-ray intensity.

The fundamental parameter method [8.3-15] is based on the physical theory of X-ray production. It requires an accurate knowledge of the shape of the excitation spectrum, the detector efficiency, and fundamental parameters such as the cross section for photoelectric absorption and fluorescence yield. The method is computationally complex because the fundamental parameter equation relates the intensity of one element to the concentration of all elements present in the sample so that again an iterative solution of a set of (integral) equations is required. The fundamental parameter method is especially interesting because it allows the semi-quantitative (5 to 10% relative standard deviation) analysis of completely unknown samples. With suitable calibration 1% accuracy can be obtained.

Trueness, precision, and detection limits

Semi-quantitative results are easily obtained, truly quantitative results (trueness better that 1%) require careful sample preparation and calibration of the instrument.

Mechanical (sample and goniometer positioning) and electric stability (high-voltage generator) can have some influence on the reproducibility of XRF measurement. In modern instruments these are minor effects. The dominant factors affecting the trueness are the sample preparation and the quantification procedure. The sample should be homogeneous to such an extent that the first few microns of the sample should represent the bulk concentration. Trueness values of 5 to 10% relative standard deviation (coefficient of variation) can be obtained with most quantification schemes for medium and high Z elements. With careful sample preparation and a calibration procedure dedicated to a specific matrix, accuracies between 0.3 and 1% relative standard deviation are possible. Determinations of low Z elements (Be–F) can easily be orders of magnitude off scale. These elements have a low fluorescence yield and emit low energetic characteristic X-rays. The few X-rays produced are difficult to detect and their attenuation in the matrix is very large.

In general, the accuracy of ED-XRF is somewhat worse than that attainable with WD-XRF because the spectrum evaluation procedure (determination of net peak areas) introduces some additional errors.

Counting statistics is an important factor affecting the precision. The fact that the X-rays are counted implies that the measurement is governed by Poisson statistics. If N is the number of X-ray photons counted during a certain time t, then the uncertainty (standard deviation) in the counts is $s_N = \sqrt{N}$. Since the concentration is proportional to the count rate, m being the sensitivity:

$$w = \frac{1}{m} I = \frac{1}{m} \frac{N}{t} \tag{8.3-10}$$

the standard deviation of the concentration becomes:

$$s_w = \frac{1}{m} \frac{1}{t} \sqrt{N} \tag{8.3-11}$$

and the relative standard deviation becomes:

$$\frac{s_w}{w} = \frac{\sqrt{N}}{N} = \frac{1}{\sqrt{N}} = \frac{1}{\sqrt{It}} \tag{8.3-12}$$

I being the count rate in cps. A determination based on a peak of 100 counts will have an uncertainty of 10%, only due to counting statistics. To determine an element with a relative precision of 0.1% one must accumulate 10^6 counts. Note that the precision is inversely proportional to the square root of the measuring time t. Measuring periods (integration times) 10 times longer improve the precision by about a factor of 3.

The fluctuation in the continuum (background) of the spectrum is one of the key factors determining the detection limit. These fluctuations are again due to counting statistics. The number of counts of an X-ray line must exceed three times the standard deviation of the continuum in order to be detectable (IUPAC $3\sigma_B$ criterium for the detection limit). If N_B is the number of counts in the continuum at the energy of the X-ray line, the detection limit in counts is $N_L = 3\sqrt{N_B}$, or expressed in concentration units:

$$w_L = \frac{1}{m}\frac{1}{t}3\sqrt{N_B} = \frac{1}{m}3\sqrt{I_B/t} \tag{8.3-13}$$

For a particular element, the detection limit depends on the sensitivity and the count rate of the continuum below the peak and is inversely proportional to the measurement time. Detection limits can be improved by increasing the sensitivity (optimization of the excitation and detection efficiency), by reducing the background (e.g., as in TXRF), or by counting for a longer period of time. The value of the attainable detection limit thus depends very much upon the sample, the element considered, and the experimental conditions, and ranges from 0.1 ppm to a few % (see below).

8.3.4 Summary

XRF is an elemental analysis technique for (mainly) solid and liquid samples. Although in principle all elements from beryllium onwards in the periodic table can be determined, low Z elements are more difficult to determine. Detection limits are in the low ppm range (0.1–10 ppm) for medium Z elements (Fe) up to 1–5% for the lightest elements (B, Be). Detection limits for ED-XRF are typically a factor 5 to 10 worse except for TXRF which has absolute detection limits in the pg range. In general, optical emission and mass spectrometric methods have better (lower) detection limits. Precision and accuracy can vary considerably, but for routine analysis they compare favourably with other instrumental techniques. Accuracies close to those obtained with wet chemical analysis are achievable if matrix-matched calibration standards are used.

The major advantage of XRF, especially WD-XRF lies in its wide dynamic range. Both elements present in the % range and elements present in the ppm range can be measured simultaneously on the same sample. Another advantage, compared to most other analytical techniques, is that solid samples can be measured as such without the need for dissolution. This enables total multielemental analysis times of a few minutes when using multichannel WD-XRF instruments (e.g., in the steel industry).

The complicated quantization procedures and the relatively high costs of the instruments, especially of WD instruments, are the major drawbacks of XRF.

XRF can be considered as a mature analytical technique. Most development today is in the improvement of the instrumentation. Research is mainly done these days on special techniques such as microbeam XRF [8.3-16], the use of synchrotron radiation [8.3-17], and polarized beam [8.3-18], and total reflection [8.3-19] setups.

References

[8.3-1] Bertin, E.P., *Introduction to X-ray Spectrometric Analysis*, New York: Plenum Press, 1978.

[8.3-2] Jenkins, R., Gould, R.W., Gedcke, D., *Quantitative X-ray Spectrometry*, New York: Marcel Dekker, 1981.

[8.3-3] Tertian, R., Claisse, F., *Principles of Quantitative X-ray Fluorescence Analysis*, London: Heyden, 1982.

[8.3-4] Van Grieken, R.E., Markowicz, A.A., (Eds.), *Handbook of X-ray Spectrometry*, New York: Marcel Dekker, 1993.

[8.3-5] Lachance, G., Claisse, F., *Quantitative X-ray Fluorescence Analysis: Theory and Applications*, Chichester: Wiley, 1995.

[8.3-6] Robinson, J.W., *Handbook of Spectroscopy, Volume I*, Cleveland: CRC Press, 1974.

[8.3-7] Helsen, J.A., Kuczumow, A., in: *Handbook of X-ray Spectrometry*, Van Grieken, R.E., Markowicz, A.A. (Eds.), New York: Marcel Dekker; 1993, pp. 75–150.

[8.3-8] Watt, J.S., in: *Handbook of X-ray Spectrometry*, Van Grieken, R.E., Markowicz A.A. (Eds.), New York: Marcel Dekker, 1993, pp. 359–410.

[8.3-9] Yoneda, Y., Horiuchi, T., *Rev. Sci. Instr.*, 1971, 42, 1069.

[8.3-10] Aiginger, H., Wobrauschek, P., *Nucl. Instr. Methods*, 1974, 114, 157.

[8.3-11] Stephenson, D., *Anal Chem.*, 1969, 41, 966.

[8.3-12] Van Espen, P.J.M., Janssens, K.H.A, in: *Handbook of X-ray Spectrometry*, Van Grieken, R.E., Markowicz, A.A. (Eds.), New York: Marcel Dekker, 1993, pp. 181–294.

[8.3-13] Lachance, G.R., Traill, R.J., *Can. Spectrosc.*, 1966, 11, 43.

[8.3-14] De Jongh, W.K., *X-ray Spectrom*, 1973, 2, 151.

[8.3-15] Criss, J.W., Birks, L.S., *Anal. Chem.*, 1968, 40, 1080.

[8.3-16] Carpenter, D.A., Taylor, M.A., *Adv. X-ray Anal.*, 1991, 34, 217.

[8.3-17] Jones, K.W., in: *Handbook of X-ray Spectrometry*, Van Grieken, R.E., Markowicz, A.A. (Eds.), New York: Marcel Dekker, 1993, pp. 411–452.

[8.3-18] Ryon, R.W., Zahrt, J.D., in: *Handbook of X-ray Spectrometry*, Van Grieken, R.E., Markowicz, A.A. (Eds.), New York: Marcel Dekker, 1993, pp. 491–515.

[8.3-19] Schwenke, H., Knoth, J., in: *Handbook of X-ray Spectrometry*, Van Grieken, R.E., Markowicz, A.A. (Eds.), New York: Marcel Dekker, 1993, pp. 453–489.

Questions and problems

1. Rewrite the Compton equation (Eq. 8.3-3) to express the wavelength (rather than the energy) of an inelastically scattered photon. What is the wavelength of an X-ray of 0.6 Å inelastically scattered over an angle of 60°?

2. Complete Table 8.3-1 for the energy levels of the N-shell.

3. Calculate the mass attenuation coefficient of Fe_2O_3 for X-rays with an energy of 6.40 and 7.47 keV.

4. How thick must a silicon layer be to stop (attenuate) 99% of the Si $K-L_{3,2}$ X-rays?

5. Verify if the following transitions are allowed or forbidden: $K-L_1$, $K-M_3$, L_2-M_4, L_1-L_2. Explain why.

6. Why is light element analysis with XRF difficult in terms of sensitivity and accuracy?

7. What is the minimum high-voltage for operating an W-anode X-ray tube if one wants to excite the sample with the W-K lines?

8. Why is it important that the anode of an X-ray tube is made from a very pure material?

9. A Rh-anode X-ray tube is operated at 20 kV. Calculate the shortest wavelength emitted by the tube and the wavelength corresponding to the maximum intensity of the continuum.

10. Explain the production of characteristic and continuum X-rays in an X-ray tube.
11. Why does a ^{109}Cd radioactive source emits Ag K X-rays?
12. Calculate the wavelength range of a WD-XRF spectrometer equipped with a LiF(220) crystal ($2d = 0.285$ nm).
13. A WD spectrum measured with a LiF(200) crystal, has a second-order ($n = 2$ in Eq. 8.3-7) diffraction peak at $2\theta = 51.7°$. What is the wavelength of the X-ray that caused this peak?
14. Why is a scintillation counter more efficient than a flow proportional counter for short-wavelength X-rays?
15. Describe the function of a complete WD-XRF spectrometer.
16. Describe the functioning of a semiconductor (Si(Li)) spectrometer.
17. Compare the WD and ED spectrometer in terms of resolution and maximum count rate.
18. Calculate the energy of the Compton peaks observable in Fig. 8.3-17 (excitation and detection done at 45°, Mo Kα = 17.44 and Mo Kβ = 19.61 keV).
19. Explain the principle of total reflection XRF.
20. Give two sample preparation methods commonly employed for the analysis of geological material. Explain why one method is preferred for major element analysis and the other for trace element analysis.
21. One wants to perform a highly accurate analysis of brass samples. What are the requirements concerning the samples?
22. What factors affect the detection limit in XRF?
23. A plastic material is analyzed with WD-XRF. The count rate of the continuum at the Co K-L$_{3,2}$ line is 4000 cps. This value is obtained after 10 s of measurement. The sensitivity of the spectrometer for cobalt is 180 cps/ppm. Calculate the detection limits. What is the value of the detection limit if one measures for 100 s?

8.4   Activation Analysis

8.4.1   Introduction

Activation analysis is a method for the determination of elements and isotopes. Although, in principle, minor and major elemental components can also be determined, its full potential is realized only when it is applied to *trace* and *ultratrace element analysis*.

The four kinds of projectile particles, i.e., slow and fast neutrons, charged particles, and photons, used for the induction of nuclear reactions to form the indicator radionuclides (IRNs) are the basis of four different activation methods: *thermal neutron activation analysis* (NAA), *fast neutron activation analysis* (FNAA), *charged particle activation analysis* (CPAA), and *photon activation analysis* (PAA). As NAA is by far the most important and the widest used activation method, it is handled in this chapter in more detail, whereas for FNAA, CPAA, and PAA, which represent rather special methods complementary to NAA, only a brief characterization is given.

Nuclear activation methods of analysis are covered in more detail in a book by Elving et al. [8.4-1] and in a newer book by Ehmann and Vance [8.4-2]. The fundamentals of radioactivity, nuclear reactions, nuclear instrumentation, and experimental techniques are excellently treated in a book by Friedlander et al. [8.4-3].

8.4.2   Principles

The nuclear activation process

Activation analysis is based on the conversion of a stable nuclide (A) of the element to be determined into a radionuclide (B), called an *indicator radionuclide* (IRN), by a *nuclear reaction*. The reaction is induced when a target material is exposed to projectiles (x) which may be neutrons, charged particles (protons, deuterons, tritons, ^{3}He, and alpha particles), or gamma photons. The nuclear reaction can be represented as:

$$A + x \rightarrow [C] \rightarrow B + y + Q \tag{8.4-1}$$

or in short form notation:

$$A(x, y)B \tag{8.4-2}$$

where y is a photon or a particle (proton, alpha particle, or neutron) emitted, and the Q value gives the amount of energy released or absorbed in the reaction. If Q is positive, energy is released and the reaction is exoergic, and if Q is negative, the reaction is endoergic, i.e., it requires energy, which must be supplied in the form of kinetic energy of the projectile particle. Some of the incident projectile energy can be used for conserving momentum of the system and, in the case of charged-particle-induced reactions, additional energy above the Q value may be required

Activation analysis is based on the measurement of radioactivity of indicator radionuclides formed from stable nuclides of the analyte elements by nuclear reactions induced in an irradiation of the samples with suitable particles, most commonly neutrons.

to overcome the Coulomb barrier that occurs between the projectile particle and the target nucleus.

In nuclear reactions used for activation analysis, especially for NAA, the predominating mechanism is that via *compound-nucleus* formation (C in Eq. 8.4-1, lifetime 10^{-20}–10^{-14} s). This mechanism consists of two steps: the capture of the projectile to form a compound nucleus in the excited state, and the de-excitation or decay of the compound nucleus via particle or gamma photon emission. A certain compound nucleus formed by capture of the incident particle may disintegrate with different probabilities in several modes:

$$A + x \rightarrow [C] \left\langle \begin{array}{l} B_1 + y_1 \\ B_2 + y_2 \\ B_3 + y_3 \end{array} \right. \tag{8.4-3}$$

etc.

Thus, absorption of the incident particle by the nucleus A can result in several different nuclear reactions, occurring with different probabilities. The number of possible reactions increases with increasing kinetic energy of the incident particles.

Activation yields

Considering the general formulation of a binuclear reaction in Eq. 8.4-1, the *production rate R* of the indicator radionuclide B (i.e., the number of its nuclei created per second) may be expressed as:

$$R = \sigma \Phi n_A \tag{8.4-4}$$

Here σ is the reaction cross section having the dimension cm^2 (its unit 1 barn (b)= 10^{-24} cm^2). The cross section is a physical constant which gives the probability that a nuclear reaction will take place and, therefore, it is similar to the rate constant of a bimolecular chemical reaction. Φ is the flux density of the projectiles, x per cm^2 per second. The number of the target nuclei in the irradiated sample portion n_A can be obtained from the mass m of the element present in the sample, the atomic weight of the element A_w, the fractional isotopic abundance of the reacting nuclide f_a, which can vary in the interval between 3×10^{-5} (^{46}Ca) and 1 (e.g., ^{27}Al), and from Avogadro's number N_A. The equation for the number of target nuclei can, therefore, be written as:

$$n_A = \frac{m}{A_w} f_a N_A \tag{8.4-5}$$

However, the radioactive product B decays during the irradiation with a decay rate, which is proportional to the *decay constant* λ having a characteristic value for each radionuclide and to the number of radioactive nuclei present, n_B. Consequently, the growth of the radioactivity is governed by the difference between the production rate R and the decay rate λn_B:

$$\frac{dn_B}{dt} = R - \lambda n_B \tag{8.4-6}$$

It is more convenient to use the *half-life* $t_{1/2}$ instead of the decay constant λ obtainable from the relationship:

$$t_{1/2} = \frac{\ln 2}{\lambda} \tag{8.4-7}$$

The functional relationship between the mass of the element in the irradiated sample and the resulting radioactivity is given by the radioactivation equation Eq. 8.4-8. Note that this equation holds only if the attenuation of the neutron flux within the sample may be neglected. In most practical situations, this approximation is well justified.

By solution of Eq. 8.4-6 and considering Eqs. 8.4-4, 8.4-5 and 8.4-7, the general *radioactivation equation* expressing the radioactivity in disintegrations per second A_t of a given radionuclide at the end of an irradiation of time t is obtained in the form:

$$A_t = \sigma \Phi \frac{m}{A_w} f_a N_A (1 - e^{-\ln 2\, t/t_{1/2}}) \tag{8.4-8}$$

The activity is always counted after a *decay time* t_d which is inevitable on account of transport, unpacking, and post-irradiation sample treatment, or, in many instances, even desired for the decay of interfering radionuclides having shorter half-lives than the indicator radionuclides. During the decay time, the activity obtained at the end of the irradiation decreases. The activity present at the beginning of counting A_c is then:

$$A_c = A_t e^{-\ln 2\, t_d/t_{1/2}} \tag{8.4-9}$$

The SI unit of radioactivity is the becquerel (Bq), defined as 1 disintegration per second.

Figure 8.4-1 illustrates the growth and decrease of the radioactivity with time during irradiation and post-irradiation decay for three indicator radionuclides (^{56}Mn, ^{76}As, and ^{182}Ta) of different half-lives. From this figure, it can be seen that the radioactivity produced increases with the irradiation time t, but as t becomes large relatively to $t_{1/2}$, the decay rate approaches the production rate, and the radioactivity achieves a limiting value called the *saturation activity*. In an irradi-

Fig. 8.4-1. Growth of radioactivity of ^{56}Mn ($t_{1/2} = 2.6$ h), ^{76}As ($t_{1/2} = 26.4$ h) and ^{182}Ta ($t_{1/2} = 114.4$ d) during the irradiation of 1 μg of Mn ($\sigma_{th} = 13.3$ b), As ($\sigma_{th} = 4.3$ b) and Ta ($\sigma_{th} = 21$ b) with neutrons of $\Phi_{th} = 4 \times 10^{13}$ cm^{-2} s^{-1}, and its decrease after the end of irradiation

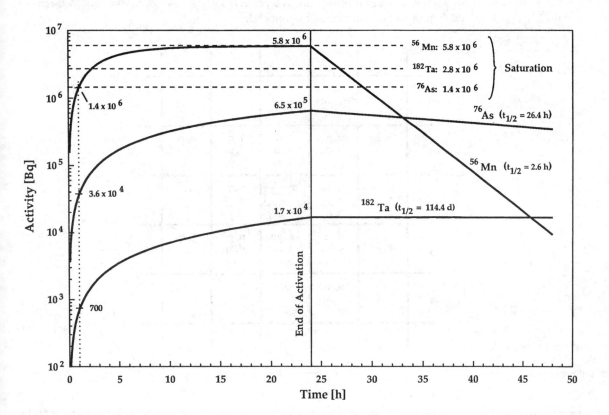

ation for $t = t_{1/2}$ and for $t = 2t_{1/2}$, 50% and 75%, respectively, of the saturation activity is produced, and with further extention of the irradiation time, no significant increase of the radioactivity is obtained. However, this consideration is not applicable to the production of long-lived radionuclides, because the maximum practicable irradiation times for a nuclear reactor is several days to some few weeks, and for an accelerator, several hours. Therefore, the achievable activities represent small fractions of the saturation activity.

Activation Eq. 8.4-8 is applicable only on condition that discrete values can be assigned to the projectile flux Φ and to the reaction cross section σ as is the case in activation with thermal and/or epithermal neutrons of a nuclear reactor (see Sec. 8.4.5), and with 14 MeV neutrons. In activation with fast neutrons produced by bombarding a suitable target with cyclotron-accelerated charged particles or with energy-rich photons, the energy and angular distribution of the projectiles have to be taken into account. Moreover, in activation with charged particles, the degradation of the projectile energy within the irradiated sample has to be considered. Thus, for these activation modes, Eq. 8.4-8 must be correspondingly modified.

Indicator radionuclides

From a stable nuclide, various indicator radionuclides can be produced, depending on the type and energy of the projectile particle used for irradiation.

Regarding the production of a suitable indicator radionuclide for the determination of an element, activation analysis offers a great variability. Various projectile particles can be used to induce nuclear reactions on a target nuclide leading to a large number of possible indicator radionuclides. The most common projectiles are neutrons, followed by charged particles, protons, deuterons, alpha and ^{3}He particles. Figure 8.4-2 shows the positioning of the indicator radionuclides produced by different nuclear reactions on the chart of the nuclides. The nuclear reaction induced, and therefore, also the indicator radionuclide produced, are determined by the type of projectile used and its energy. The larger ΔZ and/or ΔN (Z = atomic number, N = number of neutrons) between the target and product nuclide, the higher the necessary energy of the projectile particles. Most elements are polyisotopic and offer more than one target nuclide. Thus, almost for each element, one has a choice of several sensitive nuclear reactions. However, the choice is often dictated by the availability of the irradiation sources. Because of the special advantages of reactor neutron activation analysis, which are discussed in a separate section below, this is a preferred activation technique provided that it leads to a suitable indicator radionuclide.

ΔZ						
+2		$\alpha, 3n$ ^{3}He, 2n	$\alpha, 2n$ ^{3}He, n	α, n		
+1	p, 2n	p, n d, 2n	d, n t, 2n ^{3}He, pn	t, n α, pn ^{3}He, p	α, p	
0	n, 3n $\gamma, 2n$	n, 2n p, pn γ, n	Target nuclide	n, γ d, p t, pn	t, p	
-1	p, d d, αn	d, α γ, pn	n, pn t, α γ, p	n, p d, 2p	t, 2p	
-2			n, α	$\gamma, 2p$		
	-2	**-1**	**0**	**+1**	**+2**	ΔN

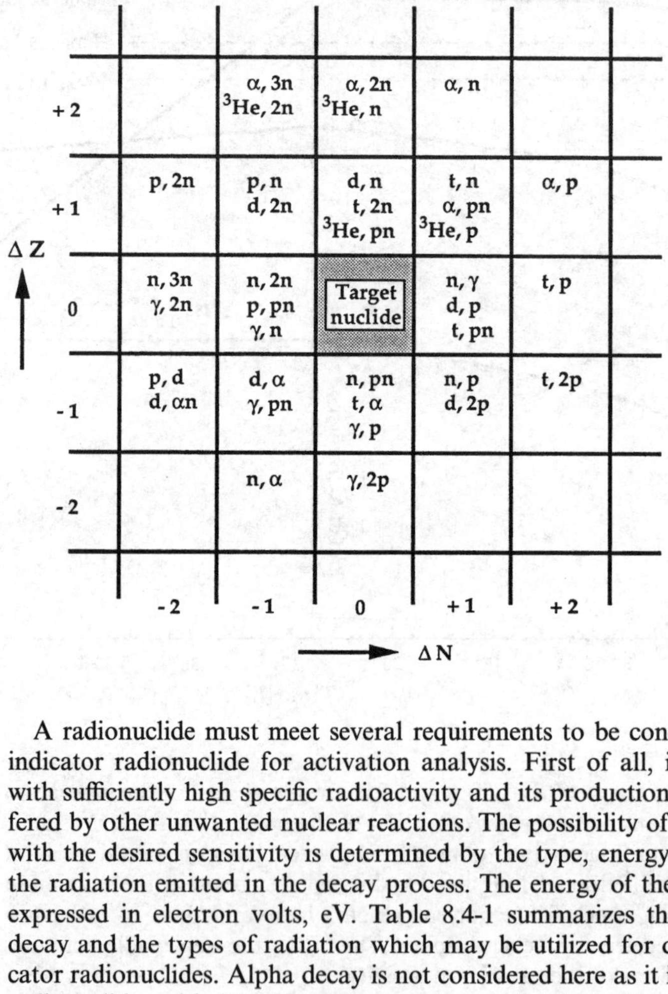

Fig. 8.4-2. Section of the chart of nuclides showing the positioning of products of nuclear reactions of different types

Main factors determining the suitability of an indicator radionuclide are: obtainable activation yield, half-life, and properties of the emitted radiation.

The electron volt eV is a unit of energy.

$1\,\text{eV} = 1.60 \times 10^{-19}\,\text{J}$

$\phantom{1\,\text{eV}} = 3.83 \times 10^{-20}\,\text{calories}$

A radionuclide must meet several requirements to be considered as a suitable indicator radionuclide for activation analysis. First of all, it must be produced with sufficiently high specific radioactivity and its production should not be interfered by other unwanted nuclear reactions. The possibility of its specific detection with the desired sensitivity is determined by the type, energy and the intensity of the radiation emitted in the decay process. The energy of the radiation is usually expressed in electron volts, eV. Table 8.4-1 summarizes the possible modes of decay and the types of radiation which may be utilized for detection of the indicator radionuclides. Alpha decay is not considered here as it is of interest only for radionuclides with $Z > 83$. Beta particles can be detected very easily. However, their continuous energy spectrum precludes specific detection of a radionuclide unless an appropriate chemical separation prior to the counting is carried out. On the other hand, gamma photons are emitted with discrete energies characteristic for each radionuclide. Characteristic X-rays associated with the electron capture and internal conversion process can also be utilized as a means of detection in some instances.

Indicator radionuclides with too short and too long half-lives are disadvantageous.

In activation analysis, it is favorable to produce indicator radionuclides with half-lives at the levels of hours to some tens of days. Although very short half-lives allow the achievement of high activation yields already with short irradiation

Table 8.4-1. Possible decay processes of indicator radionuclides and the types of emitted radiation

Decay process	Possible types of radiation emitted	
	Particles	Photons
Decay with β emission		
$^A_Z X \rightarrow ^A_{Z+1}X + \beta^- + \bar{\nu}$ $(n \rightarrow p + \beta^- + \bar{\nu})$	β-rays (e^-) with continuous energy, mono-energetic conversion electrons, antineutrino $(\bar{\nu})$ with continuous energy	mono-energetic γ-rays, bremsstrahlung with continuous energy, characteristic X-rays
$^A_Z X \rightarrow ^A_{Z+1}X + \beta^+ + \nu$ $(p \rightarrow n + \beta^+ + \nu)$	β-rays (e^+) with continuous energy, mono-energetic conversion electrons, neutrino (ν) with continuous energy	mono-energetic γ-rays, 511-keV annihilation radiation, bremsstrahlung with continuous energy, characteristic X-rays
Electron capture		
$^A_Z X + e^- \rightarrow ^A_{Z-1}X + \nu$ $(p + e^- \rightarrow n + \nu)$	mono-energetic conversion electrons, mono-energetic neutrino (ν)	mono-energetic γ-rays, bremsstrahlung with continuous energy, characteristic X-rays
Isomeric transition		
$^{Am}_Z X \rightarrow ^A_Z X + \gamma$	mono-energetic conversion electrons	mono-energetic γ-rays, bremsstrahlung with continuous energy, characteristic X-rays

times, problems occur with their rapid decay after the end of irradiation. Indicator radionuclides with very long half-lives require long radiation times to achieve the desired activity.

Overview of activation analysis methods and procedures

The sources which provide particles usable for activation analysis and their most important properties are summarized in Table 8.4-2. The projectiles of interest are of three types: neutrons, charged particles, and photons. In addition to the type

Table 8.4-2. Survey of different activation techniques

Projectiles and source	Mode of projectile production	Projectile energy	Typical analytical reaction	Projectile flux $(cm^{-2}\,s^{-1})$
Neutrons				
Nuclear reactor	fission	fission spectrum	(n, γ)	high, $10^{12}-10^{15}$
Neutron generator	$^3T(d,n)^4He$	14 MeV	$(n,p), (n,\alpha)$ $(n, 2n)$	medium, 10^8-10^{10}
Cyclotron	$d + Be \rightarrow n$	variable spectrum	$(n,p), (n,\alpha)$ $(n, 2n)$	high, $10^{10}-10^{12}$
Charged particles $\tau = p, d, t, {}^3He, \alpha$				
Cyclotron, linear accelerator	acceleration	variable	$(\tau,n), (\tau,2n)$ $(\tau,p), (\tau,\alpha)$	high, $10^{13}-10^{14[a]}$
Photons				
Electron accelerator	bremsstrahlung production	variable spectrum	$(\gamma,n), (\gamma,p)$	high, 10^{13}

[a] Number of particles per second; beam intensity of $1\,\mu A = 6.2 \times 10^{12}$ singly charged particles per second

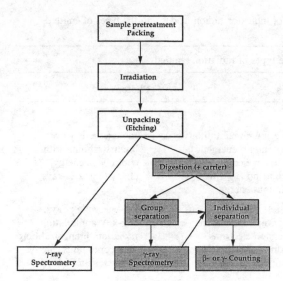

Fig. 8.4-3. Representation of the general scheme of instrumental (left course) and radiochemical (right course) activation analysis

and energy of the projectile, the projectile flux (or, in the case of charged particles, the beam intensity given in microamperes, μA) is also a very important quantity as it determines the analyte content levels to which the activation technique is applicable (see Eq. 8.4-8). The type of projectile particles also determines the maximum volume of sample which can be handled, the uniformity of the activation of the sample, the necessity of cooling the sample during irradiation, and at last the accessibility of the source and the irradiation costs. The most important activation technique, the NAA, is handled in more detail in the Sec. 8.4.5.

The instrumental mode of activation analysis is much simpler, but the radiochemical mode is more powerful and universal.

There exist two procedure modes of activation analysis, i.e., *instrumental* and *radiochemical*, which, in general, follow the scheme shown in Fig. 8.4-3. In instrumental activation analysis, the irradiated sample, after unpacking and surface etching, is directly counted, usually by a high-resolution gamma-ray spectrometer. In the radiochemical mode of activation analysis, the irradiated sample is digested and the indicator radionuclides of interest are separated from the radionuclide mixture in one or more fractions which are then counted. Due to its simpler performance, the instrumental procedure is always the first choice whenever it is adequate for solving a given problem. The elemental composition of the sample, in particular, the activation behavior of the matrix elements as well as the counting method determine whether the instrumental performance is possible or not.

Characteristic features of neutron activation analysis in relation to other trace analytical methods

As some of the features discussed below can more or less differ from one activation technique to the other, only NAA using a nuclear reactor is considered here, the application of which amounts to more than 90% of total activation analyses. In order to illustrate the role of NAA in trace element analysis, it is meaningful to compare its potential, general problems and limitations, especially the possible sources of errors and achievable accuracies, with those of the competitive methods.

The solution methods including electrothermal atomic absorption spectrometry (ETAAS), inductively coupled plasma atomic emission spectrometry (ICP-AES), mass spectrometry (ICP-MS), and some other methods normally require sample decomposition and often also analyte/matrix separation. The main limiting factor of these methods when applied to the determination of the so-called contamination risk elements is the *blank*.

Freedom of reagent blank, easy standardization, and high sensitivity together make neutron activation analysis a unique method regarding achievable limits of detection and accuracy.

Regarding the blank, activation analysis represents a unique method, even when radiochemical separations are involved. As introduction of contamination is no longer possible after the irradiation step, the analysis of solid sample pieces is completely free of blank. When applied to analysis of samples which cannot be etched, such as many environmental and biological samples, possible contamination during sampling, sample pretreatment and storage, however, must be taken into account. At any rate no blank can be introduced from reagents used for digestion and separation. At present, this can be considered as the greatest advantage of activation analysis in comparison with other methods. In NAA, highly accurate calibration can be performed by simultaneous irradiation of samples and synthetic multielement standards. This is unique for activation analysis; no other method offers the possibility of simultaneous "excitation" of sample and standard. For these two reasons, of all methods for the determination of trace elements, activation analysis best meets the stringent requirements for obtaining accurate results even at extremely low content levels – ng/g and below.

The advantages and disadvantages of NAA can be summarized as follows:

Advantages
- Freedom of blank
- Possibility to remove surface contamination
- High degree of selectivity
- Easy calibration
- Universality (about 70 elements detectable)
- Multielement character
- Nondestructive character of instrumental NAA

Disadvantages
- Dependence on large irradiation facilities
- Necessity of handling radioactive material
- In some instances long irradiation and decay times

8.4.3 Methods for radiation detection and measurement

Apart from the activation process itself, specific detection of the indicator radio-nuclides is of fundamental importance for activation analysis. As was already discussed in Sec. 8.4.2, among the different types of radiation which may be emitted in the decay process of indicator radionuclides (see Table 8.4-1), gamma rays are best suited for their identification and quantification. This has stimulated the development of *gamma-ray spectrometry* – nowadays the basic counting technique in each activation analysis laboratory. *X-ray spectrometry* can be considered as a sometimes interesting extention of the gamma-ray spectrometry into the low-energy region. Counting of beta particles requires individual separation and, therefore, use of it is made only when the indicator radionuclide emits no gamma rays and X-rays or these are of very low intensity.

All the methods applicable to the detection of gamma, beta and X-rays are based on interactions of these radiations with matter. Table 8.4-3 surveys the

A high-resolution gamma-ray spectrometer provides selectivity by its ability to discriminate among gamma-rays of different energies. Therefore, in activation analysis, it is the superior measurement technique for identification and quantification of indicator radionuclides.

Table 8.4-3. Survey of the detection techniques used in activation analysis and their typical performance parameters

Type of detector	Time resolution	Energy resolution	Common application	Importance (application frequency)
Semiconductor detectors				
Coaxial	2–10 ns	1.6–2.2 keV at 1332 keV (0.12–0.17%)	γ-ray spectrometry	high
Planar	2–10 ns	0.55–0.8 keV at 5.9 keV (9.3–13.6%)	low energy γ-ray and X-ray spectrometry	medium
NaI(Tl)-scintillation detectors	0.25 μs	50–75 keV at 1332 keV (4.5–5.5%)	γ-counting	low
Liquid scintillation detector	2–5 ns	5–8%	β-counting with water solutions	low
Geiger-Müller counter	100 μs	no	β-counting on solid samples	very low

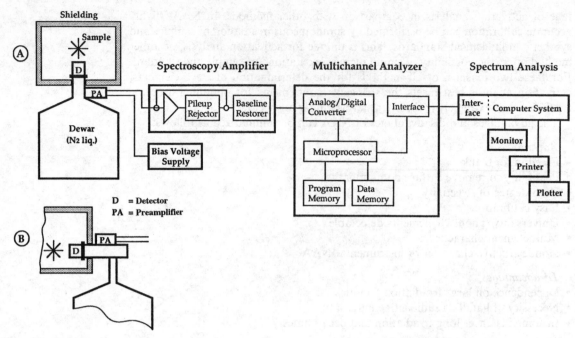

Fig. 8.4-4. Schematic diagram of a high-resolution gamma-ray spectrometer system. Detector positioning: (A) vertical, (B) horizontal

common detection techniques along with their most important performance parameters. The excellent time and energy resolution of the semiconductor detectors make their position in gamma-ray spectrometry competition-free. At present, instrumental activation analysis is performed exclusively, and radiochemical activation analysis to more than 90% with high-resolution gamma-ray spectrometers. Detailed descriptions of all methods for radiation detection and measurement are found in an excellent book by Knoll [8.4-4].

High-resolution gamma-ray spectrometry

The semiconductor detector and the electronic signal processing system are the two principal components of a gamma-ray spectrometer.

A high-resolution gamma-ray spectrometer basically consists of a semiconductor detector and an electronics system to process the pulses produced. A block scheme of a modern high-resolution gamma-ray spectrometer is shown in Fig. 8.4-4.

Germanium gamma-ray detectors

In recent years, there has been an increasing trend towards the use of *high purity* or *intrinsic germanium (HPGe) detectors* instead of *lithium-drifted germanium (Ge(Li)) detectors* which are meanwhile no longer commercially available. HPGe detectors are fabricated from germanium with impurity levels as low as 10^9 atoms/cm^3 (10^9 foreign atoms/4.4×10^{22} Ge atoms). The remaining low-level impurities act either as acceptors leading to *p-type material* or as donors with resulting *material of n type*. HPGe detectors are available in planar and coaxial configurations shown with the possibilities of their semiconductor junctions in Fig. 8.4-5. Also illustrated in this figure is the structure of the detector surface at the n$^+$ and p$^+$ contact. HPGe detectors are almost exclusively operated as *fully depleted detectors* by means of a reverse bias in which a positive voltage is applied to the n$^+$ contact with respect to the p$^+$ contact. The effect of the reverse bias is the extention of the depletion region, which is equal to the *sensitive volume* of the detector, to practically the whole detector volume. The created depletion region strongly resists the passage of electric current, and the difference in electric potential gives rise to an electric field extending over the range of the depletion region.

The active volume of the semiconductor detector determines the counting efficiency.

Under these conditions, the depletion region exhibits excellent properties for detection of gamma rays. When a gamma ray enters the depletion region, a primary electron can be produced by photoeffect, Compton scattering, or electron-pair production (for a more detailed explanation see the section "Gamma-Ray Spectra" below). In turn, each primary electron creates, by its passage within the depletion zone, electron-hole pairs which will be drawn out of the depletion region

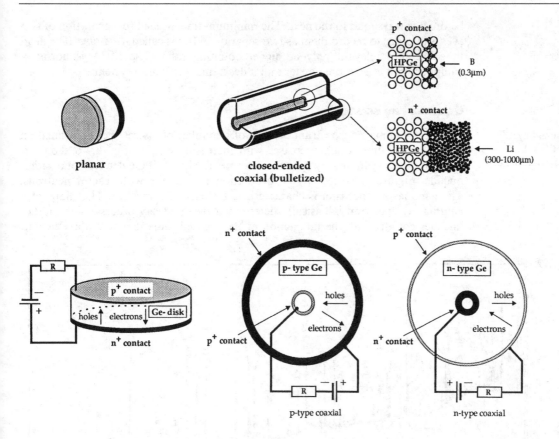

Fig. 8.4-5. Common configurations of HPGe detectors (top row) and their semiconductor junctions (bottom row)

by the electric field causing a basic electrical signal. Large sensitive volumes as needed in gamma-ray spectrometry are achievable with *coaxial geometry*, mostly fabricated in the *closed-ended coaxial configurations*, as illustrated in Fig. 8.4-5. At present, HPGe detectors with active volumes up to 600 cm^3 are commercially available providing excellent efficiencies. The maximum efficiency is obtained with the well configuration which involves placing the sample inside the detector. For weakly penetrating radiations such as low-energy gamma rays and X-rays, the best suited detectors are those with *planar configuration*.

The detectors are housed in an aluminium casing under vacuum and operated at the temperature of liquid nitrogen deposited in a Dewar, mainly to reduce the thermally induced leakage current. There are two principal positions of the detectors which are shown in Fig. 8.4-4. In order to reduce the background of external radiation, the detectors are protected from this radiation by using an effective shielding.

Electronic signal processing

The first element in the signal-processing electronic chain is the *preamplifier*, followed by the amplifier. Their main functions are to improve the signal-to-noise ratio and to provide the amplification of the small detector signal.

Another indispensible component of a gamma-ray spectrometer system is the *multichannel analyzer* (*MCA*). It consists of an *analog-to-digital converter* (ADC) and a *memory* section. The ADC converts an analog signal, i.e., the pulse height, to an equivalent digital binary number, which addresses the appropriate location (channel) of a computer-type memory. This event is registered in the memory channel as one count. In the memory part of an MCA, each channel represents a small specific incremental range of the incident gamma-ray energy, ΔE_γ. The numbers of channels can vary between 512 and 16384 (increase by a power of 2), the latter being most frequently used for gamma-ray spectrometry applied to multielement analysis. For example, if one channel represents an energy range ΔE_γ of 1 keV, a memory with 4096 channels would enable to record a gamma spectrum in the energy range of 4096 keV. The maximum content of a memory channel ranges up to about 2×10^9 counts. Each gamma-ray spectrometer system needs a certain time for processing the event induced by one incoming gamma ray before it can respond to the next. The minimum time needed for separation of two events in order to record them as two separate pulses is called *dead time*. The dead time becomes relevant only at higher counting rates. The MCA is normally equipped with a dead time meter and a dead time correction system.

Gamma-ray spectra

For a given detector, the gamma-ray spectrum is the "visiting card" of the radiation emitted by the source measured.

A typical gamma-ray spectrum consists of a number of discrete peaks situated on a background line which decreases with increasing energy. Fig. 8.4-6 shows as an example the gamma-ray spectrum measured with a HPGe detector of a radio-nuclide mixture produced by the irradiation of graphite with reactor neutrons. The gamma-ray spectrum is characteristic for each radionuclide. The shape of a gamma-ray spectrum is basically determined by the decay process of the radio-nuclides counted and the interaction of the emitted radiations with the detector.

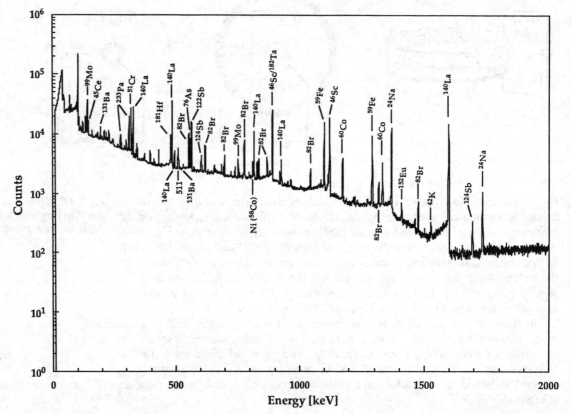

Fig. 8.4-6. Gamma-ray spectrum of a neutron irradiated high-purity graphite

The lowest energy state of a nucleus is called its ground state. In the excited state, a nucleus is promoted to a higher energy state. Most frequently, the product nucleus resulting from an alpha- or beta-decay process is in an excited state, which then returns to the ground state in one or more discrete steps by emission of gamma-rays.

Thus, for the interpretation of gamma spectra as obtained by multichannel pulse-height analyzers, knowledge of the radiations emitted in the decay and their interaction mechanisms in the detector is necessary.

Gamma-rays are emitted by excited nuclei in their transition from an *excited state* to the *ground state*, mostly via several intermediate excited levels. Due to an excess of neutrons or protons, radionuclides undergo decays by emission of neg-ative electrons (β-radiation) or positrons (β^+ radiation), respectively. The resulting daughter nuclei are mostly formed in the excited state. Only for a very few radio-nuclides, for example ^{32}P, ^{35}S and ^{204}Tl, the daughter nuclei are created in the ground state by pure beta decay. Therefore, normally, the emission of a number of gamma rays of different well-defined energies, typically in the range 0.05–4 MeV, and intensities (0–100%) accompanies the preceding beta decay. In some nuclei, prompt gamma transitions are "forbidden" and a *metastable state* or *nuclear iso-mer* is formed. This is a pure gamma emitter and decays with its own half-life. An alternative deexcitation to gamma decay is *internal conversion* whose probability increases with increasing atomic number and decreasing excitation energy. In this deexcitation mode, no gamma ray is emitted, and the excitation energy is trans-ferred to an orbital electron, preferably a K electron, which is ejected from the atom. Thus, internal conversion is always followed by emission of characteristic X-rays.

The following processes of interaction of gamma rays with the detector are essentially determining the shape of a spectrum: *photoelectric effect, Compton effect,* and *pair production*. They are illustrated in Fig. 8.4-7 along with an idealized spectrum showing the contribution of the individual types of interaction to the spectrum formation. The photoelectric effect predominates in the low-energy region and its probability decreases rapidly with increasing gamma-ray energy. The probability of the Compton effect decreases slowly with the photon energy and the probability of the pair-production process increases rapidly above the threshold photon energy of 1.02 MeV.

The photoelectric effect represents the principal interaction process in the HPGe detector. In the photoelectric effect, the incident gamma-ray photon ejects a bound electron, called a *photoelectron*, from an atom, and imparts to it the total energy less the energy with which the electron was bound (the binding energy). The binding energy is normally negligible in comparison with the gamma-ray energy. If the kinetic energy of the photoelectron is fully consumed for the secondary ionization process in the sensitive volume of the detector, then the height of the resulting pulse corresponds to the original energy of the gamma-ray photon. The signals created from gamma rays of a given energy in this way appear in a gamma spectrum as narrow peaks called *photopeaks*. However, it should be realized that, mainly but not exclusively, photoelectric events contribute to the formation of this peak. In fact, more complex processes involving the Compton effect and pair production may lead to the total absorption of the gamma-rays in the detector, too, producing pulses identical with those originating from the photoeffect. Therefore, the more correct name of this peak is *full-energy peak*.

In the Compton effect, only a part of the incident gamma-ray energy is transferred to an electron and the remainder is a scattered gamma-ray photon. The division of the original energy between the electron and the scattered photon depends on the angle at which they are emitted. The scattering angle Θ can occur between 0–180°, and consequently the energy of the electron can range from zero at $\Theta \simeq 0$ up to a maximum at a head-on collision in which $\Theta = 180°$. Recording these Compton scattered electrons results in the production of a broad continuous background called the *Compton continuum* which ranges from zero up to the

For interpretation of gamma-ray spectra, in addition to the decay schemes of the radionuclides, the interaction processes of the radiations in the detector also have to be considered.

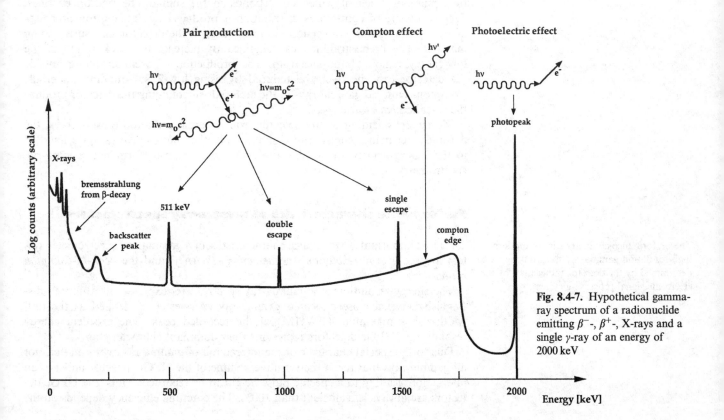

Fig. 8.4-7. Hypothetical gamma-ray spectrum of a radionuclide emitting β^--, β^+-, X-rays and a single γ-ray of an energy of 2000 keV

maximum energy on the Compton continuum at the *Compton edge*. In a spectrum of a number of gamma rays of different energies, the Compton continuum increases with decreasing spectrum energy because of the summation effect. The Compton continuum may often seriously limit the performance of instrumental activation analysis by gamma-ray spectrometry. The Compton-scattered photon can subsequently undergo further interaction, i.e., a photoelectric or Compton effect at another location within the sensitive volume of the detector. If nothing escapes from the detector in subsequent multiple interactions, the resulting pulse cannot be distinguished from the pulse caused by the photoelectric effect in terms of the time of its appearance and its height and, therefore, it contributes to the full-energy peak. The probability of this contribution increases with increasing volume of the detector.

In the *pair-production process*, the photon loses the total energy by production of an e^-–e^+ pair in the Coulomb field of a nucleus. This transformation is only possible if the energy of the incident photon exceeds 1.02 MeV which is twice the rest mass equivalent of an electron. Any energy above this threshold appears as kinetic energy of the e^-–e^+ pair. The positron created in this way will normally be slowed down and annihilated simultaneously with an electron in the detector, producing two *annihilation 0.511 MeV photons*. These photons may be absorbed in the detector or escape. The escape of one or both annihilation photons gives rise to the appearance of a *single escape peak* at the energy of $E_\gamma - 0.511$ MeV or a *double escape peak* at $E_\gamma - 1.02$ MeV. The absorption of the annihilation photons results in a spectral peak at 0.511 MeV. However, annihilation 0.511 MeV photons of much higher intensity are produced also from positions emitted from radionuclides (see Table 8.4-1). Also in this case, large detectors increase the probability for both annihilation photons to be absorbed in the detector by subsequent interactions resulting in contribution to the full-energy peak.

X-ray peaks originating from electron capture or internal conversion processes are often used for the determination of indicator radionuclides emitting no suitable gamma rays. However, it must be taken into account that beta, gamma and X-rays from the decay process may excite atoms in the sample and in material surrounding the detector to emit their characteristic X-rays.

There are also other interactions and effects which may affect the shape of the resulting gamma-ray spectrum. Additional peaks, the so-called *sum peaks*, occur due to coincident detection of two or more gamma rays of a decay cascade leading to a pulse of a height which corresponds to the sum of the photon energies. *Bremsstrahlung* of continuous distribution is produced by deceleration processes of beta particles and conversion electrons in both the detector and surrounding material. The bremsstrahlung can significantly increase the background in the low-energy region of the spectrum. The production of bremsstrahlung outside the detector can effectively be reduced by using low-Z surrounding materials. Compton-scattered gamma-rays in the materials surrounding the detector produce the so-called *backscatter peak*.

Moreover, shielding of the detector from external radiation (cosmic rays, the naturally occurring gamma-ray emitters and other radioactive samples handled in the laboratory) is an absolute necessity for a high-performance gamma-ray spectrometry.

Performance characteristics of gamma-ray spectrometers

The performance characteristics of modern high-resolution gamma-ray spectrometers are determined by the detector rather than by the electronic signal processing system.

The most important performance characteristics of a gamma-ray spectrometer system are the *energy resolution*, the *detection efficiency*, and the *peak-to-Compton ratio*.

The energy resolution of a gamma-ray spectrometer expresses its ability to distinguish between closely located gamma-ray energies. It is defined as the full-width-at-half-maximum (FWHM) of the recorded peak. The excellent energy resolution of HPGe detectors represents their dominant characteristic.

Due to the special character of the interaction of gamma rays with matter, not all gamma rays that reach the sensitive volume of the HPGe detector undergo an interaction, leading to a specific pulse height to be recorded. Thus, the HPGe detectors are always less efficient than 100%. The counting efficiency depends on the size of the detector, on the energy of the counted gamma ray, and on the geometrical arrangement of source and detector.

A further performance characteristic of the detector quoted in the specification is the peak-to-Compton ratio. It refers to the ratio of the height of the ^{60}Co 1332 keV peak to the average Compton plateau between 1040 and 1096 keV. The value of the peak-to-Compton ratio depends mainly on the energy resolution. The higher this ratio the better the detector.

Other counting methods for activation analysis

In two special and relatively rare cases in activation analysis, use is made of a complementary counting technique.

It is often advantageous to count gamma rays of either an indicator radionuclide individuum or of a simple mixture of radionuclides using a well-type NaI(Tl) scintillation detector system since this detector system provides higher counting efficiency in comparison with HPGe detectors. However, due to a rather poor energy resolution, closely spaced gamma-ray energies remain unresolved. Therefore, the NaI(Tl) scintillation detector is mainly used in connection with counting systems on the basis of a *single-channel analyzer* (*SCA*). By means of independently adjustable discriminators (a lower-level and an upper-level discriminator), the desirable window width corresponding to the wanted energy interval can be selected. Pulses passing the SCA are then recorded.

In some cases, the indicator radionuclides emit no suitable gamma or X-rays and, therefore, after a specific separation, beta rays have to be counted. For this purpose, *organic scintillators* are the preferred detectors. In these scintillators, the fluorescence process occurs as a result of π-electron transitions within a single molecule. Organic scintillators can be either liquid or solid. The *liquid scintillation counting* avoids some of the difficulties arising in other counting techniques. Liquid scintillators are prepared by dissolving an organic scintillators in an appropriate solvent. In the normally used internal source liquid scintillation counting, the sample is dissolved in the liquid scintillator and directly counted. In this counting mode, the counting efficiency is almost 100%.

Radiation measurement methods other than high-resolution gamma-ray spectrometry are used only in special situations.

Counting statistics

Radioactive decay is a random process. Consequently, any counting rate obtained by measuring the radiation emitted in radioactive decay shows some degree of statistical fluctuation. In all radioactivity measurements, this fluctuation has to be considered as an unavoidable source of uncertainty. The best estimate of the standard deviation σ of x counts is:

$$\sigma = \sqrt{x} \tag{8.4-10}$$

The relative standard deviation $\tilde{\sigma}$ (%) is given by

$$\tilde{\sigma} = \frac{\sigma}{x} \times 100 = \frac{\sqrt{x}}{x} \times 100 = \frac{1}{\sqrt{x}} \times 100 \tag{8.4-11}$$

While the absolute standard deviation increases with the number of counts, the relative standard deviation decreases. For instance, if 100, 10^3 and 10^4 counts are measured, the absolute standard deviations are ± 10, ± 31.6, and ± 100 counts, and the corresponding relative standard deviations are 10%, 3.2%, and 1%.

Usually, results of radioactivity measurements are expressed as counting rates R:

$$R = \frac{x}{t} \tag{8.4-12}$$

The standard deviations of the counting rate σ_R and $\tilde{\sigma}_R$ can be obtained from:

$$\sigma_R = \frac{\sqrt{x}}{t} = \frac{\sqrt{Rt}}{t} = \sqrt{\frac{R}{t}} \quad \text{and} \quad \tilde{\sigma}_R = \frac{\sigma_R}{R} \times 100 \tag{8.4-13}$$

Thus, for a given counting rate R, the standard deviation is inversely proportional to the square root of the counting time.

The counts recorded always include a contribution from other sources than the sample, called background. In most practical cases, the contribution of the background x_b is significant and must be subtracted from the total counts x_t in order to obtain the net counts of the sample x_n. If we assume equal counting time, then:

$$x_n = x_t - x_b \tag{8.4-14}$$

The standard deviation of the net counts is then:

$$\sigma_n = \sqrt{\sigma_t^2 + \sigma_b^2} \tag{8.4-15}$$

and, analogously, the standard deviation of the net counting rate is:

$$(\sigma_R)_n = \sqrt{(\sigma_R)_t^2 + (\sigma_R)_b^2} \tag{8.4-16}$$

or

$$(\sigma_R)_n = \sqrt{\frac{R_t}{t_t} + \frac{R_b}{t_b}} \tag{8.4-17}$$

In Eqs. 8.4-14 to 8.4-17, the indices n, t, and b refer to net, total, and background counts, respectively. The lower the radioactivity of the sample, the larger the effect of the background on the relative standard deviation in counts and count rates.

Worked example:

In counting of two samples and background, following data were recorded: sample 1 yielded 4127 counts for 10 min, sample II 24 894 counts for 5 min and the background measurements 2276 counts for 10 min. Calculate the absolute and relative standard deviations of the net counting rates of sample I and sample II.

From the recorded counts, using Eq. 8.4-12, the following counting rates in counts per second (cps) are obtained

$$\text{sample I:} \quad R = \frac{4127}{10 \times 60} = 6.88 \, \text{cps}$$

$$\text{sample II:} \quad R = \frac{24894}{5 \times 60} = 82.98 \, \text{cps}$$

$$\text{background:} \quad R = \frac{2276}{10 \times 60} = 3.79 \, \text{cps}$$

Applying Eq. 8.4-17, we obtain for sample I the absolute standard deviation

$$(\sigma_R)_n = \sqrt{\frac{6.88}{10 \times 60} + \frac{3.79}{10 \times 60}} = 0.133 \, \text{cps}$$

and the relative standard deviation

$$(\tilde{\sigma}_R)_n = \frac{0.133}{6.88 - 3.79} \times 100 = 4.3\%$$

For sample II, the absolute and the relative standard deviation is

$$(\sigma_R)_n = \sqrt{\frac{82.98}{5 \times 60} + \frac{3.79}{10 \times 60}} = 0.532 \, \text{cps}$$

$$(\tilde{\sigma}_R)_n = \frac{0.532}{82.98 - 3.79} \times 100 = 0.67\%$$

Thus, we see that the precision increases with increasing number of counts registered.

8.4.4 Radiochemical separations

The background constituted by Compton continua and sometimes also by bremsstrahlung in gamma spectra of complex radionuclide mixtures, normally produced in irradiations of real samples, may considerably limit the applicability of gamma-ray spectrometry to direct instrumental activation analysis. The major contributions to the background are usually caused by the high-activity radionuclides produced from the matrix or sometimes from one or several main impurities. Small peaks of the low-activity indicator radionuclides cannot be statistically recognized in the fluctuations of the high background. Thus, the background may drastically impair the limits of detection. For these reasons, *radiochemical separations* are often unavoidable. A radiochemical separation is always a necessity when the indicator radionuclide is a pure beta emitter.

The common criteria for the choice of a separation technique can be applied also to radiochemical separations for activation analysis. However, with short-lived indicator radionuclides, high speed becomes a decisive requirement. When matrix-produced radionuclides of very high activity are to be separated from low-activity indicator radionuclides, high separation factors, up to 10^6, are required. Although all the common separation techniques used in trace analysis, and a few special ones as well, can be applied to radiochemical activation analysis, ion exchange and some other column chromatographic techniques, and solvent extraction proved to be most important.

In some cases, activation analysis can only be used when a separation or *preconcentration prior to irradiation* is performed. In this working mode, however, activation analysis is losing its unique feature of freedom of blank and other more suitable methods are being preferred.

When radionuclides of dominating radioactivity give rise to a high background in the gamma-ray spectrum, instrumental activation analysis is not practicable. These radionuclides have to be chemically separated from the indicator radionuclides of interest.

Group separations

It is often sufficient *to remove the main-activity radionuclides* by a suitable single separation and to rely on gamma-ray spectrometers for simultaneous counting of the remaining indicator radionuclide mixture. As an example we can consider the application of NAA to the determination of trace impurities in sodium-based salts such as sodium carbonate and in biological materials with higher sodium contents. The radionuclide ^{24}Na ($t_{1/2} = 15\,\mathrm{d}$) produced in an irradiation with reactor neutrons with high activities prevents the instrumental performance of the analysis via short- and medium-lived indicator radionuclides, and must, therefore, be removed prior to counting the irradiated sample with a gamma-ray spectrometer. In many instances, the detection limits can further be improved by separating the mixture of indicator radionuclides into several groups, each of which is then separately counted. Some examples of usefulness of group separations will be given in Sec. 8.4.6 dealing with the application of NAA.

Individual separations

Maximum efficiency can be attained in activation analysis in terms of limit of detection and accuracy when the respective indicator radionuclide is selectively separated from other radionuclides. Its activity can then be measured with a simple but efficient detector system using, for example, a sodium iodide well-type crystal or liquid scintillation counter. This type of procedure is capable of exploiting to the full the extremely high sensitivity potential of activation analysis.

The separation of the desired indicator radionuclide is assisted by addition of a known amount (0.1–10 mg) of the same element in inactive but equal chemical form as carrier. Thus it is possible to separate a relatively high and fixed amount of the element in question assuming that the amount of the element in the sample is negligibly small in comparison to the amount of carrier added. After chemical equilibration, the element in question is separated from the radionuclide mixture. This separation need not be quantitative as long as the yield is determined by some conventional chemical methods such as gravimetry, titrimetry, or spectro-

Selective separation of one single indicator radionuclide allows one to closely approach the theoretical limits of detection, but with this mode, activation analysis loses its multielement character and becomes time consuming.

photometry. Activation analysis combined with individual separation and determination of the carrier recovery yields is capable of giving correct results even for very low contents of elements, at or below the ppb level.

The necessity of determining the yields can be avoided by making use of the *substoichiometric separations*. In this approach, the sample and the standard are irradiated simultaneously and are then treated in exactly the same way. After the irradiation and dissolution, the same amount of inactive carrier is added to both the sample and the standard solutions and then a fixed amount is recovered following the addition of an equal and substoichiometric amount of reagent to each solution. The count rates of the two portions are then measured under the same conditions and the amount of the element present in the original sample is calculated on the basis of the measured activities without the need for a yield determination. Most of the substoichiometric procedures involve liquid-liquid extraction of metal chelates using complexing reagents to effect the separations, but some procedures use other separation techniques including precipitation, ion exchange, and electrolytic deposition with equal success.

8.4.5 Activation with reactor neutrons (NAA)

Nuclear reactors and irradiation techniques

The most powerful and the lowest-price activation method is provided by nuclear reactors, and is based on the neutron capture (n, γ) reactions induced by thermal and epithermal neutrons. This technique is, therefore, most widely used for activation analyses.

Nuclear reactors are the most intense sources of neutrons. For the purpose of activation analysis, *research reactors* of several designs are used. Figure 8.4-8 shows a schematic cross section of a research reactor. The reactor core consists of

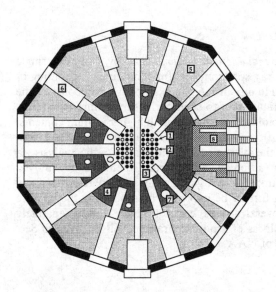

Fig. 8.4-8. Horizontal section of a nuclear reactor: reactor core consisting of fuel rods (1), and control rods (2), tank with water or heavy water (3), graphite reflector (4), biological shield (5), horizontal (6) and vertical (7) experimental channels, and a thermal column (8)

fuel rods (1) of uranium enriched in ^{235}U, which undergoes fission with thermal neutrons, and of absorber rods (2) made of a material capable of strongly absorbing neutrons (cadmium, boron) so to control the fission process via the rate of production of neutrons. In this fission process, two or three fast neutrons are released. The core is surrounded with a tank (3) filled with water or heavy water. The water serves two functions: to slow down the fast neutrons and to cool the fuel rods. A reflector (4), usually made of graphite, is installed around the core to reduce the loss of neutrons escaping from the core by reflection. The shielding (5), normally concrete, protects the reactor surroundings from the intensive radiation of the core. A number of experimental channels, which may be arranged horizontally (6) and vertically (7), lead to irradiation positions either in the reactor core or outside of it providing different neutron fluxes of different energy distributions. In many larger reactors, a thermal column (8) in a graphite block is available providing quasi-pure thermal neutrons of moderate fluxes.

Most research nuclear reactors provide neutron fluxes in the range 10^{12}–10^{14} cm^{-2} s^{-1} and some up to 10^{15} cm^{-2} s^{-1}. The energy spectrum of the fission neutrons at the usual irradiation position in or near the reactor core has a continuous character and includes *thermal neutrons* ($E_n < 0.5$ eV), *epithermal* or *resonance neutrons* (E_n between 0.5 eV and about 1 keV) and *fast neutrons* (E_n between 0.5 and 15 MeV). Normally, the ratios of thermal and epithermal neutron flux Φ_{th}/Φ_{epi} vary in the range 10–100 and those of Φ_{th}/Φ_f in the range 10–200.

In general, the low-energy thermal and epithermal neutrons undergo exoergic neutron-capture (n, γ) *reactions*:

$$^M_Z A(n, \gamma)^{M+1}_Z B$$

(M = mass number, Z = atomic number), for example:

$$^{59}Co(n, \gamma)^{60}Co \quad (Q = +7.5 \text{ MeV})$$

whereas the fast neutrons induce predominantly the threshold reactions of (n, p), (n, α) and $(n, 2n)$ type:

$$^M_Z A(n, p)^M_{Z-1} B, \quad \text{e.g.,} ^{59}Co(n, p)^{59}Fe \quad (Q = -0.8 \text{ MeV})$$
$$^M_Z A(n, \alpha)^{M-3}_{Z-2} B, \quad \text{e.g.,} ^{59}Co(n, \alpha)^{56}Mn \quad (Q = +0.3 \text{ MeV})$$
$$^M_Z A(n, 2n)^{M-1}_Z B, \quad \text{e.g.,} ^{59}Co(n, 2n)^{58}Co \quad (Q = -10.5 \text{ MeV})$$

Apart from a few exceptions, the analytical reactions induced by thermal and epithermal neutrons are of (n, γ) type. Therefore, for NAA, high flux ratios Φ_{th}/Φ_f are advantageous. Due to the existence of thermal and epithermal neutrons in the reactor neutron spectra, Eq. 8.4-4 obtains the form:

$$R = n_A(\Phi_{th}\sigma_{th} + \Phi_{epi}I) \tag{8.4-18}$$

where Φ_{th} and Φ_{epi} are the thermal and epithermal neutron flux, respectively, σ_{th} is the *thermal neutron activation cross section*, and I is the *resonance integral*. The cross sections σ_{th} and I vary considerably from one target nuclide to the next. Values for σ_{th} lie between 2.4×10^{-5} barns for the reaction $^{15}N(n, \gamma)^{16}N$ and 5300 barns for $^{151}Eu(n, \gamma)^{152}Eu$. Only a few fast neutron induced *threshold reactions* of (n, p)-, (n, α)- and $(n, 2n)$-type are utilized as analytical reactions, but they often must be considered as primary interference reactions (see "Nuclear Interferences" below). Fast neutron reaction rates can be calculated using average cross sections $\bar{\sigma}_f$ which are, in general, at the 10 millibarn level.

Sample portions usually between 50 mg and about 1 g are weighed into irradiation vials made of a high-purity material, most frequently high-pressure polyethylene and quartz. Polyethylene can be used if the irradiation time is up to about 2 h and the temperature of the irradiation position is below about 80 °C; quartz is used for longer irradiations. For medium-long and long irradiations, a larger set of vials with samples and standards are placed into an appropriate capsule for simultaneous irradiation.

The methodology of reactor neutron activation analysis is well-established [8.4-5 and 8.4-6]. If the character of the sample is roughly known, it is possible to make good estimates of the optimum performance mode, the obtainable limits of detection, and possible nuclear interferences.

Calibration in NAA

For quantitative analysis, several calibration methods can be used, which, however, provide different degrees of accuracy.

In the *absolute method*, the amount of an element is determined using slightly modified Eq. 8.4-8 and 8.4-18, where the mass of the element is the unknown. The absolute activity is obtained from the measured counting rate, the absolute detector efficiency and the absolute intensity of the gamma ray counted. The neutron flux can be measured using a suitable flux monitor such as cobalt or gold. Although the absolute method requires by far the lowest expenditure of all calibration

In neutron activation analysis, standards and samples are activated simultaneously.

methods, it is advisable to make use of it only if a lower accuracy of the result, caused mainly by the uncertainties in neutron flux and nuclear data, is acceptable.

In the *relative method*, the sample is irradiated simultaneously with multielement standards covering all elements of interest. Generally, synthetic standards prepared from high purity substances are used. Taking into consideration the decay properties of the indicator radionuclides, the elements will be combined into groups and each group given in a vial. As samples and standards are irradiated and counted under quasi-identical conditions, this method has proved to be the most reliable and provides excellent accuracy. No exact knowledge of nuclear data, neutron flux and its energy distribution, irradiation time, and detector efficiency is required. The unknown amount of an element in the sample can be calculated by the simple equation:

$$m_{sam} = m_{std} \frac{R_{sam}}{R_{std}} \tag{8.4-19}$$

where m_{sam} and m_{std} are the masses of the element in sample and standard, respectively, and R_{sam} and R_{std} are the counting rates of sample and standard. A disadvantage of this method is the relativley high time consumption for multi-element analyses.

Several approaches have been tried to develop calibration methods enabling the quantification of all determinable elements by using only one, two or three elements as comparators. These efforts have cumulated in establishing the so-called k_0-*method* that has become most widely used. Although k_0 factors are experimentally determined constants, they actually contain nuclear constants needed in the absolute method. The k_0 factors have been determined and published for practically all elements determinable by NAA. Using the k_0-method, a few additional parameters required can be determined by simultaneous irradiation of gold and zirconium with the sample. The k_0-method is excellently described in several books, e.g., [8.4-6] and papers, e.g., [8.4-7].

Worked example:

Arsenic is determined in high-purity silicon by NAA via the $^{75}As(n,\gamma)^{76}As$ reaction. For this purpose, the silicon sample and standard are irradiated in a nuclear reactor.

The experimental parameters and the physical constants are as follows:

sample mass	$= 0.2\,g$
thermal neutron flux, Φ_{th}	$= 9 \times 10^{13}\ cm^{-2}\,s^{-1}$
epithermal neutron flux, Φ_{epi}	$= 1.6 \times 10^{12}\ cm^{-2}s^{-1}$
irradiation time, t	$= 24\,h$
atomic weight of arsenic, A_w	$= 75\,g\,mol^{-1}$
isotopic abundance of ^{75}As, f_a	$= 1$
thermal neutron cross section, $\sigma_{th} = 4.3\,b$	$= 4.3 \times 10^{-24}\ cm^2$
resonance integral, $I = 60\,b$	$= 60\,b = 60 \times 10^{-24}\ cm^2$
half-life of ^{76}As, $t_{1/2}$	$= 26.3\,h$

a) Considering the detector efficiency, the background and the decay time, it was found that the minimum detectable ^{76}As radioactivity is 5.5 Bq. Estimate the limit of detection m_{LOD} of arsenic for the above given experimental conditions.

For this purpose, we use the following equation, obtained from the Eqs. 8.4-4, 8.4-8 and 8.4-18:

$$m = m_{LOD} = \frac{A_t A_w}{(\sigma_{th}\Phi_{th} + I\Phi_{epi})f_a N_A (1 - e^{-\ln 2\,t/t_{1/2}})}$$

Using the experimental parameters and physical constants, we obtain for the limit of detection in gram

$$m_{LOD} = \frac{5.5 \times 75}{(4.3 \times 10^{-24} \times 9 \times 10^{13} + 60 \times 10^{-24} \times 1.6 \times 10^{12}) \times 1 \times 6.023 \times 10^{23} \times (1 - e^{-\ln 2 \times 24/26.3})}$$

$$= 3.02 \times 10^{-12} \, g$$

Considering the sample amount of 0.2 g, the limit of detection expressed as content in g per g, c_{LOD} is

$$c_{LOD} = \frac{3.02 \times 10^{-12}}{0.2} = 1.51 \times 10^{-11} \, g \, g^{-1} = 15.1 \, ppt$$

b) The radioactivity of the irradiated sample counted via the 559.1 keV photopeak after a decay time of 5 h was 18 counts per second (cps). For the standard containing 10 ng As, a counting rate of 1036 cps was measured after a decay time of 16 h. Calculate the content of arsenic in the silicon sample.

First we have to correct the counting rates to the end of the irradiation by using the rearranged Eq. 8.4-9 (with R_t, R_c instead of A_t, A_c):

$$R_t = \frac{R_c}{e^{-\ln 2 \, t_d/t_{1/2}}}$$

Then we obtain for the sample:

$$(R_t)_{sam} = 18 \times e^{\ln 2 \times 5/26.3} = 20.54 \, cps$$

and for the standard:

$$(R_t)_{std} = 1036 \times e^{\ln 2 \times 16/26.3} = 1579.4 \, cps$$

From these corrected counting rates, the mass of arsenic m_{sam} in 0.2 g of the sample is calculated using Eq. 8.4-19:

$$m_{sam} = 10 \times 10^{-9} \times \frac{20.54}{1579.4} = 1.3 \times 10^{-10} \, g$$

Thus, the content of arsenic c_{As} in the silicon sample in g per g is

$$c_{As} = \frac{\text{mass of As in the sample}}{\text{mass of sample}} = \frac{1.3 \times 10^{-10}}{0.2} = 0.65 \times 10^{-9} \, g \, g^{-1} = 0.65 \, ppb$$

Nuclear interferences

Two kinds of *interference nuclear reactions* can occur in the activation process which can alter the proportionality between the activity of the indicator radio-nuclide and the amount of the element to be determined. These are the *primary* and *second-order* interference reactions. If they are not taken into consideration, they can give rise to systematic errors.

In primary interference reactions, the indicator radionuclide is produced in nuclear reactions, which have been induced by the original bombarding particles on the nuclei of other elements in the sample than the one being determined. In reactor NAA, the primary interference reactions are of (n, p) and (n, α) type and they are induced only by the fast neutrons of the fission spectrum. Some examples are given in Table 8.4-4. They show that, owing to relatively high Φ_{th}/Φ_f ratio and low mean fast neutron cross sections $\bar{\sigma}_f$, these interferences, in general, occur to a negligible extent if the content of the interfering element is at the trace level. However, at higher contents of the interfering elements, the contributions of the primary interference reactions to the production of the indicator radionuclides must be considered. They can easily be judged and, in the main, also quantitatively evaluated by calculation using the appropriate nuclear data, the fast neutron flux and the content of the interfering element, which is often simultaneously determined via the (n, γ)-reaction.

In neutron activation analysis, the possibility of primary interference reactions has to be considered.

Occurring less frequently, are the second-order interference reactions, which are nuclear reactions involving the products of the primary reactions. These types of interfering reactions need only be considered when very long irradiations with high neutron fluxes are applied.

The attenuation of the neutron flux within the sample itself, called *self-shielding*, has also to be taken into account as a possible source of systematic errors. This effect is significant when the matrix elements have very high total effective capture cross sections. The self-shielding effect can be estimated by calculation or by graphical correction procedures.

Table 8.4-4. Examples of the extent of primary interference reactions in NAA with fission reactor neutrons

Element determined	Analytical reaction	Primary interference reaction	Percentage contribution of the interference[a]
Na	$^{23}Na(n, \gamma)^{24}Na$	$^{24}Mg(n, p)^{24}Na$	0.02
		$^{27}Al(n, \alpha)^{24}Na$	0.01
Mn	$^{55}Mn(n, \gamma)^{56}Mn$	$^{56}Fe(n, p)^{56}Mn$	7×10^{-3}
		$^{59}Co(n, \alpha)^{56}Mn$	1×10^{-3}
Hg	$^{202}Hg(n, \gamma)^{203}Hg$	$^{203}Tl(n, p)^{203}Hg$	2×10^{-5}

[a] Percentage contribution of the primary interference reaction to the totally produced radioactivity of the indicator radionuclide related to the assumption that the concentration of the element determined and the interfering element is the same and $\Phi_{th}/\Phi_f = 10$.

8.4.6 Applications of NAA

Feasibility of instrumental NAA

Instrumental neutron activation analysis is simple and enables one to carry out the analyses nondestructively, but can only be applied when certain preconditions are fulfilled.

If *instrumental NAA* (INAA) is applicable, it is always the preferred performance mode compared with *radiochemical NAA* (RNAA). Whether the indicator radionuclides produced in an irradiation can be detected by purely instrumental means, or must first be subjected to a radiochemical separation from the other radionuclides will depend on many factors including: the matrix, the irradiation conditions, the cooling time, the decay characteristics of the activation products, and the gamma-ray spectrometer used. The feasibility of INAA requires that at least one of the following preconditions is well-fulfilled:

1. The activation cross sections of the matrix components are much lower than those of the analytes. Elements having extremely low σ_{th} include H, Be, C, O, and Pb.
2. The abundance of the matrix isotopes from which radionuclides can be produced is low, as is the case, for example, for ^2H (0.015%), ^{30}Si (3.1%), ^{34}S (4.2%) and ^{36}S (0.015%), ^{58}Fe (0.31%), and others.
3. The half-lives of the radionuclides produced from the matrix are either much shorter or much longer than those of the indicator radionuclides. In the first case, the counting is performed after the matrix radionuclides have decayed to an acceptable level. After a decay time equal to the number of half-lives n, the activity decreases by a factor $1/2^n$. For example, in the activation of an aluminium matrix, high activity of ^{28}Al ($t_{1/2} = 2.2\,min$) is produced. Already after a decay time of ten half-lives (22 min), the activity of ^{28}Al decreases by a factor of 1/1024, and after twenty half-lives (44 min) by a factor of about 10^{-6}, so that counting the indicator radionuclides with $t_{1/2} > 20\,min$ can be carried out without this interference. If the second assumption is fulfilled ($t_{1/2}(\text{matrix}) \gg t_{1/2}(\text{IRN})$), short irradiation time is sufficient for the production of the indicator radionuclides in which, according to Eq. 8.4-8, only very low activity of the long-lived matrix radionuclides is produced.
4. The feasibility of INAA is favored if the radionuclides, produced with high activity from the matrix, decay by only poor emission of gamma and beta radiation of higher energy, or by none at all. For instance, the irradiation of samples containing iron as a matrix component with reactor neutrons produces, via the reaction $^{54}Fe(n, \gamma)^{55}Fe$, a high activity of ^{55}Fe. The only radia-

tion emitted in the decay of ^{55}Fe is the characteristic low-energy X-rays of Mn, which can easily be discriminated in order to eliminate the influence on the dead time. Gamma-ray spectrometric measurements can often be carried out without considerable interference in the presence of weak beta radiation, as it contributes via bremsstrahlung to the background only in the low-energy region.

High-purity materials

In the analysis of high-purity materials, the potential of NAA can most effectively be exploited. Therefore, NAA has been widely used in this field: there is hardly a single type of high-purity material of scientific or technological interest which has not been analyzed by NAA. Below, some selected examples of analysis of some important materials in microelectronics are given.

With many materials such as diamond and graphite, silicon, and other silicon-based materials as well as organic materials used in microelectronics, e.g., polyimides, extremely low detection limits can already be achieved by INAA. In activation of carbon-based materials, no detectable activity of radionuclides from the matrix is produced. Thus, all indicator radionuclides can be counted without any matrix radionuclide interferences (for an example see Fig. 8.4-6). In NAA of silicon and silicon-based materials, due to its short half-life, the radionuclide ^{31}Si ($t_{1/2} = 2.6$ h), formed via the reaction ^{30}Si $(n, \gamma)^{31}$Si from the matrix, interferes only when counting short-lived indicator radionuclides. Moreover, the rather low σ_{th} (0.11 barns) and isotopic abundance (3.1%) of ^{30}Si as well as the fact that ^{31}Si is almost a pure beta-ray emitter further reduce the extent of the interference of ^{31}Si. Thus, INAA can be considered as the most powerful method for ultratrace characterization of silicon and silicon-based materials such as quartz, silicon nitride, and silicon carbide. With INAA using modern gamma-ray spectrometry, a neutron flux of 10^{14} cm^{-2}s^{-1} and optimized irradiation modes, extremely low limits of detection can be achieved for a large number of elements in silicon as can be seen from Fig. 8.4-9. As many as 42 elements can be assayed at content levels <1 ppb.

INAA can also be applied to analysis of aluminium and aluminium oxide; however, it is not as powerful as in the case of silicon. The reason is the formation of the relatively high activity of ^{24}Na ($t_{1/2} = 15$ h) from the matrix via the reaction ^{27}Al$(n, \alpha)^{24}$Na, induced by the fast neutrons of the fission neutron spectrum. The high Compton continuum caused by ^{24}Na seriously limits the detectability of short- and medium-lived indicator radionuclides. An essential improvement of NAA of aluminium and aluminium oxide can be achieved if the activated sample is decomposed and the highly active radionuclide ^{24}Na is specifically separated on hydrated antimony pentoxide from an HCl/HF medium [8.4-8]. This is evident from Fig. 8.4-10 showing gamma-ray spectra obtained with and without separation of ^{24}Na. Table 8.4-5 shows the improvement of the limits of detection achieved by this radiochemical procedure for some exemplary elements. In counting long-lived indicator radionuclides, the interfering effect of ^{24}Na can alternatively be eliminated by a sufficiently long decay time (5–7 days). The limits of detection can further be improved if the indicator radionuclide mixture obtained after the removal of ^{24}Na is separated into several groups by combined anion and cation exchange; for 49 elements, the detection limits are between 1 ppt and 100 ppb.

Molybdenum can be used as an example of a material which cannot be analyzed by INAA. In activation of molybdenum with reactor neutrons, six radionuclides of $t_{1/2}$ between 14 min and 10 days are produced from the matrix, so that a radiochemical separation is unavoidable. A very effective separation procedure is based on anion exchange using Dowex 1 × 8 column from 20 M HF/3% H$_2$O$_2$ medium [8.4-9]. The matrix-produced radionuclides of Mo, as well as the radionuclides of Nb and Ta are retained on the column while the indicator radionuclides of most impurities are contained in the eluate. For the majority of relevant impurities, excellent limits of detection down to the ppt level can be achieved.

The unique features of neutron activation analysis can best be exploited in analysis of high-purity solid materials. Achievable limits of detection are typically in the sub-ppb range.

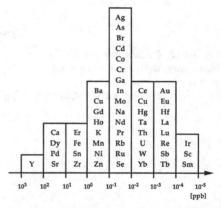

Fig. 8.4-9. Limits of detection for 51 elements in an ultra-pure silicon wafer achievable by INNA

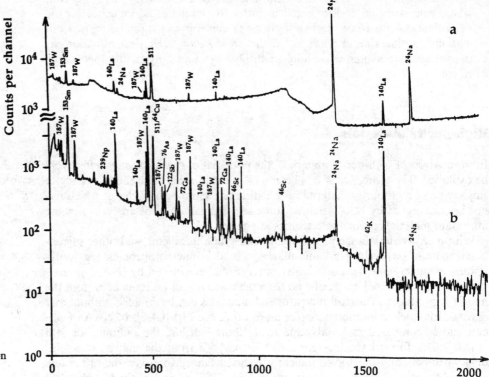

Fig. 8.4-10. Gamma-ray spectra of a high-purity Al_2O_3 sample irradiated with reactor neutrons. (a) INAA; (b) RNAA involving specific separation of ^{24}Na

Table 8.4-5. Examples of limits of detection achievable in analysis of aluminium oxide by INAA and RNAA involving a specific separation of ^{24}Na [8.4-8]

Element/IRN	Limit of detection [ng/g]		Improvement factor
	INAA	RNAA	
As/^{76}As	50	0.4	125
Ga/^{72}Ga	100	0.5	200
K/^{42}K	10 000	30	333
U/^{239}Np	30	2	15
W/^{187}W	20	0.5	40
Zn/^{69m}Zn	5 000	50	100

In materials for microelectronics, the alpha-particle-emitting natural elements, Th and U, have to be considered as impurities causing harmful effects, the so-called "soft errors" in electronic systems such as memory cells. Therefore, the maximum tolerable concentrations of these elements in different materials are in the range 0.1–1 ppb. In the analysis of several materials for Th and U, NAA provides the best limits of detection of all existing methods. In this case, the indicator radionuclides for Th and U are ^{233}Pa and ^{239}Np, respectively, which are produced by the beta decay of the short-lived reaction products:

$$^{232}Th(n,\gamma)^{233}Th \; (t_{1/2} = 22\,min) \xrightarrow{\beta^-} {}^{233}Pa \; (t_{1/2} = 27\,d) \tag{8.4-20}$$

$$^{238}U(n,\gamma)^{239}U \; (t_{1/2} = 23\,min) \xrightarrow{\beta^-} {}^{239}Np \; (t_{1/2} = 2.3\,d) \tag{8.4-21}$$

A simultaneous separation of these two indicator radionuclides by anion exchange from 12 M HCl followed by elution with 1 M NH_4F gives limits of detection for Th and U in quartz, silicon nitride, aluminium oxide, and polyimides in the range 10–100 ppt [8.4-10].

Geochemistry and cosmochemistry

The cosmochemical studies include the investigation of extraterrestrial materials such as meteorites and lunar samples. With the knowledge of the abundance of the elements in terrestrial samples, hypotheses and theories for the formation, distribution and chemical and physical interactions of the various materials can be developed. For example, analytical data from meteorites have been used by cosmochemists to formulate theories about the nuclear processes occurring in stars. The validity of these theories is strongly dependent upon the accuracy of the analytical results. Therefore, NAA has been widely used and is still one of the most important analytical tools in these areas of science.

A large variety of terrestrial and extraterrestrial samples have been analyzed for up to 50 elements by INAA and/or RNAA. The majority of the lunar rock samples brought to earth by the Apollo missions were analyzed by NAA. This method is especially well-suited for the determination of rare earth elements which are of great interest in many geochemical and cosmochemical studies. RNAA has enabled the determination of a number of elements even at the ppt level; for example, 4 ppt of Ir in the U.S. geological survey of silicate rock standard basalt BCR-1, and 9 ppt of Ir in andesite AGV-1 [8.4-11].

Environmental applications

Most environmental samples such as atmospheric aerosols, fly ash, waste incineration ash, sludge and sludge ash, plants, hair, industrial waste, automobile exhaust emission, and cigarette smoke condensate can be analyzed comprehensively by INAA.

In many instances, neutron activation analysis is an advantageous method for analyses of environmental and biological samples.

The potential of NAA has been widely explored for the analysis of atmospheric aerosol samples. The samples are usually collected over a period of time on an organic membrane or cellulose paper filter. A typical irradiation scheme involves two or three irradiations for different periods of time to assay short-, medium- and long-lived indicator radionuclides. Under these conditions, up to 50 elements can be determined in aerosols.

Several elements contained in cigarette smoke are suspected to be toxic even at very low levels of intake. Therefore, the analysis of tobacco smoke for trace elements is of great importance. Tobacco smoke condensate can be best collected in a quartz tube by electrostatic precipitation using a smoking machine. Because of high achievable detection sensitivity and the absence of reagent blank, NAA is an excellently suited technique for the analysis of tobacco smoke condensate. Since no dominant activity is produced in the irradiation of this material with neutrons, extremely low limits of detection for a large number of elements can be achieved even by INAA [8.4-12]. Table 8.4-6 gives the amounts of elements contained in the smoke condensate of one cigarette determined by INAA.

In spite of still being occasionally used for the analysis of water and other liquid samples, NAA is, in fact, not a suitable method for this purpose. NAA normally requires complicated sample preparation prior to irradiation, unlike many other methods which are, therefore, superior to NAA.

Several other methods are much better suited for analyses of liquid samples such as water than neutron activation analysis.

Table 8.4-6. Concentration of certain elements in the smoke condensate of Marlboro cigarettes determined by INAA and limits of detection (LD) [8.4-12]

Element	Concentration [ng/cig.]	LD [ng/cig.]
As	4.1 ± 0.4	0.001
Br	380 ± 4	0.5
Cd	64 ± 3	3
Co	0.08 ± 0.03	0.005
Mn	2.2 ± 0.1	0.1
Sb	0.13 ± 0.01	0.005
Se	1.5 ± 0.2	0.07

Biomedical applications

A number of trace elements including Co, Cr, Cu, Fe, Mn, Mo, Ni, Se, Zn, and V are essential for growth and health in mammals, both animal and human. Just as important, however, is the recognition of toxic effects of a number of elements already at low contents in the organisms.

NAA has proved to be a very valuable method for the determination of most of the trace elements of interest in blood, tissue, and organ samples. Blood plasma and serum are the most frequently analyzed samples. The outstanding detection sensitivity of NAA makes it the preferred method mainly in the determination of elements occurring with extremely low contents in blood plasma or serum (for example, Co, Cr, Mn, Mo, and V, which have contents of $\lesssim 1\,ng/g$), and in the analysis of biopsy samples of 2–10 mg. NAA has greatly contributed to establishing the so-called normal content levels of the trace elements in humans.

8.4.7 Nonreactor activation analysis

Fast neutron activation analysis (FNAA)

Generally, only when an analytical problem cannot be solved by reactor neutron activation, alternative activation techniques become of interest.

For the determination of a number of elements, activation with fast neutrons via threshold reactions of the (n, p), (n, α), and $(n, 2n)$ types is a well-suited technique.

The most important source of fast neutrons is the neutron generator, in which 14 MeV neutrons are produced in the reaction $^3T(d, n)^4He$. To induce this reaction, an intense beam of deuterium ions is produced, with an energy in the range of 50 to 600 keV, and focused onto a tritium target from which the neutrons are emitted with a fairly uniform spherical angular distribution. The commercially available generators provide neutron fluxes of the order of 10^8–10^{10} cm^{-2} s^{-1}.

For some specific research and industrial problems, 14 MeV NAA is the optimum method. Without doubt, the most important application is the determination of oxygen down to ppm level by the reaction $^{16}O(n, p)^{16}N$ ($t_{1/2}$ for $^{16}N = 7.2\,s$). However, 14 MeV activation analysis is limited by the relatively low neutron flux (compared with reactor fluxes) and by the invariable and, for many elements, unfavorable neutron energy of 14 MeV. High neutron fluxes of variable energy can be produced by bombarding a thick (a few cm) beryllium target with high energy deuterons [8.4-13].

Charged particle activation analysis (CPAA)

For CPAA, almost exclusively light particles, i.e., protons (p), deuterons (d), tritons (t), ^{3}He, and α particles are used. The most suitable source of these particles is the cyclotron, in which they are accelerated to a suitable energy.

On the one hand, CPAA offers unique possibilities to the analyst. As is shown in Fig. 8.4-2, a large number of nuclear reactions can be induced in each target nucleus with charged particles. Thus, for each element, several sensitive reactions yielding an indicator radionuclide with suitable nuclear properties can be chosen. Owing to high applicable beam intensities corresponding to 10^{12} to 10^{14} particles per second hitting the sample, trace element concentrations down to the ppb level can be determined. For each application, the most suitable particle and the optimum energy can be selected.

On the other hand, CPAA has also some special problems. Unlike neutrons, charged particles are quickly slowed down and stopped when they hit a "thick" sample (tenths of mm to a few mm), which makes calibration less easy and less accurate than in NAA. The activated sample thickness depends on the type and energy of the particle and on the atomic number (mean atomic number in complex materials) of the sample. It is normally low, thus, inhomogeneity errors can easily occur. The rapid deceleration of the fast particles results in the production of a large amount of heat. Therefore, the samples must be cooled efficiently during the irradiation. Even then, beam currents over 10 µA can rarely be used.

For these reasons, CPAA cannot compete with NAA, and has to be considered as an interesting complementary method to NAA. Therefore, CPAA has been most widely used for the determination of the light elements lithium, boron, carbon, nitrogen, and oxygen which are important impurities in many materials such as metals and semiconductors. For the determination of each of the light elements, there exist several extremely sensitive analytical reactions providing limits of detection at the lower ppb-level. An important aspect of the choice of the optimum reaction is, in addition to the achievable limit of detection, also the occurrence of nuclear interference reactions.

CPAA has proved to be a useful method also for the determination of medium and heavy elements in materials which cannot be analyzed by NAA because of extremely strong matrix activation. For example, while cobalt and tantalum can, for this reason, hardly be analyzed by NAA, proton activation analysis enables the determination of many elements in these matrices even with purely instrumental performance.

Fundamentals and applications of CPAA are treated in a book by Vandecasteele [8.4-14].

Determination of trace concentrations of the light elements Li, B, C, N, and O in solid materials is the application domain of charged particle activation analysis.

Photon activation analysis (PAA)

PAA represents a principally interesting, but practically seldom used activation analysis method. The reason for this contradiction is to be seen in the lack of availability of linear accelerators needed for the production of high energy electrons with which a metal target is bombarded to obtain bremsstrahlung photons of sufficiently high energy and intensity. In PAA, the determination of most elements is based on (γ, n)- and (γ, p)-type nuclear reactions. PAA has mainly been used for the determination of the light elements, carbon, nitrogen, oxygen and fluorine. PAA is described in detail by Segebade et al. [8.4-15].

References

[8.4-1] Elving, P.J., Krivan, V., Kolthoff, I.M., *Treatise on Analytical Chemistry, Part I, Theory and Practice; Vol. 14, Section K, Nuclear Activation and Radioisotopic Methods of Analysis.* New York: Wiley-Interscience, 1986.

[8.4-2] Ehmann, W.D., Vance, D.E., *Radiochemistry and Nuclear Methods of Analysis.* New York: Wiley-Interscience, 1991.

[8.4-3] Friedlander, G., Kennedy, J.W., Macias, E.S., Miller, J.M., *Nuclear and Radiochemistry.* New York: Wiley-Interscience, 1981.

[8.4-4] Knoll, G.F., *Radiation Detection and Measurement, 2nd ed.* New York: Wiley-Interscience, 1989.

[8.4-5] De Soete, D., Gijbels, R., Hoste, J., *Neutron Activation Analysis.* New York: Wiley-Interscience, 1972.

[8.4-6] Erdtmann, G., Petri, H., in [8.4-1], Chapter 7, p. 419.

[8.4-7] De Corte, F., Simonits, A., De Wispelacre, A., Hoste, J., *J. Radioanal. Nucl. Chemistry* 1987, 113, 145.

[8.4-8] Franek, M., Krivan, V., *Anal. Chim. Acta* 1993, 282, 199.

[8.4-9] Theimer, K.-H., Krivan, V., *Anal. Chem.* 1990, 62, 2722.

[8.4-10] Franek, M., Krivan, V., *Anal. Chim Acta* 1993, 274, 317.

[8.4-11] Gijbels, R., Govaerts, A., *J. Radionanal. Chem.* 1973, 16, 7.

[8.4-12] Schneider, G., Krivan, V., *Intern. J. Environ. Anal. Chem.* 1993, 53, 87.

[8.4-13] Krivan, V., Münzel, H., *J. Radioanal. Chem.* 1973, 15, 575.

[8.4-14] Vandecasteele, C., *Activation Analysis with Charged Particles.* Chichester: Ellis Horwood, 1988.

[8.4-15] Segebade, C., Weise, H.P., Lutz, G.J., *Photon Activation Analysis.* Berlin: Walter de Gruyter, 1987.

Questions and problems

1. Describe the principle of activation analysis.
2. Write the radioactivation equation, and describe the experimental parameters and nuclear constants determining the activation yields. Explain the growth of activity of an indicator radionuclide during the irradiation and its decrease during the post-irradiation decay period.
3. Calculate the minimum mass of manganese that can be determined via $^{55}Mn(n, \gamma)^{56}Mn$ under the following assumptions: minimum activity at the end of the irradiation necessary for gamma-spectrometric counting = 10 Bq; thermal neutron flux = 5×10^{13} cm^{-2} s^{-1}; irradiation time 2 h; decay time = 1 h. Physical constant: atomic weight = 54.94; natural isotopic abundance = 100%; cross section of the reaction = 13.3 barns; half-life of ^{56}Mn = 2.58 h. The contribution of manganese activation with epithermal neutrons can be neglected. (Answer: 4.3 pg)
4. Describe the possible modes of radioactive decay and the types of emitted radiation which may be utilized for detection of indicator radionuclides. Why, principally in activation analysis, is gamma-ray counting superior over beta-ray counting?
5. Describe the basic components and the working principle of a modern gamma-ray spectrometer.
6. Explain the creation of a gamma-ray spectrum on the basis of the interaction processes of gamma rays with the semiconductor detector.
7. Name and explain the most important performance characteristics of a gamma-ray spectrometer. Compare the typical performance parameters of gamma-ray spectrometers using HPGe and a scintillation detector.
8. Name the activation techniques and explain why NAA is the most important one.
9. Describe the two performance modes of NAA.
10. Discuss preconditions of INAA and situations in which radiochemical separations are inevitable.
11. Describe the unique features of NAA and explain why this method enables the achievement of high degrees of accuracy also at low analyte content levels.
12. In which ways can calibration be carried out in NAA?
13. Which nuclear interference reactions can principally occur in NAA? Discuss situations in which they become relevant.
14. Describe the major applications of NAA.
15. Explain why NAA is an extremely powerful method for analysis of high-purity materials but a disadvantageous method for analysis of water.

8.5 Inorganic Mass Spectrometry

8.5.1 Introduction

A mass spectrometer is a device which is used to separate a stream of gaseous ions into ions with different values of mass, m, divided by the charge, z. It is then necessary to use an ionization source for the production of ions. The first mass spectrometer was designed by Aston in the 1920s. Mass spectrometry gained some acceptance for use in organic analytical chemistry in the early 1940s and is currently widely used for the analysis of organic products (see Sec. 9.4). However, at that time, mass spectrometry was already being used for isotopic analysis of elements such as Pt and Pd. Improvement to the isotopic technique in terms of sensitivity, precision, and accuracy was undertaken in order to improve both the determination of low-abundance isotopes and the reliability of the atomic masses of the elements.

The first inorganic mass spectrometer (MS), which was originally developed for analytical applications, was described in the 1950s and used an rf spark as an ion source. Limits of detection were already in the sub-ppm range. The first use of a plasma as an ion source was described by Gray in 1975 and was of the dc capillary arc plasma type. Limits of detection with this device were already at the sub-ppm level. The introduction of the inductively coupled plasma (ICP) in the mid 1980s gave rise to a growing market for inorganic mass spectrometry. The large number of instrument companies which manufacture ICP-MS is evidence of the interest in this technique. Inorganic mass spectrometry is not only useful for the determination of elements in various samples, but also for the measurement of the natural isotopic abundance and the use of the isotopic dilution technique.

Basically, inorganic mass spectrometry uses the same type of mass spectrometers as for organic analysis. The major difference is that in organic analysis the mass range does not need to be extended above 300 amu. In inorganic mass spectrometry, however, it may be necessary to determine low molecular masses (<10 amu). Currently, the quadrupole mass filter is probably the most commonly used system for inorganic MS. The time-of-flight (TOF) spectrometer and the magnetic sector are also used, but to a lesser extent. Readers are referred to the section on organic mass spectrometry (Sec. 9.4) for the description of this type of mass spectrometer.

The basic task of an ion source is to atomize the sample and to subsequently ionize the atoms. Ion sources can be classified according to the form of the sample (liquid or solid) to be analyzed. The inductively coupled plasma and the heated filament are well-suited for the analysis of liquid samples, whereas the spark, the glow discharge and the laser-induced plasma are well-adapted for the direct analysis of solids [8.5-1, 8.5-2]. It is interesting to note that most of the ion sources are also used as radiation sources in atomic emission spectrometry.

It should be noted that secondary ion mass spectrometry (SIMS) can also be used for inorganic analysis. However, this method is mainly used for surface analysis and localized microanalysis. Therefore, this technique is dealt with in Sec. 10.2.

A mass spectrometer uses an ionization source to produce a stream of gaseous ions which are separated into ions with different values of mass divided by the charge.

Ionization sources can be used to ionize solids or liquids.

8.5.2 Ion sources

The heated filament

Thermoionization mass spectrometry is based on the evaporation and ionization of samples previously deposited on a refractory filament.

A single or several filaments are electrically heated in a 800–2000 °C temperature range to evaporate and to ionize liquid samples previously deposited in the form of droplets onto the filament. This is why the technique is called thermo-ionization mass spectrometry (TIMS) [8.5-1, 8.5-3]. The filament is characterized by its work function, i.e., the minimum amount of energy necessary to remove an electron from the filament metal. When a single filament configuration is used, evaporation and ionization occur from the same surface. With the use of 2 or 3 filaments evaporation and ionization steps can be separated since the gaseous sample is then transferred to another filament and adsorbed on its surface. This is useful for elements which evaporate at low temperatures but which require a high temperature for efficient ionization (e.g. Ca). The filaments are made of refractory elements such as Ta, Re, or W since their melting temperatures are 3000 °C, 3180 °C, and 3400 °C, respectively. Note that their work functions are 4.30, 4.98, and 4.58 eV, respectively. The work function of W can be lowered by adding, for instance, Th to W. The work function of thoriated W is then 2.7 eV. Elements are usually deposited in the form of nitrate or chloride salts. Ionization efficiency is particularly high for elements that exhibit a first ionization potential below 7 eV, such as alkali elements, alkaline earth elements, actinides, and lanthanides. For elements with ionization potentials higher than 7 eV (e.g., Cu, Pd, Zn), it may be necessary to add some reagents to enhance the ionization efficiency, the most popular being silica gel with or without further additives. One advantage of this type of ionization is that it produces only singly charged ions, which result in simple spectra. It should be noted that not only are positive ions observed with TIMS, but negative ions as well, particularly for nonmetals and when using low work-function filaments. Examples of negative ions include the halogens, Se, S, and Te. Positive thermal ionization theory states that the ratio of evaporated ions to neutral atoms depends exponentially on the difference between the work function of the filament and the ionization potential of the analyte, whereas negative thermal ionization depends on the difference between the analyte electron affinity and the work function. The role of the work function is therefore crucial for the production of either positive or negative ions.

The preparation of the sample, the deposition of the droplets onto the filament, and the heating stages are crucial parameters in TIMS. A typical droplet volume is 2 μL with a solid content of 2 μg. The quality of the sample preparation is of importance because of the risk of contamination, particularly during the separation of the analyte from the matrix. If an element with low ionization potential (e.g. Na) is present as a contaminant, the ionization efficiency of the analyte will be significantly modified. Separation methods such as ion exchange chromatography are thus widely used for real sample analysis.

TIMS is mostly used in combination with low resolution quadrupole filter MS and medium resolution single-focusing magnetic sector MS. However, when there is a need to measure a minor isotope in the presence of a major isotope, the abundance ratio should be better than 10^{-8}. It is then necessary to use high resolution MS, such as a triple sector system, to eliminate the tails on adjacent peaks.

Inductively coupled plasma

The inductively coupled plasma (ICP) was described in Sec. 8.1 as a radiation source in atomic emission spectrometry. A plasma is an ionized gas that is macroscopically neutral, i.e., having the same number of positive particles (ions) and negative particles (electrons). If a gas, X, is used, a plasma can be described by the following equilibrium:

$$X = \sum_{n=1}^{q} X^{n+} + \sum_{n=1}^{q} n.e \qquad (8.5\text{-}1)$$

where X^{n+} is an ion with n times the electron charge and e is the electron. A plasma is therefore well-suited for inorganic mass spectrometry since it is an efficient source of ionization. The power stability is a crucial parameter for avoiding any drift in the properties of the plasma. The generator must exhibit enough flexibility to compensate for any variation of the plasma impedance due to a change in its load, i.e., injection of different types of solutions, aqueous, or organic.

The gas that is used to generate the plasma (plasma gas) is argon. Like any noble gas, argon is an element with a high ionization energy (15.76 eV), and is chemically inert. However, some molecular species can be formed either in the plasma or during the ion transport. Although these species are not stable, their lifetime can be long enough such that they can be observed at the detector. Because of the nature of the hf field and the resultant skin effect, the energy from the generator is mainly deposited into the external part of the plasma. There is, consequently, a zone along the axis of the plasma where the viscosity is lower. This results in the creation of a central channel that facilitates the injection and the confinement of the sample. This is a unique feature among the various types of plasmas and this is a major advantage for an ionization source. Indeed, the ion zone is confined along the axis of the plasma, which facilitates an efficient extraction of the ions. The ICP is also an efficient atomization source, which explains why the ICP has rapidly become a well-established technique for inorganic MS [8.5-4–8.5-7]. Torch configuration and types of nebulizers are similar to those used in atomic emission spectrometry (Sec. 8.1).

The ICP is mostly used with low resolution quadrupole MS, although high resolution sector MS (reverse Nier-Johnson geometry) with single or multi detectors are commercially available. The crucial point of the ICP-MS is the fact that the ICP is operated both at atmospheric pressure and high temperature, whereas the MS requires high vacuum conditions and an ambient temperature. An interface is, therefore, necessary for reducing both the pressure and the temperature (Fig. 8.5-1). Currently, the interface consist of two cones usually made of Cu or Ni. The first cone is called the sampler, and the second one the skimmer. This technology arises from the 1960s. The orifices of the cones are of the order of 1 mm or less, and are aligned within the axis of the plasma. The tip of the sampler must be located in the central channel of the ICP, i.e., in a region where the ions are present. The pressure is reduced between the sampler and the skimmer by means of a rotary pump. A molecular supersonic beam is obtained behind the sampler, which ends in a Mach disk. The tip of the skimmer is located in the axis of the supersonic beam, slightly ahead of the Mach disk. The distance between the two tips is less than 10 mm. The advantage of the supersonic beam is that it can significantly reduce the temperature of the ion species because of the expansion of the plasma.

Because of a strongly divergent ion beam behind the skimmer, it is necessary to locate some ion optics. The role of ion optics is:

a) to realign the ions to obtain a colinear beam which is required for efficient ion filtering and transmission by the quadrupole filter

b) to trap the photons emitted by the plasma in order to suppress the noise contribution that they could cause if they arrived on the detector.

High vacuum at the ion optics and quadrupole mass filter level is obtained through the use of turbomolecular pumps, which have replaced the oil diffusion pumps and the cryogenic pump on most commercially available MS systems. The number of various ion optics designs which are currently commercially available illustrate the difficulty of obtaining an efficient system over the whole mass range. Moreover, the processes in the supersonic beam and at the ion optics level are not accurately known, and further work is still required to obtain a sufficient understanding of how this part of the ICP-MS system functions. The current efficiency of ion extraction and transmission results in 10^7 count s^{-1} for 1 $\mu g\,mL^{-1}$ of analyte in solution. It should be noted that, in the instance where a sector MS is used, it is necessary to accelerate the ions to a kinetic energy of a few keV before the entrance slit of the MS. In this case, the ion optics are more complex, consisting of both acceleration and x–y deflection. The deflection can be based on the use of quadrupole systems with dc voltage only.

The sample introduction systems are similar to those used in ICP atomic emission spectrometry (Sec. 8.1): pneumatic nebulization, ultrasonic nebulization, and

The argon inductively coupled plasma is well adapted to mass spectrometry because of the confinement of the sample and its high ionization efficiency.

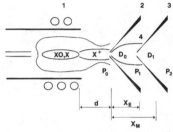

Fig. 8.5-1. Schematic drawing of the ICP-MS interface. 1: plasma coil, 2: sampler cone, 3: skimmer cone, 4: supersonic ion beam, d: sampling position, X_E: sampler-skimmer distance, X_M: location of the Mach disk, P_0: atmospheric pressure, P_1: intermediate pressure, P_2: vacuum, D_0 and D_1: orifice diameters of the sampler and the skimmer, respectively

The design of the interface and the ion optics is crucial to obtain satisfactory analytical performance in ICP-MS.

laser ablation. In particular, laser ablation ICP-MS systems are commercially available [8.5-8].

Spark source

Although an arc source (Sec. 8.1) can be used as an ionization source, only the spark source has been made commercially available [8.5-1]. Spark source mass spectrometers (SSMS) were introduced in the 1960s. The spark is of the high voltage type (Sec. 8.1). Dc sparks have been used, but commercially available systems have made use of a pulsed 1-MHz field in order to obtain trains of short pulses across the electrode gap. Since the pulse length (20–200 µs) and the repetition rate (1 Hz–10 kHz) can vary to a large extent, ionization conditions can be optimized according to the type of sample. In contrast to the spark sources used for atomic emission spectrometry, which are mostly operated at atmospheric pressure, the spark source for MS operates under vacuum conditions. The electrodes are located in a spark housing, which is also connected to the high voltage. This electrical connection prevents most of the ions from impinging onto the walls of the vacuum system, which would result in the sputtering of the housing material.

Two electrodes are generally used. The design can consist of either two identical electrodes or, as with atomic emission spectrometry, of a flat sample and a counter electrode made of a refractory material. Some systems make it possible to move the sample in order to scan the target. In atomic emission spectrometry, the gap is usually kept constant, whereas in SSMS, the gap can be modified or moved around a reference position.

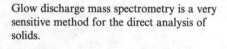

A spark source is used at reduced pressure with a photographic plate as a detector.

The spark source is mostly used with the high resolution double focusing Mattauch–Herzog design. The double focusing is provided simultaneously for all masses, and the entire spectrum is obtained on a plane, which makes it possible to use a photographic plate as a detector. The Mattauch–Herzog design has the advantage of being able to cope with the high energy distribution of the ions formed by the spark source.

Glow discharge source

Glow discharge mass spectrometry is a very sensitive method for the direct analysis of solids.

Low pressure discharges were used as ionization sources for conductive solid samples for MS due to their easy and efficient ionization. They were widely used before the introduction of the spark source. Following the use of glow discharge in atomic emission spectrometry, where an intense ion emission was observed, there was a renewed interest at the beginning of the 1970s in the utilization of this type of source in MS [8.5-9–8.5-13]. Glow discharge mass spectrometry (GDMS) has some unique features for this purpose, which can also be seen in atomic emission spectrometry (Sec. 8.1). Sample preparation is reduced to a minimum, the GD already operates at reduced pressure (0.1–10 torr), atomization is obtained by sputtering of the surface, and ionization, mainly by electron impact and Penning ionization from the metastable levels of the fill gas, is located in the glow region. The fill gas is usually high purity argon, but argon can be replaced by another rare gas such as Ne. The interface with the MS is located very closely to the glow region so to avoid the uptake of molecular ions (Fig. 8.5-2). Similar to the ICP-MS, a two stage differential pumping system is used. Ion optics are also necessary, particularly for reducing the spread of the ion energy. The adjustment of the ion optics has a crucial effect on ion extraction and transmission. The GD parameters that are used for optimization of the ionization include the nature of the gas, the gas pressure, the voltage and the current of the discharge. In some recent developments, the cell is cooled with liquid nitrogen to minimize the formation of molecular species and to permit the analysis of low melting temperature samples.

In contrast to ICP-MS where the low resolution MS systems were introduced before the high resolution systems, the first commercially available GDMS made use of a high resolution MS system. Commercially available high resolution systems are based on the reverse Nier-Johnson geometry, i.e., a magnetic sector fol-

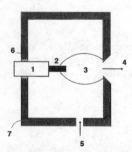

Fig. 8.5-2. Schematic drawing of a glow discharge as an ion source for mass spectrometry. 1: sample holder, 2: pin (sample) cathode, 3: glow discharge, 4: ion extraction, 5: gas inlet, 6: insulator, 7: anode

lowed by an electrostatic sector. More recently, quadrupole MS has been used with GD sources. It should be noted that at least one commercially available system offers the possibility of using alternatively GD and ICP sources on the same quadrupole MS system.

Laser-based techniques

Two main methods are based on the use of laser for ionization: laser-produced plasma mass spectrometry and resonance ionization mass spectrometry (RIMS). Laser-produced plasmas have been used since the beginning of the 1960s as radiation sources in atomic spectrometry. However, since a plasma is a source of ions, laser-produced plasmas can also be used in mass spectrometry [8.5-14]. When a laser is focused onto the surface of a target, the photons are absorbed by the material. This results in the ejection of electrons and heavy particles, which absorb the photons of the laser beam. A microplasma is formed. Its composition depends on the laser wavelength, the irradiance of the laser, and the working pressure. At the same time, partly due to the laser beam, partly due to the microplasma, ablation of the material occurs. The laser-material interaction depends on the spot size of the laser beam, which in turn depends on the type of laser, the wavelength, and the focusing system. Spot sizes as small as a few μm can be obtained with lasers working in the UV part of the spectrum. Several types of laser can be used for this purpose: the CO_2 laser (10.6 μm), the Nd:YAG laser (1064 nm, 532 nm, 355 nm, and 266 nm) and excimer lasers such as the XeCl excimer laser (308 nm) and the ArF excimer laser (193 nm). A limited number of systems are currently commercially available (LAMMA). The LAMMA system is based on the use of a low energy Q-switched Nd:YAG laser operating at the fourth harmonic (266 nm) via two frequency-doubling crystals and a TOF mass spectrometer. The use of the TOF mass spectrometer is made possible because of the short pulses of the Q-

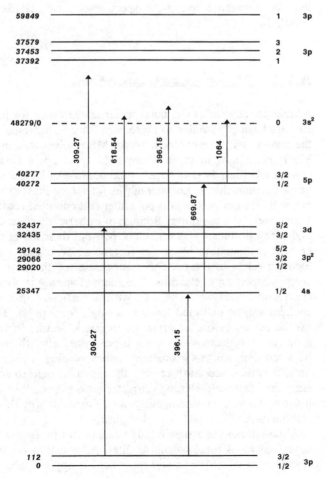

Fig. 8.5-3. Possible RIMS transitions in the case of aluminum

switched laser (15 ns). The laser beam is focused onto the target through a microscope, which results in a lateral resolution of 1 μm. Samples are usually in the form of thin sections of typically less than 2 μm thick. This requirement may be a problem for sample preparation and is a severe drawback. A more recent system, introduced in the beginning of the 1980s, was able to handle bulk samples. The mass resolution of TOF mass spectrometer in the LAMMA system is limited to 500–800 and new designs using Fourier transform (ion cyclotron resonance) are being explored.

Resonance ionization mass spectrometry is a highly selective and sensitive analytical method based on the use of laser for the ionization steps through optical transitions.

RIMS is a multistep photon absorption process which selectively ionizes free atoms using the resonant absorption of laser radiation via allowed optical transitions (see Sec. 8.1) from the ground state to the continuum of the ion [8.5-14–8.5-17] through one or several excited levels of the atom. The simplest resonance ionization process is a two-step procedure with photons at the same wavelength (λ_1, λ_1). This process is possible when the excitation energy of the excited level is higher than half of the ionization energy of the atom. A single dye laser is then required. This process can be used for most elements. However, for some elements, mostly nonmetals (e.g., halogens, rare gases, P, and S) the first excited states require a wavelength below 200 nm, which is difficult to obtain with the current laser technology. More complex schemes have been described. Two (λ_1, λ_2) or three different wavelengths, e.g., $(\lambda_1, \lambda_2, \lambda_3)$, $(\lambda_1, \lambda_2, \lambda_1)$, $(\lambda_1, \lambda_2, \lambda_2)$ can be used. In the case of Al (Sec. 8.1, Fig. 8.1-1), several schemes of type (λ_1, λ_1), (λ_1, λ_2), and $(\lambda_1, \lambda_2, \lambda_3)$ have been described [8.5-18]. Some of them are given in Fig. 8.5-3, based on the use of Al I 396.15 nm (λ_1, λ_1), 309.27 nm (λ_1, λ_1), 309.27 nm and 618.54 nm (λ_1, λ_2), 396.15 nm, 669.86 nm and 1064 nm $(\lambda_1, \lambda_2, \lambda_3)$. In the case of the (λ_1, λ_2) scheme, the laser is tuned at the 618.54 nm wavelength, and then the frequency is doubled to obtain the 309.27 nm wavelength. In the $(\lambda_1, \lambda_2, \lambda_3)$ scheme, the λ_1 and λ_2 wavelengths are obtain in a similar manner as for the previous example, whereas the 1064 nm wavelength corresponds to the fundamental wavelength of a Nd:YAG laser.

Atomization can be obtained under vacuum conditions with the use of electrothermal atomization (furnace or filament similar to TIMS), ion gun (sputtered initiated resonance ionization spectrometry or SIRIS), and laser atomization (LARIS).

8.5.3 Mass spectrometers

Mass spectrometers used in inorganic element analysis are similar to those used in organic mass spectrometry except for the mass range which is reduced to m/z values lower than 250.

As mentioned previously, mass spectrometers used for inorganic mass spectrometry are basically similar to those used for organic mass spectrometry, except for the mass range. The magnetic sector MS resolves ions in space, the time-of-flight MS resolves ions in time, whereas the quadrupole device is a mass filter. The quadrupole mass filter is largely used because of its low cost, its reliability, and it has a vacuum that is not as high as for sector mass spectrometers [8.5-19]. The major limitation of a quadrupole filter is the limited resolution, which is typically 1 amu over the mass range. Because of isobaric interferences, there is a need to use high resolution mass spectrometers. Most of these are of the double focusing type, i.e., the combination of an electrostatic analyzer and a magnetic sector. Double focusing refers to the need to compensate for the beam divergence and for the energy dispersion of the ions. The energy focusing is obtained with the use of the electrostatic analyzer. The ions with a particular m/z value but with different energies will be deflected towards a single focal point. The electrostatic analyzer can be set up before or after the magnetic sector. In the reverse Nier-Johnson geometry, the electrostatic sector is positioned after the magnetic sector. It should be noted that, for this geometry, double focusing is obtained for a single mass at time. It is then necessary to vary the magnetic field to scan over the mass range. Only the Mattauch–Herzog geometry makes it possible to obtain a double focusing for all masses simultaneously, which explains why this configuration is used in combination with a photographic plate.

Although not yet commercially available for inorganic analysis, the ion trap is a development of the quadrupole filter. An ion trap consists of a cylindrical ring electrode to which the quadrupole field is applied, and two end-cap electrodes.

The top electrode contains apertures for introducing ions, whereas the apertures of the bottom electrode serve to eject ions towards the detector. By modifying the rf field, trajectories of ions of successive m/z values are made unstable. In a standard quadrupole filter, only ions with stable trajectories are transmitted to the detector, whereas in an ion trap, only the ions with unstable trajectories are detected. It is also possible to use the ion trap to add some gases to obtain collisions within the ion trap in order to reduce the formation of molecular species. Some prototypes make use of the ion trap at the exit of a quadrupole mass filter.

The time-of-flight spectrometer is mainly used with laser-produced plasmas due to the inherent pulsed nature of the laser. Pulsing of the ions is necessary to avoid the simultaneous arrival of ions of different m/z values at the detector. Some TOF prototypes have been described in connection with an ICP. It is then necessary to pulse the extraction of the ions from the ICP, usually at right angles to the axis of the plasma. A further limitation of the TOF spectrometer is that all masses are extracted. In the case of the ICP, it is necessary to add a device to eliminate the presence of the Ar ions in order to increase the lifetime of the detector. Note that the TOF spectrometer may exhibit a rather limited resolution because of the time spread in the ion production and extraction.

8.5.4 Detectors

The most commonly used detectors are the electron multiplier (continuous or discrete dynode type) and the Faraday cup. The photographic plate is only used with the spark source. The discrete dynode type consists of a series of dynodes. The ions are converted to electrons at the first dynode and then the electron current is further multiplied by the other dynodes because of the voltage applied on each dynode. The continuous dynode or channel multiplier (channeltron) consists of a curved funnel-shaped glass tube, which is doped inside with a semiconductor such as lead oxide. For detection of positive ions, a negative high voltage is applied at the entrance of the tube. Since the potential varies with position inside the tube, the secondary electrons which are formed move to the end of the multiplier which is near ground potential. The channeltron exhibits a very low dark current, but has a relatively short lifetime, determined by the total accumulated charge. Although channeltrons are widely used in ICP-MS, there is a current trend to replace them by the discrete dynode type electron multiplier. Both analogue and counting mode are used. The counting mode is useful for low signals whereas the analogue mode is used to extend the upper end of the dynamic range of the detector. The Faraday cup (i.e., a metal hollow conductor) is a very simple device which operates independently of the energy of the ions and does not exhibit significant mass discrimination. This detector is an alternative for measuring high ion currents. In this way, high precision measurements are obtained.

The Faraday cup and the electron multiplier (with discrete or continuous detector) are the most commonly used detectors in mass spectrometry.

8.5.5 Analytical performance

TIMS is probably one of the best methods in terms of precision. This makes this method highly suitable for isotopic ratio measurements, particularly when a Faraday cup detector and a sector mass spectrometer are used. Precision can be as good as 0.1%, for instance for $^{42}Ca/^{40}Ca$, $^{25}Mg/^{24}Mg$. Depending on the application, it may be necessary to correct for the mass-dependent isotope fractionation. TIMS is relatively insensitive to polyatomic interferences, particularly those arising from the presence of Ar, such as in ICP-MS or GDMS. The sample throughput is typically at least 20 samples per day, which is better than SSMS but not as good as ICP-MS.

ICP-MS provides an efficient combination between precision (better than a few %), accuracy, number of determinable elements, limits of detection and sample throughput. The limits of detection can be as low as $1-10 \, \text{ng L}^{-1}$. However, because of a limited salt concentration (typically $0.1 \, \text{g L}^{-1}$) to avoid the blocking of the sampler, limits of detection in a solid, prior to dissolution, are only in the

One limitation of ICP-MS is the presence of isobaric interferences, particularly for m/z values below 80.

10–100 ppb range. In any case, however the limits of detection in solution are improved by a factor of 10 to 1000 compared with the use of the ICP in atomic emission spectrometry. This explains the strong interest in ICP-MS. Actually, the major drawback of the quadrupole type ICP-MS is the presence of isobaric interferences such as those arising from molecular species formed with Ar, e.g., $^{40}Ar^{16}O^+$ on $^{56}Fe^+$, $^{40}Ar^{35}Cl^+$ on $^{75}As^+$, $^{40}Ar^{40}Ar^+$ on $^{80}Se^+$. This makes the use of this technique for m/z values below 80 difficult. Isobaric interferences are also observed for elements forming refractory oxides (Ca, Ti, Zr, Mo, La, Ta, rare earth elements, etc.), e.g., $^{47}Ti^{16}O^+$ on $^{63}Cu^+$, or elements forming doubly charged ions (Ca, Ba, U, etc.), e.g., $^{138}Ba^{2+}$ on $^{69}Ga^+$. The isobaric interference problem can be partly solved by using a high resolution mass spectrometer. A resolving power of about 5000 is sufficient to separate $^{40}Ar^{16}O^+$ from $^{56}Fe^+$. ICP-MS is also used for isotopic measurements. Although the precision is not yet at the level of that obtained with TIMS, the high sample throughput of the ICP-MS makes this technique suitable for this purpose. Moreover, TIMS precludes the measurement of some isotopic ratios such as Re/Os, which can be easily determined by ICP-MS. However, it is obvious that isotopic rato measurements cannot be used for monoisotopic elements such as Be, Al, Sc, Mn, Co, As, Y, Nb, Rh, Au, Bi, and Th.

Although the ICP-MS is mainly used for the analysis of solutions, laser-produced plasmas can be used in conjunction with the ICP to provide the direct analysis of solids. The microplasma is produced at atmospheric pressure and laser ablation of the sample occurs. Fine particles are taken up by a stream of argon for further volatilization, atomization, and ionization in the ICP.

SSMS can determine a large number of elements. However, because of a relatively poor precision inherent to the instability of the ionization process and a possible inhomogeneity of the samples, a spark source is mainly used for qualitative analysis and semiquantitative panoramic analysis. Limits of detection can be in the 1–10 ppb range for many elements, the major limitation being the use of the photographic plate. Even so, the excellent limits of detection in a solid are one of the most important characteristics of the spark source. Like any inorganic mass spectrometry technique, SSMS can suffer from isobaric interferences due to the formation of molecular species. It should be noted that the sample throughput can be considered as low. This is due to the use of a photographic plate, which implies a limited dynamic range and time for development and reading.

GDMS provides more reliable results than SSMS when trace amounts are at the sub-ppm concentration. Actually, GDMS can achieve qualitative analysis of solids in a concentration range not achievable by other methods. Determination for some elements at the sub-ppb levels have been reported.

The LAMMA system exhibits limits of detection at the ppm level for easily ionized elements. This is a remarkable figure, considering the small sample size. Actually, the absolute amount which is really detected can be as low as 10^{-20} g.

RIMS is a highly sensitive and selective method, but rather complex, as RIMS implies the use of one or several lasers. Limits of detection are in the sub-ppb range and the precision is about 5–10%. Furnace RIMS can be used for both liquid and solid samples whereas SIRIS and LARIS are used for solid samples only.

8.5.6 Applications

Isotope dilution mass spectrometry provides high accuracy and precision, based on the use of stable isotope spikes.

Precision and accuracy can be significantly improved by using the method of isotope dilution mass spectrometry (IDMS). A known quantity of an isotopic spike, usually of minor abundance and being either a stable isotope or a long-lived radioisotope, is added to the sample and carefully equilibrated, which precludes the direct use of a solid. Then, the ratio of the highest abundance isotope to the spike isotope is measured to obtain an accurate value of the concentration of the analyte. Usually the ratio of the enriched sample is compared with the natural ratio, which is normally invariant. The amount of spike selected is such that the experimental ratio is near one. In this instance, the precision is maximum. It is important to verify whether any of the two isotopes suffers from isobaric interferences. Some examples of IDMS are ^{206}Pb added to ^{208}Pb, and ^{111}Cd added to ^{114}Cd. The crucial point is the sample preparation. Use of IDMS has some definite advantages: high precision and accuracy can be obtained, detection limits may be improved down to less than 1 pg, and the spike isotope acts as an ideal internal standard. The main limitation is the availability of some stable isotopes and their cost, although the among needed for the experiment is rather small. Species equilibration in the sample must be ensured if any additional sample preparation techniques are considered.

Inorganic mass spectrometry is preferably used each time extremely low concentrations and isotopic abundance are to be determined. The field of application is broad; semiconductors and ceramic materials, environmental samples, biological materials with the use of stable isotope tracers, high-purity reagents and metals, nuclear materials, geological samples, food, etc.

A growing field is also the coupling of inorganic mass spectrometry techniques such as the ICP-MS, with separation methods in order to provide information about the identification and quantitation of the chemical form of an element (speciation), particularly for elements whose toxicology depends on their chemical form: As, Se, Pb, Hg [8.5-20].

Speciation will become more and more applied based on the use of ICP-MS.

References

[8.5-1] Adams, F., Gijbels R., Van Grieken, R., (Eds.), *Inorganic Mass Spectrometry*. New York: John Wiley, 1988.

[8.5-2] Adams, F., *Spectrochim. Acta* 1983, 38B, 1379.

[8.5-3] Crews, H.M., Ducros, V., Eagles, J., Mellon, F.A., Kastenmayer, P., Luten, J.B., McGaw, B.A., *Analyst* 1994, 119, 2491.

[8.5-4] Coburn J.W., Harrison, W.W., *Applied Spectroscopy Rev.* 1981, 17, 95.

[8.5-5] Date, A.R., Gray, A.L., (Eds.), *Applications of inductively coupled plasma mass spectrometry*. Glasgow: Blackie, 1989.

[8.5-6] Jarvis, K.E., Gray, A.L., Houk, R.S., *Handbook of Inductively Coupled Plasma Mass Spectrometry*. Glasgow: Blackie, 1992.

[8.5-7] Blades, M., *Appl. Spectrosc.* 1994, 48, 12A.

[8.5-8] Denoyer, E.R., Fredeen, K.J., Hager, J.W., *Anal. Chem.* 1991, 63, 445A.

[8.5-9] Harrison, W.W., Bentz, B.L., *Progress in Anal. Spectrosc.* 1988, 11, 1.

[8.5-10] Stuewer, D., *Glow discharge mass spectrometry – Aspects of a versatile tool*, in *Applications of plasma mass source mass spectrometry*. Holland, G., Eaton, A.N., Cambridge: The Royal Society of Chemistry, 1991.

[8.5-11] Marcus, R.K., (Ed.), *Glow discharge spectroscopies*. New York: Plenum Press, 1993.

[8.5-12] Broekaert, J.A.C., *Appl. Spectrosc.* 1995, 49, 12A.

[8.5-13] Mykytiuk, A.P., Semeniuk, P., Berman, S., *Spectrochim. Acta Rev.* 1990, 13, 1.

[8.5-14] Vertes, A., Gijbels, R., Adams, F., *Laser Ionization Mass Analysis*. New York: John Wiley, 1993.

[8.5-15] Sjöström, S., Mauchien, P., *Spectrochim. Acta Rev.* 1993, 15, 125.

[8.5-16] Hurst, G.S., Payne, M.G., *Principles and Applications of Resonance Ionization Spectroscopy*. Bristol: Adam Hilger, 1988.

[8.5-17] Smith, D.H., Young, J.P., Shaw, R.W., *Mass Spectrom. Rev.* 1989, 8, 345.

[8.5-18] Saloman, E.B., *Spectrochim. Acta* 1991, 46B, 319.

[8.5-19] Miller, P.E., Bonner Denton, M., *J. Chem. Education* 1986, 63, 617.

[8.5-20] Vela, N.P., Olson, L.K., Caruso, J.A., *Anal. Chem.* 1993, 65, 585A.

Questions and problems

1. What is the major difference between mass spectrometers used in organic mass spectrometry and those used in inorganic mass spectrometry?
2. What is the principle of thermoionization mass spectrometry?
3. Could negative ions be determined with thermoionization mass spectrometry?

4. Why is the inductively coupled plasma highly suited to inorganic mass spectrometry?

5. What is the type of mass spectrometer which is the most commonly used with ICP?

6. Why is it necessary to use an interface in ICP-MS?

7. Which type of detector is commonly used with spark source mass spectrometry?

8. What are the main characteristics of the glow discharge used as an ionization source in mass spectrometry?

9. What is the role of the laser in resonance ionization mass spectrometry?

10. Why is resonance ionization mass spectrometry based on a multistep process?

11. What are the possible modes for data acquisition in ICP-mass spectrometry?

12. What is the main field of application of thermoionization mass spectrometry?

13. Why is the determination of ^{75}As difficult in the presence of Cl when using ICP-MS?

14. Selenium exhibits 6 isotopes: ^{74}Se, ^{76}Se, ^{77}Se, ^{78}Se, ^{80}Se, and ^{82}Se. In ICP-MS, Se suffers from isobaric interferences from Ar_2^+ and $ArCl^+$. Considering that Ar has the following isotopes ^{36}Ar, ^{38}Ar, and ^{40}Ar and Cl has the following ones ^{35}Cl and ^{37}Cl, what conclusions can be given as to the determination of Se?

15. What is the principle of isotope dilution mass spectrometry and its major advantages?

9 Compound and Molecule Specific Analysis

Learning objectives

- To introduce the basic principles of the main methods of molecular spectroscopy: UV-VIS, IR and Raman, ^{1}H and ^{13}C NMR, and mass spectroscopies
- To present modern instrumentation for obtaining molecular spectra from solids, liquids and gases
- To describe typical analytical applications of molecular spectroscopy including sampling, measurement and evaluation

As we learn from theory, molecular systems can be identified by their characteristic energy term schemes consisting of discrete electronic, vibrational and rotational states. At room temperature the substances are mainly in their electronic and vibrational ground states. Upon interaction with the appropriate type of electromagnetic radiation, characteristic electronic, vibrational and rotational transitions can be induced in the sample. These *excited states* usually decay to their original ground states within 10^{-8} s, either by emitting the previously absorbed radiation in all directions with the same or a lower frequency, or by "radiationless" relaxation (see Fig. 9-1).

Molecular spectra can be obtained in the absorption or emission mode from samples in the gaseous, liquid or solid state. They are images of the interactions mentioned above and contain analytical information about the sample.

This chapter introduces the principles and state of the art of today's powerful analytical molecular spectroscopy started by Bunsen and Kirchhoff in 1860. Initially only the visible spectral range was used (see. Chap. 8), but has since been expanded to utilize the ultraviolet (UV) and infrared (IR) regions of the electromagnetic spectrum for molecular systems.

The principles, the experimental setup, the analytical information, and typical applications of UV-VIS spectroscopy are treated in Sec. 9.1, and IR and Raman spectroscopy in Sec. 9.2.

Further complementary structural information about molecular systems may be obtained by investigating the *nuclear magnetic resonance* (NMR) of a sample being irradiated with radio frequencies in a high magnetic field (Sect. 9.3). More structural information may be obtained by analyzing the intensity distribution of mass fragments of a sample bombarded with free electrons, photons or ions. This technique is called *mass spectrometry* (MS) and is described in Sec. 9.4.

Ideally, all techniques of molecular spectroscopy require pure compounds for unambiguous compound identification, hence these techniques are frequently used after the purification of mixtures by classical or chromatographic methods of separation. Modern multidimensional systems are available today which link chromatography and spectroscopy on-line (see Chap. 14).

9.1 UV-VIS Spectrometry, Emission and Luminiscence

9.1.1 Principles

Most UV-VIS spectra are obtained by measuring the intensity of the absorption of monochromatic radiation across a range of wavelengths passing through a solution in a cuvette. The practical wavelength region extends from 190–400 nm (UV range) and from 400–780 nm (VIS range).

In a typical experiment, a light beam of intensity I_0 strikes a sample consisting of a quartz or glass cell containing a solution (Fig. 9.1-1).

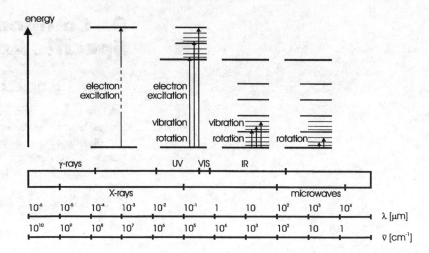

Fig. 9-1. Simplified term scheme of a polyatomic system

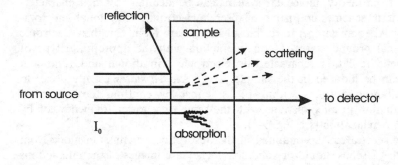

Fig. 9.1-1. Intensity loss of a light beam of intensity I_0 by reflection, scattering and absorption

Transmission measurements can be used for qualitative and quantitative analysis.

After passing through the cell, the light beam has a reduced intensity I due to reflection losses at the cell windows, absorption in the sample and, eventually, by scattering at dispersed particles. Only the absorption losses are caused by the dissolved sample. The run is repeated using an identical cell containing only solvent to compensate for reflection and scattering losses, and the transmittance T is calculated using the following equation,

$$T = \frac{I}{I_0} \approx \frac{I_{\text{solution}}}{I_{\text{solvent}}} \tag{9.1-1}$$

The presentation of transmittance T, as a function of wavelength λ, is the required *spectrum* of the sample (Fig. 9.1-2).

Using the human eye as a (qualitative) detector and sunlight as a source, the sample in Fig. 9.1-2 appears colourless because it does not absorb in the spectral region above 400 nm. Increasing the size of the conjugated-ring system shifts the absorption bands into the blue area of the visible range and so the sample looks red or yellow to us (see Fig. 9.1-7). Similarly, substances that look blue absorb strongly in the red part of the visible spectrum. This is a qualitative explanation for the many colour phenomena in our world.

The intensity of an absorption band, i.e., the *absorbance*, is proportional to the number of absorbing species in the illuminated part of the cell. Absorbance, A, is defined by the equation,

$$A = -\log T = \log \frac{I_0}{I} \tag{9.1-2}$$

and is proportional to the cell thickness, d [cm], the concentration of the solution, c [mol/L]; and a substance-specific proportionality constant ε called the *molar absorptivity*, [L mol^{-1} cm^{-1}].

$$A = \varepsilon \cdot c \cdot d \qquad \text{Beer's law}$$

For a given system, a linear relationship exists between A and the sample concentration, but usually only for dilute solution ($c \leq 0.1$ mol/L). At higher concentrations, changes of ε values may occur which lead to deviations from a linear working curve.

(a)

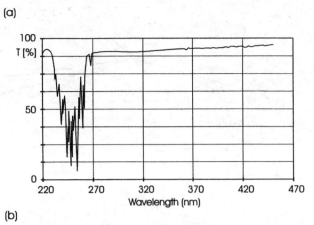

(b)

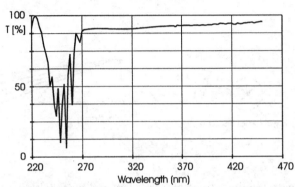

Fig. 9.1-2. UV-VIS spectrum of benzene in isooctane: (a) gaseous state, (b) liquid state

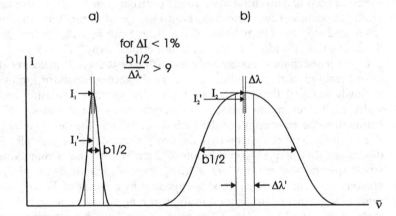

Fig. 9.1-3. Deviation from Beer's law due to nonmonochromatic radiation

A further deviation from Beer's law is observed when measurements made using polychromatic radiation are compared with calibrations done with monochromatic radiation (Fig. 9.1-3).

This observation is especially important when spectrometers with low resolutions and narrow wavelength bands are used for quantitative measurements. In such cases, when the magnitude of the spectral resolution $\Delta\lambda'$ is comparable to that of the full width at half height $b_{1/2}$ of the band (Fig. 9.1-3a), the measured intensity I_1' is much lower than the true value I_1 obtained using resolution $\Delta\lambda$. This problem is less pronounced in case where $b_{1/2} \gg \Delta\lambda$ (Fig. 9.1-3b). As a rule of thumb, one must have $b_{1/2}/\Delta\lambda > 9$ to give correct results.

Absorbance A for quantitative analysis with linear working curves.

9.1.2 Experimental

Spectrometers

In order to obtain complete spectra in the UV-VIS range, dual beam dispersive scanning systems or dispersive multichannel systems are employed. Both comply

a)

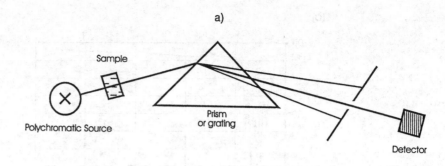

b)

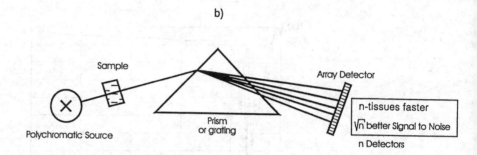

Fig. 9.1-4. Comparison of (a) dispersive monochannel and (b) multichannel spectrometers

with the limitations of Beer's law and work with monochromatic radiation. They are composed of polychromatic broad spectrum sources, dispersive units (mostly gratings), sample cells, detectors, electronics, and computers for data manipulation and storage. The position of the sample can be either before or behind the dispersive unit, if radiation damage is not critical (why?) (Fig. 9.1-4).

In the monochannel system only one detector is used. It measures the intensity of one resolution element after another as the monochromator (grating or prism) is slowly scanned through the spectrum. The spectral resolution and hence the width of the resolution element is determined by the dimension of the monochromator, the wavelength of the radiation used and the (adjustable) slit width.

The multichannel system uses an array detector (with typically 316 silicondiodes) so that a spectral resolution of 2 nm results and is maintained over the whole spectral region from 200–820 nm. Since all intensities are measured simultaneously the measurement time is improved by a factor of 316 or – assuming the same measurement time as used in the scanning system – the signal-to-noise ratio (S/N) is enhanced by $\sqrt{316}$. This property is called the *multichannel advantage* of multidetector systems. Since there are also no narrow slits required in an array spectrometer the radiation throughput is much higher, and one single deuterium source for the whole UV-VIS range from 200 to 780 nm (*throughput advantage*) is used.

In conventional spectrometers, which have a lower throughput, the deuterium lamp can only be used as a continuous source between 200 and 400 nm (UV to visible range), while tungsten-halogen bulbs serve for the range of 400–2500 nm (visible to near infrared range). Glass cuvettes with a typical cell thickness of 1 cm for trace analysis, are transparent enough for the VIS range, whereas more expensive quartz cells are available for the whole UV-VIS-NIR spectral area. In all cases, cells with plane-parallel windows and a precise sample thickness are required for quantitative work. Table 9.1-1 gives a survey of some suitable solvents for the UV-VIS range.

Filter photometers

In photometers, the expensive monochromator unit is replaced by one or more interference filters, thus making the system usable for specific applications at given

Table 9.1-1. Solvents for the UV-VIS range and their minimum wavelength in 1 cm cells

Solvent	Minimum wavelength (λ_{min})
Acetonitrile	190 nm
Water	191 nm
Cyclohexane	195 nm
Hexane	201 nm
Methanol	203 nm
Ethanol	204 nm
Diethylether	215 nm
Dichloromethane	220 nm
Chloroform	237 nm
Carbon tetrachloride	257 nm

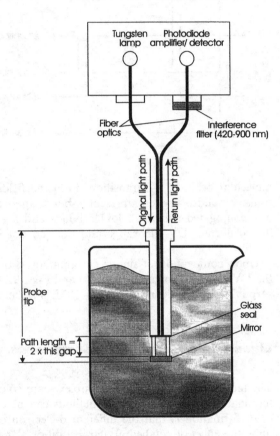

Fig. 9.1-5. Schematic of a fiber photometer

wavelengths. This old instrumental setup has gained new importance by the recent development of high-quality quartzglass fibers as used in telecommunication.

A typical experimental setup of a fiber photometer for use in the visible range is given in Fig. 9.1-5.

SiO_2 fibers can be produced today with theoretical attenuation values of 0.1 dB/km ($dB = -10\log(I_0/I)$). Hence rather than bringing the sample into a cell and to the spectrometer, the spectrometer can be brought to the sample and the measurements be carried out *in situ* (without sampling, which might change the composition of or even destroy labile samples). This new possibility is of great importance to environmental and process monitoring (see Chap. 16).

9.1.3 Analytical information in the UV-VIS range

In the UV-VIS range from 200 to 780 nm typical *chromophores* (light absorbing groups) can be observed; groups with $n \rightarrow \pi^*$, $\pi \rightarrow \pi^*$ transitions in

Non-conjugated chromophores.

Table 9.1-2. Absorption maxima of nonconjugated chromophores

Chromophore	Transition	λ_{max} (nm)
–C–C–	$\sigma \rightarrow \sigma^*$	150
–O–	$n \rightarrow \sigma^*$	185
–N<	$n \rightarrow \sigma^*$	195
–S–	$n \rightarrow \sigma^*$	195
>C=O	$\pi \rightarrow \pi^*$	190
	$n \rightarrow \pi^*$	300 (weak)
>C=C<	$\pi \rightarrow \pi^*$	190

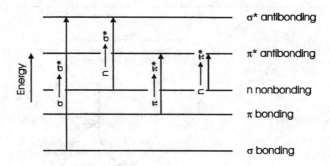

Fig. 9.1-6. Schematic of molecular orbital energy levels

Table 9.1-3. Absorption maxima of some conjugated chromophores

Substance	λ_{max} (nm)
CH₃–C–CH=CH₂ ‖ O	225
CH₂=CH–CH=CH₂	217
CH₃(CH=CH)₃ CH₃	274
CH₃(CH=CH)₅ CH₃	342
CH₃(CH=CH)₇ CH₃	401
⬡	203
⬡—CH=CH₂	248

Conjugated chromophores.

molecular orbitals, d-d transitions in ligand fields of metal chelates and charge-transfer bonds. The omnipresent σ-bonds in organic compounds and also the non conjugated (isolated) double bonds and $n \rightarrow \sigma^*$ transitions are not excited in the normal UV-VIS range and thus do not interfere (Table 9.1-2 and Fig. 9.1-6).

Upon conjugation of simple chromophores, the absorption maxima shift into the UV and VIS range (Table 9.1-3 and Figs. 9.1-2 and 9.1-7). Typical molar absorptivities are in the range of 10^3–10^5 L cm^{-1} mol^{-1}.

Ligand field and charge-transfer absorptions

The fact that transition metal complexes are colored means that trace metal determination using colorimetric methods can be carried out. The reason for this color formation is that the different d-electron orbitals of metal ions may split upon the effect of octahedral, tetrahedral or planar ligand fields (for details, read Cotton, A., Wilkinson. G., Advanced Inorganic Chemistry, New York: J. Wiley, 1966). With increasing ligand field strength the absorption maxima shift to shorter wavelengths. These weak absorption bands are usually accompanied by very strong charge-transfer bands below 400 nm. The absorption mechanism involves the transition of an electron from a donor to an acceptor orbital of the complex (Fig. 9.1-8).

Solvent effects

Solvents can interact strongly with certain solutes and thus change the observed UV-VIS spectra, either by removing vibrational fine structures, or by shifting absorption band maxima, or both. When the UV spectra of phenol in water (Fig. 9.1-9a) and in isooctane (Fig. 9.1-9b) are compared, both a blue shift and the disappearance of splitting can be observed in the water case. The blue shift is typical for the stabilization of the ground state by hydrogen bonding solvents such as water (why?), the splitting can only be observed in water-free solvents without hydrogen bonding (why?).

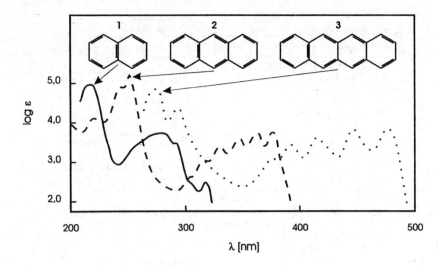

Fig. 9.1-7. UV-VIS spectra of

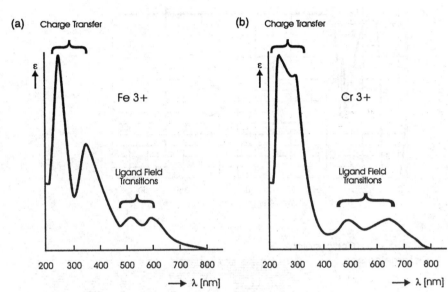

Fig. 9.1-8. Charge-transfer and ligand field spectra of (a) Fe(III) and (b) Cr(III) tetramethylene dithiocarbamates for trace metal analysis

Red shifts of band maxima can be observed in polar solvents when the excited states are more polar than the ground states.

$$\text{e.g.} \quad \underset{\overset{\|}{O}}{CH_2{=}HC{-}C{-}CH_3} \leftrightarrow \underset{\overset{|}{O}}{CH_2{-}HC{=}\overset{\oplus}{C}{-}CH_3}$$

Red or blue shifts of 10–20 nm are typical.

9.1.4 Applications of UV-VIS absorption spectroscopy

The most important use of UV-VIS absorption spectroscopy is in quantitative (trace) analysis of metals, drugs, body fluids, and food due to its sensitivity, reproducibility and ease of operation. Sensitivities in the 10^{-5} to 10^{-6} M range (with enrichment by solvent extraction), and precision of a few tenths of a percent are typical. For example, see Section 9.1.6 (metal traces, cholesterol, enzymatic samples, HIV, etc!)

Array detector systems are furthermore used as flow-through detectors for *high performance liquid chromatography* (HPLC) due to their high recording speed based on the multichannel advantage (see Sec. 9.1.2). In time intervals of 5 sec-

Application of array detectors saves time.

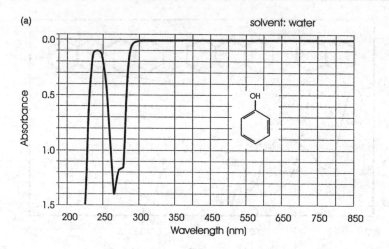

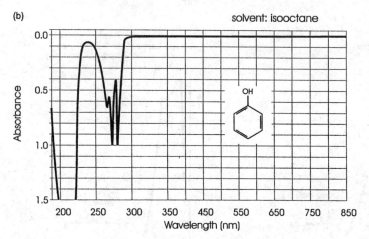

Fig. 9.1-9. UV-VIS spectra of phenol in
(a) H_2O (H-bonding with phenol) and in
(b) isooctane (no interaction with phenol)

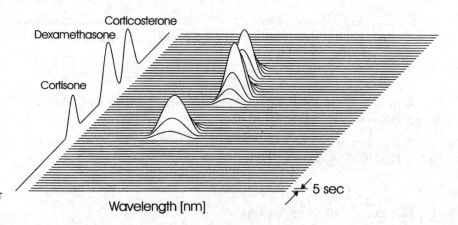

Fig. 9.1-10. LC-UV array detector signals for
a mixture of three drugs

onds, complete UV-VIS spectra of the eluted compounds can be recorded rather
than just LC peaks measured at a single wavelength (Fig. 9.1-10).

The application of UV-VIS spectroscopy for structural analysis has become
less important recently due to the increased power of IR, NMR and mass
spectroscopies.

Analysis trace metal

Nearly for every metal ion, organic chelate forming reagents are known. These
form colored and/or charge-transfer compounds which can be extracted from
aqueous into organic solutions, enriched and measured quantitatively. One important aspect, as compared to other techniques, is the selectivity for distinct oxidation states (speciation – see Sec. 7.5).

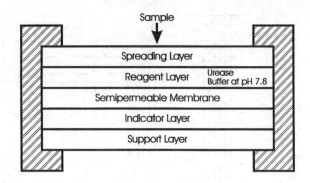

$$H_2NCONH_2 + H_2O \xrightarrow{\text{Urease}} 2NH_3 + CO_2$$

$$NH_3 + \text{Ammonia Indicator} \longrightarrow \text{Dye}$$

Fig. 9.1-11. Kodak Ektachem urea slide

Clinical analysis

Well-known colorimetric reactions – such as for urea or glucose in serum, enzymes, or cholesterol – have been adapted for rapid screening by "dry chemistry" procedures (e.g., Kodak Ektachem). In this procedure, one drop of the sample (e.g., blood serum) is applied on top of a multilayered slide (Fig. 9.1-11).

This slide contains – in distinct layers – all the reagents necessary for the enzymatic analysis of urea according to the reactions described in Fig. 9.1-11. The appearance of a characteristic coloring indicates the presence of the substance under test. The presence of other substances can be determined in a similar way, by using different enzymes, e.g., glucose oxidase for the analysis of glucose:

$$\beta\text{-D-glucose} + O_2 \xrightarrow{GO_x} \text{gluconic acid} + H_2O_2$$

$$H_2O_2 + \text{tartrazine} \xrightarrow{PO_x} \text{o-toluidine} + H_2O$$

If glucose is present in the sample, a green color develops at the lower end of the slide and can be measured photometrically in an automated way and evaluated quantitatively.

Similar tests are also available on (cheap) test strips for urine and blood serum screening, on which eventual color changes are visually compared with color scales. The test results are semiquantitative, but a rapid survey over many samples can be made in seconds.

"Dry chemistry" for clinical analysis.

9.1.5 Molecular fluorescence, phosphorescence and chemiluminescence

In addition to the absorption techniques treated in Sec. 9.1.1 to 9.1.4, this section gives an introduction to the luminescence methods of molecular spectroscopy: fluorescence, phosphorescence, and chemiluminescence. The common principle of these three techniques is the measurement of the emission spectra obtained when excited molecules decay to their ground states.

Compared to absorption techniques, luminescence methods are extremely sensitive (ppb-range) and the signal intensities are largely proportional to the concentration of the sample. The number of substances which fluoresce, phosphoresce, or chemiluminesce is, however, limited.

Fluorescence plays an important role in today's emerging fiber-optical chemical sensors (see Sec. 7.10 and 7.11). Therefore the principles of the luminescence methods and instrumentation are briefly described here.

Photoluminescence (fluorescence and phosphorescence)

The excitation of molecules to higher energy states by the absorption of energy is very rapid (10^{-15} s), and can occur to several vibrational levels of the excited electronic states (Fig. 9.1-12).

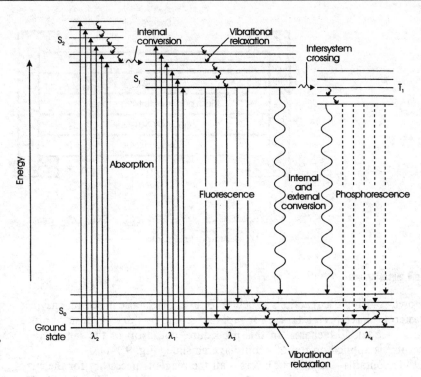

Fig. 9.1-12. Energy level diagram to describe absorption, vibrational relaxation, internal conversion, fluorescence, external conversion, intersystem crossing, and phosphorescence in molecules

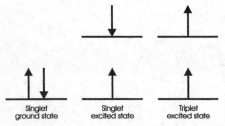

Fig. 9.1-13. Singlet and triplet states

The excited states usually decay rapidly by nonradiational "vibrational relaxation" (10^{-10} s). Internal conversion ($S_2 \rightarrow S_1$ in Fig. 9.1-12) can occur when two electronic levels are so close together that vibrations of the lower electronic state can be excited.

Fluorescence occurs when a molecule excited to a higher vibrational state of an upper electronic level returns via the lowest vibrational level of this excited electronic state to any of the vibrational levels of the ground state. In this case, radiation having a longer wavelength than the exciting radiation is emitted only after 10^{-8}–10^{-6} s.

Collision of excited species with gas molecules, such as O_2, may lead to fluorescence *quenching*. This is caused by a nonradiational transfer of energy from the excited species to other molecules (*external conversion*), and results in a decrease of the fluorescence intensity of a given sample.

During these excitation and decay processes the orientation of the electron spin is preserved. In some cases, triplet states can also be populated from ordinary singlet states by *intersystem crossing* (see Fig. 9.1-13 and the transition $S_1 \rightarrow T_1$ in Fig. 9.1-12).

A molecule in the T_1 state decays after intersystem crossing by vibrational relaxation or in some cases by "phosphorescence", a slow radiational deactivation process ($T_1 \rightarrow S_0 + h\nu$) which takes 10^{-4} s to 10 s or even longer. Phosphorescence is usually observed only if external conversion is reduced by cooling or immobilization. The wavelengths of phosphorescence are even longer than in the case of fluorescence (Fig. 9.1-14).

The most intensive fluorescence is found in aromatic compounds with low-energy $\pi \rightarrow \pi^*$ transitions (conjugated chromophores). Thus the fluorescence emission maxima shift in the homologous series from benzene (278 nm), naphthalene (321 nm), anthracene (400 nm) to naphthacene (480 nm). Phenols and aniline – compounds with acid protons – show a strong pH-dependence of their fluorescence.

Phosphorescence is not observed in compounds with strong fluorescence, but is in certain pesticides, anion acids, enzymes, and petroleum hydrocarbons. Room temperature phosphorescence can be observed for compounds absorbed on surfaces due to stabilization of the triplet state.

Instruments for measuring fluorescence and phosphorescence are similar to UV-VIS filter photometers or spectrometers, but require a reference beam and detector for background measurements (Figs. 9.1-15 and 9.1-16).

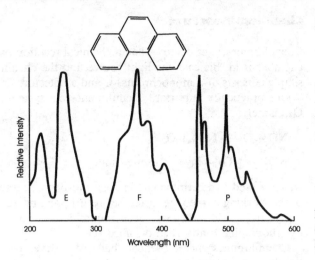

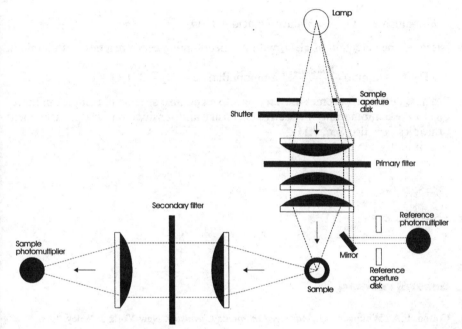

Fig. 9.1-14. Excitation (E), fluorescence (F), and phosphorescence (P) spectra of phenanthrene

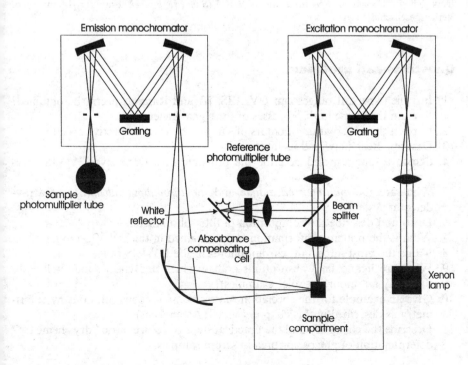

Fig. 9.1-15. Schematic of a typical fluoro photometer

Fig. 9.1-16. Schematic of a typical spectrofluorometer

Chemiluminescence

Chemiluminescence occurs when a chemical reaction produces light. The reaction chamber is in this case the light source and the chemiluminescence spectrometer simply consists of a monochromator and a detector.

One practically important chemiluminescent system is the reaction of NO with O_3, namely:

$$NO + O_3 \rightarrow NO_2^* + O_2$$

$$NO_2^* \rightarrow NO_2 + h\nu \qquad \text{*excited state}$$

which is used to determine NO in the atmosphere at trace levels. The linear section of the working curve is reported between 1 ppb and 10 000 ppm NO.

O_3 can be determined by the chemiluminescence of its reaction product with the dye rhodamine B adsorbed on silica gel.

Chemiluminescence biosensors have been developed recently for oxidase catalyzed reactions such as

$$\beta\text{-D-glucose} + O_2 \xrightarrow{\text{GO}_x} \beta\text{-gluconic acid} + H_2O_2$$

based on the detection of H_2O_2 with its chemiluminescence reaction with luminol.

$$2\,H_2O_2 + \text{luminol} \xrightarrow{\textit{Peroxidase}} \text{3-aminophthalate} + N_2 + 3\,H_2O + h\nu$$

These detection schemes are very sensitive and also specific in complex mixtures due to the combination of selective chemistry and sensitive physical measurement principles (see also Sec. 7.11).

Luminol:

General reading

Cotton, F.A., Wilkinson, G., *Advanced Inorganic Chemistry*. New York: J. Wiley, 1966.

Derkosch, J., *Absorptions-Spektralanalyse in UV, VIS und IR Bereich*. Frankfurt/Main: Akad. Verlagsgesellschaft, 1967.

Questions and problems

1. In which spectral ranges can UV, VIS, IR and Raman spectra be obtained? Explain the elementary processes of the signal generation.
2. Explain the observation of colors of samples by their respective spectra.
3. Describe Beer's law and its limitations.
4. Sketch a diagram and describe the principle of a dispersive UV-VIS spectrometer.
5. What are the *multiplex advantage* and the *throughput advantage* of array-detector systems?
6. Draw and describe the composition of filter photometers.
7. What is the principle information to be obtained in the UV-VIS range?
8. Name the most important chromophores in the UV-VIS range.
9. What are ligand field absorptions and in which analytical subdiscipline do they play an important role (examples)?
10. Give an example for the speciation of arsenic in aqueous solutions by colorimetry as described in this chapter (Marsh As-analysis).
11. Describe the chemistry and the practical steps of colorimetric "dry chemistry" determination of glucose or urea in serum samples.

12. Describe the basic principles of molecular fluorescence, phosphorescence and chemiluminescence and compare it to absorption phenomena in the UV-VIS range.
13. What is understood by "fluorescence quenching" and how can it be used for analytical purposes?
14. How can fluorescence be distinguished from phosphorescence in principle and in analytical applications? What are the detection limits?
15. Describe the spectrometer used for
    a) photoluminescence and
    b) chemiluminescence spectroscopies.
16. Describe in detail typical examples of the applications of chemiluminescence in chemical analysis.

9.2 Infrared and Raman Spectroscopy

Both IR and Raman spectroscopy belong to the group of molecular vibrational spectroscopies which, together with NMR (nuclear magnetic resonance spectroscopy), mass spectrometry and chromatography, form the supporting pillars of modern organic analysis, including structure, micro and surface analysis.

What can vibrational spectroscopy contribute to the analysis of molecular systems?

Vibrational spectroscopy is:

- molecule-specific, providing inherent information about functional groups present in the molecule, including their kind, interactions and orientations
- isomer-selective due to the "fingerprint" range
- quantitative and non-destructive, even to labile compounds, with a principal working range from 0.1% to 100%, but also usable for traces after enrichment
- universal in its sampling requirements, enabling the extraction of information from solids, liquids and gases, from surfaces, interfaces between solids and liquids or gases, microdomains, bulk samples and layered structures.

For practical reasons the spectral regions of interest in IR and Raman spectroscopy are divided into four areas according to Table 9.2-1.

The most important area for the qualitative analysis of organic systems is the MIR (mid infrared) range where most of the normal vibrations are found ("group frequencies" and "fingerprint bands" – see Sec. 9.2.3). Today, Raman spectra are taken from samples excited by use of lasers in the visible or NIR (near infrared) range (see Sec. 9.2.2). The NIR range, containing overtones from normal vibrations (see Sec. 9.2.3), is also important for the routine quantitative analysis of food or industrial mixtures, while the FIR (far infrared) range includes lattice vibrations and normal vibrations containing heavy atoms and weak bonds.

IR and Raman spectroscopy differ in their way of signal generation. While IR spectra – in a similar way to UV-VIS spectra – are obtained in the absorption mode, Raman spectroscopy is a scattering technique (see Sec. 9.2.1).

Nevertheless, since not only theory but also modern instrumentation can be described for both techniques together, and since extractable information is often complementary, analytical IR and Raman spectroscopies are treated in this chapter together. For a more detailed treatment of the subject the reader is referred to special monographs (e.g., [9.2-1 to 9.2-4]).

Spectral collections are available now that combined Fourier transform spectrometers are on the market. These clearly show, that both IR and Raman spectra

Table 9.2-1. Spectral regions for Raman and IR spectroscopy

Region	Wavelength (μm)	Wavenumber (cm^{-1})
Visible	0.4–0.8	25 000–12 500
Near IR	0.8–2.5	12 500–4 000
Mid IR	2.5–25	4 000–400
Far IR	25–1000	400–10

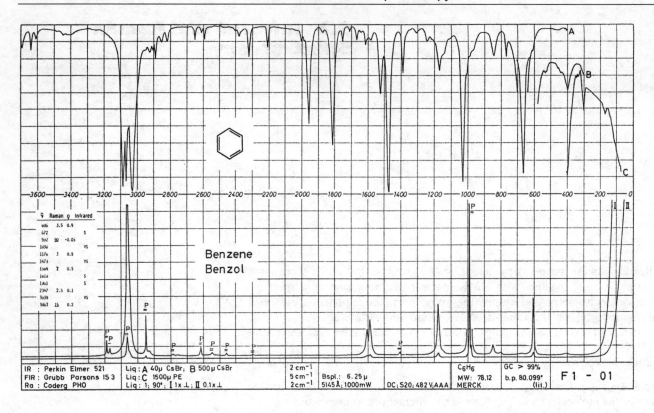

Fig. 9.2-1. IR and Raman spectra of benzene

are required to obtain complete structural information on polyatomic systems (Fig. 9.2-1).

9.2.1 Principles

Most IR spectra are recorded by measuring the absorption of the incident monochromatic radiation in the range of 2.5–25 μm (4000–400 cm^{-1}) by solid, liquid or gaseous samples. In a similar way as described in Section 9.1.1 for UV-VIS spectroscopy, the attenuation of the transmitted IR radiation can be displayed as a function of the wavelength. This is *IR spectrum* (see Fig. 9.2-1a).

In some cases, IR *emission* spectra can also be obtained in heated samples and/ or cooled detectors in special sampling modes (see Sec. 9.2.2).

Raman spectra are obtained by inelastic scattering of light at molecules (see Fig. 9.2-1b). To produce the necessary excitation of Raman spectra monochromatic laser sources in the visible or NIR range (e.g., Ar$^+$ laser at 488 nm or Nd:YAG laser at 1.06 μm) are required. This is however a very weak effect. Only approximately 10^{-8} parts of the incident radiation are elastically scattered. These form the *Rayleigh line* which has the same frequency as the exciting radiation. About 10^{-10} parts lead to the excitation of vibrational and rotational levels of the electronic ground state of the molecules. This causes an energy loss of the incident radiation and a band shift to longer wavelengths as compared to the Rayleigh line (*Stokes shift*). *Anti-Stokes* lines with higher frequencies than the incident radiation can be observed when the molecules under consideration were in excited vibrational levels (at higher temperatures) before the interaction with the laser source (Fig. 9.2-2). At room temperature these anti-Stokes lines are weaker than the Stokes lines. The intensity ratio between Stokes and anti-Stokes lines is a function of the temperature of the sample (why?).

As can be seen in Fig. 9.2-2, vibrational transitions are directly excited by IR radiation, the life time of the excited states being in the order of 10^{-6} s. With the Raman effect the molecule is excited into short-lived virtual states by the scattering of photons and returns after 10^{-12} s to the same, higher or lower vibrational states. These interactions result in the above-mentioned Rayleigh, Stokes anti-Stokes lines in the spectrum.

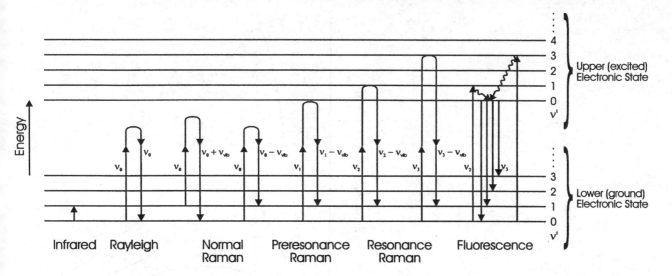

Fig. 9.2-2. Electronic and vibrational term scheme for IR, Raman and fluorescence transitions [9.2-4]

The regular Raman effect is very weak and can be analytically used only for the study of main components. Signal intensities can however be enhanced up to 10^4 times exploiting the *resonance Raman effect*. The enhancement is obtained by using an exciting wavelength that matches with electronic transitions of the sample (see Fig. 9.2-2). This is, however, not a routine technique and is not extensively treated here. Both normal and resonance Raman effects may be obscured by fluorescence, which is by a factor of 10^7 more efficient than raman scattering, when treating aromatic samples with visible or UV laser sources (see Sec. 9.1.5).

9.2.2 Experimental

The vast majority of today's IR and Raman spectrometers consist of a light source, a monochromator or interferometer, a detector and dedicated electronics and data systems. Only in IR emission spectroscopy does the sample act directly as the light source. In this section, the major components are described both for dispersive and non-dispersive systems.

Essential parts of IR- and Raman-spectrometers.

Sources

Silicon carbide rods ("globars") electrically heated to approx. 1100 °C are the most frequently used sources for MIR spectroscopy. Such sources show approximately the energy distribution of a blackbody radiator (Fig. 9.2-3).

The maximum radiant intensity at 1100 °C occurs at approximately 2 µm ($5000 \, \text{cm}^{-1}$). At lower temperatures the energy curve maximum shifts to a longer wavelength. It is important to note that the intensities fall off very rapidly at short wavelengths and also rather rapidly on the long wavelength side of the curve. These intensity changes have to be compensated for. In dispersive infrared spectrometers a slit located between the source and the sample compartment is programmed to open as the wavelength increases, resulting in an almost constant intensity level at all wavelengths. Nevertheless the low intensity of blackbody sources at short and long wavelengths requires alternatives for these spectral areas, because the spectral resolution is degraded when the slit width is increased.

Sources.

High-pressure mercury arc sources are used for the FIR. They consist of quartz-jacketed tubes containing mercury vapor which, upon passage of electricity, forms a plasma that provides continuous radiation in the FIR region. Bright UV-VIS light is also formed. This needs to be eliminated by a black polyethylene filter in order to avoid stray light and to protect the sample from eventual decomposition.

For the NIR region ordinary tungsten filament lamps are used.

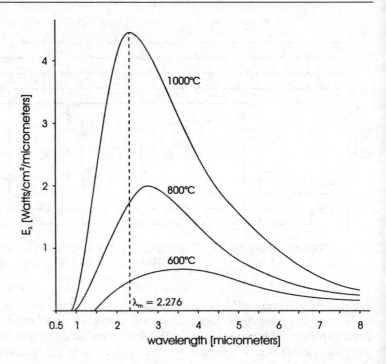

Fig. 9.2-3. Energy distribution of a blackbody radiator at 600 °C, 800 °C, and 1000 °C

Besides the blackbody-type sources, IR lasers are also used for particular applications of IR spectroscopy, where it is unnecessary to scan the whole MIR part of the spectrum. The intensity of laser sources is approximately 100 times greater than the blackbody sources mentioned above, but wavelengths are restricted to the available laser emission bands. Carbon dioxide and tunable PbS diode lasers produce radiation bands in the 9 to 11 μm (1100 to 900 cm^{-1}) range, where many organic substances absorb. Thus these laser sources are more and more used for the IR analysis of environmentally important substances, where low detection limits must be reached (e.g., monitoring of chlorinated hydrocarbons in water in the ppb range).

For Raman spectroscopy, monochromatic lasers are available for the excitation of scattered radiation in modern instruments. Visible (He-Ne, 632.8 nm or Ar, 488.0 and 514.5 nm) and IR lasers (Nd : YAG, 1.06 μm) are mainly used.

Since the intensity of the Raman lines varies with the fourth power of the exciting frequency, the 488 nm Ar-line provides scattering that is nearly three times as strong as that produced by an He-Ne laser with the same input power. VIS lasers are usually operated at 100 mW, while Nd : YAG lasers can be used up to 350 mW without causing the photodecomposition of organic samples. IR lasers are also usually incapable of causing fluorescence because they are not energetic enough to excite typical electronic transitions. Because of the relatively low Raman efficiency of this source, Nd : YAG lasers can only be used in Fourier transform spectrometers.

Detectors

Detectors.

Most common modern IR detectors are pyroelectric *triglycine sulfate* (TGS) detectors, which can be operated at room temperature, and photoconducting *mercury cadmium telluride* (MCT) detectors that have much higher sensitivity and shorter response time but can only be operated at liquid nitrogen temperature. While TGS detectors are used in the MIR and FIR ranges, MCT detectors are available for several spectral areas in the MIR.

For the FIR range, germanium bolometers are the detectors with the largest sensitivity. They are thermistors that exhibit a large change in resistance as a function of temperature. However, these temperature changes have to be measured at 1.5 K in order to suppress noise, and the detector must hence be cooled with liquid helium.

In the near IR range, lead sulfide photon detectors that an operate at room temperature are mainly used.

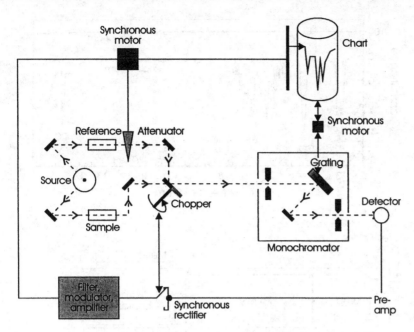

Fig. 9.2-4. Schematic of a double beam grating IR spectrometer (what are the differences to UV-VIS spectrometers?)

Raman detectors must either be sensitive in the NIR range, where cooled SiGe or InGaAs elements are required for sensitivity reasons, or in the visible part of the spectrum, where regular photodiodes can be used.

Spectrometers

IR sources emit polychromatic radiation. For spectroscopy, either monochromatic radiation is required (dispersive systems) – for the reasons discussed in the beginning of this chapter – or a complex encoding system must be used (multiplex systems). In the first case, grating or prism monochromators are applied to cover the whole spectrum. In the second case, a Michelson interferometer is typically used to modulate the IR radiation. When only narrow spectral areas are needed, optical filters or IR laser sources are sufficient (nondispersive systems).

Raman spectrometers are commercially available both in the dispersive (grating) and in the multiplex (interferometer) version. Combined IR and Raman spectrometers based on the Michelson interferometer are commercially available today.

Monochromators.

Dispersive systems

Most dispersive IR systems operate in the double beam mode. Figure 9.2-4 shows the schematic of a grating IR spectrometer. The radiation emanating from the globar is divided by aluminum mirrors into the sample and reference beams, then chopped by a sector mirror rotating at approximately 10 Hz to discriminate the signal from any background radiation, and spectrally dispersed at the grating monochromator. Finally, the modulated radiation passes through a slit system to the detector. The slit width is variable and determines the spectral resolution of the spectrometer (typically $0.1–10 \, \text{cm}^{-1}$). Several gratings have to be used in order to cover the whole IR range. The major difference of IR spectrometers as compared to UV-VIS instruments is the position of the sample compartment before the monochromator.

This is due to the nondestructive nature of the IR radiation. Consequently, the sample may be exposed to the total energy in order to efficiently eliminate the stray light produced by the sample.

The double beam arrangement has the advantage of compensating for any intensity fluctuations of the source and of optically cancelling background absorptions, such as water vapor or carbon dioxide from the air, clearly visible in the

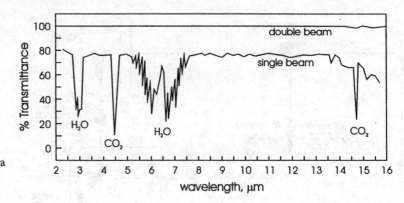

Fig. 9.2-5. Single and double beam IR spectra of atmospheric water vapor and carbon dioxide

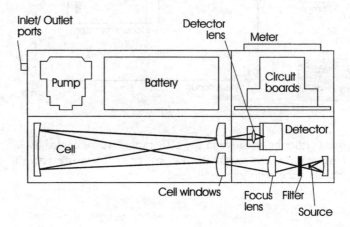

Fig. 9.2-6. A portable IR filter photometer for gas analysis

single beam spectrum (Fig. 9.2-5). This is done by rapidly switching from sample to reference beams by means of the 10 Hz chopper.

Inexpensive dispersive IR spectrometers are used for routine qualitative analysis. They operate according to the optical null principle, whereby the power of the reference beam is adjusted to the transmittance of the sample. This is achieved by introducing a comb-shaped attenuator into the reference beam; whenever a power difference between sample and reference occurs, the comb moves to compensate for it, as does the recorder pen. Unfortunately, a certain amount of overcompensation – especially at sharp peaks – and undercompensation is inevitable and so quantitative applications may not be recommendable when high precision is required.

Dispersive Raman spectrometers are similar in design to UV-VIS spectrometers (see Sec. 9.1), the difference being that double monochromators are used to eliminate spurious radiation. Unlike IR spectra, Raman spectra can be recorded in the complete MIR and FIR range from $4000-10 \, cm^{-1}$ using one single grating (why?).

Filter photometers

Small, dedicated systems using optical filters or gas-filled cells instead of monochromator units are commercially available for routine quantitative gas analysis in air. A variety of interference filters allow the range from $3000-750 \, cm^{-1}$ to be covered. The use of long path cells (path lengths up to 80 m) is possible due to the *folded ray arrangement* (Fig. 9.2-6). Detection limits in the range of parts per billion (ppb) ar reported for organic trace gases.

Some special IR photometers employ no wavelength selection device at all. Selectivity is achieved by filling the detector compartment with the analyte gas so that the detector responds only to radiation in the characteristic wavelength region of the analyte gas (Fig. 9.2-7).

The reference cell is filled with a nonabsorbing gas, and the sample flows through the sample cell. A thin metal diaphragm separates the two sections of the detector cell and serves as one plate of a capacitor. If the sample cell does not

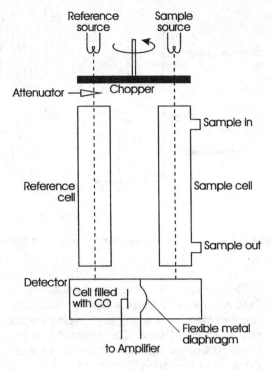

Fig. 9.2-7. A nondispersive IR photometer for the monitoring of gases (CO in this example)

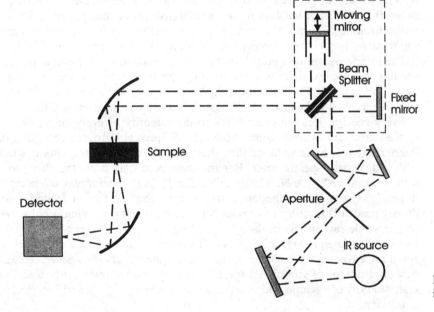

Fig. 9.2-8. Basic components of a Michelson interferometer

contain analyte gas, both compartments of the detector are heated equally by the two IR beams. If the sample cell does contain analyte gas, the beam passing through it, i.e., the right-hand beam in Fig. 9.2-7 is attenuated at the specific wavelengths were the detector gas absorbs. Thus the left-hand detector compartment becomes warmer than the right and, consequently, the diaphragm moves to the right. Therefore, the capacitance changes whenever analyte gas is present in the sample cell, resulting in a measurable IR signal.

Multiplex systems (Fourier transform spectrometers)

Instead of monochromators, multiplex systems based on the Fourier transform algorithm mainly use Michelson interferometers for coding the polychromatic spectrum of the source in both the case of IR as well as Raman spectrometers. Such a device essentially consists of a beamsplitter, a fixed mirror, a movable mirror, and a drive and positioning mechanism (Fig. 9.2-8).

The Michelson-Interferometer, a multiplex encoder.

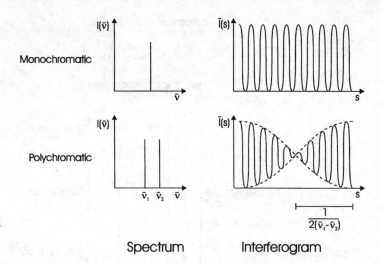

Fig. 9.2-9. Spectra-interferogram correlation
for monochromatic and polychromatic
sources

The radiation of the source is divided by the beamsplitter and directed to the
fixed and movable mirrors equally. Normally the moving mirror is scanned at a
constant velocity resulting in changing optical path differences of the two beams
as a function of time. The reflected beams interfere at the beamsplitter, from
where 50% of the radiation returns to the source, and 50% reaches the detector. At
the detector the intensity of the radiation is measured as a function of the optical
path difference of the beams in both arms of the interferometer. This signal is
called the *interferogram*. It is the Fourier transform of the spectrum. Thus for a
monochromatic source, the corresponding interferogram is a cosine function,
whereas two monochromatic liens give a more complex signal (Fig. 9.2-9).

The interferogram of a polychromatic source such as a sample in a globar light
beam is so complex, that the corresponding spectrum can no longer be recognized
by the human eye. By means of a digital computer and a special software such
interferograms can, however, be Fourier transformed almost in real time.

Every spectral element contributes to the intensity of every single data point
of the correlated interferogram. Assuming N spectral elements, recording of the
interferogram is N times faster than the recording of the corresponding spectrum
with a dispersive spectrometer. For the same recording time, the signal-to-noise
ratio is improved by $\sqrt{N}$. This is called the *Fellgett* or *multiplex advantage* over
dispersive systems. A further advantage results from the fact that interferometers
do not need slits in order to provide for a given spectral resolution but make use
of the whole ray bundle in the system (*Jacquinot or throughput advantage*). The
two result in a signal-to-noise (S/N) advantage of approx. 100 and a recording
speed improvement of 10 000 over dispersive systems. This has paved the way for
new applications of vibrational spectroscopies to micro, trace, surface and rapid-
scan analysis of organic and inorganic substances by significantly increasing the
sensitivity.

To make practical use of these Fourier advantages a very precise drive mechanism and exact He-Ne laser-based position locator for the movable mirror (in the
nm-domain – why?) and a rapid digital data system are mandatory. Enhancement
of the SN ratio can then be obtained by combining several thousand interferograms before performing the Fourier transform. A modern FTIR spectrometer
equipped with an MCT detector for the MIR range can take up to 80 interferograms per second and performs the Fourier transform for 4000 data points in less
than a second to produce routine IR spectra with a typical resolution of $1\,cm^{-1}$.
Much better resolutions can be obtained, however, when more measurement time
is given.

High resolution measurements.

The spectral resolution in Fourier spectrometers is determined by the sweep
length of the movable mirror. An optical sweep of 1 cm corresponds to a physical
displacement of the mirror of 0.5 cm and results roughly in a spectral resolution of
$1\,cm^{-1}$. Similarly, a sweep of 10 cm corresponds to a physical displacement 5 cm
and results in a resolution of $0.1\,cm^{-1}$.

High resolution interferometers with physical sweep widths of approximately
6 m and spectral resolutions of $0.001\,cm^{-1}$ are now on the market. Such instru-

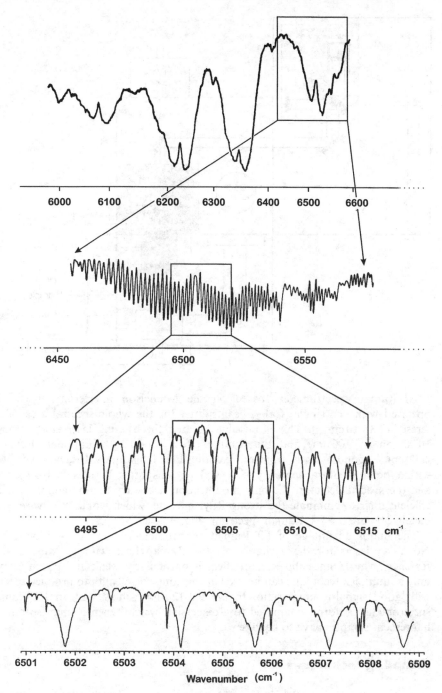

Fig. 9.2-10. Comparison of grating and FTIR spectra of gases with different spectral resolution: (a) grating (1962, Kuiper), (b) FTIR (1966, Connes et al.), (c) FTIR (1969, Connes et al.), (d) FTIR (1975, Mt. Palomar)

Table 9.2-2. Beamsplitter materials for FTIR spectrometers

Range	Material
$25000–5000\ cm^{-1}$	SiO_2 (VIS)
$15000–3000\ cm^{-1}$	SiO_2 (NIR)
$10000–1600\ cm^{-1}$	CaF_2
$7500–350\ cm^{-1}$	Ge/KBr
$550–70\ cm^{-1}$	Mylar, $6\ \mu m$
$250–50\ cm^{-1}$	Mylar, $12\ \mu m$
$120–30\ cm^{-1}$ [a]	Mylar, $23\ \mu m$
$50–10\ cm^{-1}$ [a]	Mylar, $50\ \mu m$

[a] below $100\ cm^{-1}$ with Hg lamp

Beamsplitter materials – the heart of the interferometer.

ments are needed to measure rotational fine structure of gases in the stratosphere, such as ozone or halogenated hydrocarbons, or in planetary atmospheres (Fig. 9.2-10).

Maintaining the precision of the drive mechanism is of special importance for successful applications in the visible and especially the ultraviolet range (why?). To this end, the mechanism employs air bearings for the movable Michelson mirror, and tow servo-interferometers: an He-Ne interferometer, to determine exactly the position of the mirror, and a white light-interferometer to locate the interferogram centerburst.

To cover the whole VIS, NIR, MIR and FIR range several different beam-splitters are required. Table 9.2-2 gives a survey of available beamsplitter materials. In view of the relatively narrow spectral areas to be covered by one single beam splitter, especially in the far infrared, an automatic beamsplitter changer is necessary for practical reasons.

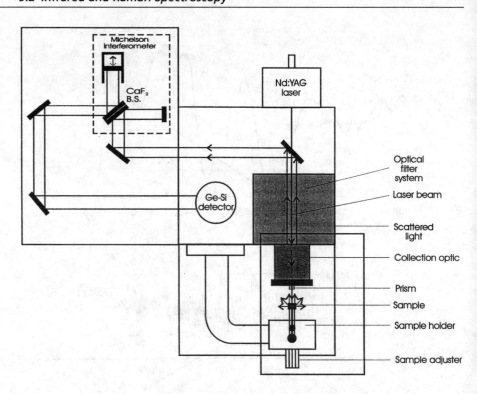

Fig. 9.2-11. Optical diagram of an FT-Raman spectrometer

FT-Raman-spectrometers.

FT-Raman spectrometers based on the Michelson interferometer can be operated with one single (CaF_2) beamsplitter for the whole spectral area of interest ($4000-10\,cm^{-1}$). This is possible because the exciting laser emits around $10\,000\,cm^{-1}$ ($1.06\,\mu m$) and the Raman effect produces vibrational shifts (of $4000\,cm^{-1}$ at most) away from this exciting line. This interval is, however, well within the range of transparency of the CaF_2 beamsplitter. Besides the laser source and the special detector, the only modification necessary for the adaption is an efficient filter to eliminate the strong Rayleigh line, which would otherwise completely obscure the FT-Raman spectrum (Fig. 9.2-11).

The major advantage of FT-Raman spectrometers is the possibility to use Nd:YAG lasers for the excitation of the Raman lines without causing fluorescence (why?). Since the Raman effect is particularly weak at $1.06\,\mu m$ (why?) only a multiplex technique can be used at this long wavelength to produce spectra with good signal-to-noise ratio. Figure 9.2-12 shows how well an FT-Raman spectrometer performs compared to a dispersive system when Raman spectra of fluorescent samples have to be taken.

Sampling techniques

Samples in the gaseous, liquid or solid state can be investigated both by IR and Raman spectroscopy.

Whereas NaCl and KBr windows for the MIR range and CsBr or polyethylene windows for the FIR range must be used, glass or quartz windows can be used in the NIR and Raman modes conveniently. Samples for Raman spectoscopy are routinely taken in glass tubes and directly excited by a laser. Since water is a very weak Raman scatterer, aqueous solutions can be readily investigated by Raman spectroscopy (in contrast to IR, where strong absorptions obscure large parts of the spectrum – which part?). Figure 9.2-13 shows a typical setup for Raman sampling.

Sample preparation for IR spectroscopy is more diversified and must be treated in more detail.

Gas samples

Sample preparation.

In order to obtain IR spectra of gases or low-boiling substances showing rotational fine structures, IR-transparent gas cells have to be filled at reduced pressure.

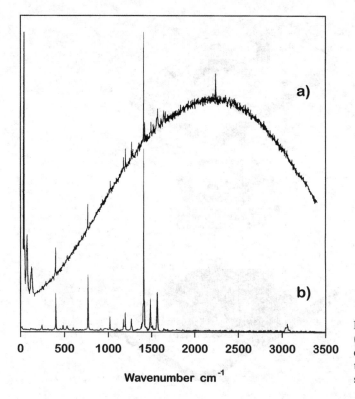

Fig. 9.2-12. Raman spectra of anthracene: (a) dispersive spectrometer, 514.5 nm excitation (fluorescence background overrides the Raman spectrum), (b) FT-Raman spectrometer, 1.06 μm excitation

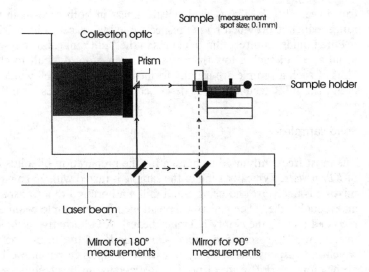

Fig. 9.2-13. Sampling device for the excitation of Raman spectra

A variety of cells with path lengths of 10 cm to 80 m are commercially available. The long path cells can be kept shorter than 1 m because they work according to the principle of multiple reflection of the radiation. Gold mirrors are used in order to keep the reflection losses low.

Solutions

No single solvent covering the entire MIR range exists. Instead, the most appropriate solvent for the given situation has to be chosen from a whole selection (which does not include water or alcohols – why not?). The solvents with the largest absorption free areas (*optical window*) are carbon tetrachloride and carbon disulfide (Table 9.2-3).

Cells used in MIR spectroscopy of solutions typically consist of two NaCl or KBr windows for nonaqueous media or CaF_2 windows for aqueous solutions. In the latter case, the strong water bands restrict the useful spectral window to the area of 1400 to 1000 cm^{-1}, where many interesting substances such as sugars in

Table 9.2-3. Useful IR-transparent solvents and their major optical windows in the MIR-range [cm^{-1}]

Solvents	Optical windows	
CCl$_4$	4000–1600	1500–850
CS$_2$	4000–1650[a]	1400–500
CHCl$_3$	4000–1250[b]	1150–850

[a] except strong bands at 2350 and 2200 cm^{-1}
[b] except strong bands at 3050 and 940 cm^{-1}

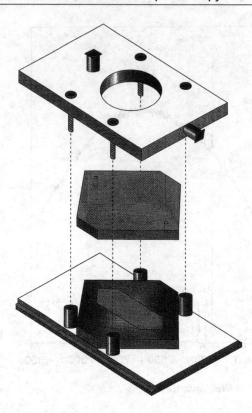

Fig. 9.2-14. Expanded view of an IR cell for liquids

fruit juices absorb. The typical cell thickness in both cases is in the 10 to 50 μm range, which is provided by the respective teflon spacer (Fig. 9.2-14).

Pure liquids require a film thickness in the 1 μm range. Since it would be difficult to fill a cell of such a low thickness and even more difficult to clean it, capillary films of such a sample are usually formed between two cell windows. As an alternative the ATR technique – as described for solid samples – can be used.

Solid samples

The most frequently used technique for the preparation of solids is the formation of *KBr pellets*. Typically 1 mg of the sample is mixed with 200 mg of dry KBr. This mixture is finely ground using a ball mill, and pressed to a transparent pellet using an evacuable die. Thee pellets are positioned in the sample beam and their spectra recorded (versus the empty reference beam). An alternative technique, which is of interest for hygroscopic or polar compounds, is the formation of *Nujol mulls*, which are suspensions of finely powdered samples in paraffin oil. Only the C–H absorptions recorded are of no diagnostic value in this technique. To circumvent this drawback, fluorolube – a highly viscous halogenated liquid – can be used instead.

Surface Analysis in the IR.

Attenuated total reflectance (ATR)

Not all solid samples of practical interest, e.g., polymers, can be finely powdered to produce homogeneous KBr pellets for IR transmission spectroscopy. Fortunately, a special sampling technique has shown itself to be of great value in the analysis of polymer surface areas and fibers, and also pastes, powders and even aqueous solutions. This technique is ATR spectroscopy (Fig. 9.2-15). Only the principles of this technique will be described here. For further details the reader is refereed to the monography by Harrick.

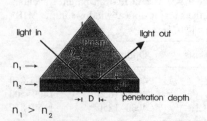

Fig. 9.2-15. Schematic of attenuated total reflectance

In the ATR technique the sample (e.g., a polymer foil) is brought into close contact with the surface of a prism made of a material with a high refractive index n, such a KRS-5, i.e, T1BrI ($n = 2.4$) or Ge ($n = 4.0$). A light beam approaching the interface from the optically denser medium at a large enough angle of incidence is totally reflected. Nonetheless it does penetrate a short distance (the pene-

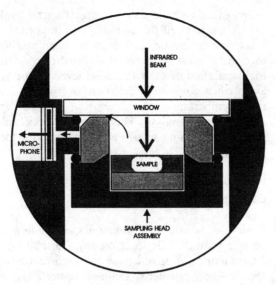

Fig. 9.2-16. Principle of a photoacoustic cell

tration depth d_p) into the optically thinner medium (the sample).

$$d_p = \frac{\lambda}{2\pi\sqrt{\sin\theta - (n_2/n_1)^2}}$$

λ: wavelength
θ: angle of incidence
n_1: refractive index in the optically denser medium (prism)
n_2: refractive index in the optically thinner medium
D: displacement of the ray

If the sample is an IR absorbed (*attenuator*), and the ATR unit is properly mounted in the sample compartment of an IR spectrometer, a transmission-like IR spectrum can be observed. The penetration depth depends on the wavelength of the radiation, the refractive indices of sample and prism material and the angle of incidence. It is – as a rule of thumb – in the order of a few tenths of the wavelength of the respective IR radiation. This is in the μm and sub-μm range. By changing the angle of incidence, depth profiling of the sample is possible to a certain extent. In commercial ATR units, multiple path cells containing trapezoidal or rod-shaped reflection elements are used in order to enhance the sensitivity.

Photoacoustic IR spectroscopy

With this technique, irregularly-shaped materials such as polymer flakes or porous beads can be easily analyzed. In principle, this device consists of a sensitive microphone in close proximity to the sample, housed in a sealed cell filled with a non-IR absorbing, inert gas (N_2 or He, Fig. 9.2-16).

When modulated IR radiation hits the sample, the sample absorbs the radiation and is heated at the frequency of this modulation in an amount corresponding to the absorption intensity. When the modulation frequency is in the audio range you can hear the sample absorbing light. Using a Michelson interferometer in the IR range produces a sound in the audio frequency range which is modulated by the IR spectral absorptions of the sample. The spectra are detected by the microphone via the inert gas in the cell. The advantage of this accessory to the FTIR spectrometer is that no sample preparation is required at all. The technique is truly "nondestructive" to the sample which means that no information can be lost due to interactions with solvents or to sample handling, such as grinding, in conventional sample preparation.

IR emission spectroscopy

Heated samples emit IR radiation at the same wavelengths at which they would absorb at room temperature. Therefore this principle can be used to replace the source of a conventional FTIR spectrometer by the hot sample, and thus to obtain the IR spectrum of this sample. Two important conditions are, however, that the spectrometer and detector must be significantly cooler than the sample and that the sample must be stable at elevated temperatures. One must bear in mind that the population of the vibrational levels of the sample is a function of its temperature when comparing emission IR spectra with conventional spectra taken at room temperatures. An interesting practical application of this principle is in environmental analysis. The potentially toxic emission of remote smoke stacks can be measured using an IR telescope without the need to enter the property for sampling.

Other techniques

Several other sampling techniques are finding more and more practical applications in industrial IR spectroscopy thus making this method even more valuable. *External* and *diffuse reflection* are important means to characterize flat and reflecting, or finely powdered samples, respectively. *IR microscopes* are commercially available accessories for conventional FTIR spectrometers that allow for the analysis of single particles or single phases with a diameter of down to 10 μm. Only the wavelength of the radiation limits the beam diameter in today's state-of-the-art FTIR instruments. Spectra of microdomains in organic samples can be readily obtained both in transmission and in reflection. In both cases, the sample thickness must not be larger than approximately 10 μm (why?). Typical detection limits are in the 10 nm thickness range, corresponding to 1 ng of a sample having a density of 1. Important applications of IR microscopes ae to be found in the pharmaceutical and electronic industries for phase and contamination analysis.

Thinner samples require the application of *grazing angle IR spectroscopy* which has monolayer sensitivity. For further details concerning this technique – which is now also commercially available for IR microscopy spectral literature, such as Harrick N.J., is recommended.

9.2.3   Analytical information

Both qualitative and quantitative information can be obtained from vibrational spectra. Key data are the number, the frequency, and the intensity of the fundamental vibrations which are observed in mainly the MIR range. Overtones which appear at approximately twice the frequency of the fundamental bands are of special value for quantitative analysis in the NIR range but are not further treated here in great detail.

Normal vibrations

IR-versus Raman-activity of vibrations.

A polyatomic molecule can be excited to discrete vibrations involving all atoms moving in phase with the same frequency but with different amplitudes. Such vibrations are called *normal vibrations*. All single normal vibrations are superimposed to the complex overall molecular motion.

For an N-atomic molecule, 3N coordinates ar required to describe its actual position in space. By subtracting three coordinates for the three translations, and another three for the three rotations of the molecule as a whole, one obtains the (3N-6) coordinates needed to describe the (3N-6) normal vibrations in nonlinear molecules. In linear molecules the rotation around the molecular axis is not active, therefore (3N-5) normal vibrations exist in such cases. Example: A diatomic (linear) molecule has of course only $3 \times 2 - 5 = 1$ vibration, a triatiomic linear molecule has $3 \times 3 - 5 = 4$, but a triatomic bent molecule has only $3 \times 3 - 6 = 3$ vibrations!

Stringent selection rules exist for the excitation of normal vibrations by electromagnetic fields. Only those vibrations that are accompanied by a change in dipole moment of the molecule can be excited by IR radiation (i.e., are *IR active*). The

Table 9.2-4. IR and Raman activity of normal vibrations in polyatomic molecules

Number of active vibrations.

Molecule	Type	Number of IR active vibrations	Number of Raman-active vibrations
S=C=S	linear (3N-5)	2	1
H–C≡N	linear (3N-5)	4	4
$\overset{O}{\underset{H \quad H}{\diagdown\diagup}}$	bent (3N-6)	3	–

condition for the *Raman activity* of a vibration is that the polarizability of the respective bond must change during the vibration.

As an example, the vibration in N_2 is not IR, but Raman active, because only the polarizability of the molecule is changed during the N–N vibration (due to the delocalization of the electrons upon interaction with the oscillating electromagnetic field). The dipole moment of the nonpolar N≡N-bond is not. Vibrations which do not lead to a change in polarizability or dipole moment cannot be excited at all by electromagnetic radiation. Further examples for larger molecules can be taken from Table 9.2-4.

Without going into theoretical details, one can see from the data in Table 9.2-4 that the symmetry of the molecule under consideration plays an important role in determining the number of IR and Raman active vibrations. Although symmetric vibrations are forbidden in the infrared, they are still Raman active. The higher the symmetry of a molecule, the smaller the numbers of vibrational bands observed with either method. This is because several normal vibrations then have exactly the same frequency. These are called *degenerate* vibrations. In highly symmetrical molecules having a center of symmetry, a normal vibration is either IR or Raman active (*exclusion principle*). Further detailed information about the IR and Raman activity of normal vibrations and their dependence on molecular symmetry can be taken from group theory (see [9.2-1]).

Qualitatively, it may be stated that the vibrational spectra of polyatomic molecules are unique. Due to the great sensitivity of the normal vibrations to mass and structural differences, every molecule has its own fingerprint-type vibrational spectrum. This is of special importance for the analysis of isomers, which cannot be differentiated by mass spectroscopy, and for many other industrial applications.

Exploiting this property of vibrational spectra, large computerized data bases have been established for IR and Raman spectra that assist with the identification of unknown compounds. Automated spectral search routines are used to compare the measured spectra of unknown compounds with stored spectra of known substances (Fig. 9.2-17).

The collection of IR standard spectra distributed by *Sadtler*, for instance, comprises more than 120 000 IR spectra obtained from commercial FTIR spectrometers, digitally stored on hard disk. The package includes also the required software for spectra search systems. These systems provide a "hit-list" of the names of the best matches. The results are only relevant, however, if the spectra of the compounds under investigation are contained in the spectra library. Several refined versions of substructure search programs are available to overcome this drawback. They rely on a second property of vibrational spectra, that is based on the concept of *group frequencies*.

Spectra collections for qualitative analysis.

Group frequencies

A second important type of information can be extracted from vibrational spectra: the presence or absence of functional groups. A comparison of the IR spectra of many different organic compounds containing the same functional group shows that absorption bands particular to this group are observed in all these spectra, independent of the backbond structure of the molecules. Such an absorption is called *group frequency*. In the normal vibrations resulting in such group frequencies, the displacement of the atoms is practically restricted to the members of the particular functional group.

"Group frequencies" are special cases within the general model of "Normal vibrations".

(a)

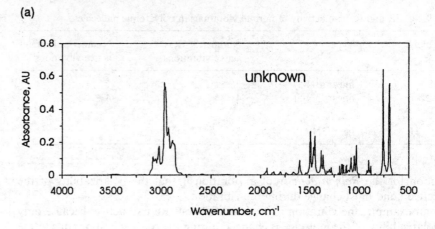

(b)

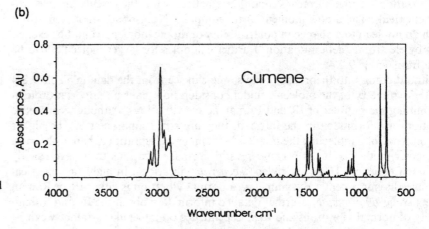

Fig. 9.2-17. Comparison of the vibrational spectra of an unknown with (a) spectra stored in a spectra library and (b) search report (Birsy™)

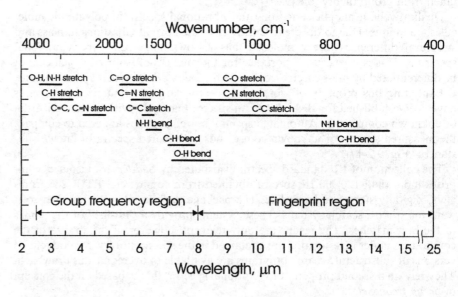

Fig. 9.2-18. Vibrational group frequencies and fingerprint area

Such behaviour is found if the atoms constituting the functional groups are their significantly lighter or heavier than the neighboring atoms or if the bond strength in the functional groups differ from those of the bonds in the vicinity. Figure 9.2-18 gives a survey of the most important functional groups which are observed in vibrational spectra along with their characteristic wavenumber ranges.

The functional groups leading to significant group frequencies contain H atoms, or isolated double or triple bonds. Group frequencies are found at wavenumber positions higher than $1300 \, cm^{-1}$. groups containing heavy atoms absorb in the FIR range below $400 \, cm^{-1}$. The area in between contains vibrational bands which cannot be easily assigned to particular functional groups. This is because both masses and bond strengths of each of the absorbing species are too similar. This

(a)

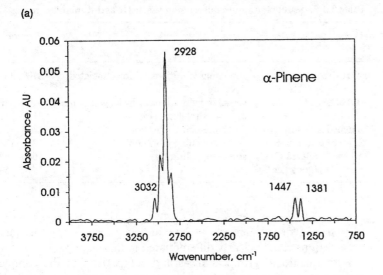

(b)

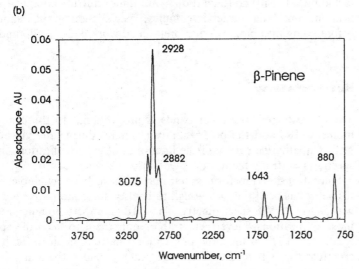

Fig. 9.2-19. IR spectra of (a) alpha and (b) beta pinene showing group frequency and fingerprint regions

area contains bands of special significance to the molecule as a whole, however, and is therefore called the *fingerprint* region. It is of high diagnostic value for compound specific analysis. A close match between two spectra especially in the fingerprint region is essential for the identification of a compound.

The *harmonic oscillator* can be used as a suitable model for the qualitative description of the frequency of diatomic functional groups. According to this spring-ball arrangement, the vibrational frequency is proportional to the square root of the force constant k of the bond, and indirectly proportional to the square root of the reduced mass μ of the atoms of masses m_1 and m_2 linked by this bond.

Hooke's law.

$$\mu = \frac{1}{m_1} + \frac{1}{m_2}$$

$$v = v = \frac{1}{2\pi c}\sqrt{\frac{k}{\mu}}$$

This simple model shows that a band shift of $x\sqrt{2}$ is obtained upon deuteration of H-containing functional groups, for example.

Very detailed empirical correlation tables for functional groups have been accumulated in the literature since vibrational spectroscopy was first used for the analysis of organic compounds (see Appendix 5, "Colthup Table", and Sec. 9.2.4). These tables include information about the vicinity of the respective groups, eventual hydrogen bonding and bond coupling.

To demonstrate the power of vibrational spectroscopy in the identification of organic compounds with similar structure, Fig. 9.2-19 shows the IR spectra of two

Isomer differentiation.

Table 9.2-5. Selected group frequencies and their IR and Raman intensities

Type		Position, cm^{-1} (Intensity)	
		IR	Raman
–OH (free)	stretching vibrations of nonassociated groups	3700–3590 (s)	(w)
–OH (assoc.)	stretching vibrations of hydrogen-bonded	3400–3200 (s)	(w)
–N=C=N	asymm. stretching vibration	2155–2130 (s)	(w)
Aldehydes (saturated)	C=O stretching vibration	1735–1725 (s)	(w)
Ketones (saturated)	C=O stretching vibration	1725–1705 (s)	(m)
Alkenes (isolated C=C)	stretching vibration	1680–1620 (m)	(s)
(conjugated)		1640–1590 (w)	(s)

isometric flavor substances – alpha and beta pinene. Group frequency and fingerprint regions can be clearly differentiated in this example.

Both functional group analysis and comparison of the complete spectrum – including intensity consideration – are important for complete substance identification. Since the matter is complex, special monographs should be consulted before going into practical interpretational work [9.2-1, 9.2-3 and 9.2-5].

Band intensity

Position and intensities of vibrational bands.

The intensity of vibrational bands is proportional to the change of the dipole moment (IR) and the polarizability (Raman) during the vibration. Table 9.2-5 gives a qualitative survey of the intensities of group vibrations in IR and Raman spectra (s = strong band, m = medium band, w = weak band).

Since Fourier transform spectrometers have been available, vibrational spectrometry has been precise enough to be used for quantitative analysis in a similar way as to UV-VIS spectrometry. The LBB (Lambert-Bouguer-Beer) law holds for most applications. Nevertheless, the requirement to keep the spectral resolution smaller than 1/9 of the band half-width is more difficult to fulfill than in UV-VIS spectroscopy, where the bands are much broader than in vibrational spectra. Linear working curves can also be obtained form Raman spectra, although scattering rather than absorption is the basis of signal generation.

Quantitative evaluation of vibrational spectra

As in UV-VIS spectroscopy, the quantitative information of vibrational spectra is contained in the peak area of the particular band. Band overlapping makes a correct evaluation of this information difficult, especially in vibrational spectra. Therefore the peak height is usually determined using the *baseline method* (Fig. 9.2-20).

With this simple method, reproducibilities in the order of $v = 1\%$ or less can be obtained.

Several chemometric techniques for the analysis of organic mixtures, employing more sophisticated evaluation methods are available for commercial FT spectrometers (see *multivariate analysis* in Sec. 12.5).

9.2.4 Application to structural analysis

In the previous chapters an introduction into the signal generation and the principal applications and evaluations of vibrational spectra have been given. This chapter deals in more detail with the most important, mainly qualitative, functional group information to be found in the different spectral areas under consideration: NIR, MIR, FIR and Raman domains.

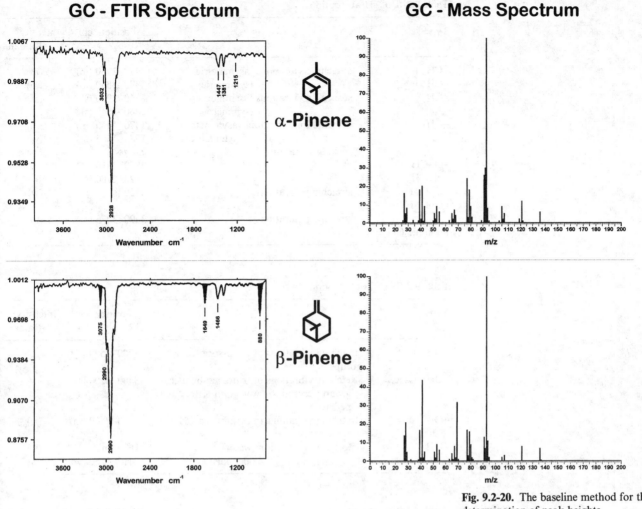

GC - FTIR Spectrum   GC - Mass Spectrum

α-Pinene

β-Pinene

Fig. 9.2-20. The baseline method for the determination of peak heights

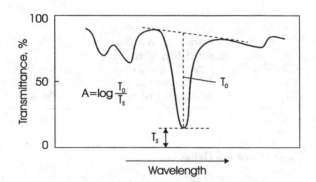

Fig. 9.2-21. NIR spectra of meats (a) Regular NIR-spectra, (b) first derivative; [9.2-6]

Near infrared (NIR)

In the NIR range, overtone and combination bands of strong normal vibrations (C–H, N–H and O–H) from the MIR area can be observed. These bands are much less intense than their corresponding ground vibrations. Nonetheless, nice the NIR is void of overlapping normal vibrations and the LBB law is well-obeyed by the weak overtones, routine quantitative analysis of the main components in food, such as proteins, starch, fat or moisture, can be performed without tedious sample preparation (Fig. 9.2-21). Samples, usually finely powdered are investigated by a special reflection technique (*diffuse reflection*) between 7000 and 5000 cm^{-1}. The detection limits are in the order of 0.1%; precision values of 1 to 2% have been reported.

Table 9.2-6. C–H absorptions

Type		Position, cm^{-1} (Intensity)	
		IR	Raman
–CH$_3$ }	stretching vibrations (asymmetric)	2970–2950 (m)	(m)
–CH }	(symmetric)	2880–2860 (m)	(m)
=CH$_2$	stretching vibrations (asymmetric)	2940–2920 (s)	(s)
	(symmetric)	2860–2840 (s)	(s)
–CH	deformation vibrations (asymmetric)	1470–1440 (m)	(m)
	(symmetric)	1385–1365 (m)	(w)
=CH$_2$	rocking vibrations	720 (w)	(–)
–O–CH$_3$		2850–2810 (m)	
=N–CH$_3$		2820–2780 (m)	
–C≡C–H	stretching vibrations	3320 (s)	(w)
=C=C=H$_2$			
	stretching vibrations	3090–3075 (m)	(m)
aromatic CH			

Table 9.2-7. O–H absorptions

Type		Position, cm^{-1} (Intensity)	
		IR	Raman
–OH (free)	stretching vibrations of nonassociated groups	3700–3590 (s)	(w)
–OH (assoc.)	stretching vibrations of hydrogen-bonded groups, crystal water (e.g., in KBr pellets)	3400–3200 (s)	(w)
–OH (assoc.)	stretching vibrations in paraffinic acids (very broad)	3200–2500 (m)	(w)
–OH	deformation vibration (water)	1620 (m)	(w)
	(alcohols)	1410–1260 (w)	(w)

Table 9.2-8. N–H Absorptions

Type		Position, cm^{-1} (Intensity)	
		IR	Raman
–NH$_2$ (free)	2 stretching vibrations }		
NH (free)	1 stretching vibration }	3500–3300 (m) 3500–3400	(w)
=NH (free)	1 stretching vibration }		
–CO–NH$_2$	2 stretching vibrations	3500–3400 (m)	(w)
–NH$_2$	deformation vibrations	1650–1560 (m)	(w)
=NH	deformation vibration	1580–1490 (w)	(w)

Mid infrared (MIR)

Characteristic bands – both frequencies and intensities.

In the MIR range most of the characteristic group vibrations and the molecular fingerprint can be observed. Some single-bonded structures such as –C–O–C– of esters or ethers (e.g. sugars) and –C–O– of alcohols absorb in the fingerprint area but can be readily identified due to their large intensity (why are C–O–C– and –C–O– vibrations so intense?). Other functional groups containing heavy, weakly bonded atoms absorb in the FIR. This paragraph lists the major functional groups of practical relevance and their absorption frequencies in the MIR range.

The intensity values in Tables 9.2-6 to 9.2-11 are relative approximations (s-strong band, m-medium band, w-weak band). For more detailed studies of spectra-structure correlations, with emphasis on the fingerprint range, (where significant vibrational coupling of group frequencies with backbone vibrations may occur and confuse inexperienced persons) the reader is referred to the special literature (e.g., [9.2-1, 9.2-3]).

Table 9.2-9. C≡C, C≡X, C=C=C and X=C=Y absorptions

Type		Position, cm^{-1} (Intensity)	
		IR	Raman
–C≡C–	stretching vibration	2260–2150 (m–w)	(s)
	(acetylene)	1961 (w)	(s)
–C≡N	stretching vibration	2250–2230 (s)	(s)
–N≡C	stretching vibration	2150–2130 (s)	(s)
=C=C=C=	asymm. stretching vibration	2000–1900 (m)	(s)
–N=C=O	asymm. stretching vibration	2275–2250 (s)	(w)
–N=C=S	asymm. stretching vibration	2140–1990 (s)	(m)
–N=C=N	asymm. stretching vibration	2155–2130 (s)	(w)
–C=C=O	asymm. stretching vibration	2195–2085 (s)	(m)
O=C=O (gas)	asymm. stretching vibration	2350 (s)	

Table 9.2-10. C=O Absorptions

Type		Position, cm^{-1} (Intensity)	
		IR	Raman
Esters (saturated) C=O	stretching vibration	1740–1730 (s)	(m)
(unsaturated)	band shift to lower wavenumbers	1680–1640 (s)	(m)
–CCl–CO–OR	band shift to higher wavenumbers	1770–1745 (s)	
a-Ketoester		1755–1740 (s)	
Aldehydes (saturated)	C=O stretching vibration	1735–1725 (s)	(w)
(unsaturated)	band shift to lower wavenumbers	1680–1650 (s)	(w)
Ketones (saturated)	C=O stretching vibration	1725–1705 (s)	(m)
(unsaturated)	band shift to lower wavenumbers	1685–1660 (s)	(m)
Carboxylic acids	coupled C=O stretch. vibration		
	(asymmetric)	1740–1660 (s)	(w)
	(symmetric)	1690–1625 (–)	(m)
Carboxylic acid anhydrides	2 sharp bands	1850–1800 and	
		1790–1740 (s)	(m)
Carboxylic acid chlorides	increased C=O double bond order	1815–1790 (s)	
Carboxylates	asymm. stretching vibration	1610–1550 (s)	
	symm. stretching vibration	1420–1300 (s)	
Amides			
–CO–NH$_2$	Amide I band (mainly C=O-stretch)	1650 (s)	(w)
	Amide II band (mainly N–H-bend. vibr.)	1600 (s)	(w)
–CO–NHR	Amide I band	1680–1630 (s)	(w)
	Amide II band	1570–1520 (s)	(w)

Table 9.2-11. C=C, C=N, N=N and N=O Absorptions

Type		Position, cm^{-1} (Intensity)	
		IR	Raman
Alkenes (isolated C=C)	stretching vibration	1680–1620 (m)	(s)
(conjugated)		1640–1590 (w)	(s)
Aromatic compounds	2–3 ring vibrations	1600, (1580), 1500 (m)	(m)
	overtone pattern	2000–1600 (w)	(–)
	out-of-plane bending pattern	880–680 (m)	(w)
=C=N–H	stretching vibration	1690–1630 (s)	(s)
–N=N–	stretching vibration in azo groups	1580–1570 (w)	(s)
R$_3$–C–N=O	stretching vibration	574 (m)	
R$_2$–N–N=O	stretching vibration	1460–1430 (s)	(s)
R$_3$–C–NO$_2$	asymm. stretching vibration	1560 (s)	(s)
	symm. stretching vibration	1350 (s)	(s)
R–O–N=O	2 stretching vibrations	1680–1650 (s)	(s)
		1625–1610 (s)	
R–O–NO$_2$	asymm. stretching vibration	1640–1620 (s)	(s)
	symm. stretching vibration	1285–1270 (s)	(s)
NO$_3$-(inorg. nitrate)	asymm. stretching vibration	1400–1380 (s)	
	symm. stretching vibration	860–810 (m)	

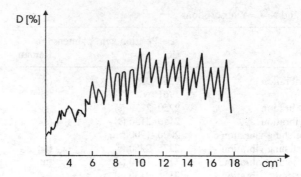

Fig. 9.2-22. FIR spectrum of CH₃CN in the gaseous state, showing the rotational fine structure

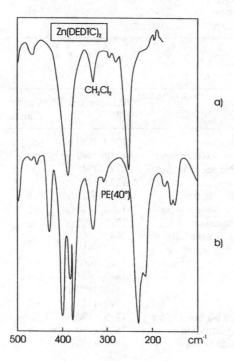

Fig. 9.2-23. FIR spectra of zinc diethyldithiocarbamate, (a) in solid sate, (b) in CH₂Cl₂ solution

Far infrared (FIR)

The far-infrared range enables the analysis of rotational transitions in gas samples that extend from below $500\,cm^{-1}$ into the microwave range. Figure 9.2-22 shows as a typical example the FIR gas spectrum of CH_3CN.

The FIR range is also very important for structural investigations of metal chelates and other compounds containing heavy or weakly bound atoms. Unlike X-ray analysis, vibrational spectroscopies are not restricted to the solid (crystalline) state, but can also analyze solutions. Thus the "real" molecular structure, undisturbed by lattice and crystal field effects can be studied in different solvents. As an example, Fig. 9.2-23 compares the FIR spectra of a metal chelate in the solid state (recorded as a nujol mull between polyethylene windows) and in methylene chloride solution.

It can be clearly seen that the higher (tetrahedral) symmetry of the complex is only stable in solution and that the splitting of the metal-ligand bands in the solid state spectrum shows a distortion from this symmetry in the crystalline state.

Furthermore, the FIR range is used for the study of crystal lattice vibrations in inorganic compounds, ions and semiconductors and skeletal bending modes of organic substances. Since water and water vapour also absorb strongly in the FIR, the instrument must be carefully purged with dry air, or better still, be evacuated to remove interferences.

Raman range

As pointed out in Section 9.2.2, the whole spectral range from 3600 to $10\,cm^{-1}$ (corresponding to the complete MIR and FIR region) can be covered by Raman spectroscopy. This may be achieved either by dispersive instruments using one grating or by FT spectrometers with one single beamsplitter made from CaF_2 (why?). Thus modern Raman instruments provide comprehensive information on the functional groups and fingerprint region contained in the Raman spectra of organic compounds. According to the selection rules, polarizable bonds such as $-C-S-$, $-S-S-$, $-N=N-$ and $-C=C-$, which are only weak IR-absorbers, produce strong absorptions in the Raman spectrum. Group frequency correlation charts are available in special monographs (e.g. [9.2-5]). Typical data have been included into Tables 9.2-6 to 9.2-11 in the MIR section.

Raman spectroscopy has been increasingly applied to biological systems due to its ability to use very small sample volumes and aqueous solutions. Conformational changes of proteins, nucleic acids and peptides in lipids and membrane can readily be followed *in situ* (that is, in the natural environment of the substances under investigation) since water is a very weak Raman scatterer. Because many biological samples are fluorescent, it is necessary to use FT Raman instruments for obtaining good quality Raman spectra (why?).

Raman spectroscopy is an important technique for the structural investigation of inorganic compounds especially metal chelates and ions in aqueous solutions. This is due to the easy access to the region of metal-ligand vibrations at wavenumbers below $500\,\mathrm{cm}^{-1}$. Together with the respective FIR bands, which are difficult to obtain (why?), a complete structural analysis of such compounds can be performed.

Besides its well-established role for the structure analysis of molecular systems together with IR spectroscopy, the Raman technique has received new importance in industrial analysis since the introduction of FT-Raman instruments and a new generation of dispersive NIR laser-powered instruments using array detectors. Such systems are used for the analysis of polymers, food and semiconductors, to name a few examples. These modern techniques are limited by the low sensitivity of the regular Raman effect.

Under certain circumstances the sensitivity of Raman measurements can be significantly increased. In *resonance Raman spectroscopy* an excitation wavelength is used which is close to that causing an electronic transition in the sample (see Fig. 9.2-2). The intensity of the vibrational modes, which show a large change in equilibrium geometry upon electronic excitation (that is to say, totally symmetric vibrations and vibrations that vibronically couple to the two electronic states), is greatly enhanced.

Enhancement factors of up to a million are reported with VIS and UV lasers. These must be tuned closely to the wavelength of the absorption maximum of the electronic transition. Resonance Raman spectra can be obtained at analyte concentrations as low as 10^{-8} M! One must consider, however, that short wavelength laser excitation may be destructive to organic samples and that only nonfluorescent samples can be analyzed successfully with this technique (why?). The most important application of resonance Raman spectroscopy to date is the analysis of biological sample; for example, the oxidation state of complex-bound iron in hemoglobin in dilute aqueous solutions. Here bands of the tetrapyrrole chromophore can be observed with little interference from the other Raman bands of the molecule, which are not enhanced due to the selective excitation.

A further possibility for the application of Raman spectroscopy to organic trace analysis exists since the discovery of the *surface-enhanced Raman scattering* (SERS) effect. Detection limits in the range of 10^{-10} M have been observed when the molecules under investigation have been absorbed onto rough metal surfaces (mainly silver, gold or copper) prior to excitation with VIS lasers. This is due to an enhancement of the respective Raman lines in the order of up to 10^6 by a mechanism that is still not yet fully understood. It is assumed that both the electromagnetic field at the surfaces and the polarizability of the adsorbed molecules with the metal islands at the rough surface. A similar enhancement of the Raman band intensities can be observed in dilute aqueous solutions containing stable Ag or Au colloids. The enhancement factor is a function of the wavelength of the exciting laser and decreases to approximately 100 when using a $1.06\,\mu m$ Nd:YAG laser. It is also a function of the correct matching of the metal particle size (in the sub-μm range) to the laser wavelength. SERS is hence not yet a routine technique and will not be further treated here.

Surface enhanced RAMAN-effect.

References

[9.2-1] Colthup, N.B., Daily, L.H., Wiberley, St.E., *Introduction to IR and Raman Spectroscopy*. New York: Academic Press, 1964.

[9.2-2] Griffiths, P.R., de Haseth, J.A., *Fourier Transform Infrared Spectrometry*. New York: J. Wiley, 1986.

[9.2-3] Hesse, M., Meier, H., Zeeh, B., *Spektroskopische Methoden in der organ. Chemie*. Stuttgart: Thieme, 1991.

[9.2-4] Grasselli, J.G., Bulkin, B.J. (ed.), *Analytical Raman Spectroscopy*. New York: J. Wiley, 1991.

[9.2-5] Lin Vien, Colthup, N.B., Fateley, W.G., Grasselli, J.G., *Infrared and Raman Characteristic Frequencies of Organic Molecules*. San Diego: Academic Press, Inc., 1991.

[9.2-6] Clancy, P., *Int. Labmale XII*, 1987, 1.

General reading

Greenler, R.G., *J. Chem. Phys.* 1966, 44, 310.

Harrick, N.J., *Internal Reflection Spectroscopy*. New York: Interscience, 1967.

Nakanishi, K., *IR Absorption Spectroscopy Practical*. San Francisco: Holden-Day, 1969.

Schrader, B., *Raman/IR Atlas*. Weinheim: VCH, 1988.

Questions and problems

1. Name and describe the features of vibrational spectroscopies that make them important for analytical chemistry.
2. What are the 4 sub-ranges of the IR region and what is their importance for analytical chemistry?
3. Describe the principle of the signal generation of Raman spectroscopies (regular Raman and resonance Raman).
4. Compare the signal generation of IR, Raman, fluorescence and UV-VIS spectroscopies.
5. Draw and describe the major components of modern IR and Raman spectrometers.
6. Which sources are available for NIR, MIR, FIR and Raman spectrometers?
7. Which detectors are used in NIR, MIR, FIR and Raman spectrometers?
8. Describe the principle of the Michelson interferometer.
9. Compare the principal components of a dispersive IR spectrometer with those of a dispersive UV-VIS system.
10. What are the major parts of a dispersive Raman spectrometer?
11. Describe the principle of IR filter photometers
12. What are the multiplex and throughput advantages of Michelson interferometer based IR spectrometers?
13. What determines the spectral resolution (a) in dispersive, (b) in Fourier transform IR spectrometers? Give examples.
14. What is the advantage of FT Raman spectrometers over dispersive systems? What are their disadvantages?
15. Describe the sampling system for raman spectroscopy.
16. Describe the sampling requirements for IR spectroscopy.
17. Describe the principles of the ATR technique.
18. What are the principal and other typical applications of photoacoustic IR spectroscopy?
19. Describe the principal and other typical applications of IR emission spectroscopy.
20. How and to what limits can small phases in organic samples be analyzed in the IR or Raman range?
21. Explain the origin of normal vibrations and group frequencies, their number and IR and Raman activity.

22. What are degenerate vibrations?
23. Name the most important group frequencies for IR and Raman spectroscopy.
24. Describe the origin and the analytical importance of the fingerprint region. Which functional groups can be found in the IR fingerprint region?
25. Under which conditions can the model for the harmonic oscillator be used for the calculation of vibrational bands?
26. Describe the basic law and the boundary conditions for the application of IR and Raman spectroscopy for quantitative analysis.
27. Describe the most important qualitative information to be obtained from spectra in the NIR.
28. Describe the most important qualitative information to be obtained from IR and Raman spectra in the MIR.
29. Describe the most important qualitative information to be obtained from IR and Raman spectra in the FIR.
30. In which spectral area do metal chelates display their metal-ligand vibrations?
31. Compare X-ray analysis and vibrational spectroscopy with respect to their strengths and weaknesses for structure analysis.
32. Name functional groups which absorb strongly in the Raman spectrum.
33. Describe the principal and other major applications of resonance Raman spectroscopy.
34. Describe the principal and other major applications of surface enhanced Raman spectroscopy.
35. Summarize briefly the main analytical information to be obtained by (a) UV-VIS, (b) IR, (c) Raman, (d) NMR, and (e) mass spectroscopies from pure organic samples.

9.3 Nuclear Magnetic Resonance (NMR) Spectroscopy

9.3.1 Introduction

The main field of application of nuclear magnetic resonance (NMR) spectroscopy is in determining the structures of molecules. Why? A high proportion of chemists are involved in preparative work. They synthesize compounds, often by methods involving many stages, and so need analytical methods to monitor and control the success of their syntheses. For several decades now, NMR spectroscopy has been the most commonly used method for this purpose. Other chemists are involved in isolating substances, for example natural products, whose molecular structure is unknown. NMR spectra, usually the proton or carbon-13 NMR spectrum or both, provide the first important indications as to the class of compounds to which the molecule belongs, and often give much more detailed information about the structure. To read all this information from an NMR spectrum some basic knowledge is essential.

The features that have led to NMR spectroscopy becoming the most important analytical method, at least for organic chemists, are not difficult to identify. Even though nuclear magnetic resonance is a quantum mechanical phenomenon, most of the basic equations can be derived from a classical model and understood by nearly everyone. Most spectra can be interpreted with a minimum of theoretical understanding and a few simple rules. Consequently, chemists very quickly adopted the method from physicists.

> Interpreting NMR spectra is relatively straightforward.

The high-resolution ^{1}H NMR spectrum of ethyl acetate ($H_3CCOOCH_2CH_3$), shown in Fig. 9.3-1, is a convenient example to illustrate the nature of the molecular structural information that can be obtained. There are three spectral parameters that we need to be acquainted with:

> Three main spectral parameters can be derived from the spectrum.

1. In this example we recognize three groups of peaks (resonance signals) at different positions in the spectrum. (The peak on the extreme right belongs not to

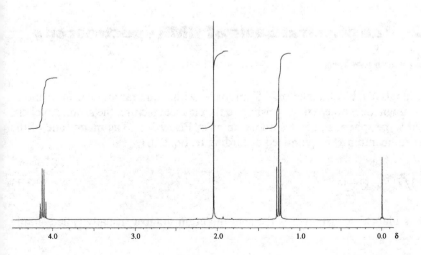

Fig. 9.3-1. 300 MHz ^{1}H NMR spectrum of ethyl acetate with integrated curve

ethyl acetate but to a small quantity of tetramethylsilane which is added to the solution to define the zero of the scale, as explained in Sec. 9.3.2, "The chemical shift"). The position of each of these groups of peaks is the first spectral parameter; we call it the *chemical shift*.

2. The groups of peaks in Fig. 9.3-1 show characteristic patterns: a quartet, a singlet, and a triplet (from left to right). In addition we see that within the quartet and the triplet the peaks are equidistant. This splitting into characteristic multiplets is caused by an interaction between nuclear magnetic dipoles separated by a small number of chemical bonds. The effect is called spin-spin coupling, and is characterized by the *coupling constants*. These coupling constants are the second spectral parameter. They contain information which helps us to elucidate the constitution, configuration and conformation of a molecule.

3. The third spectral parameter is the *integral*. This means the area under a signal. It depends on the number of equivalent nuclei that contribute to the signal. In our example (Fig. 9.3-1) the ratio of the "integrals" is measured by means of a stepped curve. We find for the quartet, the singlet, and the triplet the ratio $2:3:3$, so that we can at once assign the quartet to the protons of the methylene group, CH_2CH_3. Taking into account also the spin-spin coupling (see above), the triplet can be assigned to the protons of the methyl group of CH_2CH_3. The singlet is assigned to an isolated CH_3 group: $OCOCH_3$.

These three parameters – chemical shifts, coupling constants, and integrals – are the most important for chemists because they are directly related to the structures of molecules.

Since the discovery of nuclear magnetic resonance in 1946, NMR instruments and methods have developed at an explosive pace and no end is yet in sight. An outstanding example is the development of two-dimensional (2D) and multidimensional NMR techniques. However, the theory of these remarkably powerful methods is beyond the scope of this introductory treatment, which will be restricted to the more usual one-dimensional methods and to the uses and interpretation of some simple 2D spectra (see Sec. 9.3.5, "Two-dimensional experiments").

The nuclides ^1H and ^{13}C are especially important to chemists.

The nuclides that interest us are mainly protons (^1H) and carbon-13 (^{13}C) as their resonances are the most important for determining the structures of organic molecules. NMR resonances of other nuclides with a magnetic moment can also be observed, in many cases without difficulty, and a few examples will be mentioned. NMR spectroscopy of solids will not be treated here because the measurements are somewhat different from high-resolution NMR spectroscopy of liquid samples, and special instruments are usually required.

First we need to understand the basic theory of NMR, some properties of nuclei, their behavior in a magnetic field, and the effect of a radiofrequency pulse. After that we need to learn how the chemical shifts and coupling constants depend on the chemical structure of molecules. Finally we will look briefly at some special methods for assigning ^1H and ^{13}C resonances including the simplest kinds of 2D-correlated spectra.

9.3.2 The physical basis of NMR spectroscopy

Nuclear properties

NMR signals can be obtained only from nuclei with a nuclear angular momentum P and a magnetic moment μ. According to the classical picture, the atomic nucleus, assumed to be spherical, rotates about an axis (Fig. 9.3-2). The magnitude of the angular momentum P is quantized according to Eq. 9.3-1.

P, μ

Fig. 9.3-2. Classical description of angular momentum P and magnetic moment μ of a nucleus

$$P = \sqrt{I(I+1)}\,\frac{h}{2\pi} \tag{9.3-1}$$

Here h is Planck's constant; I is the spin quantum number, usually called simply the nuclear spin.

Associated with P is the magnetic moment μ:

$$\mu = \gamma P \qquad (9.3-2)$$

γ is a proportionality factor, a constant for each nuclide (i.e., each isotope of each element). Combining Eqs. 9.3-1 and 9.3-2 we obtain for the magnitude of the magnetic moment

$$\mu = \sqrt{I(I+1)}\,\frac{h}{2\pi} \qquad (9.3-3)$$

Nuclides with $I = 0$ therefore have no nuclear magnetic moment μ and cannot be observed by NMR spectroscopy. A fact of considerable importance to us is that ^{12}C and ^{16}O are among these nuclides with $I = 0$ (Table 9.3-1).

Table 9.3-1. Spin I, natural abundance (%), and NMR frequency (MHz) for some important nuclides in a magnetic field $B_0 = 2.3488$ Tesla

Nuclide	Spin I	Natural abundance (%)	NMR frequency (MHz) [$B_0 = 2.3488$ T]
^1H	1/2	99.985	100.00
^2H	1	0.015	15.351
^{12}C	0	98.9	–
^{13}C	1/2	1.108	25.144
^{14}N	1	99.63	7.224
^{15}N	1/2	0.37	10.133
^{16}O	0	99.96	–
^{19}F	1/2	100	94.077
^{31}P	1/2	100	40.481

Nuclei in a static magnetic field

If a nucleus with angular momentum P and magnetic moment μ is placed in a static magnetic field B_0, the angular momentum takes up an orientation such that its component P_z along the direction of the field is an integral or half-integral multiple of $h/2\pi$.

$$P_z = m\,\frac{h}{2\pi} \qquad (9.3-4)$$

Here m is the magnetic or directional quantum number, and can take any of the values $m = I, I-1, \ldots, -I$.

It can easily be deduced that there are $(2I + 1)$ different orientations of the angular momentum and magnetic moment in the magnetic field. This behavior of the nuclei in the magnetic field is called *directional quantization*.

Nuclei with a magnetic moment undergo directional quantization in an external field.

From Eqs. 9.3-2 and 9.3-4 we obtain the components of the magnetic moment along the field direction:

$$\mu_z = m\gamma\,\frac{h}{2\pi} \qquad (9.3-5)$$

In the classical representation the nuclear dipoles precess around the z-axis, which is the direction of the magnetic field. However, in contrast to the classical spinning top, for a precessing nuclear dipole only certain angles are allowed because of the directional quantization. For the proton with $I = 1/2$, for example, this angle is 54°44′ (Fig. 9.3-3). In quantum mechanics the $m = +1/2$ state is described by the spin function α, while the $m = -1/2$ state is described by the spin function β; the exact form of these functions need not concern us here.

The energy of a magnetic dipole in a magnetic field with a flux density B_0 is:

$$E = -\mu_z B_0 \qquad (9.3-6)$$

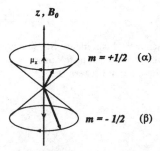

Fig. 9.3-3. Precession of nuclear dipoles with spin $I = 1/2$ around a double cone; the half-angle of the cone is 54°44′

and with Eq. 9.3-5:

$$E = -m\gamma \frac{h}{2\pi} B_0 \tag{9.3-7}$$

The pattern of energy levels of a nucleus in an applied field depends on the spin I.

For the proton and the carbon-13 nucleus, both of which have $I = 1/2$, there are two m-values and consequently two E-values. If $m = +1/2$, μ_z is parallel to the field direction, which is the energetically preferred orientation; conversely, for $m = -1/2$, μ_z is antiparallel. For nuclei with $I = 1$, such as ^{2}H and ^{14}N, there are three energy levels. Figure 9.3-4 shows the energy level schemes for nuclei with

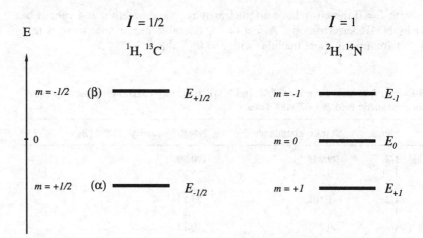

Fig. 9.3-4. Energy level schemes for nuclei with $I = 1/2$ (left) and $I = 1$ (right)

$I = 1/2$ and $I = 1$. The energy difference between two adjacent energy levels is in all cases:

$$\Delta E = \gamma \frac{h}{2\pi} B_0 \tag{9.3-8}$$

From Eq. 9.3-8 we see that $\Delta E \propto B_0$.

Our next question is: how do the nuclei in a macroscopic sample, such as that in an NMR sample tube, distribute themselves between the different energy states in thermal equilibrium? The answer to this question is provided by Boltzmann statistics. For protons – and also for all other nuclides – the energy difference ΔE is very small compared with the average energy of the thermal motion, and consequently the populations of the energy levels are nearly equal. The excess in the lower energy level is only in the region of parts per million. According to the classical picture, the nuclei precess on the surface of a double cone as shown in Fig. 9.3-5.

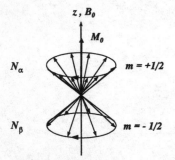

Fig. 9.3-5. Distribution of the precessing nuclear dipoles around the double cone. As $N_\alpha > N_\beta$ there is a resultant macroscopic magnetization M_0

If N is the total number of nuclei and N_α and N_β are the numbers of nuclei with $m = +1/2$ and $-1/2$ respectively there is a resultant macroscopic magnetization M_0 in the z-direction, since $N_\alpha > N_\beta$. The vector M_0 plays an important role in the classical description of all types of pulsed NMR experiments.

The NMR experiment

The resonance condition

A radiofrequency field can induce transitions between energy levels when its frequency is matched to the interval between them.

In the nuclear magnetic resonance experiment transitions are induced between different energy levels by irradiating the nuclei with a superimposed transverse radiofrequency field B_1 of the correct quantum energy, i.e., with electromagnetic waves of the appropriate frequency ν_1. Let us consider Fig. 9.3-6 (see also the left-hand side of Fig. 9.3-4). Transitions between the two energy levels can occur when the frequency ν_1 is chosen so that Eq. 9.3-9 is satisfied:

$$h\nu_1 = \Delta E \tag{9.3-9}$$

Transitions from the lower to the upper energy level correspond to an absorption of energy, and those in the reverse direction to an emission of energy. Each transition is associated with a reversal of the spin orientation. Due to the population excess in the lower level the absorption of energy from the irradiating field dominates. This is observed as a signal.

From Eqs. 9.3-8 and 9.3-9, we obtain the *resonance condition*, Eq. 9.3-10:

$$v_1 = \frac{\gamma}{2\pi} B_0 \qquad (9.3\text{-}10)$$

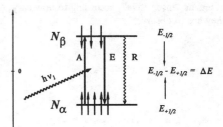

Fig. 9.3-6. Transitions between the two levels $E_{-1/2}$ and $E_{+1/2}$ will be induced by electromagnetic radiation of frequency v_1, provided that $hv_1 = \Delta E$. A = absorption, E = induced emission, R = relaxation. The small arrows represent the nuclear dipoles in the corresponding energy levels

(The term "resonance" relates to the classical interpretation of the phenomenon, since transitions only occur when the frequency v_1 of the electromagnetic radiation matches the precession frequency of the nuclei, which is called the Larmor frequency; see Fig. 9.3-3).

So far we have considered nuclei with $I = 1/2$, for which there are only two energy levels. For nuclei with $I > 1/2$ there are more energy levels, as given by Eq. 9.3-7. To understand the NMR behavior in this case we need to know that transitions are only allowed if

$$\Delta m = \pm 1 \qquad (9.3\text{-}11)$$

This *selection rule*, given by quantum mechanics, says that, for example, the transition from $m = +1$ to $m = -1$ for a ^{14}N or ^2H nucleus is forbidden (see Fig. 9.3-4, right-hand side).

Up to this point we have assumed that we are dealing with isolated nuclei. According to Eq. 9.3-10 we would expect only one resonance signal for each nuclear species. In reality the resonances are influenced in characteristic ways by the environments of the observed nuclei. In the molecules that are of interest to chemists the nuclei are always surrounded by electrons and other atoms. Consequently the nuclei are shielded to some extent, and the effective flux density B_{eff} at the nucleus is always less than the applied flux density B_0. The effect is small, only in the order of parts per million (ppm). This observation is expressed by Eq. 9.3-12:

$$B_{eff} = B_0 - \sigma B_0 = (1 - \sigma)B_0 \qquad (9.3\text{-}12)$$

σ is the shielding constant, a dimensionless quantity which is of the order of 10^{-5} for protons, but reaches larger values for heavy atoms.

With Eq. 9.3-12 the resonance condition (Eq. 9.3-10) becomes

$$v_1 = \frac{\gamma}{2\pi}(1 - \sigma)B_0 \qquad (9.3\text{-}13)$$

Thus the resonance frequency v_1 is proportional to B_0 and – more importantly – to the shielding factor $(1 - \sigma)$. From this statement we arrive at the following important conclusion: Nuclei that are chemically non-equivalent are shielded to different extents and give separate signals in the spectrum. This effect is called the *chemical shift*; we will learn more about it in Sec. 9.3.3. For practical purposes, NMR spectra are by convention recorded in such a way that the shielding constant σ increases from left to right. In our example, in the spectrum of ethyl acetate (see Fig. 9.3-1), the protons of the methylene group are the least shielded, while those of CH_3 of the ethyl group are the most shielded.

A selection rule determines which transitions are observable.

Shielding effects cause small changes to the effective field at the nucleus.

The phenomenon of small differences in the shielding constant, which gives rise to the importance of NMR spectroscopy in chemistry, is called the chemical shift.

The NMR method and the spectrometer

The resonance condition (Eq. 9.3-13) allows us to understand the principle of the method, and also the essential parts and the construction of an NMR spectrometer.

Up to about 1970 all NMR spectrometers worked in the so-called *continuous wave (CW)* mode. To get a spectrum under these conditions one must either keep v_1 constant and vary B_0 until the resonance condition is satisfied or conversely hold B_0 constant and vary v_1. Both these techniques, the *field sweep* and *frequency sweep* methods, are used in practice. At the resonance condition a signal is induced in the receiver coil. This signal is very weak and must be amplified by a large factor before the spectrum can be observed using an oscilloscope or a pen recorder.

Some nuclei are described as "sensitive" and others as "insensitive", according to how easy they are to observe.

The CW method was, and still is, suitable for recording the spectra of nuclides with the combined advantages of a large magnetic moment, spin $I = 1/2$, and high natural abundance. These are called *sensitive nuclides*; examples are ^1H, ^{19}F, and ^{31}P (see Table 9.3-1). Unfortunately neither ^{13}C nor ^{15}N belongs to this category. To make possible routine measurements on *insensitive nuclides* or on very dilute solutions, as well as for other reasons, it was necessary to develop new techniques. This involved radical changes both in the design of NMR spectrometers and in the measurement procedures.

With regard to spectrometers an important innovation was the introduction of *cryomagnets*, which provided considerably higher field strengths than could be reached with conventional permanent magnets or electromagnets. Table 9.3-2 lists some commonly used magnet types with the corresponding field strengths and resonance frequencies for ^1H and ^{13}C.

Table 9.3-2. ^1H and ^{13}C resonance frequencies at different magnetic field strengths B_0

B_0 (Tesla)	Resonance frequencies (MHz)		Types of magnet used
	^1H	^{13}C	
1.41	60	15.1	perm[a)], electro[b)]
2.35	100	25.15	electro, cryo[c)]
4.70	200	50.3	cryo
7.05	300	75.4	cryo
9.40	400	100.6	cryo
14.09	600	150.9	cryo

[a)] permanent magnet; [b)] electromagnet; [c)] cryomagnet

The pulsed method of observing NMR signals has great advantages.

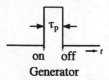

on   off
Generator

Fig. 9.3-7. Schematic representation of a pulse. A generator which works at a fixed frequency ν_1 is switched on for a short time τ_p. The pulse duration τ_p is typically several μs

The FID curve contains all the spectral information.

Fourier transformation converts the information from the time domain (FID) to the frequency domain.

However, the decisive step forward was achieved through *pulsed NMR spectroscopy*. This method uses a radiofrequency pulse to excite simultaneously all nuclei of one species in the sample. What is a *pulse*? A generator usually operates at a fixed frequency ν_1. However, if it is switched on for only a short time τ_p (Fig. 9.3-7), one obtains a pulse which contains not just the frequency ν_1 but a continuous band of frequencies symmetrical about the center frequency ν_1. The effective frequency bandwidth is proportional to τ_p^{-1}. For example, if $\tau_p = 10^{-5}$ s the frequency band is about 10^5 Hz wide. The choice of the generator frequency ν_1 is determined by B_0 and the nuclide to be observed. For example, to observe proton transitions at $B_0 = 4.7$ T (Tesla) the generator frequency is 200 MHz, whereas for observing ^{13}C resonances it must be 50.3 MHz (cf. Table 9.3-1).

If we use the pulsed method to detect NMR signals we do not directly record a spectrum like that in Fig. 9.3-1, where the abscissa is a frequency axis. Instead the spectrum looks like the example shown in Fig. 9.3-8a, with a time axis as abscissa. It is a spectrum in the so-called *time domain*. This interferogram or free induction decay curve (FID), to give it its more usual name, cannot be directly interpreted in this form. Nevertheless, it contains all the information (frequencies, multiplicities, and intensities) that is associated with an NMR spectrum. Furthermore, we can transform it into a spectrum in the *frequency domain* by a common mathematical operation called the *Fourier transformation* (FT). If we do that, we get from Fig.

Fig. 9.3-8. 300 MHz ^1H NMR spectrum of ethyl acetate. (a) Time domain spectrum, the FID; (b) frequency domain spectrum obtained by Fourier transformation of (a)

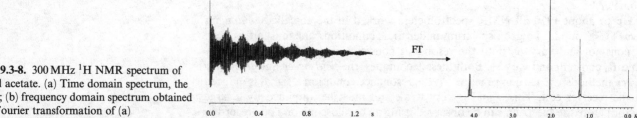

(a)

(b)

FT

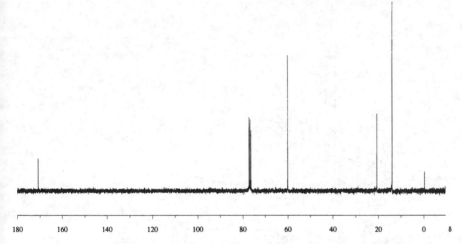

Fig. 9.3-9. 75.47 MHz ^{13}C NMR spectrum of ethyl acetate in CDCl$_3$ with ^{1}H broad-band decoupling

9.3-8a the spectrum shown in Fig. 9.3-8b which is the same as that in Fig. 9.3-1. The transformation is performed by a computer and needs less than one second! One advantage of this method is that the FIDs of many pulses can be added together, accumulated in the computer, and only then transformed as a final step. In this accumulation procedure the random electronic noise becomes partly averaged out, whereas the contribution from the signals is always positive and therefore builds up by addition. The signal-to-noise ratio increases in proportion to the square root of the number of scans NS (pulses): signal-to-noise ratio $\propto \sqrt{NS}$.

As mentioned above, the spin-1/2 isotopes ^{13}C and ^{15}N belong to the insensitive nuclides. The gyromagnetic ratio γ (and therefore also the magnetic moment μ) for ^{13}C is only 1/4 of that for ^{1}H, and that for ^{15}N is even smaller, only 1/10 of that for ^{1}H (Eq. 9.3-14).

$$\gamma(^{13}\text{C}) \approx \gamma(^1\text{H})/4 \qquad \gamma(^{15}\text{N}) \approx \gamma(^1\text{H})/10 \qquad (9.3\text{-}14)$$

In addition the natural abundances are low in both cases, 1.11% for ^{13}C, and 0.37% for ^{15}N. Without going into the theory, we note here that the signal intensity for equal numbers of nuclei is proportional to γ^3. Consequently the signals for ^{13}C and ^{15}N are, for this reason alone, much smaller than those for ^{1}H. If we also take into account the low natural abundance of ^{13}C, we can calculate that the ratio of the signal intensities for ^{1}H and ^{13}C is about 5700 : 1, assuming that we have the same numbers of carbon and hydrogen atoms, as in CHCl$_3$ for example, subject to the same experimental conditions! To overcome this problem so that ^{13}C and ^{15}N NMR spectra can be obtained routinely, it is absolutely essential to use a high field spectrometer which works in the pulsed mode.

But that is not enough. Normally the ^{13}C nuclei are coupled to protons and give multiplets because of spin-spin coupling (see Sec. 9.3.2, "The indirect spin-spin coupling"). So the already weak intensity is distributed between the several lines of the multiplet. The way out of this dilemma is to decouple the ^{13}C and ^{1}H nuclei. This can be done by irradiating all ^{1}H resonances simultaneously while observing the ^{13}C resonances; the procedure is called broad-band (BB) decoupling. This causes all the protons to change their spin orientations in the magnetic field very rapidly, so that the coupling to the ^{13}C nuclei is effectively averaged to zero. All the resonances in the ^{13}C spectrum then become singlets, recombining the individual peak intensities of each multiplet. Such a broad-band decoupled ^{13}C NMR spectrum of ethyl acetate is shown in Fig. 9.3-9. The molecule contains four chemically different sets of carbon atoms and therefore we get four signals.

Obviously, however, in modifying the spectrum by decoupling we sacrifice some information, namely that contained in the spin-spin coupling constants. Usually, therefore, the chemical shifts are the only spectral parameters measured in ^{13}C NMR spectroscopy. In contrast to ^{1}H NMR spectroscopy, these chemical shifts can be read off simply and directly from the spectra. Incidentally, the fact that this decoupling technique also distorts the signal intensities is *one* reason why integrals are not usually measured for ^{13}C NMR spectra (the difficulty that this causes is

^{13}C and ^{15}N are "insensitive" nuclides.

^{13}C spectra can be simplified and made easier to observe by broad-band decoupling of protons.

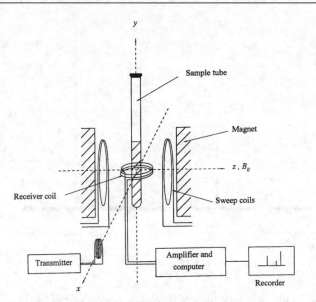

Fig. 9.3-10. Schematic arrangement of an NMR spectrometer

connected with the *nuclear Overhauser effect, NOE*; see Sec. 9.3.2, "The intensities of resonance signals" and the cited literature).

Figure 9.3-10 shows in schematic form the construction of an NMR spectrometer. In principle this diagram serves to illustrate the arrangements in both a CW and a pulsed spectrometer. It consists of a magnet, a radiofrequency transmitter, which generates the resonance frequency or the radiofrequency pulses, and a receiver to detect the signals. These signals are amplified and stored in a computer, which also serves to transform the data or process them in some other way. Finally, the spectrum is presented on a pen recorder (or an oscilloscope). The geometrical arrangement shown here is for a conventional (iron) magnet. The arrangement for a cryomagnet is slightly different, as the sample tube axis is then parallel to the magnetic field direction.

Free induction decay and relaxation

Earlier we learned that the lower level of the energy scheme is more populated than the upper. If we irradiate the sample – in the CW or pulse mode – with a radiofrequency field B_1, then transitions between the levels take place if the resonance condition is satisfied (see Fig. 9.3-6). Consequently the thermal equilibrium will be disturbed and the population ratio will change. We saw in Fig. 9.3-5 that the z-components of all the nuclear magnetic moments in the sample add to form the macroscopic magnetization M_0. If the population ratio changes, the component M_z of the magnetization vector M_0 in the z-direction also changes. But the change of population ratio under the influence of a radiofrequency field or a pulse is only one aspect of a very complex phenomenon. To understand what happens to the macroscopic magnetization M_0 under the influence of a pulse, we use a semiclassical vector representation.

Since the macroscopic magnetization M_0 is associated with a macroscopic angular momentum (made up of the nuclear spin angular momentum contributions – see Sec. 9.3.2, "Nuclear properties") it behaves rather like a spinning top or gyroscope. Nevertheless, this model borrowed from classical mechanics must not be pushed too far since we are, after all, dealing with quantum-mechanical phenomena involving individual nuclei. Therefore we will merely describe the behavior in a semiclassical way without going into theoretical details.

The motions of the magnetic vectors are best described by reference to a rotating coordinate system (rotating frame).

First we have to introduce a mathematical device which makes it much easier to discuss and visualize the otherwise very complicated motion of the vectors involved. This is the *rotating frame*, a coordinate system (x', y', z) which rotates about the z-axis at the frequency ν_1 of the transverse radiofrequency field. The effective radiofrequency field in the rotating frame is represented by a constant vector B_1, which we will assume to lie along the x'-axis. If the resonance condition is satisfied the rotating frame also keeps pace with the precessing nuclear spins, so that their motion is effectively frozen.

The effect of applying the field $\boldsymbol{B}_1$ for the duration of the pulse (τ_p) is to tilt M_0 towards the y'-axis through an angle θ which is given by the product $\gamma B_1 \tau_p$ (Fig. 9.3-11). Thus, at the end of the pulse the macroscopic magnetization vector has a transverse component $M_{y'}$. During the *acquisition time* immediately following the pulse, $M_{y'}$ (which is, of course, rotating at frequency ν_1 in the laboratory frame) induces a signal in the receiver coil, which is observed as a free induction decay (FID). If the sample contains nuclei in more than one environment, precessing at different frequencies in the static field $\boldsymbol{B}_0$, then the transverse field component will split into several components after the pulse. These components then rotate (relatively slowly) in the $x'y'$ plane and generate an interferogram-type FID such as that in Fig. 9.3-8a. Evidently the amplitude of the FID is affected by the choice of the pulse angle θ. A 90° pulse maximizes the signal; such a pulse also plays an important role in the more complex pulse sequences that will be mentioned later.

The nature of the FID can be understood by means of a vector picture.

Following the disturbance caused by the pulse, the spin system gradually returns to its original equilibrium state; this behavior is called *relaxation*. We must distinguish between two different relaxation processes, namely the return of the (reduced) longitudinal magnetization M_z to its original value M_0, and the decay of the transverse magnetization to zero.

The return of the spin system to its equilibrium condition is called relaxation.

The first of these processes is called the longitudinal or spin-lattice relaxation. The latter name comes from the fact that the energy absorbed from the radiofrequency pulse, leading to an increase in the population(s) of the upper nuclear energy level(s), becomes dissipated into the so-called "lattice" as the equilibrium population ratio is re-established. The rate at which this occurs depends on how effectively the random thermal motions of the "lattice" (a term used regardless of whether the sample is a solid or a liquid) interact with the nuclear spins to induce transitions. It can be described by a time-constant T_1, the *spin-lattice* or *longitudinal relaxation time*.

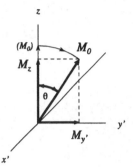

The second process is the transverse or spin-spin relaxation, the latter name indicating that it involves interactions of the nuclear spins with each other (although that is not the only mechanism in the decay of the transverse magnetization). In this process the individual precessing nuclear spins, which have become bunched to some extent to produce the transverse component of magnetization, gradually revert to their random phase distribution (see Fig. 9.3-5). Obviously this plays a role in determining the appearance of the FID curve as it dies away. No energy transfer is involved in this case, as the nuclear level populations are unaffected. The corresponding time-constant in this case is T_2, the *spin-spin* or *transverse relaxation time*.

Fig. 9.3-11. The macroscopic magnetization vector M_0 has been turned through an angle θ from its equilibrium orientation (along the z-axis of the rotating frame x', y', z) by applying a pulse. It now has a component M_z along the field direction and a transverse component $M_{y'}$

From the standpoint of chemical kinetics T_1^{-1} and T_2^{-1} are the rate constants of the relaxation processes, which are *first-order* processes. T_1^{-1} and T_2^{-1} are also important spectral parameters; they can be correlated with the flexibility and mobility of the whole molecule or parts of a molecule. For example, in a macromolecules the side-chains are more flexible than the backbone. A detailed discussion would be beyond the scope of this contribution and the interested reader is referred to more specialized treatments.

Spectral parameters: general aspects

The chemical shift

In Sec. 9.3.2, "The resonance condition," it was shown that chemically nonequivalent nuclei in a molecule are shielded differently and therefore give separate signals in the NMR spectrum. In practice the positions of signals are not reported by giving the actual σ values, nor do we measure the absolute resonance frequencies ν_1, which are typically some hundreds of MHz. Apart from the difficulties of precisely measuring such high frequencies, it is also not particularly useful, since Eq. 9.3-13 shows that ν_1 is proportional to B_0, and therefore both ν_1 and the field strength B_0 would have to be quoted. Instead we use a relative scale whereby one measures the frequency difference $\Delta\nu$ between the resonance signals of the sample and that of a reference compound. In ^1H and ^{13}C NMR spectroscopy tetramethylsilane (TMS) is usually employed for this purpose. But the $\Delta\nu$ values also depend on B_0! In practice, therefore, one defines a dimensionless quantity δ according to Eq. 9.3-15.

$$\delta = \frac{\Delta v}{\text{observing frequency}} \times 10^6 \qquad\qquad (9.3\text{-}15)$$

The factor 10^6 is introduced in order to give convenient numerical values; thus the δ-values are always given in parts per million (ppm). The δ-value for the reference compound TMS is, by definition, zero, since in this case $\Delta v = 0$.

The reference compound TMS is normally added as an *internal standard* to each sample before the spectrum is recorded (see Fig. 9.3-1, for example). In cases where TMS cannot be used (for example, TMS is insoluble in water) a different reference compound is added to the sample and the results are converted to the TMS scale. For measuring spectra of other nuclides such as ^{2}H, ^{15}N, ^{19}F, or ^{31}P, suitable reference compounds must be found. In the case of deuterium, for example, the signal of ^{2}H$_2$O (D$_2$O) is taken as internal reference. In other cases one can use an *external reference* by inserting a capillary filled with the reference compound into the sample tube. For ^{15}N an aqueous solution of NH$_4$NO$_3$ is used as (external) reference, assigning a δ-value of zero for the ^{15}NH$_4^+$ ion in this compound. External references are also used in ^{19}F and ^{31}P NMR spectroscopy; for example, the ^{19}F signal of liquid CFCl$_3$ and the ^{31}P signal of 85% H$_3$PO$_4$.

The total range of chemical shifts in ^{1}H NMR spectroscopy is approximately 10 ppm (apart from certain extreme values), whereas in ^{13}C NMR spectroscopy the range is more than 220 ppm (see Figs. 9.3-20 and 9.3-21, Sec. 9.3.3).

Before we discuss the relationship between chemical shifts and molecular structures, we will introduce the two other important spectral parameters, the *indirect spin-spin coupling constant* and the intensity, measured by the *integral*.

The indirect spin-spin coupling

Figure 9.3-1 shows the high resolution spectrum of ethyl acetate. As mentioned in the introduction, the quartet at $\delta = 4.07$ and the triplet at $\delta = 1.21$ are quite characteristic for protons in an ethyl group. The quartet can be assigned to the methylene protons, and the triplet to the methyl protons (ratio of integrals quartet:triplet = 2:3). Before explaining how these splitting patterns are caused by the indirect spin-spin coupling, we will try to understand a few simpler cases, using examples from ^{1}H NMR spectroscopy. Most of the rules that will be derived for protons can be generalized, at least for other nuclides with $I = 1/2$.

Let us consider the following three examples:

1. The spectrum of cinnamic acid (Fig. 9.3-12) shows two doublets at $\delta = 6.47$ and $\delta = 7.81$. Each doublet corresponds to one proton. They can be assigned to the olefinic protons H^A and H^X.
2. Figure 9.3-13 shows the spectrum of 2-bromoisophthalic acid dimethyl ester. We find a doublet for the two equivalent protons H^A (H-4 and H-6) at $\delta = 7.69$, and a triplet for the proton H^X (H-5) at $\delta = 7.40$. This assignment is based on the integrals; the ratio of the intensities of the two multiplets is 2:1
3. In Fig. 9.3-14 the spectrum of 2-chloropropionic acid is shown. We find a quartet for H^A at $\delta = 4.45$ and a doublet for the CH$_3$ protons at $\delta = 1.73$ (integrals in the ratio 1:3).

What is the explanation for this splitting of signals? We will consider in turn some very common coupled spin systems, starting with the simplest.

The two-spin system AX

In example 1 above, the protons H^A and H^X in cinnamic acid form a two-spin system AX. Let us confine our attention to the resonances for the proton H^A. It

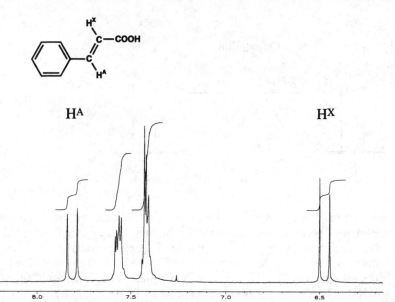

Hᴬ   Hˣ

Fig. 9.3-12. Part of the 300 MHz ^{1}H NMR spectrum of cinnamic acid in CDCl$_3$; δ(OH) $\approx$ 12. The two doublets at $\delta = 6.47$ and $\delta = 7.81$ can be assigned to the olefinic protons H^A and H^X using an empirical rule for predicting δ-values

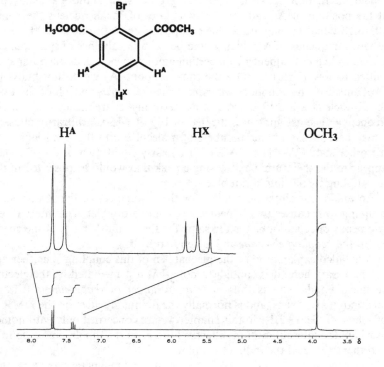

Hᴬ   Hˣ   OCH₃

Fig. 9.3-13. 300 MHz ^{1}H NMR spectrum of 2-bromoisophthalic acid dimethyl ester. The two equivalent protons H^A (H-4 and H-6) give a doublet, and the proton H^X (H-5) gives a triplet. The assignment follows from a comparison of the integrals; the ratio of the intensities of the two multiplets is 2:1

$CH_3^X CH^A ClCOOH$

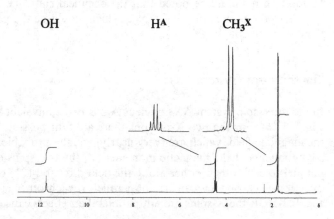

OH   Hᴬ   CH₃ˣ

Fig. 9.3-14. 300 MHz ^{1}H NMR spectrum of 2-chloropropionic acid in CDCl$_3$. The quartet (CH) and doublet (CH$_3$) are shown expanded, both by the same factor

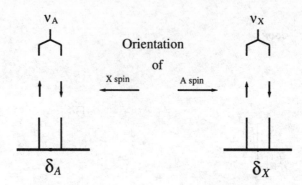

Fig. 9.3-15. Sketch to explain the splitting pattern observed for a two-spin system AX

has one neighbor proton H^X. To understand the splitting into a doublet we must distinguish between two situations. In the first, the nucleus X, coupled to A, is so orientated in the magnetic field that its magnetic moment μ has a positive component μ_z along the field direction (Fig. 9.3-15); we represent this by an arrow directed upwards (A-X↑). In the second situation the orientation of μ_z is reversed (A-X↓). These two situations correspond to the $m = +1/2$ and $m = -1/2$ states for the nucleus X, or more briefly the α and β states (Sec. 9.3.2, "Nuclei in a static magnetic field").

The interaction between A and X produces an additional field contribution at the position of A, and for the two states of the X-nucleus the contributions, although equal in magnitude, have opposite signs. In one case, therefore, the resonance frequency of A, which would be ν_A in the absence of spin-spin coupling, is shifted to higher frequency by a certain amount, whereas in the other situation it is shifted to lower frequency by the same amount. As we do not know the precise mechanism of the coupling, we cannot predict *a priori* which of the two states of the X-nucleus will shift the A-resonance to higher frequency and which to lower frequency. The assignment indicated in Fig. 9.3-15 is arbitrarily chosen. Since in a macroscopic sample the numbers of molecules with the X-nucleus in the one or the other state ((A-X↑) or (A-X↓)) are nearly equal, two peaks of equal intensity appear in the spectrum; i.e., the single peak that would be observed in the absence of coupling is split into a doublet.

An analogous situation applies for the resonances of the X-nucleus, since the coupling to A causes two X peaks, i.e., again a doublet. The interval between the two peaks of each doublet is the same for the A and X parts of the spectrum; it is called the *coupling constant* and is denoted by J_{AX}.

A detailed explanation of the mechanism of this coupling is neither appropriate nor necessary here; it is enough to note that it is mediated by the electrons of the chemical bonds, and is thus *indirect*. A *direct* or *dipole-dipole* (through-space) interaction also exists, and is normally the dominant effect on the NMR properties of solids (cf. Sec. 9.3.1). In this chapter we are concerned only with liquid samples, in which the internal motions effectively remove the dipole-dipole coupling effects and thereby reveal the indirect coupling.

Since the splitting is due solely to the nuclear magnetic moments, the value of the coupling constant J_{AX} – unlike the chemical shift – does not depend on the magnetic flux density B_0. The coupling constant is therefore always given in Hz. In addition it should be noted that the chemical shift, the δ-value, always corresponds to the middle of the doublet, which would be the position of the signal if there were no coupling.

The three-spin system AX$_2$

In the three-spin system AX$_2$ nucleus A has two equivalent neighbor nuclei X (see Fig. 9.3-16). For the two X-protons there are four possible spin configurations in the magnetic field, which are (very nearly) equally probable. Considering only the z-components of the magnetic moments, μ_z, the two spins of the X nuclei can be parallel to each other, either along the field direction (↑↑) or opposite to the field direction (↓↓), or they can be antiparallel to each other (↓↑ or ↑↓), as shown in Fig. 9.3-16. If the X-spins are antiparallel their effects cancel to zero, while if they

The indirect spin-spin coupling produces characteristic multiplet patterns.

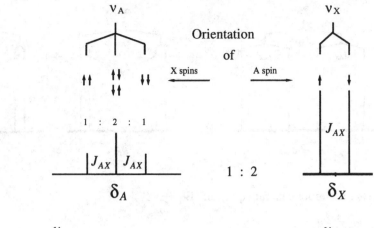

Fig. 9.3-16. Sketch to explain the splitting pattern observed for a three-spin system AX_2

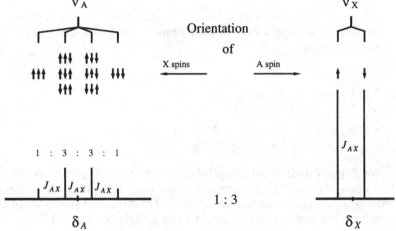

Fig. 9.3-17. Sketch to explain the splitting

are parallel ($\uparrow\uparrow$ or $\downarrow\downarrow$) the additional field contributions at the position of A are equal in magnitude but opposite in sign. Therefore we observe a triplet for the protons H^A. The intensities are in the ratio $1:2:1$, the middle signal with double intensity, since in a macroscopic sample molecules with an antiparallel arrangement of the X-spins are twice as numerous as those with a parallel arrangement. The interval between any two adjacent lines is J_{AX}.

For the two X-protons we obtain a doublet, as they are coupled to only one neighbor nucleus A. The chemical shifts δ_A and δ_X are calculated from the middle peak of the triplet and the middle of the doublet.

As an example we will consider the three-spin system of the ring protons in the symmetrically substituted aromatic compound 2-bromo-isophthalic acid dimethyl-ester, example 2 above (Fig. 9.3-13). The three spins form an A_2X system (in this case not AX_2! Normally the non-equivalent nuclei are identified by different letters of the alphabet, beginning with A and labelling them in the order of their signals from left to right in the spectrum).

The four-spin system AX_3

In the same way as shown for the two- and three-spin systems, we can construct the splitting patterns caused by three or more equivalent neighboring nuclei (Fig. 9.3-17). In the case of three coupling neighbors (AX_3 system) we obtain for A four lines, a quartet, with an intensity distribution $1:3:3:1$. A spectrum of this type is found for $CH^X_3CH^ACICOOH$ (Fig. 9.3-14) with the quartet at $\delta_A = 4.45$ and the doublet for X at $\delta_X = 1.73$.

From these few examples we can derive the rule given in Eq. 9.3-16 for the multiplicity M if all the coupled nuclei have a spin $I = 1/2$ (^1H, ^{13}C, ^{15}N, etc.).

$$M = n + 1 \tag{9.3-16}$$

Here n is the number of equivalent neighbor nuclei. For n equivalent neighbor nuclei with spin $I \geq 1$, such as ^2H or ^{14}N, the multiplicity M can be calculated

The number of lines resulting from coupling to a group of equivalent nuclei is given by simple multiplicity rules.

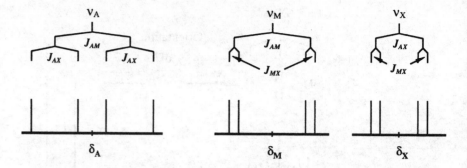

Fig. 9.3-18. Sketch to explain the splitting pattern observed for three non-equivalent nuclei AMX

from the more general equation (Eq. 9.3-17):

$$M = 2nI + 1 \qquad\qquad (9.3\text{-}17)$$

For couplings to nuclei with $I = 1/2$ the signal intensities within each multiplet correspond to the coefficients of the binomial series, which can be obtained from Pascal's Triangle:

$n = 0$						1				
$n = 1$					1		1			
$n = 2$				1		2		1		
$n = 3$			1		3		3		1	
$n = 4$		1		4		6		4		1

We can now understand the splitting patterns for ethyl acetate in Fig. 9.3-1. The two methylene protons are coupled to the three equivalent CH_3 protons of the ethyl group and give a quartet at $\delta_A \approx 4.1$. The three methyl protons of the methyl group are coupled to two protons and give a triplet at $\delta_X \approx 1.4$.

The three-spin system AMX with non-equivalent nuclei

An example of a three-spin system with non-equivalent nuclei H^A, H^M, and H^X is provided by the olefinic protons in styrene. We expect three coupling constants with different values: $J_{AM} \neq J_{AX} \neq J_{MX}$. In the spectrum we observe for each nucleus A, M, and X four peaks of nearly equal intensity, forming a doublet of doublets. Figure 9.3-18 shows how one can easily construct the splitting pattern. One begins with the signals without coupling. Then one splits each line into a doublet corresponding to one of the two coupling constants, preferably the largest in each case. This is repeated for the second smaller coupling constant, so that each line of the first doublet is further split into a doublet. The center of each such doublet of doublets corresponds to the δ-value.

With this knowledge it should be easy to analyze the olefinic part of the spectrum of styrene shown in Fig. 9.3-19.

Strongly coupled spectra

All the examples of couplings in proton spectra discussed above are of a type called *first-order*, which means that the spin-spin coupling constants are small compared with the chemical shift interval between the coupled nuclei, i.e., $\Delta\nu \gg J$. In classifying spectra this condition is conventionally denoted by using letters that are far apart in the alphabet, such as the series A, M, X. When the first-order condition is not satisfied the nuclear system is said to be *strongly coupled*. The spectrum is then more complicated, with more peaks than in the first-order case, and simple multiplet and relative intensity rules no longer apply. In all but the simplest cases (such as AB type spectra), a full analysis is only possible by a quantum-mechanical calculation, for which powerful computer programs based on iterative matching procedures have been developed. Such programs are often included in the software packages supplied with modern NMR systems. The

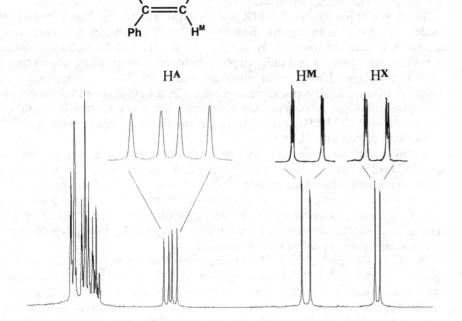

Fig. 9.3-19. 300 MHz ^{1}H NMR spectrum of styrene in CDCl$_3$

aromatic part of the styrene spectrum around $\delta = 7.3$ in Fig. 9.3-19 is an example of a strongly coupled AA′BB′C system (a system of five protons where two pairs of protons are chemically equivalent). Whether a full quantum-mechanical analysis is worthwhile depends on the nature of the problem; often the additional structural information that this would yield is insufficient to justify the extra work. One of the advantages of the very high magnetic fields nowadays available with superconducting magnets is that spectra approach the ideal first-order condition more closely as the field is increased (since $\Delta\nu$ is proportional to B_0, whereas J remains unaltered).

Couplings between equivalent nuclei

On the basis of the spin systems analyzed above, it seems surprising that for the three protons of an isolated methyl group (e.g., the signal for COCH$_3$ in Fig. 9.3-1) we observe only a singlet. The same also applies for the six protons in benzene, which also give a single peak, even though we know that in benzene derivatives (such as styrene in Fig. 9.3-19) the protons are coupled to each other. Without going into the quantum-mechanical arguments which explain this, we can say quite generally that *couplings between equivalent nuclei cannot be observed in the spectrum!*

Here we will limit the discussion to first-order spectra and summarize the results in a few general rules, which are valid for nuclei with $I = 1/2$:

Rule 1: The multiplicity is given by Eq. 9.3-16 above:

$$M = n + 1$$

Rule 2: The intensities within the multiplets can be obtained from Pascal's Triangle.

Rule 3: Couplings between equivalent nuclei cannot be observed in the spectra.

Rule 4: The coupling constants J do not depend on the magnetic flux density B_0 and can therefore be given in Hz.

Rule 5: In saturated compounds couplings between protons separated by more than three bonds are normally too small to be observed.

Five rules summarize the main features of multiplet splittings.

The last rule anticipates the later discussion of relationships between coupling constants and molecular structure.

The main applications of NMR spectroscopy are to the determination of molecular structures. In organic chemistry the NMR resonances most commonly studied are therefore those of ^{1}H and ^{13}C. In the ^{1}H NMR spectra of organic molecules one normally sees only couplings between protons, which are described as H,H couplings. However, for molecules containing fluorine, phosphorus, or other spin-1/2 nuclei, the couplings to these nuclei always appear in the spectrum too. Even couplings to nuclei which have a spin $I \geq 1$ can be seen occasionally, both in ^{1}H and ^{13}C NMR spectra. Examples are the couplings of protons or ^{13}C nuclei with deuterium, ^{2}H.

At this point special mention must be made of the couplings between protons and ^{13}C nuclei. Normally we cannot see the C,H coupling in the ^{1}H NMR spectrum because the natural abundance of ^{13}C is only 1.11%. But for very strong ^{1}H single peaks, for example the resonance of the solvent or of a methyl group, one can observe two small peaks to the left and right of the main peak with a relative intensity of 0.55%, the interval between them being the C,H coupling constant. These small peaks are called the ^{13}C satellites.

In ^{13}C NMR spectra the C,H coupling often leads to very complex splitting patterns. In Sec. 9.3.2, "The NMR method and the spectrometer", we learned how this coupling is normally eliminated in routine measurements. The primary reason for the decoupling was to get singlets instead of multiplets, so as to concentrate the whole of the intensity into one peak. This has the additional advantage of giving spectra that consist only of singlets and are therefore much easier to analyze.

The intensities of resonance signals

The area under the signal curve is referred to as the intensity of the signal. It depends on the number of nuclei that contribute to the signal. The intensity is measured by means of a stepped curve, the *integral*. By comparing the step heights in the integral trace we obtain directly the ratios of the sets of chemically non-equivalent nuclei in the molecule. We have already seen how this information was used to assign all the signals in the spectrum of Fig. 9.3-1.

In ^{13}C NMR spectra and also in spectra of many other nuclides, integrals are not normally measured because in many cases, particularly for nuclides with low natural abundance and low sensitivity compared with protons, detection methods have to be used which would distort the integrals. Such distorting effects are associated with, for example, very large chemical shifts (sometimes hundreds of ppm), long relaxation times T_1 (e.g., for ^{13}C), and differences in the magnitude of the nuclear Overhauser effect (NOE) accompanying spin decoupling.

9.3.3 Information from chemical shifts

Introduction: the shielding constant σ

In molecules the effective field at the position of a nucleus is slightly less than the externally applied field B_0, i.e., the nucleus is magnetically shielded, resulting in a small reduction in the observed resonance frequency. This effect is described by the shielding constant σ (see Eqs. 9.3-12 and 9.3-13). In Sec. 9.3.2, "The chemical shift", we learned that spectra are recorded so that the σ-values increase from left to right. In practice, however, chemical shifts are quoted as δ-values (see Eq. 9.3-15). Theoretical calculations of σ by a variety of methods are not sufficiently reliable to allow exact prediction of spectra for analytical purposes. Nevertheless, such calculations give useful insights into the causes of shielding, especially with regard to the characteristic differences between protons and the heavier nuclides, including ^{13}C.

The nuclear shielding constant σ is determined mainly by the local electron density.

Theory and experiment lead to the conclusion that the shielding constant σ is determined mainly by the distribution of electron density in the molecule. The

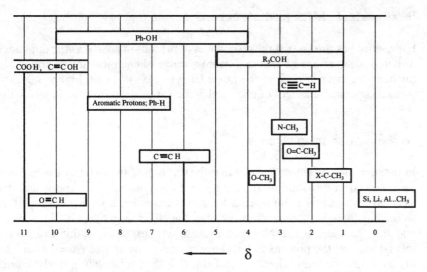

Fig. 9.3-20. Chemical shifts of ^{1}H nuclei in organic compounds

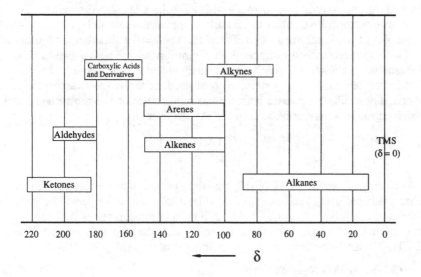

Fig. 9.3-21. Chemical shifts of ^{13}C nuclei in organic compounds

chemical shifts are therefore considerably affected by substituents which specifically influence this electron distribution.

So as to get a general idea of the positions of ^{1}H and ^{13}C NMR signals for different classes of compounds, some data are summarized in Figs. 9.3-20 and 9.3-21. It is seen that the resonances of ^{1}H and ^{13}C nuclei in similar bonding situations are grouped in characteristic regions. For example, the signals of aromatic protons are found in the region of $\delta \approx 6.5$ to 9.0, and those of aldehydic protons between $\delta = 9$ and $\delta = 11$. The corresponding ^{13}C resonances of arenes are in the range $\delta = 100$ to $\delta = 150$ and those of aldehydes between $\delta = 180$ and $\delta = 210$. Empirical guidelines such as these enable the chemist to get information about the structure of an unknown compound, to evaluate the success of a synthesis, or to quantitatively analyze the components in a mixture – to mention only a few applications.

First we will try to get an impression of the relationship between the structure of a molecule and the ^{1}H chemical shifts. As mentioned above, substituents have a large influence by affecting the electron distribution through inductive and mesomeric effects. These effects are transmitted through the bonds. But interactions through space are also possible – for example if the observed nuclei have magnetically anisotropic neighbors such as a carbonyl group, a CC double or triple bond, or a phenyl ring. Intermolecular interactions also contribute to the shielding. Because some of the effects are more important for protons than for the heavier nuclides we will first discuss these effects for the chemical shifts of protons. After that we will turn our attention to ^{13}C resonances. This part will end with a look at some examples of chemical shifts of other nuclides. –

Table 9.3-3. ^{1}H and ^{13}C chemical shifts of substituted methanes CH_3X

Substituent X	$\delta(CH_3)$	$\delta(^{13}CH_3)$
I	2.16	−24.0
Br	2.70	9.6
Cl	3.05	25.6
F	4.25	71.6
NH$_2$	2.36	28.3
OH	3.38	50.5
NO$_2$	4.33	61.2

^{1}H chemical shifts and structure

Experience has shown that the resonances of protons in nearly all the different types of bonding situations lie within a narrow range of chemical shifts, about 20 ppm; furthermore, over 95% lie in the range from $\delta = 0$ to $\delta = 10$. Figure 9.3-20 shows some characteristic ranges of chemical shifts for protons in organic molecules.

^{1}H chemical shifts in alkanes

In the previous section we learned that the position of the resonance signal depends on the electron distribution around the observed nucleus. We know also that the electron distribution is influenced by substituents. Experiments confirm this and show that there is a connection between the electronegativity of the substituents and the shielding of the observed protons. An example of this can be seen in methyl halides: the protons in CH_3I are the most strongly shielded, whereas those in CH_3F are the least shielded (see Table 9.3-3), corresponding to the increase in electronegativity from I to F. Other strongly electronegative substituents such as the NO_2 group, oxygen, or nitrogen also shift the CH_3 signal to higher δ-values, whereas electropositive substituents such as metals cause a shift to lower δ-values. Examples of the latter are Li and Si: $LiCH_3$ has $\delta = -1$ (in ether) and the δ-value of TMS, the reference substance, is by definition zero. These δ-values show that the protons in $LiCH_3$ are even more strongly shielded than those in TMS.

Inductive effects influence the electron distribution and therefore the chemical shifts.

All these examples are generally considered in terms of *inductive substituent effects*. These effects decrease with increasing distance of the substituent from the observed nucleus, as can be seen in the following series:

$$\begin{array}{ccc} CH_3Cl & CH_3CH_2Cl & CH_3CH_2CH_2Cl \\ \delta = 3.05 & 1.42 & 1.04 \end{array}$$

Shoolery's rule is just one of many useful empirical relationships.

For help in assigning ^{1}H NMR signals, approximate δ-values are in practice often predicted using various empirical rules. Only one such rule will be mentioned here, namely *Shoolery's rule*. This additivity relationship states that the chemical shift of the protons in a methylene group with the two substituents X and Y, $X\text{-}CH_2\text{-}Y$, can be estimated to a good approximation using Eq. 9.3-18:

$$\delta(CH_2) = 0.23 + S_X + S_Y \tag{9.3-18}$$

S_X and S_Y are effective shielding constants whose values for common substituents can be found in most textbooks on analytical NMR spectroscopy.

^{1}H chemical shifts in alkenes

As expected, olefinic protons, i.e., those that are linked to a double bond, are less shielded than those in alkanes. The δ-values of protons in ethylene derivatives are strongly influenced by substituent effects, which can be inductive, mesomeric, or steric, and their positions relative to the observed proton. The range of chemical shifts is from $\delta = 4$ to 7.5, and for ethylene itself, $\delta = 5.28$.

^{1}H chemical shifts in arenes

For aromatic protons, we would expect, on the grounds of bonding and electron density, to find δ-values similar to those in alkenes. However, the ^{1}H NMR signal of the protons of benzene is found at $\delta = 7.27$, compared with $\delta = 5.28$ for ethylene. This is not an isolated case; it is found generally that aromatic protons are about 1 to 2 ppm less shielded than olefinic protons. An explanation for this will emerge later.

The range of chemical shifts for aromatic protons is $\delta = 6.5$ to 9. Substituent effects, mainly the mesomeric effects of the substituents, are again responsible for the broad distribution of chemical shifts. These effects are illustrated by two examples, aniline and nitrobenzene, whose spectra are shown in Figs. 9.3-22 and

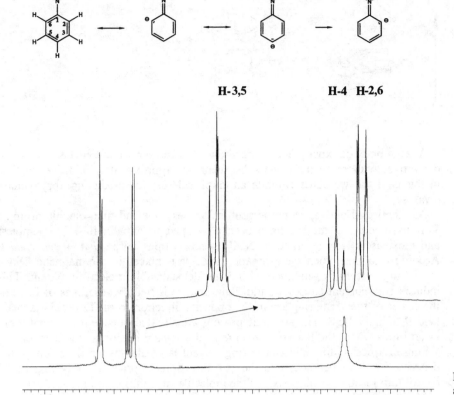

H-3,5   **H-4 H-2,6**

Fig. 9.3-22. 300 MHz ^{1}H NMR spectrum of aniline in CDCl$_3$ and mesomeric limiting structures

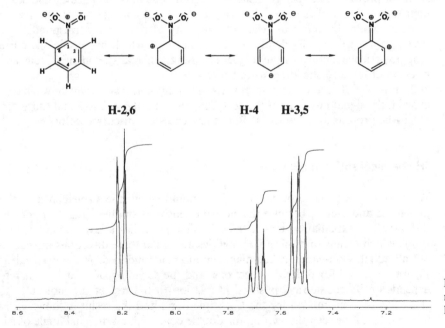

H-2,6   **H-4   H-3,5**

Fig. 9.3-23. 300 MHz ^{1}H NMR spectrum of nitrobenzene in CDCl$_3$ and mesomeric limiting structures

9.3-23. For aniline we find that the δ-values are all smaller than the value for benzene ($\delta = 7.27$). For nitrobenzene, however, all the values are greater than this. These results for aniline and nitrobenzene and also the characteristic differences between the δ-values for protons in the *ortho, meta*, and *para* positions, can be understood on the basis of the mesomeric limiting structures, which are given together with the spectra in Figs. 9.3-22 and 9.3-23. Evidently the amino group increases the electron density in the ring (+M effect) – especially at the *ortho* and *para* positions. The nitro group withdraws electrons from the ring (−M effect); as in the case of aniline, the protons in the *ortho* and *para* positions show the largest effects.

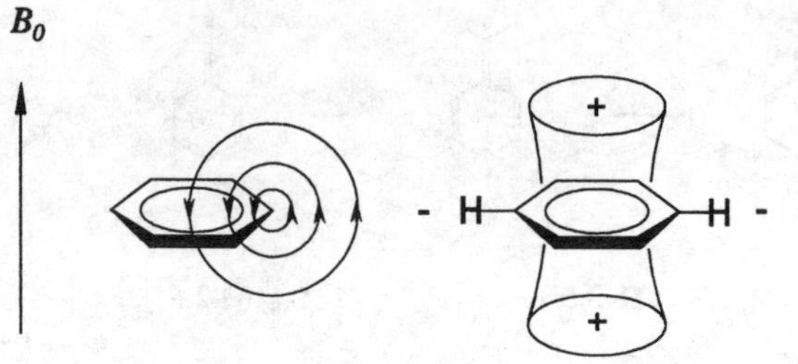

B_0

Fig. 9.3-24. The ring current effect in arenes, with zones of increased (+) and reduced (−) shielding

The effects of multiple substitution on the chemical shifts of benzene ring protons are additive.

A large body of experimental data shows that the substituent effects in benzene derivatives with more than one substituent are approximately additive, so that in this case too we can formulate an empirical rule for predicting approximate δ-values.

As mentioned above, aromatic protons are less shielded than olefinic protons. Various theoretical models have been developed to explain this. The simplest and most useful to the practical NMR spectroscopist is the *ring current model*. According to this, when the aromatic molecule is placed in the magnetic field a circulating current is generated within the delocalizable π-electron system. This induced ring current causes an additional magnetic field, whose lines of force at the center of the arene ring are in the opposite direction to the external magnetic field B_0 (Fig. 9.3-24). The aromatic protons are situated outside the current loop, at positions where the lines of force are in the *same* direction as the field B_0, resulting in a deshielding effect. The ring current is greatest when the plane of the benzene ring is perpendicular to the field direction. In practice an averaged field contribution is always observed, as the molecules in solution are in rapid motion.

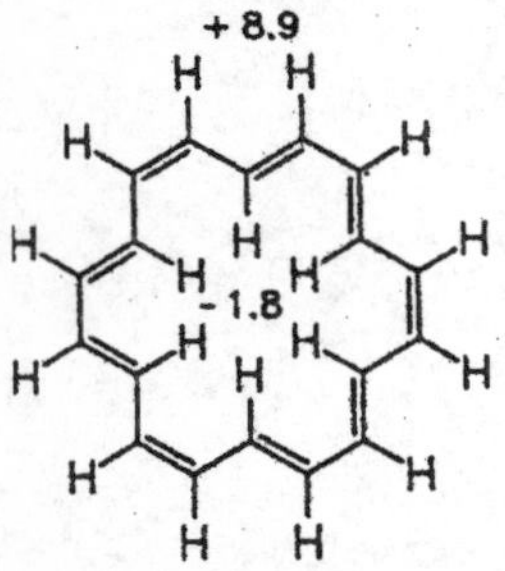

Many examples can be cited to show how useful the ring current model is. One of the most striking examples is that of [18]-annulene, which has twelve outer and six inner protons. The spectrum shows two peaks, readily assigned on the basis of their relative intensities (integrals). That at $\delta = -1.8$ corresponds to the six inner protons, while the other of twice the intensity at $\delta = +8.9$ corresponds to the twelve outer protons. Evidently the inner protons, which are inside the current loop, experience a considerably higher shielding, whereas the outer protons are in a region of reduced shielding, as in benzene.

In Fig. 9.3-20 there are two further apparent anomalies which we have to explain: the signal positions of the acetylenic protons at $\delta \approx 2$ to 3 and those of the aldehydic protons at $\delta \approx 9$ to 11. Both these cases are discussed below.

+8.9

−1.8

^{1}H chemical shifts in alkynes

We would expect that protons in alkynes should be even less shielded than those in alkenes and arenes, because the electron density is smaller. But the opposite is the case. The explanation is that the CC triple bond has a large magnetic *anisotropy*, which means that the susceptibilities along the three directions in space are not all equal. Consequently the magnetic moments induced by the external field B_0 are not equal for different directions, and the shielding of a nucleus therefore depends on its geometrical position in relation to the rest of the molecule. This effect has been calculated for the CC triple bond and the result is shown in Fig. 9.3-25a. It can be seen that within the double cone with a semiaxial angle of 54.7° the shielding is increased (positive sign, +), whereas outside this cone it is reduced (negative sign, −). In alkynes the proton attached to the triple-bonded carbon atom is in the region of increased shielding, and the signals of such protons have correspondingly low δ-values.

^{1}H chemical shifts in aldehydes

The extremely low shielding of aldehydic protons, with δ-values in the range 9 to 11, cannot be fully explained on the basis of the electronegativity of the double-bonded oxygen atom. Here again the explanation lies in the additional effect of

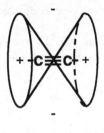 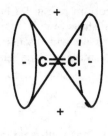 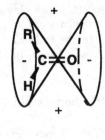

a   b   c

Fig. 9.3-25. Magnetic anisotropy of the CC triple bond (a) and the CC and CO double bonds (b, c). + increased shielding zone; − reduced shielding zone

the magnetic anisotropy of a neighbor group. Figure 9.3-25c shows the result of the calculation for the C=O group. It can be seen that the aldehydic proton lies in the zone of reduced shielding. Figure 9.3-25b shows the result of similar calculations for the CC double bond.

^{1}H chemical shifts in OH, SH and NH groups

For the signals of the protons in OH, SH and NH groups, no well-defined ranges can be given. These hydrogen atoms can form hydrogen bonds, they can undergo exchange, and they have varying degrees of acidic character. Their resonance frequencies are further influenced by concentration, temperature, and the presence of impurities such as water. The resonances of OH protons are more sensitive to these effects than those of SH and NH, and consequently OH signals can be found anywhere in the spectrum, whereas the SH and NH resonances usually fall within narrower regions. With great caution some generalizations can be made. We find the signals for OH in alcohols in the range $\delta = 1$ to 5, in phenols in the range $\delta = 4$ to 10, in acids from $\delta = 9$ to 13, and in enols from $\delta = 10$ to 17. The signals for NH protons can be found in amines within the range $\delta = 1$ to 5, in amides in the range $\delta = 5$ to 6.5, and in peptides in the range $\delta = 7$ to 10. The signals of SH protons are found in the range $\delta = 1$ to 4.

The exchange of hydrogen atoms with deuterium is of great practical importance, as the corresponding signals in the ^{1}H NMR spectrum disappear following the H–D exchange. This effect is used as an aid to assignment.

^{13}C chemical shifts in relation to molecular structure

Figure 9.3-21 gives an impression of where signals of carbon nuclei in organic molecules can be expected in the ^{13}C NMR spectrum. The reference substance is again TMS, with $\delta(\text{TMS}) = 0$. It should be noticed that ^{13}C resonances extend over a range of more than 220 ppm, which is about twenty times greater than that for ^{1}H resonances. Consequently separation of signals should be better in ^{13}C than in ^{1}H NMR spectra.

Owing to the method of observation used, the chemical shifts are usually the only spectral parameters that can be obtained from ^{13}C NMR spectra. Often the spectrum contains simply *one* signal for each non-equivalent carbon atom or group in the molecule. An example is the ^{13}C NMR spectrum of ethyl acetate (see Fig. 9.3-9). We find four signals for the four carbon nuclei. The problem is essentially that of assigning each signal to the correct carbon atom. A knowledge of general rules for the relationships between chemical shifts and molecular structure is therefore even more important in ^{13}C NMR spectroscopy than in ^{1}H NMR spectroscopy. In discussing ^{1}H chemical shifts in the previous section we introduced some special phenomena, such as the effects of ring currents and of magnetic anisotropy of neighbor groups, in order to understand the experimental results. We also mentioned intermolecular effects involving, for example, the solvent or the temperature, especially in connection with chemical shifts of OH, SH, and NH protons (H exchange and hydrogen bonding). In ^{13}C NMR spectroscopy all these effects, when expressed in ppm, are similar in magnitude to those in ^{1}H NMR spectroscopy and therefore, when considered in relation to the total shifts range of about 220 ppm, they are of minor importance. On the other hand substituent effects, which play an important role in ^1H NMR spectroscopy, continue to do so in the case of ^{13}C chemical shifts.

In the following sections we will discuss the influence of substituents on the chemical shifts of ^{13}C resonances in some selected classes of compounds.

^{13}C chemical shifts in alkanes

With alkanes one normally finds well-separated signals for all the different carbon nuclei. The range extends over more than 60 ppm from $\delta = 10$ to 85. The shifts are strongly affected by the number of neighboring carbon atoms and by steric effects such as branching of the chain. Substituents have a considerable influence on the chemical shifts. Merely replacing a hydrogen atom by a methyl group typically reduces the shielding by more than 8 ppm. In substituted alkanes there is – as already seen in ^1H NMR spectroscopy – a clear relationship with the electronegativity of the substituent X, as is evident from Table 9.3-3 for simple methane derivatives (see previous section). The "β-effect" of a substituent (through two bonds) is much smaller, and results in a deshielding of the carbon nucleus, whereas a carbon nucleus in the γ-position generally experiences an increase in shielding. For example, replacing one terminal hydrogen atom in a linear alkane by a chlorine atom alters the chemical shifts of the α, β, and γ carbon nuclei by +31, +11, and -4 ppm, respectively:

$$\begin{array}{ccc} \alpha & \beta & \gamma \\ Cl-CH_2-CH_2-CH_2- \\ \Delta\delta = \quad +31 & +11 & -4 \end{array}$$

From a theoretical standpoint these large effects cannot be explained on the basis of inductive effects alone; steric or other effects must also be invoked.

^{13}C chemical shifts in alkenes and arenes

The ^{13}C resonances of double-bonded carbon nuclei in alkenes are found in a broad range from $\delta \approx 100$ to 150. Substituents can cause large effects; in general they can reduce the shielding, but there are many exceptions.

The same range $\delta \approx 100$ to 150 also contains the ^{13}C resonances of benzene (128.5) and of substituted benzenes, so there can sometimes be problems in assignment. The inductive and mesomeric properties of the substituents influence the chemical shifts of all carbon atoms of the ring. To understand the influence of substituents with appreciable +M or −M effects, such as OH, NH_2, or NO_2, it is helpful to draw the mesomeric limiting structures, as was done in Figs. 9.3-22 and 9.3-23 for aniline and nitrobenzene.

Additivity of substituent effects on ^{13}C chemical shifts leads to useful incremental relationships.

From the large amount of experimental data available for alkanes, alkenes, and especially benzene derivatives, it is found that the substituent effects are to a good approximation additive. Based on this, convenient incremental relationships have been developed for each class of compound to predict chemical shifts in substituted derivatives. Details can be found in the cited literature.

^{13}C chemical shifts in alkynes

Comparing the observed δ-values for acetylene (71.9) and ethylene (123.5), we see that the ^{13}C nuclei in acetylene are considerably more strongly shielded. Similar values are found for the carbon nuclei in CC triple bonds of alkynes generally. The magnetic anisotropy of the CC triple bond, which accounted for the large shielding of the protons in alkynes (see Sec. 9.3.3, "^1H chemical shifts in alkynes"), is insufficient to explain the ^{13}C shielding. To find the cause we would need to know more about the theory of shielding, but that is outside the scope of this chapter. (It is in fact the so-called paramagnetic shielding term which chiefly determines the shielding of ^{13}C nuclei, and in particular this accounts for the shielding in the CC triple bond.) The substituent effects are surprisingly large in some cases, for example with the ethoxy group as substituent, C-1 (the directly bonded carbon atom) has a δ-value of 89.6, while C-2 has $\delta = 23.4$.

^{13}C chemical shifts in aldehydes and ketones

The chemical shifts of ^{13}C=O in aldehydes and ketones are quite characteristic. The carbonyl ^{13}C nuclei are among the least shielded that are found, being in the range from $\delta = 190$ to 220. For the two classes of compound the ranges overlap, and in many cases a clear decision between the two possibilities cannot be made on the basis of chemical shifts alone. In a non-decoupled spectrum (which is not normally measured) it would be easy to decide, since aldehydes have one hydrogen atom bonded to the C=O group, and therefore the signal of the corresponding ^{13}C nucleus (in ^{13}C=O) would be a doublet, whereas for ketones it would be a singlet. Various methods have been developed to overcome such problems without being forced to measure the non-decoupled spectrum. One of these, the DEPT method, will be explained later (see Sec. 9.3.5, "The DEPT experiment").

^{13}C chemical shifts in carboxylic acids and derivatives

The last important class of compounds to be mentioned here are the carboxylic acids and their derivatives. The resonances of ^{13}C nuclei in carboxyl groups are found in the region from $\delta \approx 160$ to 180; thus these nuclei are more shielded than those in aldehydes and ketones (see also Fig. 9.3-25). The resonance position varies within the stated range for derivatives such as amides, anhydrides, esters, ethers, and acyl halides. The transition from the acid to the carboxylate ion causes a reduction in the shielding of the carboxyl ^{13}C nucleus. For example, in acetic acid, $H_3C–C^1OOH$, $\delta(C^1) = 176.9$ (in D_2O), whereas for $H_3C–C^1OO^-$, $\delta(C^1) = 182.6$ (D_2O, pD 8).

Chemical shifts of "other" nuclides

Up to now we have been concerned almost exclusively with the properties of ^{1}H and ^{13}C nuclei and their NMR spectra. However, it is possible to obtain NMR spectra of nearly all the elements, although not always from the isotope with the highest natural abundance, as can be seen from the example of carbon. These "other" nuclides – together with ^{13}C – are generally referred to as *heteronuclides*. Among these we have to differentiate between nuclides with spin $I = 1/2$ and those with greater spin, as there are fundamental differences between these two cases. The group of nuclides with spin $I = 1/2$ includes, for example, ^{3}H, ^{15}N, ^{19}F, ^{31}P, ^{57}Fe, ^{119}Sn, and ^{195}Pt. Nuclides with spin $I > 1/2$ are ^{2}H, ^{10}B and ^{11}B, ^{14}N, ^{17}O, the alkali and alkaline earth metals, ^{59}Co, and many others. Nuclides with spin $I = 1/2$ behave like ^{1}H and ^{13}C, so measurements on these present no insuperable problems apart from the fact that many of them belong to the insensitive nuclides (nuclides with small γ and μ), and in addition the natural abundance is often low. Nuclides with spin $I > 1/2$ have an electric quadrupole moment eQ; they usually give broad NMR signals. Often this means that one is unable to observe any multiplet splittings due to couplings with other nuclei, or even to resolve chemical shift differences. If an element has two isotopes, one with spin $I = 1/2$ and one with spin $I > 1/2$, one always prefers to observe the isotope with spin $I = 1/2$, even if that isotope has a low natural abundance. The best examples are the two isotopes of nitrogen, ^{14}N and ^{15}N, where nearly all the work that is done uses ^{15}N resonances.

For heteronuclides, the chemical shift is often the only spectral parameter used in the analysis. The ranges of chemical shifts are generally much greater than for ^{1}H and ^{13}C; spectral widths of many thousands of ppm or several hundred kHz are not uncommon. For example, the range of chemical shifts for ^{59}Co is about 20000 ppm, which at $B_0 = 2.3488$ T corresponds to a spectral width of the order of 4×10^5 Hz! (measuring frequency 23.614 MHz). Even in ^{19}F NMR spectroscopy there is an interval of 1313 ppm between the fluorine resonance of the com-

NMR properties of heteronuclides depend critically on the spin *I*.

Nuclei with a quadrupole moment usually give broad NMR signals.

pound ClF, with the highest known shielding, and that of FOOF with the smallest shielding (at $B_0 = 2.3488$ T; ^{19}F resonance frequency 94.077 MHz). A last example from ^{195}Pt NMR: the shift difference between the two extreme values found for the analogous platinum(IV) complex ions $[PtF_6]^{2-}$ and $[PtI_6]^{2-}$ is nearly 13.4 ppm.

9.3.4 Information from spin-spin coupling constants

Introduction

Multiplet splittings are a key diagnostic feature of proton spectra, and can also occur in ^{13}C spectra.

In all the ^{1}H NMR spectra shown in the previous sections we find multiplets, caused by spin-spin coupling of neighboring nuclei in the molecule. ^{13}C NMR spectra, on the other hand, consist only of singlets because the couplings between protons and carbon-13 nuclei have been eliminated by ^{1}H broad-band decoupling. Multiplets can, however, also be observed in ^{13}C NMR spectra if the ^{13}C nucleus is coupled to another nucleus with a spin I and magnetic moment μ. This is especially common when the neighbor nuclei have spin $I = 1/2$ as in the cases of ^{19}F and ^{31}P. Deuterium, with $I = 1$, is an example of a nuclide with $I > 1/2$ whose coupling to carbon can always be seen. In Fig. 9.3-9 the triplet for the solvent CDCl$_3$ at $\delta \approx 77$ comes from such a C,D coupling. But in practice the most important types of couplings are those between protons. Therefore we will concentrate our attention almost exclusively on H,H coupling constants. A few examples of coupling constants between ^{13}C and ^{1}H will also be given, but we will leave all other heteronuclear couplings to the specialists.

Coupling constants are described by a standard notation which specifies the coupled nuclei and the number of bonds between them.

In all cases where protons are coupled, the coupling constant will be denoted by $J(H,H)$. The number of bonds between the coupled nuclei will be indicated by a superscript preceding the J; thus ^1J denotes a coupling between the nuclei of atoms directly bonded to each other (for example, H$_2$, HD, ^{13}C–^{1}H), ^2J means a geminal coupling, ^3J a vicinal coupling, and ^{3+n}J a "long-range" coupling. As shown in Sec. 9.3.2, the indirect spin-spin coupling constant J is a measure of the strength of interaction. Because the coupling constant is independent of the magnetic field strength, it can be given in Hz. Table 9.3-4 gives an approximate indication of the ranges of $J(H,H)$ and $J(C,H)$ values that can be expected for the different types of coupling constant, disregarding a few extreme cases.

As can be seen from Table 9.3-4, the values of ^1J, ^2J, and ^3J vary over quite wide ranges. Evidently, therefore, the internuclear distance alone cannot account for the values. Instead it is clear that molecular structure plays an important role. The factors influencing coupling constants include, among others:

- the hybridization of the atoms involved in the coupling
- bond angles and torsional angles
- bond lengths
- substituent effects.

In the next section we will investigate how the molecular structure influences the H,H coupling constants.

Table 9.3-4. Magnitudes and signs of H,H and C,H coupling constants

	$J(H,H)$	Sign[a]	$J(C,H)$	Sign[a]
^1J	276[b]	positive	125–250	positive
^2J	0–30	usually neg.	−10 to +20	pos./neg.
^3J	0–18	positive	1–10	positive
^{3+n}J	0–7	pos./neg.	<1	pos./neg.

a) The coupling constant can have a positive or a negative sign. The detailed implications of this cannot be discussed here; however, in first-order spectra the sign has no effect on the number and intensities of the lines (see Section 9.3.2, "The indirect spin-spin coupling").

b) For H$_2$, determined from measurements on HD.

H,H coupling constants and chemical structure

In Sec. 9.3.2, "The indirect spin-spin coupling", some spectra were presented to demonstrate the coupling patterns for two-, three-, and four-spin systems. We will now analyze the multiplets in these spectra exactly to get the coupling constants. In general we can use either the A- or X-multiplets for the determination of coupling constants. The values found for the few examples are given below. They represent typical coupling constants for the corresponding bonding situations.

1) $CH_3{}^XCH^AClCOOH$ (Fig. 9.3-14) $^3J_{AX} = 7.0\,\text{Hz}$
2) $CH_3{}^XCH_2{}^AOCOCH_3$ (Fig. 9.3-1) $^3J_{AX} = 7.0\,\text{Hz}$
3) $Ph-CH^A{=}CH^X-COOH$ (Fig. 9.3-12) $^3J_{AX} = 15.8\,\text{Hz}$ (*trans* coupling)
4) $Ph-CH^A{=}CH^MH^X$ (Fig. 9.3-19) $^3J_{AM} = 17.6\,\text{Hz}$ (*trans* coupling)
                                                   $^3J_{AX} = 10.6\,\text{Hz}$ (*cis* coupling)
                                                   $^2J_{MX} = 1.0\,\text{Hz}$ (*gem* coupling)

H,H coupling constants in saturated compounds

In saturated compounds we have to consider only *geminal* and *vicinal* coupling constants, since couplings through more than three bonds are generally smaller than 1 Hz and cannot be observed in most cases (Rule 5, Sec. 9.3.2, "The indirect spin-spin coupling").

Geminal couplings $^2J(H,H)$: First we note that the geminal coupling $^2J(H,H)$ in *symmetrical* CH_2 and CH_3 groups cannot be observed (Rule 3 in the above-mentioned section). Nevertheless, couplings between the protons of a CH_2 group do exist and can be seen in cases where the two protons are not chemically equivalent. They are non-equivalent if, for example, the CH_2 group forms part of a rigid molecule, or, more generally, if the two protons are diasterotopic. Typical values for $^2J(H,H)$ in methane derivatives are from 10 to 13 Hz (for CH_4: $(-)12.4\,\text{Hz}$). The geminal coupling constants depend on bond angle and substituents.

Vicinal coupling constants $^3J(H,H)$: More important than the geminal are the vicinal coupling constants $^3J(H,H)$. Examination of a wealth of experimental data shows that in rigid saturated molecules, for example in rigid rings, the vicinal coupling constants depend strongly and in a characteristic manner on the dihedral angle ϕ. Figure 9.3-26 shows the form of this dependence, which is called the *Karplus curve* after M. Karplus whose calculations provided the theoretical basis. Evidently the coupling constants are largest for $\phi = 0°$ or $180°$, and smallest for $\phi = 90°$.

The most important uses of the Karplus relationship are in determining the conformations and configurations of ethane derivatives and saturated six-membered rings. The ^3J-values of about 7 Hz found in examples (1) and (2) above are typical for vicinal coupling constants in ethane derivatives. The observed value corresponds to an averaged coupling, since at room temperature there is a rapid exchange between the different rotamers. As can be seen from Fig. 9.3-27, a pair of vicinal protons in an ethane derivative can have a *gauche* coupling (^3J_g) or a

Vicinal couplings depend on the dihedral angle in a characteristic way described by a "Karplus curve".

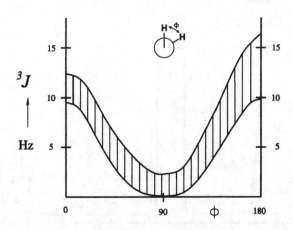

Fig. 9.3-26. Range of observed vicinal coupling constants for different values of the dihedral angle ϕ (Karplus curve)

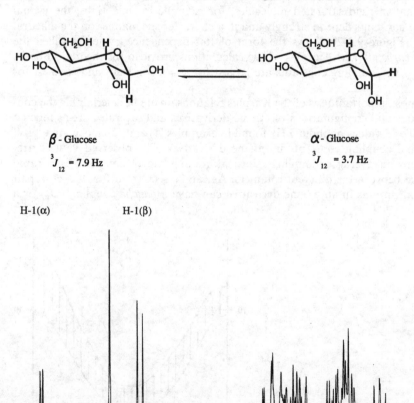

Fig. 9.3-27. Scheme of the three different rotamers of ethane derivatives. In rotamers I and III a pair of vicinal protons has a *gauche* coupling, with the dihedral angle $\phi = 60°$; in rotamer II there is a *trans* coupling and $\phi = 180°$

trans coupling (^3J_t). In one case $\phi = 60°$, and in the other $\phi = 180°$. From the Karplus curve (see Fig. 9.3-26) we read off the following values for the coupling constants at these angles: $^3J_g \approx 3$–5 Hz, $^3J_t \approx 10$–16 Hz. Provided that there is rapid rotation and the three rotamers are involved in the equilibrium in equal amounts, the vicinal coupling constant is given by Eq. 9.3-19:

$$^3J = \frac{1}{3}(2\,{}^3J_g + {}^3J_t) \approx 7\,\text{Hz} \tag{9.3-19}$$

As mentioned above, an important area of application of the Karplus curve is in determining configurations of six-membered rings, especially of carbohydrates. An example is provided by the spectrum of glucose shown in Fig. 9.3-28. The spectrum is very complex because glucose undergoes mutarotation in aqueous solution, so that we see the spectrum of a mixture of α- and β-glucose. However, for the present purpose we do not need to analyze the whole spectrum. We only need to look for the resonances of the proton on C-1, for both α- and β-glucose. These *anomeric* protons are the least shielded of the ring protons because of the two neighboring oxygen atoms; in addition they are the only protons in the molecule which give doublets, since each is coupled to only one other proton, H-2. We find two such doublets at $\delta = 4.65$ and 5.24. From these two doublets we find the vicinal coupling constants $^3J(\text{H-1,H-2})$ are 7.9 Hz and 3.7 Hz respectively. The larger value is characteristic of an axial-axial coupling (usually in the range 7 to 9 Hz), which one expects for β-glucose; the smaller value corresponds to an axial-equatorial coupling in α-glucose. Axial-equatorial and equatorial-equatorial coupling constants are usually within the same range of 2 to 5 Hz.

Fig. 9.3-28. 250 MHz ^{1}H NMR spectrum of glucose in D$_2$O. The integrated areas of the two doublets for the H-1 signals give proportions of about 40% α-glucose and 60% β-glucose. The signal at $\delta \approx 4.8$ is assigned to HDO

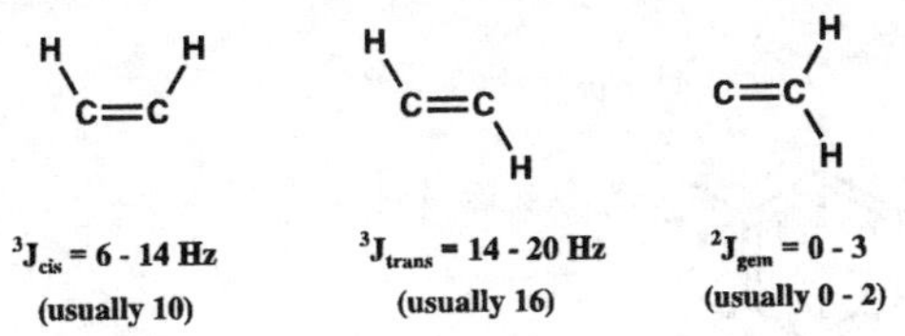

$^3J_{cis} = 6 - 14\ Hz$ $^3J_{trans} = 14 - 20\ Hz$ $^2J_{gem} = 0 - 3$

(usually 10)   (usually 16)   (usually 0 - 2)   **Scheme 9.3-1**

H,H coupling constants in alkenes

We have already encountered couplings in alkenes in two examples. The first was when we discussed the two-spin system AX, with the olefinic protons of cinnamic acid as an example (see Sec. 9.3.2, "The indirect spin-spin coupling", and Fig. 9.3-12). The second was when we analyzed the three-spin system AMX for the olefinic protons in styrene (see Fig. 9.3-19). From the analysis of the two doublets in the spectrum of cinnamic acid we find a coupling constant $^3J = 15.8\ Hz$. This value is typical for a *trans* coupling between two olefinic protons. The *cis* coupling constants in ethylene and ethylene derivatives are found to be much smaller, while the geminal coupling constants are smaller still and can only be seen under conditions of very high resolution. Typical values are shown in Scheme 9.3-1.

If we analyze the spectrum of styrene (Fig. 9.3-19), we find the three coupling constants $^3J_{trans} = 17.6\ Hz$, $^3J_{cis} = 10.9\ Hz$, and $^3J_{gem} = 1.0\ Hz$. From these values the multiplets can be assigned to the corresponding protons in the molecule without difficulty.

In principle the measurement of the vicinal coupling constant is a simple, rapid, and unambiguous method for determining the configuration at a double bond. Of course, the method cannot be used if the two coupled protons are equivalent (see Rule 3 in Sec. 9.3.2, "The indirect spin-spin coupling"). Therefore we cannot differentiate between *cis*- and *trans*-stilbene by this method.

H,H coupling constants in arenes

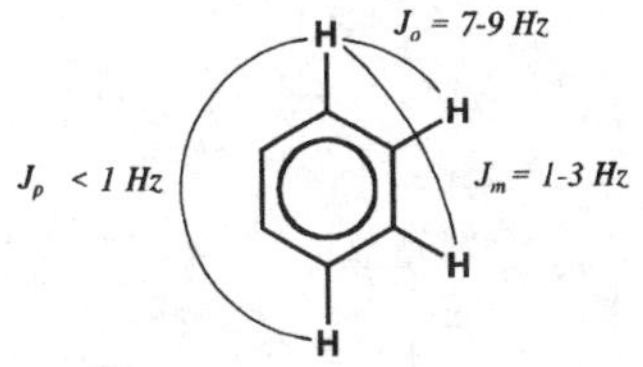

The *ortho*, *meta*, and *para* coupling constants in benzene and benzene derivatives are different. Typical values are shown in Scheme 9.3-2.

In most cases the *para* coupling is too small to be observed. As can be seen, the ranges of J_O and J_m do not overlap. From the relationship $J_O > J_m > J_p$, combined with the rules given in Sec. 9.3.2, it is possible in most cases to analyze the spectra, or at least to determine the substitution pattern. The analysis becomes particularly simple if the spectra have been recorded at a high resonance frequency. In the following we will analyze four typical spectra of benzene derivatives.

Our first example is the spectrum of a symmetrical 1,2,3-trisubstituted compound, 2-bromoisophthalic acid dimethylester (see Fig. 9.3-13), which was discussed earlier as an example of a three-spin system of type A_2X. The interesting signals are the doublet at $\delta = 7.7$ and the triplet at $\delta = 7.4$. The observed coupling constant of 7.7 Hz is characteristic of an *ortho* coupling J_O.

As our second example we will analyze the aromatic region of the spectrum of a 1,2,4-trisubstituted benzene, 2-hydroxy-4-methyl-acetophenone (Fig. 9.3-29). There are three multiplets: at $\delta = 7.5$ we have a doublet with a small *meta* coupling constant of 1 Hz, at $\delta = 7.29$ we have a doublet of doublets with coupling constants of 7.0 Hz (*ortho*) and 1.5 Hz (*meta*), and at $\delta = 6.9$ we have a doublet with the *ortho* coupling constant of 7.0 Hz as before. The *para* coupling is evidently too small to be detected. The analysis and the assignment of the multiplets to the corresponding protons are shown beside the spectrum. With the help of the coupling constants it is easy to arrive at the 1,2,4-substitution pattern, but it is not possible to say which substituent is in which position. To solve that problem we need the information from chemical shifts. In this case we would base the assignment on estimated shift values calculated from substituent increments.

The third example is acetylsalicylic acid (aspirin; Fig. 9.3-30), a 1,2-disubstituted benzene derivative. The spectrum of this compound consists of four groups of

$J_o = 7\text{-}9\ Hz$

$J_p < 1\ Hz$ $J_m = 1\text{-}3\ Hz$

Scheme 9.3-2

Ring substitution patterns can often be deduced from proton spectra.

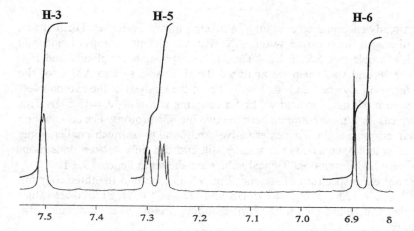

Fig. 9.3-29. Portion of the 300 MHz
^{1}H-NMR spectrum of 2-hydroxy-4-
methylacetophenone with integral curve in
CDCl$_3$. The small signal at $\delta = 7.26$
corresponds to the residual CHCl$_3$ in the
solvent

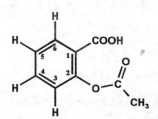

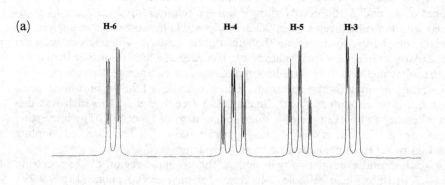

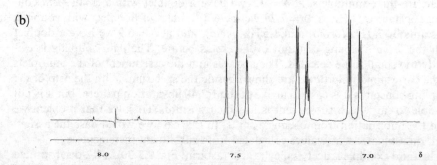

Fig. 9.3-30. Portion of the 200 MHz ^{1}H
NMR spectrum of acetylsalicylic acid
(aspirin) in CDCl$_3$. (a) Normal spectrum; (b)
H-6 decoupled (see Sec. 9.3.5)

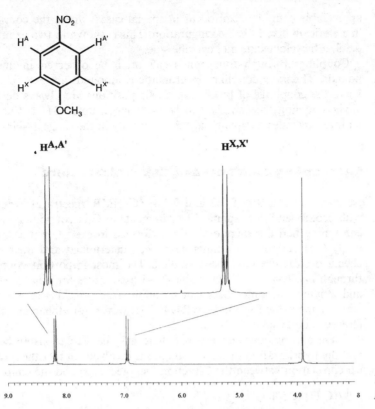

Fig. 9.3-31. 300 MHz ^{1}H NMR spectrum of *p*-nitroanisole in CDCl$_3$

peaks in the range for aromatic protons, each corresponding to one proton (from integrals), a singlet at $\delta = 2.2$ for CH$_3$ and a broad singlet at $\delta = 5.2$ for the OH proton. In Fig. 9.3-30a only the aromatic region of the spectrum is shown. Again we can neglect the *para* coupling. We find from left to right: a doublet of doublets with one large (*ortho*) coupling constant and one small (*meta*) coupling constant ($\delta = 7.96$, $J_O = 7.6$ Hz, $J_m = 1.5$ Hz; H-6), then a triplet of doublets with two (fortuitously) equal *ortho* coupling constants and one *meta* coupling constant ($\delta = 7.51$, $J'_O = J''_O = 7.8$ Hz, $J_m = 1.2$ Hz; H-4), then a (nearly) mirror image of the latter group ($\delta = 7.26$, $J'_O = J''_O = 7.8$ Hz; H-5), and lastly a doublet of doublets with one *ortho* and one *meta* coupling constant ($\delta = 7.06$, $J_O = 8.0$ Hz, $J_m = 1.0$ Hz; H-3). Because each multiplet contains at least one *ortho* coupling, the two substituents must be in the 1,2-positions. (A 1,4-substituted ring is ruled out as it would be symmetrical – see the last example below.) To assign the multiplets to the corresponding ring hydrogen nuclei 3,4,5, or 6, we must again estimate the chemical shifts using substituent increments. (Assignment: see Fig. 9.3-30a; decoupling experiment see Sec. 9.3.5, "Spin decoupling").

Lastly, we consider the uniquely characteristic case of a 1,4-disubstituted benzene, taking as an example *p*-nitroanisole. The *para*-substitution pattern is very easy to recognize, since there are two pairs of chemically equivalent hydrogen nuclei in the ring (see the structural formula).

The ring protons in 1,4-disubstituted arenes show a characteristic pattern.

Therefore there are only two chemical shift values. As can be seen from Fig. 9.3-31, the aromatic proton part of the spectrum is absolutely symmetrical, which in any disubstituted benzene derivative C$_6$H$_4$XY immediately indicates the *para* isomer. (Note, however, that in a C$_6$H$_4$X$_2$ derivative the *ortho* isomer will also give a symmetrical pattern.) In our example the ring proton spectrum resembles that of a two-spin system of the AX-type. Theory indicates that each half of the spectrum can have up to twelve lines, but often the rest of these are too weak to detect. The spectrum shown can be analyzed like an AX (first-order) pattern and the doublet splitting corresponds to the *ortho* coupling constant.

Long-range couplings

Couplings through more than three bonds are called *long-range couplings*. In saturated compounds they are normally too small to be detected (< 1 Hz). However,

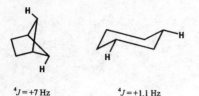

$^4J = +7\,\text{Hz}$ $^4J = +1.1\,\text{Hz}$

Scheme 9.3-3

appreciable couplings can occur in special cases, where the coupled protons are in a sterically fixed "W" configuration. This is shown in two examples in Scheme 9.3-3, a bicyclohexane and cyclohexane.

Couplings through four bonds can often be observed in unsaturated compounds. The *meta* couplings in aromatic compounds are examples of this. In alkenes the couplings of protons in allylic positions are always detectable and the allylic coupling constant ^4J can be quite large, up to 3 Hz. Where the coupling pathway includes a triple bond, ^4J is typically in the range 1–4 Hz.

C,H coupling constants and chemical structure

Table 9.3-5. Coupling constants $^1J(\text{C,H})$ in ethane, ethene, benzene, and ethyne

Compound	$^1J(\text{C,H})$
H_3C-CH_3	124.9
$H_2C=CH_2$	156.4
C_6H_6	158.4
$HC\equiv CH$	249.0

As mentioned in Sec. 9.3.2 and 9.3.4 ^{13}C NMR spectra are routinely recorded with broad-band decoupling. Measurements of C,H coupling constants are time-consuming and are therefore only performed in exceptional cases. Nevertheless, many C,H coupling constants have been determined and their relationships to chemical structure have been studied. The most important couplings are those through *one bond*, $^1J(\text{C,H})$. Table 9.3-5 lists values for ethane, ethene, benzene, and ethyne. The differences can be quite large: $^1J(\text{C,H})$ in ethane is 124.9 Hz, whereas the value for ethyne is 249 Hz! However, for ethene and benzene the difference is quite small.

These observations suggest that there may be a relationship between $^1J(\text{C,H})$ and the hybridization of the carbon atom involved. In fact there exists an empirical correlation between the *s*-fraction (denoted by s) and the coupling constant:

$$^1J(\text{C,H}) \approx 500\,s \tag{9.3-20}$$

The value of s for sp^3 hybridization is 0.25, that for sp^2 is 0.33, and for sp hybridization $s = 0.5$.

Substituents have a considerable influence on C,H coupling constants, as can be seen from the few examples below:

$$\begin{array}{ccccc} & CH_4 & CH_3Cl & CH_2Cl_2 & CHCl_3 \\ ^1J(\text{C,H}) = & 125 & 151 & 178 & 209\,\text{Hz} \end{array}$$

C,H couplings through two or more bonds are at least one order of magnitude smaller. They are used as an aid to structure determination only in special cases.

9.3.5 Special methods for assigning ^{1}H and ^{13}C signals

Introduction

Data collections are a valuable aid to assigning spectral features.

Most NMR spectra are measured to get information about the structure of molecules. This information is contained in the spectral parameters, namely the chemical shifts (δ), the coupling constants (J), and the intensities. In analyzing the spectrum to determine these parameters, the most important step is the assignment of all (or nearly all) the signals to specific nuclei or groups in the molecule. Nevertheless, a complete assignment is not always possible at first, even for an experienced NMR spectroscopist. In many cases more information is needed, and this may come from additional experiments or from data collections. In Sec. 9.3.3 we mentioned the use of empirical correlations to estimate ^{1}H or ^{13}C NMR chemical shifts. These rules are based on the observation that within a particular class of compounds the contribution of a substituent to the chemical shift is nearly constant. It would not be appropriate here to go into details of these methods, or of those based on the effects of solvent and temperature, H-D exchange, or altering the chemical structure of the molecules by derivatization. However, a few common experimental methods that have now become routine, or nearly so, in ^{1}H and ^{13}C NMR spectroscopy will be described briefly in the following sections. These are spin decoupling, the DEPT experiment, and two types of two-dimensional (2D) experiments. The DEPT and 2D experiments are introduced here with the aim of awakening the reader's interest in learning more about these procedures, and should serve to give a modest impression of the potential of modern NMR spectroscopy. However, the theory and experimental details of these procedures are beyond the scope of this short introduction.

Spin decoupling

In Sec. 9.3.2, "The indirect spin-spin coupling", it was explained that the interaction of neighboring nuclear dipoles via the indirect spin-spin coupling mechanism causes splitting of the signals, giving characteristic multiplet patterns. These coupling patterns yield information about the structures of molecules. For example, the presence of a quartet and a triplet in the spectrum indicates an ethyl group. In Sec. 9.3.2, "The NMR method and the spectrometer", we learned that ^{13}C NMR spectra are normally recorded with broad-band (BB) decoupling so that indirect spin-spin couplings are removed. This is achieved by irradiating with a high power level at the frequency corresponding to proton transitions, so that the spin orientation of the protons changes rapidly; thus the lifetime in each spin state is shortened, with the result that the coupling is averaged to zero. The signal splitting disappears; multiplets become singlets. This ^1H BB decoupling procedure is a *heteronuclear* decoupling experiment because protons are irradiated while ^{13}C resonances are observed. Homonuclear decoupling experiments are also possible; they are very important and are used routinely in ^1H NMR spectroscopy to identify signals belonging to mutually coupled protons. As an example we may take acetylsalicylic acid, whose partial spectrum in the aromatic region, which we discussed earlier, is shown in Fig. 9.3-30a. To demonstrate the method, we irradiate at the resonance frequencies of the doublet of doublets centered at $\delta = 7.95$, which we assigned to H-6, the proton in the *ortho* position relative to the carboxy group. Comparing the coupled and the decoupled spectra (see Fig. 9.3-30) we see that the doublet of triplets at $\delta = 7.25$ has simplified, so that one *ortho* coupling is now missing. Therefore we can assign that signal to H-5. However, we also see a simplification of the other doublet of triplets at $\delta = 7.5$ because the *meta* coupling $^4J(H\text{-}4/H\text{-}6)$ has been eliminated too; consequently this multiplet can be assigned to H-4.

Thus a single decoupling experiment, the irradiation of the resonance frequencies of H-6, led to the simplification of the spectrum and to an assignment of the signals of all the aromatic protons. In other cases more than one decoupling experiment may be necessary to reach a complete assignment.

In our example we *selectively* irradiated the resonances of one proton and identified the signal positions of all coupling partners. In contrast to this, the heteronuclear ^1H BB-decoupling method which results in a ^{13}C NMR spectrum consisting of singlets is *nonselective*, because all the ^1H resonances are irradiated simultaneously. However, selective heteronuclear decoupling experiments (as opposed to broad-band, BB) are also possible. Such experiments are, of course, not restricted to the combination of the nuclides ^1H and 13C; in principle we can decouple any given pair of nuclides, for example $^1H/^{19}F$ or $^{13}C/^{31}P$.

The DEPT experiment

As mentioned in the previous section and also in Sec. 9.3.2, "The NMR method and the spectrometer", ^{13}C NMR spectra normally consist only of singlets because couplings to protons are eliminated by broad-band (BB) decoupling. Consequently the only data available for the assignment of signals are the chemical shifts. In many cases it would be very helpful to know how many hydrogen atoms are bonded directly to each carbon atom. A spectrum recorded without decoupling would, of course, contain this information in the form of the multiplicities of the signals. But, as explained in Sec. 9.3.2, to measure such non-decoupled spectra is very time-consuming and the spectra are often difficult to analyze, especially when the molecule contains many carbon atoms. Up to the early 1980s, ^{13}C, ^1H coupling information was nearly always obtained by the ^1H

Heteronuclear decoupling simplifies the spectrum of one nuclide by eliminating the coupling to another.

Homonuclear decoupling is a useful technique for analyzing proton spectra.

Off-resonance decoupling reduces multiplet splittings, thus helping in assignments.

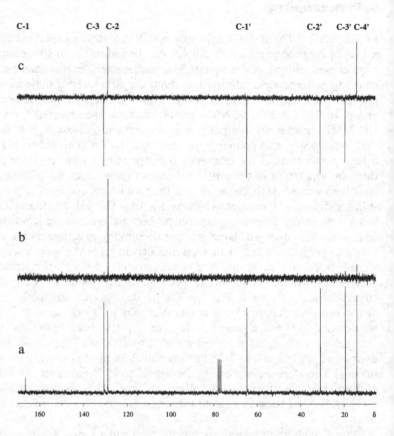

Fig. 9.3-32. Examples of DEPT experiments. (a) 50.3 MHz ^{13}C NMR spectrum of acrylic acid n-butyl ester with ^{1}H broad-band decoupling. (b) DEPT(90) spectrum; CH sub-spectrum of (a). (The small signal at $\delta = 13.9$ corresponds to the residual signal of the CH$_3$ group. A complete suppression of all CH$_2$ and CH$_3$ signals is sometimes difficult because it is necessary to make compromises in selecting the experimental conditions). (c) DEPT(135) spectrum; positive signals arise from CH and CH$_3$ groups, negative signals arise from CH$_2$ groups. (*Experimental conditions*: ≈ 50 mg in 0.5 ml CDCl$_3$ as solvent, total time approximately 15 min)

off-resonance decoupling technique. In this procedure the C,H coupling constants are not averaged to zero by decoupling, but are effectively reduced to typically one-tenth of their normal value. This means that the multiplet structures can still be recognized: quartets for CH$_3$, triplets for CH$_2$, doublets for CH and singlets for quaternary carbons (C$_q$). Couplings to protons which are two bonds or more away disappear completely. However, with the new generation of NMR experiments the DEPT technique (acronym for **D**istortionless **E**nhancement by **P**olarization **T**ransfer) has been developed as a better alternative. This technique uses a complex pulse sequence which need not concern us here; for our purpose, only the result is of importance. By carrying out two further experiments (DEPT(90) and DEPT(135)) in addition to the recording of the normal BB-decoupled ^{13}C NMR spectrum, we obtain the same information as from the off-resonance spectrum, but much more easily. Figure 9.3-32 shows the three ^{13}C NMR spectra for acrylic acid n-butyl ester. Spectrum (a) is the BB-decoupled ^{13}C NMR spectrum. Spectrum (b) is the DEPT(90) spectrum, which theoretically should contain only signals for CH groups, all with positive amplitude. In our example we find only the resonance signal for one CH group. Spectrum (c) is the DEPT(135) spectrum. As can be seen it is a sub-spectrum of (a). Besides the CH signal already seen in spectrum (b) we find one additional signal with positive amplitude and four with negative amplitudes. These additional signals can be assigned to the ^{13}C resonances of a CH$_3$ group (positive amplitude) and four CH$_2$ groups (negative amplitude). Neither the DEPT(90) nor the DEPT(135) spectrum shows a signal for the quaternary carbon, that of the carboxylic group –COO–. We therefore know that our molecule contains one CH group, four CH$_2$ groups, one CH$_3$ group, and one quaternary carbon, C$_q$.

In most practical cases it is sufficient to record only the BB-decoupled ^{13}C NMR spectrum and the DEPT(135) spectrum. It is true that we cannot then differentiate between signals of CH and CH_3 groups, because both have positive amplitudes, but one can usually distinguish between them using other criteria (for example, indications from chemical shifts). The signals of the four CH_2 groups above can be easily assigned on the basis of substituent increments such as those mentioned earlier. However, the following section describes two procedures which enable one to complete the assignments without recourse to these other sources of information.

Two-dimensional experiments

Two-dimensional (2D) NMR spectroscopy represents a new generation of NMR experiments. Whereas a one-dimensional spectrum has just one frequency axis, the abscissa, with the intensities as ordinate, in a two-dimensional spectrum both axes, the abscissa and the ordinate, are frequency axes, with the intensities constituting a third dimension. In the following we will first discuss a 2D spectrum in which ^1H chemical shifts along both frequency axes are correlated with each other. This technique has become known as H,H-COSY (from correlated spectroscopy). Next we will learn about the kinds of information that can be deduced from a 2D spectrum in which ^1H and ^{13}C chemical shifts are correlated, by the method of heteronuclear $^1H,^{13}C$-correlated spectroscopy, or H,C-COSY. Without going into the theory we can understand the results obtained by these techniques and learn how such spectra can be analyzed.

All two-dimensional methods are based on the couplings between nuclear dipoles. These interactions need not necessarily be scalar couplings (the "indirect spin-spin coupling"); dipolar interactions through space can also be involved, through their effects on nuclear relaxation (see Sec. 9.3.2, "Free induction decay and relaxation"). However, the examples considered here will be restricted to scalar coupling. In our first example we treat the homonuclear case, H,H-COSY.

Two-dimensional homonuclear (H,H)-correlated NMR spectroscopy (H,H-COSY)

In Fig. 9.3-33 the H,H-COSY spectrum of acrylic acid n-butyl ester is shown. The normal one-dimensional ^1H NMR spectrum appears at the top edge and at the left-hand edge of the 2D spectrum. The δ-scales are given below and to the right. The abscissa is called the F_2-axis and the ordinate the F_1-axis. In the 2D spectrum we find peaks on the diagonal, the *diagonal peaks*, and off the diagonal the so-called *cross peaks*. The projections of the diagonal peaks onto the two axes are equal (δ_i/δ_i; the first δ-value corresponds to the value on the F_2-axis, the second that on the F_1-axis). When we draw a vertical or a horizontal line, we come to the corresponding signal of proton i in the one-dimensional spectrum (top or left). In our example we find seven diagonal peaks, which correspond to the seven multiplets in the one-dimensional spectrum. These diagonal peaks do not give us any new information. But the cross peaks do! Their positions are (δ_i/δ_j) or (δ_j/δ_i) where δ_i corresponds to the chemical shift of proton i, and δ_j to the chemical shift of proton j. The cross peaks indicate a *correlation* between the chemical shifts of the two coupled protons (or groups) i and j. Each pair of coupled nuclei gives two cross peaks, with the diagonal and the cross peaks forming the corners of a square.

With that knowledge we can analyze the spectrum shown in Fig. 9.3-33. The best method is to begin the analysis at a diagonal peak whose assignment is known. This diagonal peak forms one corner from which we start to draw the first square. In our example we use the peak of the $OCH_2(1')$ group at $\delta = 4.16$ (on the F_2- and F_1-axes). By drawing vertical and horizontal lines we find the cross peaks, and, consequently by completing the square, also the diagonal peak of the coupled protons—here those of the $CH_2(2')$ group. Now we start again to draw the next square, beginning at the diagonal peak of $CH_2(2')$ just assigned. The cross peak in the horizontal (or vertical) direction leads us to the diagonal peak of $CH_2(3')$. By repeating this procedure we can also find the chemical shifts of CH_3. We can see

The analysis of a 2D spectrum begins with the identification of diagonal and cross peaks.

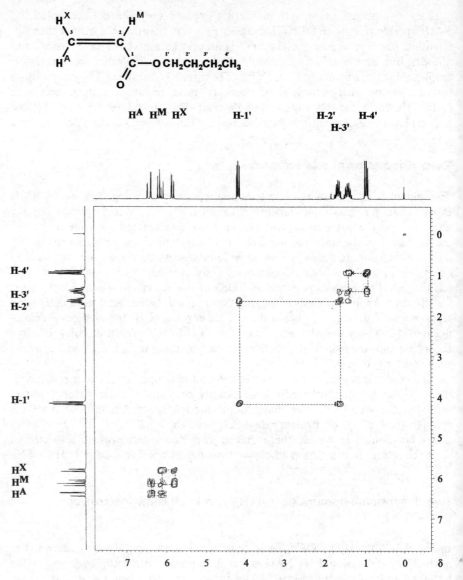

Fig. 9.3-33 200 MHz two-dimensional (H,H)-correlated NMR spectrum of acrylic acid n-butyl ester. At the left-hand edge and at the top is the one-dimensional ^{1}H NMR spectrum with assignments. The diagonal and cross peaks joined by dashed lines indicate which protons are scalar coupled; they form squares. (*Experimental conditions*: ≈ 50 mg in 0.5 ml CDCl$_3$ as solvent, total time approximately 1 h)

that diagonal peaks of protons which are coupled to more than one group of protons (here CH$_2$(2′) and CH$_2$(3′)) are corners of more than one square. In Fig. 9.3-33, the individual steps of the total analysis are drawn as dashed lines. The result of the assignment is indicated above the one-dimensional spectrum at the top edge.

The same procedure can be used to analyze the peaks of the olefinic protons in the region of $\delta = 5.8$ to $\delta = 6.5$.

The reader will recall that a similar result can be obtained by spin decoupling experiments (see Sec. 9.3.5, "Spin decoupling"). However, as was mentioned there, one may need to perform several decoupling experiments. Moreover, spin decoupling experiments are not possible when multiplets are close together, a limitation which does not apply to the two-dimensional COSY experiment (see, for example, the correlations of the olefinic protons in Fig. 9.3-33). For a fair comparison one must also take into account the measurement time for a two-dimensional COSY experiment. Although this depends on the measurement method, the spectrometer used, and the sample concentration, recording times of one hour or half an hour are fairly typical.

Finally it must be mentioned that diagonal and cross peaks can show fine structures caused by spin-spin couplings, although these cannot be recognized in Fig. 9.3-33. We ignore the fine structure here, since our aim was "only" the correct assignment of signals.

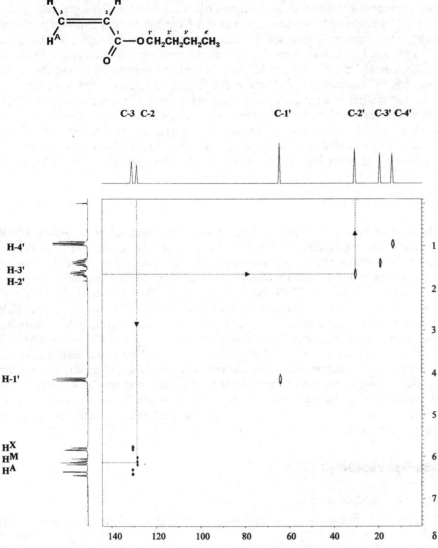

Fig. 9.3-34. Two-dimensional H,C-correlated 50 MHz NMR spectrum of acrylic acid n-butyl ester. The one-dimensional ^1H NMR spectrum is shown at the left-hand edge, while at the top edge is the projection of the two-dimensional spectrum on the F_2-axis. The dashed lines show the analysis procedure for two examples. In the first example the analysis starts at the ^1H NMR signal of CH_2; following the arrows, the ^{13}C resonance of the corresponding nucleus C-2′ can be found. In the second example we start at the signal of =CH to assign the corresponding ^1H NMR signal. (*Experimental conditions:* ≈ 50 mg in 0.5 ml $CDCl_3$ as solvent, total time approximately 1 h)

Two-dimensional heteronuclear (H,C)-correlated NMR spectroscopy (H,C-COSY)

Figure 9.3-34 shows the two-dimensional (H,C)-correlated NMR spectrum of acrylic acid n-butyl ester, the same test molecule which we used in the previous sections. The abscissa (F_2-axis) corresponds to ^{13}C chemical shifts, the ordinate (F_1-axis) to ^1H chemical shifts. As in the H,H-COSY spectrum, the 2D spectrum shows cross peaks which correlate ^1H chemical shifts with ^{13}C chemical shifts. The one-dimensional ^1H NMR spectrum is shown at the left-hand edge, while at the top edge the ^{13}C NMR spectrum is obtained by projecting the peaks of the two-dimensional spectrum onto the F_2-axis. This spectrum shows only six peaks, though the molecule contains seven carbon atoms. Theory shows that cross peaks only appear when carbon nuclei are coupled to directly bonded protons. Therefore a quaternary carbon, such as the carboxy carbon nucleus in this example (at $\delta = 166.3$; see Fig. 9.3-32), does not give a correlation peak. Consequently, in the projection of the two-dimensional spectrum onto the F_2-axis the corresponding peak is missing so that in Fig. 9.3-34 only the range from $\delta = 0$ to $\delta = 145$ is shown.

Usually some of the resonances in the ^1H and ^{13}C NMR spectra can immediately be assigned with confidence on the basis of chemical shifts and multiplicities. In our example we know the assignments of all the ^1H resonances from the H,H-COSY experiment. Also the DEPT(135) experiment discussed earlier allowed us to assign nearly all the ^{13}C signals. Thus, an H,C-COSY experiment would not be necessary in this case. However, this situation is the exception rather than the rule. Two examples will therefore be described to illustrate the analysis procedure.

In our first example, we start the analysis from the ^{1}H resonance of a proton that has been definitely assigned. For example, we use the ^{1}H signal of $CH_2(2')$ at $\delta = 1.66$ (see Fig. 9.3-34, formula, and spectrum at the left-hand edge). Drawing the horizontal line, there is no difficulty in finding the cross peak and the corresponding ^{13}C resonance of C-2' at $\delta = 30.58$ in the spectrum at the top edge.

In our second example, we now begin with the ^{13}C resonance at $\delta = 128.5$, which has been assigned on the basis of its chemical shift and the DEPT(135) spectrum to the olefinic carbon nucleus C-2 (=**CH**). By drawing the vertical line, we find the cross peak, and in the horizontal direction (following the line in Fig. 9.3-34) the corresponding ^{1}H resonance of the olefinic proton at $\delta = 6.12$.

Concluding remarks

The development of the pulsed Fourier transform method of recording NMR spectra in the 1960s later provided the basis for a remarkable variety of experimental procedures which could scarcely have been dreamed of at the time. In the space of this article it has only been possible to sketch three of the most important and useful of these, the DEPT experiment and the 2D methods H,H-COSY and H,C-COSY, without giving any experimental details. The armory of multiple pulse methods (with a plethora of acronyms that is bewildering even to experts) continues to grow. On the semiclassical picture used in Sec. 9.3.2, "Free induction decay and relaxation", all these procedures are based on manipulating the macroscopic magnetization vector by means of radiofrequency pulses, alternating with periods of data acquisition. The interested reader can find details in the extensive literature. The hectic pace of development goes on unabated, ensuring the continued fascination of NMR spectroscopy for the specialist and extending its already impressive capabilities for the chemist.

General reading

Introduction to NMR spectroscopy

H. Friebolin, *Basic One- and Two-Dimensional NMR Spectroscopy*, 3rd ed. Weinheim, VCH 1998.

H. Günther, *NMR Spectroscopy. An Introduction*, New York, John Wiley & Sons 1980.

H.-O. Kalinowski, S. Berger, S. Braun, *Carbon-13 NMR Spectroscopy*, Chichester, John Wiley & Sons 1988.

J.K.M. Sanders and B.K. Hunter, *Modern NMR-Spectroscopy. A Guide for Chemists*. Oxford, Oxford University Press 1987.

Physical basis of NMR spectroscopy

R. Freeman, *A Handbook of Nuclear Magnetic Resonance*. New York, Longman Scientific & Technical 1987.

R.K. Harris, *Nuclear Magnetic Resonance Spectroscopy. A Physicochemical View*, New York, J. Wiley & Sons 1986.

C.P. Slichter, *Principles of Magnetic Resonance*, Berlin, Heidelberg, New York, Springer Verlag 1978.

F.J.M. van de Ven, *Multidimensional NMR in Liquids. Basic Principles and Experimental Methods*, New York, VCH Publishers, Inc. 1995.

Theory

R.R. Ernst, G. Bodenhausen and A. Wokaun, *Principles of Nuclear Magnetic Resonance in One and Two Dimensions*, Oxford, Clarendon Press 1986.

NMR techniques

S. Braun, H.-O. Kalinowski and S. Berger, *100 and More Basic NMR Experiments. A Practical Course*, Weinheim, VCH 1996.

A.E. Derome, *Modern NMR Techniques for Chemistry Research*, Oxford, Pergamon Press 1987.

9.4.1 Principle

Introduction

Mass spectrometry (MS) is based on the generation of gaseous ions from analyte molecules, the subsequent separation of these ions according to their mass-to-charge (m/z) ratio, and the detection of these ions. The resulting mass spectrum is a plot of the (relative) abundance of the ions produced as a function of the m/z ratio. Nowadays, the mass spectrometer is a highly sophisticated and computerized instrument. It consists of five parts, which reflect the five important areas in analytical MS: sample introduction, analyte ionization, mass analysis, ion detection, and data handling (see Fig. 9.4-1). MS is a combined separation and detection technique.

In mass spectrometry ions are separated according to their mass-to-charge ratio.

MS is widely applied, especially in four major fields. In this chapter the analytical mass spectrometry of organic compounds is described. Other areas of interest of MS are: the analysis of surfaces, and organic mass spectrometry. The mass spectrometric analysis of inorganic compounds is discussed in a separate chapter (see Sec. 8.5). Inherent to the processes involved in MS is the extensive application of gas-phase ion chemistry. Obviously, ion chemistry can be applied for analytical purposes, but it can also be studied as such, e.g., in order to elucidate reaction mechanisms in organic chemistry. This is the field of organic mass spectrometry. Finally, MS can also be applied in the study of surfaces, e.g., in secondary ion mass spectrometry.

MS is applied analytically in a great variety of qualitative and quantitative studies. Qualitative applications comprise structure elucidation of unknown compounds such as natural substances, metabolites of drugs and other xenobiotics, and synthetic compounds. In that respect, MS is important for the determination of molecular mass, molecular formula, or elemental composition, and in structure elucidation. MS is the most sensitive spectrometric technique for molecular analysis compared to other spectrometric techniques like NMR and infrared spectrometry. In quantitative analysis, MS is applied in developing definitive and reference methods, and in the quantitation of, for example, polychlorodibenzodioxins (PCDD's) and drugs of abuse. MS is currently developing very rapidly, entering into new fields of applications, e.g., in the characterization of biomacromolecules (cf. Sec. 9.4.4).

Concepts of mass

The mass spectrometer is actually used to measure the relative molecular mass M_r of a compound in atomic units or Dalton (Da). Three different concepts of mass are used in MS. The *average* molecular mass is calculated from the elemental composition using average atomic masses (Table 9.4-1). The average molecular mass is only of importance in MS in the analysis of large molecules (see Sec. 9.4.4). The *nominal* molecular mass M_r is calculated from the elemental composi-

There are three concepts of mass: average, nominal, and exact mass.

Table 9.4-1. Relative abundance, nominal mass, exact mass, and average mass of some common elements in organic compounds

Element	Isotope nominal mass	Relative abundance (%)	Exact mass	Average mass
H	1	100	1.0078	1.008
	2	0.016	2.0141	
C	12	100	12.0000	12.011
	13	1.08	13.0034	
N	14	100	14.0031	14.007
	15	0.38	15.0001	
O	16	100	15.9949	15.999
	17	0.04	16.9991	
	18	0.2	17.9992	
F	19	100	18.9984	18.998
Si	28	100	27.9769	28.086
	29	5.1	28.9765	
	30	3.35	29.9738	
P	31	100	30.9738	30.974
S	32	100	31.9721	32.060
	33	0.78	32.9715	
	34	4.4	33.9679	
Cl	35	100	34.9689	35.453
	37	32.5	36.9659	
Br	79	100	78.9183	79.904
	81	98	80.9163	
I	127	100	126.9045	126.905

tion using the nominal atomic masses of the most abundant isotopes in nature, while for the *exact* molecular mass the exact atomic masses of the most abundant isotopes in nature are used (see Table 9.4-1). The exact atomic masses are based on the defined mass of the ^{12}C isotope of 12.0000.

Worked example

The average, the nominal, and the exact molecular mass of a compound may be calculated using the data in Table 9.4-1.

For decane, the nominal mass is calculated from the nominal mass of the most abundant C isotope (^{12}C) and the most abundant H isotope (^1H). Thus:

$$M_{r,decane} = (10 \times 12) + (22 \times 1) = 142\,Da$$

Similarly, the exact, and average mass are calculated using the exact mass of the most abundant C and H isotopes, and the average mass of C and H atoms, respectively.

The results of such calculations are given in Table 9.4-2 for three examples, i.e., the small molecule decane, the middle-mass molecule leu-enkephalin, and the large molecule bovine pro-insulin. From these examples it can be concluded that the difference between the nominal, the exact, and the average molecular mass increases with increasing mass. Below 1000 Da the nominal, and the exact molecular mass may readily be determined, while for larger molecules the average molecular mass is generally determined.

Most elements consist of a mixture of stable isotopes.

In order to achieve proper mass measurement, the mass scale of the instrument must be calibrated against well-defined masses of a calibrant (see Sec. 9.4.3).

Many of the elements commonly found in organic compounds actually consist

Table 9.4-2. Illustration of three concepts of mass with three examples

	Decane	Leuenkephalin	Bovine proinsulin
Mass	$C_{10}H_{22}$	$C_{28}H_{38}N_5O_7$	$C_{381}H_{586}N_{107}O_{114}S_6$
Nominal	142	556	8672
Exact	142.172	556.277	8676.167
Average	142.3	556.6	8681.8

of a mixture of different stable *isotopes*, e.g., 1% of a population of carbon atoms consists of ^{13}C isotopes, while the remaining 99% are ^{12}C isotopes. Data on the isotopic composition of some common elements are given in Table 9.4-1. The elements in Table 9.4-1 can be classified into four groups:

- (A) elements with only one stable isotope (e.g. F, P, and I).
- (A + 1) elements with a stable isotope peak at an m/z value of M + 1 (e.g., H, C, and N).
- (A + 2) elements with a significant stable isotope peak at m/z value of M + 2 (e.g., O, S, Cl, and Br).
- Other elements that have a more complex isotopic composition (e.g. Si).

The mass spectrometer is capable of separating the various isotopes of an element. As a consequence, the major peaks in the mass spectrum are accompanied by isotope peaks at subsequently higher m/z values. The width of the isotope cluster depends on the elemental composition and the mass of the ion.

Worked example:

Because most elements consist of a mixture of stable isotopes (cf. Table 9.4-1), the molecular ion peak in the mass spectrum actually comprises of a cluster of isotope peaks. For small, simple molecules the relative intensities of the peaks in the isotopic cluster may be estimated by simple calculations, while for larger molecules a computer program is required.

Decane, $C_{10}H_{22}$ with $M_r = 142$, contains 10 C atoms. According to Table 9.4-1, 1.1% of the C atoms is a ^{13}C. This results in an $M_r + 1$ peak with a relative intensity of $10 \times 1.1 = 11\%$ and in an $M_r + 2$ peak with negligible intensity.

The molecular ion cluster of dichloroethane, $C_2H_4Cl_2$ with $M_r = 98$, is more complex. The most important isotopic contribution is due to the presence of the ^{37}Cl isotope. It leads to peaks at $m/z = M_r + 2$ (one ^{37}Cl atom present in the molecule) and $M_r + 4$ (two ^{37}Cl present in the molecule) with estimated relativive intensities of $2 \times 32.5 = 63\%$ and $1 \times 100 \times (0.325)^2 = 10.6\%$. The ^{13}C contribution leads to a peaks at $M_r + 1$ with a relative intensity of $2 \times 1.1 = 2.2\%$, and at $m/z = M_r + 3$ with estimated relative intensities of $0.022 \times 63 = 1.4\%$.

Related to the concepts of average, nominal, and exact molecular mass is the *resolution* R of the mass analyzer, defined as:

$$R = \frac{m}{\Delta m} \qquad (9.4\text{-}1)$$

The resolution represents the ability of the instrument to distinguish between ions having masses of m and $m + \Delta m$, respectively. When Δm has to be smaller, the resolution should be increased proportionally.

Worked example:

Calculate the resolution required to separate CO, and N_2. From Table 9.4-1, the exact mass of both compounds can be calculated to be 27.9949 and 28.0062, respectively. Using Eq. 9.4-1, it can be calculated that the resolution required is:

$$R_s = \frac{28}{(28.0062 - 27.9949)} \approx 2500$$

Experimental and instrumental aspects of MS, its analytical performance, and some typical examples are discussed in the following sections.

9.4.2 Experimental

In Fig. 9.4-1 the five building blocks of a mass spectrometer are schematically depicted. The complexity in describing the instrumental aspects of MS lies in the large number of techniques that are available and important in practical applications of MS.

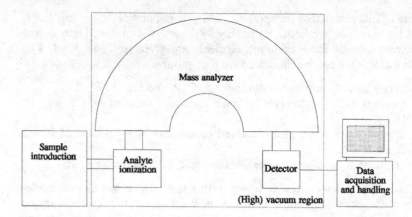

Fig. 9.4-1. Block diagram of the mass spectrometer

Mass spectrometry is performed in a high-vacuum region.

Sample introduction systems comprise controlled leaks for the introduction of analyte vapour from a reservoir, various direct insertion probes for the introduction of low volatility liquids and solids, and combinations with various chromatographic techniques, especially gas and liquid chromatography. In general, samples are introduced into a *high vacuum region*, where the ion source, the mass analyzer, and the ion detector are housed.

Ionization of the analyte can be performed in a number of ways. The most important ionization techniques can be classified into four groups: electron ionization, chemical ionization, desorption ionization, and other ionization, techniques.

After the generation of ions, these are separated according to their *m/z* value in a mass analyzer. Five types of mass analyzers are currently in use, i.e., magnetic sector, quadrupole mass filter, quadrupole ion trap, time-of-flight, and ion-cyclotron resonance instruments. The detection of ions in most cases is performed with an electron multiplier, although various other detection devices are applied as well. During operation, the mass spectrometer generates enormous amounts of data. Highly advanced computer systems and programs are used for the collection, storage, handling, and interpretation of these data.

In the next subsection a simple mass spectrometer is introduced and discussed in detail. Other approaches to sample introduction, analyte ionization, mass analysis, ion detection, and data handling are subsequently discussed in other subsections as alternatives for and/or improvements to this simple instrument.

A simple mass spectrometer

A quadrupole mass filter is a simple and widely-used mass analyzer.

The currently most widely-used mass spectrometer is a quadrupole mass filter, equipped with an electron impact ion source and an electron multiplier, and directly coupled to capillary gas chromatography (GC).

In the capillary GC the sample constituents are separated and can be consecutively transferred to the MS ion source as pure substances in gas or vapour state (cf. Sec. 5.2). The ion source, quadrupole mass filter, and electron multiplier are housed in a high vacuum manifold with a typical pressure of 10^{-3} Pa. Turbomolecular or oil diffusion pumps, backed by a mechanical pump, are used to generate the vacuum. Great caution must be taken in operation of the instrument not to destroy or contaminate the vacuum.

Electron ionization

In electron ionization the analyte is bombarded with 70 eV electrons.

The generation of ions takes place by a process called *electron ionization* (EI). In EI, the analyte vapour is subjected to a bombardment of energetic electrons from a direct electrically-heated tungsten or rhenium filament. While most electrons are elastically scattered and others, upon interaction, cause electronic excitation of the analyte molecules, a few excitations cause the complete removal of an electron from the analyte molecule. As a result, a radical cation, denoted as $M^{+\bullet}$, and two electrons are produced:

$$M + e^- \rightarrow M^{+\bullet} + 2e^- \tag{9.4-2}$$

The $M^{+\bullet}$ ion is called the *molecular ion*, since its m/z ratio corresponds to the molecular mass M_r of the analyte. Singly-charged ions are prominently produced. In the ionization process, energy is transferred to the molecular ion generated. The ions produced in EI are characterized by a distribution of internal energy, depending on the analyte and the initial energy distribution of the electrons. The maximum energy that can be transferred during ionization is the difference between the electron energy, which typically is 70 eV (6.8 MJ/mol), and the ionization energy of the analyte, which typically is between 6 and 10 eV (0.6–1 MJ/mol). In general, the internal energy distribution is centred around 2–6 eV. The excess internal energy and the radical character of the molecular ion can give rise to *unimolecular dissociations* resulting in fragment ions, which are unique for different structures. Typical fragmentation reactions of a molecule M upon electron impact comprise the formation of an ionized fragment F and the loss of either a radical $R^\bullet$ or a neutral N.

In electron ionization unimolecular dissociations lead to structure-specific fragment ions.

$$M^{+\bullet} \rightarrow F_1^+ + R^\bullet \tag{9.4-3}$$

$$M^{+\bullet} \rightarrow F_2^{+\bullet} + N \tag{9.4-4}$$

$$F_2^{+\bullet} \rightarrow F_3^+ + R^\bullet \tag{9.4-5}$$

A typical EI spectrum of diuron (N-dichlorophenyl-N′,N′-dimethyl urea) is given in Fig. 9.4-2a.

Fragmentation in EI is discussed in more detail in Section 9.4.3. EI mass spectra are highly reproducible. Extensive collections of standardized EI mass spectra are available, also for computerized evaluation (see Sec. 9.4.3).

Quadrupole mass analysis

The ions generated in the ion source are electrostatically extracted from the ion source and introduced into a *quadrupole mass filter* for mass analysis.

A quadrupole mass filter is a device consisting of four stainless-steel hyperbolic or circular rods that are accurately positioned parallel in a radial array (Fig. 9.4-3). Opposite rods are charged by either a positive or a negative direct-current (DC) potential, U, at which an alternating-current (AC) potential, $V_0 \cos \omega t$, is superimposed. The AC potential, the frequency $\omega/2\pi$ of which is in the radiofrequency region (MHz), successively reinforces and overwhelms the DC field. Ions from the ion source are introduced into this quadrupole field by means of an up to 20 V accelerating potential. The ions start to oscillate in a plane perpendicular to the rod length as they traverse through the quadrupole filter. The trajectories of the ions of one particular m/z ratio are stable; these ions are transmitted towards the detector, while ions with unstable trajectories do not pass the mass filter, because the amplitude of their oscillation becomes infinite.

The stability of the trajectory is determined by the values of the parameters A, depending on the DC potential, and Q, depending on the radiofrequent AC potential. A and Q are given by:

$$A = \frac{8eU}{mr_0^2\omega^2} \tag{9.4-6}$$

$$Q = \frac{4eV_0}{mr_0^2\omega^2} \tag{9.4-7}$$

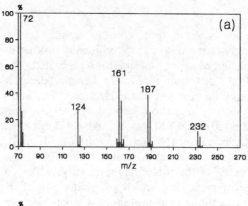

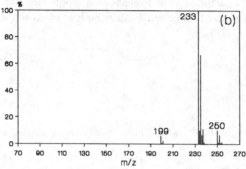

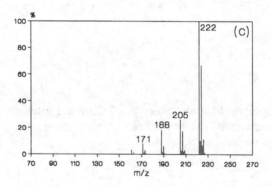

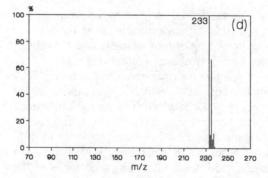

Fig. 9.4-2. Comparison of ionization methods. (a) EI mass spectrum of diuron (N-dichlorophenyl-N′,N′-dimethyl urea), (b) ammonia CI spectrum of diuron, (c) ammonia CI spectrum of DCPU (N-dichlorophenyl urea), and (d) thermo-spray mass spectrum of diuron

where m is the mass of the ion, and r_0 is the radius of the quadrupole field. A plot of A as a function of Q is a so-called stability diagram (Fig. 9.4-4), as it indicates for which values of A and Q stable trajectories can be achieved. Ions of different m/z value can be transmitted by the quadrupole filter towards the detector when the DC and the AC potentials are swept, while their ratio and oscillation frequency are kept constant. Scanning between $m/z = 1$ and 500 in a typical quadrupole requires a variation of U between 0 and 300 V, and of V_0 between 0 and 1500 V. Consecutive m/z values are positioned on the 'A/Q = constant' line in Fig. 9.4-4 and are shifted in the stable-oscillation region when the potentials are swept.

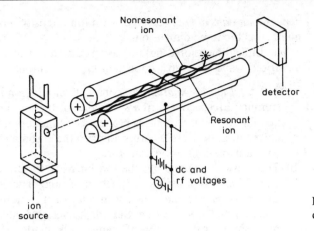

Fig. 9.4-3. Schematic diagram of a quadrupole mass filter

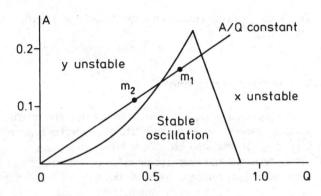

Fig. 9.4-4. Stability diagram for a quadrupole mass filter

Thus, the quadrupole mass filter acts as a band pass filter, the resolution of which depends on the ratio of DC and AC potentials, i.e., the slope of the 'A/Q = constant' line in Fig. 9.4-4. Generally, the resolution is set to 'unit-mass', indicating that, for instance, $m/z = 100$ and $m/z = 101$ can be distinguished; all ions with m/z values between 99.50 and 100.49 are attributed to $m/z = 100$. Therefore, the quadrupole mass filter is suitable for the determination of the nominal masses of a compound and its fragment ions.

Electron multiplier

The detection of ions by means of an *electron multiplier* is based on the emission of secondary electrons, resulting from the collision of energetic particles at a suitable surface. The secondary electrons can be multiplied by consecutively striking subsequent surfaces. The electron multiplier may be either of the *discrete dynode* type or of the *continuous dynode* type. A discrete dynode multiplier comprises of 12–20 beryllium-copper dynodes, electrically connected through a resistive network. The continuous dynode or channel multiplier consists of a curved lead-doped funnel-shaped tube. The voltage applied between the ends of the tube creates a uniform field along the tube length. Secondary electrons are accelerated further into the tube to cause subsequent collisions with the inner wall. The typical gain of an electron multiplier is 10^6. The current from the electron multiplier is preamplified prior to the digitization required for the data collection by the data system.

Data handling

A simple instrument, as just described, mostly operates under computer control. During acquisition, the instrument is sweeping the AC and DC potential at a fixed ratio in order to transmit ions in the range of m/z 50–500, for instance, to the detector. Such a scan is made every second or few seconds. Each scan, i.e., the ion intensities as a function of the m/z value is saved by the computer for later data handling. As a result of the acquisition during GC-MS analysis a *three-*

Computerized data handling is a prerequisite in mass spectrometry.

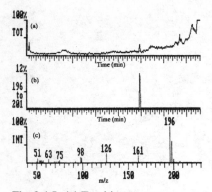

Fig. 9.4-5. (a) Total-ion current chromatogram, (b) mass chromatogram, and (c) mass spectrum of a typical GC-MS run

Selectivity may be improved by selective ion monitoring.

dimensional data-array is created, with ion intensity, m/z value, and time or scan number as the three dimensions.

The three-dimensional data-array produced during acquisition can be handled in various ways. The most important types of output are (Fig. 9.4-5):

(a) The *total-ion chromatogram*: consecutively the ion intensities in each scan are summed irrespective of the m/z value of the ion, and the summed intensities, i.e., the total-ion current, is plotted as a function of time or scan number. This chromatogram closely corresponds to the chromatogram obtained from a GC with a flame-ionization detector.
(b) The *mass chromatogram*: the ion intensity for an ion or ions with particular m/z value(s) is plotted as a function of time or scan number.
(c) The *mass spectrum*: for each scan number the ion intensity can be plotted as a function of the m/z value. Summed, averaged, and/or background-corrected mass spectra over a series of scans can be made as well.

The mass spectrum can be computer-searched against a mass spectral library (see Sec. 9.4.3).

Selective ion monitoring

In the previous sections it was assumed that the instrument was operated in the scanning mode: a series of complete mass spectra is acquired. This is obviously the way to go in the analysis of unknown compounds, but when MS is used as a highly selective and sensitive detector in the analysis of known compounds, i.e., for screening or quantitative analysis, only a limited number of ions is of importance. Then the MS can be operated in the *selective-ion monitoring* (SIM) mode. The instrument selects one particular m/z value for transmission during a preset period and then jumps to another m/z value, and so on. No complete mass spectra are acquired. The data can be displayed in terms of mass chromatograms. The main advantage of the SIM mode is that the instrument does not waste measurement time in detecting non-relevant ions. As a result, better signal-to-noise ratios and improved detection limits can be achieved.

Limitations of the simple instrument

Although the simple instrument, as just described, is widely used and can be considered as an extremely powerful tool in the analytical laboratory, the instrument suffers from a number of limitations. The major limitations are:

(a) Only volatile analytes are amenable to GC-MS analysis, thus excluding highly polar, ionic and/or macromolecular compounds.
(b) Ionization is limited to compounds with sufficient vapour pressure and stability.
(c) The unimolecular fragmentation reactions in EI may obscure the molecular ion, especially in less stable compounds, prohibiting the determination of the molecular mass.
(d) Only nominal mass information is obtained in a limited m/z range.

In order to cope with these limitations a number of other approaches in terms of sample introduction, analyte ionization, mass analysis, and ion detection have been developed. These approaches are discussed in subsequent sections. This discussion provides an elaborate overview of MS technology and instrumentation.

Soft ionization techniques

Ionization techniques currently applied in analytical MS can be classified in various ways (see Table 9.4-3). An important distinction can be made between hard and soft ionization techniques. In a hard technique a considerable amount of energy is transferred to the analyte ion during the ionization process, most likely resulting in subsequent unimolecular dissociation reactions. EI, as discussed in the

Table 9.4-3. Classification of ionization techniques for mass spectrometry

Sample state	Class of ionization technique		Examples
Gas or vapour	Electron ionization	EI	Electron ionization
Gas or vapour	Chemical ionization	CI	Chemical ionization
		DCI	Desorption chemical ionization
		FAB	Fast-atom bombardment
Liquid or solid	Desorption	SIMS	Secondary-ion MS
		FD	Field desorption
		PD	Plasma desorption
		MALDI	Matrix-assisted laser desorption/ionization
Liquid/solution	Other	TSI	Thermospray ionization
		ESI	Electrospray ionization

previous section, is the typical example of a hard ionization technique. Most other techniques are *soft ionization* techniques, generally resulting in little fragmentation and thus providing molecular mass information. Classification of the soft ionization techniques can be roughly based on the way the sample is introduced, although some apparent mixed mechanisms somewhat obscure such a classification. The most important soft ionization techniques are discussed in more detail below.

The various soft ionization methods generate the same types of ions, i.e., *cationized* molecules, as a result of attachment of a cation like H^+, Na^+ in positive-ion mode, and *deprotonated* or *anionized* molecules in negative-ion mode.

In soft ionization methods primarily protonated and deprotonated ions are generated.

Chemical ionization

Chemical ionization (CI) is performed in a relatively high-pressure ion source (0.1–100 Pa). The high pressure results in frequent intermolecular and ion-molecule collisions in the source. The ionization is based on a chemical reaction between a reagent gas ion and the analyte molecule. The reagent gas ions are generated by EI and subsequent ion-molecule reactions. with ammonia as reagent gas, ions like NH_4^+ and $(NH_3)NH_4^+$ at $m/z = 18$ and 35 are generated. The ions are even-electron species and can be described as protonated molecules. They can react with an analyte molecule M in an *ion-molecule reaction*:

Chemical ionization by gas-phase ion-molecule reactions is an important ionization mechanism.

$$NH_4^+ + M \rightleftharpoons NH_3 + MH^+ \qquad (9.4-8)$$

This proton-transfer reaction results in a protonated analyte molecule, generally with low internal energy and thus less prone to fragmentation than the molecular ion generated under EI ionization. The nominal m/z value of this ion corresponds to $M_r + 1$.

Both the thermodynamics and the kinetics of the reaction in Eq. 9.4-8 are of importance in the evaluation of the ionization process. The *proton affinity* (PA) of a molecule M is defined as the exothermicity of its protonation reaction:

$$PA_{(M)} \equiv \Delta H_{f(M)} + \Delta H_{f(H^+)} - \Delta H_{f(MH^+)} \qquad (9.4-9)$$

Some proton affinities of typical reagent gases are given in Table 9.4-4.

The reaction in Eq. 9.4-8 only takes place when the proton affinity of M exceeds that of the reagent gas. For example, the reaction between protonated ammonia and pyridine results in protonated pyridine, while no proton transfer reaction occurs between protonated ammonia and water. The excess energy of the reaction appears as internal energy of the products, with a narrow energy distribution. The internal energy of the product ion can be controlled by the choice of the reagent gas.

The CI process is an ion-molecule process (bimolecular), while fragmentation of the protonated molecule is an unimolecular process, like in EI, although based on different chemistry because an even-electron species rather than a radical cation is involved.

The proton-transfer reaction is just one of the ion-molecule reactions that can

Table 9.4-4. Proton affinities (PA in kJ/mol) of some typical reagent gases for chemical ionization

Base	PA (kJ/mole)
H_2	422
CH_4	536
H_2O	723
CH_3OH	773
CH_3CN	797
i-C_4H_{10}	823
NH_3	857
CH_3NH_2	894
Pyridine	921

take place in CI. Other possible reactions resulting in positive ions are: charge-exchange (Eq. 9.4-10) and electrophilic addition (Eq. 9.4-11).

$$M + X^{+\bullet} \rightarrow M^{+\bullet} + X \tag{9.4-10}$$

$$M + X^{+} \rightarrow MX^{+} \tag{9.4-11}$$

In the ammonia CI spectra of the herbicide diuron (N-dichlorophenyl-N',N'-dimethyl urea) and the related compound DCPU (N-dichlorophenyl urea), reproduced in Fig. 9.4-2b and c, both the protonated and the ammoniated molecule are detected. The relative abundance of the signals due to proton transfer and electrophilic addition depends on the proton affinity. The higher proton affinity of diuron is in favour of the formation of a protonated species instead of the ammoniated species. Electrophilic addition (Eq. 9.4-11) in the formation of the ammoniated molecule can, from an acid-base mechanistic point-of-view, be considered as the sharing of the proton by the analyte and reagent gas molecules, having similar proton affinities.

Besides reactions that lead to the generation of positive ions, *negative ions* can be produced by chemical ionization processes as well. Negative ions can be produced by ion-molecule reactions, e.g., proton abstraction (Eq. 9.4-12) and anion attachment, or by electron capture.

$$MH + B^{-} \rightleftharpoons M^{-} + BH \tag{9.4-12}$$

In proton abstraction the reaction is determined by the relative gas-phase acidities of the analyte and the reagent gas molecules, defined as:

$$\Delta H_{acid} \equiv \Delta H_{B^{-}} + \Delta H_{H^{+}} - \Delta H_{BH} \tag{9.4-13}$$

The reaction in Eq. 9.4-12 proceeds when gas-phase acidity of B^{-} exceeds that of M.

Negative ions might be generated via electron capture.

The electron capture process generally is classified as a CI process because of the high pressure needed in the ion source. In *electron capture*, ionization takes place by the capture of thermal electrons (with energies between 0 and 10 eV) by the analyte resulting in the generation of radical anions. The high pressure in the ion source is required both for thermalizing the electrons, and for the removal of excess energy from the anion formed upon electron attachment. Either associative (Eq. 9.4-14) or dissociative (Eq. 9.4-15) electron capture takes place.

$$AB + e^{-} \rightarrow AB^{-\bullet} \tag{9.4-14}$$

$$AB + e^{-} \rightarrow A^{-} + B^{\bullet} \tag{9.4-15}$$

Electron capture can be considered as a highly selective ionization method, as only a limited number of analytes are prone to efficient electron capture, e.g., fluorinated compounds. Often chemical derivatization to, for instance, pentafluorobenzyl derivatives is applied for improving sensitivity in negative-ion electron capture MS.

Energy-sudden ionization methods

Highly polar and/or high-mass analytes can be mass analyzed using energy-sudden ionization methods.

The desorption or energy-sudden ionization techniques (Table 9.4-5) are in principle based on the fact that at high temperatures the rate constant for vaporization are higher than the rate constants for decomposition. Although the mechanistic aspects of the various desorption methods are not completely understood, the concept of rapid sample heating is a feature common to all these methods. As indicated in Table 9.4-5, in most desorption methods a sample matrix is applied to assist in the desorption/ionization process. In general, it is assumed that the energy deposited on the sample surface can cause gas-phase ion-molecule reactions to occur near the interface of solid or liquid and the vacuum (the so-called selvedge). The formation of analyte ions in the sample matrix prior to evaporation and/or desorption is assumed to play an essential role as well. The energy deposited provides these preformed ions in the condensed phase with sufficient kinetic energy to leave the matrix. As a result, cationized molecules such as $[M + H]^{+}$ and $[M + Na]^{+}$ are generated in positive-ion mode and deprotonated molecules

Table 9.4-5. Characteristics of energy-sudden or desorption ionization methods

	Technique	Agent	Matrix
DCI	Desorption chemical ionization	Rapid heating and CI reagent ion plasma	Solid on wire
SIMS	Secondary ion MS	keV Xe^+ or Cs^+	Solids or liquids
FAB	Fast-atom bombardment	keV Ar, Xe, Cs^+	Sample in glycerol
PD	Plasma desorption	MeV ^{252}Cf fission fragments	Sample on nitrocellulose
MALDI	Matrix-assisted laser desorption/ionization	Photons	Sample in matrix, e.g., sinapinic acid
FD	Field desorption	10^9 V/m electric field	Sample on activated emitter

$[M - H]^-$ in the negative ion mode. The various methods, e.g., DCI, FAB, PD, MALDI, and FD are briefly described below.

Desorption chemical ionization

In *desorption chemical ionization* (DCI) the sample is deposited on a 50–200 μm ID wire to which direct and rapid electrical heating can be applied. The wire is placed into the CI reagent ion plasma. The combination of rapid heating and direct interaction of vaporized molecules with the reagent gas ions enables the mass analysis of many relatively nonvolatile analytes.

Fast-atom bombardment

In *fast-atom bombardment* (FAB) the analyte is dissolved in an appropriate non-volatile matrix solvent, e.g., glycerol. The solution is deposited as a homogeneous thin film on a metal target on a direct insertion probe. The target is transferred to the high vacuum of the ion source and bombarded with 8 keV particles, e.g., Ar or Xe atoms. High energy cesium ions are used for bombardment as well. An example of a FAB spectrum of a peptide Tyr-Ala-Gly-Phe-Leu is shown in Fig 9.4-6a.

Plasma desorption

The sample is dissolved in a volatile solvent and subsequently deposited on a target coated with nitrocellulose. Adsorption of the sample molecules to the target material takes place, while the solvent and possible water-soluble contaminants can be washed away. The target is transferred to the high vacuum where it is subjected to the high-energy (MeV) fission fragments from a ^{252}Californium source. Ions are generated which can be mass analyzed. In most cases, a time-of-flight mass analyzer (see "Mass analysis" and Fig. 9.4-7b) is used for this purpose. Samples with molecular masses up to 20 kDa can be analyzed in this way.

Matrix-assisted laser desorption/ionization

A sample is mixed with an appropriate matrix solution, e.g., containing sinapinic acid, and deposited onto a stainless-steel target. Upon drying, crystallization takes place. When these crystals are laser bombarded with photons, the frequency of which corresponds to the absorption maximum of the matrix molecules, sample ions are generated which can be mass analyzed by a time-of-flight mass spectrometer. Like the plasma-desorption method, the technique is especially useful for the analysis of biomacromolecules, as compounds with molecular masses up to 300 kDa can be detected. This recently introduced technique is called *matrix-assisted laser desorption/ionization* (MALDI). An example of a mass spectrum of cytochrome c obtained by MALDI is given in Fig. 9.4-6b.

Matrix-assisted laser desorption/ionization (MALDI) is an important technique in the analysis of biomacromolecules.

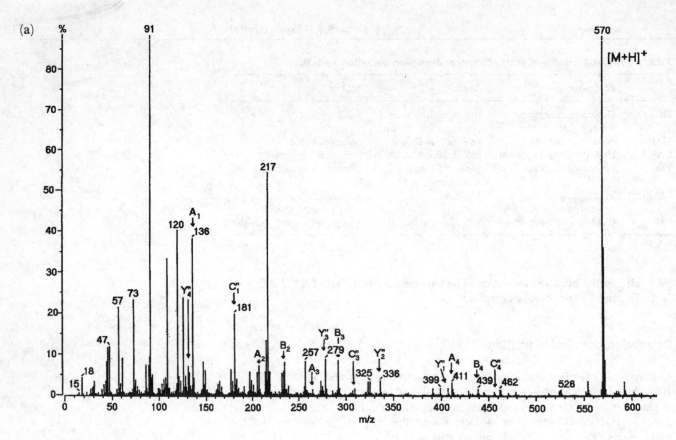

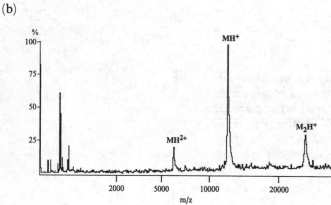

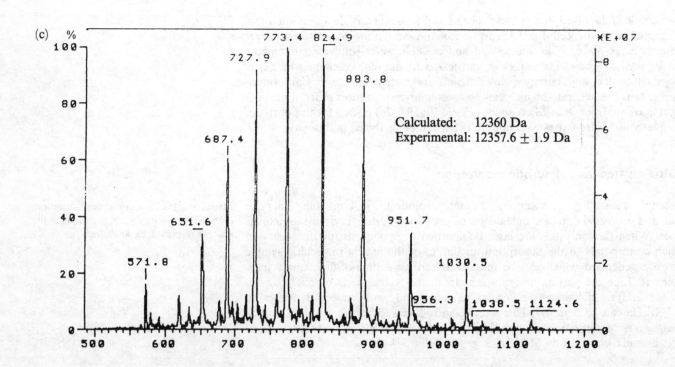

Fig. 9.4-6. Soft ionization methods. (a) FAB mass spectrum of the peptide Tyr-Ala-Gly-Phe-Leu, (b) MALDI mass spectrum of 1 pmol horse cytochrome c with 2,5-dihydroxybenzoic acid as matrix, and (c) electrospray mass spectrum of 1 pmol/μL horse cytochrome c in 50% methanol/water (1% acetic acid), showing an ion envelope of multiply-charged ions

Field desorption

In *field desorption* (FD) the sample solution is deposited on to a 10 μm tungsten wire FD emitter, which prior to the deposition of the sample on to it is activated to provide microneedles on the surface. The emitter is placed at a high electric potential, resulting in high local electrical field strengths at the tip of the micro-needles (typically 10^8 V/m). Ions are generated in various ways. The high local electrical field enables electron tunnelling from the sample to the emitter followed by desorption of a radical cation. This ion may be detected directly or react with other analyte ions in ion-molecule reactions, e.g., resulting in a protonated molecule which is subsequently detected. Furthermore, preformed ions may be extracted directly from the condensed phase as a result of the high local field strength as well.

Other ionization methods

The other important soft ionization methods, i.e., thermospray and electrospray, have been invented in the course of developing methods for the online combination of liquid chromatography and mass spectrometry. They are based on the nebulization of a liquid and the subsequent generation of ions from the liquid droplets. In *thermospray* ionization, the liquid is nebulized into a modified CI source, while in *electrospray* the nebulization and ionization are performed in a region with atmospheric pressure. The ionization mechanism is not clearly understood. It is generally assumed that preformed ions in solution are able to 'evaporate' from the small, highly-charged droplets, generated upon nebulization and subsequent evaporation of the neutral solvent molecules. However, in many instances, a significant contribution of gas-phase ion-molecule chemistry is observed as well. Both techniques are capable of extremely soft ionization of highly labile and involatile compounds, such as peptides, proteins, (oligo)nucleotides, and (oligo)saccharides without any fragmentation. The thermospray mass spectrum of diuron (N-dichlorophenyl-N',N'-dimethyl urea) is given in Fig. 9.4-2d. Compare the extent of fragmentation under thermospray with that under ammonia DI and EI (cf. Fig. 9.4-2a and b). An important feature of especially the electrospray technique in the analysis of biomacromolecules is the formation of multiply-charged ions, resulting in a so-called ion envelope (see Fig. 9.4-6c and Sec. 9.4.4), from which the molecular mass of the protein can be calculated accurately (better than 0.1%).

Thermospray and electrospray ionization are ideally suited for the on-line coupling of liquid chromatography and mass spectrometry.

Mass analysis

Four different principles of mass analysis are applied in MS instruments: a combination of magnetic and electric sectors, quadrupole filters, time-of-flight measurements, and ion-cyclotron resonance systems.

Sector instruments

The simplest mass analyzer to understand is a magnetic sector instrument. Ions with mass m and z elementary charges e are introduced into a magnetic field B. The kinetic energy E_{kin} of the ions is determined by the accelerating voltage V. Thus:

$$E_{kin} = zeV = \tfrac{1}{2}mv^2 \tag{9.4-16}$$

where v is the velocity of the ions. When the magnetic Lorentz force is counterbalanced by the centrifugal force, ions are transmitted to the detector:

$$Bzev = \frac{mv^2}{r} \tag{9.4-17}$$

where r is the radius of curvature of the path through the magnetic field. Combining these two equations leads to:

$$\frac{m}{z} = \frac{B^2 r^2 e}{2V} \tag{9.4-18}$$

From this equation it can be concluded that the separation of ions with different m/z values can be achieved in three different ways: variation of the radius of curvature ion with different m/z values are separated in space, while variation of either B or V ions of different m/z values are separated in time, i.e., they can be detected one after another by a detector at a fixed position behind a slit. The resolution of this instrument is determined by the ion entrance slit at the source side of the magnet and the ion exit slit at the detector side. A narrower slit improves the resolution but decreases the transmission of ions. Choosing the resolution for a particular application is always a compromise between the selectivity required and the signal obtained.

A double-focusing sector instrument allows high-resolution measurement for accurate mass determination.

Better performance of the sector instrument in terms of mass resolution (Eq. 9.4-1) is achieved by combining the magnetic sector with an electrostatic analyzer (ESA). The ESA provides energy-focusing: the ions with a particular m/z value but differing in kinetic energy are deflected towards one focal point. This significantly improves the mass resolution of the instrument without loss in signal. With a *double-focusing instrument*, high resolution or accurate mass determination is possible. A schematic diagram of a double-focusing mass spectrometer is shown in Fig. 9.4-7a. The order in which the magnetic sector B and the ESA are placed generally is not important: both forward (EB) and reversed (BE) geometries are in use, while some manufacturers even apply an EBE configuration.

The sector instruments are probably the most versatile instruments as they combine a relatively wide mass range, high resolution, and good quantitation capability. However, the costs are high, and operation of the instrument is more complex and demanding than that of quadrupole instruments.

Quadrupole ion traps

The quadrupole filter has been described in detail in Sec. 9.4.2. As indicated, a quadrupole instrument provides unit-mass resolution up to ca. 2000 Da, ease of operation, fast scanning, and limited costs.

A recent development in quadrupole technology is the three-dimensional *quadrupole ion trap*. An ion trap consists of a cylindrical ring electrode to which the quadrupole field is applied, and two end-cap electrodes. The top end-cap contains holes for introducing ions or electrons into the trap, while the bottom end-cap contains holes for ions ejected towards the electron multiplier. Ions that are generated inside the trap itself or in an external ion source are stored in the trap. By subsequently increasing the radiofrequency potential the trajectories of ions of successive m/z values are made unstable. These ions are ejected out of the trap where they are detected by the electron multiplier. Please note, that in the quadrupole ion trap the ions with unstable trajectories are detected, while in the quadrupole filter the ions with stable trajectories are transmitted to the detector.

Time-of-flight instruments

A *time-of-flight* (TOF) instrument consists of a pulsed ion source, and accelerating grid, a field-free flight tube, and a detector. The ions in the pulsed beam are accelerated by a potential V. The time t needed by the ions to reach the detector, placed at a distance d, is measured. That time can be related to the m/z value via:

$$t = d\sqrt{\frac{m}{2zev}} \tag{9.4-19}$$

Pulsing of the ion source is required to avoid the simultaneous arrival of ions of different m/z values at the detector. A schematic diagram of a TOF analyzer is

(a)

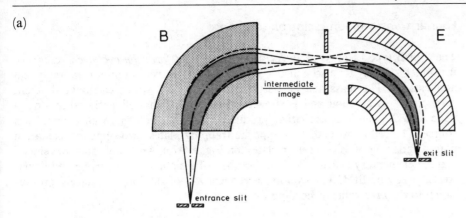

(b)

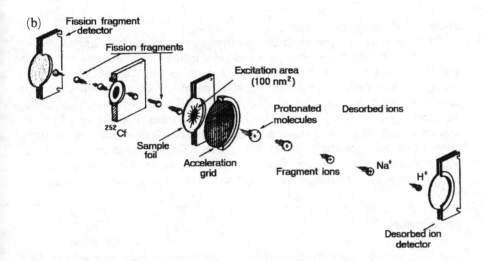

(c)

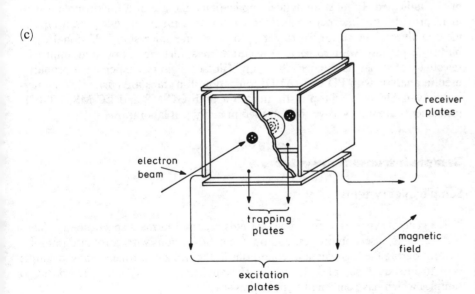

Fig. 9.4-7. Mass analyzers: (a) double-focusing sector instrument, (b) time-of-flight instrument, (c) ion-cyclotron resonance instrument

given in Fig. 9.4-7b. TOF instruments are used especially in combination with PD and MALDI ionization (see "soft ionization techniques"). In these applications, one benefits from the unlimited mass range, the high scanning speed, and the simplicity and low costs of the instrument. Although until recently the resolution of a TOF instrument was limited (generally ca. 300), new developments, viz. reflectron TOF and delayed extraction, significantly improved the resolution (5000 can now be achieved, enabling accurate mass analysis).

Fourier-transform ion-cyclotron resonance

The fourth type of mass analyzer, the *Fourier-transform ion-cyclotron resonance* instrument (FT-ICR, also called FT-MS), is not yet very common in analytical applications. The mass analysis is performed in a cell which can have different geometries, e.g., a cubic cell as shown in Fig. 9.4-7c. The cell is placed in a magnetic field B. The ions are either produced inside the cell or in an external ion source. The cell consists of two opposite trapping plates, two opposite excitation plates, and two opposite receiver plates (cf. Fig. 9.4-7c). An ion of mass m, velocity v and z elementary charges describes in the cell a circle of radius r, perpendicular to the magnetic field. The cyclotron frequency ω of the ion is inversely proportional to the m/z value, according to:

$$\omega = 2\pi f = \frac{v}{r} = \frac{Bez}{m} \tag{9.4-20}$$

where f is the frequency in Hertz.

When the ions, trapped in their cyclotron motion in the cell, are excited by means of an RF pulse the radius of the circle is increased and all ions of a particular m/z value start to move as an ensemble in phase. The coherent movement of ions generates an image current at the receiver plates. As the coherency is lost in time, a decaying image current is observed, i.e., a time-domain signal containing all frequency information of the moving ions. The time-domain signal can be Fourier-transformed into a frequency-domain signal, which in turn can be transformed to a mass spectrum via Eq. 9.4-20. The FT-ICR instrument is capable of high-resolution measurements. Disadvantages are the relatively difficult operations, which are in part due to the severe vacuum requirements, and the high costs.

Comparison

Important performance characteristics of the various instruments have been briefly indicated. In most analytical applications, the use of a quadrupole instrument provides the best perfomance/cost ratio. It is the most widely-used type of mass analyzer. The sector instrument provides some interesting additional capabilities, especially with respect to accurate mass determination and improved selectivity at somewhat higher resolution. Time-of-flight instruments are primarily used in analysis with PD or MALDI ionization. Ion traps are currently commercially available as benchtop instruments for both GC-MS and LC-MS. FT-ICR instruments are not (yet) very often used in analytical laboratories.

Sample introduction systems

Sample inlet systems

Three general types of sample inlet systems are used in mass spectrometry: (controlled) leaks, insertion probes, and on-line combinations with chromatography.

Considering the high vacuum conditions in the MS a controlled *leak* is a simple way to introduce samples. However, the application of a leak is restricted to samples which have sufficient vapour pressure.

A wider application area is provided by sample introduction via a *direct insertion probe*. The solid or liquid sample is deposited into a metal sample crucible, which is placed in the holder of a heatable insertion probe. The probe is introduced via a vacuum lock into the high vacuum and positioned against the ion-source block. Temperature-programmed heating of the probe ensures introduction of the sample. Again, the application of a direct insertion probe is limited to samples with sufficient vapour pressure at the elevated temperatures that can be reached (ca. 400 °C). A variety of modified sample insertion probes has been produced for use in combination with soft ionization techniques. Dedicated insertion probes are available and needed for DCI, FD, FAB, while special target holders are applied in PD and MALDI.

In providing samples for probe methods, considerable care is required to avoid the presence of excessive amounts of nonvolatile salts, e.g., sodium phosphates. Concentration of other contaminants, e.g., the phtalates present in many solvents, during sample preparation should be avoided as well.

Combined chromatography-mass spectrometry

The most important and versatile method for sample introduction is via a chromatograph. Especially the on-line gas chromatography mass spectrometry (GC-MS) combination is very widely used. It has developed to a routine tool in many areas of analytical mass spectrometry. In the last decade, the routine use of on-line liquid chromatography mass spectrometry (LC-MS) has become available as well. LC-MS and the ionization techniques developed in relation to it have revolutionized MS and its application areas. Considering the importance of on-line GC-MS and LC-MS some attention is paid to the experimental and instrumental aspects of these techniques.

Gas chromatography-mass spectrometry

The on-line capillary GC-MS combination can be achieved in two ways, i.e., via direct coupling or via an open-split coupling.

Gas chromatography–mass spectrometry is a powerful analytical tool in structure elucidation.

The gas flow through a 0.25 mm ID capillary GC column matches the allowable gas load of a simple MS vacuum system. Since the components eluting from the GC column are already in the vapour state, immediate introduction of the capillary column effluent into the MS ion source in electron ionization mode is possible. Although frequently applied, this *direct coupling* has some disadvantages. The colume outlet is at high vacuum, which results in changing retention times relative to other GC detectors, like a flame ionization detector, Furthermore, the gas load to the ion source changes during the temperature program of the GC analysis. This may influence the tuning of the ion source parameters. Finally, the whole sample injected on to the column is introduced into the MS, resulting in a pressure pulse with possible detuning effects when the solvent peak is eluting from the column. The problems related to the solvent load to the heated filament are easily overcome by switching off the filament at the beginning of the analysis.

An alternative to the direct coupling is the *open-split coupling*; a schematic diagram of such a device is shown in Fig. 9.4-8a. The column is connected with the ion source via a fixed restriction. A makeup gas flow is provided either to makeup the column flow to match the restrictor throughput or to rapidly remove the excess of carrier gas. As a result, the column outlet is at atmospheric pressure, just like in a conventional GC detector. By increasing the makeup flow it is possible to efficiently divert high vapour loads from the solvent away from the MS.

Liquid chromatography mass-spectrometry

The on-line coupling of LC-MS is far more difficult to accomplish than the GC-MS coupling. This is due to a number of reasons:

- The gas load to the vacuum system resulting from evaporating the mobile phase is generally too high.
- The mobile phase composition (containing nonvolatile additives like phosphate buffers) often is incompatible with MS.
- The transfer of polar and ionic analytes from the liquid to the gas phase is difficult.

In order to solve the first problem a number of LC-MS interfaces has been developed, some of which are discussed in somewhat more detail below. The second problem must be solved at the chromatography side, e.g., by changing to volatile buffers or the use of column switching. The third problem is largely solved by the advent of new soft ionization techniques, especially thermospray and electrospray, which were developed in the course of LC-MS development.

The most widely-used LC-MS interfaces at the moment span a wide range of analyte polarities: particle-beam for nonpolar to medium polarity analytes, thermospray and atmospheric pressure CI interfaces (APCI) for polar compounds, and continuous-flow fast-atom bombardment (CF-FAB) and electrospray for highly polar, ionic, and macromolecular analytes. Except for CF-FAB, all these techniques can be used at flow-rates of 0.2–1 ml/min with typical solvents used in reversed-phase LC using volatile buffers like ammonium acetate. Schematic diagrams of the LC-MS interfaces discussed below are given in Fig. 9.4-8.

In a *particle-beam* interface (Fig. 9.4-8b) the column effluent is pneumatically nebulized into an atmospheric-pressure desolvation chamber. This is connected to a momentum separator where the analytes are transferred to the MS ion source while the low molecular-mass solvent molecules are efficiently pumped away. The analyte particles hit the heated source surface, evaporate, and can be ionized by EI or CI. The evporation step obviously limits the application range of the interface to not-too-polar analytes.

In a *thermospray* interface (Fig. 9.4-8c) the column effluent is rapidly heated in a narrowbore capillary in such a way that partial (ca. 90%) evaporation of the solvent is achieved inside the capillary. As a result a mist of vapour and small droplets is formed, in which the heated droplets further evaporate and ions are generated, either by the thermospray ionization process based on ion evaporation (see "soft ionization techniques") or by solvent-mediated chemical ionization initiated by electron from a heated filament or a discharge electrode. The excess vapour is pumped away directly from the ion source.

In a CF-FAB interface a small mobile-phase stream, containing glycerol as a FAB matrix, is flowing towards a FAB target where the effluent is bombarded by fast atoms or ions.

In both atmospheric-pressure ionization systems (Fig. 9.4-8d), i.e., electrospray and APCI, nebulization of the column effluent takes place in an atmospheric-pressure region. Contrary to particle-beam, in these systems the ionization also takes place in this region and the ions are sampled into the high vacuum for mass analysis. In *electrospray*, the nebulization results from the disintegration of a liquid stream under influence of a high electric field. A potential difference of ca. 3 kV is applied between a needle through which the liquid is introduced and a counter electrode. Ions are generated via the ion-evaporation process. In an *APCI interface*, and aerosol is produced with a heated pneumatic nebulizer and ions are generated by ion-molecule reactions initiated by a corona discharge in the ion-source region.

None of the LC-MS interfaces are capable of solving all analytical problems. Therefore, the most appropriate LC-MS interface for a particular application must be selected, primarily based on analyte polarity, as indicated above. Because of the ease of operation, the sensitivity and the robustness, LC-MS interfaces based on atmospheric-pressure ionization are now the method-of-choice in most applications. Dedicated benchtop LC-MS instruments are commercially available.

Tandem mass spectrometry

Introduction

Tandem mass spectrometry allows both structure elucidation and specific screening.

In structure-elucidation problems one often wants to obtain more information on the ions generated in the ionization process. This is certainly true when soft ionization methods are applied. It must be realized that fragmentation in EI is mainly due to the excess internal energy in the analyte obtained during the ionization process. Therefore, increasing the internal energy of an ion generated by soft ionization methods will induce fragmentation of those ions as well. This can be achieved in a number of ways. The most widely-used approach is collisional activation via collisions with neutral gas molecules (collision-induced dissociation, CID), while other approaches comprise laser photodissociation and surface-induced dissociation. As a result of the increase of internal energy, fragmentation of the ion is achieved:

$$M_p^+ \rightarrow m_d^+ + m_n \tag{9.4-21}$$

(a)

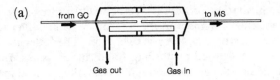

(b)

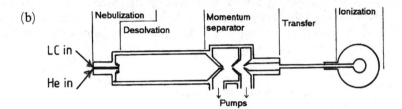

(c)

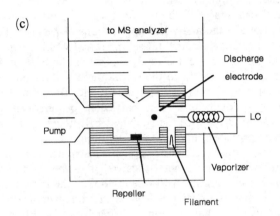

Fig. 9.4-8. Interfaces for on-line chromatography mass spectrometry. (a) Open-split coupling for GC-MS, (b) particle-beam interface, (c) thermospray interface, (d) electrospray and APCI interface

(d)

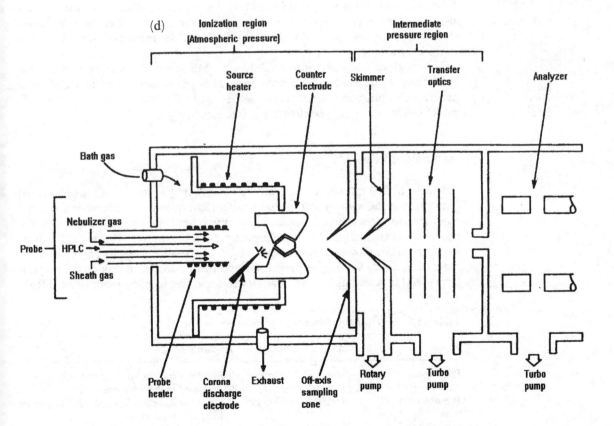

where M_p^+ is the precursor ion, m_d^+ is the product ion, and m_n represents one or more neutral species. It is important to realize that the fragmentation takes place after the presursor ion has left the ion source.

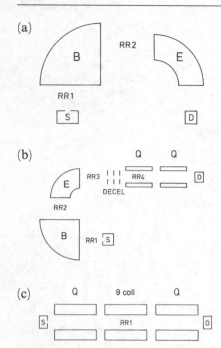

Fig. 9.4-9. Instruments for tandem mass spectrometry. (a) BE instrument, (b) hybrid instrument of BE-q_{coll}-Q configuration, (c) triple quadrupole instrument (Q-q_{coll}-Q). S – ion source; B – magnetic sector; E – electrostatic analyzer; D – detector; RR1 – field-free or reaction region; Q – quadrupole filter; DECEL – set of deceleration lenses

The fragmentation process is a probability process, meaning that part of the ionized molecules fragment in the ion source, while other ions fragment on their way to the detector, i.e., after acceleration. The latter, the so-called metastable ions, are not recognized as normal, fragment-ion peaks. In the case of CID, the fragmentation of the accelerated ions is actually induced in a collision cell with a somewhat higher pressure. This requires a special setup of the instrumentation. In principle, two mass analyzers are required: one for selecting the precursor ion from the ions generated in the ion source, and one for analyzing the product ions after the collisions. For this reason the approach is called tandem mass spectrometry (MS-MS).

Instrumentation for MS-MS

Despite the fact that MS-MS can be performed with most mass analyzers discussed in "Mass analysis" the use of sector instruments and quadrupole filters is most important in the analytical applications of MS-MS. Schematic diagrams of the most important instrumental setups for MS-MS are given in Fig. 9.4-9. In a double-focusing instrument of a BE or EB configuration, a collision cell can be placed either in the first or in the second field-free region. With a collision cell in the first field-free region of an instrument with BE geometry (cf. Fig. 9.4-9a), the detection of product ions of a particular precursor is possible by scanning with a constant B/E ratio, the so-called B/E-linked scan mode. Obviously, such an approach provides only limited resolution, i.e., ca. 1000 for the precursor and 5000 for the product ions. Other possible instruments based on sectors which can show better resolution characteristics are either the 3- or 4-sector instruments with a collision cell in the third field-free region, or the hybrid instrument, e.g., with a BE-q_{coll}-Q geometry (cf. Fig. 9.4-9b) with a quadrupole collision cell. The hybrid instrument requires that the ions passing from the BE part have to be slowed down prior to the quadrupole collision cell. An important advantage of the hybrid instrument is the possibility to select the precursor ion with increased mass resolution. This advantage obviously is also valid for the more expensive 3- and 4-sector instruments.

The most versatile and most widely-used MS-MS configuration is the triple quadrupole instrument, where mass analysis is performed in the first and third quadrupole while the second quadrupole is used as collision cell in the RF-only mode, i.e., in a Q-q_{coll}-Q configuration (see Fig. 9.4-9c).

Scan modes in MS-MS

MS-MS was first introduced as a way to achieve fragmentation of ions generated in the ion source, e.g., by soft ionization methods. In such an experiment, the first mass spectrometer selects a particular precursor ion, which is dissociated and the product ions are analyzed with the second mass spectrometer. This is the product-ion scan mode. However, other scan modes are possible as well. These are summarized in Table 9.4-6. The parent-ion and neutral-loss scan modes are particularly useful in screening (see Sec. 9.4.4), while selective reaction monitoring (SRM)

Table 9.4-6. Scan modes in tandem mass spectrometry

Mode	MS1	MS2	Application
Product-ion	Selecting	Scanning	To obtain structural information from ions produced in the ion source
Precursor-ion	Scanning	Selecting	To monitor compounds which in CID give an identical fragment
Neutral-loss	Scanning	Scanning	MS1 and MS2 are scanning at a fixed m/z difference: to monitor compounds that lose a common neutral species
Selective reaction monitoring (SRM)	Selecting	Selecting	To monitor a specific CID reaction

is important in quantitative analysis. It must be realized that the use of MS-MS, and especially SRM, is extremely useful in quantitative environmental and bio-analysis, where interfering sample matrix components may adversely influence the achievable detection limits. The monitoring of a special collision-induced reaction by SRM significantly improves the selectivity, resulting in greatly improved signal-to-noise ratios.

In acquiring product-ion mass spectra from precursors produced by soft ionization methods, the fragmentation mechanisms significantly differ from those known from EI, because even-electron species rather than odd-electron radical cations are fragmented. The fragmentation reactions of the even-electron species are not as extensively studied as those in EI.

9.4.3 Analytical performance

General aspects

In the previous section the various tools available to the mass spectrometrists for solving analytical problems have been outlined and discussed. Considering the wide variety of inlet techniques, ionization methods, and mass analyzers, the selection of an appropriate combination of these three elements for a particular application requires considerable attention, as does the proper operation of the instrument. In this section, important aspects related to the analytical performance and practical applications of MS in both qualitative and quantitative analysis are discussed.

Calibration

An important aspect related to the performance of a mass spectrometer is the calibration of the m/z axis. The mass of a compound as determined by MS is based on the calibration of the instrument with a reference compound which provides ions of known m/z values within the mass range of the measurement. The two most widely-used reference compounds for EI and CI are perfluorokerosene (PFK, $CF_3-(CF_2)_n-CF_3$) and heptacosafluoro-tributylamine (PFTBA, $(C_4F_9)_3N$). For desorption and other soft ionization methods the calibration is less standardized. Cesium iodide is ofen used in FAB, ammonium acetate clusters or polypropylene glycol in thermospray, while the latter as well as solutions of myoglobin and a smaller peptide are often used in electrospray. It is important for a calibration compound to provide a series of good intensity peaks in the m/z range of interest under the ionization conditions available. The calibration of the mass scale obviously is even more important in high-resolution measurements. While in most low- and medium-resolution applications external calibration is applied, internal calibration is required for successful high-resolution measurements. The latter means that the calibrant is introduced into the ion source simultaneously with the compound of interest. Peak matching is applied to determine the accurate mass of the unknown.

Prior to its use, the mass spectrometer must be calibrated.

Selection of technology

In general, it can be stated that the most easy solution to a problem should be preferred, e.g., GC-MS with EI on a quadrupole instrument. When the analyte volatility or stability is insufficient for GC-MS, either the application of analyte derivatization strategies or the use of soft ionization methods, when necessary in combination with off-line or on-line liquid chromatography, can be considered. When the analysis is hindered by interfering sample constituents, one should consider to either improve the chromatographic separation, or to improve the mass spectral selectivity by measuring at higher resolution or with MS-MS. High-resolution measurements are obviously also performed when accurate mass determination is required.

Dioxin analysis

The general approach outlined above can be readily illustrated by outlining strategies for the quantitative determination of polychlorodibenzodioxins (PCDD's) (Fig. 9.4-10). PCDD's are amongst the most toxic compounds known. They are, for instance, generated as an unwanted by-product in waste incineration processes, from which they may be expelled in the environment where they accumulate in the fat tissue of the animals used for food, Here, we focus our attention on the tetra-chlorodibenzodioxins (TCDD's) with a molecular mass of 320 Da.

In a straightforward GC-MS analysis these compounds show few problems since, due to their inherent stability, relatively little fragmentation is observed (see Fig. 9.4-10). However, especially considering the economic effects resulting from the detection of TCDD's in milk or animal tissues, highly reliable measurements are needed, while, due to the toxicity of the compounds, concentrations in the ppt-range should be measured.

Fortunately, the mass spectrometric analysis is somewhat facilitated due to the 4 chlorine atoms providing a distinct isotopic pattern with relative abundances of ca. $3:4:2$ at $m/z = 320$, 322, and 324 respectively. Selective ion monitoring (SIM) at these three ions is used. By checking whether the relative intensities of the three ion signals are within preset limits of the theoretical intensity values as well as coelute within a preset retention-time window more reliable result can be obtained. The use of isotopically-labelled internal standards further improve the validity of the results.

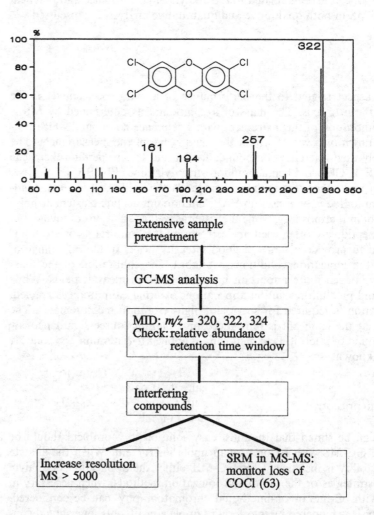

Fig. 9.4-10. Analytical strategies in the quantitative analysis of polychlorodibenzodioxins (PCDD's)

However, as TCDD's have to be analyzed at extremely low levels, the ion signals are often obscured by the signals of interfering compounds. The three possible solutions to this problem are:

- Improve the selectivity of the sample pretreatment and/or the gas chromatographic separation.
- Improve the mass spectral selectivity by increasing the resolution.
- Improve the selectivity by monitoring specific reactions in the gas phase occurring or induced in the MS-MS mode.

The first approach generally does not result in sufficient improvement. The second approach is rather powerful, since the TCDD's, having an exact mass of 319.8965, show a negative mass defect, while most compounds containing the atoms C, H, N, and O, i.e., the vast majority of possible interfering natural compounds, show a positive mass defect. Increasing the resolution to ca. 5000 results is significant improvement, while a resolution exceeding 10000 is needed to eliminate other chlorinated interferences such as chlorinated biphenyls and DDT. The selectivity can also be improved by using SRM in the MS-MS mode. In the product-ion mode a characteristic loss of 63 (COCl) is observed, enabling SRM using the ions at m/z 320 and 322 as the precursor ions selected in MS-1, and the ions at m/z 257 and 259 as the product ions transmitted in MS-2.

In practice, a combination of strategies like these is required to analyze the TCDD's at the required levels, e.g., extensive sample pretreatment, high-resolution gas chromatography, and multiple-reaction monitoring with an increased front-end resolution on a hybrid or 4-sector instrument.

In these types of applications the MS is mainly used as a highly sensitive and selective chromatography detector for a known compound. In the following section attention is paid to the type of information obtained by the MS and the way this information can be applied in qualitative and quantitative analysis.

Qualitative analysis

In qualitative analysis a wide variety of questions follows one after another: what is the molecular mass, the elemental composition, and the structure of the compound of interest.

Determination of molecular mass

Perhaps the most valuable piece of information on a compound that an MS can provide is the relative molecular mass M_r. Generally, the starting point is the EI spectrum, provided the sample is pure and the compound sufficiently volatile. The isotope cluster at the highest m/z value in the mass spectrum is likely to correspond to the molecular ion region. The assignment can be checked with the nearest fragment ions. Peaks at M-1, M-2, M-15, and M-18 tend to conform to the assignment, while losses of 3–14 amu or 21–25 amu generally indicate either the presence of impurities or an incorrect assignment. Other experiments, involving EI spectra at lower electron energies or soft ionization methods can be performed to further confirm the assignment. However, the selection of the soft ionization methods must be done with some care since these methods show considerable selectivity.

Many nonpolar compounds that cannot readily accommodate a proton, e.g., polynuclear aromatic compounds, are virtually transparent to soft ionization methods. When the application of soft ionization methods is successful, the presence of other peaks in addition to the protonated molecule, e.g., at $m/z + 17$, $+22$, and/or $+38$, due to ammoniated, sodiated, and potassiated molecules, respectively, are often helpful in assigning the correct molecular mass, as are methanol or acetonitrile adducts in thermospray LC-MS applications and glycerol adducts in FAB. Furthermore, the use of both the positive-ion mode, resulting in the m/z value of the $[M + H]^+$ ion, and the negative-ion mode, resulting in the m/z value of the $[M - H]^-$ ion, also helps in unambiguously assigning the molecular mass. In some cases, special additives, e.g., deuterated compounds to the CI reagent gas, are used to ascertain the molecular mass assignment.

In principle, assigning the molecular mass is a difficult task. In practice, however, knowledge on the nature of the compound, its physicochemical prop-

Measurement of molecular mass is the most important feature of a mass spectrometer.

erties and experience in MS in most cases lead to unambiguous molecular mass assignments.

Determination of elemental composition

In principle, there are two methods for deriving the elemental composition of an unknown from mass spectral data: (1) accurate mass measurement, and (2) isotope cluster abundance calculations.

With the use of a high-resolution mass spectrometer, measurement of the molecular mass can be made to an accuracy of 10^{-3} Da (1 mmu) or better, using an internal standard of known mass. However, even with this accuracy, a unique fit of one elemental composition is seldom obtained. The number of possible compositions increases with the increasing number of elements expected to be present and with increasing molecular mass. However, the number of reasonable elemental compositions in such a list generally is relatively small. This is illustrated with an example in Table 9.4-7: only 4 out of 15 elemental compositions for 126.05 ± 0.05 lead to reasonable structures. However, various isomers are possible in most cases. Discrimination between these isomers requires interpretation of the complete mass spectrum and possibly other techniques like NMR.

Two additional tools provide further help in the selection from the possible compositions: the nitrogen rule and the double-bond equivalent.

The *nitrogen rule* states that with an even molecular mass of a molecule the number of nitrogen atoms in the molecule is zero or even, while an odd molecular mass indicates an odd number of nitrogen atoms.

The *double-bond equivalent* (DBE) for a compound containing H, C, N, O, F, Cl, Br, and/or I can be calculated from the elemental composition using:

$$DBE = 1 + C - \tfrac{1}{2}(H + F + Cl + Br + I) + \tfrac{1}{2}N \qquad (9.4\text{-}22)$$

The DBE must have an integer value for a molecule. The DBE of a compound or an odd-electron ion has an integer value while the DBE of an even-electron ion is the true value followed by $\tfrac{1}{2}$. The DBE indicates the number of double bonds and/or rings in the compound.

Worked example:

The DBE of TCDD can be calculated to be:

$$DBE = 1 + 12 - \tfrac{1}{2}(4 + 0 + 4 + 0) + 0 = 9$$

This represents 3 rings and 2×3 double bonds in the benzene nuclei.

Table 9.4-7. Elemental composition (C, H, N, and O) for a molecular ion with $m/z = 126.05 \pm 0.05$, indicating exact mass, double-bond equivalent (DBE), and whether given composition is reasonable

Exact mass	C	H	N	O	DBE	Structure
126.0065	4	2	2	3	5	no
126.0106	9	2	0	1	9	no
126.0164	2	6	0	6	0	no
126.0178	3	2	4	2	5	no
126.0218	8	2	2	0	9	no
126.0290	2	2	6	1	5	no
126.0317	6	6	0	3	4	yes
126.0402	1	2	8	0	5	no
126.0429	5	6	2	2	4	yes
126.0470	10	6	0	0	8	no
126.0542	4	6	4	1	4	no
126.0654	3	6	6	0	4	yes
126.0681	7	10	0	2	3	yes
126.0793	6	10	2	1	3	?
126.0905	5	10	4	0	3	?

To some extent, the elemental composition can also be derived by analyzing the isotopic clusters. The presence of the so-called $(A + 2)$ elements like Cl, Br, and S is readily recognized from the $[M + 2]^+$ peak in the isotopic cluster. Monoisotopic or (A) elements like F, P, and I can be recognized from the fact that the $[M + 1]^+$ ion abundance is lower than can be explained by C, H, N, and O. Si also shows a readily recognizable isotope pattern. The maximum number of carbon atoms in the molecule can be estimated from the $[M + 1]^+/[M]^+$ percentage ratio by division by 1.1. Along these lines, the elemental composition of a peak can be calculated (see the above worked example).

Structure elucidation

The next step in the qualitative analysis after the determination of the molecular mass and the elemental composition is the interpretation of the fragment ion peaks in the mass spectrum to achieve structure elucidation. The interpretation of mass spectra requires a lot of practice and cannot be taught in a short text like this chapter. Only some basic aspects of interpretation are covered here, while a more extensive treatment can be found elsewhere.

The fragmentation of the molecular ion is a complex network of competing and consecutive reactions, the yields of which are determined by the stability of both the precursor and the product ions. The mass spectrum reflects the results of this competition. The stability and thus intensity of the molecular ion peak decreases in the series:

aromates > conjugated alkenes > cyclic compounds
> organic sulphides > short n-alkanes > mercaptans

ketones > amines > esters > ethers
> carboxylic acids, aldehydes and amides

As indicated by Eqs. 9.4-3 and 9.4-4, the fragmentation of the molecular ion may involve either the loss of a radical, thus resulting in an even-electron fragment ion:

$$\text{odd}^{+\bullet} \rightarrow \text{even}^+ + \text{radical}^\bullet \tag{9.4-23}$$

or the loss of a neutral, thus resulting in an odd-electron fragment ion:

$$\text{odd}^{+\bullet} \rightarrow \text{odd}^{+\bullet} + \text{neutral} \tag{9.4-24}$$

The former are generally more stable. The odd-electron fragment may eliminate either a radical or a neutral, while the even-electron fragment can only eliminate a neutral but not a radical unless a particularly stable odd-electron fragment ion results. As a result of the nitrogen rule, it can be stated that an odd-electron ion containing an even number of nitrogen atoms has an even m/z value.

Some important aspects that determine the ion stability are:

(a) the preference to maintain an octet of electrons, e.g., an R_4N^+ ion is more stable than an R_3C^+ ion,
(b) the sharing of an electron involving a nonbonding orbital of a heteroatom,
(c) resonance stabilization or delocalization, and
(d) charge localization on a favourable site, e.g., determined by electronegativity.

Upon fragmentation of an odd-electron ion, $AB^{+\bullet}$, two competitive pathways are possible, leaving the charge on either A or B. *Stevenson's rule* states that in such a simple bond cleavage, the fragment with the lowest ionization energy will preferentially take the charge. From Stevenson's rule some general *fragmentation rules* with respect to *simple bond cleavages* can be deduced:

The fragment ions observed may be used in structure elucidation, since they proceed from structure-specific cleavages.

- The molecular ion intensity decreases with increasing branching in the molecule.
- The molecular ion intensity in a homologous series decreases with increasing mass, with fatty acids as an exception.

- Cleavage at alkyl-substituted C-atoms is favourable due to the stability of the resulting carbenium ions:

$$R_3C^+ > R_2CH^+ > RCH_2^+ > CH_3^+$$

The largest substituent leaves most easily as a radical.
- Double bonds and (aromatic) rings tend to stabilize the molecular ion.
- Saturated rings lose their side chains at the α-position, with the charge remaining at the ring. Unsaturated rings undergo retro-Diels-Alder reactions:

- Two types of cleavage reactions occur in molecules with heteroatoms, double bonds, and aromatic rings:
  1. In the homolytic cleavages, fragmentation takes place one bond away from where the radical is localized, resulting in a carbenium ion with a delocalized charge. At double bonds allylic cleavage reactions occur, leading to a resonance-stabilized allyl carbenium ion.

$$H_2\overset{+}{C}-CH=CH_2 \leftrightarrow H_2C=CH-\overset{+}{C}H_2$$

In alkyl-substituted aromates, α-benzyl cleavage occurs, resulting in the resonance-stabilized benzyl or tropylium ion ($C_7H_7^+$) (a seven-membered ring):

  2. In the heterolytic cleavages, the bond adjacent to the charge, localized at the heteroatom, is cleaved, e.g., the C–C bond next to a heteroatom (N, O, etc.). The charge remains on the fragment containing the heteroatom due to ter-sonance stabilization via the nonbonding electrons.

$$R\!-\!\overset{+\cdot}{Y}\!-\!R \xrightarrow{i} R^+ + {}^{\cdot}YR$$

where Y can be O, N, S, or halogen. The cleavage at the α-position relative to a carbonyl group is another example of this. It results in stable acylium cations:

  3. Cleavage reactions often are accompanied by the elimination of small, stable, neutral molecules, such as CO_2, olefins, water, ammonia, etc.

Besides simple cleavage reactions, *rearrangement reactions* occur as well. Rearrangement becomes more important when the internal energy of the ions decreases. Low energy and slower reactions are involved. In 70 eV EI, only the simpler rearrangements need to be considered, involving the elimination of a stable molecule via a cyclic transition state. An example of this is the *McLafferty rearrangement*:

where Y can be H, R, OH, OR, or NR_2. A characteristic feature of this rearrangement reaction is the formation of an even-mass fragment ion from an even-mass molecular ion, which is in contrast to the simple cleavages of an even-mass molecular ion which results in the formation of an odd-mass fragment ion.

In interpreting the mass spectrum, the attention is first focused to the more important peaks, i.e., the peaks with higher intensity and/or higher m/z value. In this respect it may also be helpful to use Table 9.4-8 which summarizes some characteristic ions and Table 9.4-9 which summarizes some characteristic neutral losses observed in EI mass spectra.

Learning the interpretation of a mass spectrum can only be done by practice, as multiple strategies are applied, depending on the information available. Some approaches are briefly outlined in the worked examples below (cf. the mass spectra in Fig. 9.4-11).

Table 9.4-8. Some characteristic fragment ions in EI mass spectra

m/z	Ion	Functional group
15	CH_3^+	Methyl, alkane
29	$C_2H_5^+$ or HCO^+	Alkane, aldehyde
30	$CH_2=NH_2^+$	Amine
31	$CH_2=OH^+$	Ether or alcohol
43	$C_3H_7^+$ or CH_3CO^+	Alkane, ketone
45	CO_2H^+ or CHS^+	Carboxylic acid, thiophene
50, 51	$C_4H_2^{+\cdot}$, $C_4H_3^+$	Aryl
77	$C_6H_5^+$	Phenyl
83	$C_6H_{11}^+$	Cyclohexyl
91	$C_7H_7^+$	Benzyl (tropylium ion)
105	$C_6H_5C_2H_4^+$	Substituted benzene
	$CH_3-C_6H_4-CH_2^+$	Disubstituted benzene
	$C_6H_5CO^+$	Benzoyl

Table 9.4-9. Some characteristic neutral loss in EI mass spectra

Mass loss	Composition	Functional group
15	CH_3	Methyl
16	CH_4	Methyl
	NH_2	Amide
	O	Nitrogen oxide
17	OH	Acid, tert. alcohol
18	H_2O	Alcohol, aldehyde
19	F	Fluoride
20	HF	Fluoride
26	C_2H_2	Aromatic
27	HCN	Nitrile, heteroaromatic
28	CO	Phenol
	C_2H_4	Ether
	N_2	Azo
29	C_2H_5	Alkyl
30	CH_2O	Methoxy
	NO	Aromatic nitro
31	CH_3O	Methoxy
32	CH_3OH	Methyl ester
33	$H_2O + CH_3$	Alcohol
	HS	Mercaptan
35	Cl	Chloro compounds
36	HCl	Chloro compounds
42	CH_2CO	Acetate
43	C_3H_7	Propyl
44	CO_2	Anhydride
46	NO_2	Aromatic nitro
50	CF_2	Fluoride

Worked example:

In the first mass spectrum (cf. Fig. 9.4-11a), the peak at $m/z = 134$ is most likely the molecular ion. Two major fragments are observed at $m/z = 43$ and 91 which sum up to 134. According to Table 9.4-8, these fragments may correspond to an acetyl and a tropylium ion, respectively, leading to the structure proposal below. The structure can further be checked using the isotope peaks. No significant isotopic contributions other than from C are observed. The $m/z = 135$ to $m/z = 134$ ratio of 9% is in agreement with the proposed structure, leading to an elemental composition of $C_9H_{10}O$ and a DBE of 5 due to four double bonds and one ring.

$$\text{◯} - CH_2 - \overset{O}{\underset{\|}{C}} - CH_3$$

Worked example:

In the second example (cf. Fig. 9.4-11b), the peak at $m/z = 72$ is most likely due to the molecular ion. Major losses of 17, 27, and 45 to $m/z = 55$, 45, and 27, respectively, are in agreement with this. The ratio of $m/z = 74$ and $m/z = 72$ is small, thus excluding $(A + 2)$ elements other than O. From the ratio of $m/z = 73$ and $m/z = 72$, which is 4.2%, the presence of 3 or 4 carbon atoms is predicted. From the 0.7% ratio of $m/z = 74$ and $m/z = 72$ the presence of 1 or 2 oxygen atoms is expected. Thus, possible elemental compositions are $C_3H_4O_2$ and C_4H_8O with DBE values of 2 and 1, respectively. Possible structures which account for all this are given below. With the aldehyde at the right, a large peak at $m/z = 29$ would be expected, while the loss of 17, due to OH, is also not expected from the aldehyde. Thus, the 2-propenoic acid is the most likely structure.

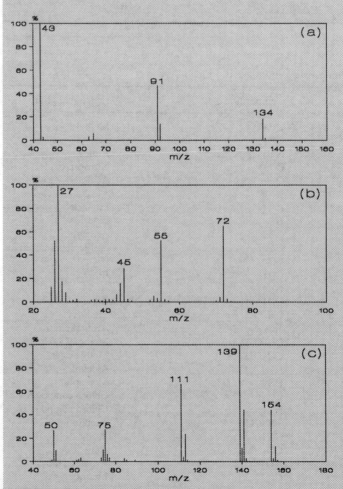

Fig. 9.4-11. Mass spectra for the interpretation examples

$$CH_2=CH-\overset{\displaystyle O}{\overset{\|}{C}}-OH \qquad H-\overset{\displaystyle O}{\overset{\|}{C}}-\!\!\!\!\!\diagdown\!\!\!\!\!\diagup\!\!-CH_3$$

Worked example:

The $m/z = 154$ peak in the third example (cf. Fig. 9.4-11c) is a good candidate for the molecular ion. The losses of 15 (due to CH_3) and 43 (due to CH_3CO or C_3H_7) give $m/z = 139$ and 111, respectively, confirm this assignment. The presence of the isotope peak at $m/z = M + 2$ is most likely due to a Cl atom (3:1 peak ratio). The Cl atom is, as expected, retained in the fragments at $m/z = 139$ and 111. The 9% $m/z = 154$ to $m/z = 155$ ratio corresponds to 8 C atoms. The relatively high intensity of the peaks at higher m/z values indicates possible aromatic character. With $1 \times$ Cl and $8 \times$ C 131 out of the 154 Da have been explained. The difference of 23 cannot be explained by H atoms only, but indicates the presence of an O atom. The elemental composition thus is C_8H_7OCl. Combining the information gathered results in the structure proposal below. The 4-position of the Cl atom on the ring obviously cannot be determined from the mass spectrum; other means, e.g., NMR, were needed for that.

$$Cl-\!\!\!\bigcirc\!\!\!-\overset{\displaystyle O}{\overset{\|}{C}}-CH_3$$

Worked example:

The final example is the EI spectrum of diuron shown in Fig. 9.4-2a. the isotope cluster at $m/z = 232$ is attributed to the molecular ion. the 10:7 ratio of $m/z = 232$ and $m/z = 234$ points to a dichloro character. Fragment peaks at $m/z = 187, 161,$ and 124 still contain two chlorine atoms. Accurate mass determination at 5000 resolution results in 203.021 Da leading to an elemental composition $C_9H_{10}Cl_2N_2O$ (calculated mass 203.017 Da, $\Delta = 4$ mmu). A computer library search identifies the compound as diuron, while again the chlorine substitution is uncertain. Some major fragment peaks further confirm the identification. The $m/z = 161$ fragment can be attributed to the dichloroaniline ion, the $m/z = 187$ peak to dichlorophenylisocyanate, while the $m/z = 72$ peak is due to $(CH_3)_2N=C=O^+$.

Fragmentation of even-electron ions

Considering the indicated importance of soft ionization methods and the use of these in combination with MS-MS, some attention must be paid to the fragmentation of protonated molecules. In general, it can be stated that this type of fragmentation is not as extensively and systematically investigated as the EI fragmentation. General rules follow from the treatment of EI fragmentation, e.g., with respect to the stability of ions. However, while in EI the odd-electron molecular ion may fragment to form either even-electron or odd-electron fragment ions (Eqs. 9.4-21 and 9.4-22), the even-electron protonated molecule generally fragments by the loss of a neutral rather than of a radical:

$$even^+ \rightarrow even^+ + neutral \tag{9.4-25}$$

Rearrangements and *hydrogen shifts* are observed more commonly than in EI. Some examples are given in Sec. 9.4.4.

Library searching

A powerful tool in structure elucidation with EI mass spectra is the use of mass spectral libraries. Most of them are on-line computer searchable. The information of a measured spectrum is reduced to a small number of the most significant peaks and then compared with the library spectra. The agreement between the measured

Mass spectra, obtained by electron ionization, may be searched against spectral libraries for analyte identification.

spectrum and the library entries is expressed in a number, usually between 0 and 1000, with 1000 representing a perfect fit. The 10 best fitting library entries are then displayed for further visual inspection by the operator.

The computerized library searching is very useful as it provides ideas in which direction to search when a completely unknown must be identified, or provides adequate confirmation when the presence of a compound is to be confirmed. However, the most frequently-used libraries contain only 20000 to 50000 mass spectra while over 12000000 compounds are known. Therefore, the results of the computer library search should not be taken for granted. In performing computerized library searching, the appropriate cast should be kept in mind: while the computer is used for rapidly matching a measured mass spectrum against the library spectra, the final decision on identity must be made by the operator after a thorough examination of the data and the computer output.

While library searching in EI is relatively powerful, also because of the reproducibility of EI spectra from day-to-day and from instrument-to-instrument, the situation is quite different for soft ionization methods and CID in MS-MS. In those cases, the mass spectra are so much determined by the experimental conditions, that the acquisition of 'universal' libraries is not possible. Starting a library within one institute or company is sometimes successful.

Quantitative analysis

Mass spectrometry also is a powerful tool in quantitative analysis.

In quantitative analysis by mass spectrometry, chromatographic inlets are used in most cases. The MS is operated either in full-scan acquisition or, and more often, in *selective ion monitoring* (SIM) mode (see Sec. 9.42). The advantage of SIM obviously is the improved signal-to-noise ratio due to the longer acquisition time at the ions of interest. However, SIM also includes a reduction of the information in the mass spectrum. Therefore, regulatory organizations often demand full-scan acquisition. A useful compromise is the use of several specific ions per component in an SIM procedure. Then, the compound of interest is only identified when the relative intensities of the selected peaks are within preset limits and the selected peaks maximize within a preset time window. When a soft ionization method has to be used in quantitative analysis, e.g., in LC-MS applications, combination with MS-MS is required since due to the absence of fragmentation only one specific ion is present for identification. By monitoring a specific reaction in MS-MS more reliability is achieved.

Like in other quantitative analysis procedures the use of an *internal standard* is highly recommended, to account not only for fluctuations during the sample pretreatment procedures, but also for fluctuations in ionization yield. In MS, the almost ideal internal standard can be selected. Since an internal standard should not be present in the sample and have a physicochemical behaviour similar to the compound of interest, the use of an isotopically labelled compound meets these requirements the most. MS is capable of differentiating between the natural and the labelled compound. In GC-MS analysis, the isotopically labelled internal standard generally shows a slightly shorter analysis time.

An important figure of merit of a quantitative method is the detection limit or minimum detectable quantity. For GC-MS, figures like 1 pg/s (the MS is a mass-flow sensitive detector) are cited. Modern quadrupole mass spectrometers are specified to provide GC-MS detection of 200 pg methyl stearate in EI and 100 pg of benzophenone in CI with a signal-to-noise ratio of 30. Double-focusing sector instruments are specified to provide signal-to-noise ratios of 200 in both EI and CI for 100 pg methyl stearate in GC-MS and the detection of only 30 fg of 2, 3, 7, 8-TCDD with a signal-to-noise ratio better than 10. However, considering the chemistry involved in the ionization processes in EI and especially in soft ionization methods, the response is highly compound dependent. Furthermore, these figures indicate little about what limits can be reached in real analysis. In real analysis, the detection limits are primarily determined by the so-called chemical noise rather than by the electronic noise of the detector and amplification circuit. The success of a method in real analysis is completely determined by the optimum mutual tuning of the various building blocks: sample pretreatment and separation, ionization, mass analysis, detection, and data handling. Furthermore, in such situation, concentration rather than absolute mass detection limits are of concern.

9.4.4 Applications

The aim of this section is to provide a brief overview of some fields of application of MS and to highlight with the use of examples some of the topics discussed in previous sections, e.g., the use of MS-MS and soft ionization methods in detection, screening, quantitation, and structure elucidation.

Numerous successful applications of MS in many fields of science are available, e.g., in petrochemistry, environmental analysis, pharmaceutical analysis, bioanalysis, and food analysis. Initially, the application in petrochemistry provided a large impetus to the development and acceptance of MS and GC-MS as an analytical technique, especially around World War II. Subsequently, the wider applicability of MS was recognized, with important breakthroughs in environmental analysis (dioxin analysis) in the late seventies and in pharmaceutical and biochemical areas more recently, especially with the advent of LC-MS.

Confirmation of synthesis products

Perhaps the most important application of MS is in the field of identification and confirmation of synthetic compounds or components isolated from (natural) products or samples. Next to NMR spectrometry, MS plays a key role in structure confirmation and elucidation in the organic synthesis laboratories. The most significant advantage of MS over NMR is the small amount of compound required, i.e., nanograms in MS contrary to micrograms in NMR. Unfortunately, MS often does not provide a complete and final answer; NMR is still needed, e.g., in solving isomerism questions. A disadvantage of MS is that it is a destructive analytical technique: the sample used can not be recovered for further analysis or chemistry.

In structure confirmation of synthetic compounds mass spectrometry plays a major role, next to nuclear magnetic resonance spectrometry.

In the applications, questioning whether the synthetic product is the compound expected or planned, the sample introduction into the instrument is achieved either by a direct insertion probe or via a gas chromatograph. The latter has the advantage that relatively smaller sample amounts can be analyzed and that preseparation of the sample is achieved prior to MS analysis. By GC-MS the acquisition of the mass spectra of several components in a mixture during one analysis cycle, and/or the on-line separation of the component of interest from possibly interfering compounds can be achieved. Relatively simple, low-cost, and easy-to-operate benchtop quadrupole or ion-trap GC-MS instruments are nowadays available for this type of work. EI is the ionization technique of choice in this area, although the organic synthetic research is increasingly directed towards more polar and labile compounds, requiring different approaches. Identification or confirmation of compounds is achieved by library searching and/or interpretation of the mass spectra, as outlined in Sec. 9.4.3.

MS has also grown in importance in determining the elemental composition of organic synthetic products using accurate mass determination with a high-resolution double-sector instrument. Again, EI is the ionization method of choice. Sometimes, low electron energies, i.e., 10–20 eV instead of the usual 70 eV are needed to achieve sufficient signal for the molecular ion. The accurate mass determination is performed by a peak-matching procedure. The compound of interest is introduced into the instrument simultaneously with a suitable reference compound, e.g., perfluorokerosene when EI is used. Mass analysis is performed by scanning the accelerating voltage over a small range, keeping the magnetic field strength constant. The mass of the analyte is calculated by linear interpolation of its peak position in between two peak positions of the reference compound, the m/z values of which are known. The measured mass is compared to the calculated value for the compound analyzed. Subsequently, the possible and/or likely elemental compositions may be calculated from the measured accurate mass. This procedure is now preferred over more conventional, chemical means of determination of the elemental composition.

This procedure of identification using GC-MS or a direct insertion probe and EI ionization obviously is not always successful. In principle, its application is restricted to compounds having sufficient vapour pressure (volatility) and thermal stability. In this respect, the direct insertion probe shows a slightly wider applicability range than GC-MS. The applicability range of GC-MS can be enlarged by the use of volatility-enhancing derivatizations, as also used in conventional GC analysis (see Sec. 5.2). In MS, these derivatization reactions can have two aims. The prime aim is to enhance the volatility of the compound by 'shielding' the polar groups, i.e., the active protons in acids, amines, alcohols, and phenols are transformed into more inert groups by, for example, esterification of acidic groups, acetylation of amine groups, silylation. Furthermore, the ionization characteristics of the compound may be improved by derivatization, e.g., the incorporation of a pentafluorobenzyl group to achieve a higher response in electron-capture negative-ion CI. Along these lines, many nonvolatile and/or thermally-labile compounds can be made amenable to GC and GC-MS analysis, e.g., steroids, (amino) acids, sugars, and a wide variety of drugs. Obviously, the derivatization procedure affects the mass of the compound of interest. In general, the shift to a somewhat higher m/z value is advantageous, because less interfering compounds will be present at higher m/z values. However, in identification of unknowns, it must be realized that the derivatization may result in unexpected artefacts. Then, the use of soft ionization methods (Sec. 9.4.2) for determination of the molecular mass is recommended.

Isotope incorporation

A related area, where MS plays a major role, is in the determination of isotope incorporation, which is important in research where stable isotopic markers, e.g., ^2H, ^{13}C, ^{15}N, or ^{18}O, are used, for instance, to determine the fate of a compound in the body or to elucidate biosynthetic routes in plants or animals. MS is the standard technique for those measurements, principally enabling the determination of both the extent and site of isotope incorporation. Proper measurements require highly reproducible measurement conditions, a reasonable abundance of the ions to be studied, and the absence of interfering ions as well as of reactions that may obscure the extent of incorporation, e.g., proton/deuterium exchange. The $[M + H]^+$ and $[M - H]^-$ ions are renowned interferences, since the isotopic forms of these species have the same nominal mass as the isotopic forms of $M^{+\bullet}$.

The approach of the determination of the isotope incorporation involves obtaining the spectra of both the labelled and the unlabelled compound from multiple slow scans at low resolution. Comparison of the natural isotopic abundances, as found in the spectrum of the unlabelled compound, and the isotopic abundances in the labelled compound allows calculation of the extent of incorporation. When the $[M + H]^+$ and $[M - H]^+$ ions can be removed, accuracies of typically 1% in the isotopic incorporation can be achieved.

Structure elucidation with soft ionization methods

In structure elucidation with soft ionization methods, the use of tandem mass spectrometry plays a vital role.

An increasingly important application area of MS is concerned with the analysis of polar, nonvolatile, and/or thermally-labile compounds. When the derivatization procedures outlined above are not successful, and/or the type of application does not favour the sometimes elaborate derivatization, MS analysis can only be achieved by the use of soft ionization methods (Sec. 9.4.2). In terms of structure elucidation, the soft ionization methods have the disadvantage that, while the molecular mass can readily be determined, in general, no structural significant fragmentation is obtained. In that case, the soft ionization method must be combined with MS-MS (Sec. 9.4.2). Fragmentation of the even-electron species generated by soft ionization methods can be achieved by collision-induced dissociation (CID).

The combination of soft ionization methods and MS-MS is nowadays quite frequently applied, especially owing to the relatively simple operation of triple-quadrupole instruments. This approach is highly suitable for studies like the identification and quantitation of drugs and drug metabolites, and the screening of food or environmental samples for contaminants. This is briefly demonstrated by discussing a few examples, which also demonstrate how the analytical procedure is developed.

The example chosen is related to the fabrication of sausage. A method was developed to detect sulphadimidine (SDM), a common antibiotic and growth-stimulating compound used in pig farming, and some of its possible conversion products (SOH and DAS) in meat products especially sausage.

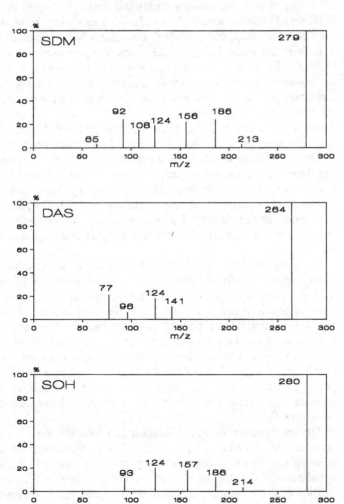

$$\text{SDM} \quad R = NH_2 \quad m/z = 279$$
$$\text{SOH} \quad R = OH \quad m/z = 280$$
$$\text{DAS} \quad R = H \quad m/z = 264$$

The compounds of interest are thermally-labile at the sulphonyl link $(-NH-SO_2-)$ but can readily be protonated in a soft ionization method such as thermospray ionization. No fragmentation is observed in the thermospray mass spectrum. In order to develop the analytical procedure, the product-ion mass spectra of SDM and its analogies are acquired (Fig. 9.4-12). An important and striking feature of the production mass spectra is the absence of isotope peaks. Because the fragment ions are produced in CID from a selected monoisotopic precursor ion, no isotope peaks are detected in the product-ion MS-MS spectrum.

The product-ion mass spectra to some extent can be used to elucidate the structural differences between the three compounds. The common peak at $m/z = 124$ corresponds to the protonated amino-dimethyl pyrimidine, being complementary to the peak due to the phenyl-SO_2^+ group, found at $m/z = 141$ for DAS,

Fig. 9.4-12. Product-ion mass spectra of sulfadimidine and two of its analogues obtained by thermospray LC-MS-MS. (Samples were provided by Dr. N. Haagsma and R. Smit, University Utrecht, The Netherlands)

at $m/z = 156$ for SDM, and at $m/z = 157$ for SOH, which indicates that the difference between the three compounds is located in the phenyl ring. The mass shift of 15 and 16 Da relative to DAS most likely corresponds to substitution of the proton by an amino and a hydroxy group respectively. The same mass shift is observed between the peaks at $m/z = 77$ in DAS, $m/z = 92$ in SDM, and $m/z = 93$ in SOH, also reflecting the differences in the phenyl ring. Two interesting peaks are found at $m/z = 213$ in SDM and $m/z = 214$ in SOH, corresponding to a loss of 66 Da. The peaks are due to a rearrangement of the ion in which H_2 and the SO_2 group are removed from the molecule. The corresponding analogue is absent in the DAS spectrum, because a stable resonance structure enabling the removal of the SO_2 in SOH and SDM cannot be achieved in DAS.

The aim of the study was the screening of food samples for the presence of SDM and its analogues. Inspection of the product-ion mass spectra shows that all three compounds show a common neutral loss of 123 u, corresponding to amino-dimethyl pyrimidine. According to Table 9.4-6, a common neutral loss can be applied in setting up a neutral-loss scan, in which MS1 and MS2 are both scanned at a fixed mass difference of 123 u. This provides a highly selective screening method for SDM-related compounds: only those compounds in the sample that show the neutral loss of 123 u are detected by the MS. When occasionally during screening of real samples an additional compound next to the expected SDM, SOH and DAS is detected, the product-ion mode can be used to further elucidate the structure of this compound; it most likely will be another SDM analogue.

Besides the neutral loss of 123 u, another common characteristic of the mass spectra is the detection of a peak at $m/z = 124$, which is due to the protonated amino-dimethyl-pyrimidine ring. According to Table 9.4-6, such a common fragment may be used in a precursor scan, in which MS1 is scanned and MS2 is selecting the common fragment ion ($m/z = 124$). Peaks will only be detected by the detector at the end of MS2 when MS1 passes a ion which upon CID results in a fragment ion at $m/z = 124$. Again, a highly selective screening method is obtained.

The situation in which either a precursor-ion scan or a neutral-loss scan involving the same fragmentation reaction can be used is relatively rare. In most cases, there is a stronger preference between the two parts of the molecule for accommodating the charge. After fragmentation by CID, the part of the molecule with the highest proton affinity or lowest ionization energy will be observed as an ion peak in the mass spectrum, while the complimentary part is lost as a neutral and thus not detected.

As indicated in Sec. 9.4.2, the detection limits in the analysis may be enhanced by the application of selective reaction monitoring (SRM) to detect only a few ions involved in the reaction without spending time on measuring nonrelevant parts of the mass range. It must be emphasized that the use of MS-MS screening methods, as described for SDM, provides improved detection limits. The MS-MS method results in a significant improvement in the selectivity of the method: only a specific reaction, characteristic of the compound class investigated, is monitored. Obviously, the specificity of the reaction selected in setting-up the procedure is of great importance for achieving optimum results; monitoring the loss of water, for instance, is generally not very specific since many other compounds show a loss of water in CID.

The combination of soft ionization and MS-MS, possibly in combination with on-line liquid chromatography, is also frequently used in the identification of drug metabolites. This is an important application area of MS, because during the early development of a drug only little metabolite material is available and the identification of drug metabolites is of utmost importance in pharmaceutical studies of a potential drug. While originally GC-MS was most often used in combination with extensive chemical derivatization procedures, in the last few years the application of LC-MS has become increasingly important in this field, especially in combination with MS-MS.

Identification of drug metabolites is not a search for complete unknowns. Modern computer-aided molecular modelling (CAMM) procedures provide good insight into the type of metabolites to be expected. The application of soft ionization methods is both important and useful in this field, because on the one hand labile compounds like glucuronide and sulphate conjugates have often to be analyzed, and on the other hand many common biotransformation routes like hydroxylation, conjugation, hydrolysis of esters and amides, and demethylation can readily be identified from molecular mass determination.

A nice example is the identification of metabolites of the H_1-antagonist Temealastine (relative molecular mass 441). Temelastine is extensively metabolized with Phase I hydroxylation and Phase II glucuronidation being the major routes. Prior to thermospray LC-MS analysis, the sample is treated with β-glucuronidase to convert the glucuronides into hydroxyls. Four isometric hydroxylated metabolites and an N-oxide are detected (cf. Table 9.4-10).

Two of these compounds were directly identified from the thermospray mass spectra: the N-oxide (#1) and the #2-hydroxy compound (reference compounds were available). Differentiation of the other three isomers can only be achieved by the use of LC-MS-MS. The product-ion mode was used, selecting the ^{81}Br-containing protonated molecule at $m/z = 460$ in MS1 and scanning MS2, while the isomers were separated by LC. The results obtained are summarized in Table 9.4-10. The method was subsequently applied to the analysis of temelastine metabolites in human faecal extracts.

While with thermospray LC-MS the routine analysis of drug conjugates was not possible, the advent of electrospray ionization enabled the analysis of intact glucuronide and sulphate conjugates. Sensitive and selective screening for glucuronides and sulphates during drug metabolism studies can be achieved by applying neutral loss scans with a loss of 176 u, owing to the cleavage of the C–O-glycosidic bond, and 80 u, due to the loss of SO_3, respectively.

Table 9.4-10. Elucidation of the Phase I metabolism of temelastine by thermospray LC-MS and LC-MS-MS (based on data from Blake and Beattie, *Biomed. Environ. Mass Spectrom.*, 18, 860, 1989)

	LC-MS data	LC-MS-MS data	Comment
N-oxide #1	$m/z = 458/460$ (weak) $m/z = 442/444$ $m/z = 242/244$	$m/z = 228$	Unambigious identification by LC-MS
Hydroxy #2	$m/z = 440/442/444$ $m/z = 337/339$	$m/z = 228$	Unambigious identification by LC-MS
Hydroxy #3	$m/z = 458/460$ $m/z = 442/444$ (weak)	$m/z = 228$	Unambigious identification by LC-MS-MS
Hydroxy #4	$m/z = 458/460$	$m/z = 224$	Unambigious identification by LC-MS-MS
Hydroxy #5	$m/z = 458/460$	$m/z = 224, 226$	Unambigious identification by LC-MS-MS

Characterization of biomacromolecules

Current research in analytical mass spectrometry concerns the characterization of biomacromolecules.

One of the exciting new fields of application of MS, which is developing at the moment, is the field of biochemistry; more specifically, the field of protein characterization. This is due to the recent advent of techniques like MALDI and electrospray which allow a rapid and accurate determination of the average protein molecular masses requiring only small amounts of material (pmol range or below). The average mass of the protein is determined, because a mass spectral resolution in excess of 10 000 would be required to separate the various isotope peaks present. In comparison with other more conventional, biochemical methods for molecular mass determination of biomacromolecules, such as SDS-PAGE and gel permeation chromatography, MS provides a fast and easy measurement using small amounts of material and providing superior accuracy. However, MS is a destructive method: the material used cannot be recovered for further experiments.

MALDI and electrospray differ considerably in the appearance of the information they offer. The MALDI spectrum is relatively simple as it contains an intense peak of the protein analyzed at the m/z value of a singly-charged ion (cf. Fig. 9.4-6b). Further, some less intense peaks owing to multiply-charged ions, like $[M + 2H]^{2+}$, and proton-bound dimers, i.e., $[2M + H]^+$, might be observed. With the use of appropriate (internal) standards, mass accuracies of ca. 0.1% are reached. The *electrospray* spectrum contains an envelope of multiply-charged peaks (cf. Fig. 9.4-6c), from which the molecular mass can be calculated. Assuming that adjacent peaks in the ion envelope differ by only one charge and that the charging is due to protonation, the relation between a multiply-charged peak at $m/z = p_1$ and the relative molecular mass M_r is:

$$p_1 z_i = M_r + m_H z_i \tag{9.4-26}$$

where m_H is the mass of the proton and z_i is the number of charges. The adjacent peak at the high m/z side of p_1 is described by:

$$p_2(z_i - 1) = M_r + m_H(z_i - 1) \tag{9.4-27}$$

This set of equations can be solved for z_i resulting in:

$$z_i = \frac{(p_2 - m_H)}{(p_2 - p_1)} \tag{9.4-28}$$

The molecular mass is subsequently calculated by substituting the integer value of the calculated z_i in Eq. 9.4-26. More accurate computer algorithms have been developed which mathematically transform the measured mass spectrum to a transformed spectrum containing one peak corresponding to the molecular mass of the protein. The determination of a protein molecular mass is important in itself, e.g., because the molecular mass can readily indicate the identity of a recombinant product.

Both MALDI and electrospray can easily be combined with enzymatic digestion of the proteins for further characterization. When a protein is thus digested the complete mixture can be deposited onto the MALDI target and analyzed. In favourable cases the mass of over 90% of the peptide fragments can be determined. This approach can be used to elucidate changes in a protein, e.g., for characterization of recombinant proteins, or to identify covalently-bound modifications in the protein. Electrospray MS, as it is readily combined with both LC-MS and MS-MS, can provide even further information on the amino acid sequence in a peptide. Upon CID a peptide will fragment into two complementary series of ions, which contain the amino acid sequences from the C- and N-terminals of the peptide. The electrospray MS-MS method is faster and more universally applicable than conventional biochemical methods like Edman degradation.

This type of application is possible for only a few years. Therefore, substantial progress and importance in this field is expected for the future. Considering the present speed of development in these areas, it was decided to pay some attention to these relatively recent developments, as it may be expected that this type of application will further grow in importance and play a prominent role in the future of MS.

References

[9.4-1] Karasek, F.W., Clement, R.E., *Gas Chromatography Mass Spectrometry, Principles and Techniques*. Amsterdam, The Netherlands: Elsevier Science Publishers, 1988.

[9.4-2] Niessen, W.M.A., van der Greef, J., *Liquid Chromatography – Mass Spectrometry*. New York, NY, USA: Marcel Dekker Inc., 1992.

[9.4-3] McLafferty, F.W., Turecek, F., *Interpretation of mass spectra*. 4th edition. Mill Valley, CA, USA: University Science Books, 1993.

[9.4-4] M.L. Gross, *Mass Spectrometry in the Biological Sciences: A Tutorial*. NATO ASI Series C, Vol. 353, Dordrecht, The Netherlands: Kluwer Academic Publishers, 1990.

[9.4-5] Chapman, J.R., *Practical Organic Mass Spectrometry*. 2nd edition, London, UK: Wiley & Sons, 1993.

Questions and problems

1. Explain the differences between the average, the nominal, and the exact molecular mass of a molecule.
2. Calculate the average, the nominal, and the exact molecular mass of (a) propane, (b) 2-propanol, (c) 1-chloropropane, and (d) 1-chloro,2-bromopropane.
3. Calculate the mass spectrometric resolution required to resolve the following pairs of ions:
   (a) $C_{10}H_8N^+$ and $C_{11}H_{10}^+$ with mass 142.06567 and 142.07825, resp.
   (b) an ion with mass 17 497 and an ion with mass 17 498.
   (c) $^{116}Sn^+$ (mass 115.90219) and $^{232}Th^{2+}$ (mass 232.03800).
   (d) $C_3H_7N_3^+$ (mass 85.0641) and $C_5H_9O^+$ (mass 85.0653).
4. Explain the differences between hard ionization methods such as electron impact, and soft ionization methods.
5. Discuss the advantages and limitations of selective ion monitoring in qualitative and quantitative analysis.
6. Explain why primarily $[NH_4]^+$ ions and no $[CH_3OH_2]^+$ ions are detected in a chemical ionization reagent gas generated from a mixture of methanol vapour and ammonia.
7. Explain the use of glycerol as a matrix in fast-atom bombardment mass spectrometry.
8. Derive Eq. (9.4-19), describing the relation between the flight time, the acceleration potential, and the mass and charge of the ion.
9. Discuss the advantages and limitations of sector relative to quadrupole instruments.
10. Explain why no isotope peaks are observed in product-ion MS-MS spectra.
11. Collision induced dissociation of protonated phthalates affords the protonated anhydride at m/z 149. Select an MS-MS scan mode to selectively screen a sample for the presence of phthalates.
12. In which way discrimination of tetrachlorodibenzodioxin and hydroxytetrachlorodibenzofuran, both having a molecular formula of $C_{12}H_4O_2Cl_4$ with $M_r = 321.8936$, is possible in GC-MS?
13. In the thermospray mass spectrum of a certain compound peaks are detected at $m/z = 369$, 386 and 391. What is the molecular mass of the compound?
14. Calculate the double-bond equivalent for diuron (see Fig. 9.4-2a).

10 Microbeam and Surface Analysis

Learning objectives

■ To introduce the main methods of surface and interface analysis: photon probe techniques, electron probe techniques, ion probe techniques, field probe techniques, and scanning probe techniques

■ To describe single techniques with special emphasis on their analytical figures of merit

■ To describe different analytical applications in order to show the scope and limitations of the techniques used for solving specific analytical problems in different fields of research and technology

The report of the "Committee to Survey the Development of Chemical Sciences" of the US National Academy of Sciences [10-1] states that, "surface science is one of the most rapidly growing areas of physical sciences at this time". Thus, the analytical characterization of surfaces and interfaces ("inner surfaces") has become an important task for modern analytical chemistry.

The significance of surface and interface analysis is based on their potential to gain information about important fundamental chemical processes like corrosion, adsorption, chemisorption, reactivity, oxidation, passivation, diffusion, or segregation, and to provide a substantial input for the development of processes, materials, and devices for high technology.

Surface and interface analyses are most important for materials science, especially for the development and production of catalysts, semiconductors and microelectronic devices, metals, ceramics, glass, thin film structures, polymers, and composites.

From the perspective of materials science, surface analysis covers the characterization of surface layers at a thickness from less than 1 atomic layer up to several micrometers (this latter task often being called "thin film analysis").

For practical reasons, it is convenient to distinguish between a physical surface – the outermost layer(s) of a material whose thermodynamic properties are different from the bulk; and a technical surface – the outer zones of a material with different properties than the inner domains of the material (Fig. 10-1).

The significance of the physical surface is based on the fact that the bonding forces of the atoms in the outermost atomic layer are not saturated – thus, these atoms experience structural reorganization (relaxation, reconstruction) and a high chemical reactivity. The physical surface determines the surface reactivity of materials.

Surface reactions, like corrosive processes, also proceed into the material, often to a large depth. Defined material structures extending well beyond the dimensions of a physical surface are generated to achieve specific mechanical, optical, or electrical properties. Thus, analytical characterization must include not only the outermost atomic layer but also "near-surface" zones of a much larger thickness.

Not only do surfaces have different thermodynamic properties than the bulk material, but they also have different interfaces, which are defined as boundaries between two condensed phases [10-2]. In materials science these boundaries are mostly between two solid phases, and sometimes, as in electrochemistry, between a solid and a liquid phase.

Interfaces are "inner surfaces" which define many material properties like corrosion potential and mechanical strength. The characterization of interfaces is intimately connected with surface analysis. To some extent the methods used in interface analysis are the same as those applied in surface analysis. But, in addition, high (lateral) resolution techniques are needed (micro and nanoprobes) which use the same or similar processes for generating analytical signals as in surface analysis. Thus, surface and interface analyses are treated together in this chapter.

In selecting the methods for surface and interface analysis we have to consider

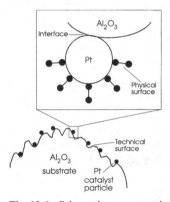

Fig. 10-1. Schematic representation of surface and interface

Surfaces and interfaces strongly determine the properties of materials.

not only the type of information sought, but also the specific constraints given by the small spatial domains to be characterized. Generally, these features of a surface or an interface determine the properties of a material and must therefore be characterized:

- topology, morphology
- elemental composition
- chemical bonding of elements
- structure (geometric and electronic)

This information must often be obtained from small spatial domains: one atomic layer in the characterization of physical surfaces, nanometer extensions in interface analysis, and nanometer lateral and depth resolution in the analysis of technical surfaces (thin films). Consequently, the number of atoms or molecules to be characterized is small: one atomic layer (AL) contains atoms in the order of 10^{15} cm^{-2}. With an analytical area of, e.g., 1 μm^2 the total number of atoms or molecules available for analysis is only 10^7. Furthermore, we have to consider that trace components often have to be analyzed at a surface: e.g., for the analysis of 10 ppm of a component in 1 AL with 1 μm^2 lateral resolution only about 100 atoms or molecules are available.

High requirements of surface and interface analysis can be met only by high-performance physical techniques.

These requirements – high information content, high spatial resolution, and high absolute and relative detection power – can only be met by physical techniques based on the interaction of photons, electrons, ions, and electrical fields with the material investigated. Chemical methods play only a minor role – for example, in the etching of surfaces.

The interaction of these physical reagents generates a variety of analytical signals from which *chemical* information about surfaces and interfaces can be gained.

In order to obtain relevant chemical information for small spatial domains and even lower concentrations, a substantial number of different techniques is needed due to the specific potential and limitations of each particular method.

Due to very different requirements a large number of different techniques with different figures of merit is needed.

Presently more than 30 methods are actually in use for surface and interface analysis. Among these about 15 can be considered as major techniques. The most important methods being applied on a large scale in industry are XPS, AES, SIMS, and RBS for compositional surface analysis, SEM for characterizing surface morphology, AEM for characterization of interfaces, and IR and Raman Spectroscopy for molecular surface and interface analysis.

In this chapter an attempt will be made to present the fundamentals, instrumental features and applications of the major techniques for surface and interface analysis in an introductory manner. The methods will be classified according to the reagent applied: photon probes, electron probes, ion probes, and electrical field probes.

It is not purpose of this chapter to discuss chemistry and physics of the solid state and of surfaces, for which background information can be found in [10-3, 10-4]. For a more in-depth treatment of the fundamentals of the methods, two books [10-5, 10-6] are particularly recommended. Furthermore, for nearly each technique (or group of techniques), monographs have been written describing in great detail fundamentals, instrumentation, and application [10-7 to 10-21].

Research reports dealing with methodological developments and applications of methods for surface and interface analysis can mostly be found in the following journals: *Surface Science, Surface and Interface Analysis, Journal of Vacuum Science and Technology, Nuclear Instruments and Methods B, Journal of the Electrochemical Society, Thin Solid Films, Vacuum, Journal of Applied Physics*.

Furthermore, the various journals for analytical chemistry contain articles dealing with surface and interface analysis which are often of great interest to the practising materials analyst.

10.1 Photon Probe Techniques

Photons in the IR, VIS, UV, or X-ray region are applied for surface and interface analysis. The major interactions between the photon beam and the sample, the resulting techniques, and their typical area of application are listed in Table 10.1-1.

Table 10.1-1. Survey of photon probe techniques

Fundamental process	Mechanism of signal generation	Spectral range of photons	Technique	Applications
Inelastic processes:				
Scattering	Transfer of vibrational energy	VIS	Raman spectroscopy	Molecular and structural surface probe
Absorption	Excitation of vibrational states	IR	Infrared spectroscopy	Molecular and structural surface probe
	Excitation of electronic states	X-ray	EXAFS, SEXAFS, NEXAFS, S-NEXAFS	Bond lengths, atomic distances, coordination numbers, oxidation states, orientation of molecules at surfaces
Emission	Photoelectron emission	UV	UPS	Electronic structure of surfaces, adsorbed species, elemental surface analysis, oxidation states, chemical bonding
		X-ray	XPS	
	Photodesorption	VIS UV	LAMMS	Trace elements at surfaces, molecular microprobe

Raman and IR spectroscopy are presented in Chap. 9, and thus are not further discussed here.

10.1.1 Emission spectroscopy

Photoelectron spectroscopy (PES)

As described already in the previous paragraph, absorption of electromagnetic radiation of sufficient energy (X-rays or UV light) leads to the emission of a photoelectron. Only these photoelectrons which are generated in the outermost layers of a solid can emerge from the surface into the vacuum and be detected. Photoelectron spectroscopy is therefore a surface-sensitive technique. The measurement of the kinetic energy of the emitted photoelectrons allows the binding energies of electrons in the emitting atom or molecule to be determined, and the intensity function (number of photoelectrons vs. kinetic energy) allows conclusions to be made about the number of atoms of a particular species involved in the emission process.

Since only the electrons from the outermost surface layers can leave the sample, photoelectron spectroscopy is very surface sensitive.

X-ray photoelectron spectroscopy (XPS) introduced by Siegbahn in 1949 (Nobel prize 1981) is a major technique for qualitative and quantitative elemental analysis of surfaces yielding additional information about the chemical bonding of the excited atoms.

XPS is a major technique for qualitative and quantitative elemental analysis.

UV photoelectron spectroscopy (UPS) is one of the major techniques for electronic structure analysis of solid surfaces and for characterization of adsorbates.

The basic process for XPS and UPS is illustrated in Fig. 10.1-1. The kinetic energy of the photoelectrons as measured with an X-ray electron spectrometer (Fig. 10.1-2) is given by:

$$E_{kin}(PE) = h\nu - E_B - \phi$$

where E_B = binding energy of the PE in the atom; and ϕ = work function (for conducting specimen in contact with the spectrometer, this is the spectrometer work function [10-5].

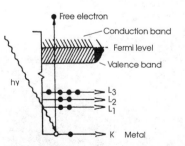

Fig. 10.1-1. Photoelectron spectroscopy: electronic structure of a metal and the process of generating photoelectrons

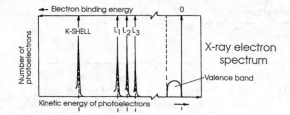

Fig. 10.1-2. Photoelectron spectroscopy: schematic representation of a photoelectron spectrum

Bonding information can be obtained with XPS.

Since ϕ can be determined with reference materials, measurement of E_{kin} allows a direct determination of the binding energy. In XPS this can be used for qualitative elemental analysis and for obtaining bonding information, if the atomic energy levels are sufficiently influenced by chemical bonding. In UPS, molecular levels of adsorbed species can be measured, but more often the shape of a PE peak originating from the valence band is evaluated for the determination of the density of states.

Each photoelectron spectrum contains peaks on a relatively intense electron background. The peaks originate from electrons which have escaped from the solid without experiencing any energy loss. The continuous background signal consists of photoelectrons having lost some of the kinetic energy during emergence and of secondary electrons (see Sec. 10.2.1).

Since only those electrons, which have not undergone any energy loss, can be evaluated for analytical purposes, the analytical volume is therefore confined to very thin surface layers. The (loss-free) escape depth of the photoelectrons is determined by their mean free path, which is in the range between 0.5 and 10 nm, depending on the kinetic energy of the photoelectrons. Typical values for XPS are in the order of 1 nm (approximately 5 atomic layers). The mean free electron path is a statistical value designating the distance in a solid which electrons can travel until a fraction of $1/e$ ($\sim 35\%$) still have not suffered an energy loss due to a collision with an atom in the solid.

The escape function of photoelectrons from a solid is given by $e^{-x/\lambda}$ with x designating the depth scale. When x equals λ then for a normal take-off angle ($\alpha = 90°$, the spectrometer collects electrons which emerge perpendicular to the sample surface) 65% of the photoelectrons originate up to a depth equal to λ. 35% still come from lower layers with the probability of emergence dropping off with the escape function. Depth resolutions can be increased by reducing α, thus increasing the distance the electrons have to travel. The mean escape depth x (measured perpendicular to the surface) is then $x \sin \alpha$.

After the presentation of the fundamentals common to both types of photoelectron spectroscopies the individual techniques should now be discussed.

XPS

In XPS, photons in the keV range generated by an X-ray anode (usually Al K_α, $E = 1.49$ keV or Mg K_α, $E = 1.25$ keV) or a synchrotron are used to produce photoelectrons. In modern instruments the photon beam is monochromated with a crystal disperser which selects a wavelengh from the emission spectrum according to Bragg's law (Fig. 10.1-3):

$$n \lambda = 2d \sin \Theta$$

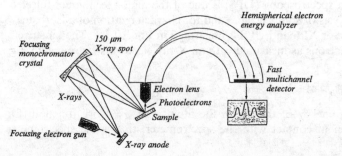

Fig. 10.1-3. XPS: scheme of instrumental configuration (Surface Science Labs)

where n = order of diffraction, here $n = 1$ is selected; d = lattice spacing of diffraction crystal; and Θ = diffraction angle.

The full-width-at-half-maximum (FWHM) of such monochromated radiation is typically 0.3 eV allowing the determination of the kinetic energy of the photoelectrons with high accuracy. Monochromatization has the additional advantage that the X-ray beam can be focused onto a rather small spot on the sample (typically 10–100 µm in diameter) allowing spatially resolved surface analysis.

The energy of the emitted electrons is measured with an electron spectrometer – either a hemispherical or a cylindrical mirror analyzer. Both operate on the basis of electrostatic or magnetic deflection principles like a mass spectrometer. After passing through the analyzer, electrons of the selected kinetic energy are detected using an electron multiplier or channel plates.

Qualitative elemental analysis with XPS is straightforward through determination of E_B from E_{kin} (see Fig. 10.1-2). Quantitative evaluation of the peak intensities is based on the determination of the relative concentrations of elements within the excited surface layer [10-5]:

$$\frac{C_A}{C_B} = \frac{I_A}{I_B} \cdot \frac{\sigma_B^K}{\sigma_A^K} \cdot \frac{\lambda_B}{\lambda_A} \cdot \frac{\eta_B}{\eta_A}$$

where C_A, C_B = atomic concentrations of A and B; I = measured photoelectron intensity; σ = cross section for ejection of photoelectrons from a given orbital (here K-orbital); λ = electron mean free path; and η = spectrometer efficiency.

If intensity ratios of lines with similar energy are evaluated, then the quantitative relation is reduced to:

$$\frac{C_A}{C_B} = \frac{I_A}{I_B} \cdot \frac{\sigma_B^K}{\sigma_A^K}$$

Values for σ are in the range between 10^{-18}–10^{-21} cm² and can be obtained from the tables of [10.1-1].

XPS is, however, not only suited for qualitative and quantitative elemental analysis of surfaces, but also for gaining information about chemical bonding, since the exact values of E_B depend on the chemical environment of an excited atom. Figure 10.1-4 shows a pronounced example for this effect – the silicon $2p$ spectra of silicon and silicon dioxide. The binding energy of the silicon $2p$ electron is shifted towards higher values in SiO_2 by 4.25 eV exhibiting the reduced screening of the positive nuclear charge by the lower valence electron density as compared to silicon. This so-called "chemical shift" can be used to determine the oxidation state or the stoichiometry of atoms in a sample. Usually, however, the chemical shifts encountered are smaller (typically less than 1 eV), thus peak deconvolution techniques must often be used to resolve overlapping peaks originating from different oxidation states. With peak deconvolution, chemical shifts as small as 0.1 eV can be evaluated.

The possibility of determining chemical shifts with high accuracy makes XPS a specially valuable tool for surface analysis, since compound identification through quantitative elemental analysis is often strongly limited by an inhomogeneity of the surface layers analyzed and by surface contaminations.

Applications of XPS include virtually all the problems involved in the change of surface composition by processes such as oxidation, corrosion, segregation, adsorption, chemisorption, etc. Figure 10.1-5 shows, as an example, the spectra of chromium and nickel from a nickel-chromium alloy in its initial stage of oxidation. It can be observed that most of the chromium is oxidized (being present as Cr^{3+}) while nickel has reacted only to a small extent [10.1-2].

XPS also plays a significant role in catalysis and thin film technology (e.g., for optical or microelectronic devices).

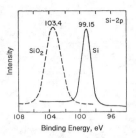

Fig. 10.1-4. XPS: Si-2p photoelectron spectra of silicon and silicon dioxide showing a chemical shift of 4.25 eV. (From *Handbook of X-ray Photoelectron Spectroscopy*. Eden Prairie, MN: Perkin Elmer Corp., Physical Electronics Division, 1992)

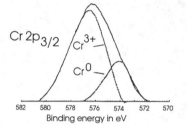

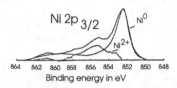

Fig. 10.1-5. XPS: Cr and Ni $2p_{3/2}$ spectra of nickel-chromium alloy in initial stage of oxidation (from exposure to 20 L O_2). Peak deconvolution was applied to obtain the contributions of the various oxidation states to the photoelectron peaks of chromium and nickel [10.1-2]

Chemical shifts can be used to determine oxidation states.

XPS can be used to study, e.g., oxidation, corrosion, segregation, or adsorption.

UPS

UPS being the major technique to study the electronic structure of surfaces and fundamental parameters of adsorption exhibits major differences to XPS:

UPS is a major technique to study electronic structures of surfaces.

- Due to the low energy radiation (10–40 eV) used, only outer electronic levels are ionized (valence band, and molecular orbitals of adsorbed species).
- High resolution spectra can be obtained due to the narrow line width of the photon source. The source is usually a helium discharge lamp with lines at 21,21 (HeI) and 40,82 (HeII) eV whose width is several meV. Alternatively, monochromatized synchrotron radiation is applied.

The instrumental configuration is similar to XPS. UV photoelectron spectra of clean metal surfaces represent the electronic structure (density of states) of the valence bands. It must be emphasized that such spectra represent a mixture of surface and bulk states, since the photoelectrons originate from several atomic layers at the surface of the material. Surface states become more pronounced when low angle measurements are employed, reducing the analytical information depth.

Absorption and chemisorption of substances at the surface causes a significant change in the shape of the spectra, and thus can be used to follow such processes.

Information about the orientation of adsorbed molecules can be obtained by angle resolved UPS experiments (ARUPS) using polarized photon beams. Another important application of UPS is the determination of the coverage of the surface with adsorbed species. Due to the high surface sensitivity, fractions of a monolayer of an adsorbate can be detected (by measuring photoelectrons emitted by the absorbed species). The intensity of the photoelectron peaks is evaluated in order to determine the coverage. Thus, UPS is especially important for fundamental studies in catalysis.

Laser micro mass spectrometry (LAMMS)

The transfer of energy from intense photon sources such as laser beams to the surface of a material can result in desorption of the surface species with a fraction of the species in an ionized state. Depending on the laser power, non-thermal desorption processes (below $10^8 \, W \, cm^{-2}$) yielding molecules of adsorbed species, or thermal desorption processes (above $10^8 \, W \, cm^{-2}$) dominate [10-9]. The latter process is characterized by thermal volume evaporation leading to atomization of the surface material. This technique is mainly suited for elemental microanalysis, but some molecular information can be obtained from fragmentation patterns.

In order to achieve microanalysis the laser beam (typically a frequency quadrupled Nd-YAG laser with $\lambda = 266 \, nm$ and a pulse duration of 10–20 ns) is directed onto the sample surface through an optical microscope illuminating an area of 1–2 µm in diameter. The atomic and molecular ions generated by such a laser pulse are analyzed with a time-of-flight mass spectrometer. Recent instruments make use of ion cyclotron resonance (Fourier transform) mass spectrometry.

LAMMS is a rather sensitive tool for microanalysis – detection limits for elements in the range between 1 and 100 µg/g can be achieved. Thus, the technique is often used for distribution analysis of trace elements, especially in biological specimens. Evaluation of the ion intensities is rather qualitative, since the ionization probabilities vary strongly for different elements and matrices.

For molecular analysis a fingerprint evaluation is usually performed. The technique seems to be particularly useful for the characterization of organic layers on substrates with a lateral resolution of 10–100 µm.

10.2 Electron Probe Techniques

The interaction of electrons in the energy range between some eV and 1000 keV with the atoms of a specimen provides the basis for several techniques which are most important for surface and interface analysis. A survey of these fundamental processes of interaction, i.e., the major signals, the techniques, and their typical applications, is given in Table 10.2-1. Basically, the techniques can be classified

Table 10.2-1. Survey of electron probe techniques

Fundamental process	Mechanism of signal generation	Typical energy of electrons	Technique	Application
Elastic processes: Scattering	Deflection of transmitted electrons	100–1000 keV	TEM	Imaging of inner structure in thin specimens
	Deflection of reflected electrons	100 keV 20 eV 20 keV	REM LEEM BSE imaging	Imaging of surface structure
	Bragg diffraction	100 keV 1–5 keV 10 keV	THEED LEED RHEED	Crystal structure, lattice spacings
Inelastic processes: Absorption	Excitation of vibrational states	5 eV	REELS	Molecular surface analysis
	Excitation of electronic states and plasmons	100 keV	TEELS	Elemental nanoanalysis
Emission	Decay of excited electronic states by:			
	photons	20 keV	EPXMA	Elemental microanalysis, elemental surface analysis with nanometer resolution
	Auger electrons	5 keV	AES	
	Secondary electrons	20 keV	SEM	Morphological and topographical imaging

into two groups. The first group are those techniques which exhibit a high surface sensitivity (low depth of analytical volume); such as AES, LEED, RHEED, and REM. These techniques are used for surface characterization of (bulk) specimens. The second group comprises those methods which have actually a poor depth resolution, but a high lateral resolution, such as SEM, EPXMA, TEM, THEED, and TEELS. These techniques allow the characterization of micro- and nanodomains of a material, and are therefore suited for thin film and interface analysis.

Compared with photon probe techniques these electron probes show several distinct differences:

- Electrons usually have a high cross section of interaction with the target atoms leading to intense analytical signals. This also has the consequence that all measurements must be performed in vacuum ($p < 10^{-5}$ mbar).
- Electron beams can be focused to diameters in the range between 1 nm and 100 nm (depending on the beam intensity), thus high lateral resolution analysis is possible.

Several techniques listed in Table 10.2-1 are often combined (e.g., in one instrument). As a result, the fundamental processes of interaction of electrons with matter and the resulting analytical signals will first be presented in this section, and how these are utilized in analytical methods will then be discussed. The following methods will be discussed:

- EPXMA, SEM
- Analytical electron microscopy (AEM) combining the analytical techniques TEM, THEED, TEELS, and X-ray analysis
- AES

Electrons have a high cross section of interaction with target atoms, usually leading to intense analytical signals. Electron beams can be focused to very small diameters between 1 and 100 nm.

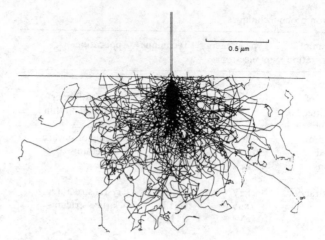

Fig. 10.2-1. Penetration of electrons into a solid: Monte Carlo calculations of electron trajectories, initial kinetic energy = 20 keV, material: Fe, normal incidence [10.2-1]

10.2.1 Fundamentals

Penetration of electrons in a solid state material

When electrons impinge on the surface of a solid state material they experience elastic and inelastic processes as shown in Table 10.2-1. Elastic scattering leads to deflection of electrons and diffraction phenomena. Inelastic processes mainly lead to excitation of electronic states and plasmons (collective excitation of conduction electrons). In this case, the energy transfer is also accompanied by a change in direction of the impinging electron. Elastic and inelastic processes therefore cause a lateral broadening of a finely focused electron beam. Figure 10.2-1 shows the result of Monte Carlo calculations for 20 keV electrons impinging on iron perpendicular to the surface. It is clearly seen that the electron trajectories extend deeply into the bulk of the material undergoing the interactions for signal generation until their kinetic energy has fallen below the critical value necessary to generate a particular signal.

The range of an electron is defined as the total distance that an electron travels along a trajectory and is expressed by:

$$R = \int_{E_0}^{0} \frac{dE}{dE/dx}$$

where R = range; E_0 = initial kinetic energy; and dE/dx = energy loss (due to inelastic processes).

For practical purposes an approximate expression for R has been developed [10.2-1]:

$$R = 0.064 \frac{E_0^{1.68}}{\rho}$$

where ρ = mass density (g cm^{-3}); E_0 = initial kinetic energy (keV); and R = range (µm).

For the calculation of the range of production of a signal (e.g., X-rays) integration goes from E_0 to E_c (the critical energy needed to excite a state), leading to the following expression for R [10.2-1]:

$$R = 0.064 \frac{(E_0^{1.68} - E_c^{1.68})}{\rho}$$

The range of electrons in a solid is of the order of micrometers.

This expression shows that the range of electrons in a solid is in the order of micrometers, which also means that the volume in which the analytical signals originate is in this order of magnitude. A higher spatial resolution as is necessary for surface and interface analysis can be achieved by:

(a) Evaluating signals which originate only from the outermost atomic layers, e.g., Auger electrons and secondary electrons.

(b) Analyzing thin (e.g., 50 nm) specimens with high energies. Then the lateral electron diffusion is negligible. This approach is chosen in AEM.

Elastic interactions of electrons with matter

In elastic processes, energy and momentum are conserved. This means that the kinetic energy of the electrons is not transferred into other forms of energy, like in the excitation of electronic states.

The elastic interaction of electrons with atoms occurs in the form of scattering at the atomic nuclei due to Coulomb interactions. Multiple small angle scattering events (2–3°) dominate, but large angle scattering up to 180° also takes place, although with a much lower probability. Elastic scattering is the major effect for the broadening of the electron beam when penetrating the specimen, and is the cause for a fraction of the impinging electrons to be backscattered from the specimen after having experienced multiple scattering events. The cross section for elastic scattering is proportional to Z^2 of the target atom. Thus, a specimen composed of different phases experiences a different amount of scattering in micro areas of different composition.

> The cross section for elastic scattering is proportional to Z^2 of the target atom.

This effect is used analytically for backscattered and transmitted electrons. Measurement of the *backscattered electron* intensity as a function of the position of a finely focused electron beam scanning scross a sample surface yields an image whose contrast is dominated by differences in the (mean) atomic number in the various micro areas of the sample. The BSE (backscattered electron) image allows the microstructure of a specimen according to the atomic number distribution to be characterized (Fig. 10.2-2). This is a valuable mean for microanalytical purposes and is frequently used in EPXMA, and SEM.

For electron transparent specimens of a certain thickness (e.g., 100 nm) the intensity of *transmitted electrons* measured under a particular nanodomain of the sample is inversely proportional to Z^2 (if deflected electrons are eliminated in imaging). Thus, the transmitted electron micrograph (obtained either by scanning a finely focused beam across the sample or by illuminating a larger area and applying microscope imaging) shows a Z-dependent contrast and allows nanodomains of different mean atomic number to be distinguished. This effect forms a part of the image formation mechanism in TEM (Fig. 10.2-3). Another major contribution to the image contrast in TEM originates from diffraction effects. Due to these, the image intensity varies strongly with changes in crystal orientation around the angles near Bragg incidence (crystal lattice planes). Strain fields around dislocations and other extended image defects give characteristic contrast patterns. Thus, the typical TEM micrographs are obtained showing the nanostructure of a material.

This simple description of the scattering processes must be extended to explain lattice images or atomic (column) images obtained with TEM (as shown in Fig.

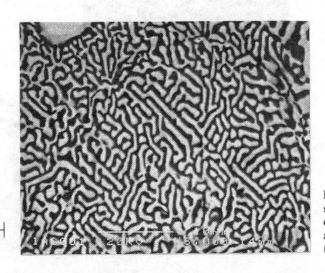

10 μm

Fig. 10.2-2. Elastic electron scattering: backscattered electron micrograph of eutectic chromium carbide/iron alloy showing the atomic number contrast in the microstructure (dark: chromium carbide, bright: iron)

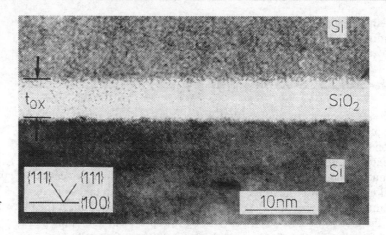

Fig. 10.2-3. Elastic electron scattering: transmitted electron micrograph of a cross section of a thin layer structure consisting of silicon and silicon dioxide

Diffraction patterns can be used to determine lattice spacings in a solid.

10.2-3): In atomic resolution imaging, the coherent interference between the electron wave on the specimen and the scattered electron wave leads to an intensity pattern which exhibits a systematic correlation between atom positions and the scattered electron intensity distribution.

Scattering under Bragg-angle conditions is responsible for obtaining *electron diffraction* patterns which can be measured either in reflection (LEED, RHEED) or transmission mode (THEED). A point pattern is obtained for single crystals when the Bragg condition is fulfilled (Fig. 10.2-4):

$$n\lambda = 2d \sin \Theta$$

where n = order of diffraction; λ = electron wavelength; and d = interplanar lattice spacing.

The diffraction patterns of polycrystalline materials are concentric rings. Amorphous materials show only diffuse circular halos reflecting variations in atomic distances typical for an amorphous material.

From point and ring patterns lattice spacings can be obtained by evaluation of the distances between the diffraction spots or rings using the geometric relation between Θ and the projection distance:

$$\lambda L = Rd$$

(a)

(b)   (c)

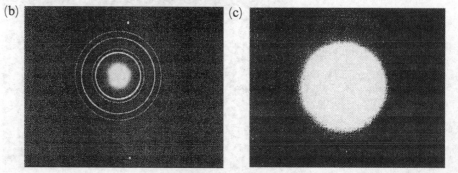

Fig. 10.2-4. THEED: Transmission electron diffraction patterns of (a) single crystal, (b) polycrystalline material, (c) amorphous material [10.2-2]

where $L =$ distance between sample and observation screen; and $R =$ radius of diffraction ring.

In a similar manner LEED patterns are evaluated.

Transmission electron diffraction patterns serve mainly as a means of determining the crystal structure of a nanodomain in a material. The lattice parameters obtained are also valuable supplementary information for analytical identification of phases. The minimum phase size for obtaining a diffraction pattern is in the order of 20 nm.

Reflection electron diffraction patterns allow the atomic spacing of the topmost monolayer [10-5] to be determined. For a two-dimensional array of atoms with interatomic spacings a and b the diffraction conditions for both spacings, when considering only 1st order diffraction spots, must be met simultaneously:

$$\lambda = a \sin \Theta_a$$

$$\lambda = b \sin \Theta_b$$

This means that a two-dimensional point array is formed in the imaging device revealing the periodicity of atoms on the surface and the overall symmetry. Different real space patterns yield different diffraction patterns.

LEED is a most valuable tool for the study of adsorption or segregation processes. It is often combined with AES in one instrument.

Inelastic interactions of electrons with matter

In inelastic processes part of the kinetic energy of the impinging electrons is transferred to potential energy by excitation of vibrational or electronic states of the target atoms. At low energies (eV) dominant inelastic processes are excitation of vibrational states and plasmons. At higher energies (keV) ionization and plasmon excitation are mainly responsible for the energy loss of the electrons penetrating a material. Excitation of vibrations and plasmons, and ionization lead to discrete energy losses of the electrons, causing distinct "absorption features" in the energy spectrum of an originally monoenergetic electron beam. These processes form the basis of *electron energy loss* spectroscopy (EELS).

The energy of plasmon excitation depends on the force between the oscillating electron cloud and the (positive) ions in the lattice of a material. For bulk plasmons it is, e.g., 10.6 eV for Mg, 15.3 eV for Al, or 16 eV for Si. The energy of surface plasmons is $E_{p(vol)} \cdot 1/\sqrt{2}$, e.g., 10.3 eV for Al [10-5]. Since valence levels are involved, the excitation energy of plasmons is dependent on the chemical composition and structure of alloys. Sometimes this can be evaluated for analytical purposes.

The energy loss due to ionization is determined by the ionization energy (being practically equivalent to E_B) of the particular core level. Evaluation of these features for qualitative elemental analysis is therefore straightforward. Consequently, TEELS is a valuable tool for the analysis of nanodomains in materials.

The number of ionizations taking place in a material per incident electron is determined by the ionization cross section σ_e:

$$\sigma_e = \frac{\pi e^4}{E E_B}$$

where $e =$ charge of electron; $E =$ kinetic energy of electron; and $E_B =$ binding energy.

Ionization cross sections show a maximum for $E \sim 4 E_B$. Typical values for K and L shell ionizations are in the range $10^{-17} - 10^{-20}$ cm^{-2} [10-5].

These cross section values are important for quantitative evaluation of core level energy loss spectra, since:

$$\frac{N_A}{N_B} = \frac{I_A}{I_B} \cdot \frac{\sigma_B}{\sigma_A}$$

where $N_A/N_B =$ atomic concentration ratio of elements A and B in analyzed volume; and $I_A/I_B =$ intensity ratio of energy loss peaks.

Ionization of electronic levels generates electrons. These are ejected from the

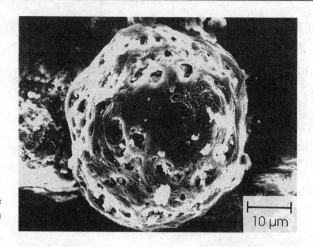

10 µm

Fig. 10.2-5. SEM: Secondary electron image of airborne carbon particle showing the high spatial resolution and depth of field

Secondary electrons are important for topographical and morphological imaging of samples.

atom during primary electron impact and travel through the solid losing kinetic energy by the processes described. Those produced near the surface can be emitted into the vacuum and collected by secondary electron detectors. These secondary electrons allow topographical and morphological imaging of bulk samples when the primary ion beam is scanned across the surface of a material and the secondary ion current is measured as a function of beam position (Fig. 10.2-5). The image contrast is mainly determined by the angle between the impinging electron beam and the surface plane in the particular area. This means when rough specimens are scanned, the angle changes from position to position, and thus also secondary electron emission. One obtains an image of the surface which looks very much like the image of an object we are able to see with a stereo microscope. There are, however, two major differences:

- The lateral resolution of secondary electron images is about 3 nm (as compared to 1 µm with the light microscope) allowing high magnifications (100 000) in imaging. The reason is that the volume from where the secondary electrons originate is only about 10 nm in depth. Within this depth there is only marginal beam spreading, thus the lateral resolution is practically equal to the beam diameter.

In SEM, the depth of field is several orders of magnitude larger than for light microscopy.

- The depth of field is several orders of magnitude larger than for the light microscope because a nearly parallel electron beam is used in image formation. This allows imaging of rough samples. SEM has become one of the most valuable tools in micro and surface analysis extending the range of classical microscopy by nearly three orders of magnitude in spatial resolution.

The electronic excitation of core levels produces ions in a highly excited state, since the hole is located in an inner orbital. The ionized atom, therefore, decays immediately (that is within 10^{-8}–10^{-13} s) into a less energetic state by transfer of an electron from an upper (lower E_B) to a lower (larger E_B) level. The energy difference gained in this transition $\Delta E = E_{B(1)} - E_{B(2)}$ is removed from the atom by either emission of an X-ray photon (Fig. 10.2-6a) or an Auger electron (Fig. 10.2-6b). The energy of the photon being equivalent to the difference in binding energies of the two levels involved:

$$h\nu = E_{B(1)} - E_{B(2)}$$

Electronic transition must obey selection rules.

is in the (lower) keV range and can be measured to determine the identity of the emitting atom (qualitative X-ray analysis). All those transitions are possible, and take place with largely different probabilities, which are allowed by the selection rules:

$$\Delta l = \pm 1, \Delta j = 0, \pm 1.$$

The kinetic energy of the Auger electron is determined by the binding energies of the three levels involved (and by the work function ϕ if measured in vacuum):

$$E_{\text{kin}(\Delta E)} = E_{B(1)} - E_{B(2)} - E_{B(3)} - \phi$$

These energies are also in the (low) keV range and are typical for the identity

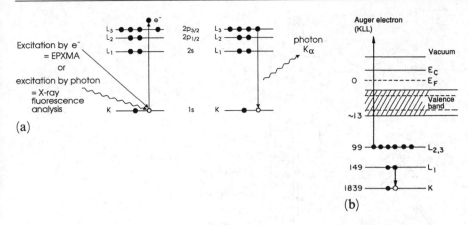

(a)

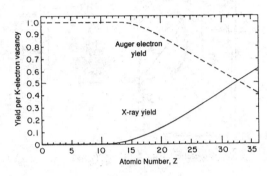

(b)

Fig. 10.2-6. Decay of excited core level states (after excitation by electrons or photons) by (a) X-ray, (b) Auger electron emission

Fig. 10.2-7. Dependence of probability for X-ray and Auger electron emission by decay of 1s core level states on atomic number [10-13]

of the emitting atom. For Auger transitions, selection rules do not apply as for transitions leading to X-ray emission. X-ray and Auger electron emission are competing processes and they provide the only way for energy dissipation from core-level excited atoms to occur. Thus, the sum of the probability of Auger electron emission (a) and X-ray emission (ω) is unity:

$$a + \omega = 1$$

where a = Auger yield; and ω = fluorescence yield.

The Auger and fluorescence yield strongly depend on the atomic number for a given core level ionization:

$$\omega_x = \frac{Z^4}{a_x + Z^4}$$

a_x = constant for a particular core level ionization

$$a_K = 10^6$$

$$a_L = 10^8$$

Figure 10.2-7 shows the graphical representation of this function for K ionizations, indicating that for $Z < 33$ Auger electron emission is the dominant process. Both Auger electron and X-ray signals can be used for quantitative elemental analysis which will be discussed with the description of the methods.

X-ray fluorescence and Auger electron emission are competitive processes.

10.2.2 Electron probe x-ray microanalysis (EPXMA) and scanning electron microscopy (SEM)

Although still traditionally designated as two separate techniques, in today's reality the distinction is vanishing: SEM refers to imaging, EPXMA to elemental analysis, but usually the same instrument (with various attachments) is used for both purposes. In this section EPXMA will mainly be used to designate the method, which uses secondary and backscattered as well as X-ray signals.

SEM and EPXMA are among the most important physical techniques used routinely.

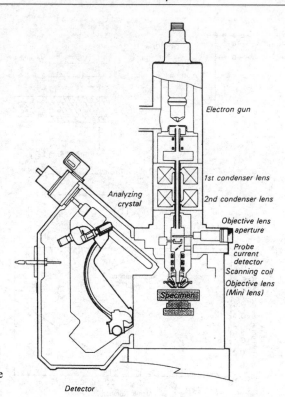

Fig. 10.2-8. EPXMA: Scheme of a scanning electron microprobe for secondary and backscattered electron imaging and X-ray analysis. Electromagnetic condenser lenses are used to focus the electron beam

Figure 10.2-8 shows the configuration of a modern scanning electron microprobe. An electron beam produced by emission from a tungsten or LaB_6 filament is accelerated to an energy between 1 and 50 keV. Carrying a beam current in the range of 10 pA to 1 µA, it is focused to a diameter between 1 and 100 nm with electrostatic lenses and directed onto the surface of the sample. Apart from point analysis, line and area scans can also be performed with the aid of a beam scanning generator.

The emerging secondary electrons have a low kinetic energy of about 10 eV and are therefore accelerated by a potential of several 100 volts for detection with a scintillator crystal. The backscattered electrons whose energy is close to the primary electron energy are detected without acceleration using either a scintillator or a semiconductor device.

For the measurement of the energy and intensity of the characteristic X-ray spectrum, energy dispersive and wavelength dispersive spectrometers are used (Fig. 10.2-9). Energy dispersive X-ray spectrometers allow the simultaneous registration of the X-ray intensities of all elements from Be to U if windowless detectors are used. The spectrometer consists of a semiconductor detector (silicon doped with Li) which converts the photons into electrical pulses whose voltage is proportional to the energy of the photon. Thus, energy discrimination of the photons is achieved. The energy dispersive X-ray spectra show a resolution of approximately 140 eV at medium X-ray energies (Fe K_α). Spectral interferences occur between K_α and K_β lines of adjacent elements and K and L lines of elements with a larger difference in atomic number.

Wavelength dispersive spectrometers have a higher resolution and a better signal-to-noise ratio than energy dispersive spectrometers.

The wavelength dispersive spectrometer consists of a diffraction crystal which reflects a particular wavelength of the spectrum according to the Bragg condition. The intensity of this radiation is then measured with (gas) ionization or scintillation detectors. Wavelength dispersive spectrometers have a much higher resolution ($\sim$5 eV) and a better signal-to-noise ratio than energy dispersive spectrometers, but allow only sequential analysis of various elements. Also several diffraction crystals must be used to cover the whole wavelength range. In practice, analytical microprobes are equipped with one energy dispersive and one to five wavelength dispersive spectrometers.

X-ray spectrometry allows an easy elemental identification, since the number of lines is rather small with the electronic transitions $2p_{3/2,1/2} \rightarrow 1s$ ($K_{\alpha_{1,2}}$ line),

(a)

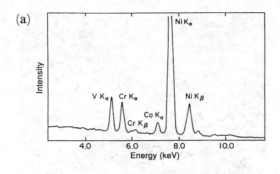

(b)

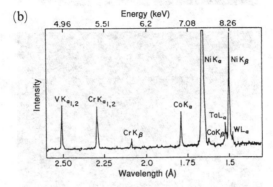

Fig. 10.2-9. EPXMA: Energy dispersive (a) and wavelength dispersive (b) X-ray spectra of a nickel-based alloy. The unspecific continuous background intensity is Bremsstrahlung [10.2-1]

$3d_{5/2,3/2} \to 2p_{3/2}$ ($L_{\alpha_{1,2}}$ line) and $4f_{7/2,5/2} \to 3d_{5/2}$ ($M_{\alpha_{1,2}}$ line) being dominant for low, medium, or high Z elements, respectively.

Quantitative evaluation is always based on a comparison of measured X-ray intensities of a particular line and element (e.g., Cu K_α) in a specimen and a reference material. Pure elements or stoichiometric compounds can be used for first order quantifications:

$$C_A = C_A^\star \cdot \frac{I_A}{I_A^\star}$$

$\star$ denotes reference material.

This simple quantitation algorithm yields an analytical accuracy between 1 and 10% relative depending on how closely matched in composition are the specimen and reference material.

Correction procedures have been developed accounting for the differences in ionization (energy loss), backscattering of electrons, absorption of X-rays generated below the surface and secondary fluorescence effects (ZAF-correction). Applying these, one usually obtains quantitative analytical results with an accuracy of better than 1% relative (with the exception of low Z elements where the matrix effects are rather high). Thus, EPXMA belongs to the most accurate analytical methods available for analysis of solid state materials. The detection limits for X-ray analysis are in the order of 0.01–0.1% of the element in the analyzed volume of a few μm^3.

Lateral distribution analysis is possible by performing line or area scans. This is a valuable tool for the microstructural chemical characterization of materials. An example is shown in Fig. 10.2-10.

The combination of secondary electron signals for morphological imaging, backscattered electrons for Z mapping and qualitative and quantitative analysis makes EPXMA the most important method for microanalysis of solids. It is used on a routine basis for all kinds of problems and materials – e.g., identification of precipitates in metals, phases in geological materials, toxic dust particles, and asbestos fibers. Its major limitation is the size of the analytical volume for X-ray generation – typically 1–3 μm in diameter and depth. Thus, nanophases cannot be analyzed quantitatively with X-ray spectroscopy, although they can be imaged using the secondary or backscattered electron signals. Also thin surface layers can be detected if they are typically thicker than a few nanometers, but selective analysis is not possible. Obviously different techniques have to be applied – analytical

The analytical accuracy is typically in the range of 1–10% relative.

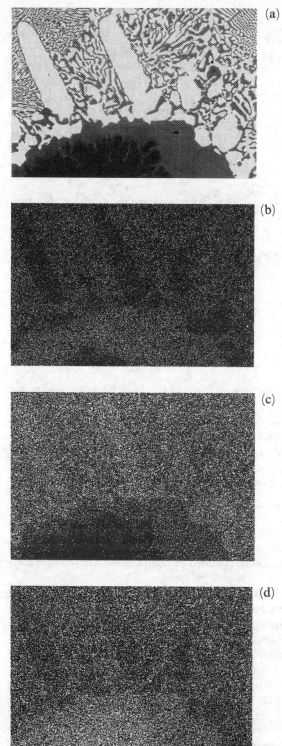

Fig. 10.2-10. EPXMA: microstructural characterization by backscattered electrons and by X-ray mapping. (a) BSE-micrograph showing a metallic iron phase (bright), chromium carbide (grey) and chromium oxide (dark), (b) X-ray mapping of carbon, (c) X-ray mapping of iron, (d) X-ray mapping of chromium

electron microscopy for nanoanalysis and Auger electron spectroscopy for microanalysis with a high depth resolution (a few nanometers).

The major use of EPXMA for surface analysis is:

- high resolution topographical imaging with secondary electrons
- thin film compositional analysis evaluating X-ray ratios
- determination of the thickness of thin films ($t > 10$ nm), since $I_A = kNt$

where I_A = measured intensity of element A; k = proportionality factor; N = area density [atoms/cm^2]; t = film thickness. The factor k is determined from reference materials or with (calculated) depth distribution functions for X-ray generation.

Despite this rather limited applicability EPXMA is a most valuable tool for surface and interface studies because the method provides an extensive basic material characterization in interesting spatial domains.

10.2.3 Analytical electron microscopy (AEM)

AEM is the most important method for nanoanalysis of materials. It combines transmission (TEM) and reflection (SEM) imaging with electron diffraction (THEED) for structural characterization and elemental analysis evaluating X-ray emission and electron energy loss spectra (TEELS) (Fig. 10.2-11).

In order to utilize the high spatial resolution potential either small particles or thin specimens on electron transparent substrates (thin carbon foils on copper grids) must be analyzed. Preparation of thin specimens is done by cutting and ion milling or electrochemical etching and demands optimized procedures for each material. The thickness of the specimens in the analytical area is in the range between 10 and 100 nm. The energy of the primary electrons in AEM is in the range between 40 and 400 keV. Lower energies are advantageous for X-ray analysis, higher energies for high resolution imaging. Maximum beam intensities must be obtained for small diameters, since virtually all the signals produced are proportional to the beam current. With high brightness electron sources (field emission guns) a beam current of 1 nA can be achieved at a diameter of only 1 nm. This is the basis for sensitive nanoanalysis and extensive characterization of interfaces.

Analytical modes

Transmission electron imaging is the most important microscopic technique for characterizing the nanostructure of a material down to 1 Å. Precipitates, physical defects like dislocations, stacking faults, etc., and grain boundaries can be imaged with high resolution.

Transmission high energy electron diffraction (often also called selected area electron diffraction (SAED)) is one of the most common means of identification of such precipitates. Increasingly, X-ray and energy loss spectroscopy are used for the analytical characterization of nanoprecipitates, because elemental identity and composition can be obtained directly, not indirectly through structural parameters which are often ambivalent.

X-ray analysis is the major technique for elemental analysis. The typical advantages are: simple quantitation, and high signal-to-noise ratios (see also Fig. 10.2-10). The limitations for X-ray analysis in AEM are set however by the extremely small interaction volume – e.g., for a sample of 10 nm thickness and a beam diameter of 10 nm the excited volume is only $10^{-6}\,\mu m^3$ corresponding to an analytical mass of roughly 10^{-17}–10^{-18} g. Furthermore, the collection yield for X-rays is determined by the solid angle which is observed by the detector. Due to the isotropic emission of X-rays only a fraction of 10^{-2}–10^{-4} of the photons generated in the interaction volume can be detected. This sets the detection limit for X-ray analysis at $\sim 10^{-20}$–10^{-19} g, if large area energy dispersive detectors are positioned close to the position of electron impact. The lateral resolution (e.g., for concentration profiles across interfaces) is in the order of 10–20 nm.

Quantitative evaluation of X-ray intensities uses intensity ratios:

$$\frac{C_A}{C_B} = k\,\frac{I_A}{I_B}$$

where k = relative sensitivity factor, depends on lines and elements, must be determined from reference materials.

The analytical accuracy is mainly determined by counting statistics and is typically a few percent only at analyzed volumes of $10^{-4}\,\mu m^3$.

X-ray analysis in AEM has not only the principal disadvantage of a low collection yield, but also that of a low fluorescence yield for elements of low atomic numbers. Both of these limitations are less severe for TEELS. The collection yield

AEM is the most important technique for analysis of nanometer-sized domains in a solid.

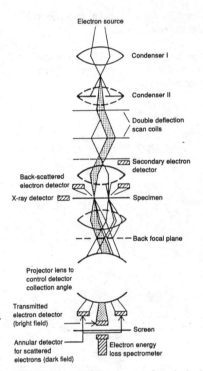

Fig. 10.2-11. AEM: scheme of analytical electron microscope for transmission and reflection electron imaging, electron diffraction, X-ray analysis and electron energy loss spectrometry [10-12]

of transmitted electrons is very high. Since the number of ionizations in the analytical volume determines the signal, low-Z elements can be analyzed with good sensitivity too. The big disadvantage of energy loss spectra is, however, the poor signal-to-noise ratio, because the transmitted electrons also lose energy in multiple scattering events leading to a continuous electron background. The signal-to-noise ratio can be improved by analyzing very thin (10–20 nm) specimens. Quantitative evaluation of energy loss spectra with ionization cross sections exhibits a typical accuracy of 10–20% relative.

10.2.4   Auger electron spectroscopy (AES)

AES combines low information depth with high lateral resolution.

High primary beam intensities are needed to achieve usable Auger peaks with sufficient signal-to-noise ratio.

AES is the most widely used technique for surface analysis due to the combination of low information depth and high lateral resolution. While the information depth is determined by the mean free electron path (Sec. 10.1.1), which is in the range between 0.5 and 10 nm, high lateral resolution is achieved by excitation of the Auger electron signal with a finely focused electron beam ($E_0 = 3$–10 keV). The analytical area of interest can easily be selected using the secondary electron image. The minimum diameter of the beam is limited to about 100 nm by the need to use a high primary beam intensity in order to achieve Auger peaks of sufficient signal-to-noise ratio. The Auger electron peaks are superimposed with a rather high continuous electron background resulting from multiple scattering events (Fig. 10.2-12). Often the derivative spectra are recorded to allow a better evaluation of the spectrum peaks. Registration of the integral peaks with pulse counting techniques is preferred for quantitative analysis and elemental mapping.

Figure 10.2-13 shows an instrumental configuration. The determination of the kinetic energy of the emitted electrons is usually performed with a cylindrical mirror analyzer (at a given potential between inner and outer cylinder only electrons of a certain energy can reach the detector) which has a constant spectral resolution over the whole range. Auger instruments are equipped with a small ion (Ar^+) gun to remove contaminants from technical samples and for sputter depth profiling. Often in-situ sample preparation is possible to generate surfaces which are not contaminated by the ambient atmosphere. Such attachments allow sample manipulations like fracturing, annealing, and thin film deposition.

As in every true surface analytical instrument the analysis chamber must be at ultra high vacuum (UHV) with residual gas pressures below 10^{-9} mbar. The reason is that every fresh surface (generated, e.g., by sputtering) reacts strongly with the residual gas. Assuming that every gas molecule which hits such a surface is adsorbed (which is a good approximation for reactive gases like oxygen or H_2O) the surface will be covered with 1 monolayer within 1 s if the residual gas pressure is 10^{-6} mbar. By reducing the pressure to 10^{-9} mbar the time needed for adsorption of 1 monolayer will be 1000 s – leaving enough time to perform analysis under uncontaminated conditions.

Quantification can be performed with external reference materials or sensitivity factors.

The designation of the Auger peaks is determined by the three levels involved in the transition, e.g., KL_2L_3 means ionization in 1 s orbital, electron decay from $2p_{1/2} \rightarrow 1$ s and ejection of an Auger electron from $2p_{3/2}$ orbital. The most intense lines are KLL and LMM. Auger lines originating from the valence band reflect

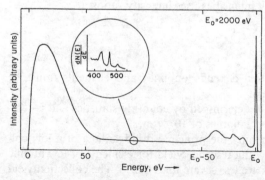

Fig. 10.2-12. AES: spectrum of electrons emitted from a solid at bombardment with 5 keV primary electrons. The high intensity peaks at low and high energies are due to secondary electron and backscattered electron emission. Auger electron peaks are superimposed on a continuous electron background resulting from multiple scattering events. The insert shows a differentiated part of the spectrum with Auger peaks

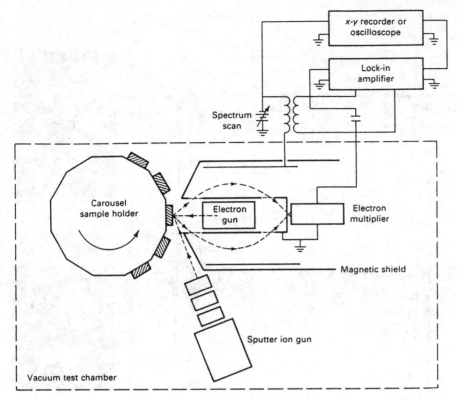

Fig. 10.2-13. AES: scheme of instrument [10-15]

the density of states and carry information about oxidation state and chemical bonding. It must be emphasized however that the chemical shifts are less pronounced than in XPS, since the kinetic energy of the Auger electrons is determined by the energy differences between various levels, which partially compensates chemical bonding effects. The accessible element range for AES covers Be to U. Qualitative analysis is based on the measurement of the kinetic energy of the Auger electrons which does not depend on the primary electron energy. Quantitative evaluation of the Auger peaks can be performed with external reference materials using this relation:

$$\frac{C_A}{C_{A^*}} = \frac{I_A}{I_{A^*}} \cdot \frac{\lambda^*}{\lambda} \cdot \left[\frac{1 + R^*}{1 + R}\right]$$

where C_A, C_A^* = atomic concentration of element A in sample and reference material (*); I_A, I_A^* = Auger peak intensities; λ, λ^* = mean free electron path (escape depth); and R, R* = back-scatter coefficient (accounts for generation of Auger electrons by backscattered electrons).

When sample and reference have a similar composition, $C_A/C_A^* = I_A/I_A^*$ serves as a good first approximation for quantitative analysis.

An alternative common approach to quantitation is the evaluation of the peaks in a spectrum with sensitivity factors for the various elements:

$$\frac{C_A}{\sum_i C_i} = \frac{I_A/S_A}{\sum_i I_i/S_i}$$

where S_i = sensitivity factors obtained from pure element or compound reference materials (tabulated in handbooks).

The analytical accuracy of AES is in the order of 5% for polished samples under optimum conditions. However, at small analytical areas the statistical errors become dominant increasing the inaccuracy.

Also a variety of artefacts, such as electron-induced decomposition of materials at high beam densities or compositional changes due to preferential sputtering in depth profiling, can severely reduce analytical accuracy.

Analytical accuracy can be reduced by artefacts.

The relative detection power depends on lateral resolution and the sensitivity of the Auger line. At a lateral resolution of 100 nm the relative detection limit is in the order of 1% for sensitive lines, at 1 μm resolution detection limits of about

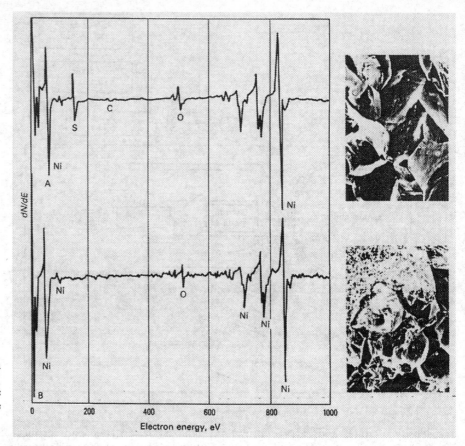

Fig. 10.2-14. AES: Auger electron spectrum of fracture surface of nickel heat-treated at 600 °C showing surface segregation of sulfur. The Auger intensity of sulfur corresponds to a surface coverage of 0.2 monolayers. The bulk concentration of S in the alloy was <5 atomic ppm. The peaks of O and C are probably due to a small amount of surface contamination. A = Ar from sputtering Ar^+ ions [10.2-3]

AES is an important method for distribution analysis of elements.

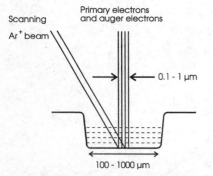

Fig. 10.2-15. AES: depth distribution analysis by combination of sputtering and Auger analysis of the freshly exposed surface

AES is frequently used in combination with depth profiling.

0.1% can be achieved. This is sufficient to detect fractions of a monolayer of a substance at the surface. Figure 10.2-14 shows, as an example, the Auger electron spectrum of a fracture surface of nickel heat-treated at 600 °C [10.2-3]. Segregation of sulfur to the grain boundary can be detected leading to an embrittlement of the alloy. The Auger intensity of sulfur corresponds to 0.2 monolayers of sulfur at the grain boundary (this means that only 20% of the atoms at the interface are covered with one sulfur atom). This demonstrates, on the one hand, the surface sensitivity of AES and, on the other hand, the large influence of trace impurities (like sulfur, phosphorus) on the mechanical properties of metals.

Distribution analysis of elements and phases at surfaces and interfaces is one of the major tasks for AES. For lateral distribution analysis a finely focused electron beam (100–1000 nm beam diameter) is applied. By area and line scanning, secondary electron imaging, elemental mapping, and (quantitative) line profiling is possible. This technique is widely used for the characterization of grain boundaries, nanoprecipitates and surface inhomogeneities for all kinds of materials except highly insulating targets (e.g., oxidic-ceramics) which practically cannot be analyzed with AES due to the build-up of a negative surface charge which can hardly be compensated.

Apart from point analysis and high resolution lateral distribution characterization, depth profiling by combining sputter removal of surface atoms (see Sec. 10.3.2) with Auger analysis of the freshly exposed surface is most important for thin film and surface analysis. The principle of this measurement technique is shown in Fig. 10.2-15. A focused Ar^+ beam of 3–10 keV kinetic energy and a typical diameter between 1 and 100 μm is scanned across the analytical area in a rectangular or square pattern removing surface atoms by collisional transfer of energy. Sputter rates depend on the ion beam density and are typically in the order of 0.1–10 nm/s. Thus, a crater is eroded from the material. The freshly exposed surface in the crater bottom is analyzed by focusing the electron beam in the center of the crater and measuring the Auger electrons emitted. By this, one measures in the flat central part of the crater achieving a high dynamic range of the depth profiles (steep profiles which are not distorted by the slanted crater walls). Sput-

(a)

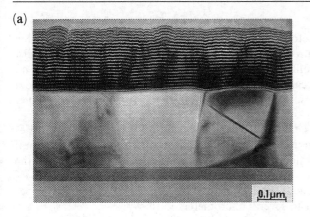

(b)

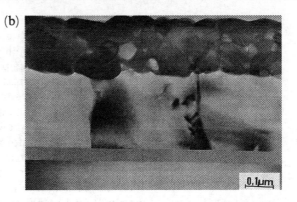

(c)

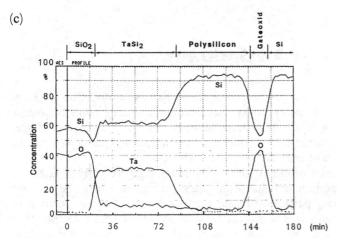

Fig. 10.2-16. AES: characterization of a thin film structure for microelectronic devices by combination of transmission electron microscopy of cross sections and depth profiling with AES. (a) TEM micrograph of layer system before annealing showing the following sequence from top to bottom: approximately 20 Ta and 20 Si layers each 5 nm thick produced by sputtering, polysilicon 275 nm thick, SiO_2 approximately 50 nm, silicon substrate. (b) TEM micrograph after annealing at 900 °C showing a newly formed layer on the top: polycrystalline tantalum silicide approximately 200 nm, polysilicon approximately 250 nm, SiO_2 approximately 50 nm, silicon substrate. (c) Quantitative Auger depth profiles of oxygen, silicon and tantalum indicating the formation of a silicon oxide layer on top of stoichiometric $TaSi_2$ [10.2-4]

tering and analysis can be done simultaneously or in an intermittent manner. A "raw profile" is obtained in which the Auger intensities are plotted versus sputter time. Conversion of the Auger peak intensities to concentrations is achieved by means of sensitivity factors. Calibration of the depth scale can be achieved with knowledge of the sputter rates, or better, by measuring the depth of the crater after analysis with a profilometer. Since the sputter coefficients are dependent on chemical composition, calibration problems occur when there is a strong change in composition during depth profiling, as is the case for systems containing several layers. Then each layer must be calibrated individually.

Another severe artefact is preferential sputtering, which designates a sputter-induced change in the surface composition when the various elements in the surface zone have different sputter coefficients – e.g., if the sputter coefficient for A is larger than for B in an alloy A–B, then the surface becomes enriched in B. Calibration correction is necessary by a factor taking into account the ratio of the sputter coefficients.

Further problems arise when laterally heterogeneous materials (inclusions in a matrix leading to different surface compositions within the sputtered area) are depth-profiled: if the sputter coefficient of the inclusion is different from the matrix, a surface profile (height differences) between different areas will develop leading to a decrease in the depth resolution ("selective sputter effects"). Depth resolution is dependent on the mass of the primary ions, their energy, and their angle of incidence. In ideal specimens (e.g., a film of Ta_2O_5 on Ta) a depth resolution of the profile of 5–10 nm is obtainable under typical analytical conditions.

Despite the often severe problems in the quantitation of depth profiles this measurement technique is of greatest importance because for many questions in materials science the depth distribution of components must be known. Also often a qualitative or semiquantitative evaluation is sufficient to solve a problem in materials science. Depth-profiling applications include virtually all types of materials, but are particularly important for thin film technology, in metallurgy, and microelectronics. Figure 10.2-16 shows, as an example, the cross section of a multilayer structure of a microelectronic device consisting of a $TaSi_2$ layer/polysilicon/ silicon dioxide and silicon before and after annealing (TEM micrographs) and the quantitative Auger depth profile of the annealed specimen [10.2-4]. The TEM micrographs and Auger profiles show that the sputtered layers of tantalum and silicon (Fig. 10.2-16a) had reacted to stoichiometric $TaSi_2$ during annealing and a protective surface layer of silicon oxide had formed on top of the $TaSi_2$ layer under the particular process conditions applied.

This example also shows that the combination of AES with TEM provides a particularly large information content, since TEM offers the superior lateral resolution (and depth resolution for cross section specimens), but AES yields the chemical composition of the layers.

Depth profiling as a means of thin layer analysis usually covers surface zones ranging from 1 atomic layer to several micrometers in thickness.

10.3 Ion Probe Techniques

The common principle of ion probe techniques is that the sample (the target) is bombarded with an ion beam (primary particles) in the energy range between a few hundred eV and several MeV. Depending on the primary particle/target combination and the kinetic energy, different interactions dominate and can be exploited for generation of analytical signals (Table 10.3-1).

If a target is bombarded with particles of low mass, e.g., He^{2+} (M = 4), elastic nuclear-nuclear Coulomb interaction dominates leading to elastic large angle scattering of the primary particles. Measurement of the energy loss spectrum of the backscattered primary ions forms the basis of scattering spectrometries. The interaction between the nuclei is most pronounced at high primary ion energies (1–3 MeV): Rutherford backscattering spectrometry (RBS). In the energy range of a few hundred keV the incoming ion interacts with a nuclear potential which is screened by the electrons of the atom at which backscattering occurs: medium energy ion scattering – MEIS. Screening effects get more pronounced in the low keV range leading to neutralization of incoming ions beneath the top atomic layer. Thus, in ion scattering spectrometry (ISS) only the outermost atomic layer contributes to the backscattering signal.

During the elastic collision between the incoming ion and the target atom a

Table 10.3-1. Survey of ion probe techniques

Fundamental process	Mechanism of signal generation	Typical energy of ions	Technique	Application
Elastic processes:				
Scattering	Back-scattering of light ions (He^{2+}, Ne^+)	1–3 MeV 200–600 keV 0.5–6 keV	RBS MEIS ISS	Quantitative thin film analysis, position of foreign elements, qualitative monolayer analysis
	Forward recoil emission	1–30 MeV	ERD	Quantitative thin film analysis of H, C, N, and O
Inelastic processes:				
Nuclear reactions	Emission of nuclear particles	MeV	NRA CPAA	Quantitative thin film analysis of light elements
Sputtering	Sputtering and surface ionization	1–30 keV	SIMS	Trace analysis at surfaces, micro- and nanoanalysis of trace elements, isotopic analysis
	Sputtering and post-ionization of neutrals	0.2–20 keV	SNMS	Quantitative thin film analysis, trace analysis

significant part of the kinetic energy of the primary particle is transferred to the target atom leading to its displacement ("recoil"). This effect can be utilized analytically with MeV ions knocking out light elements (like H, C, N, O) at the surface of the sample. Detection of those atoms which are ejected as ions is the basis of elastic recoil detection (ERD).

For specific combinations of primary particles and target atoms and energies high enough to overcome the Coulomb-barrier of the nuclei (typically several MeV) nuclear reactions can be obtained. Measurement of particles emitted during the decay of the unstable nucleus formed provides the basis for nuclear reaction analysis (NRA)/charged particle activation analysis (CPAA).

When a target is bombarded with ions of medium mass, e.g., O^+ ($M = 16$), Ar^+ ($M = 40$), and energies in the low keV range inelastic nuclear-electronic interactions dominate leading to disruption of chemical bonds of the surface atoms and sputter removal of these atoms. A fraction of the sputtered atoms is ionized at the surface during ejection. Measurement of these forms the basis of secondary ion mass spectrometry (SIMS). The majority of the ejected target particles are, however, in a neutral charge state. These can be analyzed if a postionization process (ionization of neutral species after ejection from the surface) is applied: sputtered neutrals mass spectrometry (SNMS).

Ion probe techniques are of great importance in surface analysis with RBS and SIMS belonging to the most important techniques. Therefore these are described in this section.

10.3.1 Techniques based on scattering

Rutherford backscattering spectrometry (RBS)

The experimental setup for RBS is shown schematically in Fig. 10.3-1. A helium ion beam (He^{2+}), energy 1–3 MeV, collimated to a few mm in diameter, impinges on a planar sample and the spectrum of the backscattered helium ions is measured with an energy dispersive surface barrier detector. Such a nuclear particle spectrometer consists of a silicon solid state detector covered by a thin layer of gold. The charge generated by an incoming helium ion is proportional to its kinetic energy, thus the energy spectrum of the backscattered ions is measured in an energy dispersive mode with a resolution of 10–20 keV (see also energy dispersive Si (Li) X-ray detector which operates by the same principle).

The scattering mechanism is Coulomb repulsion between the nuclei of the helium ion (which is actually an α-particle, He^{2+}) and the target atom occurring when the helium ion gets sufficiently close to the nucleus – typically within less than 10^{-3} Å (the K-shell radius is 10^{-2} Å!). In this elastic nuclear-nuclear collision the scattered helium ion transfers a certain amount of its kinetic energy to the target atom (which is consequently displaced) (Fig. 10.3-2). The energy loss depends on the masses of the atoms involved and the scattering angle. It is usually expressed as the kinematic factor $k_{M_2}(\Theta) = E_1/E_0$.

Since a purely elastic collision takes place, energy and momentum are conserved, thus:

$$k_{M_2}(\Theta) = \left[\frac{(M_2^2 - M_1^2 \sin^2 \Theta)^{1/2} + M_1 \cos \Theta}{M_2 + M_1}\right]^2$$

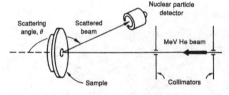

Fig. 10.3-1. RBS: experimental setup. A collimated helium ion beam is directed on to a planar sample, and primary particles backscattered at an angle of Θ are measured by an energy dispersive detector [10-5]

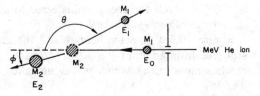

Fig. 10.3-2. RBS: elastic collision process between a projectile of mass M_1 and energy E_0 with a target atom of mass M_2. After backscattering at an angle of Θ ("scattering angle") the initial projectile has the energy E_1 [10-5]

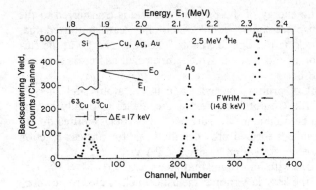

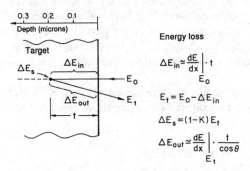

Fig. 10.3-3. RBS: back-scattering spectrum obtained from 1 monolayer of Au, Ag, and Cu on a silicon substrate showing the dependence of the energy loss on the target mass ($E_o = 2.5$ MeV) [10-5]

Fig. 10.3-4. RBS: energy loss processes for a projectile that is backscattered from depth t. ΔE_{in} and ΔE_{out} are due to inelastic electronic interactions, ΔE_s is due to elastic nuclear backscattering [10-5]

For Θ near 180° (a conventional experimental arrangement) and $M_1 = 4$:

$$\frac{E_1}{E_0} = k_{M_2}(\Theta) = \frac{(M_2 - 4)^2}{(M_2 + 4)^2}$$

This means that the energy of the backscattered helium ions is simply determined by the mass of the target atom at which backscattering takes place.

Figure 10.3-3 shows this relationship for the case of backscattering from a thin layer of about 1 monolayer of Au, Ag, and Cu on a light element substrate (silicon). It can be seen that the energy loss ($E_0 - E_1$) due to this elastic collision process is very different for the three films due to the different atomic masses. This is the basis for separating the contributions of different atomic species to the backscattering intensity. The mass resolution of RBS decreases with increasing mass of the target atoms and is about 2 for lighter elements, meaning that a ΔM of 2 is necessary to be separated on the energy scale.

For the example presented in Fig. 10.3-3 the backscattering peaks of Au, Ag, and Cu are narrow – their width is determined by detector resolution. The reason is that these metals are present in a thickness of only 1 monolayer.

The backscattering spectrum of silicon is however broad and extends from a steep edge down to (near) zero energy of the backscattered ions. This can be explained by the fact that helium ions penetrating a solid state material are passing through a "sea of electrons" between the nuclei. Due to the Coulomb interaction between He^{2+} and the electrons of the target the incoming ion loses some of its kinetic energy before the elastic collision. This energy loss also occurs when the helium ion, backscattered from a target atom at some distance from the surface, travels back to the surface. The energy loss due to inelastic electronic interactions is rather small compared with the energy transfer during an elastic backscattering collision. E.g., for 2 MeV He^{2+} ions in aluminum, the inelastic energy loss dE/dx is 30 eV/Å, while the energy transferred to an aluminum atom in an elastic backscattering collision is 900 keV.

Figure 10.3-4 shows schematically the energy balance of a projectile backscattered from an atom at a depth t from the surface. When the impinging particle reaches the atom at which scattering takes place it has lost the energy $\Delta E_{in} \simeq dE/dt|_{E_0} \cdot t$.

At the scattering location this particle does not have the energy E_0 as at the surface, but:

$$E_t = E_0 - \Delta E_{\text{in}}$$

During the elastic scattering process it loses the energy:

$$\Delta E_s = E_t - E_t \cdot \mathrm{K}_{M_2} = (1 - \mathrm{K}_{M_2})E_t,$$

thus, after scattering it has the energy:

$$E_1 = E_0 - \left.\frac{\mathrm{d}E}{\mathrm{d}x}\right|_{E_0} \cdot t - (1 - \mathrm{K}_{M_2})E_t.$$

The energy loss on its way out is then:

$$\Delta E_{\text{out}} \simeq \left.\frac{\mathrm{d}E}{\mathrm{d}x}\right|_{E_1} \cdot \frac{t}{\cos\Theta}$$

The energy which is measured for a particle backscattered from a depth t in direction $\sim180°$ ($\cos\Theta \sim 1$) is:

$$E_1(t) = \mathrm{k}_{M_2}(\Theta)\left[E_0 - t \cdot \left(\frac{\mathrm{d}E}{\mathrm{d}x}\right)_{E_0}\right] - t \cdot \left.\frac{\mathrm{d}E}{\mathrm{d}x}\right|_{E_1}$$

This means that backscattering from a bulk material – like silicon in Fig. 10.3-3 does not lead to peaks, but areas whose high energy edge is determined by the energy loss of the projectiles at the surface.

The electronic energy loss feature allows the determination of the depth of a target atom at which the backscattering process takes place, provided that dE/dx is known. Since these parameters can be calculated with high accuracy or measured with reference materials (thin films), depth distribution information can be extracted directly from Rutherford backscattering spectra: e.g., the thickness of a film can be determined accurately from the energy widths of the corresponding peaks.

Figure 10.3-5 shows an illustrative example of the elastic and inelastic energy loss features. The backscattering energy of the top gold layer corresponds to the elastic loss during backscattering ($\Delta E_s = 250$ keV). The kinetic energy of the helium ions backscattered from the bottom layer is given by the inelastic energy losses ΔE_{in} (22 eV/Å) + ΔE_{out} (23 eV/Å) and the elastic energy loss during scattering (250 keV). Thus the backscattering peak of the bottom Au layer appears at $E_1 \sim 2.570$ MeV.

The high energy edge of the backscattering signal of aluminum is at 1.65 MeV since $\Delta E_s = 1.35$ MeV. The position of the low energy side is determined by the inelastic energy loss of the incoming (22 eV/Å) and outgoing (29 eV/Å due to lower kinetic energy) particle plus the elastic energy loss (experienced at the bottom atomic layer of the aluminum film). Accounting for these losses the lower edge of Al in the backscattering spectrum is at 1.36 MeV.

Depth profiling with RBS is successfully applied for the study of heavy dopant atoms in semiconductors (e.g., Sb in silicon). The depth resolution is in the range

The depth resolution of RBS is in the range of 20–50 nm.

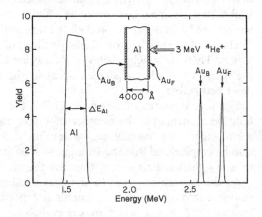

Fig. 10.3-5. RBS: back-scattering spectrum for 3.0 MeV helium ions incident on a 4000 Å Al film with gold markers (3 monolayers) on front and back surface ($\Theta = 170°$) [10-5]

of 20–50 nm and not as good as achievable with sputter-based techniques like SIMS.

After discussing the energy of the backscattered ions we now want to consider the intensity of the peaks in the RBS spectrum. The probability that the impinging ion is backscattered in the elastic collision is determined by the scattering cross section $d\sigma/d\Omega$.

The measured intensity is given by:

$$I = N \frac{d\sigma}{d\Omega} \Delta\Omega\, Q$$

where I = (integrated) intensity of backscattering signal; N = area density of backscattering target atom [at cm^{-2}] (equals the product of atomic concentration [at cm^{-3}] times thickness); $\Delta\Omega$ = acceptance angle of detector; and Q = number of incident projectiles.

Since $d\sigma/d\Omega$ can be calculated from fundamental parameters or determined from the measurement of reference materials quantitative evaluation is straightforward and simple. $\Delta\Omega$ can be determined experimentally for the detector geometry applied, and the incident ion current can be measured during the experiment. At this stage it is important to emphasize that the backscattering events being nuclear collisions are not influenced by the chemical bonding of the elements to be determined, thus RBS is a quantitative surface analysis technique which does not exhibit matrix effects.

Quantitation becomes even more simple for the typical application of RBS, the determination of the composition (stoichiometry) of thin films on a light substrate, e.g., metal silicides on silicon. Then the intensity ratio of (usually) two elements is measured, which leads to:

$$\frac{I_A}{I_B} = \frac{N_A}{N_B} \cdot \frac{\sigma_A}{\sigma_B}$$

Since σ is proportional to Z^2, it follows that:

$$\frac{N_A}{N_B} = \frac{I_A}{I_B} \cdot \left(\frac{Z_B}{Z_A}\right)^2$$

The integrated intensities I are the product of peak width (ΔE) and height (H). For a thin film, ΔE is constant for both elements, thus:

$$\frac{N_A}{N_B} \simeq \frac{H_A}{H_B} \cdot \left(\frac{Z_B}{Z_A}\right)^2$$

With this simple approach an analytical accuracy of 5% relative can be achieved. For a higher accuracy, peak integration and exact cross section values must be used. Then the composition of thin films can be determined to 1–2% relative.

If depth and quantitative information is evaluated from RBS spectra, (nondestructive) quantitative depth profiling and thin film characterization (with depth) is possible. Figure 10.3-6 shows an example in which the reaction of a nickel film (1000 Å) on silicon during annealing forming nickel silicide is characterized by RBS. After annealing, an increase of ΔE of the nickel peak is observed showing that the surface zone which contains nickel has been broadened. Then the silicon backscattering signal has moved to a higher energy to $k_{Si} \cdot E_0$, showing that silicon has diffused into the nickel and all the way to the surface. From ΔE_{Si} and ΔE_{Ni} the thickness of the newly formed silicide film can be determined using the energy loss values. Then the stoichiometry of the silicide film can be determined by evaluating the intensity of silicon (shaded area) in the film and the intensity of nickel as described before.

For investigations of this type RBS is the method of choice, because it is not only nondestructive (the number of target atoms being displaced by the elastic scattering event is usually negligible), but also fast and accurate. It must however also be emphasized that the strength of RBS lies in the analysis of heavy atoms on or in a lighter matrix – only then are the analytically interesting peaks separated from the substrate signal. In the case of films consisting of light elements on a heavy matrix the analytical signals are positioned on top of a large substrate (background) signal, which makes evaluation much more difficult and inaccurate.

Quantification is straightforward, since backscattering events are independent of chemical bonding of the target atom (no chemical matrix effects).

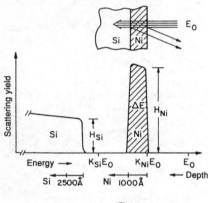

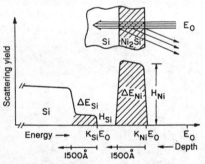

Fig. 10.3-6. RBS, back-scattering spectra of thin film system. Top: 1000 Å Ni on silicon. Bottom: 1000 Å Ni on silicon after annealing leading to formation of a thicker Ni$_2$Si film. Evaluation of ΔE yields the film thickness and of the integrated peak intensities (shaded areas) the stoichiometry of the silicide layer formed during annealing [10-5]

RBS is a nondestructive method.

The sensitivity for light element analysis is also much lower since σ is proportional to Z^2. Another limitation for thin film analysis is the thickness of the films – usually a maximum thickness of 0.5–1 μm can be analyzed because the energy losses due to inelastic scattering events become too dominant for thicker structures.

Low energy ion scattering spectrometry (ISS)

ISS has a unique position among surface analytical methods as the backscattering signal originates exclusively at the first atomic layer of a target. This is due to the fact that low energy noble gas ions are neutralized in inelastic electronic interactions when penetrating into the solid due to their high ionization potential. Thus, ions can only be detected when the elastic collision occurs immediately at the surface of the solid. ISS is the only surface analysis technique which is selective for monolayers irrespective of the identity of the atoms underneath. Monolayer selectivity with other techniques can only be achieved if the top monolayer is chemically different from those underneath (e.g., a film of adsorbed molecules on a metal surface).

ISS is the only method that provides information exclusively from the top monolayer of a solid irrespective of the substrate composition.

In ISS not only helium but also other noble gas ions (often Ne^+) with energies in the range 0.5–5 keV are used. For the determination of the kinetic energy, electrostatic energy analyzers are used. For qualitative analysis, the kinematic relations described for RBS are applicable (energy loss in elastic scattering processes with target nuclei). Quantitative analysis is practically not possible, since neither scattering cross sections nor the probability of neutralization of the ions during reflection are well-described.

ISS is largely used for fundamental studies (e.g., of adsorption) due to its unique surface selectivity.

10.3.2 Secondary ion mass spectrometry (SIMS)

Fundamental aspects of sputtering

Sputtering describes the emission of target particles under ion bombardment. Sputtering becomes the dominant process for ion-solid interactions when ions of medium or higher mass and energies in the keV range impinge on a target. The sputter process can be described using nuclear-nuclear interactions with screened potentials. This means that interactions of the ions with the electrons of the target atoms play a significant role in the process.

The actual mechanism of sputtering is seen as a transfer of energy from the incoming ion to target atoms in inelastic collisions. The kinetic energy is deposited in several hundred collisional encounters during about 10^{-13} s. In the collision cascade recoiled atoms are generated some of which can leave the surface. The kinetic energy of most of these particles is around 5 to 20 eV only. The depth of origin is for most particles (~90%) the topmost atomic layer. The probability of ejection of an atom under ion impact is expressed as the *sputter yield Y*, which is the number of emitted atoms per incident primary ion. Y is in the order between 1 and 10 under typical analytical conditions (sputtering with Ar^+, Xe^+, Kr^+, O^-, O_2^+, Cs^+, Ga^+ in the energy range of 0.2–30 keV).

Sputtering in all cases causes a change in the composition of the surface of the target material, since the primary ions are implanted into a surface zone of 5–10 nm thickness. Equilibrium concentrations of the implanted species ranging from a few percent to 50 percent are encountered. If reactive ion bombardment is used (e.g., with oxygen, gallium, or cesium ions) the surface chemistry, and consequently, the ionization probability for ejected particles is changed drastically, which has a great influence in SIMS (see "Principles and Applications of SIMS").

Sputtering always causes changes of the surface composition of a specimen.

When sputter removal is used in an analytical technique, the sputter induced changes in the specimen have to be considered. The major effects are:

1. Change of surface chemistry due to primary ion implantation.
2. Preferential sputter effects: When the various elements of a multielement specimen (e.g., an alloy) have different sputter yields then a change of surface

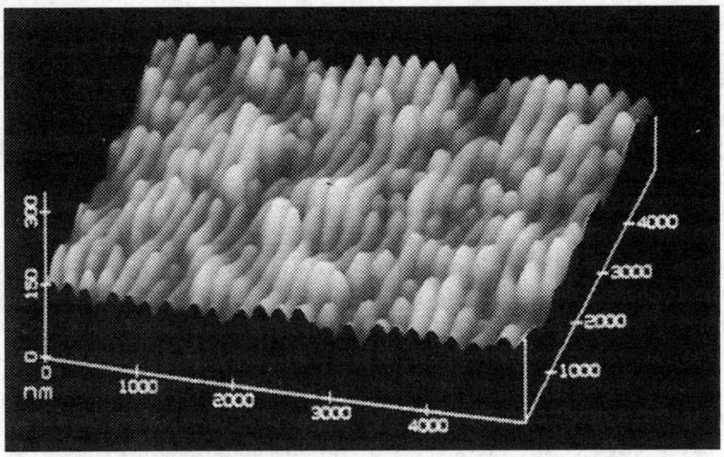

Fig. 10.3-7. AFM image of a SIMS sputter crater in an AlGaAs/GaAs multilayer material. The ripple structure has formed through bombardment with O_2^+ ions with an energy of 12 keV [10.3-1]
Image size: 5 μm × 5 μm,
Depth scale: 300 nm from black to white

composition is encountered. The atoms which have a lower sputter yield are enriched at the surface. Consequently, the composition of the remaining surface does not have the original composition. A correction is necessary for XPS and AES, taking into account the different sputter yields. For those techniques which analyze the removed particles (SIMS and SNMS) this effect will be compensated by a continuous surface enrichment in the element with the smaller sputter coefficient. In equilibrium this element will be enriched to the extent that the removed particle flux (being the product of area density and sputter coefficient) represents the original composition. Thus, corrections for preferential sputtering are not necessary for SIMS and SNMS for analysis in the sputter equilibrium, which is reached after removal of a surface layer whose thickness is approximately 2 R_p (R_p = projected range of implanted particles, typically 2–10 nm).

3. Selective sputtering: Specimens containing phases of different composition (e.g., an alloy with precipitates) exhibit different sputter yields for the different phases. In addition, the sputter coefficient at grain boundaries is usually larger than in the center of a phase. As a consequence the different components are removed at a different rate, thus yielding a particle flux not proportional to their average content in the specimen, and a roughening (step and cone formation) in the sputtered area can be observed. Problems occur in quantitation of analysis of heterogeneous specimens. Also the depth resolution is severely reduced due to the roughening effect. Figure 10.3-7 shows as an example for sputter-induced surface roughening an AFM image (see Sec. 10.5.2) of the bottom of a SIMS sputter crater (depth 1.2 μm) in an AlGaAs/GaAs multilayer material. While the initial wafer surface was flat with an average roughness (peak to valley) or 14 nm, Fig. 10.3-7 shows a characteristic ripple structure with an average roughness of 105 nm.

4. Sputter induced mixing effects: Due to the transfer of kinetic energy from the primary ions to the target atoms the target atoms are displaced. Consequently, the crystallographic surface structure is largely destroyed within the implantation zone and recoiling of target atoms into greater depths of the target occurs. The ion collision effects leading to a material transport have the consequence that an originally sharp interface between 2 layers is broadened in depth profiling (recoil and cascade mixing). The amount of broadening of an initially sharp layer is characterized by the so-called "decay length", which is the distance between position of the maximum in the profile and its 1/e intensity value. Recoil and cascade mixing pose the physical limits for depth resolution in sputter profiling and cannot be eliminated, only reduced (e.g., by grazing angle incidence or reducing primary ion energy). At optimized conditions for maximizing depth resolution, values for the decay lengths of 2–3 nm can be achieved.

Evaluation of raw data from sputter techniques is difficult.

But it must always be kept in mind that every sputter-generated profile does show a certain amount of broadening compared to the real distribution. In spite of these artefacts, sputtering is the only feasible way of controlled removal of surface layers for analysis – it is not only a good method for cleaning surfaces before analysis, but also the basis for destructive depth profiling. Three of the major surface analysis techniques (AES, XPS, and SIMS) rely on sputtering in surface analysis. The consequence of the fact that the sputter processes are complex and not well-described is that in every analysis the results have to be checked very carefully for possible sputter-induced influences. These have to be evaluated at least on a semiquantitative basis. Empirical knowledge and combination of sputter-based techniques with other methods using different principles of signal generation is of great importance.

Principles and applications of SIMS

SIMS is based on the bombardment of a surface with an ion beam in the energy range from 0.2–30 keV and measurement of the emitted ions of the target material. The most common primary ions are Ar^+, Ga^+, O_2^+, Cs^+, and O^-.

It is common to distinguish between two methods of SIMS:

1. Static SIMS: The primary current density is very low. Thus, only a fraction of a monolayer is removed during analysis [10.3-2]. Static SIMS is applied for analysis of extremely thin surface layers and organic materials (to reduce molecule fragmentation during analysis).
2. Dynamic SIMS: The primary current density is high. Consequently, material is removed at a rate of 0.1 to 100 atomic layers per second. This technique is the basis for depth profile measurements.

Instrumentation is optimized for these operational modes, e.g., static SIMS needs UHV conditions ($< 10^{-9}$ mbar), whereas dynamic SIMS, being less surface-contamination dependent can be used in moderate vacuum conditions of, e.g., 10^{-6} mbar.

Static SIMS is usually used for analysis of extremely thin surface layers (organic materials), dynamic SIMS for depth profiling and three-dimensional distribution analysis.

Static SIMS

Figure 10.3-8 shows the instrumental configuration for static SIMS. A pulsed Ar^+ beam containing about 10^3 ions per pulse and delivering about 10 000 pulses per second of < 1 ns width bombards a rather large area (10–1000 μm in diameter). With a primary ion density of 10^9–10^{13} ions per cm^2 s, only a very small fraction of an atomic layer is removed during analysis. The secondary ions are measured with a high-resolution ($M/\Delta M \leq 10\,000$) TOF (time-of-flight) mass spectrometer exhibiting a very high transmission (~20%), which is the basis for accurate determination of molecular masses and the high absolute detection power necessary for monolayer analysis.

Static SIMS seems to provide a substantial potential for molecular surface analysis [10-17, 10.3-3, 10.3-4]. Sputtering of an organic material under these conditions leads to a mass spectrum, which usually allows identification of the surface species by evaluation of the molecular fragments. Figure 10.3-9 shows an example which is particularly illustrative – the spectra of polycarbonate films with different chemical surface structures. The spectrum of the original surface of the injection moulded polymer contains mainly fragmentation peaks, which are typical for a polycarbonate. Also the end group of the polymer chain (i-octylphenolate) and residues of the release agent (palmitate and stearate anions) are found.

The spectrum of this polycarbonate surface changes dramatically if the polymer is exposed to an oxygen plasma: mainly fragmentation peaks from chain scission can be found showing the breaking of molecular bonds in the polymer due to exposure to the oxygen plasma. When this surface is rinsed with water the spectrum resembles that of the original surface, indicating that the fragments formed by plasma etching are water soluble.

This example shows that static SIMS is obviously able to describe chemical changes at organic surfaces. The most interesting features being its high surface sensitivity and depth resolution (1 monolayer, since virtually all detected ions originate from the top surface layer under static sputtering conditions).

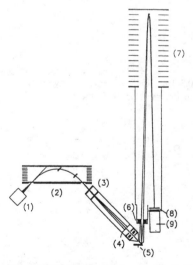

Fig. 10.3-8. SIMS: scheme of a static SIMS instrument. (1) primary ion source; (2) 90° deflector; (3) buncher (for focussing of the ion beam); (4) mass separation aperture; (5) target; (6) Einzel lens (electrostatic lens basically consisting of charged metal plates with holes for the ion beam); (7) reflector (reflects the ion beam at an angle of approximately 180° by means of an electrical field); (8) channel plate and scintillator; (9) photomultiplier [10.3-4]

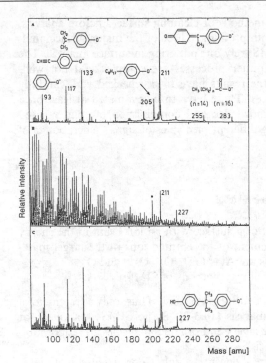

Fig. 10.3-9. SIMS: secondary ion mass spectra of polymer surface obtained in static sputtering mode (Ar^+) [10.3-5].
(a) Original surface of injection moulded polycarbonate
$M/e = 93, 117, 133, 211$ – fragments of polymer chain
$M/e = 205$ – i-octylphenolate (endgroup)
$M/e = 255$ – palmitate anion
$M/e = 283$ – stearate anion
(b) After treatment in oxygen plasma
(c) After treatment in oxygen plasma and rinsing with water

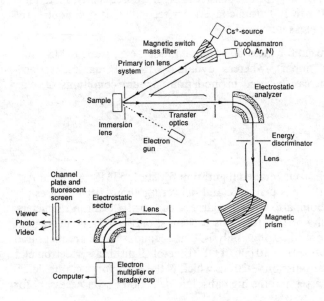

Fig. 10.3-10. SIMS: scheme of a dynamic SIMS instrument (from Cameca)

Dynamic SIMS

Figure 10.3-10 shows a dynamic SIMS instrument based on a double focusing mass spectrometer.

An ion beam generated by a duoplasmatron (e.g., O_2^+, O^-, or Ar^+ ions are generated by electrical discharging in a chamber purged with the corresponding gases) or a liquid metal ion source (e.g., Cs^+ ions are generated by thermal evaporation and ionization) is focused onto the surface of the sample. The diameter of the primary ion beam is in the range from about 100 nm to several hundred μm. The secondary ions generated are extracted with a voltage of 4500 V and directed into the entrance slit of an electrostatic analyzer of a double-focusing mass spectrometer. After mass separation with the magnetic analyzer, the secondary ions are detected using an electron multiplier or a Faraday cup. Secondary ion imaging can be achieved in two ways:

Lateral distribution information can be obtained either by scanning of the primary ion beam or by microscopic imaging.

• By scanning a finely focused beam across the sample surface and monitoring by means of the electron multiplier the secondary ion emission as a function of beam position. A lateral resolution of about 100 nm can be achieved, but depth profiling is slow due to the sequential analysis of each point at the surface.

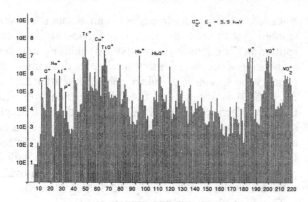

Fig. 10.3-11. SIMS: Mass spectrum of a WC–TiC–TaC–NbC–Co hard metal containing phosphorus as a trace constituent (bulk value = 2–3 μg/g) obtained with oxygen primary ions (O_2^+) at 5.5 keV [10.3-6]

- By making use of the microscopic imaging capabilities of this type of ion micro-analyzer. The sample surface is illuminated with a large primary ion beam (e.g., 300 μm in diameter) and the lateral distribution of the secondary ion emission is projected onto an imaging screen (e.g., channel plate with fluorescence screen). This microscopic imaging mode has a lateral resolution of approximately 1 μm, but a higher depth profiling speed, since atomic layers extending over the whole image area are removed simultaneously. Magnetic sector field spectrometers offer a higher mass resolution ($M/\Delta M < 20\,000$) compared with quadrupole spectrometers ($M/\Delta M < 1000$), which are also often used. This is a great advantage for the separation of interfering species in the complex mass spectrum.

Figure 10.3-11 shows, as an example, the positive secondary ion mass spectrum of a technical material – a WC–TiC–TaC–NbC–Co hard metal obtained under primary ion bombardment with oxygen. The survey (low resolution) mass spectrum shows a peak at virtually each mass number above 5. The large number of peaks originates from the emission of atomic ions (e.g., Ti$^+$ at m/e = 48), cluster ions (e.g., Ti$_2^+$ at m/e = 96), oxide ions generated by oxygen bombardment (e.g., TiO$^+$ at m/e = 64), other "molecular" species (e.g., TiC$^+$ at m/e = 60), and multiply-charged species (e.g., Ti^{2+} at m/e = 24). Then nearly every element has several isotopes which causes a further increase in the number of peaks in the mass spectrum.

Dynamic SIMS instruments also make use of a quadrupole instead of the sector field mass spectrometer. These instruments are more simple, but have only a mass resolution of approximately 800 and a lower transmission. The major potential of mass spectrometry lies in its large detection power, which is obvious from the dynamic range of the mass spectrum which can be up to 10^{10}. This means that signals from major and ultra trace constituents can be obtained in one spectrum.

Quantitative evaluation of secondary ion mass spectra is based on the following equation describing signal generation:

$$I_A^{+(-)} = I_p \cdot C_A^i \cdot Y \cdot \alpha_A^{+(-)} \cdot f(\Delta E) \cdot f(\Delta\Omega) \cdot T \cdot \beta$$

where $I_A^{+(-)}$ = measured ion intensity (ions per second); I_p = primary ion intensity (ions per second); C_A^i = isotopic concentration of analyzed element ($C_A^i = C_A \cdot \eta_A^i$; η_A^i = isotopic abundance); Y = (total) sputter yield (atoms per primary particle); $\alpha_A^{+(-)}$ = ionization yield; $f(\Delta E)$ = fraction of ionized particles within the energy range ΔE (energy window) that can enter the mass spectrometer; $f(\Delta\Omega)$ = fraction of ionized particles ejected from the surface at angles within $\Delta\Omega$ (solid angle acceptance window) that can enter the mass spectrometer; T = transmission of mass spectrometer; and β = yield of ion detector.

Typical values for these parameters are:

$I_p < 10^{13}$ ions/s;

$Y = 0.5$–20;

$\alpha_A^{+(-)} = 10^{-1}$–$10^{-4}$;

$f(\Delta E) \cdot f(\Delta\Omega) \cdot T \cdot \beta \sim 10^{-1}$–$10^{-3}$ (depending on mass resolution)

The major problem in quantitation of SIMS data is that sputter and secondary ion yields are strongly dependent on the chemical composition of the sample. This

SIMS spectra are very complex.

Chemical matrix effects can change signal intensities by several orders of magnitude.

so-called "chemical matrix effect" can amount to several orders of magnitude for sputtering with noble gas ions. A reduction of this effect is possible by reactive ion sputtering. The ions used are oxygen, gallium, and cesium, which are capable of influencing surface chemistry in a defined manner through implantation of the primary ion species: e.g., under noble gas sputtering, the secondary ion intensity of B^+ in SiO_2 is about 1000 times higher than in Si (for the same boron concentration). When sputtering with oxygen under a 60° angle of incidence (measured to sample surface), the signal in silicon is increased by a factor of about 50 due to the effect of implanted oxygen. Thus, the ratio of B^+ in SiO_2 and Si is only about 20. If normal incidence bombardment of the sample with oxygen is used then the implanted oxygen concentration in silicon can be equivalent to that of SiO_2, thus the same intensities of B^+ in SiO_2 and Si are obtained.

This example shows two aspects of SIMS:

- Secondary ion yields can be strongly influenced by reactive ion bombardment. Oxygen primary ions increase the yield for positive secondary ions by about 2 orders of magnitude (reaching a maximum under conditions of saturation of the surface with the oxygen being equivalent to the oxidic state). Cesium primary ions increase the yield for negative secondary ions by several orders of magnitude. In practice, oxygen and cesium primary ions are both used to achieve a maximum sensitivity: electropositive elements are analyzed as positive secondary ions generated usually under oxygen bombardment, and electronegative elements are analyzed as negative secondary ions generated under cesium bombardment.
- Chemical matrix effects can be eliminated or reduced by reactive ion bombardment.

The major consequence of the dependence of the secondary ion yield on surface chemistry (matrix composition) for quantitation is that relative sensitivity factors obtained from either an internal calibration or an external calibration with closely matrix-matched reference materials must be determined [10.3-6]. The internal calibration procedure is applicable for depth profiling when the total amount of the element to be determined is known, which is the case, e.g., for ion implanted dopants in silicon (from fluence measurements, which determine the total number of implanted ions per area). Then the relative sensitivity factor is obtained from integrating the measured depth profile. This value is proportional to the fluence (usually expressed as the total number of ions per cm^2) of the element.

External reference materials must be very similar in composition to the sample to be analyzed. Thus, ion implantation or doping of the same matrix is often used to prepare reference materials. For doped materials an extensive analytical characterization of the reference material is necessary. Often there is a lack of reference materials or reference analytical techniques. All these problems make quantitative analysis with SIMS a difficult task which needs a thorough methodological development for every particular element and matrix. Thus, SIMS should primarily be used for problems which cannot be solved by another less complex technique. Or, in other words, the analyst should deliberately make use of the unique potential of this method.

The outstanding features of SIMS which distinguish this method from AES, XPS, RBS, or other methods, are these:

- All elements can be analyzed. SIMS, for example, is particularly useful for the analysis of hydrogen.
- Isotopes can be distinguished. SIMS has the potential for isotope ratio measurements in microdomains (important for geochemistry and cosmochemistry) and in surface zones (important for the determination of diffusion coefficients of atoms in solid-state materials).
- SIMS has a large detection power. The absolute detection limit for most elements is in the range between 100 and 10^6 atoms, the relative detection limit typically in the ng/g range. Thus, SIMS is particularly suited for distribution analysis of trace elements.
- SIMS provides lateral resolution for microanalysis and high depth resolution. SIMS has the potential for three dimensional stereometric characterization of materials (Fig. 10.3-12).

SIMS is primarily used for problems that cannot be solved by less complex methods.

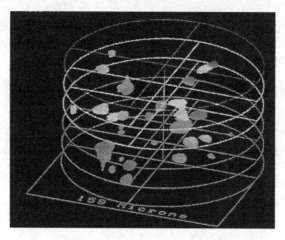

Fig. 10.3-12. SIMS: Three dimensional distribution analysis of oxygen in silicon by recording secondary ion images during depth profiling and reconstruction with image processing techniques. Three-dimensional representation of the distribution of SiO_2 precipitates.
Analytical signal: O^-, diameter of analyzed area $= 150\,\mu m$, depth $= 2.4\,\mu m$ [10.3-7]

SIMS is applied for virtually all types of solid-state materials, particularly for the characterization of isotopic effects and trace elements. The technique is of particular value for trace analysis of ultrapure materials due to the large detection power, and for the study of trace elements at grain boundaries and trace elements in thin films. Interface analysis is often performed by ion microscopy and lateral distribution characterization.

The most important task for SIMS is the characterization of semiconductors, particularly the (depth) distribution analysis of dopants. Since the distribution of dopant elements (e.g., B, P, As, Sb in silicon, and Si, Be in GaAs) determines the electrical and other properties, knowledge of this parameter is essential for the development and production of microelectronic devices.

Detection limits for depth profiling are in the order of 10^{14}–10^{15} atoms/cm^{-3}. The analytical accuracy of the concentration scale in depth profiles is about 5–20% rel., and that of the depth scale is 5–10 nm.

Insulators show basically a positive charging under secondary ion bombardment due to the emission of electrons from the sample surface. This charging effect can be reduced by the use of negative primary ions (O^-) and can be eliminated in practically all cases by simultaneous electron bombardment during analysis. SIMS offers, in this respect, a substantial advantage over AES. Some of the important applications of SIMS for insulators are the compositional analysis of ceramic coatings (produced, e.g., by chemical vapor deposition) or glasses (for the study of ion exchange phenomena on the surface).

10.4 Field Probe Techniques

High local electrical fields at the surface of a material can lead to ionization processes of either gases in contact with the surface or of atoms of the material itself. These processes are the basis for *field ion microscopy* and *atom probe analysis*. Another effect of high local fields is the stimulation of electrical currents which is the basis of *scanning tunneling microscopy* (STM). Moreover, in principle, the different *scanning force microscopy* techniques – the most important one being *atomic force microscopy* (AFM) – also belong to this group, since the measured forces can be seen as a result of localized electrical fields, too. Table 10.4-1 shows a survey of field probe techniques. However, due to the unique common properties of STM and AFM, these techniques will be described separately in Sec. 10.5 (Scanning Probe Microscopy Techniques).

Field ion microscopy (FIM) and atom probe analysis

FIM, invented by Müller [10.4-1], is based on the ionization of helium atoms on the surface of a fine metal tip (radius some nanometers) with a positive potential

FIM can achieve atomic resolution in real space.

Table 10.4-1. Survey of field probe techniques

Fundamental process	Mechanism of signal generation	Technique	Application
Field ionization	Imaging of surface ionized noble gas atoms	FIM	Imaging with atomic resolution
	Desorption of specimen atoms	atom probe analysis	Quantitative nanoanalysis of monolayers
Electron tunneling	Measurement of tunneling current	STM	Imaging with atomic resolution at surfaces, single atom spectroscopy
Interatomic force interaction	Measurement of forces	AFM	Surface imaging with atomic resolution (also insulators!)

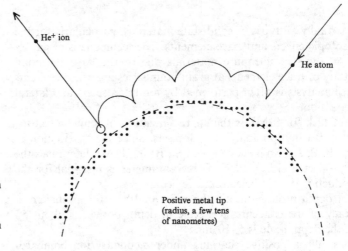

He⁺ ion

He atom

Positive metal tip (radius, a few tens of nanometres)

Fig. 10.4-1. FIM: Field ionization of helium atoms at the surface of a fine metal tip at high positive potential. Helium atoms (pressure ∼ 10^{-5} mbar) diffuse to the surface, are polarized, and are eventually ionized at an atomic row [10-4]

Atom probe analysis can obtain distribution information of elements with atomic resolution and a detection limit of only 2 atoms.

applied, such that a field in the order of 5×10^8 Vcm⁻¹ is present at the tip surface. The ionized helium atoms are accelerated towards an imaging screen due to a positive potential of several hundred volts between tip and screen. The helium ions impinge on the screen at different locations depending on their place of ionization on the tip yielding an image of the tip. Ionization of the helium atoms takes place at positions of highest fields; that is, on atomic rows on the tip surface (Fig. 10.4-1). Consequently, the atoms on the screen form an image of the atom positions on the tip. The magnification of this image is about 10^6, thus individual atoms on the tip are resolved in the image.

While FIM is exclusively concerned with atomic imaging of tip surfaces from which interesting structural information can be derived, atom probe analysis is directed towards identification and quantitation of surface atoms within a locally, extremely confined area. This is achieved by high voltage pulse or laser pulse induced desorption of surface atoms of the tip and measurement of the ionized species in a TOF mass spectrometer. Due to the high local electric field all desorbed atoms are ionized.

In a high voltage pulse atom probe the sample shaped to a tip is mounted in a UHV-sample chamber opposite an imaging screen (two-dimensional array of multichannel plate ion detectors). Atomic imaging of the tip of the specimen is performed with a potential of several hundred volts between tip and screen with helium ions at a pressure of approximately 10^{-5} mbar. Imaging allows an analytical area for analysis to be selected. Desorption of target atoms is achieved by a high voltage pulse (approximately 4000 V) of about 10 ns width. The ions (singly and multiply-charged) are accelerated towards the screen and can be detected there after a particular flight time. An elemental distribution image which has a lateral resolution of approximately 1–2 nm is obtained by accurate determination of the flight times of the different ions. The mass resolution is about 2 and is limited by the extremely short discrimination of flight times necessary, since the drift length is only 20–30 nm.

Qualitative and quantitative measurement of the target atoms desorbed from a selected nanoarea on the tip is achieved by directing these ions into a TOF mass spectrometer through a small aperture in the channel plate array. Selection of the analytical area is performed in the field ion imaging mode (with helium ions) by moving the tip of the sample or the screen. Due to the high magnification of the field ion microscope, the entrance aperture of the TOF mass spectrometer can define a very small area on the tip of the specimen. Lateral resolution for qualitative and quantitative analysis is 2 nm. Such an area contains only about 100 atoms per atomic layer. But due to the fact that the ionization yield is unity, the transmission of a TOF is above 50%, and single ions can be detected, the absolute detection limit is about 2 atoms. The relative detection limit for an analytical area of 2 nm in diameter and consumption of only 1 atomic layer is 2%.

Another advantage of the field ionization principle is that the ionization yield is unity for all elements and not dependent on matrix composition. Thus, quantitative analysis consists simply of determining the number of the various ions of a species and putting this number in relation to the total number of ions (of all atomic species) counted:

$$C_A = \frac{\sum I_A^i}{\sum I_i^i}$$

where C_A = atomic concentration of element A, $\sum I_A^i$ = sum of all ions of various charge states and mass of element A, $\sum I_i^i$ = total number of ions in mass spectrum.

The analytical error in quantitative results is purely based on counting statistics. Since often a very low number of atoms are desorbed, the counting error can be considerable.

These figures of merit show the great potential of atom probe analysis for ultrasensitive nano and surface analysis. The method has, however, two severe disadvantages: insulators cannot be analyzed, and the sample has to be shaped to an extremely fine tip. This is achieved mainly by pulsed electrochemical etching. Sample preparation becomes especially tedious when a particular feature in a material, e.g., a grain boundary, has to be prepared in such a way that it can be analyzed selectively. Repetitive polishing with intermittent observation of the nanostructure of the tip with a transmission electron microscope is necessary.

This obvious disadvantage limits the practical use of the method and confines it to the research laboratory.

Spectra generated by only about 100 atoms can be evaluated quantitatively.

10.5 Scanning Probe Microscopy (SPM) Techniques

The term scanning probe microscopy stands for a group of techniques in which a fine probe (usually a sharp tip) is scanned across a sample surface by means of piezoelectric translators. The signal measured by that local probe is recorded as a function of the lateral position. The signal generated by the probe can be, e.g.,

- the tunneling current (scanning tunneling microscopy, STM)
- different forces (scanning force microscopy, SFM; e.g., atomic force microscopy, magnetical force microscopy, electrical force microscopy, van der Waals force microscopy)
- the ionic current (scanning ion conductance microscopy, SICM) or
- electromagnetic radiation (scanning near-field optical microscopy, SNOM)

Since STM and AFM have gained up to now the highest importance in that field, they will be discussed in more detail in this section.

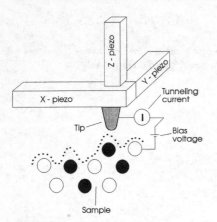

Fig. 10.5-1. Principle of STM

STM produces real-space images of surfaces with atomic lateral resolution, and a depth resolution in the sub-Å range.

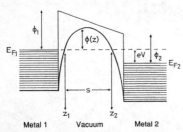

Fig. 10.5-2. STM: Energy level diagram of a metal-vacuum-metal junction with a voltage V applied across the barrier of a thickness s. The barrier height ϕ is lowered with respect to the work functions ϕ_1 and ϕ_2. In this situation electrons tunnel from the occupied states below E_{F1} (metal 1) to the empty states above E_{F2} (metal 2) [10.5-7]

At low bias voltages the STM produces images of the local density of states near the Fermi level (LDOS_{E_F}).

10.5.1 Scanning tunneling microscopy (STM)

STM, invented by Binnig and Rohrer in 1981 [10.5-1] (Nobel prize 1986) is based on measurement of the electron tunneling current between a fine metal tip and the surface of a material as a function of lateral position. Information is obtained on topographical surface structure and electronic properties with atomic spatial resolution [10.5-2 to 10.5-6].

Figure 12.5-1 shows the principle of STM. A sharp metal tip (e.g., Pt/Ir or W) is mounted on a piezoelectric translator which allows the tip to be moved in the x, y, and z directions. If the tip is close to the sample surface (approximately 1 Å), a small voltage (typically a few mV) can generate a tunneling current (typically a few nA) between tip and sample. Since the tunneling current is an extremely sensitive measure (exponential relation) for the tip-sample separation, the recording of this current as a function of the lateral position enables surface topography with high resolution in the sub-Ångstrom range to be obtained. When interpreting STM images, it must be considered however that electronic densities are monitored, which, on the atomic level, do not necessarily reflect the positions of the atomic nuclei.

STMs can be built to be compact and small, so that they can be easily mounted on flanges for UHV instruments and used in combination with AES, LEED, and sputtering, etc for surface studies under controlled conditions. It must be emphasized however that STMs can also be operated in air and even under liquids, thus opening up an exciting potential for studying surface chemistry. Elaborate image processing units complement the basic STM instrument.

The major significance of STM is that it produces real-space images of the surface, investigated with atomic lateral resolution, and a vertical resolution in the sub-Ångstrom range. The physical basis of the microscope is the tunneling effect, which describes the finite probability for an electron to penetrate an energy barrier which is higher than the kinetic energy of the electron.

Figure 10.5-2 shows a schematic representation of the physical phenomena in terms of an energy-level diagram of a junction between the tip and the surface separated by vacuum. The barrier height ϕ is lowered with respect to the individual work functions and electrons can tunnel through that barrier from atoms at negative potential to atoms at positive potential. These electrons will come from the occupied states of one atom and tunnel to the unoccupied states of the other atom. The direction of flow of electrons is given by the polarity of tip and surface.

The electron tunneling current between a tip and a solid surface at low voltages (mV) is in a first order approximation given by:

$$I \propto U e^{-A\phi^{1/2}s}$$

where I = electron tunneling current; U = potential between tip and surface; A = constant $(2(\sqrt{2m_e}/\hbar)$; ϕ = effective barrier height (local work function); and s = distance.

At low sample bias, the tip follows lines of constant density of states at the Fermi level (E_F) when scanned across the surface under a constant tunneling current, which is achieved by feedback control of the z-piezo which controls the motion of the tip perpendicular to the sample surface (z-direction). At high voltages not only stats in the vicinity of the Fermi level contribute to the tunneling current, but also states below or above E_F (depending on the polarity of the voltage). The amount of these contributions can be varied by changing the bias. Thus, information on the density of states of the valence band (when the tip is at positive potential and electrons tunnel from the surface to the tip) or the conduction band (tip at negative potential, electrons tunnel into empty states of the substrate) can be obtained.

In any case the measured electron tunneling current – typically in the range between 1 and 10 nA – is a function of the distance between tip and surface, of the voltage applied and of the effective barrier height. Thus, different information about the surface can be obtained with different operational modes [10.5-3, 10.5-6].

1. The most widely used technique is to measure the tunneling current across the surface at constant voltage. As the local distance between the tip and the

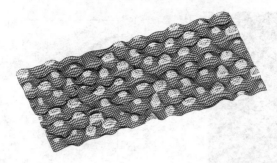

Fig. 10.5-3. STM: Topographic image of silicon (111) showing the (7×7) unit cells originating from surface reconstruction [10.5-8]

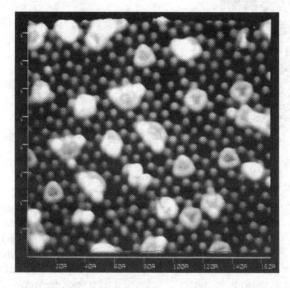

Fig. 10.5-4. STM: Atomic resolution image of silicon covered with 1/3 monolayer of silver. The adsorbed Ag atoms can be distinguished clearly from the silicon atoms due to topographic and local barrier height contrast [10.5-9]

surface changes, a varying tunneling current is obtained. In practice, the measurement technique is such that the tip moves along lines of constant electronic density of states by adjustment of the distance between tip and sample in each lateral position so as to achieve a constant tunneling current (*constant current topography* [CCT] mode). Thus, a two dimensional topographical image (strictly speaking, electronic density at the Fermi level) of the surface with atomic resolution is obtained. Figure 10.5-3 shows, as an example, the image of the silicon (111) surface. The individual atoms at the surface are clearly separated and the surface reconstruction $(7 \times 7$ structure) of silicon can be observed. When the surface contains adsorbed atoms the local work function (*effective barrier height*) is changed yielding a higher or lower tunneling current for these atoms.

Figure 10.5-4 shows, as an example, the topographic image obtained in the CCT mode of the silicon (111) surface with 1/3 monolayer of silver on top. The large contrast of the Ag atoms with respect to the silicon atoms is also due to a change in the effective barrier height. Thus, for systems where the surface chemistry changes, both topography (distance) and local electronic structure must be considered for interpretation of the images.

2. To record the local barrier height distribution, the distance between tip and surface is modulated at each location while scanning across a sample and the function dI/ds is measured. Since dI/ds is proportional to $\phi^{1/2} \cdot I$, information about the local barrier height is obtained at each position.

Since local barrier heights are influenced by chemical composition (e.g., adsorbed species), a two-dimensional measurement of this property enables information to be obtained about the species present.

3. Measurement of the dependence of the tunneling current on the bias voltage (dI/dU) for each spatial position of the tip at the surface at constant distance yields the spatial distribution of the density of states. Since different electronic states are accessible by variation of the bias voltage, a tool for electron spectroscopy with high spatial resolution is available (often called scanning tunneling spectroscopy, STS).

Generally, STM images cannot be directly interpreted in terms of atomic arrangements. The electronic (chemical) structure of the surface must be considered.

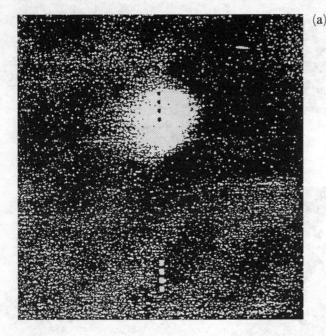

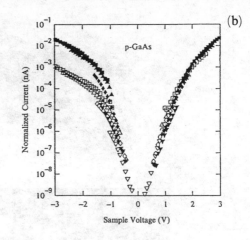

Fig. 10.5-5. STM: Micrograph of one adsorbed oxygen atom on the surface of gallium arsenide (a). $\mathrm{d}I/\mathrm{d}V$ curves taken from oxygensite and an oxygen-free area (b). [10.5-10]

Due to this exciting and large potential, STM is developing at a tremendous speed. The majority of applications of the technique is still in high resolution topographical imaging where STM extends SEM by more than two orders of magnitude in spatial resolution. Furthermore, STM is a much better tool for measuring vertical features on the surface of a sample, e.g., steps at surface planes. High-resolution topographical imaging is of great value for many problems and questions in material science and basic research – e.g., etching, polishing of metal surfaces, electrode processes, surface reconstruction, and terrace formation.

The evaluation of the spectroscopic features seems to be the center of interest for future work with atomic resolution being the ultimate goal for chemical surface analysis. The major problem encountered here is that the local potential barrier is not specific for a certain element. Thus, analysis is still largely restricted to known components with some qualitative distinctions possible – e.g., between adsorbed carbon or oxygen atoms by the shape of the dI/dV curves. Figure 10.5-5 shows an example. One single oxygen atom adsorbed at the surface of gallium arsenide as clearly observable in the STM (CCT) micrograph yields a $\mathrm{d}I/\mathrm{d}U$ curve which is significantly different from the clean gallium arsenide surface.

The major analytical drawback of STM is its limitation to conductive samples.

One of the major limitations of STM is that good insulators cannot be characterized. Another limitation is the possibility of field-induced effects, particularly when operating at higher voltages. Both limitations can be overcome with atomic force microscopy (AFM), which will be described in the next section.

10.5.2 Atomic force microscopy (AFM)

Besides the STM, the AFM is the most important scanning probe microscopy technique.

AFM belongs to the group of scanning force microscopy (SFM) techniques, a term which is used for several techniques which are based on the measurement of different forces (e.g., attractive, repulsive, magnetic, electrostatic, and van der Waals). Besides STM, the AFM, which was introduced in 1986 [10.5-11], is the most important and widest-used technique among the scanning probe microscopy (SPM) techniques.

In the AFM, a sharp tip mounted on a soft lever is scanned across the sample surface by means of piezoelectric translators, while the tip is in contact with the surface. The force acting on the tip changes according to the sample topography, resulting in a varying deflection of the lever (see scheme in Fig. 10.5-6), which in most commercial instruments is measured by means of laser beam deflection off a microfabricated cantilever and subsequent detection with a double segment photodiode. Figure 10.5-7 shows a schematic view of this detection concept. The laser

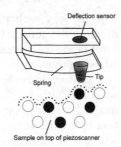

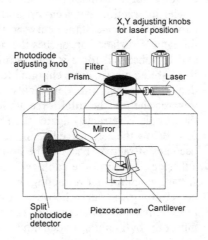

Fig. 10.5-6. Principle of AFM. The sample symbolized by the white and black circles is scanned by means of a piezoelectric translator

Fig. 10.5-7. Schematic view of the deflection sensing system as used in the NanoScope III AFM (Digital Instruments, Santa Barbara, Calif.). The deflection of the triangularly shaped cantilever is amplified by a laser beam focused on top of the cantilever and reflected towards a split photodiode detector

beam can be focused and positioned on the rear of the cantilever by means of adjusting knobs. The reflected laser beam is directed towards a double segment photodiode. As the cantilever deflection – and thus the mirror plane for the laser beam – changes, the position of the laser beam on the photodiode – and thus the difference signal between the two segments – changes, too. Therefore the difference signal is a sensitive measure for the cantilever deflection (resolution = 0.01 nm). Other options for sensing the cantilever deflection (e.g., tunneling contact, interferometry, capacitance) are described by Sarid [10.5-12].

Today, cantilevers with extremely low force constants of less than 0.1 N/m and resonance frequencies up to more than 100 kHz can be produced, which allow imaging with forces in the nanoNewton and sub-nanoNewton range. These forces are about 10 000 to 100 000 times lower than the force of gravity introduced by a fly (1 mg) sitting on a surface. But high local pressures (MPa up to GPa) are encountered, which may cause artefacts during measurement especially on soft systems (e.g., biological samples).

Silicon nitride cantilevers with integrated pyramidal tips (base: 4 μm × 4 μm, height: 4 μm) are commonly used in commercial instruments. The nominal radius of curvature of the tip apex is typically 20 to 50 nm. In the ideal case the tip should be terminated by a single atom (Fig. 10.5-8). In the AFM, the tip is always touching the sample surface, which we call the contact mode. As a consequence there is always an interatomic repulsive force at the contact area due to penetration of the electronic shells of the tip and substrate atoms. In addition to this short-range force long-range forces (e.g., coulomb forces between charges, dipole-dipole interactions, polarization forces, van der Waals dispersion forces, capillary forces due to adsorbate films between tip and substrate), which can be attractive or repulsive, are also encountered (Fig. 10.5-8). Although both types of forces contribute to the total force acting on the cantilever, only the varying repulsive interatomic force allows high-resolution imaging of surfaces. Since long-range attractive forces pull the tip towards the sample surface, they give rise to an increase in the local repulsive force which can deteriorate the experimental conditions. Thus, it is important to minimize those long-range forces (e.g., by imaging under liquids) in order to achieve very low repulsive forces (nanoNewton and less) in the contact area between tip and sample. This is especially important when looking at soft materials, which can be deformed or destroyed easily by the load of the tip.

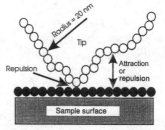

Fig. 10.5-8. Schematic view of forces encountered when the tip touches the sample surface. Bright circles symbolize tip atoms, dark circles symbolize symbolize sample atoms

In AFM, short-range, repulsive interatomic interaction forces are utilized for signal generation.

The force load in AFM is typically in the n-N range.

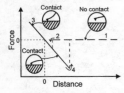

Fig. 10.5-9. Force-distance curve depicting the interaction of the AFM tip with the sample surface. The operation range of the AFM is between 3 and 4. For explanation refer to the text

The interaction between the sample surface and the tip can be described by force–distance curves.

AFM can be used to image features with dimensions at the atomic scale up to approximately 100 µm.

AFM reveals the surface topography with direct depth information.

With AFM insulators can be imaged, too.

In analogy with the STM, the AFM can be operated in two different with modes – the constant force mode and the constant height mode. In the constant force mode the cantilever deflection – and thus the force – is kept constant by re-adjusting the sample in vertical direction following the topographic features on the surface. In this mode comparatively large and rough sample areas can be imaged without destroying tip and/or sample surface ("tip crash"). In the constant height mode the vertical position of the sample is kept constant and the varying deflection of the cantilever is recorded. In this mode higher scan rates can be achieved which is advantageous for eliminating thermal drifts in high-resolution imaging; however, large scan sizes should be avoided, because "tip crashes" are possible.

The interaction between tip and sample can be described by force-distance curves. The force-distance curve in Fig. 10.5-9 shows how the force changes when the sample surface approaches the tip. At large separations there is no interaction and the observed force is zero (straight line between 1 and 2, if we assume that there are no electrostatic charging forces). At position 2 the tip jumps into contact due to attractive van der Waals interaction. As the sample is further moved towards the tip, the total force acting on the cantilever becomes repulsive. When the sample is retracted again, the force is reduced along the line from position 3 to 4. Below the zero force line in the diagram the net force acting on the cantilever becomes attractive, because the tip is held at the surface due to adhesion. In 4 the adhesion force and the cantilever load are just balanced and the tip flips off the surface when further retracting the sample. For AFM measurements the force can be set along the curve between position 3 and 4, preferably close to 4 in order to minimize the contact force. The value of the pull-off force can be reduced significantly by imaging under liquids due to elimination of capillary forces, which pull the tip towards the sample. For a general overview on force interactions see [10.5-13].

The *Analytical Potential of AFM* is based on the following possibilities:

1. Atomic resolution can be achieved, but imaging can also be performed on areas larger than 100 µm square. This capability is able to provide overview images, which can be used to zoom in on details with high resolution without changing the sample or the instrumental setup, which would make it impossible to find the identical sample position again. Figure 10.5-10 shows, as an example, an image of a PVD gold film on silicon, which gives a good overview on size and distribution of the single crystallites, which are typically 100 nm in diameter. The rms-roughness (rms stands for root-mean-square and reflects the standard deviation of all height values within a considered area) determined from that image is 3 nm. From XRD measurements, it is known that the surface of the gold film is mainly a (111) surface. Atomic resolution images of the position marked with the arrow in Fig. 10.5-10a revealed a gold (111) surface for that specific crystallite (Fig. 10.5-10b). Although the corrugation in one lattice direction is more prominent, caused by convolution with an asymmetric tip, hexagonal symmetry can be observed and the spacings are in good agreement with the expected values (0.29 nm).

2. A further important advantage of AFM is the fact that images contain direct depth information witch is important for roughness determinations. Figure 10.5-11 shows the topography of SrS films on glass produced with an atomic layer epitaxy process at different deposition temperatures. Such films are gaining importance in the production of electroluminescence displays (ELDs). Since light emission strongly depends on surface topography, interesting parameters are roughness and grain size of the thin films grown in the gas phase. Figure 10.5-11a shows that deposition at 300 °C leads to a layer with an rms-roughness of 15 nm and with a grain size in the range of 50 to 100 nm, while a temperature of 400 °C (Fig. 10.5-11b) leads to a coarser structure with an rms-roughness of 33 nm and a grain size of 150 to 200 nm. This kind of information, which is not accessible in such a straightforward way with other techniques, is useful for the optimization of the deposition process.

3. In addition to conducting samples, insulators can also be imaged readily with AFM without the need to coat them with a conductive film. This means that sample preparation is facilitated in many cases, and that artefacts introduced by the coating process can be avoided. Therefore, a wide variety of organic,

(a)

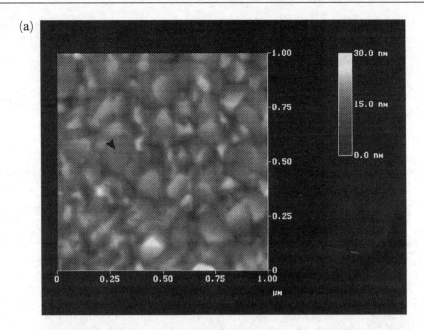

(b)

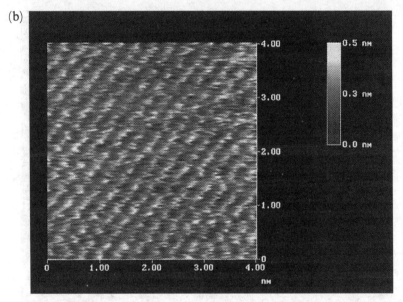

Fig. 10.5-10. (a) AFM image of a PVD gold film on silicon. Image size: 1 μm × 1 μm, depth scale: 30 nm from black to white. (b) Atomic resolution image of the crystallite marked by the arrow in (a). Image size: 4 nm × 4 nm, depth scale: 0.5 nm from black to white [10.5-14]

biological, but also inorganic insulators can be studied in a straigthforward manner.

As an important group of samples, which can be investigated readily with AFM and which are of high technological importance, glasses should be mentioned. An example is shown in Fig. 10.5-12 and will also be discussed in context with the in-situ imaging capabilities of AFM. Here, it should be emphasized that such topographical information on these samples is not accessible by other techniques in such a straightforward manner, while this image could be taken practically without any special sample preparation.

4. An important potential of AFM for surface chemistry is its capability to perform in-situ measurements under liquids and in air, since it allows direct observation of surface processes by operating the microscope with a liquid cell. In this cell, formed by the sample as bottom and a glass cover as top with a silicone o-ring seal between, both sample and cantilever can be immersed in the liquid while measuring. Such studies can also be performed under potential control which opens up valuable opportunities for electrochemistry. Moreover, imaging under liquids opens up the possibility to protect sensitive surfaces by in-situ preparation and imaging under an inert fluid. Figure 10.5-13 shows two AFM images taken on a sodium chloride crystal under acetic acid [10.5-15]. The time inter-

In-situ measurements in gases and under liquids allow direct observation of surface processes with high resolution.

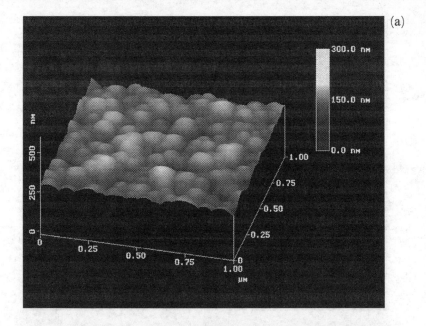

(a)

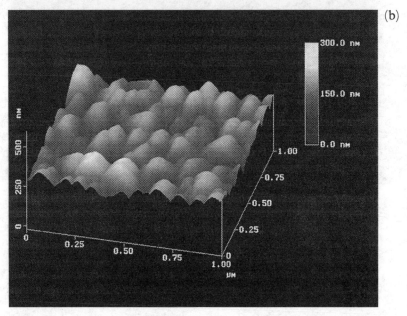

(b)

Fig. 10.5-11. Surface roughness and grain structure of SrS-EL-layers produced with atomic layer epitaxy (film thickness: 500 nm). Image size: 1 μm × 1 μm, depth scale: 300 nm from black to white. (a) Layer deposition at a substrate temperature of 300 °C; rms-roughness: 15 nm, grain size: 50 to 100 nm. (b) A deposition temperature of 400 °C leads to a coarser structure of the EL-film; rms-roughness: 33 nm, grain size: 150 to 200 nm [10.5-14]

AFM can greatly enhance analytical capabilities in many fields of research.

val between the images is 6 seconds. A motion of monoatomic steps across the atomically flat cleavage surface can be observed as the material dissolves in the liquid.

Although the field of AFM is still rapidly growing, its great potential for a wide variety of applications ranging from materials science to biology is already well-documented in literature. From the analytical point of view the in-situ imaging capabilities are of special interest, due to their great potential for investigating technologically relevant processes, e.g., corrosion.

Within the frame of this section only one example should be shown, the in-situ observation of the corrosion process on a potash-lime-silica glass [10.5-16]. Figure 10.5-12 shows the surface of a cleavage cross section through such a glass taken under humid air at different times after cleaving the sample. Immediately after cleaving, a flat surface with an rms-roughness of 1 nm can be observed. in the course of time, the surface topography changes due to dissolution of the sensitive network in the humid air. This process obviously leads to the formation of unstable corrosion products, which can be scratched away by the scanning AFM tip, leaving those pronounced craters (diameters and depths of order 1 μm and 5 to 10 nm, respectively). This kind of investigation bears a great potential for studying

(a)

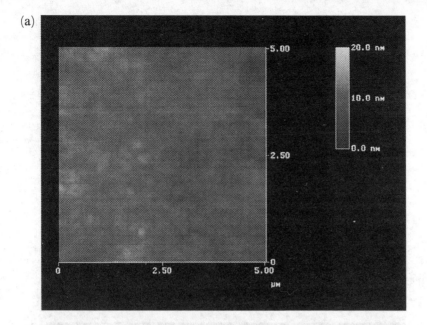

(b)

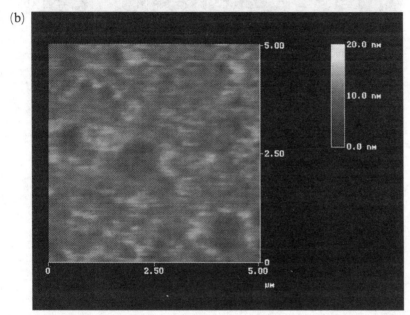

Fig. 10.5-12. AFM images of a potash-lime-silica glass surface produced by cleaving the material. Image size: 5 μm × 5 μm, depth scale: 20 nm from black to white. (a) Image taken shortly after cleaving the glass shows a smooth topography with an rms-roughness of approximately 1 nm. (b) Image taken at the same position after 100 minutes of exposure to humid air shows pronounced pitting [10.5-16]

surface properties, since such experiments can be performed under a variety of liquids and gases of controlled composition.

Although there is no doubt about the high potential of AFM in analytical sciences, it should also be mentioned that possible artefacts like convolution of the surface topography with the tip geometry (e.g., structures sharper than the tip result in an image of the tip) or destruction of sensitive samples by the load of the tip have to be considered when using this technique. However, developments for improving the method are still underway at a high pace, since AFM is still a comparatively young technique.

One improvement, the so-called tapping mode AFM, should be mentioned here. In this technique, the cantilever is vibrated above the sample surface in such a way that the tip is just slightly tapping the surface while scanning across it. In this way the force load can be reduced significantly and shear forces that could scratch off material can be practically eliminated. In the tapping-mode AFM, the surface topography is accessible by monitoring the oscillation amplitude of the cantilever, which is influenced by the damping due to tip-sample interaction. As an example for the benefits of tapping-mode AFM, Fig. 10.5-14 shows an image of the same glass sample as shown in Fig. 10.5-12b. While in Fig. 10.5-12b craters

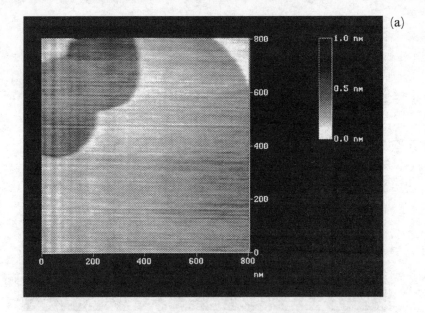

(a)

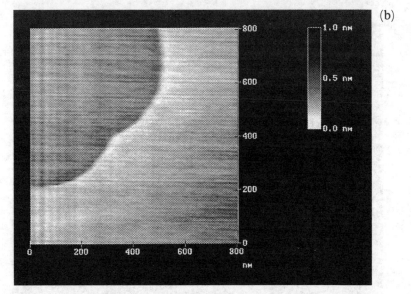

(b)

Fig. 10.5-13. AFM images of NaCl under acetic acid. The two images show the motion of a monoatomic step (from upper left to lower right) caused by dissolution of the crystal. The image in b) has been recorded 6 seconds later than the one in a). The step heights are 0.28 nm. Image size: 800 nm × 800 nm, depth scale: 1 nm from black to white [10.5-15]

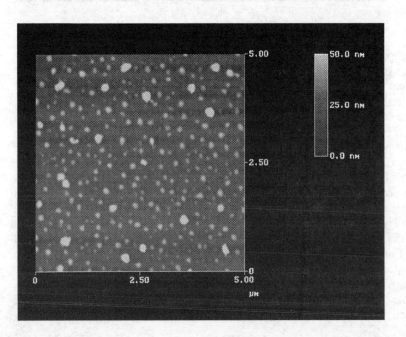

Fig. 10.5-14. Tapping mode AFM image of the same glass sample as shown in Figure 10.5-12b. Image size: 5 μm × 5 μm, depth scale: 50 nm from black to white [10.5-17]

produced due to the removal of unstable corrosion products by the AFM tip can be seen, the tapping mode AFM allows these corrosion products to be imaged as protrusions (Fig. 10.5-14).

Although the field of force microscopy is still developing rapidly, AFM has already become a powerful tool for characterizing surface morphology and for studying surface processes, and an ever-increasing importance of the technique for analytical chemistry is foreseeable.

References

[10-1]   Pimentel, G.C., Coonrad, J.A., *Opportunities in Chemistry: Today and Tomorrow.* Washington D.C.: US National Academy of Sciences, 1985.

[10-2]   Freiser, H., Nancollas, G.H. (Eds.), *IUPAC Compendium of Analytical Nomenclature.* Oxford: Blackwell Scientific Publishers, 1987.

[10-3]   Kittel, Ch., *Introduction to Solid State Physics.* New York: Wiley, 1976.

[10-4]   Prutton, M., *Surface Physics.* Oxford: Oxford University Press, 1983.

[10-5]   Feldman, L.C., Mayer, J.W., *Fundamentals of Surface and Thin Film Analysis.* New York-Amsterdam-London: North-Holland, 1986.

[10-6]   Woodruff, D.P., Delchar, T.A., *Modern Techniques of Surface Science.* Cambridge: Cambridge University Press, 1988.

[10-7]   Oechsner, H. (Ed.), *Thin Film and Depth Profile Analysis.* Berlin: Springer, 1984.

[10-8]   Ertl, G., Küppers, J., *Low Energy Electrons and Surface Chemistry.* Weinheim: Verlag Chemie, 1985.

[10-9]   Adams, F., Gijbels, R., Van Grieken, R. (Eds.), *Inorganic Mass Spectrometry.* New York: Wiley, 1988.

[10-10] Heinrich, K.F.J., *Electron Beam X-Ray Microanalysis.* New York: Van Nostrand-Reinhold, 1981.

[10-11] Benninghoven, A., Rüdenauer, F.G., Werner, H.W., *Secondary Ion Mass Spectrometry. Basic Concepts, Instrumental Aspects, Applications and Trends.* New York: Wiley, 1987.

[10-12] Joy, D.C., Romig, A.D., Goldstein, J.I. (Eds.), *Principles of Analytical Electron Microscopy.* New York: Plenum Press, 1986.

[10-13] Siegbahn, K. et al., *ESCA – Atomic, Molecular and Solid State Structure Studied by Means of Electron Spectroscopy.* Uppsala: Almquist and Wicksell, 1967.

[10-14] Ibach, H., Mills, D.L., *Electron Energy Loss Spectroscopy and Surface Vibrations.* New York: Academic Press, 1982.

[10-15] *Metals Handbook: Materials Characterization.* Metals Park, OH: American Society for Metals, 1986, Vol. 10.

[10-16] Briggs, D., Seah, M.P. (Eds.), *Practical Surface Analysis by Auger and X-Ray Photoelectron Spectroscopy.* New York: Wiley, 1983.

[10-17] Briggs, D., Brown, A., Vickerman, J.C., *Handbook of Static Secondary Low Mass Spectrometry.* Chichester: Wiley, 1989.

[10-18] Vanselow, R., Howe, R. (Eds.), *Chemistry and Physics of Solid Surfaces IV–VII.* Heidelberg: Springer, 1982, 1984, 1988, 1989.

[10-19] Chu, W.K., Nicolet, M.A., Mayer, J.W., *Back Scattering Spectrometry.* New York: Academic Press, 1978.

[10-20] Clark, R.J.H., Hester R.E. (Eds.), *Spectroscopy of Surfaces.* Chechester: Wiley, 1988.

[10-21] Werner, H.W., Garten, R.P.H., *Rept. Progr. in Physics,* 1984, 47, 221.

[10.1-1] Scoffield J.H., *J. Electron. Spectrosc.* 1976, 8, 129.

[10.1-2] Steffen, J., Hofmann, S., *Fresenius Z. Anal. Chem.* 1987, 329, 250.

[10.2-1] Goldstein, J.I., Newbury, D.E., Echlin, P., Joy, D.C., Fiori, C., Lifshin E., *Scanning Electron Microscopy and X-ray Analysis.* New York: Plenum Press, 1981.

[10.2-2] Goodhew, P.J, Humphreys, F.J., *Electron Microscopy and Analysis.* London: Taylor and Francis, 1988.

[10.2-3] Joshi, A., *Auger Electron Spectroscopy*, in: *Metals Handbook; Materials Character-ization*. Metals Park, OH: American Society for Metals; 1986, Vol. 10; p. 549.

[10.2-4] Van Criegern, R., Hillmer, T., Huber, V., Oppolzer, H., Weitzel, I., *Fresenius Z. Anal. Chem.* 1984, 319, 861.

[10.3-1] Friedbacher, G., Schwarzbach, D., Hansma, P.K., Nickel, H., Grasserbauer, M., Stingeder, G., *Fresenius J. Anal. Chem.*, 345, 615, 1993.

[10.3-2] Benninghoven, A., *Z. Phys.* 1970, 230, 403.

[10.3-3] Niehuis, E., Van Velzen, P.N.T., Lub, J., Heller, T., Benninghoven, A., *Surf. Inter-face Anal.* 1989, 14, 135.

[10.3-4] Van der Wel, H., Lub, J., Van Velzen, P.N.T., Benninghoven, A., *Mikrochim. Acta* 1990, 2, 3.

[10.3-5] Grasserbauer, M., Stingeder, G., Friedbacher, G., Virag, A., *Surf. Interface Anal.*, 1989, 14, 623.

[10.3-6] Grasserbauer, M., Wilhartitz, P., Ortner, H.M., Kny, E. *Int. J. Hard Refract. Met.*, 1986, 5, 30.

[10.3-7] Gara, S., Hutter, H., Stingeder, G., Tian, C., Führer, H., Grasserbauer, M., *Mikro-chim. Acta*, 1992, 107, 149.

[10.4-1] Müller, E.W., *Science* 1965, 149, 591.

[10.5-1] Binnig, G., Rohrer, H., Gerber, Ch., Weibel, E., *Phys. Rev. Lett.*, 1982, 49, 57.

[10.5-2] Golovchenko, J.A., *Science*, 1986, 232, 48.

[10.5-3] Hansma, P.K., Tersoff J., *J. Appl. Phys.*, 1987, 61, R 1.

[10.5-4] Frommer, J., *Angew. Chem.*, 1992, 104, 1325.

[10.5-5] Magonov, S.N., *Appl. Spectrosc. Rev.* 1993, 28, 1.

[10.5-6] Bonnell, D.A., *Scanning Tunneling Microscopy and Spectroscopy*. New York: VCH Publishers, 1993.

[10.5-7] Van der Walle, G.F.A., Van Loenen, E.J., in: *Analysis of Microelectronic Materials and Devices*: Grasserbauer, M., Werner, H.W. (Eds.), Chichester: Wiley, 1991.

[10.5-8] Binnig, G., Rohrer, H., *Spektrum d. Wissenschaft*, 1985, 10, 62.

[10.5-9] Neddermeyer, H., *Trends Analyt. Chem.* 1989, 8, 230.

[10.5-10] Stroscio, J.A., Feenstra, R.M., *J. Vac. Sci. Technol.*, 1988, B6, 1472.

[10.5-11] Binnig, G., Quate, C.F., Gerber, Ch., *Phys. Rev. Lett.*, 1986, 56, 930.

[10.5-12] Sarid, D., *Scanning Force Microscopy*. New York: Oxford University Press, 1991.

[10.5-13] Israelachvili, J., *Intermolecular and Surface Forces*. London: Academic Press, 1991.

[10.5-14] Friedbacher, G., Prohaska, T., Grasserbauer, M., *Mikrochim. Acta*, 113, 179, 1994.

[10.5-15] Prohaska, T., Friedbacher, G., Grasserbauer, M., *Fresenius J. Anal. Chem.*, 1994, 349, 190.

[10.5-16] Schmitz, I., Prohaska, T., Friedbacher, G., Schreiner, M., Grasserbauer, M., *Frese-nius J. Anal. Chem.* 1995, 353, 666.

[10.5-17] Schmitz, I., Schreiner, M., Friedbacher, G., Gasserbauer, M., *Fresenius J. Anal. Chem.*, 1997, 69, 1012.

Questions and problems

1. Name a few aspects of surface analysis which are important and significant in various fields of science and technology.
2. Name some criteria required for selecting a surface analytical technique for a given problem.
3. What is the typical range of sample volumes, masses, and concentrations for surface analytical techniques?
4. (a) Calculate the thickness of a layer (in multiples of the mean free path λ) from which 99% of the photoelectrons are originating assuming a .normal take-off angle. (b) Calculate the portion of photoelectrons originating from a depth between $10\,\lambda$ and $11\,\lambda$.

5. What are the typical applications for XPS and UPS, and which properties determine their range of applications?

6. We have divided electron probe techniques into two groups: techniques with high surface sensitivity and others with poor surface sensitivity but high spatial resolution. Which phenomena are responsible for these different properties?

7. In electron–electron interaction, which phenomena are caused by elastic and which by inelastic processes, and which techniques utilize these phenomena?

8. Name the different physical processes occurring in EPXMA and the related analytical techniques.

9. Describe the advantages and disadvantages of wavelength dispersive spectrometers as compared with energy dispersive systems.

10. Name the different physical processes occurring in AEM and the related analytical techniques.

11. What are the reasons for preparing thin specimens for AEM?

12. Compare the advantages and disadvantages of X-ray versus energy-loss analysis in AEM.

13. What are the most prominent analytical figures of merit of AES compared with other surface analytical techniques?

14. For which elements and problems is AES especially important and why?

15. What is the depth resolution of RBS? Why is RBS still a valuable technique especially when considering aspects of quantification?

16. Which analytical figure of merit is unique for ISS? What is the reason for that?

17. Which experimental parameters have an influence on the depth resolution of SIMS?

18. What aspects determine the choice of primary ions for SIMS measurements?

19. How can the roughness development in SIMS sputter craters be reduced?

20. Why is quantification of SIMS data not straightforward and what does this mean with respect to problems SIMS is usually used for?

21. What does the term static SIMS mean and what are typical applications for it?

22. What are chemical matrix effects and how can they be reduced in SIMS measurements?

23. Name the outstanding features of SIMS compared with other techniques such as AES, XPS, or RBS?

24. What are the common features of scanning probe microscopy techniques?

25. What kind of information do STM images primarily show?

26. What are the unique features that make STM an interesting tool for surface science?

27. What are the problems associated with interpretation of STM images especially on the atomic scale?

28. What can be done in STM experiments to distinguish between electronic and topographical information – for example, in order to decide whether a bright spot in the image is caused by a bump or a different atom?

29. Describe the constant current and the constant height mode, and compare their advantages and drawbacks for specific applications.

30. Which specification (force constants, resonance frequencies) should be fulfilled by AFM cantilevers and what are the consequences for practical designs?

31. Which measures can be taken in AFM experiments to reduce the force acting on the sample surface?

32. Which effects determine the lower limit of possible forces applicable for AFM measurements?

33. What information can be extracted from force-distance curves?

34. What are the advantages of performing AFM measurements under liquids?

35. Describe the constant force and the constant height mode and compare their advantages and drawbacks for specific applications.

36. Describe the analytical figures of merit for AFM and compare them with STM.

37. Put together the analytical figures of merit (e.g., sample volume, resolution, absolute and relative detection limits, dynamic range, aspects of quantification, possible artefacts, and so on) for several techniques described in this chapter. By comparing these data draw your conclusions about the usefulness of the single methods for different analytical problems.

11 Structural Analysis

Learning objectives

■ To provide a general appreciation of the diffraction and spectroscopic methods available for the determination of molecular structures

■ To introduce the basic physical principles of X-ray diffraction by crystals

■ To present modern instrumentation for the study of X-ray diffraction patterns from powder samples or single crystals

■ To describe the analytical applications of powder diffractometry

■ To introduce the basic principles of crystal structure analysis

Crystals, with their long-range ordered structure, provide an excellent diffraction grating for X-rays, giving rise thereby to a discontinuous scattering pattern that is dependent on the relative positions and scattering powers of the individual atoms. It is not possible to focus the scattered waves physically, so as to measure the electron density distribution in the crystal lattice in a direct manner. This task is performed in X-ray structural analysis by Fourier transformation of the observed diffraction pattern of a single crystal.

11.1 General Philosophy

At its basic level, the goal of structural analysis is restricted to the determination of atom connectivity in molecular or solid-state compounds. With a selection from today's powerful array of spectroscopic techniques, discussed in Chap. 9, it is generally possible to establish the constitution of newly synthesized organic molecules in a more or less routine manner. This may often not be the case for inorganic compounds, which often exhibit greater structural diversity and which are often less amenable to the major analytical methods, such as NMR and mass spectrometry.

It is apparent that a three-dimensional visualization of the molecular structure, as provided by an optical microscope for samples with dimensions in the range 1–100 µm, would be capable of delivering the required information in a direct manner. However, the resolution of such an instrument (that is its ability to distinguish between neighboring objects) is limited by the wavelength of the radiation or particles employed. As distances between chemically bonded atoms typically lie between 0.9 and 3 Å ($1\,\text{Å} = 10^{-8}\,\text{cm}$), X-rays with wavelengths within or close to this range might be expected to be appropriate to view molecular structures. Unfortunately, such a direct approach is not feasible, as no material has as yet been found that is capable of focusing X-rays in a manner analogous to a glass lens in an optical microscope. In contrast, high-energy electrons, which have an appropriate wavelength (as given by the de Broglie relationship), can be focused by an electrostatic field. Although electron microscopy does indeed allow the visualization of larger molecules and in favourable cases individual atoms, it cannot yet achieve the resolution of X-ray diffraction (Sec. 11.2) and so is not available as a routine method of structural analysis.

In addition to absorption accompanied by fluorescence (Sec. 8.3), the interaction of X-rays with atoms can lead to scattering, which can be both elastic (Rayleigh effect) or inelastic (Compton effect). During the former process, electrons in the atom involved are accelerated by the incident X-ray beam and become, themselves, a source of radiation with the same energy and wavelength as the original X-ray beam. In contrast, the Compton effect reflects the particle nature of electromagnetic radiation and can be treated as a collision between a photon and an electron, which leads to loss in energy and increase in wavelength of the X-ray radiation – in accordance with the laws of conservation of energy and momentum. Fortunately, inelastic scattering only plays a subordinate role for wavelengths,

Elastic scattering of X-rays produces radiation of the same wavelength as the original source.

such as CuKα (1.5418 Å) or MoKα (0.7107 Å), commonly employed in X-ray experiments. It does, however, lead to a relatively high background level of scattering. In the elastic (coherent) scattering process, the accelerated electrons give rise to scattered waves in all directions.

For amorphous materials or liquid samples with a very restricted degree of internal order (covering a range of only a few atoms or molecules), interference effects between the wavefronts from neighboring atoms lead to a continuous scattering pattern of limited use. Crystals, with their long-range ordered structure (Sec. 11.2.1) provide an excellent diffraction grating for X-rays, and therefore give rise to a discontinuous scattering pattern that is dependent on the relative positions of the individual atoms. As has been discussed, it is not possible to focus these scattered waves physically. In X-ray structural analysis this task is performed by appropriate calculations using the method of Fourier transformation (Sec. 11.2.3). As a result of the advent of high-performance computers and appropriate computer programs, what used to be a very arduous assignment – even up to the late 1960s – has now developed into a routine analytical method. More than 10000 crystal structures worldwide are now published per annum.

X-ray diffraction dominates the methods of structural analysis at the level of molecule visualization. However, a number of other diffraction and spectroscopic techniques are available which can provide metrical information, These include the follow methods:

- Neutron diffraction
- Gas-phase electron diffraction
- Extended X-ray absorption fine structure (EXAFS)
- Rotation spectroscopy
- Liquid-crystal NMR spectroscopy

The ordered internal arrangement of atoms/molecules in crystals acts as a diffraction grating for X-rays.

X-ray and neutron diffraction are complementary methods, as X-rays are scattered by electrons, neutrons by atom nuclei.

Neutron diffraction, like X-ray diffraction, requires crystalline samples. It also turns out to be complementary to X-ray diffraction: X-rays are scattered by electrons; neutrons by atom nuclei. The combination of high-quality data from these diffraction techniques yields information on the electron-density distribution for molecules in crystals.

EXAFS can provide interatomic distances for both amorphous and gaseous samples.

Since molecules are present in all orientations in the gas phase, electron diffraction for this state of aggregation yields a continuous scattering pattern and is only capable of providing one-dimensional radial distribution curves dependent on interatomic distances and the scattering power of the atoms involved. This technique is, therefore, restricted to relatively small molecules that must, of course, exhibit an adequate vapor pressure. EXAFS, which may be used in the investigation of both amorphous and gaseous samples, is a related method in which photoelectrons are liberated from atoms under investigation by high-energy X-rays. The back-scattering at neighbouring atoms of the resultant electron waves leads to modification of the observed absorption for X-rays with energies just above an absorption edge such as Kα. By analogy to gas-phase electron diffraction, EXAFS provides one-dimensional radial distribution cures that have, for instance, proved useful in the investigation of the active sites in metalloenzymes.

Rotation spectroscopy of gas-phase molecules allows the determination of moments of inertia (which are dependent on interatomic distances and atomic masses). The use of isotopomers for molecules with more than three independent geometrical parameters means the technique is restricted to very small molecules – preferably of relatively high symmetry.

Liquid-crystal NMR spectrometry measures the direct through-space coupling of nuclear magnetic moments in orientated molecules.

The direct coupling of two nuclear magnetic moments i and j through space, D_{ij}, is dependent on r_{ij}^{-3}, where r_{ij} is the internuclear distance. Due to molecular tumbling, the average direct coupling in solution is zero. Although modern NMR methods enable the observation of direct spin-spin coupling in single crystals and amorphous solids, the extraction of structural information is complicated by the presence of both inter- and intramolecular interactions. The use of liquid crystals as orientating solvents removes the disadvantage of solid samples by averaging out the intermolecular direct coupling while still retaining a nonzero value for the intramolecular direct coupling Liquid-crystal NMR spectrometry of orientated molecules now provides an efficient method of structure determination for small molecules having up to ten magnetic nuclei.

11.2 X-Ray Diffraction

11.2.1 Diffraction by crystals

Crystals are solids in which a basic arrangement of atoms or molecules is repeated periodically by translation vectors in three nonplanar directions to generate a three-dimensional structure. This description leads to the crystallographic concept of a small building block, the *unit cell*, which contains the full structural information of the crystal as a whole. The unit cell (Fig. 11.2-1) is a parallelepiped, characterized by three vectors **a**, **b**, and **c** that define its edges, and three angles α, β, and γ between these vectors. The eight corners of the unit cell, all of which must exhibit an identical environment within the crystal are referred to as *lattice points*. Implementation of the translation vectors of the periodic internal crystal structure yields an infinite three-dimensional array of evenly spaced lattice points denoted as the *crystal lattice*

The *unit cell* with dimensions a, b, c, α, β, γ is the repeating unit of the *crystal lattice*.

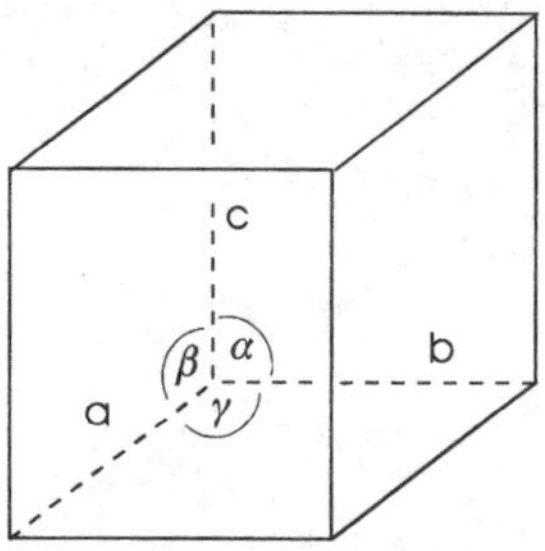

Fig. 11.2-1. The unit cell

Symmetry operations

As was established by mineralogists in the nineteenth century, two classes of symmetry operation are required to describe the internal arrangement of atoms or molecules in crystals. *Proper symmetry operations* such as rotations or translations retain the chirality of an object. In contrast, *improper symmetry operations* convert an object into its mirror image, e.g., an change in configuration from R to S for a chiral tetrahedral atom. Symmetry operations are performed on points, axes, or planes referred to as *symmetry elements*. In a crystal they transfer atoms or molecules to sites with identical environments. For instance, a crystal structure exhibiting *n*-fold rotation axes will appear indistinguishable from the starting view after rotation by an angle of $2\pi/n (360°/n)$ about such an axis. As a result of internal periodicity, only the rotation axes $n = 1$ (monad), 2 (diad), 3 (triad), 4 (tetrad), and 6 (hexad) are possible for crystals. The crystallographic symbols for these axes and the symmetry-equivalent positions generated by their use are depicted in Fig. 11.2-2. Translation describes the transfer of an object in a given direction and will, of course, leave the chirality of the object unchanged. In crystals, rotations of $2\pi/n$ can be combined with translations of $(r/n) \cdot \mathbf{x}$ ($r = 1, 2, \ldots, n - 1; \mathbf{x} = \mathbf{a}, \mathbf{b}, \mathbf{c}$) to yield so-called *screw axes* n_r.

The *symmetry operations* of a crystal are performed on points, axes or planes known as *symmetry elements*, and transfer atoms or molecules to positions with identical environments.

Worked example: Symmetry operations

A tetrad rotation axis *4* in direction *c* can be combined with translation vectors $c/4$, $c/2$ and $3c/4$ ($r = 1, 2, \ldots, n - 1; n = 4$) in a crystal lattice to afford the respective screw axes 4_1, 4_2 and 4_3. The symbols for these combination symmetry elements are depicted below together with the sites generated from an original atom at height z/c (given as + for atom 1).

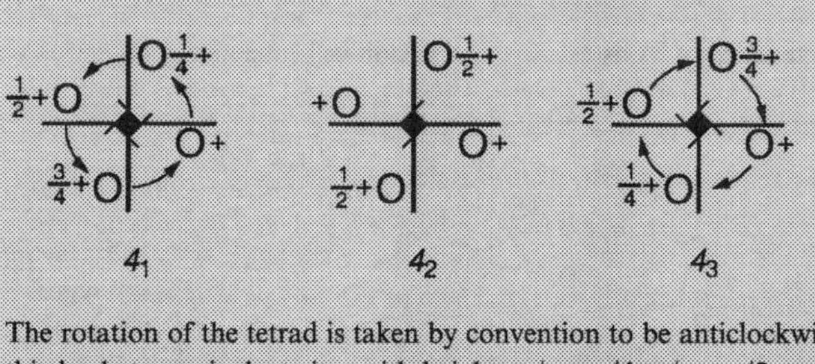

The rotation of the tetrad is taken by convention to be anticlockwise. For the 4_1 screw axis this leads to equivalent sites with heights $z/c + c/4$, $z/c + c/2$ and $z/c + 3c/4$. The fourth operation yields a position with the same coordinates as the original atom in directions **a** and **b** but a height of $z/c + 1$, which means that this atom occupies a site identical to that of atom 1 in the next unit cell in the **c** direction. Only the height of the translation-related atom in the original cell is, therefore, given in the crystallographic projection. An analogous procedure affords the displayed atom sites for the 4_2 and 4_3 screw axes. Comparison of the diagrams for 4_1 and 4_3 indicates that these axes generate sets of equivalent positions that are mirror images and cannot be superimposed. Such a pair of screw axes is described as *enantiomorphous*. In contrast the 4_2 screw axis, like the tetrad itself, is without a sense of rotation.

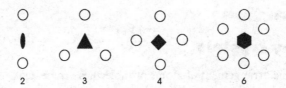

Fig. 11.2-2. Crystallographic rotation axes with symmetry-equivalent positions

The improper symmetry operations in crystals can be described by use of rotary inversion axes, $\bar{n}$, and glide planes. Rotary inversion axes correspond to a rotation by $2\pi/n$, followed by inversion through a point (center of symmetry) lying on the axis involved. This leads to inversion of configuration, which is denoted by a comma in the circle for the position involved. As may be seen from Fig. 11.2-3, the symmetry operation $\bar{2}$ is equivalent to reflection in a mirror plane (m) perpendicular to the rotation axis. This has led to a conventional use of the symbol m in the description of crystal structures. *Glide planes* can result from the coupling of a mirror plane with a translation $\mathbf{a}/2$, $\mathbf{b}/2$ or $\mathbf{c}/2$ (or combinations of two of these). These symmetry elements are denoted as a, b, c or n respectively, depending on the direction of translation or whether a combination of translations is involved. The coupling of a mirror plane with a translation $(\mathbf{a} + \mathbf{b})/4$, $(\mathbf{a} + \mathbf{c})/4$ or $(\mathbf{b} + \mathbf{c})/4$ yields a *diamond glide plane* represented by the symbol d.

Crystal symmetry

The characteristic symmetry elements of a crystal structure determine the shape of the unit cell and allow its assignment to one of the seven crystal systems.

There are six basic shapes for unit cells (Table 11.2-1) which reflect the characteristic symmetry of a crystal structure. These correspond to seven *crystal systems*, because triad and hexad axes lead to unit cells with identical geometries. In addition to the primitive lattices (type P) already discussed, which just have lattice points at the corners of the unit cell, additional points can be added either at the center of the cell (type I), at the center of opposite faces (type A, B or C), or at the centers of all faces (type F). In some cases, however, it may still be possible, after centering, to choose a simpler lattice; alternatively, the new lattice may not be compatible with the symmetry of the crystal system. As a result there are only 14 independent three-dimensional lattices. These are known as the *Bravais lattices*.

Miller indices hkl are used to characterize both the external faces and internal lattice planes of a crystal.

The external appearance of crystals is controlled by their internal periodic structure. Characteristic faces develop by addition of atoms or molecules to a crystal nucleus in an ordered stepwise manner. An example of crystal growth is depicted in Fig. 11.2-4 for an arrangement of spherical atoms. Similar growth rules also apply to the packing of molecules. The external faces of a crystal are parallel to a set of lattice planes, which can be drawn through the lattice points of

Table 11.2-1. The seven crystal systems and their unit cells

Crystal system	Characteristic symmetry	Axis	Unit cell constants
Triclinic	1		$a \neq b \neq c;\ \alpha \neq \beta \neq \gamma$
Monoclinic	2	$\mathbf{b}$	$a \neq b \neq c;\ \alpha = \gamma = 90°,\ \beta > 90°$
Orthorhombic	2	$\mathbf{a, b, c}$	$a \neq b \neq c;\ \alpha = \beta = \gamma = 90°$
Trigonal	3	$\mathbf{c}$	$(a = b) \neq c;\ \alpha = \beta = 90°,\ \gamma = 120°$
Tetragonal	4	$\mathbf{c}$	$(a = b) \neq c;\ \alpha = \beta = \gamma = 90°$
Hexagonal	6	$\mathbf{c}$	$(a = b) \neq c;\ \alpha = \beta = 90°,\ \gamma = 120°$
Cubic	3($\times$4)	cube diagonals	$a = b = c;\ \alpha = \beta = \gamma = 90°$

Fig. 11.2-3. Crystallographic rotary inversion axes with symmetry-equivalent positions (+ and − denote respectively above and below the plane of the page)

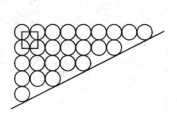 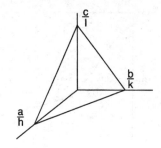

Fig. 11.2-4. (a) Development of the crystal face (210) shown in two dimensions. (b) Miller indices (*hkl*) of a crystal face

the structure. Both the faces and their respective lattice planes may be characterized by their *Miller indices hkl*, provided by the intercepts a/h, b/k and c/l of the first lattice plane from the origin. Consideration of the conditions of face development suggest that the values of the integers h, k and l should be small and, indeed, values greater than 3 or 4 are rarely observed.

Only crystallographic operations of the type n and $\bar{n}$ are admissible for the description of symmetry relationships between the external faces of crystals. These can be combined to give 32 crystallographic point groups; these are known as *crystal classes*. The periodic internal arrangement of atoms in a crystal structure requires the implementation of translation vectors, which can also be coupled with rotation axes and mirror planes, as discussed above. Inclusion of the derived symmetry operations, screw axes and glide planes, leads to the 230 *space groups* permissible for combinations of symmetry elements in the unit cell. These are listed in the *International Tables for Crystallography* [11.2-1]. It is interesting to note, in this context, that approximately 75% of all organic and organometallic compounds crystallize in only 5 space groups and that 12 space groups, all belonging to the triclinic, monoclinic, and orthorhombic crystal systems, cover about 87% of such derivatives. These space groups all allow a relatively good close packing of organic molecules with their typically low symmetry.

Whereas only 32 *crystal classes* (point groups) are possible for crystals themselves no less than 230 *space groups* are permissible for combinations of *symmetry elements* in the *unit cell*.

Bragg reflections

As mentioned in Sec. 11.1 crystal lattices provide an excellent diffraction grating for X-rays. When registered on a photographic film, an X-ray diffraction pattern of a single crystal consists of a series of regularly spaced spots, whose positions depend on the size of the unit cell and the crystal orientation. Bragg demonstrated in 1913 that the angular distribution of such scattering maxima may be calculated by considering the diffraction process to require the geometrical conditions for reflection on a series of lattice planes (Fig. 11.2-5). Under these conditions, constructive interference can only occur when the parallel diffracted waves $(1', 2', 3',$ etc.) all display a path difference of $\lambda(2\pi)$ to their nearest neighbours. The Bragg equation:

$$\lambda = 2d_{hkl} \sin \theta \qquad (11.2-1)$$

relates the wavelength λ to the interplanar distance d_{hkl} and the Bragg angle θ. Although diffraction is involved, the geometrical requirements have led to the use of the expression *Bragg reflection hkl* for such scattering maxima.

The *Bragg equation* relates the angular distribution of scattering maxima, known as *Bragg reflections hkl*, to the *unit cell* dimensions and orientation of a crystal.

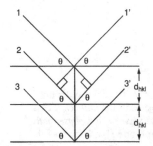

Fig. 11.2-5. Bragg reflection on a series of lattice planes

Structure factors

The intensity I_{hkl} of a Bragg reflection is determined by the relative positions and scattering powers of the atoms in the unit cell. For the elastic scattering involved, the contributions of the individual atoms may be added vectorially. An atom j will give rise to a scattered wave with amplitude f_j known as the *atomic scattering factor*, and a phase angle ϕ_j with respect to that of an atom placed at the origin of the unit cell. The atomic scattering factor f, which measures the scattering power of an atom relative to that of single point electron at the center of the atom, is given by the expression:

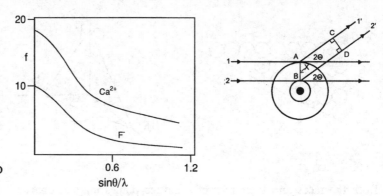

Fig. 11.2-6. (a) Atomic scattering factors for F^- and Ca^{2+}. (b) Path differences AC and BD for scattering of X-rays within an atom

$$f = f_0 \exp\left(\frac{-B(\sin^2 \theta)}{\lambda^2}\right) \qquad (11.2\text{-}2)$$

Atomic scattering factors f measure the scattering power of individual atoms in the unit cell.

B is known as the temperature factor (Eq. 11.2-3). For a zero Bragg angle θ, stationary atoms exhibit a scattering power f_0 proportional to the total number of electrons (Z) in the atom. As the effective size of the scattering atom (as determined by the orbital distribution of electrons) and the wavelength of the X-rays employed are similar, interference between radiation scattered from the various parts of the atom will lead to a rapid exponential decrease in f as the Bragg angle increases (Fig. 11.2-6a). This phenomenon is exemplified in Fig. 11.2-6b, where two incident waves 1 and 2 are scattered by electrons at positions A and B separated at a distance x. The resulting waves $1'$ and $2'$ then exhibit a path difference (BD-AC) at positions C and D of $x \cdot \sin 2\theta$. Interference between all such scattering waves leads to the dependence of the scattering power f on 2θ depicted in Fig. 11.2-6a. In contrast f shows effectively no decrease on going to higher 2θ values for neutron diffraction, as the neutron wavelength is much larger than the radii of the atom nuclei responsible for scattering. In typical X-ray structural analyses, diffraction data are collected for a $\sin\theta/\lambda$ range 0.0–0.7 Å^{-1}.

Molecules in a crystal lattice constantly display translational, rotation and generally, to a lesser extent, internal vibrational motions. These movements, which will, of course, increase with temperature, lead to an increase in the effective atomic sizes, as viewed by X-ray diffraction over the average of about 10^{15} unit cells in a typical single crystal. The resultant smearing of the electron density distribution is represented in Eq. 11.2-2 by the temperature factor B (or Debye-Waller factor), where B or alternatively U is related to the mean-square amplitude of atomic movement $\langle u^2 \rangle$:

$$B = 8\pi^2 \langle u^2 \rangle = 8\pi^2 U \qquad (11.2\text{-}3)$$

Typical values of B are low for minerals and solid-state compounds at room temperature (1–3 Å^2), but often lie in the range 5–10 Å^2 for organic compounds or metal complexes.

Superposition of individual wavefronts produced by atoms in the unit cell yields a total scattered wave with a phase angle ϕ_{hkl} and an amplitude F_{hkl}, known as the structure factor.

Relative phase angles ϕ_{hkl} for atoms in a unit cell may be determined by use of the Bragg condition (Eq. 11.2-1), as exemplified for the Bragg reflection 120 in Fig. 11.2-7. The positional coordinates x_j, y_j, z_j for the j-th atom are expressed as fractions of the unit cell constants a, b and c. For atoms in neighboring lattice planes, such as O and A in Fig. 11.2-7, constructive interference will only be possible when the diffracted waves (here $1'$ and $2'$) exhibit a phase difference (here ϕ_{120}) of 2π after Bragg reflection. Inspection of Fig. 11.2-7 indicates that atoms B and C, which are displaced from the plane through the origin by two and three lattice planes respectively, will similarly be associated with phase differences of respectively 4π and 6π. For a generally placed atom j with fractional coordinates x_j, y_j, z_j, the phase angle with respect to the origin (0,0,0) will be given by the expression:

$$\phi_{hkl} = 2\pi(hx_j + ky_j + lz_j) \qquad (11.2\text{-}4)$$

Superposition of the individual wavefronts (Fig. 11.2-8a) produced by N atoms in the unit cell yields a total scattered wave with an amplitude F_{hkl}, known as the *structure factor*:

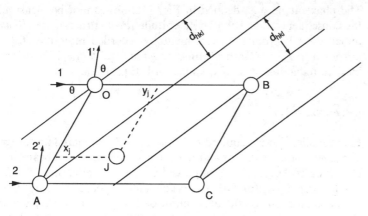

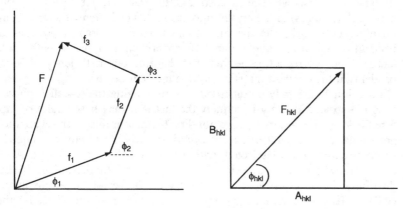

Fig. 11.2-7. Phase angles ϕ for atoms in the unit cell

Fig. 11.2-8. (a) Vectorial addition of scattered waves. (b) Structure factor F_{hkl} and phase angle ϕ_{hkl}

$$A_{hkl} = \sum_{j}^{N} f_j \cos\{2\pi(hx_j + ky_j + lz_j)\} \qquad (11.2\text{-}5)$$

$$B_{hkl} = \sum_{j}^{N} f_j \sin\{2\pi(hx_j + ky_j + lz_j)\} \qquad (11.2\text{-}6)$$

$$F_{hkl} = \sum_{j}^{N} f_j \exp\{2\pi i(hx_j + ky_j + lz_j)\} \qquad (11.2\text{-}7)$$

Worked example: Structure factors

CsCl exhibits a cubic crystal structure with $a = b = c = 4.123$ Å, $\alpha = \beta = \gamma = 90°$ and only two atoms in the unit cell. The Cl^- anion is sited at the center $(1/2, 1/2, 1/2)$ of the unit cell and is coordinated by 8 Cs^+ cations $(0,0,0)$. The Bragg reflection 001 corresponds to a d_{001} spacing of c and a $\sin\theta/\lambda$ value of 0.12 (Eq. 11.2-1), for which the atomic scattering factors of Cs^+ and Cl^- are respectively 50 and 15.
Using Eq. 11.2-5 and 11.2-6

CsCl

$$A_{001} = f_{Cs} \cdot \cos\{2\pi(1.z_{Cs})\} + f_{Cl} \cdot \cos\{2\pi(1.z_{Cl})\}$$

$$= f_{Cs} - f_{Cl} = 35$$

$$B_{001} = f_{Cs} \cdot \sin\{2\pi(1.z_{Cs})\} + f_{Cl} \cdot \sin\{2\pi(1.z_{Cl})\}$$

$$= 0$$

This gives a structure factor F_{001} of 35 with a phase angle ϕ_{hkl} of 0°. B_{hkl} is always zero for crystal structures such as CsCl that contain centres of symmetry.

The phase angle ϕ_{hkl}, defined in Fig. 11.2-8b, cannot be measured directly, with the consequence that only the magnitude $|F_{hkl}|$ (*structure amplitude*) can be determined directly from the experimentally recorded intensities I_{hkl}, which are proportional to F_{hkl}^2. *Observed structure amplitudes* $|(F_{hkl})_0|$ (or $|F_0|$) may be obtained by data reduction of the I_{hkl} values with Eq. 11.2-8:

$$|(F_{hkl})_0| = \mathrm{K}\sqrt{\frac{I_{hkl}}{\mathrm{A \cdot L \cdot P}}} \qquad (11.2\text{-}8)$$

Observed *structure amplitudes* $|F_0|$ are obtained by data reduction of experimentally recorded intensities I_{hkl}.

K is a scaling factor required to bring the experimental data (on an arbitrary scale dependent on crystal size and the intensity of the X-ray beam) onto the absolute scattering scale (f values) employed in the determination of *calculated structure amplitudes* $|(F_{hkl})_c|$ (or $|F_c|$) from known atom positions x_j, y_j, z_j using Eq. 11.2-7. A is a correction for the absorption of X-rays in accordance with the Lambert-Bougner-Beer law, which must also take the size and habit (distribution of symmetry-related faces) of the crystal into account. The Lorentz factor L compensates for differences in the effective measurement time for Bragg reflections and is dependent on the Bragg angle θ and the experimental set-up. P is the polarization factor, which corrects for the fact that the diffraction efficiency of X-rays is dependent on the polarization status of the incident beam.

The electron density distribution $\rho(xyz)$ in a crystal is given by an inverse Fourier transform with coefficients F_{hkl}, the calculation of which requires phase angles from a structure model.

Equation 11.2-8 may be regarded as summarizing the so-called phase problem of X-ray structural analysis, namely the fact that the phase angles ϕ_{hkl} and therefore the structure factors F_{hkl} are not directly measurable. If this were the case, the determination of crystal structures would be a trivial procedure – requiring only a Fourier summation at selected grid points (xyz) in the unit cell. As discussed above, the values of the atomic scattering factors f are dependent on the average distribution of electron density in the crystal lattice for the atoms involved. The electron density function, $\rho(xyz)$, in a crystal is therefore given by the inverse Fourier transform (Eq. 11.2-9) of the structure factor

$$\rho(xyz) = \frac{1}{V_c} \sum_h \sum_k \sum_l F_{hkl} \exp\{-2\pi i(hx + ky + lz)\} \qquad (11.2\text{-}9)$$

where V_c is the volume of the unit cell. An alternative formulation (Eq. 11.2-10) is provided by use of the readily accessible observed structure amplitudes $|(F_{hkl})_0|$ (or $|F_0|$) and the not directly measurable phase angles ϕ_{hkl} as coefficients:

$$\rho(xyz) = \frac{1}{V_c} \sum_h \sum_k \sum_l |(F_{hkl})_0| \cdot \exp(i\phi_{hkl}) \cdot \exp\{-2\pi i(hx + ky + lz)\}$$

$$(11.2\text{-}10)$$

As will be discussed in Sec. 11.2.3, structure solution methods are available that enable the atom sites in the unit cell to be proposed. These positions may then be employed to estimate values of the phase angles ϕ_{hkl} using Eqs. 11.2-5 and 11.2-6 (Fig. 11.2-8b). After successful solution and refinement of the crystal structure, electron density maps may be calculated by using Eq. 11.2-10 with calculated phase angles $(\phi_{hkl})_c$ and observed structure amplitudes $|(F_{hkl})_0|$ as Fourier coefficients.

The diffraction pattern for a substance is characterized by the angular position (2θ), symmetry relationships and intensities of the Bragg reflections hkl. It is convenient to distinguish between two experimental techniques for the registration of diffraction data:

- Powder diffraction
- Single crystal diffraction

In the former method (Sec. 11.2.2) microcrystalline samples containing approximately 10^6 randomly orientated individual crystallites (size $5 \times 10^{-5} - 5 \times 10^{-4}$ cm) provide one-dimensional intensity/diffraction angle distribution diagrams that can be used for: the identification of substances; the determination of physical properties; the measurement of crystal size; and, to a limited extent, the elucidation of crystal structure. Single crystals with dimensions in the range 0.1–0.6 mm are generally required for the latter method, which is the standard technique for X-ray structural analysis (Sec. 11.2.3).

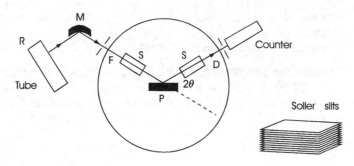

Fig. 11.2-9. Powder diffractometer with Bragg-Brentano geometry

11.2.2 Powder diffraction

Diffraction studies on powder samples are performed with monochromated radiation provided by a conventional sealed X-ray tube (e.g., CuKα radiation) fitted with a crystal monochromator (e.g., graphite). The large number of crystallites in the irradiated sample guarantees that these will be present in all possible orientations, so that Bragg reflections *hkl* will therefore, in principle, be capable of being registered for all possible lattice plane spacings d_{hkl} corresponding to 2θ values (Eq. 11.2-1) within the angular range of the instrument. The diffracted X-rays lie on a series of cones coaxial with the path of the incident beam. Computer-controlled automatic diffractometers are in widespread use, both for routine analytical measurements and for individual structural or physical studies. However, relatively inexpensive focusing cameras with photographic registration (e.g., Guinier cameras) still play a role in both qualitative and quantitative studies of powder samples [11.2-2].

Powder diffraction requires microcrystalline samples.

Powder diffractometer

The geometry of a powder diffractometer with *Bragg-Brentano* focusing is illustrated in Fig. 11.2-9. The powder samples are compressed on a metal sample holder (P), which can be rotated about an axis normal to its plane during exposure so as to further increase the randomness of orientation of the crystallites. Use is made of the parafocusing effect, in which the line focus (F) of the X-ray tube (R) and the exit slit of the diffractometer (D) are constrained to lie on a circle, so as to be equidistant from the sample holder (P), the surface of which is tangential to the circle. A bent crystal monochromator (M) – adjusted to fulfill the Bragg condition $\lambda = 2d_{hkl} \sin\theta$ for a strong *reflection hkl* (λ being the wavelength of Kα radiation) – is employed to focus the X-rays onto the entrance slit F. The geometrical construction of the powder diffractometer then ensures that the diffracted X-rays are effectively focused onto the detector slit D. Divergence of the incident and diffracted beams within the diffractometer is restricted by passing these through a series of thin metal plates (S), known as Soller slits.

The diffraction pattern is registered by rotating the detector at a constant speed ω about the center of the specimen, with the sample holder rotating about the same axis at a speed $\omega/2$ to maintain the geometry of the focusing circle. Scintillation counters with Tl-doped NaI crystal scintillators or proportional counters are employed as detectors in conventional powder diffractometers at scan speeds ω commonly not exceeding $2° \cdot min^{-1}$. A drastic reduction in measurement time can now be achieved for powder diagrams with a new generation of transmission diffractometers which employ position-sensitive detectors. For instance, a proportional counter with a wire curved to cover all 2θ values between 0 and 60° on the scanning circle can simultaneously register all Bragg reflections *hkl* in this range. Such detectors enable powder patterns to be recorded within 1–2 min.

Modern powder diffractometers employ position-sensitive detectors.

Analytical applications

A powder diffraction pattern delivers the following information, which is characteristic either of the substance (d_{hkl}, I_{hkl}) or the crystallites ($\beta_{1/2}$) measured:

- lattice plane spacings d_{hkl}
- intensities I_{hkl}
- line widths characterized by FWHM values $\beta_{1/2}$ (full-width-at-half-maximum)

The powder pattern of the cubic compund BaS is provided as an example in Fig. 11.2-10. Within the scope of this chapter, it is not possible to offer an exhaustive list of all potential applications of X-ray powder diffraction.

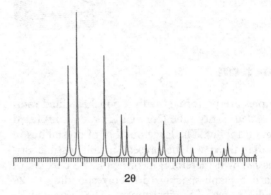

2θ

Fig. 11.2-10. Powder pattern for BaS

Table 11.2-2. Some analytical applications of powder diffraction

No	Experimental data	Area of application
1	d_{hkl}	unit cell constants $a, b, c, \alpha, \beta, \gamma$ density D_x
	for solid solutions	composition
	as function of temperature	coefficients of thermal expansion
	as function of pressure	internal stress
2	I_{hkl}	symmetry elements, space groups
		Bravais lattices
	for solid mixtures	quantitative analysis
3	$d_{hkl} + I_{hkl}$	identification
		kinetic studies
		solid-state reactions
		phase diagrams
4	$\beta_{1/2}$	crystallite size
5	$d_{hkl} + I_{hkl} + \beta_{1/2}$	crystal structure analysis (Rietveld method)

Table 11.2-2 summarizes a variety of typical uses, which will now be briefly discussed:

1. Computer programs are available for the automatic indexing of lattice plane spacings, thereby enabling the determination of unit cell constants (to an accuracy of 0.001 Å or better) and densities, if the chemical formula of the substance is known. For solid solutions, the values of the unit cell constants vary linearly with the atomic percentages of the components, e.g., in the cubic Cu-Au system a increases from 3.608 Å for pure copper to 4.070 Å for pure gold. Measurement of the lattice plane spacing for a high angle Bragg reflection will, therefore, yield the composition of the alloy. The changes in d_{hkl} values with temperature or external pressure allow, respectively, the determination of coefficients of thermal expansion or the internal stress in a crystal. The former will be anisotropic for all crystal systems other than cubic.

2. The presence of symmetry elements in the unit cell leads to equivalence between I_{hkl} values, which are characteristic of the various crystal systems. Screw axes and glide planes, both of which exhibit a translational component, are associated with systematic destructive intereferences (*systematic absences*) for certain types of Bragg reflection, e.g., $0k0$, $k = 2n + 1$, for a 2_1 screw diad in direction **b**. Systematic absences are also produced by centred Bravais lattices, e.g., hkl, $h + k + l = 2n + 1$, are absent reflections for an I lattice. For homogeneous samples containing two or more microcrystalline substances, characteristic powder lines can be used for the quantitative analysis of individual components. The intensity I_{hkl} of such a powder line should, in principle, be directly proportional to

the quantity of the constituent responsible for its appearance. However, absorption of X-rays by other substances present in the sample can lead to systematic errors. Use of the internal standard method is therefore advisable; in this method a calibration curve is constructed by addition of known quantities of the substance in question to the original sample. It is also important that the crystallites are randomly orientated with a particle size between 5×10^{-5} and 5×10^{-4} cm.

3. The lattice plane spacings d_{hkl} with their intensities I_{hkl} are characteristic for a crystalline substance and generally allow a straightforward identification of a sample. More than 50 000 sets of powder diffraction patterns have been archived in the Powder Diffraction File (PDF) of the Joint Committee on Powder Diffraction Standards (JCPDS). Identification can be achieved by use of reference books (e.g., Hanawalt or Fink indices) with systematically tabulated data or automatically with computer search programs. The following entry for silicon (Nr. 27–1402) in the *Hanawalt index* exemplifies the registration of powder patterns in a reference file:

> Powder diffraction provides a powerful nondestructive analytical method for identifying solid-state compounds.

$$3.14_x \quad 1.92_6 \quad 1.63_3 \quad 1.11_1 \quad 1.25_1 \quad 0.86_1 \quad 0.92_1 \quad 1.36_1 \quad 27 - 1402$$

In this index the lattice plane spacings for the eight strongest lines are listed (the first three in bold text) in order of decreasing intensity, with the position in the file being determined by the d_{hkl} value (3.14 Å) for the strongest line. The subscripts refer to the relative intensities I_{hkl} with respect to the first line ($x = 10$). Identification of the constituents of a mixture is also relatively straightforward for two or three substances, provided these each make up more than ca. 5% of the specimen. An important aspect of the powder method is its nondestructive nature.

Worked example: compound identification by powder diffraction

The following lattice plane spacings d and observed intensities I_0 were measured by powder diffraction for Bragg reflections of an unknown compound.

d(Å)	I_0	$10 \cdot I_0/I_{max}$
3.123	8600	100
2.705	857	10
1.912	4386	51
1.633	2582	30
1.561	172	2
1.351	516	6
1.240	774	9
1.209	168	2
1.103	770	9
1.040	433	5

Identification of the substance may be performed using the Hanawalt index of the Powder Diffraction File (PDF) in the following way. Firstly, the observed intensities are normalized relative to the powder line with the highest intensity ($d = 3.123$ Å). Then the d values of the eight strongest lines are listed in order of decreasing relative intensity as follows:

$$3.12_x \quad 1.91_5 \quad 1.63_3 \quad 2.71_1 \quad 1.24_1 \quad 1.10_1 \quad 1.35_1 \quad 1.04_1$$

The position of the unknown compound in the Hanawalt file is determinded by the d value for the line of strongest intensity, 3.12 Å. Comparison with the listed entries in the file suggests that the compound is sphalerite β-ZnS. Modern powder diffractometers are equipped with software for the automatic identification of a number of components in substance mixtures.

4. Powder line widths are inversely proportional to the size of crystallites. Measurement of FWHM values $\beta_{1/2}$ allows an estimation of particle size and is of particular interest in the pharmaceutical industry. A direct analogy to *small angle scattering* is apparent. In this method the broadening of the undeflected X-ray beam is used to investigate the size and shape of noncrystalline macromolecules.

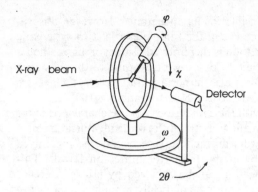

Fig. 11.2-11. Four circle diffractometer

The *Rietveld method* is used to determine the crystal structure of compounds that are only available in microcrystalline form.

5. The experimental data provided by powder diffraction is one-dimensional. As 100–200 measurable independent Bragg reflections *hkl* can be expected per atom in a typical crystal structure, overlapping of powder diffraction lines will be extreme for all but the simplest structures. Although the powder method was, indeed, employed for the determination of small structures in the early days of X-ray diffraction, its limitations in comparison with single-crystal studies (Sec. 11.2.3) led to a long period of relative dormancy. Profile-fitting structural refinement of the one-dimensional intensity data of powder patterns by the *Rietveld method* has, however, led to a renaissance in powder diffraction in the past 20 years. Although limited, at present, to structures with less than 200 parameters (see Sec. 11.2.3), the Rietveld method is now of great importance in materials science, where many compounds of technological interest are only available in a microcrystalline form [11.2-2, 11.2-3].

11.2.3 Crystal structure analysis

Four-circle diffractometer

Computer-controlled diffractometers are used for the collection of three-dimensional intensity data I_{hkl} from single crystals, which should typically have dimensions in the range 0.1–0.6 mm. Photographic methods of registration are still employed for initial crystal control. The geometry of a four-circle diffractometer is depicted in Fig. 11.2-11. A scintillation counter is used as detector and may be rotated about the 2θ circle to intercept the diffracted X-ray beam; this beam is provided by an X-ray tube fitted with a crystal monochromator. Standard normal focus sealed tubes can be run at 2–2.4 kW and allow the collection of 100–200 reflections per hour. Rotating anode generators, previously only employed for macromolecular (protein) crystallography, are beginning to become more popular for small-molecule studies (up to 3–4×10^3 Da). These provide a 7–10 fold increase in intensity of the X-ray source and enable a significant increase in the speed of data collection or the use of smaller crystals of masses down to about 100 ng. The remaining three circles ω, ϕ, and χ allow the crystal to be orientated in the diffracting position for the Bragg reflection under study. The crystal itself is glued to a thin glass fiber and mounted on a goniometer head that enables orientation and centering in the X-ray beam. If crystals are air-sensitive, they may be contained in thin-walled glass capillary tubes instead. Intensity data are collected by scanning a diffraction peak in either the $\omega - 2\theta$ or the ω scan mode. In the former mode, the ω and 2θ circles are moved simultaneously so as to retain the condition $\omega = \theta$ throughout the scan. In the latter mode, the ω circle is driven through a range of typically 1–1.5° centered on $\omega = 2\theta$, while the detector is held at a fixed position. It is advantageous to lower the temperature of the crystal by bathing it in a stream of cold nitrogen gas, in order to reduce the thermal motions of the atoms in the structure. Such cooling also permits the X-ray structural analysis of compounds that are unstable or liquid at room temperature.

Area detectors now greatly reduce the measurement time for single crystal data sets.

Conventional scintillation counters are relatively inexpensive and exhibit excellent long-term stability. They are, however, limited to a maximum count rate of about $5 \times 10^4 - 10^5$ counts s^{-1} and exhibit no positional sensitivity. Recent technological developments have resulted in the use of three types of area detector in X-ray crystallography:

- multiwire proportional counters
- television area detectors (fluorescent phosphor)
- imaging plates (storage phosphor)

The imaging plates (Eu-doped BaFBr) can store a latent X-ray image, which is then exposed to radiation from a He-Ne-laser to emit light proportional to the number of X-ray photons received. Such detectors exhibit extremely high count rates and enable the collection of data sets within a few hours, thus reducing the time required by up to a factor of ten in comparison with conventional counters.

Structure solution

Data reduction of the registered integrated intensities I_{hkl} yields observed structure amplitudes $|F_0|$ (Eq. 11.2-8). As discussed in Sec. 11.2.1, the phase problem prevents the immediate calculation of an electron density map of the crystal structure (Eq. 11.2-9). Phase angles ϕ_{hkl} may generally be determined for small molecules through one of two strategies:

> As *phase angles* ϕ_{hkl} cannot be measured directly by standard X-ray techniques it is necessary to generate a structure model to provide initial estimates of their values.

- Patterson synthesis (heavy-atom method)
- direct methods

The *Patterson synthesis* (Eq. 11.2-11) employs the experimentally accessible coefficients F_{hkl}^2 and may be shown to represent vectors (uvw) between atoms in the unit cell:

$$P(uvw) = \frac{1}{V_c} \sum_h \sum_k \sum_l F_{hkl}^2 \{\cos 2\pi(hu + kv + lw)\} \qquad (11.2\text{-}11)$$

The height of a Patterson peak is proportional to the product of the atomic numbers of the atoms at the end of the vector peaks. For molecules containing one or two heavy atoms, (e.g., the Br derivative of an organic molecule), the vector peaks between these atoms will dominate the Patterson map and enable their positions to be located. It is then possible to calculate *an electron density synthesis* (Eq. 11.2-10) by using the observed structure amplitudes $|F_{hkl}|_0$ and the phases $(\phi_{hkl})_c$ calculated for the heavy atom position(s). As this requires the summation of trigonometrical Fourier series (sine and cosine functions from Eqs. 11.2-5 and 11.2-6) it is also commonly referred to as a *Fourier synthesis*. Although such a map will be biased towards the heavy atoms, it will also provide low peaks for some, if not all, of the lighter atoms (apart from hydrogen) which contribute, of course, to the magnitude of the structure amplitude. Inclusion of these light atom positions in the structure model leads to an improvement in the calculated phase angles, with the result that a subsequent Fourier synthesis will often allow all remaining non-hydrogen atoms to be located. An example of an electron density map is provided by Fig. 11.2-12. It is, however, usually easier to interpret maps provided by *difference Fourier syntheses*, which use the differences between the observed and calculated structure amplitudes as coefficients:

> *Difference Fourier syntheses* are used to locate lighter atoms missing in the initial structure model.

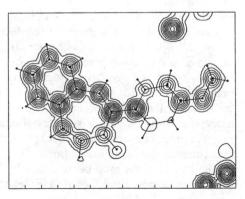

Fig. 11.2-12. Electron density map for $C_{18}H_{14}O_2$

$$\Delta_\rho(xyz) = \frac{1}{V_c} \sum_h \sum_k \sum_l (|F_0| - |F_c|) \cdot \exp(i\phi_c) \cdot \exp\{-2\pi i(hx + ky + lz)\}$$

$$(11.2\text{-}12)$$

Solution of macromolecular structures involves the method of *isomorphous replacement*.

Such a difference map charts only the unknown part of the crystal structure.

Protein crystallography makes use of the method of *isomorphous replacement*, in which crystals of a native protein are soaked with the solution of a metal salt to allow heavy metal atoms to occupy sites in the crystal structure. The positions of the lighter atoms remain virtually unaffected – with the result that such heavy atom derivatives are *isomorphous* with the original macromolecule, i.e., they display the same space group and virtually unchanged unit cell constants. Differences in intensities I_{hkl} of isomorphous crystals are produced solely by the contributions from the various heavy atoms, whose sites may be located by Patterson syntheses. Such information from at least two additional isomorphs enables phase angles and consequently electron density distributions to be calculated for protein crystals.

Direct methods of phase determination are based on relationships between the phases of intense Bragg reflections.

Electron density is localized at atom sites and must be positive or zero. These restrictions lead to mathematical constraints on possible phase angles, which are the basis of the so-called *direct methods* of phase determination, initially explored by Karle und Hauptman in the 1950s. The simplest case is provided by centrosymmetric crystal structures, which always contain atoms with coordinates x, y, z and $-x, -y, -z$. Inspection of Eq. 11.2-6 indicates that components B_{hkl} will be zero for such crystals, i.e., the phase angles are restricted to 0 and 180° with respective signs "s" of plus (for $\phi_{hkl} = 0$) and minus (for $\phi_{hkl} = 180°$) for the A_{hkl} terms. In *triplet relationships*, the signs of three intense Bragg reflections with indices hkl, $h'k'l'$, and $h - h', k - k', l - l'$ are related with a high degree of probability by the following equation:

$$s(hkl) = s(h'k'l') \cdot s(h - h', k - k', l - l')$$

$$(11.2\text{-}13)$$

With such relationships, it is usually possible to derive phase angles for the strongest Bragg reflections (typically about 10 per nonhydrogen atom), thereby enabling the calculation of a good approximation of the electron density map. Similar methods are available for noncentrosymmetric crystals. The development of high-performance computer programs coupled with the advent of automatic diffractometers and high-speed computers led in the 1970s to the breakthrough of X-ray diffraction as a routine method of structural analysis. Today it is commonplace for the first report of the synthesis of a novel compound to be accompanied by an X-ray structural analysis.

Structure refinement

Structure refinement is performed by the least-squares method to deliver the best fit to the observed $|F_0|$ or F_0^2 values.

Atomic coordinates (x, y, z), localized by direct methods and/or difference Fourier syntheses, may be improved by *least-squares refinement*, which delivers the best fit for the structural model ($|F_c|$) to the set of experimental observations ($|F_0|$). Traditionally, most crystal structures have been refined against $|F_0|$, with observed structure amplitudes less than $m \cdot \sigma(|F_0|)$ being treated as unobserved. σ is the experimental standard deviation; m is an arbitrary value between 2 and 4. Rejection of such weak reflections is common practice as it leads to an improvement in the final figures of merit (Eqs. 11.2-13 to 11.2-15). It has no justification and may bias the structural results obtained. The refinement method involves minimization of the function $\sum w(|F_0| - |F_c|)^2$, with individual weights w for the structure amplitudes. This procedure is generally satisfactory for good data sets but less reliable for weakly diffracting crystals because up to 60–70% of the experimental data may be considered as unobserved. In such cases, it is clearly better to include all $|F_0|$ values. Recently there has been a marked increase in the number of structures being refined for the complete set of intensity data against F_0^2, involving minimization of the sum $\sum w(F_0^2 - F_c^2)^2$. Such a strategy also avoids the problems associated with calculation of observed structure amplitudes $|F_0|$ (Eq. 11.2-8) from negative intensities, caused by background counts being higher than peak counts as a result of statistical fluctuation.

Thermal motions may be modelled by ellipsoids (Fig. 11.2-13) defined by six anisotropic temperature factors. This leads to a total of nine refinable parameters

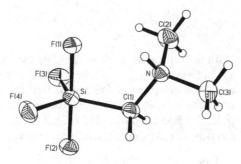

Fig. 11.2-13. Molecular structure of $C_3H_9NF_4Si$ with thermal ellipsoids (50% probability)

per atom. Their low scattering power ($f_0 = 1$) may prevent hydrogen atoms being located in difference syntheses for heavy atom structures or weakly diffracting crystals. In such cases hydrogen atom sites may be calculated geometrically at the expected distance from the bonding partner. X-ray structures are massively over-determined, e.g., for good centrosymmetric data sets the number of experimental $|F_0|$ values will typically exceed the number of refined parameters by 10 : 1 or more. This generally leads to rapid convergence of the structure model within a few cycles of refinement. However, difficulties may be experienced in cases of pseudosymmetry or disordered structures and for solvent molecules of crystallization with pronounced thermal motion.

Accuracy

The final atom coordinates from the least-squares refinement allow the calculation of bond lengths and angles together with estimated standard deviations (esds), which should always be quoted in structural discussions. For high-quality diffraction data esds should be about 0.001–0.002 Å for distance to heavy atoms and about 0.005 Å or less for C–C distances in organic compounds. A series of criteria are available for assessing the correctness of an X-ray structural analysis.

The correctness of an X-ray structural analysis should always be assessed through use of figures of merit and other parameters.

1. Figures of merit: R, R_w, S

The traditional crystallographic R factor is defined as:

$$R = \frac{\sum |F_0 - F_c|}{\sum |F_0|} \qquad (11.2\text{-}14)$$

For good data sets R should lie between 0.02 and 0.05. However, systematic measurement errors (e.g., absorption, crystal decay) or inadequacies in the structural model, such as problems resulting from disorder or the presence of large thermal motions, may lead to R factors between 0.05 and 0.10. Although such analyses provide, in virtually all cases, chemically correct structures (i.e., correct connectivities), bond lengths and angles will be less accurate. Further figures of merit for refinement on $|F_0|$ are the weighted factor R_w and the goodness of fit, S, which has an expected value of 1, with n being the number of Bragg reflections and ρ the total number of parameters refined:

$$R = \sqrt{\frac{\sum w|F_0 - F_c|^2}{\sum w|F_0|^2}} \qquad (11.2\text{-}15)$$

$$S = \sqrt{\frac{\sum w|F_0 - F_c|^2}{n - p}} \qquad (11.2\text{-}16)$$

2. Number of reflections/number of parameters: n/p

As mentioned previously n/p should be 10 or more for good data sets. This quotient also reflects the data resolution of the X-ray diffraction study, which is defined as the smallest lattice plane spacing $(d_{hkl})_{min}$ for which significant intensity data can be recorded. Use of the Bragg equation gives:

$$(d_{hkl})_{min} = \frac{\lambda}{2 \sin \theta_{max}} \qquad (11.2\text{-}17)$$

With the typical value of $\theta_{max} = 30°$ for small molecule crystallography with Mo-Kα radiation, $(d_{hkl})_{min}$ will be 0.7107 Å. The rapid decrease in intensity with Bragg angle for macromolecules means that the resolution for such structures will generally be restricted to values above 2 Å. Although atom resolution is then not possible, known molecular fragments (e.g., amino acids) may be fitted to the electron density map to afford protein structures.

3. Convergence of the refinement: Δ/σ

This may be assessed from the largest and average values of Δ/σ, where Δ is the last shift for a parameter with standard deviation σ. In well-behaved structures Δ/σ should be 0.01 or less. If shifts larger than 0.10 are observed (e.g., for hydrogen atoms or disordered fragments), it is advisable to constrain the atoms involved.

4. Residual electron density: $\Delta\rho$

$\Delta\rho$ is calculated with a difference Fourier synthesis after completion of the refinement. Both the highest peak (positive) and lowest hole (negative) are usually quoted. For R factors between 0.02 and 0.05 $|\Delta\rho|$ should not exceed about $0.5\,e\,\text{Å}^{-3}$. Positive values above approximately $1.0\,e\,\text{Å}^{-3}$ at van der Waals or hydrogen-bonding distances from refined atoms positions can indicate the presence of solvent molecules. Such $\Delta\rho$ values at distances within about 1 Å from heavy atoms may be caused by systematic errors such as absorption or the incorrect treatment of thermal motion.

Worked example: Assessment of X-ray structural analyses

The crystal structure analysis of the compound $C_3H_9F_4NSi$ depicted in Fig. 11.2-13 could be summarized for a chemical journal in the following typical manner.

$C_3H_9F_4NSi$ crystallizes in the triclinic space group $P\bar{1}$ with the cell constants ($-110°C$, standard deviations in brackets refer to the last digit) $a = 7.449(2)$, $b = 8.653(4)$, $c = 11.728(5)$, $\alpha = 108.08(3)$, $\beta = 96.90(3)$, $\gamma = 101.58(3)$, $V_c = 690.3(5)\,\text{Å}^3$, Z (molecules in the unit cell) $= 4$, D_c (calculated density) $= 1.570\,g \cdot cm^{-3}$. The structure was refined (236 parameters for two independent molecules) against F_0 to $R = 0.036$, $R_w = 0.031$ for 3396 reflections with $F_0 > 4\sigma(F_0)$ (from 4064 independent measured reflections) collected with graphite-monochromated MoKα ($\lambda = 0.7107$ Å) radiation for $2\theta \le 60°$. The residual electron density of a final difference Fourier synthesis was $+0.31$ (maximum) and $-0.45\,e\cdot\text{Å}^{-3}$ (minimum).

An assessment of this X-ray structural analysis can be performed with the criteria discussed in this chapter

1. Figures of merit (reliability indices)

   $R = 0.036$, $R_w = 0.031$

2. Number of reflections/number of parameters

   $n/p = 3396/236 = 14.39$

3. Convergence of the refinement

   $\Delta/\sigma < 0.01$ for all nonhydrogen atoms (not stated)

4. Residual electron density (max./min)

   $\Delta\rho = +0.31/-0.45\,e\cdot\text{Å}^{-3}$

These parameters are typical for a good data set of a well-behaved compound. Hydrogen atoms were located in difference Fourier syntheses and refined freely. Standard deviations for the Si-F bonds fall in the range 0.001–0.002 Å.

Even very careful X-ray analyses, in which systematic measurement errors (absorption, extinction) are reduced to a minimum, are not capable of delivering R factors below 0.015–0.02. This is because of inadequacies in the structural model used for refinement, in particular the treatment of thermal motion (as an ellipsoid) and the electron density distribution (as being about discrete spherical atoms). The latter assumption is clearly inadequate for the description of valence electrons in molecules. It leads to the well-known systematic shortening of the lengths of bonds to hydrogen atoms by 0.10–0.15 Å in X-ray determinations; this is caused by distortion of the density distribution of the single valence electron of a hydrogen at this atom towards its more electronegative partner (e.g., C, N, O). As neutron diffraction yields the positions of atom nuclei, combination of the two diffraction techniques (*X-N-syntheses*) for high quality data can yield experimental valence electron distributions. For a more detailed description of X-ray structural analysis the reader is refered to modern textbooks [11.2-4, 11.2-5].

Combination of X-ray and neutron diffraction data $(X - N)$ can provide experimental valence electron distributions.

References

[11.2-1] Hahn, T. (Ed.), *International tables for crystallography*, Vol. A. *Space-group symmetry*. 2nd ed. Dordrecht: Reidel Publishing Co., 1987.

[11.2-2] Krischner, H., Koppelhuber-Bitschnau, B., *Röntgenstrukturanalyse und Rietveldmethode*. Braunschweig/Wiesbaden: Vieweg, 1994.

[11.2-3] Young, R.A., *The Rietveld Method*. New York: Oxford University Press, 1993.

[11.2-4] Giacovazzo, C. (Ed.), *Fundamentals of Crystallography*. New York: Oxford University Press, 1992.

[11.2-5] Glusker, J.P. *Crystal Structure Analysis for Chemists and Biologists*. New York: VCH, 1994.

Questions and problems

1. Which methods are available for the determination of bond lengths and angles:
   a) in the gas phase
   b) in the solid state?
2. List the seven crystal systems together with their characteristic rotation axes and unit cell constants.
3. How are the Miller indices *hkl* of a lattice plane determined? Draw the first lattice planes 110, 120, and 010 from the origin for a two-dimensional projection of the orthorhombic unit cell $a = 10$ Å, $b = 20$ Å, $c = 5$ Å.
4. Describe the Bragg law. Explain the definition of resolution in X-ray crystallography and determine the best resolution possible for:
   a) CuKα radiation with $\lambda = 1.5418$ Å
   b) MoKα radiation with $\lambda = 0.7107$ Å.
5. Define the atomic scattering factor f and discuss its dependence on:
   a) atomic number
   b) Bragg angle θ
   c) X-ray wavelength λ
   d) temperature T
6. What is the phase problem in X-ray structural analysis and which strategies are available for the estimation of phase angles ϕ_{hkl}?
7. Describe the geometry of a powder diffractometer with Bragg-Brentano geometry.
8. Discuss typical analytical applications of powder diffraction.
9. Discuss the criteria available for assessing the correctness of an X-ray structural analysis.
10. What are the chief sources of:
    a) systematic measurement error and
    b) inadequacies in the structural model
    in X-ray structural analysis?

Part IV
Computer-Based Analytical Chemistry (COBAC)

12   Chemometrics

Learning objectives

- To understand the meaning of terms such as: *discrete and random variable, population, probability density* and *probability distribution function, moments of a probability distribution function, mean and variance, sample size, sampling distribution, sample parameters*
- To understand the properties of some important distributions such as: *normal, chi-squared, Student's t and F*
- To recognize that uncertainties associated with the measurement process lead to a distribution of error in the experimental results
- To recognize that the probability of obtaining a result in a given range is given by the corresponding area under the normal curve
- To recognize that sample mean and sample variance are *estimates* of the corresponding (usually unknown) population parameters
- To predict a confidence interval for the population mean μ
- To understand the meaning of terms such as: *decision making criteria, hypothesis testing, null hypothesis, alternative hypothesis, confidence level, type I error, type II error, power of a test*
- To perform a significance test to decide whether or not the population mean has a given value
- To compare two sample means (e.g., means obtained by two different analytical methods, as required in testing for accuracy)
- To compare two sample variances (e.g., the precision of two different analytical methods)
- To recognize the presence of an *outlier* in a given set of experimental results, and to decide whether or not it must be rejected
- To decide on a statistical basis if a given set of experimental results fits a given probability distribution function
- To recognize the importance of the sampling procedure in the context of the overall analytical process
- To understand the meaning of: *random sampling, sampling plan, lot, sampling unit, sample increments, gross sample, sub-sample, test-portion.*

12.1   Analytical Quality Criteria and Performance Tests

12.1.1   Basic statistics

In order to evaluate experimental data on a statistical basis, knowledge of the basic ideas of *probability theory* is necessary. A detailed presentation of statistical principles is beyond the scope of this discussion, but can be found in many specialized books. Here only some basic concepts and equations will be described.

Discrete and continuous random variables

A variable is called *random* if a certain probability is associated with each value; any function of a random variable is a random variable itself. Alternative terms tolrresponding to a given outcome of the experiment, is called *realization*.

If the random variable can assume only a finite number of values then it is regarded as *discrete*; if not, it is regarded as *continuous*. Particle counts from a radioactive source offer an example of a *discrete random variable*; examples of *continuous random variables* are temperature, volume, concentration, etc. Of

(a)

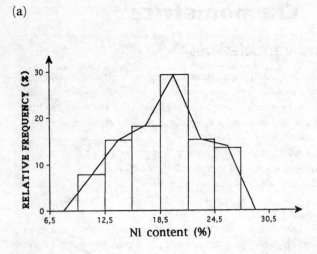

(b)

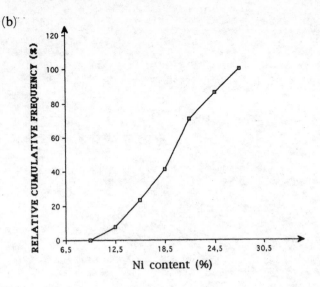

(c)

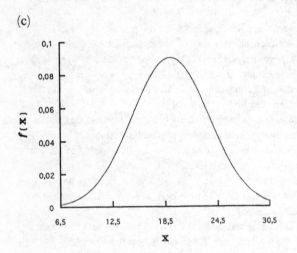

(d)

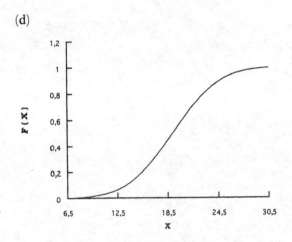

Fig. 12.1-1. (a) Histogram and frequency plot constructed using data in the following worked example; (b) relative cumulative frequency plot for data in (a).
(c, d) probability density, $f(x)$, and probability distribution, $F(x)$, functions, respectively

course, a continuous variables can assume all values from $-\infty$ to $+\infty$, but in some cases only positive values have a physical meaning. In practice, it is often difficult to experience a continuous variable because of the limitations of experimental devices. However, since most variables the analytical chemist encounters are continuous, only these will be considered in the following discussion.

Probability density and distribution functions

When an experiment is replicated n times, data can be grouped into classes of a given width and the number of measurements belonging to each class counted. After grouping into classes, the data can be visually inspected through a *histogram* constructed from rectangles having a width equal to the class width, and area proportional to the relative frequency in the given class (the total area is 1). By joining the points with coordinates corresponding to the central value and relative frequency of the class, a frequency plot is obtained (see the following worked example and Fig. 12.1-1a).

Alternatively, the *cumulative relative frequency* can be plotted against the *upper boundary* of the class to give a *cumulative frequency distribution* plot (see the following worked example and Fig. 12.1-1b). On increasing the number of samples analyzed and decreasing the class width, plots such as a and b in Fig. 12.1-1a and b tend (as n tends to ∞) to continuous curves such as 12.1-1c and d. These curves represent the so-called *probability density*, $f(x)$, and *probability distribution*, $F(x)$, *functions*, respectively. Note that:

$$f(x) = \frac{dF(x)}{dx} \tag{12.1-1}$$

and

$$\int_{-\infty}^{+\infty} f(x)\,\mathrm{d}x = 1 \tag{12.1-2}$$

This last equation means that it is certain that x is in the interval $(-\infty, +\infty)$.

The standard probability density and standard probability distribution functions for a Gaussian distribution will be discussed extensively later on.

Worked example:

Sixty-five alloy samples are analyzed for their percentage content of nickel. Results are grouped as follows

Category or class	Ni content (%) LL UL	Number of samples
1	10–12	5
2	13–15	10
3	16–18	12
4	19–21	19
5	22–24	10
6	25–27	9

LL, UL: lower and upper *limits of the class*; $UL–LL$ = class interval; Note that the first class, for example, includes all values between 9.5 and 12.5 which represent the *lower boundary* (LB) and the *upper boundary* (UB), respectively. The difference ($UB - LB$) gives the *class width*. If all classes have the same width, the width itself can be calculated as the difference between the LL (or UL) of two adjacent classes.

Represent these data in the form of a histogram, a plot of relative frequencies, and a plot of relative cumulative frequencies.

The relative frequency is obtained by dividing the number of observations in each class by the total number of observations, and can be conveniently expressed as a percentage. The relative frequencies for classes 1–6 are: 7.7, 15.4, 18.5, 29.2, 15.4, and, 13.8% respectively. A histogram is constructed of rectangles with area proportional to the relative frequency, and widths equal to the class width. Each rectangle is centred on the mid-value of each class, i.e., $(UL - LL)/2$. If all classes have the same width, the height of each rectangle is proportional to the class frequency and is usually taken to be equal to the relative frequency. The resulting histogram is shown in Fig. 12.1-1a. A relative frequency plot is obtained when the points with coordinates $\{(UL - LL)/2, \text{relative frequency}\}$ are joined by a straight line (see Fig. 12.1-1a).

The total number of observations less than or equal to the UB of a class is called *cumulative frequency*. *Relative cumulative frequencies* can be calculated as above and can also be expressed as percentages. In this case, one has:

Category or class	Ni content (%)	Cumulative frequency	Relative cumulative frequency (%)
1	≤12.5	5	7.7
2	≤15.5	15	23.1
3	≤18.5	27	41.6
4	≤21.5	46	70.8
5	≤24.5	56	86.2
6	≤27.5	65	100.0

Data in the 2nd and 4th columns of the table are used to draw a plot of relative cumulative frequencies (see Fig. 12.1-1b).

Expectation

The expectation (or expectancy) E of any function g of a random variable X is defined as:

$$E\{g(x)\} = \int_{-\infty}^{+\infty} g(x)f(x)\,dx \tag{12.1-3}$$

Some important examples are listed below.

- The r_{th} (noncentral) moment of X (or of the distribution $F(x)$) is defined as:

$$E(x^r) = \int_{-\infty}^{+\infty} x^r f(x)\,dx \quad r = 1, 2, 3, \ldots \tag{12.1-4}$$

Mean and variance are the first (non central) and the second (central) moment, respectively.

Note that the first noncentral moment $E(x) = \int_{-\infty}^{+\infty} xf(x)\,dx$ represents the *distribution mean* of X, indicated in the following by μ, and gives a measure of the *central location* of a distribution.

- The r_{th} central moment of X, or of $F(X)$, is defined as:

$$E\{(X - \mu)^r\} = \int_{-\infty}^{+\infty} (x - \mu)^r f(x)\,dx, \quad r = 1, 2, 3, \ldots. \tag{12.1-5}$$

In statistics a finite number n of observations is called a "*sample of size n*" and represents a small fraction of an infinite number of observations that could in principle be made, representing what is called a "*population*". Unfortunately confusion may arise between "*statistical sample*" and "*sample*" as ordinarily intended in analytical chemistry. For example, five 1 mL *aliquots* taken from a urine *specimen* and individually analyzed represent a (*statistical*) *sample of size* 5. Note how the use of terms *specimen/aliquot* may help to avoid confusion.

where μ stands for $E(X)$. Note that when the distribution is symmetrical about μ all the odd central moments are zero. The *second central moment*, commonly indicated as σ^2 or $V(X)$ is the *distribution variance* of X:

$$V(X) = \sigma^2 = \int_{-\infty}^{+\infty} (x - \mu)^2 f(x)\,dx \tag{12.1-6}$$

and gives a measure of the spread of the distribution about the mean. The square root of the variance is the *standard deviation* σ. Table 12.1-1 summarizes the definitions of mean and variance for a population and a sample of size n.

Table 12.1-1. Mean and variance in sample and population

	Mean	Variance
Sample of size n	$\bar{X} = \sum_i X_i/n$	$s^2 = \sum_i (X_i - \bar{X})^2/(n-1)$
Population[a]	$\mu = \int_{-\infty}^{+\infty} xf(x)\,dx$	$\sigma^2 = \int_{-\infty}^{+\infty} (x - \mu)^2 f(x)\,dx$

[a] Continuous variable

- The *covariance* of two random variables X and Y is defined as:

$$C(X, Y) = E\{(X - \mu_x)(Y - \mu_y)\} \tag{12.1-7}$$

where μ_x and μ_y are the means of X and Y.

Note that:

$$C(X, X) = V(X) = E\{(X - \mu_X)^2\} \tag{12.1-8}$$

i.e. the variance is a special case of covariance.
The *distribution correlation coefficient* $r(X, Y)$ is defined as:

$$r(X, Y) = \frac{C(X, Y)}{\{V(X)V(Y)\}^{0.5}} \tag{12.1-9}$$

and can assume values in the range ± 1. It is a measure of the degree to which X and Y are related.
Some useful properties of expectations are listed in Table 12.1-2.

Table 12.1-2. Useful properties of expectations

$E(a) = a$	$E(aX + b) = aE(X) + b$
$V(X) = E\{(X - \mu)^2\}$	$V(aX + b) = a^2 V(X)$
$\quad = E(X^2) - \{E(X)\}^2$	
$E(X + Y) = E(X) + E(Y)^*$	$C(X, Y) = E(XY) - \mu_x\mu_y$
$V(X + Y) = V(X) + V(Y) + 2C(X, Y)$	$V(X + Y) = V(X) + V(Y)^{**}$
$V(X - Y) = V(X) + V(Y) - 2C(X, Y)$	$V(X - Y) = V(X) + V(Y)^{**}$

a, b are constants.
* When X and Y are independent $E(X, Y) = E(X)E(Y)$ which means that in this particular case $C(X, Y) = 0$ and $r(X, Y) = 0$.
** Valid only if $C(X, Y) = 0$, i.e., if X, Y are independent. Similar equations can be derived for more than two random variables.

Sampling distributions

By performing n replicates, a *set* of $X_1, X_2, \ldots, X_n$ independent, identically distributed observations of the random variable X will be produced (the subscript indicates the particular experiment in which the random variable is observed). A function of a random variable is a random variable itself; if $Z(X)$ is a function of the X_i observations, a new random variable is obtained whose probability distribution is "induced" by that of X; this distribution is called the *sampling distribution of Z*. Two important examples of $Z(X)$ are the sample mean $\bar{X}$ and the sample variance s^2. Note that it is important to make a clear distinction between a *sample parameter* ($\bar{X}$ or s^2) and the corresponding *distribution parameter* (μ or σ^2).

Distribution parameter: *a fixed, predetermined quantity associated with the probability distribution of X.*

Sample parameter: *a random variable i.e. a function of the particular observation which varies from a sample to another and affords an* estimator *of the corresponding distribution parameter which is often unknown.*

It is possible to derive some useful information concerning certain parameters of the distribution of $\bar{X}$ even without prior knowledge of the distribution of X. Making use of the proper expectation properties (*see third equation in the first row and second equation in the second row of Table 12.1-2*) one can write:

$$E(\bar{X}) = E\left(\frac{\Sigma_i X_i}{n}\right) = \frac{\{E(X_1 + X_2 + \cdots + X_n)\}}{n}$$

$$= \frac{\{E(X_1) + \cdots + E(X_n)\}}{n} = \frac{n\mu}{n} = \mu \qquad (12.1\text{-}10)$$

i.e., the expectation of $\bar{X}$ is the distribution (or population) mean μ. With a similar approach it can be shown that:

$$E(s^2) = \sigma^2 \qquad (12.1\text{-}11)$$

i.e., the expectation of the sample variance is the distribution variance.

Taking into account that the X_i are independent, the variance of $\bar{X}$ can be easily calculated, as follows:

$$V(\bar{X}) = V\left(\frac{\Sigma_i X_i}{n}\right) = \frac{[V(X_1) + \cdots + V(X_n)]}{n^2}$$

$$= \frac{n\sigma^2}{n^2} = \frac{\sigma^2}{n} \qquad (12.1\text{-}12)$$

In other words, if several sets of n replicates are performed different sample means (i.e., different estimates of μ) will be obtained which are less spread around μ than the original X_i, having a standard deviation of $\sigma/\sqrt{n}$ instead of σ (see Fig. 12.1-2). The distribution of the sample mean is called the *sampling distribution of the mean.*

A sample parameter affords an estimate of the corresponding population parameter (fixed, predetermined value). Estimation is the first and most important purpose of statistics.

The random variable $\bar{X}$ (the *sample mean*) is an estimator of μ (*the population mean*).

The random variable s (the *sample standard deviation*) is an estimator of σ (the *population standard deviation*).

If the sample parameter P is the estimator of the population parameter Π and it is verified that $E(P) = \Pi$ (which means that the distribution of P is centered just at Π) then P is an *unbiased estimator*.

$\bar{X} = \Sigma_i X_i/n$ and $s = \{\Sigma_i(X_i - \bar{X})^2/(n - 1)\}^{1/2}$ are unbiased estimators of μ and σ, respectively.

As n increases $V(\bar{X})$ tends to zero and $\bar{X}$ tends to μ which, in the absence of systematic error, is also the *true value* for the measured quantity.

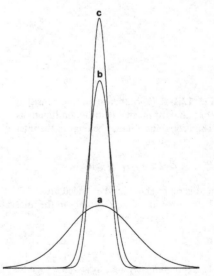

Fig. 12.1-2. Sampling distributions of the mean (curves b, c) for different sample sizes n compared with that of the original X_i (curve a) having variance σ^2 and mean μ. The sample size for curves b and c is 9 and 16, respectively

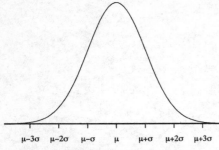

Fig. 12.1-3. Probability density function for a normal (Gaussian) distribution with mean μ and variance σ^2. The area under the normal curve can be divided into areas defined by μ and σ. Areas under the curve represent probabilities that an observation will fall within defined ranges. Areas in the ranges $\mu \pm \sigma$, $\mu \pm 2\sigma$, and $\mu \pm 3\sigma$ are 68.3, 95.4, and 99.7%, respectively

Normal (Gauss) distribution

A random variable X is normally distributed with mean μ and variance σ^2 if it possesses the following probability density function (see also Fig. 12.1-3):

$$f(x) = \frac{1}{\sigma\sqrt{2\pi}}\exp\left\{-\frac{(x-\mu)^2}{2\sigma^2}\right\} \quad -\infty \leq x \leq +\infty \tag{12.1-13}$$

In short notation one can write $X \sim N(\mu, \sigma^2)$. For practical reasons, e.g., to remove the effect of scale and measurement units, it is convenient to reduce any possible normal distribution, with its own mean μ and variance σ^2, to a unique normal distribution with predetermined, known values of mean and variance. This can be achieved by the so-called *z-transformation*, i.e., defining a new random variable Z (*the standard normal variate*) as:

$$Z = \frac{X - \mu}{\sigma} \tag{12.1-14}$$

The standard normal density $f(z)$ and normal distribution $F(z)$, functions can then be written as:

$$f(z) = (2\pi)^{-1/2}\exp\left(-\frac{z^2}{2}\right) \quad -\infty \leq z \geq +\infty \tag{12.1-15}$$

and

$$F(z) = \int_{-\infty}^{+\infty}(2\pi)^{-1/2}\exp\left(-\frac{z^2}{2}\right)dz \tag{12.1-16}$$

respectively, and are represented in Fig. 12.1-4. It can be shown that the mean and the variance of the Z distribution are 0 and 1 respectively (in short notation $Z \sim N(0,1)$). The normal density function is symmetric around the mean, is bell-shaped and the $f(z)$ practically approaches zero at z values of about ± 3. The area under the curve is given by:

$$\int_{-\infty}^{+\infty} f(z)\,dz = 1 \tag{12.1-17}$$

(a)

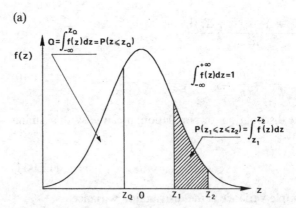

Fig. 12.1-4. (a) Standard normal density. (b) Standard normal distribution functions. The probability that Z belongs to the interval (z_1, z_2) is given by:

$$P(z_1 \leq Z \leq z_2) = \int_{z_1}^{z_2} f(z)dz$$

and is represented by the shaded area. The probability that Z belongs to the interval $(-\infty, z_Q)$ is given by:

$$P(Z \leq z_Q) = F(z_Q) = \int_{-\infty}^{z_Q} f(z)dz$$

The Q *quantile* z_Q, i.e., the z-value such that $F(z_Q) = Q$ with $0 \leq Q \leq 1$ is also shown; this quantity is also called the $100Q$ *percentile*. For instance $z_{0.95}$ indicates the 0.95 quantile (or the 95 percentile) and is such that the probability of observing a value of $z \leq z_{0.95}$ is 0.95. Note that the point q has coordinates (z_Q, Q)

(b)

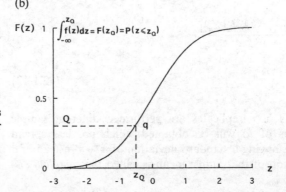

The normal distribution is the most widely used distribution for random variables and/or derived statistics since practical experience proves that, at least as a good approximation, it is valid for most physicochemical parameters. As a consequence of the *central limit theorem*, many derived statistics can be treated as normally distributed, irrespective of the original distribution of data, provided a large sample size is considered. For example, even if the original data are not normally distributed the *sampling distribution of the mean* tends to a normal one with mean μ and variance σ^2/n as n increases. A further example of the consequences of the central limit theorem is the following: the variable $S = X_1 + X_2 + \cdots + X_n$ (sum of n independent variables X_i with mean μ_i and variance σ_i^2) tends to be normally distributed with mean $\sum_i \mu_i$ and variance $\sum_i \sigma_i^2$ as n increases, irrespective of the original distribution of the X_i.

Fluctuations in an analytical measurement originate from several sources, each contributing to a certain extent to the random error. For a well-designed analytical system and properly performed analysis the measurement error can be (approximately) considered as a *linear combination* of independently distributed component errors each giving a *small contribution to the sum*, with no single source of error dominating all the others. This is the reason why errors are normally distributed in these cases.

It has also been proven that many statistical techniques based on the assumption of a normal distribution of data are *robust*, i.e., they remain valid as long as the departure of the original data from normality is reasonably small. In any case the assumption of normality of data must be done with proper care; techniques for assessing the goodness-of-fit of an experimental set of data to a model distribution, e.g., χ^2-test, must be used whenever possible.

Analytical errors are usually expected to be normally distributed because of the cumulation of several, small, and independent component errors.
Note that an outlier would originate a single large error dominating all the others.

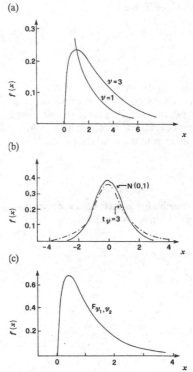

Fig. 12.1-5. Typical shapes for the χ^2, Student's t and F probability density functions $f(x)$.
(a) χ^2-distribution at 1 and 3 degrees of freedom; (b) Student's t-distribution at 3 degrees of freedom (*dashed line*) compared with a normal distribution (*solid line*); (c) F-distribution at 4, 10 degrees of freedom

Chi-squared, Student's t and F distributions

The properties of these important distributions are briefly summarized here. The use of χ^2, t, and F distributions in solving statistical problems in chemical analysis such as the evaluation of accuracy (see, e.g., tests on mean, paired t-test), estimation of confidence intervals, comparison of the precision of two methods (see F-test), and assessment of the goodness-of-fit test (see χ^2-test) will be described in the following sections with the aid of several numerical examples.

The *chi-squared* distribution, at v degrees of freedom, is described by the following probability density function:

$$f(x) = \frac{x^{\frac{v}{2}-1}e^{-x/2}}{2^{v/2}\Gamma(v/2)} \quad v > 0; \ 0 \le x < +\infty \tag{12.1-17}$$

where Γ is the gamma function. Typical shapes of this function, at different v, are shown in Fig. 12.1-5a; it is not necessary to remember the equation describing this probability density since the corresponding distribution function is widely tabulated. Mean and variance of the distribution are $E(X) = v$ and $V(X) = 2v$. Examples of random variables distributed as chi-squared are given; it can be shown that if $X_1, X_2, \ldots, X_n$ are independently, normally distributed as $N(0,1)$, the variable $\sum_i X_i^2$ is distributed as χ^2_{n-1}. Analogously, if $X_1, X_2, \ldots, X_n$ are independently distributed as $N(\mu, \sigma^2)$, the variable $\sum_i(X_i - \bar{X})^2/\sigma^2 = (n-1)s^2/\sigma^2$ is distributed as χ^2 at $n-1$ degrees of freedom. As a consequence of the *additivity property*, if X_1 and X_2 are independently distributed as $\chi^2_{v_1}$ and $\chi^2_{v_2}$ the variable $X = X_1 + X_2$ is distributed as $\chi^2_{v_1+v_2}$.

The *Student's t* distribution is quite important and very useful in describing small ($n < 30$) samples, i.e., in describing a sampling distribution rather than a population. A *Student's t* distribution at v degrees of freedom is mathematically described by the following probability density function:

$$f(x) = \frac{(\pi v)^{-1/2}\Gamma\{(v+1)/2\}}{\Gamma(v/2)}\left(1 + \frac{x^2}{v}\right)^{-(v+1)/2} \quad -\infty < x < +\infty \tag{12.1-18}$$

Again it is not necessary to remember the above equation since the corresponding distribution function is widely tabulated (see Table 12.1-3 and Fig. 12.1-6). The *Student's t* distribution is symmetric around $x = 0$ with variance $V(x) = v/(v-2)$ and tends to normality for $v \to \infty$ (in practice for $v > 30$ the t distribution collapses

Table 12.1-3. Probability points for a t distribution at different degrees of freedom v. Tabulated values give the 90, 95, and 97.5 percentiles, i.e., those values of t_p, such that the area from $-\infty$ to t_p is 0.90, 0.95, and 0.975, respectively

v	$t_{0.90}$	$t_{0.95}$	$t_{0.975}$
1	3.08	6.31	12.71
2	1.89	2.92	4.30
3	1.64	2.35	3.18
4	1.53	2.13	4.78
5	1.48	2.01	2.57
6	1.44	1.94	2.45
7	1.42	1.89	2.37
8	1.40	1.86	2.31
9	1.38	1.83	2.26
10	1.37	1.81	2.23
20	1.32	1.72	2.09
30	1.31	1.70	2.04
60	1.30	1.67	2.00
120	1.29	1.66	1.98
∞	1.28	1.645	1.96

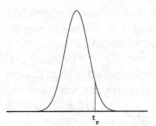

Fig. 12.1-6. Probability density function for a Student's *t*-distribution. t_p represents the *p* percentile of the distribution, i.e., the value of *t* such that the probability $P(-\infty \leq t \leq t_p)$ is equal to *p* (see Table 12.1-3)

to a normal one – see Fig. 12.1-5b). If X and Y are independently distributed as $N(0,1)$ and χ_v^2, respectively then the variable $Z = X/(Y/v)^{1/2}$ is distributed as a Student's *t* distribution at v degrees of freedom; it follows that since:

$$\frac{\bar{X} - \mu}{\sigma/\sqrt{n}} \sim N(0,1), \quad \text{and} \quad \frac{(n-1)s^2}{\sigma^2} \sim \chi_{n-1}^2$$

Equation 12.1-19 gives the theoretical foundation for estimating the population mean from the sample mean $\bar{X}$ by calculating the relevant confidence interval as $\bar{X} \pm ts/\sqrt{n}$ as described in more detail later in this section.

the random variable:

$$\frac{(\bar{X} - \mu)/(\sigma/\sqrt{n})}{\{(n-1)s^2/\sigma^2(n-1)\}^{1/2}} = \frac{\bar{X} - \mu}{s/\sqrt{n}} \tag{12.1-19}$$

is distributed as t_{n-1}.

The F distribution is useful in the comparison of the precision of two methods.

The *F distribution* is of considerable importance since it allows statistical tests to be performed on random variables following distributions other than the normal one (e.g., chi-squared). The probability density function describing an F distribution at v_1, v_2 degrees of freedom can be found in specialist books; an example of a F distribution is given in Fig. 12.1-5c.

Examples of variables distributed as an F distribution are given; if $X \sim \chi_{v_1}^2$ and $Y \sim \chi_{v_2}^2$ then the variable $(X/v_1)/(Y/v_2)$ is distributed as Fv_1, v_2.

If $X \sim N(\mu_1, \sigma_1^2)$ and $Y \sim N(\mu_2, \sigma_2^2)$, then $(n_1 - 1)s_1^2/\sigma_1^2 \sim \chi_{n_1-1}^2$ and $(n_2 - 1)s_2^2/\sigma_2^2 \sim \chi_{n_2-1}^2$; it follows that the variable $(s_1^2/\sigma_1^2)/(s_2^2/\sigma_2^2)$ is distributed as F_{n_1-1,n_2-1}. In the particular case when $\sigma_1 = \sigma_2$, the variable s_1^2/s_2^2 is distributed as F_{n_1-1,n_2-1}.

12.1.2 Performance tests

Estimation and confidence interval

Estimation is one of the most important goals of statistics. Population parameters are estimated through analyzing a random sample of a given size.

Rarely is the whole material under investigation (population in statistical terms) analyzed. Suppose one is interested to know the phosphate content in soil; for practical reasons it is necessary to collect and analyze random samples (see Sec. 12.1.3). In other words a *random sample* is extracted from a *population* in order to *estimate* some *properties of the population* itself. This approach relies on the assumption that the information content of the data obtained on a sample reasonably reflects that of the whole population.

An unbiased estimator is such that its expectancy equals the population parameter it estimates.
A confidence interval gives more information than an estimate because it tells us how close to τ we are.

If a sample of *size "n"* is analyzed a set of *"n"* independent random observations $X_1, X_2, \ldots, X_n$ will be obtained which fit a certain probability distribution. A parameter τ (e.g., mean or variance) of the distribution can be *estimated* by a function $T(X)$ of the observations, called *estimator*, which is itself a random variable possessing its own probability distribution, mean, and variance. An example of $T(X)$ is the sample mean, already described in Sec. 2.4. Of course for each different set of observations a particular value (or *realization*) of T will be obtained which is called the *estimate*. Evidently a reliable estimator should have a high probability of being close to the unknown parameter which it estimates; in the ideal situation the distribution of T should be centred just at τ, i.e. $E(T) = \tau$. An estimator satisfying this condition is called *unbiased*. It has already been demonstrated that $E(\bar{X}) = \mu$ and $E(s^2) = \sigma^2$, i.e., the sample mean and the sample variance are *unbiased estimators*.

Note that the probability that a confidence interval contains the true value of τ can only take the values of 1 or 0 according to whether it includes the true value, or not. This means that we cannot say that we have a probability of $(1 - \alpha)$ that a given confidence interval $t \pm k$ (where t is a realization of T) includes the true value. What we can say is that we have a $(1-\alpha)100\%$ confidence that the interval $(t - k, t + k)$ contains the true value of τ. In other words, if we draw a sufficiently large number of samples of size n and for each of them calculate the confidence interval, then a proportion of $(1 - \alpha)$ will contain the true value of τ.

It is possible to quantify how close an estimate is to the population (unknown) parameter τ by evaluating the probability that T lies in a certain interval $\tau \pm k$ where k is a constant. Note that the probability that T lies in the interval $\tau \pm k$ is the same that τ lies in the interval $T \pm k$; the interval $T \pm k$ is random (because T is random) and a realization of it is called the *confidence interval*; the extreme values are called *confidence limits*. The probability $P = (1 - \alpha)$ that the random interval can include τ is called the *confidence coefficient* (or *level*) while α is called the *significance level* and typically takes the value of 0.05, 0.01, or 0.001 in decreasing order of usage. So if $\alpha = 0.05$, one has a 95% confidence that the interval $T \pm k$ includes the true value. The greater the confidence level, the greater the interval, the absolute certainty requiring an infinitely wide interval.

For example, one can calculate the *confidence limits on the mean* in the ideal case of a normal distribution with known variance. Since:

$$\bar{X} \sim N\left(\mu, \frac{\sigma^2}{n}\right)$$

where μ, *in the absence of systematic errors*, is equal to the true value, the standard normal variate is:

$$z = \frac{\bar{X} - \mu}{\sigma_{\bar{X}}} = \frac{\bar{X} - \mu}{\sigma/\sqrt{n}} \qquad (12.1\text{-}20)$$

and is distributed as $N(0, 1)$. To calculate the confidence interval associated with a 95% confidence level it is necessary to know the value of z such that the area from $-z$ to $+z$ equals 0.95 and the area of each tail equals 0.025 (see Fig. 12.1-7 and Table 12.1-4). In other words we require the 0.975 quantile (or the 97.5 percentile of the normal distribution, see also Fig. 12.1-2) that is the value $z = z_{0.975}$ such that $F(z) = 0.975$.

From a normal integral table (see also Table 12.1-4) one can easily find that $z_{0.975} = 1.96$ so that the 95% confidence interval can be expressed by the following inequality:

$$\mu - \frac{1.96\sigma}{\sqrt{n}} \leq \bar{X} \leq \mu + \frac{1.96\sigma}{\sqrt{n}} \qquad (12.1\text{-}21)$$

Clearly the size of the confidence interval depends on the confidence coefficient $(1 - \alpha)$ or the significance level α chosen. Then the general expression is:

$$\mu - \frac{z_{(1-\alpha/2)}\sigma}{\sqrt{n}} \leq \bar{X} \leq \mu + \frac{z_{(1-\alpha/2)}\sigma}{\sqrt{n}} \qquad (12.1\text{-}22)$$

By rearranging this inequality, one can easily estimate the population mean from the sample mean $\bar{X}$ by calculating its confidence interval as follows:

$$\bar{X} - \frac{z_{(1-\alpha/2)}\sigma}{\sqrt{n}} \leq \mu \leq \bar{X} + \frac{z_{(1-\alpha/2)}\sigma}{\sqrt{n}} \qquad (12.1\text{-}23)$$

As can be realized, a confidence interval gives more information than an estimate because the interval width, which is completely defined by the chosen confidence coefficient, gives an indication of the closeness of the estimate to the unknown population parameter. It is also clear that, for a given α, the size of the confidence interval is determined by the number of replicates performed. However due to the $1/\sqrt{n}$ dependence, increase of the sample size over a certain value is not advisable because it is unlikely that the corresponding increase of the analysis cost and analysis time could be compensated for by the reduction in the size of the confidence interval.

Note that the above expressions have been derived assuming knowledge of σ and normality of data. This is rarely the case in practice. A more realistic situation is represented by any distribution with finite (unknown) variance and large n (typically > 30). Due to the central limit theorem $\bar{X}$ is approximately distributed as $N(\mu, \sigma^2/n)$ and:

$$\frac{\bar{X} - \mu}{\sigma/\sqrt{n}} \sim N(0, 1)$$

so that the above considerations are still valid. The σ^2 value is unknown, but it can be reasonably substituted by the sample variance s^2 since $(\bar{X} - \mu)/(s/\sqrt{n})$ is also approximately distributed as $N(0, 1)$. Thus the confidence interval for the population mean at a confidence level $1 - \alpha$ is given by:

$$\mu = \bar{X} \pm \frac{z_{(1-\alpha/2)}s}{\sqrt{n}} \qquad (12.1\text{-}24)$$

However, in most situations, the sample size is small due to, e.g., a limited amount of the available material or the high cost of a single analysis. In this case, assum-

Confidence limits on the mean define an interval around the sample mean $\bar{X}$. We have a confidence of $(1 - \alpha)100\%$ that this interval includes the population mean μ.

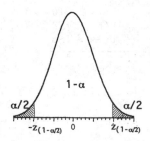

Fig. 12.1-7. Illustration of the significance of the various quantities involved in the calculation of a confidence interval at a significance level α (see Table 12.1-4)

Table 12.1-4. Area in one tail of the normal distribution (see Fig. 12.1-17), and $z_{(1-\alpha/2)}$ percentile values required in the calculation of the confidence intervals at different confidence levels. As can be seen from this Table and Eq. 12.1-23 the width of a confidence interval doubles on increasing the confidence level from 90 to 99.9%

Confidence level	Area in one tail	$z_{(1-\alpha/2)}$
90%	5%	1.64
95%	2.5%	1.96
99%	0.5%	2.58
99.9%	0.05%	3.29

In the most common situation involving a small sample size and unknown population variance, the σ^2 value, is replaced by sample variance s^2 and the normal distribution by a Student's t distribution.

Table 12.1-5. Values of $t_{(1-\alpha/2)}$ at confidence levels, $(1-\alpha)$, of 0.90, 0.95, and 0.99

Degrees of freedom	$t_{0.95}$	$t_{0.975}$	$t_{0.995}$
1	6.31	12.71	63.66
2	2.92	4.30	9.92
3	2.35	3.18	5.84
4	2.13	4.78	4.60
5	2.01	2.57	4.03
6	1.94	2.45	3.71
7	1.89	2.37	3.50
8	1.86	2.31	3.55
9	1.83	2.26	3.25
10	1.81	2.23	3.17
20	1.72	2.09	2.85
30	1.70	2.04	2.75
60	1.67	2.00	2.66
120	1.66	1.98	2.62
∞	1.645	1.96	2.58

ing that $X_1, X_2, \ldots X_n$ are independently, identically distributed as $N(\mu, \sigma^2)$, the statistics $T = (\bar{X} - \mu)/(s/\sqrt{n})$ is distributed as t_{n-1} (see paragraph on Student's t distribution) and then the confidence interval for the population mean μ are given by:

$$\mu = \bar{X} \pm \frac{t_{(1-\alpha/2),n-1}s}{\sqrt{n}} \qquad (12.1\text{-}25)$$

where $t_{(1-\alpha/2),n-1}$ represents the $(1-\alpha/2)$ quantile of the t distribution at $n-1$ degrees of freedom, i.e., the value of t such that the area from $-\infty$ to t is equal to $(1-\alpha/2)$.

Worked example:

The lead content in a water sample is determined by anodic stripping voltammetry. The following values (µg/L) are obtained: 1.2, 1.8, 1.4, 1.6, 1.3, 1.5, 1.4, 1.3, 1.7, and 1.4. Calculate the 95% and 99% confidence limits for the lead concentration.

The mean and standard deviation of the above values are 1.46 and 0.19, respectively. There are 10 measurements and 9 degrees of freedom. From a t table (see Table 12.1-5) it can be found that $t_{0.975,9}$ is 2.26 and $t_{0.995,9}$ is 3.25; from Eq. 12.1-25 the 95% confidence limits are:

$$\mu = 1.46 \pm 2.26 \times \frac{0.19}{\sqrt{10}} = 1.46 \pm 0.14$$

Analogously the 99% confidence limits are:

$$\mu = 1.46 \pm 3.25 \times \frac{0.19}{\sqrt{10}} = 1.46 \pm 0.20$$

Worked example:

An analyst is asked to determine lead in a consignment of fruit juice. The client specifies that the lead content is of the order of 100 µg/kg (ppb) and that he requires an "accuracy" of 5 µg/kg and accepts a 95% confidence level. Calculate the sample size necessary to satisfy this request assuming that, at the specified concentration level, the precision of the analytical method used is known to be ± 8 µg/kg.

The client's request for a 95% confidence level means that μ lies in the range $\bar{X} \pm 1.96\sigma/\sqrt{n}$ where $1.96\sigma/\sqrt{n} = 5$; it follows that: $\sqrt{n} = (1.96 \times 8)/5 = 3.14$ and $n \sim 10$. Ten samples must be analyzed to meet the client specifications; note, however, that the underlying assumption in the above approach is that the analytical method used is *free from systematic error.*

Evaluation of accuracy

As pointed out in Table 2.1-1 of Sec. 2.4, careful working can minimize random errors; such errors fluctuate randomly and are indeterminate. Systematic errors, on the other hand, can occur irrespective of how carefully the work is carried out and, in principle, can be detected, quantified, and in some cases completely removed; this is the reason why they are also called *determinate errors* (note that systematic errors cannot be revealed by repeated measurements).

As pointed out in Sec. 2.4, there can be several sources of systematic errors: *methodological*, e.g., sample contamination, interferences, loss of analyte due to improper sample handling, incomplete washing of a precipitate in gravimetric analysis, indicator error in volumetric analysis, and so on; *instrumental*, e.g., noncalibrated instruments, and *personal* bias, such as color blindness, astigmatism, tendency to prefer even (or odd) numbers in reporting the results. Systematic errors can be *constant*, i.e., concentration independent, or *proportional*, i.e., concentration dependent. For example a constant systematic error can originate from "blank" contamination, while a wrong calibration can be the origin of a proportional error. Of course, they can occur simultaneously.

As for random errors, the first defence against systematic errors is their prevention, so that, before performing a determination, the analyst must attempt to eliminate any possible source of errors, for instance, by properly designing the experiment once the crucial steps of the analytical procedure have been identified. For example, the bias introduced in a titration by the indicator can be removed by a *blank correction*. However, not all systematic errors can be so easily detected and eliminated. The final action for checking the presence of systematic errors and evaluating the accuracy of an analytical procedure, involves the use of *reference methods or standard reference materials*, e.g., those proposed by institutions such as the American Society for Testing and Materials (*ASTM*), le Bureau Communitaire de Reference (*BCR*), the International Atomic Energy Agency (*IAEA*), and the National Institute of Standards and Technology (NIST). In the latter case, a certified material identical to the one to be analyzed, or of very similar composition (matrix and analyte concentration) is analyzed; the agreement between *certified* and *found* values, ascertained by *statistical significance tests* – see below, is taken as evidence of the absence of systematic errors.

This procedure of evaluating accuracy has the obvious disadvantage of only being valid for the particular *reference material* used; moreover, a suitable standard material is, in some cases, not readily available or not available at all. In such circumstances the test method can be compared, on a statistical basis, with a *reference method* which is assumed *free of method or laboratory bias*. In the view of the previous discussion on bias and accuracy (see Sec. 2.4) it should be emphasized that the above assumption may be hazardous so that the final evaluation of a method should be, preferably, carried out by an interlaboratory trial. If the test and the reference method are compared using a certain sample, the conclusion on the accuracy of the method will be, of course, restricted to the particular sample analyzed; clearly this approach can be extended to different samples in order to draw more general conclusions. It is of paramount importance that the comparison of methods covers the concentration range over which the analytical procedure will be used. The results obtained with both methods can be compared on a statistical basis (e.g., *t*-test, regression analysis) to ascertain whether the observed difference is significant or not, i.e., to decide whether or not it can be explained by random variations only.

Hypothesis testing

Chemists are constantly faced with problems such as: Is a given material of the purity declared on the label? Is an optical method as reliable as an electrochemical one for a particular analysis? Is a given method of comparable precision to another? Is the test method free from systematic error?

For example, a way to test this last hypothesis is, to analyze a standard reference material whose analyte concentration is known with high accuracy and precision. The difference between the *certified* value and the mean of replicate determinations (i.e., the *found* value), performed by the test method, originates from random variations and, if present, from systematic variations too. Because of the unavoidable occurrence of random errors, it is unlikely that the *found* will equal the *certified* value even if systematic errors are absent. This is why in such a situation one must be able to decide whether the observed *difference is significant* or not, i.e., if it can be accounted for by random variations alone. Such a decision can be taken on a statistical basis by using a *significance or hypothesis test*. It is clear that statistical techniques for accepting or rejecting a hypothesis are of fundamental importance for a correct interpretation of analytical data. The fundamental steps in performing the test are outlined in Fig. 12.1-8. The basic principles underlying the test will be briefly described in the following.

In general one supposes that the observations come from a distribution of known form, but with one or more unknown parameters: typically mean and/or variance. The hypothesis to be tested, called *null hypothesis* H_0, is usually formulated, in a way that implies no significant difference between the quantities to be compared. So, in testing for accuracy, the null hypothesis is that systematic error is absent; this is expressed as H_0: *found = certified*. Thus, under the null hypothesis, the sampling distribution of the mean is symmetric around the true (certified) value with variance σ^2/n. The null hypothesis is tested against an *alter-*

```
┌─────────────────────────────────────────────────────────┐
│            SELECTION OF THE APPROPRIATE TEST              │
│                         ⇓                                │
│           SELECTION OF THE SIGNIFICANCE LEVEL            │
│                         ⇓                                │
│   FORMULATION OF THE NULL AND ALTERNATIVE HYPOTHESIS    │
│                         ⇓                                │
│   CALCULATION OF THE TEST STATISTICS AND COMPARISON     │
│   WITH THE CRITICAL VALUE AT THE CHOSEN SIGNIFICANCE    │
│                       LEVEL                             │
│                         ⇓                                │
│                      DECISION                           │
└─────────────────────────────────────────────────────────┘
```

Fig. 12.1-8. Fundamental steps in the elaboration of a hypothesis test

native hypothesis H_1; H_0 and H_1 are mutually exclusive. H_1 can be formulated in several ways, but only a particular couple H_0,H_1 is the most appropriate for the given problem. So in the above situation, for instance, H_1 can be reasonably formulated as H_1 : *found* $\neq$ *certified* unless there is an indication that the *found* can only lie on one side of the *certified* value. In general the following possibilities exist:

$H_0 : T = T_0$ (*simple null hypothesis*)

$H_1 : T \neq T_0$ (*two-sided test*)

$H_1 : T > T_0$ (*one-sided test*)

$H_1 : T < T_0$ (*one-sided test*)

In a significance test a null and an alternative hypothesis must be formulated. The choice of H_0 and H_1 will depend on the nature of the question to be answered.

where T is the unknown parameter of the distribution from which our data are drawn, and T_0 is a predetermined value. According to the formulation of H_1 one can have a *two-sided test* ($H_1: T \neq T_0$) or a *one-sided test* ($H_1: T > T_0$ or $T < T_0$).

The significance level α represents the probability of incorrectly rejecting the null hypothesis, i.e., the probability of rejecting H_0 when it is true.

In order to *reject*/accept H_0 (*accept*/reject H_1) one must decide, on a statistical basis, if an estimate of T is too different, too large or too small compared with T_0. Since analytical measurements are in most cases carried out on samples extracted from the population, the risk of taking a wrong decision must be taken into account when performing a hypothesis test. Then one must fix a *significance level* α which represents the *maximum probability of rejecting* H_0 *when it is true*. Figure 12.1-9 helps to understand the meaning of the various quantities. If a realization t of the test statistics T falls within the shaded area of Fig. 12.1-9 one rejects H_0 (accept H_1) at the given significance level.

Type I and type II errors

The *type I* error, α, represents the probability of incorrectly rejecting the null hypothesis. The probability of correctly accepting the null hypothesis is represented by $(1 - \alpha)$.
The *type II* error β represents the probability of incorrectly accepting the null hypothesis, i.e., the probability of accepting H_0 when H_0 is false.

In using the procedure outlined above, it is possible that the null hypothesis may be rejected when it is in fact true (*type I error*); that is to say that the probability, $P(I)$, of the *type I* error is equal to α.

The power of a test, $1 - \beta$, represents the probability of correctly rejecting the null hypothesis.

Acceptance of H_0 when it is false (or rejection of H_1 when it is true) is known as *type II error* and its probability, $P(II)$, is represented by the shaded area β in Fig. 12.1-10. Note that if $P(I)$, for a given situation, is reduced by decreasing the significance level α (e.g., passing from 0.05 to 0.01), then $P(II)$ increases. Conversely as T moves far away from T_0, $P(II)$ decreases while, as T moves toward T_0, $P(II)$ increases up to $(1 - \alpha)$ when $T = T_0$; that is to say that β is a function of T. Since $H_0(H_1)$ may be either true(false) or false(true) the possible outcomes in a hypothesis test can be summarized as in Fig. 12.1-11. In order to avoid possible misunderstandings, one must consider that in performing a significance test at a given significance level (say 5%) *we just evaluate the probabilities of rejecting the null hypothesis given that it is true or false* so that the acceptance of H_0 *does not imply* that the probability for H_0 of being true is 95%.

For a given significance level, the probability of correctly rejecting the null hypothesis (power of the test) increases on increasing the sample size n. For a given sample size n, the power of the test $(1 - \beta)$ decreases on decreasing the significance level α.

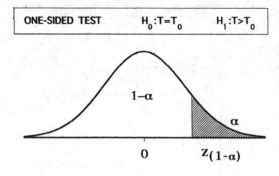

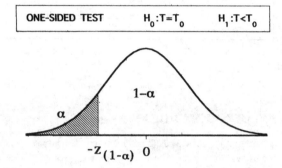

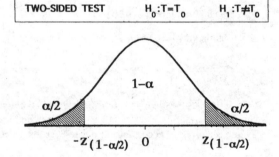

Fig. 12.1-9. Illustration of the rejection criteria for one- and two-sided tests (see also Table 12.1-6). The test statistics $T = (\bar{X} - T_0)/(\sigma/\sqrt{n})$ is distributed as $N(0,1)$ under H_0, i.e., when the null hypothesis is true. Shaded areas represent the region of rejection of the null hypothesis; nonshaded areas represent the region of acceptance of H_0

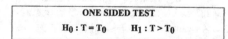

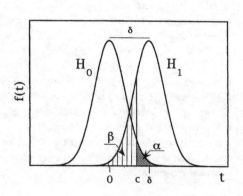

Fig. 12.1-10. Illustration of significance level, and *type I* and *type II* errors. Suppose the X values come from a normal population with variance σ^2; hence $\bar{X} \sim N(T, \sigma^2/n)$ and the statistics $T = (\bar{X} - T_0)/(\sigma/\sqrt{n})$ is distributed as $N(0,1)$ under H_0 and as $N(\delta, 1)$ under H_1. α represents the probability, $P(T \geq c)$, that the test statistics T is greater than or equal to a *critical value c* under the assumption that H_0 is true. One rejects H_0 at the significance level α if $t \geq c$ (where t is a realization of the test statistics). β represents the probability of the *type II* error.
$\delta = (T - T_0)/(\sigma/\sqrt{n})$; thus when T tends toward T_0, β tends to $(1 - \alpha)$
Curves labelled H_0 and H_1 represent the probability density functions under H_0 and H_1, i.e., when the null and alternative hypotheses, respectively, are true

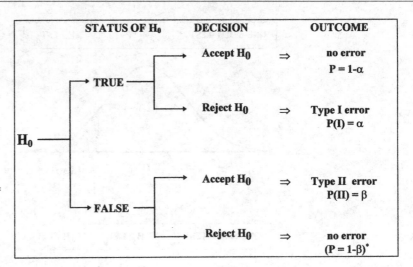

Fig. 12.1-11. Summary of the possible outcomes (and relevant probabilities) in a hypothesis test
* Note that P(rejecting H_0 when H_0 is false) = $1 - P$ (accepting H_0 when H_0 is false) = $1 - \beta$. This last probability, i.e., $1 - P(II)$ is known as the *power of a test* which is useful especially when the performances of different statistical test are compared

Worked example:

A manufacturer claims that the content of acetylsalicylic acid in his aspirin tablets is 0.30 g, with a standard deviation of 0.02 g. A sample of 9 tablets is taken from the production line and analyzed; the mean content is found to be 0.285 g. Test the manufacturer's assertion at 95 and 99% confidence levels (i.e., at 0.05 and 0.01 significance levels).

The null and alternative hypothesis are formulated as:

$$H_0 : \mu = 0.30\,g \quad H_1 : \mu \neq 0.30\,g$$

The test statistics $T = (\bar{X} - \mu)/(\sigma/\sqrt{n})$ is distributed as $N(0, 1)$. The realization of T is:

$$t = \frac{0.285 - 0.300}{0.02/\sqrt{9}} = -2.25$$

At $\alpha = 0.05$ the critical value is $z_{0.975} = 1.96$ (H_1 requires a two-sided test).

Since $|t| \geq z_{(1-\alpha/2)}$ we reject H_0 at the given significance level α. Alternatively we can calculate the range of acceptance of our test results, calculating the 95% confidence interval as:

$$\bar{X} \pm \frac{z_{(1-\alpha/2)}\sigma}{\sqrt{n}} = 0.30 \pm 1.96 \times \frac{0.02}{\sqrt{9}} = 0.30 \pm 0.013.$$

The confidence limits are 0.287 and 0.313. Since the test result lies outside this range H_0 must be rejected.

At $\alpha = 0.01$ the critical value (the 99.5 percentile of the distribution) is $z_{0.995} = 2.57$ so we accept the manufacturer's assertion $H_0 : \mu = 0.30\,g$. The same conclusion is reached by considering the range of acceptance since the test result lies inside the 99% confidence limits 0.283 and 0.317.

As can be seen, *applying different significance levels different conclusions are drawn*: H_0 has been rejected in the first case and accepted in the second one. *Note, however that in both cases the right decision has been taken.* In reality the manufacturer's claim (H_0) is either true or false, but, unfortunately, we do not know which is the real situation (otherwise there would be no need to carry out the test); so we can only discuss the probability of rejecting H_0 given that it is true or given that it is false. At $\alpha = 0.05$, H_0 has been rejected; so if H_0 is true we have a 5% probability of making a *type I* error, while if it is false we make no error. At $\alpha = 0.01$ H_0 has been accepted: in this case we cannot make a *type I* error, but if H_0 is false, we make a *type II* error. We cannot predict $P(II)$ unless the true value for the population mean is known (but in this case there is no need for the test). Moreover we can realize that on decreasing the significance level, $P(I)$ is reduced, since the range of acceptance for our test results is increased. This conversely increases the probability of false acceptance i.e., of *type II* error (decreasing α increases β). In other words, we must define our decision criterion according to the cost associated with the *type I* error (see also the following example).

Worked example:

For the same problem of the previous example calculate the probability of the type II error and the power of the test for sample sizes of 9 and 27. Choose a significance level of $\alpha = 0.05$ and suppose that the true population mean is 0.31 g (remember that in a real situation the population mean remains unknown unless an infinitely large number of measurements is performed).

The null hypothesis is the same as before $H_0 : \mu = 0.30$ g, but now we know that the true value for the population mean is 0.31 g. So for $n = 9$ the situation can now be sketched as in Fig. 12.1-12.

The shaded area β, i.e., $P(\text{II})$ can be easily calculated as follows:

$$z = \frac{\overline{X} - \mu}{\sigma/\sqrt{n}} = \frac{0.313 - 0.310}{0.02/\sqrt{9}} = -1.05$$

The area corresponding to this value of z for a standard normal distribution is 0.634 or 63.4%. The power of the test is $1 - \beta = 36.6\%$.

For $n = 27$, the situation can be sketched as in Fig. 12.1-13.

The critical value changes from 0.313 to 0.307 and the corresponding z value is -2.33; the area β is now 0.218 (or 21.8%) and the power of the test is now 78.2%. As can be seen, a threefold increase in the sample size more than doubles the power of the test. The importance of increasing the sample size; is now clear for a given significance level, increasing n decreases the probability of *type II* error and increases the *power of the test*, i.e., *the probability of correctly rejecting the null hypothesis*.

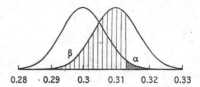

Fig. 12.1-12. Illustration of the significance of type I and type II error in the situation described in the worked example for a sample size $n = 9$ (see text)

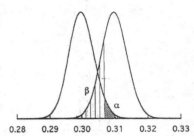

Fig. 12.1-13. Illustration of the significance of type I and type II error in the situation described in the worked example for a sample size $n = 27$ (see text)

Test on one sample mean

Table 12.1-6 summarizes the tests on the mean of a single sample referred to in different cases.

Tests on two samples means

In addition to the use of a certified reference material, other approaches can be adopted to evaluate accuracy. For instance, one can compare the procedure under evaluation with a well-established reference method. The same sample is analyzed by the two methods and the sample means $\overline{X}_A$ (an estimate of μ_A) and $\overline{X}_B$ (an

Table 12.1-6. Summary of hypothesis tests on the mean

Case	H_0, H_1	Test statistics and its distribution under H_0 (type of test)	Rejection criteria
1	$H_0: \mu = \mu_0$	$T = (\overline{X} - \mu_0)/(\sigma/\sqrt{n}) \sim N(0, 1)$	
	$H_1: \mu > \mu_0$	(one-sided)	$t \geq z_{(1-\alpha)}$
	$H_1: \mu < \mu_0$	(one-sided)	$t \leq z_{(1-\alpha)}$
	$H_1: \mu \neq \mu_0$	(two-sided)	$\lvert t \rvert \geq z_{(1-\alpha/2)}$
2	$H_0: \mu = \mu_0$	$T = (\overline{X} - \mu_0)/(s/\sqrt{n}) \sim N(0, 1)$	
	$H_1: \mu > \mu_0$	(one-sided)	$t \geq z_{(1-\alpha)}$
	$H_1: \mu < \mu_0$	(one-sided)	$t \leq z_{(1-\alpha)}$
	$H_1: \mu \neq \mu_0$	(two-sided)	$\lvert t \rvert \geq z_{(1-\alpha/2)}$
3	$H_0: \mu = \mu_0$	$T = (\overline{X} - \mu_0)/(s/\sqrt{n}) \sim t_{n-1}$	
	$H_1: \mu > \mu_0$	(one-sided)	$t \geq t_{(1-\alpha),n-1}$
	$H_1: \mu < \mu_0$	(one-sided)	$t \leq t_{(1-\alpha),n-1}$
	$H_1: \mu \neq \mu_0$	(two-sided)	$\lvert t \rvert \geq t_{(1-\alpha/2),n-1}$

μ_0 is the test value for the mean; $z_{(1-\alpha)}$ and $z_{(1-\alpha/2)}$ are the $(1 - \alpha)$ and the $(1 - \alpha/2)$ quantile, respectively, of the standard normal distribution;
$t_{(1-\alpha),n-1}$ and $t_{(1-\alpha/2),n-1}$ are the $(1 - \alpha)$ and the $(1 - \alpha/2)$ quantile, respectively, of a t-distribution at $n - 1$ degrees of freedom; the other symbols have their usual meaning (see text).
case 1: observations distributed as $N(\mu, \sigma^2)$ with known σ^2
case 2: any distribution with finite variance and n large
case 3: normal distribution with unknown variance (n small); in this case the test statistics is approximately distributed as t at $n - 1$ degrees of freedom.

estimate of μ_B) are compared on the basis of a significance test. An example of how to derive the test statistics to be used is given in the following. If $\bar{X}_A$ and $\bar{X}_B$ are independently and identically distributed, with the same variance and means μ_A and μ_B, respectively then $\bar{X}_A \sim N(\mu_A, \sigma^2/n_A)$ and $\bar{X}_B \sim N(\mu_B, \sigma^2/n_B)$. Due to the additivity property of the normal distribution, (i.e., if $X_1 \sim N(\mu_1, \sigma_1{}^2)$ and $X_2 \sim N(\mu_2, \sigma_2{}^2)$ a linear combination $X_1 \pm X_2$ is distributed as $N(\mu_1 \pm \mu_2, \sigma_1{}^2 + \sigma_2{}^2)$), the difference $(\bar{X}_A - \bar{X}_B)$ is distributed as $N(\mu_A - \mu_B, \sigma^2(1/n_A + 1/n_B))$. It follows that the standard normal variate (see Eq. 12.1-26) is distributed as $N(0,1)$:

$$\frac{(\bar{X}_A - \bar{X}_B) - (\mu_A - \mu_B)}{\sigma(1/n_A + 1/n_B)^{1/2}} \sim N(0,1) \tag{12.1-26}$$

We also know (see the Chi-squared distribution) that $(n-1)s^2/\sigma^2$ is distributed as χ^2_{n-1}; due to the additivity property of the chi-squared distribution:

$$\frac{[(n_A - 1)s_A^2 + (n_B - 1)s_B^2]}{\sigma^2} \sim \chi^2_{n_A + n_B - 2} \tag{12.1-27}$$

Hence (see the Student's t distribution):

$$\frac{(\bar{X}_A - \bar{X}_B) - (\mu_A - \mu_B)}{\left(\dfrac{1}{n_A} + \dfrac{1}{n_B}\right)^{1/2} \left[\dfrac{(n_A - 1)s_A^2 + (n_B - 1)s_B^2}{n_A + n_B - 2}\right]^{1/2}} \sim t_{n_A + n_B - 2} \tag{12.1-28}$$

Note that the term in square brackets is a pooled standard deviation s_p; so the above equation may be rewritten as:

$$\frac{(\bar{X}_A - \bar{X}_B) - (\mu_A - \mu_B)}{\left(\dfrac{1}{n_A} + \dfrac{1}{n_B}\right)^{1/2} s_p} \sim t_{n_A + n_B - 2} \tag{12.1-29}$$

The obvious null hypothesis is that no systematic difference exists between methods A and B, i.e., $\bar{X}_A$ and $\bar{X}_B$ come from the same population; H_0 is then formulated as $H_0 : \mu_A - \mu_B = 0$. The alternative hypotheses can be formulated as usual. Table 12.1-7 summarizes significance tests for comparisons of two-sample means valid in the specified typical situations.

Worked example:

In the comparison of two methods for the determination of methyl mercury content in marine organisms, the following results (mg/kg) were obtained:

	Mean	Standard deviation	Sample size
Reference method	1.25	0.05	10
Test method	1.40	0.09	11

Do the two means differ significantly at a significance level of 0.05?
The null and alternative hypothesis are formulated as:

$H_0: \mu_A - \mu_B = 0$ and $H_1: \mu_A - \mu_B \neq 0$.

Since the two variances are not significantly different (see the following example) the situation corresponds to case 3 in Table 12.1-7. Then:

$$s_p = \left[\frac{(n_A - 1)s_A^2 + (n_B - 1)s_B^2}{n_A + n_B - 2}\right]^{1/2} = 0.07_3$$

and

$$t = \frac{(\bar{X}_A - \bar{X}_B) - (\mu_A - \mu_B)}{\left(\dfrac{1}{n_A} + \dfrac{1}{n_B}\right)^{1/2} s_p} = 4.70$$

The 97.5 percentile of a t distribution with 19 degrees of freedom is 2.09; the null hypothesis must be rejected. In other words the test method is affected by systematic errors.

Table 12.1-7. Tests on two-sample means

Case	H_0, H_1	Test statistics and its distribution under H_0	Rejection criteria
1	$H_0: \mu_A - \mu_B = 0$ $H_1: \mu_A - \mu_B > 0$ $H_1: \mu_A - \mu_B < 0$ $H_1: \mu_A - \mu_B \neq 0$	$T = \dfrac{\overline{X}_A - \overline{X}_B}{(\sigma_A{}^2/n_A + \sigma_B{}^2/n_B)^{1/2}} \sim N(0,1)$	$t \geq z_{(1-\alpha)}$ $t \leq z_{(1-\alpha)}$ $\lvert t \rvert \geq z_{(1-\alpha/2)}$
2	$H_0: \mu_A - \mu_B = 0$ $H_1: \mu_A - \mu_B > 0$ $H_1: \mu_A - \mu_B < 0$ $H_1: \mu_A - \mu_B \neq 0$	$T = \dfrac{\overline{X}_A - \overline{X}_B}{(s_A{}^2/n_A + s_B{}^2/n_B)^{1/2}} \sim N(0,1)$	$t \geq z_{(1-\alpha)}$ $t \leq z_{(1-\alpha)}$ $\lvert t \rvert \geq z_{(1-\alpha/2)}$
3	$H_0: \mu_A - \mu_B = 0$ $H_1: \mu_A - \mu_B > 0$ $H_1: \mu_A - \mu_B < 0$ $H_1: \mu_A - \mu_B \neq 0$	$T = \dfrac{\overline{X}_A - \overline{X}_B}{s_p(1/n_A + 1/n_B)^{1/2}} \sim t_v$	$t \geq t_{(1-\alpha),v}$ $t \leq t_{(1-\alpha),v}$ $\lvert t \rvert \geq t_{(1-\alpha/2),v}$
4	$H_0: \mu_A - \mu_B = 0$ $H_1: \mu_A - \mu_B > 0$ $H_1: \mu_A - \mu_B < 0$ $H_1: \mu_A - \mu_B \neq 0$	$T = \dfrac{\overline{X}_A - \overline{X}_B}{(s_A{}^2/n_A + s_B{}^2/n_B)^{1/2}} \sim t_v$	$t \geq t_{(1-\alpha),v}$ $t \leq t_{(1-\alpha),v}$ $\lvert t \rvert \geq t_{(1-\alpha/2),v}$

case 1: normal distribution with known variances. Extension to the case of equal variances is obvious.
case 2: any distribution with finite variances and n_A, n_B large.
case 3: normal distribution with the same but unknown variances (n_A, n_B small); the hypothesis of equal variances must be verified by a F-test. s_p is a pooled standard deviation as defined in the text; the degrees of freedom v are given by $v = n_A + n_B - 2$
case 4: normal distribution with unequal, unknown variances and n_A, n_B small; inequality of variances must be ascertained on the basis of an F-test on s_A and s_B.
The degrees of freedom v can be calculated as:

$$\frac{1}{v} = \frac{1}{(n_A - 1)}\left(\frac{s_A^2/n_A}{s_A^2/n_A + s_B^2/n_B}\right) + \frac{1}{(n_B - 1)}\left(\frac{s_B^2/n_B}{s_A^2/n_A + s_B^2/n_B}\right)$$

Paired *t*-test

In laboratory practice, one can be faced with the necessity of comparing two methods under particular situations such as:

- the quantity of material available is sufficient for only one determination by each method
- only samples from different sources, and/or containing slightly different amounts of analyte, are available
- test samples are analyzed over an extended period of time

In such cases tests on sample means described in the previous section are useless because effects caused by differences in analyte concentrations or variations in environmental conditions can obscure small differences between the methods. In such cases a *paired t-test* is more appropriate. In this test *n paired* measurements must be performed (i.e., one measurement by each method on each sample) and the difference D_i between each pair is calculated. Under the assumptions that no systematic error exists, i.e., that both methods give the same results (in other words the observed differences arise from random variations alone), the D_i value should be distributed with mean zero. Null and alternative hypotheses can be formulated as $H_0 : \mu_D = 0$ and $H_1 : \mu_D \neq 0$, respectively. The set of D_i values is then used as a single sample of size n to perform the test described in case 3 of Table 12.1-6 with $\mu = 0$:

$$t = \frac{\overline{D}}{(s_D/\sqrt{n})} \sim t_{n-1} \tag{12.7-30}$$

The null hypothesis is rejected if $\lvert t \rvert \geq t_{(1-\alpha/2),n-1}$. This test loses some power compared with the corresponding two-sample means test (case 3 in Table 12.1-7) since the test statistics has $n - 1$ degrees of freedom, in the present case, and twice this value in the two-sample case. However, the use of a test with less power is more acceptable than the possible effects arising from other sources of variation which could completely invalidate the conventional *t*-test on the two-sample means. Note that the paired *t*-test relies on the assumption that the errors (random or systematic) are independent of the concentration. Evidently this assumption may no longer be justified when the concentration of the test samples varies over a wide range; in such a situation the paired *t*-test is no longer appropriate and the preferred statistical method is linear regression (*see below*).

Worked example:

Ten samples of human plasma are analyzed for their glucose content by a routine enzymatic-photometric test and a flow injection analysis (FIA) method based on ampero-metric detection of H_2O_2 at a biosensor with immobilized glucose oxidase. The following results (expressed in mg/100 mL) are obtained:

Sample	Photometric	FIA	D_i
1	75	70	+5
2	100	103	−3
3	82	83	−1
4	85	82	+3
5	93	94	−1
6	78	77	+1
7	80	83	−3
8	90	88	+2
9	84	86	−2
10	95	94	+1

Do the two methods give different results at a significance level of 0.05?

The null hypothesis is that the two methods give the same results; furthermore the question, as formulated, requires a two-sided test. Then $H_0 : \mu_D = 0$ and $H_1 : \mu_D \neq 0$.

$$\bar{D} = \frac{\sum_i D_i}{n} = 0.20 \quad \text{and} \quad s_D = \frac{\sum_i (D_i - \bar{D})^2}{n-1} = 2.66$$

$$t = \frac{\bar{D}}{(s_D/\sqrt{n})} = 0.24$$

The critical value obtained from a *t* table at $\alpha = 0.05$ and 9 degrees of freedom is $t_{0.975,9} = 2.26$. Since the calculated value is smaller than the critical one, H_0 is accepted at 95% confidence level. In other words in the present experiment ($n = 10$) and at the chosen α no significant difference between the two methods can be detected. Remember that the validity of this test relies on the assumption of normality of data.

Comparison of two methods by least-squares fitting

The previously described statistical tests are not appropriate for comparison of methods when the analyte concentration in the test sample varies over a wide range. In such instances, the results obtained by the test procedure are plotted against those obtained by a reference procedure; in the absence of errors a straight regression line $y = \beta_0 + \beta_1 x$ should be obtained with a slope equal to 1, a zero intercept, and a correlation coefficient $r = 1$. Of course this is an *ideal* situation since, even in the absence of systematic errors, a scatter of the results around the best-fit line will always be observed due to the unavoidable presence of random variations. One is faced again with the problem of taking decisions; in particular one must be able to decide if the experimental estimates b_0, for β_0, and b_1, for β_1, obtained e.g., by a least-squares regression method (see Sec. 12.5.4), are significantly different from 0 and 1, respectively. This can be done by using significance tests; for example to test whether b_0 is significantly different from a given value β_0 (e.g., $\beta_0 = 0$) one uses the statistics:

$$T = \frac{(b_0 - \beta_0)}{s_{b_0}^2} \tag{12.1-31}$$

which is distributed as t_{n-2}; $s_{b_0}^2$ is the standard deviation on the intercept. For more details on this matter see Sec. 12.5.4.

If slope and intercept are found not to differ significantly from 1 and 0, respectively, one can conclude that systematic errors are absent. The presence of a proportional systematic error will be revealed by a slope significantly different from 1, while a constant systematic error is revealed by an intercept significantly different from zero. Of course, these two type of errors could also occur simultaneously. More complex situations can also occur, especially in the presence of speciation problems.

In spite of its extensive use, there are several theoretical objections to this approach, the most important being that the conventional technique for deriving the regression line of *y* on *x*, assumes that the random error on *x* is zero which, clearly, is not the case here. However it has been shown that the consequences of this approach are minimal provided:

- that a sufficiently high number of samples (typically > 10) with analyte concentration evenly spaced throughout the range of interest is used
- the method of higher precision is plotted on the *x* axis

Moreover, if an "unweighted" regression line is used, a second objection arises from the assumption of constant errors in the *y*-values (i.e., independent of analyte concentration).

Comparison of the precision of two methods (*F*-test)

To evaluate the performances of different methods, precision, other than accuracy, is one of the most important criteria. Comparison of precision can be performed by using tests on variances such as the *F-test* which compares two variances. Tests on variance assume that data are normally distributed and are not robust toward departure from normality.

As an example, one has to compare variances obtained in two sets of observations performed by *methods A* and *B*, in order to decide whether they differ significantly in precision or if *method A* is more or less precise than *method B*. In the first case, the null hypothesis $H_0: \sigma_A^2 = \sigma_B^2$ must be tested against the alternative hypothesis $H_1: \sigma_A^2 \neq \sigma_B^2$ (two-sided test) while in the second case it must be tested against $H_1: \sigma_A^2 > \sigma_B^2$ or $H_1: \sigma_A^2 < \sigma_B^2$ (one-sided tests). If the null hypothesis is true, the ratio between the estimates s_A^2 and s_B^2 should not differ significantly from unity. The test statistics (see the *F*-distribution) is:

$$T = \frac{s^2_{\text{num}}}{s^2_{\text{den}}} \sim F_{\nu_{\text{num}}, \nu_{\text{den}}} \tag{12.1-32}$$

The largest variance among s_A^2 and s_B^2 is allocated in the numerator in order to have values of $T \geq 1$; ν_{num}, ν_{den} are the degrees of freedom associated with the estimates appearing in the numerator and denominator of T, respectively. In a two-sided test H_0 is accepted, at the significance level α, if the realization $t < F_{\nu_{\text{num}}, \nu_{\text{den}}}(1 - \alpha/2)$. Other possible alternative hypothesis ($H_1: \sigma_A^2 < \sigma_B^2$ or $H_1: \sigma_A^2 > \sigma_B^2$) can also be tested according to a one-sided test in order to decide whether *method A* is more or less precise than *method B*.

Worked example:

In the worked example in Sec. "Test on two sample means" it has been assumed that the two methods do not differ significantly in their precision. Verify this assumption at a 0.05 significance level.

We have to compare two variances on the basis of an *F*-test with $H_0: \sigma_A^2 = \sigma_B^2$ and $H_1: \sigma_A^2 \neq \sigma_B^2$. The realization of the test statistics is:

$$t = \frac{(0.09)^2}{(0.05)^2} = 3.24$$

Since the critical value of F for a two-sided test at $\alpha = 0.05$ is $F_{10,9} = 3.96$ we must accept the null hypothesis and conclude that the two variances are not significantly different.

Outliers

Rejection of data is a process requiring extreme caution.

An extensive tabulation of the Q-values can be found in: D.B. Rorabacher, *Anal. Chem.*, *1991, 63, 139–146*.

When a series of replicate measurements is performed it can happen that one (or more) of the individual results appear substantially different from the others. The obvious question arising in such a situation is whether these *outlying* results must be rejected or not before doing further operations (calculation of mean and standard deviation, hypothesis tests, and so on). As will be seen, removal of an outlier has important practical consequences, especially when only a small data set is available (a very common situation). When facing such a situation the most obvious approach (if possible) is to increase the number of observations, since any statistical test will have more power, and the decision of whether to eliminate the suspected value will produce a smaller effect on the calculated mean, standard deviation, etc. The most widely used parametric test for outlier rejection, which assumes that data belong to a population with a Gaussian error distribution, is the *Dixon's Q-test*. For a *single outlier* the test statistics is calculated as:

$$Q = \frac{|\text{suspected outlier} - \text{nearest value}|}{\text{range}} \tag{12.1-33}$$

and compared with the critical value taken from the appropriate table at the desired confidence level (e.g., 95%) and with n degrees of freedom (where n is the number of observations).

As an example, one performs four replicate titrations, obtaining the following titrant volumes (mL)

$$20.85; 20.80; 20.95; \mathbf{21.35}$$

with the last value appearing suspect. The calculated Q-value is 0.727 which is less than the critical value at the 95% confidence level (0.831); *the suspected value thus must be retained*. The message is that when n is small, as is often the case, a suspected value must be very different from the others to be rejected on the basis of a *Dixon's Q-test*. This also suggests the importance of having as large a data set as possible. Suppose a further three results are added to the above set:

$$20.85; 20.80; 20.95; \mathbf{21.35}; 20.70; 20.90; 20.82$$

The calculated Q-value is now 0.615; for $n = 7$ the tabulated critical value is 0.570 so that *the suspected outlier must be rejected* at the 95% confidence level. After outlier rejection, the mean and the standard deviation change from $\bar{X} = 20.91$ and $s = 0.21$ to 20.84 and 0.09, respectively (note the dramatic change in the standard deviation); since s is routinely used in significance tests the consequences of incorrect rejection of a suspected value are now clear. Note, however, that as n increases, there is an increasing possibility of obtaining two (or even more) outliers. In this case the test statistics Q (as defined above) will be of no use because of *masking effects* (to convince yourself calculate the Q-value for the following set of data $20.85; 20.80; 20.95; \mathbf{21.35}; \mathbf{21.25}; 20.90; 20.82$). A possibility of circumventing the problem is to consider the two suspected results as a pair (the so-called *block method*) calculating a more appropriate test statistics; the disadvantage of this approach is obvious: both values must be retained or discarded, even in cases where rejection of only one value is most appropriate. The situation is evidently more complicated when the two suspected outliers are at the two extremes of the data range. A more extensive treatment of outliers (including the use of nonparametric tests) can be found in specialist texts.

Goodness-of-fit test (χ^2-test)

In most situations described in the preceding part of this section, it has been assumed that the experimental observations fit a particular probability distribution. The validity of the conclusions is, in most cases, strictly related to the validity of the assumptions involved in the model distribution of the original data. The goodness of fit of the experimental observations to a certain model distribution can be verified by means of statistical tests such as the χ^2-test. In this test, the null hypothesis can either completely specify the distribution or specify only the form of the distribution (i.e. distribution parameters such as the mean and the variance

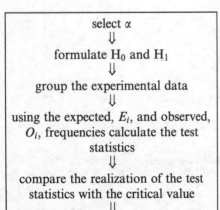

Fig. 12.1-14. Basic steps in performing a goodness-of-fit test

remain unknown). In this last case the unknown parameter needs to be estimated both in order to perform the test and to use the model distribution once H_0 has been accepted. The basic steps in performing a χ^2-test can be summarized as follows:

1. The significance level α is fixed and the null and alternative hypotheses are formulated, e.g., $H_0 : F = F(x, \tau)$; H_1 : not H_0, the null hypothesis stating that the data are extracted from a given distribution function with F known and the parameter τ which can be either known or unknown.
2. The range covered by the observations is divided into k categories (or groups) having a convenient width and the frequencies O_i observed in the ith category $(i = 1, 2, \ldots k)$ are calculated.
3. The expected frequencies, E_i, under H_0 are calculated as well as the test statistics $T = \sum_i (O_i - E_i)^2 / E_i$; if the number of observations is sufficiently large (typically $n > 50$) and none of the E_i is too small (typically not less than 5) the statistics T is approximately distributed as χ_v^2 under H_0. The degrees of freedom v are calculated as $v = k - 1 - h$ where k is the number of categories and h is the number of distribution parameters which need to be estimated.
4. The realization t of the test statistics T is, as usual, compared with a critical value, i.e., with $\chi_{v,(1-\alpha)}^2$ the $(1 - \alpha)$ quantile of a χ^2-distribution at v degrees of freedom. The rejection/acceptance criteria are as usual:

   if $t \geq \chi_{v,(1-\alpha)}^2$, H_0 is rejected at level of significance α;

   if $t < \chi_{v,(1-\alpha)}^2$, H_0 is accepted at level of significance α.

Worked example:

Prove at a level of significance α, that (grouped) data of worked example of page 711 fit a normal distribution.

The null hypothesis is: $H_0 : F = N(\mu, \sigma^2)$; the alternative hypothesis can be formulated as: H_1 : not H_0.

In order to calculate the expected frequencies E_i we need to standardize the upper boundary (UB) of each class calculating the following quantity $z = (UB - \mu)/\sigma$. First of all we calculate the estimate of μ and σ ($\bar{X}$ and s, respectively). For grouped data mean and variance can be calculated as:

$$\bar{X} = \frac{\sum_i f_i X_i}{n} \quad \text{and} \quad s^2 = \frac{\sum_i f_i (X_i - \bar{X})^2}{(n - 1)}$$

where n is the total number of observations, f_i is the frequency in the ith class, and X_i is the mid-value in the ith class.

In the particular case $\bar{X} = 19.1_{23}$ and $s = 4.3_{96}$.

Data can be arranged as follows:

Class	Interval LB	UB	O_i	$z = (UB - \mu)/\sigma$	$F(z)$	E_i	$(O_i - E_i)^2 / E_i$
1	$-\infty$	12.5	5	-1.51	0.0647	4.20	0.152
2	12.5	15.5	10	-0.82	0.2061	9.19	0.0714
3	15.5	18.5	12	-0.14	0.4443	15.48	0.782
4	18.5	21.5	19	0.54	0.7054	16.97	0.243
5	21.5	24.5	10	1.22	0.8888	11.92	0.309
6	24.5	∞	9	∞	1	7.23	0.433

An example of how to calculate E_i for the second class is given. Note that the probability $P(12.5 \leq X \leq 15.5)$ is given by: $\{F(-0.82) - F(-1.51)\} = 0.2061 - 0.0647 = 0.1414$. This probability is represented by the shaded area in Fig. 12.1-15 and has the meaning of a relative frequency. Since $n = 65$, the expected frequency in the given class is calculated as $E_i = 65 \times 0.1414 = 9.19$.

The realization of the test statistics: $T = \sum_i (O_i - E_i)^2 / E_i$ is $t = 1.99$. Since we have used μ and σ to calculate the E_i values, the degrees of freedom are:

$$v = k - 1 - h = 6 - 1 - 2 = 3$$

The tabulated value of $\chi_{3,0.95}^2$ is 7.82; thus, we accept the null hypothesis at the given significance level.

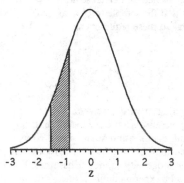

Fig. 12.1-15. Graphic representation of the probability $P(12.5 \leq X \leq 15.5)$ see text

12.1.3 Statistics in sampling

The *quality of an analytical result* cannot be better than that of the *sample* it is extracted from.

We have seen in the previous sections that analytical results can be affected by different sources of error due to reagent contamination, interferences, wrong procedures, data treatment, and so on. Most of these error sources can be controlled by proper use of blanks, standards, and reference materials. A similar approach cannot be used for error sources associated with wrong sampling. This is why sampling uncertainty is usually treated separately from the uncertainties associated with other steps of an analytical procedure. The sampling step will of course be of crucial importance when the bulk material is heterogeneous.

The total variance s^2_{total} is the sum of the individual variances related to sampling, $s^2_{sampling}$ and the other analytical operations, $s^2_{analytical}$:

$$s^2_{total} = s^2_{sampling} + s^2_{analytical} \qquad (12.1\text{-}34)$$

If the sampling variance is large and cannot be reduced, the use of very sophisticated instrumentation as well as any effort to refine the intermediate steps of the analytical procedure are, of course, of no use in reducing the overall variance. When $s^2_{analytical}$ is known, $s^2_{sampling}$ can be evaluated from s^2_{total}; otherwise a series of replicate measurements on a certain sample and on replicate samples must be performed in order to estimate $s^2_{analytical}$ and $s^2_{sampling}$.

Random sampling

Rarely an *object* is, or can be, analyzed as a whole; a small fraction of it, **the** *sample*, is analyzed instead. An *object* represents the entity (*the bulk material or population*) which needs to be described in terms of certain parameters that can be obtained, e.g., by chemical analysis. An *object* is for example a blood specimen, a biological tissue, a lake, a shipload of fuel, the daily production of a chemical plant, etc.

As mentioned, analysis of samples is mandatory, whenever it is impossible to analyze all the bulk material under investigation. Since these samples are processed to deduce the properties of the population from which they have been drawn, their collection must be done by a random process which ensures that all members of the population have an equal chance of being selected. The way to perform a random sampling is to divide the bulk material into real (or imaginary) cells, assigning to each cell a number, composed of a given number of digits; selection of the cells from which sample increments must be taken is done by a table of random numbers entered at an arbitrary point. The cells are chosen according to criteria, based on preconsideration of the nature of the bulk material and its consignment. For example, one can choose a cell every 2, 3, or *n* units until the pre-established number of cells is collected.

The *sample* must be representative of the analyzed object and must preserve the properties which the analyzed object had at the time of sampling.
Sample deterioration must be clearly avoided. Finally, the sample size must be suitable for analysis.

A possible alternative to random sampling consists of drawing at regularly spaced intervals, e.g., from a conveyor belt or a production line. Note, however, that this procedure, convenient for its simplicity, is more prone to bias than random sampling because errors from periodicity in the material composition can be introduced.

Sampling is the procedure (i.e., a series of steps performed in a given sequence) which ensures that the sample to be analyzed conforms to these requirements.

Sampling plan

For a nonhomogeneous bulk material, the sampling plan (see also Fig. 12.1-16) must take into account the following points:

Sampling of *homogeneous objects* is clearly a very straightforward procedure.
Homogeneous objects have the same, time-invariant, composition at any point. Note, however, that they are quite rare; moreover, object homogeneity cannot be assumed, but needs to be experimentally verified with the aid of statistical techniques, e.g., ANOVA.

- *size of the sample increment* (a sample increment is an individual portion of material collected by a single operation of a sampling device)
- *number of sample increments*
- *opportunity to prepare a gross sample* or to analyze individually the single increments (this last procedure can give information on the eventual heterogeneity of the material)
- *reduction of a gross sample* to a *sub-sample* from which *test portions* can be taken for analysis

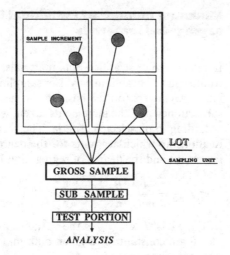

Fig. 12.1-16. Sampling for chemical analysis
A *lot* identifies a quantity of bulk material of similar composition whose properties are under study. For a packaged product, a *lot* identifies a collection of primary containers or units produced under identical conditions; a "single-day" production may also form a lot. *Sampling units* represent parts of a lot which are separated in space or in time. A *sample increment* identifies a portion of material collected by a single operation of the sampling device from a sampling unit. *Increments* can be analyzed individually or as a whole, after they have been composited (mixed). A *gross sample* identifies one or more sample increments taken from a lot. A *subsample* or *laboratory sample* identifies a sample, specifically intended for analysis, prepared from a gross sample and retaining its composition. A *test portion* (or *specimen* or *aliquot*) identifies a portion of material of a size suitable for carrying out the measurement of a parameter of interest

Size of sample increments

The between-sample variance ($s^2_{sampling}$ in Eq. 12.1-34) decreases as the sample size increases; considering this fact it can be shown that the equation:

$$M(\mathrm{RSD})^2 = K_s \qquad (12.1\text{-}35)$$

is valid in many situations. M is the mass of the sample analyzed; RSD the relative standard deviation (expressed in percent) of the sample composition, and K_s the sampling constant, defined as the sample mass required to limit the sampling uncertainty to 1% with 68% confidence. From Eq. 12.1-35 the minimum mass M required for a given RSD can be easily calculated, provided K_s is known. The dependence of M on RSD for different K_s values is shown in Fig. 12.1-17. It can be seen, for example, that a sample mass of 20 g will give a RSD% of 0.5, 1.4, and 1.9 for K_s values of 5, 35 and 70 g. If K_s is not known it must be determined experimentally by a series of replicate determinations on samples of different mass value. For example, a K_s value of about 35 g was found for the determination of sodium in human liver. It is clear from the above that the K_s value reflects in practice the degree of heterogeneity of the bulk sample.

The *purpose of sampling* must be clearly defined in order to design the sample procedure, optimizing the sampling parameters (size and number of sample increments, subsampling, mixing, frequency of sampling, etc.). Sampling aims typically to give a *description* (gross or detailed) of the object or to control, e.g., an industrial process. In this last case the task is to *decide*, on statistical bases, whether variations observed arise from random fluctuations of process conditions, or from something going definitely wrong (process out of control). In case of loss of control, suitable actions can be taken to regain control. These aspects are described in more detail in other sections where "control charts" are described.

Minimum number of increments

When the bulk material is not homogeneous a number of replicate sample increments must be analyzed. If a prior estimate of the sampling variance $s^2_{sampling}$ is available the number of increments n can be estimated by the equation:

$$n = \frac{t^2 s^2_{sampling}}{(MAD)^2} \qquad (12.1\text{-}36)$$

where:
MAD is the *maximum allowable difference* between the estimate to be made from the sample and the actual value, t is the Student's t value at n degrees of freedom and at a given confidence level (see also problem 22).

If $s^2_{sampling}$ is unknown, it must be determined by carrying out a sufficiently high number of preliminary measurements on the same lot to ensure a valid and stable estimate. Note that in Eq. 12.1-36 t must be taken from the appropriate table at a given n (which is also unknown); so if a value of $\alpha = 0.05$ is selected, t can be initially set at 1.96 (Gaussian distribution) and this preliminary value of n can be substituted in Eq. 12.1-36 and the system iterated to constant n.

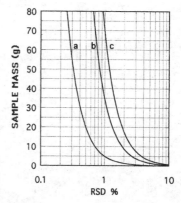

Fig. 12.1-17. Plots of the *sample mass* vs *RSD*% for different values of the *sampling constant* K_s: 5g (a), 35g (b) and 70g (c).

Minimum number of individual increments in sampling segregated materials

In segregated or stratified bulk materials, the average amount of a substance is distributed in a nonrandom way. For sampling, the material must be divided into real or imaginary layers (on the basis of the suspected pattern of segregation), which are subsectioned and the increments taken, with the use of a table of random numbers, according to preselected criteria. The variance in sample composition is given by a relationship which accounts for the degree of homogeneity within a given sample increment and the degree of segregation between sample increments, according to:

$$s^2_{sampling} = \frac{A}{mn} + \frac{B}{n} \tag{12.1-37}$$

In Eq. 12.1-37, $s^2_{sampling}$ is the variance of the average of n samples of mass m; A and B are constant for a given bulk material and must be determined by preliminary experiments.

In conclusion, sampling is often not a simple matter and requires extra effort to reduce the associated variance; quantities such as $\bar{X}$, $s^2_{sampling}$, K_s, A, B, n are in most cases unknown and must be determined by preliminary experiments. This must be clear to any analytical chemist, as well as the fact that sampling theory, certainly a useful tool, cannot completely replace experience, common sense, and "chemical feeling". Furthermore, all the necessary procedures for sample preservation and storage, which are as important as sampling itself, can be dictated only by a thorough knowledge of the chemistry of the analytes and of the analyzed matrix.

Questions and problems

1. Give examples of *discrete* and *continuous* random variable.
2. How would you define the rth central and noncentral moments of a probability distribution function? Are you able to define mean and variance in terms of moments?
3. Comment on the meaning of terms such as: *population, sample size, aliquot, specimen*.
4. Among the following situations, distinguish between examples of "measurement" and "population" variability:
   • different titrant volumes in an acid–base titration
   • different weights of crisp bags having a nominal content of 100 g
   • glucose concentration measured in a given blood specimen
   • glucose concentrations in 100 randomly collected blood specimens
5. Comment on the meaning of *sample* and *population parameters*.
6. What is meant by an *"unbiased estimator"*? Can you give examples of *unbiased estimators*?
7. Why is the Gaussian distribution considered so important?
8. Which probability density function would you recommend to describe a sampling distribution rather than a population?
9. How would you define the 95th percentile of a distribution? Do you envisage any relationship between the median and the 50th percentile?
10. Define the confidence interval of the mean.
11. Five replicate titrations give the following values for the titrant volume (mL) at the equivalence point 14.10, 14.20, 14.15, 14.25, 14.20:
    • calculate mean and standard deviation
    • calculate the 90% and 95% confidence intervals
    • calculate the 90% and 95% confidence intervals in the case where the population standard deviation σ is 0.03 mL; observe the difference between cases 2 and 3 and comment on this
12. A plasma specimen was analyzed for its glucose content by an enzymatic-colorimetric method based on the conversion of glucose to gluconic acid and H_2O_2, catalyzed by the enzyme glucose oxidase. Five replicate measurements gave the following results (mmol/L): 5.40; 5.35; 5.60; 5.45; 5.50:

- calculate sample mean and sample standard deviation
- calculate the 90 and 95% confidence intervals for the plasma glucose concentration
- how many replicate measurements would be necessary to reduce the calculated 95% confidence interval to $\pm 0.08\,\text{mmol/L}$?
- how would you report the 95% confidence interval if you had a previous knowledge that the precision of the method, expressed as RSD, is 1.5%?

13. What is a hypothesis test and how it is performed? Explain the meanings of the following terms: *significance level, null and alternative hypothesis, type I error, type II error, power of a test*.

15. Suppose you wish to buy a batch of a disinfectant solution only if its content of active chlorine conforms to the producer's declared value of 5.6% wt/vol; Would you adopt a one- or two-tailed test? How would you formulate the null and alternative hypothesis? How would you formulate your decision-making criterion? What would you do in practice (suppose you have a well-equipped lab)? Do you think the number of replicate measurements you perform will influence the power of the test?

16. A certified reference material with a protein content of 5.65% is analyzed by a new method to check for the possible presence of a bias. Five replicate determinations gave the following results: 5.60, 5.55, 5.64, 5.70, and 5.52%.
- how would you formulate the null and alternative hypotheses?
- what decision should you make concerning the null hypothesis (assume $\alpha = 0.05$)?
- which of the following statements concerning the null hypothesis, as previously formulated, are correct?
  - there is no bias
  - there is no evidence of bias
  - there is a 95% probability that the null hypothesis is true
  - the alternative hypothesis cannot be rejected
  - the null hypothesis cannot be rejected

17. Ten randomly selected blood samples were analyzed for creatinine content by an enzymatic colorimetric (Jaffè) method and an ion-pair HPLC method. The following results (μmol/L) were obtained:

Sample	Jaffè method	ion-pair HPLC
1	65.1	64.3
2	80.5	81.0
3	75.8	74.3
4	90.2	88.0
5	79.4	78.5
6	99.0	97.2
7	98.2	95.3
8	85.3	84.2
9	81.3	82.1
10	72.4	70.1

Is there a bias between the two methods?

18. Which test (one sample, two sample, paired and one- or two-tailed) would you use to assess whether:
- the Hg content of a batch of tuna for human consumption is above the legal limit (suppose five replicates are performed in the same lab by the same method)
- two different methods for Hg determination in tuna give results which are not significantly different (suppose one sample is analyzed once by the two methods in n different laboratories)
- a batch of tuna from Mediterranean has a Hg content higher than a batch from Atlantic (suppose five replicate determinations are performed on each batch in the same lab by the same method)
- two different labs give different results in the determination of Hg in tuna (suppose the same sample is repeatedly analyzed in both labs)

- mineralization of tuna by a H_2SO_4/HNO_3 mixture in the presence of $K_2Cr_2O_7$ gives:
  - the same recovery than the H_2SO_4/HNO_3 mixture alone
  - a recovery higher than that obtained by the H_2SO_4/HNO_3 mixture alone

19. Suppose two different methods (A and B) give the following results for the determination of the Ni content (%) in a stainless steel sample:

$$\bar{X}_A = 18.0 \quad s_A = 0.52 \quad n_A = 10$$

$$\bar{X}_B = 18.3 \quad s_B = 0.21 \quad n_B = 10$$

In order to check whether the two sample variances are not significantly different (assume $\alpha = 0.05$ and $F_{9,9} = 3.14$):
- how would you formulate H_0 and H_1?
- would you accept or reject H_0?

Based on the above results, which of the following statements are correct?
- the precision of the two methods is not significantly different
- the hypothesis of method A being less precise than method B cannot be rejected
- the two methods have the same precision

20. A water specimen is analyzed for its nitrate content at the site of collection. The same sample is reanalyzed in the laboratory by a different method. The obtained results (mg/L) are summarized as follows:

field method : $\bar{X} = 5.50 \quad s = 0.45 \quad n = 8$

laboratory method : $\bar{X} = 4.80 \quad s = 0.12 \quad n = 8$

- test whether the precision of the laboratory method is significantly greater than that of the field method? (assume $\alpha = 0.05$ and $F_{7,7} = 3.787$).
- test whether the mean results obtained by the two methods differ significantly (assume $\alpha = 0.05$). *Before doing the calculation check which of cases 1–4 in Table 12.1-7 conforms to the present situation.*

21. The manufacturer of your preferred brandy claims that the ethanol content is 40.0% wt/vol. Ten replicate determinations, carried out on a random sample, give a sample mean and a sample standard deviation of 38.75 and 0.53 wt/vol, respectively.
- test the manufacturer's assertion at the 95% confidence level
- the rejection of a batch of brandy which is "*off specification*" is, in your opinion, associated with:
  - type I error
  - type II error
  - power of the test
- the acceptance of a batch of brandy which is "*off specification*" is, in your opinion, associated with:
  - type I error
  - type II error
  - confidence level adopted in the test
- the acceptance of a batch of brandy which *conforms to the specification* is, in your opinion, associated with:
  - type I error
  - type II error
  - significance level α
  - $(1 - \alpha)$

22. Calculate the number of sampling increments necessary to ensure, at a 95% confidence level, that the maximum allowable difference between the estimate of the mean and the true value is not greater than 5%. Assume that $s_{sampling}^2$ is known (from previous experience) to be 6.1%. (note that Eq. 12.1-36 remains valid if $s_{sampling}$ and MAD are both expressed as relative percentages of the mean).

12.2 Calibration

Learning objectives

- To get acquainted with some basic terminology on calibration
- To understand calibration as a comparison exercise for physical quantities (wavelengths, time, mass/charge ratio) or amounts of substances with known content (standards)
- To memorize important factors contributing to the quality of calibration
- To distinguish between absolute and relative methods of chemical analysis
- To recognize the titer factor as a special type of calibration factor
- To study the general sequence of steps in setting up a calibration experiment
- To recognize frequently encountered shapes of calibration curves
- To learn to use least-squares procedures for deriving calibration factors
- To understand assumptions made in calibration by a least-squares procedure
- To understand the different modes of calibration and their particular advantages and disadvantages

12.2.1 Calibration is comparison

Most measurements become useful only through comparisons: a measured value by itself can not usually be interpreted without a reference point. Moreover, the raw data in most chemical measurements are recorded in units of potential difference (volts, V) or current (amperes, A) that can be actuated by many different phenomena not necessarily the quantity we need to measure. Such a quantity is technically called a *measurand*. In analytical chemistry the term *analyte* is also commonly used if one refers to a certain chemical species, such as an element, ion, molecule, or radical. Before an electrical signal becomes readily understood and interpretable it is compared with a similar one recorded under identical experimental conditions, but on a sample with well-defined values of the measurand. Such a sample with known properties regarding the quality under investigation is called a *standard*. The higher the signal in the measurement of the standard, the higher is the expected signal in the measurement of the same property on a sample with a (previously) unknown value of the measurand.

The signal can increase or decrease with increasing value of the measurand (Fig. 12.2-1). In the first case, increasing signal with increasing value of the measurand, and a larger signal in the standard than in the unknown relates to a smaller value of the measurand in the unknown. For signals showing the reverse behavior (smaller signals with increasing value of the mesurand), a smaller signal in the unknown relative to the signal from the standard points to a higher value of the measurand in the unknown sample. This dependence of the signal on the value of the measurand can be utilized for the *quantification* of this measurand. If the signal depends also on other concomitant species these are regarded as *interferents* and, if not properly dealt with, this (secondary) dependence can lead to inaccurate measurements.

There is also a similar type of comparison useful for *identification* of chemical species (Fig. 12.2-2). In order to identify a species, one frequently resorts to spectroscopic measurements, where one measures a spectrum as a function of wavelength (or wavenumber). The property of interest in these cases is not only the size of the signal but also the wavelength axis. To pinpoint a certain wavelength, e.g., 280 nm, along the wavelength axis it is useful to have a standard with one or more distinct features around this wavelength. If a feature expected at 250 nm shows up at the (uncalibrated) nominal wavelength of 275 nm, then the wavelength scale can be adjusted (calibrated) by a factor of 1.1 to give the true

Quantity: an attribute of a phenomenon, body or substance, which may be distinguished qualitatively and determined quantitatively.

Measurand: a quantity subjected to measurement.

Calibration: set of operations that establish, under specified conditions, the relationship between values of quantities indicated by a measuring instrument or measuring system, or values represented by a material measure or a reference material, and the corresponding values realized by standards.

Primary standard: standard that is designated or widely acknowledged as having the highest metrological qualities and whose value is accepted without reference to other standards of the same quality.

Secondary standard: standard whose value is assigned by comparison with a primary standard of the same quantity.

Reference standard: standard, generally having the highest metrological quality available at a given location or in a given organization, from which measurements made there are derived.

Working standard: standard that is used routinely to calibrate or check material measures, measuring instruments, or reference materials.

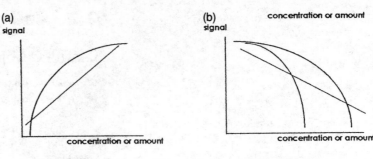

Fig. 12.2-1. Some signals increase with increasing value of the measurand (a), others decrease with increasing value of the measurand (b). Typical measurands in analytical chemistry are amount of substance (mol) and concentration (mol/L, mol/kg)

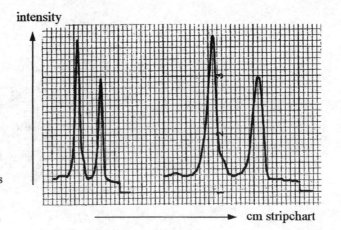

Fig. 12.2-2. Two different energy calibrations for the same measurement: on the left the calibration factor is 1 cm per degree, on the right the calibration factor is 2 cm per degree

physical wavelength. This type of calibration is automated and is unnoticed by the operator in many modern instruments so that the reliability of the operation is increased.

Such a comparison of a wavelength scale is also required for electrical measurands (potential, current), but this calibration can often be taken for granted if the measuring instruments (e.g., voltmeter) are properly attended. In liquid chromatography there is the need to calibrate the time axis in terms of the eluent stream (mL/min) in order to facilitate the correlation between elution times (elution volumes) and chemical substances.

Here we concern ourselves mainly with quantitative measurements of chemical species and the role of calibration in this process. One calibrates the reading of a measurement system in order to give it a quantitative interpretation. Another type of comparison will be dealt with later (library search, pattern recognition) where the aim is to recover qualitative information regarding the nature of a chemical species.

Three factors contribute to the quality of calibration:
(a) repeatability of measurement
(b) trueness of standards
(c) validity of comparison.

Traceability: property of the result of a measurement or the value of a standard whereby it can be related to stated references, usually national or international standards, through an unbroken chain of comparisons all having stated uncertainties.

Trueness: the closeness of agreement between the average value obtained from a large series of test results and an accepted reference value.

Accuracy: the closeness of agreement between a test result and the accepted reference value.

12.2.2 Quality of calibration

No quantitative measurement can therefore be better than the intrinsic quality of this comparison process called calibration, and there are a couple of ways to improve this comparison process as shall be seen later. What then are the major factors determining (and limiting) the quality of calibration?

One factor is the *repeatability* of the measurement itself (Fig. 12.2-3). Suppose the measurement gives highly variable readings, it will then tend to do so both for the standard and the unknown. The signal size is therefore poorly defined leading to poor predictions of the (value of the) measurand. In practice the variability for the standards may be different (and often smaller) than the variability of the measurement of the sample, thus the latter is limiting the quality of calibration.

The other factor relates to the *value of the measurand in the standard* (Fig. 12.2-4). Although ideally this should be known with ultimate accuracy, there is always some uncertainty remaining since its value is also determined by measurements, albeit of higher quality than routinely obtainable. Even with perfectly reproducible measurements this uncertainty regarding the value of the standard leads to imperfections in the comparison process. The value assigned to a standard usually carries a smaller (relative) uncertainty if the standard is produced from pure chemicals with well-defined stoichiometry. It has a larger uncertainty if the standard is a reference material with a matrix similar to the samples. This does not need to be a disadvantage as the calibration is then particularly relevant to the problem and one needs to worry less about matrix effects.

The final factor limiting the quality of calibration, and in many real life analytical systems the overwhelming one, is the *validity of comparison* itself. In calibrating an instrument one must be reasonably sure that the instrument will respond to the measurand in the standard in the same way it does to the measurand in a different environment, the unknown. Physical differences between sample and standard also frequently hamper the calibration by invalidating the comparison process. This problem is considerably more likely to occur if calibration is based on measurements of standards prepared from chemicals of high purity, and advantages can be expected from the use of in-type *standard reference materials* for calibration, if available. Which particular physical properties can affect the results and need to be identical (or at least very similar) in standards and unknowns depends on the principle of the measurement and the physical state of the sample. Among them are temperature, pressure, viscosity, turbidity, particle size, surface roughness, and thickness.

Differences in response can also be caused by concomitants and the measurement is then said to be of insufficient *selectivity*. Differences could also be traced to contamination or loss whereby a higher or lower signal might be obtained. The judgement regarding the validity of a comparison is one of the most difficult aspects of analytical work and can ultimately only be done by checking the accuracy of results on reference materials. Alternatively, results obtained on several samples by procedure A compared with the results obtained by another procedure B that is based on a completely different measuring principle can show the accuracy of the data. For instance it might be possible to assess the accuracy of a spectrophotometric determination of phosphate by one based on ion chromatography. Figure 12.2-5 shows such a comparison: the tighter the points lie to the line, the closer the slope to unity, and the smaller the intercept, the better the correspondence between the two methods.

12.2.3 Frequency of calibration and recalibration

Although great efforts are being made to produce analytical instruments that give stable readings ("response") over time, the long-term stability is usually limited. This is partly the reason for the importance of calibration in practice: while for an analytical balance it might be sufficient to calibrate it once every two months (weighing serves to determine a physical measurand, mass). Most chemical measurements are calibrated daily or even more frequently than that. In this way chemists attempt to reduce the day-to-day (and even sample-to-sample) variability of their measurements as much as possible. Thus, on each working day every chemist in the world produces his/her own scale that may or may not be significantly different from all other ones.

12.2.4 Absolute vs. relative methods of analysis

Measurements that are evaluated by comparison with those made on standards are called *relative methods*. *Absolute methods* can be evaluated without comparative measurements, but they rely instead on fundamental physical and chemical constants and principles.

Certain electrochemical measurements rely on a definite oxidation state of a chemical substance before and after the measurement, and therefore the transferred charge is directly convertible to the amount to substance (in mol), the conversion factor being Faraday's constant. Similarly, volumetric measurements (e.g.,

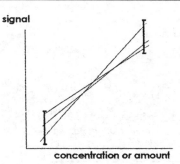

Fig. 12.2-3. Some calibrations are limited by poor repeatability that leads to a highly variable relation between concentration/amount and signal

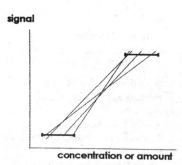

Fig. 12.2-4. Other calibrations are limited by significant uncertainties regarding the concentration/amount in the standards

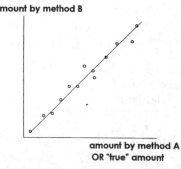

Fig. 12.2-5. The validity of a calibration can be assessed by a method comparison exercise if both procedures are based either on independent principles or on measurements on standard reference materials, where reliable values of the measurand are available ("true" value). The validity is the ultimate and crucial property in any calibration that limits its usefulness. This curve is commonly called a "recovery function"

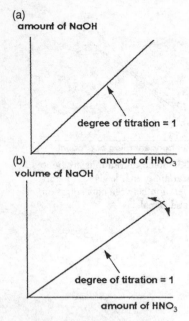

(a)
amount of NaOH

degree of titration = 1

amount of HNO$_3$

(b)
volume of NaOH

degree of titration = 1

amount of HNO$_3$

Fig. 12.2-6. Titrations are based on the amounts of substances in stoichiometric relations to each other. They can therefore be regarded as absolute methods. This is, however, only true for amounts (a); if volumes are measured, the relation is calibrated by titer factors (b) that vary from day to day

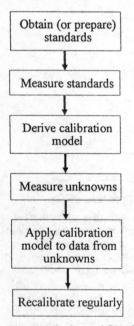

Obtain (or prepare) standards

↓

Measure standards

↓

Derive calibration model

↓

Measure unknowns

↓

Apply calibration model to data from unknowns

↓

Recalibrate regularly

Fig. 12.2-7. General flowchart of calibration

titrations) rely on the presence of well-defined chemical species, before and afterwards, that are related to each other by stoichiometric coefficients from a reaction equation. The knowledge of the particular reaction equation and of atomic masses is therefore sufficient for the quantitation of the measurand.

In these (and similar) instances, no calibration for the actual measuring step is necessary provided the titrant is available in high purity. If, however, other operations are also involved in the analytical procedure, e.g., decomposition, extraction, or clean-up, it may be advisable to carry a standard through the entire procedure in order to check for contamination, losses, and degradation of the measurand. This is why in practice very few procedures are operated completely without standards. Another complication for such absolute methods is the fact that the actual measurement is done in terms of volume (of the titrant), the exact concentration of which is rarely known with sufficient accuracy (Fig. 12.2-6). Therefore in spite of the validity of the fundamental stoichiometric relation, the actual volume is calibrated by comparison with the result of the measurement of a standard.

An example is the titration of a strong acid with a strong base, e.g., NaOH. An NaOH solution is made up to have a molarity of 1.0 mol/L. Instead of relying on the dilution process, a measurement is done on a very well-defined amount of strong acid, e.g., HNO_3. If 0.0100 mol HNO_3 is taken, one expects a volumetric consumption of NaOH solution of 10.0 mL; instead, 10.3 mL are consumed, and the relation between the volume of titrant and the amount of acid is adjusted by the factor 1.03. This proportional adjustment corresponds to the construction of a separate calibration function for this particular batch of titrant using a single standard. It is necessary only because in practice the volume of the basic solution is measured instead of the amount of base, and so does not invalidate the fundamental stoichiometric equation underlying this volumetric measurement.

The jargon "standard-less" procedure also often relates to relative methods that do not require calibration by the operator in the laboratory or in the field. Generally, these procedures are characterized by a fairly stable relationship between the measurand and the instrumental response leading to great intervals between calibrations. In the extreme, the calibration might be valid for the lifetime of an instrument and therefore carried out prior to shipping in the factory. If this is possible it is a great convenience for the chemist, but the responsibility for adequate calibration rests still with him/her. It is therefore wise to check regularly whether the system still operates according to the same response characteristics (see Sec. 3.3.4). The sequence of experimental steps in calibration are summarized in Fig. 12.2-7.

12.2.5 Calibration protocols and calibration models

The simplest calibration is one that utilizes only one standard. One may be forced into such a protocol when only one sufficiently reliable standard is available. In this instance, it is supposed that for a zero measurand (e.g., zero wavelength, retention time, mass, amount, or concentration) the analytical system gives zero (or a very well-known value other than zero) response (signal, reading). Between these two values at intermediate mass of the analyte, one resorts to a linear interpolation, but no extrapolation beyond the value of the standard is recommended. The corresponding model for measuring concentration is:

signal = constant × concentration

$$y = b_1 c \tag{12.2-1}$$

The constant b_1 is called the *sensitivity* and can be regarded as a proportionality constant between the signal and the concentration, or more generally, between the signal and the measurand. This proportionality is usually valid over a restricted range of values of the measurand; below this range the signal is too small to be useful, and above there are generally deviations from this proportionality, so that the value of b_1 is no longer the same. The region of validity of b_1 is called the *working range* of an analytical system; it can often be extended to higher values of the measurand by suitable dilutions, while for reaching lower values, more involved chemical operations may be necessary, e.g., preconcentration of the analyte. At higher values of the measurand the curve tends often to bend towards the x-axis. As long as there is a noticeable change in signal as the concentration changes, one operates within the *dynamic range* of the method although the sensitivity decreases continuously for such a bent curve.

Worked example: Determination of sensitivity b_1 for a titration of a strong acid with a strong base

The titration of hydrochloric acid with 0.5 mol/L aqueous sodium hydroxide proceeds according to $NaOH(aq) + HCl(aq) \rightarrow NaCl(aq) + H_2O(liq)$. What is the sensitivity, b_1, and the titer factor, f, if the volume of $NaOH(aq)$ is recorded as 20.25 mL for 10 mmol $HCl(aq)$.

Strategy: From the reaction equation we see that there is a 1:1 stoichiometric relation between NaOH and HCl. The volume consumed in the titration must therefore contain exactly the amount of HCl that was titrated. The sensitivity b_1 should be expressed in units of mL NaOH/mmol HC. From this, the titer factor can be found as an appropriate relation to the actual molarity close to 0.5 mol/L.

Solution: There were 20.25 mL of base consumed for 10.00 mmol of the acid. The sensitivity $b_1 = 20.25$ mL $NaOH/10.00$ mmol $HCl = 2.025$ mL/mmol. The molarity of the base is the inverse of this value $1/2.025$ mL mmol$^{-1} = 0.4938$ mol/L $= 0.9876 \times 0.5$ mol/L. The titer factor, f, is 0.9876.

Exercise: In the titration of KOH with H_2SO_4, 20.30 mL of the 0.5 molar acid were consumed. What is the value of b_1 and the titer factor f for the (diprotonic) acid?

(Answer: $b_1 = 1.015$ mL/mmol; $f = 0.9852$)

If the response at zero concentration is not known beforehand it is measured in the course of calibration. A measurable signal at zero (or negligible) concentration of analyte is either due to contamination of the sample or due to inadequate recovery of the net signal from the background, or both. The simplest model for this two-point calibration that can also be extended to multiple points is:

$$\text{signal} = \text{blank} + \text{constant} \times \text{concentration (or amount)}$$

$$y = b_0 + b_1 c \tag{12.2-2}$$

The term b_0 thus indicates the magnitude of the blank whereas b_1 is again a measure of sensitivity. More formally, a test for statistical significance will reveal the necessity to include the b_0 term in the calculations.

For a curvilinear dependence of signals from measurand, other models must be employed, such as the quadratic model:

$$\text{signal} = \text{blank} + (\text{constant}_1 \times \text{concentration}) + (\text{constant}_2 \times \text{concentration}^2)$$

$$y = b_0 + (b_1 c) + (b_{11} c^2) \tag{12.2-3}$$

The more involved the model the greater the required number of standards. Another reason for using a larger number of standards is in multivariate calibration when more than one analyte is considered simultaneously, or if – due to the lack of selectivity – corrections of the signal for the contributions of the concomitant species are applied.

This not only requires the use of multiple standards, but also the knowledge of the concentrations of all species contributing to the signal in any of the standards. In mathematical terms the simplest model involving k species corresponds to:

$$y = b_0 + b_1 c_1 + b_2 c_2 + b_3 c_3 + \cdots + b_k c_k \tag{12.2-4}$$

but much more involved ones (e.g., hyperbolic or sigmoid) are also in use. For instance, in X-ray fluorescence spectrometry it is not uncommon to measure 30–60 standards. A good strategy in calibration, however, is to use only as parsimonious a model as possible and to test for the most parsimonious model consistent with the data by a formal statistical testing procedure, thereby avoiding any overfitting of the experimental results.

Once the *calibration models* are derived by one of the methods presented in the next part of this chapter, they are employed to predict the measurand in a sample other than a standard. In order to do so, the models need to be reformulated in terms of their inverse. These inverse functions are called *analytical models* (or *analytical functions*). Such an analytical model for Eq. 12.2-1 is:

$$c = \frac{y}{b_1} \tag{12.2-5}$$

For Eq. 12.2-2 the analytical model is:

$$c = \frac{(y - b_0)}{b_1} \tag{12.2-6}$$

Analytical models for the more complicated calibration models (Eqs. 12.2-3 and 12.2-4) are either obtained by algebraic methods or – equally important – by numerical approaches in an iterative manner.

12.2.6 Calibration functions from least-squares procedures

While in principle it is possible to bypass the computational procedures in calibration by appropriate graphical methods, it is advisable to resort to more formal procedures to improve the traceability and reproducibility of this important operation. The calibration functions are thus derived from statistical computations that yield the coefficients b_0, b_1, b_{11}, etc., along with some figures-of-merit useful for judging the quality of the calibration. The strategy is such that one of the models from Eqs. 12.2-1 to 12.2-4 (or even a more complex one) is adopted and – once the coefficients are computed – is tested for its adequacy. If not suitable, another somewhat more complex model is chosen.

For the model in Eq. 12.2-2 we can estimate the coefficients by the following computations:

$$b_1 = \frac{\sum_{i=1}^{N}(c_i - \bar{c}) \cdot (y_i - \bar{y})}{\sum_{i=1}^{N}(c_i - \bar{c})^2} \tag{12.2-7}$$

$$b_0 = \bar{y} - b_1\bar{c} \tag{12.2-8}$$

where $\bar{c}$, $\bar{y}$ are the arithmetic mean values of given c values (concentrations of standards) and measured y values (signals). While the obtained coefficients are optimal in the least-squares sense there is no complete reproduction of the measured values possible due to scatter in the data. The (statistical) quality of calibration is evaluated by applying the derived model (in our case $\hat{y} = b_0 + b_1c$) to reproduce all experimental data y measured earlier and is used to obtain b_1 and b_0; the measured y and the predicted $\hat{y}$ are taken to compute:

$$s_y = \sqrt{\frac{1}{(N-2)}\sum_{i=1}^{N}(y_i - \hat{y}_i)} = \sqrt{\frac{1}{(N-2)}\sum_{i=1}^{N}(y_i - (b_0 + b_1c_i))} \tag{12.2-9}$$

This s_y is called the *residual standard deviation* and is given in the signal domain. Thus its units are the same as those of the original signal (mV, nA, s^{-1}, ...). The *standard deviation of the procedure* is derived from it by computing:

$$s_c = \frac{s_y}{b_1} \tag{12.2-10}$$

which is now given in the same units as the measurand (mmol, ng/g, mg/mL) and is therefore easier to interpret than the residual standard deviation.

Before moving on to applying the calibration curve a couple of essential comments are in order. While the above computations are straightforward and provided by many computer programs even in hand-held calculators, their application in the chemical laboratory is tricky nevertheless, as a couple of assumptions made in these computations are not always met in practice. The analyst, therefore, has to check the assumptions first, or at least verify that the violation of these assumptions does not do any harm analytically whereby the results of analytical measurements are invalidated.

Here are the assumptions:

(a) The uncertainties of the standards are negligible compared with the standard deviation of the measurements. It is simpler with standards prepared from pure chemicals to meet this assumption; if calibration is done using standard reference material the described mathematical procedure is usually not applicable.

(b) All measurements are statistically independent of each other. Any trend over time or any cross-contamination would invalidate the measurements.

(c) The measurements have an identical standard deviation which does not depend on the value of the measurand and, therefore, higher signals have the same magnitude of the standard deviation as smaller signals (Fig. 12.2-8). This is a particularly critical assumption for measurements made over a large concentration range, such as molecular fluorescence, atomic emission, mass spectrometry, ECD-GC data.

(d) All measurements are normally distributed.

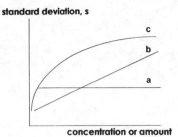

standard deviation, s

concentration or amount

Fig. 12.2-8. It is fairly common – particularly in instrumental analysis – that the standard deviation is NOT constant over the calibration range (as in a), but either proportional (b) or it shows a more complicated rise as concentration or amount increases (c). Special precautions need to be made to obtain valid calibration lines in cases (b) and (c)

It is fairly frequent that assumption (c) is violated to a degree that measurements at low levels of the measurand carry significant systematic errors. This needs to be checked and can then be avoided by restricting the range of calibration to a range considerably smaller than that provided by the measuring principle. Alternatively, one can produce multiple calibration curves, one for the low, another one for the intermediate, and a third one for the high range. If this also fails, one can resort to the so-called weighted least-squares approach that is dealt with in [12.2-1, 12.2-2].

If all assumptions are met satisfactorily one can move on to predicting concentrations of the measurand in the samples. This is done using the analytical model. If the calibration model corresponds to Eq. 12.2-2 then the analytical model is provided by Eq. 12.2-6. The first step is to estimate the quality of the model itself by computing a confidence band. This is constructed on the basis of:

$$\Delta y' = t_{\alpha/2, N-2} \cdot s_y \cdot \sqrt{\frac{1}{N} + \frac{(c' - \bar{c})^2}{\sum (c_i - \bar{c})^2}} \qquad (12.2\text{-}11)$$

One has, however, to take into account that there is some variability involved in this prediction that comes not just from imperfect measurements of standards, but also from imperfect measurements of the sample itself. The confidence interval of the prediction of the measurand in a sample is therefore larger and is given by:

$$\Delta c' = t_{\alpha/2, N-2} \cdot \frac{s_y}{b_1} \cdot \sqrt{\frac{1}{M} + \frac{1}{N} + \frac{(c' - \bar{c})^2}{\sum (c_i - \bar{c})^2}} \qquad (12.2\text{-}12)$$

where $M = 1$ if just one measurement is made on the sample, $M = 2$ for duplicate measurements, $M = 3$ for triplicate measurements, and so on; t is the appropriate factor from the Student's t-distribution and $\bar{c}$ is the mean concentration covered by all standards.

This equation tells us that large sensitivities (relative to the residual standard deviation) are advantageous; also that multiple measurements of the sample and a large number of calibration samples all help to keep the confidence interval small.

Worked example:

A calibration curve is constructed for the determination of formaldehyde by a spectrophotometric procedure. The signal has an absorbance at 570 nm and it is plotted against concentration of formaldehyde on the abscissa. There are seven points in this calibration and the order of the measurement is randomized. The results of the measurements are given in the table in columns 3 and 4.

Experimental	Order of measurement	Concentration	Absorbance		Residuals
		c_i	y_i	$\hat{y}_i$	$(y_i - \hat{y}_i)$
1	5	0.1	0.086	0.0939	−0.0049
2	2	0.3	0.269	0.2691	−0.00001
3	1	0.5	0.445	0.4443	0.007
4	6	0.5	0.452	0.4443	0.0077
5	4	0.6	0.538	0.5319	0.0061
6	7	0.7	0.626	0.6195	0.0065
7	3	0.9	0.782	0.7947	−0.0127

Using least-squares estimation, the estimated dependence of the signal from the concentration is found to be $\hat{y} = 0.00633 + 0.876c$, and the calibration line (Fig. 12.2-9) is inspected.

It should always be apparent that this curve is an estimation by putting a circumflex on top of the y (read as "y hat"). A more in-depth view, however, is provided if not the data and the curve themselves, but the *difference* between the curve and the data is plotted (Fig. 12.2-10). This difference is given in the last column of the table and is called the *residual* as it is a measure of the variability that remains around the plotted curve. Ideally, this residual is small and unstructured. In the example it is small as it should be, but there is some structure that points to deviations from Beer's law.

In making a judgment whether the deviations are too large to be ignored, one has to look at a concentration value that is measured twice, such as the one at 0.5 mL of formaldehyde. If the variation between these replicated values is much smaller than the whole spread of the curvature, a bent curve (such as a quadratic polynomial) has to be considered. Here we estimate the deviation of the two measurements to be 0.007 abs. units (from comparison

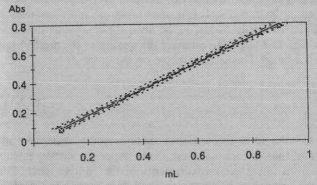

Fig. 12.2-9. Calibration line of formaldehyde data with confidence bands from Eqs. 12.2-11 and 12.2-12

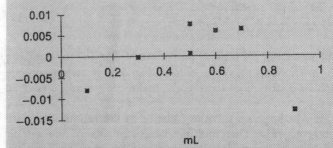

Fig. 12.2-10. The residuals $(y_i - \hat{y}_i)$ of formaldehyde calibration show some structure hinting at possible curvature

of data in the table, column 4, experiments 3 and 4) and the whole spread of the diagram to be from −0.015 to +0.010 in the same order of magnitude. We therefore conclude that the estimated curve fits well enough the data without resorting to a more formal (statistical) test.

The residual standard deviation calculated from Eq. 12.2-9 is 0.0085 abs. units and the standard deviation of the procedure from Eq. 12.2-10 is 0.0097 mL formaldehyde.

Another way of judging the quality of calibration is the correlation coefficient that varies between −1 and +1. Values very close to +1 or −1 point to a very tight fit of the curve. A disadvantage of the correlation coefficient is, however, that a good fit in on part of the curve can obscure a bad fit in another part of the curve. It is therefore advisable to always resort to the graphical inspection and to Eq. 12.2-9 for assessment of calibration curves. This inspection can also spot nonlinearities with better success than the correlation coefficient.

12.2.7 Calibration modes and protocols

While calibration is generally practiced in analytical chemistry the exact mode of operation varies depending on the analytical method. In some methods it is enough to measure a single standard; this is called single-point calibration. Sometimes a two-point calibration is practiced, one is a standard at a low concentration (or amount) value, another is at a high concentration (or amount) value. The two are chosen to bracket the values of the unknowns; the lower one might be the blank value. These protocols are all versions of a calibration mode that is called *external calibration* because the samples containing the unknown amount, and the standards containing the known amount are separate throughout.

In addition to the external calibration some other modes are also important: internal calibration, standard additions method, and isotope dilution techniques.

If prior to the determination, a fixed amount of a different, but similar substance is added to the sample, this substance is called the *internal standard* and the mode of operation is called *internal calibration*. The addition serves the purpose of controlling a critical step that would otherwise introduce a large element of uncertainty. Therefore it is wise to add the surrogate substance as early as possible in the analytical procedure. Care has to be taken, however, that the added substance is well mixed with the sample and is in a physical state (particle size, surface, etc.) and binding state comparable with that of the measurand. The assumption in the internal standard mode is that the measurand shows the same behavior in all the critical steps; thus, the ratio between the data of the measurand and the surrogate substance constitutes more reliable information than the data of the measurand itself.

A related mode is the *method of standard additions* since one uses the measurand itself for internal calibration. In order to do so, the sample is split into several subsamples before the analysis. One subsample is treated as usual, while one adds increasing amounts of the measurand to the other subsamples. The analytical determination is then carried out on all subsamples and the resulting data is plotted as shown in Fig. 12.2-11. The purpose of this mode is generally to correct for

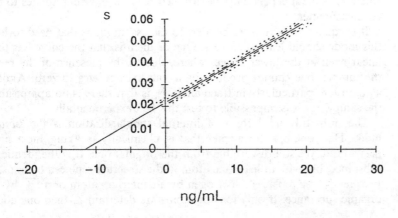

Fig. 12.2-11. Standard additions for lead in a water supply reservoir by graphite-furnace atomic absorption. Three aliquots were spiked with 0, 10, and 20 ng/mL respectively and measured in duplicate. The resulting concentration value is 11.5 ng/mL

Table 12.2-1. Overview of calibration protocols

Mode of calibration	Operation and protocol	Typical area of application	Limitations
External	a set of standards is measured independently, but usually immediately preceding the determinations of unknowns	widely used	equal standard deviations over the entire range, adequate model
Internal	a substance similar in chemical behavior and analytical response to the measurand (surrogate substance) is added to each sample and to all standards; all measurements are ratios to the response of this "internal standard"; this supplements, but usually does not replace external standardization	applied if a particular operation in the analytical procedure is highly variable from measurement to measurement, e.g., extraction yield or injection volume	existence of a suitable surrogate substance that behaves similarly with respect to the critical step
Standard additions	each sample is split into several (at least two) subsampels, and a known amount of measurand is added to all subsamples, but one	if signal enhancement or signal depression due to the action of the matrix is likely to hamper the determination	no (or extremely well-characterized) blank; straight-line curve, equal standard deviation over the range
Isotope dilution	a spike of known amount, but with a different isotopic pattern is homogeneously mixed with the sample prior to the analytical procedure	highly accurate measurements if losses of measurand in the procedure are likely	isotope specific measurement, such as mass spectrometry, availability of isotopes, chemical state of spike identical to native substance, well-mixed spike, accuracy of added amount, no contamination

proportional errors that stem from the differences in response between standards in clean solution (e.g., water or ethanol) and response in the matrix loaded sample. In effect one attempts to correct for different slopes between the two curves. In order to have a good estimate of the sensitivity it is necessary to add enough spike so that the signal is at least doubled. As one frequently prepares two or three spiked samples this leads to multiple work on single samples, this being an obvious disadvantage in terms of throughput.

The measurement on the unknown is evaluated by:

$$c = \frac{b_0 - bl}{b_1}$$

where bl is the independently determined blank signal that applies to the particular measurement.

It is equally important to be alert to the assumptions that need to be fulfilled if this mode should lead to success. One of them is that the curve has to be strictly linear even in the lower range, where, due to the presence of the measurand in the sample, one cannot produce data, but extrapolates to zero. Another crucial assumption, particularly in trace analysis, is that there is no appreciable blank in the sample for it is impossible to test this fact experimentally.

The most advanced form of internal standardization is the *isotope dilution* mode. Here one opts for a spike that is, chemically speaking, most similar to the measurand, yet still discernible from the original one. It is the identical chemical substance, but with at least one atom in the structure replaced by another isotope. Frequently one replaces a hydrogen by a deuterium atom or a ^{12}C by a ^{13}C in an organic substance. If only ions or atoms are determined, then one adds a known amount of the ion/atom in different isotopic abundances so that the isotopic ratio of the sample is altered: the more of the measurand originally present, then the smaller the resulting overall change.

For this technique an isotopically selective detector is required, so that in practice the detection is generally done by mass spectrometry. The technique is then termed isotope dilution mass spectrometry to highlight the essential feature of this operation. In order to give the best accuracy some prerequisites are necessary:

(a) The isotopically discernible spike needs to be added in a similar amount to that present in the sample. This generally requires a rough preanalysis.
(b) The spike must be in the same chemical state as the measurand is in the sample. For example, if the substance is $Cr(H_2O)_6^{3+}$, then the spike must also be the hexahydrate of $Cr(III)$, and no CrO_4^-, or some other form of chromium.
(c) One must be certain that prior to any chemical pretreatment of the sample the spike is homogeneously mixed with the native measurand.

However, if viable, this technique corrects for very incomplete recoveries that moreover may vary from sample to sample. Care must be taken not to contaminate the sample after addition of the spike.

References

[12.2-1] Draper, N., Smith, H., *Applied Regression Analysis*, 2nd ed., New York: J. Wiley & Sons, 1981.

[12.2-2] Taylor, P.D.P., Schutyser, P., *Spectrochim. Acta* (1986) 41B, 1055, *on weighted regression*.

General reading

Mandel, J., *The Statistical Analysis of Experimental Data*, New York: J. Wiley & Sons, 1964.

Miller, J.C., Miller, J.N., *Statistics for Chemical Analysis*, Chichester: Horwood, 1988.

ISO 8466, *Water quality – Calibration and evaluation of analytical methods and estimation of performance characteristics*, Geneva 1993.

Draft ISO Guide 32, *Calibration of Chemical Analyses and Use of Certified Reference Materials*, Geneva 1993.

International Vocabulary of Basic and General Terms in Metrology, BIPM/IEC/IFCC/ISO/IUPAC/IUPAP/OIML, Geneva 1993.

Questions and problems

1. At trace levels the following calibration data were measured for Pb in water by graphite furnace atomic absorption spectrometry:

Pb (ng/mL)	0	0.25	0.50	0.75	1.0	2.0	3.0
Integrated signal (s)	0.0017	0.0031	0.0053	0.0072	0.0091	0.0156	0.0221

   What is the calibration model on the basis of Eq. 12.2-2; what is the residual standard deviation (Eq. 12.2-9) and the standard deviation of the procedure (Eq. 12.2-10)? Plot the curve, the residuals, and the confidence interval of the prediction (Eq. 12.2-12). Is the curve acceptable? Does the residual show marked structure?
   [Answer: $b_0 = 0.00183\,s$; $b_1 = 0.00683\,s/(ng/mL)$; the curve looks fine, but the residual has some structure pointing to potential curvature]
2. Compute the calibration curve, the residual standard deviation, and the standard deviation of the procedure for the determination of *o*-xylene by headspace GC:

o-xylene (ng/mL)	0.0	4.0	10.0	20.0	40.0	80.0	100.0	120.0	140.0
peak area (a.u.)	0.061	0.132	0.262	0.525	0.770	1.031	1.284	1.534	1.790

[Answer: $b_0 = 0.0070$ a.u.; $b_1 = 0.01275$ a.u./(ng/mL); $s_y = 0.00454$ a.u.; $s_c = 0.356$ ng/mL]

3. In the HPLC calibration of chlorophenols in water (after extraction) the following raw data were obtained:

chlorophenol (ng/mL)	0.5	1.0	6.2	12.2	35.0	56.0	75.1	100.8
peak area (a.u.)	45	86	615	1121	3046	4838	6331	8173

Compute the calibration line, draw the curve and the residual plot. What is your assessment of the validity of the model based on an inspection of the residual plot and the curve itself?
[Answer: $b_0 = 94.7$ a.u.; $b_1 = 81.93$ a.u./(ng/mL); $s_y = 121$ a.u.; the calibration line is curvilinear and should be computed on the basis of Eq. 12.2-3; one obtains $b_0 = 8.7$ a.u.; $b_1 = 92.0$ a.u./(ng/mL); $b_{11} = -0.108$ a.u./(ng/mL)^2; $s_y = 32.8$ a.u.; now, the residual plot does not show structure any longer]

4. What is the purpose of calibration? Do you know of a case when calibration is not required in analytical chemistry?

5. What is the general sequence of steps one takes in a calibration process?

6. Define the desirable properties of a standard?

7. Describe in your own words the three factors contributing to the quality of calibration?

8. What is meant by "validity of calibration"?

9. What is an advantage and what a disadvantage of using standard reference materials as standards in calibration?

10. How can one determine the right frequency of recalibration?

11. When would you favor a calibration model of the type $y = b_1 c$ over $y = b_0 + b_1 c$? Discuss.

12. When would you favor a calibration model of the type $y = b_0 + b_1 c + b_{11} c^2$ over $y = b_0 + b_1 c$? Discuss.

13. What is sensitivity?

14. What do you have to consider regarding the sensitivity of a system if the proper calibration model is $y = b_0 + b_1 c + b_{11} c^2$?

15. What are the residuals and which components contribute to their size?

16. Explain the difference in operation between internal standardization and the method of standard additions?

17. Describe the potential and the limitations of isotope dilutions mass spectrometry.

12.3   Signal Processing

Learning objectives

■ To get an overview of signal enhancement techniques in analytical chemistry

■ To understand an analytical signal as consisting of various frequency components

■ To define the Nyquist frequency and the aliasing effect

■ To compute a simple convolution of an analytical signal with a point-spread function

■ To understand the importance of deconvolution

■ To list the advantages of digital over analog filters based on hardware realisations

■ To apply general rules for choosing appropriate filters for maximum signal enhancement with minimal distortion

■ To understand the potential of second-derivative filters in analytical chemistry

12.3.1   Extraction of information

Signals are the original data obtained from an analytical process. They can be spectra, chromatograms, voltammograms, pH-meter traces as a function of time, voltages from an ion-selective electrode, and many more. All of these consist of a series of data spread over a certain time, wavelength range, or distance (as in spatial data). Usually, not every data point or every fluctuation is equally important to the analyst, some of the information is just regarded as noise or baseline. This is particularly easy to see in a chromatogram, where some regions contain the peaks whose area is ultimately related to the amount of a substance and in between there is a flat or curved baseline. Over both of these components, baseline and peak, there is spread out noise that can more or less dominate the overall signal to the point where it is hardly recognizable. The quality of such raw data is thus frequently described by the signal-to-noise ratio, SNR. It is one of the prime purposes of *signal processing* to distinguish between the useful part of the signal and the noise by discriminating against noise, thus favoring the extraction of a "clean" signal and thereby enhancing the analytical information. The useful part of the analytical signal is frequently peak-shaped, occasionally it may take the shape of a step function.

The enhancement of the useful part of the signal must capitalize on the different nature of this signal component from the noise, the most important factor is the difference in frequency between the two. Electronic noise often has a much higher frequency than the signal. At still lower frequencies another undesired feature of an analytical process is observed: trends, drifts, or shift of the baseline constitute signal components with lower frequency than the information-bearing part of the signal (Fig. 12.3-1)

All signal enhancement techniques discussed here capitalize on the dissimilarities regarding phase and frequency between the useful and the undesirable parts of the raw signal. The analytical system and its hardware components, however, need

Analytical signals are made up of relevant parts, background, and noise. The relevant parts are frequently peak-shaped.

Frequency	Component
Intermediate	Peaks, structure, useful part
Low	Drift
High	Noise

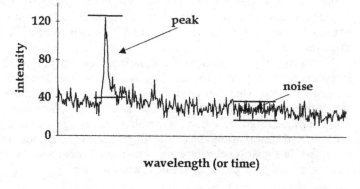

Fig. 12.3-1. An analytical signal displaying a peak, noise, and a slightly downward-sloping baseline. The SNR of this peak is roughly 5

to be selected with the aim of minimizing the amplitude of the interfering features because it is in principle much simpler to extract a useful signal from low-noise raw data with a high SNR than from those with a low SNR.

The effective distinction between desirable and undesirable features in a signal therefore depends on extracting information regarding the useful parts of the signal. This is typically done at high SNR, realized at high analyte concentrations. The "shape" of the signal is characterized in terms of frequency and phase information. Optimal extraction of analytical information is therefore guided by this "shape" information and discriminates against the undesirable features inter alia by enhancing the desirable ones.

The most important means of extracting information are transforms, smoothing, correlation, convolution, differentiation, and integration. For all of these it is nowadays typical for the continuous analog information (as registered on a chart recorder) to be digitized in order to store it in a computer; the information extraction is then effected by numerical procedures.

12.3.2 Discretization and the Fourier transform

Any signal can be represented in discretized form by either reading in original data at constant intervals or as averages (integrals) over some small interval. The delay between successive readings can be longer or shorter, depending on the purpose. A longer delay leads to fewer data points that need to be registered, stored, and processed, but it also results in some loss of information regarding rapidly changing signals. A wider spacing between readings provides less information about rapidly changing signals, thereby also sacrificing information on the high-frequency part of the signal.

Alternatively, one can shorten the observation time by recording densely spaced data in time (or wavelength, as it may be). While this may lead to the same amount of data as obtained from sparsely recorded signals, one loses information regarding low frequencies; longer cycles and trends can no longer be observed. The extraction of information from signals thus involves incorporation of some prior knowledge on the part of the analytical chemist as a guide in making appropriate choices regarding the density of observations in time and space, and the total length of a recording. It is, however, a rule rather than an exception that several measurements are required for good averaging and/or for recovering good estimates of the signal shape.

Fast changes in a signal (corresponding to high first derivatives of the signal) can be observed only if there are at least two data points recorded per cycle. This is called the *Nyquist frequency* constituting a lower limit:

$$f_{max} = \frac{1}{2\Delta x} \tag{12.3-1}$$

with f_{max} the maximum frequency that can be observed when the data are recorded with a spacing of Δx. If high frequencies might be of interest to the analytical chemist, Δx needs to be reduced. Once Δx is fixed, a choice is also made regarding the highest frequency of interest, just as the run of the observation fixes the longest cycle (or drift) that might be detectable. A direct consequence of the lack of detectability of frequencies higher than the Nyquist frequency is that any number of frequencies may exist in the signal at the same time. These are said to be *aliased* with each other. This is shown in Fig. 12.3-2 for two arbitrary frequencies sampled at an interval Δx.

Averaging over neighboring observations corresponds to the observation of integrals of the original data, and has two effects. The first is that it decreases the highest frequency that can be detected according to Eq. 12.3-1. The second is a smoothing, as averaging will eliminate short-term spikes and noise, both in the positive (higher signals) and in the negative (lower signals) direction. This smoothing will be the more pronounced the longer the averaging interval and the smaller the spike. In signal processing many other and more sophisticated procedures for smoothing are practiced and will be discussed later.

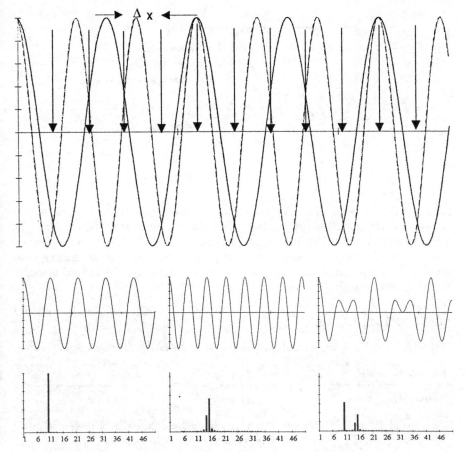

Fig. 12.3-2. Δx is too large to distinguish between the two frequencies (here cosine functions have been chosen for demonstration). When the signal is sampled at the time indicated by the arrows, the two frequencies are said to be "aliased"

Fig. 12.3-3. Simple signals (top) and their main frequencies (bottom) derived from a Fourier transform; the signal at the right is the sum of the other two and therefore the frequencies of the other two show up at the lower right

An alternative way to picture raw data is to convert them to their Fourier transform (Fig. 12.3-3). While the original data from dispersive instrumentation are expressed in terms of their time (or wavelength) behavior, the Fourier transform gives the identical information in terms of their frequency behavior. This is important in practice, particularly when data are recorded by interferometric procedures (FTIR, FTNMR, etc) where the original information is in terms of frequency and the wavelength information can be recovered only after back-transformation. Another reason for the importance of the Fourier transform is the fact that it makes readily visible those frequencies which actually contribute to a signal. This knowledge is relevant for the design and improvement of analytical instrumentation, as certain frequencies can be assigned to specific sources. It is well known that a frequency of 50 Hz is observed if the electronic circuit is "transparent" to the ac mains frequency.

Wavelength can always be converted to frequency, but in a Fourier transform one obtains information on frequencies measured in units of measurement^{-1} while in dispersive instrumentation the unit is Hz or cm^{-1}

12.3.3 Convolution

No signal as recorded has exactly the same shape as it had when it was generated. This can be regarded as a kind of blurring of the original signal by some property of the analytical recording step. A typical property of this type is the slit function in spectroscopy which leads to a distortion (broadening, reduction in intensity) of the originally emitted signal.

Mathematically, this distortion can be regarded as a *convolution* of two functions, the original signal and the slit function, that renders only the resultant of these two. While the recovery of the original signal – albeit never perfect – can be accomplished by the reverse mathematical process, *deconvolution*.

Convolution is expressed as:

$$g[x] = \sum_{i=0}^{m} f(x)h[x(i) - x] = f[x] * h[x] \qquad (12.3-2)$$

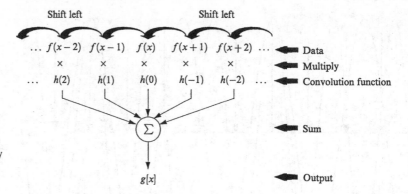

Fig. 12.3-4. Schematic respresentation of the convolution process. The result $g[x]$ is greatly dependent on the choice of the convolution function, thus on the actual numbers $h(k)$

where $f(x)$ is the original function (e.g., the spectrum as generated), $h(x)$ the distorting function (e.g., the slit function), and $g[x]$ the resultant function (e.g., the spectrum as recorded). This process, as explained in Fig. 12.3-4, is very important as it occurs frequently, not only when recording spectra, but also in imaging or as the effect of dead volume on dispersion in chromatography. The following example will help to clarify the concept.

Worked example:

Consider a signal as it is generated (but not recorded) as a function of wavelength, $f(x)$:

wavelength (a.u.)	1	2	3	4	5	6	7	8	9	10	11	12
original signal (a.u.)	0	1	2	3	6	8	10	8	6	3	2	1

On registration, the signal becomes somewhat distorted, the distortion being described by the function $h(x)$ to account for the imperfection of the recording devices. This function is often called the slit function or, generally, the point-spread function:

k	-1	0	$+1$
$h(k)$	0.2	0.6	0.2

and at all other values of k we have $h(k) = 0$. We note that while both functions are symmetric in this example, the analytical signal, typically a spectrum, a chromatographic peak, or a continuously recorded trace, is not necessarily so.

The basic operation of convolution is easy to understand by considering the signal at wavelength 1; while the original spectrum as physically generated does not show a (non-zero) signal at this wavelength, this is not true of the recorded (distorted) signal after convolution, as this operation gives:

$$g[1] = f(1)h(1-1) + f(2)h(2-1) + f(3)h(3-1) + \cdots$$
$$= 0 \times 0.6 + 1 \times 0.2 + 2 \times 0 + \cdots = 0.2$$

$$g[2] = f(1)h(1-2) + f(2)h(2-2) + f(3)h(3-2) + \cdots$$
$$= 0 \times 0.2 + 1 \times 0.6 + 2 \times 0.2 + \cdots = 1.0$$

$$g[3] = f(1)h(1-3) + f(2)h(2-3) + f(3)h(3-3) + f(4)h(4-3) + \cdots$$
$$= 0 \times 0 + 1 \times 0.2 + 2 \times 0.6 + 3 \times 0.2 + \cdots = 2.0$$

$$g[4] = \cdots + f(3)h(3-4) + f(4)h(4-4) + f(5)h(5-4) + \cdots$$
$$= 2 \times 0.2 + 3 \times 0.6 + 6 \times 0.2 + \cdots = 3.4$$

$$g[5] = \cdots + f(4)h(4-5) + f(5)h(5-5) + f(6)h(6-5) + \cdots$$
$$= 3 \times 0.2 + 6 \times 0.6 + 8 \times 0.2 + \cdots = 5.8$$

$$g[6] = \cdots + f(5)h(5-6) + f(6)h(6-6) + f(7)h(7-6) + \cdots$$
$$= 6 \times 0.2 + 8 \times 0.6 + 10 \times 0.2 + \cdots = 8.0$$

$$g[7] = \cdots + f(6)h(6-7) + f(7)h(7-7) + f(8)h(8-7) + \cdots$$
$$= 8 \times 0.2 + 10 \times 0.6 + 8 \times 0.2 + \cdots = 9.2$$

$$g[8] = \cdots$$

A comparison of the spectrum generated with that actually recorded (Fig. 12.3-5) highlights some of the general results of convolution:

wavelength (a.u.)	1	2	3	4	5	6	7	8	9	10	11	12
original signal (a.u.)	0	1	2	3	6	8	10	8	6	3	2	1
recorded signal (a.u.)	0.2	1.0	2.0	3.4	5.8	8.0	9.2	9.0	...	...	...	...

The recorded spectrum appears distorted – broader and flatter than the original one. While the reduction in height may not seem significant in this case, the fact that a real spectrum is always accompanied by some noise leads to a deterioration of the SNR, and thus potentially to problems in distinguishing the signal from the noise when the signal is small, near the limit of detection.

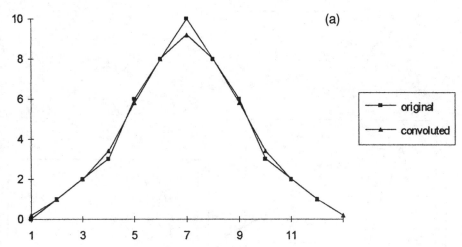

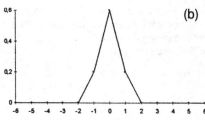

Fig. 12.3-5. (a) A comparison of the original (squares) with the convoluted (triangles) function shows that the latter gives a lower and broader peak. (b) The slit (or point-spread) function

12.3.4 Deconvolution, cross-correlation and signal restoration

Deconvolution is clearly the reverse operation of convolution and has as its aim the recovery of the original signal. This operation is easier the more is known about the shape and position of the original signal components such as peak position, peak shape, noise, slit function, etc. Also deconvolution is in analytical practice a more important operation than convolution, as it offers the possibility of recovering a signal from other, interfering signals and from noise, both of which are a frequent nuisance in analytical measurement.

One simple approach is *cross-correlation*, where the sought-for signal is enhanced by convoluting it with its own shape: the better the shape is known the better the signal is restored. To illustrate this operation, consider the Gaussian peak in Fig. 12.3-6a, which is blurred by substantial noise. We subject this signal to cross-correlation and form, as in Eq. 12.3-2:

$$g[x] = f[x] * h[x]$$

where $g[x]$ is the resulting restored signal, $f[x]$ the original signal, and $h[x]$ the (best estimate of the) true signal.

In Fig. 12.3-6b $h[x]$ is defined as in the inset, not very similar to a Gaussian peak and the result is an improved signal. However, if the function is closer to the original one, as in the inset of Fig. 12.3-6c, the restored signal resembles the original much more closely. This operation also has a smoothing effect that will be the focus of our attention in the following discussions.

A somewhat different problem arises, if one supposes more than one relevant feature in a peak-like structure, as is common in vibrational spectroscopy or in chromatograms with overlapping peaks. In general, one is then faced with a case of insufficient resolution, this being distinctly different from the former case of low

The best convolution function for cross-correlation is the actual peak shape without noise.

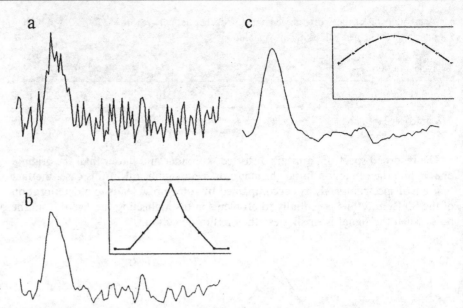

Fig. 12.3-6. The noisy signal (a) is cross-correlated with the functions in the insets; the function giving (b) is clearly less suitable than the function giving (c)

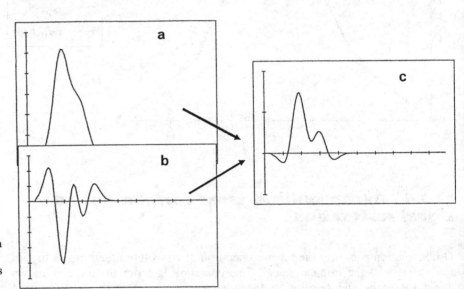

Fig. 12.3-7. From the overlapping signal (a) a fraction of its second derivative (b) is subtracted to yield better resolution (c). In this example the fraction $\alpha = 0.3$

SNR. This is exemplified in Fig. 12.3-7, where two Gaussian peaks overlap each other. For better visualization, the peak structure can be enhanced by subtracting its second derivative (or a fraction $1 - \alpha$) from it:

$$g[x] = \alpha \cdot f[x] - (1 - \alpha) \cdot f''[x] \quad with\ 0 < \alpha < 1 \tag{12.3-3}$$

For quantification of the relative intensities of the two peaks, the convolute can be described by a *least-squares fit* with the resulting coefficients a_1 and a_2 giving the relative intensities:

$$f[x] = a_0 + a_1 \cdot peak_1 + a_2 \cdot peak_2 \tag{12.3-4}$$

and a_0 being the background. This is shown in Fig. 12.3-8 for a convolute of two noisy Gaussian peaks, where the peak structure is composed of a higher peak at the left and a lower peak at the right, the sum of which make up the signal. Such a least-squares procedure also has some smoothing properties, but can only be employed reliably if the number, the shape, and potentially also the positions of the peaks are known beforehand.

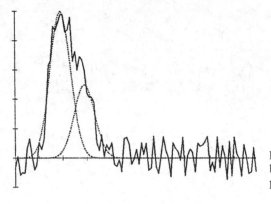

Fig. 12.3-8. The noisy peak is deconvoluted by a least-squares procedure into two separate peaks with different heights a_1 and a_2

12.3.5 Digital filters

One of the most common types of signal processing is digital filtering of spectra and of other (quasi)continuous recordings. It is a standard operation in practically all modern instrumentation and has its precursor in analog filters based on hardware. There are several reasons why digital filters are gaining more and more acceptance:

- They can be implemented in software on almost any computer and are therefore easy to "build" and to test.
- They are based purely on simple arithmetic, such as addition and multiplication, and, in contrast to their analog counterparts, do not change in space and time.
- They are easy to adapt to specific situations.
- They are simple to understand.

Before considering the different techniques, the purpose of digital filters requires some discussion. The primary aims of applying digital filters are the following:

1. Enhancement of the signal over the noise (smoothing)
2. Numerical differentiation with the aim of improving visual resolution or subtracting the background
3. Integration of signals

It is important to note at this point, however, that averaging and filtering cannot increase the information content of a signal, but information that already exists is made more accessible and visually enhanced.

The simplest digital filter involves averaging (arithmetic average): successive data are summed and described by their average. As is known from statistics (Sec. 12.1) this average exhibits less noise (usually described as standard deviation) than the raw data.

A problem common to all filters can also be discussed for the arithmetic mean: for nonstationary data (those that change with time and space) the change cannot be properly described by the average; the structure of data that bounce up and down (a peak!) cannot be represented by an average.

A remedy exists in that the width for averaging is chosen in a way that enhances the signal and suppresses the noise. This width is called the *filter width*, and is one of the most important properties of a digital filter. Too broad a width suppresses the structure of the signal, too narrow a width does not remove noise well enough. The simplest averaging filter is called the *boxcar average* (or *moving average*) filter, and its application is shown in Fig. 12.3-9. While in Fig. 12.3-9c the structure is obviously enhanced, the noise level is clearly lower, as seen from a comparison of Fig. 12.3-9b and d. The inset in Fig. 12.3-9c shows the particular boxcar filter. All boxcar filters are defined by filter coefficients of $1/n$ for a filter support of n. Thus, if the filter support is 8 data points wide, each coefficient assumes the value $1/8 = 0.125$.

One of the most useful filter types is obtained if the coefficients are chosen such that the resultant approximation $g[x]$ corresponds to a least-squares estimate of the signal $f(x)$, where x is the center of the filter support. The principle then is that for each data point a local least-squares solution is sought to represent the signal at this point. Fortunately, these coefficients have all been tabulated for various widths of the filter support and for different polynomials. In practice it suffices to consider the polynomials of order two or three, the parabola and the cubic function.

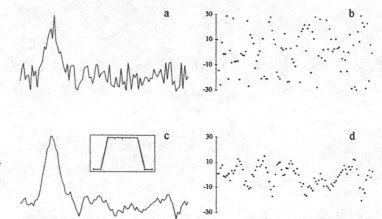

Fig. 12.3-9. The original signal (a) exhibits a higher noise level (b) than the signal after boxcar averaging (c) which gives the noise level (d). The shape of the boxcar filter is shown in the inset

Such coefficients are given in Table 12.3-1 and can be viewed as a special type of point-spread function $h[x]$ leading to a polynomial approximation and, if the filter support is chosen properly, also to a substantial smoothing. These coefficients serve as weights for producing a suitable weighted average. The width varies as $(2n + 1)$ and the sum of the coefficients divided by the NORM (last line in Table 12.3-1) is always 1.0.

Table 12.3-1. Coefficients for quadratic and cubic smoothing (Savitsky and Golay, 1964)

Points	25	23	21	19	17	15	13	11	9	7	5
−12	−253										
−11	−138	−42									
−10	−33	−21	−171								
−09	62	−2	−76	−136							
−08	147	15	9	−51	−21						
−07	222	30	84	24	−6	−78					
−06	287	43	149	89	7	−13	−11				
−05	322	54	204	144	18	42	0	−36			
−04	387	63	249	189	27	87	9	9	−21		
−03	422	70	284	224	34	122	16	44	14	−2	
−02	447	75	309	249	39	147	21	69	39	3	−3
−01	462	78	324	264	42	162	24	84	54	6	12
00	467	79	329	269	43	167	25	89	59	7	17
01	462	78	324	264	42	162	24	84	54	6	12
02	447	75	309	249	39	147	21	69	39	3	−3
03	422	70	284	224	34	122	16	44	14	−2	
04	387	63	249	189	27	87	9	9	−21		
05	322	54	204	144	18	42	0	−36			
06	287	43	149	89	7	−13	−11				
07	222	30	84	24	−6	−78					
08	147	15	9	−51	−21						
09	62	−2	−76	−136							
10	−33	−21	−171								
11	−138	−42									
12	−253										
NORM	5175	8059	3059	2261	323	1105	143	429	231	21	35

Worked example:

We examine a part of a goniometer scan on a wavelength-dispersive X-ray fluorescence spectrometer where the SNR is rather low. The spectrum is treated by repeated smoothing and the effect of this operation is studied with respect to the change of peak height and peak position.

As initial filter support we choose five data points and read from Table 12.3-1 the following coefficients: −3/35, 12/35, 17/35, −3/35. We note the symmetry of this and all other filters in Table 12.3-1 and the fact that the sum of all coefficients is equal to $(-3 + 12 + 17 + 12 - 3)/35 = 35/35 = 1.0$. This ensures that there is very little change in peak height and peak area to be expected unless the filter support is chosen too wide, in which case neighboring background would "spill over".

The particular peak that will be scrutinized is marked with an arrow in Fig. 12.3-10a. This peak has a FWHM of about 7, and this makes the filter support with 5 points somewhat narrower than the FWHM. The effect of repeated smoothing on peak area and peak height is shown in Fig. 12.3-11. A general result is that the peak height is more affected by filtering than is the peak area, but nevertheless the peak height decreases only slowly with this 5-point filter and 7 data points for FWHM (5/7 = 0.71 FWHM).

The situation is different if the filter support width is changed instead. As can be seen from Fig. 12.3-12, the reduction of noise is concurrent with a decrease in resolution, caused by peak broadening, and a decreased peak height much more pronounced than that caused by multiple application of a filter of appropriate width. Fig. 12.3-13 clearly shows a sharper decrease in peak height and also in area than that shown in Fig. 12.3-11.

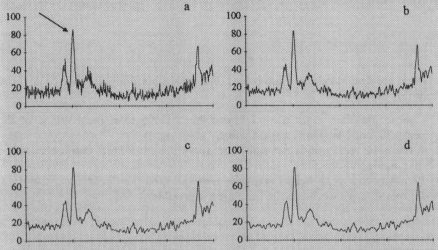

Fig. 12.3-10. Effect of 5-point polynomial smoothing on a spectrum. (a) Original spectrum; (b) one-time smoothed spectrum; (c) three-times smoothed spectrum; (d) eight-times smoothed spectrum

Fig. 12.3-11. Effect of repeated smoothing on peak area (shaded bars) and peak height (open bars)

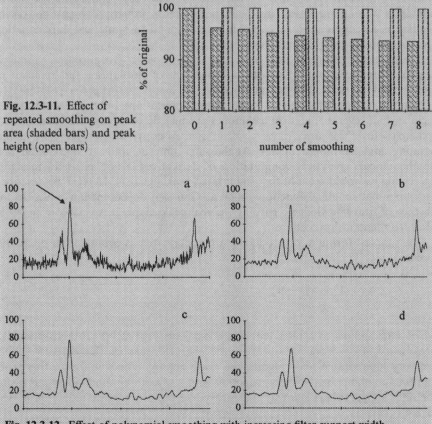

Fig. 12.3-12. Effect of polynomial smoothing with increasing filter support width. (a) Original spectrum; (b) 7-point smoothed spectrum; (c) 13-point smoothed spectrum; (d) 19-point smoothed spectrum

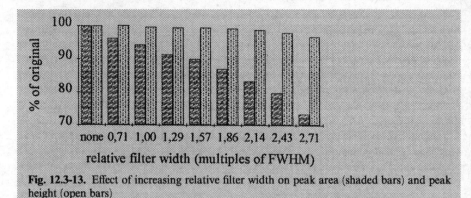

Fig. 12.3-13. Effect of increasing relative filter width on peak area (shaded bars) and peak height (open bars)

There are several rules of thumb for polynomial smoothing that can be deduced from this example:

1. Choose the filter width $(2n + 1)$ somewhat smaller than the FWHM of the peak; 70% of FWHM is a good value.
2. If the noise reduction is not sufficient, apply this filter more than once, rather than increasing the filter width.
3. Be aware that peak height is always more affected than peak area; therefore, for quantitative work use the peak area if possible.

12.3.6 Numerical differentiation and integration

Another important function of digital filters is that they produce derivatives of signals as well as integrals. This again presupposes properly chosen filter coefficients $h[x]$ as convolution functions that correspond to numerical differentiation and integration in applied mathematics.

In analytical chemistry these operations mainly serve two purposes: visualization of overlapped peaks through enhancement of resolution, and removal of background for estimating net intensities.

In order to appreciate the value of these applications fully, one needs to remember that several analytical techniques produce derivatives as they are applied. Examples for these are Auger electron spectroscopy, differential pulse polarography, and thermogravimetry. At the same time, it needs to be realized that (mathematical) analytical computation of derivatives is not as useful in signal processing as would seem at first sight; most peak signals *CANNOT* be described as simple mathematical functions, such as Gaussian or Lorentzian peaks. In the absence of firm knowledge regarding the real peak shape, it is very convenient to resort to numerical procedures.

The simplest formula for producing a first derivative is computation of the difference:

$$x'(k) \approx \frac{x(k+1) - x(k-1)}{2}$$

which corresponds to a filter with the coefficients $(1/2, 0, -1/2)$. Unfortunately, because of error propagation, the noise characteristics of this derivative filter are very unfavorable and enhance the noise greatly. The solution to this problem is to combine the derivative function with a smoothing function by increasing the filter support the same way that proved so useful for noise reduction in Sec. 12.3.4.

For estimation of the first derivative, the following filters have better noise properties:

order two: $x'_k \approx 0.2x_{k+2} + 0.1x_{k+1} - 0.1x_{k-1} - 0.2x_{k-2}$

order three: $x'_k \approx \dfrac{-x_{k+2} + 8x_{k+1} - 8x_{k-1} + x_{k-2}}{12}$

order five: $x'_k \approx \dfrac{-x_{k+3} + 9x_{k+2} - 45x_{k+1} + 45x_{k-1} - 9x_{k-2} + x_{k-3}}{30}$

order seven: $x'_k \approx \dfrac{1}{280}(-x_{k+4} + 10.666x_{k+3} - 56x_{k+2} + 224x_{k+1}$

$$- 224x_{k-1} + 56x_{k-2} - 10.666x_{k-3} + x_{k-4})$$

Filters for producing second derivatives are even more important in analytical practice as they produce (negative) peaks, thus the visual impression is analogous for the operator to the directly recorded signal, and they remove sloping and (moderately) curved background. The coefficients of these filters are given in Table 12.3-2.

A powerful application of this principle is shown in Fig. 12.3-14 where two peaks not visible in direct registration are distinguishable as second derivatives. Again, a balance must be sought between good smoothing properties, which call for a wide filter support, and good resolution enhancement, which requires narrow filter support. A 7-point second derivative filter is a good compromise in this case.

In a similar manner one can produce integrals numerically. Integration can be performed pointwise if the pieces are then summed within the desired limits of the

Table 12.3-2. Coefficients for computing second derivatives (Savitsky and Golay, 1964)

Points	25	23	21	19	17	15	13	11	9	7	5
−12	92										
−11	69	77									
−10	48	56	190								
−09	29	37	133	51							
−08	12	20	82	34	40						
−07	−3	5	37	19	25	91					
−06	−16	−8	−2	6	12	52	22				
−05	−27	−19	−35	−5	1	19	11	15			
−04	−36	−28	−62	−14	−8	−8	2	6	28		
−03	−43	−35	−83	−21	−15	−29	−5	−1	7	5	
−02	−48	−40	−98	−26	−20	−48	−10	−6	−8	0	2
−01	−51	−43	−107	−29	−23	−53	−13	−9	−17	−3	−1
00	−52	−44	−110	−30	−24	−56	−14	−10	−20	−4	−2
01	−51	−43	−107	−29	−23	−53	−13	−9	−17	−3	−1
02	−48	−40	−98	−26	−20	−48	−10	−6	−8	0	+2
03	−43	−35	−83	−21	−15	−29	−5	−1	7	5	
04	−36	−28	−62	−14	−8	−8	2	6	28		
05	−27	−19	−35	−5	1	19	11	15			
06	−16	−8	−2	6	12	52	22				
07	−3	5	37	19	25	91					
08	12	20	82	34	40						
09	29	37	133	51							
10	48	56	190								
11	69	77									
12	92										
Norm	26910	17710	33649	6783	3876	6188	1001	429	462	42	7

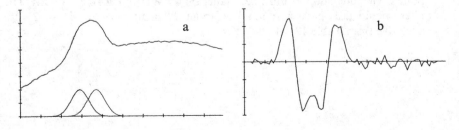

a

b

Fig. 12.3-14. Signal (a) consists of two Gaussian peaks separated by about 0.6 FWHM on a sloping and curved background with 5% noise. The second derivative (b) removes the background and achieves enough resolution for the two peaks to be identified as separate minima. The 7-point filter of Table 12.3-2 was applied in this case

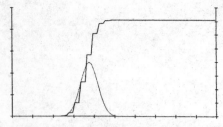

Fig. 12.3-15. Application of Simpson's rule for integration of a Gaussian peak

integral. For three points we have:

$$\int_0^{N-1} x_k \, dt = \sum_{k=0,2,4,\dots,N-3} \frac{x_{k+2} + 4x_{k+1} + x_k}{3}$$

this is generally known as Simpson's rule (Fig. 12.3-15).
  For a cubic polynomial we take

$$\int_k^{k+3} x_k \, dt \approx \frac{3x_{k+3} + 9x_{k+2} + 9x_{k+1} + 3x_k}{8}$$

this is known as the 3/8-th rule.

General reading

M. Abramovitz, I. Stegun, *Handbook of Mathematical Functions*, New York: Dover, 1972.

C.G. Enke, T.A. Nieman, Anal. Chem. *48*, 705A (1976).

D.L. Massart, B.G.M. Vandeginste, S.N. Deming, *Chemometrics. A Textbook*, Amsterdam: Elsevier, 1988.

A. Savitsky, M.J.E. Golay, Anal. Chem. *36*, 1627 (1964).

C.S. Williams, *Designing Digital Filters*, New York: Prentice Hall, 1986.

Questions and problems

1. What are the components that make up a signal?
2. What are the high-frequency and low-frequency parts of a signal called?
3. Explain the SNR.
4. Explain the limit expressed by the Nyquist frequency.
5. Use a sketch to illustrate the aliasing effect.
6. What simple rule leads to the optimal choice of the convolution function in cross-correlation?
7. What are the advantages of digital over analog filters?
8. Two main reasons exist for applying digital filters in analytical chemistry. What are these?
9. What does the convolution function that is used in boxcar averaging look like?
10. What is a good compromise for choosing the "right" filter support for smoothing?
11. Why are second derivative filters popular in analytical signal processing?
12. Compute the missing values of $g[x]$ in the Example in Sec. 12.3.3. Answer:

wavelength (a.u.)	1	2	3	4	5	6	7	8	9	10	11	12
original signal (a.u.)	0	1	2	3	6	8	10	8	6	3	2	1
recorded signal (a.u.)	0.2	1.0	2.0	3.4	5.8	8.0	9.2	9.0	**5.8**	**3.4**	**2.0**	**1.0**

13. A Gaussian peak described by 12 points per standard deviation ($s = 12$ points) has to be smoothed. Make a reasonable choice for the filter support with the additional information that 1 FWHM corresponds to 2.35 standard deviations. Answer: One FWHM corresponds to 2.35×12 point $= 28.2$ points. The filter support must be smaller than FWHM, e.g., 0.7 FWHM. The filter should thus have a support of about 20 points – either 19 or 21 are available from Table 12.3-1].

12.4 Optimization and Experimental Design

Learning objectives

■ To provide an introductory course into systematic optimization methods in analytical chemistry
■ To select the most important factors that influence a given analytical problem based on statistical approaches of experimental design as well as on evaluating the factor effects and their interactions by means of statistical tests
■ To discuss the design of experiments for modeling the relationship between factors and to apply response surface methods for locating the optimum
■ To search for the optimum by sequential methods, i.e. by means of the Simplex method.

12.4.1 Introduction

The aims of applying optimization and design methods in analytical chemistry can be numerous. The influence of different factors on an analytical signal might be investigated in order to judge the robustness of a method or to evaluate the most important interfering factors. Very often the performance of an analytical procedure is optimized with respect to analytical quality criteria, such as precision, trueness, sensitivity, detection limit, or signal-to-noise ratio.

Another problem in this field would be optimization of the composition of a sensitive layer that should be used for recognition of heavy metal ions in a chemical sensor.

Thus, typical *factors* in analytical chemistry comprise the pH value, reagent concentration, temperature, flow rate, solvent, elution strength, composition, irradiation, atomization time, or sputtering rate. Typical *responses* are the analytical figure-of-merits as introduced in Chap. 2 as well as objective functions that consist of combinations of different quality criteria.

Systematic optimization procedures are carried out in the following sequence:

1. *Choice of an objective function.* Very often the optimization criterion is simply an analytical signal or the analysis time. In more complicated situations, however, objective functions that are composed of several criteria, such as selectivity, sensitivity, and precision have to be considered. Therefore, combination of objective criteria to a single function is an important topic in analytical chemistry.

Objective function Z obtained by aggregation of quality criteria z_i by a weighted sum:

$$Z = \sum_{i=1}^{p} w_i z_i$$

where w_i-weight of criterion i.

2. *Selection of the most important factors.* Potential factors that affect a given objective function are best selected by an expert in the particular analytical field. The test for significance of the factors' influence has to be performed on the basis of a simple experimental design, a screening design, by means of statistical tests.

3. *Optimization.* To find the most suitable factor combinations we can distinguish between simultaneous and sequential optimization approaches. With *simultaneous* strategies the relationship between responses and factors is studied by running an experimental design, constructing a mathematical model, and investigating the relationship by so-called response surface methods (RSM). Very often RSMs are aimed at judging this relationship graphically and the consequences are drawn from those plots. If the optimal point is desired it can be found by calculating the partial derivative with respect to the individual factors or by applying a grid search over the whole response surface. *Sequential* strategies of optimization are based on an initial design of experiments followed by a sequence of further measurements in the direction of the steepest ascent or descent. That is, no quantitative relationship between factors and responses is evaluated but the response surface is searched along an optimal (invisible) path. The two strategies are exemplified in Fig. 12.4-1.

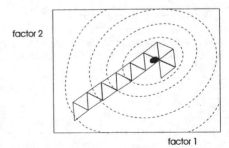

Fig. 12.4-1. Response vs. factors plot: In the case of response surface methods the response is described by a mathematical model (dotted contour lines of the response surface). By using search methods the response is measured along a search path, here along a Simplex path

12.4.2 **Experimental design**

Basic principles

Ceteris-paribus principle

Optimization of an analytical problem or of an analytical device has to be carried out by studying a limited number of factors. Very often it is easy to select the most important factors from deep knowledge about a given problem, e.g., for developing a new spectrophotometric method the factors pH value and reagent concentration would have to be studied, or in HPLC important factors are represented by the constituents of the mobile phase and their actual concentrations.

Sometimes the effect of a factor can only be presumed and its effect would have to be assured by suitable screening experiments. Because of the complexity of most of the analytical problems there will be additional factors that are either unknown or that cannot be controlled by the experimenter. Uncontrolled factors might be the impurity of reagents, intoxication of an electrode surface, the instability of a plasma source, or the changing quality of a laboratory assistant's work.

Since the study of all potential factors is usually prohibitive the effect of selected factors will be investigated and the remaining factors should be kept as constant as possible. This general principle is known as the *ceteris-paribus principle*.

Replication

Replication of measurements at a given factor combination is necessary for estimation of the experimental error. Furthermore, the error can be decreased with repetitive measurements by averaging, i.e., by a factor of $1/\sqrt{n}$ if n repetitive measurements have been carried out.

Randomization

Randomization means running the experiments in a random order. Randomized experiments are obligatory if systematic errors (bias) cannot be avoided and should be detected. Imagine the construction of a calibration graph. If the concentrations are measured in ascending order it would not be possible to detect a systematic error caused by a positive signal drift correlated with time. All experimental observations might lie on a straight line but the measured slope of the calibration graph is shifted systematically from the true slope (Fig. 12.4-2a). On the other hand, if the concentrations are measured in a random sequence, then you will notice that something is wrong because the variation of observations will be exceptionally high (Fig. 12.4-2b).

In the present example we have studied only one factor – the component concentration. In the case of studying several factors randomization is likewise necessary. In multifactor experiments the experimental design will have to be run in a randomized order as we will see below.

One of the basic prerequisites of all statistical tests is the existence of random and independent data. This assumption will also be valid for statistical tests in connection with experimental designs. Therefore, randomized experimentation is obligatory in this context. Carrying out the experiments in a random sequence will be one of the suppositions for being able to measure independent, uncorrelated, and usually normally distributed data.

Random sequences should be read from random number tables or nowadays from a random number generator. This guarantees genuine randomness of the sequence instead of subjective selections.

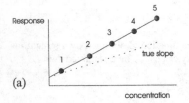

(a)

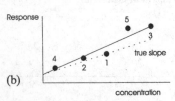

(b)

Fig. 12.4-2. Experimental observations for constructing a calibration graph in systematic (a) and randomized order (b)

Blocking

Uncontrolled factors lead to higher experimental errors. This increased error will also decrease the sensitivity of the experiments with respect to the factors to be studied. Therefore, the experiments should be designed such that uncontrolled factors can be detected and in a further step can be kept constant or eliminated.

There are two categories of uncontrolled factors. Uncontrolled influences might arise from either *unknown factors* or from *known factors that cannot be controlled*. The eventual impurity of a reagent is an example of an unknown factor. The changing quality of a laboratory assistant's work is an agreed factor that is difficult to control. Completely unknown factors cannot be accounted for at all.

Known or presumably known factors can be detected by blocking the experiments. The idea is to run the experiments in blocks that show a minimum experimental variance within one block. For example, if a systematic investigation requires 12 experiments and you could run only 4 experiments a day the experiments should be arranged in 3 blocks with 4 experiments each day. Day-to-day effects could then be detected by considering the block effects with an adequate mathematical model as given below.

Of course, the experiments, within a block should be run at random. Randomized experimentation with respect to some factors, and blocking of the experiments with respect to some other factors exclude each other. Therefore, in practice a compromise between randomized runs and blocked experiments has to be found.

Factorial experiments

Factorial experiments are based on varying all factors simultaneously at a limited number of factor levels. This kind of experimentation is especially important in the beginning of an experimental study, where the most influential factors, their ranges of influence and factor interactions are not yet known. Factorial experiments allow experiments to take place in the whole range of the factors space. They reveal high precision at a minimum experimental effort, and they enable factor interactions to be detected, such as the dependence of enzyme activity on both pH-value and co-enzyme concentration.

Confounding

Confounding of parameter estimations for different factors occurs if the factor combinations are highly correlated and, therefore, no difference between the factor effects can be detected. Confounding depends highly on the concrete experimental design. If, for example, the levels of two factors are changed in a constant ratio it would not be possible to distinguish between the effects of those two factors.

Symmetry

Factorial experiments should be partitioned in the whole factor space in a balanced manner. The same is true for replications in the experimental space. One reason for performing symmetric experiments is the avoidance of confounded factor effects. In addition, symmetric experiments might simplify data evaluation.

Important designs

Screening designs

Designs on the basis of two levels for each factor are called screening designs. The most general design is a *full factorial design* at two levels. These designs are described as 2^k-designs where the base 2 stands for the number of factor levels and k expresses the number of factors. As an example, a two-level three factorial design, 2^3, is given in Table 12.4-1 and Fig. 12.4-3. The factor levels are scaled here to -1 for the lower level and $+1$ for the higher level. Other coding schemes

Table 12.4-1. Full factorial design at two levels, 2^3 design. The star labelled experiments represent the half-fraction factorial design of Fig. 12.4-4

Experiment	Factors			Response
	x_1	x_2	x_3	
1	-1	-1	-1	y_1
2*	$+1$	-1	-1	y_2
3	$+1$	$+1$	-1	y_3
4*	-1	$+1$	-1	y_4
5*	-1	-1	$+1$	y_5
6	$+1$	-1	$+1$	y_6
7*	$+1$	$+1$	$+1$	y_7
8	-1	$+1$	$+1$	y_8

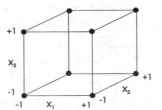

Fig. 12.4-3. Full factorial design at two levels, 2^3 design. x_1, x_2, and x_3 represent the three factors

Table 12.4-2. Plackett and Burman fractional factorial design for estimating the main effects of 11 factors at two levels

Run	x_1	x_2	x_3	x_4	x_5	x_6	x_7	x_8	x_9	x_{10}	x_{11}	Response
1	+	+	−	+	+	+	−	−	−	+	−	y_1
2	−	+	+	−	+	+	+	−	−	−	+	y_2
3	+	−	+	+	−	+	+	+	−	−	−	y_3
4	−	+	−	+	+	−	+	+	+	−	−	y_4
5	−	−	+	−	+	+	−	+	+	+	−	y_5
6	−	−	−	+	−	+	+	−	+	+	+	y_6
7	+	−	−	−	+	−	+	+	−	+	+	y_7
8	+	+	−	−	−	+	−	+	+	−	+	y_8
9	+	+	+	−	−	−	+	−	+	+	−	y_9
10	−	+	+	+	−	−	−	+	−	+	+	y_{10}
11	+	−	+	+	+	−	−	−	+	−	+	y_{11}
12	−	−	−	−	−	−	−	−	−	−	−	y_{12}

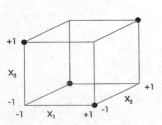

Fig. 12.4-4. Fractional factorial design at two levels, 2^{3-1} half-fraction design. x_1, x_2, and x_3 represent the three factors

Table 12.4-3. 2×2 Latin square design in different representations

Conventional representation:

Blocking variable 1	Blocking variable 2	
	low	high
low	A	B
high	B	A

Factorial representation:

Blocking variable 1	Blocking variable 2	Blocking variable 3
low	low	A
low	high	B
high	low	B
high	high	A

4×4 Latin square

Blocking variable 1	Blocking variable 2			
	1	2	3	4
1	A	B	D	C
2	D	C	A	B
3	B	D	C	A
4	C	A	B	D

are also used, e.g., 0 and 1, or − and +, or A and B for the low and high levels, respectively. One advantage of working with scaled factor levels is the feasibility of applying the same design to several investigations. Another advantage can be seen if we model the relationship between responses and factors quantitatively. With coded levels the size of the parameters will become comparable. Of course, the original variables can also be used in a study.

As long as the number of factors is small full factorial designs can easily be run. At high factor numbers, however, the number of experiments will increase dramatically. For example, for the study of 7 factors in a 2^7-design 128 experiments in total may be necessary. At this point we have to discuss the objectives of running a factorial experimental design. One reason is the estimation of factor effects. We know already from the section on calibration by linear models (Sec. 12.2) that it will not be necessary to evaluate the effects of 7 factors by running 128 experiments. Another aim is to model the responses in dependence on the factors. For modeling dependences of responses on two factor levels we would use a polynomial of first order (cf. Eq. 12.4-5). So in our case we would have to estimate 7 parameters linked to the effects of the 7 factors and perhaps an additional parameter that models the shift at the ordinate axis. For computing this statistical model we deduce $128 - 8 = 120$ degrees of freedom which are obviously too many to test for the adequacy of a simple first-order polynomial model.

The number of experiments can be reduced if *fractional factorial designs* are used. For fractional factorial designs the number of experiments is reduced by a number p according to a 2^{k-p} design. In the case of $p = 1$, so-called half-fraction designs result. For the example of three factors at two levels (cf. Fig. 12.4-3) the half-fraction design consists of $2^{3-1} = 4$ experiments as given in Fig. 12.4-4. In Table 12.4-1 the points of this design are labeled by a star. The experiments can be run either at the given points or at the complementary corner points.

Apart from saturated designs, numerous other fractional factorial designs can be used if the effect of many factors is to be studied. Concrete designs can be taken from tables and can be generated with most of statistical software packages. Special designs for estimating only the main effects have been tabulated by Plackett and Burman. As an example in Table 12.4-2 their fractional factorial design for 11 factors at two levels is represented.

A special case of a two-level factorial design is the *Latin square design*, which was introduced very early on to eliminate more than one blocking variable. A Latin square design for 2 variables is given in Table 12.4-3 along with the representation as a fractional factorial design.

Estimation of factor effects. Screening designs are mainly used to estimate the effects of factors in an analytical investigation on a statistical basis. To understand the test procedure we will consider an example from kinetic-enzymatic determinations.

Worked example: Factor effects estimation

The determination of the enzyme ceruloplasmin based on spectrophotometric measurements of the initial rate of p-phenylenediamine oxidation is investigated at a constant enzyme concentration of 13.6 mg/L. In the first step, a screening 2^3-design is used to study the effects of the factors pH value, temperature, and the substrate concentration p-phenylenediamine.

Table 12.4-4. 2^3-screening design and factor levels for estimation of the factors pH-value, temperature (T), and p-phenylenediamine concentration (PPD)

Run	Coded factor level						y, min^{-1}
	main effects			interaction effects			
	T	PPD	pH	$T \cdot PPD$	$T \cdot pH$	$PPD \cdot pH$	
1	+1	−1	−1	−1	−1	+1	6.69
2	+1	+1	−1	+1	+1	−1	11.71
3	+1	+1	+1	+1	−1	+1	14.79
4	+1	−1	+1	−1	+1	−1	8.05
5	−1	−1	−1	+1	−1	+1	6.33
6	−1	+1	−1	−1	+1	−1	11.11
7	−1	+1	+1	−1	−1	+1	14.08
8	−1	−1	+1	+1	+1	−1	7.59

Factor	Level	
	−1	+1
T, °C	35	40
PPD, mM	0.5	27.3
pH	4.8	6.4

The factor levels are given in Table 12.4-4 along with the experimental design and the measured initial rates, y. Based on the 2^3-design both *main* factor effects as well as *interactions* can be studied. The levels for factor interactions are calculated as products of the actual factor level combinations.

The factor effects are calculated as the absolute difference, $|D|$, between the responses of a factor at high and low level. These differences are then tested against the experimental error expressed by the standard deviation s multiplied by the Student's t-value, i.e., $|D| \geq t(P, f)s$:

$$D_T = \frac{y_1 + y_2 + y_3 + y_4}{4} - \frac{y_5 + y_6 + y_7 + y_8}{4} \qquad (12.4\text{-}1)$$

$$= \frac{6.69 + 11.71 + 14.79 + 8.05}{4} - \frac{6.33 + 11.11 + 14.08 + 7.59}{4} = 0.53$$

$$D_{PPD} = \frac{11.71 + 14.79 + 11.11 + 14.08}{4} - \frac{6.69 + 6.33 + 8.05 + 7.59}{4} = 5.76$$

$$D_{pH} = \frac{14.79 + 14.08 + 8.05 + 7.59}{4} - \frac{6.69 + 6.33 + 11.71 + 11.11}{4} = 2.17$$

$$D_{T \cdot PPD} = \frac{11.71 + 14.79 + 6.33 + 7.59}{4} - \frac{6.69 + 8.05 + 11.11 + 14.08}{4} = 0.123$$

$$D_{T \cdot pH} = \frac{14.79 + 8.05 + 6.33 + 11.11}{4} - \frac{6.69 + 11.71 + 14.08 + 7.59}{4} = 0.053$$

$$D_{PPD \cdot pH} = \frac{6.69 + 14.79 + 6.33 + 14.08}{4} - \frac{11.71 + 8.05 + 11.11 + 7.59}{4} = 0.858$$

With a standard deviation of 0.24 and a degree of freedom of $f = 3$ measured at factor level of run 3 we calculate for the experimental error: $t(0.95, 3)s = 3.18 \cdot 0.24 = 0.76$.

The comparison of the experimental error with the absolute differences reveals that the main factors pH and PPD concentration show a significant effect (D_{PPD} and D_{pH} are higher than 0.76) while the effect of the temperature can be neglected in the studied range between 35 and 40 °C ($D_T < 0.76$).

Graphical inspection of the main factor effects can be carried out in Fig. 12.4-5. As found by the calculations, there is minimal influence by the temperature. If the enzymatic reaction is run at a common 37 °C the method will be rugged against temperature fluctuations. Notice that all measured initial rates are slightly higher at 40 than at 35 °C. Compared with the general experimental error, however, this effect has been found to be statistically insignificant.

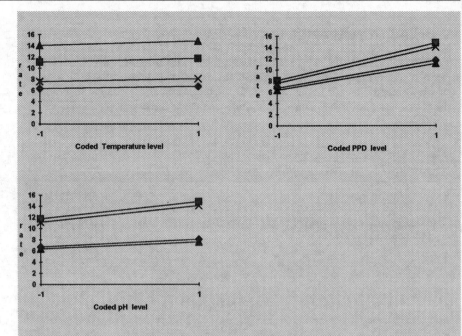

The enhancing effect of the substrate concentration PPD and of the pH value on the rate can also be seen in Fig. 12.4-5. As a general rule main effects will give parallel straight lines in the factor effects plot. In the case of factor interactions the slope of the straight line will differ if the levels of the alien factors change. The latter is the case for interactions of the pH and PPD factors. As can be seen from Fig. 12.4-5 the changes of rates between the lower and higher levels of PPD are more pronounced, if the factor pH is at high level (points ■ and x). A similar effect is observed if the rate changes in dependence on pH are compared at high (points ◆ and x) and low (points ◆ and ▲) PPD concentrations.

The calculated interaction effects are significant for the interaction of the substrate PPD and the pH value ($D_{PPD \cdot pH} > 0.75$).

In contrast to the simultaneous factorial design study, experimentation by variation of one variable at a time is limited to the estimation of main effects, and no interactions, as are common in analytical chemistry, are to be found. What cannot be evaluated with screening designs are curved dependences, i.e., for more complicated relationships between responses and factors designs at 3 or more factor levels are needed.

Response surface designs

In order to describe the relationship between responses and factors quantitatively we will use below mechanistic (physicochemical) or empirical models. e.g., polynomial models. These mathematical models should be able to describe linear and curved response surfaces likewise. Curved dependences can be modeled if the factor levels have at least been investigated at three levels.

Three-level factorial designs are known, therefore, as response surface designs. Full factorial three-level designs can be formalized in the same way as known for two-level designs, i.e., a 3^k-design means k factors at 3 factor levels. Figure 12.4-6 gives an example for a 3^2-design.

Full factorial three-level designs are sometimes used for investigating few factors (two or three) although their statistical properties with respect to symmetry or confounding of parameter estimates are less favorable than those known for the two-level designs. In the case of many factors the same problem as with two-level designs arises, i.e., the number of experiments gets very high. These disadvantages led to the development of so-called optimal designs of which the *central composite design* and the *Box-Behnken design* are the most important ones.

Central composite design. Composite designs consist of a combination of a full or fractional factorial design and an additional design, often a star design. If the

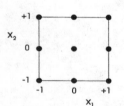

Fig. 12.4-6. Full factorial three-level design for two factors, 3^2-design

The degree of freedom of a factorial design is the number of runs minus the number of independent factor combinations.

center of both designs coincide they are called central composite designs. Consider a design that consists of a full factorial two-level design linked to a star design. For the number of runs r we obtain:

$$r = 2^{k-p} + 2k + n_0 \tag{12.4-2}$$

where k is the number of factors, p is the number for reduction of the full design, and n_0 is the number of experiments in the center of the design.

Worked example: Runs in a central composite design

For a design with three factors one gets for the number of experiments r according to Eq. 12.4-2:

$$r = 2^3 + 2{*}3 + 1 = 15$$

A complete three factor central composite design is depicted in Table 12.4-5 and Fig. 12.4-7. The distance of the star points α from the center can be differently chosen. For a uniformly rotatable design $\alpha = 2^{(k-p)/4}$, e.g., for 3 factors $\alpha = 2^{(3-0)/4} = 1.682$.

To estimate the experimental error, replications of factor combinations are necessary. Usually the center point is run thrice. The total number of runs in a central composite design with three factors amounts then to 17. Apart from good statistical properties of the central composite design there is one experimental disadvantage. Because of the star points outside the hypercube the number of levels that have to be adjusted for every factor is actually 5 instead of 3 in a conventional three level design. If the adjustment of levels is difficult to achieve an alternative response surface design would be the design introduced by Box and Behnken.

Box-Behnken design. In a Box-Behnken design the experimental points lie on a hypersphere equidistant from the center point as exemplified for a three factor design in Fig 12.4-8 and Table 12.4-6. In contrast to the central composite design the factor levels have only to be adjusted at three levels. In addition, if two replications are again performed in the center of the three factor design the total number of experiments is 15 compared to 17 with the central composite design.

In general, the Box-Behnken design requires few factor combinations as given in Table 12.4-7. The usage of the design will be further explained in the section on response surface methods below.

One disadvantage of Box-Behnken designs might be the representation of responses in dependence on a single factor. This is because the corner points of the cube have not been measured but have to be computed by an appropriate response surface model.

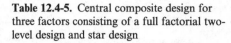

Table 12.4-5. Central composite design for three factors consisting of a full factorial two-level design and star design

Run	Factors			Response
	x_1	x_2	x_3	
2^3-kernel design				
1	-1	-1	-1	y_1
2	$+1$	-1	-1	y_2
3	$+1$	$+1$	-1	y_3
4	-1	$+1$	-1	y_4
5	-1	-1	$+1$	y_5
6	$+1$	-1	$+1$	y_6
7	$+1$	$+1$	$+1$	y_7
8	-1	$+1$	$+1$	y_8
$2k$-star points				
9	$-\alpha$	0	0	y_9
10	$+\alpha$	0	0	y_{10}
11	0	$-\alpha$	0	y_{11}
12	0	$+\alpha$	0	y_{12}
13	0	0	$-\alpha$	y_{13}
14	0	0	$+\alpha$	y_{14}
center point				
15,16,17	0	0	0	y_{15}, y_{16}, y_{17}

Replication here means carrying out a given factor combination several times.

In a Box-Behnken design the experimental points lie on a sphere rather than on a cube.

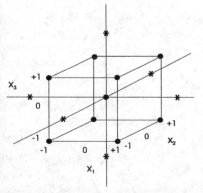

Fig. 12.4-7. Central composite design for three factors

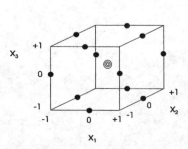

Fig. 12.4-8. Box-Behnken design for 3 factors

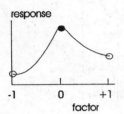

response

factor

-1   0   +1

Table 12.4-6. Box-Behnken design for three factors

Run	Factors			Response
	x_1	x_2	x_3	
1	+1	+1	0	y_1
2	+1	−1	0	y_2
3	−1	+1	0	y_3
4	−1	−1	0	y_4
5	+1	0	+1	y_5
6	+1	0	−1	y_6
7	−1	0	+1	y_7
8	−1	0	−1	y_8
9	0	+1	+1	y_9
10	0	+1	−1	y_{10}
11	0	−1	+1	y_{11}
12	0	−1	−1	y_{12}
13,14,15	0	0	0	y_{13}, y_{14}, y_{15}

Table 12.4-7. Number of factors and experimental points for the Box-Behnken design with three replications in the center of each design

Number of factors	Number of experiments
3	15
4	27
5	46
6	54
7	62

Mixture designs. A special problem arises if additional relationships hold between the factors in an analytical investigation. This is true if mixtures, such as eluents in liquid chromatography or formulations in the pharmaceutical and textile industry, are under investigation. The constituents of a mixture given in portions of weight, volume, or moles are confined to the assumption that the amounts of the N constituents sum up to 100% or (normalized) to 1, i.e.,

$$\sum_{i=1}^{N} x_i = 1 \qquad \text{for} \quad x_i \geq 0 \tag{12.4-3}$$

The most important mixture design for an analyst is based on a so-called (k, d)-lattice. For the k factors the lattice describes all experimental points having the factor levels $0, 1/d, 2/d, \ldots, (d-1)/d$ or 1. In total the (k, d)-lattice has the following number of points:

$$\binom{d+k-1}{d} \tag{12.4-4}$$

Water

10

8   9

7   6   5

1   2   3   4

THF   ACN

Ternary mixtures of water, THF (tetrahydrofuran) and ACN (acetonitrile) as mobile phase compositions in HPLC

Worked example: Lattice design

Consider a (3,3)-lattice design as given in Fig. 12.4-9. The number of points is calculated by:

$$\binom{3+3-1}{3} = \binom{5}{3} = \frac{5!}{(5-3)!3!} = \frac{5 \cdot 4 \cdot 3 \cdot 2 \cdot 1}{2 \cdot 1 \cdot 3 \cdot 2 \cdot 1} = 10$$

The factor levels are $0, 1/3, 2/3, 1$, i.e., 0, 0.33, 0.66, 1.00, respectively. In HPLC the three constituents could be the solvents methanol, acetonitrile, and water of the mobile phase.

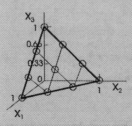

Fig. 12.4-9. Mixture design for 3 factors at 10 levels based on a (3,3)-lattice design

12.4.3 Response surface methods (RSM)

Response surface methods are very useful in order to quantify and interpret the relationships between responses and factor effects. In analytical chemistry the relationships can be based on physical or physicochemical models that are generalized by statisticians as so-called *mechanistic* models. Another way is empirical modeling where the parameters have no mechanistic meaning. General empirical models are polynomials of second order, where the response y is related to the variables (factors) x as follows:

$$y = b_0 + \sum_{i=1}^{k} b_i x_i + \sum_{1 \le i \le j}^{k} b_{ij} x_i x_j + \sum_{i=1}^{k} b_{ii} x_i^2 \qquad (12.4\text{-}5)$$

with k the number of variables (factors), b_0 the intercept parameter, and b_i, b_{ij}, b_{ii} regression parameters for linear, interaction and quadratic factor effects.

To estimate all parameters in Eq. 12.4-5 the experiments have to be carried out at three factor levels as discussed above for the response surface designs. The responses of experiments based on two factor levels can also of course be explored by RSM. However, no parameter estimates can then be obtained for curved (quadratic) factor effects, but only linear and interaction effects can be visualized.

The estimation of the empirical parameters is a general problem of least-squares estimation by linear models. The basics are introduced in Sec. 12.5.4. With the parameters in hand the model can be used to plot the response in dependence on the individual factors. The response surface method is explained here by further exploring the enzymatic determination of ceruloplasmin from the previous section about screening designs (cf. Table 12.5-4).

In the above study the factors pH value and PPD were found to significantly influence the enzyme-catalyzed oxidation of the substrate PPD. To study the relationship between the response (initial reaction *rate*) and the significant factors quantitatively a design at three levels, a Box-Behnken design, is run at a temperature of 37 °C. The concentration of the enzyme ceruloplasmin (CP) is included as a third factor in order to investigate the response characteristics of the analyte ceruloplasmin. Details of the Box-Behnken experiments are outline in Table 12.4-8. The three levels for each factor are given along with the concrete experimental design and the results of rate measurements.

> Small deviations from the factor levels -1, 0 and $+1$ will not significantly alter the statistical properties of a design.

Parameter estimation with a polynomial of second order reveals the following final model:

$$\text{rate} = 22.1298 + 5.48\,pH + 2.57\,PPD + 12.35\,CP + 4.73\,PPD \times CP$$
$$- 6.32\,PPD^2 - 5.02\,pH^2$$

Table 12.4-8. Factor levels and Box-Behnken design for studying the ceruloplasmin determination by RSM

Run Systematic	Run Randomized	PPD	pH	CP	Rate y, min^{-1}	Factor	-1	Level 0	$+1$
1	2	$+1$	$+1$	0.02	20.67	PPD, mM	0.5	14.3	27.3
2	12	$+1$	-1	0.02	12.67	pH	4.8	5.6	6.4
3	13	-1	$+1$	0.02	8.21	CP, mg L^{-1}	0.7	13.6	26.0
4	4	-1	-1	0.02	6.58				
5	6	$+1$	0	$+1$	37.2				
6	7	$+1$	0	-1	5.27				
7	1	-1	0	$+1$	14.95				
8	9	-1	0	-1	1.87				
9	11	0.03	$+1$	$+1$	33.63				
10	14	0.03	$+1$	-1	4.4				
11	10	0.03	-1	$+1$	26.02				
12	15	0.03	-1	-1	1.06				
13	5	0.03	0	0.02	23.86				
14	3	0.03	0	0.02	24.43				
15	8	0.03	0	0.02	23.29				

Coding of factor level:

$$x_i^* = \frac{x_i - M}{H}$$

where mid-range

$$M = \frac{high + low}{2}$$

and half-range

$$H = \frac{high - low}{2}$$

Decoding of factor level:

$$x_i = x_i^* \cdot H + M$$

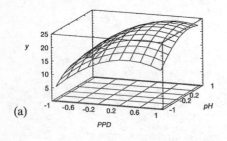

(a)

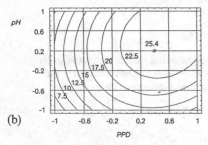

(b)

Fig. 12.4-10. Surface (a) and contour (b) plots of enzymatic rate versus the factors PPD concentration and pH. The contour lines in B represent iso-rate lines

From the empirical model, it can be concluded that the three factors pH, PPD, and the enzyme concentration (CP) show main and quadratic effects (cf. Eq. 12.4-5) on the rate of the reaction. In addition, there is a statistically significant interaction between the factors substrate (PPD) and enzyme (CP). In analytical terms this means that for determinations of the enzyme the substrate concentration should be kept constant with high precision.

Based on the mathematical model the response surfaces can be explored graphically. An example plot of the response rate in dependence on PPD concentration and pH is seen in Fig. 12.4-10a. The curved dependences in the direction of both factors lead to a maximum rate at coded levels of PPD of about 0.4 and of pH at 0.2. This relates to decoded levels of 16.6 mM PPD and a pH value of 5.95. Maxima are best found from the contour plots as represented in Fig. 12.4-10b.

Factor effects versus regression parameters

Although we have considered factor effect calculations and regression parameter estimation independently it is important to understand that both concepts are linked together. More exactly, the following relationship holds:

$$\text{regression coefficient} = \frac{\text{factor effect}}{\text{factor range}} \qquad (12.4\text{-}6)$$

Worked example: Comparison of factor effects and regression coefficients

Consider our enzyme example. Modeling the rate dependence by a polynomial that accounts for the same factors as in the screening design in Table 12.4-4 we will obtain the regression equation:

$$y = 10.04 + 2.88\,PPD + 1.084\,pH + 0.4288\,PPD \times pH$$

The insignificant regression parameters have been eliminated in this equation. If you now compare the factor effects D in Eq. 12.4-1 with the regression coefficients you will find that the latter are half as large as the factor effects, i.e.,

$$D_{PPD} = 5.76$$

$$D_{pH} = 2.17$$

$$D_{PPD \times pH} = 0.858.$$

This is because the coded factor range is for all three factors equal to 2 (from -1 to $+1$) and is therefore within rounding-off errors. In this example, the factor effects should be two times larger than the regression parameters.

Blocking of experiments

If a large number of experiments is to be carried out it will be difficult to run the experiments under identical conditions. During experimentation reagent charges might change or the activity of an enzyme might deteriorate. Often it will be necessary to interrupt experimentation during the night so that important changes in the experimental conditions may result.

To reflect systematic changes in such situations the sequence of experiments has up to now been randomized. Strong systematic changes, however, will increase the overall experimental error to a large extent. Elimination of these systematic changes can be accounted for if the changes are taken as a discrete event and the estimation of the time-dependent effects are confounded with the estimation of unimportant interactions such as a three factor interaction.

The experimental design is then divided into blocks. Considering the 2^3-design of Table 12.4-1 the blocks could be designed as the half-fraction designs given in Table 12.4-9.

Estimation of parameters for, e.g., the main effects would be performed by the following polynomial:

$$y = b_0 + b_1 x_1 + b_2 x_2 + b_3 x_3 + b^* x^* \qquad (12.4\text{-}7)$$

Table 12.4-9. Full factorial 2^3-design arranged in two blocks

Experiment		Factors			Response
	x_1	x_2	x_3	$x(x_1x_2x_3)$	
Block 1:					
1	−1	−1	−1	+1	y_1
2	+1	+1	−1	+1	y_2
3	+1	−1	+1	+1	y_3
4	−1	+1	+1	+1	y_4
Block 2:					
5	+1	−1	−1	−1	y_5
6	−1	+1	−1	−1	y_6
7	−1	−1	+1	−1	y_7
8	+1	+1	+1	−1	y_8

Because of the orthogonality of the experimental design the changes between the blocks will not influence the estimation of the parameters b_0 and b_i. In addition, an averaged measure for the changes between the two blocks can be derived from the size of $2b^*$.

12.4.4 Sequential optimization: simplex method

"Change and hope" is the name sometimes given to an optimization procedure where the individual factors are changed independently of each other. As long as no interaction effects of factors are valid the single-factor-at-a-time approach will succeed. In Fig. 12.4-11a this situation is explored for a response surface including curved factor effects. If one starts with variation of factor 2 keeping factor 1 at the coordinate value labeled (1) an optimal value will be found at the coordinate (2) for factor 2. In the next step factor 1 would be investigated at a constant factor 2 value at label (2) and the optimum would be found.

However, if interaction effects become valid the single-factor-at-a-time approach does not guarantee that the optimum is reached. As seen in Fig. 12.4-11b the change of factor 2 will result in an optimal coordinate at label (2) that would be fixed for changing factor 1. In this case the real optimum will never be found and the result remains suboptimal. The reason is that the ridge in Fig. 12.4-11b does not lie parallel to the factor axis. So changes of factor 1 are not independent on factor 2. Instead both factors interact and have to be considered simultaneously.

The most common sequential optimization method is based on the simplex method by Nelder and Mead. A simplex is a geometric figure having a number of vertices equal to one more than the number of factors. Therefore a simplex in one dimension is a line, in two dimensions a triangle, in three dimensions a tetrahedron, and in multiple dimensions a hyper-tetrahedron.

Fixed-size simplex

To find the steepest path along a response surface by means of the simplex method an algorithm has to be followed that consists of designing an initial simplex, running the experiments at the initial vertices and calculating the new vertex point by reflection of the vertex with the worst response. Movement with steps of fixed size is called the fixed-size simplex.

The algorithm works as follows (cf. Fig. 12.4-12):

- Generate the initial simplex according to the coded levels of the factors as given in Table 12.4-10
- Run the experiments at the initial simplex coordinates
- Decide from the responses which vertex represents the best response (vector b), the next-to-best (n), and the worst response (w)
- Calculate the new experimental point by:

$$r = p + (p - w) \tag{12.4-8}$$

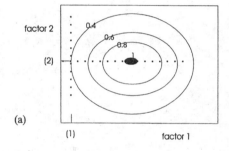

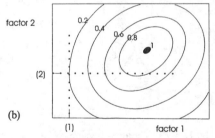

(a)

(b)

Fig. 12.4-11. Sequential optimization with single-factor-at-a-time strategy in case of a response surface without factor interactions (a) and for a surface with interactions (b)

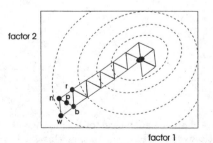

Fig. 12.4-12. Fixed-size simplex according to Nelder and Mead along an unknown response surface

Table 12.4-10. Choice of initial simplexes for up to 9 variables coded in the interval between 0 and 1

Experiments	x_1	x_2	x_3	x_4	x_5	x_6	x_7	x_8	x_9
1	0								
2	1	0							
3	0.50	0.87	0						
4	0.50	0.29	0.82	0					
5	0.50	0.29	0.20	0.79	0				
6	0.50	0.29	0.20	0.16	0.78	0			
7	0.50	0.29	0.20	0.16	0.13	0.76	0		
8	0.50	0.29	0.20	0.16	0.13	0.11	0.76	0	
9	0.50	0.29	0.20	0.16	0.13	0.11	0.094	0.75	0
10	0.50	0.29	0.20	0.16	0.13	0.11	0.094	0.083	0.75

where p is the centroid of the face remaining if the worst vertex x has been eliminated from the full simplex.

The centroid is calculated according to:

$$p = \frac{1}{n} \sum_{j \neq i}^{n} v_j \tag{12.4-9}$$

where n is the number of factors, i is the index of worst vertex to be eliminated, and j is the index of the vertex considered.

These steps are repeated as long as the simplex begins to rotate around the optimum or the response satisfies the experimenter's needs. The fixed step width of the fixed-size simplex may reveal problems if the step width is chosen either too large or too small. In the first case, the optimum might be missed and in the latter case the number of required experiments becomes very large. These disadvantages can be circumvented if the step width is tunable as with the variable-size simplex.

Variable-size simplex

With the variable-size simplex the step width is changed by expansion and contraction of the reflected vertices. The algorithm is modified as follows (cf. Fig. 12.4-13):

Reflexion: $r = p + (p - w)$ (12.4-10)

(a) if r is better than b: expand the simplex

$$e = p + \alpha(p - w) \tag{12.4-11}$$

with $\alpha > 1$, for example 1.5 for all directions or α is chosen different for each direction; e – new vertex after expansion

(b) if r lies between b and n: keep the simplex bnr

(c) if r is worse than n: contract the simplex according to:

　　1. r is worse than n but better than w, contract in "positive" direction:

$$c_+ = p + \beta(p - w) \quad \text{with} \quad 0 < \beta < 0.5 \tag{12.4-12}$$

　　2. r is worse than w, contract in "negative" direction:

$$c_- = p - \beta(p - w) \quad \text{with} \quad 0 < \beta < 0.5 \tag{12.4-13}$$

(d) At the experimental boundaries the simplex is reflected into the space of the experimental variables.

(e) Stop the simplex if the signal change is less than the experimental error or if the step width is less than a given threshold.

In practice, the simplex method is the most used experimental optimization algorithm. The main advantages are its simplicity, speed, and good convergence properties. Problems with the simplex method arise if multimodal response sur-

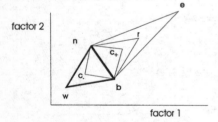

factor 2

factor 1

Fig. 12.4-13. Variable-size simplex

Local optima are typical for optimization of selectivity in HPLC separations. This is caused by changes of elution order of peaks in different mobile phases.

faces are investigated, i.e., if several local optima exist. In this case the simplex will climb the nearest local maximum or minimum and the global optimum might be missed. Mathematical theory provides more efficient optimization methods, such as the conjugate gradient method or Powells method. These methods, however, are mainly used in locating optima of mathematical functions and are scarcely used in experimental optimization.

Worked example: Simplex optimization

In this example the performance of the variable-size simplex is demonstrated for the enzyme determination based on the problem in Table 12.4-8. For a fixed enzyme concentration of $13.6\,mg\,L^{-1}$ ceruloplasmin (coded 0) the conentration of the substrate PPD and the pH value is sought for the maximum rate of the reaction, y. Since the simplex searches for a minimum the rate as the objective criterion has to be transformed. Here we use the difference to $100\,min^{-1}$, i.e., the objective criterion is $100 - y$.

The *initial simplex* is chosen according to the scheme in Table 12.4-10 in coded levels. The responses for the initial simplex are as follows:

| Vertex | Coded variable | | $100 - y$ |
	PPD	pH	
1	0.0	0.0	76.547
2	1.0	0.0	77.550
3	0.5	0.87	76.450

Fig. 12.4-14 demonstrates the initial simplex by bold lines.

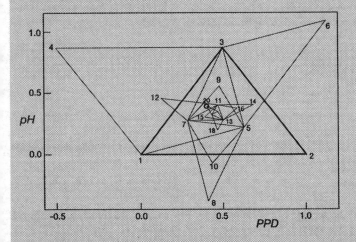

Fig. 12.4-14. Simplex search for optimum PPD concentration and pH value for the enzymatic determination of ceruloplasmin (cf. Table 12.4-8)

The best (minimum) response is found for vertex 3 (best response *b*), the next-to-best response is 1 (vertex *n*) and the worst response is found for vertex 2 (*w*). The latter vertex is eliminated and a new (reflected) vertex 4 is calculated based on the centroid (Eq. 12.4-9) according to Eq. 12.4-8:

centroid: $\quad p = \frac{1}{2}[(0,0) + (0.5, 0.87)] = (0.25, 0.435)$

reflexion: $\quad r = p + (p - w) = (0.25, 0.435) + [(0.25, 0.435) - (1, 0)] = (-0.5, 0.87)$

The new simplex consists now of the vertices 1, 3 and 4 having the following responses:

| Vertex | Coded variable | | $100 - y$ |
	PPD	pH	
1	0.0	0.0	76.547
3	0.5	0.87	76.450
4	−0.5	0.87	83.317

The vertex 4 (r) produces a worse response than the next-to-best vertex 1 and is even worse than w. Therefore, the simplex is contracted according to Eq. 12.4-13. At first the centroid is calculated without the worst vertex 4. Here the centroid is the same as in the first step:

centroid: $p = \frac{1}{2}[(0.0) + (0.5, 0.87)] = (0.25, 0.435)$

contraction: $c_- = p - \beta(p - w) = (0.25, 0.435) - 0.5[(0.25, 0.435) - (-0.5, 0.87)]$

$$= (0.625, 0.218)$$

The new simplex is 1, 3 and 5. The coordinates of this simplex and the measured responses are:

Vertex	Coded variable		$100 - y$
	PPD	pH	
1	0.0	0.0	76.547
3	0.5	0.87	76.450
5	0.625	0.218	75.131

After contraction the simplex is reflected to point 6 and again contracted to vertex 7 (cf. Fig. 12.4-13). If the calculation is continued the simplex moves in the direction of the optimum as given in the figure. After 20 iterations the following optimum is found in coded levels:

$PPD = 0.46$

$pH = 0.316$

$100 - y = 74.88$

Thus, the maximum rate $y = 100 - 74.88 = 25.12 \, \text{min}^{-1}$. This optimum compares well with the results from the response surface study in Fig. 12.4-10.

General reading

Box, G.E.P., Draper, N.R., *Empirical Model-building and response surfaces*. New York: Wiley, 1987.

Box, G.E.P., Hunter, W.G., Hunter, J.S., *Statistics for experimenters: An introduction to design, data analysis and model building*. New York: Wiley, 1987.

Deming, S.N., Morgan, S.L., *Experimental design: a chemometric approach*. Amsterdam: Elsevier, 1987.

Massart, D.L., Vandeginste, B.G.M., Deming, S.N., Michotte, Y., Kaufmann, L., *Chemometrics: a textbook*. Amsterdam: Elsevier Science Publishers, 1987.

Nelder, J.A., Mead, R., *A simplex method for function optimization*. Comput. J., 1965, 7, 308.

Questions and problems

1. Specify the following characteristics with a standard statistical software package
   - number of experiments
   - concrete design
   - randomized run sequence
   - possible blocking
   - alias structures
   for two-level designs: 2^9, 2^{7-3}, 2^{10-5} as well as
   for central composite designs based on 3 and 6 factors.
2. Give definitions for the following terms: unimodal, multimodal, replication of experiments, factor level, coding of factors, local and global optimum, randomization, screening designs, factor effects.

3. Graphical inspection of response surfaces is restricted to 3D-plots. How do you plot response surfaces if more the two factors are included in the study?
4. How can one avoid factors and their parameter estimates being confounded?
5. How can one account for interactions of factors in a polynomial model?
6. What difficulties may arise if experimental factors are changed one by one?
7. Why do chemists run their experiments mostly by one-factor-at-a-time methods?

12.5 Multivariate Methods

Learning objectives

- To evaluate and interpret analytical data from full chromatograms, spectra, depth profiles or electroanalytical records, from multidimensional detectors, and from samples for which concentrations of several chemical constituents or other properties have been measured
- To learn about methods for data preprocessing and for calculating distances and similarity measures
- To introduce grouping of analytical data based on unsupervised learning methods, i.e., projection methods and cluster analysis
- To handle multivariate data for which their class membership is determined by means of supervised pattern recognition approaches
- To model analytical relationships by multiple linear regression analysis, such as in multicomponent analysis or by target transform factor analysis

12.5.1 Fundamentals

Modern analytical instrumentation generates a vast amount of data. A digitized spectrum in the IR-spectral range consists of about 2000 wavenumber points. In a GC-MS experiment, it is not difficult to provide in a single run 600 000 data that amount to about 2.4 Mbytes of digital information. There are different methods for dealing with this extensive amount of information. One approach is ignorance. This means that quantitative analysis in spectroscopy is restricted to data evaluation at a single wavelength or the GC-MS trace is followed at a single mass unit. However, with the advent of computers in the analytical laboratory most of the multidimensional data are stored in instrument computers or external data bases so that important information might be wasted if only small fractions of data are evaluated. The extensive use nowadays of chemometrics enables the evaluation and interpretation of these data to be carried out efficiently as will be examined in the present chapter.

The main objectives for application of multivariate methods in analytical chemistry are aimed at the grouping and classification of objects (samples, compounds, or materials) as well as at modelling relationships between different analytical data. Some typical examples include:

(a) *Grouping or clustering* of rock samples according to their similar elemental pattern, or of material samples with respect to comparable chemical composition and technological properties.
(b) *Classification* of samples, such as rocks, materials, or chemical compounds, by means of analytical data (spectrum, chromatogram, or elemental pattern) on the basis of known class membership of those objects.
(c) Calibration of a single chemical constituent by means of full spectrum, or calibration of several components by means of mixture calibration techniques. In mathematical terms these are problems of *parameter estimation* where the parameters represent the calibration coefficients.

To understand the multidimensionality of analytical problems let us consider the interpretation of results from a clinical analytical laboratory. The calcium content of serum can be used as an indicator for diagnosing different diseases. At high calcium levels there might be kidney insufficiency, and at extraordinary low levels of calcium resorption problems may occur. The comparison of calcium levels is a one-dimensional problem that can be solved by direct comparison of the individual values. As a second parameter, one might wish to include the actual phosphate levels of the patients into the comparison. Here we know that low calcium levels correspond with high phosphate levels, and that high calcium levels correspond with low phosphate levels as shown in Fig. 12.5-1.

Clustering and classification methods are often summarized by the notion **pattern recognition**.

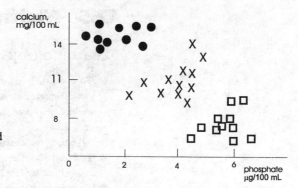

Fig. 12.5-1. Joint visualization of calcium and phosphate levels in blood serum for normal patients (x), for kidney insufficiency (●), and for resorption problems (□)

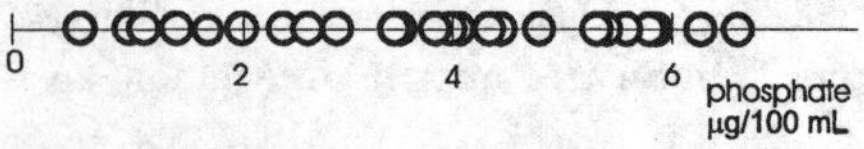

Fig. 12.5-2. "Discrimination ability" of a single variable for different groups of patients

Multivariate data constitute a row of a data matrix.

Graphically we could add a third parameter, e.g., the blood pH-value. Visualization, however, of additional parameters that are to be considered simultaneously, e.g. the glucose, copper or bilirubin levels, visualization is hampered by our 3-dimensional world, i.e., one cannot see the higher dimensions. Thus, one of the main goals will be to use multivariate methods to enable visualization of data for which more than three variables have been measured. The mathematical basis will be representation and manipulation of data by vectors and matrices. Distances and similarities in multidimensional space have to be defined, and finally methods for projection of multivariate data onto two or three-dimensional space are needed.

Another reason for multivariate measurements is the separability of data clusters on the basis of the feature variables. At the beginning of a study it is not known how to discriminate between different diseases. If the phosphate level has been studied first as shown in Fig. 12.5-2, then by exploring this single variable, no complete discrimination will be possible. By considering a second variable, here the calcium level, we arrive at Fig. 12.5-1, where a clear distinction between different groups of patients can be observed even without knowing the membership of samples to a certain cluster. Thus, measuring more variables can improve the *discrimination ability* (Fig. 12.5-2).

In general, the analytical data can be arranged as a data matrix X of N objects (rows) and K features (columns). The objects might be samples, molecules, materials, findings, or fertilizers. Typical features or variables of those objects will be elemental patterns, spectra, structural features, or physical properties. The $N \times K$ data matrix X can be written as follows:

Object	Features			
	1	2	...	K
1	x_{11}	x_{12}	...	x_{1K}
2	x_{21}	x_{22}	...	x_{2K}
3	x_{31}	x_{32}	...	x_{3K}
⋮				
N	x_{N1}	x_{N2}	...	x_{NK}

For some problems these data are divided into dependent and independent variables, e.g., for calibrating concentrations on spectra. The dependent variables are then renamed, for example, by the character y.

A *class* comprises a collection of objects which have similar features. A *pattern* of an object is its collection of characteristic features. For multivariate data evaluation, not all objects and features are necessarily used. On the other hand, some of the available data cannot be used as they are reported. Therefore, pretreatment of data is a prerequisite for efficient multivariate data anlaysis.

Preprocessing of data

Missing data, centering, and scaling

In the first step, the data have to be reviewed with respect to completeness. *Missing data* do not hinder mathematical analysis. Of course, missing data should not be replaced by zeros. Instead the vacancies should be filled up either by the column/row mean or in the worst case by generating a random number in the range of the considered column/row. *Features* can be removed from the data set if they are highly correlated with other features, or if they are redundant or constant.

To eliminate a constant offset the data can be translated along the coordinate origin. The common procedure is *mean centering*, where each variable x_{ik} is centered by subtracting the column mean $\bar{x}_k$ according to:

$$x_{ik}^* = x_{ik} - \bar{x}_k \qquad (12.5\text{-}1)$$

where i is the row index, and k column index.

Very often the features represent quite different properties of a sample or of an object, so that the metric might differ from column to column to a great extent. This may imply different absolute values of the variables as well as different variable ranges (variances). Both types of distortions will affect most of the statistically based multivariate methods. Elimination of these differences can be carried out by *scaling* the data to similar ranges and variances. Two scaling methods are important, i.e., scaling the data by range or by standard deviation (autoscaling):

Range scaling:

$$x_{ik}^* = \frac{x_{ik} - x_k(min)}{x_k(max) - x_k(min)} \quad 0 \le x_{ik}^* \le 1 \qquad (12.5\text{-}2)$$

Autoscaling:

$$x_{ik}^* = \frac{x_{ik} - \bar{x}_k}{s_k} \quad \text{with} \quad s_k = \sqrt{\frac{\sum\limits_{i=1}^{N}(x_{ik} - \bar{x}_k)}{N - 1}} \qquad (12.5\text{-}3)$$

where N is the number of objects.

Since autoscaling reveals data with zero mean and unit variance this method is usually the method of choice for scaling the data. Figure 12.5-3 gives a graphical impression for centering and autoscaling.

Rotation

Transformation of the original data to a new coordinate system is another possibility of data pretreatment. The methods are based on principal component analysis or factor analysis. The aim is to expand the original data set into as many components as correspond to the number of rows and columns of the data matrix. The original coordinates are then expressed in new orthogonal variables. Since this objective is closely related to the projection of the original data set onto orthogonal variables that span a space of lower dimensionality, the methods concerning unsupervised pattern recognition are introduced in Sec. 12.5.2.

12.5.2 Unsupervised methods

Grouping of analytical data is possible either by means of clustering methods or by projecting the high dimensional data onto lower dimensional space. Since there is no supervisor in the sense of known membership of objects to classes these methods are performed in an unsupervised manner.

Projection methods

These methods are aimed at projecting the original data set from a high dimensional space onto a line, a plane, or a 3D-coordinate system. Perhaps the best way would be to have a mathematical procedure that allows you to sit before the computer screen pursuing the rotation of the data into all possible directions and stopping this process when the best projection, i.e., optimal clustering of data

Coded data are obtained by multiplying, dividing, adding and/or subtracting a constant in order to convert the original data into more convenient values.

In some cases, normalization of a data vector to length 1 is an important preprocessing procedure:

$$x_{ik}^* = \frac{x_{ik}}{\|x_k\|}$$

where $\|x_k\| = \sqrt{x_{1k}^2 + x_{2k}^2 + \cdots x_{Nk}^2}$

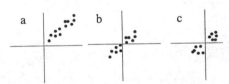

Fig. 12.5-3. Demonstration of translation and scaling procedures: the original data (a) are centered in (b) and autoscaled in (c). Notice that the autoscaling decreases the between-group distance in the direction of greatest within-group scatter, and increases it in the perpendicular direction in the sense of sphericization of groups

The *dispersion matrix* describes the scatter of multivariate data around the mean. For centered data the dispersion matrix equals $X^t X$.

groups, has been found. In fact, such methods of projection pursuits already exist in statistics and are tested within the field of chemometrics.

At present, data projection is performed mainly by methods called principal component analysis (PCA), factor analysis (FA), singular value decomposition (SVD), eigenvector projection, or rank annihilation. The different methods are linked to different science areas. They also differ mathematically in the way the projection is computed. i.e., which dispersion matrix is the basis for data decomposition, which assumptions are valid, and whether the method is based on eigenvector analysis, singular value decomposition, or on other iterative schemes.

An explanation of projection methods is based here on PCA in comparison to SVD. Similar methods, such as factor analysis, can be found in Sec.12.5.4.

The key idea of PCA is to approximate the original matrix X by a product of two small matrices – the score and loading matrices – according to:

$$X = TP^t \tag{12.5-4}$$

$$_N\boxed{X}^K = {_N}\boxed{T}^A {_A}\boxed{P^t}^K + {_N}\boxed{E}^K$$

where X is the original data matrix consisting of N rows (objects) and K columns (features); T is the scores matrix with N rows and A columns (number of principal components); P is the loading matrix with K columns and A rows; E is the error matrix that is $N \times K$; and t is the transpose of a matrix.

Principal components in PCA or common factors in factor analysis are sometimes called *latent variables*.

In other words, the projection of X down on an A-dimensional *subspace* by means of the projection matrix P^t gives the object coordinates in this plane, T. The columns in T are the score vectors and the rows in P are called loading vectors. Both vectors are orthogonal, i.e., $p_i^t p_j = 0$ and $t_i^t t_j = 0$, for $i \neq j$.

The simplest method for PCA is based on the NIPALS algorithm. Details of the algorithm are given in the following worked example.

Worked example: Iterative principal component analysis by the NIPALS algorithm

Determine principal components for the following data matrix based on the iterative NIPALS algorithm:

$$X = \begin{pmatrix} 2 & 1 \\ 3 & 2 \\ 4 & 3 \end{pmatrix}$$

1. Standardization of data:

$$X = \begin{pmatrix} -\dfrac{1}{\sqrt{2}} & -\dfrac{1}{\sqrt{2}} \\ 0 & 0 \\ \dfrac{1}{\sqrt{2}} & \dfrac{1}{\sqrt{2}} \end{pmatrix}$$

2. Estimation of the loading vector p^t. Usually the first of the X-matrix is used:

$$p^t = \begin{pmatrix} -\dfrac{1}{\sqrt{2}} & -\dfrac{1}{\sqrt{2}} \end{pmatrix}$$

3. Computation of the new score vector t:

$$t = Xp = \begin{pmatrix} -\dfrac{1}{\sqrt{2}} & -\dfrac{1}{\sqrt{2}} \\ 0 & 0 \\ \dfrac{1}{\sqrt{2}} & \dfrac{1}{\sqrt{2}} \end{pmatrix} \begin{pmatrix} -\dfrac{1}{\sqrt{2}} \\ -\dfrac{1}{\sqrt{2}} \end{pmatrix} = \begin{pmatrix} 1 \\ 0 \\ -1 \end{pmatrix}$$

Comparison of the new t-vector with the old one. If the deviations of the elements of the two vectors are within a given threshold of 10^{-z}, e.g., $z = 5$, then continue at step 6, otherwise go to step 4.

4. Compute new loadings p^t:

$$p^t = t^t X = (1 \quad 0 \quad -1) \begin{pmatrix} -\dfrac{1}{\sqrt{2}} & \dfrac{1}{\sqrt{2}} \\ 0 & 0 \\ \dfrac{1}{\sqrt{2}} & \dfrac{1}{\sqrt{2}} \end{pmatrix} = \left(-\dfrac{2}{\sqrt{2}} \quad -\dfrac{2}{\sqrt{2}} \right)$$

Normalize the loading vector to length 1:

$$p^t = \dfrac{p^t}{\|p^t\|} = \left(-\dfrac{1}{\sqrt{2}} \quad -\dfrac{1}{\sqrt{2}} \right)$$

5. Continue at 3 if the number of iterations does not exceed a redefined threshold, e.g. 100; otherwise go to step 6.
6. Determine the matrix of residuals:

$$E = X - tp^t = \begin{pmatrix} -\dfrac{1}{\sqrt{2}} & \dfrac{1}{\sqrt{2}} \\ 0 & 0 \\ \dfrac{1}{\sqrt{2}} & \dfrac{1}{\sqrt{2}} \end{pmatrix} - \begin{pmatrix} 1 \\ 0 \\ -1 \end{pmatrix} \left(-\dfrac{1}{\sqrt{2}} - \dfrac{1}{\sqrt{2}} \right) = \begin{pmatrix} 0 & 0 \\ 0 & 0 \\ 0 & 0 \end{pmatrix}$$

If the number of principal components is equal to the number of previously fixed principal components or of cross-validated components go to step 8. Otherwise continue at 7.
7. Use the residual matrix E as the new X-Matrix and compute additional principal components t and loadings p^t at step 1.
8. As a result, the matrix X is represented by a principal component model according to Eq. 12.5-4, i.e.,

$$X = TP^t = \begin{pmatrix} 1 \\ 0 \\ -1 \end{pmatrix} \left(-\dfrac{1}{\sqrt{2}} \quad -\dfrac{1}{\sqrt{2}} \right)$$

The actual two-dimensional data can be described by just one principal component.

With real data, more principal components are necessary. Therefore, there are more columns of scores in the T-matrix and more rows in the P^t-matrix representing the loadings.

The NIPALS method represents one algorithm for matrix diagonalization. Others are SVD, the power method or bidiagonalization algorithms, such as the PLS method.

To outline some different approaches let us compare it with SVD. The square roots of eigenvectors are called singular values. Thus SVD is basically a kind of eigenvector analysis. The formulation for SVD is as follows:

$$X = UWV^t \tag{12.5-5}$$

$$
\begin{bmatrix}
x_{11} & x_{12} & \cdots & x_{1K} \\
x_{21} & x_{22} & \cdots & x_{2K} \\
\vdots & & & \\
x_{N1} & x_{N2} & \cdots & x_{NK}
\end{bmatrix}
$$

$$
= \begin{bmatrix}
u_{11} & u_{12} & \cdots & u_{1A} \\
u_{21} & u_{22} & \cdots & u_{2A} \\
\vdots & & & \\
u_{N1} & u_{N2} & \cdots & u_{NA}
\end{bmatrix}
\times
\begin{bmatrix}
w_{11} & 0 & 0 \\
0 & w_{22} & \cdots & 0 \\
\vdots & & & \\
0 & 0 & \cdots & w_{AA}
\end{bmatrix}
$$

$$
\times
\begin{bmatrix}
v_{11} & v_{21} & \cdots & v_{K1} \\
v_{12} & v_{22} & \cdots & v_{K2} \\
\vdots & & & \\
v_{1A} & v_{2A} & \cdots & v_{KA}
\end{bmatrix}
$$

In SVD the columns of U and V are the left and right singular (eigen) vectors. In the case of $N = K$ (square matrix) the singular values coincide with the eigenvalues of X, and $U = V$, containing the eigenvectors of X.

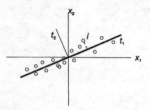

Fig. 12.5-4. Projection of a swarm of objects from the original two dimensions onto one dimension, i.e., the score vector t_1

Table 12.5-1. Elemental contents of hair samples given in ppm (parts per million)

Hair-No.	Cu	Mn	Cl	Br	I
1	9.2	0.30	1730	12.0	3.6
2	12.4	0.39	930	50.0	2.3
3	7.2	0.32	2750	65.3	3.4
4	10.2	0.36	1500	3.4	5.3
5	10.1	0.50	1040	39.2	1.9
6	6.5	0.20	2490	90.0	4.6
7	5.6	0.29	2940	88.0	5.6
8	11.8	0.42	867	43.1	1.5
9	8.5	0.25	1620	5.2	6.2

where U contains the same column vectors as does T in Eq. 12.5-4 but normalized to length one; W is a diagonal matrix containing the square roots of the eigenvalues or singular values; if small singular values are not truncated then $A = N$; and V^t is identical to P^t.

Geometric interpretation of the projection onto principal components is given in Fig. 12.5-4. The original data exemplified by the variable coordinates x_1 and x_2 are represented by a swarm with N points in one, two, in general, K-dimensional space.

The objects are projected orthogonally onto the new principal component axes, which in the present case, is one dimension. The distance of the projected object i from the centroid (after centering from the coordinate origin) is the score value t_{i1} of the object i.

The principal component axes are orthogonal to each other. Most of the variance of the data is contained in the first principal component. In the second component there is more information than in the third one, etc. For interpretation of the projected data both the score and loading vectors are plotted. In the score plots, the grouping of objects can be recognized. A loading plot reveals the importance of the individual variables with respect to the principal component model.

As an example, consider the elemental data of hair (features) measured on 9 subjects (objects) in Table 12.5-1.

These data constitute the X-matrix, i.e.,

$$X = \begin{bmatrix} 9.2 & 0.30 & 1730 & 12.0 & 3.6 \\ 12.4 & 0.39 & 930 & 50.0 & 2.3 \\ 7.2 & 0.32 & 2750 & 65.3 & 3.4 \\ 10.2 & 0.36 & 1500 & 3.4 & 5.3 \\ 10.1 & 0.50 & 1040 & 39.2 & 1.9 \\ 6.5 & 0.20 & 2490 & 90.0 & 4.6 \\ 5.6 & 0.29 & 2940 & 88.0 & 5.6 \\ 11.8 & 0.42 & 867 & 43.1 & 1.5 \\ 8.5 & 0.25 & 1620 & 5.2 & 6.2 \end{bmatrix} \quad (12.5\text{-}6)$$

Graphical representation of these data is not possible since 5 dimensions would be necessary. In the first step the variables are usually autoscaled to unit variance and zero mean (cf. Eq. 12.5-3) as demonstrated above in the NIPALS algorithm.

Estimating the number of principal components
In a first step of PCA the number of components in the PCA model has to be decided. Several criteria exist:

- percentage of explained variance
- eigenvalue-one criterion
- Scree-test
- cross validation

The percentage of *explained variance* is applied in the sense of a heuristic criterion (cf. Eq. 12.5-91). It can be used if enough experience is gained by analyzing similar data sets. If all possible principal components are used in the model the variance can be explained by 100%. Usually a fixed percentage of explained variance is specified, e.g., 90%.

In our example of hair data in Table 12.5-1, 90.7% of the data variance can still be explained by two principal components (Table 12.5-2).

The *eigenvalue-one criterion* is based on the fact that the average eigenvalue of autoscaled data is just one. In this case only eigenvalues greater than 1 are considered important. According to this criterion the eigenvalues of the hair data in Table 12.5-2 reveal two significant principal components.

The *Scree-test* is based on the phenomena that the residual variance levels off when the proper number of principal components is obtained. Visually the residuals or more often the eigenvalues are plotted against the number of components in a Scree-plot. The component number is then derived from the leveling-off in this

Table 12.5-2. Eigenvalues and explained variances for the hair data in Table 12.5-1

Component	Eigenvalue λ	Explained variance %	Cumulative variance %
1	3.352	67.05	67.05
2	1.182	23.65	90.70
3	0.285	5.70	96.40
4	0.135	2.70	99.10
5	0.045	0.90	100.00

dependence. Figure 12.5-5 demonstrates the Scree-plot for the hair data. The slope can be seen to flatten between the second and third component.

The fourth method for deciding on the number of principal components is *cross validation*. In the simplest case, every object of the *X*-matrix is removed from the data set once, and a model with the remaining data is computed. Then the removed data are predicted by means of the PCA model and the sum of the square root of residuals over all removed objects is calculated. In case of large data sets, the leave-one-out method can be replaced by leaving out groups of objects.

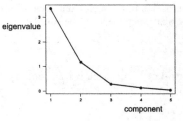

Fig. 12.5-5. Scree-plot for the principal component model of the hair data of Table 12.5-1

Graphical interpretation of principal components
Interpretation of the results of a principal component analysis is usually carried out by visualization of the component scores and loadings. Sometimes the data can be interpreted from a single component. Commercial software provides two- or three-dimensional plot facilities. In order to recognize groups or clusters in the data the PC scores are plotted against each other. Figure 12.5-6 demonstrates the score plot for the first two principal components of the hair data of Table 12.5-1.

In the score plot, the linear projection of objects is found, representing the main part of the total variance of the data. As can be seen, there are three clusters with three objects (hair) each. These objects belong to the hair of different persons.

Correlation and importance of feature variables is to be decided from plots of the PC loadings. Figure 12.5-7 demonstrates the loading plot of the first two components for the hair data of Table 12.5-1.

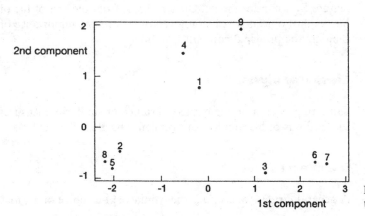

Fig. 12.5-6. Principal component scores for the hair data in Table 12.5-1

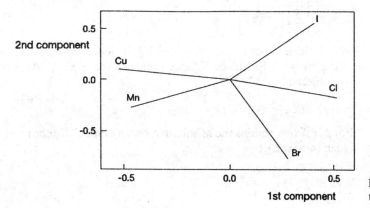

Fig. 12.5-7. Principal component loadings of the hair data in Table 12.5-1

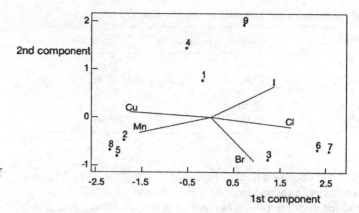

Fig. 12.5-8. Biplot for the simultaneous characterization of the scores and loadings of two principal components of the hair data in Table 12.5-1

Karhunen-Loeve expansion is synonymous with principal component analysis.

The loading plot provides the projection of the features onto the principal components. From this plot, information about the *correlation* of feature variables can be deduced. The correlation of features is described by the cosine of the angle between the loading vectors. The smaller the angle, the higher is the correlation between features. Uncorrelated features are orthogonal to each other. If variables are highly correlated then it is sufficient to measure only one out of the correlated variables.

The size of the loadings in relation to the considered principal component is a measure of the *importance* of a feature for the PC model. Loadings in the origin of the coordinate system represent unimportant features.

In our hair data example (Fig. 12.5-7) the elements Cu and Mn are correlated higher than the halogens Br, Cl, and I. For Cu and I anticorrelation is observed. All elements are important for describing the first principal component. The second component is mainly characterized by the elements I and Br.

A joint interpretation of scores and loadings is possible if the loadings are properly scaled and superimposed on the score plots. The so-called *biplot* is given for the present example in Fig. 12.5-8. From the loading direction the discriminating ability of variables can be deduced. In our example the features Cu, Mn, I, and Cl separate the object clusters into the groups (2,8,5) and (3,6,7) whereas the feature Br separates the left cluster (2,5,8) from the rest of the objects. The neighborhood of objects to a loading vector reflects the importance of that variable for building the principal component model.

Cluster analysis

Clustering of data on the basis of distance or similarity measures is carried out by the methods of hierarchical or nonhierarchical cluster analysis.

Distance measures

A general *distance measure* is the Minkowski-distance or L_p-metric:

$$d_{ij} = \left[\sum_{k=1}^{K} |x_{ik} x_{jk}|^p \right]^{1/p} \tag{12.5-7}$$

with K the number of variables, and i, j the index for objects i and j, respectively.

Typically the Euclidian distance is chosen where $p = 2$. As an example consider two objects:

$$d_{12} = [(x_{11} - x_{21})^2 + (x_{12} - x_{22})^2]^{1/2}$$

For $p = 1$ one obtains the Manhatten or 'city block' distance, i.e., the way round a block in Manhatten:

$$d_{ij} = \sum_{k=1}^{K} |x_{ik} - x_{jk}| \tag{12.5-8}$$

Graphical demonstrations for the two distance measures are given in Fig. 12.5-9. A disadvantage of all L_p-metric measures is the dependence of the calculated distances on the units of measures. Therefore, scaling of variables is often unavoidable if these measures can be used.

A measure that considers the different scales of variables and takes into account the correlation among those variables by definition is the *Mahalanobis-distance*. This invariant measure has the following formulation:

$$D_{ij}^2 = (x_i - x_j)^t C^{-1} (x_i - x_j) \tag{12.5-9}$$

where C is the covariance matrix as explained in Sec.12.5.4, and x_i, x_j are column vectors for objects i and j, respectively.

Thus, normalization of data becomes unnecessary if the Mahalanobis-distance is used as the distance measure. In addition, distortion of ratios that arise from correlations of features or feature groups can be avoided. For example, if the Euclidian distance is used, two highly correlated variables are counted even though they described the same information.

Complementary to distance measures are similarity measures. If we base the similarity measure on the Minkowski measure we can formulate the following similarity measure S_{ij}:

$$S_{ij} = 1 - \frac{d_{ij}}{d_{ij}(max)} \tag{12.5-10}$$

where $d_{ij}(\text{max})$ is the maximum distance in the whole distance matrix.

A distance measure that considers the standard deviation of a variable j, s_j, is the Pearson distance:

$$d_{ij} = \sqrt{\frac{\sum_{k=1}^{K}(x_{ik} - x_{jk})^2}{s_j^2}}$$

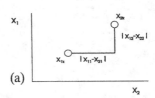

(a)

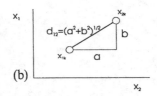

(b)

Fig. 12.5-9. City block distance (a) versus Euclidian distance (b) for two variables x_1 and x_2

Hierarchical clustering

Formation of object cluster can be performed in a hierarchical way, i.e., the objects are aggregated according to their distance of similarity in ascending order. To understand the whole procedure we will follow a clustering example based on the clinical data on calcium and phosphate. A selection of data for six objects is given in Table 12.5-2.

In the first step, the distance matrix for the whole data set is computed based on a distance measure. In principle, all the distance measures can be used as described above. Here the Euclidian distance according to Eq. 12.5-7 with $p = 2$ is applied. As an example, the distance between object number 1 and 2 is calculated by:

$$d_{12} = [(8 - 8.25)^2 + (5.5 - 5.75)^2]^{1/2} = 0.354$$

In the distance matrix every object is compared to all remaining objects by computing the Euclidian distance. Of course, the distance between an object and itself will be zero:

Table 12.5-2. Example data for diagnosing a disease based on the calcium and phosphate levels of blood serum

Object (sample)	Features	
	calcium, mg/100 mL	phosphate, mg/100 mL
1	8.0	5.5
2	8.25	5.75
3	8.7	6.3
4	10.0	3.0
5	10.25	4.0
6	9.75	3.5

Object	1	2	3	4	5	6
1	0	0				
2	0.354	0				
3	1.063	0.711	0			
4	3.201	3.260	3.347	0		
5	2.704	2.658	2.774	1.031	0	
6	2.658	2.704	2.990	0.559	0.707	0

Reduction of the distance matrix is carried out by aggregation of objects. As a rule the objects with the shortest distances are aggregated first. Several aggregation operators are used as will be discussed below. For our example, we shall use the method of average linkage, where the objects are combined by averaging the calculated distances. The following steps demonstrate the stepwise aggregation procedure.

First reduced matrix: The shortest distance in the distance matrix is between objects 1 and 2, where $d_{12} = 0.354$. The two objects are combined to a new object

1* with a distance set to zero. The new distance values are calculated by averaging the single distances according to:

$$d_{1 \cdot 3} = \frac{d_{13} + d_{23}}{2} = \frac{1.063 + 0.711}{2} = 0.887$$

$$d_{1 \cdot 4} = \frac{d_{14} + d_{24}}{2} = \frac{3.202 + 3.260}{2} = 3.231$$

$$d_{1 \cdot 5} = \frac{d_{15} + d_{25}}{2} = \frac{2.704 + 2.658}{2} = 2.681$$

$$d_{1 \cdot 6} = \frac{d_{16} + d_{26}}{2} = \frac{2.658 + 2.704}{2} = 2.681$$

For the reduced matrix we obtain:

Object	1*	3	4	5	6
1*	0				
3	0.887	0			
4	3.231	3.347	0		
5	2.681	2.774	1.031	0	
6	2.681	2.990	0.559	0.707	0

Second reduced matrix: The shortest distance in the remaining matrix is now between objects 4 and 6 with $d_{46} = 0.559$. The objects 4 and 6 are joined to object 4* and set to zero. Averaging the d-values of that row reveals the following distances and a new distance matrix:

$$d_{54^*} = \frac{d_{54} + d_{56}}{2} = \frac{1.031 + 0.707}{2} = 0.869$$

$$d_{4^* 3} = \frac{d_{43} + d_{63}}{2} = \frac{3.547 + 2.990}{2} = 3.269$$

$$d_{4^* 1^*} = \frac{d_{41^*} + d_{61^*}}{2} = \frac{3.231 + 2.681}{2} = 2.956$$

Object	1*	3	4*	5
1*	0			
3	0.887	0		
4*	2.956	3.269	0	
5	2.681	2.774	0.869	0

Third reduced matrix: The minimum distance value is now $d_{54}^* = 0.869$ so that a new object 5* is formed and the following new distances are calculated:

$$d_{1 \cdot 5^*} = \frac{d_{51^*} + d_{4^* 1^*}}{2} = \frac{2.681 + 2.956}{2} = 2.819$$

$$d_{35^*} = \frac{d_{4^* 3} + d_{53}}{2} = \frac{0.887 + 2.774}{2} = 1.831$$

Object	1*	3	5*
1*	0		
3	0.877	0	
5*	2.819	1.831	0

Fourth reduced matrix: The objects 1* and 3 now give the lowest distance value, i.e., $d_{1*3} = 0.887$. These objects are aggregated to object 3* with the new distance to the remaining object 5*:

$$d_{3*5*} = \frac{d_{5*1*} + d_{5*3}}{2} = \frac{1.831 + 2.274}{2} = 2.325$$

Object	3*	5*
3*	0	
5*	2.547	0

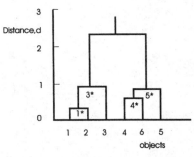

Graphically the distances can be represented in a *dendrogram* as shown for the example in Fig. 12.5-10. Decision about the number of clusters can be made for different reasons. Often a fixed number of clusters is to be assumed. This would be the case in our example from clinical chemistry, where the classes are dictated by the number of diseases to be diagnosed. Sometimes a distance measure or an allowed difference between classes is used for evaluating the cluster numbers.

Fig. 12.5-10. Dendrogram for the clinical analytical data of Table 12.5-1

For aggregating the clusters we have used the average linkage method up to now. The general form for calculating the distances between the new objects or cluster k (in the example labeled by a star) and another object i is carried out by averaging the individual distances of objects A and B to object i:

(a) *Average linkage*

$$d_{ki} = \frac{d_{Ai} + d_{Bi}}{2} \tag{12.5-11}$$

Other aggregation methods exemplified for combining two clusters are the following:

(b) *Single linkage*
Here the shortest distance between the clusters is computed according to:

$$d_{ki} = \frac{d_{Ai} + d_{Bi}}{2} - \frac{|d_{Ai} - d_{Bi}|}{2} = min(d_{Ai}, d_{Bi}) \tag{12.5-12}$$

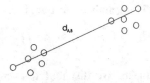

(c) *Complete linkage*

$$d_{ki} = \frac{d_{Ai} + d_{Bi}}{2} + \frac{|d_{Ai} - d_{Bi}|}{2} = max(d_{Ai}, d_{Bi}) \tag{12.5-13}$$

This method is based on the widest distance between the points of the clusters:

(d) *Weighted average linkage*
The number of objects is used for weighting the distances of the clusters:

$$d_{ki} = \frac{N_A}{N} d_{Ai} + \frac{N_B}{N} d_{Bi} \quad \text{with } N = N_A + N_B \tag{12.5-14}$$

(e) *Centroid*

The centroid calculated as the mean of a cluster is used as the basis for aggregating:

$$d_{ki} = \frac{N_{Ai}}{N} d_{Ai} + \frac{N_B}{N} d_{Bi} - \frac{N_A N_B}{N^2} d_{AB} \tag{12.5-15}$$

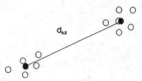

(f) *Median*

The centroid can also be computed as the median value according to:

$$d_{ki} = \frac{d_{Ai}}{2} + \frac{d_{Bi}}{2} - \frac{d_{AB}}{4} \tag{12.5-16}$$

(g) *Ward's method*

This method combines those two clusters that reveal by their combination the minimum increase in the total error or squared error within the individual groups:

$$d_{ki} = \frac{N_A + N_i}{N + N_i} d_{Ai} + \frac{N_B + N_i}{N + N_i} d_{Bi} - \frac{N_i}{N + N_i} d_{AB} \tag{12.5-17}$$

Which of the aggregation methods provides the best clustering of data cannot be stated in advance. In practice, the different methods should be tested on the concrete multivariate data.

Nonhierarchical clustering

For nonhierarchical clustering an initial distribution of the different objects is assumed, the membership of each object to the clusters is checked, and the objects are redistributed again. As a general procedure, which can likewise be applied to classical cluster analysis as well as to fuzzy clustering, we consider the *c-means algorithm*.

The objects are partitioned in subsets S_i with $i = 1, \ldots, c$, where c is the number of clusters. The membership of an object with feature vector x_k to cluster i can be expressed by a membership function

$$m_{ik} = m_{Si}(x_k) \tag{12.5-18}$$

where, for *crisp* sets the membership value is either 0 or 1 and for *fuzzy* sets the membership value could assume values in the interval [0,1].

A matrix $M = [m_{ik}]$ is called *c-partition* if the following conditions are fulfilled:

(a) The membership of the N objects to the clusters can be either crisp or fuzzy, i.e.,

$$m_{ik} \in \{0, 1\} \; or \; [0, 1] \qquad 1 \le i \le c, 1 \le k \le N \tag{12.5-19}$$

(b) The sum of memberships of objects to a specific partition is equal to one for crisp sets, and in the case of fuzzy sets the sum is normalized to one, i.e.,

$$\sum_{i=1}^{c} m_{ik} = 1 \quad 1 \le k \le N \tag{12.5-20}$$

(c) Within a particular partition the available objects are distributed over all clusters, i.e., each cluster of a partition contains at least one object. On the other hand there will be $N - 1$ objects at maximum in a single cluster.

$$0 < \sum_{k=1}^{N} m_{ik} < N \quad 1 \le i \le c \tag{12.5-21}$$

Consider for the three objects x_1, x_2, and x_3 some 2-partitions:

$$M_1 = \begin{bmatrix} x_1 & x_2 & x_3 \\ 1 & 1 & 0 \\ 0 & 0 & 0 \end{bmatrix} \quad M_2 = \begin{bmatrix} x_1 & x_2 & x_3 \\ 1 & 1 & 0 \\ 0 & 1 & 1 \end{bmatrix} \quad M_3 = \begin{bmatrix} x_1 & x_2 & x_3 \\ 1 & 0 & 0 \\ 0 & 1 & 1 \end{bmatrix}$$

$$M_4 = \begin{bmatrix} x_1 & x_2 & x_3 \\ 1 & 1 & 1 \\ 0 & 0 & 0 \end{bmatrix} \quad M_5 = \begin{bmatrix} x_1 & x_2 & x_3 \\ 1 & 0 & 1 \\ 0 & 1 & 0 \end{bmatrix}$$

Each row represents a particular cluster and the columns characterize the objects as assigned to different clusters. There are only two genuine 2-partitions, i.e., M_3 and M_5. In partition M_1 object x_3 is not partitioned at all. In M_2 the object x_2 is distributed twice and in M_4 the second partition (second row) contains no object at all.

As an example of fuzzy c-partitions we have the following partitions for three objects:

$$M_1 = \begin{bmatrix} x_1 & x_2 & x_3 \\ 1 & 0.5 & 0 \\ 0 & 0.5 & 1 \end{bmatrix} \quad M_2 = \begin{bmatrix} x_1 & x_2 & x_3 \\ 0.8 & 0.5 & 0.2 \\ 0.2 & 0.5 & 0.8 \end{bmatrix}$$

$$M_3 = \begin{bmatrix} x_1 & x_2 & x_3 \\ 0.8 & 0.9 & 0.3 \\ 0.2 & 0.2 & 0.7 \end{bmatrix} \quad M_4 = \begin{bmatrix} x_1 & x_2 & x_3 \\ 0.8 & 1 & 0.9 \\ 0.2 & 0 & 0.1 \end{bmatrix}$$

As can be seen there is one wrong partition, i.e., M_3. Here the sum of membership values for object x_2 does exceed 1. To find the genuine partitions the following scheme is worked out:

Description of clusters by their centroids:

$$v_i = \frac{1}{\sum\limits_{k=1}^{N} m_{ik}} \sum_{k=1}^{N} m_{ik}^q x_k \tag{12.5-22}$$

with v_i the centroid of cluster i, and q the exponent for expressing the degree of fuzziness; for $q = 1$ we obtain the classical c-means algorithm.

Compute the differences between objects and the cluster centroid:

$$\|x_k - v_i\|^2 = \left[\sum_{j=1}^{p} (x_{kj} - v_i)^2 \right]^{1/2} \tag{12.5-23}$$

where p is the number of variables.

Finally a distance function is minimized as follows:

$$min\, z(M, V) = \sum_{i=1}^{c} \sum_{k=1}^{N} m_{ik} \|x_k - v_i\|^2 \tag{12.5-24}$$

Minimization of the function $z(M, V)$ is a computational problem. The number of partitions that have to be checked is calculated according to:

$$\#\,partitions = \frac{1}{c!} \left[\sum_{j=1}^{c} \binom{c}{j} (-1)^{c-j} j^N \right] \tag{12.5-25}$$

For example, for 10 clusters with 25 objects we would need to test 10^{18} different partitions. Fortunately, not all the partitions have to be computed because there exist algorithms for finding an optimum partition iteratively according to appropriate criteria. Often a threshold is chosen for judging the change on going from a partition M^l to M^{l+1}. If the threshold is small enough the procedure can be stopped.

The results of a c-means clustering can be represented by an object-cluster distance plot. For the data in Table 12.5-2 one obtains the following plot:

cluster 1

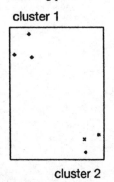

cluster 2

12.5.3 Supervised methods

If the membership of objects to particular clusters is known in advance the methods of supervised pattern recognition can be used. Here the following methods are explained: linear learning machine, linear discriminant analysis, *k*-nearest neighbor, and SIMCA methods.

Linear learning machine (LLM)

The first analytical application of a pattern recognition method dates back to 1969 when classification of mass spectra with respect to certain molecular mass classes was tried with the *linear learning machine* (LLM). The basis for classification with the LLM is a discriminant function that divides the *N*-dimensional space into category regions that can be further used to predict the category membership of a test sample.

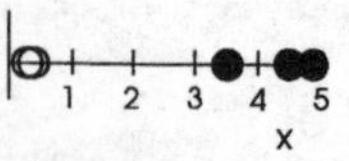

Fig. 12.5-11. Representation of iodine data from hair samples in Table 12.5-2; the two groups are labeled as open and full circles

Consider the data in Table 12.5-3 that represent the iodine content of hair samples from five different patients belonging to two categories. Since only one feature has been measured the data can be represented in 1-dimensional space as given in Fig. 12.5-11. To find a decision boundary that separates the two groups the data vectors have to be augmented by adding a $(N + 1)$ component equal to 1.0. This ensures that the boundary for separating the classes passes through the origin. If more than two categories are to be separated several linear discriminant functions would have to be constructed.

Table 12.5-3. Iodine content of hair samples from different patients

Hair sample	Iodine content, ppm	Augmented component
1	0.29	1.0
2	4.88	1.0
3	0.31	1.0
4	3.49	1.0
5	4.46	1.0

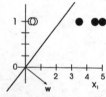

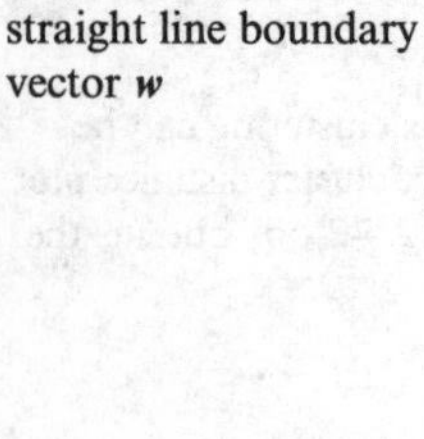

Fig. 12.5-12. LLM: representation of iodine data of Table 12.5-3 augmented by an additional dimension and separated by a straight line boundary with the normal weight vector *w*

The boundary that separates the two categories is found iteratively by adjusting the elements of a weight vector, w, which is normal to the boundary, such that the dot product of w and any vector representing the full circles is positive, while that of w and the empty circles is negative (Fig. 12.5-12). The decision boundary s is expressed by:

$$s = w_1 x_1 \tag{12.5-26}$$

or in general:

$$s = w^t x = \|w\| \cdot \|x\| \cos\theta \tag{12.5-27}$$

where $\|.\|$ is the vector norm, i.e. $[\sum_i x_i^2]^{1/2}$, x the augmented data vector, s the scalar variable, and θ the angle between w and x.

If the angle θ is less than 90° it is obvious that the objects represented as full circles are categorized and that $s > 0.0$. Conversely, if the angle θ is greater than 90° the scalar variable will be $s < 0$ and the empty circle objects are described.

To find the weight elements, w, they are set initially to random numbers. The objects are then classified by computing s and checking against the correct answer. If all classifications are correct the training process can be stopped and the LLM can be used for further classification purposes. However, if the response is incorrect new weights have to be calculated by updating the old ones, e.g. by:

$$w^{(new)} = w^{(old)} + cx \quad \text{with } c = \frac{-2s}{x^t x} \tag{12.5-28}$$

Here the constant c is chosen such that the boundary is reflected to the correct side of a data point the same distance as it was in error. The update of weights is repeated as long as all objects are correctly classified.

Of course, the LLM will work properly only if the data are linearly separable. One should also remember that the solution for positioning the boundary is not unique so that different solutions will emerge if the order for presenting the training objects is changed.

Linear discriminant analysis (LDA)

A more formal way of finding a decision boundary between different classes is based on LDA as introduced by Fisher and Mahalanobis. The boundary or hyperplane is calculated such that the variance between the classes is maximized and the variance within the individual classes is minimized. There are several ways to arrive at the decision hyperplanes. In fact, one of the routes Fisher described can be understood from the principles of straight-line regression.

Consider again the iodine data in Table 12.5-3. If you create an augmented variable y with values $+1$ for objects from one group and -1 for objects from the other group and perform a linear regression (LR) of y on x for the five training objects then the regression equation for y will have coefficients similar to the LDA weights, i.e.,

$$y = -1.082 + 0.447x$$

Comparison with the LLM reveals that the boundary is not constrained to pass through the origin but a cut-off point is included into the model. Application of LDA gives the discriminant function

$$s = -4.614 + 1.718x$$

Both boundaries are given in Fig. 12.5-13. To arrive at the (nonelemental) LDA solution an eigenvalue problem has to be solved. To generalize the problem we have to consider a data matrix X with N objects and K feature variables. There are G different groups or classes indexed by G_1 to G_{Nj}:

$$\begin{bmatrix} x_{11} & x_{12} & \cdots & x_{1K} \\ x_{21} & x_{22} & & x_{2K} \end{bmatrix} \left. \vphantom{x} \right\} G_1 \\ \begin{matrix} x_{31} & x_{32} & \cdots & x_{3K} \\ x_{41} & x_{42} & & x_{4K} \end{matrix} \left. \vphantom{x} \right\} G_2 \\ \vdots \\ \begin{matrix} x_{j1} & x_{j2} & \cdots & x_{jK} \\ x_{N1} & x_{N_2} & & x_{NK} \end{matrix} \left. \vphantom{x} \right\} G_{N_j}$$ (12.5-29)

The weights of the linear discriminant functions are found as the eigenvectors of the following matrix:

$$G^{-1}Hw = \lambda w \qquad (12.5\text{-}30)$$

where λ is the eigenvalue.

The matrix G is derived from the covariance matrix C of the different classes or groups G as follows:

$$G = (N - G)C = (N - G)\frac{1}{N - G}\sum_{j=1}^{G}(N_j - 1)C_j \qquad (12.5\text{-}31)$$

$$C_j = \frac{1}{N_j - 1}\sum_{l \in G_j}(x_{li} - \bar{x}_{ji})(x_{lk} - \bar{x}_{jk}) \qquad (12.5\text{-}32)$$

where N is the total number of objects, N_j the number of objects in group j, and the index $l \in G_j$ is an element of the jth group G_j.

Matrix H describes the spread of the group means $\bar{x}_j$ over the grand average $\bar{x}$, i.e.,

$$H = \sum_{j=1}^{G}N_j(\bar{x}_j - \bar{x})(\bar{x}_j - \bar{x})^t \qquad (12.5\text{-}33)$$

$$\bar{x} = \frac{1}{N}\sum_{j=1}^{G}N_j\bar{x}_j \qquad (12.5\text{-}34)$$

A decision boundary separates two or more groups of data.

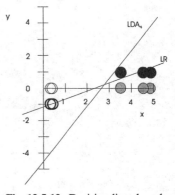

Fig. 12.5-13. Decision lines based on LR and LDA for separating the objects represented by full and empty circles

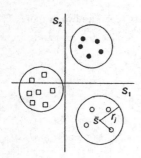

The eigenvector, w, that is found on the basis of the greatest eigenvalue λ_1 provides the first linear discriminant function, s_1, by:

$$s_1 = w_{11}x_1 + w_{12}x_2 + \cdots + w_{1K}x_K \tag{12.5-35}$$

With the residual x-data the second largest eigenvalue is computed and with the new eigenvectors the second discriminant function, s_2, is obtained:

$$s_2 = w_{21}x_1 + w_{22}x_2 + \cdots + w_{2K}x_K \tag{12.5-36}$$

The procedure is continued until all discriminant functions are found for solving the discrimination problem. By plotting pairs of discriminating functions against each other the best separation of objects into groups after the linear transformation of the initial features can be visualized. The projection of a particular object onto the separating line or hyperplane is called its *score* on the linear discriminant function.

Classification of unknown objects is carried out by inserting the feature data into the discriminant functions in order to transform its coordinates in the same way as for the original data set. Then, the object is assigned to that class for which its centroid has the smallest Euclidian distance:

$$\min_j[w^t(x_{unknown} - \bar{x}_j)] \quad \text{with} \quad j = 1, \ldots, G \tag{12.5-37}$$

A disadvantage with this method of unique decision is the fact that simultaneous membership of an object to several classes is not detected and that outliers, which do not belong to any of the classes, will always be categorized. Therefore, the unique categorization is often replaced by assigning the object to all classes within a fixed variance range, e.g. 95%. If the object lies outside of any of those variance ranges it will not be categorized at all.

The calculation of the variance radius, r_j, is done by:

$$r_j = \frac{D(N - G)}{N - G - D + 1} \frac{N_j + 1}{N_j} F_{(D, N-G-D+1; \alpha)} \tag{12.5-38}$$

where D is the number of discriminant functions used, and F is Fisher's F-statistics with risk α.

Assignment of an object to a particular class is performed if:

$$\sum_{d=1}^{D}(s_d - \bar{s}_{dj})^2 \leq r_j \tag{12.5-39}$$

Here s_d represents the new coordinates of the unknown object and $\bar{s}_{dj}$ is the jth class centroid.

Necessary *assumptions* of LDA are the normality of data distributions, the existence of different class centroids, as well as the similarity of variances and covariances among the different groups. Therefore, classification problems arise if the variances of groups differ substantially or if the direction of objects in the pattern space is different as depicted in Fig. 12.5-14.

Better discriminations are often obtained by means of regularized linear discriminant analysis (RDA). This method uses biased covariance estimates and is based on Baysian classification.

k-Nearest neighbor method (*k*-NN)

A simple nonparametric classification method is the nearest neighbor method as introduced by Fix and Hodges in 1951. For classification of an unknown object, its distance, usually the Euclidian distance, is computed to all objects. The minimum distance is selected and the object is assigned to the corresponding class.

The k-MN method is also used for filling in missing values or in library searches.

Typically the number of neighbouring objects k is chosen to be 1 or 3. With the k-NN method, very flexible separation boundaries are obtained as exemplified in Fig. 12.5-15.

Unfortunately, classification is dependent on the number of objects in each class. In case of overlapping classes, the object will be assigned to the class with the larger number of objects. This situation can sometimes be handled if no single criterion is used but one allows alternative counting of neighborhood, e.g., for a class A with less objects there must be found 5 neighbors, whereas for another class B with more objects 7 neighbors would have to be considered.

Soft independent modelling of class analogies (SIMCA)

Apart from discrimination methods, class membership of objects can also be determined by description of the individual classes by means of a separate mathematical model independent on the model for the other classes. In terms of geometry the model describes an envelope or a 'class box' around the class so that unknown objects can be classified according to their fit to a particular class model.

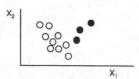

Fig. 12.5-14. Try to find a boundary which separates the differently spread and differently directed objects of the full and empty circled classes after projection onto that boundary

An early developed method uses multivariate normal distribution to model the classes on the basis of their data variances. Although this model has been sometimes used in analytical chemistry it lacks general application because the method is based on the covariance matrix where it is tacitly assumed that many data exist and that the ratio between objects and variables is favorably about 6 to 1, respectively.

More often the SIMCA method is used which finds separate principal component models for each class. By using SIMCA, the object variable number ratio is less critical and the model is constructed around the projected rather than the original data. The basic steps of principal component calculations as needed for SIMCA have been outlined in the chapter on projection methods with the NIPALS algorithm (Sec. 12.5.2).

Fig. 12.5-15. Separation boundary for classification of objects into two classes with $k = 1$

For each class q a separate model is constructed that reveals for a single x-observation:

$$x_{ij}^q = \bar{x}_{qj} + \sum_{a=1}^{A_q} t_{ia}^q p_{ja}^q + e_{ij}^q \tag{12.5-40}$$

where $\bar{x}_{qj}$ is the mean of variable j in class q, A_q is the number of significant principal components in class q, t_{ia}^q is the score of object i on component a in class q, p_{ja}^q is the loading of variable j on principal component a in class q, and e_{ij}^q is the residual error of object i and variable j.

The principal components are found by the iterative NIPALS algorithm. Each separate model may reveal a different number of significant principal components A_q. Thus, the class models may represent lines, planes, boxes, or hyperboxes as demonstrated in Fig. 12.5-16. The models can then be used to eliminate outlying objects, to estimate the modeling power of a particular variable, and to classify new objects.

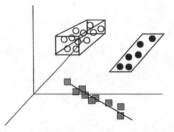

Fig. 12.5-16. SIMCA models for different numbers of significant principal components

Elimination of alien objects

Objects that do not fit the estimated principal component model can be eliminated by testing the total residual variance of a class q against the residual variance of that object. The two variances are calculated as follows:

Total residual variance of class q

$$s_0^2 = \sum_{i=1}^{N} \sum_{j=1}^{K} \frac{e_{ij}^2}{(N - A_q - 1)(K - A_q)} \tag{12.5-41}$$

where N is the number of objects, and K is the number of variables.

Residual variance for object i

$$s_i^2 = \sum_{j=1}^{K} \frac{e_{ij}^2}{K - A_q} \tag{12.5-42}$$

If both variances are of the same order of magnitude the object is assigned as a typical object to the class. If $s_i^2 > s_0^2$ then the object should be eliminated to enable a more parsimonious class model to be described.

Modeling power

The residual variance of a variable j of class q is used for estimating the modeling power of a particular variable if related to the so-called meaningful variance – the familiar expression for the variance:

Residual variance of variable j

$$s_j^2(error) = \sum_{i=1}^{N} \frac{e_{ij}^2}{N - A_q - 1} \qquad (12.5\text{-}43)$$

Meaningful variance of variable j

$$s_j^2(x) = \sum_{i=1}^{N} \frac{(x_{ij} - \bar{x})^2}{N - 1} \qquad (12.5\text{-}44)$$

The comparison of the two variances reveals a measure for the noise-to-signal ratio of this variable. The modeling power for variable j, R_j, is derived from the following expression:

$$R_j = 1 - \frac{s_j(error)}{s_j(x)} \qquad (12.5\text{-}45)$$

If the modeling power approaches values of 1 then the variable will be highly relevant, because the ratio between the residual error for the variable is small compared to its meaningful variance.

Classification

An unknown object with data vector x_u is checked to belong to a particular class by regression of the vector x_u on the q class models. Multiplying the data vector by the loading matrix P reveals an estimation for a new score vector $\hat{t}$. With the score vector the residuals are computed and used to decide on the membership of the object to a class:

$$\hat{t} = x_u P \qquad (12.5\text{-}46)$$

$$e = x_u - \hat{t} P^t \qquad (12.5\text{-}47)$$

As the residual variance for object u we get:

$$s_u^2 = \sum_{j=1}^{K} \frac{e_{uj}^2}{K - A_q} \qquad (12.5\text{-}48)$$

The object is assigned to class q if the variances s_u^2 and s_0^2 are of similar order of magnitude. In case s_u^2 is greater than s_0^2 the object is not a member of class q.

12.5.4 Multivariate modeling

Modeling of analytical dependences is usually based on linear models. Throughout the book these linear models have already been used for calibration of analytical procedures as well as for optimization based on RSM techniques. Calibration and response surface methods are indeed the most important applications where multivariate modeling of dependent variables on independent variables is needed. Another important field is environmental analysis, where receptor models are based on multivariate relationships, or environmental pollution can be specified on the basis of certain pollution patterns.

Linear models consist of additive terms, each of which contains only one multiplicative parameter.

In spite of the fact that our environment does not seem to be a simple linear world, in analytical chemistry one can show that by using multivariate approaches most of the relationships can be described by linear models. Of course, with linear models we mean models as introduced in Sec. 12.2, i.e., models that are linear in the parameters. These models can also be used to model curved dependences if appropriate transformations of the variables, e.g., quadratic, cubic, or exponential transformations are performed.

Linear regression analysis as used up to now is based on ordinary least-squares regression (OLS). To broaden the view in this chapter we will formalize the OLS method and add some more soft modeling techniques, i.e., principal component regression (PCR), the partial least-squares method (PLS), as well as target transform factor analysis (TTFA).

Ordinary least-squares regression (OLS)

The general least-squares problem that relates a matrix of dependent variables Y to a matrix of independent variables X can be stated as follows:

$$
\begin{bmatrix}
y_{11} & y_{12} & \cdots & y_{1M} \\
y_{21} & y_{22} & & y_{2M} \\
\vdots & & & \\
y_{N1} & y_{N2} & \cdots & y_{NM}
\end{bmatrix}
=
\begin{bmatrix}
x_{11} & x_{12} & \cdots & x_{1K} \\
x_{21} & x_{22} & \cdots & x_{2K} \\
\vdots & & & \\
x_{N1} & x_{N2} & \cdots & x_{NK}
\end{bmatrix}
\begin{bmatrix}
b_{11} & b_{12} & \cdots & b_{1M} \\
b_{21} & b_{22} & \cdots & b_{2M} \\
\vdots & & & \\
b_{K1} & b_{K2} & \cdots & b_{KM}
\end{bmatrix}
$$

In matrix notation we get:

$$Y = XB \qquad (12.5\text{-}49)$$

$$_N\overset{M}{\boxed{Y}} = {}_N\overset{K}{\boxed{X}}{}_K\overset{M}{\boxed{B}} + \text{Residuals}$$

where Y is the $N \times M$ matrix of dependent variables, X is the $N \times K$ matrix of independent variables, and B is the $K \times M$ matrix of regression parameters. Residuals are the differences between measured and modeled data, i.e., $Y - XB$.

For example, in multivariate calibration this equation will be used to model the concentrations of M constituents in N samples (Y-matrix) on the N spectra recorded at K wavelengths (X-matrix).

In OLS the number of columns in the X-matrix is maintained. As an example, the set of linear algebraic equations for the first column of Eq. 12.5-49 looks like this:

$$y_{11} = b_{11}x_{11} + b_{21}x_{12} + \ldots b_{K1}x_{1K}$$

$$y_{21} = b_{11}x_{21} + b_{21}x_{22} + \ldots b_{K1}x_{2K}$$

$$\vdots \qquad\qquad (12.5\text{-}50)$$

$$y_{N1} = b_{11}x_{N1} + b_{21}x_{22} + \ldots b_{K1}x_{NK}$$

> OLS is synonymous with the following terms: least-squares regression, linear least-squares regression, multiple least-squares regression, multivariate least-squares regression.

Parameter estimation

Usually the matrix of the independent variables, X, is not square so that the regression parameters B have to be estimated by the generalized inverse method. B is given by:

$$B = (X^t X)^{-1} X^t Y \qquad (12.5\text{-}51)$$

In principle this equation could be solved by directly inverting the matrix $X^t X$. This, however, will only work if no linear dependences are valid and the system is, in a mathematical sense, well-conditioned. The conditioning of the system is given by the **condition number**:

> OLS provides the *best linear unbiased estimator* (BLUE) that has the smallest variance among all linear and unbiased estimators.

$$cond(B) = \|B\|\|B^{-1}\| \qquad (12.5\text{-}52)$$

where $\|B\|$ is the norm of matrix B.

The matrix norm of B is computed as its largest singular value (square root of eigenvalue λ) and the norm of B^{-1} as the reciprocal smallest singular value of B i.e.,

$$cond(B) = \sqrt{\lambda_{max}} \cdot \frac{1}{\sqrt{\lambda_{min}}} \qquad (12.5\text{-}53)$$

This definition holds for exactly determined systems where the number of rows in X is equal to the number of columns, i.e., $N = K$ (cf. Eq. 12.5-49). In the case of overdetermined systems, with $N > K$, the condition number is obtained from:

$$cond(\boldsymbol{B}) = [cond(\boldsymbol{B}^t\boldsymbol{B})]^{1/2} \tag{12.5-54}$$

Well-conditioned systems have condition numbers near to one. A matrix is singular if its condition number is infinite, and it is ill-conditioned if its condition number is too large, that is, if its reciprocal approaches the machine's floating point precision.

Linear dependences between rows or columns lead to singularities. This might even happen if there is no exact linear dependency and roundoff errors in the machine render some of the equations linearly dependent.

Typical procedures to solve the OLS problem are Gaussian elimination or Gauss–Jordan elimination. More efficient solutions are based on decomposition of the X-matrix by algorithms, such as LU-decomposition, Housholder reduction, or singular value decomposition (SVD). Here one of the most powerful methods, SVD, will be outlined below.

Prediction

In regression analysis the dependent variable y is also called a *response* variable and the independent variable x is denoted as *predictor* variable or *regressor*.

Modeling of the analytical relationships by estimating the regression parameters in OLS is one of the objectives. Most often in a second step the model parameters are used to predict some unknown x- or y-values from the measured y- or x-values, e.g., in multivariate calibration the concentrations are predicted from the recorded spectra.

Prediction of a single $\boldsymbol{y}_0$-vector from an $\boldsymbol{x}_0$-vector is easily performed by Eq. 12.5-49. For predicting an $\boldsymbol{x}_0$-vector (dimension $1 \cdot K$) from a $\boldsymbol{y}_0$-vector ($1 \cdot M$) the following expression is used:

$$\boldsymbol{x}_0 = \boldsymbol{y}_0\boldsymbol{B}^t(\boldsymbol{B}\boldsymbol{B}^t)^{-1} \tag{12.5-55}$$

To estimate the upper bound error for prediction the following relationship holds:

$$\frac{\|\delta\boldsymbol{x}_0\|}{\|\boldsymbol{x}_0\|} = cond(\boldsymbol{B})\left(\frac{\|\delta\boldsymbol{y}\|}{\|\boldsymbol{y}\|} + \frac{\|\delta\boldsymbol{B}\|}{\|\boldsymbol{B}\|}\right) \tag{12.5-56}$$

where $\|\delta\boldsymbol{x}_0\|/\|\boldsymbol{x}_0\|$ – relative error for prediction, $\|\delta\boldsymbol{y}\|/\|\boldsymbol{y}\|$ – relative error of measurements y, $\|\delta\boldsymbol{B}\|/\|\boldsymbol{B}\|$ – relative error of parameter estimation.

From Eq. 12.5-56 it is obvious that the prediction error will be small if the error in the dependent variable y and the modeling error can be kept low.

Principal component regression (PCR)

PCR is best performed by means of SVD (singular value decomposition). With this method the matrix X is decomposed into two orthonormal matrices U and V that are joined by a diagonal matrix W of singular values (cf. Eq. 12.5-5):

$$\begin{bmatrix} x_{11} & x_{12} & \cdots & x_{1K} \\ x_{21} & x_{22} & \cdots & x_{2K} \\ \vdots & & & \\ x_{N1} & x_{N2} & \cdots & x_{NK} \end{bmatrix} = \begin{bmatrix} u_{11} & u_{12} & \cdots & u_{1A} \\ u_{21} & u_{22} & \cdots & u_{2A} \\ \vdots & & & \\ u_{N1} & u_{N2} & \cdots & u_{NA} \end{bmatrix} \begin{bmatrix} w_{11} & 0 & \cdots & 0 \\ 0 & w_{22} & \cdots & 0 \\ \vdots & & & \\ 0 & 0 & \cdots & w_{AA} \end{bmatrix}$$

$$\times \begin{bmatrix} v_{11} & v_{21} & \cdots & v_{K1} \\ v_{12} & v_{22} & \cdots & v_{K2} \\ \vdots & & & \\ v_{1A} & v_{2A} & \cdots & v_{KA} \end{bmatrix}$$

$$\boldsymbol{X} = \boldsymbol{U}\boldsymbol{W}\boldsymbol{V}^t \tag{12.5-57}$$

where A is the rank of matrix X, i.e. for full rank $A = K$.

Computation of the regression coefficients b vector-wise is carried out by formation of the pseudo-inverse matrix X^+ (Moore-Penrose matrix) according to:

$$X^+ = V\left(\text{diag}\left(\frac{1}{w_{ii}}\right)\right)U^t \tag{12.5-58}$$

$$b = X^+ y \tag{12.5-59}$$

In the case of full rank, all singular values will be obviously different from zero and the SVD solution equals that of OLS. However, one often comes up with several small singular values because of ill-conditioned systems. Therefore, the main goal of PCR is not to keep all singular values for an exact representation of the Moore-Penrose matrix but to select a subset of singular values that best guarantee predictions of unknown cases.

Partial least-squares regression (PLS)

Regression of Y on X by OLS and PCR is based on solving the linear equations column-wise in order to estimate the regression coefficients in the columns of the matrix B In Eq. 12.5-49. The decomposition of the X-matrix is performed independently on the Y-matrix. A method for using the information from the Y-matrix is the PLS-algorithm as developed by H. Wold and propagated by his son S. Wold. Each PLS latent variable direction of the X-matrix is modified so that the covariance between it and the Y-matrix vector is maximized. The PLS method is based on a bilinear model with respect to the objects and the variables of the X- or Y-matrix. Both the X- and Y-matrices are decomposed into smaller matrices according to the following scheme:

$$X = TP^t + E \tag{12.5-60}$$

$$_N\boxed{X}^K = {}_N\boxed{T}_A \, {}_A\boxed{P^t}^K + {}_N\boxed{E}^K$$

$$Y = TQ^t + F \tag{12.5-61}$$

$$_N\boxed{Y}^K = {}_N\boxed{T}_A \, {}_A\boxed{Q^t}^M + {}_N\boxed{F}^M$$

where X, Y, N, K, M, A have the same meaning as given in Eq. 12.5-49; T is the $N \times A$ scores matrix containing orthogonal rows; P is the $K \times A$ loading of the X-matrix; E is the $N \times K$ error matrix of the X-matrix; Q is the $M \times A$ loading of the Y-matrix; and F is the $N \times M$ error matrix for the Y-matrix.

To compute the B-coefficients for the general model of Eq. 12.5-49 the matrices P, Q, and W are required:

$$B = W(P^t W)^{-1} Q^t \tag{12.5-62}$$

with W the $K \times A$ matrix of PLS-weights.

The meaning and estimation of the weight matrix W can be understood from the PLS algorithm in the following example.

Orthonormal means a set of vectors that are orthogonal and have their norm equal to one.

PCR and PLS are examples of *biased* regression methods where the expected estimated value is different from the true value of the parameter. Another powerful biased regression method is *ridge regression*.

Worked example: PLS algorithm

The first dimension A (index a) is computed from the column mean vectors of the X- and Y-matrix as follows (centering):

$a = 0$:

$$X = X - \bar{x} \tag{12.5-63}$$

$$Y = Y - \bar{y} \tag{12.5-64}$$

Next the dimensions $a = 1$ to $a = A_{opt}$ are computed based on a suitable stopping criterion, usually the standard error of prediction due to cross-validation (SEP_{CV}):

Loop for the number of dimensions: $\qquad a = a + 1$

The principal components are estimated iteratively, e.g., by the NIPALS algorithm. Iteration is stopped if the computer precision is reached. Iteration loop for NIPALS:

1. Use as a starting vector for the y-score vector u the first column of the actual Y-matrix:

$$u = y_1 \qquad (12.5\text{-}65)$$

2. Compute the X-weights:

$$w^t = \frac{u^t X}{u^t u} \qquad (12.5\text{-}66)$$

3. Scale the weights to a vector of length one:

$$w^t = \frac{w^t}{(w^t w)^{1/2}} \qquad (12.5\text{-}67)$$

4. Estimate the scores of the X-matrix:

$$t = X w^t \qquad (12.5\text{-}68)$$

5. Compute the loadings of the Y-matrix:

$$q^t = \frac{t^t Y}{t^t t} \qquad (12.5\text{-}69)$$

6. Generate the y-score-vector u:

$$u = \frac{Y q}{q^t q} \qquad (12.5\text{-}70)$$

Compare u(old) with u(new). If $\|u(\text{old}) - u(\text{new})\| < \|u(\text{new})\| \times$ THRESHOLD, convergence is obtained. Otherwise iteration is continued at (1). The threshold can be chosen on the basis of computer precision.

7. Compute the loadings of the X-matrix:

$$p^t = \frac{t^t X}{t^t t} \qquad (12.5\text{-}71)$$

8. Form new residuals for the X- and Y-matrix:

$$E = X - T P^t \qquad (12.5\text{-}72)$$

$$F = Y - T Q^t \qquad (12.5\text{-}73)$$

Compute the SEP_{CV}. If the SEP_{CV} is greater than for the actual number of factors then the optimum number of dimensions has been found. Otherwise the next dimension is computed. In general, the SEP is obtained by:

$$\text{SEP} = \left[\frac{\sum_{j=1}^{M} \sum_{i=1}^{N} (y_{ij}^{estimated} - y_{ij}^{true})^2}{\sum_{j=1}^{M} \sum_{i=1}^{N} (y_{ij}^{true})^2} \right] \qquad (12.5\text{-}74)$$

The B-coefficients are finally computed according to Eq. 12.5-62, i.e.,

$$B = W (P^t W)^{-1} Q^t$$

One can show that the matrix $P^t W$ is an upper *bidiagonal* matrix so that the PLS algorithm represents just a variation of diagonalizing a matrix before its inversion.

To estimate the error of the model there are two possibilities. Either the objects or samples are predicted by *resubstitution* into the model equation Eq. 12.5-49. This gives an estimate of the standard error of modeling. Or the model is built by leaving out objects randomly and predicting the left-out samples from the reduced model. The latter prediction is known as *cross-validation* and reveals the standard error of prediction from cross-validation, SEP_{CV} (cf. Eq. 12.5-74).

The PLS-algorithm is one of the standard methods used for *two-block* modeling, e.g., for multivariate calibration as given below.

Target transform factor analysis (TTFA)

One of the methods for the decomposition of a data matrix into factors has been explored by means of principal component analysis (Sec. 12.5.2). The ways for this sort of mathematical analysis, however, are numerous. In addition, representation of the X-matrix might vary since in PCA we have associated the objects with samples and the features with their properties whereas in factor analysis the objects usually represent properties, and the feature variables describe cases or samples. One additional objective is to test analytical data vectors as targets in relation to the obtained abstract factors, e.g., in order to verify an absorption spectrum under a spectrochromatogram or to detect real sources of airborne contaminants in the analysis of environmental data.

The general aim of factor analysis (FA) is to separate a data matrix X into two matrices, a matrix of *common* factors and another of *unique*, also called *specific*, factors E. The matrix of common factors is modeled by the scores matrix S, that consists of the values of properties for each of the A underlying causative factors and the matrix L that gives the amount of each factor or loading:

$$X = SL + E \qquad (12.5\text{-}75)$$

$$_N[\overset{K}{X}] = {}_N[\overset{A}{S}]_A[\overset{K}{L}] + [E]$$

The factor matrices are found from diagonalizing a symmetric matrix derived from the original data matrix X. Several real, symmetric matrices exist that describe the interrelationship within the data, the most important are the covariance and correlation matrix.

The *variance-covariance matrix* can be obtained from either:

$$C_Q = X^t X \quad \text{or} \quad C_R = XX^t \qquad (12.5\text{-}76)$$

where X is a $N \times K$ matrix, and Q, R label for the so-called Q- and R-modes, respectively.

Usually the centered data are used, so that we obtain for the covariance matrix in *Q-mode* in detail:

$$C_Q = \begin{vmatrix} s_{11}^2 & cov(1,2) & \ldots & cov(1,K) \\ cov(2,1) & s_{22}^2 & & cov(2,K) \\ \cdot & & & \\ cov(K,1) & cov(K,2) & \ldots & s_{KK}^2 \end{vmatrix} \qquad (12.5\text{-}77)$$

In general, Q- and R-mode analysis is used to describe the relationship among n objects by p variables or the relationship among p variables determined by n objects, respectively. Typically, clustering of objects is Q-analysis but correlating variables is R-analysis.

where

$$s_{jj}^2 = \frac{1}{K-1} \sum_{i=1}^{K} (x_{ij} - \bar{x}_j)^2 \quad \text{for } j = 1 \ldots K$$

$$cov(j,k) = \frac{1}{K-1} \sum_{i=1}^{K} [(x_{ij} - \bar{x}_j)(x_{ik} - \bar{x}_k)] \quad j,k = 1, \ldots K \quad \text{and} \quad j \neq k$$

The covariance matrix is used in cases where the metric of the variables is comparable. If widely different metrics are valid, e.g., in the case of simultaneous data treatment of main and trace constituents, the variables are scaled by the variable variances giving the *correlation matrix* as follows:

$$R_Q = (XV_Q)^t(XV_Q) \quad \text{or} \quad R_R = (V_R X)(V_R X)^t \qquad (12.5\text{-}78)$$

where

$$v_Q(ij) = \cfrac{1}{\cfrac{1}{K-1}\sqrt{\sum_{i=1}^{K}(x_{ij}-\bar{x}_j)^2}} \qquad v_R(ij) = \cfrac{1}{\cfrac{1}{N-1}\sqrt{\sum_{i=1}^{N}(x_{ij}-\bar{x}_j)^2}}$$

Again exemplified for the Q-mode the correlation matrix looks as follows:

$$R_Q = \begin{vmatrix} 1 & r_{12} & \cdots & r_{1K} \\ r_{21} & 1 & & r_{2K} \\ \cdot & & & \\ r_{K1} & r_{K2} & \cdots & 1 \end{vmatrix} \qquad (12.5\text{-}79)$$

where

$$r_{jk} = \frac{cov(j,k)}{s_j s_k} = \frac{\sum\limits_{i=1}^{K}(x_{ij} - \bar{x}_j)(x_{ik} - \bar{x}_k)}{\left[\sum\limits_{i=1}^{K}(x_{ij} - \bar{x}_j)^2 \sum\limits_{i=1}^{K}(x_{ik} - \bar{x}_k)^2\right]^{1/2}}$$

$$s_j = \sqrt{\frac{\sum\limits_{i=1}^{K}(x_{ij} - \bar{x}_j)^2}{K-1}}; \quad s_k = \sqrt{\frac{\sum\limits_{i=1}^{K}(x_{ik} - \bar{x}_k)^2}{K-1}} \quad \text{and} \quad j \neq k$$

Diagonalization of the dispersion matrix is carried out by solving an eigenvalue problem. We will demonstrate this for a dispersion matrix based on the correlation matrix R.

An eigenvector u is obtained for a real, symmetric matrix, R, by

$$Ru = u\lambda \qquad (12.5\text{-}80)$$

where λ is some unknown scalar, the eigenvalue. This eigenvalue is to be determined such that the vector Ru is proportional to u. For this, Eq. 12.5-80 is rewritten as:

$$Ru - u\lambda = 0 \quad \text{or} \quad (R - \lambda I)u = 0 \qquad (12.5\text{-}81)$$

where I is the identity matrix and the vector, u, is orthogonal to all of the row vectors of the matrix $(R - \lambda I)$. The equation obtained is equivalent to a set of A equations where A is the rank of R.

Unless u is a null vector, Eq. 12.5-81 only holds if:

$$R - \lambda I = 0 \qquad (12.5\text{-}82)$$

A solution to this set of equations exists only if the determinant on the left side of Eq. 12.5-82 is zero:

$$|R - \lambda I| = 0 \qquad (12.5\text{-}83)$$

Computing the determinant reveals a polynomial in λ of degree A. Then the roots, λ_i with $i = 1, \ldots, A$, of those equations have to be found. There will be an associated vector u_i such that:

$$Ru_i - u_i\lambda_i = 0 \qquad (12.5\text{-}84)$$

or in matrix notation:

$$RU = U\Lambda \qquad (12.5\text{-}85)$$

The matrix U is a square and orthogonal matrix of the eigenvectors. Its dimension is either $K \times K$ or $N \times N$ depending on which mode, Q or R, is used. From the different mode of forming the dispersion matrix we also obtain different sets of eigenvectors. If we denote the matrix of eigenvectors derived from R_Q by V and the eigenvector matrix derived from R_R by U we obtain a relationship as known from SVD:

$$X = U\Lambda^{1/2}V^t \qquad (12.5\text{-}86)$$

The Q- and R-mode factor solutions are interrelated as follows:

$$X = L_Q S_Q \quad \text{or} \quad X = S_R L_R \qquad (12.5\text{-}87)$$

where

$$X = \overbrace{U}^{L_Q} \overbrace{\Lambda^{1/2} V^t}^{S_Q} \quad \text{or} \quad X = \overbrace{U\Lambda^{1/2}}^{S_R} \overbrace{V^t}^{L_R}$$

The direction of the solution is less dependent on the mode, Q- or R-mode, but is directed by the scaling procedures applied. The scaling may be based on scaling the columns or rows or on the scaling of both.

The number of significant eigenvalues is determined in an additional step (cf. Sec. 12.5.2). In the absence of experimental error, the number of eigenvalues will equal the number of true components or sources apart from calculational error. With real experimental data the choice becomes more difficult. Functions of general usefulness account for the data variance in relation to the experimental uncertainty, such as the chi-squared measure:

$$\chi^2 = \sum_{ij} \frac{(x_{ij} - \bar{x}_{ij})^2}{s_{ij}^2} \tag{12.5-88}$$

where x_{ij} is the reconstructed data point based on A factors and s_{ij} represents the uncertainty in the value of x_{ij}. A similar measure for estimating the number of factors is the Exner function:

$$E^p = \left[\sum_{ij} \frac{(x_{ij} - \bar{x}_{ij})^2}{(x^0 - \bar{x}_{ij})^2} \right]^{1/2} \tag{12.5-89}$$

with x^0 the-grand mean.

For both measures the number of factors is decided from their minimum in dependence on the number of factors.

An important measure for the adequacy of the factorial model is the fraction of explained variance in relation to the total variance of the system. The *total variance* of data is calculated from the sum of all eigenvalues (Q-mode):

$$\text{total variance} = \sum_{a=1}^{K} \lambda_a \tag{12.5-90}$$

The fraction of *explained variance* is calculated as the ratio of the sum over all significant eigenvalues and the sum over all eigenvalues according to:

$$\text{explained variance} = \frac{\displaystyle\sum_{a=1}^{A} \lambda_a}{\displaystyle\sum_{a=1}^{K} \lambda_a} \quad \text{for} \quad A \leq K \tag{12.5-91}$$

Depending on the actual problem the variance of the data should be explained to a certain degree, e.g., by 90%.

Target rotation

Decomposition of the original data matrix into the matrices S and L is not unique, i.e., there is an infinity of mathematically equivalent matrices that will diagonalize the correlation matrix. To relate the abstract matrices to one with physical significance it is necessary to realign the factor axes of S with axes that represent, e.g., spectra, or as in environmental analysis, source emission concentration profiles. This can be done by target transformation of the abstract factor space.

Target transformation involves the computation of a rotation vector, t, which aligns a column of the S matrix with the input target vector, b. To minimize the difference between the rotated axes of S and the test vector the least squares estimate is computed:

$$t = (S^t S)^{-1} S^t b \tag{12.5-92}$$

The rotation vector, t, can then be used to predict the assumed spectrum or source profile by:

$$\hat{b} = St \tag{12.5-93}$$

Without a target vector rotation of factors is possible by applying different criteria. Orthogonal rotation is carried out by, e.g., the varimax or quartimax criterion. Oblique rotation criteria also exist, e.g. oblimax.

Comparison of the predicted vector $\hat{b}$ with the assumed vector b reveals the contribution of the target to the measured data. Typically the relative deviation between both vectors is computed:

$$\text{relative deviation} = \frac{\sum_{i=1}^{K} |\hat{b}_i - b_i|}{\sum_{i=1}^{K} |b_i|} \tag{12.5-94}$$

Finally a threshold is to be chosen in order to decide on the presence of the target in the data matrix X.

Multicomponent analysis

As an example for multivariate modeling we consider the simultaneous determination of several components in low selective analytical systems (multicomponent analysis). These components can be elements, compounds or chemical/physical properties. By means of multicomponent analysis, constituents of pharmaceutical formulations can be determined in the UV-range, the water and protein content of cereal can be estimated from NIR-spectra or chemical elements, and technological parameters of coal are predictable on the basis of IR-spectra. The limited selectivity of chemical sensors can also be overcome by applying the principles of multicomponent analysis.

The principles will be explained on the basis of spectroscopic multicomponent analysis because this is at present the dominating area of application.

Spectroscopic multicomponent analysis can be based on Beer's law that is formulated for a single component as (cf. Sec. 9.1):

$$A_\lambda = \varepsilon_\lambda \, dc \tag{12.5-95}$$

where A_λ is the absorbance at wavelength λ, ε_λ the molar absorptivity, L $\text{mol}^{-1}\,\text{cm}^{-1}$, d the cell thickness, cm, and c the concentration, mol/L.

The absorbances can be normalized to a constant thickness of the cuvette revealing a simplified Beer's law:

$$A_\lambda = k_\lambda c \tag{12.5-96}$$

where k_λ is the normalized molar absorptivity.

In multicomponent systems the absorbance at a specific wavelength λ is assumed to be the result of the sum of the absorbances of the M individual constituents according to:

$$A_\lambda = k_{\lambda 1} c_1 + k_{\lambda 2} c_2 + \cdots + k_{\lambda M} c_M = \sum_{j=1}^{M} k_{\lambda j} c_j \tag{12.5-97}$$

In multiwavelength spectroscopy the spectrum is acquired at K wavelengths leading to exactly determined linear equation systems $(K = M)$ or overdetermined systems $(K > M)$:

$$A_1 = k_{11} c_1 + k_{12} c_2 + \cdots + k_{1M} c_M$$

$$A_2 = k_{21} c_1 + k_{22} c_2 + \cdots + k_{2M} c_M$$

$$\vdots$$

$$A_K = k_{K1} c_1 + k_{K2} c_2 + \cdots + k_{KM} c_M$$

where A_i is the absorbance at wavelength λ_i, k_{ij} the absorptivity of the jth component at wavelength λ_i, and c_j-concentration of the jth component.

In matrix notation we obtain:

$$a = Kc \tag{12.5-98}$$

where a is a $K \cdot 1$ dimensional vector that represents the spectrum. K a $K \times M$ matrix of normalized molar absorptivities, and c a $M \cdot 1$ dimensional vector of concentrations.

Direct calibration method

This method is used if all the absorptivities are known. This implies that the pure component spectra can be measured or can be obtained elsewhere, that there is no interaction between the different components in the sample or between constituents and the solvent, as well as that no unknown matrix constituents interfere with the determination.

Analysis of unknown samples is based on the sample spectrum, a_0, and the known absorptivities K according to (cf. prediction Eq. 12.5-55):

$$c_0^t = a_0^t K^{-1} \quad \text{for} \quad K = M \quad \text{or} \tag{12.5-99}$$

$$c_0 = K^t (KK^t)^{-1} K^t a_0 \quad \text{for} \quad K > M \tag{12.5-100}$$

where c_0 is the vector of predicted concentrations.

Worked example:

Two constituents in a sample are to be determined from their absorbances at either two or three wavelengths. The following K-matrix for the absorptivities (arbitrary units) is given:

Wavelength	Constituent	
	1	2
1	3.00	2.00
2	3.00	4.00
3	2.00	6.00

The absorbance data measured on the sample are:

$$a_0 = \begin{pmatrix} 0.71 \\ 1.09 \\ 1.41 \end{pmatrix}$$

The true concentrations of that sample are:

$$c_0 = \begin{pmatrix} c_1 \\ c_2 \end{pmatrix} = \begin{pmatrix} 0.100 \\ 0.200 \end{pmatrix}$$

Determination at the *two wavelengths* 1 and 2 (exactly determined linear system, $K = M$) reveals:

$$c_0^t = a_0^t K^{-1} = (0.71, 1.09) \begin{bmatrix} 3.0 & 3.0 \\ 2.0 & 4.0 \end{bmatrix}^{-1} = (0.71, 1.09) \begin{bmatrix} 0.666 & -0.5 \\ -0.333 & 0.5 \end{bmatrix} = (0.110, 0.190)$$

or expressed for the single concentrations:

$$c_1 = 0.110 \quad \text{and} \quad c_2 = 0.190$$

As a result, the relative deviation between estimated and true concentrations is obtained in percent to be:

$$\delta c_1 = \frac{c_1^{predicted} - c_1^{true}}{c_1^{true}} \cdot 100 = \frac{0.110 - 0.100}{0.100} \cdot 100 = 10\%$$

$$\delta c_2 = \frac{c_2^{predicted} - c_2^{true}}{c_2^{true}} \cdot 100 = \frac{0.190 - 0.200}{0.200} \cdot 100 = 5\%$$

This relatively high error can be decreased if all *three wavelengths* are used in the analysis:

$$c_0 = K^t(KK^t)^{-1}a_0 = \begin{bmatrix} 3 & 3 & 2 \\ 2 & 4 & 6 \end{bmatrix} \left(\begin{bmatrix} 3 & 2 \\ 3 & 4 \\ 2 & 6 \end{bmatrix} \begin{bmatrix} 3 & 3 & 2 \\ 2 & 4 & 6 \end{bmatrix} \right)^{-1} \begin{bmatrix} 0.71 \\ 1.09 \\ 1.41 \end{bmatrix}$$

$$= \begin{bmatrix} 3 & 3 & 2 \\ 2 & 4 & 6 \end{bmatrix} \left(\begin{bmatrix} 13 & 17 & 18 \\ 17 & 25 & 30 \\ 18 & 30 & 40 \end{bmatrix} \right)^{-1} \begin{bmatrix} 0.71 \\ 1.09 \\ 1.41 \end{bmatrix}$$

$$= \begin{bmatrix} 3 & 3 & 2 \\ 2 & 4 & 6 \end{bmatrix} \left(\begin{bmatrix} 0.125 & 0.0479 & -0.0967 \\ 0.0479 & 0.0209 & -0.0309 \\ -0.0967 & -0.0309 & 0.0889 \end{bmatrix} \right) \begin{bmatrix} 0.71 \\ 1.09 \\ 1.41 \end{bmatrix}$$

$$= \begin{pmatrix} 0.0995 \\ 0.2002 \end{pmatrix}$$

For the relative deviations from the true concentrations we obtain now:

$$\delta c_1 = \frac{c_1^{predicted} - c_1^{true}}{c_1^{true}} \cdot 100 = \frac{0.0995 - 0.100}{0.100} \cdot 100 = 5\%$$

$$\delta c_2 = \frac{c_2^{predicted} - c_2^{true}}{c_2^{true}} \cdot 100 = \frac{0.2002 - 0.200}{0.200} \cdot 100 = 0.1\%$$

Notice that the error has been decreased by adding just one wavelength to the analysis scheme. With higher floating point precision the result could be even further improved. Therefore multiwavelength spectrometry can be a powerful alternative to systems where, for every component, only a single wavelength is applied.

In practice, some or even all of the above mentioned prerequisites for direct calibration on the basis of OLS regression are often not obeyed. Therefore, more sophisticated calibration procedures have to be carried out based on mixture or multivariate calibration.

Indirect calibration methods

Indirect methods are based on estimating the calibration parameters from calibration mixtures. These methods offer the following advantages:

(a) Interactions between constituents or between constituents and the sample matrix can be accounted for in the calibration. Thus, the validity of Beer's law, i.e., the additivity of spectra for every single component and linear response-concentration relationships, is not a prerequisite any more.
(b) Modeling of background in a principal component becomes feasible.
(c) Systems of highly correlated spectra can also be used for multicomponent analysis.

Collinearity of data refers to approximate linear dependence among variables.

The different methods for multivariate calibration differ by the mathematical model that is either based on Beer's law, i.e., the spectra are regressed on concentrations as with the *K*-matrix approach or on inverse models where the regression of concentrations on spectra is carried out.

(a) *K-matrix approach*

The *K*-matrix approach is based on an extension of Eq. 12.5-97 to matrix form:

$$\begin{bmatrix} a_{11} & a_{12} & \cdots & a_{1K} \\ a_{21} & a_{22} & \cdots & a_{2K} \\ \vdots & & & \\ a_{N1} & a_{N2} & \cdots & a_{NK} \end{bmatrix} = \begin{bmatrix} c_{11} & c_{12} & \cdots & c_{1M} \\ c_{21} & c_{22} & \cdots & c_{2M} \\ \vdots & & & \\ c_{N1} & c_{N2} & \cdots & c_{NM} \end{bmatrix} \begin{bmatrix} k_{11} & k_{12} & \cdots & k_{1K} \\ k_{21} & & & k_{2K} \\ \vdots & & & \\ k_{M1} & k_{M2} & \cdots & k_{MK} \end{bmatrix}$$

or in matrix notation:

$$A = CK \tag{12.5-101}$$

where A is the $N \times K$ matrix of absorbances, C is the $N \times M$ matrix of concentrations of constituents, K is the $M \times K$ matrix of absorptivities, N is the number of samples, K is the number of wavelengths, and M is the number of components.

In the present notation it is assumed that the absorbance data are centered and that, therefore, there is no intercept at the absorbance axis. If uncentered data are used the first column in the concentration matrix should consist of 1s, and in the K-matrix the intercept coefficients would have to be introduced as the first row.

Calibration is based on a set of N samples of known concentrations for which the spectra are measured. By means of the calibration sample set, estimation of absorptivities is possible by solving for the matrix K according to the general least-squares solution:

$$K = (C^t C)^{-1} C^t A \qquad (12.5\text{-}102)$$

The *analysis* is then based on the spectrum $a_0 (1 \cdot K)$ of the unknown sample by:

$$c_0 = a_0 K^t (K K^t)^{-1} \qquad (12.5\text{-}103)$$

where c_0 is the $(1 \cdot M)$ vector of sought-for concentrations.

A great advantage of the K-matrix approach is the fact that the elements of the K-matrix represent genuine absorptivities with reference to the spectra of the individual constituents. Also, the general assumption in least-squares regression analysis is valid, such that only the dependent variable, here the absorbance, is error prone.

In the K-matrix approach, all absorbing constituents of a sample must be explicitly known to be included into the calibration procedure. As we will see below, with more soft modeling techniques it will also be possible to account for unknown constituents without their explicit calibration.

Another disadvantage of the K-matrix approach results from the fact that calibration *and* analysis are connected to the inversion of a matrix. Although this is not a problem from the point of view of computational time, it might become a problem if ill-conditioned (less selective) systems are applied, where the spectra of the constituents are very similar. Then in the analysis step (Eq. 12.5-103) a badly conditioned matrix of absorptivities has to be inverted that might be almost singular, i.e., all singular values or eigenvalues are zero. To overcome this difficulty, powerful algorithms for solving the linear equations, such as SVD, should be used in connection with reduction of the dimensionality of the problem.

An alternative to the K-matrix approach is to calibrate the concentrations directly on the spectra. These methods are called *inverse calibration methods*.

(b) *Inverse calibration methods*

P-matrix approach. The P-matrix approach is based on the following model:

$$\begin{bmatrix} c_{11} & c_{12} & \cdots & c_{1M} \\ c_{21} & c_{22} & \cdots & c_{2M} \\ \vdots & & & \\ c_{N1} & c_{N2} & \cdots & c_{NM} \end{bmatrix} = \begin{bmatrix} a_{11} & a_{12} & \cdots & a_{1K} \\ a_{21} & a_{22} & \cdots & a_{2K} \\ \vdots & & & \\ a_{N1} & a_{N2} & \cdots & a_{NK} \end{bmatrix} \begin{bmatrix} p_{11} & p_{12} & \cdots & p_{1M} \\ p_{21} & & & p_{2M} \\ \vdots & & & \\ p_{K1} & p_{K2} & \cdots & p_{KM} \end{bmatrix}$$

and in matrix notation:

$$C = A P \qquad (12.5\text{-}104)$$

The calibration coefficients are now the elements of the P-matrix that are estimated by the generalized least-squares solution (OLS) according to:

$$P = (A^t A)^{-1} A^t C \qquad (12.5\text{-}105)$$

Analysis is carried out by direct multiplication of the measured sample spectrum a_0 by the P-matrix:

$$c_0 = a_0 P \qquad (12.5\text{-}106)$$

A disadvantage of this calibration method is the fact that the calibration coefficients (elements of the P-matrix) have no physical meaning, since they do not reflect the spectra of the individual components. The usual assumptions about errorless

The concentration matrix for uncentered data is:

$$\begin{bmatrix} 1 & c_{11} & c_{12} & \cdots & c_{1M} \\ 1 & c_{21} & c_{22} & \cdots & c_{2M} \\ \vdots & & & & \vdots \\ 1 & c_{N1} & c_{N2} & \cdots & c_{NM} \end{bmatrix}$$

independent variables (here the absorbances) and error-prone dependent variables (here concentrations) are not valid. Therfore, if this method of inverse calibration is used in connection with OLS for estimating the *P*-coefficients there is only a slight advantage over the classical *K*-matrix approach because a second matrix inversion is avoided. However, in connection with more soft modeling methods, such as PCR or PLS, the inverse calibration approach is one of the most frequently used calibration tools.

Soft modeling. The methods of soft modeling are based on the inverse calibration model where concentrations are regressed on spectral data:

$$C = AB \tag{12.5-107}$$

where *C*, *A* – are again the $N \times M$ concentration and $N \times K$ absorbance matrices, respectively, and *B* is the $K \times M$ matrix of regression or *B*-coefficients.

PCR approach. The method of PCR was outlined earlier in this section on the basis of singluar value decomposition (SVD). For simultaneous spectroscopic multicomponent analysis the decomposition of the absorbance matrix *A* can be written as (cf. Eq. 12.5-57):

$$A = UWV^t \tag{12.5-108}$$

Estimation of the matrix of regression coefficients *B* is performed column-wise by:

$$b = A^+c \tag{12.5-109}$$

with A^+ being the pseudo-inverse of the absorbance matrix *A* (cf. Eq. 12.5-58).

The main advantages of PCR-calibration are as follows:

(a) Decomposition of the absorbance matrix into smaller orthogonal matrices enables reduction of the dimensionality of the problem in the case of ill-conditioned systems. So, if highly correlated spectra are to be investigated, one will always obtain the best solution even in the case of nearly singluar matrices.
(b) Additional unknown components or background components can be automatically modeled as principal components if the concentrations of those components vary within the different calibration samples.

Problems may occur if small principal components are eliminated in the process of reducing the number of significant principal components/singluar values, because it might happen that one of the eliminated singluar values is important for the prediction of a certain constituent concentration. The decomposition of the absorbance matrix *A* does not consider relationships between the concentrations and the absorbances. Therefore the decomposition might be not optimal with respect to further use of the calibration model for prediction of concentrations in unknown samples.

A method that accounts for the concentration-spectra relationships during decomposition is the PLS approach.

PLS approach. Details of the PLS method were given earlier in this section. In multicomponent analysis we obtain the following equations for the decomposition of the absorbance matrix *A* (the former *X*-matrix) and the concentration matrix *C* (formerly the *Y*-matrix) according to the inverse calibration model in Eq. 12.5-107:

$$C = AB \tag{12.5-110}$$

$$A = TP^t + E \tag{12.5-111}$$

$$C = TQ^t + F \tag{12.5-112}$$

$$B = W(P^tW)^{-1}Q^t \tag{cf. Eq. 12.5-62}$$

The meaning of the additional matrices is the same as in Eqs. 12.5-60 to 12.5-62.

The main advantage of the PLS method is based on the interrelated decomposition of the concentration matrix *C* and the absorbance matrix *A*, so that with this algorithm the most robust calibrations at present can be obtained.

Diagnostic statistics

Leading vendors of software for multicomponent analysis provide nowadays a great variety of tools for diagnosing the suitability of the calibration model, for detecting outliers and influential samples or for estimating realistic prediction errors. It is not unusual that for a calibration set consisting of 30 standard samples, about 5000 different diagnostic plots could be generated.

Visual inspection should be possible from plots of predicted versus measured concentrations, from principal component plots of loadings and scores in case of soft modeling techniques, and by plotting the standard error of calibration (SEC) or the standard error of prediction (SEP) from cross-validation in dependence on the number of eigenvalues or of principal components.

Very important to diagnostic statistics is the study of the residuals. Let us return to the general least-squares model given in Eq. 12.5-49. We rewrite it here for a single y-variable as follows:

$$y = Xb + e \qquad (12.5\text{-}113)$$

where e is the vector of residuals of y-values, i.e., the difference between the measured y-value, y, and the y-value estimated by the model, $\hat{y}$; for a single y-value j $e_j = y_j - \hat{y}_j$.

In case of an inverse calibration model one can interpret the model in Eq. 12.5-113 as regressing the concentrations of a single component, y, on the spectra of the calibration samples collected in the matrix X with N rows (samples) and K columns (wavelength) according to Eq. 12.5-107. The regression coefficients b are then in a $K \cdot 1$ vector.

The relationship between the estimated and measured y-values can be described by a fundamental matrix, the *hat-matrix*, H. As explained by Eq. 12.5-51 the regression parameters are estimated by the general inverse as:

$$b = (X^t X)^{-1} X^t y \qquad (12.5\text{-}114)$$

The fitted model has the form:

$$\hat{y} = X\hat{b} \qquad (12.5\text{-}115)$$

Substitution of Eq. 12.5-114 into 12.5-115 reveals:

$$\hat{y} = X[(X^t X)^{-1} X^t y] = X(X^t X)^{-1} X^t y = Hy \qquad (12.5\text{-}116)$$

where H is the $N \times N$ hat-matrix defined by:

$$H = X(X^t X)^{-1} X^t \qquad (12.5\text{-}117)$$

The hat-matrix transforms the vector of measured y-values into the vector of fitted $\hat{y}$-values. The element of H, denoted by h_{ij}, is computed by:

$$h_{ij} = x_i^t (X^t X)^{-1} x_j \qquad (12.5\text{-}118)$$

Many special relations can be found with the h_{ij}. The relationship of the hat-matrix with the residuals can be understood from the following equations:

$$\hat{e} = y - \hat{y} = y - X(X^t X)^{-1} X^t y = [I - X(X^t X)^{-1} X^t] = [I - H]y \qquad (12.5\text{-}119)$$

The **rank** of the matrix X is easily found from the diagonal elements of the hat-matrix by the formula

$$rank(X) = \sum_{i=1}^{N} h_{ii} \qquad (12.5\text{-}120)$$

Also the elements of the hat-matrix are important for estimating the standard error of prediction. In general, the prediction error is calculated as the predictive residual sums of squares by:

$$PRESS = \sum_{j=1}^{N} \hat{e}_j^2 = \sum_{j=1}^{N} (y_j - \hat{y}_j)^2 \qquad (12.5\text{-}121)$$

Diagnostic plots in multi-component analysis:

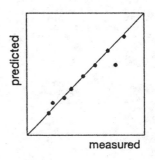

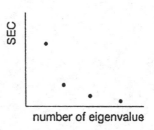

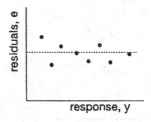

The influence of observations on a regression model can be assessed by its *leverage*.

If we have to estimate the prediction error on the basis of the calibration sample set we can leave out samples i to calculate new prediction residual errors, $\hat{e}_{(i)}$. By using the elements of the hat-matrix, sometimes also called *leverages* of case i, a good approximation of the estimated prediction error from cross-validation, e_{cv}, can be obtain according to:

$$e_{cv} = \frac{\hat{e}_{(i)}}{1 - h_{ii}} \tag{12.5-122}$$

For the new PRESS we obtain:

$$\text{PRESS} = \sum e_{cv}^2 \tag{12.5-123}$$

Low prediction errors go along with good models. Samples that do not follow the same model as the rest of the data are called *outliers*. Testing for outliers can also be based on the leverage values, h_{ii}. Suppose that the ith sample is an outlier, then a new model is calculated after deleting the ith sample from the data set. Based on the new estimation for the regression parameters $\hat{b}_{(i)}$, new residual values, $\hat{e}_{(i)}$, are obtained where the ith sample was not used. To test for the significance of the outlier a Student's t-test can be applied. If y_i is not an outlier the nul hypothesis can be assumed, such that there is no difference in predicting y_i with the full model or with the model estimated without the potential outlier i. If y_i is an outlier then the t-value should exceed the critical value at a certain risk level. The t-value can be approximated by the leverage value by formula:

$$t_i = \frac{\hat{e}_i}{s_{(i)}\sqrt{1 - h_{ii}}} \tag{12.5-124}$$

where t_i is called externally Studentized residual since case i is not used in computing $s_{(i)}$; it has $N - K - 1$ degrees of freedom; $s_{(i)}$ is derived from the residual mean square:

$$s_{(i)}^2 = \frac{\sum\limits_{j=1, j \neq 1}^{N} e_j^2}{N - K - 1}$$

Outliers should not be confused with *influential points*. Up to now we have used the residuals in order to find problems with a model. If we want to study the robustness of a model to perturbations we do an influence analysis. This kind of study is done as though the model were correct. Influential points cannot be detected by large residuals. Their removal, howver, may cause major changes in subsequent use of the model. The difference can be understood from Fig 12.5-17. A straight-line model that includes the influential point will give a different slope if that point is deleted. On the other hand, if the obvious outlier is included in the model we will estimate larger residuals for all of the cases.

To measure the change of the influential point the model has to be built by including or deleting it. From the two models we obtain different estimations for the y-values that can be used to compute a measure, the so-called *Cook's distance* D_i:

$$D_i = \frac{(\hat{y}_{(i)} - \hat{y})^t (\hat{y}_{(i)} - \hat{y})}{K s^2} \tag{12.5-125}$$

where $\hat{y}_{(i)}$ is the vector of y-values estimated from the model without case i, $\hat{y}$ is the vector of y-values estimated with the full model, and s^2 is the variance computed as in Eq. 12.5-124 for the full model:

$$s^2 = \frac{\sum\limits_{j=1}^{N} e_j^2}{N - K}$$

Again the leverage value can be used to estimate the D-value as follows:

$$D_i = \frac{1}{K} r_i^2 \left(\frac{h_{ii}}{1 - h_{ii}} \right) \quad \text{with} \quad r_i = \frac{\hat{e}_i}{s\sqrt{1 - h_{ii}}} \tag{12.5-126}$$

r_i is called the internally Studentized residual because s is estimated by including all of the data.

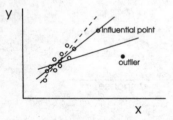

Fig. 12.5-17. Straight-line modeling in the presence of an outlier and an influential point

Large D_i-values reflect the substantial influence of case i. Therefore, samples or cases with the largest D_i-values will be of interest. In practice, those samples should be deleted and the model should be recomputed in order to understand which changes will happen.

Transformations

In spectroscopy the measurement of absorbances is only one type of response that is recorded and which obeys Beer's law. Many other principles are used such as the measurement of optical remission spectra or of X-ray fluorescence spectra that do not obey Beer's law. As long as the measured responses can be transformed to give linear dependences with the concentrations multiwavelength spectroscopy is applicable in the same way as outlined above.

Transformations can be based on mechanistic, e.g., physical models, or on empirical models, such as polynomials. A typical mechanistic model is in remission spectroscopy the Kubelka-Munk function that reveals the following linear concentration relationship after transformation of the measured reflectance:

$$F(R) = \frac{(1-R)^2}{2R} = \frac{K}{S} = \frac{kc}{S} \qquad (12.5\text{-}127)$$

where R is the remission in the sense of diffuse reflectance at infinite sample thickness, K is the absorbance coefficient, k is the absorptivity, S is the coefficient for stray light, and c is the concentration.

In case of a constant stray light coefficient the following relationship holds for M different components:

$$F(R) = \frac{1}{S} \sum_{j=1}^{M} k_j c_j \qquad (12.5\text{-}128)$$

The absorption coefficients k_j can be estimated by least-squares analysis in the same way as described for the Beer's law calibration models or by one of the methods of inverse calibration. The same is true for other specroscopic transformations, such as the Saunderson correction or the William-Clapper transformation.

Another way for empirical transformation can be based on polynomial models as we have used in case of response surface techniques. Consider an inverse calibration model where apparent absorbances, a, are to be related to a curved concentration dependence. This could be modeled by transforming the concentrations, e.g., by a quadratic model:

$$c = b_0 + b_1 a + b_2 a^2 \qquad (12.5\text{-}129)$$

Because this is still a linear model with respect to the calibration parameters, b_i, the same chemometric methods as described can be used. As an alternative to transformation of either the response or concentration variables nonlinear modelling techniques are currently investigated, such as the alternating conditional expections (ACE) algorithm or nonlinear PLS approaches. In practice, however, nonlinear dependences are seldom found. This is because spectroscopic knowledge very often allows judicious choice of a suitable transformation and, in addition, the data reduction at many wavelengths usually eliminates the need for nonlinear models in multivariate spectroscopic calibration.

General reading

Henrion, R., Henrion, G., *Multivariate data analysis*. Berlin: Springer, 1994.

Massart, D.L., Brereton, R.G., Dessy, R.E., Hopke, P.K., Spiegelman, C.H., Wegscheider, W. (Eds.). *Chemometrics Tutorials*, Collected from Chemolab, Amsterdam: Elsevier, 1990; Vol. 1–5.

Massart, D.L., Vandeginste, B.G.M., Deming, S.N., Michotte, Y., Kaufmann, L., *Chemometrics: a textbook*. Amsterdam: Elsevier Science Pubishers, 1987.

Sharaf, M.A., Illman, D.A., Kowalski, B.R., *Chemometrics*. New York: Wiley, 1986.

Weisberg, S., *Applied Linear Regression*. New York: Wiley, 1980.

Questions and problems

1. Explain the following methods for data preprocessing: centering, range scaling, autoscaling, scaling to variance one, normalization, FT-transformation, principal component projection, linear transformation, logarithmic transformation.
2. Specify the following terms in multivariate analysis: principal component, eigenvector, common and unique factor, score, loading, target vector, latent variable.
3. Which methods can be used to decide on the number of abstract component in a princial components analysis?
4. What is the difference between principal component and factor analysis?
5. The content of heavy metals in soil samples has been determined at 9 different locations. The following table provides the results of the analysis for the elements Zn, Cd, Pb, and Cu (ppm):

Sample	Zn	Cd	Pb	Cu
1	35.3	0.08	0.25	6.5
2	20.2	1.20	0.52	3.2
3	34.2	0.05	0.28	5.8
4	22.2	1.50	0.48	2.9
5	33.8	0.07	0.26	4.9
6	25.3	0.09	0.60	3.6
7	38.1	2.10	1.20	3.0
8	39.2	1.90	1.50	2.5
9	37.8	2.80	1.40	2.6

(a) What are the eigenvalues of the data matrix
(b) How many significant principal components/factors are in the data set if more then 98% of the total variance is explained?
(c) Which sample numbers belong to the found classes?
(d) Assign the following unknown sample to the appropriate class of origin: Zn 21.8 ppm; Cd 1.3 ppm; Pb 0.5 ppm; Cu 3.3 ppm.

6. Summarize the advantages and disadvantages of direct and inverse multivariate calibration.

13 Computer Hard- and Software and Interfacing Analytical Instruments

Learning objectives

- To introduce the principles of representation, conversion, and storage of analytical data
- To learn how to count with bits and how to perform arithmetic or logical operations in a computer
- To understand the principal terminology for computer systems and the meaning of robotics and automation
- To code spectra and chemical structures in analytical data bases and to learn about library search methods and the simulation of spectra

13.1 The Computer-Based Laboratory

Nowadays the computer is an indispensable tool in research and development. The computer is linked to analytical instrumentation; it serves as a tool for acquiring data, for word processing or for handling data bases and quality assurance systems. In addition, the computer is the basis for modern communication techniques, such as electronic mail or video conferences. In order to understand important principles of computer usage some fundamentals are considered here, i.e., coding and processing of digital information, the main components of a computer, programming languages, computer networking, and automation processes.

Analog and digital data

The use of digital data provides several advantages compared with the use of analog data. Digital data are less noise sensitive. The only noise arises from round-off errors due to finite representation of the digits of a number. They are less prone to, e.g., electrical interferences and they are compatible with digital computers.

As a rule, primary data are generated as analog signals either in a discrete or a continuous mode (Fig. 13.1-1). For example, monitoring the intensity of optical radiation by means of a photocell provides a continuous signal. Weak radiation, however, could be monitored by detecting individual photons by a photomultiplier.

Usually the analog signals generated are converted into digital data. This is carried out by an analog-to-digital converter as explained below.

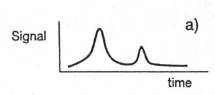

Fig. 13.1-1. Signal dependence on time for an analog (a) and a digital detector (b)

Binary versus decimal number system

In a digital measurement the number of pulses occurring within a specified set of boundary conditions is counted. The easiest way to count is to have the pulses represented as binary numbers. In this way only two electronic states are required. To represent the decimal numbers from 0 to 9 one would need 10 different states. Typically, the binary numbers 0 and 1 are represented electronically by voltage signals of 0.5 and 5 V, respectively. Binary numbers characterize coefficients of the power of 2, so that any number of the decimal system can be described.

Worked example:

The decimal number 77 is expressed as a binary number by 1001101, i.e.,

1	0	0	1	1	0	1	
1×2^6	0×2^5	0×2^4	1×2^3	1×2^2	0×2^1	1×2^0	$=$
64	$+0$	$+0$	$+8$	$+4$	$+0$	$+1$	$= 77$

Table 13.1-1. Relationship between binary and decimal numbers

Binary number	Decimal number
0	0
1	1
10	2
11	3
100	4
101	5
110	6
111	7
1000	8
1001	9
1010	10
1101	13
10000	16
100000	32
1000000	64

Table 13.1-1 provides further relationships between binary and decimal numbers. Every binary number is composed of individual bits (*bit* for binary digit). The digit lying farthest to the right is termed the *least significant* digit and the one on the left is the *most significant* digit.

How are calculations done using binary numbers? Arithmetic operations are similar, but simpler than those for decimal numbers. For addition, e.g., four combinations are feasible:

```
 0    0    1    1
+0   +1   +0   +1
─────────────────
 0    1    1   10
```

Notice that for addition of the binary numbers 1 plus 1, a 1 is carried over to the next higher power of 2.

Worked example:

Consider addition of 21 + 5 in case of a decimal (a) and a binary number (b):

```
(a)   21      (b)   10101
     + 5           +  101
     ────          ──────
      26            11010
```

Apart from arithmetic operations in the computer logical reasoning is necessary too. This might be in the course of an algorithm or in connection with an expert system. Logical operations with binary numbers are summarized in Table 13.1-2.

It should be mentioned that a very compact representation of numbers is based on the *hexadecimal number system*. However, hexadecimal numbers are easily converted to binary data so the details need not be explored here.

Table 13.1-2. Truth values for logical connectives of predicates p and q based on binary numbers. 1 relates to *true* and 0 represents *false*

p	q	p AND q	p OR q	IF p THEN q	NOT p
1	1	1	1	1	0
1	0	0	1	0	–
0	1	0	1	1	1
0	0	0	0	0	–

Digital and analog converters

Analog-to-digital converters (ADC)

In order to benefit from the advantages of digital data evaluation the analog signals are converted into digital ones. An analog signal consists of an infinitely dense sequence of signal values in a theoretically infinite small resolution. The conversion of analog into digital signals in the ADC results definitely in a reduction of information. For conversion, signal values are sampled in a predefined time interval and quantified in a n-ary raster (Fig. 13.1-2). The output signal is a code word consisting of n-bits. Using n-bits, 2^n different levels can be coded, e.g., an 8-bit-ADC has a resolution of $2^8 = 256$ amplitude levels.

Digital-to-analog converters (DAC)

Converting digital into analog information is necessary if an external device is to be controlled or if the data have to be represented by an analog output unit. The resolution of the analog signal is determined by the number of processed bits in

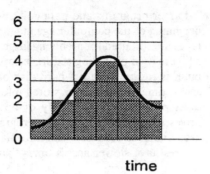

Fig. 13.1-2. Digitization of an analog signal by an analog-to-digital converter (ADC)

the converter. A 10-bit DAC provides $2^{10} = 1024$ different voltage increments. Its resolution is then 1/1024 or approximately 0.1%.

Computer terminology

Representation of numbers in a computer by *bits* has already been considered. The combination of eight bits is called a *byte*. A series of bytes arranged in sequence to represent a piece of data is termed a *word*. Typical word sizes are 8, 16, 32, or 64 bits, or 1, 2, 4, and 8 bytes.

Words are processed in *registers*. A sequence of operations in a register enables *algorithms* to be performed. One or several algorithms make up a *computer program*.

The physical components of a computer form the *hardware*. Hardware includes disk and hard drives, clocks, memory units as well as registers for arithmetic and logical operations. Programs and instructions for the computer including the tapes and disks for their storage represent *software*.

Components of computers

CPU and buses

The heart of a computer is the central processing unit or *CPU*. In a microprocessor or minicomputer this unit consists of a highly integrated chip.

The different components of a computer, its memory and the peripheral devices, such as printers or scanners, are joined by *buses*. To guarantee rapid communication among the various parts of a computer information is exchanged on the basis of a definitive word size, e.g., 16 bits, simultaneously over parallel lines of the bus. A data bus serves the exchange of data into and out of the CPU. The origin and the destination of the data in the bus is specified by the address bus. For example an address bus with 16 lines can address $2^{16} = 65536$ different registers or other locations in the computer or in its memory. Control and status informations to and from the CPU are administrated in the control bus. The peripheral devices are controlled by an external bus system, e.g., an RS 232-interface for serial data transfer or the IEEE-488 interface for parallel transfer of data.

A *bus* consists of a set of parallel conductors that forms a main transition path in a computer.

Memory

The microcomputer or microprocessor contains typically two kinds of memory – the RAM (*random access memory*) and ROM memory (*read only memory*). The term RAM is somewhat misleading and historically reasoned, since random access is feasible for RAM- and ROM-memory likewise. The RAM-memory can be used to read and write information. In contrast, information in a ROM is written once, so that it can be only read but not reprogrammed. ROMs are needed in micromputers or pocket calculators in order to perform fixed programs, e.g., for calculation of logarithms or standard deviations.

Larger programs and data collections are stored in *bulk storage devices*. In the beginning of the computer age magnetic tapes were the standards here. Nowadays tapes are still used for archiving large data amounts. Routinely 3.5″ disks (formerly $5\frac{1}{4}″$) are used providing a storage capacity of 1.44 Mbytes. In addition, every computer is equipped with a hard disk of at least 20 Mbyte up to several Gbyte. The access time to retrieve the stored information is of the order of a few milliseconds.

At present the availability of optical storage media is increasing. CD-ROM drives serve for reading large programs or data bases. Optical hard disk can be used either to read or write information. Although optically based bulk storage devices have slower access times than magnetic bulk storage media their storage capacity is larger.

I/O systems

Communication with the computer is carried out by input-output (I/O) operations. Typical input devices are the keyboard, magnetic tapes and disks, or the signals of an analytical instrument. Output devices are screens, printers, plotters, as well as tapes and disks. To convert analog information into digital or vice versa the above mentioned AD- or DA-converters are used.

Programs

Programming a computer at 0 and 1 states or bits is possible by *machine code*. Since this kind of programing is rather time-consuming higher level languages have been developed where whole groups of bit-operations are assembled. However, these so-called *assembler languages* are still difficult to handle. Therefore, high-level, algorithmic languages, such as FORTRAN, BASIC, PASCAL, or C, are more common in analytical chemistry. With high-level languages the instructions for performing an algorithm can easily be formulated in a computer program. Thereafter, these instructions are translated into machine code by means of a *compiler*.

For logical programing additional high-level languages exist, e.g., LISP (**List** **P**rocessing language) or PROLOG (**Pro**graming in **Log**ic). Further developments are found in so-called *Shells*, which can be used directly for building expert systems.

Networking

A very effective communication between computers, analytical instruments, and data bases is based on networks. There are local nets, e.g., within an industrial laboratory as well as national or worldwide networks. Local Area Networks (LAN) are used to transfer information about analysis samples, measurements, research projects, or in-house data bases. A typical local area network is demonstrated in Fig. 13.1-3. It contains a Laboratory-and-Information-Management System (LIMS), where all informations about the sample or the progresses in a project can be stored and further processed (cf. Sec. 13.2).

Wordwide networking is feasible, e.g., via Internet or Compuserve. These nets are used to exchange electronic mail (E-mail) or data with universities, research institutions, or industry.

Robotics and automation

Apart from acquiring and processing analytical data the computer can also be used to control or supervise automatic procedures. To automate manual procedures a *robot* is applied. A robot is a reprogrammable device that can perform a task cheaper and more effectively than a person. Typical geometric shapes of a robot arm are sketched in Fig. 13.1-4. The anthropomorphic geometry (Fig. 13.1-4a) is derived from the human torso, i.e., there is a waist, shoulder, elbow, and

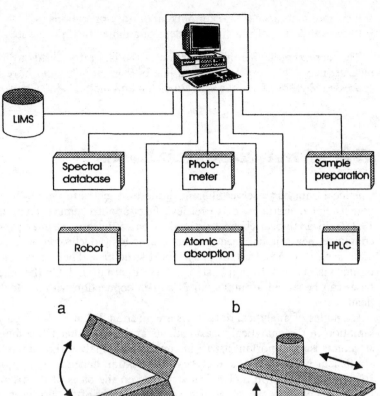

Fig. 13.1-3. Local Area Network (LAN) to connect analytical instruments, a robot, and a LIMS

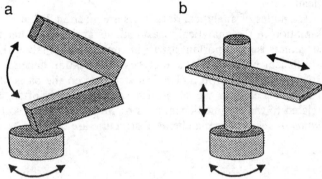

Fig. 13.1-4. Anthropomorphic (a) und cylindrical (b) geometry of robot arms

wrist. Although this type of robot is mainly found in the automobile industry it can also be used for manipulation of liquid or solid samples.

In the chemical laboratory the cylindrical geometry dominates (Fig. 13.1-4b). The revolving robot arm can be moved in horizontal and vertical directions. Typical operations of a robot are:

- *Manipulation* of test tubes or glassware around the robotic work area
- *Weighing* for determination of a sample amount or for checking unit operations, e.g., addition of a solvent
- *Liquid handling* in order to dilute or add reagent solutions
- *Conditioning* of a sample by heating or cooling
- *Separations* based on filtration or extraction
- *Measurements* by analytical procedures, such as spectrophotometry or chromatography
- *Control and supervision* of the different analytical steps

Programming of a robot is based on software dedicated to the actual manufacture. The software consists of elements to control the peripheral devices (robot arm, balance, pumps), to switch the devices on and off as well as to provide instructions on the basis of logical structures, e.g., IF-THEN rules.

Alternatives for automation in a laboratory are *discrete analyzers* and *flowing systems*. By means of discrete analyzers, unit operations can be automatized, such as dilution, extraction, or dialysis (cf. Sec. 7.4). Continuous flow analyzers or flow injection analyses serve similar objectives for automation, e.g., for the determination of clinical parameters in blood serum.

The transfer of manual operations to a robot or an automated system provides the following advantages:

- High productivity and/or minimization of costs
- Improved precision and trueness of results

- Increased assurance for performing laboratory operations
- Eased validation of the different steps of an analytical procedure

The increasing degree of automation in the laboratory leads to more an more measurements that are available on-line in the computer and have to be further processed, e.g., by chemometric data evaluation methods (cf. Chap. 12).

13.2   Analytical Databases

Chemical data bases serve different purposes, such as to search the scientific and patent-related literature or to retrieve facts about chemical compounds [13.2-1, 13.2-2]. In analytical chemistry the data bases that are of interest are those that contain either original measurements (spectra and chromatograms) or derived data, such as concentrations or chemical structures. These data can be retrieved on-line via network from a host, e.g., STN International. On the other hand, data bases can be stored at individual PCs or in connection with an analytical instrument.

Examples of analytical data bases are given in Table 13.2-1. Apart from representation of the analytical measurements in the computer the coding of chemical structures is an important aspect in constructing analytical data bases.

Efficient retrieval of the analytical information depends on appropriate *search strategies*. To confirm a chemical structure on the basis of its spectrum the data base must contain the sought-for spectrum. Very often, however, no spectrum related to the assumed chemical structure is available, so then methods for *simulation of spectra* from a chemical structure are needed.

Table 13.2-1. Some analytical data bases

Method	Data base/supplier	Data	Compounds	Remarks
NMR	SpecInfo (CC)	120 000 spectra	80 000	Host/workstation
	Bruker	19 000 spectra		Spectrometer-PC
MS	NIST/EPA/MSDC	50 000 spectra	50 000	Host/PC
	John Wiley&Sons	125 000 spectra	110 000	PC
IR	Sadtler	60 000 spectra		PC
	Aldrich-Nicolet	>100 000 spectra	10 600	Spectrometer-PC
Atomic emission	Plasma 2000 (PE)	50 000 atomic lines	60 elements	PC
GC	Sadtler	retention indices		PC

13.2.1   Representation of analytical information

Type of information

Different sorts of information for analytical data bases exist:

- numerical
- alphanumerical, e.g., text
- topological
- graphical

Numerical data are in question, if spectral, chromatographic or electroanalytical data have been measured and if concentrations, errors or analysis costs are to be stored. Typical alphanumeric data concern descriptions of sample identity or analytical procedures. Chemical structures are represented topologically. Graphical information includes plots of spectra or calibration curves as well as data of imaging procedures, e.g., from electron microprobe analysis.

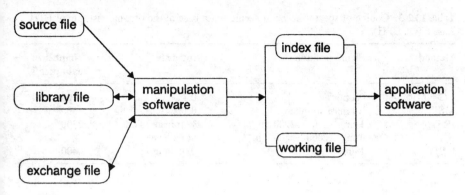

Fig. 13.2-1. Structure of analytical data bases

Structure of data bases

The demands for storing and processing information dictates the format of a data base. Usually the content of the data base is acquired in different steps.

Source and library files

The raw analytical data are stored in the source file (cf. Fig. 13.2-1). Elimination of unimportant data, filtering, transformations or compression of data leads to the library file, which is archived. The library files of an analytical data base consist of a header and a collection of data blocks. The header contains information about the file organization as well as control parameters. Stored in the data block are different sets of data that contain information about the analytical data, such as spectra, about chemical structures, and additional remarks.

Exchange files

To transfer data between different bulk storage media, files of fixed exchange formats are created. The most important exchange format in spectroscopy is the JCAMP/DX-format [13.2-3]. This format has been elaborated by the Joint Committee on Atomic and Molecular Data with the following objectives:

1. Different sorts of spectra should be describable, e.g., FT-IR, Raman, UV/VIS spectrometry, X-ray diffraction spectrometry, NMR or MS.
2. The text of the file should be readable by computer, humans, and telecommunication systems alike. Therefore, only ASCII-characters are allowed.
3. Descriptive information about the sample should be compact.
4. The format must be flexible enough to guarantee later extensions.

Table 13.2-2 demonstrates an example of a JCAMP/DX data file for storing the IR-spectrum of epichlorhydrin vapor. Here the minimum information is given. Further items concern information about the compound, e.g., the molecular mass or the Chemical Abstract number, about sample preparation, the instrument used or about measuring conditions and data processing methods, such as smoothing or derivatives.

The format is general enough that it can be exploited for similar purposes. Thus, a convention exists for describing chemical structures (JCAMP/CS-format where CS stands for Chemical Structure).

Working and index file

In Fig. 13.2-1 two additional files are mentioned, i.e., the working and index file. The working file contains the data that are needed in an actual application. Judicious organization of this file guarantees fast access to the data. The organization is determined by the type of information, i.e., whether there are numerical, alpha-

Table 13.2-2. Important information of a JCAMP/DX exchange file for an IR-spectrum

1.	##Title = Epichlorhydrin vapor
2.	##JCAMP-DX = 4.24
3.	##DATA TYPE = INFRARED SPECTRUM
4.	##ORIGIN = Sadtler Research Laboratories
5.	##OWNER = EPA/Public Domain
:	rows 6–24 are optional
25.	##XUNITS = 1/CM
26.	##YUNITS = ABSORBANCE
27.	##XFACTOR = 1.0
28.	##YFACTOR = 0.001
29.	##FIRSTX = 450.
30.	##LASTX = 4000.
31.	##NPOINTS = 1842
32.	##FIRSTY = 0.058
33.	##XYDATA = (X + +(... Y))
34.	450 58 44 34 39 26 24 22 21 21 19 16 15 15 17
35.	etc.
36.	3998 15 15 14
37.	##END =

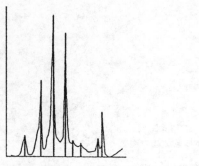

Fig. 13.2-2. Evaluation of the peak list of a spectrum

Table 13.2-3. Coding of spectra in the molecular data base of the organic institute of ETH Zurich (OCETH)

Method	Signal position	Intensity	Number of data points
^{13}C-NMR	0.1 ppm	0.1%	100
MS	0.1 mass number	0.1%	500
IR	$1\,cm^{-1}\,(200\ldots2000\,cm^{-1})$	% transmission	2300
	$4\,cm^{-1}\,(2000\ldots4000\,cm^{-1})$	at 63 levels	
UV/VIS	1 nm	0.01 in lg ε	400

numerical, topological or graphical data. Also the searching algorithm is responsible for the data representation.

Access to other data or files is organized via the index file. For processing databank informations additional software will be necessary.

Coding of spectra

The limited storage capacity of bulk storage media hindered for a long time the complete storage of spectra and chromatograms in an analytical data base. Therefore, many spectroscopic data bases contain only features of the spectra. For example UV-spectra are based on maximum absorbances or IR-spectra are represented as peak lists (cf. Fig. 13.2-2). At present one attempts to store the complete analytical information, i.e., full spectra, complete chromatograms or even spectrochromatograms that are generated by hyphenated systems, such as HPLC with diode array detection. Usually the digitization rate is determined by the measuring conditions. Table 13.2-3 exemplifies the representation of signal position and intensity of molecular spectra as stored in the beginning of the 1980s in a data base of the ETH Zurich.

As a rule the data bases contain additional information, e.g., about the starting point of a measurement, resolution or multiplets in NMR-spectroscopy.

Apart from the storage of full spectra at certain digitalization levels several algorithms are known to represent the original data in a compact, and for further processing, efficient way. As an extreme case the measured data or chemical structures are encoded as binary vectors of constant length.

As a rule the user does not need to know about the details of data compression. In addition, because of the increasing performance of computers the necessity for compressing the original data decreases, at least for analytical data banks.

Table 13.2-4. Symbols for the HOSE-substructure code ordered by priority

Symbol	Meaning
R	Ring
%	triple bound
=	double bond
*	aromatic bond
C	C
O	O
N	N
S	S
X	Cl
Y	Br
&	ring closure
,	separator
(//)	sphere separator

Coding of chemical structures

Fragmentation codes

An easy possibility to convert a chemical structure into a computer readable format is based on fragmentation codes. Typical fragmentations are aromatic rings, structural skeletons, the OH-group or azo-group (–N=N–). The fragmentations are numbered and stored in a fragmentation list. More generally, a fragmentation is formed on the basis of a freely eligible centre of the molecule and is described by its 1st to 4th spheres by means of hierarchically ordered symbols.

Consider a conventional encoding of chemical structures in ^{13}C-NMR-spectroscopy as introduced with the so-called HOSE-Code (Hierarchically Ordered Spherical Description of Environment) [13.2-4]. Table 13.2-4 contains some symbol descriptions of this code.

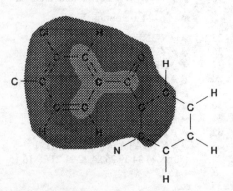

Fig. 13.2-3. Spheres around a carbon atom (bold face) as the basis for encoding the structure by the HOSE code

Worked example:

The molecule given in Fig. 13.2-3 is to be coded by the HOSE-code. The carbon atoms around the centre of the molecule (bold face C-atom) in the 1st sphere are described according to Table 13.2-3 by:

    ˆC˙CC(

Brackets symbolize the end of the first sphere. The 2nd sphere reads then as:

    ˆC,ˆC,= OC/

Here the end of the sphere is characterized by the symbol /. Analogously the 3rd and 4th sphere is obtained and the HOSE code for all of the four spheres of the molecule in Fig. 13.2-3 is:

    ˆC˙CC(ˆC,ˆC,=OC/ˆCX,ˆ&,,CC/ˆ&C,,CN,C)

Although with this kind of code a fragmentation can be represented up to the 4th sphere, in many applications the description of the 1st sphere is a good approximation. Rings are characterized by special fragmentation codes, e.g., the HORD code (**H**ierarchically **O**rdered **R**ing **D**escription). In principle, the assignment of fragmentations is easy and unique. Thus, fragmentation codes are also used in structure elucidation in IR-spectroscopy, although the vibrations of a molecule are coupled and a decomposition of the molecule into individual fragmentations is sometimes misleading for the interpretation process.

The easy assignment of fragmentations does not provide the structure of the unknown compound. In a next step the fragmentations have to be connected and the hypothetical structure has to be compared with candidate structures atom-by-atom. Of course, only those molecules that contain all of the found fragmentations have to be considered as candidates.

Generation of a molecule from its fragments is carried out by a structure generator.

Matrix representation of chemical structures

A mathematical description of a chemical structure can be derived by means of *graph theory*. Fig. 13.2-4 demonstrates as an example the graph of the molecule phosgene.

The atoms of the molecule form the nodes of the graph and the bonds the edges (cf. Table 13.2-4). The coding of the graph is performed by the so-called *adjacency matrix A*. For every element a_{ij} of this matrix the following is valid: if the node K_i is connected with another node K_j, then its value is 1, otherwise it is 0. For the phosgene molecule we obtain:

$$A = \begin{matrix} & \begin{matrix} 1 & 2 & 3 & 4 \end{matrix} \\ \begin{matrix} 1 \\ 2 \\ 3 \\ 4 \end{matrix} & \begin{pmatrix} 0 & 0 & 1 & 0 \\ 0 & 0 & 1 & 0 \\ 0 & 0 & 1 & 0 \\ 0 & 0 & 1 & 0 \end{pmatrix} \end{matrix} \qquad (13.2\text{-}1)$$

Fig. 13.2-4. Chemical structure of phosgene represented as an undirected graph

The additional numbers above and at the side of the brackets correspond to the nodes in the graph as numbered in Fig. 13.2-4.

In the adjacency matrix no bonds are considered. This would be possible by the analogous representation of a molecule based on the *bond electron matrix*, BE matrix for short. In the latter the *g*-fold connection of two nodes as well as the number of *n* free electrons of an atom is accounted for. For the phosgene molecule

Table 13.2-4. Representation of a chemical structure based on undirected graphs

Chemical term	Graph theoretical term	Symbol
Molecular formula	molecular graph	*G*
Atom	node	*K*
Covalent bond	edge	*g*
Free electrons	loop	*n*
Topological map	adjacency matrix	*A*

Table 13.2-5. Connection matrix for the molecule phosgene (cf. Fig. 13.2-4).

	1	2	3	4
1	Cl	0	1	0
2	0	Cl	1	0
3	1	1	C	2
4	0	0	2	O

Table 13.2-6. Bond atoms and bonds in a connection table of phosgene (cf. Fig. 13.2-4); 1 = single bond, 2 = double bond

Atom number	Atom symbol	Atom 1	Bond	Atom 2	Bond	Atom 3	Bond
1	Cl	3	1				
2	Cl	3	1				
3	C	1	1	2	1	4	2
4	O	3	2				

the BE-matrix $\boldsymbol{B}$ is given by:

$$
\boldsymbol{B} = \begin{array}{c} \\ 1 \\ 2 \\ 3 \\ 4 \end{array}
\begin{array}{cccc} 1 & 2 & 3 & 4 \\ \left(\begin{array}{cccc} 6 & 0 & 1 & 0 \\ 0 & 6 & 1 & 0 \\ 1 & 1 & 0 & 2 \\ 0 & 0 & 2 & 4 \end{array}\right) \end{array}
\qquad (13.2\text{-}2)
$$

The advantage of matrix representation consists of the fact that all of the matrix operations can be applied to the encoded chemical structures. As can be deduced from Eqs. 13.2-1 and 13.2-2 the quadratic matrices are symmetric around the main diagonal. However, even in the case where the matrices are stored as triangular matrices a disadvantage arises from the need for high storage capacity that accompanies representation of a chemical structure in this way. In addition, the graphs are sparsely connected. Therefore many matrix elements are equal to zero and many parts of the matrix are redundant.

Connection tables

Table 13.2-7. A nonredundant connection table for phosgene (cf. Fig. 13.2-4): 1 = single bond, 2 = double bond

Node no.	Atom	Connected to	Bond
1	Cl		
2	Cl	3	1
3	C	1	1
4	O	3	2

In practice, the node-oriented connection table is applied. This table can be derived from the connection matrix of atoms (cf. Table 13.2-5) and contains only the numbered chemical elements, the bonds connected to the atoms, and the kind of the actual bond. In Table 13.2-6 the connection table for the phosgene molecule is written down. A less redundant connection table representation is given in Table 13.2-7.

As a general rule, in the representation of chemical structures the hydrogen atoms are not coded at all. If necessary, e.g., for graphical representation of a molecule, hydrogen atoms can be added by a suitable algorithm automatically.

Connection tables can be easily extended in their rows, e.g., by having information about alternating bonds, cyclic and noncyclic bonds, stereochemistry, or by description of variable positions or generic groups in a molecule. Generic groups are represented by *Markush*-structures.

Markush-structures

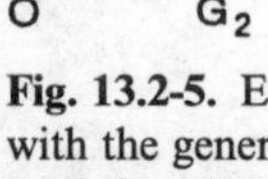

Fig. 13.2-5. Example of a Markush-structure with the general groups G_1 = phenyl, naphthenyl, N-pyridinyl and G_2 = cycloalkyl

Searching for chemical structures is often related to searching for a whole family of structures rather than for a single compound. The description of general, so-called generic classes, is feasible by means of Markush-structures (cf. Fig. 13.2-5). Such a structure consists of a core with well-defined atoms and bonds at at least one general group. The general groups are additionally specified. Substructure search is then performed for a whole compound class. This problem is especially important in the field of patent-literature where the most general claim for a compound is envisaged. Eugene A. Markush was the first to apply for a patent in the U.S.A. in 1923 that claimed for the generic class of a chemical compound.

Canonization

Important for the representation of a chemical structure as matrix or table is the unique assignment (canonization) of atoms in the structure. This can be calculated, e.g., by Morgan's algorithm [13.2-5, 13.2-6].

LIMS: Laboratory information and management systems

Among analytical data bases there are also systems for organizing laboratory work, for exchanging information and for communicating within a company. They are termed laboratory information and management systems (LIMS).

Typical performance characteristics of a LIMS are:

- Sample identification
- Design of analytical procedures
- Compilation of analytical reports
- Release of analytical results
- Acquisition of raw data and data reduction
- Archiving of analytical results
- Data base functions for chemicals, reference materials, suppliers, specifications, personnel and bibliographies

Organization of data in a LIMS can be carried out by different models of data bank theory:

- Entity relationship model
- Network data model
- Hierarchical data model
- Relational data model

Today the relational data model is the dominating one. In contrast to a telephone directory this model enables related lists to be represented. It is based on combinations of keys. The individual list can be kept short and the data structure can be extended any time if required. Thus, in the development of a LIMS not all options of the applicant have to be known in advance.

Table 13.2-8 demonstrates an example of a relation in a LIMS, where the primary key serves here the sample identification. The origin and matrix of the sample form the attributes of the relation. Every row represents a tupel of the relation.

Manipulation of entities in the data base is performed either by relation-oriented operations, such as projection, connection, and selection, or by set-oriented operations, i.e., union, intersection, and negation.

Table 13.2-8. Example relation for characterization of an analytical sample

Sample identification	Origin	Matrix
P1	final control	alloy
P2	plant 1	steel
P3	plant 2	fly ash
P4	supplier 4	ore
P5	customer 007	sewage

13.2.2 Library search

Spectra and chemical structure searches are based on distance and similarity measures as introduced in Sec. 12.5. Different strategies are known: sequential search, search based on inverted lists, and hierarchical search trees. The strategies are explained for searching of spectra.

Search strategies

Sequential search

This kind of search is based on comparing the measured spectrum with candidate spectra of the library bit by bit. Sequential search is only useful if a small data set is to be treated or if it is obligatory to retrieve every individual data set. A more efficient way is to sort the entities in a data base by deriving appropriate keys.

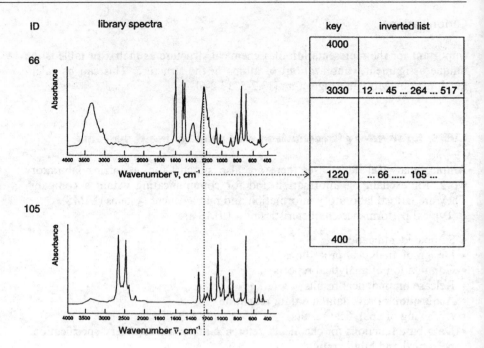

Fig. 13.2-6. Inverted list for an IR-spectral library. ID – identity number of spectrum

Inverted lists

With this method, selected data keys are defined and the data are arranged in new files that contain the information on the individual data sets. Consider an inverted list for a spectral library in the IR-range (Fig. 13.2-6). The key consists here of the numbered spectral features, i.e., in this example, the wavenumbers representing absorption maxima.

Every feature appears in the list of keys together with the identity numbers (IDs) of all spectra that contain the actual feature. After collection of all features of the unknown spectrum a rather short file can be generated on the basis of the keys that consist of all the candidate spectra.

Problems may arise if the length of the inverted lists differs. This may be because certain wavenumbers are more typical than others or certain chemical structures appear more often, e.g., –C–C– is more frequent than –C=C–. The solution to this problem is the application of Hash coding algorithms: the key is coded by a random number that is then stored in a random access file.

Hierarchical search trees

Hierarchical arrangements of spectra or chemical structures are based on grouping of data by means of some similarity measure. The fundamentals have been introduced with cluster analysis in Sec. 12.5.2. The main problem in library search is in deciding the metric to be chosen in order to describe the similarity of spectra or chemical structures. In addition, clustering of large data amounts may still be limited by the computer resources available at present.

Similarity measures for spectra

Comparison of a measured spectrum with a candidate library spectrum is feasible by different principles:

- Correlation of spectra
- Similarity and distance measures
- Logical operations

Correlation measures

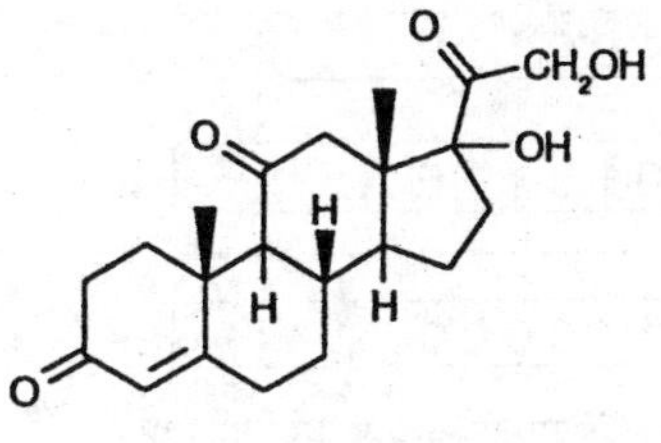

Comparison of full spectra prove a success by applying correlation or similarity measures. In the case of correlation the coefficient of correlation between the spectra to be compared is computed (cf. Eq. 12.5-79 in Sec. 12.5.4). Ranking the comparisons by the size of correlation coefficients provides a *hit list* that describes the quality of comparison between the unknown and candidate library spectrum (Table 13.2-9). In principle, the spectrum with the highest correlation coefficient is the sought-for spectrum. In order to ensure that a certain degree of similarity is reached for the top spectrum of the hit list a threshold for assigning the library spectrum should be specified.

cortisone

Compound	Correlation coefficient
Cortisone	0.999
Dexamethason	0.965
Betamethason	0.962
Prednisolon	0.913

Table 13.2-9. Hit list for comparison of UV-spectra of corticoids in methanol/water (70%/30%) eluents based on the correlation coefficient (Eq. 12.5-79 in Sec. 12.5.4)

Typically, the correlation coefficient is used for comparison of UV-spectra, e.g., as is common in HPLC with diode array detection.

Similarity and dissimilarity measures

Comparisons of spectra with similarity or distance measures are based on the same definitions as given in Eqs. 12.5-7 to 12.5-10 in Sec. 12.5.2.

In the case of full spectra or of other analytical signal curves both the Euclidian distance and the Manhattan distance are used as similarity measures. In the case of the Manhattan distance the differences between the unknown and the library spectrum are summed. As a result of comparison, a hit list ranked according to distances or similarities of spectra is again obtained.

Grouping and feature selection

One possibility to speed up the search is to sort the data sets as a preliminary. Here the methods of unsupervized pattern recognition are used, e.g., principal components and factor analysis, cluster analysis or neural networks (cf. Sec. 12.5.2). The unknown spectrum is then compared with every class separately.

To improve spectral comparisons the selection of features will very often be necessary. For example, in mass spectrometry the original spectra are scarcely used for spectral retrievals. Instead a collection of features is derived, such as the modulo-14-spectrum.

Logical operations

Comparison of spectra is also possible by using logical connectives (cf. Table 13.1-2). A prerequisite for logical comparison is the conversion of spectra into a bit format. Bit-wise conversion can be performed with complete spectra. More frequently, however, the bit vectors are formed from features derived from the raw spectral data. The logical operations can also be considered as distances of the derived vectors in bit space.

In the simplest case the unknown spectrum is compared with the candidate library spectrum by an AND-connective (cf. Table 13.1-2).

A typical dissimilarity measure is the so-called Hamming-distance based on the exclusive OR (cf. Fig. 13.2-7) calculated as follows:

$$\text{Hamming-distance} = \sum_{i=1}^{p} \text{XOR}(y_i^A - y_i^B) \qquad (13.2\text{-}1)$$

$0\ 0 \rightarrow 0$

$0\ 1 \rightarrow 1$

$1\ 0 \rightarrow 1$

$1\ 1 \rightarrow 0$

Fig. 13.2-7. Connection of bits by exclusive OR (XOR)

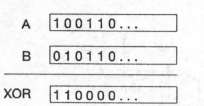

A   1 0 0 1 1 0 . . .

B   0 1 0 1 1 0 . . .

XOR 1 1 0 0 0 0 . . .

Fig. 13.2-8. Comparison of an unknown spectrum (A) with a candidate library spectrum (B) by the exclusive OR-connective (XOR)

where y_i^A is the bit vector for the unknown spectrum at point i, y_i^B is the bit vector of the candidate library spectrum, and p is the number of points (wavelengths).

Fig. 13.2-8 demonstrates the distance calculation. For identical spectra the Hamming-distance would be zero.

A combination of different logical operations can be found in mass spectrometry with the following dissimilarity measure S':

$$S' = 2 + \text{XOR} - 2 \sum_{i=1}^{p} \text{AND}(y_i^A - y_i^B) \tag{13.2-2}$$

In the examples discussed the comparison is based on bit *vectors*. Another type of logical operation can be based on *set*-oriented comparisons. The latter procedure is necessary if the length of data sets differs from spectrum to spectrum. Typical examples are the peak list in IR-spectrometry or capillary chromatograms. Fig. 13.2-9 demonstrates the set-oriented comparison of two spectra.

The peak list of the library spectrum is assumed as errorless or crisp and the peak positions of the measured unknown spectrum are characterized by intervals. Comparison is performed usually by the AND-connective, i.e., by the intersection of both sets.

In case of fuzzy intervals the spectra have to be compared on the basis of fuzzy set theory.

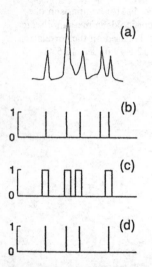

Fig. 13.2-9. Set-oriented comparison of two spectra: (a) Unknown original spectrum; (b) signal positions for the unknown spectrum; (c) library spectrum with intervals for signal positions; (d) comparison of (b) and (c) by intersection of the two sets (AND-connective)

Similarity measures for chemical structures

If the chemical structures are encoded in *fragmentation codes* a preselection of substructures is feasible. The comparison between the coded vectors of the unknown and library structure is possible by means of AND-connection.

A comparison of structures *atom-by-atom* is based on the connection tables. Consider the classical example presented by E. Meyer in 1970 as given in Fig. 13.2-10.

Worked example:

Molecule:

$$H_2N - \underset{1}{\overset{\overset{5}{\overset{O}{\|}}}{C}} - \underset{2}{O} - \underset{3}{CH_2} - \underset{4}{\overset{\overset{7}{H_2C} - \overset{8}{NH_2}}{C}} \equiv \underset{5\ \ 9}{CH} - \underset{10}{\overset{\overset{11}{\overset{O}{\|}}}{C}} - \underset{12}{CH_2} - \underset{13}{OH}$$

Connection table:

Atom no.	Symbol	Atom1	Bond	Atom2	Bond	Atom3	Bond
1	N	2	1				
2	C	1	1	3	1	5	2
3	O	2	1	4	1		
4	C	3	1	6	1		
5	O	2	2				
6	C	4	1	7	1	9	2
7	C	6	1	8	1		
8	N	7	1				
9	C	6	2	10	1		
10	C	9	1	11	2	12	1
11	O	10	2				
12	C	10	1	13	1		
13	O	12	1				

Sought-for substructure:

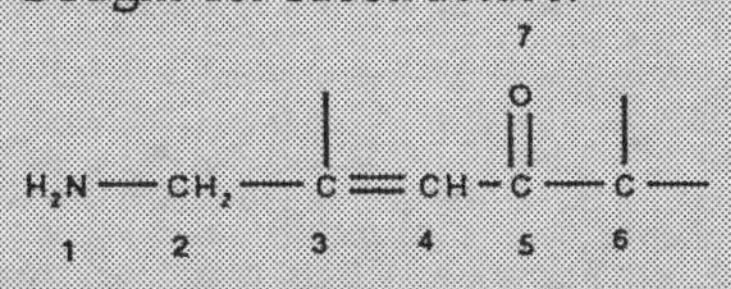

$$H_2N \underset{1}{---} CH_2 \underset{2}{---} \underset{3}{C} \equiv CH \underset{4}{-} \underset{5}{C} \underset{6}{-} \underset{}{C} ---$$

with O (7) double-bonded to atom 5

Atom no.	Symbol	Atom1	Bond	Atom2	Bond	Atom3	Bond
1	N	2	1				
2	C	1	1	3	1		
3	C	2	1	4	2		
4	C	3	2	5	1		
5	C	4	1	6	1	7	2
6	C	5	1				
7	O	5	2				

Match

Substructure:	1	2	3	4	5	6
Molecule structure:	8	7	6	9	10	12

Fig. 13.2-10. Example of substructure search based on connection tables of the chemical structures (for representation of the connection table cf. Table 13.2-7)

The task is to check whether the specified substructure is contained in the given molecule. Starting with the atom-wise comparison in the molecule at the nitrogen atom no. 1 matches the attached C-atom. The next atom, however, is oxygen in the molecule and carbon in the substructure. At this point the search is stopped and back-tracked. A new trial starting at the N-atom no. 8 in the molecule will match the substructure step by step.

13.2.3 Simulation of spectra

In case there is no spectrum in the library to elucidate a chemical structure interpretative methods are needed. The methods of pattern recognition and of artificial intelligence have to be used then. As a result different chemical structures will be obtained as candidates for the unknown molecule. To verify an assumed structure, simulation of spectra becomes important. In a final step the simulated spectrum could be compared with the measured one.

High-performance methods for routine simulations of IR- and mass spectra are not available yet. In IR-spectroscopy the best simulations are obtained on the basis of quantum chemical approaches.

More successful simulations are for NMR-spectroscopy (Fig. 13.2-11). Simulation of chemical shifts, δ, are based on increments that are derived from investigations of a set of well-characterized compounds by means of multiple linear regression analysis (cf. Sec. 12.5.4):

$$\delta = b_0 + b_1 X_1 + b_2 X_2 + \cdots + b_n X_n \tag{13.2-3}$$

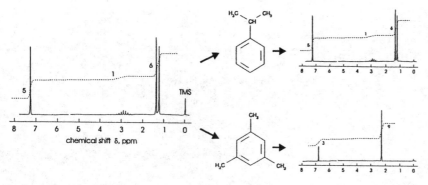

Fig. 13.2-11. Simulation of the ^{1}H-NMR-spectrum for cumene (above right) and 1,3,5-trimethyl benzene (below right) to verify the measured spectrum (shown left) as being cumene

where X_i is a numerical parameter describing the environment of the structure, b_i are regression coefficients, and n are number of descriptors.

References

[13.2-1] Ash, J.E., Hyde, E. (Eds.), *Chemical Information Systems*. Chichester: Ellis Horwood, 1975.

[13.2-2] Warr, W.A., Suhr, C., *Chemical Information Management*. Weinheim-New York: VCH, 1992.

[13.2-3] McDonald, R.S., Wilks, P.A., *JCAMP-DX: A Standard Form for Exchange of Infra-red Spectra in Computer Readable Form*. *Applied Spectroscopy*, 1988, **42**, 151–162.

[13.2-4] Bremser, W., HOSE – *A Novel Substructure Code*, *Anal. Chim. Acta*. 1978, **103**, 355–365.

[13.2-5] Morgan, H.L., *The generation of unique machine description for chemical structures – a technique developed at Chemical Abstract Service*. *J. Chem. Doc.*, 1965, 107–113.

[13.2-6] Zupan, J., *Algorithms for Chemists*. Chichester: Wiley&Sons, 1989.

Questions and problems

1. Calculate the resolution for the following 10, 16, and 20-bit analog-to-digital converters.
2. How many bits are stored in an 8-byte word?
3. Draw the chemical structure of the molecule given by the following connection table according to Table 13.2-7:

Node no.	Atom	Connected to	Bond
1	C		
2	C	1	1
3	C	2	1
4	C	3	2
5	N	3	1

4. Which substructures describe the two fragmentations based on the HOSE-codes =OCC(,*C*C,//) and *C*CC\$(*C,*C,C//)?
5. How does one derive a peak list from a full spectrum?
6. Which spectral features can be used as keys in sorting spectra in an inverted list?
7. Mention common similarity measures in retrieving spectra and chemical structures.
8. Which methods can be used to group spectra and structures?

Part V
Total Analysis System

14 Hyphenated Techniques

Learning objectives

■ To introduce the most important hyphenated techniques
■ To describe the interfaces necessary to couple chromatographic to spectroscopic techniques
■ To discuss the analytical information provided by hyphenated techniques and their application to complex analytical problems

14.1 Introduction

"GC-MS, LC-MS, GC-IR, LC-IR, TLC-IR.... As the rising tide of alphabet soup threatens to drown us, it seems appropriate to look at the common denominator behind all of these new techniques. The hyphen which is the single common constituent of all of these acronyms is also the symbol of their common principle, the marriage (sometimes a shotgun one) of two separate analytical techniques via appropriate interfaces, usually with the backup of a computer tying everything together." With this pointed remark, Hirschfeld began his milestone essay on hyphenated techniques [14.1-1], thereby coining, or at least redefining, this term.

In principle, we can understand under the term "hyphenated techniques" the coupling of two (or more) analytical techniques with the goal of obtaining a faster and more efficient analytical tool. This spans the range from the combination of two spectroscopic techniques (e.g., MS and MS), two chromatographic techniques (e.g., LC and GC) to the coupling of separation techniques with spectroscopic detection techniques (e.g., GC-MS) which expresses a closer understanding of hyphenated techniques.

Hyphenated techniques couple a separation technique with a spectroscopic detection technique.

The motivation to do so stems from the ever-increasing demands posed on analytical methods to cope with complex samples. Even for high-performance separation techniques such as gas or liquid chromatography, the number of substances that can be separated is amazingly small if there is no restriction on randomness. Hirschfeld has demonstrated that this number can be as small as 14 for a column with 1000 theoretical plates. In practice, this limitation is circumvented by choosing nonrandom conditions, e.g., by optimizing the temperature program in GC or gradient elution in HPLC. Another way to get around this limitation is the practice of applying multidimensional separations. Here, the requirement for successful method development is that the coupled techniques are orthogonal, i.e., they provide different information. In many types of chromatography (GC, HPLC, SFC, TLC) this multidimensional approach is already well established, often as part of the sample preparation procedure. When, e.g., analyte enrichment and clean-up is done by solid-phase extraction, this is basically a simple liquid chromatography prior to gas chromatographic analysis. The combination of two orthogonal techniques, however, is not the only criterion for a hyphenated system. If the combination is done in two separate steps, we call it sample preparation. In hyphenated systems, the coupling is done on-line. The combination of separation techniques and spectroscopic techniques for detection is the focus of this chapter.

The multidimensionality of the data provides more information than either of the separate techniques. This can be understood from the fact that chromatographic techniques can provide excellent separation and quantitation, while identification of the separated compounds is often impossible on the basis of the chromatographic data alone. Spectroscopic techniques, on the other hand, have superior identification capabilities. Identification from spectroscopic data is often unambiguous, either via comparison with library spectra or by interpretation. The feasibility of analyte identification, however, depends critically on the purity of the compounds in the chromatographic peak. This is where the hyphenation of

The advantage of hyphenated techniques is the additional dimensionality of the data obtained.

Table 14.1-1. The state of the art in hyphenated methods (adapted from [14.1-1])

	Mass spectrometry	Infrared	Optical emission	Atomic absorption	Nuclear magnetic resonance	Fluorescence	UV-VIS
Gas chromatography	***	***	**	*		**	*
Liquid chromatography	***	**	*	**	**	***	***
Supercritical fluid chromatography	*	*	*				
Thin layer chromatography	*	**				***	***
Capillary electrophoresis	**	*				***	***

*** commercially available and widely used
** commercially available, but less frequently used
* at the research stage, not commercially available

Most hyphenated techniques require an interface between the separation technique and the detector which is decisive for their performance.

separation and spectroscopic techniques demonstrates its strength and efficiency. Sections 14.2 and 14.3 present the major hyphenated techniques, their instrumental requirements, and their strengths and shortcomings.

The main problem of a hyphenated system is the adaptation of one set of parameters, i.e., chromatographic running conditions, to the operating conditions of the second technique. In general, this requires an interface. It is in the nature of the development of hyphenated systems that, in the early stages, this interface is rather complicated and restricting, whereas at a later stage of instrumental development the interface either becomes simpler and eventually even obsolete, or it becomes an integral part of the analytical method, opening new possibilities and widening the scope and the strength of the hyphenated technique. Table 14.1-1 gives an overview of the possibilities of hyphenation and their present status.

14.2 Hyphenated Gas Chromatographic Systems

Gas chromatography (GC) is by far the most important separation technique for volatile organic compounds. A variety of detectors for gas chromatography has been developed since the mid 1950s, some of them showing remarkable sensitivity and/or selectivity (see Sec. 5.1). However, the driving force for the development of hyphenated GC-MS and GC-FTIR systems has been the need for unambiguous detection of the separated compounds.

14.2.1 Gas chromatography-mass spectrometry (GC-MS)

Principles

Both techniques coupled in GC-MS, i.e., gas chromatography (GC) and mass spectrometry (MS) had already been used separately in analytical chemistry before the first succesful hyphenation was reported in 1952 [14.2-1]. Since both GC and MS were at an early developmental stage at that time, it took five more years until the first GC-MS instrument was commercially available.

Nowadays, GC-MS is the most popular of all hyphenated techniques, and the combination of one of the most powerful separation techniques with the high degree of structural information provided by mass spectrometry has made GC-MS the workhorse of trace organic analytical laboratories. The fact that modern (bench-top) GC-MS instruments are simple to use, small, robust, allow flexible operation, and are reasonably priced has greatly contributed to their success.

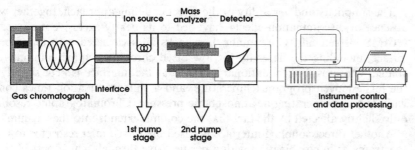

Fig. 14.2-1. Basic components of a GC-MS-system

Mass spectrometric detection for gas chromatography offers a number of unique advantages, e.g., the use of isotope-labeled compounds as standards to increase analytical accuracy, the determination of the elemental composition of a compound if a high-resolution instrument is used, and the possibility of separating chromatographically unresolved peaks based on the difference in their mass spectra.

Instrumentation

Only a brief account will be given here of the instrumentation for GC-MS, as instrumental aspects of both gas chromatography (Sec. 5.1) and mass spectrometry (Sec. 9.4) have already been covered in full detail. Special emphasis is laid on instrumental features relevant to the coupling of both techniques, i.e., the interface.

The principal difficulty of coupling gas chromatography to mass spectrometry is to introduce a compound at atmospheric pressure (at which the chromatographic separation is carried out) into a high vacuum (the analyzer part of the MS is usually operated at a pressure of approximately $10^{-5}-10^{-8}$ Torr). Different GC-MS interfaces have been developed, however, to overcome this apparent incompatibility. The basic components of a GC-MS system are shown in Fig. 14.2-1.

Coupling GC to MS detection requires the introduction of the analytes at atmospheric pressure into the high vacuum of the mass analyzer.

GC-MS interfaces

At the time when GC-MS was developed, GC was performed with packed columns where the flow rate was of the order of 60 mL/min and higher. This flow rate was not compatible with the high vacuum of the MS system. The most decisive point for the commercial success of hyphenated GC-MS systems was thus the development of suitable interfaces to overcome this limitation. The requirements for an interface include: the ability to reduce the volumetric flow of the GC column to the extent that the high vacuum in the mass analyzer can be maintained; selective separation of the carrier gas; and maintenance of the chromatographic separation.

Meanwhile capillary GC (CGC) has become the most prominent mode of GC. In CGC the flow rate is typically in the low and sub-mL/min range which practically eliminates the need for a special interface with its associated problems.

Of the different coupling schemes and interfaces, only the direct coupling interface will be discussed in detail, since the devices designed for interfacing packed-column GC to a mass spectrometer, such as molecular separator interfaces, are becoming less important.

Direct coupling interface. The simplest of all ways of introducing the chromatographic effluent into the MS is by direct coupling, i.e., to insert the chromatographic column directly into the ion source of the mass spectrometer via a vacuum-tight flange. However, this can be realized only with narrow-bore capillary columns with typical flow rates of 1–2 mL/min. This gas flow is still compatible with modern MS vacuum systems and also close to the optimum flow rate for open tubular capillary columns.

Direct-coupling GC-MS is suitable for capillary columns up to a flow rate of ca. 2 mL/min.

The simplicity and versatility of the direct coupling approach, together with the absence of discrimination effects, have led to its nearly exclusive use in capillary GC-MS instruments. Only minor drawbacks are associated with this type of interface, including the necessity for low bleed chromatographic columns (usually coated with cross-linked stationary phases) as the interface is kept slightly above the highest oven program temperature, and a shift in retention times compared with detectors operating at atmospheric pressure. Chromatographic resolution is only slightly affected by the fact that the column extends into the vacuum.

Another direct coupling interface is based on a fixed inlet restrictor to the mass spectrometer in combination with a needle valve through which part of the chromatographic effluent is bypassed. This kind of interface provides more flexibility in the column flow rates as variable fractions can be split and bypassed from the mass spectrometer.

Sample ionization

For a comprehensive coverage of the different ionization techniques, the reader is referred to Sec. 9.4. Although various ionization methods are available, electron impact (EI) and chemical ionization (CI) are the most common for general use in GC-MS analysis. Of these two techniques, electron impact ionization is, not only for historical reasons, by far the most widely used ionization technique (>90% of all aplications). A short discussion of the reasons is given.

Electron impact (EI) ionization produces fragment-rich mass spectra that provide structural information.

Electron impact ionization.   In electron impact ionization (EI), sample molecules, entering the ion source from the gas chromatographic column, are ionized by thermal electrons emitted from a tungsten or rhenium filament (the cathode) and accelerated toward the anode. Collision of the electrons with sample molecules, upon which part of the kinetic energy of the electrons is transferred to the molecules, leads to their excitation, fragmentation, and ionization. As the distribution of the internal energy directly affects the appearance of the mass spectra, and is strongly dependent on the electron beam energy E_{el}, it is usually fixed at a standard value of $E_{el} = 70\,eV$.

The reasons for this are:

- This is the electron energy at which the highest ion formation occurs
- The mass spectrum appearance changes only slightly with changes in E_{el}
- Both a relatively large molecular peak and intense fragment ion peaks may be formed (depending on the molecular structure)
- The internal energy distribution of the ions formed is practically the same for different instruments, making spectra virtually instrument-independent

Spectra are compiled in libraries and used for identification of compounds via a search procedure (see Sec. 9.4).

Chemical ionization (CI) produces very simple mass spectra that usually provide molecular mass information.

Chemical ionization.   In chemical ionization (CI), the analytes are ionized by gas-phase ion–molecule reactions. To achieve this, the reagent gas (usually methane, *iso*-butane, ammonia, or water) is introduced into the ion source at comparatively high pressure (0.01–2 Torr) and reagent gas ions are generated by electron impact ionization. The analyte molecules are ionized indirectly via a series of reactions with the reagent gas, by which only a small amount of energy, with a narrow distribution, is transferred to the analyte molecules via collisions. This explains why CI is often termed a "soft" ionization technique. Soft ionization leads to less fragmentation and thus more abundant molecular ions are obtained in CI compared with EI. While the low degree of fragmentation increases the sensitivity of CI-MS, it yields only limited structural information. A valuable feature of CI, however, is that its selectivity can be tuned by the choice of reagent gas (Fig. 14.2-2). If a very mild reagent gas such as ammonia is used, only very basic compounds, e.g., amines, are protonated and can be detected. The dependance of CI spectra on experimental conditions, notably on the ion source pressure, precludes CI spectra from different instruments from being compared or library searched. Depending on the reagent gas, positive or negative ions are obtained.

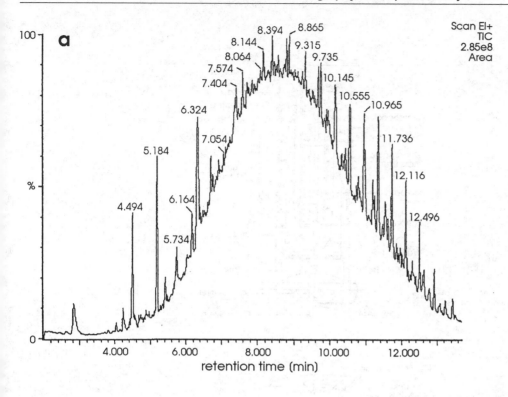

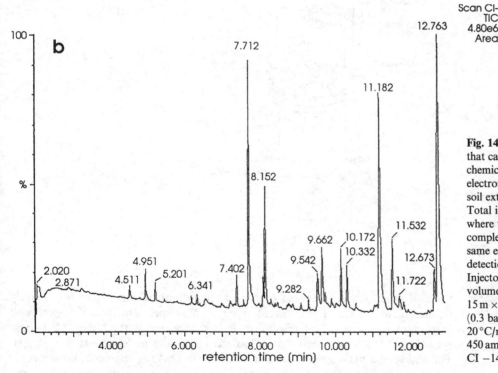

Fig. 14.2-2. Demonstration of the selectivity that can be achieved by (negative ion) chemical ionization versus (positive ion) electron impact ionization for the analysis of a soil extract for polychlorinated biphenyls. (a) Total ion chromatogram (TIC) in EI mode where the huge interferences from the complex matrix are evident: (b) TIC of the same extract in methane-CI with negative ion detection. Chromatographic conditions: Injector temperature: 250 °C, Injection volume: 1 µl (splitless), Column: DB 5ms, 15 m × 0.25 mm × 0.25 µm, Carrier gas: He (0.3 bar), Oven program: 60 °C (1 min) → 20 °C/min → 280 °C (10 min), Scan range: 34–450 amu, Source temperature: EI −250 °C; CI −140 °C [14.2-2]

Types of mass analyzer

The different types of mass spectrometers have been covered in Sec. 9.4. We therefore restrict ourselves to the description of features relevant for GC-MS systems, such as sensitivity, linear dynamic range, resolution, mass range, and scan speed. The scan speed of a mass spectrometer is the time needed to scan an order of magnitude on the mass axis (e.g. from m/z 50 to 500). Owing to the small peak width in capillary column GC-MS, fast scan speeds are necessary (≤ 1 s/decade) in order to obtain at least 3–5 spectra per peak in full-scan mode. The limited mass range of some mass analyzers does not present a problem, since the molecular

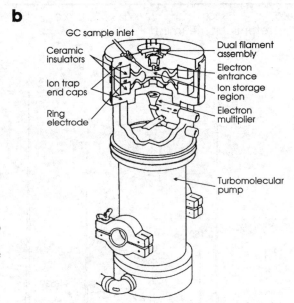

Fig. 14.2-3. Schematic view of (a) a bench-top quadrupole mass spectrometer which is fully integrated with the ion source, the quadrupole analyzer, and the detector in the diffusion pump housing, leading to a very compact design; (b) an ion-trap detector [14.2-3]

Table 14.2-1. Resolution of ions with similar nominal mass

Ion	Nominal mass	Exact mass
$C_5H_{11}O_2$	103	103.07595
$C_4H_{11}N_2O$	103	103.08718
$C_6H_{15}O$	103	103.11236
$C_5H_{15}N_2$	103	103.12359

mass of compounds amenable to gas chromatographic separation is generally smaller than 600 Da. The different types of mass spectrometer differ greatly in their resolving power. The resolution R is a measure of the ability of a mass spectrometer to resolve two ions of different m/z and is defined as $R = m/\Delta m$. The ability of a mass spectrometer to resolve two peaks of adjacent unit masses is called unit mass resolution. Quadrupole instruments are usually operated at unit mass resolution. In contrast to this, double-focusing instruments are capable of high mass resolution ($R \geq 10\,000$). This is important, as the elemental composition can often be deduced directly from the exact mass of a fragment ion. To Separate the ions $C_5H_{11}O_2$ and $C_4H_{11}N_2O$ (Table 14.2-1) with $\Delta m = 0.01123$ Da, a resolution of at least $R = 9172$ is required.

Quadrupole mass filters allow high scanning speeds up to a transmission range of ca. m/z 2000.

Quadrupole mass filters. The most popular mass spectrometers for coupling with gas chromatography are quadrupole instruments (Fig. 14.2-3a). They are comparatively inexpensive, compact, and robust isntruments. Their operation and maintenance takes less skill than that of magnetic sector instruments. They allow high scanning speeds (<1 s/decade) and automatic switching from positive to negative ion detection. The drawbacks of this type of instrument are that they show mass discrimination and they are less sensitive than magnetic sector instruments, especially in the higher mass range (>500 Da). The upper transmission limit of common quadrupole instruments is between m/z 1000 and 2000. This is usually not a critical point since most compounds separated by GC have molecular masses less than this.

Magnetic sector (double-focusing) instruments. Double-focusing instruments consist of a magnetic sector analyzer that disperses the ions according to their momentum (which depends on the m/z ratio) and an electrostatic analyzer (ESA) where the ions of different kinetic energy are focused again. This double-focusing effect allows one to reach a much higher resolving power than with a quadrupole instrument. With modern magnetic sector instruments, resolving powers of 100 000 and more can be realized, and compounds of up to 20 000 Da be transmitted. When operated at low resolution, magnetic sector instruments are much more sensitive than quadrupole instruments and still equally sensitive at a resolving power of 10 000. At this high resolving power, the elemental composition of a compound can be be calculated directly from its exact molecular mass. Since this ability is more dependent on the mass accuracy than on the mass resolution, it will not yield unambiguous information except for the lowest molecular mass compounds (300 Da or less).

> Double-focusing instruments can reach a resolution of up to ca. 100 000 and transmit molecules up to ca. 20 000 Da.

Being much more complex than quadrupole instruments, they are also significantly more expensive. As they are more difficult to operate and to maintain, they are used only where their high resolution, high sensitivity, or high mass range are essential.

Ion trap detectors (ITD). Although a fairly recent development, the ion trap detector is gaining great popularity as a gas chromatographic detector. It is derived from the quadrupole mass analyzer, but has undergone a complete rearrangement. The rf voltage which controls scanning is applied to a circular ring electrode situated between two end-caps at ground potential (Fig 14.2-3b). Sample molecules entering the detector directly from the GC column are ionized by electron impact. Both ionization and introduction of the sample into the ion trap are pulsed by switching the acceleration voltage on and off for appropriate intervals (gating). The ions are then stored in the ion trap on stable trajectories. Analysis of the stored ions is accomplished by increasing the rf voltage stepwise. This makes the trajectories of the ions successively unstable, and leads to their expulsion from the trap into the electron multiplier where they are detected. In contrast to quadrupole instruments where ionization, separation, and detection of the ions occurs in separate compartments, in ion traps these steps take place in the same chamber at comparatively high pressures 10^{-2}–10^{-3} Torr). A unique feature of the ion trap detector is its automatic gain control, which adjusts the actual ionization time according to the product of sample size and ionization efficiency to obtain the optimum amount of ions in the ion trap. This is achieved by a prescan of approximately 0.2 ms during which all ions above a certain threshold value are detected. From this preliminary scan, the ionization time and the full-scan acquisition time are estimated. It is mainly this feature that contributes to the remarkable sensitivity of the ion trap detector, making it possible to obtain full-scan mass spectra of less than 5 pg substance, and providing a dynamic range of ca. 10^6. The very compact design, its simple operation and comparatively low price, as well as the possibility of using both EI and CI ionization and carrying out MS-MS experiments, make the ion trap a very attractive detector for gas chromatography.

> Ion trap detectors are very compact quadrupole-like mass filters in which ionization, mass filtering and detection takes place in the same chamber.

Other types of mass spectrometer. To a lesser extent, other mass spectrometric detectors have been used for gas chromatography. These include Fourier transform mass spectrometers (FT-MS), time-of-flight mass spectrometers (TOF-MS) and tandem mass spectrometers (MS-MS). In most cases, the considerably higher costs and experimental effort have prevented the wider use of these techniques. While very popular for HPLC detection, MS-MS is less frequently applied in GC. The advantage of the very high selectivity of the MS-MS arrangement in its different modes of operation is very attractive and can be decisive for the determination of analytes at low levels in complex matrices (e.g., dioxins in environmental samples).

Fig. 14.2-4. Representation of the three-dimensional data structure obtained by GC-MS operated in the full-scan mode. The data types that can be obtained are: (a) Mass spectra; (b) Ion chromatograms and selected ion monitoring traces; (c) Total ion chromatograms

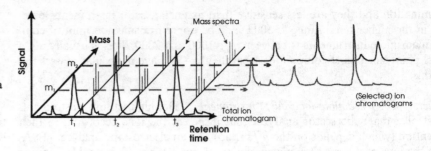

Operational modes of GC-MS

In the full-scan mode, mass spectra are continuously acquired during a chromatographic run.

Full-scan mode. The popularity of GC-MS owes a lot to the power of modern computer systems used for instrument control, acquisition, and evaluation of the huge amount of data that is produced within a chromatographic run with mass spectrometric detection. In general, GC-MS operated in the scan mode yields a three-dimensional data array where the intensity is recorded as a function of both time (containing the chromatographic information), and mass (spectroscopic information). Different types of data with consequently different information content can be extracted from this three-dimensional data array (Fig. 14.2-4).

Mass spectra. A section through the data array parallel to the mass axis at a certain time t_1 during the elution of a peak produces a mass spectrum of the respective compound (the plot of the relative abundance of ions as a function of m/z). With its characteristic fragmentation pattern, the mass spectrum is the "fingerprint" of the molecule and can serve to identify the unknown substance either via library search or by interpretation of the spectrum.

Ion chromatograms. Taking a section through the above-mentioned data array parallel to the time axis at a certain mass m_1 results in an ion chromatogram, a plot of the relative abundance of an ion of fixed m/z ratio as a function of the retention time. This type of information can be useful to indicate the presence or absence of either individual compounds (if the molecule peak is extracted from the full-scan data array) or compound groups (if the signal of a characteristic fragment ion common to all compounds to be monitored is extracted). To a limited degree, this mode of operation allows the retrieval of functional group chromatograms.

In the selected ion monitoring (SIM) mode, sensitivity is enhanced by monitoring only few selected m/z ratios and proportionally increasing the registration time.

Selected ion monitoring mode. Allowing the most flexible data evaluation, the full-scan mode as presented above suffers from the drawback of rather low sensitivity as usually a fairly wide range (e.g., $m/z = 50$–500) is scanned and the signal of the mass of interest is only extracted post-run. Thus, most of the data acquisition time is wasted on masses that are analytically irrelevant. This can be overcome by defining before the run the mass(es) to be monitored. Instead of scanning the whole range in a scan cycle, the instrument now spends the whole cycle time acquiring data at the preset mass(es). This is called single-ion monitoring (SIM) or multiple-ion detection (MID) as shown in Fig. 14.2-5. Obviously, the SIM mode is far more sensitive, owing to the improved S/N ratio (being able to detect as few as 10 pg abs. compared with ≥ 10 ng abs. in the full-scan mode), but spectral information has to be sacrificed. In order to find a compromise between sensitivity and specificity, generally two to four characteristic ions of the suspected compounds have to be measured simultaneously and the intensity ratios of the signals at different m/z ratios have to fall within defined limits for positive identification.

Total ion chromatograms (TIC) are similar to the signals obtained by a flame ionization detector.

Total ion chromatograms. The total ion chromatogram (TIC, also called the reconstructed ion chromatogram, RIC) is a chromatogram obtained by summation of the abundance of all m/z ratios in a mass spectrum (recorded either in full-scan or MID mode) and plotting the results against retention time or scan number. In its appearance, and also in its sensitivity it resembles the signal obtained by a flame-ionization detector (Fig. 14.2-5).

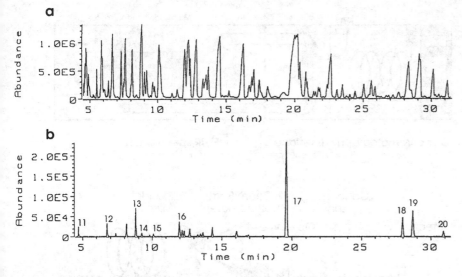

14.2.2 Gas chromatography-Fourier transform infrared detection (GC-FTIR)

Principles

While the mass spectrometer has been considered in many ways the ultimate GC detector, it still has some shortcomings. The most important are:

- The destructive nature of the MS detector
- Its general inability to distinguish between structural isomers
- The lack of direct functional group information

This is where infrared detection can be an attractive alternative, as the information provided is complementary to that obtained by mass spectrometric detection. As described in Sec. 9-2, an infrared spectrum contains information characteristic both of the functional groups contained in the molecule (in the group frequency region, 1500–4000 cm^{-1}) and the individual molecule (in the so-called fingerprint region, below 1500 cm^{-1}). As in mass spectrometry, infrared spectra can therefore be considered characteristic, or even specific for a given molecule.

GC-FTIR provides both functional group and molecule specific information of the separated compounds.

Instrumentation

For the basic principles and instrumentation of infrared spectrometry, the reader is referred to Sec. 9-2. Here, the focus will be on instrumental requirements of infrared spectrometry only as far they are essential in combination with gas chromatography.

Gas chromatography with infrared detection nowadays implies the use of a Fourier transform infrared (FTIR) spectrometer. Although in early work [14.2-5] grating instruments were used as GC detectors, their wider use for trace organic analysis was hampered by severe shortcomings. Basically, grating instruments were too slow and too insensitive to be successfully used as GC detectors in trace organic analysis.

A significant improvement has been achieved by the use of FTIR instruments as GC detectors. As explained in detail, in Sec. 9-2, they are superior to grating instruments in scanning speed, optical throughput, and thus sensitivity and wavelength accuracy. Only with these features is the versatility of IR detection in gas chromatography achieved.

GC-IR interfaces

To the same extent as with other hyphenated techniques, the interface for coupled GC-IR is of crucial importance for optimum performance. Since the mobile phases

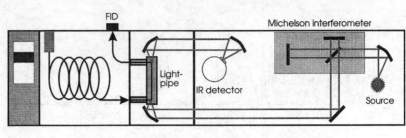

Fig. 14.2-6. Schematic diagram of a GC-FTIR system with a lightpipe interface

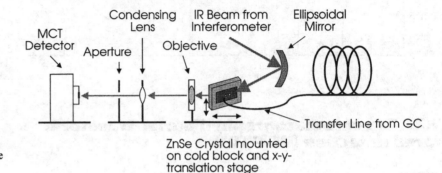

Fig. 14.2-7. Schematic diagram of a cold-trapping GC-FT-IR interface (based on the Tracer, BioRad) [14.2-7]

commonly used in gas chromatography (He, Ne, or H_2) are transparent in the mid-IR range, direct coupling via a flow-through cell is feasible. Mobile phase elimination techniques are applied mainly to increase sensitivity. Thus, a number of interfaces have been developed for GC-FTIR, two of which have gained greater importance: the flow-cell approach, and the cold-trapping approach.

The lightpipe interface is a multiple-reflection flow through cell that provides only moderate sensitivity for FTIR detection.

Flow-cell (lightpipe) interface. The flow-cell approach, the feasibility of which was shown in the early 1980s by Azzaraga [14.2-6] represents the simplest interface for coupled GC-FTIR: The chromatographic column is connected via a heated transfer line to a flow-through cell (a "lightpipe"). This is a heated, internally gold-coated glass tube with IR-transparent windows made from KBr or ZnSe on both ends, which is located in the optical path of the spectrometer (Fig. 14.2-6). Typical lightpipe dimensions are 1 mm ID and 10–20 cm length (corresponding to a lightpipe volume V of ca. 50–200 µL) for use with capillary columns, and 1–3 mm ID and 20–100 cm length ($V = 0.8–5$ ml) for packed columns. The lightpipe volume has to be carefully matched to the chromatographic peak width. This usually results in a compromise between maximum sensitivity (achieved by a larger flow-cell volume) and maintenance of chromatographic resolution (requiring smaller lightpipe volumes). The simplicity of the lightpipe interface for GC-FTIR is one of its main advantages. Measurements can be carried out in real-time, but gas-phase spectra are obtained, requiring the use of special gas-phase libraries for identification. The principal limitation of this approach is its comparatively low sensitivity, 5–100 ng of a substance, depending on the absorption strength, are required to obtain a spectrum.

The cold-trapping interface offers higher sensitivity for GC-FTIR by trapping the separated analytes on a liquid-nitrogen-cooled ZnSe plate for data acquisition.

Cold-trapping interface. The cold-trapping interface (Fig. 14.2-7) was developed as a more sensitive alternative to lightpipe GC-FTIR by Wilkins et al. in the late 1980s [14.2-8]. The concept was subsequently adopted by Griffiths et al. [14.2-9] and is based on cryotrapping of the analytes prior to analysis. The chromatographic effluent is continuously deposited through a heated, small diameter capillary onto a ZnSe plate that is cooled by liquid nitrogen to 77 K. The ZnSe plate is moved to transport the condensed sample into the focus of a dedicated FTIR microscope that focuses the IR beam diameter to match the sample spot dimensions (ca. 100 µm). Thus, near real-time detection is possible. This approach pro-

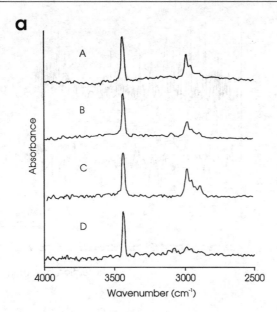

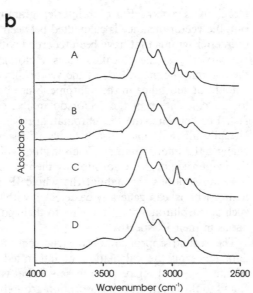

Fig. 14.2-8. Comparison of (a) lightpipe and (b) direct deposition FTIR spectra obtained from a GC separation of different barbiturate derivatives: (A) barbital; (B) aprobarbital; (C) butabarbital; and (D) phenobarbital in the region from 4000 to 2500 cm^{-1}. 12.5 ng and 375 pg of each component were injected for the lightpipe and direct deposition spectra, respectively [14.2-7]

vides a sensitivity similar to GC-MS (spectra can be obtained from as little as 20–50 pg) (Fig. 14.2-8). The IR spectra can be compared with available condensed-phase spectra libraries. A limitation is, however, that highly volatile compounds are not trapped efficiently at the temperature applied.

Data handling and processing in GC-FTIR

The success of commercial GC-FTIR systems would have been impossible without recent developments in computer technology to ever faster microprocessors and larger storage capacities. During a 30-min chromatographic run with a spectral acquisition rate of 4 scans/s, 7200 scans with, e.g., 2048 data points per scan must be acquired, stored, and processed in real-time. This requires extensive storage and computing capabilities. Also, the generation of real-time chromatograms from GC-FTIR data is not as straightforward as with GC-MS, since the data are obtained primarily in the form of interferograms rather than spectra. Two different ways to reconstruct chromatograms from the spectral data have gained importance.

The first is based on the Gram–Schmidt (GS) vector orthogonalization algorithm. In short, this method treats the information-containing parts of an inter-

The Gram–Schmidt algorithm allows a fast reconstruction of functional group or total absorbance chromatograms.

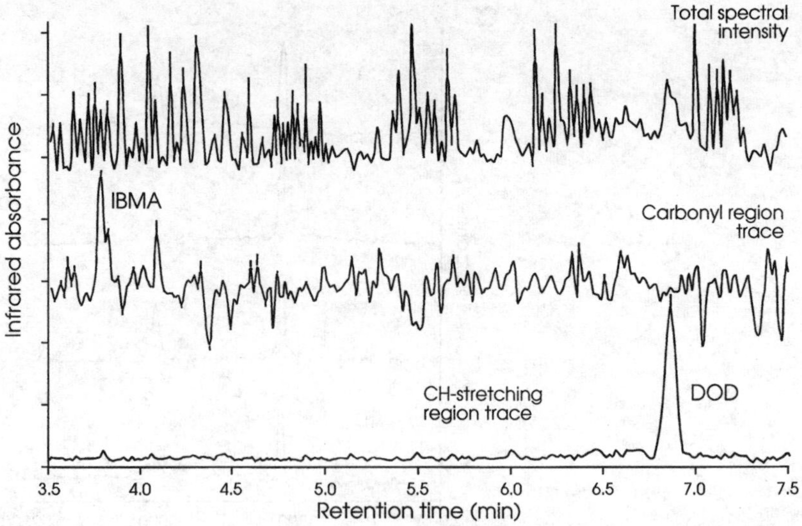

Fig. 14.2-9. Traces from a GC-IR separation of less than 70 pg each of isobutyl methacrylate (IBMA) and dodecane (DOD). The top trace is the total spectral intensity from which individual compounds can hardly be detected. The middle and bottom traces are from the carbonyl and C-H stretching regions, respectively, and clearly show both IBMA (RT ~3.8 min) and DOD (RT ~6.8 min) [14.2-10]

ferogram as vectors. For each interferogram measured during a chromatographic run, the vector distance is calculated between this particular vector and a number of interferograms that have been recorded with nothing in the lightpipe except the He carrier gas (the so-called basis vectors). When an analyte is eluting from the column, the magnitude of the vector difference is roughly proportional to the amount of substance in the lightpipe. Since only part of the interferogram is used for the calculation of the vector difference, computation of the GS trace is very fast. Furthermore, the GS chromatogram is effectively a universal indicator, as virtually all molecules absorb at least in some part of the IR spectrum, which changes the interferograms. The sensitivity of the GS trace naturally depends on the magnitude of the absorption by the individual molecule. Molecules that show weak absorptions throughout the whole IR spectrum (e.g., polycyclic aromatic hydrocarbons) can generally be detected with much less sensitivity than substances such as barbiturates which, owing to their polarity, have several strong sbsortion bands in their IR spectra.

Functional group chromatograms indicate the elution of a compound that contains the functional group considered.

The second approach to construct chromatograms from the interferometric data relies on the calculation of integrated absorption traces in one or more specified spectral regions. As these spectral windows are usually chosen to correspond to the characteristic absorption group frequencies of the functional group(s) of interest, the name functional group (FG) chromatograms has come into use. FG chromatograms are clearly more selective than GS chromatograms. This selectivity, however, is not complete, owing to overtone vibrations or combination bands of the molecules. In the case of strongly absorbing functional groups, as is the case for the C=O stretching vibration of carbonyl compounds, the FG trace can be even more sensitive than the GS chromatogram. In most other cases the GS trace provides a higher sensitivity (Fig. 14.2-9).

While chromatograms are obtained in either of the two reconstruction modes on-line (at least with the lightpipe interface), the IR spectra are obtained by post-run data processing. Spectral identification is usually achieved by library search. The goodness of fit of the library entry and the measured spectrum is expressed in the form of a hit quality index. This is the Euclidean distance of the two spectra taken as vectors. The better two spectra match, the closer to zero is the value of the hit quality index.

Combined GC-MS-FTIR

Combined GC-MS-FTIR allows the differentiation of both homologs and isomers.

There are well-founded arguments for operating both a mass spectrometer and FTIR spectrometer as parallel detectors for GC. As pointed out above, the information from MS and IR is complementary to a certain degree. Both are molecule-specific detectors. MS is capable of distinguishing homologs, while differentiation between isomers is often impossible. IR can provide isomer-specific detection while differentiation between homologs represents a problem. Different possibilities exist to couple both detectors either in parallel or in series [14.2-11]. If the parallel arrangement is used, the flow is usually split so that 90–99% of the flow is directed to the IR spectrometer. This accounts for both the different sensitivities and the different flow requirements of infrared and mass spectrometric detectors. Serial coupling of both detectors requires makeup gas to be added first, to match the optimum conditions for IR lightpipe detection. The exit flow has to be split, frequently using a jet separator or an open split interface, to reduce the gas flow introduced into the mass spectrometer. The use of the open split interface is usually preferable, since it allows variation of the flow rate over a wider range to maximize chromatographic efficiency and sensitivity with lightpipe IR detection.

14.2.3 Gas chromatography-atomic emission detection (GC-AED)

Principles

Although the first developments directed toward element-specific detection for gas chromatography by plasma optical emission spectroscopy were undertaken in the mid-1960s by McCormack and co-workers [14.2-12] and Bache and Lisk [14.2-13], a commercial instrument employing this hyphenated technique was not introduced until 1989 [14.2-14], making the atomic emission detector (AED) the most recent addition to the group of spectroscopic gas chromatographic detectors.

In contrast to GC-MS or GC-FTIR, GC-AED provides *element-specific* gas chromatographic detection rather than molecule-specific detection, and can thus be considered an ideal complement to these detectors.

Atomic emission detection is based on the fact that the chromatographic effluent is directed into a noble-gas-sustained plasma where it is completely atomized and the atoms and ions formed in the plasma are excited to emit light. Different types of plasma have been used as excitation sources with varying success. Among these are the microwave-induced plasma (MIP) sustained in helium or argon, the direct current argon plasma (DCP), the inductively coupled argon plasma (ICP), the capacitively coupled plasma (CCP), or the stabilized capacitive plasma (SCP). Of all these, the microwave-induced helium plasma has gained the widest acceptance for several reasons. The plasma is operated at atmospheric pressure which makes interfacing to the GC very simple. Flow rates are in the range 30–300 mL/min and thus considerably lower than, e.g., with an ICP. Using He as plasma gas is convenient, as it is also commonly used as GC carrier gas, and particularly advantageous, as it provides a simpler spectral background and a significantly higher excitation energy than argon (ionization energy 24.59 eV versus 15.76 eV), allowing effective excitation even of nonmetallic elements. The power coupled into the plasma by a ca. 60 W radio frequency generator operated at 2.45 GHz is significantly lower than in an ICP, leading to lower plasma temperatures (3000–4000 K). This is by no means a problem as long as care is taken not to overload the plasma (by introducing too large amounts of analytes or solvent into the plasma).

Different reactions have been proposed to explain the excitation and the combined ionization/excitation or fragmentation/excitation mechanisms in the plasma, taking into account the abundant number of low-energy electrons that are especially effective in recombination excitation:

$$e^- + He + A^+ \longrightarrow He + A^* + h\nu \text{ (continuum)} \tag{14.2-1}$$

or

$$e^- + A^+ \longrightarrow A^* + h\nu \text{ (continuum)} \tag{14.2-2}$$

where A is an analyte atom. High-energy, fast electrons sustain the plasma according to:

GC-AED provides element-specific detection based on the elemental emission excited in the microwave-induced He-plasma.

The high first ionization energy of He leads to the efficient excitation of metals and non-metals in the plasma.

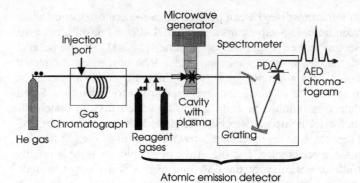

Fig. 14.2-10. Scheme of a GC-atomic emission detection instrument

$$e^- + He \rightarrow He^+ + 2\,e^- \qquad (14.2\text{-}3)$$

but can also be involved directly in excitation processes such as

$$e^- + A \rightarrow A^* + e^- \qquad (14.2\text{-}4)$$

or

$$e^- + A^+ \rightarrow A^{+*} + e^- \qquad (14.2\text{-}5)$$

Both ions (He^+) and metastables (He_m) of the plasma gas can lead to excitation of the analyte:

$$He^+ + A \rightarrow He + A^{+*} \qquad (14.2\text{-}6)$$

$$He_m + A \rightarrow He + A^+ + e^- \qquad (14.2\text{-}7)$$

$$He_m + A^+ \rightarrow He + A^{2+} + e^- \qquad (14.2\text{-}8)$$

$$He_m + A \rightarrow He + A^* \qquad (14.2\text{-}9)$$

$$He_m + A^+ \rightarrow He + A^{+*} \qquad (14.2\text{-}10)$$

Reactions (14.2-7) and (14.2-8) are the well-known Penning ionization processes. Dissociative excitation reactions are also of importance in the plasma, e.g.:

$$He_m + AB \rightarrow He + A + B^* \qquad (14.2\text{-}11)$$

where AB is an analyte molecule.

The excited atoms, ions, or molecules return to their ground state by emitting light of characteristic wavelength which is detected by a spectrometer.

Instrumentation

GC-AED with diode array detection allows for simultaneous multielement detection.

In one commercial AED instrument of widespread use, the analytes eluting from the chromatographic column are introduced directly into the plasma by leading the end of the chromatographic column directly into the discharge tube in which the plasma is located (Fig. 14.2-10). As the stable operation of the plasma and the sensitive and selective detection of the different elements requires a flow of typically 30–200 mL He/min, additional He makeup gas is used. Reagent or scavenger gases (oxygen, or hydrogen, or a combination of both for the detection of most elements, but, for the detection of oxygen a CH_4/N_2 mixture) are also added before entering the plasma in order to increase selectivity and to prevent the formation of carbonaceous deposits on the discharge tube wall. The plasma is powered by a low-power (60 W) microwave generator and sustained in a quartz discharge tube of approximately 1 mm ID which is housed in the center of the microwave cavity. As the plasma cannot tolerate the introduction of larger amounts of carbon compounds, a valve device is integrated to vent the solvent

Table 14.2-2. Figures of merit for helium microwave-induced plasma detection for GC [14.2-15]

Element	Wavelength (nm)	Detection limit (pg/s (pg))	Selectivity vs. C	Linear dynamic range
Carbon	247.9	2.7	1	>1000
Carbon	193.1	2.6	1	21000
Hydrogen	656.3	7.5 (22)	160	500
Hydrogen	486.1	2.2	variable	6000
Deuterium	656.1	7.4	200	500
Boron	249.8	3.6 (27)	9300	500
Chlorine	479.5	39	25000	20000
Bromine	470.5	10	11500	>1000
Fluorine	685.6	40	30000	2000
Sulfur	180.7	1.7	150000	20000
Phosphorus	177.5	1	5000	1000
Silicon	251.6	7.0	90000	40000
Oxygen	777.2	75	25000	4000
Nitrogen	174.2	7.0	6000	43000
Aluminum	396.2	5.0	>10000	>1000
Antimony	217.6	5.0	19000	>1000
Gallium	294.3	ca. 200	>10000	>500
Germanium	265.1	1.3 (3.9)	7600	>1000
Tin	284.0	1.6 (6.1)	36000	>1000
Tin	303.1	(0.5)	30000	>1000
Arsenic	189.0	3.0	47000	500
Selenium	196.1	4.0	50000	>1000
Chromium	267.7	7.5	108000	>1000
Iron	302.1	0.05	3500000	>1000
Lead	283.3	0.17 (0.71)	25000	>1000
Mercury	253.7	0.1	3000000	>1000
Vanadium	292.4	4.0	36000	>1000
Titanium	338.4	1.0	50000	>1000
Nickel	301.2	1.0	200000	>1000
Palladium	340.4	5.0	>10000	>1000
Manganese	257.6	1.6 (7.7)	110000	>1000

Detection limits are calculated as three times the signal-to-noise ratio. The values in brackets indicate the absolute detection limits. Selectivity is calculated as the ratio of peak area response for 1 mol analyte element by 1 mol carbon.

before entering the plasma. At plasma temperatures of more than 3000 K, the analytes are completely atomized, excited, and emit characteristic radiation. This element-specific emission is viewed from the open end of the discharge tube (to avoid interferences by deposits on the discharge tube walls) and is led through the transfer optics to a holographic grating that disperses the polychromatic light. Located along the focal plane of the grating, a movable 211-pixel photodiode array can be positioned to allow detection of the element-specific emission. As the diode array covers only approximately 25 nm of the whole accessible spectrum (165–800 nm), the only elements that can be detected simultaneously are those that have emission lines lying so close together that they can be detected with one diode array position. For this reason and the restrictions due to the use of different scavenger gases, only a limited set of elements can be monitored at the same time, e.g., carbon at 495.8 nm (second order of the 247.9 emission line), hydrogen at 486.1, Br at 470.5, and Cl at 479.5 nm. The result of a GC run with atomic emission detection is a three-dimensional data matrix where the emission intensity has been recorded as a function of both retention time and wavelength (similar to HPLC-DAD). Element-specific chromatograms are calculated in real-time from this data matrix by a special algorithm that corrects for background emission and spectral interferences.

The sensitivity of the AED is strongly element dependent. It is excellent for, e.g., sulfur, carbon, and metals, but less satisfactory for the halogens, especially when compared with the ECD (Table 14.2-2).

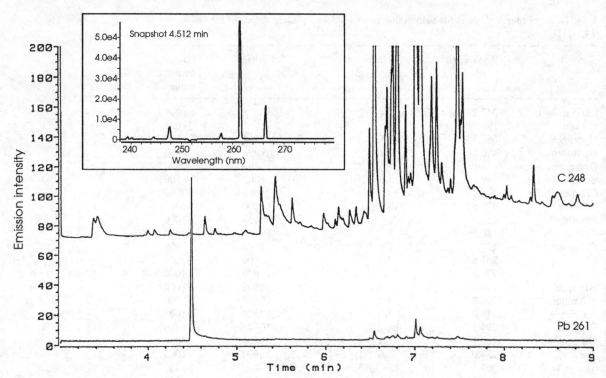

Fig. 14.2-11. Element-specific chromatogram of an extract of a road dust sample on the carbon (top) and the lead channel (bottom). The insert with the spectrum of the peak at 4.5 min clearly indicates the presence of an organolead compound (trimethyllead, derivatized by ethylation)

As element-specific detector, GC-AED provides complementary information to GC-MS and GC-FTIR.

The success of the AED is mainly due to the following three facts:

- The unique selectivity. The selectivity of the heteroelements versus carbon often equals or exceeds 10^3 which is hardly achieved by other element-specific detectors such as ECD or NPD. Furthermore, the presence of the detected element can be verified by comparing the emission spectrum for the compound in question (Fig. 14.2-11).
- The possibility of compound-independent calibration, which is particularly valuable in quantitative analysis if authentic standards for the analytes are not available. The independence of the elemental response of the molecular structure also allows one to calculate the elemental composition of the analyte. Although the accuracy of this determination cannot compete with that obtained from conventional elemental microanalysis, results can be obtained directly from chromatograhic peaks at sample amounts up to six orders of magnitude smaller.
- The possibility of detecting almost every element that can be excited by He and also some of the stable isotopes, such as ^{2}D, ^{13}C, and ^{15}N based on the slightly shifted emission spectrum of the molecular bands compared with the most abundant isotope.

The use of GC-AED is most successful in combination with other hyphenated techniques in environmental analysis, where the formation provided by techniques such as GC-MS or GC-FTIR is dieally complemented, and often allows the unambiguous identification of organic pollutants, via information on the presence or absence of a suspected element.

Worked example:

A pulp mill effluent was analyzed for chlorinated and halogenated compounds. Since the number of compounds present in the sample prevented identification of all compounds GC-AED was used as a complementary analytical tool. Figure 14.2-12 gives both the TIC from the GC-MS analysis (upper left-hand corner) and the element-specific chromatograms for C, H, Cl, S, and O. The element-specific response simplified the identification of the unknown compounds as: (1) 1,1-dichlordimethyl sulfone; (2) trichlorothiophene; (3) 1,1,2-trichlorodimethyl sulfone; (4) tetrachlorothiophene; (5) trichloromethoxythiophene; (6) 1,1,2,2-tetrachlorodimethyl sulfone; (7) dichloroacetylthiophene; (8) trichloroacetylthiophene. The concentration of the identified compounds was between 3 and 118 µg/L.

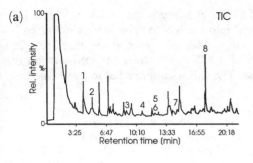

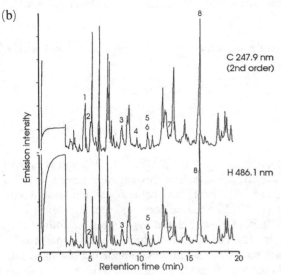

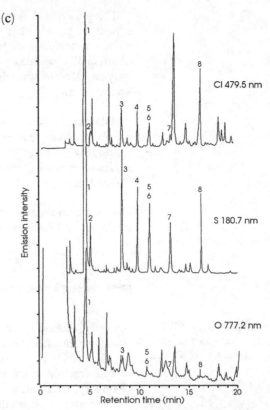

Fig. 14.2-12. GC-MS total ion chromatogram (upper left-hand corner) and GC-AED element-specific traces for C, H, Cl, S, and O of an alkaline extraction liquor from a bleach plant [14.2-15]

14.3 Hyphenated Liquid Chromatographic Systems

While gas chromatography is the key separation technique for the analysis of volatile compounds, liquid chromatography (LC) is its equivalent for polar and high molecular mass compounds. However, unlike gas chromatography, LC has suffered from the lack of detectors that are both sensitive and selective, or even specific (see Sec. 5.2). Most of the commonly used detectors are either sensitive, but not specific (e.g., refractive index, light scattering, or fluorescence detectors) or specific to a certain degree at the cost of sensitivity (e.g., diode array detection). This has called for the development of hyphenated LC techniques combining both features.

14.3.1 Liquid chromatography-mass spectrometry (LC-MS)

Principles

Liquid chromatography coupled to mass spectrometric detection has long been considered an "impossible marriage". The mismatch between the flow rates encountered in conventional HPLC systems (0.5–2 mL/min of normal or reversed-phase solvent) and the vacuum requirements of the mass spectrometer seemed far too great. Still, the lack of a sensitive, selective, and universal detector for HPLC was the driving force for research directed toward the coupling of HPLC and MS. Several different interfaces have been developed over the past 20 years to overcome the problems arising from this apparent incompatibility.

The problem of introducing liquid flow rates of up to a 2 mL/min into the vacuum of the MS requires the use of a suitable interface.

The principal difficulty of the HPLC-MS interface is that either as much LC effluent as possible has to be introduced into the MS to achieve maximum sensitivity, or efficient enrichment of the analyte must take place at the interface. The differentially pumped vacuum system of a MS can tolerate the introduction of only ca. 50 nL/s of liquid mobile phase. The various approaches to overcome this limitation are:

- Enlargement of the pumping capacity of the MS vacuum system
- Solvent elimination prior to the introduction into the vacuum system
- Splitting the effluent stream at the cost of sensitivity
- Use of micro-LC columns that allow efficient operation at substantially lower flow rates.

One or a combination of these strategies is used in LC-MS interfaces. They will be discussed in the following section.

Instrumentation

In the direct liquid introduction (DLI) LC-MS interface, only a small fraction of 10–20 µL/min of the column flow is allowed to enter the MS.

An early solution for the introduction of liquid samples into the MS was the *direct liquid introduction* (DLI) interface. In this interface, developed in the early 1970s by McLafferty and co-workers [14.3-1], the HPLC column effluent is split and a small fraction of ca. 1–4% or 10–20 µL/min is allowed to enter the desolvation chamber through an orifice or a diaphragm where it forms a liquid jet, followed by disintegration into small droplets. The larger part of the solvent is separated by evaporation, while the other part serves as reagent gas, supporting chemical ionization (CI), before the ionized compounds are introduced into the mass analyzer (Fig. 14.3-1).

Although it was one of the first interfaces to demonstrate the feasibility of LC-MS, the DLI interface has some disadvantages which eventually led to its disappearance. The sensitivity is comparatively low, as only small effluent volumes can be introduced in the MS and no separation of the analytes from the mobile phase takes place. Only chemical ionization is possible, using the reversed-phase solvent as reagent gas. Finally, operation is difficult, owing to frequent clogging of the small diameter orifices or diaphragms (typically <5 µm ID).

In thermospray (TSP-)LC-MS, the solvent is evaporated by spray formation in a heated capillary.

Based on experience from the DLI, the *thermospray* (TSP) interface was developed later by Vestal and co-workers [14.3-2]. In the TSP interface, the effluent from the HPLC column, often containing a volatile buffer, is passed through a resistance-heated vaporizer capillary where exactly as much heat is transferred to the mobile phase as is needed to yield partial evaporation of the solvent. At the end of this stainless steel capillary, a very fine aerosol is formed with a droplet size of less than 1 µm which expands directly into the heated ionization chamber (Fig. 14.3-2). A very high-throughput pumping system is necessary to remove up to 2 mL/min of aqueous solvent introduced into the MS system.

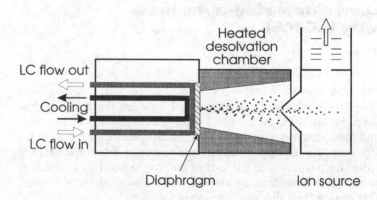

Fig. 14.3-1. Scheme of the direct liquid introduction interface

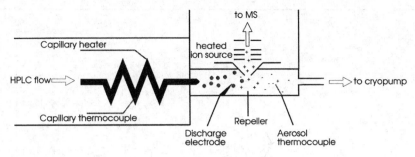

Fig. 14.3-2. Scheme of the thermospray interface

Ionization of the analytes can take place either by solvent-mediated chemical ionization or by the thermospray ionization process. The first is achieved by using electrons from a spray electrode or a filament to create the reagent gas from the solvent molecules, which in turn initiates a charge transfer to the analytes. The latter is based on an ion evaporation mechanism from the droplets in which the volatile buffer is involved. Depending on whether the discharge is used, the ionization mechanism changes, and this greatly influences the sensitivity. Ion evaporation usually results in $[M + H]^+$ ions for samples of high proton affinity. Otherwise, $[M + NH_4]^+$ ions are detected, where the NH_4^+ stems from the buffer, e.g., ammonium acetate. When negative ions are detected, either $[M - H]^-$ or negative cluster ions of the molecule and a solvent or a buffer anion can be found. However, either mode of ionization is soft, and results in only limited fragmentation. Nevertheless, to obtain characteristic fragmentation spectra from TSP-LC-MS analyses, a tandem quadrupole instrument is often used. Unlike single quadrupole instruments, a MS/MS instrument allows one to obtain fragmentation spectra of molecular ions selected by the first quadrupole (Fig. 14.3-3). The ions are sampled through a skimmer with a small orifice that reaches directly into the ionization chamber. This permits one to attain the high vacuum required for ion separation.

The TSP interface has been very successful over the past few years, owing to the fact that normal HPLC methods can easily be adapted, since normal flow rates and most of the common buffers may be used, provided they are volatile. It seems, however, that the newer methods using atmospheric pressure ionization are gradually replacing the TSP, owing to their greater robustness and often higher sensitivity.

The *particle beam* (PB) interface, also called MAGIC (monodisperse aerosol generator interface for chromatography) by its developers Willoughby and Browner [14.3-3], produces an aerosol under atmospheric pressure (pneumatically assisted). The aerosol expands into a heated desolvation chamber where the small particles, contained in an additional gas flow, are separated from the majority of the solvent molecules in a momentum separator (Fig. 14.3-4). The small particles are transported to the ion source where they disintegrate upon collision with the heated source chamber walls. The residual solvent vaporizes and the released gaseous analyte molecules can be ionized by the method of choice (mostly EI or CI).

The possibility of obtaining gas-phase-like EI spectra which can be library searched is the most valuable advantage of the PB interface. Its efficiency is highly dependent upon the uniformity of aerosol generation, as the momentum separator works better as the size distribution of the aerosol becomes narrower. Although used extensively, among other applications, in environmental analysis, the PB interface has some shortcomings. These are mainly the inadequate detection limits (in the ng abs. range), as well as wide variations in sensitivity, even for structurally similar compounds, and the lack of linear response over a wide concentration range. Furthermore, the co-elution of two compounds has rather unpredictable effects on the response. LC-MS with the PB interface has been a useful extension of GC-MS to more polar and higher molecular mass compounds (Fig. 14.3-5), but it seems to have been supplanted by the other interfaces.

All the interfaces discussed have in common the fact that ionization takes place in a low pressure region (typically 1–1000 Pa in TSP or when CI is used and

Ionization is soft in TSP-LC-MS and the degree of fragmentation is small, which provides only limited structural information.

In the particle beam interface, analyte molecules are separated from the solvent molecules by a momentum separator.

PB-LC-MS produces gas-phase-like, library searchable EI spectra.

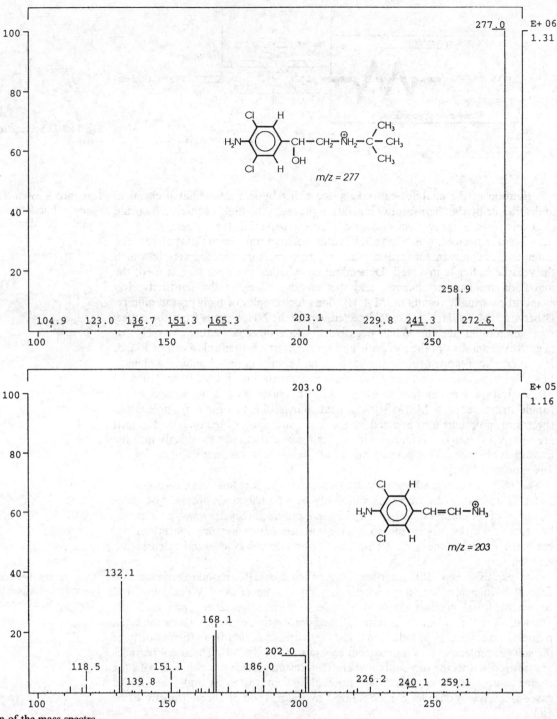

Fig. 14.3-3. Comparison of the mass spectra of clenbuterol (an illegal growth promotor) obtained by: (a) Thermospray MS; (b) Thermospray MS/MS. The MS/MS spectrum was obtained by mass selecting the $[M + H]^+$ ion at m/z 277 (quadrupole 1), collisional activation with argon (24 eV) in quadrupole 2, and then scanning quadrupole 3 to obtain the product ion spectrum

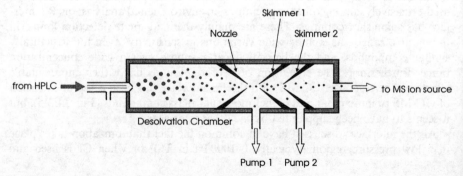

Fig. 14.3-4. Scheme of the particle beam interface

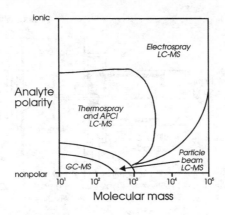

Fig. 14.3-5. Representation of the application ranges of LC-MS interfaces [14.3-4]

$\leq 10^{-2}$ Pa with EI). This clearly distinguishes them from the techniques presented in the following discussion, where ionization takes place at atmospheric pressure.

Although it was one of the first techniques developed for interfacing LC and MS [14.3-5], it took more than 20 years until the group of *atmospheric pressure ionization* (API) techniques developed into probably the most versatile group of interfaces and sample introduction techniques. The term API denotes three different techniques which differ mainly in their nebulization principle: the heated nebulizer, the electrospray, and the ionspray interface.

The first technique to be discussed here is the *heated nebulizer* or APCI interface. In APCI, *atmospheric pressure chemical ionization*, ionization mechanisms are identical to those in medium pressure chemical ionization. Reagent gas ions are generally formed by a corona dicharge. Positive ions can be formed either by proton transfer, the formation of adducts, or charge extraction reactions. Negative ions, in contrast, can be generated by proton abstraction, anion attachment, or electron-capture reactions. Mass spectra obtained from conventional (medium pressure) CI and APCI differ somewhat, which can be explained by the fact that ion formation in APCI is an equilibrium process while it is kinetically controlled in medium pressure CI. An important advantage is also the theoretically achievable sensitivity in APCI versus medium pressure CI, owing to the much higher reaction efficiency of ion–molecule interaction at higher pressure. APCI however fails to deliver the expected gain in sensitivity of three to four orders of magnitude, owing to significantly less effective transmission of the ions through the mass analyzer at higher pressure.

A typical form of the heated nebulizer interface consists of a concentric pneumatic nebulizer from which the aerosol directly enters a heated quartz or stainless steel tube (Fig. 14.3-6). The aerosol is swept through the heated tube with the aid of an additional makeup gas flow, and is introduced into the API source, where APCI takes place, initiated by a corona discharge. Alternative designs include the use of a drying gas, led countercurrently. These systems are said to tolerate the use of nonvolatile buffers, as the uncharged (non)volatile material is swept away by the countercurrent drying gas flow and, even if the source has to be cleaned frequently, cleaning of the atmospheric pressure ionization chamber can take place without switching off the vacuum system. Even when compared with the thermospray interface, the APCI interface is remarkably robust and simpler to operate since it can handle aqueous eluent flow rates of up to 2 mL/min.

In all atmospheric pressure ionization (API) techniques, ionization takes place in an atmospheric pressure region.

In APCI, a corona discharge supports the chemical ionization through the solvent molecules.

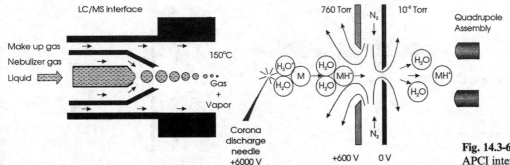

Fig. 14.3-6. Scheme of the heated nebulizer/ APCI interface [14.3-6]

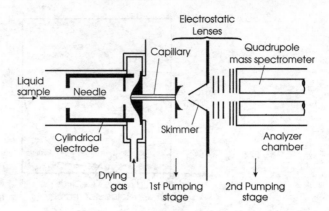

Fig. 14.3-7. Scheme of the electrospray interface [14.3-6]

The electrospray interface desolvates and ionizes the analytes by applying a strong electric field, and usually produces multiply charged ions of large molecules.

The *electrospray interface* (ESI) operates at essentially lower flow rates of typically 1–10 µL/min. The process of electrospray ionization includes both the nebulization of the liquid stream into an aerosol of highly charged droplets and the ionization of the analyte molecules after desolvation of the charged droplets. ESI is subsumed under the API interfaces, as the sample is introduced, after an appropriate split from the chromatographic column or by direct introduction via an infusion apparatus, through a stainless steel hypodermic needle into an atmospheric pressure desolvation chamber (Fig. 14.3-7). While the needle is at ground potential, a strong electric field (2–5 kV) is applied to the cylindrical counter electrode which charges the surface of the liquid emerging from the needle, creating a fine spray of charged droplets. Driven by the electric field, the droplets pass through a curtain of nitrogen drying gas. The gas curtain has the function of supporting evaporation of the solvent and also preventing the uncharged material from entering the ion source. By expansion of the curtain gas, the ions are carried through a capillary into the vacuum of the first pumping stage, and after passing further pumping stages and the lens system, finally into the mass analyser.

The mechanism of ion formation in the ESI is still under discussion. A simple explanation is that the aerosol is charged to such an extent that the droplets virtually explode by Coulomb repulsion into smaller droplets which eventually evaporate to yield highly charged molecules.

Electrospray LC-MS has attracted much attention since it was observed that biomolecules (peptides and proteins) form multiply charged ions. It is thus possible to measure proteins with molecular masses up to 100 000 Da or more with a simple quadrupole instrument that transmits ions up to a m/z ratio of, e.g., 2000.

The ionspray interface is a pneumatically assisted electrospray interface that can accommodate higher flow rates.

The main drawback of the ESI, that it can accept only very small liquid flow rates (1–10 µL/min) was overcome by the development of the *ionspray interface* (ISP) in 1987 [14.3-7]. The ISP interface (Fig. 14.3-8) combines electrospray nebulization with pneumatic nebulization (and is thus often also referred to as pneumatically assisted or high flow electrospray). In the first ionspray instruments, the flow range could be extended to only about 50 µL/min. Additional improvements, e.g., the use of a drying gas curtain, made possible the introduction of up to 2 mL/min. The ISP interface is able to handle mobile phases with high water content and can also be operated with gradient elution systems.

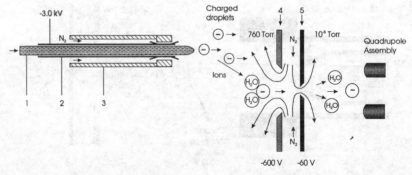

Fig. 14.3-8. Scheme of the ionspray interface. (1) 50 µm I.D. fused-silica capillary; (2) 0.20 mm ID stainless steel capillary; (3) 0.8 mm ID Teflon tube with narrow bore insert; (4) ion focusing lens, serving as counter electrode for ion spray; (5) orifice holding plate with 100 µm ID conical orifice [14.3-6]

Table 14.3-1. Comparison of LC-MS interfaces in terms of allowable flow rate and mobile phase composition [14.3-4]

Interface	Maximum flow rate (mL/min)	Mobile phase composition
Direct liquid introduction	0.05	Reversed-phase solvents, no buffers
Thermospray	2	Reversed-phase solvents, with volatile buffers
Particle beam	0.5	Reversed-phase solvents (low water content), with volatile buffers
Electrospray	0.005–0.5 ⎫	Reversed-phase solvents, with volatile buffers
Ionspray	0.05 ⎬	
Heated nebulizer/APCI	2	Reversed-phase solvents, with volatile buffers

Table 14.3-2. Comparison of single-ion detection limits in the LC-MS analysis of *N*-methyl carbamates with different interfaces

Interface	Detection limit (ng)				Flow rate (mL/min)
	Methomyl	Aldicarb	Carbofuran	Carbaryl	
Direct liquid introduction			50	40	0.02
Thermospray	2.8	0.9	0.8	0.8	1.0
Ionspray	0.4	1.5	1.5	1.0	0.05 after 20:1 split
APCI	0.06	0.07	0.05	0.05	1.0
Particle beam EI	250	500	55	10	0.4 after 3:2 split

Comparison of the different LC-MS interfaces is difficult. No general advice can be given on the use of LC-MS interfaces; the choice of interface depends strongly on the particular application (see Table 14.3-2). If maximum sensitivity is required, the APCI (heated nebulizer) interface is often the best choice, probably in just a few cases being matched or exceeded by electrospray. CI techniques are, however, scarcely able to provide any structural information. For this purpose, a technique with EI ionization, such as the particle beam interface should be chosen. The gain in chemical information (when the typical fragmentation pattern can be obtained) can compensate for its significantly lower sensitivity. A comparison of the different interfaces in terms of the tolerable flow rates and their detection limits is given in Tables 14.3-1 and 14.3-2.

14.3.2 Liquid chromatography-Fourier transform infrared detection (LC-FTIR)

Principles

The use of an FTIR detector for liquid chromatography is a further hyphenated technique, yielding on-line molecule-specific information on the separated compounds. In comparison with LC-MS, the interfaces are simpler since the sample does not need be introduced into a spectrometer operating under high vacuum. Additionally, functional group information can be obtained in a very straightforward way be monitoring the absorption in a characteristic frequency window.

Compared with a GC-FTIR instrument, however, the operation of a LC-IR system is impaired by the strong absorption that both normal and reversed-phase solvents show in the mid-IR region. This either calls for a solvent elimination approach or restricts the application of LC-FTIR to monitoring the absorption at selected frequency windows in order to obtain functional group information when used with a flow-cell. Scanning of the whole spectrum is hardly possible, owing to the prominent solvent absorption bands which, notably with FT instruments, dominate the spectra and thus make detection of the analyte bands difficult in spectral regions of strong solvent absorption.

Instrumentation

The flow-cell interface for LC-FTIR requires very small pathlengths and is thus rather insensitive.

The *flow-cell* approach is the most straightforward for LC-IR operation. The chromatographic effluent is passed through a flow-cell directly after the column, and interferograms are continuously recorded through the whole run. The use of the Gram–Schmidt algorithm, as in GC-FTIR for the real-time calculation of total absorption traces is not feasible, because the mobile phase absorbs strongly and the small variations in absorbance due to the elution of the analytes can hardly be detected. Data processing is thus usually carried out at the end of a run after subtracting the mobile phase absorption spectrum. In order to prevent total absorption by the solvent bands, short pathlengths are necessary, generally less than 0.2 mm for organic mobile phases and less than 0.03 mm for aqueous mixtures. Together with the fact that absorption coefficients in the mid-IR are rather small compared with the UV/VIS spectral region, this accounts for the comparatively low sensitivity of this technique which is in the 0.1–1 µg range. Additional disadvantages of this interface are that usually no information on analyte absorption can be obtained in the spectral regions where the solvent absorbs, as correct spectral subtraction is very difficult, especially with reversed-phase solvent mixtures. Furthermore, background subtraction cannot be performed in a satisfactory way when gradient elution is necessary, limiting applications to those ideal cases where isocratic separation can be achieved. Consequently, the flow-cell approach for LC-FTIR does not have a wide range of application, although its simplicity and cost advantages justify the interest in this interface.

Solvent elimination LC-FTIR techniques are more sensitive, owing to the removal of the interfering solvent.

The *solvent elimination* approach has three major advantages over the flow-cell interface. First, the full spectral information from the analyte can be exploited. Second, the separation of chromatography and detection allows one to optimize the conditions of chromatographic separation without imposing restrictions on mobile phase composition or mode of separation. Third, storage of the chromatographically separated compounds allows one to improve sensitivity by off-line accumulation of data and signal averaging without time constraints. Still, the mobile phases commonly used for HPLC separation pose problems, since the solvent elimination interface requires the solvent to be significantly more volatile than the analytes.

In practice, solvent elimination is achieved by dropping the eluting compounds onto KCl pellets, evaporating the solvent under a gentle stream of nitrogen, and finally transferring the analytes deposited on the KCl powder into the optical path of an FTIR spectrometer where spectra are recorded in diffuse reflectance.

The micro-LC-FTIR approach deserves special mention. The use of micro-LC is attractive for the following reasons. Microscale columns (with typically 1 mm ID compared with 4.6 mm for conventional HPLC columns) are operated at significantly higher peak concentration, which leads to a proportionally higher detectability. The low flow rates optimal for separations on micro-HPLC columns also allow IR-absorbing solvents (which are easier to eliminate) or expensive solvents (e.g., deuterated) to be used. A solvent elimination interface that has been designed especially for micro-LC-FTIR is presented in Fig. 14.3-9. This *buffer memory technique* is similar to the solvent elimination approach, with the difference that the chromatographic effluent is continuously deposited in a narrow line along a slowly translated KBr crystal plate. In order to maintain chromatographic resolution, the effluent on the KBr plate has to be measured with either a conventional IR spectrometer with a beam condenser, or directly under an IR microscope.

Although these interfaces for LC-FTIR based on solvent elimination offer the potential for significantly more sensitive detection than the flow-cell approach,

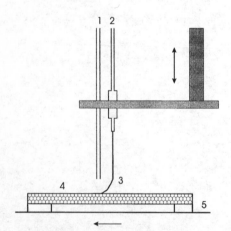

Fig. 14.3-9. Interface device for micro LC-FT-IR via the buffer memory technique: (1) nitrogen gas for elimiating solvents; (2) LC effluent from the microcolumn; (3) stainless steel capillary; (4) KBr crystal plates; (5) plate support [14.3-8]

they are still at the developmental stage and will require further improvement before finding wider acceptance.

14.3.3 Liquid chromatography-nuclear magnetic resonance detection (LC-NMR)

Principles

While nuclear magnetic resonance (NMR) spectroscopy is one of the most powerful and widely used techniques for structure elucidation in organic chemistry (Sec. 9.3), its use as a detector for liquid chromatography has, in contrast to LC-MS or LC-FTIR, long been precluded for the following reasons [14.3-9, 14.3-10].

- NMR is a rather insensitive detection technique, especially compared with MS, normally requiring sample amounts in the low µg range
- Receivers and electronics were not able to handle the large dynamic range required for LC detection
- Suitable flow-cells and probes were not commercially available
- The quality of solvent suppression was not satisfactory.

Significant improvements in the 1980s [14.3-11] have largely overcome these problems and have finally led to the introduction of commercially available LC-NMR instruments.

The driving force behind these developments was to collect NMR spectra and to perform 2D-NMR experiments with compounds separated by liquid chromatography on-line, and ideally also on-flow, or at least with the flow stopped.

The most common nuclei for direct detection are ^1H, ^{19}F, and ^{31}P, while (owing to the lower sensitivity) for ^{13}C or ^{15}N the inverse mode with ^1H detection is preferred.

LC-NMR allows one to carry out on-line NMR experiments with minute sample amounts separated by LC.

Instrumentation

In order to cope with the above-mentioned limitations of stand-alone NMR instruments, they had to be substantially modified and improved to allow their incorporation into flow systems. A modern LC-NMR setup is shown in Fig. 14.3-10. The chromatographic system consists of a HPLC pump, with injector and the analytical column. A simple detector (UV absorbance or RI) serves to indicate the elution of peaks and to obtain quantitative information. The signal from this detector can also be used to trigger data acquisition or pump operation, or to collect

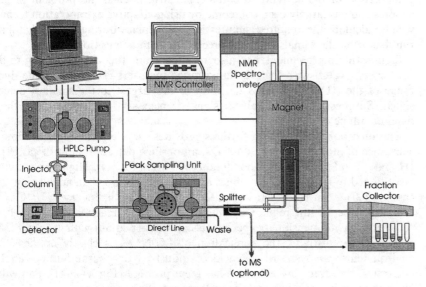

Fig. 14.3-10. Schematic view of a hyphenated LC-NMR system [14.3-12]

the peaks (either intermediately in storage loops before detection or after detection in a fraction collector). The chromatographic effluent is routed from the LC-NMR interface, either through the flow-cell MNR probe or directly to waste. Following passage through the NMR probehead, the flow is led to a fraction collector for storage and further investigation of the different fractions analyzed by NMR.

A dedicated NMR flow-probe had to be designed to maintain chromatographic resolution and, at the same time to achieve maximum sensitivity. Important parameters to be considered are the flow-cell diameter and its volume, which have to be chosen according to the expected sample quantities and LC peak volumes, the amount of back-mixing that can be tolerated, but also to the field strength and the nucleus to be detected by direct or inverse experiments.

LC-NMR flow-cells have to be dimensioned with respect to chromatographic resolution, flow rate, and NMR magnet field strength.

Typical flow-cells have internal diameters ranging from 2 to 4 mm (3–5 mm OD) and an active volume of ca. 60–250 µL. While, from the chromatographic point of view, a flow-cell diameter as small as possible is desirable to avoid back-mixing (especially in stopped-flow measurements), the 5 mm probeheads are best suited for LC-NMR at 7 T (300 MHz for ^{1}H). The 4 mm OD probeheads are a reasonable compromise between good sensitivity and low back-mixing when working at higher frequencies. It is evident that sample size, flow-cell volume, and flow rate have to be matched to achieve maximum sensitivity. The 3 mm OD probeheads perform best for stopped-flow detection of the smaller LC peak volumes obtained from micro-LC columns.

The sample is required to spend a certain time in the magnetic field so that the nuclei are polarized before entering the flow cell. As a result of the limited residence time of a nucleus in the flow-cell, however, both the spin–lattice and the spin–spin relaxation times are reduced compared with static measurements. This leads to a signal which increases with flow rate. This effect is counterbalanced, however, by the increasing spectral line-broadening at higher flow rates, which requires that compromise conditions be found.

As a further measure to achieve maximum sensitivity, the receiver coil is attached directly to the flow-cell. This also provides superior stability of the lock signal.

LC-NMR requires either the use of deuterated or aprotic solvents or solvent suppression techniques.

The choice of a suitable solvent for LC-NMR is critical, since solvents commonly used in NMR experiments are either deuterated and thus (with the exception of D_2O) too expensive to be used in HPLC separations, or they are aprotoic ($CHCl_3$, Freons) and lack universal applicability in normal-phase operation. The use of protonated solvents makes suppression of the solvent signals necessary. Although a variety of different techniques exist for solvent suppression in NMR, based on either differences in chemical shifts (e.g., selective presaturation, selective excitation, or composite pulse techniques) or differences in relaxation times (e.g., progressive saturation, spin echo techniques), neither is completely suitable for LC-NMR. While this is less pronounced in isocratic separations, it is especially important in gradient elution, where the resonance frequencies change with the composition of the solvent. In commercial instruments, the problem of solvent suppression has largely been overcome by using adaptive extrapolation techniques which calculate the required suppression frequencies during a chromatographic run, based on the signals from the preceeding data acquisition blocks.

Significant improvement has also resulted from the use of analog-to-digital converters (ADC) with high dynamic range digitizers (16–19 bit). The dynamic range of the ADC determines the ratio of the largest to the smallest detectable signal. Suppression of the solvent signal is necessary to benefit from the high dynamic range of the ADC.

The incorporation of these features provides, in a typical configuration for a substance of molecular mass ≤400 Da, an on-flow detection limit of ca. 10 µg for ^{1}H and ^{19}F (at 11.7 T, corresponding to 500 MHz for ^{1}H). Higher sensitivity can be obtained by prolonged scan times during stopped-flow data acquisition. In this case, ≥200 ng can be detected in ca. 3 h for ^{1}H and ^{19}F. Inverse ^{1}H/^{13}C correlation experiments still require overnight scanning to reach a detection limit of ca. 15 µg. Two-dimensional experiments can be carried out with ≥1 µg in 3 h.

With this sensitivity, the application of LC-NMR is clearly not feasible for environmental analysis. In the field of industrial or pharmaceutical analytical chemistry, however, this technique has great promise (Fig. 14.3-11), and will gain importance if its sensitivity can be further increased.

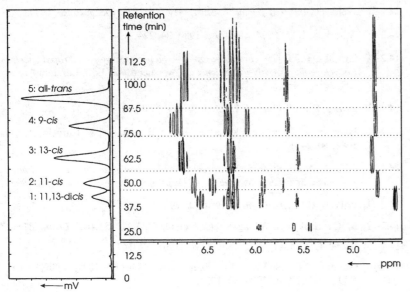

Fig. 14.3-11. UV-chromatogram, detected at 325 nm (left), and contour plot of a LC-NMR separation (right) of vitamine A acetate isomers in n-heptane. (Flow rate: 0.2 mL/min, Column: cyanopropyl-modified silica). The contour plot shows the ^{1}H NMR chromatogram in the olefinic region (at 400 MHz) recorded with on-flow detection. The peak labels in the chromatogram indicate the different isomers of vitamine A acetate, characterized by the conformation at the double bond [14.3-13]

14.4 Other Techniques

An account of hyphenated techniques would not be complete without at least mentioning the recent progress in coupling two further, generally used separation techniques, thin-layer chromatography (TLC) and supercritical fluid chromatography (SFC), to spectroscopic detection.

Succesful interfaces have already been proposed for the hyphenation of TLC with MS and FTIR detection. They have not yet found general application, however, probably due to the fact that the simplest and approximately least expensive separation technique in the analytical laboratory is coupled to probably the most sophisticated and expensive of detection techniques. However, the separation power and versatility of modern high-performance thin-layer chromatography (HPTLC) should not be underestimated.

SFC has also been successfully coupled to mass spectrometric, FTIR, and atomic emission detection. Owing to the nature of the mobile phase used in SFC (usually supercritical carbon dioxide, often with the addition of small amounts of a modifier e.g., methanol) the interfacing requirements are intermediate between those of gas and liquid chromatography. Thus, existing GC and LC interfaces can be adapted with minor changes for successful operation with the various types of spectroscopic detectors.

Looking into the future of hyphenated techniques, it is not difficult to predict that novel hyphenated techniques are likely to appear. Some have already achieved a certain degree of maturity, e.g., capillary electrophoresis (CE) coupled to FTIR and MS detection. Others are just on the horizon, such as GC with UV absorbance detection. The driving force behind all these developments is the need for highly sensitive and specific detectors for the various separation techniques, capable of solving increasingly demanding and complex analytical problems; their success will finally depend on their ability to satisfy these needs.

References

[14.1-1]   Hirschfeld, T., *Anal. Chem.* 1980, **52**, 297A.

[14.2-1]   Jones, A.J., Martin, A.J.P., *Analyst* 1952, **77**, 915.

[14.2-2]   *Fisons-Chromatography News*, 3/94, p. 6.

[14.2-3]   Poole, C.F., Poole, S.K., *Chromatography Today*, Amsterdam: Elsevier, 1991.

[14.2-4]   Janson, O., *Analytik von Spurenstoffen in Deponiegasen*, in: *Buch der Umweltanalytik*, Weber E., Weber R. (Eds.), Darmstadt: GIT-Verlag, 1992; Vol. 4, p. 130.

[14.2-5]   Welti, D., *Infrared Vapour Spectra*, New York: Heyden/Sadtler, 1970.

[14.2-6]   Azzaraga, L.V., *Appl. Spectrosc.* 1980, **34**, 224.

[14.2-7]   Griffiths, P.R., *Gas chromatography – Fourier transform infrared spectrometry*, in: *Encyclopedia of Analytical Sciences*, Townshend, A. (Ed.), London: Academic Press, 1995.

[14.2-8]   Brown, R.S., Wilkins, C.L., *Anal. Chem.* 1988, **60**, 1483.

[14.2-9]   Haefner, A.M., Norton, K.L., Griffiths, P.R., Bourne, S., Curbelo, R., *Anal. Chem.* 1988, **60**, 2441.

[14.2-10] Gilbert, A.S., *Infrared spectroscopy – Hyphenated techniques*, in: *Encyclopedia of Analytical Sciences*, Townshend, A. (Ed.), London: Academic Press, 1995.

[14.2-11] Wilkins, C.L., *Anal. Chem.* 1994, **66**, 295A.

[14.2-12] McCormack, A.J., Wong, S.S.C., Cooke, W.D., 1965, *Anal. Chem.*, **37**, 1470.

[14.2-13] Bache, C.A., Lisk, D.J., *Anal. Chem.* 1965, **37**, 1477.

[14.2-14] Quimby, B.D., Sullivan, J.J., *Anal. Chem.* 1990, **62**, 1027; Sullivan, J.J., Quimby, B.D., *Anal. Chem.* 1990, **62**, 1034.

[14.2-15] Uden, P.C., J. Chromatogr. A 1995, 703, 393.

[14.2-16] Pedersen-Bjergaard, S., Asp, T.N., Vedde, J., Carlberg, G.E., Greibrokk, T., *Chromatographia* 1993, **35**, 193–198.

[14.3-1]   Baldwin, M.A., McLafferty, F.W., *Org. Mass Spectrom.* 1973, **7**, 1111.

[14.3-2]   Blakley, C.R., McAdams, M.J., Vestal, M.L., *J. Chromatogr.* 1978, **158**, 264.

[14.3-3]   Willoughby, R.C., Browner, R.F., *Anal. Chem.* 1984, **56**, 2625.

[14.3-4]   Niessen, W.M.A., Tinke, A.P., *J. Chromatogr. A.* 1995, **703**, 37–57.

[14.3-5]   Carrol, D.I., Dzidic, I., Stilwell, R.N., Horning, M.G., Horning, G.C., *Anal. Chem.* 1974, **46**, 706.

[14.3-6]   Slobodnik, J., van Baar, B.L.M., Brinkman, U.A.Th., *J. Chromatogr. A*, 1995, **703**, 81–121.

[14.3-7]   Bruins, A.P., Covey, T.R., Henion, J.D., *Anal. Chem.* 1987, **59**, 2642.

[14.3-8]   Jinno, K., *FTIR Detection*, in: *Detectors for Liquid Chromatography*, Yeung, E.S., (Ed.) New York: Wiley, 1986; p. 82.

[14.3-9]   Dorn, H.C., *Anal. Chem.* 1984, **56**, 747A.

[14.3-10] Albert, K., Bayer, E., *Trends Anal. Chem.* 1988, **7**, 288.

[14.3-11] Laude, D. A., Jr., Wilkins, C.L., *Anal. Chem.* 1984, **56**, 2471.

[14.3-12] Courtesy: Bruker Analytische MeBtechnik GmbH, Rheinstetten, Germany.

[14.3-13] Albert, K., *J. Chromatogr. A*, 1995, **703**, 123–147.

General reading

Hill, H.H., McMinn, D.G., (Ed.), *Detectors for Capillary Chromatography*, New York: Wiley, 1992.

Niessen, W.M.A., van der Greef, J., *Liquid Cromatography-Mass Spectrometry*, New York: Marcel Dekker, 1992.

Uden, P.C., *Element-Specific Chromatographic Detection by Atomic Emission Spectroscopy*, ACS Symposium Series, vol. 479, Washington, DC: American Chemical Society, 1992.

White, R., *Chromatography/Fourier Transform Infrared Spectroscopy and its Applications*, New York: Marcel Dekker, 1990.

Yeung, E.S., (Ed.), *Detectors for Liquid Chromatography*, New York: Wiley, 1986.

J. Chromatogr. A, 703 1995: A symposium volume with reviews on all aspects of hyphenated techniques

Questions and problems

1. What are the basic advantages of coupling a separation technique to a spectroscopic detection technique?
2. How can peak purity be checked by spectroscopic detection for a separation technique?
3. What is the advantage of the selected ion monitoring mode in GC/MS detection?
4. Why has capillary column GC-MS almost completely replaced packed column GC-MS applications?
5. What are the requirements for a mass spectrometer to be useful as a GC detector?
6. What is the usual way to obtain a chromatogram in GC-FTIR? And in GC-MS?
7. What complementary information can be obtained by combined GC-MS-FTIR?
8. What are the advantages and limitations of the cold-trapping interface for GC-FTIR?
9. Describe the application of the Gram–Schmidt algorithm in GC-FTIR.
10. What is the function of adding reagent (scavenger) gases in GC-AED?
11. Why does either the grating have to rotate or the photodiode array be movable along the focal plane in multielement GC-atomic emission detection?
12. How can interferences be eliminated in GC-AED?
13. What are the limitations of LC-FTIR performed with the flow-cell interface?
14. Why is ionization in LC-MS more favorable at atmospheric pressure than at reduced pressure and which interfaces provide atmospheric pressure ionization?
15. What is the typical application range of the thermospray, APCI, electrospray, and particle beam interface for LC-MS?
16. Compare the characteristics of spectra obtained by particle beam- and thermospray LC-MS.
17. Calculate the gas volumes that are introduced into the ion source of a LC-MS instrument by the vaporization of (a) 1 mL/min of water (b) 1 mL/min of acetonitrile/water 50/50 (v/v) and (c) 1 mL/min cyclohexane at 373 K and 1.015×10^5 Pa.
18. Which parameters affect the choice of the internal volume and diameter of a LC-NMR flow cell?
19. Which considerations apply for the choice of a suitable solvent in LC-NMR?

15 Miniaturized Analytical Systems

Learning objectives

- To provide an overview on the principles and developments in miniaturization of chemical analysis systems
- To highlight the advantages of miniaturization in comparison to conventional systems
- To introduce the fabrication method
- To give examples for successful chip designs

15.1 Principles

Chemical analysis nowadays is mostly performed with bench-top systems, roughly the size of a large television set. As we have seen in the previous chapters, there exist a variety of analytical chemistry methods for laboratory analysis of a given sample with respect to component identification, quantification, and possible structural elucidation. The analysis consists of sample collection, pretreatment, sample injection, separation, and detection. Originally, all these steps had to be performed manually with different instruments.

However, for on-line analysis a more automated approach is necessary. In most modern on-line systems, sample handling, separation, and detection stages are integrated into a single instrument, with most stages automated and computer controlled. Examples of these so-called *total analysis systems* (TAS, Fig. 15.1-1b) are flow injection analysis (FIA), electrophoresis, chromatography (Chap. 5) and mass spectrometry (Sec. 9.4). These methods operate ex-situ, i.e., a sample has to be taken and transferred to the laboratory, usually requiring some more or less complicated sample pretreatment prior to analysis. These systems offer the advantages of a high degree of automation, the possibility of inbuilt calibration and, owing to the sample pretreatment, the need for only moderately sensitive detector systems. The main disadvantages are the reagent consumption, slow transport of samples in solution, and the slow separation speed of liquid chromatographic methods.

As with electronics some decades ago, there is a strong move toward the miniaturization of chemical analysis systems. The potential applications for small, portable analyzers are numerous, for example:

- Environmental measurements
- Pollution control, directly at the sources of contamination
- Pharmaceutical and agrochemical research
- Biomedical screening and other high-throughput medical applications
- Industrial process control, e.g., in the food industry
- Forensic studies
- Artificial senses for "smart machines", e.g., the "artificial nose"
- Domestic applications in the medical and hygiene field

Size reduction in analytical systems has several distinct advantages:

- Faster analysis, due to the reduced transport lengths, optimized mass transport for chemical reactions and separations
- Reduced consumption of reagents
- Less waste production
- Smaller sample volumes
- Portability

Many of the analytical systems described in previous chapters, such as chromatography, electrophoresis, or flow injection analysis, have been subject to this miniaturization, leading to the concept of *miniaturized total analysis systems* (μ-TAS, Fig. 15.1-1c).

Total analysis systems are analytical systems where sample handling, separation and detection are integrated into a single instrument.

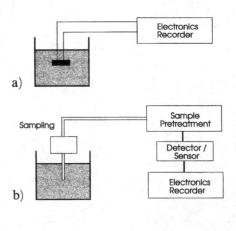

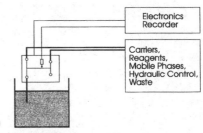

Fig. 15.1-1. Concept comparison between (a) a chemical sensor; (b) a total analysis system and (c) a miniaturized total analysis system

In order to define μ-TAS, one has to take a look at competing concepts in analytical chemistry. On one hand, there is the chemical sensor (Fig. 15.1-1a, see Sec. 7.8 and 7.9). An ideal sensor shows a high sensitivity toward the substance to be detected while at the same time suppressing response to any other substance. It has a large dynamic (e.g., concentration) range, and displays reproducible signals and low noise over an extended period of time. Ideally, it can be used in situ, e.g., the sensor should be small enough to be operated at the place of interest, possibly immersed in the liquid or the gases under consideration. In on-line applications, signal acquisition must be fast and continous. So far, problems remain in obtaining selectivity (partly overcome by the use of sensor arrays) and with the life time of the devices. Furthermore, the development of a market-ready sensor is very time- and labor-consuming.

Miniaturized total analysis systems are TAS, where the analytical functions take place at the location of the measurement itself.

If a TAS can be reduced in size, leading to the advantages listed above, we can define it as a miniaturized total analysis system (μ-TAS), and it should be able to perform all the necessary steps of sampling, handling, and pretreatment at the same location as the measurement itself. This approach, illustrated in Fig. 15.1-1, combines the advantages of chemical sensors with the resolving power of modern bench-top analytical systems. These properties make μ-TAS an ideal concept for high-throughput applications, such as clinical analysis or process control of simple mixtures.

15.2 Microfabrication

The fabrication process for μ-TAS makes use of integrated circuit technology, developed originally for the microelectronics industry. Chip fabrication relies on well-established processes, including photolithography, wet etching, or advanced gas-phase technologies such as reactive ion etching (RIE). Figure 15.2-1 shows a typical fabrication process. A substrate, usually silicon, glass, or quartz, but potentially also polymers, is covered with a metal film (usually chromium or gold with a thin chromium layer to promote adhesion) and a photoresist layer. With a photomask, which contains the structural information for the device, produced by standard lithography processes, the resist film is exposed and then developed, removing the photoresist from the exposed areas. The substrate is then placed in an etching bath, which etches through the metal film in the areas not covered by photoresist. In a second etch step, the substrate itself is etched, typically in HF/ HNO_3 or KOH. Depending on the etchant and the substrate, the microchannels have different profiles. In glass or other amorphous materials, one usually has

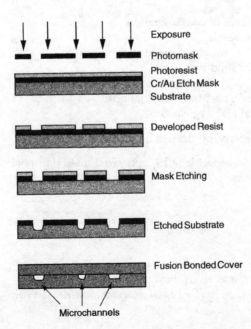

Fig. 15.2-1. Microfabrication process for the production of microchannels

isotropic etching conditions, yielding the same etch rate in any direction, and therefore forming rounded channel edges. On a monocrystalline silicon or quartz substrate, anisotropic etching is possible with suitable etchants, leading to a channel profile which depends on the direction of the exposed crystal planes. In a final step, a cover lid is fusion-bonded on top of the substrate, closing the channels. Another possibility is to grow a top layer of SiO_2 onto the substrate. Further process steps, such as the metallization of certain areas to act as electrical contacts, or the formation of oxide or nitride layers for insulation, etc., can easily be included.

15.3 Examples and Experimental Results

15.3.1 Chromatography

The credit for the very first fully integrated gas chromatography system on a silicon chip, including valves and a detector system, goes to Terry and co-workers in 1979. But only in recent years have new designs, particularly for liquid chromatography, emerged. A design of an open-capillary liquid chromatography system is shown in Fig. 15.3-2. Typical dimensions of the column are: width 5–50 μm, depth 1–10 μm, length 5–15 cm, yielding a total column volume between 1.5 and 10 nL. The detection is achieved by amperometric or conductometric methods, measuring the current and resistance across the column outlet. The detector volume can be as little as 1.2 pL.

Another design with an optical detection unit is shown in Fig. 15.3-1. It basically consists of a split injector, controlling the portion of the sample entering the separation column, the column itself, a frit at the end of the column (which holds back packing material from the column), and the detector cell. The optical detection, based on the excitation of fluoresence in the sample, demands an optical path-length of the order of 1 mm to achieve satisfactory sensitivity. The light is therefore guided along the whole length of the detector groove, where it is reflected in and out of the groove by the aluminum covered side walls, and transferred to a photomultiplier via an optical fiber. The detector volume is 2.3 nL, with a column volume of 490 nL. Figure 15.3-3 shows a separation of the two fluorescent dyes fluorescein and acridine orange, obtained within less than 1 min.

The main obstacle to these HPLC chips becoming a complete μ-TAS is the current lack of integratable high-pressure pumps and valves suitable for the range between 10 and 400 bar. Although recent years have seen commercially available micromachined silicon valves and pumps, they can operate only at pressures up to a few bar, producing a pulsating flow. New approaches, such as piezoelectric pumps used in ink-jet printers, may be able to overcome this problem. So far, the available HPLC chips give far from optimal performance and require further improvement.

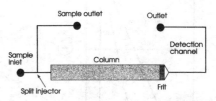

Fig. 15.3-1. Design for a packed column HPLC-chip with optical detection cell

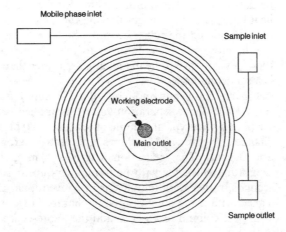

Fig. 15.3-2. Design for an open tubular liquid chromatography system on a chip

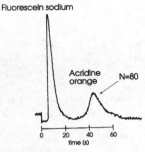

Fig. 15.3-3. Example of a separation of two fluorescent dyes obtained with the HPLC-chip

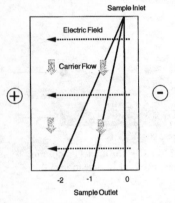

Fig. 15.3-4. The principle of free-flow electrophoresis (FFE)

Electrophoretic separation is the separation of individual molecules under the influence of an electric field applied along the capillary.
Electroosmotic flow is the movement of the complete liquid column in a capillary under the influence of an electric field applied along the capillary.

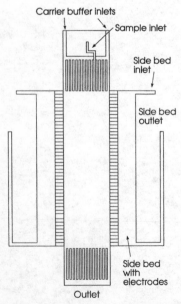

Fig. 15.3-5. Design of an FFE system on a silicon chip

15.3.2 Free-flow electrophoresis

Free-flow electrophoresis (FFE) is a very useful technique for sample pretreatment. The principle can be explained by Fig. 15.3-4. The sample is fed into a carrier stream. An electric field is applied perpendicular to the flow of the carrier. Ions are then deflected from the flow direction according to their electric charge and electrophoretic mobility. The angle of deflection increases with the applied electric field, the ion charge and mobility, and decreases with the flow speed of the carrier. After the separation process, the different species can be observed as they exit the outlet channel.

Figure 15.3-5 shows the layout of a FFE device fabricated on silicon. The sample is fed into the system at the location marked 'sample inlet' and is transported by the carrier, accessing the chip through the carrier buffer inlet channels. The voltage is applied by two platinum arrays in the side beds, separated by a spacer, consisting of 2500 V-shaped grooves. A device of this type can be used for continuous sample fractioning where specific species are selected for a subsequent analysis step.

15.3.3 Capillary electrophoresis (CE)

This electrophoretic technique has also been used in μ-TAS. The sample and the buffer liquid are fed into small capillaries. If a voltage is applied at either end of the capillary, two phenomena can be observed. The first effect, the electrophoretic separation, is simply the motion of individual ions, positive or negative, in the fluid under the influence of the applied field. The second effect is called electroosmotic flow and pumps the fluid through the capillary. It arises from the existence of an electric double layer (Helmholtz layer) close to the capillary walls, consisting of a negative immobile charge at the walls (ionized silanol groups) and a layer of positive ions from the liquid which are attracted by this negative charge. If an electric field is applied along the capillary, the mobile positive ions in the fluid start to move, owing to the electrostatic force. Because of the viscosity of the liquid, the complete fluid column is dragged along by the moving ions.

An important parameter for the performance of a separation system is its efficiency. It can be shown theoretically that the separation efficiency, the number of theoretical plates N, is given by:

$$N \propto \frac{L}{d},$$

where L is the capillary length and d its diameter. The analysis time t is given by:

$$t \propto L \cdot d,$$

so a decrease in the capillary diameter leads to a reduction in analysis time as well as an increase in performance. This property makes CE particularly suitable for miniaturization. The limiting factor for the performance of CE chips is the Joule heat because of the current flowing through the capillary. Good thermal coupling of the fluid and the substrate and a high heat dissipation in the substrate material are therefore necessary. In order to achieve sufficiently high electric filed strength, these devices have to be made on glass rather than on a silicon substrate, as silicon is a semiconductor.

Figure 15.3-6 shows a layout of a CE chip. After feeding in the sample between ports 1 and 4, the separation voltage is applied between ports 2 and 5 and a sample plug with a typical volume of less than 100 pL is driven through the separation channel. Figure 15.3-7 shows an electropherogram of six fluorescein-labeled amino acids. The separation length is 24 mm, the field strength 1060 V/cm. A separation efficiency with plate numbers up to 200000 has been achieved within seconds, rather than minutes as in conventional CE. Several chips have been designed with a pre-column mixing chamber to monitor fast chemical reactions or with post-column reactors to add the fluorescence labeling after the separation.

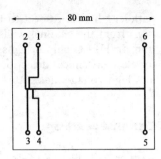

Fig. 15.3-6. Layout of a CE chip. The numbers denote fluid reservoirs. *1* sample inlet; *2* carrier electrolyte inlet; *3* modifier electrolyte inlet; *4* sample outlet; *5, 6* outlets

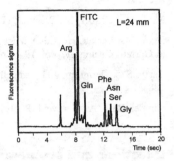

Fig. 15.3-7. Electropherogram of six fluorescein labeled amino acids recorded with a system shown in Fig. 15.3-6

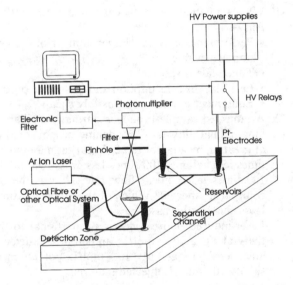

Fig. 15.3-8. Experimental setup for a CE experiment on a chip

15.3.4 Experimental setup

An example of a typical μ-TAS experiment, a capillary electrophoresis device, is shown in Fig. 15.3-8. Fluid reservoirs, usually plastic vials or pipet tips, are glued onto the chips. Through holes drilled into the cover plate they are connected to the microchannels of the CE chip. Platinum electrode wires are placed into these reservoirs and connected to a high-voltage power supply via a relay. For the sake of clarity, the electrodes connected to the lower reservoirs are omitted from the diagram. Detection is usually carried out with laser-induced fluorescence, where light from an Argon ion, He–Ne, or He–Cd laser is coupled into the detection zone via an optical fiber or other optical arrangement. Recently, blue LEDs have also been used as a light source for fluorescence excitation, indicating the possibility of integrating the detection system onto the chip. The excited fluorescence is sampled, either directly with a microscope or indirectly by an optical fiber, filtered to suppress the light of the laser, and detected in a photomultiplier tube or a CCD camera. Another detection method is optical absorbance, used frequently in conventional HPLC systems. As the sample volumes in CE devices are of the order of 1 nL or less, care has to be taken to optimize the coupling of the light in and out of the sample volume. Special flow-cells have been designed to maximize the optical

Laser induced fluorescence is the most sensitive and widely used method for on-chip detection. It enables single molecule detection.

path-length with respect to the detection volume, either by using multiple reflections from the channel walls or by probing the capillary in a longitudinal direction. In HPLC chip designs, platinum electrodes have been included in the chip for electrochemical detection. The requirement of floating the detector electronics at a high voltage makes this approach more difficult to use in CE.

General reading

Harrison, D.J., Fluri, K., Seiler, K., Fan, Z., Effenhauser, C.S., Manz, A., *Science* 1993, 261, 895–897.

Manz, A., Graber, N., Widmer, H.M., *Sensors and Actuators* 1990, B1, 244–248.

Manz, A., Becker, H. (Eds.), Micro System Technology in Chemistry and Life Science, Topics in Current Chemistry, Vol. 194, Springer 1997.

Terry, S.C., Jermann, J.H., Angell, J.B., *IEEE Trans. Electron. Devices*, 1980, ED-26, 1880–1886.

van den Berg, A., Bergveld, P. (Ed.), *Micro Total Analysis Systems*, Dordrecht, Kluwer 1995.

Questions and problems

1. Why is a miniature analytical system attractive?
    Answer:
    *   Performance is better: larger number of theoretical plates and better resolution achievable
    *   Time scale is shorter
    *   Less reagent consumption and less waste production
    *   Equipment is small and possibly cheap
2. An injected sample broadens during a capillary electrophoresis separation by longitudinal diffusion (broadening proportional to square root of time). Miniaturization by a factor of 10 (linear) at maintained voltage would result in 1000 times less volume, 100 times less cross-section, 10 times shorter capillary, and 100 times faster linear flow rate. How is the peak maximum of an injected component affected? Which of the following three detector schemes seem most favorable for this case?
    Laser-induced fluorescence is proportional to the amount of material, sensitivity at $1\,\mu L$, $10^{-12}\,mol/L$; amperometric detection is proportional to surface area, sensitivity at $1\,\mu L$, $10^{-9}\,mol/L$; potentiometric detection is constant, sensitivity $10^{-6}\,mol/L$ injected material.
    Answer:
    Compare the known system at $1\,\mu L$ with the new system at $1\,nL$:
    *   Detection volume 1000 times smaller
    *   Concentration in detection volume = peak maximum, becomes 10 times larger (square root of time)
    *   Amount of compound in detection volume 100 times smaller
    *   Fluorescence: 100 times smaller, $10^{-19}\,mol/nL$
    *   Amperometric: 10 times smaller, $10^{-17}\,mol$
    *   Potentiometric 10 times larger, $10^{-16}\,mol$
    Conclusion: fluorescence is still a lot more sensitive at the $1\,nL$ level.
    Note: at the $1\,pL$ level, all three detection methods show about the same sensitivity. (If you have answered this question correctly, or if these kinds of question interest you, consider doing a PhD thesis in this area: e-mail a.manz@ic.ac.uk.)

16 Process Analytical Chemistry

Learning objectives

■ To introduce the student to the differences in process and laboratory instrumentation

■ To acquaint students with representative sampling, chromatographic, and spectroscopic apparatus for process analysis

16.1 What is Process Analysis?

Throughout this text book, you have learned the principles of analytical chemistry:

- how to obtain a sample representative of the material being analyzed;
- how to treat samples to prepare them for analysis (e.g., dissolution, adjusting solution conditions such as pH);
- how to perform necessary separation steps;
- the types of measurement systems available;
- the treatment and interpretation of the analytical data;
- how to assure the accuracy of the results through quality assurance and quality control.

You have learned of the use of computers, to control and automate some of the steps, and chemometrics, to optimize methods and extract the most useful information from measurements.

In process analysis, the goal is to perform many or all of these functions as part of the chemical process in real time, and to use the information to control or optimize the process. Because of the harsh environment of many chemical processes, and the use of untrained personnel, the measurement systems must meet higher standards of automation, ruggedness, and simplicity than do laboratory instruments. In this chapter, we shall explore the unique measurements for process analysis, how they are met, and give some examples of process measurement systems and their applications. For more detail on process analysis see [16.1-1 to 16.1-4].

Process analysis involves real-time measurements, and includes sampling, pretreatment, interpretation, and use of the results for controlling the chemical process.

16.2 Why do Process Analysis?

There are many reasons for performing real-time measurements. In today's competitive and regulated world, manufacturers are concerned with quality of product, increased productivity, optimal consumption of feed chemicals, and minimization of wastes generated [16.2-1]. Real-time measurements are needed to optimize and control the chemical process, especially during scale-up in a pilot plant phase.

The types of process measurements that may be performed are shown in Fig.16.2-1 [16.2-2]. Transport of samples from the chemical plant to a laboratory for measurement (off-line measurements) has the advantage of the availability of sophisticated measurement systems and trained laboratory personnel. But the transport and measurement are generally slow, requiring hours to days, yielding historical data rather than data that can be used for immediate process adjustments. Hence, off-line measurements are really quality control measurements, i.e., they are used to determine whether the product meets certain specifications of purity, quantity, etc. At-line measurements, in which the instrument is brought into the plant, are more efficient, but still require trained personnel. They are also subject to the harsh environment of the plant, and still may not provide sufficiently rapid measurements.

Process analysis optimizes chemical processes in terms of use of chemicals, waste produced, and quantity and purity of product.

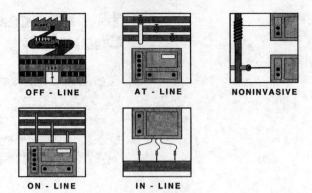

OFF - LINE   AT - LINE   NONINVASIVE

ON - LINE   IN - LINE

Fig. 16.1-1. Types of process analytical techniques. (Adapted from [16.2-2])

A better approach is to interface the instrument to the chemical process to automatically sample the process and make intermittent measurements. The instrumentation requirements will differ from those of laboratory instrumentation. In-line sensors inserted into the process stream are convenient for obtaining continuous measurements; for example, a glass pH electrode with a reference electrode. The electrode design must be such to protect the sensor in the process environment. There are a limited number of such sensors that possess the required selectivity for the analyte of interest in the presence of the process matrix. Noninvasive measurements can sometimes be performed. For example, near infrared spectroscopy may be used to detect the analyte by passage of radiation through the chemical system via a transparent window in the process pipe.

16.3 How Does Process Analysis Differ from Laboratory Analysis?

Process analyzers are automated and must be rugged for long-term operation in harsh environments.

Samples for laboratory analysis are handled under strictly controlled conditions and may be pretreated prior to measurements to provide improved selectivity or sensitivity. Laboratory instrumentation is not subjected to harsh environments or corrosive samples, and can be designed for sophisticated measurements such as high resolution spectral measurements. It is often complicated to use and requires trained analytical chemists for operation. Process analyzers, on the other hand, must be able to withstand the environment of the chemical plant, with changes in temperature and humidity, and must be able to sample and analyze materials, under high pressure or temperature, that may be viscous, contain particulate matter, or may be nonaqueous. Furthermore, the operations of sampling, sample pretreatment, measurement, and data collection and processing must be automatic, with unattended operation for days or weeks at a time. The instrumentation should remain in calibration for long periods of operation, and be capable of autocalibration. It should be capable of rapid measurements to allow appropriate control of the process. Process analyzers will generally be less versatile than laboratory instruments and be dedicated to specific measurements in order to increase their ruggedness and reliability. They are constructed using durable parts that do not require frequent maintenance and with as few moving parts as possible; for example, by using light-emitting diodes or laser diodes as radiation sources of selected wavelengths, or array detectors to eliminate the need to scan a spectrum. Fiber optics may be used as optical light pipes to allow connection to a remote spectrometer or for multipoint measurements with multiplexing to a single instrument.

16.4 Process Analytical Techniques and Their Applications

Traditional process measurements performed include physical measurements of temperature, pressure, refractive index, density, specific gravity, viscosity, turbidity, boiling point, flash point, cloud point, and fluid flow. These parameters can influence the progress of a reaction or are a measurement of its progress. But they provide little in the way of chemical information that may be used to control the process in a more timely and efficient manner. Chemical measurements of reactants, intermediates or products provide information about the process of a reaction, its efficiency or product yield, and can be used to alter reactant feeds to maximize yields and purity. Measurement of pH is common since this influences many reactions.

Measurements may be continuous or intermittent, depending on the information needed and the type of measurement systems. The frequency of measurement will depend on the kinetics of the process, defined by the time constant, which may be several minutes or more. There are two variables, θ and k, which describe the time response of the reaction. Consider, for example, a stirred continuous flow batch reactor in which reactants flow into the reactor and product exits. The mean residence time in the reactor is given by [16.4-1]

Process measurements may be continuous or intermittent.

$$\tau_R = \text{volume/volumetric flow rate} \qquad (16.4\text{-}1)$$

Thus, for a 1 liter reactor in which the flow rate into the reactor is 100 mL/min, the residence time is 10 min. τ_R is known as the hydrodynamic time constant of the reaction and is often referred to as the hold-up time or residence time by engineers. The time constant, τ, of the reaction is

$$\tau = \frac{\tau_R}{(1 + k\tau_R)} \qquad (16.4\text{-}2)$$

where k is a chemical reaction rate constant. The value of k represents a first-order rate constant in time^{-1}. While many reactions are not first order, many processes are operated under conditions of quasi first-order reactions so that they are readily controllable. From the equation, either the flow or the reaction time will be the limiting time sensitive variable. The time constant is usually on the order of minutes. If the time response of the analytical measurement is T, then the effective overall time constant is:

$$\tau = \frac{\tau_R}{(1 + k\tau_R)} + T \qquad (16.4\text{-}3)$$

The measurement interval should be rapid compared to the process time constant, but not so short as to emphasize short term variance (noise) in the process. It is advisable to use a running average measurement, with a minimum of two measurements per time constant, and typically four measurements are used. Very often, the time constants are several minutes or more and so measurements every few minutes may suffice. The following examples illustrate the application of the time constant.

Worked example:

In the manufacture of corn syrup, high and lower concentrations of syrup are frequently mixed to make the final target "Dextrose Equivalent" (DE) product. This mixing can even take place as the product feed stocks are being loaded into rail cars. Thus, it is an important measurement and control problem to assure proper mixing and ratios. What is the minimum measurement time required?

A customer has ordered a load of 42 DE corn syrup. The two available feed stocks are nominally 40 and 55 DE. The analyzer is located at the exit of a 2 meter high × 1 meter diameter stirred cylindrical surge tank that is kept full. The surge tank is located directly after the blend valve where these two feeds are mixed. The flow is controlled by a pump that puts out 200 L/min. The engineers assure you that the mixing rate is much faster than the feed rate. Because this is a blending operation, the chemical reaction is technically $A \Rightarrow A$ and k is zero because the rate is instantaneous.

Solution: The volume of the reaction vessel ($V = \pi r^2 l$) is $(3.14)(50\,\text{cm})^2(200\,\text{cm}) = 1.57 \times 10^6\,\text{cm}^3 = 1570$ L. The problem is easier to solve by putting everything into centimeters and then using 1000 cubic centimeters in a liter as a conversion factor. Given these parameters, θ is equal to 1570 L/200 L min^{-1} or 7.85 min. θ times k is approximately 0. The response time of the system is 7.85/(1 + 0) or 7.85 minutes. The analyzer should be capable of reporting results at least every 3 min and 55 s (twice per resident time). Going much faster than one reading every 2 min adds very little to the information. If the analyzer is much faster than one reading every 2 min, the readings should be averaged to reduce noise.

Worked example:

Ethylene glycol is made commercially by adding ethylene oxide to water. The process uses excess water and is run catalyst free at 140–230 °C. Because water is in large excess, this is a pseudo first-order reaction. Laboratory reaction tests measure the rate constant of this reaction at the temperature used as approximately $5 \times 10^{-5}\,s^{-1}$. How fast should an analyzer make measurements to monitor this reaction in a 20 L vessel with a feed rate of 20 L/min? How fast should the analyzer make measurements for a reaction rate of $5 \times 10^{-2}\,s^{-1}$? How fast should the analyzer respond for a reaction rate of $5 \times 10^{1}\,s^{-1}$?

Here $\theta = 10\,L/20\,L\,min^{-1}$ or 30 s. The response time is $30\,s/(1 + 30s \times 5 \times 10^{-5}\,s^{-1})$ or $30\,s/(1.0015)$ or approximately 30 s. The analyzer should respond at least once every 15 s, with speeds faster than 7.5 s not adding much information. If the reaction is much slower $(k = 5 \times 10^{-2}\,s^{-1})$, the response time is $30/(1 + 30 \times 5 \times 10^{-2})$ or $30/(2.5) = 12$ s. The analyzer should respond at least every 6 s. If the reaction is quite slow $(k = 50\,s^{-1})$, the response rate is $30/(1 + 30 \times 50) = 0.02$ s. The analyzer needs to make about 100 readings per second. Notice that the slower the chemical reaction, the more it dominates the response time and the more sensitive the control system must become. This might seem contrary to intuition. However, it is easy to manage a reaction that goes to completion and harder to manage one that is substantially incomplete. This is one reason why most processes are designed with θ much more important than k.

Common analytical instruments may be adapted for process measurements.

Most classes of instruments used in the laboratory may be used for process measurements, but they are not designed to meet the rugged requirements of process analyzers. Absorption spectrometers (visible, ultraviolet, or infrared), X-ray fluorescence spectrometers, and gas and liquid chromatography are examples. Probe-type sensors include pH electrodes, oxidation-reduction probes (ORP's) and fiber-optic sensors. Fiber-optic sensors may be designed for absorbance or luminescence measurements.

On-line instruments will periodically collect a sample, which will be conditioned prior to presentation to the instrument. This conditioning process is called sample pretreatment. It is usually accomplished by diverting a small fraction of the reaction mixture (e.g., in a flowing stream) into a side test-stream where reagents may be merged and mixed with it. The sample may undergo prefiltration at the instrument-solution interface, dilution or concentration, and temperature adjustment prior to being introduced into the measurement instrument.

Technology for process analytical instrumentation that provides chemical compositional information is generally developed in the analytical laboratories first. There is usually a long lag time between the invention of the technology in the laboratory and its transfer to the process environment as a process analytical tool. This is because the technique first needs to be ruggedized and made suitable for the process environments. This lag time depends on the nature of the technique in terms of its suitability to modifications. Among the first analytical tools that were transferred to process environments were pH sensors in the late forties and fifites. Gas chromatography and wet chemical analysis techniques such as titration followed the pH sensors in the late sixties and early seventies [16.4-2]. Since then, technology has been expanding in both the numbers of applications and the numbers of techniques used. We shall investigate several of these important techniques below.

16.4.1 Separations (chromatography)

Gas chromatography

Process gas chromatographs are widely used.

One of the earliest and most widely used separation techniques is gas chromatography (GC). The popularity of this technique is due to the adaptability of its instrumentation to process environments and its wide range of applicability. It has been heavily implemented in the petroleum industry and is popular in chemical and environmental industries. For a comprehensive theory of chromatography, the student should refer to reference [16.4-3] and Chap 5.

Various types of detector are used to detect the species eluting from the column. Four have gained popularity in process gas chromatography over the years. These are thermal conductivity, flame or photoionization, electron capture, and flame photometric detectors. The thermal conductivity detector is the most frequently used for process gas chromatographs (GCs) due to its simplicity.

There are two general types of gas chromatography columns: packed columns and open tubular (capillary) columns. Historically, most process gas chromatography was done with packed columns, but these are being replaced by some form of open tubular column chromatography since open tubular columns provide increased efficiency and much faster analysis times than conventional packed columns.

Process GC has been applied in the oil industry more than any other area. Determination of boiling points of aliphatic or aromatic hydrocarbons, octane number, and calorific value of natural gas are a few of the example from the oil and related industries [16.4-4–16.4-6]. The use of process GC in environmental analysis for both ambient air and water quality monitoring are other important areas of applications [16.4-7, 16.4-8]. Ambient air leak detection, such as detection of aromatic hydrocarbons, sulfur dioxide, hydrogen sulfide, and carbonyl sulfide in emission from coal combustion are a few recently reported examples of the use of process GC in environmental monitoring [16.4-9]. The development of techniques such as column switching, or the incorporation of cryogenic trapping, spray extraction or membrane separators, has provided wider applicability for GC in the chemical industry [16.4-10,16.4-11].

Liquid and other chromatographic techniques

The application of liquid chromatography (LC) to process analysis has occurred much more slowly than the application of GC. One of the reasons is the increased maintenance requirements due to the liquid mobile phase needs of LC. The reproducible preparation of mobile phase in production plants difficult and so the system requires calibration after changing each batch of mobile phase. Hence, process LC systems require highly trained personnel to operate and service them. Recycling of the mobile phase is one way of reducing the maintenance, but this may not be possible to implement in every application.

Process LC instruments are more maintenance intensive.

Before process LC can contribute significantly to process analysis, improvements must be made in the reliability and in the level of maintenance required. One way to reduce maintenance is to use micropacked LC columns that have a mobile phase flow rate of 1–10 µL/min. At 5 µL/min flow rate, a liter of mobile phase would last more than 4 months, while at conventional flow rates of 2 mL/min, a liter lasts a little more than 8 hours. Recently reported studies in this area are encouraging for the future of process LC [16.4-12].

Other chromatographic techniques such as supercritical fluid chromatography (SFC) and size exclusion chromatography (SEC) have been applied in process analysis [16.4-13, 16.4-14]. However, they require even more attention than LC, which prevents the wide application of these useful techniques.

16.4.2 Spectroscopic techniques

Optical spectroscopy

Optical spectroscopic techniques used in process analysis can be grouped in two major categories: electronic spectroscopy (ultraviolet (UV) and visible (VIS) spectroscopies), and vibrational spectroscopy (infrared (IR), near infrared (NIR), and Raman spectroscopies). In the UV and visible regions of the spectrum, absorption

Absorption spectrophotometry is widely used in process analysis.

is due to transitions between atomic or molecular electronic energy levels. Transitions between electronic energy levels can only occur when the energy of the incident photon is the same as the difference between the atomic or molecular energy levels. These energy levels for near ultraviolet and visible regions occur in isolated atoms, certain inorganic ions, organic compounds that contain conjugated double bonds, and numerous miscellaneous molecular species. Absorption in the ultraviolet and visible regions tends to be very strong, thus, measurements of concentrations in the parts-per-million range are often possible. Nonetheless, the absorption bands are typically very broad compared with those in the infrared region, so that selective measurement of one component in the presence of another may be difficult.

Diode-array analyzer for measurements in the UV-VIS region are becoming important in process analysis [16.4-15]. Filter-based photometric devices are also popular due to their simplicity. Measurements of SO_2 and H_2S in the petroleum industry and of chlorine and phenol in the chemical industry are a few examples of the uses of this technology in process analysis [16.4-16,16.4-17].

Absorption in the IR region results from changes in the vibrational energy levels of a molecule. In particular, these absorptions are almost exclusively due to overtones and combinations of vibrations involving hydrogen atoms. Bands can be observed for C–H, N–H, O–H and other functional groups containing hydrogen. Absorption bands in the NIR and IR ranges are several orders of magnitude weaker than bands in the ultraviolet and visible regions; NIR and IR are therefore not as useful for trace level analysis. NIR and IR bands are relatively narrow and are quite characteristic of the functional groups causing the absorption. As a result, these measurements offer much better selectivity than ultraviolet or visible measurements when analyzing mixtures of organic compounds. Filter-based NIR instruments have been used extensively in process analysis.

There are numerous IR applications reported from the chemical and food industries in the literature [16.4-18–16.4-20]. For analysis of complex organic mixtures, the use of Fourier Transform infrared (FTIR) or scanned grating IR instruments is essential. FTIR has been implemented in process analysis widely. Its applications have spread to all industries such as chemical, oil, and food industries, and even into environmental and biotechnology areas [16.4-21–16.4-25].

The principle of Raman spectroscopy is based on the *Raman effect*, which is described as follows: When light of a single wavelength interacts with a molecule, the light scattered by the molecule contains wavelengths different from that of the incident light. These wavelengths are characteristic of the structure of the molecule, and their intensities depend on the concentration of the molecule. Therefore, the identities and concentrations of various molecules in a substance can be determined by illuminating the substance with light of a single wavelength and then measuring the different wavelengths, and their intensities, of the scattered light.

In the past five years, Raman spectroscopy has received increasing attention as a process analytical tool for in situ real time analysis [16.4-26–16.4-28]. This is due in part to the excellent throughput of optical fibers in the visible and near-infrared regions, where small, portable, air-cooled lasers may be used for excitation, and to the availability of highly efficient detectors such as silicon charge-coupled devices (CCD). Raman spectra can now be acquired in seconds using CCD detectors, as compared with the many minutes required with a scanned grating and a photomultiplier tube detector. Moreover, Raman spectra are composed of vibrational fundamentals and the spectra are usually quite simple, obviating the need for sophisticated mathematical treatments (although, recently, there has been much activity in incorporating chemometrics in data treatment).

Optical fibers are useful for transmitting light between the sample and the spectrometer.

The use of optical fibers in optical spectroscopy eliminates the need for sampling and enables in-line analysis for obtaining chemical information in process analysis. Neither fiber-optic sensors which utilize the intrinsic properties of fibers for detection nor fibers modified, for example, by immobilized reagents at the fiber tip to create chemical specificity, are yet reliable and rugged enough to be used routinely in process analysis. These type of fiber-optic sensors are outside the interest of this chapter.

Fiber-optic sensors or probes used to provide a simple interface between the spectrometer and the sample have facilitated the implementation of many of these spectroscopic techniques in process analysis. In these types of spectrometers, light

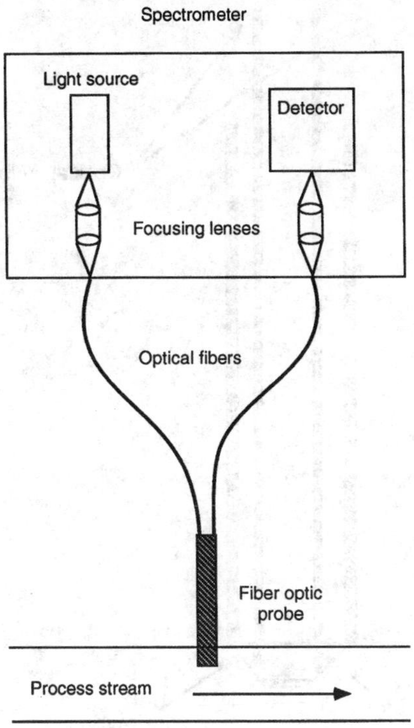

Spectrometer

Light source

Detector

Focusing lenses

Optical fibers

Fiber optic probe

Process stream

Fig. 16.4-1. An optical spectrometer coupled with fiber-optic probe. (Courtesy of the Dow Chemical Company)

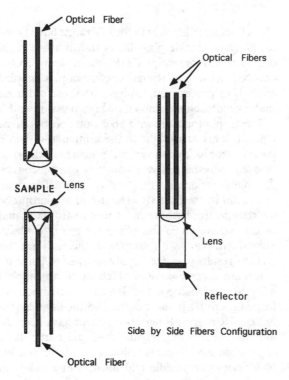

Optical Fiber

Optical Fibers

SAMPLE Lens

Lens

Reflector

Side by Side Fibers Configuration

Optical Fiber

Facing Fibers Configuration

Fig. 16.4-2. Transmission probe configurations. (Courtesy of the Dow Chemical Company)

from the light source of the spectrometer is carried via a fiber-optic cable into the sample. After interacting with the sample, the light is carried back to the detector via fiber optics. Figure 16.4-1 illustrates the component of an optical spectrometer coupled with a fiber-optic probe. Two different types of light/sample interface (probe) are popular in UV, VIS, and NIR spectroscopies: transmission probes, and attenuated total reflection (ATR) probes. Configurations of these probes are illustrated in Figs. 16.4-2 and 16.4-3.

Optical fibers transmit best in the visible and near-IR ranges. Recently, quartz

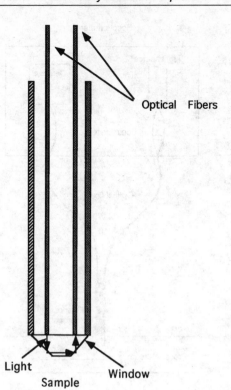

Optical Fibers

Light

Sample   Window

Fig. 16.4-3. Principle of the Attenuated Reflectance Probe (ATR). (From McLachlan, R.D., Jewett, G.L., Evans, J.C. (1986), *U.S. Patent* 4,573,761; courtesy of the Dow Chemical Company)

fibers for short distances in the UV range have become commercially available. The use of chalcogenide glass fibers transmitting in the min-IR region also has been demonstrated for remote FTIR spectroscopy [16.4-29]. The applications of UV-VIS and NIR as well as Raman spectroscopies coupled with optical fibers for remote analysis of process streams have been useful for particularly hostile environments such as radioactive samples, or high pressure and temperature processes [16.4-30].

Spectrophotometers with fiber-optic probes do not have a sample compartment. Optical fibers carry light to the sample and then return sample-modified light to the analyzer for measurement. A variety of detectors, gratings, filters, and slits can be used with the analyzer and they can be conveniently changed. Depending on the configuration, the analyzer can be used from approximately 250 nm in the UV to 220 nm in the near IR. The use of a microprocessor is essential as a controller for the spectrophotometer, as well as for data analysis. Often, a reference scan is taken and stored in the computer prior to placing a sample in the light path. Subsequent scans with samples in the path are then ratioed to the reference scan and the results presented in absorbance or percent transmittance format [16.4-31].

Raman spectroscopy is much more amenable to fiber-optic technology than infrared spectroscopy. Yet, Raman spectroscopy contains a similar amount of information to IR in most cases, and the light source requirements are in the visible or near-IR ranges. Fiber-optic probes have been developed for light scattering and luminescence measurements; these are particularly suitable for Raman spectroscopy when used in conjunction with a laser as a light source. These probes consist of a fiber-optic bundle and an optical window at the end of the fibers. In the bundle, optical fibers are arranged such that there is one single fiber at the center that delivers the light (usually a 600 micrometer silica core with doped silica cladding and a polyamide buffer), and several other similar fibers that surround the center fiber collect the scattered light (Fig. 16.4-4). This type of arrangement provides maximum collection efficiency for the scattered light. Lenses are used to focus the light from the laser into the center fiber, and the returning light is passed through a holographic notch filter to remove the laser line.

Mass spectrometry

A mass spectrometer separates the molecules of the sample according to their masses and measures their quantities. One way to separate molecules of different

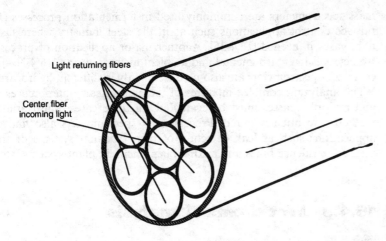

Fig. 16.4-4. Arrangements of the optical fibers in Raman optic probe. (Courtesy of the Dow Chemical Company)

masses for analysis is to first ionize (or apply an electrical charge to) the molecules, and then send them into an electric and/or magnetic field. These fields interact with the ions (charged molecules) thereby separating them on the basis of their charge and mass. The separated ions, when they reach the detector, create an electrical current, which is detected by sensitive current measurement devices. The magnitude of the measured current correlates with the concentration of the molecules in the sample. The current and mass information are typically recorded in a data storage device such as a computer. There are five parts of a typical mass spectrometer: sample introduction chamber, ionization chamber, mass analyzer, detector, and data recorder.

In a typical mass spectrometer, samples are introduced into a vacuum chamber in the gas phase. Therefore, solids or very high boiling liquids (b.p. > 250 °C) generally cannot be analyzed using a conventional process mass spectrometer. The pressure inside a mass spectrometer is approximately a billion times lower than normal atmospheric pressure, so connecting an atmospheric pressure sample to a mass spectrometer which is being installed for on-line analysis is a considerable technical challenge. To maintain the mass spectrometer's low pressure without overloading the spectrometer vacuum pumps, special flow restrictors must be used. There are four common methods for connecting mass spectrometers to process streams: capillary inlet, molecular leak, porous frit, and membrane interfaces. After the sample has been introduced into a mass spectrometer, it is ionized in the ionization chamber. The most common method of ionization is electron impact (EI) ionization. After ionization of the sample molecules, the next step is to separate the charged particles according to their mass. This step is performed in the mass analyzer portion of the instrument. There are two common types of mass analyzers used in process mass spectrometers: magnetic sector analyzers and quadrupole analyzers [16.4-32, 16.4-33]. Magnetic sector analyzers generally provide the most stable readings. Process mass spectrometers that are capable of measuring ions with greater than 200 mass units typically have quadrupole analyzers because they are generally less expensive and more compact than magnetic sector analyzers for the 200–600 mass unit range.

Two types of detector are available in process mass spectrometers: the Faraday cup and the secondary electron multiplier (SEM). The Faraday cup has traditionally been the most common detector used in process mass spectrometers. In general, Faraday cup detectors are useful down to approximately 10 parts per million (ppm). The Faraday cup detector is simple, rugged, stable, relatively inexpensive, and is the preferred detector – provided that it has sufficient sensitivity and speed for the application. A second commonly employed detector is the SEM. This detector works by multiplying the electrons produced when an ion strikes a surface that easily emits electrons. The increase in the number of charges makes the SEM a very sensitive detector. The detector is also very fast and can measure changes as fast as 10^{-9} seconds. However, it is not as stable as the Faraday cup detectors.

The analysis of light gases such as hydrogen, oxygen, nitrogen, carbon dioxide, carbon monoxide, argon and water vapor can be performed by process mass spectrometry. Because of the suitability of mass spectrometers for these gases, process

Magnetic sector and quadrupole mass analyzers are used in process analysis.

The gases H_2, O_2, N_2, CO_2, Ar, and H_2O vapor are measured by mass spectrometry.

mass spectrometers are commonly used in fermentation processes [16.4-34], and furnace control applications such as in the steel industry where fast analysis of light gases in needed [16.4-35]. Another major application of process mass spectrometers is in environmental and ambient air monitoring [16.4-36–16.4-38]. Process mass spectrometers are also used frequently in different hydrocarbon analyses. When analyzing complex mixtures of these chemicals, interferences are common and multiple masses must be evaluated. In these cases, multivariate calibration methods are often required. Process mass spectrometers also find other special applications such as leak detection of plant vacuum systems or tracer tests to measure residence times and mixing phenomena in plant systems [16.4-39].

16.4.3 Wet chemical analysis

The most popular automated wet chemical process analytical techniques include process flow injection analysis (PFIA), titration, and continuous flow analysis (CFA). On-line titration analysis is one of the oldest wet chemical analysis techniques. Its process analytical use is mostly for total acid-base determinations and other well-established laboratory titration techniques such as chloride determination by silver nitrate. The instrumentation required for titrations is considered more complicated than FIA or CFA instrumentation, thus it is prone to higher maintenance. However, there are numerous successful examples of this technique [16.4-40].

Process FIA (Fig.16.4-5) was often confused with other automated wet chemical analysis techniques. The most significant characteristic of FIA is the concentration gradient caused by dispersion of an injected sample in the carrier stream: in other automated-wet chemical analysis techniques, such an continuous flow analysis, sample integrity is conserved by segmenting the flow stream with air bubbles. The concentration gradient caused by dispersion produces chromatography-like peaks in FIA. Quantification of the analyte concentration is based on a peak profile, rather than deviation from a continuous signal. This leads to more accurate results because of the ability to correct for baseline drift. Sequential injection analysis, born from flow injection analysis and based on the same principles, differs in operational aspects [16.4-41, 16.4-42]. However, sequential injection analysis has to be proven by demonstrating long-term operation in process environments.

The popularity of flow injection analysis is mostly attributed to its simplicity and the adaptability of its instrumentation to different environments. The versatility of flow injection analysis is another advantage of the technique in process monitoring. This feature also enables the technique to be used in combination with other analytical techniques as a solution handling system, e.g., for sample introduction, concentration, or dilution.

A variety of analysis techniques have been implemented in process flow injection analysis. The flow-injection-gradient-dilution technique [16.4-43, 16.4-44] has been used in monitoring of dye processes. Flow injection titration techniques are based on measurements of the peak-width, and have found several applications in industrial process monitoring [16.4-45, 16.4-46]. Silicon membrane separators have been incorporated into process flow injection analysis in order to increase selectivity [16.4-47]. These membrane separators have found use in fermentation monitoring where the culture media is brought in contact with buffer solutions via membranes [16.4-48, 16.4-49]. Gas diffusion-FIA systems enable the analysis of many volatile compounds such as ammonia, carbon dioxide, acetic acid, ozone, chlorine, and amines [16.4-50, 16.4-51].

A variety of detection techniques have been used in the laboratory for flow injection analysis. However, many are too complex and expensive for practical process flow injection detectors. The popular detection techniques in process flow injection are spectrophotometric and electrochemical. The primary spectrophotometric techniques employ UV, visible, and fluorescence-based detectors. Recently, diode-array detectors have been used successfully in flow injection analysis for multicomponent monitoring coupled with quantitative chemometrics [16.4-52]. Fluorescence detection has been implemented in FIA to follow enzymatic reactions for the determination of fermentation products [16.4-53, 16.4-54].

Continuous flow and flow injection analyzers are used in process analysis.

Flow injection can be used to introduce samples into various instruments.

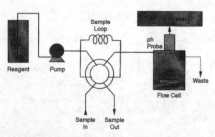

Fig. 16.4-5. Block flow diagram of a typical flow injection analyzer based on titration (gradient dilution) principle. (Courtesy of the Dow Chemical Company)

The most common electrochemical detection techniques used in process flow injection analysis are conductometric and potentiometric (pH probes, ion-selective and metal electrodes). A flow-injection-dual-end-point-titration technique for the determination of sodium carbonate and caustic in alkaline process streams has been demonstrated [16.4-55]. The determination of sodium thiosulfate by a flow-injection-oxidation/reduction-titration technique is an example of the use of metal electrodes in process flow injection analysis [16.4-56]. Ion-selective electrodes have been used for cyanide determination in the monitoring of process streams from cyanide-leaching plants in the mining industry [16.4-57].

Reagent simplicity, stability, and cost are important factors which are carefully considered when selecting the detection technique in process flow injection analysis. Electrochemical detection techniques seem to have an advantage in this area over colorimetric detection techniques. High reagent consumption rates in PFIA contribute to the maintenance requirements of the analyzers. Miniaturization of the instrumentation is expected to minimize this problem in future applications [16.4-58].

Although there are innovative ways of increasing selectivity, insufficient selectivity remains one of the limiting factors in the implementation of process flow injection analysis for on-line determination of complex matrices in the chemical industry. The sampling and maintenance of sample lines present an additional problem for PFIA. The filtration of particulates in the sampling of mixed phases further complicates process flow injection applications.

The instrumentation for process flow injection analysis is an important issue when considering how to provide reliable, low maintenance analyzers in process environments. Different types of pumps, valves, and detectors are being used in construction of the analyzers. The separation of wet components and electronics is described as an advantage since it enables the installation of wet components closer to the sampling points, which in turn shortens the troublesome sample lines [16.4-59]. Overall, the lack of commercially available rugged process flow injection analyzers seems to be the most important limiting factor in the implementation of process flow injection analysis in process monitoring and control.

16.4.4 Other techniques

There are several other analytical techniques, well-established in the laboratory environment, that find niche applications in process analysis, such as nuclear magnetic resonance spectroscopy (NMR), X-ray fluorescence, and atomic spectroscopic techniques – either flame or inductively coupled plasma (ICP). These techniques are not widely applied, mainly due to the complexity of the instrumentation. Laboratory robotics are becoming established in process analysis, for instance for the automated sampling of solid materials in chemical processes.

In case the analysis technique does not give sufficient selectivity or sensitivity, an emerging useful technique in process analysis is chemometrics. On the whole, chemometrics is the application of statistics to chemical data analysis. The value of implementing statistical analysis has long been recognized in chemical engineering and in process control. In process analysis, the implementation of chemometrics has improved selectivity as well as precision and limits of detection, for example, by examining a whole section of an NIR spectrum. Sophisticated statistics software packages, when incorporated into data analyses, can recognize variations from established concentrations for efficient process reactions. These types of pattern recognition techniques can be quite useful in process control since they can warn of the occurrence of abnormalities of the process.

Chemometrics applies statistics to enhance the selectivity of measurements in complex media.

16.5 Sampling Strategies (Analyzer/Process Interface)

Sampling strategies will depend on the nature of the process analytes and matrix, the information required, and the type of measurement device [16.5-1]. Interfacing of the analyzers with the process is achieved by removing a representative sample from the process for on-line analyzers, by insertion of in-line sensors in the process for in-line analysis, and by providing a suitable interface for noninvasive techniques. In this section, each of these procedures will be discussed separately and examples will be given to demonstrate the concepts.

The process analyzer must be interfaced to the chemical process.

Interfacing of the analyzers with the process is one of the most important issues in process analysis. The successful application of an analysis for process control is highly dependent on its associated sampling system. No matter how sophisticated and accurate the analyzer is, the analysis is limited by the sampling system. Maintenance of these sampling systems sometimes outweighs the maintenance of the analyzer itself. Design, construction, and installation of the sampling system – and its maintenance – must be considered very carefully at the beginning of each application.

16.5.1 Sampling for on-line analysis

The sampling of liquids and gases both involve the transfer of sample in a pipe. Solids present much more challenging problems for automatic sampling due to the difficulty of obtaining representative samples and to the transportation of the non-fluid material.

In general, a good sampling system must perform the following functions:

A representative sample must be taken, conditioned, and transported to the analyzer.

1. Take an adequate sample of the material that is representative of the bulk being analyzed. Since liquid and gas materials are more homogeneous than solids, this issue gains more importance for solid sampling.
2. Treat or condition the sample (e.g., clean, vaporize, condense, adjust pressure or temperature, dilute, etc.) in a way that it is compatible with the analyzer without destroying the integrity of the analyzer to be measured.
3. Transport the representative sample to the analyzer with a minimum lag time.
4. Remove the effluent sample from the analyzer to an appropriate waste container, or return it to the process without adversely affecting it.
5. Must be safe, free of leaks, and must not present any other danger to the surrounding environment.
6. Must not interfere with or present any danger to the process being monitored.

The location of the sampling point and the length of the sample lines are interdependent. In order to minimize the length of the sample lines, all possible sampling points in the process must be considered carefully. Sometimes, multiple sampling points are necessary to analyze the material at different points in the process with one or more analyzers. This task must be considered very carefully when using the same analyzer since each analysis will contribute to the delay time for process control. While the use of sampling systems eliminates the need for manual sampling, spot samples requiring manual sampling points next to the analyzer sampling points may still be needed for calibration or occasional cross-check on the analysis system [16.5-2].

Sampling lines should be short.

Sample lines between the sampling point and the analyzer should generally be as short as possible in order to minimize the time lags in sampling and to eliminate potential problems such as plugging or leakage. If the analyzer can not be located at a short distance from the sampling point, fast sample loops can be used in which the material to be sampled is piped close to the analyzer location and then returned to the process following withdrawal of sample at the take-off point to the analyzer.

Sample conditioning is necessary if the analyzer specifications and sample conditions are not in agreement. Sample treatment can involve temperature and pressure adjustments, filtration and the general removal of contaminants in the stream. When needed, chemical treatment and concentration or dilution of the sample is also undertaken. Homogeneity of the sample is considered at two levels of sampling: micro-level – where dispersed phases or discrete particles (e.g., dissolved gases, liquid droplets, or solid particles) are present; and macro-level – dealing with multiphase systems (e.g., liquid or solid or both, dispersed in gas;

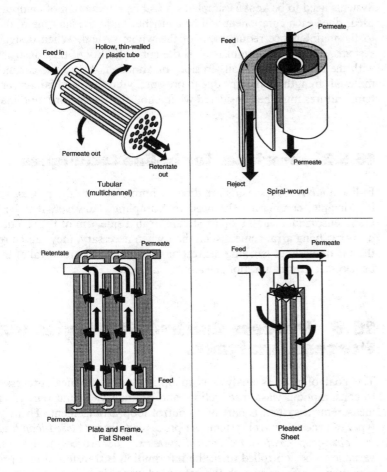

CROSS-FLOW FILTERS ARE GROUPED IN FOUR MAJOR CATAGORIES

Fig. 16.5-1. Types of cross-flow filters. (Adapted from Michaels, S.L., *Chem. Eng.* 8, 84–91, June 1989, with permission)

liquid or solid or any combination thereof, dispersed in liquid; gas or liquid in solid). When required, filters are used for micro-level phase separation. For macro-scale phase separation, other types of phase separators, such as cyclones, coalescers, or condensers/evaporators, may be needed. Removal of micro-particulate matter from gas phase samples can be accomplished with through-flow filters. On the other hand, cross-flow filtration is favored for removal of solids or biological cells from liquid samples since cross-flow filters are by nature self-cleaning filters, thus requiring less maintenance. There are four main types of cross-flow filter design: tube and shell, plate and frame, pleated sheet, and spiral-wound. Figure 16.5-1 shows the differences between these cross-flow filters schematically. A generic two stage sampling system for liquid sampling for on-line analyzers, including a cross-flow and a through-flow filter, is illustrated in Fig 16.5-2. One of the important

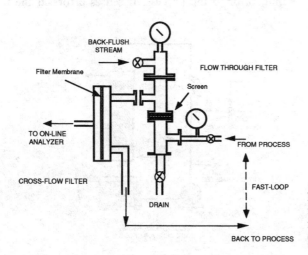

Fig. 16.5-2. A generic liquid sampling system for on-line analyzers showing two stage filtration. Courtesy of the Dow Chemical Company

criteria in designing sample conditioning systems is the simplicity requirement. Systems need to be kept as simple as possible; a minimum of components must be used, since each component will be contributing to the lag time of the sample and to the maintenance requirement of the whole system. When designing sampling systems, another important issue is the compatibility of the construction material with the sample. Corrosion, erosion, or abrasion, due to the chemical reactivity; material strength weakening due to pressure, vibration, and fatigue; and the sample temperatures must be considered when choosing the construction material.

16.5.2 Interfaces for in-line techniques

In-line analyzers or sensors are directly immersed into the process stream – thus, in principle, eliminating the need for sampling. However, it is better to install these sensors in a small by-pass stream or in a side arm of the reaction vessel with proper valving arrangement such that, when necessary, they can be removed from the stream without harming the rest of the process. The removal of the sensor may be necessary for calibration, cleaning, or repair.

16.6 Process Control Strategies via Process Analyzers

Process analysis is performed to better control and optimize the chemical process. Feedback and feedforward control are used.

The goal of process analysis is to optimize the chemical process with respect to consumption, waste production, and product yield and purity. The analytical measurement system is part of a control loop arrangement. There are two main types of process control systems via process analyzers: *closed-loop feedback control and open-loop feedforward control systems*. The process analyzer measures the variable to be controlled, then the information is transferred to a process control computer which compares the measured variable against a reference set point value. Based on this difference, other process parameters are controlled – such as flow rates, mixing rate, cooling water flow rate, etc. Figure 16.6-1 shows the block flow diagram of packed bed, countercurrent, gas-liquid absorber and sample lines to an on-line process analyzer. The concentration of the compound of interest is determined in the effluent stream of the absorber (gas scrubber), then the information is sent to the process control computer. The process control computer, using this information, adjusts the addition of the feed to the scrubber by changing the setting in an automated flow controller.

There is a lag time associated with these types of process control loops. The lag time is the time interval between alteration of the variable and when the analyzer senses a consequent change. It is minimized by installing the sampling points appropriately and utilizing increased flow rates between the input and the analyzer.

In an *open-loop feedforward control* system, the analyzer is placed at or near the input, prior to the process. It senses errors in the input variable and corrective

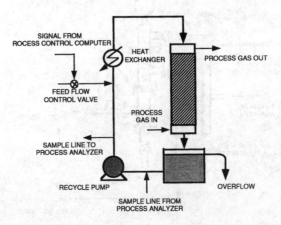

Fig. 16.6-1. Closed-loop feedback control of a gas-liquid absorber via an on-line analyzer. (Courtesy of Dow Chemical Company)

action is taken before the process reaction begins. Hence, it directly controls the input to the process, maintaining reactants at predetermined levels, as opposed to closed-loop feedback systems that maintain a product at desired levels or purity by adjusting the reactant inputs.

16.7 Future of Process Analysis

Process analyzers are not only useful in providing real-time analysis for process control, but they are also used in research organizations to decrease the optimization and development time of new processes. This area will probably enjoy a big expansion in the future, since process analyzers have the potential of contributing significantly to the productivity of research, optimization, and scale-up of the processes. In production areas, process analysis will have increasing importance such that it will eventually replace all off-line analyses, even for final product quality control analysis. However, current process analyzer technology is not yet sufficiently advanced to meet the requirements of this type of task. High speed separation techniques such as capillary electrophoresis may replace liquid chromatography. One advance that is needed in the area is in the miniaturization of the equipment.

Miniaturization of techniques like LC, FIA, GC, and MS would provide numerous advantages such as reduction in reagent consumption, maintenance, and cost of the analyzer. Future process analyzers will also incorporate self-function controllers in order to check on the proper operation of the analyzers, i.e., flow rate controllers to raise alarms about the operation of the analyzer. The trend will be towards the increased use of in-line sensors and their development for routine operation – such as the development of fiber-optic technology to couple all optical spectroscopic techniques with probe-type sensors. Development of more noninvasive techniques would also be ideal for elimination of sampling problems. Current trends, such as the development of remote detection techniques and micro-analyzer/sensor technology, are in the right direction and offer clues about the look of future process analytical technology.

References

[16.1-1] Mix, P.E., *The Design and Application of Process Analyzer Systems*. New York: Wiley Interscience, 1984.

[16.1-2] Clevett, K.J., *Process Analyzer Technology*. New York: John Wiley & Sons; 1986.

[16.1-3] Nichols, G.D., *On-Line Process Analyzers*. New York: John Wiley & Sons; 1988.

[16.1-4] Considine, D.M., *Process Instruments and Controls Handbook*, 3rd ed. New York: McGraw-Hill; 1985.

[16.2-1] Riebe, M.T., Eustace, D.J., *Anal. Chem.* 1990, 62, 65a.

[16.2-2] Callis, J.B., Illman, D.L., Kowalski, B.R., *Anal. Chem.* 1987, 59, 624a.

[16.4-1] Honigs, D.E., *Am. Lab.* 1987, 48.

[16.4-2] Denn, N.N., *Process Modeling*, Marshfield, Mass.: Pitman Publishing, 1986.

[16.4-3] Annino, R., Villalobos R., *Process Gas Chromatography: Fundamentals and Applications*. ISA Press: Research Triangle Park, NC, 1992.

[16.4-4] Crandall, J.A. et al., *Adv. Instrum. Control* 1990, 45(2), 605.

[16.4-5] Durand, J.P., Boscher, Y., Dorbon, M., *J. Chromatogr.*, 1990, 509(1), 47.

[16.4-6] Kenter, R., Struis, M., Smith, A.L.C. *Proc. Control Qual.* 1991, 1, 127.

[16.4-7] Pau, J.C., Knoll, J.E., Midgett, M.R., *JAPCA* 1988, 38(12), 1528.

[16.4-8] Villalobos, R., Annino, R., *Adv. Instrum. Control* 1991, 46(1), 707.

[16.4-9] Clayton, R. et al., *Report, EPA/600/2-89/006*; Order No. PB89-166623. Avail. NTIS from: *Gov. Rep. Announce. Index (U.S.)* 1989, 89(12).

[16.4-10] Melcher, R.G., Morabito, P.L., *Anal. Chem.* 1990, 62, 2183.

[16.4-11] Mouradian, R.F. et al., *J. Air Waste Manage. Assoc.* 1991, 41(7), 1067.

[16.4-12] Cortes, H.J., Larson, J.R., McGowan, G.M., *J. Chromatogr.* 1992, 607(1), 131.

[16.4-13] Renn, C.N., Synovec, R.E., *Anal. Chem.* 1988, 60, 200.

[16.4-14] Lee, M.L., Markides, K.E., *Analytical Supercritical Fluid Chromatography and Extraction* Copyright by Chromatography Conferences Inc., 1990.

[16.4-15] Small, J.R., Hassel, R.L., *Am. Lab.* 1988, 20(11), 88.

[16.4-16] Saltzman, R.S., Bilinski, J.L., *ISA Trans.* 1987, 26, 65.

[16.4-17] Mooney, E.F., *Anal. Div.* 1989, 23, 37.

[16.4-18] Kunikawa, K., Hatsutori, M., Yuzaki, M., Jpn. Kokai Tokkyo Koho JP 01 13 5761, 1989.

[16.4-19] Walling, P.L., Dabney, J.M., *J. Soc. Cosmet. Chem.* 1988, 39(3), 191.

[16.4-20] McDermott, L.P., *Cereal Foods World* 1988, 33(6), 498.

[16.4-21] Chauvel, J.P., May, L., *Proc. Contr. Qual.* 1992, 2, 199.

[16.4-22] Friedrich, J.B. et al., *Anal. Chim. Acta* 1989, 222(2), 221.

[16.4-23] Wilks, P.A., *Tech. Q. Master Brew. Assoc. Am.* 1988, 25, 113.

[16.4-24] Cronin, J.T., *Spectroscopy* 1992, 7, 33.

[16.4-25] Hess, C., *Ger. DD*, 256525, 1988.

[16.4-26] Leugers, M.A., McLachlan, R.D., *Proc. SPIE-Int. Soc. Opt. Eng.* 1989, 990, 88.

[16.4-27] Roberts, M.J. et al., *Proc. Control Qual.* 1991, 1, 281.

[16.4-28] McCreery, R.L., *Proc. SPIE-Int. Soc. Opt. Eng.* 1990, 1439, 25.

[16.4-29] Beebe, K.R. et al., *Anal. Chem.* 1993, 65, 199R–216R.

[16.4-30] Nave, S.E. et al., *Proc. Control Qual.* 1992, 3, 43.

[16.4-31] Schirmer, R.E., Gargus, A.G., *Am. Lab.* 1986, 18(12), 37–43.

[16.4-32] Scrivens, J.H., Ramage, J.C., *Int. J. Mass Spectrom. Ion Proc.* 1984, 60, 299.

[16.4-33] Heppner, R., Hertel, R., Niu, W., Sarrao, J., *Perkin Elmer Technical Note* 1986.

[16.4-34] Berecz, I. et al., *Vacuum* 1987, 37(1–2), 85.

[16.4-35] Schuy, K.D., Reinhold, B., *Measurement and Control* 1971, 4, T84.

[16.4-36] Hemberger, P.H. et al., *Int. J. of Mass Spectrom. Ion Proc.* 1991, 106, 299.

[16.4-37] Sinha, M.P., Gutnikov, G., *Anal. Chem.* 1991, 63(18), 2012.

[16.4-38] Ketkar, S.N., Penn, S.M., Fite, W.L., *Anal. Chem.* 1991, 63(5), 457.

[16.4-39] Walsh, M.R., LaPack, M.A., *49th Symp. Instrument. Process Industr.* 1994.

[16.4-40] Weiss, M.D., *Control* 1989, 2, 72.

[16.4-41] Gubeli, T., Christian, G.D., Ruzicka, J., *Anal. Chim. Acta* 1991, 63, 2407.

[16.4-42] Ruzicka, J., Marshall G.D., Christian, G.D., *Anal. Chem.* 1990, 62, 1861.

[16.4-43] Garn, M.B. et al., *17th Annual Meeting of the Federation of Analytical Chemistry and Spectroscopy Societies* 1990, paper no. 212.

[16.4-44] Garn, M.B., Thommen, C., Frenk, K., *18th Annual Meeting of the Federation of Analytical Chemistry and Spectroscopy Societies* 1991, paper no. 162.

[16.4-45] Yalvac, E.D., *17th Annual Meeting of the Federation of Analytical Chemistry and Spectroscopy Societies*, 1990, paper no. 209.

[16.4-46] Swaim, P., *17th Annual Meeting of the Federation of Analytical Chemistry and Spectroscopy Societies* 1990, paper no. 217.

[16.4-47] Melcher, R.G., Morabito, P.L., Bakke, D.W., Yalvac, E.D., *16th Annual Meeting of the Federation of Analytical Chemistry and Spectroscopy Societies* 1989, paper no. 540.

[16.4-48] Ogbomo, F., Prinzing, U., Schmidt, H.L., *J. of Biotech.* 1990, 14, 63.

[16.4-49] Ludi, H. et al., *J. of Biotech.* 1990, 14, 71.

[16.4-50] Risinger, L., Johansson, G., Thorneman, T., *Anal. Chim. Acta.* 1989, 224, 13.

[16.4-51] Cornham, J.S., Gordon, G., Pacey, G.E., *Anal. Chim. Acta.* 1988, 209, 157.

[16.4-52] Lukkari, L., Lindberg, W., *Anal. Chim. Acta.* 1988, 211, 1.

[16.4-53] Schrader, U. et al., *GBF Monogr.* 1991, 17 (Biosens.: Fundam., Technol. Appl.), 119.

[16.4-54] Spohn, U. et al., *GBF Monogr.*, 1991, 14 (Flow Injection Analysis (FIA) Based Enzymes Antibodies), 51.

[16.4-55] Chung, S. et al., *Anal. Chim. Acta.*, 1991, 249, 77.

[16.4-56] Yalvac, E.D., Bredeweg, R.A., Albers, D.R., *18th Annual Meeting of the Federation of Analytical Chemistry and Spectroscopy Societies* 1991, paper no. 66.

[16.4-57] Robert, R.V.D. et al., *Inorg. Anal. Chem.* 1988, 79.

[16.4-58] Luedi, H. et al. *GBF Monogr.*, 1991, 17 (Biosens.: Fundam., Technol. Appl.), 511

[16.4-59] Yalvac, E.D., *16th Annual Meeting of the Federation of Analytical Chemistry and Spectroscopy Societies* 1989, paper no. 536.

[16.5-1] Cornish, D.C., Jepson, G., Smurthwaite, M.J., *Sampling Systems for Process Analysers.* London: Butterworths, 1981.

[16.5-2] Huskins, D.J., *On-Line Process Analysers.* New York: John Wiley & Sons, 1981.

Appendix

1 Key to Literature

Complementary analytical and instrumental textbooks

Christian, G.D., *Analytical Chemistry*, 5th edn., Wiley, New York, 1994.

Harris, D.C., *Quantitative Chemical Analysis*, 4th edn., Freeman, New York, 1995.

Skoog, D.A., West, D.M., Holler, F.J., *Fundamentals of Analytical Chemistry*, 7th edn., Saunders College Publishing, New York, 1996.

Skoog, D.A., Leary, J.L., *Principles of Instrumental Analysis*, 4th edn., Saunders College Publishing, New York, 1992.

Strobel, H., Heineman, W.R., *Chemical Instrumentation: A Systematic Approach*, 3rd edn., Addison-Wesley, Boston, 1989.

Official methods of analysis and nomenclature

ASTM-Book of Standards, Vol. 3.06, *Chemical Analysis of Metals and Related Materials*, American Society for Testing Materials, Philadelphia.

Official Methods of Analysis, Association of Official Analytical Chemists, 15th edn., Washington, DC, 1990.

Watson, C.A. (Ed.), *Official and Standardized Methods of Analysis*, 3rd edn., Royal Society of Chemistry, London, 1994.

International Union of Pure and Applied Chemistry, Analytical Chemistry Division, in Pure and Applied Chemistry, Great Britain.

Periodicals

Accreditation and Quality Assurance
American Laboratory
Analytical Biochemistry
Analytical Chemistry
Analytica Chimica Acta
Analytical Instrumentation
Analytical Letters
Applied Spectroscopy
Chemometrics and Intelligent Laboratory Systems
Chromatographia
Clinical Chemistry
Fresenius' Journal of Analytical Chemistry (former Zeitschrift für Analytische Chemie)
Journal of Analytical Atomic Spectrometry
Journal of the Association of Official Analytical Chemists
Journal of Chemometrics
Journal of Chromatographic Science
Journal of Chromatography
Journal of Electroanalytical Chemistry and Interfacial Electrochemistry
Microchemical Journal
Mikrochimica Acta
Separation Science
Spectrochimica Acta
Talanta
The Analyst
Trends in Analytical Chemistry

Review serials

Analytical Chemistry, Fundamental Reviews (published biannually in the April issue of Analytical Chemistry in even-numbered years)

Analytical Chemistry, Applicational Reviews (published biannually in the April issue of Analytical Chemistry in odd-numbered years)

CRC Critical Reviews in Analytical Chemistry (published quarterly providing in-depth articles on biochemical analysis)

Table compilations

Martell, A.E., Smith, R.M., *Critical Stability Constants*, Plenum Press, New York, 1974–1989, 6 volumes.

Meites, L. (Ed.), *Handbook of Analytical Chemistry*, McGraw-Hill, New York, 1963.

Treatises

Kolthoff, I.M., Elving, P.J. (Eds.), *Treatise on Analytical Chemistry*, Wiley, New York, 1961–1986.

Svehla, G., Wilson, C.L., Wilson, D.W. (Eds.), *Comprehensive Analytical Chemistry*, Elsevier, New York, 1959–1992.

Winefordner, J.D., Kolthoff, I.M., *Chemical Analysis. A Series of Monographs on Analytical Chemistry and its Applications*, Wiley, New York, about 100 volumes have appeared.

Sources on WWW

Many research journals are now offered in electronic versions. For example, take a look at *Analytical Chemistry* on the Web:
http://pubs.acs.org/journals/ancham

Fresenius' Journal of Analytical Chemistry you find at:
http://link.springer.de

Educational hypermedia on analytical chemistry are linked to the Chemistry Hypermedia Project:
http://www.scimedia.com

2   List of SI Units

1. Base SI Units and Physical Quantities

Physical Quantity	Symbol for Quantity	Name of SI Unit	Symbol for SI Unit
length	l	metre	m
mass	m	kilogram	kg
time	t	second	s
electric current	I	ampere	A
thermodynamic temperature	T	kelvin	K
amount of substance	n	mole	mol
luminous intensity	I_v	candela	cd

2. SI Prefixes

Multiplication factor	Prefix	Symbol
10	deca	da
10^2	hecto	h
10^3	kilo	k
10^6	mega	M
10^9	giga	G
10^{12}	tera	T
10^{15}	peta	P
10^{18}	exa	E
10^{-1}	deci	d
10^{-2}	centi	c
10^{-3}	milli	m
10^{-6}	micro	μ
10^{-9}	nano	n
10^{-12}	pico	p
10^{-15}	femto	f
10^{-18}	atto	a
10^{-21}	zepto	z

3. Examples of SI Derived Units with Special Names and Symbols

Physical Quantity	Name of SI Unit	Symbol for SI Unit	Expression in Terms of SI Base Units
frequency	hertz	Hz	s^{-1}
force	newton	N	$m\,kg\,s^{-2}$
pressure, stress	pascal	Pa	$m^{-1}\,kg\,s^{-2}\ (= N\,m^{-2})$
energy, work, heat	joule	J	$m^2\,kg\,s^{-2}\ (= N\,m = Pa\,m^3)$
power	watt	W	$m^2\,kg\,s^{-3}\ (= J\,s^{-1})$
electric charge	coulomb	C	$s\,A$
electric potential	volt	V	$m^2\,kg\,s^{-3}\,A^{-1}\ (= J\,C^{-1})$
electric capacitance	farad	F	$m^{-2}\,kg^{-1}\,s^4\,A^2\ (= C\,V^{-1})$
electric resistance	ohm	Ω	$m^2\,kg\,s^{-3}\,A^{-2}\ (= V\,A^{-1})$
electric conductance	siemens	S	$m^{-2}\,kg^{-1}\,s^3\,A^2\ (= \Omega^{-1})$
magnetic flux	weber	Wb	$m^2\,kg\,s^{-2}\,A^{-1}\ (= V\,s)$
magnetic flux density	tesla	T	$kg\,s^{-2}\,A^{-1}\ (= V\,s\,m^{-2})$
inductance	henry	H	$m^2\,kg\,s^{-2}\,A^{-2}\ (= V\,A^{-1}s)$
Celsius temperature*	degree Celsius	°C	K
plane angle	radian	rad	1 (rad and sr may be
solid angle	steradian	sr	1 included or omitted in expressions for the derived units)
activity (of radionuclides)	becquerel	Bq	1/s

*The Celsius temperature is defined by $\theta/°C = T/K - 273.15$

4. Examples of Commonly Used Units Outside the SI

Physical Quantity	Unit	Symbol for Unit	Value in SI Units	
time	minute	min	60	s
time	hour	h	3600	s
volume	litre	L	10^{-3}	m^3
energy	electronvolt*	eV	1.602 18	$\times 10^{-19}$ J

* Defined in terms of best values of certain physical constants

3 Collection of Data

Table I. Acid Dissociation Constants at 25 °C

Name	pK_{A1}	pK_{A2}	pK_{A3}	pK_{A4}
$Al(H_2O)_6^{3+}$	4.85			
$Al(OH)_3$	12.2			
H_3AsO_3	9.20	13.50		
H_3AsO_4	2.22	6.98	11.53	
$H_3BO_3 \cdot H_2O$ (20 °C)	9.28	12.75	13.80	
$H_2B_4O_7$	4.00	9.00		
HBr	−6.00			
HBrO	8.60			
Bromocresol green	4.68			
Bromocresol purple	6.30			
p-Bromophenol	9.24			
Bromophenol blue	3.86			
Bromothymol blue	7.10			
$CO_2 \cdot H_2O$	6.35	10.25		
HCOOH	3.75			
CH_3COOH	4.75			
C_2H_5COOH	4.87			
CH_2FCOOH	2.57			
CH_2ClOOH	2.87			
$CHCl_2COOH$	1.29			
CCl_3COOH	−0.08			
$CH_2BrCOOH$ (20 °C)	2.89			
CH_2ICOOH (20 °C)	3.16			
$CH_2CNCOOH$	2.47			
C_6H_5COOH (benzoic acid)	4.19			
$C_6H_5CH_2COOH$	4.31			
Phenol	9.89			
Phenol red	7.90			
Phenol phthaleine	9.4			
1,10-Phenanthroline $\cdot$ H^+	9.40			
Cresol purple (acidic range)	1.51			
Cresol purple (alkaline range)	8.32			
Cresol red	8.20			
Thymol blue	8.90			
Thymol phthaleine	10.0			
Oxalic acid	1.25	4.28		
Succinic acid	4.19	5.48		
Tartaric acid	3.04	4.37		
Phthalic acid	2.95	5.41		
Citric acid	2.94	4.14	5.82	
Ascorbic acid	4.30	11.82		
Salicylic acid	3.00	12.38		
Glycine $\cdot$ H^+	2.35	9.77		
N-tris(hydroxymethyl)methylglycine	8.08			
Cysteine $\cdot$ H^+	1.96	8.36	10.28	
Cystine $\cdot$ $2\,H^+$	1.04	2.05	8.00	
Ethylenediaminetetraacetic acid	2.07	2.75	6.24	10.34
Hydroquinone	10.0	12.0		
8-Hydroxyquinoline $\cdot$ H^+	5.01	9.90		
Methyl orange	3.40			
Methyl red	4.95			
Vanillin	7.40			
HCN	9.32			
HOCN	3.66			
HSCN	0.85			
HCl	−3.00			
HClO	7.53			
$HClO_2$	2.00			
$HClO_3$	0			
$HClO_4$	−9.00			
Chlorophenol red	6.00			
$Cr_2O_7^{2-} \cdot H_2O$ ($HCrO_4^-$)	6.50			
HF	3.17			

Table I. (cont.)

Name	pK_{A1}	pK_{A2}	pK_{A3}	pK_{A4}
$Fe(H_2O)_6^{3+}$	2.22			
$H_4[Fe(CN)_6]$			2.22	4.17
HI	−8.00			
HIO	12.3			
HIO_3	0.79			
H_5IO_6	1.64			
NH_4^+	9.25			
$CH_3NH_3^+$	10.64			
$(CH_3)_2NH_2^+$	10.99			
$(CH_3)_3NH^+$	9.83			
$(C_2H_5)_2NH_2^+$	10.59			
$(C_2H_5)_3NH^+$	10.75			
$C_6H_5NH_3^+$	4.54			
$Pyridine \cdot H^+$	5.10			
$Imidazole \cdot H^+$	6.83			
$H_3N-NH_3^{2+}$	0.27	7.24		
$H_3N-CH_2-CH_2-NH_3^{2+}$	6.85	9.93		
$(C_2H_4OH)NH_3^+$	9.50			
$(C_2H_4OH)_3NH^+$	7.76			
$CH_3-CH_2-NH_3^+$	10.63			
$Triaminopropane \cdot 3\,H^+$	3.72	7.95	9.59	
Tris(hydroxymethyl)aminomethane	8.08			
HN_3	4.72			
HNO_3	−1.32			
H_2O_2 (20 °C)	11.65			
PH_4^+	0			
H_3PO_2 (18 °C)	2.00			
H_3PO_3 (18 °C)	1.8	6.16		
H_3PO_4	2.23	7.21	12.32	
$H_4P_2O_7$	1.52	2.36	6.60	9.25
H_2S	7.00	12.92		
H_2SO_3	1.76	7.20		
H_2SO_4	−3.00	1.99		
NH_2SO_3H	0.988			
N-2-Hydroxyethylpiperazine-N′-2-ethanesulfonic acid	7.55			
2-(N-morpholin)ethanesulfonic acid	6.15			
H_2Se	3.89	11.00		
H_2SeO_3	2.54	8.02		
H_2SeO_4	−3.00	2.00		
H_4SiO_4 (30 °C)	9.70	11.70	12.00	12.00
$Zn(H_2O)_6^{2+}$	9.60			

Table II. pH-Indicators

Indicator	pH-range	Acid-base color	Preparation (g in 100 ml)[1]	
Methyl violet 1st Step	0.1 ... 1.5	yellow-blue	0.05	
Cresol red 1st Step	0.2 ... 1.8	red-yellow	0.04	
m-Cresol purple 1st Step	1.2 ... 2.8	red-yellow	0.04	90% A
Thymol blue 1st Step	1.2 ... 2.8	red-yellow	0.1	20% A
Tropaeolin 00	1.3 ... 3.2	red-yellow	1.0	W
Methyl violet 2nd Step	1.5 ... 3.2	blue-violet	0.05	W
2,4-Dinitrophenol	2.6 ... 4.0	colorless-yellow	0.05	W
Dimethyl yellow	2.9 ... 4.0	red-yellow	0.1	A
Bromophenol blue	3.0 ... 4.6	yellow-blue	0.04	A
Methyl orange	3.0 ... 4.4	red-orange	0.1	W
Congo red	3.0 ... 5.2	blue-red	0.1	W
α-Naphthol red	3.7 ... 5.7	violet-yellow	0.1	70% A
2,5-Dinitrophenol	3.8 ... 5.2	colorless-yellow		
Bromocresol green	3.8 ... 5.4	yellow-blue	0.04	90% A
Methyl red	4.4 ... 6.2	red-yellow	0.2	60% A
Bromocresol purple	5.2 ... 6.8	yellow-purple	0.04	90% A
Litmus	5.0 ... 8.0	red-blue	0.1	W
Bromothymol blue	6.0 ... 7.6	yellow-blue	0.1	20% A

Table II. (cont.)

4-Nitrophenol	6.0...8.0	colorless-yellow			
Phenol red	6.4...8.0	yellow-red	0.1		W
Neutral red	6.8...8.0	red-yellow	0.1	60%	A
Cresol red 2nd Step	7.2...8.8	yellow-red	0.04		W
α-Naphtholphthalein	7.3...8.7	rose-colored-green	0.1	70%	A
m-Cresol purple 2nd Step	7.4...9.0	yellow-purple	0.04		W
3-Nitrophenol	7.0...9.0	colorless-yellow			
Thymolblue 2nd Step	8.0...9.6	yellow-blue	0.1	20%	A
Phenolphthalein	8.2...10.0	colorless-red	0.1	70%	A
Thymolphthalein	9.4...10.6	colorless-blue	0.04	90%	A
β-Naphthol violet	10.0...12.0	orange-violet	0.04		W
Alizarin yellow	10.0...12.1	yellow-orange	0.1		W

[1] A...Alcohol; W...Water

Table III. Solubility Products (at 25 °C)

Compound	pK_{sp}	Compound	pK_{sp}	Compound	pK_{sp}
$AgCl$	9.75	CdS (red)	28.0	MgC_2O_4	4.1
$AgCl$ (50)	8.88	$Ce(IO_3)_3$	9.5	$MgNH_4PO_4$	12.6
$AgBr$	12.28	$Ce_2(C_2O_4)_3$	28.4	$Mn(OH)_2$	14.2
AgI	16.08	$Co(OH)_2$	16	$MnCO_3$	10.1
$AgCN$	14.2	$CoCO_3$	12.0	MnS (green)	12.6
$AgSCN$	12.0	CoS	22.1	MnS (pink)	9.6
$AgBrO_3$	4.28		(22...26)	MnS	15
$AgIO_3$	7.52	$Cr(OH)_3$	30	$NaHCO_3$	2.9
$AgOH$ (18)	7.9	$CuCl$ (20)	6.0	$Ni(OH)_2$	13.8
$AgOOCCH_3$	2.40	$CuBr$ (20)	7.4	$NiCO_3$	6.87
AgN_3	8.54	CuI (20)	11.3	NiS	20.7
Ag_2CO_3	11.2	$CuSCN$ (18)	10.8		(21...26)
Ag_2CrO_4	11.95	Cu_2S	48.0	PbF_2 (9)	7.5
$Ag_2Cr_2O_7$	6.7	$Cu(IO_3)_2$	7.13	$PbCl_2$	4.77
$Ag_2C_2O_4$	11.30	$Cu(OH)_2$	19.3	$PbBr_2$	4.4
Ag_2S	49.6	$CuCO_3$	9.86	PbI_2	8.06
Ag_2SO_4	4.80	CuC_2O_4	7.54	$Pb(OOCH)_2$	6.70
Ag_3PO_4	17.8	$CuCrO_4$	5.44	$Pb(IO_3)_2$	12.59
Ag_3AsO_4	19		(37...44)	$PbCO_3$ (18)	13.5
Ag_3AsO_3	18.5	CuS	35.1	PbC_2O_4 (18)	10.56
Ag_3VO_4 (20)	6.3	$Fe(OH)_2$ (18)	15.3	$PbCrO_4$	13.75
$Al(OH)_3$	32.7	$FeCO_3$ (20)	10.6	$PbSO_4$	7.80
BaF_2	5.76	FeC_2O_4	6.7	PbS	26.6
$Ba(IO_3)_2$	9.2	FeS	17.3		(27...29)
$BaCO_3$	9.26		(18...21)	$RaSO_4$ (20)	10.37
BaC_2O_4 (18)	6.79	$Fe(OH)_3$ (18)	37.4	$RbClO_4$	2.4
$BaCrO_4$	9.93	$[Hg_2]Cl_2$	17.88	$Sn(OH)_2$	25.3
$BaSO_4$	9.96	$[Hg_2]Br_2$	20.9	SnS	25.0
$BaSO_4$ (50)	9.70	$[Hg_2]I_2$	27.9		(25...28)
$Be(OH)_2$	18.6	$[Hg_2](CN)_2$	39.3	SrF_2	8.54
$[BiO]Cl$	6.15	$[Hg_2]CO_3$	16	$SrCO_3$	9.96
Bi_2S_3	97	$[Hg_2]CrO_4$	8.70	SrC_2O_4 (18)	7.25
CaF_2	10.40	$[Hg_2]SO_4$	6.17	$SrCrO_4$ (18)	4.44
$Ca(IO_3)_2$ (18)	6.13	HgS (black)	51.8	$SrSO_4$	6.55
$Ca(OH)_2$ (18)	5.26	HgS (red)	52.4	$TlCl$	3.72
$CaCO_3$ (Calcit)	8.18	$KClO_4$	2.05	$TlBr$	5.41
CaC_2O_4	8.07	$K_2[PtCl_6]$	5.85	TlI	7.24
$CaCrO_4$ (18)	1.64	$La(IO_3)_3$	11.21	$TlIO_3$	5.66
$CaSO_4$	4.32	$La(OH)_3$	20	TlN_3	3.66
$Ca[HPO_4](H_2O)_2$	6.57	$La_2(C_2O_4)_3$	27.7	Tl_2CrO_4	12.01
$Ca_3(PO_4)_2$	26.0	Li_2CO_3	0.5	$Zn(OH)_2$	17
$Cd(OH)_2$ (18)	13.9	MgF_2	8.18	α-ZnS	23.8
$CdCO_3$	13.6	$Mg(OH)_2$	10.74	β-ZnS	21.6
CdS (yellow)	26.1	$MgCO_3(H_2O)_3$	4.25		(22...25)

Table IV. Complex Formation Constants

Complex	Temperature in °C	Ionic strength	lg K	lg β
$[Cu(NH_3)]^{2+}$	30	0.5…5	4.15	4.15
$[Cu(NH_3)_2]^{2+}$	30	0.5…5	3.50	7.65
$[Cu(NH_3)_3]^{2+}$	30	0.5…5	2.89	10.54
$[Cu(NH_3)_4]^{2+}$	30	0.5…5	2.13	12.67
$[Cu(NH_3)]^{+}$			6.18	6.18
$[Cu(NH_3)_2]^{+}$			4.69	10.87
$[Ag(NH_3)]^{+}$	30	0.5…5	3.20	3.20
$[Ag(NH_3)_2]^{+}$			3.83	7.03
$[Zn(NH_3)_4]^{2+}$				9.46
$[Cd(NH_3)_4]^{2+}$				7.12
$[Ni(NH_3)_4]^{2+}$				7.95
$[Ni(NH_3)_6]^{2+}$				8.90
$[AlF_4]^{-}$	25	0.5		17.75
$[AlF_5]^{2-}$	25	0.5	1.62	19.37
$[AlF_6]^{3-}$	25	0.5	0.47	19.84
FeF_3	25	0.5		12.06
$[FeF_5]^{2-}$	25	0.5		15.4
$[Zn(OH)_4]^{2-}$	25			15.44
$[Al(OH)_4]^{-}$			−1	32
$[AgCl_2]^{-}$	25	0		4.75
$[AgCl_4]^{3-}$	25	5		5.32
$[AuCl_4]^{-}$	18			21.30
$[HgI_4]^{2-}$	25	0.5		29.83
$[CdI_4]^{2-}$	25	5		6.05
$[PbI_4]^{2-}$	25	0		3.85
$[Fe(CN)_6]^{4-}$	18			35.00
$[Fe(CN)_6]^{3-}$	18			42
$[Co(CN)_6]^{4-}$				29.5
$[Co(CN)_6]^{3-}$				48
$[Ni(CN)_4]^{2-}$				15.5
$[Cu(CN)_4]^{3-}$	25	0		30.3
$[Zn(CN)_4]^{2-}$	25			16.9
$[Cd(CN)_4]^{2-}$	25	3		18.45
$[Hg(CN)_4]^{2-}$	25			41.4
$[Au(CN)_2]^{-}$				38
$[Ag(CN)_2]^{-}$	18	0.3		21.1
$[Ag(SCN)_2]^{-}$	25	2.2		7.6
$[Ag(SCN)_4]^{3-}$	25	2.2		10.1
$[Hg(SCN)_4]^{2-}$	25	0.3		21.3
$[Ag(S_2O_3)]^{-}$	20	0		8.8
$[Ag(S_2O_3)_2]^{3-}$	20	0		13.5
$[Ni(en)_2]^{2+}$	30	1.2		14.1
$[Ni(en)_3]^{2+}$	30	1.2		18.6
$[Cu(en)_2]^{2+}$	30	0.5		19.6
$[Zn(en)_2]^{2+}$	30	1		10.4
$[Ag(en)_2]^{+}$	18			7.84
$[Cd(en)_2]^{2+}$	30	0.5		10.0
$[Cd(en)_3]^{2+}$	30	0.5		12.1
$[Co(en)_3]^{3+}$	30	1		48.7
$[Fe(phen)_3]^{2+}$				21.3
$[Fe(dipyr)_3]^{2+}$	25	0.33		17.58
$[Cu(dipyr)_2]^{2+}$	25	0.1	3.65	14.2
$[Cu(dipyr)_3]^{2+}$	25	0.5		17.85
$[Cu(citr)]^{+}$	25			14.20
$[Cu(OH)_2(citr)_2]^{-}$	25			18.77
$[Cu(OH)(citr)]^{2-}$	25			16.35
$[Cu(OH)_2(citr)_3]^{6-}$				19.3
Ca-Erio T				5.4
Mg-Erio T				7.0

Table V. Logarithms of side reaction coefficients $f_Y = [Y']/[Y^{4-}]$ for the anion $[Y^{4-}]$ of ethylenediaminetetraacetic acid (EDTA)

pH	$\lg f_Y$	pH	$\lg f_Y$	pH	$\lg f_Y$	pH	$\lg f_Y$
0	21.4	3.5	9.65	7.0	3.51	10.0	0.50
0.5	19.41	4.0	8.61	7.5	2.86	10.5	0.23
1.0	17.44	4.5	7.60	8.0	2.35	11.0	0.09
1.5	15.51	5.0	6.61	8.5	1.85	11.5	0.03
2.0	13.70	5.5	5.65	9.0	1.36	12.0	0.01
2.5	12.12	6.0	4.78	9.5	0.90	12.5	0.00
3.0	10.79	6.5	4.03				

Table VI. Standard Electrode Potentials at 25 °C

Reducing agent	$\rightleftharpoons$	Oxidizing agent		$E°$ in V	pK
Ag	$\rightleftharpoons$	Ag^+	$+e^-$	0.79996	13.5
$Ag + Cl^-$	$\rightleftharpoons$	$AgCl$	$+e^-$	0.2221	3.75
$Ag + Br^-$	$\rightleftharpoons$	$AgBr$	$+e^-$	0.0713	1.20
$Ag + I^-$	$\rightleftharpoons$	AgI	$+e^-$	-0.1519	-2.57
$Ag + 2CN^-$	$\rightleftharpoons$	$[Ag(CN)_2]^-$	$+e^-$	-0.395	-6.7
$2Ag + S^{2-}$	$\rightleftharpoons$	Ag_2S	$+2e^-$	-0.7051	-23.8
Ag^+	$\rightleftharpoons$	Ag^{2+}	$+e^-$	1.987	33.6
Al	$\rightleftharpoons$	Al^{3+}	$+3e^-$	-1.67	-84.6
$As + 3H_2O$	$\rightleftharpoons$	$H_3AsO_3 + 3H^+$	$+3e^-$	0.2475	12.5
$H_3AsO_3 + H_2O$	$\rightleftharpoons$	$H_3AsO_4 + 2H^+$	$+2e^-$	0.58	19.6
$AsO_3^{3-} + 2OH^-$	$\rightleftharpoons$	$AsO_4^{3-} + H_2O$	$+2e^-$	-0.08	-2.7
Au	$\rightleftharpoons$	Au^+	$+e^-$	1.46	24.8
Au	$\rightleftharpoons$	Au^{3+}	$+3e^-$	1.42	72.0
$Au + 4Cl^-$	$\rightleftharpoons$	$[AuCl_4]^-$	$+3e^-$	1.0	50.7
$Au + 2CN^-$	$\rightleftharpoons$	$[Au(CN)_2]^-$	$+e^-$	-0.60	-10.1
Ba	$\rightleftharpoons$	Ba^{2+}	$+2e^-$	-2.90	-98
Be	$\rightleftharpoons$	Be^{2+}	$+2e^-$	-1.70	-57.4
Bi	$\rightleftharpoons$	Bi^{3+}	$+3e^-$	0.277	14.0
$Bi^{3+} + 3H_2O$	$\rightleftharpoons$	$BiO_3^- + 6H^+$	$+2e^-$	1.73	58.4
$2Br^-$	$\rightleftharpoons$	Br_2 (aq)	$+2e^-$	1.087	36.7
$Br^- + H_2O$	$\rightleftharpoons$	$HBrO + H^+$	$+2e^-$	1.33	44.9
$Br^- + 3H_2O$	$\rightleftharpoons$	$BrO_3^- + 6H^+$	$+6e^-$	1.44	146
$Br^- + 2OH^-$	$\rightleftharpoons$	$BrO^- + H_2O$	$+2e^-$	0.75	25.3
$Br^- + 6OH^-$	$\rightleftharpoons$	$BrO_3^- + 3H_2O$	$+6e^-$	0.61	61.8
$HCHO + H_2O$	$\rightleftharpoons$	$HCOOH + 2H^+$	$+2e^-$	0.056	1.89
HCOOH	$\rightleftharpoons$	$CO_2 + 2H^+$	$+2e^-$	-0.196	-6.62
Ca	$\rightleftharpoons$	Ca^{2+}	$+2e^-$	-2.76	-93.2
Cd	$\rightleftharpoons$	Cd^{2+}	$+2e^-$	-0.402	-13.6
Cd(Hg)	$\rightleftharpoons$	Cd^{2+}	$+2e^-$	-0.3519	-11.9
Ce^{3+}	$\rightleftharpoons$	Ce^{4+}	$+e^-$	1.713	28.9
$2Cl^-$	$\rightleftharpoons$	Cl_2	$+2e^-$	1.3583	45.89
$Cl^- + H_2O$	$\rightleftharpoons$	$HClO + H^+$	$+2e^-$	1.498	50.6
$Cl^- + 2H_2O$	$\rightleftharpoons$	$HClO_2 + 3H^+$	$+4e^-$	1.57	106
$Cl^- + 3H_2O$	$\rightleftharpoons$	$ClO_3^- + 6H^+$	$+6e^-$	1.45	147
$Cl^- + 4H_2O$	$\rightleftharpoons$	$ClO_4^- + 8H^+$	$+8e^-$	1.36	184
$Cl^- + 2OH^-$	$\rightleftharpoons$	$ClO^- + H_2O$	$+2e^-$	0.88	29.8
$Cl^- + 4OH^-$	$\rightleftharpoons$	$ClO_2^- + 2H_2O$	$+4e^-$	0.77	52
$Cl^- + 6OH^-$	$\rightleftharpoons$	$ClO_3^- + 3H_2O$	$+6e^-$	0.62	62.8
$Cl^- + 8OH^-$	$\rightleftharpoons$	$ClO_4^- + 4H_2O$	$+8e^-$	0.53	71.6
Co	$\rightleftharpoons$	Co^{2+}	$+2e^-$	-0.277	-9.4
Co^{2+}	$\rightleftharpoons$	Co^{3+}	$+e^-$	1.842	31.1
Cr	$\rightleftharpoons$	Cr^{3+}	$+2e^-$	-0.0557	-18.8
Cr^{2+}	$\rightleftharpoons$	Cr^{3+}	$+e^-$	-0.40	-6.76
$2Cr^{3+} + 7H_2O$	$\rightleftharpoons$	$Cr_2O_7^{2-} + 14H^+$	$+6e^-$	1.36	138
$Cr(OH)_3 + 5OH^-$	$\rightleftharpoons$	$CrO_4^{2-} + 4H_2O$	$+3e^-$	-0.12	-6.1
$Cr^{3+} + 8OH^-$	$\rightleftharpoons$	$CrO_4^{2-} + 4H_2O$	$+3e^-$	-0.72	-36.5
Cs	$\rightleftharpoons$	Cs^+	$+e^-$	-2.923	49.4
Cu	$\rightleftharpoons$	Cu^+	$+e^-$	0.522	8.82
Cu	$\rightleftharpoons$	Cu^{2+}	$+2e^-$	0.3460	11.69
Cu(Hg)	$\rightleftharpoons$	Cu^{2+}	$+2e^-$	0.3511	11.86
Cu^+	$\rightleftharpoons$	Cu^{2+}	$+e^-$	0.170	2.87
$2F^-$	$\rightleftharpoons$	F_2	$+2e^-$	2.87	96.9

Table VI. (cont.)

Reducing agent	⇌	Oxidizing agent		$E°$ in V	pK
Fe^{2+}	⇌	Fe^{3+}	$+ e^-$	0.7704	13.0
$[Fe(CN)_6]^{4-}$	⇌	$[Fe(CN)_6]^{3-}$	$+ e^-$	0.36	6.1
Ferroin	⇌	Ferriin	$+ e^-$	1.04	17.6
H_2	⇌	$2 H^+$	$+ 2 e^-$	0.0000	0
$2 H^-$	⇌	H_2	$+ 2 e^-$	-2.24	-76
Hg	⇌	Hg^{2+}	$+ 2 e^-$	0.852	28.8
$2 Hg$	⇌	Hg_2^{2+}	$+ 2 e^-$	0.7986	27.0
$2 Hg + 2 Cl^-$ (sat. KCl)	⇌	Hg_2Cl_2	$+ 2 e^-$	0.2412	8.15
$2 Hg + 2 Cl^-$	⇌	Hg_2Cl_2	$+ 2 e^-$	0.2677	9.04
$2 Hg + 2 Br^-$	⇌	Hg_2Br_2	$+ 2 e^-$	0.1396	4.72
$2 Hg + 2 I^-$	⇌	Hg_2I_2	$+ 2 e^-$	-0.0405	-1.37
$2 Hg + SO_4^{2-}$	⇌	Hg_2SO_4	$+ 2 e^-$	0.6151	20.8
$2 I^-$	⇌	I_2	$+ 2 e^-$	0.5355	18.1
$I^- + H_2O$	⇌	$IOH + H^+$	$+ 2 e^-$	0.99	33.4
$I^- + 3 H_2O$	⇌	$IO_3^- + 6 H^+$	$+ 6 e^-$	1.085	110
$I^- + 2 OH^-$	⇌	$IO^- + H_2O$	$+ 2 e^-$	0.49	16.6
$I^- + 6 OH^-$	⇌	$IO_3^- + 3 H_2O$	$+ 6 e^-$	0.26	26.4
K	⇌	K^+	$+ e^-$	-2.9241	-49.4
Li	⇌	Li^+	$+ e^-$	-2.9595	-50
Mg	⇌	Mg^{2+}	$+ 2 e^-$	-2.375	-80
Mn	⇌	Mn^{2+}	$+ 2 e^-$	-1.18	-39.9
Mn^{2+}	⇌	Mn^{3+}	$+ e^-$	1.51	25.5
$Mn^{2+} + 2 H_2O$	⇌	$MnO_2 + 4 H^+$	$+ 2 e^-$	1.23	41.6
$Mn^{2+} + 4 H_2O$	⇌	$MnO_4^{2-} + 8 H^+$	$+ 4 e^-$	1.74	118
$Mn^{2+} + 4 H_2O$	⇌	$MnO_4^- + 8 H^+$	$+ 5 e^-$	1.51	127
$Mn + 2 OH^-$	⇌	$Mn(OH)_2$	$+ 2 e^-$	-1.52	-51
$MnO_2 + 4 OH^-$	⇌	$MnO_4^{2-} + 2 H_2O$	$+ 2 e^-$	0.58	19.6
$MnO_2 + 4 OH^-$	⇌	$MnO_4^- + 2 H_2O$	$+ 3 e^-$	0.58	29.4
Na	⇌	Na^+	$+ e^-$	-2.7131	-45.8
Ni	⇌	Ni^{2+}	$+ 2 e^-$	-0.23	-7.77
$2 NH_4^+$	⇌	$N_2H_5^+ + 3 H^+$	$+ 2 e^-$	1.28	43.2
$NH_4^+ + H_2O$	⇌	$NH_3OH^+ + 2 H^+$	$+ 2 e^-$	1.32	44.6
$2 NH_4^+$	⇌	$N_2 + 8 H^+$	$+ 6 e^-$	0.27	27.4
$2 NH_4^+ + H_2O$	⇌	$N_2O + 10 H^+$	$+ 8 e^-$	0.65	87.8
$NH_4^+ + H_2O$	⇌	$NO + 6 H^+$	$+ 5 e^-$	0.84	70.9
$NH_4^+ + 2 H_2O$		$NO_2^- + 8 H^+$	$+ 6 e^-$	0.86	87.2
$NH_4^+ + 2 H_2O$	⇌	$NO_2 + 8 H^+$	$+ 7 e^-$	0.89	105
$NH_4^+ + 3 H_2O$	⇌	$NO_3^- + 10 H^+$	$+ 8 e^-$	0.88	119
$2 NH_3$		$3 N_2 + 2 H^+$	$+ 2 e^-$	-3.09	-104
$2 NH_3 + 2 OH^-$	⇌	$N_2H_4 + 2 H_2O$	$+ 2 e^-$	0.10	3.38
$NH_3 + 2 OH^-$	⇌	$NH_2OH + H_2O$	$+ 2 e^-$	0.42	14.2
$2 NH_3 + 6 OH^-$	⇌	$N_2 + 6 H_2O$	$+ 6 e^-$	-0.73	-74
$2 NH_3 + 8 OH^-$	⇌	$N_2O + 7 H_2O$	$+ 8 e^-$	-0.42	-56.8
$NH_3 + 5 OH^-$	⇌	$NO + 4 H_2O$	$+ 5 e^-$	-0.10	-8.45
$NH_3 + 7 OH^-$	⇌	$NO_2^- + 5 H_2O$	$+ 6 e^-$	-0.16	-16.2
$NH_3 + 7 OH^-$	⇌	$NO_2 + 5 H_2O$	$+ 7 e^-$	-0.013	-1.54
$NH_3 + 9 OH^-$	⇌	$NO_3^- + 6 H_2O$	$+ 8 e^-$	-0.12	-16.2
$2 H_2O$	⇌	$O_2 + 4 H^+$	$+ 4 e^-$	1.229	83.0
H_2O_2	⇌	$O_2 + 2 H^+$	$+ 2 e^-$	0.682	23.0
$2 H_2O$	⇌	$H_2O_2 + 2 H^+$	$+ 2 e^-$	1.77	59.8
$4 OH^-$	⇌	$O_2 + 2 H_2O$	$+ 4 e^-$	0.401	27.1
$2 HO_2^-$	⇌	$2 O_2 + 2 H_2O$	$+ 4 e^-$	-0.075	-5.07
$P + 3 H_2O$	⇌	$H_3PO_3 + 3 H^+$	$+ 3 e^-$	-0.50	-25.3
$P + 4 H_2O$	⇌	$H_3PO_4 + 5 H^+$	$+ 5 e^-$	-0.41	-34.6
$PH_3 + 3 OH^-$	⇌	$P + 3 H_2O$	$+ 3 e^-$	-0.87	-44
$P + 2 OH^-$	⇌	$H_2PO_2^-$	$+ e^-$	-2.05	-34.6
$P + 5 OH^-$	⇌	$HPO_3^{2-} + 2 H_2O$	$+ 3 e^-$	-1.57	-79.6
$P + 8 OH^-$		$PO_4^{3-} + 4 H_2O$	$+ 5 e^-$	-1.49	-126
Pb	⇌	Pb^{2+}	$+ 2 e^-$	-0.1263	-4.27
$Pb^{2+} + 2 H_2O$	⇌	$PbO_2 + 4 H^+$	$+ 2 e^-$	1.46	49.3
$PbSO_4 + 2 H_2O$	⇌	$PbO_2 + SO_4^{2-} + 4 H^+$	$+ 2 e^-$	1.685	56.9
Rb	⇌	Rb^+	$+ e^-$	-2.9259	-49.4
H_2S	⇌	$S + 2 H^+$	$+ 2 e^-$	0.141	4.76
$H_2S + 4 H_2O$	⇌	$SO_4^{2-} + 10 H^+$	$+ 8 e^-$	0.30	40.5
$2 S + 3 H_2O$	⇌	$H_2S_2O_3 + 4 H^+$	$+ 4 e^-$	0.50	33.8
$2 S + 4 H_2O$	⇌	$H_2S_2O_4 + 6 H^+$	$+ 6 e^-$	0.63	64

Table VI. (cont.)

Reducing agent	$\rightleftharpoons$	Oxidizing agent		$E°$ in V	pK
$S + 3 H_2O$	$\rightleftharpoons$	$H_2SO_3 + 4 H^+$	$+4 e^-$	0.45	30.4
$S + 4 H_2O$	$\rightleftharpoons$	$SO_4^{2-} + 8 H^+$	$+6 e^-$	0.356	36.1
$2 S_2O_3^{2-}$	$\rightleftharpoons$	$S_4O_6^{2-}$	$+2 e^-$	0.06	2.0
S^{2-}	$\rightleftharpoons$	S	$+2 e^-$	-0.48	-16.2
$S + 6 OH^-$	$\rightleftharpoons$	$SO_3^{2-} + 3 H_2O$	$+4 e^-$	-0.61	-41.2
$S + 8 OH^-$	$\rightleftharpoons$	$SO_4^{2-} + 4 H_2O$	$+6 e^-$	-0.72	-73
$S_2O_3^{2-} + 10 OH^-$	$\rightleftharpoons$	$2 SO_4^{2-} + 5 H_2O$	$+8 e^-$	-0.76	-103
$Sb + H_2O$	$\rightleftharpoons$	$SbO^+ + 2 H^+$	$+3 e^-$	-0.212	-10.7
$H_3SbO_3 + 3 H_2O$	$\rightleftharpoons$	$H[Sb(OH)_6] + 2 H^+$	$+2 e^-$	0.75	25.3
Sn	$\rightleftharpoons$	Sn^{2+}	$+2 e^-$	-0.1496	-4.75
Sn^{2+}	$\rightleftharpoons$	Sn^{4+}	$+2 e^-$	0.15	5.1
Sr	$\rightleftharpoons$	Sr^{2+}	$+2 e^-$	-2.89	-97.6
Tl	$\rightleftharpoons$	Tl^+	$+ e^-$	-0.336	-5.7
Tl^+	$\rightleftharpoons$	Tl^{3+}	$+2 e^-$	1.247	42.2
V^{2+}	$\rightleftharpoons$	V^{3+}	$+ e^-$	-0.253	-4.3
$V^{3+} + H_2O$	$\rightleftharpoons$	$VO^{2+} + 2 H^+$	$+ e^-$	0.337	5.7
$VO^{2+} + 3 H_2O$	$\rightleftharpoons$	$V(OH)_4^+ + 2 H^+$	$+ e^-$	1.00	16.9
Zn	$\rightleftharpoons$	Zn^{2+}	$+2 e^-$	-0.7628	-25.8

4   Laser Principles and Characteristics

A laser (light amplification by stimulated emission of radiation) is a device consisting of a gaseous, liquid, or solid medium whose molecules or atoms are excited so as to obtain an inversion of population of the electrons. Stimulated emission is then obtained, resulting in a nearly monochromatic and coherent (i.e., spatially and temporally in phase) light beam at the same frequency as the stimulating photons. The so-called *pump* energy needed to obtain and maintain population inversion can be provided in the form of photons (e.g., with a flash lamp or another laser) or in the form of an electrical discharge. Lasers, with the exception of some gas lasers, make use of an *optical cavity* resonator to increase the laser power. This resonator consists of two parallel mirrors, one being slightly transparent. The photons are reflected back and forth between the two mirrors and only a small percentage of the beam emerges through the slightly transparent mirror. Depending on the media, a single line or several lines can be stimulated. Another possibility is the emission of a range of wavelengths, which makes possible the continuous tuning of the wavelength to a given value (*tunable* laser, e.g., *dye lasers, Ti:sapphire laser*). The fundamental wavelength emitted by the laser can be changed using *nonlinear optics*, e.g. a fixed frequency change crystal or *an optical parametric oscillator*. In the former case, the wavelength can be divided by two, three, or four, while in the latter case, the wavelength can the tuned over a given wavelength range.

The beam can be *circular* or *rectangular*. An important parameter is the distribution of energy through the beam profile. For instance, the energy can exhibit a Gaussian profile. The operation of the laser may be *continuous wave* (cw) or *pulsed*. Two pulse modes are currently used: *Q-switched* mode and *mode-locked*. *Pulse duration* and *repetition rate* are used to describe pulsed lasers. To express the amount of photons, *power* (W or mW) is usually used for cw laser, while *energy* (J, mJ, or μJ) per pulse is usually used for pulsed lasers.

If the beam is further focused, important parameters are the *irradiance* ($W\,cm^{-2}$) and the *flux* ($J\,cm^{-2}$) at the focal spot. The irradiance and the flux are related to the power and the energy, respectively, and to the size of the spot, which in turn depends on the focusing system and the resulting beam waist.

Analytical applications of the lasers are based on at least one of the following properties: *monochromaticity, coherence*, or *high irradiance*, (or flux). Examples of the use of monochromaticity are resonance ionization mass spectrometry (RIMS, see Sec. 8.5) and Raman spectrometry (see Secs. 9.2 and 10.5). High flux is used for laser ablation (see Secs. 8.1 and 8.5).

A summary of the principles and characteristics of currently available lasers is given in the table below.

Principle	Laser	Pump	Wavelength	Line(s)/wavelength range	Beam size (mm)	Operation	Repetition rate	Pulse duration	Energy/power
Gas	He–Ne	discharge	single	632.8 nm	Ø0.5	cw			a few 10 mW
	He–Cd	discharge	several	442 nm	Ø0.5	cw			<200 mW
				325 nm					<60 mW
	argon ion	discharge	multiline	454 nm; 514 nm	Ø0.3–1.0	cw			25 mW–25 W
				488 nm					mW–8 W
				514.5 nm					mW–10 W
	krypton ion	discharge	multiline	multiline	Ø0.5–1.0	cw			a few W
				IR: 752–800 nm					1.6 W
				red: 647–677 nm					4.6 W
				green/yellow: 520–568 nm					3.3 W
				blue/green: 476–482 nm					3.5 W
				violet: 406–468 nm					3.0 W
				UV: 337–356 nm					2.0 W
	nitrogen	discharge	single	337.1 nm	5–25 × 2	pulsed	1–100 Hz	5–12 ns	0.3–3.5 mJ
					3–4 × 8–10		1–30 Hz	3 ns	<0.3 mJ
	carbon dioxide			10.6 µm					
Gas-excimer	F$_2$	discharge	single	157 nm	6–14 × 20–23	pulsed	10–500 Hz	9–30 ns	10–30 mJ
	ArF		single	193 nm					150–700 mJ
	KrCl			222 nm					100 mJ
	KrF			248 nm					300–1200 mJ
	XeCl			308 nm					150–600 mJ
	XeF			351 nm					150–400 mJ
Solid state	Nd:YAG	flash lamp	single	1064 nm[1]	Ø0.5	cw			12–20 W
					Ø3–5	Q-switched	2–100 Hz	3–14 ns	200–2000 mJ
						Q-switched	1–40 KHz	60–300 ns	0.3–2 mJ
					Ø3.5	locked	10–20 Hz	20 ps	75 mJ
		diode laser	single	1064 nm	Ø0.2–0.5	cw			25 mW–3 W
					Ø0.1–0.3	Q-switched	1 Hz–50 KHz	10–20 ns	10–80 µJ
					Ø0.5		1 Hz–5 KHz	<3 ns	250 µJ
	Ti:sapphire	argon ion	tunable	675–1100 nm	Ø0.5	cw			1–5 W
		argon ion		710–1000 nm	Ø0.4–1	locked	75–100 MHz	2 ps	0.27–1.3 W (790 nm)
		Q-switched Nd:YAG		695–910 nm	Ø2	Q-switched	1–10 Hz	5–9 ns	100–180 mJ (800 nm)
Dye		argon/krypton ion	tunable	370–960 nm	Ø0.25–0.3	cw			0.1–1.4 W
		Q-switched Nd:YAG	tunable	340–920 nm[2]	1 × 3	Q-switched	1–20 Hz	5–6 ns	50–250 mJ
		locked Nd:YAG	tunable	575–880 nm	Ø0.5	locked	10 Hz–100 KHz	0.5–7 ps	0.6–4 nJ
		nitrogen	tunable	360–700 nm	2 × 3	pulsed	1–20 Hz	3 ns	60 µJ (500 nm)
		XeCl excimer	tunable	320–950 nm	1 × 2	pulsed	1–250 Hz	15 ns	
Diode laser			tunable	780–980 nm		cw			up to 15 W
Optical parametric oscillator		Q-switched Nd:YAG at 355 nm	tunable	410–2200 nm (signal + idler)	Ø3–5		10–20 Hz	1–6 ns	5 mJ (250 nm), 60 mJ (500 nm)
				213–426 nm + frequency doubling					

[1] 532 nm, 355 nm and 266 nm can be obtained by using frequency doubling, frequency tripling, and frequency quadrupling, using nonlinear optics (crystals)
[2] Can be extended to 205–432 nm (4–50 mJ) with nonlinear optics

5 Colthup Table

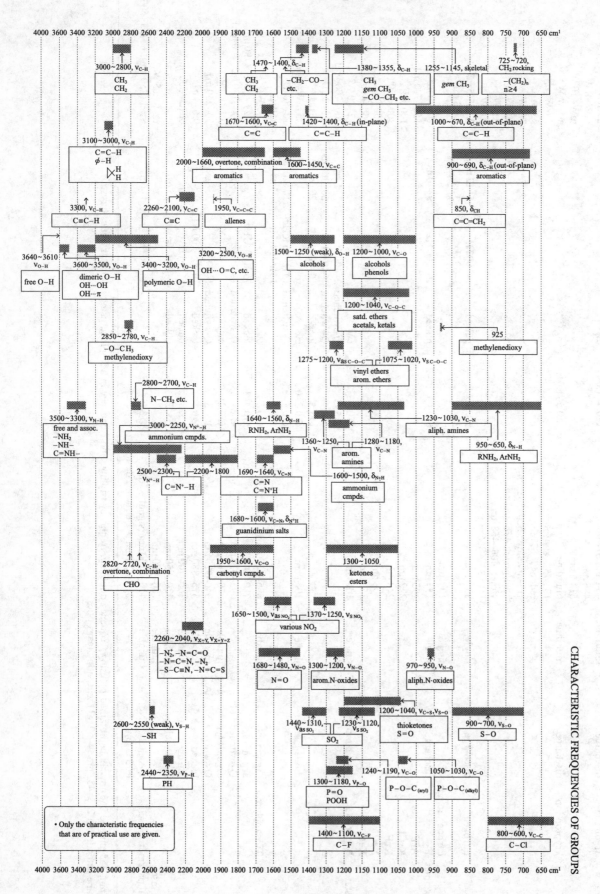

6 Statistical Tables

Two-sided and one-sided **Student t-distribution** for different α-levels at degrees of freedom from $f = 1$ to $f = 20$

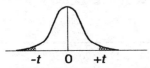

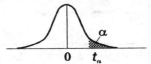

Two-sided Student t-distribution:			One-sided Student t-distribution:		
f	$\alpha = 0.05$	$\alpha = 0.01$	f	$\alpha = 0.05$	$\alpha = 0.025$
1	12.706	63.657	1	6.314	12.706
2	4.303	9.925	2	2.920	4.303
3	3.182	5.841	3	2.353	3.182
4	2.776	4.604	4	2.132	2.776
5	2.571	4.032	5	2.015	2.574
6	2.447	3.707	6	1.943	2.447
7	2.365	3.499	7	1.895	2.365
8	2.306	3.355	8	1.860	2.306
9	2.262	3.250	9	1.833	2.262
10	2.228	3.169	10	1.812	2.228
11	2.201	3.106	11	1.796	2.201
12	2.179	3.055	12	1.782	2.179
13	2.160	3.012	13	1.771	2.160
14	2.145	2.977	14	1.761	2.145
15	2.131	2.947	15	1.753	2.131
16	2.120	2.921	16	1.746	2.120
17	2.110	2.898	17	1.740	2.110
18	2.101	2.878	18	1.734	2.101
19	2.093	2.861	19	1.729	2.093
20	2.086	2.845	20	1.725	2.086

F-distribution for risk levels $\alpha = 0.05$ (lightface type) and $\alpha = 0.01$ (boldface type) at the degrees of freedom f_1 and f_2

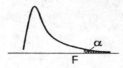

f_2	$f_1 = 1$	2	3	4	5	6	7	8	9	10	12	20	50	∞
1	161	200	216	225	230	234	237	239	241	242	244	248	242	254
	4052	**4999**	**5403**	**5625**	**5764**	**5859**	**5928**	**5981**	**6022**	**6056**	**6106**	**6208**	**6302**	**6366**
2	18.51	19.00	19.16	19.25	19.30	19.33	19.36	19.37	19.38	19.39	19.41	19.44	19.47	19.50
	98.49	**99.00**	**99.17**	**99.25**	**99.30**	**99.33**	**99.36**	**99.37**	**99.39**	**99.40**	**99.42**	**99.45**	**99.48**	**99.50**
3	10.13	9.55	9.28	9.12	9.01	8.94	8.88	8.84	8.81	8.78	8.74	8.66	8.58	8.53
	34.12	**30.82**	**29.46**	**28.71**	**28.24**	**27.91**	**27.67**	**27.49**	**27.34**	**27.23**	**27.05**	**26.65**	**26.35**	**26.12**
4	7.71	6.94	6.59	6.39	6.26	6.16	6.09	6.04	6.00	5.96	5.91	5.80	5.70	5.63
	21.20	**18.00**	**16.69**	**15.98**	**15.52**	**15.21**	**14.98**	**14.80**	**14.66**	**14.54**	**14.37**	**14.02**	**13.69**	**13.46**
5	6.61	5.79	5.41	5.19	5.05	4.95	4.88	4.82	4.78	4.74	4.68	4.56	4.44	4.36
	16.26	**13.27**	**12.06**	**11.39**	**10.97**	**10.67**	**10.45**	**10.29**	**10.15**	**10.05**	**9.89**	**9.55**	**9.24**	**9.02**
6	5.99	5.14	4.76	4.53	4.39	4.28	4.21	4.15	4.10	4.06	4.00	3.87	3.75	3.67
	13.74	**10.92**	**9.78**	**9.15**	**8.75**	**8.47**	**8.26**	**8.10**	**7.98**	**7.87**	**7.72**	**7.40**	**7.10**	**6.88**
7	5.59	4.74	4.35	4.12	3.97	3.87	3.79	3.73	3.68	3.63	3.57	3.44	3.32	3.23
	12.25	**9.55**	**8.45**	**7.85**	**7.46**	**7.19**	**7.00**	**6.84**	**6.71**	**6.62**	**6.47**	**6.16**	**5.87**	**5.65**
8	5.32	4.46	4.07	3.84	3.69	3.58	3.50	3.44	3.39	3.34	3.28	3.15	3.03	2.93
	11.26	**8.65**	**7.59**	**7.01**	**6.63**	**6.37**	**6.19**	**6.03**	**5.91**	**5.82**	**5.67**	**5.36**	**5.08**	**4.86**
9	5.12	4.26	3.86	3.63	3.48	3.37	3.29	3.23	3.18	3.13	3.07	2.93	2.80	2.71
	10.56	**8.02**	**6.99**	**6.42**	**6.06**	**5.80**	**5.62**	**5.47**	**5.35**	**5.26**	**5.11**	**4.81**	**4.53**	**4.31**
10	4.96	4.10	3.71	3.48	3.33	3.22	3.14	3.07	3.02	2.97	2.91	2.77	2.64	2.54
	10.04	**7.56**	**6.55**	**5.99**	**5.64**	**5.39**	**5.21**	**5.06**	**4.95**	**4.85**	**4.71**	**4.41**	**4.13**	**3.91**
12	4.75	3.88	3.49	3.26	3.11	3.00	2.92	2.85	2.80	2.76	2.69	2.54	2.41	2.30
	9.33	**6.93**	**5.95**	**5.41**	**5.06**	**4.82**	**4.65**	**4.50**	**4.39**	**4.30**	**4.16**	**3.86**	**3.58**	**3.36**
20	4.35	3.49	3.10	2.87	2.71	2.60	2.51	2.45	2.39	2.35	2.38	2.12	1.96	1.84
	8.10	**5.85**	**4.94**	**4.43**	**4.10**	**3.87**	**3.70**	**3.56**	**3.46**	**3.37**	**3.23**	**2.94**	**2.65**	**2.42**
50	4.04	3.19	2.80	2.57	2.41	2.30	2.23	2.14	2.08	2.04	1.96	1.80	1.61	1.45
	7.20	**5.08**	**4.22**	**3.74**	**3.45**	**3.21**	**3.04**	**2.91**	**2.81**	**2.72**	**2.58**	**2.29**	**1.98**	**1.70**
∞	3.84	3.00	2.60	2.37	2.21	2.10	2.01	1.94	1.88	1.83	1.75	1.57	1.36	1.00
	6.63	**4.61**	**3.78**	**3.32**	**3.02**	**2.80**	**2.64**	**2.51**	**2.41**	**2.32**	**2.18**	**1.88**	**1.53**	**1.00**

7 Matrix Algebra

A point in n-dimensional space $\mathbf{R}^n$ is represented by a *vector*, i.e.,

$$
x = \begin{pmatrix} x_1 \\ x_2 \\ \vdots \\ x_n \end{pmatrix} \quad \text{or in transposed form} \quad x^t = (x_1, x_2, \ldots, x_n)
$$

The sum of two vectors $x, y \in \mathbf{R}^n$ reveals:

$$
x + y = \begin{pmatrix} x_1 + y_1 \\ x_2 + y_2 \\ \vdots \\ x_n + y_n \end{pmatrix} \quad \text{example:} \quad \begin{pmatrix} 1 \\ 2 \\ 3 \end{pmatrix} + \begin{pmatrix} 2 \\ 4 \\ 7 \end{pmatrix} = \begin{pmatrix} 3 \\ 6 \\ 10 \end{pmatrix}
$$

Multiplication of a vector x with a scalar $l \in \mathbf{R}^n$ provides the following vector:

$$
l \cdot x = \begin{pmatrix} lx_1 \\ lx_2 \\ \vdots \\ lx_n \end{pmatrix}
$$

A *matrix* of elements of real numbers consisting of n rows and m columns, i.e., a $n * m$-matrix, is defined by:

$$
A = \begin{pmatrix} a_{11} & a_{12} & \cdots & a_{1m} \\ a_{21} & a_{22} & & a_{2m} \\ \vdots & & & \vdots \\ a_{n1} & a_{n2} & \cdots & a_{nm} \end{pmatrix} \quad \text{example:} \quad A = \begin{pmatrix} 2 & 4 & 6 \\ 3 & 1 & 5 \\ 5 & 8 & 9 \end{pmatrix}
$$

A *square matrix* has the same number of rows and columns, i.e., its dimension is $n * n$.

In a transposed matrix A^t, the rows and columns are interchanged giving for the matrix A:

$$
A^t = \begin{pmatrix} a_{11} & a_{12} & \cdots & a_{n1} \\ a_{12} & a_{22} & & a_{n2} \\ \vdots & & & \vdots \\ a_{1m} & a_{2m} & \cdots & a_{nm} \end{pmatrix} \quad \text{example:} \quad A^t = \begin{pmatrix} 2 & 3 & 5 \\ 4 & 1 & 8 \\ 6 & 5 & 9 \end{pmatrix}
$$

If the transpose of a matrix is identical to the original matrix in every element, i.e., $A^t = A$, it is called a *symmetric matrix*.

A *diagonal matrix* is a special case of a symmetric matrix. In a diagonal matrix only the diagonal contains values different from zero and all off-diagonal elements are zero:

$$
A = \begin{pmatrix} a_{11} & 0 & \cdots & 0 \\ 0 & a_{22} & & 0 \\ \vdots & & & \vdots \\ 0 & 0 & \cdots & a_{nm} \end{pmatrix} \quad \text{example:} \quad A = \begin{pmatrix} 2 & 0 & 0 \\ 0 & 1 & 0 \\ 0 & 0 & 9 \end{pmatrix}
$$

The diagonal matrix that has all 1's on the diagonal is termed the *identity matrix*:

$$
I = \begin{pmatrix} 1 & 0 & \cdots & 0 \\ 0 & 1 & & 0 \\ \vdots & & & \vdots \\ 0 & 0 & \cdots & 1 \end{pmatrix}
$$

The following examples describe *matrix addition* and *matrix subtraction*:

$$A + B = \begin{pmatrix} 2 & 4 \\ 1 & 3 \end{pmatrix} + \begin{pmatrix} -1 & 2 \\ 5 & -3 \end{pmatrix} = \begin{pmatrix} 1 & 6 \\ 6 & 0 \end{pmatrix}$$

$$A - B = \begin{pmatrix} 2 & 4 \\ 1 & 3 \end{pmatrix} - \begin{pmatrix} -1 & 2 \\ 5 & -3 \end{pmatrix} = \begin{pmatrix} 3 & 2 \\ -4 & 6 \end{pmatrix}$$

Multiplication of a $n * n$-matrix A and a $n * n$-matrix B gives the $n * n$-matrix C:

$$C = AB = \begin{pmatrix} a_{11} & \cdots & a_{1k} \\ \vdots & & \\ a_{n1} & \cdots & a_{nk} \end{pmatrix} \begin{pmatrix} b_{11} & \cdots & b_{1m} \\ \vdots & & \\ b_{k1} & & b_{km} \end{pmatrix} = \begin{pmatrix} c_{11} & \cdots & c_{1m} \\ \vdots & & \\ c_{n1} & & c_{nm} \end{pmatrix}$$

where $c_{ij} = \sum_{l=1}^{k} a_{il} b_{lj}$ for $1 \leq i \leq n$ and $1 \leq j \leq m$.

Example: $\quad C = \begin{pmatrix} 2 & 3 & 1 \\ 3 & 4 & 1 \end{pmatrix} \begin{pmatrix} 7 & 4 \\ 1 & 3 \\ 5 & 0 \end{pmatrix} = \begin{pmatrix} 22 & 17 \\ 30 & 24 \end{pmatrix}$

The *rank* of a matrix is the maximum number of linearly independent vectors (rows or columns) in a $n * p$ matrix X denoted as $r(X)$. Linearly dependent rows or columns reduce the rank of a matrix.

The *determinant* of a matrix is calculated by:

$$D = \begin{vmatrix} a_{11} & a_{12} & \cdots & a_{1m} \\ a_{21} & a_{22} & & a_{2m} \\ \vdots & & & \\ a_{n1} & a_{n2} & \cdots & a_{nm} \end{vmatrix} = \sum_{i=1}^{n} (-1)^{i\text{th}} a_{ik} \det(M_{ik})$$

Here M_{ik} is the $(n-1)(n-1)$ matrix where the ith row and kth column has been deleted.

Example: $\quad D = \begin{vmatrix} 2 & 4 & 6 \\ 3 & 1 & 5 \\ 7 & 8 & 9 \end{vmatrix} = 2(1 \cdot 9 - 5 \cdot 8) - 4(3 \cdot 9 - 5 \cdot 7) + 6(3 \cdot 8 - 1 \cdot 7)$

$$= 72$$

For *inversion* of a matrix A we get for A^{-1}:

$$A = \begin{pmatrix} a_{11} & a_{12} \\ a_{21} & a_{22} \end{pmatrix} \qquad A^{-1} = \begin{pmatrix} \dfrac{a_{22}}{D} & -\dfrac{a_{12}}{D} \\ -\dfrac{a_{21}}{D} & \dfrac{a_{11}}{D} \end{pmatrix}$$

where D is the determinant of the matrix.

Example: $\quad A = \begin{pmatrix} 4 & 2 \\ 3 & 1 \end{pmatrix} \qquad A^{-1} = \begin{pmatrix} -0.5 & 1 \\ 1.5 & -2 \end{pmatrix}$

In this example we have inverted a $2 * 2$ matrix. Perhaps an inversion by head could also be performed in case of a $3 * 3$ matrix. For larger matrices, however, a computer algorithm is necessary. In addition, matrix inversion is a very sensitive procedure, so that powerful algorithms, such as singular value decomposition (cf. Sec. 12.5.2) is to be applied.

A *linear transformation* from R^n to R^m (case a) or from R^m to R^n (case b) is possible by:

(a) multiplication of a n-dimensional column vector by an $n * m$-matrix forming a m-dimensional vector:

$$x^t A = (x_1, x_2, \ldots, x_n) \begin{pmatrix} a_{11} & \cdots & a_{1m} \\ a_{21} & & a_{2m} \\ \vdots & & \\ a_{n1} & \cdots & a_{nm} \end{pmatrix} = \left(\sum_{i=1}^{n} x_i a_{i1}, \ldots, \sum_{i=1}^{n} x_i a_{im} \right)$$

(b) multiplication of a $n * m$-matrix by a m-dimensional row vector:

$$Ax = \begin{pmatrix} a_{11} & \cdots & a_{1m} \\ a_{21} & & a_{2m} \\ \vdots & & \\ a_{n1} & \cdots & a_{nm} \end{pmatrix} \begin{pmatrix} x_1 \\ x_2 \\ \vdots \\ x_m \end{pmatrix} = \begin{pmatrix} \sum\limits_{i=1}^{m} x_1 a_{1i} \\ \sum\limits_{i=1}^{m} x_i a_{2i} \\ \vdots \\ \sum\limits_{i=1}^{m} x_i a_{ni} \end{pmatrix}$$

Important are vectors that do not change their direction during a linear transformation. They are termed *eigenvectors* of the matrix A. For every eigenvector of A there exists a real number λ, the eigenvalue, for which the following equation holds:

$$Ax = \lambda x$$

Eigenvector analysis is needed, e.g., in Sec. 12.5.4 for projection of multidimensional data.

Index